Vegetable Crops

— Volume 3 —

THE EDITORS

Professor T.K. Bose obtained his M.Sc. (Ag.) and Ph.D. degree from Calcutta University, West Bengal, India. Trained at the Royal Botanic Garden, Kew, UK. Former Secretary of the Agri-Horticultural Society of India; Project Coordinator, All India Coordinated Floriculture Improvement Project (ICAR); Professor and Head, Department of Horticulture; Dean, Faculty of Agriculture; Director of Research, Bidhan Chandra Krishi Viswavidyalaya (BCKV), West Bengal; Member of Scientific Advisory Committees of ICAR, CSIR (Govt. of India) and Government of West Bengal. Awarded (D.Sc.) Honoris Causa from OUAT, Bhubaneshwar and UBKV, Coochbehar, India. Guided around 45 Ph.D. students, published more than 250 research papers and authored 30 books on horticultural crops.

Prof. M.G. Som served as former Vice-chancellor, Dean, Faculty of Agriculture and Professor of Vegetable Science, BCKV, West Bengal. Obtained his M.Sc. and Ph.D. degrees from ICAR-IARI, New Delhi. Received fellowship in 1997 and Plaque in 1998 from the Indian Society of Vegetable Science (ISVS). Served as FAO expert in Bangladesh for increasing the vegetable production. Acted as a member of governing body of the ICAR, New Delhi; member of Advisory Research Council, ICAR-IIVR, Varanasi, and ICAR-CIARI, Port Blair, Andaman. Published more than 100 research papers, written 8 books, 2 monographs and few bulletins and several popular articles. Guided 14 Ph.D. students and more than 50 M.Sc (Ag.) students. In 2019, ISVS felicitated Prof. Som for his outstanding contribution in the field of vegetable research and education.

Prof. Arup Chattopadhyay obtained his M.Sc. and Ph.D. degrees from BCKV, West Bengal. He has 23 years of experience in Research, Teaching and Extension. Acting as the Officer-in-Charge, AICRP on Vegetable Crops, and served as Professor for more than 8 years. He has guided 15 M.Sc. and 9 Ph.D. students in Vegetable Science. Published more than 150 research papers, authored 05 books, 07 book chapters, and many popular articles, folders, leaflets. Developed nine trait specific varieties/hybrids of different vegetable crops. First recipient of 'Dwarikanath Young Scientist Gold Medal Award' for the best Ph.D. thesis in Vegetable Science; 'Fellows' of ISVS, Varanasi, and CWSS, West Bengal. BCKV has been felicitated 'Lt. Amit Singh Memorial Award' by the ICAR for the Best coordinating centre on Vegetable Crops in India for outstanding contribution in vegetable research under his able leadership.

Prof. Tapan Kumar Maity has acclaimed professional career of more than 35 years in the Department of Vegetable Science, BCKV, West Bengal. Acted as Nodal Officer of Quality Control Laboratory for Horticultural Produce, and PI of AINRP on Onion and Garlic (ICAR). Served as the Head of the Department of Vegetable Science and the Dean, Faculty of Horticulture. Published more than 100 original research papers, contributed 13 book chapters, edited 3 books. Supervised 12 Ph.D. and 32 M.Sc. students. Acted as Member of the Working Group of the Agriculture Commission, Government of West Bengal, Member of the Monitoring Team of AINRP on Onion and Garlic for different centers; acting as consultant of National Horticulture Board, Government of India.

Prof. Jahangir Kabir obtained his M.Sc.(Ag) and Ph.D.(Horticulture) degrees from BCKV) West Bengal. He has 33 years of varied experience in Teaching and Research in Vegetable Crops and Post Harvest Physiology and Technology of Horticultural Crops. Acted as former Head of the Department of Post Harvest Technology, , and served as Professor for more than 15 years. He has guided more than 30 M.Sc. and 10 Ph.D. students, authored more than 120 publications including books (edited), research papers, book chapters, review/technical and popular articles. Acted as reviewer of research papers, paper setter, and member of selection committee of Public Service Commission and several State and Central Universities.

Vegetable Crops

— Volume 3 —

— *Senior Editor* —

M.G. Som

Department of Vegetable Science
Faculty of Horticulture
Bidhan Chandra
Krishi Viswavidyalaya
West Bengal 741 252, India

— *Editor Emeritus* —

T.K. Bose

Faculty of Horticulture
Bidhan Chandra
Krishi Viswavidyalaya
West Bengal 741 252, India

— *Edited by* —

A. Chattopadhyay • T.K. Maity • J. Kabir

Faculty of Horticulture
Bidhan Chandra Krishi Viswavidyalaya
West Bengal 741 252, India

2022

Daya Publishing House®
A Division of
Astral International Pvt. Ltd.
New Delhi – 110002

Fourth Revised & Illustrated edition 2022

Publisher's Note:

Every possible effort has been made to ensure that the information contained in this book is accurate at the time of going to press, and the publisher and author cannot accept responsibility for any errors or omissions, however caused. No responsibility for loss or damage occasioned to any person acting, or refraining from action, as a result of the material in this publication can be accepted by the editor, the publisher or the author. The Publisher is not associated with any product or vendor mentioned in the book. The contents of this work are intended to further general scientific research, understanding and discussion only. Readers should consult with a specialist where appropriate.

Every effort has been made to trace the owners of copyright material used in this book, if any. The author and the publisher will be grateful for any omission brought to their notice for acknowledgement in the future editions of the book.

Cataloging in Publication Data--DK

Courtesy: D.K. Agencies (P) Ltd. <docinfo@dkagencies.com>

Vegetable crops / senior editor, M.G. Som ; editor emeritus, T.K. Bose ; editors, A. Chattopadhyay, T.K. Maity, J. Kabir. -- Fourth revised & illustrated edition.

volumes cm

Includes bibliographical references and index.

ISBN 9789390435227 (HB)

1. Vegetables. 2. Vegetables--Pictorial works. I. Som, M. G., editor. II. Bose, T. K., editor. III. Chattopadhyay, A. (Arup), editor.

LCC SB320.8.V44 2021 | DDC 635.0954 23

Published by : **Daya Publishing House®**
A Division of
Astral International Pvt. Ltd.
– ISO 9001:2015 Certified Company –
4736/23, Ansari Road, Darya Ganj
New Delhi-110 002
Ph. 011-43549197, 23278134
E-mail: info@astralint.com
Website: www.astralint.com

Digitally Printed at : Replika Press Pvt. Ltd.

From the Desk of Editor Emeritus

The undersigned had initiated the publications entitled 'Fruits: Tropical and Sub-tropical, Vegetable Crops and 'Commercial Flowers' in one volume each in the 1980's in collaboration with co-editors, humbly expresses from the core of his heart that the updated revised editions of the above mentioned publications with colour illustrations in thirteen volumes which was planned in June, 2020, are being printed from this year, a rare event in Horticultural Science has been possible by the blessing of Almighty God, with the cooperation of dedicated horticulturist as editors and contributors.

Kolkata

T.K. Bose
Editor Emeritus

From the Desk of Senior Editor

The book entitled "Vegetable Crops in India" was first published as a single volume in 1986, afterwards revised and published in 1993 and 2003 in three volumes by the name "Vegetable Crops". The previous three editions proved to be very useful to the students, teachers and researchers across the world. Concerted research in production and productivity of vegetable crops during last decades necessitates further revision of the book with coloured illustrations as the fourth edition in four volumes.

The third volume of this edition includes chapters on Solanaceous, Salad and Leafy Vegetable crops. I am extremely happy to acknowledge the contributors for their dedicated efforts to make this volume very effective and meaningful. I would like to thank all the co-editors of this volume who tirelessly and sincerely participated in editing and proof reading of the manuscript. It is expected that this revised edition will act as a platform to enrich and strengthen the students and the researchers over their knowledge on vegetable crops.

I would like to specially appreciate Prof. T. K. Bose, the Editor Emeritus, who took active initiative and guided us to complete the fourth edition of this book. It would have been impossible for us to put this book into table without his encouragement and guidance.

Despite of our best efforts, the editors are conscious about the lapses that might have crept in the manuscript, however, we would like to emphasis on the fact that these lapses or errors are unintentional but circumstantial.

M.G. Som

Kolkata *Senior Editor*

Preface to the Fourth Edition from Editors

The first edition of the book entitled 'Vegetable Crops in India' published in 1986 was greatly appreciated and proved valuable to the students, teachers, researchers and extension specialists in horticulture in general and vegetable science in particular, in different parts of the world. Then the book was revised twice in 1993 and 2003 with the title 'Vegetable Crops' in three volumes due to its increasing demand. All the previous editions, which contained updated scientific information of all commercially grown vegetable crops, were considered as valuable contribution on the subject. After about 17 years of the last revised publication in 2003, voluminous information pertaining to vegetable crop improvement, production, protection and postharvest management has been generated necessitating further revision in four volumes.

This particular edition has been enlarged by incorporation of up-to-date information on various aspects of vegetable crop production, protection and improvement particularly through biotechnological interventions with meaningful illustrations, and published in four volumes. Efforts have been made to cover as much information as possible. Third volume in this edition includes chapters on Solanaceous, Salad and Leafy Vegetable crops. Most of the contributors have taken care to review up-to-date research works and prepared the manuscript. The revision has been made by the editors and in few cases new contributors have been associated with some chapters.

We express our sincere thanks to all contributors, the Editor Emeritus Prof. T. K. Bose and the Senior Editor Prof. M.G. Som who sincerely participated in revision, editing and proof-reading of the manuscript and we appreciate their commendable efforts to bring out this publication. We express our sincere thanks to Prof. P. Hazra for correction of first draft. We express our sincere gratitude to Dr. Praveen Kumar Maurya, Mr. Tridip Bhattacharjee, Mr. Swadesh Banerjee, Mr. Soham Hazra, Mr. Subhrajyoti Chatterjee, and SK. Masudul Islam, Ph.D. scholars, Bidhan Chandra Krishi Viswavidyalaya, India for their help and cooperation for arranging illustrations, necessary typing and correction of citations.

Apart from own collection of photographs, editors have also taken some photographs which are in public domain from different websites as far as possible. Therefore we acknowledge with enormous thanks to all State Agricultural Universities and Research Institutes under Indian Council of Agricultural Research, New Delhi, India; Public and Private International Agricultural Institutes across the world for taking photographs from their websites to make this book a valuable knowledge for the readers across the world. We think images can be powerful teaching tools, as illustrations to in-class lectures, or for studying concepts outside of the classroom. We also think identification of vegetable crops, their package of practices, cultivars, utilization, diseases and pests with the help of suitable illustrations in a consolidated manner will make a dent for the students and teachers of vegetable science.

Our thanks are due to Shri Prateek Mittal, Director of Astral International (P) Ltd., New Delhi for his keen interest in bringing out this publication.

A. Chattopadhyay • T.K. Maity • J. Kabir

Kolkata ***Editors***

Contents

SALAD CROPS

LEAFY VEGETABLES

List of Contributors

Akhtar, S.

Department of Horticulture (Vegetable and Floriculture), Bihar Agricultural University, Sabour, Bhagalpur 813210, Bihar, India

Banerjee, S.

Department of Vegetable Science, Bidhan Chandra Krishi Viswavidyalaya, Mohanpur 741252, Nadia, West Bengal, India

Bhattacharjee, T.

College of Agriculture, Lembucherra 799210, Tripura (West), Tripura, India

Bhutia, N.D.

College of Horticulture and Forestry, Central Agricultural University, Pasighat 791102, Arunachal Pradesh, India

Chatterjee, R.

Department of Plantation, Spices, Medicinal and Aromatic Crops, Bidhan Chandra Krishi Viswavidyalaya, Mohanpur 741252, Nadia, West Bengal, India

Chatterjee, S.

Department of Horticulture, Centurion University of Technology and Management, Paralakhemundi 761211, Odisha, India

Chattopadhyay, A.

Department of Vegetable Science, Bidhan Chandra Krishi Viswavidyalaya, Mohanpur 741252, Nadia, West Bengal, India

Choudhury, B.

Division of Vegetable Science, ICAR–Indian Institute of Agricultural Research, Pusa 110012, New Delhi, India

Devi, Jyoti

Division of Crop Improvement, ICAR–Indian Institute of Vegetable Research, Varanasi 221305, Uttar Pradesh, India

Dhua, R.S.

Department of Postharvest Technology, Faculty of Horticulture, Bidhan Chandra Krishi Viswavidyalaya, Mohanpur 741252, Nadia, West Bengal, India

Hazra, P.

Department of Vegetable Science, Bidhan Chandra Krishi Viswavidyalaya, Mohanpur 741252, Nadia, West Bengal, India

Hazra, S.

Department of Genetics and Plant Breeding, Bidhan Chandra Krishi Viswavidyalaya, Mohanpur 741252, Nadia, West Bengal, India

Irulappan, I.

Faculty of HorticultureTamil Nadu Agricultural University, Coimbatore 641003, Tamil Nadu, India

Islam, Sk Masudul

Department of Vegetable Science, Bidhan Chandra Krishi Viswavidyalaya, Mohanpur 741252, Nadia, West Bengal, India

Kalloo, G.

Division of Vegetable Crops, Haryana Agricultural University, Hissar 125004, Haryana, India

Karmakar, P.

Division of Crop Improvement, ICAR–Indian Institute of Vegetable Research, Varanasi 221305, Uttar Pradesh, India

Lyngdoh, Y.A.

Division of Vegetable Science, ICAR–Indian Agricultural Research Institute, Pusa 110012, New Delhi, India

Lal, S.

Division of Vegetable Crops, Haryana Agricultural University, Hissar 125 004, Haryana, India

Maity, T.K.

Department of Vegetable Science, Bidhan Chandra Krishi Viswavidyalaya, Mohanpur 741252, Nadia, West Bengal, India

Mallick, S.C.

Department of Horticulture, Faculty of Agriculture, Bidhan Chandra Krishi Viswavidyalaya, Mohanpur 741252, Nadia, West Bengal, India

Mishra, M.

Regional Plant Resource Centre, Bhubaneswar 751015, Orissa, India

Mohapatra, P.P.

College of Agriculture, Central Agricultural University, Kyrdemkulai, Meghalaya, India

Muthukrishnan, C.R.

College of Agriculture, Tamil Nadu Agricultural University, Coimbatore 641003, Tamil Nadu

Palai, S. K.

Regional Plant Resource Centre, Bhubaneswar 751015, Orissa

Pandita, M.L.

Division of Vegetable Crops, Haryana Agricultural University, Hissar 125 004, Haryana

Parthasarathy, U.

ICAR–Indian Institute of Spices Research, 32/482 C Narmada Nilayam, Bharathan Bazar, Chelavoor, Calicut 673571, Kerala, India

Parthasarathy, V.A.

ICAR–Indian Institute of Spices Research, 32/482 C Narmada Nilayam, Bharathan Bazar, Chelavoor, Calicut 673571, Kerala, India

Sagar, V.

Division of Crop Improvement, ICAR–Indian Institute of Vegetable Research, Varanasi 221305, Uttar Pradesh, India

Sharangi, A.B.

Department of Plantation, Spices, Medicinal and Aromatic Crops, Bidhan Chandra Krishi Viswavidyalaya, Mohanpur 741252, Nadia, West Bengal, India

Singh, B.K.

Division of Crop Improvement, ICAR–Indian Institute of Vegetable ResearchVaranasi 221305, Uttar Pradesh, India

Som, M.G.

Department of Vegetable Science, Bidhan Chandra Krishi Viswavidyalaya, Mohanpur 741252, Nadia, West Bengal, India

Thangaraj, T.

College of Agriculture, Tamil Nadu Agricultural University, Coimbatore 641003, Tamil Nadu

Tiwari, R. N.

Division of Vegetable Science, Indian Institute of Agricultural Research, Pusa 110012, New Delhi, India

SOLANACEOUS CROPS

1

TOMATO

S. Akhtar, P. Hazra, B. Choudhury, R.N. Tiwari, M. Mishra and S.K. Palai

1.0 Introduction

Tomato is one of the most popular and widely grown vegetables in the world ranking second in importance to potato in many countries. The fruits are eaten raw or cooked. Tomato in large quantities is used to produce different value added products; it supplies vitamin C and adds variety of colours and flavours to the food. Tomato and tomato products made its attention even in terms of value of micronutrients existing at low concentration. Apart from contributing nutritive elements, colour and flavour to the diet, tomatoes are also a valuable source of antioxidants, or chemo-protective compounds, and may thus be termed a "functional food" (Ranieri *et al.*, 2004). The antioxidant potential of tomato is derived from a mixture of antioxidant biomolecules, including lycopene, ascorbic acid, phenolics, flavonoids and vitamin E, and is especially high in cherry tomatoes (Kaur *et al.*, 2004; Mukherjee *et al.*, 2020). Because of their importance to human health (Madhavi and Salunkhe, 1998), antioxidants may be considered a valuable quality attribute of tomatoes and it is important to minimize losses of these compounds during the postharvest period. Tomato ranks 11th as the possible sources of vitamin A and 25th for vitamin C among the vegetables grown in India (Gopalan *et al.*, 1989). Its many forms are adapted to wide range of soils and climates extending from the tropics to almost the Arctic circle. By virtues of its many attributes, tomato is considered as a favourite crop for research in physiology and cytogenetics, all over the world.

2.0 Area and Production

Because of its wider adaptability and versatility, tomato is grown throughout the world either in outdoors or indoors. The estimated world production of tomato is about 182.26 million tonnes from an area of about 4.76 million hectares (FAO, 2020). China is the leading producer with a production of 61.63 million tonnes followed by India (19.38 million tonnes) and USA (12.61 million tonnes) (FAO, 2020). A list of major tomato producing countries with their area and production is presented in Table 1.

Table 1: Area and Production of Major Tomato Producing Countries

Country	*2016*$^{+}$		*2018*^{F}	
	Production (million tonnes)	*Area (million hectares)*	*Production (million tonnes)*	*Area (million hectares)*
China	56.42	1.00	61.63	1.04
India	18.40	0.76	19.38	0.79
USA	13.04	0.14	12.61	0.13
Turkey	12.60	0.19	12.15	0.18
Egypt	7.94	0.20	6.62	0.16
Italy	6.44	0.10	5.80	0.10
Iran	6.37	0.16	6.58	0.16
Spain	4.67	0.05	4.77	0.06

Country	2016+		2018F	
	Production (million tonnes)	Area (million hectares)	Production (million tonnes)	Area (million hectares)
Brazil	4.17	0.06	4.11	0.06
Uzbekistan	2.65	0.06	2.28	0.06
World	177.04	4.78	182.26	4.76

+ Anon.(2018).

F = FAO estimate (FAO website accessed on July 15th, 2020)

3.0 Composition and Uses

3.1.0 Composition

Tomato is known for its outstanding nutritive value as indicated in Table 2 below:

Table 2: Composition of Tomato Fruit (per 100 g of edible part)*

Moisture	93.1 g	Vitamin A	320 I.U.
Protein	1.9 g	Thiamine	0.07 mg
Fat	0.1 g	Riboflavine	0.01 mg
Minerals	0.6 g	Nicotinic acid	0.4 mg
Fibre	0.7 g	Vitamin C	31 mg
Carbohydrates	3.6 g	Calcium	20 mg
Sodium	45.8 mg	Magnesium	15 mg
Potassium	114 mg	Oxalic acid	2 mg
Copper	0.19 mg	Phosphorus	36 mg
Sulphur	24 mg	Iron	1.8 mg
Chlorine	38 mg		

* Aykroyd (1963).

3.2.0 Uses

Tomato is one of the important crops used as fresh vegetable as well as a variety of processed products such as juice, ketchup, sauce, canned fruits, puree, paste, *etc.* (Chattopadhyay *et al.*, 2013). The desirable qualities for a tomato cultivar to be used for processing includes high total soluble solids (4-8° Brix), acidity not less than 0.4 per cent, pH less than 4.5, uniform red colour, pear shaped, smooth surface, free from wrinkles, small core, firm flesh and uniform ripening (Adsule *et al.*, 1980). Green tomatoes are also used for pickles and preserves. It has many other uses; tomato seeds contain 24 per cent oil used as salad oil and in the manufacture of margarine.

The major pigments present in tomatoes are the carotenoids that impart the red, orange or yellow colour to the ripe fruits. The major carotenoid lycopene followed by β-carotene has been found to be associated with varied health benefits (Khachik

Uses of Tomato

Value Added Products from Tomato

et al., 2002), owing to their antioxidant properties as quencher of reactive oxygen species (ROS) (Jomova and Valko, 2013). Consumption of fresh and processed tomato products has been proposed to reduce the risk of several chronic diseases including cardiovascular diseases and even cancer, where the protective role is attributed to the carotenoids (Fraser and Bramley, 2004; Palozza *et al.*, 2011; Burton-Freeman and Sesso, 2014; Niranjana *et al.*, 2015). The role of lycopene has been recently reviewed where the pigment has been documented to reduce blood pressure, lower the risk of artery blockage, and prevent cholesterol oxidation, lower risk of lung, prostrate, breast and uterine cancer and have positive effects on the skeletal system and on neurodegenerative diseases, including Alzheimer's and Parkinson's (Przybylska, 2019). Apart from the carotenoids, a few tomato mutants have been reported to biosynthesize and accumulate flavonoids comprising of anthocyanins in ripe fruits. Anthocyanins exert potential health benefits (Tsuda, 2012; Gerardi *et al.*, 2018), particularly, anti-inflammatory and anti-atherosclerotic effects (Amin *et al.*, 2015; Olejnik *et al.*, 2016; Blando *et al.*, 2018). Several experiments on animals and even clinical trials on human have confirmed the role of anthocyanins in the prevention of cardio-vascular diseases and cancer (Nunez and Magnuson, 2014; Wallace *et al.*, 2016; Lin *et al.*, 2017).

Tomato is also rich in medicinal value. The pulp and juice are digestible, mild aperient, a promoter of gastric secretion and blood purifier. It is reported to have antiseptic properties against intestinal infections. It is said to be useful against cancer of the mouth, sore mouth, *etc.* Dried tomato juice retains vitamin C. It stimulates torpid liver and is good in chronic dyspepsia. It is one of the best vegetables which keep our stomach and intestine in good order.

According to Kyzlink *et al.* (1981), tomatine content in small hard tomato is 300 mg per 100 g of edible part and 150 mg per 100 g larger green fruits. Light cooking slightly decreases the alkaloids but 4 week storage of the freeze-dried product reduces 5 to 10 per cent of the original level of the alkaloids. The solanine content of the pulp made from unripe tomato is much lower, similar to peeled potato (5 mg/100 g of edible part).

4.0 Origin, Taxonomy and Botany

4.1.0 Origin

The first reference of the tomato appeared in the writings of the 16th century herbalists, but they were brief and only identified as newly introduced plant into Europe. Evidences about the first tomato cultivation was compiled and evaluated by Jenkins (1948). Two hypotheses were put forward with regard to its place of origin. Most authorities, notably Sabine (1819), De Candolle (1884), Muller (1940a, 1940b) and Luckwill (1943) suggested that tomato was transported to Europe from Peru sometime in 1535. Only two areas namely Peru and Central Mexico were indicated as sites of early domestication. The cited evidences and the early names 'Mala peruviana' and 'Pomi de! Peru' suggested a Peruvian origin, on the other hand, some considered the possibility of Mexican origin of tomato. It was also mentioned among a list of plants that might probably be considered as Mexican, but both the hypotheses were frequently quoted.

In 1544, Mathiolus gave the common name 'Pomid'oro' with its latin equivalent 'Malaaurea' for the first time, the name 'Pomid'oro' still persists in Italy as the common name for tomato. Dodonaeus, in his Flemish edition of *Herbal* in 1554 described tomato in a separate chapter entitled 'Ofamorous apples or golden apples'. Neither Mathiolus nor Dodonaeus speculated on the origin of the plant though both of them clearly stated that it was grown only in the gardens of the herbalists. They gave the impression that tomato was not considered as food plant, rather as a curiosity with probable medicinal properties.

In Anguillara's Semplici (1561), a short note entitled '*Lycopersico di galeno*' was given. On Anguillara's edition, two new names 'Pomide Peru' and 'Lycopersicon' were introduced into the synonymy and became firmly established through repetition in subsequent works. In fact, the *Lycopersicon* has persisted as the scientific name of the genus.

Despite earlier wide acceptance of Peru as the centre of domestication, the bulk of the historical, linguistic, archaeological and ethnobotanical evidences favour Mexico region as the origin of cultivated tomato that was first transported to the old world and that tomato was already well developed in cultivation in the new world before it was discovered. It attained fairly advanced stage of domestication before being transported to Europe and from Europe it was introduced to many other countries of the world. Since this genus is found in nature in other areas of South America, the archaeological data does not positively favour any particular region as the place of its origin.

Jenkins (1948) described the morphological variability in the cultivated tomatoes of Mexico. Similar variability was also observed in Central America (Lehmann and Schwanitz, 1965) and coastal Peru (Rick, 1958). However, the genetic variability *per se* was not a diagnostic feature for determining centres of origin. Rick (1976) reported overall similarity in allele frequency in the common cultivars and var. *cerasiforme* from Mexico and Central America. The differing alleles found in the Ecuadorean and Peruvian cultivars were also common to var. *cerasiforme* and *Lycopersicon pimpinellifolium* from their respective regions. Whatever might be the geography of domestication of the tomato, its immediate ancestor was probably var. *cerasiforme* as concluded by Jenkins (1948) and it is still widely accepted. He also suggested that it is reasonably certain that Mexico and not Peru was the source of the cultivated tomatoes of the old world and probably the only centre of domestication of the tomatoes because of the widespread distribution of var. *cerasiforme* in Mexico. The var. *cerasiforme* also bears greater genetic resemblance with the cultivated tomato than *L. pimpinellifolium* probably a by-product rather than a member of the same line of the crop.

In the evolution of cultivars, the fear of toxicity restricted its consumption. Such notion was based upon the presence of poisonous glycoalkaloids in the family and this notion continued upto the 19th century. Later, the predominant alkaloid in tomato 'tomatine' was found to be non-toxic. The history of utilization of cultivated tomato after its introduction to Europe, could be traced to the very complete account given by McCue (1952). In a foreword to McCue's (1952) bibliography, Anderson wrote that tomato was extensively used in many different ways in Turkey and suggested that Italians perhaps acquired tomato from the Turks or at least from the people of the Levant.

Whatever might be its early history of cultivation, the tomato gained wide popularity from the middle of 19th century to the present time and in respect of its climatic tolerance it has proved to be one of the most versatile cultivated plants.

4.2.0 Taxonomy

Tomato belongs to the family Solanaceae with chromosome number 2n = 2x = 24 (Jenkins, 1948; Peralta *et al.*, 2008). It is a native of South America. Tomato was distinguished from the genus *Solanum*, that comprised of eggplant (*Solanum melongena*) and potato (*Solanum tuberosum*) as well as 1400 other species, and classified as a separate genus *Lycopersicon* characterized by the absence of spines in the pinnatifid leaves, narrow anther tips and exhibit longitudinal dehiscence of anthers. The scientific name *Lycopersicon esculentum* was given to tomato in the 1760s by a Scottish botanist Philip Miller.

However, with rapid progress in DNA sequencing technology in the 1990s, a large amount of DNA sequence data was produced and accumulated in public databases. Molecular phylogenetic analyses were accelerated using sequence data. The obtained results suggested that it was necessary to reclassify the genus *Solanum* with special emphasis on molecular phylogenetic studies at the DNA level, tomato was renamed as "*Solanum lycopersicum*" (Peralta *et al.*, 2008; Asamizu and Ezura, 2009). Thus, phylogenetic classification of the family Solanaceae was recently revised and the genus *Lycopersicon* reintegrated into the genus *Solanum* with its new nomenclature (Peralta *et al.*, 2008). As a result, the genus Solanum was integrated with the clade Lycopersicon.

Müller (1940a) and Luckwill (1943) produced the two most complete taxonomic treatments of wild tomatoes based on morphological concepts, and treated them under *Lycopersicon*. Müller (1940a) divided the genus into two subgenera:

1. Eulycopersicon C. H. Müll., with two species, the cultivated tomato *L. esculentum* and *L. pimpinellifolium* possessing glabrous, and red- to orange- to yellow-coloured fruits, flat and silky pubescent seeds, bractless inflorescences, and leaves without pseudostipules. The typical characters of this sub-genus were red-fruited edible species with carotenoid pigmentation and annual nature of plants.
2. Eriopersicon C. H. Müll., with four species, *L. peruvianum*, *L. hirsutum*, *L. glandulosum* and *L. cheesmanii*, bearing pubescent green or greenish white to yellowish and purple tinged fruits, frequently with a dark green, lavender, or purple stripe, thick glabrous seeds or pilose only at the apex, bracteate inflorescences, and leaves usually with pseudostipules. Müller (1940a) also described *L. glandulosum* C. H. Müll. and classified the highly polymorphic *L. peruvianum* into two varieties: var. *dentatum* Dunal and var. *humifusum* C. H. Müll.

Luckwill (1943) adopted the two subgenera recognized by Müller (1940a) and proposed a phylogeny of *Lycopersicon*. He hypothesized that the two subgenera might have evolved from an ancestral simple form characterized by imparipinnate leaves with 5 to 7 entire leaflets, few interjected leaflets, and probably no secondary leaflets, unbranched inflorescences, and undeveloped pseudostipules. Two different lineages diverged from the ancestral forms, one characterized by fruits with carotenoid pigments and the other by green fruits with anthocyanin pigments. Luckwill (1943) agreed with Müller (1940a) in the circumscription of subgenus Eulycopersicon, but within Eriopersicon he considered *L. pissisi* a distinct species, with *L. peruvianum* var. *humifusum* C. H. Müll. as its basionym. He also proposed different intraspecific categories in *L. esculentum*, *L. cheesmanii*, and *L. peruvianum*. *Lycopersicon cheesmanii*, with yellow to orange fruits, bractless inflorescences, and leaves without pseudostipules, apparently was misclassified in the subgenus Eriopersicon by Müller (1940a) and Luckwill (1943).

Later, Rick and Lamm (1955) recognized the importance of *chilense* as a separate species. Two distinct species, *L. esculentum* and *L. pimpinellifolium* were recognized by most authorities, with 4 or 5 botanical varieties in the former (Thompson and Kelly, 1957).

Bailey (1949) classified tomatoes into two species, *L. esculentum* and *L. pimpinellifolium*, with 5 botanical varieties in *L. esculentum* var. commune (common tomato), var. *grandfolium* (large-leafed tomato), var. *validum* (upright tomato), var. *cerasiforme* (cherry tomato) and var. *pyriforme* (pear tomato). The chromosome number of all the species were reported to be 2n = 2x = 24. Warnock (1988) reviewed taxonomy and phylogeny of the genus *Lycopersicon* and recognised 9 species (*L. esculentum, L.pimpinellifolium, L. cheesmanii, L. hirsutum, L. pennellii, L. chmielewskii, L. parviflorum, L. peruvianum and L. chilense*) comprising of various botanical varieties and forms.

The immediate ancestor of cultivated tomato was probably *L.esculentum* var. *cerasiforme* found throughout tropical and subtropical America (Jenkins, 1948). *L. esculentum* and near relatives are self-fertile but the former is out-crossed to a considerable extent in certain parts of subtropical areas and in its native region. However, in other areas, it is almost completely self-pollinating (Rick, 1950). The other species are self-incompatible with showy flower and consequently out-crossed. Cultivated tomato can be crossed with other species with varying degrees of difficulty.

D'Arcy (1972) treated tomatoes in the genus *Lycopersicon* but did not address subgeneric relationships. More recently, Child (1990) placed tomatoes in *Solanum* subg. Potatoe (G. Don) D'Arcy, sect. Lycopersicon (Mill.) Wettst., subsect. Lycopersicon, mainly characterized by anthers with sterile appendages and laterally connivent forming a tube, and classified them in three series: Lycopersicon, Eriopersicon (C. H. Müll.) Child, and Neolycopersicon (Correll) Child. The first two series correspond to the subgenera Eulycopersicon and Eriopersicon used by Müller (1940a) and Luckwill (1943). Series Lycopersicon includes herbs with few or no pseudostipules, ebracteate inflorescences, and fruits with carotenoid pigments, and Eriopersicon includes herbs to subshrubs with pseudostipules, bracteate inflorescences, and fruits with no carotenoid pigments (fruits pure green or greenish white, dark green to purple striped). Series Neolycopersicon only includes

Variability in Fruit Shape, Size and Colour of Tomato

Variability in Cherry, Cocktail and Slicing Tomato

Species of *Lycopersicon*

L. piminellifolium **(Left),** ***L. cheesmanii*** **(Middle),** ***L. hirsutum*** **(Right)**

L. pennellii **(Left),** ***L. peruvianum*** **(Middle),** ***L. chilense*** **(Right)**

L. esculentum **(Left),** ***L. chmielewskii*** **(Middle),** ***L. parviflorum*** **(Right)**

S. pennellii (Correll, 1958), a species with curved, markedly unequal length anthers, loosely coherent and without sterile appendages. Child (1990) also proposed *Solanum* sect. *Lycopersicon* (Mill.) Wettst., subsection Lycopersicoides Child, and section *Juglandifolium* (Rydb.) Child as the closest relatives of subsection Lycopersicon.

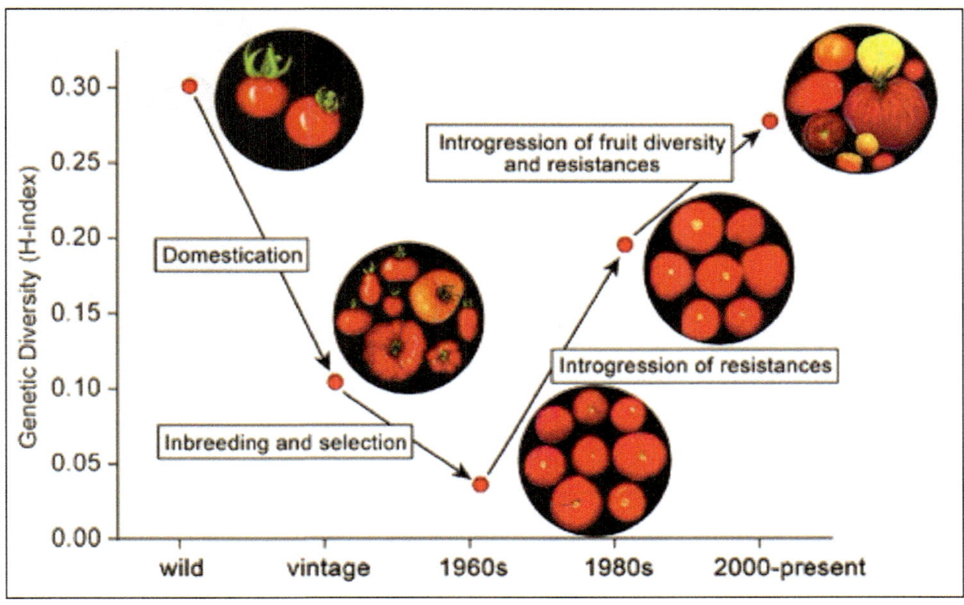

The Evolution of Genetic Diversity in Tomato.

The upper left group represents the ancestors (*S. lycopersicum* var. *lycopersicum*, *S. lycopersicum* var. *cerasiforme*, and *S. pimpinellifolium*), which gave rise to the vintage types including landraces through a process of domestication (Blanca, 2015)

Müller, 1940	Luckwill, 1943	Rick, 1979	Child, 1990
Subgenus *Eulycopersicon*	Subgenus *Eulycopersicon*	Esculentum complex	Series *Lycopersicon*
L. esculentum (3)	*L. esculentum* (8)	*L. esculentum* (2)	*S. lycopersicum*
L. pimpinellifolium	*L. pimpinellifolium*	*L. pimpinellifolium*	*S. pimpinellifolium*
		L. cheesmaniae	*S. cheesmaniae*
		L. pennellii	Series *Neolycopersicon*
		L. hirsutum	*S. pennellii*
		L. chmielewskii	Series *Eriopersicon*
		L. parviflorum	*S. habrochaites*
Subgenus *Eriopersicon*	Subgenus *Eriopersicon*	Peruvianum complex	*S. chmielewskii*
L. peruvianum (3)	*L. peruvianum* (5)	*L. chilense*	*S. chilense*
	L. pissisi — ?	*L. peruvianum*	*S. peruvianum*
L. cheesmaniae (2)	*L. cheesmaniae* (2)		*S. neorickii*
L. hirsutum (2)	*L. hirsutum* (2)		
L. glandulosum	*L. glandulosum*		

Comparison of Classifications of *Solanum* sect. *Lycopersicon*.
The numbers in parentheses represent numbers of infraspecific ranks (subspecies, varieties, and forms). The lines connect synonymous taxa.

Most recently, a new orange-fruited species of tomato was recognized, *S. galapagense* S. Darwin and Peralta (Darwin *et al.*, 2003). Within tomatoes it shares its orange fruit colour only with *S. cheesmaniae*, but is clearly distinguished from it by a number of leaf and flower traits. This species, formerly recognized as *S. cheesmaniae* f. *minor*, is clearly related to the other species with carotenoid pigments.

Peralta and Spooner (2001) examined interspecific relationships of 10 wild tomato species, with a concentration on the highly polymorphic and widespread species *Solanum peruvianum*, with DNA sequence data of the single-copy nuclear encoded GBSSI gene. Their outgroup results were concordant with the cpDNA restriction fragment length polymorphism (RFLP) phylogeny of Spooner *et al.* (1993), supporting *S. juglandifolium*, *S. lycopersicoides*, and *S. sitiens* as outgroups. The ingroup of wild tomatoes showed a basal polytomy and a terminal clade. The basal polytomy was composed of the self incompatible green-fruited species *S. chilense*, the central to southern Peruvian populations of *S. peruvianum*, *S. habrochaites*, and *S. pennellii*. The terminal clade contained the northern Peruvian populations of *S. peruvianum* (also self-incompatible, green fruits), *S. chmielewskii*, and *S. neorickii* (self-compatible, green fruits), and the self-compatible and red to orange- to yellow-fruited species *S. cheesmaniae*, *S. galapagense*, *S. lycopersicum*, and *S. pimpinellifolium*. The results supported allogamy, self-incompatibility, and green fruits as plesiomorphic in tomatoes.

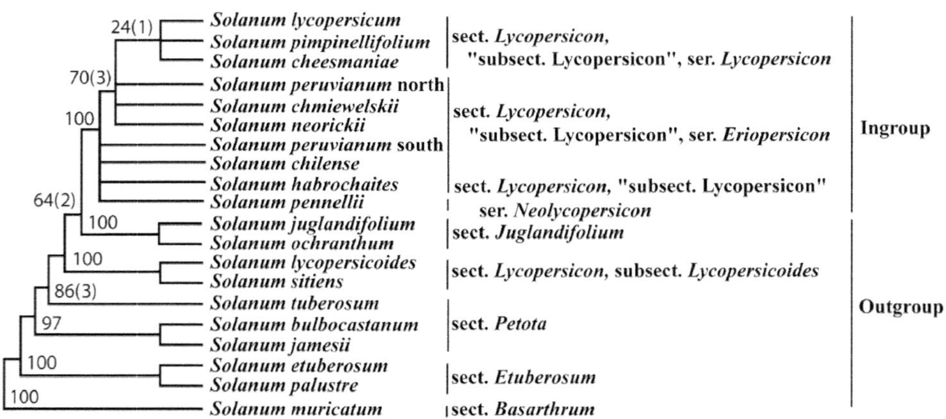

Abstracted Cladistic Results of the 65 Accessions of the 9 Tomato Species and 14 Outgroup Taxa Examined in the Phylogenetic Analysis of the GBSSI Gene Sequences by Peralta and Spooner (2001).

Nomenclature of *Solanum* subg. Potatoe according to Child (1990). They have informally labelled wild tomatoes in the clade containing all members of "subsect. Lycopersicon" (*S. lycopersicum* to *S. pennellii*) to provide a coordinate name for their formally described outgroups in section Juglandifolium (Rydb.) Child and subsection Lycopersicoides Child. Numbers indicate bootstrap values, and decay values are indicated between parentheses.

Darwin *et al.* (2003) recognized 13 species of wild tomatoes (*Solanum* section *Lycopersicon*) based on morphological characters, phylogenetic relationships, and geographic distribution, including *S. galapagense* and *S. cheesmaniae* that are endemic to the Galápagos Islands and the cultivated tomato (*Solanum lycopersicum*) and its weedy escaped forms that are distributed worldwide.

Peralta *et al.* (2005) described and illustrated two new wild tomato species segregated from *Solanum peruvianum sensu lato, Solanum arcanum* and *S. huaylasense,* placed in a key with two other segregates of *S. peruvianum sensu lato*: *S. peruvianum sensu stricto* and *S. corneliomulleri,* and the morphologically similar species *S. chilense.*

Peralta (2007) described two species in *Solanum* section *Juglandifolia, S. juglandifolium* and *S. ochranthum,* closely related to wild tomatoes, distributed in Colombia, Ecuador, and Peru, and two species of *Solanum* section *Lycopersicoides, S. lycopersicoides* and *S. sitiens,* distributed in southern Peru and northern Chile.

Table 3: Species List for Tomatoes along with Characteristic Fruit Colour, Distribution and Habitats*

Name	*Fruit Colour*	*Distribution and Habitats*
Solanum lycopersicoides Dunal	Green-yellow when maturing, black when ripe	Southern Peru to northern Chile on the western slopes of the Andes on dry rocky hillsides, 2800-3700 m.
Solanum sitiens I.M. Johnst.	Green-yellow when maturing, brown when ripe	Northern Chile, western Andean slopes on rocky hillsides and dry quebradas, 2350-3500 m.
Solanum juglandifolium Dunal	Green to yellow-green	Northeastern Colombia to southern Ecuador, on the edges of forest clearings, open areas and roadsides, 1200-3100 m.
Solanum ochranthum Dunal	Green to yellow-green	Central Colombia to southern Peru, in montane forests and riparian sites, 1400-3660 m.
Solanum pennellii Correll	Green	Northern Peru to northern Chile, in dry rocky hillsides and sandy areas, from sea level to 2850 m.
Solanum habrochaites S. Knapp and D.M Spooner	Green with darker greenstripes	Central Ecuador to Central Peru. Inpremontane forests to dry forests on the western slopes of the Andes, occasionally in lomas formations in northern Peru, 400-3600 m.
Solanum chilense (Dunal) Reiche	Green to whitish green with purple stripes	Southern Peru to northern Chile. On western slopes of the Andes, hyper-arid rocky plains, dry river beds, and coastal deserts, from sealevel to 3000 m.
Solanum huaylasense Peralta	Typically green with darkgreen stripes	Northern Peru (Department of Ancash). On the rocky slopes along rivers, 1700-3000 m.
Solanum peruvianum L.	Typically green to greenish-white, sometimes flushed with purple	Central Peru to northern Chile. In lomas formations and occasionally in coastal deserts from sea level to 600 m, sometimes growing as a weed at field edges in coastal river valleys.

Name	Fruit Colour	Distribution and Habitats
Solanum corneliomuelleri J.F.Macbr. (1 geographic race: Misti, Arequipa)	Typically green with darkgreen or purple stripes, sometimes flushed with purple	Central to southern Peru. On western slopes of the Andes, (400) 1000-3000 m, and on lower slopes on the edges of landslides towards the southern range of the species distribution.
Solanum arcanum Peralta (4 geographic races: 'humifusum', lomas, Marañon, Chotano-Yamaluc)	Typically green with darkgreen stripes	Northern Peru. Coastal and inland Andean valleys, on dry rocky slopes, 100 - 2500 m.
Solanum chmielewskii (C.M. Rick,Kesicki, Fobes and M. Holle) D.M. Spooner, G.J. Anderson and R.K.Jansen	Typically green with darkgreen stripes	Southern Peru to northern Bolivia (Sorata). In high dry Andean valleys, 2300-3000 m.
Solanum neorickii D.M. Spooner, G.J. Anderson and R.K. Jansen	Typically green with dark green stripes	Southern Ecuador to southern Peru. In dry Andean valleys, 1950-3000 m, often growing over rocky banks and roadsides. Sometimes found in sympatry with *S. chmielewskii*
Solanum pimpinellifolium L.	Red	Apparently native to coastal areas from central Ecuador to southern Peru, although populations are found in Vallenar, Chile, 0-500 m. Grows in humid places and on the edges of cultivated fields throughout its native range and has apparently escaped from cultivation in the Galápagos.
Solanum lycopersicum L.	Red	Apparently native to Peru; the domesticated form of *S. lycopersicum* now occurs worldwide. The cherry tomato, *S.lycopersicum* var. *cerasiforme,* is the possible ancestor of cultivated tomato and can often be found growing as a weed in temperate habitats and the edges of cultivated fields, where it is not necessarily native.
Solanum cheesmaniae (L. Riley)Fosberg	Yellow, orange	Endemic to the Galápagos Islands (Ecuador) from sea level to 1350 m.
Solanum galapagense S.C. Darwinand Peralta	Yellow, orange	Endemic to the Galápagos Islands, particularly the western and southern islands, mostly occurring on coastal lava and on volcanic slopes, sea level to 650 m, exceptionally up to 1,500 m in Fernandina and Santiago Islands

* Peralta (2007).

Four informal groups are recognized within the *Solanum* section *Lycopersicon* (Peralta *et al.,* 2008):

1. 'Lycopersicon group' with *Solanum lycopersicum, Solanum cheesmaniae, Solanum galapagense,* and *Solanum pimpinellifolium* (the 'red/orange fruit' clade);
2. 'Arcanum group' with *Solanum arcanum, Solanum chmielewskii,* and *Solanum neorickii* (the 'green fruit' clade);

Table 4: Species List for *Solanum* Section *Lycopersicon*, with Equivalents in the Previously Recognized Genus *Lycopersicon*, now part of a Monophyletic Genus *Solanum**

Solanum Name	*Lycopersicon Equivalent*
S. arcanum Peralta	Part of *L. Peruvianum* (L.) Miller
S. cheesmaniae (L.Riley) Fosberg	= *L. cheesmaniae* L. Riley (incorrectly published as *L. cheesmanii*)
S. chilense (Dunal) Reiche	= *L. chilense* Dunal
S. chmielewskii (C.M. Rick, Kesicki, Fobes and M. Holle)D.M. Spooner, G.J. Anderson and R.K. Jansen	= *L. chmielewskii* C.M. Rick, Kesicki, Fobes and M. Holle
S. corneliomuelleri J.F. Macbr.	Part of *L. peruvianum* (L.) Miller; also known as *L.glandulosum* C.F. Mull.
S. galapagense S. Darwin and Peralta	Part of *L. cheesmaniae* L. Riley (previously known as forma or var. *minor*)
S. habrochaites S. Knapp and D.M Spooner	= *L. hirsutum* Dunal
S. huaylasense Peralta	Part of *L. peruvianum* (L.) Miller
S. lycopersicum L.	= *L. esculentum* Miller
S. neorickii (C.M. Rick, Kesicki, Fobes and M. Holle) D.M. Spooner, G.J. Anderson and R.K. Jansen	= *L. parviflorum* C.M. Rick, Kesicki, Fobes and M. Holle
Solanum pennellii Correll	=*Lycopersicon pennellii* (Corre l) D'Arcy
S. peruvianum L.	= *L. peruvianum* (L.) Miller
S. pimpinellifolium L.	= *L. pimpinellifolium* (L.) Miller

* Peralta *et al.* (2005).

3. 'Eriopersicon group' with *Solanum huaylasense, Solanum chilense, Solanum corneliomulleri, Solanum peruvianum*, and *Solanum habrochaites;*
4. 'Neolycopersicon group' containing only *Solanum pennellii*, which was considered to be sister to the rest of the section by (Peralta *et al.*, 2008) based on its lack of the sterile anther appendage that occurs as a morphological synapomorphy in *S. habrochaites* and the rest of the core tomatoes.

Mating systems have played an important role in the evolution of wild tomato species, varying from allogamous self-incompatible, to facultative allogamous and self-compatible, to autogamous and self-compatible (Rick, 1963, 1979, 1982b, 1986b). The self-incompatibility system in tomatoes is gametophytic and controlled by a single, multiallelic S locus (Tanksley and Loaiza-Figueroa, 1985). The self-incompatibility system has shown strong relationships between the extent of outcrossing and allelic polymorphisms, floral display, and degree of stigma exsertion in wild tomatoes (Peralta and Spooner, 2005). Rick (1982b) studied the genetic bases of self-compatibility, self-incompatibility, and flower characters, using interspecific hybrids between the self-compatible (SC) *S. pimpinellifolium*, used as recurrent parent, and the two self-incompatible (SI) species, *S. habrochaites* and *S. pennellii* and concluded that the evolution of the mating system in wild tomatoes has occurred from self-incompatibility, as ancestral condition, to self-compatibility, and probably never reversed to self-incompatibility, and these changes from self-incompatibility to self-compatibility are expected to arise frequently and independently. In the self-compatible species, the extent of outcrossing and genetic variation is also related to floral display and degree of stigma exsertion (Peralta and Spooner, 2005).

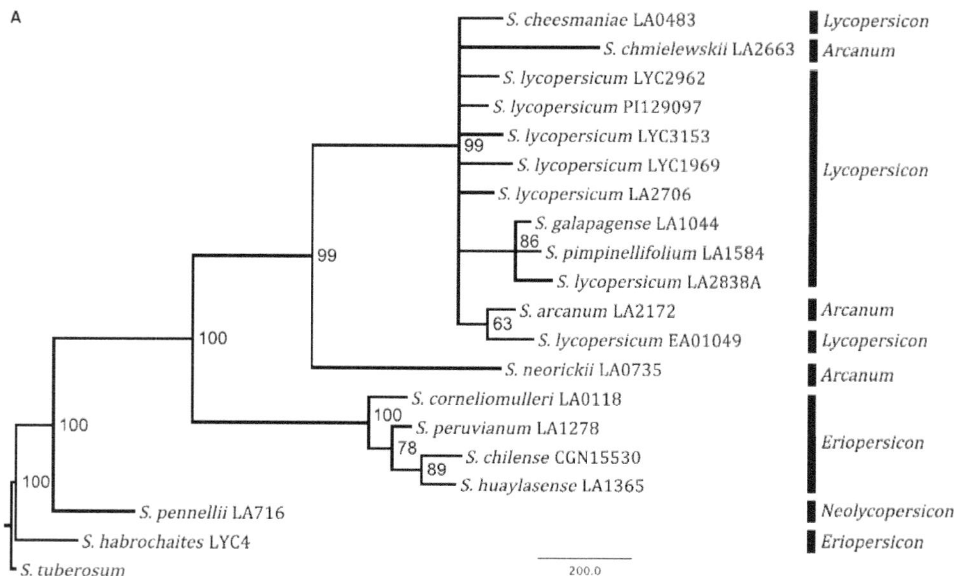

Phylogenetic Relationships in *Solanum* Section *Lycopersicon.*

A, the single most parsimonious tree topology is shown based on abundance values of the 1000 most abundant repeats identified in Illumina/454 next-generation sequencing runs. A total of 0.2 per cent of the genome for each accession was used. Bootstrap values are shown for each node (10 000 symmetric bootstrap replicates). Branch lengths are proportional to numerical step changes in repeat abundances (scale bar). Accession numbers are given for each sample (Peralta *et al.*, 2008)

Marginal populations of *Solanum pimpinellifolium* are highly autogamous with little or no genetic variation, small flower parts, and little or no stigma exsertion, while the central facultative allogamous populations have high genetic variation, larger corollas, and marked stigma exsertion (Rick *et al.*, 1977). A comparison of different *S. pimpinellifolium* genotypes in experimental plots in Peru showed that different outcrossing rates could be largely attributed to differences in floral characters, especially the level of stigma exsertion, rather than to differences in numbers and types of pollinators (Rick *et al.*, 1978). Two closely related self-compatible species, *S. chmielewskii* and *S. neorickii*, illustrate another example of changes in flower characters associated with outcrossing and genetic variation. *Solanum neorickii* is exclusively autogamous with low intra-population genetic variation and small flowers with stigmas included in the anther tube, while the facultative allogamous *S. chmielewskii* exhibited higher levels of heterozygosity, larger flower parts, and exserted stigmas (Peralta and Spooner, 2005). Rick *et al.*

(1976) postulated that *S. neorickii* evolved from *S. chmielewskii*. All populations of *S. chilense* are self-incompatible, and the related species *S. lycopersicoides*, *S. sitiens*, *S. ochranthum*, and *S. juglandifolium* are exclusively self-incompatible (Peralta and Spooner, 2005).

Rick (1963, 1979, 1986b) hybridized numerous accessions of all tomato species in all possible combinations and found two major crossing complexes. The Esculentum complex included seven species, mainly self-compatible and intercrossable. *Lycopersicon hirsutum* and *L. pennellii* typically are self-incompatible, but with some self-compatible populations. Three species of the Esculentum complex have mostly glabrous, red- to orange- to yellow-colored fruits (*L. cheesmaniae*, *L. esculentum*, *L. pimpinellifolium*), while the others have pubescent, green fruits (*L. chmielewskii*, *L. hirsutum*, *L. parviflorum*, *L. pennellii*). Hybrids were obtained in all combinations, in spite of the presence of some unilateral incompatibility. The Peruvianum complex, on the other hand, included the self-incompatible species *L. chilense* and *L. peruvianum*, both with pubescent green fruits. The barrier between these two complexes can be broken only by embryo rescue (Rick, 1979).

4.3.0 Botany

Cultivated tomato is an annual herb, 0.7 to 2 m tall, erect with thick solid stems or spreading, coarsely hairy with strong characteristic odour. Strong tap root with dense fibrous and adventitious roots are formed.

According to the growth habit, tomato is characterized by two types of plant:

1. *Determinate type*–inflorescence occurs more frequently in almost every internode until terminal ones are formed and elongation ceases at this point; in other words, it may be defined as self-topping and the main stem terminates with a flower cluster.
2. *Indeterminate type*–inflorescence cluster occurs at every third internode and the main axis continues to grow indefinitely.

The leaves are compound pinnatifid with small leaflet. The petiole is long clasping. The main leaflet is shortly stalked, 5 to 7 cm long, ovate to oblong with margin irregularly toothed, simple leaves with smooth margin also occur.

Flowers are borne in small forked raceme cyme. They vary in numbers from 5 to 12. Each flower is borne on short pedicel which is constricted at the middle and it marks the position of abscission layers. Flowers are pendent, perfect and hypogynous. There is short calyx tube bearing 5-7 persistant enlarged lobes. Corollas rotate with a short tube of 5-7 wrinkled lobes, yellow to pale yellow in colour. Stamens are 5-7 in number and inserted on the throat of corolla tube, filament bright yellow. Carpels 5-7, united and the basal ovary typically 5-7-celled with a central fleshy placenta. The style is first short and later elongated with flattened stigma which protrudes beyond the androecium in some cases. Dehiscence of anther is longitudinal, and remain viable till 1-2 days after opening of corolla. If the pollen is shed as the style grows up through anther tube, self-fertilization occurs and when the stigma protrudes, chances of cross-pollination through bees increase. The optimum temperature for pollination is around 21°C.

Rick (1976) reported change in style-length in the evolution of cultivated tomato, *L. esculentum*. Due to intensive breeding, floral structure evolved so as to improve the fruit setting ability and position of the stigma. Initially, the cultivars from Latin America tended to have well exserted stigmas and fruit set was greatly hampered due to lack of pollinating insects, especially in glass house crop where insects and wind-induced vibrations were excluded. As a result of continuous selection, the stigma of most cultivars became fixed at the mouth of the anther tube. Rick and Dempsey (1969) reported that stigma position was further shifted to even lower levels well within the anther tube. This change of style-length greatly increased self-pollination, fruit set and practically eliminated out-crossing in California cultivars.

So far as cross-pollination in tomato was concerned, pollinators like honey bees and bumble bees played a vital role, the latter being more effective in transferring the pollen (Amoako and Yeboah-Gyan, 1991; Palumbo *et al.*, 1996). In the field as well as in green houses, such insect pollination resulted in a high (84-100 per cent) fruiting rate with larger fruits and almost no puffy fruit (0-7 per cent). This was reported by Ruiter *et al.* (1988); Ravestijn (1990); Ravestijn and Sande (1991); Banda and Paxton (1991); Asada and Ono (1996, 1997); Dogterom *et al.* (1998); Pressman *et al.* (1999).

5.0 Growth and Development

Tomato seed germinates when adequate moisture, aeration and suitable temperature are provided. Both the cotyledon leaves store and assimilate food, develop chloroplast and increase in size with emergence. The area of cotyledon leaf was the largest when grown at 15.5°C (Calvert, 1959). The temperature for optimum dry matter increase of cotyledons varied between 18 and 20°C and the cotyledon of 4-day-old seedling expanded more in area and showed greater dry matter increase. Tomato seedling was very sensitive to temperature soon after cotyledons expanded. The rate of initiation of leaf primordia is faster at 25°C day and night temperature.

An investigation on seed treatment in magnetic fields of different intensity (150 to 900 oersted) on growth, development, earliness and productivity revealed best results by treatment at 450 oersted (Lukyan, 1988).

The development of axillary shoot was influenced by mineral nutrients, hormones and other chemicals, photoperiod, moisture and temperature (Verkerk, 1964; Abdel-Hafeez and Verkerk, 1969; Tucker, 1976). Depending upon the cultivar and the cultural practices, the proportion of assimilating leaf area of the axillary shoots in relation to whole plant constituted one third or half of the entire plant weight or leaf area and their contribution to yield were considerable. It was suggested by Baboy and Sabiti (1994) that plant height had the strongest effect on the productivity *i.e.*, fruit yield/plant.

Two additional aspects of correlation on the influence of axillary shoots on growth of main stem were (i) removal of young axillary bud prior to unfolding enhanced flower production and fruit ripening of field grown tomato (Asahira and Kelly, 1966) and (ii) the development of root system was restricted by pruning the axillary shoots (Thompson, 1927).

Root system constituted an integral part of the plant and was important in contributing to the structural and functional integrity of the shoot (Davis and Lungle, 1961; Asahira, 1974). The roots of field grown tomatoes penetrated 1 m depth with a spread of 1.3 m (Thompson, 1927). Reid (1926) observed that ability of tomato stem cutting to regenerate root and shoot depended upon and varied with the carbohydrate and nitrogenous substances in the cutting. Both exogenous and endogenous hormones were connected with tomato rooting (Asahira *et al.*, 1975).

The development of the root system among tomato cultivars differed and the night temperature was an important factor which regulated the size. A day temperature of 26.5°C and night temperatures between 16-22°C resulted in the heaviest root system.

Flower initiation and development in tomato began soon after the expansion of cotyledonary leaves. The development stage was critical in the seedlings, which responded to low temperature for induction of early flowering. Generally, seed vernalization at 5°C or less had no effect on flowering, in contrast, tomato seedlings were thermosensitive and could be vernalized to flower earlier and produce greater number of flowers per inflorescence. Tomato seedlings exposed to 14°C after cotyledon expansion increased the number of flowers on first inflorescence (Lewis, 1953). The commercial significance of this treatment was early flowering and fruiting and greater fruit yield (Wittwer and Teubner, 1956).

There is temperature sensitive period for each inflorescence during which low temperature promoted greater flower production. The sensitive period on Kondine Red cultivar for the first inflorescence was between the 8th and the 12thday after cotyledon expansion (Lewis, 1953) and this varied with cultivars. Differential temperature exposure of tomato seedlings during most sensitive stages indicated that the shoot temperature between 10 and 13 °C determined the morphological position of the first inflorescence. The root temperature controlled the number of flower on the inflorescence (Phatak *et al.*, 1966). In plants grown in regimes with decreasing temperatures during each growth stage upto the 6-leaf stage and decreasing the temperature at the 4-leaf stage to 10°C gave more abnormalities in fruits of the 1st and 2nd trusses than temperatures of 15°C or 20°C at that stage (Mametsuka *et al.*, 1991).

After initiation of flower, the ovarian tissues constituting the fruit organ of tomato, already possessed the majority number of cells to be reflected in a mature fruit (Houghtaling, 1935). The growth and nature of the fruit following pollination and fertilization was predominantly due to the cell enlargement accompanied by massive influx of water, organic and inorganic nutrients.

The rate of dry weight accumulation was higher in the early than the late-set fruits and was unaffected by the water regime but water stress shortened the duration of fruit growth and accelerated ripening (Wolf and Rudich, 1988). Temperature did not affect the dry matter production directly, which meant that temperature effect on the generative sink strength was not much different from that of the vegetative sink strength (Heuvelink, 1995). Under green house conditions, regulation of the water regime helped in the reduction of fruit cracking (Peet and

Willits, 1995). It was also suggested that plant water absorption could be suppressed by polyethylene glycol (PEG) treatment resulting in a decrease in fruit cracking (Ohta *et al.*, 1995). Also differences in the rate of fruit cracking were detected between irrigation treatments, more cracking occurring in the drier treatments at higher temperatures (Pascual *et al.*, 1999).

According to Ayuso *et al.* (1987), non-parthenocarpic varieties had much longer development periods than parthenocarpic varieties for fruits of similar weight.

The change in chemical composition during growth and development of fruit was studied by Boe *et al.* (1967). The acid content was found to be lower in immature fruit and it was highest at the stages when colour appeared and decreased rapidly with the advancement of ripening of the fruit. The total soluble solids and reducing sugars increased throughout the development of fruit. Citric acid was the major constituent of total acid. Malic acid occurred in small concentration and decreased as the fruit ripened. Both, β-carotene and lycopene increased in concentration as the fruits developed from mature green to ripening stages. The main route of carbohydrate accumulation by tomato fruit during the period of rapid fruit growth could be by cleavage of sucrose by apoplastic acid invertase prior to hexose transport across the plasma membrane (Brown *et al.*, 1997). Duden and Schlarzus (1982) found that the amino acids which increased markedly during ripening were glutamic acid and glutamine. Cabibel and Ferry (1980) suggested that lycopene and β-carotene content could be used to determine the stage of maturity. Endogenous calcium level in cell wall increased during fruit development to fully grown immature stage but declined just before ripening. It was also reported that calcium was solubilized during early ripening stage (Rigney and Wills, 1981). Salinity shortened the time of fruit development by 4-15 per cent and fruits from the salt treated plants were smaller and tasted better due to higher dry weight, TSS, acidity, the reducing sugar but pH was lower (Mizahi, 1982). However, the occurrence of blossom end rot increased but the incidence of puffy fruits decreased with increasing salinity (Cho *et al.*, 1996).

5.1.0 Pigmentation of Fruits

Three flesh colours in tomato, namely red (lycopene pigment), yellow (carotenoid pigment) and tangerine (prolycopene pigment) were found. Genetically, two pairs of alleles gave 4 colours as red, yellow, dark tangerine and light orange. The intensity of colour depended upon the balance between yellow and red pigments. Environment also regulated the colour development. At temperature below 10°C, tomato did not develop red or yellow colour whereas at temperatures between 10-25°C, red and yellow pigments developed and finally red colour was supressed above 30°C. If temperatures lower than 30°C was restored, red colour again developed. At temperatures above 40°C, lycopene was destroyed and no red colour development took place. Robertson *et al.* (1995) suggested that IAA was substantially more effective than 2,4-D in promoting lycopene formation and sucrose inhibited lycopene synthesis. The optimum temperature for lycopene accumulation was between 18-26°C. Mehta and Bhavamaryana (1981) found that red light and far red light promoted carotenoid synthesis, particularly lycopene in ripening fruits of tomato cv. Marglobe. Red light enhanced chlorophyll degradation, whereas far red light retarded it. Chlorophyll-a degraded more rapidly than chlorophyll-b, Chlorophyllase activity was enhanced by red and inhibited by far red light. If fruits were harvested at maturity stage 1 (10 per cent ripening colour) and kept at 10-20°C for 12 days, the carotenoid pigments continued to increase for 7 to 10 hours at 20°C and 5 to 6 hours at 10°C. The proportion of lycopene in total carotenoids was up to 75 per cent in fruits kept at 20°C. Chlorophyll content of fruit decreased during the first 4 to 20 hours at both temperatures (Dobreanu, and Indachescu, 1980).

Pigmentation in Tomatoes.

Furui *et al.* (1997) revealed that tomatine, a steroidal glycoalkaloid was localized in the locular tissue and epidermis. The tomatine content of leaves and flowers was higher than that of fruits and the wild tomatoes contained larger amounts of tomatine than the cultivated types. According to Czapski and Saniewski (1995), methyl jasmonate vapour increased the β-carotene content in the outer part (peel with 3 mm pericarp tissue) of the fruit, while it had no effect in peeled fruit pericarp. Methyl jasmonate treatment decreased the lycopene content in the outer part and pericarp tissues, but at a higher rate in the outer part of the fruits. It also increased the tomatine content 4-fold in the outer part of the fruits and by 62 per cent in the peeled fruit pericarp compared to the untreated controls.

6.0 Cultivars

The cultivars developed and recommended by various Research Institutes and State Agricultural Universities of India with their chief characteristics are given below:

Pusa-120 (Indian Agricultural Research Institute, New Delhi)

Plants are semi determinate, spreading, late maturing with dark green foliage. Fruits are flattish round, attractive, medium to large, uniform red, less acidic, less seeded, resistant to nematode and suitable for winter and summer seasons. Average yield is 300-320 q/ha. It is recommended for cultivation all over India

Pusa Ruby (Indian Agricultural Research Institute, New Delhi)

This is an early maturing cultivar. Plants are indeterminate with a height of 80-85 cm and spreading habit with few hardy branches. Fruits are flat to round, slightly lobed (4-5 locules), medium sized, uniform red at maturity with slightly acidic pulp, suitable for fresh consumption and processing. It is ideal for autumn, winter and spring summer seasons. Average yield is 280-300 q/ha. It is recommended for cultivation all over India

Pusa Sheetal (Indian Agricultural Research Institute, New Delhi)

Plants are determinate, fruit set successfully under low night temperature (up to 8°C) and suitable for early spring season, fruits are flattish round with yellow stem end, smooth, attractive, medium size, red colour and uniform ripening. Harvesting starts from early March in northern plains. Average yield is 350 q/ha. It is recommended for cultivation in Bihar, Punjab, Rajasthan, U.P., West Bengal, Delhi, Haryana, H.P. and J&K.

Pusa Gaurav (Indian Agricultural Research Institute, New Delhi)

Plants are dwarf, bushy with moderate foliage cover. Fruits are smooth, elliptical (egg-shaped) and borne in clusters. The unripe fruits are firm with thick flesh (0.6 cm) with two well filled locules facilitating easy transportation over long distances. Whole fruit is suitable for processing and canning as it is without neck constriction and has higher TSS (6 per cent) and better keeping quality at room temperature. It has average yield potential of 330-350 q/ha. It is recommended for cultivation in Punjab, U.P., Bihar, Rajasthan, Gujarat, Haryana and Delhi.

Pusa Uphar (Indian Agricultural Research Institute, New Delhi)

Plants are indeterminate, prolific bearer, upright, compact with dark-green foliage; fruits in bunches with 4-6 fruits per bunch, attractive, round, medium in size, thick skinned and uniform in ripening. Average yield is 370 q/ha. It is recommended for cultivation in Punjab, U.P., Bihar, Rajasthan, Gujarat, Haryana and Delhi.

Pusa Rohini (Indian Agricultural Research Institute, New Delhi)

Fruits are having thick pericarp and it is suitable for long distance transportation. It matures in 80 days. Average yield is 415 q/ha. It is recommended for cultivation in National Capital Regions.

Pusa Sadabahar (Indian Agricultural Research Institute, New Delhi)

Determinate, thermo insensitive, fruit set takes place almost round the year, both hot and cold set, maturity in 60 days. Average yield is 350 q/ha. It is recommended for cultivation in National Capital Regions.

Arka Vikas (Indian Institute of Horticultural Research, Bengaluru)

It is a pureline selection from American cultivar Tiptop. Fruits are medium large (80-90 g), oblate with light green shoulder, which become deep red on ripening. Developed for fresh market Adapted to both rainfed and irrigated conditions. Average yield is 35 t/ha. It is recommended for cultivation in H.P., J&K, Hills of U.P., Punjab, Tarai region of U.P., Bihar, Jharkhand, Chhattisgarh, Odisha, A.P., Rajasthan, Gujarat, Haryana, Delhi, M.P., Maharashtra, Karnataka, Tamil Nadu and Kerala.

Arka Abha (Indian Institute of Horticultural Research, Bengaluru)

Fruits are oblate with light green shoulder. Fruits have stylar end scar with average fruit weight of 75 g. Develops deep red colour on ripening. Resistant to bacterial wilt caused by *Ralstonia solanacearum*. Developed for fresh market and average yield 43 t/ha. It is recommended for cultivation all over India.

Arka Saurabh (Indian Institute of Horticultural Research, Bengaluru)

It is a pureline selection from Canadian breeding line V-685. Fruits are medium large, (70-80 g), round with light green shoulder, deep red in colour, firm with nipple tip. Developed for both fresh market and processing. Average yield 30-35 t/ha. It is recommended for cultivation in H.P., J&K, Hills of U.P., Punjab, Tarai region of U.P., Bihar, Jharkhand, Chhattisgarh, Odisha, A.P., Rajasthan, Gujarat, Haryana, Delhi, M.P., Maharashtra, Karnataka, Tamil Nadu and Kerala.

Arka Alok (Indian Institute of Horticultural Research, Bengaluru)

It is a cultivar developed through pure line selection from IIHR-719-1-6 (CL-114-5-1-0) from AVRDC, Taiwan. Fruits on the lower clusters are square round, large in size (120g), while in later clusters oblong, medium (80g), firm with light green shoulder. It is resistant to bacterial wilt. The cultivar is developed for fresh market. Average yield is 46 t/ha. It is recommended for H.P., J&K, Hills of U.P., Punjab, Tarai region of U.P., Bihar, Jharkhand, Karnataka, Tamil Nadu and Kerala.

Kashi Vishesh (Indian Institute of Vegetable Research, Varanasi)

This cultivar is resistant to TLCV and has been developed using *L. hirsutum* f. *glabratum* B'6013' as donor parent following backcross pedigree selection method. Plants are determinate, dark green, fruits are red, spherical, medium to large size. It is ready for first harvest at 70-75 days after transplanting. Average yield is 400-450 q/ha. It is recommended for cultivation in J&K, H.P., Uttarakhand, Punjab, U.P., Bihar, Jharkhand, Chhattisgarh, Odisha, A.P., Karnataka, Tamil Nadu and Kerala.

Kashi Amrit (Indian Institute of Vegetable Research, Varanasi)

This is a determinate cultivar derived from inter-specific cross *L. esculentum* (cv. Sel. 7) x *L. hirsutum* f. *glabratum* (accession B6013) through backcross pedigree method. Fruits are round, attractive red and fleshy with an average weight of 108 g. Suitable for cultivation during ToLCV infested period. Average yield is 620 q/ ha. It is recommended for cultivation in U.P., Bihar and Jharkhand.

Kashi Hemant (Indian Institute of Vegetable Research, Varanasi)

This cultivar has been developed through pedigree selection from a cross Sel-18 x Flora Dade. The plants are determinate, fruits are attractive red and round, weight varies from 80 to 85 g. Average yield is 400-420 q/ha. It is recommended for cultivation in Chhattisgarh, Odisha, A.P. and M.P.

Kashi Sharad (Indian Institute of Vegetable Research, Varanasi)

This cultivar has been developed through pedigree selection from a cross MTH-6 x Kalyani Eunish. Plants are indeterminate, leaves broad, fruits are attractive red, slightly oval, firm, thick pericarp, longer shelf life, weight 90 to 95 g. Average yield is 400-500 q/ha. It is recommended for cultivation in J&K, H.P. and Uttarakhand.

Kashi Anupam (Indian Institute of Vegetable Research, Varanasi)

This is a cultivar derived from inter-specific cross *L. esculentum* (cv. Sel. 7) x *L. hirsutum* f. *glabratum* (accession B6013) through backcross pedigree method. Plants are determinate, fruits are large, flattish round (slightly indented at blossom end of fruit), attractive red with 5-6 locules, medium maturity in 75-80 days after transplanting. Average yield is 500-600 q/ha. It is recommended for cultivation in Rajasthan, Gujarat and Haryana.

Kashi Aman (Indian Institute of Vegetable Research, Varanasi)

The fruits of this cultivar are round and very firm with a pericarp thickness of 0.52-0.57cm. Average fruit weight ranges from 80-110 g with an average locule number of 3-4. The fruits are attractive red in colour with an average total soluble solid content of 4.6 °Brix at red ripe stage. Average yield is 50-60 tonnes/ha. It is resistant to ToLCVD. The cultivar carries an allele of *Ty-3* gene. It has shown high level of resistance in field tests and to infection by both monopartite and bipartite viruses in agro inoculation experiments. It is recommended for cultivation in Punjab, Uttar Pradesh, Bihar and Jharkhand in kharif and rabi seasons.

Kashi Adarsh (Indian Institute of Vegetable Research, Varanasi)

The fruits of this cultivar are round and very firm with a pericarp thickness of 6 mm. Average fruit weight ranges from 80-115 g with 3-4 locules. The fruits are attractive red in colour. Cultivar has an average yield potential of 60 tonnes/ha. It is resistant to tomato leaf curl virus disease. Cultivar carries an allele of *Ty-3* gene. It has shown high level of resistance in field tests and artificial screening conducted by agroinoculation. It is recommended for cultivation in Madhya Pradesh and Maharashtra in kharif and rabi seasons.

Kashi Amul (Indian Institute of Vegetable Research, Varanasi)

The cultivar has an average yield potential of 50-60 tonnes/ha. The fruits of this cultivar are round and firm with a pericarp thickness of 6-6.5mm and an average fruit weight of 90-115 g in the initial three pickings. The fruits mature in 85-90 days post transplanting. This cultivar has also shown high level of ToLCV resistance in artificial screens and field tests conducted over years in disease hot spot at Indian Institute of Vegetable Research, Varanasi. This cultivar has been recommended for Karnataka, Tamil Nadu and Kerala.

HS-101 (Chaudhary Charan Singh Haryana Agricultural University, Hisar)

Plant are determinate sturdy, multi-branched, fruit in clusters of 2-3, round, small to medium in size, red at ripening and suitable for winter season. Average yield is 250-275 q/ha. It is recommended for cultivation all over India.

HS-102 (Chaudhary Charan Singh Haryana Agricultural University, Hisar)

Plant are determinate, fruit in clusters of 3-4, small to medium in size, uniform in ripening and suitable for winter and summer seasons. Average yield is 250-275 q/ha. It is recommended for cultivation all over India.

Hisar Lalit (Chaudhary Charan Singh Haryana Agricultural University, Hisar)

Plants are determinate, fruit maturity in 65-70 days after planting, average fruit weight 50 g, resistant to root knot nematode. Average yield is 250-300 q/ha. It is recommended for cultivation all over India.

Hisar Arun (Chaudhary Charan Singh Haryana Agricultural University, Hisar)

Plants are determinate, heavy bearer, early and high yielding. Average yield is 275-300 q/ha. It is recommended for cultivation all over India.

Hisar Lalima (Chaudhary Charan Singh Haryana Agricultural University, Hisar)

Plants are determinate, extremely early to medium large fruited. Average yield is 275-300 q/ha. It is recommended for cultivation in Haryana.

HS-110 (Chaudhary Charan Singh Haryana Agricultural University, Hisar)

Plants are determinate, potato leaf type, late maturing, fruits are large fleshy, suitable for table and processing. It is recommended for cultivation in Haryana.

Gujarat Tomato-2 (Anand Agricultural University, Anand)

Plant growth habit of this cultivar is determinate type. Immature fruit colour is green with dark green shoulder and at mature stage fruits are attractive red in colour with hard pericarp. Average yield is 340 q/ha. It is recommended for cultivation in Middle Gujarat.

Anand Tomato-3 (Anand Agricultural University, Anand)

The fruits of this cultivar are big in size with high pulp to juice ratio and very good firmness. Fruits are round in shape, dark red in colour and possess long shelf life. Average yield is 326 q/ha. It is recommended for cultivation in Middle Gujarat.

Pant Bahar (Govind Ballabh Pant University of Agriculture and Technology, Pantnagar)

The plants are bushy and profusely branched. Fruits are flattish round, medium in size with 5-6 locules slightly ridged and become red on maturity. The first picking can be taken in 75-80 days after transplanting. The yield potential of this cultivar is about 250 q/ha. It is recommended for cultivation in H.P., J&K, Hills of U.P., Punjab, Tarai region of U.P., Bihar and Jharkhand.

Pant T-3 (Govind Ballabh Pant University of Agriculture and Technology, Pantnagar)

The fruit weight is about 70 g. After ripening, the fruits attain uniform red in colour. It is suitable for processing. The yield potential of this cultivar is about 300 q/ha. It is recommended for cultivation in H.P., J&K, Hills of U.P., Punjab, Tarai region of U.P., Bihar and Jharkhand.

Pant Poly House Tomato-1 (Govind Ballabh Pant University of Agriculture and Technology, Pantnagar)

It is suitable for poly house cultivation. Number of fruits per bunch are 5-6. Average fruit weight is 202 g and yield potential is 1400 q/ha. It is recommended for cultivation in H.P., J&K, Hills of U.P., Punjab, Tarai region of U.P., Bihar and Jharkhand

Dhanashree (Mahatma Phule Krishi Vidyapeeth, Rahuri)

Fruits are medium to small and round with orange red in colour, suitable for *kharif* and *rabi* seasons. Average yield is 600-800 q/ha. It is recommended for cultivation in Maharashtra.

Phule Raja (Mahatma Phule Krishi Vidyapeeth, Rahuri)

Fruits are orange red in colour, medium size with attractive thick skin; suitable for *kharif* and *rabi* seasons. Average yield is 650 q/ha. It is recommended for cultivation in Maharashtra.

CO 1 (Tamil Nadu Agricultural University, Coimbatore)

Fruits are round, light green when unripe and red at ripening, bears in clusters. Average yield is 36 t/ha. It is recommended for cultivation in Tamil Nadu.

CO 2 (Tamil Nadu Agricultural University, Coimbatore)

Fruits are ovate, large and flat. Average yield is 42 t/ha. It is recommended for cultivation in Tamil Nadu.

CO 3 (Tamil Nadu Agricultural University, Coimbatore)

The cultivar is having dwarf plants, fruits are globular and red in colour, bears in clusters of 4-5 fruits. It is suitable for summer. Average yield is 50 t/ha. It is recommended for cultivation in Tamil Nadu.

PKM 1 (Tamil Nadu Agricultural University, Coimbatore)

The cultivar is having determinate plants, fruits are flattish with ridges. It is suited for long distance transport. Average yield is 32 t/ha. It is recommended for cultivation in Tamil Nadu.

Paiyur 1 (Tamil Nadu Agricultural University, Coimbatore)

A hybrid derivative suitable for irrigated and rainfed conditions of Dharmapuri district. Fruits are round, smooth fleshy and suitable for long distance transport. Average yield is 32 t/ha. It is recommended for cultivation in Tamil Nadu.

Azad T-5 (Chandra Shekhar Azad University of Agriculture and Technology, Kanpur)

The cultivar is indeterminate, round in shape, red in colour, bold and having better shelf life. Average yield is 400-450 q/ha. It is recommended for cultivation in Punjab, Tarai region of U.P., Uttarakhand, Bihar, Jharkhand.

Azad T-6 (Chandra Shekhar Azad University of Agriculture and Technology, Kanpur)

The cultivar is determinate, round, red in colour, medium and smooth. Average yield is 450-500 q/ha. It is recommended for cultivation in Punjab, Tarai region of U.P., Uttarakhand, Bihar, Jharkhand.

Punjab Ratta (Punjab Agricultural University, Ludhiana)

It gets ready for first harvest in 125 days after transplanting. The average yield is 225 q/acre. It is suitable for processing.

Punjab Tropic (Punjab Agricultural University, Ludhiana)

Plant height is about 100 cm. It becomes ready to harvest in 141days. Fruits are of large size and round shape, they borne in cluster. Gives average yield of 90-95q/acre.

Punjab Upma (Punjab Agricultural University, Ludhiana)

It is suitable for cultivation in rainy season. Fruits are oval shape, medium size and of firm deep red colour. The average yield is 220q/acre.

Punjab NR-7 (Punjab Agricultural University, Ludhiana)

It is a dwarf cultivar having medium size juicy fruits. It is highly resistant to fusarium wilt and root knot nematodes. Gives average yield of 175-180q/acre.

Punjab Red Cherry (Punjab Agricultural University, Ludhiana)

These cherry tomatoes are used in salads. These are of deep red colour. Sowing is done in August or September and plant is ready to harvest in February and gives yield up to July. Its early yield is 150 q/acre and total yield is 430-440 q/acre.

Punjab Varkha Bahar 1 (Punjab Agricultural University, Ludhiana)

The cultivar gets ready to harvest in 90 days. It is suitable for sowing in rainy season. It gives resistance to leaf curl virus. It gives average yield of 215q/acre.

Punjab Varkha Bahar 2 (Punjab Agricultural University, Ludhiana)

It gets ready to harvest in 100 days after transplanting. It is resistant to leaf curl virus. The average yield is 215 q/acre.

Punjab Swarna (Punjab Agricultural University, Ludhiana)

It was released in 2018. It has dark green colour foliage. It has oval shaped fruits which are orange in colour and medium in size. First picking should be done after 120 days of transplanting. It gives an average yield of 166q/acre till end-March (5months) and gives total yield of 1087q/acre in a span of 8-9 months. The cultivar is suitable for table purpose.

Punjab Sona Cherry (Punjab Agricultural University, Ludhiana)

It was released in 2016. It gives an average yield of 425q/acre. The fruits are yellow in colour and bears in bunches. The average weight of the fruit is approximately 11gm. It contains 7.5 per cent sucrose content.

Punjab Kesari Cherry (Punjab Agricultural University, Ludhiana)

It was released in 2016. It gives an average yield of 405q/acre. The average weight of the fruit is approximately 11gm. It contains 7.6 per cent sucrose content.

Punjab Varkha Bahar-4 (Punjab Agricultural University, Ludhiana)

It was released in 2015. It gives an average yield of 245q/acre. It contains 3.8 per cent sucrose content.

Punjab Gaurav (Punjab Agricultural University, Ludhiana)

It was released in 2015. It gives an average yield of 934q/acre. It contains 5.5 per cent sucrose content.

Punjab Sartaj (Punjab Agricultural University, Ludhiana)

The cultivar was released in 2009. It has round shape fruit, moderate and hard. It is suitable for rainy season. It gives an average yield of 898q/acre.

TH-1 (Punjab Agricultural University, Ludhiana)

It was released in 2003. The fruits are dark red in color, round hard and approximately 85gm weight. It gives an average yield of 245q/acre.

Swarna Lalima (ICAR-RCER, Ranchi)

It was developed through pure line selection. The fruits are deep red, round, weight (120-125 g) with TSS 4 ^{o}Brix. It is resistant to bacterial wilt. It gets ready for first harvest in 55-60 days after sowing. The average yield is 60-70 t/ha.

Swarna Naveen (ICAR-RCER, Ranchi)

It was developed through pure line selection. The fruits are about 40-50g in weight, oblong, deep red with TSS 5 °Brix. It is resistant to bacterial wilt. It gets ready for first harvest in 60-65 days after sowing. The average yield is 60-65 t/ha.

6.1.0 Cultivars Suitable for Growing in the Hills

Sioux (Indian Agricultural Research Institute, New Delhi)

It is having indeterminate growth habit, fruits are medium to large with yellow stem end, round fruits, sub-acidic in flavour, suitable for short distance market.

Best of All (Indian Agricultural Research Institute, New Delhi)

The cultivar is having indeterminate growth habit, fruits are round with green stem end, borne in bunches, fruits are firm and good yielder.

Marglobe (Indian Agricultural Research Institute, New Delhi)

The cultivar is late in maturity, indeterminate, fruits are large, round with green stem end, smooth and juicy.

6.2.0 Cultivars Suited for Processing

Roma (Indian Agricultural Research Institute, New Delhi)

It is prolific bearer, determinate in growth habit, having good foliage cover, excellent bearing, fruit shape is elliptical with yellow stem end, fruits are thick at stem end.

Punjab Chhuhara (Punjab Agricultural University, Ludhiana)

The cultivar is having determinate growth habit, good foliage cover, prolific bearer, fruits elliptical in shape with yellow stem end. Fruits are firm with thick pericarp and skin. It is very less seeded. It is suitable for long distant market. The average yield is 325q/acre.

S-152 (Indian Agricultural Research Institute, New Delhi)

It is prolific bearer, determinate, good coverage, fruits egg-shaped with yellow stem end, borne in bunches, suitable for canning and long distance transportation. It is an all-purpose cultivar and known as Pusa Gaurav.

Besides the cultivars mentioned above, some more cultivars and hybrids were recently reported from India: **Pusa Sheetal** can set fruit at low temperature (up to 8°C) and has been recommended as an early spring crop in North India (Tiwari *et al.*, 1990). Studies on tomato cultivars for processing industry revealed that **NDT 120, Pant Bahar and NDT 101** were highly suitable for the production of quality juice, ketchup and chutney, respectively (Nainwal *et al.*, 1992). Saikia *et al.* (1990) assessed a few cultivars growing under arid region of N.E. India and concluded that **Pant-I** and **Hybrid-10** showed higher TSS and percentage of reducing sugars.

Xinfan 4 (originally called Xinhong 718) is a processing tomato F_1 hybrid developed in China by crossing inbred lines from Introducing-6 and Meidong.

The fruits contain 5.65 per cent soluble solids, 12.63 mg/100 g pectin and yields 75-105 t/ha. It is resistant to tobacco mosaic tobamovirus and tolerant of cucumber mosaic cucumovirus (Wan g*et al.*, 1997). **Hongza 18** is a processing tomato hybrid developed by crossing Hongmanao 213 a female parent with Huangmiao as a male parent in China. It is suitable for storage and long distance shipment besides having good processing characters (Xu *et al.*, 1997). In China, **Hongza 20**, a tomato hybrid developed by crossing 8753 as female parent with Red Agate 144 as male parent, is suitable not only for processing but also for fresh consumption (Xu *et al.*, 1999).

Melka Shola and **Melka Salsa** have been reported to be cultivars superior in yield, quality and suitable for processing industries (Dessalegne *et al.*, 1999).

6.3.0 Hybrids in India

6.3.1 Public Sector Hybrids

Pusa Divya (Indian Agricultural Research Institute, New Delhi)

It is a F_1 hybrid between Long style x Roma. Plants are indeterminate, profusely branched; fruits are thick skinned, round to oval; first picking in 80 days after transplanting. Average yield is 350 q/ha. Suitable for Punjab, U.P. and Bihar

Arka Ananya (Indian Institute of Horticultural Research, Bengaluru)

It is a high yielding F_1 hybrid with combined resistance to *Tomato Leaf Curl Virus* (ToLCV) and bacterial wilt. Fruits are oblate-round, medium (65-70 g), deep red, medium firm, suitable for fresh market. Average yield is 65-70 t/ha in 140 days. This can be preferably grown in Punjab, Tarai region of U.P., Bihar, Jharkhand, Chhattisgarh, Odisha, Arunachal Pradesh, Rajasthan, Gujarat, Haryana, Delhi, M.P. and Maharashtra.

Arka Vardan (Indian Institute of Horticultural Research, Bengaluru)

It is a high yielding F_1 hybrid with root knot nematode resistance. This variety was developed for fresh market. Average yield is 75 t/ha in 160 days. This can be cultivated in H.P., J&K, Hills of U.P., Punjab, Tarai region of U.P., Bihar, Jharkhand, Karnataka, Tamil Nadu and Kerala.

Arka Vishal (Indian Institute of Horticultural Research, Bengaluru)

It is a high yielding F_1 hybrid for fresh market. Average yield is 75 t/ha in 160 days. This can be cultivated in Punjab, Tarai region of U.P., Bihar, Jharkhand, Karnataka, Tamil Nadu and Kerala.

Pusa Hybrid-2 (Indian Agricultural Research Institute, New Delhi)

Plants are compact, semi determinate with good foliage cover having prolific bearing. Fruits are round to flattish round, firm, smooth and attractive with uniform red colour at maturity. Plants are highly tolerant to root knot nematode. Average yield 600-625 q/ha.

Pusa Hybrid-4 (Indian Agricultural Research Institute, New Delhi)

Plants are determinate, compact with good dark green foliage. Fruits are attractive, round, smooth, average fruit weighing 70-80 g with three locules, thick pericarp and uniform ripening. It has field resistance to root knot nematode. Average yield potential is 425-450 q/ha.

Pusa Hybrid-8 (Indian Agricultural Research Institute, New Delhi)

Plants are determinate, with green compact foliage and heavy fruit bearer. Fruits are round, medium (75-80 g) with uniform ripening. Average yield is 430-450 q/ha.

Arka Abhijit (Indian Institute of Horticultural Research, Bengaluru)

It is high yielding F_1 hybrid with bacterial wilt resistance. It is developed for fresh market. Average yield is 65 t/ha in 140 days. This can be cultivated in Chhattisgarh, Odisha, Arunachal Pradesh, Rajasthan, Gujarat, Haryana, Delhi, Karnataka, Tamil Nadu and Kerala.

Arka Rakshak (Indian Institute of Horticultural Research, Bengaluru)

It is a high yielding F_1 hybrid developed by crossing IIHR-2834 x IIHR-2833. It is the first F_1 hybrid with triple disease resistance to ToLCV, BW and early blight. Fruits are square round, large (90-100g), deep red colored and firm. It is suitable for fresh market and processing. Average yield is 75-80 t/ha.

Arka Samrat (Indian Institute of Horticultural Research, Bengaluru)

It is a high yielding F_1 hybrid developed by crossing IIHR-2835 x IIHR-2832. It is another F_1 Hybrid with triple disease resistance to ToLCV, bacterial wilt (BW) and early blight. Fruits are oblate to high round, large (90-110g), deep red and firm. This variey is suitable for fresh market and yields 80-85 t/ha.

Arka Abhed (Indian Institute of Horticultural Research, Bengaluru)

It is a high yielding F_1 hybrid with multiple disease resistance, *i.e.*, Tomato leaf curl disease (*Ty2+Ty3*), bacterial wilt, early blight and late blight (*Ph2* + *Ph3*). Plants are semi-determinate with dark green foliage, fruits are firm, oblate round and medium large (90-100g). This hybrid is suitable for summer, *kharif* and *rabi* cultivation. This hybrid is bred for fresh market and yield ranges from 70-75 t/ha.

Arka Apeksha (Indian Institute of Horticultural Research, Bengaluru)

It is a high yielding hybrid developed by crossing IIHR 2834 and IIHR 2918. It has triple disease resistance to Tomato leaf curl disease (*Ty1+Ty2*), bacterial wilt and early blight. Plants are semi-determinate with dark green foliage. Fruits are firm, deep red, oblong, medium large (90-100g) with jointless (*j2*) pedicel. Fruits are suitable for processing with TSS 4.7° Brix, acidity 0.36 per cent, lycopene 14.15mg/100g fresh weight. It is recommended for summer, *kharif* and *rabi* cultivation. It has a yield potential of 43 to 90 t/ha depending on the season.

Arka Vishesh (Indian Institute of Horticultural Research, Bengaluru)

It is a high yielding F_1 hybrid developed by crossing IIHR 2834 and IIHR 2917. It has triple disease resistance to Tomato leaf curl disease (*Ty1+Ty2*), bacterial wilt and early blight. Plants are semi-determinate with dark green foliage and joint-less pedicel. It is recommended for summer, *kharif* and *rabi* cultivation. It has a yield potential of 43.3-90 t/ha depending on season. Fruits are firm, deep red, oblong and medium large (90-100g). Fruits have a TSS of 4.6° Brix, acidity 0.36 per cent and lycopene content of 14.14 mg/100 g fresh weight.

Kashi Abhiman (Indian Institute of Vegetable Research, Varanasi)

This is a determinate tomato hybrid recommended for J&K, H.P., Uttarakhand, Punjab, U.P., Bihar and Jharkand during mid September – March. It has high yield potential up to 93 t/ha. Fruits are deep red in colour and mature uniformly. The fruits are firm (pericarp thickness of 0.6cm) and are ideal for long distance transportation.

KTH-2 (Chandra Shekhar Azad University of Agriculture and Technology, Kanpur)

It is round shaped, red in colour. Average yield is 550-600 q/ha.

KTH-1 (Chandra Shekhar Azad University of Agriculture and Technology, Kanpur)

This hybrid produces oval shaped fruit. Fruits are red in colour. Average yield is 500-550 q/ha.

COTH 1 (Tamil Nadu Agricultural University, Coimbatore)

This F_1 hybrid has been developed by crossing IHR 709 x LE 812. Plants are semi determinate. Fruits are medium in size, bright red in colour, smooth, round to slightly oblong and borne in clusters of 4 - 5. Average yield is 95 t/ha.

COTH 2 (Tamil Nadu Agricultural University, Coimbatore)

This is a F_1 hybrid developed by crossing of LCR2 x CLN 2123 A. It is semi determinate in nature. Fruits are smooth, round to slightly oblong, medium in size. It is resistant to leaf curl virus. Average yield is 90 t/ha.

TNAU Tomato Hybrid CO 3 (Tamil Nadu Agricultural University, Coimbatore)

It is a F_1 hybrid developed by crossing of HN2 x CLN 2123 A. It is semi determinate in nature. Fruits are medium in size, smooth, round, borne in clusters of 3-5. This hybrid is resistant to leaf curl virus and moderately resistant to root knot nematode. Average yield is 95 t/ha.

Pant Poly House Tomato Hybrid-1 (Govind Ballabh Pant University of Agriculture and Technology, Pantnagar)

It is suitable for poly house cultivation. Number of fruits per bunch are 7-8. Average fruit weight is 265 g and yield potential is 1800 q/ha.

Swarna Vaibhav (ICAR-RCER, Ranchi)

It is a determinate F_1 hybrid. The fruits are round, deep red colour, firm, weight (140-150 g), contain high pulp. It is suitable for long distance transportation and processing. It gets ready for first harvest in 65-70 days after transplanting. The average yield is 90-100 t/ha.

Swarna Samridhi (ICAR-RCER, Ranchi)

It is an indeterminate F_1 hybrid, developed through heterosis breeding. Fruits are borne in cluster (8-10), round, red, weight (40-50 g), firm, TSS 5.0-6.0°Brix, and contain high pulp. It is resistant to bacterial wilt and early blight disease. It gets ready for first harvest in 55-60 days after sowing. The average yield is 100-105 t/ha.

Swarna Sampada (ICAR-RCER, Ranchi)

It is a determinate F_1 hybrid, developed through heterosis breeding. The fruits are Round, red, weight (70-80 g), firm, TSS 4.5-5.0 °Brix, acidity 0.35-0.40 per cent and contain high pulp. It is resistant to bacterial wilt and early blight disease. It becomes ready for first harvest 55-60 days after sowing. The average yield is 100-105 t/ha.

Swarna Vijaya (ICAR-RCER, Ranchi)

It is a determinate F_1 hybrid, developed through heterosis breeding. The fruits are about 80-90 g in weight, round, red, firm with TSS 4-5°Brix and acidity (0.35-0.40 per cent), and borne in cluster of 6-8. It is resistant to bacterial wilt and early blight disease. It gets ready for first harvest in 55-60 days after sowing. The average yield is 100-105 t/ha.

Swarna Deepti (ICAR-RCER, Ranchi)

It is a determinate F_1 hybrid, developed through heterosis breeding. Fruits are round, deep red in colour, firm, weight is 120-130 g with TSS 4-5°Brix and acidity (0.35-0.40 per cent) and borne in cluster of 4-5. It is resistant to bacterial wilt and early blight disease. This variety is suitable for table and processing. Average yield is 100-105 t/ha.

Swarna Anmol (ICAR-RCER, Ranchi)

It is a F_1 hybrid suited for protected cultivation. Swarna Anmol was developed by heterosis breeding. The fruits are round, red, with light green shoulder and borne in clusters of 5-6. The average fruit weight is 60-70 g, TSS 4.5-5.0 °Brix, acidity 0.30-0.35 per cent and ascorbic acid 40-42 mg/100 g fruit. It is indeterminate in growth habit (5.0-5.5 m), high yielding (160-180 t/ha), resistant to bacterial wilt and can be cultivated round the year.

Bidhan Tomato Hybrid-4 (BCTH-4) (Bidhan Chandra Krishi Viswavidyalaya, West Bengal)

It is a semi-determinate F_1 hybrid, developed by crossing between BCT-15 and BCT-59. Fruits are flattish-round, medium size (75 g), TSS 4.0° Brix; lycopene 3.13 mg/100 g fresh; total carotenoids 4.27 mg/100 g fresh; acidity 0.51 per cent ; vitamin C 40.5 mg/100 g, highly tolerant to ToLCV. Average yield is 50-55 t/ha. It is recommended for cultivation in West Bengal and Assam.

Bidhan Tomato Hybrid-62 (BCTH-62) (Bidhan Chandra Krishi Viswavidyalaya, West Bengal)

It is a semi-determinate F_1 hybrid, developed by crossing of BCT-15 x BCT-59. Fruits are oblong-heart shaped, with green shoulder, weighing 110-125 g with 4 locules and light ribbing on fruit surface. The fruits show somewhat slow ripening character and suitable for long distance transportation. Processing quality of the fruits is marked by its thick pericarp (0.62 cm), high TSS (4.4° Brix), lycopene (4.91 mg/100 g fresh) and total carotenoid (7.26 mg/100 g fresh) contents. Less severity to early blight and tomato leaf curl virus disease has been observed. Average yield is 50-55 t/ha. It is recommended for cultivation in West Bengal and Assam.

6.3.2 Private Sector Hybrids

Indo-American Hybrid Seeds (IAHS) at Bangalore developed and commercially released several tomato hybrids which were widely accepted by the growers in different parts of the country.

Rashmi - round, firm fruits, suitable for processing, tolerant to fusarium and verticillium wilt and alternaria stem canker, **Rupali** - fruits round, widely adapted, can set fruit at high temperature, **IAHS-88.2** -large fruits, tolerant to fusarium and verticillium wilt, root-knot nematodes, recommended for fresh market, **Naveen**-medium-sized, round fruits with good colour, high yielding, fresh market and processing cultivars, **IAHS-88.3**, **Karnataka**, **Rajani**, and **Vaishali** were also promising hybrids; **Suttongram Wonder, Maruraj and Sweetheart** were high yielding hybrids from Suttons India. **Gourav, Jagat, Tomard, Soma, NTH-337, HTH-245 and NTH-416** are some other hybrids marketed by different seed companies in India.

Some recent tomato hybrids marketed by different companies throughout India are as follows:

Abhilash, Saksham (Seminis)

NS-526 (Namdhari Seeds)

Lakshmi-5005 (BASF)

Rocky, To-1458, To-1156, Heemshikhar (summer hybrid), Heemsona (summer hybrid), To-3038, Amlik, Hard Rock (Syngenta)

Abhimanyu (UPL-Advanta Seeds)

Samriddhi F_1, Siddhi F_1 (East West Seed)

Tanuja (summer hybrid) (Known-you Seed)

Pan-1286 (Pan Seeds)

US-440 (US Agri Seeds)

Jigar-204 (Acsen Hyveg Pvt. Ltd.)

Ajeet Icon, Ajeet Paramveer (Ajeet Seeds Pvt. Ltd.)

6.4.0 Cultivars in some Countries of the World

Some important characters with tomato cultivars grown in different parts of the world were reported. Michalik *et al.* (1986) in Poland assessed few cultivars and lines for processing characters and reported that **Pulawski Przemyslowy** and line **PH 1703** produced high fruit extract (5.1 per cent), closely followed by **Przemyslowy**

Cultivars of Tomato

Arka Alok

Arka Abha

Kashi Sharad

Kashi Hemant

Kashi Amrit

Kashi Vishesh

Swarna Lalima

Swarna Naveen

Cultivars of Tomato

Arka Saurabh

Arka Vikas

Arka Ashish

Arka Ahuti

Kashi Amul

Kashi Adarsh

Bidhan Purple

Pusa Gaurav

Pusa Rohini

Cultivars of Tomato

Kashi Abhiman

COTH-1

Arka Shrestha

Arka Abhijit

Pusa Hybrid-2

Pusa Hybrid-4

NS 501

US 618

Hybrids of Tomato

Arka Rakshak  **Arka Samrat**  **Arka Ananya**

Abhinava: ToLCV resistant hybrid

Swarna Vaibhav **Swarna Samriddhi** **Swarna Sampada**

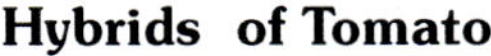

Hybrids of Tomato

Bidhan Tomato Hybrid-4 (BCTH-4) and Bidhan Tomato Hybrid-62 (BCTH-62)

PAN-1286 (Left), and Heemshikhar (Right)

Arka Ralshak

IW (5.0 per cent). Grand had the highest fruit sugar content (2.7 per cent) and best sugar to acid ratio (7.8) was found in cultivars 01355 and VF92-12.

Prima, a processing tomato cultivar of Hungary, was grown either by direct sowing or transplanting and could be harvested once-over or consecutively. The cultivar had a high dry matter content, very good processing quality and was suitable for both commercial and garden cultivation (Gyori, 1986). Derived as a single plant selection from 5737M. **Processor 278**, a small, determinate processing variety was suitable for high density planting adapted well to machine harvesting. It exhibits fruit cracking resistance along with resistance to *Sclerotium rolfsii, Fusarium oxysporum* f.sp. *lycopersici* and *Stemphylium solani* (Leeper *et al.*, 1993).

In the Netherlands, round-fruited tomato cultivars, **Counter** and **Selection 661** were reported as promising for interplanting (Cools and Jansen, 1986); **Cherita** remains the principal cherry tomato cultivar (Buitelaar, 1987). It was also suggested that the composition and consumers' acceptability of cherry tomatoes like **Sugar Lump, Gardeners Delight** and **84 FLI** were better than normal cultivars in UK. Some other established cherry tomato cultivars are **Prairie Schooner** and **Elitsa** (Summers, 1996; Stancheva, 1997).

Poncu (1987) from Romania reported that **Vidra**, a high early-season cultivar was suitable for canning or paste production while **Romex 554,** a mid-late cultivar of determinate growth habit was also suitable for tomato paste.

Verlioka an early determinate hybrid was reported from Moscow. It sets fruit under low temperature and high humidity conditions, ripened uniformly and had multiple disease resistant characters (Gavrish and Sysina, 1990).

Thirteen new hybrids of processing tomato were evaluated in the western Po Valley, Italy in 1999. Highest commercial fruit yields were obtained from mid-early hybrids **Snob** (83.5 t/ha), **ES 96/100** (81 t/ha) and **Platone** (79.1 t/ha) and the early hybrid **EPT x 690** (79.5 t/ha) (Siviero *et al.*, 2000).

In Belgium, Adriaenssens (1997) studied storage life duration and found that among long life types, cultivar **Atletico, DRW3503** and **DRW400** showed great promise.

Mountain Delight: a mid-season, determinate F_1 hybrid from U.S.A. was suitable for fresh market. Fruits were deep oblate to globate, resistant to fruit cracking and few races of fusarium wilt (Gardner, 1990). Cultivar **Monte Verde** is an inbred line derived from the cross Flora-Dade x Summit in USA. It is a fresh market cultivar with a jointless pedicel for easier harvest and improved marketable yield for mature-green production (Gardner, 1999a). Hybrid **Mountain Fresh** was derived from the cross NC 84173 PVP x NC 109 in USA. It is a fresh market hybrid with a desirable combination of disease resistance (*Verticillium* and *Fusarium oxysporum*) and plant and fruit characters (Gardner, 1999b). **Mountain Gold** is a determinate type, had multiple resistance to fruit angularity, puffiness, *Fusarium oxysporum* f. sp. *lycopersici* and *Verticillium dahliae* (Gardner, 1993).

An export grade new cultivar **Red Rose** was selected through six generations from hybrid fruit acquired in Hongkong. It has excellent taste and resistance to bursting and cracking *i.e.*, excellent transport quality.

Some other tomato cultivars grown in different parts of the world are:

Burpee's Big Boy: Indeterminate hybrid of USA origin; beefsteak type, fruits are 10 oz. of red colour.

Hillbilly: Heirloom tomato, beefsteak type, open pollinated, indeterminate, potato leaf type, fruits are yellow with red streaks of 12oz.

Brandywine: Heirloom tomato, beefsteak type, open pollinated of USA origin, leaves of potato leaf type, pink coloured fruits of 12-16 oz. for fresh purpose.

Italian Giant Beefsteak: Heirloom tomato, beefsteak type, indeterminate, open pollinated of Italian origin, red coloured fruits of 16 oz. for fresh purpose.

White Wonder: Heirloom tomato, beefsteak type, indeterminate growth habit, open pollinated, having white flesh and white skin and mild sweet taste.

Striped Stuffer: Speciality tomato, indeterminate plants with bell shaped orange coloured fruits of about 5oz.

Hard rock: Garden tomatoes, determinate growth habit, leaves narrow, fruits blocky of 3oz, red coloured, firm with acidic flavour, having resistance to Verticillium wilt, Fusarium wilt Race 1, Nematodes

Cordova: It is a hybrid of determinate growth habit, blocky fruits of 4oz. red coloured, having resistance to Verticillium wilt, Fusarium wilt Race 1 and Race 2, Nematodes.

Indigo Rose: Speciality tomato rich in anthocyanin, indeterminate growth habit, fruits round of 1oz., flesh colour red, skin colour black.

Isis Candy: Open pollinated heirloom tomato, cherry type, fruits 1.5oz., very sweet flavour, red with yellow bicoloured fruits.

7.0 Soil and Climate

7.1.0 Soil

Tomato is grown in many types of soils from sandy to heavy clay. A well-drained, fairly light fertile loam with a fair moisture holding capacity is ideal for growing a good crop of tomato. Good texture of the soil is of primary importance. Even poor and medium quality land produces good early crop, if managed properly. Tomato crop prefers a soil reaction ranging from pH 6.0 to 7.0. In acidic soils, liming was beneficial.

7.2.0 Climate

Tomato being a warm season crop, required relatively long season to produce a profitable crop. It is highly susceptible to frost. High temperature and high humidity favour development of foliar diseases. Dry wind results in dropping of flowers. Effects of acid rain and dustfall on tomato yield was studied by Yang and Mu (1998) and they reported that in the calcareous purple soil the acid rain increased yield by 21.6 per cent and resulted in Hg accumulation in the leaves as compared to the control. In the acid soil with a relatively low fertility, dust fall increased yield by 19.2 per cent and Hg concentration in the roots. Perez *et al.* (1997) studied the physiological and biochemical changes in tomato cultivars under flooding and reported significant increase in peroxidase activity in roots of stressed plants of cv. Inca9 (1) and explained in relation to its greater tolerance to flooding. Dell' Amico *et al.* (1996) observed that tomato plants exposed to average dew fall of 1.79 m^3/ ha for 38 days had better growth and yield than plants protected from dew fall by polyethylene covers.

Environmental factors such as light intensity, temperature and moisture markedly influenced the process of fruit set in tomato and subsequent fruit development and yield (Calvert, 1959; Sugiyama, 1962; Coyne, 1968; Kemp, 1968; Abdel-Hafeez, 1971; Kinet, 1977). Liptay *et al.* (1998) in the analysis of vapour pressure deficit (VPD) identified 2 years *i.e.*, 1994 and 1995 as having high frequency of very low daily vapour pressure deficits during the growing season. The site of experiment where low VPD was recorded had a normal, increased yield response to fertigation treatment. Leonardi *et al.* (2000) reported that fruit growth and transpiration rates varied greatly during daylight hours and these variations were enhanced under high VPD conditions. The increase in VPD produced a significant reduction in fruit fresh weight and in fruit water content and an increase in soluble solids, while fruit dry weight was not affected. The study showed that during summer, as VPD increased, effects both on tomato growth and quality characteristics were observed. Fruit quality and shelf-life were adversely affected by high relative humidity (Bakker, 1989). Therefore, moisture loss was considered useful.

7.2.1 Light

Nightingale *et al.* (1928) concluded that the length of daily illumination along with varying nitrate supply markedly affected the reproduction in tomato. Plants grown under 7 hours day with nitrate in the nutrient solution were highly vegetative and unfruitful, while at 14 hours day, they set fruit abundantly. Under short day, in absence of nitrate, flowers formed abundantly but did not set fruits. Under long day condition, without nitrate there was no flower and fruit set. Watt (1931) showed that temperature had an effect similar to photoperiod. He reported that increase in temperature from15.5 to 21.1°C produced results similar to that obtained by increasing the length of photoperiod. Hamner *et al.* (1945) showed that light intensity was a very important factor for ascorbic acid content in tomato fruits. Under low light intensity, vitamin C was much lower than the higher intensity. Ellis and Hamner (1943) showed that carotene content of tomato fruit was less when produced in green house than in the open.

Janes and McAvoy (1991) demonstrated in green house culture that harvesting date was inversely correlated with the amount of light received by the crop during the seedling phase, while fruit weight was positively correlated with light received after transplanting. Thus, supplementary lighting was most effective in promoting fruit development when provided between 15 and 45 days after flowering. Baevre (1990) reported that reducing the incoming light by 30 and 60 per cent resulted in significant reduction in the number of flowers, percentage of fruit set and yield. The percentage of hollow fruits (puffiness) classified by external appearance, was not affected by light reduction, but the volume of hollow spaces in each fruit increased with light reduction.

7.2.2 Temperature

Temperature represented a working thermal condition essential for metabolic and cellular function of the tomato plants. The rates of many physiological processes of the tomato plants were determined by temperature which was manifested in subsequent morphological changes. The vegetative and reproductive responses were strongly modified by temperature alone or in conjunction with other environmental factors like light, mineral nutrients and moisture (Abdalla and Verkerk, 1970; Watenbe *et al.*, 1977). The temperature which favours high percentage of germination within a prescribed time differed between cultivars. Suitable temperature for germination of seeds was 24°C (Brooks, 1969) and 18.5 to 21.0°C (Wittwer, 1960). Many tomato cultivars germinated rapidly at a temperature range of 26 to 32°C (Thompson, 1974). Singer (1990) noticed that the minimum, optimum and maximum temperatures for seed germination were 10, 25 and 30°C, respectively. Liptay and Moore (1989) stated that the adverse effects on seed vigour at high temperature could be overcome by healthy foliage cover, or artificial or by partial shading of the plants with a fast growing vegetation sown on both sides of the rows 3 weeks after transplanting tomato in the field. It was noted that the seed vigour from fruits that matured at temperature more than or equal to 35°C decreased as the temperature increased.

The number of flowers per inflorescence in plants grown at 14°C was higher than those grown at 20°C. Flowering intensity of tomato reduced at relatively higher temperatures. A day temperature of 26°C and night temperature of 22°C under phytotron condition significantly reduced flower numbers in the first inflorescence (Lewis, 1953). Studies in phytotron on tomato plants revealed that reduction in night temperature markedly increased the biological and commercial yields (Petruchuk, 1987). Results from growth chamber studies in which tomato plants were grown at 3 air temperatures (high, normal and low) indicated that small differences in air temperature altered the time to first flower (Sauser *et al.*, 1998).

Temperature had a marked effect on fruit set in tomato. Although tomato grew under a wide range of temperature, fruit set was usually poor either at low or high temperatures. Watt (1931) found that fruit set was greater at 24°C than at 16°C. Went (1944) and Went and Cosper (1945) reported that night temperature was the critical factor in setting of fruit. The optimal range was 15 to 20°C. Fruits failed to set at 13°C or below. When the average maximum day temperature was above 32°C and average minimum night temperature was about 21°C, fruit set was low (Moore and Thomas, 1952). High light intensity accompanied by high temperature was harmful for fruit set. Reduction in the light intensity by shading increased fruit set significantly at high temperature; when temperature was optimum, decrease in light intensity had no beneficial effect. Schaible (1962) also noticed that fruit set was generally poor when night temperature exceeded 23°C. Howlett (1962) pointed out that both night and day temperatures were important factors limiting fruit set. Sugiyama (1962) attributed poor fruit set and yield to high day temperature. Ranges between 30 and 45°C caused marked reduction in fruit set (Iwahori and Takahashi, 1963). El-Abd *et al.* (1986) reported that high or low day/night temperatures (35/24°, 10/15°C) had marked deleterious effects on flowering and fruit set compared to the controls at 25/18°C. High temperature increased flower abscission, reduced fruit set and pollen grain viability and stimulated heterostylosis, whereas low temperature increased flower size and reduced pollen grain viability. Kim (1988) studied the effects of night temperatures, during the early stages on growth, flowering and fruiting of tomatoes and reported that plants kept at 15°C showed moderate growth but the best fruit quality while plants kept at 25°C or 5°C produced small malformed fruits with lower yields. The earliest flower bud differentiation and flowering took place at 15 °C and latest at 5°C; fruitset was minimum at 25°C and maximum at 15°C. Growth rate was the highest at 25°C initially but slowed at 15°C later. Also net assimilation rate was maximum at 15°C initially and then at 5°C. Varietal differences in the ability to set fruit at high temperature were reported by Went (1944) and Daubney (1961). Curme (1962) and Kemp (1965) also found differences among cultivars in their ability to set fruit at low temperature. Rick and Dempsey (1969) showed that it was possible to select cultivars capable of setting fruit under relatively high temperature condition. The reason of low fruit set was studied and they found that style elongation relative to the antheridial cone was one of the characters that reduced fruit set at high temperature. High temperature also produced high abscisic acid level in tomato which favoured premature senescence and abscission of reproductive organ (McGlasson and Adata, 1977). Endogenous ABA concentration increased at high and decreased at low temperatures, whereas IAA concentration was reduced both at high and low temperatures. Application of IAA, ACC (aminocyclopropane-1-carboxylicacid), GA_3, ABA or BA to flowers hastened fruit set in almost all cases but the subsequent development of the flower cluster varied in relation to the growth substance applied (El-Abd *et al.*, 1986). It has been reported that some cultivars, capable of setting fruit under low night temperature, were also able to set fruit at relatively high night temperature. On the other hand, some cultivars which set fruit under low night temperature were unfruitful at high night temperature (Aung, 1978).

Various attempts have been made by using hormones and other chemicals to counteract the adverse effect of unfavourable temperature on tomato fruitset (Odland *et al.*, 1950; Marre and Murneek, 1953; and Rappaport, 1960). In Tucson, Arizona, U.S.A., Satti and Oebker (1986) observed that application of BA (10 ppm) alone or BA in combination with gibberellin (GA_4+7) substantially increased fruit set and yield under high temperature (day/night temperatures of 32-42/15-21°C). Arora *et al.* (1981) studied the performance of tomato cultivars under high temperature condition and found H.S.102 tolerant to high temperature upto 35°C. At Taiwan, parthenocarpic fruit in tomato cultivar Severianin from U.S.S.R. was found to develop at 20°C. The increase in percentage of parthenocarpy in 3 parthenocarpic tomato lines-Severianin, PSET-1 and RP 75/59 at 30° and 36°C was regarded to be due to reduced pollen germination and tube growth (Splittstoesser, 1988). Charles and Harris (1972) conducted extensive investigations on fruit set at high or low temperature. They observed that low fruit set at 10 and 13°C was primarily due to poor pollen viability and germination. Maisonneve and Philouze (1982) studied the effect of low night temperature on some cultivars of tomato and observed that low temperature (below 10°C) affected pollen formation about 2 weeks before anthesis. The cultivars which were susceptible to cold, produced only 20-25 per cent pollen grain as compared to optimum temperature. Gautam *et al.* (1981) evaluated tomato genotypes for fruit set under low temperature condition. They found highest fruit set in Ostankinsk cultivar. In Sudan, under arid conditions artificial vibration of inflorescences using a battery-operated vibrator increased fruit set in cultivars Chico 111, Pearson A-1, Moneymaker and Strain B by 10-17 per cent (Satti, 1986).

The effect of temperature on fruit development was investigated by Charles and Harris (1972). They observed that fruit size was reduced in all the cultivars at 18.3°C and at lower temperature the size became larger. Thompson *et al.* (1960) found that temperature had pronounced effect on many parameters of physical quality and major reduction in quality resulted from exposure to low temperature. Ryloski (1979) noted deformation in tomato fruit to be a common occurrence in winter due to low temperature at flower development stage. Sun *et al.* (1999) determined proline bound and free water as indicators of cold resistance and observed that cold resistance was best in the seedlings grown at the lowest temperature and the least in seedlings grown at the highest temperature. The degree of cold resistance, however, declined rapidly when seedlings were held at 25°C.

High temperature injuries were also studied by Iwahori (1965). Under normal temperature condition, meiosis of macro- and micro-spore mother cells was found to take place about 8-9 days before anthesis.

Elevated ozone levels produced wilting and drying, pink colouration, necrosis, pale green and chlorotic leaves with leaf area and dry matter production of plants being severely affected (Seresinhe, 1998). Ozone stress caused a mean increase in stomatal conductance by 18 per cent while in drought stress, it decreased by 22 per cent and both ozone and drought stress significantly suppressed the biomass and fruit yield (Hassan *et al.*, 1999). Studies on recovery of tomato plants from ozone injury revealed extensive foliar injury after three exposures at the vegetative stage; there was no significant decrease in fruit yield or change in fruit quality at the final harvest which was in contrast to plants exposed to increasing doses of three concentrations at flowering showing progressively reduced yield and fruit titrable acidity (Tenga *et al.*, 1990). Varshney and Rout (1998) in their effort to protect tomato plants from ozone injury concluded that soil drenching of 400 ppm of ethylene diurea (EDU) at 12day intervals gave some protection to cv. Pusa Ruby under Delhi condition.

8.0 Cultivation

8.1.0 Seed Treatment

According to Kunavin (1982) seed treatment with hydrogen peroxide (0.2to 0.6 percent) for 2 to 4 days at 21 °C gave an early emergence of seedlings by 7 to 8 days and also yielded 22.8 per cent higher yield. Heat treatment at 60°C for 6 hours also proved effective in increasing the yield by 33.3 per cent (Vikmane *et al.*, 1981). Treatment at high temperature retarded germination in cv. Intermesh but did not affect the total percentage of emergence while germination was faster with 3 per cent KNO_3 at 12°C (Petrikova, 1991). The effect of temperature on tomato seed germination was investigated by Sakthivel and Thamburaj (1998). They found that germination in cv. Naveen was completely inhibited at low (2-5°C) as well as at high temperature (40°C). The germination was, however, the highest at 25°C and 30°C, *i.e.*, 82 and 86 per cent, respectively. It is reported that the seeds stored at 5 °C in poly ethylene bags and at -20°C in aluminium foil laminated pouches maintained a high percentage of germination and vigour after 5 years of storage.

Odell and Cantliffe (1987) from USA observed that priming seeds in an aerated solution of 1.5 per cent K_2PO_4 + 1 per cent KNO_3 and then drying to 6 per cent moisture content before storage (at 10°C and 45 per cent RH) accelerated germination at 25 and 35°C. Prolonged storage (up to 19 months) of primed seeds reduced germination rates, especially at a high temperature (35°C). Bano *et al.* (1987) concluded that seeds pre-soaked in choline chloride, especially at 2 mM concentration germinated better. In general, priming with KNO_3 or PEG 8000 solution consistently improved germination at 35°C and priming for 6-8 days gave better results than priming for 4-5 days (Cantliffe and Abebe, 1994). The effects of priming and ageing on resistance to deterioration was studied by Argerich *et al.* (1989) who stated that enhancement of seed germination rates by priming treatments simultaneously lowered the resistance of seeds to deterioration and that primed seeds were considered as vigorous seeds with a reduced storage life. Germination was found to be adversely affected by GA_3 especially in relation to light/dark conditions (Schwabe, 1986). Pre-germination seed treatment with 2-(3,4-dichlorophenoxy) triethylamine (10 ppm) doubled the dry weight of leaves, stems, roots and yield at harvest (Keithly *et al.*, 1990). Studies on effect of seed treatment with growth regulators revealed that the highest fruit yield (379.5 q/ha) was obtained with cv. Pusa Early Dwarf and that the yield increased by 12, 8, 6 and 5 per cent over control by application of NAA, GA_3,IAA and indolepropionic acid, respectively (Mozarkar *et al.*, 1991). The preservation of seeds of cv. Arka Saurabh by immersing in $CaCl_2$, glycerol or hexylene glycol was a simple and cost-effective method for improving short- or medium-term longevity at ambient temperatures (Pandey, 1995). Abscisic acid and auxins inhibited root elongation while gibberllic acid promoted root elongation. Adenine, a precursor of cytokinins promoted root elongation (Bhore *et al.*, 1999).

Mikhov and V'lchev (1995) suggested that covering with plastic sheet improved the microclimate for germination. Effect of gamma irradiation on seed germination and vigour was investigated by Nargis *et al.* (1998) who concluded that seeds of cv. PKM-1, exposed to 10 and 20 kr (lower dose) increased the speed of germination and vigour index while treatment with vitamin C or E increased the seedling length even at 30 kr. According to Neelamathi and Karivaratharaju (1998) the seed germination and seedling vigour in cv. CO 1 were the best in seeds exposed to red or white light and the lowest in those exposed to blue or violet light.

De Souza *et al.* (2005) studied the effects of pre-sowing magnetic treatments on the growth and yield of tomatoes (cv. Vyta), cultivated late in the season by exposing tomato seeds either to a 120 mT dynamic magnetic field (induced by an experimental electromagnet) for 10 min or to a 80 mT field for 5 min, while non-treated seeds were used as controls. They reported that the magnetically treated seeds produced plants with significantly more fruits (17.9-21.3 per cent), with a significantly greater mean fruit weight (22.3-25.5 per cent), and with a greater fruit yield per plant (47.3-51.7 per cent) and per area (48.6-50.8 per cent) as compared to the control plants. Besides, the root length, fresh and dry root weight, stem length, fresh and dry stem weight, leaf area and foliage dry weight in nursery stage of the magnetically treated plants were higher in nursery stage and during the vegetative stage, the leaf, stem and root relative growth rates of plants derived from magnetically-treated seeds were greater than those shown by control plants, as well as the relative growth rate of the fruits belonging to the 'magnetically treated plants' was greater than that of control plant fruits during fruiting stage.

Ahammed *et al.* (2012) reported that brassinosteroid treatment of tomato seeds could improve germination and early seedling development under environmental pollutant stress condition (exerted by three-ring polycyclic aromatic hydrocarbon namely phenanthrene (PHE)) and seed treatment with 1.0 nM 24-epibrassinolide (EBR) was most effective followed by 100 and 0.01 nM for the improvement of germination and seedling growth under PHE stress in tomato.

Rao *et al.* (1996) observed that 0.2 per cent serpentine, alkaloids isolated from the roots of *Catharanthus roseus* and 0.2 per cent citronellol, essential oil of *Pelargonium graveolens* significantly increased growth of tomato and decreased *M. incognita* (root knot nematode) infestation, while at 0.5 per cent both significantly reduced the root-knot nematode infestation, but did not increase the growth of tomato.

Lewis and Miller (2005) demonstrated the efficacy of hot water treatment of tomato seeds and reported that seedlings from hot water treated seeds did not show incidence of bacterial diseases *viz.*, bacterial speck, spot and canker in both green house and open field, and also produced bigger fruits and higher yield as compared to plants developed from non-treated seeds.

8.2.0 Raising of Seedlings

About 400 to 500 g seeds are normally needed for raising crop in one hectare. The seed quantity varies with season and region of cultivation.

In North India for spring-summer crop, the seeds are sown in late November and transplanted in second fortnight of January. In area where frost does not occur, only one sowing is done in July-August with transplanting in August-September. For autumn crop in North India, seeds are sown in July-August and transplanted in August-September. In hilly areas, seeds are sown in March-April with transplanting in April-May.

Nandpuri *et al.* (1971) found that seedlings transplanted in early February gave higher yield than those transplanted in March under North Indian conditions. Time of sowing mainly depends upon climatic conditions. In Eastern India, especially in West Bengal, autumn-winter crop is transplanted in September-October and spring summer crop is transplanted in late December.

Hossain *et al.* (2014) studied the effect of sowing three sowing dates, 1st October, 15th October and 30th October on tomato yield and attributing traits in Bangladesh and reported that flowering and fruit harvest were early for the 1st October planted crop and it also produced the highest number of fruits per plant and highest yield while yield was lowest in the 30th October crop.

Abdalbagi *et al.* (2010) in Sudan also studied the effect of three sowing dates, *viz.*, 1st October, 15th October and 1st November, on tomato productivity and found that sowing at October 1st increased the productivity of tomato as it positively influenced the plant height, stem diameter, days to 50 per cent flowering, fruit yield and marketable yield.

In Brazil, Pereira and Reisser (1998) working in plastic green houses reported that earliest sowing date resulted in the highest early and total yield. Sowing on or around 15th December was recommended for good out-of season yields. Tillmann *et al.* (1994) also from Brazil stated that the rate of emergence and the final stand were the best with seeds when sown at a depth of 1.5cm. Seeds are sown in rows in raised nursery beds of 60 to 75 cm wide. The seedlings were protected against heavy rains and extreme weather condition. For early germination during winter season, seed beds were covered with polythene film. Ali and El-Said (1987) suggested that virus-free tomato transplants could be produced in tunnels 1 m high covered with fine white cloth before being transplanted to plastic houses. Highest quality tomato transplants were obtained by El-Beltagy *et al.* (1986) using growing media of peat moss + sand or peat moss + vermiculite. Tesi and Tosi (1986) from Italy reported best growth in 6-week-old seedlings with peat substrates having a high content of organic matter and a high water retaining capacity. Growth was the maximum when seedlings were supplied with NPK fertigation (20-20-20 at 1.5g/litre) three times a week. According to Silva and Giorgi (1992), the best transplants were obtained on substrates containing 50 or 75 per cent FYM while high rates of poultry manure retarded transplant development and the poorest results were obtained with pine needles. Seedling age had a greater significance so far as the establishment as well as the yield of tomato crop was concerned. Rahman *et al.* (1994) while experimenting with cv. Manik concluded that plants grown from younger seedlings flowered and were ready for harvest earlier than those grown from older seedlings. The number of fruits/plant and average fruit weight were maximum when seedlings were 40 days old at transplanting. According to Lee and Kim (1999) the highest marketable yields (1699-1849 g/plant) were, however, obtained from 60 days old seedlings of cv. Momotaro T-93.

In China, plantlets in large number were obtained from axillary and adventitious buds of shoot tips of tomato cultures on MS media with various growth regulators. They formed vigorous root systems when transferred onto White's medium with NAA or IAA followed by hardening and transplanting in the field (Deng *et al.*, 1988).

8.2.1 Hardening of Seedlings

For better survival in the field, hardening of the seedlings was suggested. The hardened plants withstood the extremes of temperature. Jemakinde and Onuveme (1980) studied the heat hardening (40°C for 6 hours daily for 5 to 12 days) for drought conditions. Plants were allowed to nearly wilt for 2 to 3 days before watering and this practice was repeated 2-3 times. The percentage of seedling survival in field

Filling of Pot Mixture and Seed Sowing in Tomato

Mixing Coconut Husks (Left) with Soil in a 1: 2 Ratio for Raising Seedlings.

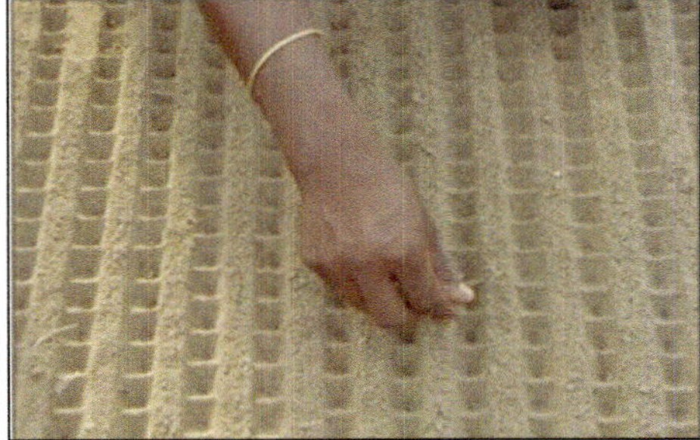

Making holes 0.5 inches deep prepared by wooden teeth bar in rows 2 inches apart (Left) for seedling raising and one seed per hole is sown (Right) at the commercial vegetable nursery.

A dibbling board can be used to make 0.5 cm deep holes (Left); Sow 2 seeds per hole and cover the seeds with fine compost (Right).

Cover the Seeds with Potting Mixture (Left) or Fine Compost (Right) after Sowing.

Different Stages of Seedling Production of Tomato

Soil Solarisation of Nursery Bed (Left) and Preparation of Poly Tunnel (Right).

Sowing of Seeds in Line.

Seedling Raising in Raised Bed (Left) and Plugtray (Right).

Raising of Seedlings under Insect Proof Net to Protect against White Fly.

under stress condition is increased by the hardening treatment. The fruit weight and number of' fruits per plant also increased. Pardossi *et al.* (1988) in their trials with cv. Marmande obtained better hardened plants by withholding water. Reduction of watering increased the chilling resistance without affecting the growth rate in the nursery and fruit yield of mature plants in a cold green house.

Ghanbari and Sayyari (2018) subjected tomato seedlings (cv. C.H Falat) to pretreatment of 0, 10 and 20 per cent Polyethylene glycol (PEG) for 7 days followed by subjecting them to chilling stress at 3 °C for 6 days and 6 h per day. The results showed that PEG-induced drought improved growth rate of tomato seedlings subjected to chilling stress and enhanced their antioxidant enzyme activity, abscisic acid (ABA), anthocyanin accumulation, potassium (K^+) and proline content compared with the control (0 per cent PEG) at the end of chilling stress period. PEG pretreatment provided significant protection against chilling stress and reduced poly phenol oxidase (PPO) activity, electrolyte leakage (EL) and H_2O_2 content in root and leaf of chilled seedlings. Moreover, when PEG pretreatment was applied with chilling stress, alleviation was observed in the growth impeding and a decrease in chilling symptoms as compared to control seedlings. The highest chilling tolerance was induced by application 20 per cent of PEG. Thus PEG-induced drought could be effectively used to protect tomato seedlings from the adverse effects of low temperatures stress.

Hardening of transplants was also recorded by treatment with CCC (chlormequat) at 0.2 per cent on seedlings at 4 to 5-leaf stage or by removing 30 per cent of cover for 15 days (Shulgina and Andrienko, 1980). Moussa and Doss (1986) in Egypt studied the methods for hardening of seedlings and stated that soaking of seedlings for 24 hours in glucose solution (20 g/l) or glucose + kinetin (30 ppm) gave the best results with regard to early and total yields. Seedlings treated at the 4 to 5-leaf stage with hydrel (ethephon) at 0.015 per cent, dextrel at 0.02 per cent, sulpha at 0.001 per cent, ethrel (ethephon) at 0. 0 1 per cent or CCC (chlormequat) at 0.1 per cent produced short well-rooted transplants, and showed advanced flowering and improved fruitset (Zhukova *et al.*, 1987). In Iran, Ramin (1998) reported that ethephon and CCC applied at 50, 100 and 200 mg/l in cv. Early Urbana and Red Clode seedlings significantly decreased the stem length but increased stem diameter. Also ethephon at 50 and100 mg/l produced transplants with more adventitious roots; 200 mg/l produced plants with massive adventitious roots, but CCC had little effect on root initiation in both the cultivars. Foliar application of sucrose (10 per cent with small quantity of detergent), three times at one day interval before planting markedly reduced post-planting shock and mortality. Role of physical impedance and brushing for mechanical conditioning was elaborately studied by a number of workers and it was found that such treatments produced transplants with desirable qualities without adverse effects on field performance (Samimy, 1993; Gamer and Bjorkman, 1997, 1999). Gamer and Bjorkman (1999) observed that mechanically conditioned transplants of processing tomatoes resumed growth after transplant shock as quickly as did untreated plants, and subsequent canopy development was also equal, besides yield was not reduced by mechanical conditioning. There were no adverse effects on field performance of either processing or fresh-market tomatoes as a result of reducing stem elongation by mechanical conditioning before transplanting. Thinning of seedlings before they become crowded also helped in hardening of the seedlings.

Tomato seed priming with shikimic acid (60 ppm) has been found to improve plant growth, and fruit quality of tomato plants grown under field conditions through enhancement of photosynthetic pigments, transpiration rate, and the fruits were found to contain higher amounts of vitamin C, lycopene and total carotenoid and sugars in fruits, apart from increased nutrient contents of tomato plants, and also increased the fruit yield.

Seedlings are ready for transplanting in 4 to 5 weeks time. Harbaoni and Yellod (1980) observed that seedlings 5 mm in diameter were better for field setting. The seedlings of about 15 cm in height were more suitable for transplanting.

8.3.0 Direct Sowing

Direct seeding was compared with transplanting by Green (1980). He reported that transplanting showed earlier ripening and gave higher yield. Similar effects were observed by Singh *et al.* (1995b).

El-Hassan (1986) from Sudan reported that there were no significant differences in the marketable yields or percentage of defective fruits from the direct sown and transplanted crop, but direct sown plants gave larger fruits than transplanted ones. From Cuba, it was reported that for direct sowing, seeds pelleted with a mineral/ organic mixture (containing peat and superphosphate) at a depth of 2 cm and covered with peat at sowing produced high yield (23.2 t/ha) (Martinez, 1987). For direct sowing, the seeds of cv. Campbell soaked for 2 hours in water at 60°C prior to sowing resulted in faster germination and earlier flowering and fruiting. Surlekov *et al.* (1995) reported that direct sowing of tomato seeds between 25th February and 1st March advanced the fruit earliness by 8 per cent as compared to the controls sown between 20th March and10th April, without reducing the yield. Yields were the highest in plants raised from treated seeds and sown at 1 or 2 cm depth (Martinez and Careaga, 1987). The complex fertilizer (10-30-00) when applied at 16 ml/m row gave the best result in case of direct sown tomato cv. Petomech but became phytotoxic at 32 ml (Churata-Masca *et al.*, 1987).

Maboko and Du Plooy (2014) reported that direct seeding of hydroponically grown tomatoes reduced the cost of cultivation. No significant differences in marketable and total yield were observed in direct sown and transplanted crop, and direct sown crop was early.

However, the studies of Javanmardi and Moradiani (2017) showed that direct seeded tomato crop produced lower yield and late harvest.

8.4.0 Grafting

Extensive researches were carried out with grafting mechanisms, choice of rootstock, time of grafting and other related aspects in tomato with the aim to enhance the yield as well as fruit characteristics. This method was effectively exploited in maintaining vigorous seedlings to attain desirable characters.

Grafting of Tomato

Grafting of Tomato to Combat Soil Borne Pathogens.

Grafted Tomato Plants with Silicone Clip.

Grafted Tomato Production in Soil Borne Pathogen Prone Areas.

There are a variety of techniques to graft tomatoes (Lee, 2003). Tube grafting has been adopted as the commercial method for tomato grafting worldwide as a result of high-throughput production capability and typical success rates of 85 per cent to 90 per cent (Oda, 1995). This technique is carried out when the rootstock and scion seedlings are very small, and the two are attached with a small silicon tube or clip (Rivard and Louws, 2006).

Potato (cv. Kufri Sindhuri) stocks were cleft-grafted with tomato (cv. Pusa Ruby) scions to investigate the effect of time of grafting on yield and the highest yield of tomato (870 g/plant) and that of potato tubers (120 g/plant) were obtained from the plants grafted in December (Maharana *et al.*, 1995).

Nobuoka *et al.* (1996) studied the conditions required to promote the healing of graft unions in tomato seedlings. They attributed the decline of transpiration rate not only to stomatal closure but also to the decrease in cuticular transpiration rate by increasing water potential in scions. Therefore, high humidity and low light intensity prevented wilting of scions, favouring healing of the graft union.

Oda *et al.* (1994) designed a new grafting plate for simultaneous horizontal grafting of multiple young tomato plants. By applying pressure between the cut surfaces of the scions and rootstocks, they increased the survival rate from 56 to 93 per cent but the productivity of plants produced by horizontal grafting tended to be lower than those produced by conventional cleft grafting. Jaksch and Kell (1997) while comparing the yields between grafted and non-grafted tomato plants obtained the highest early yields (first 4 weeks) by top grafts and total yields of class I fruits.

Grafting has been important in tomato production to manage bacterial wilt incidence in solanaceous crops (Peregrine and Binahmad, 1982). CRA 66 is a breeding line that has been identified as a resistant rootstock genotype for grafted tomato production in Germany and India (Grimault *et al.*, 1994; Tikoo *et al.*, 1979). Hawaiian lines (Hawaii 7996 to 7998) have been identified as suitable candidates for resistance to bacterial wilt (Oda, 1999; Tresky and Walz, 1997; Nakaho *et al.*, 2004).

Rivard and Louws (2008) evaluated different grafted tomato plants against soil borne diseases in organic production system and reported that scion grafted on to Hawaii 7996 and CRA 66 rootstocks showed no symptoms of bacterial wilt while the scion grafted on to rootstock Maxifort showed no symptoms of Fusarium wilt. Thus, grafting could be an important component of integrated disease and pest management system, particularly, in organic production system.

8.5.0 Soil Preparation and Transplanting

Tomato is planted in well pulverized and levelled field. Tomato is normally planted in raised beds of 60 to 75 cm width.

Sharma and Tiwari (1996) studied the performance of cv. Pusa Ruby under Pantnagar conditions (subtropical) and found that individual fruit weight and diameter and the total and the marketable yields were the greatest (48.94 g, 15.45 cm and 62.89 kg/21.6 m^2 plot and 56.81 kg/plot, respectively) when transplanting was done on 5th March; the number of days from sowing to harvesting was the least (143 days) when transplanting was done on 25th March. In Jammu however,

Pusa Ruby produced the maximum yield following transplanting on 30th January (Samnotra *et al.*, 1998). Under Rahuri condition (arid tropical climate), the highest yield of 0.911kg/plant in cv. Pusa Ruby was obtained when planting was done on 15th November and the fruit size declined as planting was delayed while the fruit size increased on staked plants (Kadam and Deore, 1990). According to Singh *et al.* (1995a) cv. Pant Bahar produced the highest fruit yield at Pantnagar when planted on ridges.

Transplanting is usually done during late afternoon and the seedlings are placed on side of the beds. This provides ample moisture for the plants to survive.

In two years trial, Hanna *et al.* (1997) observed that transplanting tomatoes to a depth of 15.0 cm significantly increased marketable yield in both years and total yield in one year. Mean fruit mass was not influenced by transplant depth, but plant dry mass was significantly increased by deeper transplanting in one year. Root zone temperature was significantly lower in 15 cm than at the 7.5 cm depth in both years.

8.6.0 Spacing

Many experiments were conducted on the effect of spacing on growth and fruiting of tomato. Most of the workers observed that proper spacing resulted in higher yield. Roy *et al.* (1954) observed that a spacing of 4 x 2 ft (1.2 x 0.6 m) produced the highest yield and increased spacing resulted in slight increase in fruit weight and number of fruits per plant. Samundar (1964) recorded the highest yield with a row spacing of 3 ft (0.94 m). These results were confirmed by Thoke (1966). He found that a spacing of 2 ft (0.6 m) gave 3400 lbs higher total yield per acre with higher vitamin C content than in wider spacings. Roy (1966) also reported that vitamin C content increased with increased spacing of 3 x 2 ft (0.94 x 0.6 m). Sethi and Kuksal (1977) obtained the highest yield in two mutants of Chaubattia Laghu, when planted at 45 cm between plants and rows. The cultivar CO 1 gave the highest yield at closer spacing of 75 x 45 cm at Coimbatore in India (Kamlanathan and Thamburaj, 1970). Higher acid or ascorbic acid content was associated with closer spacing of 60 x 45 cm and higher TSS and juice content with wider spacing of 60 x 75 cm (Pandita and Bhatnagar, 1981). Joshi *et al.* (1980) obtained the highest yield of 826 q/ha with a spacing of 30 x 30 cm in cultivar H.S. 102. A field experiment in Himachal Pradesh (hill region) showed maximum fruit yields of 38.38 t/ha and 42.92 t/ha with a spacing of 60 x 30 cm for cv. Roma and 60 x 45 cm for Sioux, respectively (Jamwal *et al.*, 1995). Effect of high density planting with different cultivars was also studied by Verma and Sharma (1981). They found that Russian- 700 was the most suitable cultivar due to its short ripening period. Shukla *et al.* (1999) from Himachal Pradesh concluded from their evaluation trials that high density planting in May increased the yields in cvs. Karnataka, Vaishali, MTH-3 and Mangala.

In Greece, Tsambanakis (1984) observed with cultivars Dembo and Dombito that a spacing of 0.8 m between rows and 0.25 m within the rows produced the highest yield of 15.8 kg/m^2 but fruit quality was better at a wider spacing of 0.8 m between rows and 0.5 m within rows. Hamar and Olabi (1985) from Hungary reported that quantitative distribution of yield was best for cultivars of indeterminate growth (Gala and Primset) with 4 plants/m^2 and for determinate cultivars (Luca and Balca) with 5 plants/m^2. In Egypt, Midan *et al.* (1985) obtained the highest yield with cv. Pritchard at a spacing of 30 cm between hills with 2 plants/hill (6.67plants/m^2). Moccia and Katcherian (1999) from Argentina reported that yield per unit area increased linearly with the increase in planting density, while yield per plant decreased. Increasing plant density increased the number of trusses, flowers and fruits harvested, but decreased the average fruit weight.

Ara *et al.* (2007) observed that a spacing of 60 x 50 cm gave the higher marketable yield (82.39 t/ha) while a spacing of 60 x 40 cm produced lower marketable yield (68.32 t/ha) in tomato indeterminate variety BARI Tomato-4 at Bangladesh.

Awas *et al.* (2010) in Ethiopia reported that a spacing of 40 x 30 cm gave higher marketable yield (607.9 q/ha) followed by 60 x 30 cm and 50 x 30 cm (570.4 q/ha and 568 q/ha, respectively), while 100 x 30 cm inter -row spacing showed the least marketable yield (475.85 q/ha) when planted in double row system.

For the cherry tomato group grown under field conditions, there is no reported information on the most adequate spacings (de Azevedo *et al.*, 2010). In green house, Alvarenga (2004) suggested spacings of 1.0 to 1.1 m between rows and 0.5 to 0.7 m between plants.

De Azevedo *et al.* (2010) reported that a spacing of 0.6 x 1.5 m resulted in lower number of fruits in cherry tomatoes (hybrid 'Super-Sweet' and a local self-pollinated variety 'Perinha') when compared to the 0.4 x 1.5 m, however, producing fruits with higher individual average mass, which resulted in similar final yield, but these yields were higher than that obtained from the 0.8 x 1.5 m spacing.

8.7.0 Irrigation

Tomato being very sensitive to soil water regime, the water table plays a major role in meeting the irrigational needs of the crop. To strike a balance between yield and quality, a water table depth of 0.6 and 0.8 m was necessary for tomato production on sandy loam soils (Dodds *et al.*, 1997). It is necessary to maintain even moisture supply. Over watering is harmful, so also insufficient irrigation. Singh and Joon (1950) studied the effect of irrigation on root development and yield of tomato. In cultivar Best of All, half an inch of irrigation every fortnight caused the maximum penetration of roots which resulted in the highest yield. Quality of fruits was also influenced by moisture supply. Flooding was harmful for tomato crop (Kuo *et al.*, 1982). Salinity had detrimental effect on flowering, fruit set, yield and fruit quality (Rajasekharan and Shanmuzadin, 1981). But Caruso and Postiglione (1993) from Italy reported that the marketable yields were highest when 0.125 per cent saline water was used for irrigation. Fruit TSS ranged from 5-7 per cent with fresh water to 6.3 per cent with 0.5 per cent saline water, the sugar/TSS ratio was the highest with 0.125 per cent saline water which decreased with the increasing in salinity. The fruit quality based on chemical constituents (mainly glucose contents), taste and shelf life improved when the crop was irrigated with water containing salts (Plant, 1997). Production and accumulation of dry weight in fruits also increased. However, the fruit size decreased with salinity although it had much less effect on fruit number. Flower development in tomato cv. Roma was arrested and flowers at all stages dropped and fruit growth ceased when water was withheld for a few days. When water stress was relieved, small fruits developed but partially developed fruits did not recover fully. A relatively dry period followed by sudden heavy watering during fruiting period caused cracking of the fruits (Chiaranda *et al.*, 1982). Excessive water stress in tomato after pinching resulted in skin browning, sour flavour and hard texture (Ito and Kawai, 1994).

Drip Irrigation in Tomato.

Quality Production under Drip System of Irrigation.

In New South Wales, Australia, Hermus (1986) obtained the highest yield from trickle-irrigation with 100 per cent evapotranspiration soil water replacement. In France, Branthome and Ple (1987) conducted trials with a single-harvest cv. Petogro and concluded that single-harvest cultivars needed a water supply close to the evapotranspiration (ETP) level during flowering and fruit setting. It was suggested to reduce 50 per cent of ETP level during the last phase (fruit development and ripening) until irrigation was stopped at the 30-50 per cent maturity stage. According to Alvino *et al.* (1986) high irrigation levels did not increase the yield but depressed quality; about 30 cm of depth of irrigation water gave the best yield and quality. Rudich *et al.* (1977) reported that frequent irrigation at the time of fruit expansion had an adverse effect on quality and reduced TSS, viscosity and increased acidity. Similar observations were also made by Vittum and Flocker (1967).

Irrigation with cold water at 5-15°C reduced stem length of tomato by 28-32 per cent in comparison with water at room temperature. These results demonstrated that use of cold water provided an effective means for improving the quality of plug-sown seedlings (Chen *et al.*, 1999).

Hanna *et al.* (1997) reported that morning irrigation increased the marketable and total yields and mean fruit mass in both years (1994 and 1995) of their investigation and plant dry mass in 1995, while morning irrigation provided only a minor reduction in soil temperature.

Zhai *et al.* (2010) recorded highest fruit yield when the tomato plots in green house in Jiangsu Province, China were irrigated at an interval of 3 days with 1.5 dm irrigation water through drip irrigation system while the lowest fruit yield was recorded when irrigated at an interval of 4 days with 1.2 dm water. This suggested that the fruit yield would be further lower if the irrigation interval was set longer and the irrigation water amount was set lower at the same time.

Li *et al.* (2020) studied the impact of subsurface drip irrigation and furrow irrigation on tomato root distribution and rhizosphere in an organic system and observed that biomass allocation to different aboveground plant tissues was significantly affected. In comparison to furrow irrigation, marketable tomato yield decreased by 28.3 per cent in the subsurface drip irrigation system, while the biomass of non-marketable fruits and stems and leaves increased. Subsurface drip irrigation substantially reduced the amount of irrigation water applied, leading to 66.4 per cent increase in irrigation water productivity compared to the furrow irrigation system.

8.8.0 Manuring and Fertilization

8.8.1 Macronutrients

In high yielding cultivars, 65 to 75 per cent of total dry matter was accumulated in the fruit end. Extensive investigations on nutrient uptake were carried out on tomato. Yawalkar *et al.* (1962) reported that a crop of tomato yielding 378 q/ha removed 104 kg N, 22 kg P and 141 kg K. On the other hand, More (1978) observed that for producing a tonne of tomato fruits, 2.90 kg N, 0.40 kg P, 4.00 kg K and 0.45 kg Mg were utilized. Cholakov (1987) from Bulgaria reported that tomato plants yielding 65.01 to 83.24 t/ha removed about 187.8 to 200.2 kg N, 58.4 to 70.6 kg P_2O_5 and 226.6 to 262.0 kg K_2O from meadow soil. He (1988) also observed that for producing 1 tonne fruits, average uptakes of N, P and K were 3.82 kg, 1.01 kg and 3.65 kg, respectively. The cultivars Arka Alok, Pusa Ruby and Arka Saurabh were found to be efficient P absorbers and were recommended for soils with marginal phosphorus content (Anjaneyulu *et al.*, 1998). As regards method of fertilization, dry matter accumulation was always higher in fertigated plants than those conventionally irrigated and fertilized. Fruit production and quality parameters such as size, firmness and soluble sugars were consistently enhanced by fertigation (Alcantar *et al.*, 1999).

Fifth leaf from the top was usually used as the index organ of the plant for estimation of the nutrient status of the crop. The different estimates of N status corresponding to maximum yield were: 5.52 per cent (Ward, 1963) and 4.5-5.1 per cent (Adams *et al.*, 1978a). Roorda Van Eysinga (1971) considered 0.44 per cent P in the leaf dry matter to be satisfactory for optimum tomato yield. According to Adams *et al.* (1978b) potassium concentration of more than 6 per cent would ensure good fruit quality and flavour.

Nitrogen, phosphorus and potash nutrition influenced the tomato crop in many ways. Nitrogen promoted vegetative growth, flower and fruit set but excess of nitrogen delayed maturity and decreased fruit size. Rajgopal and Rao (1974) found that nitrogen deficient plants had lower level of endogenous auxins and reduced gibberellin activity. Severity of root rot disease decreased by addition of nitrogen but blossom-end rot appeared to increase with increasing levels of nitrogen. Adequate nitrogen increased fruit quality, fruit size, keeping quality, colour and taste and acidity also increased by excess doses of nitrogen (Sharma, 1971).

The effect of NO^-_3/NH_4^+ ratio on the mineral nutrition of tomato was studied by many workers. High ammonium reduced leaf and root growth and developed symptoms of ammonium toxicity. Under these conditions, uptake of Ca and Mg was depressed. In trials with cultivars Marmande and Precorder F_1, Graifenberg *et al.* (1986) reported highest yield in both the cultivars by using urea + NO_3 + NH_4 at 2: 1: 1. Borowski (1995) concluded that feeding with NH_4-N had the greatest inhibitory effect on plant growth in sand, to a lesser extent in loam and none in the garden soil as compared to NO_3-N. In the garden soil, NH_4-N underwent complete nitrification in comparison with 33 and 9 per cent nitrification in loam and sand, respectively.

High level of phosphorus throughout root zone was essential for rapid root development and good utilization of water and other nutrients by the plants. Phosphorus had pronounced effect on the developing flowers. It also affected cytokinin activity. Phosphorus in combination with N and K improved peel colour, taste, hardiness and vitamin C content, and hastened maturity (Su, 1974).

Potassium deficient plants usually had dark green leaves with short internodes; in extreme cases the leaves showed marginal necrosis. It also affected fruit size and fruit quality. Potassium tended to increase fruit size and rectified many disorders like blotchy ripening, vascular browning, white wall and grey wall (Hayslip and Iley, 1967). Ozbun *et al.* (1967) showed that blotchy ripening was caused by low doses of K. Increased level of K helped in decreasing incidence of irregular shaped fruits (Gallagher, 1972). According to Trudel and Ozbun (1970), potassium played an important role in the process of tomato fruit pigmentation; it increased carotenoids particularly lycopene and decreased chlorophyll. Potassium level also exerts a strong influence on acid metabolism in tomato fruits. Fruits from potash deficient plants were low in total solids, sugar, acid, carotene and lycopene contents. Chiesa *et al.* (1997) suggested that dry matter and soluble-solids increased as K rate increased and titrable acidity was lower in fruits harvested at the firm red stage and decreased during storage although there were no significant differences in pH and titrable acidity at different K rates. It was also noted that application of K fertilizer increased plant resistance to tomato early blight and decreased the disease incidence and disease index (Huang *et al.*, 1999). The work of Ortas (2013) indicated that increase in N and K uptake and constant level of P contributes to increased yield of fresh market tomato and high fruit quality is influenced by high levels of potassium.

From various parts of the world, many reports on optimum N, P, K fertilizer combinations for yield maximization were available. Some of the effective rates of N, P, K were: NPK (2: 2: 3) at 350 kg/ha (Borna, 1969); at 90: 120: 60 kg/ha (Mursanov and Otreper, 1977) and 240: 240: 120 kg/ha (Belichki, 1983). Adams (1986) thoroughly reviewed the nutritional aspects in tomato. Kamaev (1996) recommended application of NPK fertilizers at 90-180 kg N, 120 kg P, 120 kg K and organic matter 40-60 t/ha annually.

Sharma and Singh (1963) recommended application of 60 lb N per acre as ammonium sulphate, 60 lb P as super phosphate with a basal dose of 10 cart load FYM. Seth and Dhandar (1970) obtained the highest yield of fruits and seeds in Pusa Ruby when 90 kg nitrogen/ha was used. Doses of nitrogen, however, did not influence the quality of tomato (Dastane *et al.*, 1963). Maschio *et al.* (1982) observed the highest yield with a basal dose of 60 kg N per hectare without any top dressing, but Sharma *et al.* (1978) noted that application of 60 kg N/ha in split doses (half before and half after transplanting) gave the highest yield in tomato cultivar Angurlata. Effects of different N (100 or 150 kg/ha), P (0, 50, 100 or 150 kg/ha) and K (0, 50 or 100 kg/ha) ratios on yield and quality of tomato were studied by Pansare *et al.* (1994). They concluded that yields in tomato cv. Pusa Ruby under Rahuri conditions were highest with a 3: 1: 3 N: P: K ratio while TSS was the highest at 3: 1: 2 ratio.

As regards application of organic matter, Chinaswamy (1967) found best results with a combination of farmyard manure (FYM) and groundnut cake in an organic mixture. Marked interaction between N and P was observed by Kuksal *et al.* (1977). A combination of 120 kg N along with either 60 kg or 90 kg P, or 90 kg N + 60 kg P per ha gave the best results. When P was applied alone, it resulted in early yield. However, the most economical application was 110 kg N, 60 kg P and 95 kg K/ha (Bandopadhyay *et al.*, 1972). In Punjab, Randhawa and Nandpuri (1977) recorded high yield in cultivar Punjab Chuhhara with the application of 100 kg N, 260 kg each of P and K per hectare. Trials conducted in Orissa (cv. Punjab Chhuhara) however, recorded the highest fruit weight (58.0 g) and total yield (341.9 q/ha) with the application of 90 kg N/ha and 40 kg K/ha (Hossain and Mohanty, 1999). Mortley *et al.* (1991) found that total fruit yield by band application of 112 kg/ha N was 24 per cent higher than that for the same rate as broad cast and similar to the yield from 224 kg/ha as broadcast.

Sharma and Mann (1971, 1972) reported that N and P application increased the juice and ascorbic acid content of the fruit. The application of N decreased total sugar and number of ripe fruits; enhanced plant height, branching and number of leaves per branch. The percentage of fruit set and number of fruits per plant were increased by phosphatic fertilizers. Leaf P at the pre-flowering stage was found to be positively correlated with application of P up to 100 kg/ha. Pandey *et al.* (1996) reported overall yield of 499.50 q/ha from Acc.90 when applied with 80 kg/ha each of N and P. Application of 200 kg N/ha resulted in significantly greater fruit size, fruit weight and plant height and a significant improvement in plant height, fruit size and total fruit yield was observed with the application of phosphorus from 60 to 180 kg P_2O_5/ha (Sharma *et al.*, 1999).

Studies were conducted at Quincy and Gainesville, Florida by Locascio *et al.* (1997) to compare the effects of time of N and K application on production of drip irrigated tomato. At both sites, total N was applied at 196 kg/ha. At Quincy, no K was applied but at Gainesville 112 kg/ha was applied. The sources of N and K nutrients were ammonium nitrate and potassium chloride, respectively. At Quincy, the early marketable yields, total yields of extra large fruit and total marketable yields in 1994 were similar with 0 per cent, 40 per cent and 100 per cent preplant applied fertilizer. However, the total marketable yield at Quincy in 1993 was significantly higher with 100 per cent preplant compared with no preplant or 40 per cent preplant applied fertilizer. At Gainesville, on a coarse-textured soil, application of 40 per cent and 100 per cent preplant fertilizer resulted in higher yields of early extra large and total early marketable fruit than with no preplant fertilizer. Yields of total extra large and total marketable fruit were significantly higher with 40 per cent of the fertilizer applied preplant than with no preplant fertilizer or with 100 per cent preplant fertilization. Total seasonal extra large and marketable fruit at Gainesville were affected by treatments in the following order: 40 per cent > 100 per cent > 0 per cent.

Yaffa *et al.* (2000) examined the effects of tillage [no-till (NT), chisel ploughing (CP), and mouldboard ploughing (MP)], cover cropping [hairy vetch (*Vicia villosa* Roth) vs. winter weeds], N fertilization (0, 90 and 180 kg/ha N), and date of sampling on tomato yield, N uptake and soil inorganic N in a Norfolk sand loam, in Fort Valley, USA for 2 years. Yield was greater with CP and MP than with NT in 1996 and was greater with 90 and 180 than with 0 kg/ha (Nin, 1996 and 1997). Similarly, aboveground tomato biomass (dry weight of stems + leaves + fruits) and N uptake was greater with CP and MP than with NT from 40 to 118 days after transplanting (DAT) in 1996; greater with hairy vetch than with winter weeds at 82 DAT in 1997; and greater with 90 or 180 than with 0 kg/ha Nat 97 DAT in 1996 and at 82 DAT in1997. They suggested that minimum tillage, such as CP and 90 kg/ ha N can better sustain tomato yield and reduce potentials for soil erosion and N leaching than can conventional tillage, such as MP, and 180 kg/ha N, respectively. Because of increased vegetative cover in the winter, followed by increased mulch and soil N in the summer, hairy vetch can reduce the potential for soil erosion and the amount of N fertilization required for tomato

Foliar spray of urea was beneficial. The size and number of fruits per plant and yield per acre were found to increase with increased uptake of nitrogen. This was confirmed by Tan *et al.* (1999) who also suggested urea to be the most appropriate foliar N sources due to its low ability to injure foliage because of the rapid absorption and translocation, fast assimilation and the wide and suitable range of solution pH. Hence, foliar application of N was usually recommended on the lower leaf surface. Foliar spray of 7500 ppm of urea had better effect on vegetative growth but delayed maturity (Bajpai and Bajpai, 1963). Choudhury and De (1972, 1975) observed that half soil and half foliar applications of 120 kg N and 60 kg P/ha with basal dressing of 60 kg K were better than basal application alone. Foliar applications of N and P were on an average 2.02 and 1.63 times more efficient than corresponding soil application. Verma *et al.* (1970) applied 1.25 per cent urea as foliar spray and noticed improved growth and yield in cv. Marglobe. Treatment with phosphorus generally resulted in increased amino acid concentration. Sharma and Shukla (1972) and Dhuria and Shukla (1972) reported that 125 kg N and 120 kg P/ha applied 3/4th in soil and 1/4th as foliar spray gave the maximum yield.

The effects of potassium fertilization and season on quality of fresh market tomato were reported by Picha and Hall (1982). Reducing sugar, dry weight and pH were higher in autumn crop with increasing K application upto 744 kg/ha. Potassium nutrition during plant development was studied by Widder and Lorenz (1982). The maximum rate of potassium accumulation occurred in plants at 90 days which coincided with the period of maximum K accumulation in the fruit. About 70 per cent of K accumulated in the plant was utilized by the fruits.

In field trials, Hartz *et al.* (1999) determined the relationship between soil K availability and fruit colour disorders. Soil application of either K or gypsum reduced yellow shoulder (YS) and total colour disorders. Multiple foliar K applications were effective in reducing fruit colour disorders at only one of two sites. In no field trial did K application improve yield, soluble solids, or juice colour.

A number of physiological and ripening disorders were reported to be related with inadequate supply of calcium. Blossom-end rot (BER) was closely associated with Ca-deficiency of the fruit. The percentage of yield affected by blossom-end rot was the lowest with the composted manure treatment given at the start of cultivation (Salminen, 1997). Application of $CaSO_4$ at 1.0 or 2.0 kg/m^2 reduced the incidence of blossom-end rot by altering the conditions which were conducive to this disorder (Anon., 1998). Pasture (1971) found fruit cracking due to calcium deficiency. Magnesium deficiency caused vascular browning and it was corrected by using 1.5 per cent $MgSO_4$ as foliar spray (Winsor *et al.*, 1965).

Application of N: P: K @ 180: 80: 75 (kg/ha) improved the growth, yield and quality of tomato cultivar Azad T6 under Lucknow condition (Kumar *et al.*, 2013).

Increasing the levels of N and K over the control increased cherry tomato height and yield, and reduced the days to flowering and maturity, but when the levels of N and K were increased beyond (100 kg/ha), reduction in tomato height and yield sets in and tomato flowering and maturity periods are shortened (Ddamulira *et al.*, 2019). Based on their study, application of 100, 60 and 100 kg per hectare of N, P and K, respectively could be adopted by cherry tomato farmers to increase their yields in central Uganda where the study was conducted.

The combined use of 150 kg/ha of NPK 15-15-15 fertilizer + poultry manure @10 t/ha increased tomato yield and could be recommended for sustainable productivity in south-western Nigeria (Adekiya and Agbede, 2009).

8.8.2 Micronutrients

Among the micronutrients, boron, zinc and manganese deserve special attention. A positive correlation was observed between boron and flower number, number of aborted flower and fruit weight. Boron deficiency caused reduction in root growth, swollen hypocotyl and cotyledons, brittle leaves and necrosis of shoot apex. Dunger (1962) reported that cracking of tomato fruits was also due to boron deficiency. He suggested that both seedlings and plants should be sprayed with 0.3 to 0.4 per cent boron solution. First spray in nursery bed before planting and second spray 3 to 4 weeks after transplanting was suggested. Datta (1961) recommended soil application of 20 to 30 kg of borax per hectare. Like boron, zinc deficiency was found to occur in high pH soils. Zinc deficiency was aggravated by phosphorus application. El-Sherif *et al.* (1993) concluded that mixing of $ZnSO_4$ with the soil prior to sowing was the most effective method and that foliar application could be used as an emergency treatment if and when Zn deficiency was detected in a crop. Mallick and Muthukrishnan (1980) studied the effect of micronutrients on the quality of tomato fruits. They found that micronutrients, especially zinc at high doses (500 ppm) increased ascorbic acid content by 2 to 7 mg/100 g. Maximum increase of 22 per cent was recorded in sugar/acid ratio using manganese at 10 ppm. The yield was highest when zinc was applied at the rate of 10 ppm.

Multiple regression analysis suggested that yield was consistently depressed by deficiencies of boron and copper, and responses to iron, and manganese were also found. Increasing levels of Fe in the nutrient solution increased the chlorophyll content of leaves and Fe uptake by plants (More and Shinde, 1991).

Chand and Prasad (2017) observed that combination of boron, copper sulphate and zinc sulphate @ 100 ppm each with recommended dose of NPK enhanced the yield, fruit diameter, dry matter content and TSS (Brix), while ascorbic acid content of fruits could be enhanced with 250 ppm boron, copper sulphate, zinc sulphate at 10 and 20 days after transplanting.

Spray of boron @ 100 ppm + Zn @ 100 ppm + Cu @ 100 increased the vegetative growth of the tomato plants enhancing the plant height, plant spread and number of branches per plant while fruit formation was earlier with spray of boron @ 250 ppm + Zn @ 250 ppm + Cu @ 250 ppm (Chand *et al.*, 2018).

Sivaiah (2012) studied the effects of boron, zinc, molybdenum, copper, iron, manganese and mixture of all on tomato varieties Utkal Kumari and Utkal Raja. All the micronutrients except manganese (@ 50 ppm) were applied @100 ppm in three sprays at an interval of 10 days starting from 30 days after transplanting. There was improved vegetative growth, enhanced number of tender and mature fruits per plant as well as higher fruit and seed yield and quality and the best results were obtained from the micronutrients mixture.

Sultana *et al.* (2016) recommended integrated use of foliar application of micronutrients and soil application of macronutrients is recommended to enhance tomato yield and suggested that foliar applications of micronutrients were more suitable than the soil application, due to the rapid overcoming on deficient, easy to use, reducing the toxicity caused by accumulation and preventing of elements stabilization in the soil. They found that foliar application of Zn (0 05 per cent) in form of zinc sulphate ($ZnSO_4.7H_2O$) + B (0.03 per cent)) in the form of boric acid (H_3BO_3) at three different stages of plant growth, *viz.*, before flower initiation, after fruit set when it becomes approximately marble sized and at 20 days interval of second spray improved the tomato yield and its attributing traits.

Two foliar sprays of mixture of $FeSO_4$ @ 0.2 per cent, Calcium nitrate @ 0.2 per cent, Boron @ 0.1 per cent and $ZnSO_4$ @ 0.2 per cent at 15 days and 21 days after transplanting in tomato hybrid Arka Rakshak had a very positive influence on plant growth and increased yield (Dixit *et al.*, 2018).

Foliar application of 12.5 ppm $ZnSO_4$ + 12.5 ppm H_3BO_3 was effective in increasing yield, fruit size and earliness and reduced the percentage of diseased plants in summer season in tomato cv. BARI hybrid tomato-4 in Bangladesh (Ali *et al.*, 2015).

Aminochelates are relatively new fertilizer formulations with improved efficiency for agricultural applications. Many plant responses to these amino acid-based fertilizers have not yet been well studied. Souri *et al.* (2017) evaluated the effects of soil (six split applications at a final volume of 5 mL) and foliar application (six applications at 0.2 per cent concentration) of a commercial aminochelate (liquid fertilizer consisting of 2.0 per cent N-amino acid, 2.5 per cent Zn, 2.0 per cent Fe, 1.5 per cent Mn and 0.4 per cent Cu) on growth, yield, and quality traits in tomato under field conditions in lime soil. Plant growth, yield, and quality parameters significantly increased in response to foliar application, and to lesser extent soil application of aminochelate fertilizer. Plants treated with foliar application of aminochelate had a significantly higher chlorophyll index, number of lateral shoots, shoot dry weight, number of fruits, plant yield, and leaf N concentration compared to the control. Soil application of aminochelate resulted in significantly higher fruit vitamin C content and total soluble solids content versus the control. However, for most traits, there was no significant difference between soil application of aminochelate versus chemical fertilizer (N, P, K mixture). Nevertheless, the results indicate that the application of aminochelate especially foliar application, is a preferred method for fertilization under lime soil conditions to improve plant growth, productivity, and quality.

8.8.3 Biofertilizers

Biofertilizers not only played an important role in maintaining good health of the plants, but also served as a natural source of plant nutrient to increase productivity. Application of plant growth promoting rhizobia increases absorption of nutrients due to the bacterium's ability to produce hormones such as IAA (Wibowo, 2007). The increased of nutrient absorption due to PGPR application such as *Pseudomonas* sp. and *Bacillus* sp. bacteria can release organic acids such as formic acid, acetate, and lactate which are able to solubilize phosphate forms which in normal condition is difficult to dissolve and become available for the plant and the organic acids released by these bacteria can form chelation with cations binding P in the soil (Han and Lee, 2005). The beneficial effects of biofertilizers were manifold. Sasai (1991) reported that VAM fungi promoted phosphate uptake in low phosphate soils during the early stages of plant growth. According to Terry *et al.* (1995) *Azospirillum brasilense* strain UAP-154 + 30 kg N/ha applied to the seedbed + 45 kg N/ha at transplanting resulted in a higher fruit yield, and a fertilizer saving of 45 kg N/ha. Mishra (1996) reported that pre-establishment of *Glomus fasciculatum* in soil increased plant growth, decreased the size and number of galls (caused by *Meloidogyne incognita*) and improved uptake of NPK. In another experiment Hameedunnisa (1998) observed that seeds of cv. Arka Vikas, treated with suspension culture of *Azotobacter* and subsequent application of 150 kg N/ha produced maximum fruit yield. Barbolina and Arkhipchenko (1999) reported that application of Bamil (produced from wastes obtained from an animal breeding farm) in combination with mineral fertilizers increased the yield and stimulated growth as compared to tomatoes receiving mineral fertilizer alone. Bamil application also suppressed nematode production and infection as compared to control.

Altuhaish *et al.* (2014) reported that biofertilizers had positive effect on vegetative growth as well as yield of tomatoes in Indonesia which could be attributed to the enhanced nutrient absorbing capacity favoured by the biofertilizers.

Dinu *et al.* (2015) reported that foliar application of a combination of biofertilizers humic acids extracted from brown coal in combination with polyphenolic extract from the seeds of *Vitis vinifera* and boron enhanced the yield and had positive effects on fruit quality in three tomato cultivars (Antalya, Cemil and Lorely) grown in a protected space.

Angadi *et al.* (2017) observed that application of *Azospirillum* and phosphate solubilizing bacteria 2.5 kg/ha each along with recommended fertilizer dose (60: 50: 30 NPK/ha) improved vegetative growth, earliness and yield and its attributing traits in tomato.

Seed and foliar application of microalgae *Acutodesmus dimorphus* aqueous extracts in lower concentrations (50 per cent) had a major positive effect on seed germination and plant growth tomato seedlings treated with dry *A. dimorphus* biomass (50 g/plant) 22 days prior to transplant resulted in the greatest plant growth, the highest number of floral development, and early fruit setting, which could theoretically mean a greater crop yield (Garcia-Gonzalez and Sommerfeld, 2016).

Azolla biofertilizer extract having a concentration of 20 per cent expressed the highest effect on all vegetative growth parameters and also average fruit weight, fruit numbers and ultimately yield (Hanafy and Gehan, 2018).

Application of dried cyanobacteria *Anabaena* strain E-3 as biofertilizer showed superior fresh shoot and root weight, dry shoot, root weight, leaf area, and number of branches, which was at par with the treatment with liquid cyanobacteria biofertilizer inoculation and gave better yield also in Ethiopia (Abuye and Achamo, 2016).

Spore-forming industrial *Bacillus amyloliquefaciens* FZB42 (trade name Bioeffector, produced by Rhizovital) as single inoculums on tomato var. Mobil improved soil microbial counts, phosphorus availability, yield and fruit quality, such as total soluble solids (TSS) content and sugars (glucose, fructose) (Biró *et al.*, 2018).

Application of biofertilizers both *Azotobacter* and *Mycorrhiza* in combination with compost 15 t ha^{-1} showed a positive effect on tomato growth components including plant height, shoot fresh and dry weight, and root fresh and dry weight (Marajan *et al.*, 2017).

8.9.0 Weed Control

Frequent shallow cultivation was recommended to control weeds. According to Lugo *et al.* (1988) critical period of weed competition in Puerto Rico was between 24 and 36 days after transplanting. For controlling the weeds, the role of weedicides in tomato crop was studied by many workers. Leela *et al.* (1972) recommended Tillam 6E for monocot weeds and TOK E 25 for dicot weeds. Tok E 25 was, however, banned in many countries. Lasso at 1. 0 kg per hectare was the most effective in controlling weeds. *Orobranche ramosa*, a notorious saprophytic dicot weed of tomato could be controlled by the application of chlorsulfuron (pendimethalin) at 2.44 g a.i. when thoroughly mixed with the soil prior to transplanting (Qasem, 1998). Sandyan and Banerjee (1981) observed Basalin at the rate of l.25 l/ha as most effective in controlling weeds. Metribuzin at the rate of 1.0 kg/ha was found effective in transplanted crop of tomato. About 0.56 kg was applied as preplant incorporation and rest of the quantity was applied as post-planting treatment (Sanok *et al.*, 1981; Kalia *et al.*, 1980). According to Sandhu *et al.* (1993), pendimethalin at 0.75 kg, metribuzin at 0.37-0.5 kg, fluchloralin at 1.2 kg or fluchloralin and pendimethalin followed by one hoeing resulted in effective weed control and increased tomato yields. However, Rana and Barevadia (1995) concluded from their trial in Gujarat that the highest fruit yield was obtained with pendimethalin at 1.0 kg/ha along with one hand weeding at 45th day after transplanting. Therefore, in general, herbicide application with one handweeding increased weed control efficacy compared to the herbicide application alone. This was again confirmed by Pandey and Padhiar (1998) by pre-emergence application of pendimethalin at 1 kg/ha followed by one hand weeding at 45th day after transplanting which resulted in a higher fruit yield of tomatoes as compared to either the herbicide application alone or with one handweeding. For directly drilled tomatoes, Dymid at the rate of 10 kg/ha or Devrinol @ 4 kg/ha was the most effective (Acosta, 1981). In direct sown tomatoes in Bulgaria effective weed control was achieved with pre-emergence application of 1.5 kg Devrinol and 0.105 kg Sencor/ha and post-emergence applications of 0.25 kg Fusilade and 0.26-0.35 kg Sencor/ha to the extent of 86.0-91.8 per cent and 85.8-94.4 per cent control of weeds in terms of weed number and weed biomass, respectively. Banded herbicide application resulted in 50 per cent reduction of costs (V'lchev, 1987). From Italy, Pimpini *et al.* (1986) found effective control of weeds by presowing application of 4 kg Diphenamid + 0.4 kg Metribuzin/ha or Diphenamid alone followed by hoeing. Incorporation of Trifluralin at 0.5 kg and Napropamide at1.0 kg/ha before sowing followed by colour plastic mulch gave better weed control than clear mulch (El-Sayed and El-Fadaly, 1991).

Pre-transplant application of Fomasafen 840 g a.i. per hectare controlled annual grasses, common purslane, redroot pigweed in tomatoes till 42 DAT (days after transplanting) with minimal crop injury at 7 and 14 DAT in Fremont, Ohio (Mohseni-Moghadam and Doohan, 2016).

In Nigeria, Daramola *et al.* (2020) observed that pre-emergence application of propaben® (metolachor+prometryn) and butachlor, each at 2 kg/ha, plus hand weeding at 40 days after transplanting reduced weed density and biomass, and increased weed control efficacy (57- 64 per cent) less than the weed-free check (100 per cent) and better than two hand weedings (45 per cent), or either herbicide applied alone (34 - 38 per cent). Propaben® or butachlor plus hand weeding increased number of leaves and branches, number of fruit and fruit yield of tomato similar to the weed-free check and better than two hand weedings, or either herbicide applied alone. Although the weed-free check gave similar fruit yield, propaben® or butachlor plus hand weeding was 37- 44 per cent higher in net return. Application of propaben® or butachlor alone gave comparable net returns to two hand weedings and the weed-free check, but had a 61- 88 per cent higher cost-benefit ratio.

The use of cover crops for allelopathic and physical weed control is among the effective strategies to control weeds in organic and conventional vegetable production systems. Cover crops like *V. villosa*, Egyptian clover (*T. alexandrinum* L.), *A. sativa*, *T. meneghinianum*, wheat (*Triticum aestivum* L.), *S. cereale*, *V. sativa*, *L. multiflorum* have been tested for their weed control efficacy in tomato; the cover crops were killed before transplanting tomato seedlings and out of the different cover crops, *L. multiflorum* and *S. cereale* decreased the weed biomass by more than 75 per cent while, *V. sativa* and *V. villosa* were also effective for weed control in tomato (Mennan *et al.*, 2009). *V. villosa* as a cover crop in tomato produced a biomass of 7.49 t/ha, and the mulch from this cover crop decreased the biomass of some weed species like *A. retroflexus*, *D. sanguinalis*, *P. oleracea* by 40 per cent, and their density by >70 per cent in tomato (Campiglia *et al.*, 2010a, 2010b). *S. cereale* was grown as a cover crop and dried and kept as mulch one week before planting tomato, along with additional weed control measures to achieve satisfactory weed control (Caamal-Maldonado *et al.*, 2001). *T. incarnatum* and *V. villosa* were found to control weeds effectively in tomato production but did not improve the tomato yield in a separate study conducted by Teasdale and Abdul-Baki (1998).

8.10.0 Mulching

Mulching with organic mulches such as straw or jute felt or polythene mulches improved yield in tomato as compared to non-mulched crop (Kundu *et al.*, 2019).

Clear plastic mulch gave a higher yield and a lower percentage of defective fruits than either the non-mulched control or straw mulch (El-Hassan, 1986). Schalk and Robbin (1987) reported from USA that aluminium mulches (aluminium and aluminized plastic films) lowered soil temperature, reduced heat stress for young transplants, increased their survival and height and improved yield. Yield of large fruits were enhanced by application of mulch. Aphids were repelled by the aluminium mulches, while fruit injury increased due to attack by tomato pin worm and tomato fruit worm which appeared to be attracted to the reflective surfaces. Castellane *et al.* (1994) also observed that the incidence of thrips and aphids was lower in mulched areas and total fruit yield increased by pest control in spring-summer in Brazil. Mulching with black or transparent polyethylene decreased levels of *Bemisia tabaci* infestation and tomato yellow leaf curl by geminivirus infection below the aluminium-coloured polyethylene sheets (Mauromicale *et al.*, 1996). Mulch colour also affected the yield and growth of plants. Plants grown with red mulch generally had the highest early marketable yields and produced the least amount of foliage. Mulch surface colour influenced the plant microclimate to act through the natural regulatory systems within the growing plant to affect growth

Mulching and Staking of Tomato

Straw Mulching (Left) and Agrimulching (Right) in Tomato.

Black Polythene Mulching in Tomato.

Staking of Tomato Plants.

and fruit production (Decoteau *et al.*, 1989). This was in conformity with the findings of Almasoum (1998).

In Meerut, India during the summer seasons, tomato cv. Pant Bahar were mulched with *kans* (*Saccharum spontaneum*), dry sugarcane leaves, rice straw, rape straw, dry poplar leaves or nothing (control). All mulches effectively reduced weed populations and the percentage of sunscalded fruits, thereby increasing the marketable yield as compared to the control, but the increase was only significant with kans mulch (Singh, 1994).

In general, plastic mulches increased the soil temperature during daytime whereas organic mulches decreased it (Ravinder Kumar and Srivastava, 1997). But use of polyethylene mulch resulted in greater increase in growth and yield of tomato plants than rice straw or water-hyacinth mulch. Black and red polyethylene mulches, however, most effectively increased the yield (Chakraborty and Sadhu, 1994). According to Ravinder Kumar and Srivastava (1998a) the quality parameters such as fruit diameter, volume per fruit, total soluble solids and ascorbic acid were highest in cv. Pant Bahar, under 400 gauge polyethylene with 100 perforations/m^2.

Marketable and total yields of the heat tolerant tomato cultivars tested by Hanna *et al.* (1997) were significantly higher when tomatoes were grown with white than with black-surface polyethylene mulch. Fruits also were heavier in both the years of their study (1994-95) and plant dry mass was higher in 1995 with white mulch. White-surfaced mulch significantly reduced root zone temperature.

Mulching with rice husk increased the yield of tomato by 17.93 per cent over non-mulched crop and also the number of fruits per plants and TSS in BARI Tomato-14 grown in Chittagong, Bangladesh and the yield obtained was 79.74 t/ha while TSS was 6.67°Brix (Tipu *et al.*, 2014)

Use of black plastic mulch combined with appropriate planting time improved earliness and total yields of tomato in areas where the production is constrained by suboptimal temperature conditions (Dhaliwal *et al.*, 2017). In their study conducted in Punjab, black polyethylene mulch increased fruit number under the poly house as well as in the open field, while fruit weight was improved by mulch treatments only in the open field. Highest early, marketable, and total yields were from planting in the poly house in the third week of October using black polyethylene mulch, which could be attributed to increased root zone temperature, air temperature, and net radiation.

Black polythene mulches produced the highest yield in cultivar Rocky in two consecutive years in coastal belt of West Bengal (60.3 t/ha and 58.7 t/ha, respectively) whereas the non-mulched crop produced 38.7 t/ha and 36.0 t/ha; straw and jute felt mulches also gave significantly higher yield than the non-mulched plots (Kundu *et al.*, 2019).

8.11.0 Intercropping

Besides enhancing the net returns from tomato plants, intercropping with suitable species could add to the income from the same plot of land. Some of the recommended intercrops for tomato were cowpea, beans, maize and marigold

(*Tagetes erecta*). According to Subhan (1989) from Indonesia, tomato yield was the best *i.e.*, 543 g fruit/plant with 1 row of beans between the tomato rows planted 2 weeks after the tomatoes. Pilatti *et al.* (1994) reported that in green house tomato yield was significantly greater in alternate rows of capsicum and tomato as well as with two rows of capsicum alternating with one row of tomato. The differences in the yield were due to increased fruit numbers rather than increased fruit size.

According to Zavaleta-Mejia (1995) from Mexico, all treatments intercropped with *Tagetes erecta* independent of planting date and plant spacing, showed a reduction in infection of the tomato roots by the nematode *Nacobbus aberrans* as compared to the control treatment with tomato alone. Tomato foliar as well as fruit damage caused by *Alternaria solani*, was significantly reduced in all the intercropped plots. A significant increase of total (69 per cent) and commercial fruits (120 per cent) per plant was obtained over control when *T. erecta* was planted 30 days before the tomatoes and with a planting distance of 50 cm. It was thus concluded that by manipulating the *Tagetes* planting dates and spacing between plants, it was possible to provide phytosanitary protection to tomato and a significant increase in fruit production. Similarly, maize was also protective for tomato under stress as suggested by Pino *et al.* (1994). Maize cv. Francisco was intercropped with tomato cv. INCA-17 in Cuba and it was noticed that the incidence of pests and diseases was lower in the intercropped tomato crop. They also stated that three rows of tomatoes alternated with one row of maize resulted in the highest tomato yield as well as the cob yields. The economic analysis showed that the best return for an intercropping system in which single row of *Vigna unguiculata* was alternated with double rows of tomato (Omotude and Adeniyi, 1995).

The productivity of tomato was not hampered by intercropping lettuce with tomato by planting lettuce 0, 10, 20 and 30 days after transplanting tomatoes in poly house in Sao Paolo, Brazil and the land use efficiency was 1.57 to 2.22 (Filho *et al.*, 2011).

Tomato intercropped with common beans showed no decrease in yield compared to sole tomato crop in Egypt and 1 tomato: 3 common bean plants gave the highest land equivalent ratio and the best economic returns (Abd El-Gaid *et al.*, 2014).

Tomato intercropped with garlic reduced the total yield of spring tomato but produced a higher net income than tomato monoculture under plastic tunnel organic medium cultivation in China, besides enhancing the microorganism populations and the enzyme activity of the medium, which together reduce the obstacles of continuous tomato production (Liu *et al.*, 2014).

Intercropping with potato onion (*Allium cepa* L. var. *aggregatum* G. Don.), also called multiplier onion, promoted the growth and P uptake of tomato in phosphorus rich soil and affected the community structure and function of phosphobacteria in tomato rhizosphere, besides improving soil quality by lowering levels of soil acidification and salinization (Wu *et al.*, 2016).

Persian clover (*Trifolium resupinatum* L.) intercropped with tomato as 1 tomato: 2 clover and 2 tomato: 2 clover exhibited lower densities of eggs and larvae of tomato fruit borer and higher numbers of predatory bugs, coccinellids and *Chrysoperla carnea* (Stephans), while the highest value of land equivalent ratio was found in 1 tomato: 2 clover planting system (Abad *et al.*, 2020).

Intercropping in Tomato.

8.12.0 Effect of Growth Substances

Many workers studied the effect of various plant growth substances on seed germination, growth, flowering, fruit set, yield and quality of fruits. Seed treatment, sprays on seedling, flower cluster or whole plant were found to be effective in increasing the production of tomato (Choudhury and Singh, 1960; Choudhury, 1964; Singh and Upadhya, 1967). Balasubramaniam (1988) opined that seed treatment with GA_3 enhanced percentage of seed germination and shoot growth, while IAA inhibited both. Biomass production was, however, significantly promoted by IAA as compared to GA_3, both as seed pre-treatments and as foliar sprays.

Singh and Choudhury (1966) and Choudhury (1967) found that whole plant and flower clusters treated with 2, 4-D and GA_3 improved the fruit set while seed and seedling treatment did not show any effect in this regard. It was also noticed that 2,4-D at 5 ppm as a whole plant spray, at 30, 45 and 60 days after transplanting increased the plant height, earliness and fruit yield in cv. Pusa Ruby but reduced the leaf area, plant dry weight and percentage of sunscalded fruits (Sharma and Tiwari, 1995). Kaushik *et al.* (1974) observed that NAA, kinetin and morphactin at low concentration (1mg/l) increased the number and weight of fruits per plant, while GA_3 had best effect at higher concentration (100 mg/l). High concentration of kinetin and morphactin showed adverse effect. Mehta and Mathai (1975) found best response of 2, 4-D at 5 ppm followed by NAA at 0.2 ppm in respect to maximum fruit set, early and total yield. Saimbhi *et al.* (1976) observed hastening of fruit ripening in plants sprayed with ethrel. Application of growth regulators considerably influenced the fruiting pattern in tomato. Shittu and Adeleke (1999) stated that GA_3 promoted fruit set and plants treated with GA_3 at 500 ppm had the highest number of fruits 12 weeks after sowing. Similarly, triacontanol treatment at 7.5 ppm, 4 weeks after transplanting gave the maximum fruit and seed yields/ha and improved the seed quality.

In Italy, treatments with IAA, BNOA and 2,4-D at 10 and 50 ppm each on tomato cultivars at the basal cluster gave the best yield as compared to whole plant treatment

(Lipari, 1982). Multiple ethephon (500 or 1000 ppm) applications significantly increased early marketable yields and number of fruits of late tomatoes (cv. Claudia Raf) in plastic houses in the Jordan Valley (Suwwan, 1986). Ripening was enhanced by ethephon, especially at 1000 ppm but average fruit weight was reduced. The increase in early yield caused by ethephon was of considerable commercial value. Parafen [Na salt of *p*-chlorophenoxy acetic acid (4-CPA)] applied in winter and early spring when natural light was poor improved fruit formation and ripening and increased fruit yield, especially at the start of picking. A single application of 30 ppm produced the highest yield per plant (Drozdov *et al.*, 1988; Ozguven *et al.*, 1998). Ozguven *et al.* (1997) reported that application of 4-CPA at 40 ppm, twice on opened flowers of F-144 (Fantastic) tomato cultivar grown under green house produced highest yield per plant and good quality fruits. In the green house, under high temperature (with day/night temperatures of 30/18°C) conditions, application of benzyladenine (BA) 10 ppm alone or in combination with gibberellin (GA_4) at 25 ppm increased fruit set and yield (Satti and Oebker, 1986). The highest yield per inflorescence was 714 g in green house compared with 385 g in the field. Generally, the mean fruit weight decreased when the number of fruits increased at both the locations. Plants treated with Atonik (sodium 5- nitroguaiacolate + sodium 1-nitrophenolate + sodium 4-nitrophonolate) increased the number of fruits/ha, yield and hence resulted in increased income (Shi, 1999).

In an experiment in Cuba, the plants (cv. Rossoi) were sprayed with 2,4-D (1.0 or 2.0 ppm) prior to anthesis of the first flower cluster and again 14 days after anthesis. Spraying with 1.0 ppm 2, 4-D gave the maximum increase in yields from 22 to27.4 t/ha (Francisco *et al.*, 1987). In China, application of growth regulator SD-100 at 25 ppm on potted tomato plants at the beginning of flowering and at peak flowering increased fruitset and yield by 20.28 and 32 per cent, respectively and fruit ripening was 12-15 days earlier in treated plants compared to controls (Chenandsu, 1991). In plants exposed to high temperature, ABA treatments advanced fruit maturity with a concomitant increase in fruit sugar content. ABA treatment (0.1 ppm) tended to increase fruit yield while at 0.5 ppm, cold hardiness of tomato seedlings increased (Takahashi *et al.*, 1993). Green house trials involving drought stress indicated improved seedling tolerance in paclobutrazol treated seedlings and earliness was equated to 7-10 days advanced maturity in paclobutrazol plots only, compared to control (Souza Machado *et al.*, 1999). Grunzweig *et al.* (2000) studied on involvement of endogenous gibberellins in the regulation of increased tomato shoot growth in solarized soil. They concluded that increased quantities of GA_1 and eventually GA_3 played a role in the increased growth of shoots in solarized soil as early as 7 days after transplanting.

Srivastava and Srivastava (1966) observed changes in ascorbic acid content in presowing treatments. The acidity of fruits was reduced by whole plant spray with GA_3 and 2,4-D, while ascorbic acid content increased with higher concentrations of 2, 4-D and PCPA. Pandita *et al.* (1976) reported that plants treated with NAA at 100 ppm gave the highest fruit acidity followed by 2,4-D at 5ppm. Total soluble solids increased with NAA at 100 ppm and boron at 2 ppm increased reducing sugars in fruits. Bisaria and Rastogi (1988) investigated on the effect of kinetin on composition of tomato fruit and concluded that fruit yield, TSS, sugars, protein and ascorbic acid content increased on spraying kinetin (100 ppm), one month after transplanting but it had no effect on fruit acidity.

Mehta and Bhavamaryana (1981) found that ascorbic acid at 25 ppm or sucrose 10000 ppm stimulated respiration in unripe and red ripe fruits, whereas cycocel at 10 ppm inhibited it. In field and indoor trials, calcium carbide applied as an ethylene source directly to the plants, or to the soil with irrigation water, increased yields, especially of early crops. It had a growth regulating effect on the plants and also increased the resistance of the fruits to *Macrosporium solani* infection. Optimum rates were 40-80 kg/ha applied in split doses at the early stages of plant development (Karpenko *et al.*, 1989).

Tomato fruit set was promoted by GA_3 at low concentration (Khan *et al.*, 2006). Synthetic auxin 4-CPA (4-chloro phenoxy acetic acid) reduced pre-harvest fruit drop with increased number of fruits per plant and yield (Sasaki *et al.*, 2005). Application of 4-CPA is more effective during anthesis period than one week after anthesis (Poliquit *et al.*, 2007). Combined application of 4-CPA and GA_3 each at a concentration of 20 ppm has been used to improve yield, fruit number, average fruit weight in summer season in BARI hybrid-4 and BARI hybrid-8 in a study in Bangladesh (Rahman *et al.*, 2015).

Choudhury *et al.* (2013) reported that GA_3 (30 ppm) applied at 21 days after transplanting and 4-CPA (30 ppm) applied during anthesis stage showed an increased fruit set, fruit diameter, single fruit weight, number of fruits and yield as compared to the plants treated with 4-CPA and GA_3 alone or untreated plants.

Ding *et al.* (2013) could induce parthenocarpic fruit in tomato cv. Microtom by 1-(2-chloro-4-pyridyl)-3-phenylurea (CPPU, an active cytokinin). The fresh weight of CPPU-induced parthenocarpic fruits was comparable with that induced by GA_3. Importantly, CPPU-induced parthenocarpy was found to be compromised by simultaneous application of paclobutrazol (a GA biosynthesis inhibitor), and this effect could be restored by exogenous GA_3. Like pollination, CPPU-induced fruit showed enhanced accumulation of GA_1 + GA_3 and indole-3-acetic (IAA), which were accompanied by elevated expression of GA biosynthesis genes like *SlGPS*, *SlGA20ox1*, *SlGA20ox2* and *SlGA*$_3$*ox1*, and IAA biosynthesis gene *ToFZY*. Elevated GAs level in CPPU-induced fruits was also associated with down-regulation of GA inactivation genes, namely *SlGA2ox1,2,3,4,5* in comparison with untreated control. These results suggested that cytokinins may induce parthenocarpy in tomato partially through modulation of GA and IAA metabolisms.

Tiwari and Singh (2014) observed that in Uttarkhand conditions, on tomato variety Pant T-3, highest fruit yield were obtained by application of CIPA 20 ppm, 2 4-D 5ppm and Alar 50 ppm which showed 71.47, 40.1 and 33.2 per cent higher yield over control, respectively. The lowest leaf curl virus infection was recorded in the plot treated with 2, 4-D 50 ppm (40 per cent), 2, 4-D 10 ppm (45 per cent) and NAA 40 ppm (68.3 per cent). CIPA 20 ppm and Ethephon 100 ppm showed early maturity of fruits. More number of fruits per plant was recorded in CIPA 20 ppm, 2-4 D 5 ppm and NAA 40 ppm. The percent fruit set was higher in CIPA 20 ppm, 2,

4-D 5 ppm and GA_3 10 ppm than control. Equatorial fruit diameter was greater in 2, 4-D 5 ppm, CIPA 20 and 10 ppm whereas polar diameter was higher in Ethephon 100 ppm, Alar 100 ppm and CIPA 10 ppm. More number of locules per fruit was higher in 2, 4-D 5 ppm and CIPA 20 ppm on the other hand, NAA 40 ppm, 2, 4-D 10 ppm and paclobutrazol 20 ppm showed higher TSS and pericarp thickness was greater in Ethephon 50 ppm, NAA 40 ppm, 2, 4-D 5 ppm and CIPA 20 ppm.

9.0 Harvesting and Yield

9.1.0 Harvesting

The stage of maturity for harvest of tomato depends upon the purpose for which they are used and distance covered to reach the market. The following stages of maturity for harvesting are recognized (according to USDA).

- (i) Green stage (Stage 1): Surface of fruit is completely green, the shade varying from light to dark green, fruits are fully developed.
- (ii) Breaker stage (Stage 2): There is definite break in colour from green to tannish yellow or pink or red, not exceeding 10 per cent of fruit surface.
- (iii) Turning stage (Stage 3): More than 10 per cent but less than 30 per cent of the fruit surface shows changed colour from green to tannish yellow or pink or red.
- (iv) Pink stage (Stage 4): More than 30 per cent but less than 60 per cent of fruit surface shows pink or red colour.
- (v) Light red stage (Stage 5): More than 60 per cent but less than 90 per cent of the total fruit surface shows pink or red colour.
- (vi) Red stage (Stage 6): More than 90 per cent of the fruit surface is red in colour.

Tomato is harvested at different pickings. The interval between picking depends upon the weather. Tomatoes are normally picked at 4-day intervals during warm season and at weekly intervals when weather is cool. Tomatoes should not be pulled from vines but should be picked with a twisting motion of hand to separate fruits from stem. Harvesting schedule also differed with the genotypes. Ziauddin and Rukshana (1999) from Bangladesh reported that the optimum harvest interval for tomato cv. Roma during early harvest, peak harvest and late harvest were obtained at 6, 2 and 1 day, respectively, while the same for cv. Ratan were 4, 2 and 2 days, respectively.

Fruits harvested at immature green stage were usually tough leathery and developed poor colour when ripened artificially. Fruits picked in mature green stage developed good colour when ripened under favourable conditions.

Lutz (1944) found that immature fruits which did not develop good colour, were not high in ascorbic acid content. Similar observations were made by Hamner *et al.* (1945). Fryer *et al.* (1954) found that fruits harvested in yellow red stage and ripened at 23.8°C out of direct sunlight, developed ascorbic acid found in normal red ripe tomatoes. Anju Kumari *et al.* (1993) observed that the longest shelf-life

Ripening and Harvesting of Tomato

Stage I Stage II Stage III

Stage IV Stage V Stage VI

Ripening Stages of Tomato (USDA).

Manual and Mechanical Harvesting Methods of Tomato.

Harvesting of Tomato Grown under Open Field and Protected Structure

was attained by harvesting tomato fruits at the mature green stage which resulted in the lowest ascorbic acid content after storage and patchy colour development on ripening. This was also confirmed by Mallik and Bhattacharya (1996). As the largest fruits often fetch a premium price in the market, it was, therefore suggested to harvest fruits when in pink stage to have higher price. Green fruits on the other hand were generally firmer, more crack-resistant and required fewer harvests than pink fruits (Davis and Gardner, 1994).

Tomato grown for processing was harvested when fully ripe to ensure desired quality and red colour of the product. Quality characteristics such as flavour, texture and colour in fruits were superior to those harvested at earlier stage of maturity.

Green house grown tomato fruits were harvested at the breaker to turning stage (80-143 days after planting) and held at 21°C up to 8 days. The later fruits had lower fruit weight with higher resonance frequency and stiffness factor and the rate of colour development was slower. The principal axis of shelf-life changes was described by the decrease in stiffness factor or the increase in the percentage of red on the fruit surface (Schrevens *et al.*, 1995). Fruit shape and firmness were found to be the best when plants were grown at constant temperature of 23°C. Higher temperature resulted in enhancement of time of harvest as well as time for harvest to fruit ripening. The incidence of skin cracking was the highest at 17°C and decreased at higher temperature upto 23°C (Buitelaar and Janse, 1987).

Davis and Gardner (1994) observed that the total yield and average fruit size were reduced when fruit were harvested at the green stage, and total yields for pink harvested fruit were higher than green harvested fruit. However, fruit cracking was more severe for pink than for green fruit that lowered the marketable yield. But since bigger sized fruits fetched a premium price, harvesting fruit when pink probably will result in a higher price per kilogram than harvesting fruit when green. Fruits harvested at green were generally firmer, more crack resistant, and fewer harvests were required for green harvest stage.

Moneruzzaman *et al.* (2009) reported that tomato cultivar Roma VF harvested at mature green stage has the highest shelf life, but weight loss was also maximum in it. Half ripened tomato possessed highest value of vitamin C and titrable acidity, while total sugar content was highest in full ripe tomato, which also exhibited the maximum rotting.

Tomato fruit harvesting robot has been manufactured and tested in Japan. Robotic hands comprising of rubber fingers to grasp fruits without causing damage to the peel are attached to the wrist of an articulated robot (Kawamura *et al.*, 1987a, b). New and important technico-functional improvements were made to the 1998 series of Polaris evolving a self-harvester for tomatoes. The 1998 Polaris harvester had efficient harvesting head with a driver's seat and maintained productive yield by harvesting quality products (Maresca, 1998).

Population based approach to optimize fruit harvesting strategies (harvest date and harvest frequency) with regard to fruit quality and its derived economic value under Vietnamese conditions was studied and a kinetic fruit growth model was successfully calibrated which then was used quantify the population variation in terms of the physiological maturity of the tomatoes has been developed for tomato cultivar Saviour (Tran *et al.*, 2017). The applied model does not pretend to be a physiological model as its level of detail seemed to be fit for the intended purpose of optimizing the postharvest economic value of the crop taking into account pre-harvest biological variation.

9.2.0 Yield

The yield depended upon various agroclimatic factors, including soil fertility, and cultivar and cultural practices. On an average in India, under field conditions, an open pollinated variety can produce fruits of 16-25 t/ha. The hybrids often produce 60-80 t/ha.

Productivity of tomato in different countries of the world is presented in Table 5. It shows that productivity of some of these countries is much above the world's average. This is mainly due to the fact that hybrid tomato cultivars are mostly grown under protection with high-tech culture in these countries.

Table 5: Productivity (tonnes/hectare) of Tomato in different Countries

Country	*2002* +	*2006* +	*2012* +	*2014* +	*2016* +	*2018* F
China	34.61	37.67	50.71	52.63	56.20	59.25
India	16.28	17.98	20.49	21.24	24.21	24.65
USA	75.41	72.18	88.07	88.85	90.29	96.81
Turkey	37.05	43.08	37.83	37.13	66.93	68.87
Egypt	35.45	38.96	39.23	38.73	39.77	40.97
Italy	47.11	51.97	57.00	54.51	61.94	59.72
Iran	31.85	34.34	37.50	37.54	40.05	41.37
Spain	67.15	67.04	81.00	89.29	86.19	84.96
Brazil	58.42	57.09	64.50	66.85	65.14	71.94
Uzbekistan	21.74	26.18	35.33	35.14	43.34	37.85
World	29.06	31.04	39.39	33.99	37.02	38.27

+ Anon. (2018).

F = FAO estimate (FAO website accessed on July 15th, 2020)

10.0 Growing Tomato under Protection

Tomato being a very popular vegetable is in constant demand throughout the year. However, in cold countries it is not possible to grow tomato in the open field conditions during the winter months. Results of experiments conducted at different places revealed successful cultivation grown under different structures like glass house, hotbed, cold frame, plastic tunnel and temporary structures, *etc.* The commercial cultivation of table tomato in glass houses has become a profitable enterprise in countries like Northern Europe, Canada and USA. However, during the last 10 years, the growers of glass house tomato were under extreme financial pressure, fuel input for a heated crop being high, the production cost rose approximately fifteen-fold. The crop was also labour intensive and, therefore, became much more expensive. This was somewhat compensated by the high yield obtained under controlled conditions. The yield per hectare went even upto 200-250 tonnes stretched to 10-11 months. McAvoy and Giacomelli (1986) through high density sequential planting (90,000 plants/ha) obtained a continuous yield on a predictable schedule throughout the year in green house.

The advantages of glass house are manifold as it provided protection against unfavourable conditions. It ensures desired light conditions, temperature and humidity for specified periods. The carbon dioxide content of the air can also be increased assuring better photoperiod activity. In countries like Netherlands, Denmark, Canada and USA many institutes have been working on the cost-benefits of different glass house structures. Earlier glass houses were constructed on timber frames. In recent years iron or aluminium frames were in use on extensive scale with computerized environment control. Tanaka and Yasui (1986b) suggested growing of tomatoes in green house with floor area of 1000 or 200 m^2 with virtually no heating (frost free) or with heating provided when the air temperature fell to 2 or 8°C. All the houses had N-S orientation and were lined with 0.05 mm gauge PVC sheeting. On the other hand, Bartkowski and Nowosielski (1988) suggested the use of a tray container technique for commercial tomato production in a green house. Containers of fibre glass 30 cm wide, 10 cm high and 535 cm long, were placed in 340 m^2 units. The trays in the containers consisted of perforated 15 litres PVC square baskets, wider at the top than the bottom, filled with a substrate and planted with 4 plants in each basket. The nutrient solution in the containers was continuously maintained between 5-7 cm level with the aid of a valve. The electrical conductivity was controlled automatically by the injector. Plants were grown up to 7 clusters in a spring cycle and higher yields were recorded. In yet another case, the roof of a plastic green house was utilized to catch water in a specially designed ground reservoir along the green house. This water was later used to irrigate a green house tomato crop using a solar system to operate an irrigation pump. Tomato yield obtained from this green house was 8 times greater than that obtained under open field rainfed conditions. It was suggested that the feasibility of this technique might be applicable in marginal and arid zones if sufficient catchment area was allocated to each green house. The use of a drip irrigation system was essential with this technique (Snobar *et al.*, 1988).

Stoppani *et al.* (1994) suggested growing of tomato in a green house 10 m wide and 15 m long heated by circulating hot water in an aerial pipe and ventilated by opening windows. The use of cheap green house for tomato production was in practice in Greece as reported by Tzouramani *et al.* (1995). Green houses in particular, the wooden prefabricated or round arch types, were found to be highly economical. Mayno (1998) recommended growing of tomatoes in plastic film tunnels without heating for obtaining higher yields. On the other hand, Barroso *et al.* (1999) found that a single tunnel was as effective as a double-span green house (both round arched and covered with a 200 mm-thick thermal polythene film) for cultivation of tomato crop. Venlo type glass house with automated ventilation was found to be as good as standard Colombian plastic green houses with fixed open ridge and manual side wall ventilation for tomato growing. During the night, a plastic thermal screen was required in the green house (Cooman *et al.*, 1999).

10.1.0 Green House Environment

10.1.1 Temperature

Glass house has to be equipped with heating and cooling systems and maintained interior temperature of 15-22°C. Tesi (1983) concluded that economic returns of tomato under glass house were better when less heat was applied. Moss (1982) also reported the lower temperature regime for glass house tomato. Gosselin and Trudel (1982) observed that in lower temperatures (13.8 and 18.6°C), heaviest roots were produced; while shoot production was best at 23.9°C. A rise in soil temperature from 14.0 to 21.8°C increased total yield by 47 per cent in the spring under warm air-conditions. Drews and Heissner (1982) found that earliness was delayed when night temperature fell below16°C but total yield increased slightly. A low night temperature regime (either based on light conditions during the preceding day, minimum 10-12°C falling to 8°C, or a set of minimum of 8°C) produced larger fruits than the standard programme (18° falling to 16°), especially during the first 4 weeks. The yield was also higher (Erlandsson, 1987). The minimum night temperature of 13°C produced fruits with higher yield than the plants grown at 5°C (Abak *et al.*, 1994). Malfa *et al.* (1995) also found that tomato plant grown in a heated green house (minimum temperature at 13 °C) produced fruits at a similar time and with higher yields than those grown in unheated green house. Investigations carried out by Baytorun *et al.* (1999) revealed significantly higher yield in green houses with a minimum night temperature of 13°C; crop yield was higher in heated green houses than in unheated green houses and also the fruits were much larger in heated green houses during the cold season. However, it was also found that low minimum night temperature (2-5°C) in the early stages of plant growth greatly retarded stem and leaf development and delayed flower opening. When flowering and fruiting started, development again was slowed down by low air temperature. If plants at the fruit expansion stage of growth were exposed to low soil temperature (8-12°C) for 30 days; growth abnormalities and a drop in fruit yield resulted (Tanaka and Yasui, l986a).

Higher temperature is found to be detrimental for plant growth and development. High plant temperature during hot, sunny days is a primary cause of poor fruit quality and crop growth in green house tomatoes. A temperature of about 33°C and 30°C on the sunny and shaded sides of the plants, respectively and 27°C on the lower leaves in mid-summer was found to be detrimental. The most effective method of reducing the temperature was the use of sprinklers mounted on the roofs as it provided a combination of cooling and an increase in RH (Holsteijn, 1990). Malfa (1993) observed that higher temperature of 34°C is detrimental to tomato plants as it affected the growth of vegetative organs namely leaves, stems and shoots. The carbon exchange rate and stomatal conductance were also reduced. A temperature of 30.3°C was found to be congenial for tomato growing. However, Buitelaar and Janse (1990) reported an optimum temperature of 21°C for growing tomato. In an experiment conducted by them, wherein tomato plants were maintained at an average 24-h temperature of 17°C, 19°C, 21° or 23°C from 11 Jan to 1 May. Fruit shape improved with increasing temperature until mid-May, but later on there was little difference. Fruit firmness increased until mid-April, as temperature rose to 21°C, declining at highest temperature. Time for best colour development of fruits after harvest decreased with increasing temperature. Fruit flavour also improved as temperature rose. Shelf-life of fruit was the maximum at 21°C.

Two methods are commonly used for heating of glass houses, (i) hot water or steam circulation, the cost is too high in this system. Tuzel (1994) reported that water tubes used for frost protection in green house resulted in an increase in the soil temperature as well as overall yield of the crop. Hence, this was recommended for heating of green houses (ii) warm air, this system is more suitable for small glass houses. Conventional heating systems (fuel oil) for heating of green house air were used for keeping the preset temperature of 10 and 15°C. Higher yields, earlier harvests and better quality produce were obtained when preset temperature was 15°C. Fuel cost was, however, more for15°C but this was compensated by the extra income obtained from higher yield (Vakis and Photiades, 1989). Gent and Malerba (1994) recommended heating of soil in green houses by blowing hot air from a forced air heater or through drainage pipes buried beneath raised beds. Heating increased the early yields as well as total yields of tomato crop. The quality of fruit was also enhanced. This method of soil heating was found to be economical. Green house heating by gas oil furnace also gave higher yields and better fruit quality and hence was recommended for green house heating (Traka-Mavrona *et al.*, 1995). Another system of green house heating *i.e.*, bench top heating which too was economical for heating as reported by Sachs *et al.* (1992). Fresh and dry weight, plant height and development of tomato crop are found to be at par with both the bench top and perimeter heated green houses. However, soil temperature was 3-5°C higher with bench-top system and this system also used 50 per cent less energy.

Traka-Mavrona *et al.* (1995, 1997) have recommended heating of green houses by passive solar system (PSS) or a gas-oil-air furnace to maintain temperature of 11-12°C (in the green house). The mean number of flowers produced by plants and average weight of individual fruits were found to be significantly higher with the PSS and gas-oil-air furnace. Solar energy systems were used for interiors of gable even span green houses (Al-Amri, 1997). The solar heating process during the night was needed to maintain an optimum air temperature (18°C). Heating enhanced the rate of growth and increased the yield of fresh tomatoes in comparison with commercialgreen houses (with conventional heating system). The area of the solar water heater was ascertained according to solar radiation and temperature differences prevailing inside and outside the green houses. Automatic control of heating was more useful.

For cooling of the green houses for raising tomato crop, shading, misting and static ventilation were used (Boulard *et al.*, 1991). Misting and ventilation together significantly decreased transpiration, water vapour deficit and leaf and air temperature. The decrease in transpiration was roughly proportional to the degree of misting. Shading reduced transpiration rate due to decrease in natural radiation transmission but decreased leaf and air temperature only slightly. Fernandez and

Bailey (1994) found that number and position of air recirculation fans were very important for regulation of environmental conditions in the green houses. Uniform distribution of the fans decreased the temperature, humidity and CO_2 concentration. Willits and Peet (1994) suggested the use of shade cloth misting for cooling green house tomatoes. This comprised of an external black polyethylene cloth providing 55 per cent shade fitted in the green house and was misted every other day; mist being applied for 30 sec every 3 min whenever solar radiation exceeded 400 W/ m^2. With misting there was a reduction in the air temperature and leaf temperature thus having a cooling effect on the crop.

A zone cooling system was reported by Nishina *et al.* (1996). In this system the growing space to be cooled around the tomato plants was surrounded by transparent vinyl film. An air-to-water heat exchanger with fans was installed in the cooled space. A heat pump system supplied the heat exchanger with cold water. Air from the cooled space was circulated by fans for cooling and dehumidification was done by heat exchanger.

Hayashi *et al.* (1998) recommended fogging in green house for cooling. Fogging was produced by intermittent spraying (1 min spray at 3 or 4 min interval) from nozzles installed 2 m above the floor. The side and roof ventilators of the green house remained open to increase air circulation and fog evaporation during cooling. This resulted in decrease in air temperature in the green house within 1 min after fog production started.

Proper ventilation is important for glass house. The ventilation is necessary to reduce air temperature which is raised by solar radiation. Hot air is removed and replaced by cool air to keep down temperature. The process of evaporation combined with transpiration causes humidity of air to reach saturation. Flap ventilation performs well although automatic operation of ventilation equipment is necessary for commercial glass houses. According to Gustafsson and Weich (1991), ventilation was very important for removal of moisture as well as for cooling. Meneses *et al.* (1994) from their investigations in Portugal found that earliness of the crop was affected in ventilated green house. They also observed that night ventilation dramatically reduced the incidence of *Botrytis* from 10-12 per cent in the closed green house to <1 per cent. Green house with vents open at night (NOG) was found to be warmer than vents closed at night (NCG), with a lower RH. This also reduced the incidence of *Botrytis cinerea* development (Abreu and Meneses, 1994; Abreu *et al.*, 1994). The best ventilation was affected by both roof and lateral vents which had chimney-like effect and was particularly suited to light wind conditions (Feuilloley *et al.*, 1994).

10.1.2 Light

Boivin *et al.* (1987) from Canada reported that the number of leaves formed below the first inflorescences reduced significantly with supplementary lighting which also reduced the incidence of flower abortion. Supplementary lighting increased early marketable yields for the December sowing by 100 per cent and total yields by 10 per cent.

McAvoy and Giacomelli (1986) showed that using supplemental HPS lighting (100 micromoles $m^{-2}s^{-1}$ for 18 hours/day) on tomato plants grown as a single cluster crop in green houses, a saleable yield of 9.3 kg/m^2 was produced during the midwinter cropping period from 28 December to 18 January. The unlit control crop produced 3.9 kg/m^2 saleable fruit during this period of low natural radiation. Using an eight-week cropping cycle (transplant to final harvest) five crops could be produced per unit area of green house in 10 months period from 1 October to 30 July in New Jersey, USA (McAvoy and Giacomelli, 1986). Yield significantly increased as plant population increased and supplemental lighting from a HPS source was employed. An additional light of 6.48 moles $m^{-2}s^{-1}$ was given for increased yield (McAvoy and Janes, 1988). Plants grown at high density of 3.5 plants/m^2, when exposed to supplementary lighting of 150 mmol $m^{-2}s^{-1}$ produced the highest yield of 84 kg/m^2 (Dorais *et al.*, 1988, 1991). Vezina *et al.* (1991) reported that prolongation of the photoperiod over 14 h did not increase yield. Plants exposed to continuous lighting were damaged even if they had been previously exposed to long photoperiods (17 h and 20 h). For photoperiods of 17 h and 20 h, lighting during the night, which generated two dark periods per day, damaged the plants and reduced yield. Trials conducted in Canada revealed that continuous light increased leaf starch and sugar contents (Demers *et al.*, 1998). The reductions on growth and yield and leaf chlorosis in plants grown under 24 h photoperiod were possibly due to starch and sugar accumulation in the leaf.

Shading was effective in reducing the light penetration and temperature inside the green houses, thereby creating a better microclimate inside the green house for production of higher yields and quality fruits. Different types of covering materials and screens were used to reduce the temperature of green houses as well as to monitor the intensity of light so as to produce an optimum crop. Koning (1988) recorded loss in yield when overhead screens of aluminum slats with 42 per cent light penetration were closed and the light intensity in the glass house attained 450, 550 or 650 W/m^2. It was also observed that card board caps having a surface area of 17 x 18 cm and upper surface aluminized and attached by clips at weekly intervals above the tomato trusses, greatly enhanced the fruit colour. In addition, incidence of star cracks, yellow shoulder and gold fleck disorders were highly reduced. Shelf-life was also slightly extended (Holsteijn and Janse, 1990). Protective overhead screens of aluminum slats with 45 per cent light penetration resulted in lowering of the green house temperature thereby enhancing the fruit quality. The unshaded fruits had a temperature of about 5.5°C higher than the shaded ones. Evapotranspiration was also much less with consequent reduction in stress (Holsteijn, 1988). According to Janse (1988) the optimum packhouse temperature for uniform coloration of harvested fruits was about 20°C. On application of shading screens, the average percentage of multi-colored fruits was reduced and uniform coloration was achieved as compared to tomatoes grown without shade. Holsteijn (1989) also recommended summer shading to lower temperature for obtaining maximum yield. Summer shading also reduced the percentage of unevenly coloured fruits from 13 to 6 per cent and increased fruit shelf life from 4-6 days. In case of beef steak tomatoes, the percentages of fruits with green or yellow shoulders and star-shaped cracks were reduced as the degree of shading increased. Placing reflective (card board) caps over the trusses reduced the incidence of yellow shoulders from 27 per cent to nil and markedly reduced the incidence of star cracks. Shelf-life was also enhanced and the fruits had a lower incidence of gold fleck disorder (Janse, 1990). Jimenez (1991) suggested the use of shelter nets for cultivation of tomatoes in the winter in the Canary Islands because of low structural cost, wind resistance, easier climate control, savings in irrigation and improved pollination but the crop development was slower due to reduced light transmission. Other investigators observed that shading reduced average fruit size and hence reduced the proportion of fruit in the larger size grade. Heavy shade (23.4 per cent shading) reduced the incidence of uneven ripening in summer while shading with 20 per cent shadow plastic mesh placed on the roof reduced the leaf, air and soil temperature as well as incidence of blossom end rot which often caused water stress conditions and decreased yield. However, shading significantly increased the number of fruits per plant and the total yield. Highest yields were obtained under 35 per cent shading (wire wetting). The physical (weight, length, diameter and volume) as well as chemical (titrable acidity) characteristics also improved. Shading reduced the percentage of fruits affected by sunscald (Cockshull *et al.*, 1992; El-Gizawy *et al.*, 1993 and Francescangeli *et al.*, 1994).

Shading by lime white wash (50 g/m^2) or plastic mesh providing 20 or 65 per cent shade were effective in reducing the high summer temperature (air, leaf and soil) along with reduction in light transmission (Francescangeli *et al.*, 1994). They were able to reduce temperature by shading with agro textile mesh (horizontally 15 cm above the top of the tunnel or leading on it). Plant transpiration was also reduced. White knitted weave cloth was found to be effective in reducing the temperature in green house. Misted SR55 (black flat weave cloth with manufacturers shade rating of 55 per cent and measured SR of 54 per cent) and 60 per cent cloth (manufacturer shading rate of 60 per cent) were found to be more effective in reducing leaf temperature and energy consumption in green houses (Willits, 1994).

Increase in total yield was obtained by growing tomato crop inside a double-skinned green house with polyethylene (PE) screening. The highest yield was obtained with PE screened single skinned green house. The double skin plus LS-17 screen were effective in reducing and delaying frost damage in autumn (Abak *et al.*, 1994; Bascetincelik *et al.*, 1994). In yet another experiment it was found that green house covered with polyethylene film (PE) containing UV resistant (PEI), UV + IR thermal trapping (PE2) or UV + IR plus antifogging (PE3) agents resulted in higher yield. Fruit set was slightly lower under PE3 while yield was little lower under PEI (Baytorun *et al.*, 1994). Cerne *et al.* (1994) obtained vigorous growth and higher yield using LDPE-IR (low density polythene) thermal film and EVA (ethyl vinyl acetate) film in green house. The air and soil temperatures were also maximum under the films as compared to normal LDPE film. Recycled multiplayer film made from waste plastic was found to be as effective as conventional virgin monolayer film. Even though this recycled film was viable only for 7 months (one growing season), it was found to be cost effective (Kasrawi *et al.*, 1994). Double and single clad green houses were also found to be effective screens for obtaining optimum yield as reported by Vargues *et al.* (1994). Double cladding increased the soil temperature thereby reducing *Botrytis* attack on tomato crop. Cho *et al.* (1997)

found that 5 month old polyethylene (PE) film, 4-year old PC (polycarbonate) or glass were equally effective as covering materials. Double inflated polythene film (D-poly) and rigid twin wall acrylic panels (acrylic) as cladding material were found to be very effective cover materials in the spring season. Marketable yield was found to be higher under the covers as compared to single layer glass house. The covers also saved 30 per cent in heating energy apart from having a low incidence of blossom end rot (Papadopoulos and Hao, 1997). Use of an infrared absorbing new glass designated floraser glass resulted in greater energy saving by reducing transmission of photosynthetically active radiation (Cevri *et al.*, 1999). Plants grown under double layer heat + fluorescent film produced the highest yields followed by ethylene vinyl acetate (EVA) films as reported by Park *et al.* (1999). The light transmission was the highest under heat+ fluorescent film (86.4 per cent of solar radiation).

10.1.3 CO_2 Concentration

The enrichment of atmosphere by two or three-fold CO_2 increased tomato yield. Kim *et al.* (1982) were of the view that at normal light intensity total fruit yield was 40 per cent higher with CO_2 enrichment and fruit size was greatest at the lowest night temperature. Plants enriched with CO_2 in glass houses showed significant increase in fresh and dry weight and total yield (Nilsen *et al.*, 1983). Under normal condition, atmosphere contained 300 ppm of CO_2, the glass house plants utilized 3-4 times more CO_2 (1000 ppm) for its normal photosynthesis process. CO_2 was obtained from cylinders containing solid CO_2 known as dry ice or by combustion of hydrocarbon fuel such as butane or paraffin

Supplementary lighting in the glass house exerted a synergistic effect when combined with CO_2 treatments and the yields increased by 32, 73 and 122 per cent for 3 CO_2 concentrations (300, 900 and 1500 ppm), respectively, in comparison with plants grown under natural day light. CO_2 at 900 and 1500 ppm increased water use efficiency, apparently due to an increase in stomata resistance and transpiration rate (Yelle *et al.*, 1987). Higher fruit yields were obtained with intermittent enrichment of CO_2 than with enrichment in the morning and after-hours only (Zipori *et al.*, 1988). CO_2 supplementation from the flue-gas (minimum concentration 415 ppm) during the first 5 weeks of harvest resulted in higher yields. The CO_2 supplementation varied with different light intensities as mentioned above (Nederhoff and Rijsdijk, 1990). CO_2 uptake by a tomato crop was found to be linearly related to the intensity of solar radiation (Gustafsson and Weich, 1991). Lee and Lee (1994a) found that long-term CO_2 enrichment (800 ppm) increased resistance to water vapour diffusion thereby decreasing transpiration especially in leaves in the middle portion. CO_2 enrichment enhanced leaf starch while chlorophyll and soluble protein contents were reduced (Lee and Lee, 1994b). On the other hand, CO_2 enrichment at 1200 cm^2/m^3 for 10 days to plants at 5 leaf stage resulted in an increase in net assimilation rate, leaf area and dry weight as observed by Khavari (1996). Zwart (1998) recommended continuous CO_2 supply from exhaust gases during warm periods under Dutch conditions for effective heating of the green house.

Application of 700 and 1000 µl CO_2 at 22/16°C and 25/16°C resulted in lowering of transpiration rates and enhanced the photosynthetic rates. Biomass production was higher when 700 µl of CO_2 was applied at a temperature of 25/16°C while better growth was observed when plants were exposed to 8 h of CO_2 concentration (Behboudian and Lai, 1994). Hand and Slack (1988) recommended an optimum level of 455 ppm. CO_2 for obtaining optimum yield. They recorded highest yield by growing tomato plants with low levels of summer CO_2 enrichment (350, 375, 450 or 525 ppm) combined with ventilation at 21°C. The effect of preharvest CO_2 enrichment at 1000 µl on postharvest quality of tomato was studied by Behboudian and Tod (1995). They observed better fruit desirability in terms of slower postharvest ripening and higher concentrations of soluble sugar and total soluble solids. The earliness and total yields increased by 37 per cent and 43 per cent respectively, in the green house when the CO_2 concentration was maintained at 1000 ppm and the temperature less than 27°C (Özcelik and Akilli, 1999).

10.1.4 Humidity

It is very important to maintain the humidity level in the green houses. Dehumidifiers are used for regulating the RH inside the green houses. The humidity affects the crop growth, development and yield and quality of fruits. The increased humidity level from VPD 0.1 to 0.8 Kpa, was found to be congenial for good growth and yield (Jolliet *et al.*, 1993).

10.2.0 Cultivation in Green House

In green houses, plants were placed in borders and trenches along the heating pipes. Plants were planted in benches made of asbestos or corrugated iron with a layer of peat or sand. Angle iron components were excellent for benches. Tanaka and Komochi (1982) studied the relationship between plant density and topping on the growth and yield of tomato in green house. With increasing plant density, leaf size, stem diameter, weight of leaves and stem per plant decreased and flowering as well as ripening was delayed up to 7 days but yield per unit area increased.

10.2.1 Cultivars Suited for Green House

Suitable cultivars contributed to high yields in glass houses. The cultivars suited for protected cultivation should ideally have indeterminate growth habit. In New Zealand, Nemarex, E 6208 and Angela were the best cultivars for glass house cultivation (Brundell *et al.*, 1982). Scott and Farley (1983) reported that Ohio CR.6 was the best adopted cultivars for glass house cultivation in the USA. In Italy, tomato cv. Carmelo gave early and higher yields under a cover of corrugated 0.9 mill PVC than under reinforced 1 mm corrugated polyester or 10 mm polycarbonate (Tesi and Tosi, 1986). Two hybrid cultivars, Karlson TmC_3F_1 and Rusich TmC_4 bred in Moscow province of USSR were promising for protected cultivation (Tarakanov *et al.*, 1987). Kusainova (1985) from Russia identified nematode resistant tomato hybrids, Nemato, Belkanto, Grenada and Vainmon for green house cultivation in winter along with earliness, good yields and high marketable fruits. Hybrids Arletta, Carmello, Tresor and the line Gr-p were found to be the earliest and the most productive when grown under green houses heated either by passive solar system or a gas-oil-air furnace (Traka-Mavrona *et al.*, 1995). In India, some tomatoes that can perform well under protected structures are Pusa Ruby, Pant Poly house

Tomato Cultivation under Climate-Controlled Green House

Tomato-1, Pant Polyhouse Tomato Hybrid-1, TNAU Tomato Hybrid CO-3, Swarna Samridhi, Swarna Anmol, Punjab Swarna, Punjab Red Cherry, Punjab Sona Cherry and Punjab Kesari Cherry.

10.2.2 Soilless Culture

The prospects of soilless culture of tomato, using substrates like sand, gravel, peat bags, rockwool blocks and mats and agrofoam (polyurethane) mats were studied (Adamson and Mass, 1971, 1976; Verwer, 1975; Bleyaert, 1987). Hydroponics, aeroponics and nutrient film technique (NFT), a modified form of hydroponic, were successful in many countries for production of tomatoes in green houses. Benoit and Ceustermans (1986) found that when tomato plants (cv. Dombito) in rockwool blocks were placed in NFT channels and after the appearance of the second truss some of the blocks were placed on larger rockwool or polyurethane (PU) blocks (20 x 15 x 4 cm or 20 x 15 x 2 cm) the large blocks enhanced earliness but reduced total yield. Yield was higher with thin PU blocks than rockwool blocks. They also (1988) compared hydro culture (NFT) with aeroculture and the Israeli Ein-Gedi system. Early yields were greater using the Ein-Gedi system but total yields and fruit grade size were little affected by the growing system. Yields of tomato cultivars were increased when grown in Dokki and Cooper's nutrient solution as compared to the plants grown in soil. Growing in NFT system also increased the total number of flowers and fruits produced. Plants were also very vigorous when grown in NFT system (Abou-Hadid *et al.*, 1989).

Blain (1998) reported that Agro-bag containing 24 litres of a sphagnum peat-based substrate measuring 90 x 20 x 12 cm and weighing 7 kg, with an air-filled porosity of 20 per cent and an available moisture capacity of 25 per cent gave higher yields. On the other hand, Koning (1989) found good growth and optimum yields in tomato grown on rock wool in an environmentally controlled green house. The light utilization efficiency was also found to be good. Other workers also obtained good yield on rockwool (Tanis, 1990; Abou-Hadid *et al.*, 1993; Starck and Wojcik, 1995; Pivot *et al.*, 1999). Starck and Wojcik (1995) also recorded good growth and yield using an artificial foam Dynamite while Pivot *et al.* (1999) obtained good yields when tomato plants were grown on substrates coir, pozzolano (volcanic ash) or pumice as compared to soil.

Growing media varied from place to place. Peat was mixed with small dose of chemicals. In this mixture, risk of pollution by soil-borne pests was negligible. Abou-Hadid *et al.* (1993) recommended a mixture of peat, sand and vermiculite at 1: 1: 1 and stated that rockwool cubes gave better results than plastic trays. Varis and Ozuyaman (1994) found that the early fruit yield of green house grown tomato was the highest with top watered coarse perlite medium (930 g/plant). This top watered coarse perlite technique produced as good a yield as the best conventional methods, without any fermentation or the need for sterilization and also had the lowest labour requirements. Cocopeat was found to be as good as Agrofoam (polyurethane) and hence was recommended for tomato production under controlled environment (Pek *et al.*, 1996). Sahin *et al.* (1998) recommended that peat alone or growing media consisting of 50 per cent peat to grow under glass gave better growth, yield and quality of tomato fruits.

Tomato plants were also found to grow well in NFT or plant plane hydroponic (PPH) culture. Fruits obtained in hydroponic culture had low acidity but higher sugar contents. But plants grown in NFT produced better quality fruits than PPH culture (Averswald *et al.,* 1996). The pH of the nutrient solution in NFT is very important. Papadopoulos and Pararajasingham (1998) suggested the use of HCl instead of H_3PO_4 to regulate the pH of nutrient solution in NFT based green house tomato production. The total marketable yield was found to be better when HCl was used, probably due to high fruit production during the early part of the harvesting period. In contrast to NFT, Yang *et al.* (1989) observed better root growth of seedlings in aeroponic culture. They (1989b) also found that plant growth, fruit yield and size were the highest in aeroponic culture. Solutions sprayed frequently at an interval of 5 minutes gave the best results in aeroponic culture. Olympios *et al.* (1994) observed that under hydroponic culture, substrates like natural pumiceous perlite (CKP), row perlite (YP) and pumice (EL) gave better performance than horticultural perlite (PE). Plant growth parameters (height and number of leaves) and total yield (number and weight of fruits) were also better on natural pumiceous perlite and pumice.

In trials with soilless cultivation in green house substrate like black peat, white peat, pumice, Grodan (rockwool), Green-Mix, Agrofoam were used singly or in various combinations for tomato cultivars Candela and Acor tomatoes. It was grown in Agrobags, with fertilizer supplied as NFT type solution for the artificial substrates (Pardossi *et al.,* 1989). Pumice with or without Green-Mix gave the highest fruit yield, but there were no significant differences between different substrates indicating possible good yields through the use of inexpensive materials. Jensen (1989) while working with inert growth media recorded the earliest cropping on standard Grodan matting (rockwool, 900 x 200 x 75 mm) and Plantex (polyurethane). Fruit weight was the greatest and with overall good quality on Grodan. The age of polyurethane (PU-Agrofoam) mats made little difference to crop earliness, total production or average fruit size (Benoit and Ceustermans, 1987).

The use of 30 litre bags of perlite with a shallow reservoir of nutrient solution, was the most widely adopted hydroponics system for glass house crops in Scotland. In comparative trials over a number of seasons, crops grown in the perlite system consistently out-yielded those grown in rockwool or in NFT. Being inert, physically stable and easy to handle, use of perlite is commercially attractive. The used perlite was subjected to range of pasteurization treatments by steam, dry heat and fungicide to grow tomato successfully (Szmidt *et al.,* 1988). Yield increments in terms of quantity, quality and earliness were found to be enhanced when plants were grown in a recirculating sand culture system in plastic houses (Ibrahim *et al.,* 1989). Calabretta *et al.* (1994) obtained higher yields when tomato plants were grown in soilless culture systems comprising white polyethylene bags containing 2 to 4 mm diameter expanded clay granules. Properly matured, composted commercial municipal solid waste product could also be safely used as soilless medium for tomato seedling production. Xing *et al.* (1997) observed faster rates of plant growth, higher dry matter productivity and higher fruit yield in a new multilayer soilless culture system and suggested as a replacement of the soil based production system to achieve significant yield improvement in green house grown tomatoes.

10.2.3 Spacing

Ihsan *et al.* (1988) recommended growing plants with single stem and topping above the 5th truss and growing the plants with single stem at a spacing of 115 x 25 cm or 2 stems at a spacing of 115 x 40 cm for short production cycles. For longer cycles, topping the plants above the 8th or 10th truss with single stems (at 115 x 25 cm) was recommended. According to Papadopoulos and Ormrod (1988) closer spacings decreased the leaf area thereby reducing the net photosynthetic area but upper leaves were not affected. They (1988, 1991) also opined that as the carbohydrate supply was less in the spring than the autumn, it led to more intense competition for light in the spring. For maximum yield, plants were spaced more closely (23, 30 and 38 cm) corresponding to 113,000, 74,000 and 53,000 plants/ha in the autumn than the spring (45, 53 or 60 cm) corresponding to 40,000, 31,000 or 25,000 plants/ha. A wider spacing of 60 cm produced plants with a thinner head and less leaf area than those plants at 30 cm. At closer spacing higher CO_2 (510 ppm) increased the yield of harvested fruits as well as those remaining on the plants (Rijsdijk *et al.*, 1991). In Kenya, Kirimi *et al.* (2011), from their study of indeterminate tomatoes planted in green house, reported that spacing significantly affected fruit size, and a spacing of 40 cm x 30 cm produced the smallest fruits, while the largest fruits were borne on plants spaced at 50 cm x 40 cm, whereas highest yield and economic returns were obtained when the plants were planted at a spacing of 50 cm x 30 cm.

Liptay and Hoffman (1988) obtained satisfactory growth and agronomic performance in the field when seedlings were grown at a plant density of 3300/m^2 inside the green house, were transplanted outside. Papudopoulos and Ormrod (1990) found that marketable yields could be increased by increasing plant density from 25,000/ha to a maximum of 74,000 to 113,000/ha, but increase in plant density resulted in decrease in fruit size. In yet another report, highest yield of good quality fruit was obtained when tomato plants were planted at 70,000/ha and retaining 10 trusses/plant (Apahidean *et al.*, 1991).

For seedlings transplanted in July, early fruit yield and total marketable yield was the highest at 50,000 plants/ha but for seedlings planted in February, fruit yield was highest at 40,000 plants/ha (Streck *et al.*, 1996). They (1998) also obtained highest yield when plants were grown at 80,000 to 10,000 plants/ha with 3 clusters. Good yield was also obtained in plants grown at 30,000 or 40,000 plants/ha pruned to 7 clusters.

10.2.4 Planting Time

Best yields were obtained when tomato was planted in late March or early April (Apahidean *et al.*, 1989). According to Simidchiev and Kanazirska (1995), seedlings were planted when the first raceme or the 3rd- 4th raceme was initiated. Delayed planting increased the yield.

10.2.5 Fertilization

The nutrients are supplied in the form of chemicals. It is possible to supply nutrients in regulated amount with irrigation. Phosphorus, calcium, iron and magnesium were added to peat before planting; basal dressings to the crop grown in soil were rarely necessary. The pH of the soil was maintained at 6.0-7.0. Nitrogen and potassium were supplied from potassium nitrate, ammonium nitrate and urea. Basic solution was prepared by adding 1.5 kg of potassium nitrate in 10 litres of water. Before application, this solution was diluted to 800 ppm. In tomato cultivar Moneymaker grown in peat compost under glass, Varis and George (1985) recommended high level of N (100 g m^{-2}) for increased flower number, fruit and seed yield and early flowering and ripening. The most favourable combination for high fruit and seed yield was 100 g N, 43.69 g P and 166 g K. Voican *et al.* (1989) obtained the highest yields with 240 kg N +160 kg P + 200 kg K/ha.

Fruit yield and quality could be controlled through fertigation in the green house (Bar-Yosef, 1988). Fertigation was applied at regular intervals with a solution of fixed N: P: K ratio (6: 1.1: 5) by trickle irrigation at an emitter discharge rate of 2 litres/h. Careful control of nutrient solution and substrate conditions were essential for production of high yields of good quality fruits (Musard, 1988). The composition of nutrient solution varied with the growth stages of the plants. A test system for fertilization studies in green house crops was described by Garedal and Lundegardh (1997).

Chen *et al.* (1998) observed that tomatoes required higher Ca, Mg and S, and relatively low N; the nutrient formula also needed adjustment at different growth stages. The nutrient formula could also be adjusted by analyzing various nutrients in the drainage water. Satisfactory yields and quality tomato were obtained following reuse of the nutrients solution although yield was reduced and there was an increase in salinity (Lozano *et al.*, 1998). Sahin *et al.* (1998) reported that during fruit development plants were irrigated with a 20-20-20 solution (800 ppm) and at harvesting plants were irrigated with a 15-10-30 solution (1200 ppm) through drip (trickle) irrigation. This resulted in better yields with fully recycled nutrient solution. Fully recycled irrigation saved 42 per cent water.

10.2.6 Irrigation

Tomato requires an ample supply of moisture in glass house. Water stress leads to stunted growth of plants and dropping of flowers. Veken *et al.* (1982) conducted an irrigation experiment during 4 successive growth seasons in Belgium with non-controlled standard watering and 2 tensiometer controlled watering. The yield in both tensiometer controlled treatments was consistently better than in the non-controlled ones. Pavlov *et al.* (1983) reported that irrigation to the cultivars Extase and Virosa in Bulgaria with magnetized water increased the early and total yield by 22.8 and 45.8 per cent, respectively and improved fruit quality compared to the control plants.

Irrigation via a central inlet into the canals was found to enhance yield of tomato grown on rockwool mats. Incidence of skin cracking was greater with drip irrigation but yields were almost at par with controlled inlet system. Average fruit weight was found to be greater under drip irrigation; central inlet irrigation resulted in higher Mg deficiency. Hence drip irrigation was found to be better for tomato (Tanis, 1990; Tiwari *et al.*, 1998). However Abou-Hadid *et al.* (1993) found that sub-irrigation was more effective than sprinkler irrigation. Tomato plants in green house frequently suffered from excessive vegetative growth, resulting in malformed fruits and delayed maturity. By impulsing a mild water stress on the plants, a balance between vegetative and reproductive growth could be restored in favour of better fruit development. Soil water management based on leaf water potential was found to be very useful in the forcing cultivation of tomato in green houses. Water potential of -0.1 MPa in the early growing stage (from planting to flowering on the third cluster) and -0.5 MPa thereafter were recommended as indicators for irrigation, with a soil depth of 40 cm (Araki, 1994). Norrie *et al.* (1994) reported that PET (potential evapotranspiration) and electronic tensiometer technology were used jointly to improve irrigation management for tomatoes grown in peat based substrates by responding more accurately to the crop's water and nutrient needs. Irrigation to the green house crop was scheduled every 3 days or daily on the basis of pan evaporation coefficient. The best pan coefficient for maximum early and total yields was found to be 1.2 and irrigation was scheduled accordingly (Tuzel *et al.*, 1994). Xu *et al.* (1994) reported that oil water deficit resulted in decrease in photosynthesis, thereby resulting in decreased yield. According to Pulupol *et al.* (1996) deficit irrigation (DI) in which plant water potential was allowed to decrease from -0.5 to -1.2 MPa, reduced plant growth, fruit yield, size, count and caused blossom end rot. But the fruits had higher concentration of sucrose, glucose and fructose thus enhancing the fruit desirability. Guistiniani *et al.* (1997) observed that increasing salinity in irrigation water (734 mm) retarded emergence of seedlings and the seedling growth declined significantly. Higher crop yield was obtained by Benoit and Ceustermans (1998) with deep flow technique (DFT) when plants were grown in plastic containers in the green house. According to Chen *et al.* (1998) the most important factor affecting water supply is the light intensity and the irrigation volume is determined from the lighting intensity and electrical conductivity of drainage water from the substrate.

10.2.7 Mulching, Pinching and Training

It was reported from Italy that in an unheated green house mulching of tomatoes with polyethylene, grown in double rows at 40,000 plants/ha did not affect date of flowering or fruit set, but significantly increased fruit yields for cultivars Count (90 t/ha) and Duke (75 t/ha) (Marchesi, 1986). Pimpini *et al.* (1987) reported that the use of 1.6 m wide and 0.6 m high clear polyethylene tunnels of polypropylene sheets placed directly on the plants, and of mulching with clear polyethylene, advanced earliness of the cultivars Arado 320 and Pakela grown in a green house.

Tomato plants produced axillary shoots rapidly. The recommended practice was to remove such shoots as early as possible. When the plants reached a height of 120 cm, the leaf area was considered to have maximum potential for transpiration. It was suggested to remove leaves. Gradually removal of leaves was suggested to provide air circulation around the whole plant. Pinching above the 1st and 2nd trusses together with close spacing advanced earliness and shortened the harvesting period. Wien and Minotti (1988) observed increase in yield of tomatoes in the green house with plastic mulch and apex removal. As reported by Ankara and Arin (1998), total yield was found to be highest in plants pruned from the 8th truss and mulched with wheat straw. The highest early yield was obtained from plants pruned from the 4th truss and mulched with any one of the used mulches *i.e.*, transparent polythene, black polythene and wheat straw under low tunnel.

Training of plants was essential for glass house cultivation of tomato. The tomato plant was unable to support itself on its own stem and crept on the ground. The usual method is to put the strings twined all along the stem to be hung from supports above. Arch method of training is extensively used. Many wires were placed over the rows of plants 180 cm above soil.

Khoshkam *et al.* (2014) studied the impact of different plant training systems on quantitative and qualitative parameters of green house tomato cultivars. They took three levels of training system, one stem, two stem and four stem training system. They found that significantly maximum fruit length (34.8 cm), diameter (36.3 cm), total soluble solid (2.8°Brix), vitamin C (85.3 mg/100 g), acidity (0.79 per cent) and yield (137.5 t/ha) in two stem training systems.

10.2.8 Weeding

Weeding is to be done to keep the crop in a healthy condition. Of different weeds, *Solanum nigrum* was very rampant. Black night shade (*Solanum nigrum*) emerged at the 6-, 4-, and 2-leaf stages of tomato crop and simultaneously grew with it (Gonzalez *et al.*, 1996). The weed competed for light and nutrients with tomato when its height exceeded with the beginning of flowering of the main crop and reduced fruit yield, number of fruits and the N, P and K uptake.

10.3.0 Fruit Growth and Yield

Cold treatment (15°C) accelerated reproductive growth, lowered the node position of the truss, increased the number of flowers and decreased the occurrence of fruit puffiness. Although vegetative growth was delayed for 2 weeks by cold treatment, the average weight of fruit on the first and second fruit truss was superior in cold treated plants; 90 to 100 per cent fruit set was obtained after treatment with 4-CPA, vibration or a combination of both. As 4-CPA was associated with lower seed production, it was suggested to use 4-CPA to enhance fruit set in plants (Vorayos *et al.*, 1992). Higher yields and more number of tomatoes per unit area were obtained from plants growing in green house as compared to the outside raised plants (Ahluwalia and Singh, 1966). Foutes *et al.* (1997) also observed that crop growth in high tunnel covered in plastic (0.1 mm thickness) gave better growth and yield as compared to those grown in the open fields. Martinez *et al.* (1998) observed that defoliation increased the correlation between fruit weight and the grades of fruit (C1: fruits over 150 g, C2: weight of fruits between 100 and 150 g, C3: weight of fruits between 100 and 120 g). Defoliation also decreased the negative correlation between fruit weight and fruits per cluster, thereby enhancing the total yield of tomatoes.

Under protected cultivation depending upon cultivar and cultural method, the yield may vary from 80 to 400 t/ha. The average yield in Denmark and Netherland is as high as 272 tonnes and 466 t/ha respectively (Anon., 1998a).

11.0 Grading

The grading of tomato is recommended for specialized city markets in India. Indian Standard Institution specified 4 grades, *viz.*, Super A, Super, Fancy and Commercial for tomato fruits.

Because of the high cost and unreliability of manual tomato sorting, a computer vision-based sorting system was developed in USA. It classified tomatoes according to size, shape, colour and surface defects (Sarkar, 1985). A photoelectric system based on 4 light sensors was used to explore the possibility of colour sorting for ripeness in Japan. A colouring index was calculated which covered the change from green to pink or red (Mohri *et al.*, 1987). In Italy, Giametta (1989) described an electronic tomato sorting machine in which image processing techniques classified tomatoes on their colour characteristics. Colour sorters from Italy were fitted to harvesters. This machine helped the industry to stay competitive as these colour sorters improved the quality of the product delivered to the processor. A 3-axis robotic manipulator with a vacuum grab end effector was developed for gripping and carrying tomatoes during sorting and it was observed that 97.3 per cent of tomatoes were grabbed and delivered successfully at a rate of 33.7 seconds per tomato (Lin and Lan, 1997). Laykin *et al.* (1999) from Israel reported an automatic quality sorting system based on two sensors, vision and impact and designed in a way to sort the fruits into 10 classes according to the information of the fused sensors. This overall system yielded 88 per cent exact classification with 95 per cent of the rejected fruits correctly classified.

12.0 Ripening

Transformation of the mature green tomato to red ripe fruit involved many complex sequences of biochemical reactions. In developing fruits, chlorophyll-a and b contents increased 15 days after the anthesis till 55th day and then decreased to zero at 55th day (McMohan *et al.*, 1990).

Fruits took 66 d to ripen in a 15°C night temperature compared with 74 days at a nominal 11°C night (Hurd and Graves, 1985). At 15°C there was an approximately linear delay in fruit maturation time down the truss from the proximal to the distal end, such that the eighth fruit took a mean of 11.2 days longer to ripen compared with the first fruit (Adams *et al.*, 2001). Adams *et al.* (2001) also reported that tomato fruits ripened 95, 65, 46 and 42 days after flower opening when plants were grown under controlled environmental conditions at 14, 18, 22 and 26°C, respectively. There was a tendency towards small parthenocarpic fruits at both high (26°C) and low (14°C) temperature regimes which, combined with low flower numbers and poor fruit set at 26°C, resulted in low fruit yields.

In the process of ripening, the green chlorophyll pigment degraded and accumulation of lycopene started. The biosynthesis of these pigments was light and temperature dependent (Sayre *et al.*, 1953; Khudairi, 1972; Raymundo *et al.*, 1976). Carotene and lycopene development were promoted by the red and inhibited by the far red light, indicating that the process was under phytochrome control.

The ripening of tomato was accelerated as the internal temperature of fruit rose but development of red colour (lycopene) in the fruit slowed down and ceased as fruit temperature approached 26.6°C. On the other hand, the yellow colour carotene continued to develop at higher temperatures. This was the reason why tomato ripening at temperature above 29.4°C became yellowish red. It was reported that fruits of Bonny Best and Marglobe maintained at 24 to 28°C developed deep orange red colour and become yellow when kept at 32 to 36°C. At 40°C or higher, fruits remained green and failed to ripen. Masarirambi *et al.* (1996) reported that tomato fruit ripening is reversibly inhibited at high temperature (> 30°C). Thomas *et al.* (1982) advocated the use of durometer for assessing firmness of ripe fruits and hence predicting long-term storage.

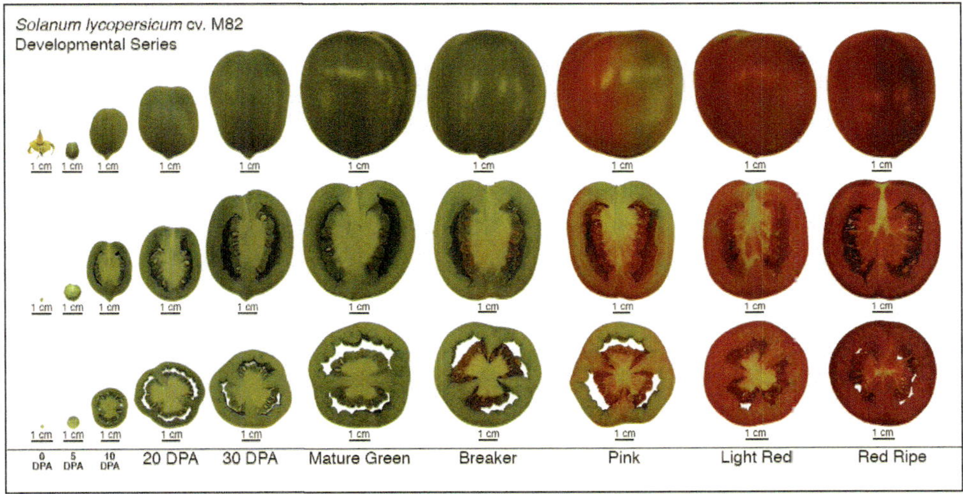

Fruit Development and Ripening Stages of Tomato.

The interplay of ethylene and polygalacturonase (PG) in tomato fruit ripening and practical possibilities for the control of tomato ripening were reviewed by Hobson and Harman (1986). Thakur and Pandey (1999) assayed PG and pectin methyl esterase (PME) in fruits of 10 tomato cultivars during ripening and observed that activities of both enzymes increased after harvest until the red-ripe stage. Changes in texture during ripening of a normal tomato cv. Ivan (USA) and the long shelf-life cv. Graziella (Israel) raised in green house were studied by Chiesa *et al.* (1997). PG activity differed significantly with cultivar, maturity stage and storage temperature. PG activity increased during storage particularly at 20°C and activity was higher in cv. Ivan than in cv. Graziella fruits after storage at both temperatures (10 or 20°C). Graziella fruits were firmer than Ivan fruits at similar maturity stages and storage temperatures.

12.1.0 Pre-harvest Treatments

The effect of chemicals on tomato ripening was first reported by Russo *et al.* (1968). They found that 14 days after treatment with 5000 ppm ethephon, the ripeness increased from 62 to 84 per cent. Read and Fieldhouse (1970) applied Alar, a growth retardant at 2500 ppm at first and 4-leaf stage and observed increase in concentration of harvest and early yield. Splittstoesser and Vandenmark (1971) treated Eniz 1350, Manalucie and Roma cultivars with ethephon 500 ppm when 10 per cent fruits were pink. It promoted ripening in 10 days and induced 55 to 65 per cent fruits to ripen at second harvest, while 25 to 30 per cent fruits ripened in untreated plants. Dostal and Wilcox (1971) observed increase in total yield and percentage of ripe fruits significantly following single application of ethephon at various stages of fruit ripening and reduced the percentage of green fruits. Iwahori and Lyon (1970) reduced the ripening period by 7 days in a green house crop by 500 ppm ethrel applied 15 to 35 days after anthesis.

Sims and Kashmere (1972) observed accelerated fruit ripening and concentrated fruit harvest by use of ethephon to green tomatoes 42 days after pollination. Ethylene was also used to hasten ripening but Heinze and Craft (1953) found that rate of ripening of mature green tomato in small container was not affected by addition of ethylene. Baranov (1982) found that treatment of ethrel at the rate of 1.6 to 2.0 kg/ha induced ripening 6 to 8 days earlier than control and amount of red fruits increased by 28 to 34 per cent. Adriaenssens (1998) investigated the effects of ethephon (480 g/1W.S.), applied by spraying at 2 or 4 l/ha (in solution at pH 3 or 7) or by smearing (100 per cent or 50 per cent in water) on ripening of tomatoes (cv. Fausto). Spray did not exceed the residue limit (3 ppm), whilst application by smearing did. Harvesting at later stage and application at pH 3 resulted in more rotted fruits. The faster ripening occurred following application of ethephon by smearing. The ripening delay in ethrel-treated fruits (applied 2 days after anthesis) may be caused by longer time required for the increased cell number to reach maturation (Atta-Aly *et al.*, 1999).

Powell and Stoffella (1998) reported that in autumn, low (14.5 mg a.i./plant), medium (24 mg a.i./plant) and high (24 or 43.5 mg a.i./plant) rates of imidacloprid reduced the percentage of tomato fruits with external irregular ripening symptoms from 33.4 per cent in the untreated control to 1.4, 0.3 and 0.7 per cent, respectively. The percentage of fruits with internal symptoms was reduced from 83.8 per cent in the untreated control plot to 14.2, 4.0 and 6.1 per cent for low, medium and high imidacloprid applications, respectively.

Pre-harvest treatment of tomato fruits with 4.0 and 6.0 g/litre Chitosan combined with 2.0 g/litre of salicyclic acid increased fruit firmness, weight, length, diameter, juice volume, colour intensity, carotenoids, chlorophyll, total and reducing sugars, soluble sugar content, acidity, ascorbic acid content besides an extended fruit shelf life and delayed ripening and there was decrease in fruit decay incidence, electrolyte leakage and pH as well as, unmarketable fruits (Almunqedhi *et al.*, 2017).

Individual application of ethephon (1.6 ml/litre) and iodine (5 ml/litre) significantly increased fruit weight (133.71 g) and firmness (3.26 N) respectively, with respect to the control (80.36 g and 0.95 N), while those sprayed with 125 mg/litre of sodium selenite showed the highest titrable acidity (0.34 per cent citric acid) (Martínez-Damián *et al.*, 2019).

12.2.0 Post-harvest Treatments

Park *et al.* (1994) observed that tomatoes at the breaker and pink stages, coated with corn-zein film and stored at 21°C showed a delay in colour change and loss of

Post-harvest Handling Operations in Tomato

Harvested Tomatoes (Left) should be Washed and Wiped (Right) with Moist Clean Cloth.

Packing in Bamboo Baskets (Left) during Transport Lead to Damage on the Fruits, and Exposure of Newly Harvested Tomatoes will Result in Heat Build-up and Accelerated Ripening.

Sorting of Tomatoes using a Sorting Table (Left) and Packing in Rigid Plastic Crates (Right) Provide Adequate Protection.

Packaging of Tomato for Long Distance Transport (Left, Middle) and Selling in Supermarket (Right).

firmness and weight and enhanced their shelf-life by 6 days. Maharaj *et al.* (1999) reported that photochemical treatment might have potential for preservation of fresh tomato fruits. The delay in senescence was attributed, in part to the maintenance of a high level of putrescine (anti-senescence agent exerting opposite physiological effect to ethylene).

Semperfresh fruit coating reduced weight loss compared with controls, but the difference between coated and uncoated fruits was not significant. Compared with controls, the shelf-life of pink tomatoes was increased by 3 days at 23 ± 2°C and 6 days at 12°C in air by coating. Fruits could be stored for 40 days in controlled atmosphere (Tasdelen and Bayindirli, 1998).

Effect of anti-senescent regulators on the shelf-life of tomato were studied by Kumar *et al.* (1998) and it was noted that fruits treated with GA_3 at 10 ppm or BA at 25 ppm showed the minimum physiological weight loss and incidence of decay. Sharma and Dashora (1999) observed that weight loss, rotting, volume reduction, visual appearance deterioration and decrease in TSS content were reduced by 0.5 per cent $KMnO_4$ treatment as compared to the control. The $KMnO_4$ (0.5 per cent) treatment also resulted in better retention of acidity.

Yanuriati *et al.* (1996) noticed the ethanol vapour treatment for 6 hours at 20°C prior to storage significantly slowed down the colour changes and softening. Smid *et al.* (1996) observed that tomatoes (cv. Chaser) treated with an aqueous solution of trans-cinnamaldehyde reduced the number of bacterial and fungi infection and increased the storage life.

Hong and Lee (1999) reported that infiltrated glucose delayed ethylene climacteric rise by 2-3 days in mature green tomato fruits (cv. Rutgers) compared with control fruits. However, there is no significant difference in ethylene production rate between treated and control fruits. The tissue injected with glucose retained its green colour for upto 10 days after treatment, while red colour developed in untreated control.

Tomato treated with chlorine, packed in perforated (0.25 per cent) polyethylene bag and kept at ambient (temperature 20-25°C and relative humidity 70-90 per cent) condition resulted in substantial reduction in losses caused by decay and weight loss (Nasrin *et al.*, 2008).

Cherry tomato cv. Marilee red (*Lycopersicon esculentum* Mill) harvested at pink stages treated with 2 per cent $CaCl_2$ was effective in controlling storage loss as well as in maintaining the quality of the produce during storage, though the ascorbic acid decreased during storage (Gharezi *et al.*, 2012).

13.0 Storage

13.1.0 Storage at different Temperature

Tomato is kept in store either at mature green or breaker stage of maturity. Storage temperature has been reported to influence the fruit firmness in cultivars like Florida MH-1 and Floradade. Fruit remained firm up to 21 days when kept at 20°C. There was no relationship between fruit colour and firmness. Risse *et al.* (1980) and

Storage of Tomato under different Environments

Hall and Augustine (1981) found that fruit developed moderate chilling injury and moderate to severe *Alternaria* rot when stored at 7.2 to 10°C. Tomato stored at 15°C had less decay. Thompson *et al.* (1960) showed that keeping fruits for 6 days at 6°C before ripening delayed colour development by 6 days as compared to fruit stored at 20°C. Inaba and Crandall (1987) from USA reported that when hand-harvested mature green tomatoes cv. Sunny, in plastic bags were immersed in ice cold water (0°C) or ice cold water with $CaCl_2$ (-2°C) for 60 or 120 minutes and then stored at room temperature (20°C), the colour development was significantly delayed and shelf-life extended without increasing decay or chilling injury. Treatment at 0°C for 60 minutes delayed colour development by 4-5 days and -2°C treatment for 120 minutes extended shelf-life by 2-3 weeks.

Lurie *et al.* (1993) reported that pre-storage heating of tomato can extend their storage life. They also observed a correlation between the maintenance of heat-shock protein mRNA during storage and the inhibition of chilling injury in heated tomatoes. Artes and Escriche (1994) and Hakim *et al.* (1997) found that intermittent warming reduces chilling and decay of tomato fruit. Kluge *et al.* (1998) however, recorded that there was no chilling injuries if tomato cv. Santa Clara fruits were stored at 5°C and 90-95 per cent RH with intermittent warming at 25°C for 96 hours. Quality and shelf-life of tomatoes are also improved by intermittent warming (Artes *et al.*, 1998a). Intermittent warming of fruits of long life tomatoes (cv. Durinta) in cold storage condition increased pectolytic enzyme activity accompanied with a fall in respiration rate and C_2H_4 production (Artes *et al.*, 1998b).

Kluge *et al.* (1996) observed that after storage of tomato fruits (cv. Santa Clara) at 13° ± 0.5°C and 80 ± 10 per cent RH for 21 days, fruit decay (mainly *Rhizopus* and *Botrytis*) was significantly higher in fruits harvested at the pink and red maturity stages and packed in non-perforated sealed polyethylene.

Tomato cvs. ES-58, H-2274, Tobol, Rio-Grande harvested at green mature and breaker stage could be stored at 12°C for 40 days. Fruit firmness and titrable acidity decreased and soluble solids increased and weight loss (3-8 per cent) was more severe at an early stage (Kayanas and Surmeli, 1995).

Fruits from 3 tomato hybrids, Pioneer 5692 F_1 (normal ripening), Pioneer 6353 F_1 (ripening inhibitor, *Rin*) and Pioneer 4127 F_1 (non-ripening, *Nor*) were picked at mature green and pink stages of maturity and stored at 10°C and 85-90 per cent RH or picked when red and stored at 20°C to assess shelf-life by Agar *et al.* (1994). *Rin* fruits had the firmnest fruits over all the treatments and had the highest titratable acidity but fruits from this hybrid had the lowest TSS. Nor fruits had the best shelf life, *Rin* fruits were the most suitable for cold storage. Green and pink fruits of normal and *Rin* types could be stored in cold storage for upto 30 days.

Fully ripe tomatoes of cocktail tomato cultivars Amoroso and Brioso and truss tomato cultivars Roterno and Capricia were stored for 10 days at two temperatures: 4°C as simulation of home-refrigerator storage and 15°C as an optimal storage temperature by Farneti *et al.* (2010). There were no apparent symptoms of chilling injury, but 4°C temperature stimulated firmness decay in both cocktail tomato cultivars, increasing fruit susceptibility to mechanical injury. Moreover, already after

5 days of 4°C storage, tomatoes generally showed decreased sugar and increased acid content (especially in cocktail tomatoes) compared to 15°C stored fruit, indicating a loss of sensory quality at 4°C.

Colour values *"L"*, *"a"*, and *"b"* depicted change in colour properties during storage and an electronic nose comprising of 18 semi-conductor oxide metallic sensors pure and doped semiconductor (MOS), coupled with a mass spectrometer system could be used to discriminate odours of the fruits and these could be used for monitoring short-term storage of tomato (Messina *et al.*, 2012).

Cold room storage was best for retaining shelf life and quality attributes in harvested tomatoes but, evaporative cooling system was also a cost effective means of extending shelf life in rural areas with minimum resources (Nkolisa *et al.*, 2019).

Tomato fruits harvested ripe and initially stored at 12.5°C for 1 day, at 20°C for 2 days (depicting the storage conditions in phase of harvest to transport and retail), and afterward at either 20 or 7°C (depicting the storage conditions at consumers' place) for another 4 days did not show much change in flavour and the variation in flavour and taste attributes varied according to varieties (Kanski *et al.*, 2020).

13.2.0 Modified Atmosphere Storage

In an investigation by Yang and Chinnan (1988) tomatoes were stored under 20 gas environments involving O_2 and CO_2 concentration of 5-20 per cent. Two models were developed to predict the respiration rate of tomatoes stored under controlled atmospheric conditions and the simplicity of use and the accuracy were assessed. Oxygen concentration had a more pronounced effect than CO_2 concentration on firmness. Tomatoes stored at 21°C and 5 per cent O_2 and 5-15 per cent CO_2 gas environment up to 45 days had no significantly different off-flavour than tomatoes stored at 12°C under ambient conditions. Ripening of 'Improved Rope' tomatoes was delayed for one week under ambient conditions (29-33°C) in 0.05 mm polyethylene bags with zero to 4 pinholes for gas diffusion, maintaining high CO_2 (7-12 per cent) and low CO_2 (2-7 per cent) concentrations (Esquerra and Bautista, 1990). Batu and Thompson (1998) reported that ripening time of tomato fruits (cv. Criterium) harvested at green mature stage, exposed to 60 per cent CO_2 for 1 day was 18 days (without CO_2 injury), whereas it was 14 days for fruits exposed to 20 per cent CO_2 for 5 days and 15 days for fruits exposed to 40 per cent CO_2 for 1 or 3 days.

Controlled and modified atmosphere storage (initial gas composition of 5 kPa O_2 and 3 kPa CO_2 for both and RH 85-90 per cent for controlled atmosphere storage) significantly retarded the ripening process compared to conventional cold storage and tomato quality, measured by firmness, colour, TSS, TA, ripening index, was preserved in controlled atmosphere storage for almost 90 days, better than MAP and superior to conventional cold storage (Majidi *et al.*, 2014).

14.0 Cherry Tomatoes

Cherry tomato [*Solanum lycopersicum* var. *cerasiforme* (Dunnal) A. Gray] is a highly priced culinary as well as ornamental vegetable. One of the most popular high value exotics, it is a favorite among chefs who cook for high profile restaurants and hotels. Nevertheless, it is becoming increasingly popular among common people, who are now interested in garnishing their dishes and diversifying their nutritional intake. Cherry tomatoes look not only attractive in kitchen gardens but are commercially valuable horticultural commodity and have impressive nutritional and pharmaceutical properties. According to the USDA nutritional information, one cup of cherry tomatoes (149 g) provides 26.8 calories, 1.3 g protein, 4.5 mg omega-3 fatty acids, 119 mg omega-6 fatty acids, 1241 IU of vitamin A, 18.9 mg vitamin C, 22.3 mcg folic acid, 11.8 mcg vitamin C, 353 mg potassium, 35.8 mg phosphorus and 14.9 mg calcium. All these nutrients and additives put cherry tomato into the category of functional foods. Whole cherry tomatoes are sold afresh as well as preserved in brine. The fruit shape ranges from spherical to oblong to pear and the colors include red, pink, yellow, grape green and brown. Cherry tomato can be grown in open or protection in poly or glass or shade houses. However, yield and fruit quality are higher when cultivated under protection.

The cultivation practices of cherry tomatoes have been detailed by Malik *et al.* (2017).

14.1.0 Varieties

Many varieties of cherry tomato have been popularized by private and public sector. Brief description of varieties released and/or popularized by public organizations is given in Table 6.

Table 6: Varieties of Cherry Tomatoes

Name	*Brief Description*	*Source*
Punjab Red Cherry	Red fruited variety, suitable for protected cultivation, tolerant to leaf curl virus, yields 110 t/ha	Punjab Agricultural University, Ludhiana
Punjab Sona Cherry	Yellow fruits borne in clusters having TSS of 7.5°Brix, 13 mg carotene/100 g fruits, yields 105 t/ha	Punjab Agricultural University, Ludhiana
Punjab Kesar Cherry	Orange fruits borne in clusters having TSS of 7.6 per cent, 13 mg carotene/100 g fruits, yields 100 t/ha	Punjab Agricultural University, Ludhiana
Pusa Cherry Tomato-1	Red spherical fruits, suitable for protected cultivation, yields 150 – 180 t/ha, first harvesting takes 70-75 days, and crop lasts for 9-10 months, tolerant to root knot nematode	Indian Agricultural Research Institute, New Delhi

14.2.0 Climate and Soil

Like slicing tomato, cherry tomato is a warm season crop that thrives during summer months in Kashmir. Optimum temperature for its growth and development is 20-30 °C with low humidity and plenty of sunshine. Temperature during March is congenial for seed sowing and transplanting is successful by the time temperature elevates in the month of April. Fruit setting, development and ripening takes place from May to August. However, under protected conditions, nursery production can be advanced by 15-20 days. Crop can also be produced in green houses, wherein the production window is stretched to a large extent. Pickings can continue till November-December compared to open cultivation where crop lasts till July-August only. Soil for cherry tomato production should be rich in organic matter, sandy loam to loam with pH 6.5 to 7.5, well aerated and well drained.

14.3.0 Fertilizer Requirements

General guideline for basal fertilizer dosage is 10 tonnes FYM, 60 kg DAP, 45 kg MOP, 50 kg secondary nutrient mixture and 10 kg micronutrient mixture per acre of field. Urea is not required if above mentioned dose of DAP is applied, as the requirement of nitrogen will be met out of DAP itself. However, precise requirement of fertilizers in a particular field should be calculated according to soil health card and on the basis of this general guideline. Half of the DAP and MOP may be applied as basal dose and remaining half at 15-20 days intervals throughout the crop life cycle. Biofertilizers like *Azospirillum* and Phosphobacteria can also be mixed with FYM to increase and sustain nutrient uptake for longer duration.

14.4.0 Nursery Raising

Cherry tomato seeds are minute and delicate. So, their nursery should be raised very carefully. It is best to do plug tray nursery production using mix of loamy soil, sand, and cocopeat at 2: 1: 1 ratio by volume (Mukherjee *et al.,* 2020). Plug trays ensure proper care of each seedling and least damage to root zone due to retention of soil ball while transplanting. Nursery should be protected from water logging and over nutrition. Seeds can be treated with captan or thiram @ 2-3 g/kg seed before sowing to protect from seed borne pathogens. Alternatively, the medium or soil can be drenched with 1-2 per cent captan/thiram suspension. Nursery should be well protected from cold. For the purpose, small polythene tunnels and poly houses can be used. If not using plug trays, nursery should be sown on 15 cm raised or flat beds depending on water retention capacity of the soil, free of clods, stones, pebbles, plant debris *etc.* Soil should be well ploughed, levelled and nourished with FYM or vermicompost (10 kg) and neem cake (1 kg) and chemical fertilizers like enriched super- phosphate (100 g) per square meter of bed. In case of varieties, 300-400 g seed is required per hectare while in hybrids 100-150 g is enough. The seeds should be sown in shallow lines dug parallel to each other about 5-7 cm apart. After sowing, the lines should be well covered lightly with soil or vermicompost and immediately irrigated with sprinkler. Weed and insect-disease management of the nursery should be done on regular basis.

14.5.0 Transplanting

When seedlings reach 4-5 true leaf stage after about 30-35 days of sowing, irrigation should be withheld about 2-3 days prior to transplanting. Few hours before transplanting, light irrigation may be done to ensure safety of roots while pulling the seedlings. If there is intense sunshine and high temperature, transplanting may be done in the evening to minimize environmental shock to the establishing seedlings. Plough the soil thoroughly and finely incorporate FYM into the soil. Apply basal

Different Cultivars and Growing of Cherry Tomato under Protected Condition

dose of fertilizers. Make ridges and furrows or raised beds or keep the field flat depending on moisture retention capacity of soil and availability of man power. In general, for open conditions, cherry tomato seedlings can be transplanted on ridges or raised or flat beds at 90 cm row to row and 60 cm vine to vine spacing. Under protection, spacing can be reduced to 60 cm row to row and 45-60 cm vine to vine for optimum space economy. The spacing should be optimized on the basis of experience with varieties, as they differ for vegetative growth, plant structure and shape.

14.6.0 Plant Architecture

Whether the crop is grown in open or protected condition, vines may require pruning and will definitely require staking with bamboos/wooden/iron rods and jute ropes to support the indeterminate growth habit. 2 or 3 main branches just below where first flower has blossomed should be retained. All other branches should be removed and well ventilated and strong plant architecture should be ensured. The thumb rule is that more the number of branches, smaller the fruit size and *vice-versa*.

14.7.0 Harvesting and Yield

Harvesting of the fruits should be done in bunches or individually, preferably with calyx intact in the early morning to avoid excess heat build up that will increase the respiration rate in fruits during transportation and marketing. The increased respiration shortens shelf life, which affects both grower and consumer. Harvesting is done at different developmental stages of the fruit depending on distance of the market. For local market, cherry tomatoes are best harvested at pink or light red stage and for distant market at colour break or turning stage. Harvesting at ripe and fully ripe stages is best suited to local processing markets.

The yield depended upon various agroclimatic factors, including soil fertility, cultivar and method of cultivation. On an average in India, under protected conditions, an open pollinated variety can produce fruits of 100-150 t/ha.

14.8.0 Post-harvest Handling

After harvesting, cherry tomatoes should be properly washed with tap water and dried in shade before packaging. Washing removes any chemical residue and reduces microbial load on the fruit skin, if there is any. This enhances its shelf life and wholesomeness. Proper grading according to fruit size, shape, color and quality is a necessity. Fruits can be minimally packaged in plastic pouches/bags or more ornately in plastic or PVC bowls, punnets and trays. Cardboard/corrugated cartons provide more protection during long distance transportation. Both these practices protect commodity, enhance its attractiveness and fetch higher prices in market. The changes in pericarp thickness, fruit firmness and polygalacturonase (PG) activity were observed in parents and hybrids at different harvesting stages (Mukherjee *et al.*, 2019). They observed that the enhanced activity of PG, leading to fruit softening, was found maximum at the ripe stage of cherry tomato.

15.0 Diseases, Pests and Disorders

15.1.0 Diseases

15.1.1 Fungal Diseases

15.1.1.1 Damping Off and Root Rot

These diseases are caused by *Pythium* spp., *Rhizoctonia solani, Phytopthora parasitica* and a few other organisms. These are very serious diseases affecting plants at the nursery stage.

Pre-emergence damping off consisted of killing the seedlings at initial stages of seed germination and emergence. Post-emergence damping off consisted of rapid shrinking and darkening of cortical tissue of the hypocotyls. As the secondary thickening of seedlings got underway, the incidence of early post-emergence damping off declined.

The soil inhabitant fungus was generally, active at low temperatures contrary to *Rhizoctonia* which ramified at higher temperatures. Tripathi and Grover (1975) observed that virulence of *Pythium* was conditional by nutritional status of its substrate glucose. Salinity also affected the incidence of the pathogens. Snapp (1994) reported reduction in *Phytophthora* root rot incidence under salt stress. Therefore, saline condition was congenial for nursery seedlings and prevented root rot at nursery stage.

Control measures consisted of soil treatment with formaldehyde 2 weeks before sowing and seed treatment with 1 per cent mercuric chloride. Spraying of seedlings with fytolan or captan was effective. Hot water seed treatment at 52°C for 30 minutes was recommended along with 5 year crop rotation. *Phytophthora* root rot (*P. nicotianae* var. *nicotianae*) in processing tomatoes could be effectively controlled by application of metalaxyl as soil drench and fosetyl-Al as foliar spray under green house conditions (Flett *et al.*, 1991). Ayub *et al.* (1998) also recommended soil drench with ridomil (metalaxyl) and captan for control of seed rot and damping off in tomato nursery. Wadi (1999) found that soil solarisation with polythene film effectively reduced the population of *Fusarium* spp., *Pythium* spp., *Rhizoctonia* spp., and *Verticillium* spp.

Thiram (0.2 per cent), metalaxyl (0.2 per cent), fosetyl-Al (0.2 per cent), propeconazole (0.05 per cent and 0.1 per cent) were cent percent effective against *Pythium aphanidermatum*, while carbendazim (0.25 per cent) was less effective (Zalte *et al.*, 2013).

Of late, biological control of different pathogens was advocated in order to reduce the cost on chemical fungicides and its residual effects on plants. Good control over phytopathogens (*Fusarium, Alternaria, Botrytis,* and *Pythium*) was achieved by oil inoculation and seed treatment with *Trichoderma viridae, T. harzianum* and *Laetisaria arvalis* (Krishnamoorty and Bhaskaran, 1990; Dinakaran and Ramakrishnan, 1996; Gromovikh *et al.*, 1998). Buysens *et al.* (1993) found that *Pseudomonas aeruginosa* str.7NSK2, a bacterium beneficial to plant growth, gave successful control of *Pythium* spp. Seed treatment with *P. aeruginosa* significantly reduced damping off and additional protection was obtained when the bacterium was added to the nutrient solution. Damping off incidence and its severity (caused by *Rhizoctonia solani*) could be reduced by inoculation with rhizobacteria like *Azospirillum* spp., *Azotobacter chroococcum* and *Pseudomonas fluorescens* (Gupta *et al.*, 1995). Biocontrol of soil borne diseases by *Rhizoctonia solani* and *Pythium ultimum* by treating seeds with *Gliocladium virens* (G 1-3) and *Burkhoderia cepacian* (Bc-F) were recommended by Mao *et al.* (1998). Shoda (1999) recommended *Bacillus subtilis* as a potential biocontrol agent to replace chemical pesticides against damping off in tomato. *B. subtilis* exhibited broad suppressive spectrum against pathogens mainly by producing a lipopeptide antibiotic. Iturins *et al.* (1999) recommended an integrated control of damping off disease caused by *Phytophthora* spp. combining fungicide metalaxyl along with biocontrol agent, binucleate *Rhizoctonia* isolate. This integrated control method was also effective against other damping off fungi *viz.*, *Pythium* spp. and *Rhizoctonia solani*.

The native bacterial isolates of *Klebsiella oxytoca* (one), *Exiguobacterium indicum* (one) and *Bacillus cereus* (two) isolated from Osmani soil were effective in control of *Pythium aphanidermatum* and the two isolates of *Bacillus cereus* particularly reduced the damping off incidence by 27 per cent relative to infected control (Al-Hussini *et al.*, 2018).

Damping off could be controlled by the seed treatment of *Pseudomonas fluorescens* @ 10 g/kg + metalaxyl @ 2 g/kg seed and combination of antagonists *Pseudomonas fluorescens* and *Bacillus subtilis* were found effective in increasing seed emergence, shoot length, root length and vigour index (Zalte *et al.*, 2013).

15.1.1.2 Late Blight

Payne recorded late blight in tomato caused by *Phytophthora infestans* as early as 1848. High rainfall and cool temperature favoured development of this disease. The disease appeared on the foliage at any stage of growth. Brown to purple black lesions began at any point on the leaflet, rachis, petiole or stem and advanced rapidly to cause a severe blight when conditions were favourable. Lesion growth was checked with low humidity. When lesion advanced to outermost zone it was pale yellow. On lower side of leaves, the white fructification of the fungus appeared where the pale and purple areas joined. In fruits, symptoms were mostly at stem end. Grey water-soaked spots enlarged to indefinite size and shape.

The fungus survived between the crop seasons on other host plants and refuge like potato. Fungus also lived in soil. From potato it might be transmitted to tomato seed beds or both the crops could be attacked at the same time.

Most effective control measure consisted of the use of resistant cultivars. Andrus (1946) classed 17 cultivars as resistant. Bonde and Murphy (1952), Walter and Conover (1952), Gallegly (1952) and Khalid *et al.* (1995) reported resistance in *Lycoperscicon* species and cultivars. However, by the identification of physiological races, the problem became more complicated. First evidence of dominant gene resistance was presented by Bonde and Murphy (1952) and later it was labelled as *TR-1* by Gallegly and Marval (1955). West Virginia-700 carried duplicate dominant genes and this accession was known to carry resistance to two races, 0 and 1. Multiple gene resistance was noted in Southland and Wisconsin cultivars. Gallegly (1962) listed Geneva-II and New Hampshire Surecrop as homozygous for *TR-1*. In 1962, cv. Rockingham was released as resistant to *TR-1* (Rich *et al.*, 1962).

Fungicide sprays of Dithane Z-78 and partate were effective for controlling the disease. Other fungicides found effective were Dithane M-45, difolatan, daconil 2787 (Gupta, 1988) and metalaxyl (Barrantes, 1989). Szekely *et al.* (1990) and Khalid *et al.* (1995) recommended a mixture of fungicides like bravo (chlorothalonil), dithane (mancozeb), polihom (metiram zinc + copper oxychloride), ridomil plus 48 (metalaxyl + copper oxychloride), Temoal 52 WP (cymoxanil + metalaxyl + copper oxychloride) for best control of late blight. Under green house conditions, slow release formulations like tebuconazole as soil application gave good control of late blight (Bickers *et al.*, 1999). Ravinder and Srivastava (1998b) observed that plants covered with plastic of any gauge *i.e.*, those grown under low plastic tunnels were less affected by late blight than uncovered plants. Plant extracts from *Potentilla erecta* and *Salvia officinalis* were also found to be highly effective in controlling *P. infestans* because of their antifungal properties (Blaeser and Steiner, 1999).

Krilaxyl (metalaxyl 8 per cent + mancozeb 64 per cent) and Dithane M-45 (mancozeb 80 per cent) were effective in control of late blight in Nepal and resulted in 180.97 per cent and 117.12 per cent increase in yield over control, besides neem, *Trichoderma* and *Artemisia* could also control the disease to some extent and the yield increase were 36.49 per cent, 41.55 per cent and 26.13 per cent, respectively with their application (Shrestha and Ashley, 2007)

Garita *et al.* (1999) reported that a species of *Pencillium, Pencillium* (067) either alone or in sequence with 2 applications of chlorothalonil was effective in reducing the disease. Extracts from bryophytes like *Bazzania trilobata* and *Diplophyllum albicans* were also found to have high efficacy against *P. infestans* pathogens. The antifungal metabolites from these liverworts acted as resistance inducers (Mekuria *et al.*, 1999).

Gudero *et al.* (2018) reported that two times Ridomil MZ gold 68.5 per cent WG foliar sprays for resistant and moderately resistant varieties and three times for susceptible varieties at 10 days interval were found effective in reducing tomato late blight epidemics and increasing fruit yields and they recommended this spray frequency gave the highest protection against late blight and the highest monetary benefit as compared to the other treatments.

15.1.1.3 Buckeye Rot (Fruit Rot)

It is caused by *Phytophthora parasitica* and the disease appeared on lower fruits. The pathogens did not affect the foliage and thus the disease was distinct from late blight. The disease appeared as a brownish spot, often at the point of contact between the fruit and the soil. As the spot enlarged, the surface of lesion assumed a pattern of concentric rings of narrow, dark brown and wide, light brown bands.

This has wide host range and is a soil inhabitant. Plenty of soil moisture and soil temperatures between 18 and 30°C are congenial for development of the disease.

Control measures consisted of sprays with copper fungicides and avoidance of poorly drained soils. Prophylactic sprays of non-systemic fungicides like dithiocarbamates have been used for its control (Dodan *et al.*, 1994). Rattan and Saini (1969) linked association of ascorbic acid with the resistance. In susceptible conditions, ascorbic acid content ranged from 17 to 22 mg/100 g. Spraying of difolatan 4 times at 10 day intervals was effective. A few cultivars like Money-Maker, Flat Large Red, Red Cherry were resistant in the field (Sharma *et al.*, 1974). Two sprays of copper oxychloride 50 at 0.25 per cent proved effective in reducing the incidence of buck eye rot (*P. nicotiana* var. *nicotiana*). Staking also reduced the incidence of the disease (Bhardwaj *et al.*, 1995).

Metalaxyl + mancozeb was reported to be the most effective fungicide against buck eye rot with suppression of disease to 3.82 per cent compared to 41.50 per cent in control, while cyamoxil + mancozeb (moximate) @ 0.35 per cent and 0.30 per cent and cyamoxil + mancozeb (curzate) @0.30 per cent were next in order with control to 6.71 per cent, 9.30 per cent and 9.78 per cent, respectively and the yield was also high in these respective treatments (Gupta and Bharat, 2008).

15.1.1.4 Early Blight

The early blight, one of the most common foliage blights of tomato is caused by *Alternaria solani*. The pathogen was first recorded in 1882 in New Jersey. Disease appeared first as spots on leaflets. They were circular to angular dark brown to black and ranged from pinhead to 4 mm in diameter in size. Concentric ridges often formed leathery necrotic tissues. Usually a narrow chlorotic zone around the necrotic spot faded into normal green. Numerous spots lead to premature senescence and dropping of leaves. On the stem dark spot occurred which lead to collar rot. Fruits were infected both at the green and ripe stage. At stem end, dark black or brown lesions usually developed with sunken spots which became leathery.

The mycelium remained viable in dry infected leaves for more than a year. Contamination of seed might occur during seed extraction. The fungus was transmitted between successive crops through the debris and the seed. Heavy dew with frequent rains seemed to be necessary for sporulation. High soil fertility tended to reduce the severity of the infection.

There were many early reports on resistance to early blight (Andrus *et al.*, 1942; Moore and Reynard, 1945; Reynard and Andrus, 1945 and Alexander and Hoover, 1955). Seven pathogenic races were reported and several accessions of *L. esculentum* like Floradel, Floralon, Indian River, Strobelle, Tropics and Walter, *L. pimpinellium*, *L. peruvianum* were found as resistant. Joi and Khade (1981) found resistance in cultivars S-12, Manale, Manalucie, Kalyanpur No. 1, Poona 142-14 to *A. solani*. Dithane Z-78 is reported to be effective for control of early blight and it also imparted attractive fruit colour. Mancozeb, chlorothalonil, copper oxychloride and their mixtures were found to be effective against early blight (Mathur and Shekhawat, 1986; Mohammad, 1988; Bhardwaj *et al.*, 1995; Devanathan and Ramanujam, 1995; Keinath *et al.*, 1996 and Dillard *et al.*, 1997). Okasha *et al.* (1989) recorded good control of *A. solani* by treatment with propineb, metiram and dichlofluanid. Sattar and Kassem (1991) got best control with iprodione followed by metalaxyl (ridomil 5 ml) both under the green house as well as field conditions. Spraying plants with cysteine at 150 mg/l during vegetative growth phase reduced the level of infection by early blight (Reda *et al.*, 1986). Alas and Bustamante (1993) obtained greater resistance to *A. solani* by application of limestone to infected green house tomato plants.

Vakalounakis (1991, 1992) reported that early blight of green house tomato could be effectively controlled by inhibiting sporulation of *A. solani* with ultra-violet absorbing vinyl film and infrared vinyl film. Disease incidence in the Ira-vinyl green house and UVA-vinyl green house was reduced to less than 50 per cent of that of common agricultural green house. To combat this disease, Gleason *et al.* (1995) developed a disease warning system, FAST (Forecasting *A.solani* on tomatoes) model for processing tomatoes so that the disease could be checked well in time. Another method developed by Stevens *et al.* (1998) to combat the disease was integration of plasticculture methodologies *i.e.*, solarised soil, solarised base soil mulched with black plastic film, non-solarised soil mulched with black agriplastic mulch and biological strategies (biofungicide). This was found to give better control of early blight than chemical fungicides plus non-solarised bare soil.

Ridomil Plus 50 per cent WP (15 per cent metalaxyl + 35 per cent copper oxychloride) @ 2g/litre and plant extracts of *Ocimum basilicum*, *Azadirachta indica*, *Eucalyptus chamadulonsis*, *Datura stramonium*, *Nerium oleander* and *Allium sativum* @ 1 per cent and 5 per cent each were applied as foliar spray at every 15 days interval till 60 days on inoculated plants in green house condition in Egypt by Sallam (2011) and it was observed that Ridomil Plus led to maximum disease reduction over control (82.8 per cent), followed by 5 per cent extract of *A. sativum* (71.7 per cent) and 5 per cent *D. stramonium* extract (68.2 per cent). These treatments also increased yield over control by 85.7 per cent, 66.7 per cent and 76.2 per cent.

Abu El-Samen (2015) suggested that application of azoxystrobin might be beneficial in reducing disease severity under high disease pressure.

Sreenivasulu *et al.* (2019) reported that tebuconazole 25.9 per cent EC @ 0.1 per cent was most effective followed by propiconazole 25 per cent EC @ 0.1 per cent for the management of early blight disease along with high yield, besides foliar spray of fungal bioagent *Trichoderma viridae* @ 0.25 per cent was highly effective and bacterial biocontrol agent *Pseudomonas fluorescens* @ 0.25 per cent was moderately effective in controlling the disease.

15.1.1.5 Fusarium Wilt

Fusarium wilt, caused by *Fusarium oxysporum* f.sp. *lycopersici* was first reported from Italy by Saccards in 1886; in India, this disease was first mentioned by Butler in 1918.

In young plants, symptoms like clearing of veinlets and drooping of petioles occurred. In the field, yellowing of the lower leaves appeared first and the affected leaflets wilted and died. Other leaves gradually got affected along with the vascular system resulting in stunted growth and ultimately death of the plants. Samson *et al.* (1942) found that it occurred rarely in seed extracted from infected plants. Chief means of dissemination was by transplants, wind borne soil and drainage water. Clayton (1923) studied soil-moisture and air- and soil-temperature relations. He found that the optimum soil temperature for spread of the disease was 28°C. The disease was inhibited when temperature went above 33°C and below 21 °C. Walker and Foster (1946) found that the low potassium and high nitrogen nutrients increased disease incidence, while high potassium and low nitrogen decreased the disease. Chand and Thakur (1969) found that wilt caused by *F. oxysporum* developed most in sandy soil and the least in clay soil. The percentage of infection rose with the increase in sandy soil and decreased from 94 per cent at pH 5.4 in sandy soil to 24 per cent in pH 7.0 in clay soil.

Control measures, therefore, pertained to soil sterilization and its amendment in order to check the pathogen population. Soil solarization using polythene mulches (transparent and black) considerably reduced the pathogen population from the soil especially under irrigated conditions (Raj and Kapoor, 1993). Amendment with oil cakes also reduced the activity of *F. oxysporum* f. sp. *lycopersici*. Groundnut oil cake was particularly very effective as it not only reduced a higher disease index but also improved plant growth (Raj and Kapoor, 1996). Pathogen population was also considerably reduced by more than 70 per cent by soil amendments with banana leaves, mushroom and bagasse composts (Raj and Kapoor, 1997). Chemical control measures also included seed treatment with fungicides like agrocit 50 SD and chinoin-fundozo 150 WP (Iloba and Achor, 1989). Mladenov *et al.* (1989) recommended the use of methyl bromide and ditrapex (dichloro-propane with 1,3-dichloropropene) as soil infectant for controlling *F. oxysporum* f sp. *lycopersici* and other *Fusarium* spp. under polythene cover.

First work on development of resistant cultivars was done by Essary in Tennessee (1912). Edgerton was next to carry breeding for wilt resistance introducing the cultivars Louisiana Pink and Louisiana Red. Pritchard (1922) produced a number of resistant cultivars including Marglobe, Norton and Pritchard. Resistance in the cultivars was not completely fixed and behaved as a quantitative genetic character comparable to type B yellow resistance in cabbage. Bohn and Tucker (1940) recommended high type of immunity to wilt, comparable to type A yellow resistance in cabbage which was controlled by a single dominant gene in red currant tomato. First commercial tomato with type A resistance was introduced in 1941 as Pan America and second cultivar as Southland. Two races, 1 and 2 were reported. Cultivars with immunity to race-1 might be susceptible to race-2. The existence of race-3 was reported by Alexander and Hoover (1955). It was found that gene effective against race-2, was independent of race-1. Gilbert and McGuire (1956) reported that resistance to *Fusarium* derived from *L. peruvianum* appeared to be governed by number of dominant genes. Choudhury *et al.* (1973) found Sel. 673 and EC 37192 were highly resistant, whereas Moneymaker and Marglobe were moderately resistant. Out of the related species, *L. pimpinellifolium* and *L. hirsutum* were highly resistant. Under green house conditions, infection of tomatoes due to *Fusarium oxysporum* f.sp. *lycopersici* was prevented by grafting the plants on resistant rootstock, KNVF (*L. esculentum* x *L. hirsutum*) (Dufour and Taillens, 1994). Kuwata *et al.* (1994) found BF-Okitsu No.101and LS-89 as resistant rootstocks for susceptible cv. Momotaro (susceptible to *F. oxysporum* race J2).

Many commercial cultivars resistant to Fusarium wilt were developed by various workers. These include Homestead (Andrus and Ruehle, 1953), Indian

River (Hayslip *et al.*, 1958), Manale (Kelbert and Walter, 1954), Manalucie (Walter and Kelbert, 1953), Manapal (Walter and Hayslip, 1961) and Florilon (Whitner *et al.*, 1962). McFerran *et al.* (1982) found Ark 7985 as resistant to *Rhizoctonia solani* and *Fusarium oxysporum* race-1. For control of Fusarium wilt, Iizuka *et al.* (1988) recommended an integrated method comprising of application of chloropicrin and the use of grafting stock having resistance to wilt.

Role of different strains of *Fusarium* in controlling wilt disease was reported by various workers. Yamaguchi *et al.* (1992) found effective control of Fusarium wilt by non-pathogenic *Fusarium oxysporum* MT0062. These Fusarium strains could be used to induce systemic resistance in tomato plants (Suszkiw, 1997). Another *Fusarium* species, isolated from roots of faba beans (non-pathogenic to tomato) was shown to be an effective inhibitor of the tomato wilt pathogen under *in vitro* as well as green house conditions (Attitalla *et al.*, 1998).

Useful pathogens were incorporated in integrated biological control programmes for *Fusarium oxysporum* f. sp. *lycopersici*. *Trichoderma harzianum* was found to be most effective against *Fusarium equisetti* while *Trichoderma aureoviride* was found to be effective against *F. oxysporum*. *Trichoderma* spp. was applied to seeds sown in soil infected with pathogens (Monaco *et al.*, 1991). Fungi producing lytic enzymes and the enzyme complexes were found to significantly reduce incidence and the severity of the disease. The most effective antagonists were *Aspergillus nidulans*, *Penicillium oxalicum* and a mixture of *Fusarium moniliforme* (*Gibberella fujikorai*) and *G. fujikorai var. subglutans*. Active enzyme complexes were obtained from *A. nidulans*, *Mucor rouxii*, *P. oxalicum* and *F. oxysporum* f. sp. *lycopersici* (Cal *et al.*, 1994). The most effective biological control agent was *P. oxalicum* (Cal *et al.*, 1999). Other bio-control agents found effective were isolates of *Gliocladium virens*, *Trichoderma hamatum*, *Pseudomonas fluorescens* and *Burkholderia cepacian* (Larkin and Fravel, 1998 and Mao *et al.*, 1998).

Song *et al.* (2004) observed that tomato *Fusarium* wilt disease could be well controlled by low toxicity and systemic fungicides added in a hydroponic system at their appropriate concentration. Prochloraz and carbendazim were the most effective fungicides in inhibiting mycelial growth with a preventive effect of 69.6 per cent after 0.4 µg/ml prochloraz was added to the liquid media for 2 weeks with a curative effect of 50.0 per cent, while preventive effect was 87.0 per cent after 5 µg/ml carbendazim was added to the liquid media for 2 weeks with a curative effective of 34.4 per cent.

Prochloraz and bromuconazole at the concentration of 10 µg/ml were the most effective, followed by carbendazim that had good preventive and curative effects on tomato wilt caused by *F. oxysporum* f. sp *lycopersici* both *in vitro* and in glass house (Amini and Sadovich, 2010).

15.1.1.6 Crown and Root Rot

Crown and root rot is caused by *Fusarium oxysporum* f. sp. *radici-lycopersici*, which is a soil borne pathogen (Jarvis, 1988). The infected plants showed rotting of the crown portion of the shoot and root. Soil fumigation with methyl bromide and chloropicrin (Apodaca-Sanchez *et al.*, 1994; Chellemi *et al.*, 1994a) or application of 935 litres metam-sodium (sodium N-methyldithio-carbonate) per hectare to the soil surface prior to bed formation controlled the disease (McGovern *et al.*, 1998).

Hymexazol, hydroxyquinolin, sodium tetraborate decahydrate, oxyquinoleine, flutriafol + thiabendazole, have all controlled *Fusarium oxysporum* f. sp. *radici-lycopersici in vitro*, and all these fungicides have limited its growth in green house on cultivar Rio Grande and the disease incidence was below 24 per cent compared to 80 per cent in untreated plants (Hibar *et al.*, 2007).

Under green house conditions, *Trichoderma harzianum* was found to be most effective in controlling *F. oxysporum* f. sp. *radici-lycopersici* when the antagonist was added to the nutrient solution in a recirculating artificial substrate system and it reduced the number of wilted plants by 70 per cent (Rattink, 1993; Datnoff *et al.*, 1995). Tu and Zheng (1994) enumerated four biological control agents, *Gliocladium roseum, G. virens, Bacillus subtilis* and *Pseudomonas fluorescens* for controlling the pathogens in green house tomatoes. In tomatoes grown on recirculated nutrient system, yet another biological control agent, *F. oxysporum* str. Fo 47 was found to prevent the spread of the disease and increase yield (Reist *et al.*, 1995). Menzies and Ehret (1997) found that fungal isolates, isolate rf 18 of *F. culmorum*, isolate rf 34 of *Penicillium brevicompactum* and isolate rf 41 of *P. crustosum* reduced the infection and severity of the pathogens.

Yu *et al.* (1997) found that sugi (*Cryptomeria japonica*) bark greatly reduced the incidence of Fusarium crown and root rot. The inhibition was caused by volatile oils, phenolics and acidic substances in the bark. Takacs and Dolej (1998) treated tomato plants infected by *F. oxysporum* f. sp. *radici-lycopersici* with plantactivatorbion (a product from Ciba Geigy) at 0.01-1.0 per cent and obtained good results.

15.1.1.7 Powdery Mildew

Powdery mildew is caused by *Erisiphe* spp., *Oidium lycopersici, Sphaerotheca fuliginea, Leveillula taurica, etc.* White pustules appeared on the upper surfaces of leaves of affected plants and only rarely on the undersurfaces. Stems were severely affected. Correll *et al.* (1988) observed that powdery mildew occurred at an early stage of crop development with each successive planting date and was not related to the physiological age of the crop or any specific meteorological conditions. More lesions were observed on older leaves than younger leaves. In the field, both immature and mature host tissues were susceptible to infection by *Leveillula taurica*. Powdery mildew affected both green house as well as field grown tomatoes (Sobolewski and Robak, 1994). Stravato (1993) first reported about powdery mildew in green house in Albenga, Italy.

Low concentration of fenarimol was very effective against the pathogen (Fletcher *et al.*, 1988). Patemotte (1988) obtained good control by 2 post-inoculation application of bitertanol, bupirimate, fenarimol, pyrazophos and triforine. Forsberg (1989) recommended application of bayleton (triadimefon), euparen M (tolylfluanid), funaflor (imazil) and triforine for control of powdery mildew caused by *Oidium lycopersici*. Myclobutanil, fenarimol, sulfur triadimefon and triforine were suggested by Bains *et al.* (1999). For controlling *Erisiphe polygoni* affecting green house tomatoes, sodium bicarbonate in combination with a wetting agent Agra 90 was recommended by Bourbos *et al.* (1999).

Hexaconazole 5 EC @ 1 ml/l and difenoconazole 25 EC @ 0.05 ml/l were found highly effective against powdery mildew caused by *Oidium neolycopersici* with 91.0 and 89.2 per cent disease control and 42.0 and 39.0 per cent increase in yield as compared to check, respectively sprayed at an interval of 10 days in poly house in Kangra, H.P. (Arushi *et al.*, 2018). Besides, *in vivo* evaluation of bioagents *viz.*, *Trichoderma harzianum*-1 (TH-1), *T. harzianum*-2 (TH-2), *T. viride*-1 (TV-1), *T. viride*-2 (TV-2) and *Pseudomonas fluorescence*-1 (PF-1) @ 10 g/l showed 63.9-75.41 per cent disease control. Among these, both strains of *T. harzianum* (TH-1 and TH-2) were found highly effective in controlling the disease. *In vivo* evaluation of three botanicals *viz., Eupatorium adenophorum, Melia azedarach* and *Azadirachta indica* showed that aqueous extract at 100 per cent concentration provided more than 50 per cent disease control, being maximum *i.e.* 65.1 per cent was provided by *E. adenophorum*.

Biological control of powdery mildew caused by *Sphaerotheca fuligiena* included a combination of *Bacillus brevis* (bacterial biocontrol agent) and plant extracts from *Reynoutria sachalinensis*. The plant extract was found to act indirectly by inducing resistance in the plant (Seddon and Schmitt, 1999).

Compost tea applied as foliar spray reduced powdery mildew disease incidence produced by the foliar pathogen *Erysiphe polygoni* by 19 per cent when used as a preventive treatment and eradicated the pathogen on the leaves when applied as a curative treatment in tomato cv. Roma grown in perlite in an unheated green house probably due to direct effects on the pathogen and indirect effects through improvement in plant resistance (Segarra *et al.*, 2009)

15.1.1.8 White Mold

White mold disease is a soil-borne fungus caused by *Sclerotinia sclerotiorum* (Lib.) De Bary. Under mild temperatures and high soil moisture content, environmental conditions commonly found in tomato crops, *S. sclerotiorum* sclerotia germinate in the infested soil, producing hyphae, which may directly infect hosts or, in most cases, apothecia, which in turn release ascospores colonizing senescent flowers (Jones *et al.*, 1991). A cotton-like white mycelium is found on the affected shoots, which is easily broken later on, and originate numerous black sclerotia structures that are used for long-term survival in the soil, externally and/or within the stem.

Trichoderma harzianum + *T. viride* (1 x 10^9 CFU/ml), associated with fungicide fluazinam applied by drip irrigation showed control of white mold and improved yield, pulp content and TSS of processing tomato hybrid Heinz 7155 in Brazil (de Aguiar *et al.*, 2010).

Carbendazim, thiophanate methyl and phenylpyrrole had completely inhibited the growth of *S. sclerotiorum* at all tested concentrations *in vitro* (Shivpuri *et al.*, 2001). Ridomil gold, Benlate, Tecto-60 and Topsin-M were the most effective fungicides in inhibiting the sclerotial germination of *Sclerotinia* spp. (Iqbal *et al.*, 2003). Tebuconazol, Andrachite, Metalaxyl, Roko, and Bavistin were found effective against pathogen under *in vitro* and *in vivo* conditions (Upadhyay *et al.*, 2019). Combined use of compost, chemical inducers (like salicylic acid at 8mM and 4mM concentration) and fungicides might be useful as a good tool for controlling tomato white mold disease caused by *Sclerotinia sclerotiorum* under green house and field conditions (Gomaa *et al.*, 2016).

15.1.1.9 Grey Mold

Grey mold is caused by *Botrytis cinerea*, which damaged stems, leaves and fruits. The key initial infection sites on tomato fruits were the residual petal and stigma from which the infection extended to the fruit stalk and finally to other parts of the plant.

Control measures included both chemical as well as integrated programme involving biological agents. Locke and Fletcher (1988) found that fungicides iprodione with dichlorofluanid gave effective control of the pathogens. Takeuchi (1991) recommended 2-3 sprays of mixture of diethofencarb and thiophanate-methyl at an interval of 2-4weeks. Gullino *et al.* (1991) found that disease incidence was higher under conditions of high nitrogen fertilization and reduced ventilation. Under these conditions only procymol-thiram sprays used alone or alternately with carbendazim, diethofencarb satisfactorily controlled grey mould. In presence of low nitrogen fertilization, alternation of chemical and biological control (*Trichoderma* spp.) gave good control of grey mold. Biological control by *Trichoderma harzianum* in alternation with dicarboximide fungicides like iprodione or procymidone gave best control of the pathogens (Elad *et al.*, 1995). In vinyl houses, a single treatment of dichlorofluanid and a mixture of diethofencarb + carbendazim gave effective disease control (Park and Yu, 1998). Seddon and Schmitt (1999) reported that combination of bacterial biocontrol agent, *Bacillus brevis* and plant extracts from *Reynoutria sachalinensis* effectively controlled the pathogens in green house tomatoes. Li *et al.* (1999) suggested a new method for controlling grey mold *i.e.*, removal of the residual petal and stigma responsible for the spread of the disease to other plant parts, 7-15 days after treating flowers (10-20 mm diameter in young fruits) with 2,4-D. Vakalounakis (1992) observed that incidence of the disease was much less in IRA (infrared absorbing) vinyl green house as compared to CA-vinyl green house.

The tomato plant protection against *Botrytis* gray mold can be done with iprodione 50 per cent (100g/l) or procymidone 50 per cent (100g/l) by one or two foliar sprays before fruit setting, if necessary, and after the fruit setting 4 kg/ha of trichodex (*Trichoderma harzianum*) controlled grey mould effectively in green house tomatoes (Fiume and Fiume, 2006).

Fenhexamid was effective against gray mold and resistant isolates were not detected (Study *et al.*, 1999; Haensler *et al.*, 2000). Low-dose fenhexamid (12.5 ml a.i./l) combined with four strains of *Pseudomonas fluorescens* (Pf52/16, Pf163, Pf9/11 and Pf141) reduced the disease incidence by an average of 78 per cent on tomato plants inoculated with *B. cinerea* in green house trials (Yildiz *et al.*, 2007).

15.1.1.10 Leaf Spot

Septoria leaf spot (*Septoria lycopersici* Speg.) is an important soilborne fungal disease of fresh market and processing tomatoes. Tomato plants become infected during the production season when infested soil carrying spores is splashed onto lower leaves during rainfall and overhead irrigation. Periods of high relativity, high temperatures, and leaf wetness favor the development of *Septoria* leaf spot during the production season (Jones *et al.*, 1991) and if not managed properly, defoliation of tomato plants can be as high as 100 per cent (Wyenandt *et al.*, 2008).

Leaf spot caused by *Septoria lycopersici* was easily controlled by seed treatment with agrocit 50 SD and chinoin-fundazol 50 WP (Iloba and Achor, 1989). Best control of leaf spot pathogens was obtained by application of bravo 500 (daconil), dithane Flowable (maneozeb), polihom (metiram zinc + copper oxychloride), ridomil plus 48 (metalaxyl + copper oxychloride), temoal 52 WP (cymoxanil + matelaxyl + copper oxychloride) and experimental products A6, A7, A8 (20 per cent copper oxychloride), IAMN56 and IAMNSN220 (Szekely *et al.*, 1990). Spraying of chlorothalonil or mancozeb at 7 or 10 day intervals, effectively controlled the pathogen (Bhardwaj *et al.*, 1995; Dillard *et al.*, 1997).

Use of leaf antagonists like yeast isolate Y236 (*Cryptococcus laurentii*) and bacterial isolate BTL (*Pseudomonas putida*) significantly reduced the incidence or the severity of *Septoria* leaf spot and could be an alternative to reduce fungicide applications in tomato (Blum, 2000).

Kumar and Sugha (2000) showed that single row or two rows of maize intercropped with tomato developed lesser *Septoria* leaf spot compared to single crop or intercrop with beans, which in fact enhanced the incidence. They also observed that staking and defoliation could reduce the disease incidence.

Cover crop mulch and weekly fungicide sprays reduced the development of *Septoria* leaf spot and increased marketable yield in processing tomato production in Ohio (Wyenandt *et al.*, 2008). Fall sown winter rye or vetch cover crop killed through chemical or mechanical means left on the soil surface before tomato transplanting to act as a physical barrier and prevent spore dispersal of the soilborne fungal pathogen, through soil splashing. The prevention of soil splashing may be critical for reducing the amount of potential inoculum from reaching the tomato plant canopy and inciting a disease such as *Septoria* leaf spot. The cover crop showed lesser disease incidence than no treatment even with no fungicide application.

15.1.1.11 Southern Blight and White Rot

Southern blight or white rot is caused by *Sclerotium rolfsii* (*Corticum rolfsii*) in tomato. Tu *et al.* (1987, 1991) found that the pathogens failed to survive in soil covered with transparent plastic sheet. The addition of green manure gave increased control compared to plastic sheet with the covering alone. Among different biocontrol agents, *Trichoderma koningii* and *Gliocladium virens* were found to be the most effective in controlling the blight pathogens (Monaco *et al.*, 1991; Ristaino *et al.*, 1991 and Latunde-Dada, 1993). *Sclerotium rolfsii* could be controlled by spraying carboxin + thiram (vitavax-200) and quintozene (PCNB). It was effective in inhibiting both mycelial growth and sclerotia formation at low concentrations (Rondon *et al.*, 1995). Ramirez *et al.* (1998) recommended a combined application of vinclozolin, chlorothalonil, liquid humus and warm humus which resulted in

91.56 per cent healthy plants and 86 per cent reduction in the number of sclerotia. Mao *et al.* (1998) suggested biological control of the pathogen by treating seeds with *Gliocladium virens* (Gl-3) and *Burkholderia cepacia* (Bc-F) both under green house as well as field conditions.

An eco-friendly approach to control the disease has been through antagonistic microbes like plant growth promoting rhizobacteria *Paenibacillus lentimorbus* B-30488 that controlled *S. rolfsii* in *in vitro* and *in vivo* conditions and improve plant growth through ACC deaminase production that led to mechanisms for development of defense enzymes and other biochemical attributes to counter the pathogen (Dixit *et al.*, 2016).

15.1.1.12 Leaf Mold

Leaf mold is caused by *Passalora fulva* (previously called *Cladosporium fulvum* or *Fulvia fulva*). Effective fungicides for controlling the pathogens were perenox (cuprous oxide), bravo (chlorothalonil), rovral (iprodione), bunginex (triforine), bayleton (triadimerfon) and captan (Gerlach *et al.*, 1986). Okasha *et al.* (1989) recommended an integrated control measure by treatment with antagonists like *Penicillium pinophilum, P. stipitatum* and *Trichoderma harzianum* along with chemical fungicides like dichlorofluanid, propineb and metiram. Vakalounakis (1992) observed that leaf mold incidence was low under IRA-vinyl green house as compared to CA-vinyl green house. Control over leaf mold was achieved with the sterol biosynthesis inhibitor fungicides fenarimol (rubigan), hexaconazole (anvil) and cyproconazole (atemi 10 peptitc), sprayed once or twice after the appearance of the first symptoms of the disease (Cobelli *et al.*, 1995).

Difenoconazole + cyprodinil, difenoconazole + mandipropamid, cymoxanil + famoxadone were very effective in controlling leaf mold (Johnson, 2015).

15.1.1.13 Grey Leaf Spot

Grey leaf spot in tomato is caused by *Stemphylium* spp. *Stemphylium solani, S. floridanum*, and *S. botryosum* f. sp.*lycopersici* differ in some taxonomic details but cause similar symptoms on tomato foliage, the host-free season is assumed to be passed in debris, and warm zones are preferred to temperate zone and host age does not affect susceptibility to *S. solani* or *S. floridanum*,, while 6 to 8 week-old plants are most susceptible to *S. botryosum* f. sp. *lycopersici* and also require short wettings for infection (Rotem and Bashi, 1977).

Fungicidal application of chlorothalonil, captafol and mancozeb gave good control over the disease (Mathur and Bhatnagar, 1993). Kim *et al.* (1997) suggested the use of resistant varieties and/or spraying of fungicides like imazalil, flusilazole and prochloraz (ergosterol biosynthesis inhibitors) and polyoxin-B (chitin synthesis inhibitor).

An early recognition method of tomato leaf spot based on MobileNetv2-YOLOv3 model through statistical analyses of images captured under four different conditions was developed and a good balance between the accuracy of the model and real-time detection of tomato gray leaf spot was observed (Liu and Wang, 2020).

15.1.1.14 Leaf Blight and Black Mold

Leaf blight and black mold are caused by *Alternaria alternata*. Leaf blight was effectively controlled by spraying of carbendazim (Chandravanshi *et al.*, 1994). Black mold was effectively controlled by application of chlorothalonil and single application 6 weeks before harvest was the best (Davis *et al.*, 1997).

Khan *et al.* (1995a) observed that mycelial growth of *A. alternata* was completely inhibited by propiconazole and tridemorph under *in vitro* conditions. Mancozeb + metalaxyl also effectively reduced the mycelial growth.

As an alternative to chemical fungicides, plant extracts have also been tried to provide environmental friendly way of checking the disease. The application of mentha essential oil (40 µl/ml) exhibited maximum inhibition in the mycelial growth (93.6 per cent) and conidia germination (90.6 per cent) *in vitro* and also significantly checked the *Alternaria alternata* incidence *in vivo* (3.5 per cent in the untreated plants, 0.93 per cent for the mentha essential oil treated plants and 0.08 per cent in the carbendazim treated plants), besides increasing the plant height (84.6 per cent), fresh weight (81.5 per cent) and dry weight (80.0 per cent) when compared to the untreated tomato plants (Perveen and Bokhari, 2020).

15.1.1.15 Verticillium Wilt

Verticillium wilt is caused by *Verticillium dahliae*. Solarization of soil for 30 to 50 days is found to reduce the incidence of *Verticillium* wilt. Solarization and treatment with methyl bromide also reduced the pathogen population considerably (Ghini *et al.*, 1993). Solarization also effectively controlled *verticillium* wilt of green house grown tomatoes (Bourbos and Skoudridakis, 1996). Mycostop®, a commercial formulation of strain K61 of *Streptomyces griseoviridis*, isolated from *Sphagnum* peat can control *Verticillium* wilt by colonizing the rhizosphere prior to pathogens and there may be a potential additive effect of the commercial biofungicide and solarization in increasing tomato yield (Minuto *et al.*, 2006).

Integrated control measures including proper field selection, heat treatments, solarization, sanitation, tillage, use of plant residues, weed control, resistant rootstocks, cultivars, fertilization, irrigation, chemical treatments and use of microbial antagonists can be effective in control of the disease (Bubici and Cirulli, 2008).

15.1.1.16 Anthracnose Fruit Rot

It is caused by *Colletotrichum coccodes*. Cultivation of resistant cultivars along with 3-4 sprays of chlorothalonil or mancozeb at 7-10 days intervals controlled the disease (Fulling *et al.*, 1995; Byrne *et al.*, 1997; Dillard *et al.*, 1997).

Since anthracnose occurs more when soil moisture is high, thus soil should be well drained and staking of plants can check the disease and rotten fruits should be disposed off to avoid build up of inoculum in the soil; besides fungicides like chlorothalonil, mancozeb, azoxystrobin can effectively control the disease at when sprayed at regular intervals (Li, 2013).

15.1.1.17 Fruit Rot

Fruit rot is caused by *Aspergillus niger*. Flower extract of *Lantana camara* was found effective in reducing fruit rot damage in storage. The extract of garlic and onion was most effective in inhibiting spore germination and growth of the pathogens (Sinha and Saxena, 1990).

Alternative current (AC) heating treatment used to inactivate in microorganisms in tomato was studied and *A. niger* inactivation was increased with increase in electric field strength and increased temperature and its concentration (cfu/g) was determined to be <10 at 108, 68, 48, and 36 V/cm electric field strengths for the treatment times of 6, 12, 20 and 30 s, respectively (Yildiz and Baysal, 2006).

Citrus essential oil vapour has been successful in controlling the mycelial growth of *Aspergillus* by 34 per cent and could be used as an alternative measure to check the pathogen (Philips *et al.*, 2012).

15.1.1.18 Black Leaf Mold

Black leaf mold is caused by *Pseudocercospora fuligena*. *Lycopersicon esculentum* was more susceptible to black leaf mold than *L. hirsutum*, *L. pimpinellifolium* and *L. glandulosum* (Wang *et al.*, 1994). Use of resistant cultivars, proper phytosantitary measures and fungicidal sprays could reduce the incidence of this disease (Subedi *et al.*, 2015).

15.1.1.19 Corky Root

Corky root is caused by fungal pathogen, *Pyrenochaeta lycopersici*. Iizuka *et al.* (1988) suggested that effective control of the pathogen could be obtained by spraying chloropicrin and using resistant rootstocks. According to Shiskoff and Campbell (1990), *P. lycopersici*, an ecologically obligate parasite, survived in the root tissues as dormant propagules. Cool temperature, particularly during the first few weeks of growth of seedlings, had significant effect on increasing disease severity and this effect could not be overcome as seasonal warming occurred. D'Amore *et al.* (1999) found that row covers (low tunnels of non-woven polyester fabric) and soil fumigation (with methyl bromide) reduced the corky root symptoms significantly. Soil solarization using a polythene film greatly reduced the population density of the pathogen as observed by Wadi (1999).

In vitro dual culture of the fungal pathogen *P. lycopersici* with *Trichoderma*, *Streptomyces* and *Bacillus* showed inhibition of the pathogen ranged up to 81.2 per cent (caused from *T. viride* 18/17 SS), 75.6 per cent (from *Streptomyces* spp. AtB42) and 66.8 per cent (from *B. subtilis* M51 PI) and the green house trials also checked the disease efficiently in the same order (Fiume and Fiume, 2008).

A compost with low NH_4-N concentration and high Ca concentration reduced corky root disease severity in green house experiments in Sweden (Hasna *et al.*, 2009).

15.1.1.20 Target Leaf Spot

Target leaf spot of tomato, caused by the fungus *Corynespora cassiicola* (Berk. and Curt.) Wei., is one of the most serious foliar diseases in the winter tomato production

Diseases of Tomato

Early Blight (Left), and Damping Off (Right) Infection.

Late Blight Infection.

Fusarium Wilt (Left), and Crown and Root Rot (Right) infection.

Diseases of Tomato

Powdery Mildew Infection.

Leaf Buckeye Rot (Left) and Septoria Spot (Right) Infection.

Grey Mold (Left), and Southern Blight (Right) Infection.

Leaf Mold (Left), and Grey Leaf Spot (Right) Infection.

Diseases of Tomato

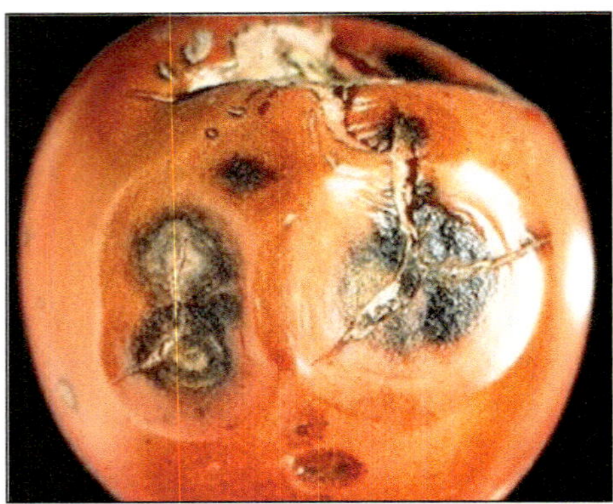

Anthracnose Fruit Rot (Left), and *Verticillium* Wilt Infection.

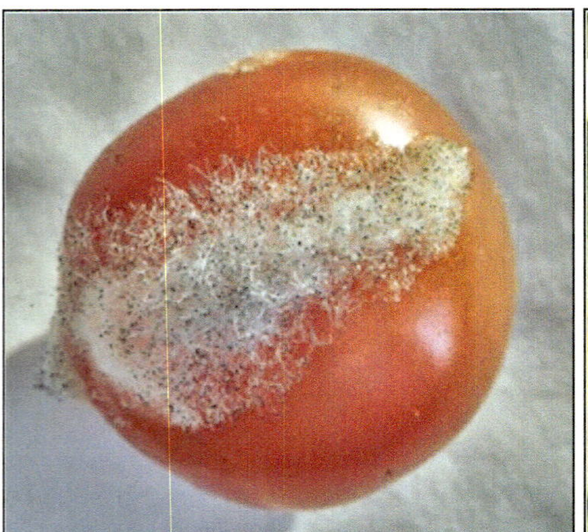

***Aspergillus niger* (Left), and Black Leaf Mold (Right) Infection.**

Corky Root (Left), and Target Leaf Spot (Right) Infection.

areas of Florida, USA, and the Gangetic plains of eastern India. Leaf symptoms of target spot include necrotic lesions with light-brown centers. These lesions often coalesce and result in large blighted areas on leaves that lead to premature defoliation. Perhaps the most serious aspect of target spot is the development of lesions on fruit. These fruit symptoms range from small, brown, sunken flecks to large, deeply pitted areas that render the fruit totally unsuitable for market. Fruit are often predisposed to target spot by injury from sand and soil particles impinging the fruit surfaces during stormy weather (Volin *et al.*, 1989).

Host-plant resistance and cultural methods have not been effective strategies for management of this disease (Blazquez, 1977). Chlorothalonil is a broad-spectrum, protectant fungicide that has been generally accepted as control of choice for target leaf spot for a number of years (Jones and Jones, 1984). However, several newer compounds provided outstanding control of target leaf spot. Azoxystrobin, one of the strobilurin fungicides with a chemistry based on a natural product from a mushroom, provided excellent control of target spot. Systemic acquired resistance (SAR) activator compound, such as acibenzolar-S-methyl, control plant target leaf spot disease in an effective way (Pernezny *et al.*, 2002). When applied to plant surfaces, they elicit a plant defense reaction, analogous to the immune reaction in animal systems (Stricher *et al.*, 1997).

15.1.2 Bacterial Diseases

15.1.2.1 Bacterial Wilt

Bacterial wilt, caused by *Ralstonia solanacearum* (*Pseudomonas solanacearum*) is one of the most serious diseases in tropical humid region. Lower leaves drop before wilting starts, the vascular system turns brown and development of adventitious roots along the stem enhances. Slimy bacterial material exudes from the bundles when the stem is cross-sectioned.

The bacterium is a soil inhabitant, persisting for long periods may be for years in some soils. It invades through wounds, through root cortex and travels through xylem vessels and in succulent stem and infection extends through the intercellular space. Relatively high soil moisture and soil temperature favours the disease. Gallegly and Walker (1949) found that wilt development decreased with increase of salt concentration. Soil treatment was found to reduce the pathogen population considerably. Yao *et al.* (1994) recommended a soil amendment comprising of bagasse, rice husk powder, oyster shell powder, urea, KNO_3, $Ca(H_2PO_4)_2$ + $CaSO_4$, CaO, $Na_2B_4O_7$, $ZnSO_4$, mineral ash and chicken manure for controlling bacterial wilt in the field. They were of the view that the mixture might stimulate proliferation of saprophytic microorganisms especially *Actinomycetes* and *Bacillus* sp. which suppressed the growth of pathogens. Soil amendments with soyabean flour, rice husk + urea along with recommended dose of fertilizers also proved to be beneficial (Miah *et al.*, 1995; Michel *et al.*, 1997). Soil drenching with a mixture of asafoetida, turmeric powder and water (ATW) after transplanting of tomato plants was effective for disease control in the field. Spraying 5 per cent ATW also inhibited growth of *R. solanacearum in vitro* (Bora, 1995). Yi and Sul (1998) observed that the wilt pathogens showed 105 times population reduction when the bacterium was cultured in acidified nutrient solution (pH 3.5-4.0) and suggested that control of wilt might be possible in acidified nutrient solutions. Crop rotation is also very useful in checking the incidence and spread of wilt pathogens. Growing of susceptible cultivars in rotation with maize, okra, cowpea or resistant tomato cultivars is very helpful in suppressing the disease (Adhikari and Basnyat, 1998). Resistant cultivars are recommended for controlling the disease. However resistance to bacterial wilt varied with cultivars (Badekar, 1977).

Buddenhagen *et al.* (1962) defined three pathogenic races, race-1 affecting tobacco and tomato, race-2 affecting banana and race-3 infecting tomato and potato. Roque (1936) in Puerto Rico developed a cultivar resistant to bacterial wilt from a cross of Lousiana Pink and native resistant type. Attempts were made to develop resistant cultivars in various parts of the world. One line CRA-6 from Hawaii was reported to be resistant to this disease. MooSeong *et al.* (1996) found that the cultivars, Naebyongchangsu, Kwangmyong and Seojin were mildly resistant to wilt pathogens out of 31 cultivars tested.

Different strategies were developed to control *R. solanacearum*, such as agrochemicals, soil disinfection, antibiotics, antimicrobial plant extracts, resistant cultivars, genetic modification, crop rotations, organic amendments, lytic bacteriophages, and bacterial antagonists (reviewed by Yuliar *et al.*, 2015).

Goode and Sasser (1980) found that hot water treatment at 50°C for 25 minutes was effective in controlling the disease. In Indonesia, tomato crops sprayed with 200 ppm streptomycine sulfate (oxytetracycline) or agromin at 4 and 7 days intervals gave high degree of control of *Pseudomonas solanacearum* (Gunawan, 1988, 1989; Mazumder, 1998).

Biological control of *R. solanacearum* has been achieved using several beneficial microorganisms which include *Bacillus* spp. (Ji *et al.*, 2008; Akintokun *et al.*, 2019), *Pseudomonas* spp. (Ramesh *et al.*, 2009; Vanitha *et al.*, 2009), avirulent mutants of *R. solanacearum* (McLaughlin and Sequeira, 1988; Frey *et al.*, 1994), *Acinetobacter, Enterobacter* spp. (Xue *et al.*, 2009), *Stenotrophomonas maltophilia* (Messiha *et al.*, 2007) and *Actinomycetes* (Tan *et al.*, 2006) under laboratory and/or green house conditions.

Abdalla *et al.* (1999) found antagonist bacteria like *Bacillus subtilis* (B1), *Pseudomonas* sp. (106, 115) and *Pseudomonas cepacia* (*Burkholderia cepacia*) had a highly inhibitory effect against *Ralstonia solanacearum* both in the laboratory and the green houses. Thus, these bacteria were promising for the biological control of *P. solanacearum*.

Application of FYM and poultry manure to soil at rates of 4 per cent and more suppressed the bacterial wilt and 8 per cent FYM and 6-8 per cent poultry manure were most effective, but showed inhibitory effect on germination and plant growth, thus 4 per cent could be used for control of the disease without hampering plant growth (Islam and Toyota, 2004).

Bacillus velezensis (B63) and *Pseudomonas fluorescens* (P142)-inoculated plants showed drastically reduced wilt disease symptoms, accompanied by significantly lower abundance of the *R. solanacearum* (biovar 2, race 3) population compared to the non-inoculated pathogen control (El-Sayed *et al.*, 2020).

Volatile plant essential oils thymol (0.7 per cent) could control bacterial wilt of tomato effectively by containing the disease incidence to 33.1 per cent in cultivar Equinox in 2002, compared to 92.5 per cent in untreated plot and to 12 per cent in cultivar Solar Set in 2003 compared to 65.5 per cent in untreated plot (Ji *et al.*, 2005).

The extracts of garlic, *Datura* and *Nerium* exhibited antibacterial activity against bacterial wilt *in vitro* and *in vivo* when added to the soil at the time of inoculation, two days before inoculation and two days after inoculation of the pathogen, and significantly reduced the disease index of bacterial wilt on Super Marmande tomato cultivar and when applied at the same time as inoculation the highest reduction of disease index was observed and garlic was most efficient followed by *Datura* and then *Nerium* (Abo-Elyousr and Asran, 2009).

Biochar (2 per cent) application significantly reduced the disease severity of bacterial wilt, increased soil total organic carbon, total nitrogen, C: N ratio, organic matter, available P, available K, pH, and electrical conductivity, and there was increased relative abundance of *Chitinophaga, Flavitalea, Adhaeribacter, Pontibacter, Pedobacter* and *Ohtaekwangia* at the genus level of *Bacteroides* and *Pseudomonas* at the genus level of Proteobacteria in soil at the phylum level under *R. solanacearum* inoculation that shifted the soil bacterial community structure contributing to the resistance of tomato plants against bacterial wilt (Gao *et al.*, 2019).

15.1.2.2 Bacterial Canker

Bacterial canker, caused by *Corynebacterium michiganense* (*Clavibacter michiganensis* subsp. *michiganensis*) is recognized for temporary or permanent wilting of leaflets at the extremities of plants of any size. The symptoms are light streaks beginning at the juncture of leaf petiole and stem, extending down the internode and up the petiole. At later stage, longitudinal canker-like opening appears in stem, petiole and midrib resulting in stunting of plants. In fruits, blister-like exterior lesions may occur and the discoloration extends to the placenta and the seeds.

The bacterium is soil inhabitant and remains in seed-coat. Fermentation of pulp is helpful in destroying the bacteria. The seed-coat is attached to cotyledon and transmits above ground after germination. Invasion of the cotyledon through stomata is the locus of infection. The organism after getting established in the phloem invades cortex and causes lysigenous cavities. The disease is influenced by soil and air factors. The optimum soil and air temperature for disease development is about 28°C. Shirakawa *et al.* (1991) observed that the density of *C. michiganensis* sub sp. *michiganensis* required to cause bacterial canker of tomato was $>10^6$ c.f.u./g tomato leaves. No disease was observed at a density of <103-105 c.f.u./g. The disease was more severe in an open field than in a field sheltered from rain by a plastic film. It was concluded that the multiplication and dissemination of *C. michiganensis* subsp. *michiganensis* increased in wet weather and that cultivation under plastic film helped in controlling the disease.

The disease is also controlled by fermentation of seed and pulp at room temperature for 72 hours before extraction. Naumann (1980) recommended a 4-year crop rotation, plant hygiene and soil sterilization for controlling the disease. Under laboratory conditions good protection of tomato plants in hydroponic culture was obtained by acidification of the nutrient solution to pH 3.0 for 12 hours once a week. In green house trials on possible control agents, disease incidence was greatly reduced by application of NTC (a new antibiotic) and 10 to 30 antagonistic isolates of *Bacillus subtilis* to the nutrient solutions.

Seed treatment with *Bacillus subtilis* (strain GBO3), *Bacillus amyloliquefaciens* (strain IN937a) and *Brevibacillus brevis* (strain IPC11) enhanced the seed quality parameters like seed germination and seedling vigor and high control of bacterial canker disease under green house conditions and there was significant increase in the level of PAL and total phenol contents increased significantly upon the PGPR treatment, which was directly proportional to the rate of reduction in the bacterial canker disease incidence (Girish and Umesha, 2005).

Biological seed treatment with antagonistic *Pseudomonas fluorescens* improved seed quality (p=0.05) under laboratory conditions and drastically reduced the bacterial canker incidence in field (Umesha, 2006).

Seed treatment with 0.1M HCl or 0.05 per cent (w/v) o-hydroxydiphenyl by soaking them for 1 hour reduced seed infection to < 1 per cent and field incidence of primary canker to <2 per cent (Dhanvantari and Brown, 1993). Seed treatment with copper hydroxide (3 g/litre), 10 minutes immersion in acidified nitrite (300 ml/litre) and certain strains of *Bacillus* effectively controlled the disease (Kasselaki *et al.*, 2009).

Soil solarization (approx. 6 weeks of soil mulching with transparent polythene sheets) significantly reduced disease incidence throughout the cropping season (Antoniou *et al.*, 1995).

Composts based on tomato or pepper residues combined with cattle or chicken manure reduced bacterial canker disease by 79 per cent and 100 per cent under both natural infection of mature plants and intentional inoculation and within 15-20 days the bacterial population of *C. michiganensis* subsp. *michiganensis* declined to undetectable levels within 15– 20 days, and therefore plant residue based composts can serve as a component in integrated-management programs (Yogev *et al.*, 2009).

15.1.3 Viral Diseases

The cultivated tomato suffers from numerous viral diseases. Most of the works were carried out in respect of conspicuously damaging viruses. In India, tobacco mosaic virus and leaf curl virus were widely encountered.

15.1.3.1 Tobacco Mosaic Virus (TMV)

This virus was first reported in 1899 by Clinton (1909). It had wide host range of solanaceous as well as other crops like spinach, beet, turnip, *etc.*

The disease was characterized by mild mottling of the foliage, sometimes leaf blade reduced to fern-like appearance, yellow chlorosis predominant, acute chlorotic spots, necrosis and distortion of fruits occurred. Many strains brought about streak symptoms. Mukhopadhyay and Som (1969) reported that this TMV disease substantially reduced the number of branches, inflorescences and flowers in all the susceptible cultivars of tomato, although the degree of reduction varied.

Diseases of Tomato

Bacterial Wilt Infection

Bacterial Canker Infection

Tobacco Mosaic Virus Infection

TYLCV Infection

The extent of reduction was found to be maximum in branches, intermediate in inflorescences and minimum in case of flowers.

Sources of primary inoculation consisted of debris of infected plants which remained in soil and also from certain infected manufactured tobacco product. Dissemination from plant to plant took place through mechanical means, contact by man, animal and machinery during cultivation.

Doolittle *et al.* (1946) reported high resistance to TMV in certain lines of *L. hirsutum*. He also reviewed the use of wild species *L. hirsutum, L. peruvianum, L. chilense, L. glandulosum* for resistance to TMV. Watson *et al.* (1954) studied the hybrid between *L. esculentum* and *L. hirsutum* and concluded that two dominant genes were involved in resistance. Kikuta and Frazier (1947), however, found in the crosses of *L. chilense* with cultivated species a single dominant gene responsible for TMV resistance. Various symbols like *Tm 1, Tm 2* were used. Pecaut and Laterot (1962) compared *Tm 1* and *Tm 2* and found that *Tm 2* with high level of resistance. Philips *et al.* (1964) reported multigenic resistance in the cross 235673 x Fireball. Davis and Webb (1965) reported a new virescence-free TMV resistant line. According to Choudhury *et al.* (1973) breeding lines B 2247 and Holmes PI showed high degree of resistance to TMV. Among other species, *L. peruvianum* and *L. hirsutum* were categorized as resistant to TMV. It was found that resistance in B 2247 was dominant and governed by a single gene, whereas resistance in other lines was complex in nature.

Christova and Schuster (1991) reported that 3 treatments of tomato cv. Rivermoon with DHT (2,4-dioxyhydro-1,3,5-triazine) 2 days before and 2 and 7 days after inoculation reduced severity of TMV (tomato mosaic tobovirus). This virus could also be effectively controlled by slows and filtration in closed soilless growing systems according to Runia *et al.* (1997) and Os *et al.* (1998). The filters eliminated the root pathogens.

15.1.3.2 Leaf Curl Virus

Before the perspective and history of tomato leaf curl virus (ToLCV) or tomato yellow leaf curl virus (TYLCV), the symptoms of leaf curl virus was first described on tobacco by Storey (1931). In India causal virus has first been identified as tobacco leaf curl virus or Nicotiana virus 10 (Pruthi and Samuel, 1939). Symptoms of the disease were first described by Vasudeva and Samraj (1948).

History of tomato leaf curl was started from Middle East. A TYLCV-like disease was first reported in Israel in 1939-40 associated with outbreak of *Bemisia tabaci* (Pico *et al.*, 1996). Twenty years later, in 1959, the entire tomato crop was destroyed by a disease with TYLCV like symptoms in Jordan Valley (Cohen and Antignus, 1994). Cohen and Harpaz (1964) published the first description of this new disease transmitted by the white fly, *Bemisia tabaci*. It has since become an economically important disease in many tomato growing countries of the world. There are strain variations in TYLCV. In particular, the virus from India (Padidam *et al.*, 1995), from sub-Saharan Africa Nigeria, Senegal (Deng *et al.*, 1994) and from southeast Asia may prove to be different from the Mediterranean virus, which in turn is referred

Diseases of Tomato

Different Stages of ToLCV Infection

Tomato Spotted Wilt Virus

Diseases of Tomato

Symptoms of Groundnut Bud Necrosis Virus (GBNV) in Tomato Plants.

(a) Necrosis of young leaf and flower buds, (b) petiole and stem necrosis, and (c) chlorotic concentric circular patches on fruit under field conditions. (d) Concentric chlorotic and necrotic rings on leaf at 10–12 days post inoculation (dpi), (e) sunken necrotic streaks on stem and leaf petioles at 14–15 dpi, and (f) necrosis of growing buds at 20–22 dpi under glass house conditions following mechanical sap inoculation by the isolate GBNV-Tom-Del (Basavaraj *et al.,* 2017). to be included

Symptoms of Emerging Virus [Tomato brown rugose fruit virus (ToBRFV)] in Tomato Plants.

A, Necrotic spots at the leaflet base induced by Tomato torrado virus; B, leaf deformation, yellowing, and stunting induced by Tomato yellow leaf curl virus; C, fruit marbling induced by Pepino mosaic virus; D, chlorotic rings and line patterns on leaves induced by Pelargonium zonate spot virus; E, fruit necrosis induced by Tomato marchitez virus; F, interveinal leaf chlorosis induced by Tomato chlorosis virus; G, necrotic leaf spots induced by Tomato necrotic spot virus (Hanssen *et al.,* 2010)

Diseases of Tomato

Symptoms in tomato plants agro-inoculated with infectious clones of tomato mottle leaf curl virus (ToMoLCV - A and J), tomato severe rugose virus (ToSRV - B and K), and tomato leaf curl purple vein virus (ToLCPVV - C and L). Tomato plants agroinoculated with the pCAMBIA empty vector as a negative control (D, I and M). Symptoms in plants co-agroinoculated with ToLCPVV + ToSRV + ToMoLCV (E and N), ToLCPVV + ToMoLCV (F and O), ToSRV + ToMoLCV (G and P), and ToLCPVV + ToSRV (H and Q) (Macedo *et al.,* 2017).

Tomato Infectious Chlorosis Closterovirus

to as western Mediterranean, Sardinian, Israeli and eastern Mediterranean strain (Nakhla *et al.*, 1993).

The geminivirus causing leaf curl and yellowing in tomato has a wide spread distribution in many tropical countries (Czosnek *et al.*, 1990; McClean, 1940; Nour-Eldin *et al.*, 1969; Retuer *et al.*, 1977; Yassin and Nour, 1965) and it is the most serious disease of tomato in India (Pruthi and Samuel, 1939; Sastry and Singh, 1973; Butter and Rataul, 1981). Apart from TYLCV, confinement of ToLCV is mainly oriented in the tropical and subtropical Asian countries and parts of Africa (Hayati, 1978). Zauba *et al.* (1993) designated ToLCV as the main enemy of tomato crop in Sultanate of Oman. Reports were there that Gemini viruses transmitted by whiteflies were believed to be responsible for the devastating epidemic in tomato crop and one of the viruses from three isolates was Sinaloa tomato leaf curl virus (Rojas *et al.*, 2000).

In India the disease has risen to alarming proportions in the plains of India and has become a limiting factor in tomato cultivation particularly during summer crop (February to May) in southern Indian states (Saikia and Muniyappa, 1989; Sadashiva *et al.*, 2002) and autumn crop (August to December) in northern plains (Som, 1973; Mayee *et al.*, 1974; Banerjee and Kalloo, 1987) and both early-autumn and autumn-winter (September to February) in Eastern India, particularly in West Bengal (Nath, 2003; Pandiarana *et al.*, 2015 a, b; Acharya *et al.*, 2018). The presence of large level variability among different strains present in different parts of the country make the disease problematic as well as breeding for disease resistance more complicated (www.danforthcenter.org/iltab)

Three types of leaf curl symptom were differentiated by Thung (1932), while Pal and Tandon (1937) differentiated four types of tobacco leaf curl symptoms namely, A, B, C and X. The type and severity of symptom vary according to virus isolate, the host genetic background, the environmental conditions and the growth stage and physiological condition of tomato plant at the time of infection (Pico *et al.*, 1996). In South Africa, McClean (1940) recorded that a distinct strain of this virus affected petunia plants. In the year 1964, Singh and Lal described 3 strains on tomato on the basis of symptoms, namely the typical strain (Nicotianavirus 10A), severe strain (Nicotianavirus 10B), and golden yellow mottle strain (Nicotianavirus 10C). In Sudan, Yassin and Nour (1965) also reported 2 strains of leaf curl which did not interfere with each other in combined infections on tomato. In Israel, Cohen and Nitzany (1966) reported a new strain of this virus designated as the tomato yellow leaf curl virus (TYLCV). Nariani (1968) from India, reported another strain causing curling, twisting and rolling of the leaves and dark green vein enations on the under surface of the leaflets in tomato. It was designated as Nicotianavirus 10A or tomato enation leaf curl virus. Seth and Dhanraj (1972) identified another new strain of tobacco leaf curl virus which was isolated from *Capsicum annuum* showing leaf curl disease. It produced severe leaf curl symptoms accompanied by enations on tomato and chilli. Chermiti *et al.* (1993) observed that incidence of TYLCV was influenced by cultivar, date of sowing, planting and frequency of vector treatment in nursery and field. Late sowing and planting were favourable for less disease incidence. Insecticidal treatment at an interval of 5-10 days in seed beds and field crops, reduced infection frequency considerably (Kegler, 1994). A delayed infection was observed in plastic houses protected by nylon net/screens owing to low invasion of *Bemisia tabaci* (Polizzi and Asero, 1994; Rivas *et al.*, 1994).

ToLCV symptoms is characterized by stunting and chlorosis of plant with slight yellowing of young leaves followed by crinkling, reduced leaflet size and rolling of leaf margins, both upward and downwards. Flowers drop and fruit setting is drastically reduced (Swarup, 1999, Nath, 2003). Sometimes the areas between the veins become completely yellowish. The plants have been found to have a greater tendency to produce stunted lateral branches, which impart a bushy appearance particularly when the plants are infected at young stage. In older leaves, purple patches may appear. The infection induces partial or complete sterility of infected plant depending upon the stage of infection. In early infection, symptoms are associated with deformed small flowers, which seldom open and soon drop off. At late infection, infected plants produce fewer and smaller fruits that usually drop (Sastry and Singh, 1973; Sakalani and Mathai, 1977; Muniyappa, 1980; Capoor and Saikia, 1981). However, type of symptom varied according to genetic background of the host plant (Pico *et al.*, 1996; Nath, 2003; Anonymous, 2007). Symptoms produced by TYLCV are very much similar to that expressed due to the infection of ToLCV. Leaves that appear soon after infection have leaflets cupped downward and inward in a hook like shape. Leaves developing later are misshapen and smaller showing interveinal and marginal chlorosis and upward curling of the leaflet margin and purpling can also appear on the lower surface of the leaflets. Infected plants show severe symptoms consisting of marked yellowing, puckering and severe size reduction in the top leaves and is severely stunted exhibiting an erect growth habit with many small branches and shortened internodes. Young and early infected plants are usually unfruitful because of severe flower shedding (Nitzany, 1973; Credi *et al.*, 1989; Jorda, 1993).

Choudhury *et al.* (1973) and Som and Choudhury (1975) found P-13, XXXII-354-A- Silvestra and Sci. 498 as resistant to leaf curl virus. They also found that two lines each of *L. peruvianum* (Lp-6, Lp-8) and *L. chilense* as resistant to leaf curl virus disease. The resistance to leaf curl virus was found to be incompletely dominant over susceptibility and governed by polygenes. Five lines of *L. esculentum* (PI 390513, PI 390658, PI 390659, LA 386 and LA 1777) and 2 of *L. peruvianum* (PI 127830 and PI 127831) were found to be resistant to ToLCV (Saikia and Muniyappa, 1989). In the host and environment interaction study by Hayati and Varma (1984), Red Cherry exhibited severe symptoms but *Lycopersicon glandulosum* (EC 66003) did not show leaf curl. Tomato variety EC 104395 also showed mild or less severe symptoms. Banerjee and Kalloo (1987) screened 122 entries consisting of promising varieties, lines and wild species. They found all the wild relatives were resistant and in *L. esculentum* assemblage, 'ACC-99' was moderately susceptible. Raghupati and Naranjanswamy (2000) screened 160 germplasm entries and found H-24 only from *L. esculentum* germplasm as moderately resistant. Kalloo and Banerjee (2000) made a conclusive study with H-24, Sel-7 and Punjab Chhuhara for 4 consecutive years and found that Punjab Chhuhara was the most susceptible and H-24 the moderately resistant cultivars to leaf curl virus. Numbers of tomato lines from wild accessions were found resistant against this disease by many workers as given below.

Wild Species	*Name of Line*	*Reference*
L. perruvianum	—	Nariani and Vasudeva, 1963
L. chilense	414-2 x 414-1 SIB, 986-22 x 986-24 SIB	Joshi, 1970
L. glandulosum	B66003	Hayati and Varma, 1984
L. hirsutum, L. gladulosum, L. peruvianum, L. hirsutum	B6013A1904	Saikia and Muniyappa, 1986Banerjee and Kalloo, 1987
L. hirsutum	LA 386, LA 1777, PI 390513	Muniyappa *et al.*, 1991
L. hirsutum f. *glabratum*	LA 1223	Raghupathi and Narayanswamy, 2000
L. pimpinellifolium	—	Nath, 2003

Phenols were very often implicated in the defence mechanism of plants against diseases caused by different types of pathogens including viruses. In some cases, a good correlation was reported between the degree of resistance and phenol level in plants, particularly after infection (Farkas and Krialy, 1962). Som and Choudhury (1976) reported that due to infection by leaf curl virus in tomato, total phenol content in tomato plants generally increased and extent of increase following inoculation in resistant plants was significantly higher than that in susceptible plants. Phenolic substances are oxidized by the enzymes, polyphenol oxidases to produce the toxic product quinones, protective melanin pigments and other oxidation products which play an important role in imparting resistance in the host (Hung and Rhode, 1973; Singh *et al.*, 1989). The exact mode of action of phenolic compounds in reducing ToLCV incident in the host is not yet known. However, high phenol content inactivates the virus (Patil *et al.*, 1964). Higher average ascorbic content in the leaves of ToLCV tolerant genotypes as observed by Hazra and Nath (2008) suggested the possible role of ascorbic acid synthesis in manifesting tolerance against this biotic stress. Higher ascorbic acid content in tolerant genotypes could be result from decreased turnover or increased biosynthesis in mitochondria (Smirnoff *et al.*, 2001). It appeared that definite proteins translated from resistant gene(s) was involved in higher metabolism of ascorbate through triggering the ascorbic acid biosynthesis genes in the host. The parent H-24 possessing definite resistance genes introgressed from *L. hirsutum* f. *glabratum* (Kalloo and Banerjee, 1990b) contained the maximum of 24.80 mg ascorbic acid in 100 g fresh leaves and also showed maximum resistance [Coefficient of infection (C.I.) = 0.85]. In some *L. peruvianum* and *L. hirsutum* accessions, ToLCV tolerance operates indirectly, preventing vector-feeding by means of physical barriers, such as leaf hairs or by secretion or presence of antifeeding chemicals in sap that reduce feeding time (Muniyappa *et al.*, 1991; Channarayappa *et al.*, 1992).

Cohen and Nitzany (1966), Caciagli *et al.* (1995) and Muniyappa *et al.* (2000) confirmed that females of *B. tabaci* were more efficient than male as vectors of tomato yellow leaf curl virus. Kobatake *et al.* (1981) found that wild population of *Lonicera japonica* near tomato field served as reservoir for white fly. The whole process of virus transmission goes on in the following manner: the insect acquires the virus while feeding on the plants, reaching the phloem cells with its stylet. The virus passes through specific cells within the gut wall into the body cavity to enter the haemocoel, where it is accumulated before passing out of the insect during feeding via the salivary glands. The groEL protein of the white fly interacts with the coat protein of the virus which implicates circulative transmission of the virus. The virus circulation inside the insect accounts for the latent transmission period that ends once the virus reaches the salivary glands and the insect become viruliferous (Markham *et al.*, 1994; Caciagli *et al.*, 1995).

Most of the control measures suggested was related to the control of insect vector. Sastry and Singh (1974) suggested use of systemic insecticides for control of white fly. The use of parathion containing insecticide was also recommended by Capoor (1958). Zidan *et al.* (1994) found that a mixture of the juvenile hormone analogue pyriproxyfen and the carbamate fenobucarb gave significant reduction in the population density of immature stages of *Bemisia tabaci*. Control of *B. tabaci* by cultural techniques like intercropping with cucumber and late sowing was advocated by Conde and Connelly (1995). Leaf extracts of *Clerodendron aculeatum* applied as a powder to the soil or in liquid form as a foliar spray resulted in delayed incidence of virus infection as observed by Baranwal and Nasim (1997).

The border cropping of sunhemp or maize with application of insecticides and delayed planting proved effective in checking insect vector (Vasudeva and Lal, 1948; Mandal *et al.*, 2017). Spraying of mineral oil at 10 day intervals applied 4 times reduced population of white fly and increased yield (Singh *et al.*, 1973; Saklani and Mathai, 1977). Mandal *et al.* (2017) suggested an integrated approach which includes nursery management [Seed priming with Seed Pro (*Bacillus subtilis*,IIHR strain) @4 g/kg of seed followed by soil application of Seed Pro @10 g/kg of soil while filling of plug trays and soil drenching of Seed Pro @5 per cent after seed germination followed by covering of nursery bed with 50-mesh nylon net] and main field management [Border row planting (2 rows) of maize at least 30 days before transplanting of seedlings in the main field followed by seedling dip with 0.1 per cent of carbendazim 12 per cent + mancozeb 63 per cent WP) at the time of transplanting and sequential spraying with acetamiprid 75 per cent WP @1.5 g/l on 10 days after transplanting (DAT), fipronil 5 per cent SC @1.5 ml/l on 20 DAT, copper hydroxide 77 per cent WP (2.0 g/l) on 25 DAT, imidacloprid 70 per cent WG @2 g/15l on 40 DAT, fenamidone 10 per cent + mancozeb 50 per cent WDG (0.25 per cent) two to three times from 45 DAT at 10 days intervals] for the better management of tomato diseases particularly leaf curl virus in the Gangetic plains of West Bengal, India.

Attempts were made to control virus by stimulating growth with the help of growth regulating substances cycocel, IBA, NAA and GA_3 (Raychoudhury and Lele, 1966). Hayati and Varma (1985) reported that foliar application of 0.02 per cent Validamycin-A suppressed tomato leaf curl virus symptoms in tomato and significantly increased yields.

Antignus *et al.* (1996) found that ultraviolet-blocking green house covers had an impressive effect, greatly reducing the population of key insect pests in green houses and correspondingly reducing the incidence of viral diseases carried by these insects.

Artificial Inoculation Technique Using White Flies in Screen House

Tomato Cultivation with Maize Border

15.1.3.3 Spotted Wilt Virus

This disease is of minor importance in India. The symptoms of spotted wilt virus consisted of bronze coloured marking on upper side of young leaflet which was accompanied by a slight downward curling of the leaves. Growth was checked temporarily or permanently, upward marginal rolling and stiffening of leaflets followed and small, circular, necrotic spots appeared on the leaves under optimum conditions. The virus perpetuated on host plant and disease was transmitted by thrips. The control of disease caused by tomato spotted wilt virus, including transmission and epidemiology, was reviewed by Reddy and Wightman (1988). They identified four main vector species *Thrips tabaci, Frankliniella schultzei, F. ocidentalis* and *F. fusa,* and some lines of red currant tomato were highly resistant to spotted wilt virus. Pearl Harbour cultivar from Hawaii was highly resistant. Argentinian cultivar Manzana was reported as resistant to this disease. Control of thrips might reduce the disease incidence. Guimaraes *et al.* (1997) observed that the incidence of TSWV was more near maize barrier which was used as a cold windbreak. This was due to greater occurrence of viruliferous thrips (*Frankliniella schultzei*) which acted as vector. Pentangelo *et al.* (1999) used a protective covering tissue called agri-fabric tissue for direct crop cover to prevent viral diseases like TSWV.

15.1.3.4 Other Viral Diseases

Resende and Cupertino (1996) enumerated different viral diseases affecting tomato. They were tomato mosaic tobamo virus, tomato spotted wilt tospo virus complex, potato Y potyvirus, potato leaf roll luteovirus, peppering spot tobravirus, tobacco streak virus, eggplant mosaic tymo virus gemini viruses and cucumber mosaic cucumo-virus. For the control of the viral diseases, Dyakunchak and Bolakhonenkova (1997) recommended measures including the use of resistant tomato cultivars, presowing treatment of seed with trace elements, the use of biological agents such as rhizoplan, crop rotation, the use of fertilizers, *etc.*

Tomato infectious chlorosis closterovirus (TICV) was identified in field and green house grown tomatoes. TICV was transmitted by the white fly, *Trialeurodes vaporariorum* in a semi-persistent manner (Wisler *et al.,* 1997).

Cucumber mosaic virus was transmitted by aphids namely *Aphis fabae, A. gossypii, Aulacorthum solani, Macrosiphum euphorbiae* and *Myzus persicae* (Contangelo *et al.,* 1994). Soil fumigation (with methyl bromide) and row covers (low tunnels of non-woven polyster fabric gave good control of the disease (D' Amore *et al.,* 1999). For prevention of CMV, covering the crop with a protective covering called agri-fabric tissue, proved to be very effective (Pentangelo *et al.,* 1999).

Tomato brown rugose fruit virus (ToBRFV) was reported by Hanssen *et al.* (2010), infecting both tomato and pepper. The symptoms include chlorosis, mosaic and mottling, occasionally narrowing of leaves, necrotic spots on peduncles, calyces and petioles, Yellow or brown spots on fruit resulting deformed or irregular fruit.

15.1.4 Post-harvest Diseases

Alternaria, Botrytis, Colletotrichum, Fusarium, Geotrichum, Mucor, Stemphyllium, Rhizopus and *Penicillium* were the most frequently isolated fungi and *Acetobacter,*

Gluconobacter, Klebsiella, Leuconostoc and *Pectobacterium* were the prevalent bacteria on fresh market tomatoes (Ahmed *et al.*, 2017). *Botrytis* and *Pectobacterium* were some of the most important pathogens causing spoilage decay on tomato (Fillinger and Elad, 2015; Etebu *et al.*, 2013; Akbar *et al.*, 2013).

Alternaria alternata, Aspergillus niger, Geotrichum candidum, Alternaria solani, Mucor racemosus, Aspergillus flavus, Fusarium oxysporum, Fusarium moniliforme, Penicillium digitatum,Rhizopus stolonifer, alternaria alternata, Colletotrichum lycopersici, Sclerotium rolfsii, Myrothecium roridum, Phoma destructiva and *Trichothecium roseum* have been reported to be the most common pathogens infesting tomatoes postharvest in Indian markets (Sajad *et al.*, 2017).

For managing postharvest diseases of tomato, proper harvesting and handling techniques besides sanitation is important. During harvesting care should be taken for gentle handling to avoid mechanical injury, cuts and bruises, since these become the potential sites of pathogen attack. Besides, sanitation of people involved in harvesting and handling or other packhouse operations is also necessary. The field bins and crates in which harvested fruits are to be kept should be cleaned and disinfected. The bruised or damaged fruits should be sorted and discarded, before loading to trucks, that have already been cleaned for keeping the postharvest diseases away.

Sodium hypochlorite solution has been used to sterilise tomato fruits in order to reduce the incidence of fungal infection before any other postharvest treatment was applied (Genanew, 2013). Tomatoes treated with chlorinated water after washing reduced the microbial load prior to packaging (Workneh *et al.*, 2012). Dipping of tomato fruits in thiabendazole solution also reduced the microbial load on the fruits (Batu and Thomson, 1998).

Cladosporium oxysporum caused storage rot of tomato and 10 per cent of fruits were affected by this fungus alone. Disease intensity was low at 10-20°C but progressed rapidly at temperatures above 25°C, with a maximum infestation at 30°C and RH 95 per cent. Pre-infection treatments with miltox (zineb + copper) and difolatan (captafol) prevented 75 per cent of fruits from rotting (Singh *et al.*, 1986).

Akhtar *et al.* (1994) identified 7 fungal species which caused postharvest losses to ripe tomato. They were *Alternaria alternata, Fusarium oxysporum, Rhizopus stolonifer; Geotrichum candidum, Aspergillus niger A. flavus* and *A. famigatus*. Thiabendanole (Tecto-60) gave the best control of the diseases occurring after harvest of tomato. Loss in marketable yield was found to be mostly due to direct damage of fruits by target spot fungus (*Corynespora cassicola*) which was easily controlled by spraying chlorothalonil before harvesting (Pernezny *et al.*, 1996).

Tomatoes were particularly vulnerable to microbial spoilage on the fruit surface. Disinfection extended the quality and shelf life of the fruits. Trans-cinnamaldehyde, a compound occurring naturally in plants was used for disinfection with good results. Treatment of the fruits for 30 min with cinnamaldehyde, delayed visible mould growth by 7 days during storage at 18°C (Smid *et al.*, 1996).

Akhtar *et al.* (1999) reported that fungi causing damage after harvest *viz., Fusarium oxysporum, Rhizopus stolonifer, Geotrichum candidum, Aspergillus niger* and

A. flavus grew best at a temperature range of 25-35°C and 80-100 per cent RH. Hence under low temperature and pressure, these fungi could be effectively controlled to prevent postharvest losses.

Tomato fruits treated with chitosan has negative effect on gray mold (due to *Botrytis cineraria*) and blue mould (due to *Penicillium oxalicum*) in tomato fruit due to direct fungitoxic property against the pathogens, and the induction of biochemical defense responses in fruit (Liu *et al.*, 2007).

15.2.0 Pests

Tomato crop is attacked by a number of insect pests, some important insect pests of tomato crop are:

15.2.1 Tomato Fruit Worm (*Heliothis armigera*)

Moths are brown to yellowish brown upto 4 cm long with slight stripes. The caterpillars were greenish. Immediately after hatching, the caterpillar crawls over the leaves and feeds on vegetative parts. It eventually finds its way to the fruit into which it cuts holes and burrows. The life cycle is completed in 28-60 days. The period of activity commences from October and continues up to March.

Flubendiamide 40 SC@ 0.21ml/l and Emamectin benzoate (0.625g/l) are best chemicals for controlling the pest (Regmi *et al.*, 2018). Chatterjee and Mondal (2012) also suggested use of these two chemicals for effective control of tomato fruit borer. Indoxacarb 14.5 SC (0.01 per cent) was also very effective against fruit borer besides novaluron 10 EC (0.01 per cent) and acephate 75 SP (0.037 per cent) (Singh *et al.*, 2017).

The pyrethroides were found to be persistent on the plant for a longer time in contrast to methomyl and *Bacillus thuringiensis* which resulted in high mortality 4-8 days after application. Thus, pyrethroids could prevent the plant from *H. armigera* infestation for a longer time (Walgenbach *et al.*, 1991; Zehnder *et al.*, 1995). Spraying 2 rounds of Eco neem (a neem based insecticide) was found to be the best means of controlling the pest (Mahadevan, 1998). Neem products act as deterrents of oviposition for *H. armigera* (Rijal and Dahal, 2019).

Srinivasan *et al.* (1994) recommended *T. erecta* as trap crop for controlling fruit worm infestation in the field. *Solanum nigrum* was observed to be heavily infested by *H. armigera* and was also recommended for use as a trap crop for *H. armigera* to prevent damage to tomato crop (Talckar *et al.*, 1999).

Biological control using various agents was recommended for control of this pest. *Trichogramma chilonsis*, Helionucleo polyhedrosis virus and neem based pesticide are best biological method for the pest management (Rijal and Dahal, 2019). Sivaprakasham *et al.* (1986) observed that *Eucelatoria bryani* was very effective against *H. armigera* under field conditions, especially for controlling adult fruit worm. Xing *et al.* (1988) suggested control of *H. armigera* with a strain of (*Bacillus thuringiensis* sub sp. *kurstakis*) [Bt (HD-1)] in tomato fields. *Bacillus thuringiensis* K. @ 3 g per liter of water is effective for management of *Helicoverpa* (Paneru and Aryal, 2004). Hoffmann *et al.* (1990) identified different species of *Trichogramma* (*T. pretiosum, T.thalense, T.deion* and *T.brevicapillum*) parasitizing *H. zea* for use as biocontrol agents for control of *H. zea, T. brasiliensis* and *T. pretiosum* and egg larval parasitoid *Chelonus armigera* were recommended for biocontrol of *H. armigera* (Rawatand Pawar, 1993). The augmented release of *T. chilonsis* @ 30 cards/ha was found effective against *Helicoverpa* (Gouli 2008). *Hyposoter exiguae* is a parasite of *H. armigera* (Vaishampayan and Singh, 1995).

In recent years, there has been an integrated approach to control the fruit worm affecting both field grown as well as green house tomatoes. Best control of *H. armigera* was achieved after treatment with different strains and formulations of *B. thuringiensis* in combination with pyrethroids at different stages of plant growth (Dibyantoro and Siswojo, 1988; Lutwama and Matanni, 1988; Cameron *et al.*, 1995; Gupta and Babu, 1998). A spray programme involving 4 sprays of NPV (Nuclear Polyhedrosis Virus) @ 300 l/ha, applied at an interval of 8 days, beginning from the sighting of 2 eggs or larvae, or during peak flowering, was recommended for the field management of *H. armigera* on tomato (Mohan *et al.*, 1996; Gopalakrishnan and Ashokan, 1998). NPV + chemical insecticide spray was found to give good control against *H. armigera* (Sivaprakasham, 1998; Pokharkar *et al.*, 1999; Satpathy *et al.*, 1999).

Germplasm screening revealed high levels of TFW resistance only in the wild *Lycopersicon* species, particularly *L. hirsutum* (accessions 76W-PI134417-A and 77W-PI134417-1) and *L. pennellii* (accession LA 716) (Talekar *et al.*, 2005). In India, screening of tomato germplasm for *H. armigera* revealed that genotype 122775 and the cross Rick x Solan Gola were resistant (Kakar *et al.*, 1990). Pusa Red Plum was the least affected cultivar to fruit worm insect (Srinivasan and Basheer, 1961). They also found hybrids Altair F_1 (90-5225), Mercur F_1 (90-5223), F_1 958930, Rupali and commercial cv. Roma to be highly resistant to *H. armigera*. The application of molecular genetics to incorporate *Bacillus thuringiensis* genes into varieties and the selection of varieties with specific leaf morphologies and biochemical compositions were proposed to develop resistance against *H. armigera* (Shukla and Yadav, 1998). Recently, insecticidal crystal protein genes of *Bacillus thuringiensis* (*viz., Cry1Ac, Cry1Ab, Cry2A, Cry2Ax1*) have been successfully engineered into tomato for resistance against tomato fruit borer.

15.2.2 Epilachna Beetles (*Epilachna vigintioctopunctata*)

Both larvae and adult feed on the leaves. Feeding is irregular and gives the plant a characteristic lace-like appearance. Spraying of chlorpyriphos at 0.25 per cent controls this pest. Sprays of quinalphos EC were also found to be effective for 20 days (Rajagopal and Trivedi, 1989). Emamectin benzoate 5 SG @ 10 g a.i./ha and thiodicarb 75 WP @70g a.i./ha were very effective with highest reduction in pest population of 75.3 and 73.9 per cent, respectively under field conditions (Kodandaram *et al.*, 2014). Use of plant extracts (10 per cent, 20 per cent and 30 per cent) of *Ricinus communis* and *Calotropis procera* reduced oviposition, egg-hatch, prolonged larval duration, and inhibited pupae formation and adult emergence in Epilachna beetles on tomato (Islam *et al.*, 2011).

15.2.3 Jassids (*Empoasca devastans*)

Jassids suck the sap from the leaves causing curling of the leaves. Imidacloprid 30.5 SC (0.005 per cent), Flonicamid 50 WG (0.015 per cent), Clothianidin 50 WDG

(0.025 per cent) and Dinotefuran 20 SG (0.01 per cent) provided effective check over the jassid population in tomato (Bambhaniya *et al.*, 2018).

15.2.4 Tobacco Caterpillar (*Spodoptera littoralis*)

The insects are greyish brown with white marking on upper wings. The thorax and abdomen are light brown. Young caterpillars feed on tender leaves, shoots and fruits at night (Ayyar, 1963). The pest is generally confined to nursery beds and is classed as cutworm. Their life cycle is completed in 32 to 49 days.

It could be controlled by spraying 1ml nuvan-100 mixed in 2 litres of water at the rate of about 800 l/ha.

Another species, *Spodoptera exigua* was also found to affect tomato crop. Jansson *et al.* (1996) found solid formulation of emamectin benzoate effective to control the pest. Shorey and Gerber (1996) suggested the disruption of pheromone communication for control of *S. exigua*.

Flubendiamide, emamectin benzoate, indoxacarb, bifenthrin and chlorantraniliprole were effective in controlling the larval population of *Spodoptera* (Nagal *et al.*, 2016).

Binary mixtures of plant extracts from rhizomes of *Acorus calamus* and fruits of *Piper retrofractum* showed synergistic effects, and these mixtures could serve as an acute toxicant or antifeedant to *Spodoptera litura* (Yooboon *et al.*, 2019)

15.2.5 White Fly (*Bemesia tabaci*) and Green House White Fly (*Trialeurodes vaporariorum*)

They are minute white insects and suck the sap of the plants. It acts as vector for transmitting leaf curl disease.

Chemical control of white fly included spraying of different pesticides and pyrethroids. Spraying (with pyrethroids) and cooling of green house (with fans which forced outside air into the green house) resulted in reduction of population of white fly (Zipori *et al.*, 1988). Hankawa *et al.* (1988) used cold fogging with Supracide (methidathion) for control of *Trialeurodes vaporariorum* (TV) on tomatoes in plastic houses. Yasui *et al.* (1991) reported that 1 or 2 treatments of insect growth regulator, buprofezin was sufficient to keep levels of green house white fly (TV) below the tolerable pest density. Salguero and Morales (1994) found mixtures of fenpropathrin + butocarboxim and methamidophos + cyfluthrin to be effective against white fly on tomatoes. White fly was effectively controlled by spraying of methomyl and cypermethrin (Chirinos and Geraud-Pouey, 1996). Soil drenching of systemic insecticide imidacloprid protected the transplanted seedlings from egg deposition by *B. argentifolii* in green house while in the field, application of imidacloprid at the time of transplanting provided protection from white fly and was better than or equal to the protection obtained with weekly sprayings of organophosphate/pyrethroid insecticide mixtures (Stansly *et al.*, 1998). Oudard (1999) found that insect growth regulator Admiral alone or in alternation with Applaud FL or Adulticide gave good control of white fly larvae (TV) without any damage to crop.

Srinivasan (2010) suggested an overall control measure for white flies described as follows. When the crop is grown in open field at early stage the crop may be protected by using a floating row cover. Nitrogen fertilizer and manure application should be limited to check vigorous vegetative growth that may harbour more population of the insect, whereas phosphorus and magnesium deficiencies should be avoided as they lead to higher infestation. Silver mulches repel white fly population and may be used. Azadirachtin (Nimbecidine EC available formulation 0.03 per cent, dosage @ 5 ml/l) spray is acceptable for use on organically certified produce. Yellow sticky traps at 30 cm above ground may be used for monitoring the adult white flies, 50 traps/hectare is optimum. 5 per cent solution of neem seed kernel extract is also useful mixed with neem oil. LASTRAW, a special concoction produced by Pest Control of India applied when the insects are noticed and repeated 2-3 times also gives efficient control of white flies.

15.2.6 Thrips (*Thrips tabaci* and *Frankliniella intonsa*)

White swelling spots appear on the fruit due to oviposition of thrips. Murai (1988) reported that *Frankliniella intonsa* adults occurred from the beginning of May to end of November; numbers peaked in mid-June to end of July. He found that thrips over-wintered in a state of reproductive diapause. Sprays of 36 per cent DMTP wettable powder (methionine) and 50 per cent MEP emulsifiable concentrate or soil treatment with granular formulation of 5 per cent acephate effectively controlled the insect. Cyantraniliprole (HGW 86) 10 per cent OD @ 90 g ai. ha^{-1} and Spinosad 45 SC 56 g a.i. ha^{-1} may be recommended as an alternative of other conventional insecticides for the management of tomato thrips and it was safe for Coccinellid population (Mishra, 2012).

Successful protection from thrips was achieved using UV absorbing vinyl film house. Silver strip mulching was also effective. Biological control agents included anthocorid predators *Orius sauteri*, the thomisid predator (*Misumenopstricus pidatus*) and eulophid parasitoid (*Ceranisusbrui*). Labour saving methods like use of dust formulations and fogging and smoking of insecticides were recommended for control of *F. intonsa* in green house in contrast to high volume spraying which was more expensive (Hankawa, 1988).

Use of entomopathogenic *Beauvaria bassiana* granules to kill soil-dwelling stages of thrips is an effective strategy to manage *F. occidentalis* populations in tomato (Lee *et al.*, 2017).

15.2.7 Leaf Miner (*Liriomyza trifolii* and *Liriomyza bryonii*)

Martin (1988) found that abamectin and cyromazine were able to control *L. trifolli* attacking tomato plants. Cryomazine was found to be effective against *L. trifolii.* (Patel *et al.*, 1998).

Among various cultural practices followed to control the pest population, Keularts and Lindquist (1989) found that polyethelene soil covering combined with irrigation (polyethylene cover wetted periodically) was the most practical method for management of *L. trifolii*. Scholz-Dobelin (1997) recommended the use of a neem seed extract for reducing leaf infestation by *L. trifolii.*

Many workers investigated on the biological control of the leaf miner. Viere (1988) and Scholz-Dobelin (1997) recommended braconids (*Dacnusa sibrica* and *Diglyphus isaea*) as bioagents for *L. trifolii* and braconids (*Opius pallipes*) for *L. bryoniae*. Biological control of *L. trifolii* by *D. isaea* in green house tomato was also found to be effective by Ushchekov (1994) and Ozawa *et al.* (1999). Effective control of *L. bryonii* was obtained using *Opius pallipes* (Schelt and Altena, 1997).

Aqueous extracts from two plants, *Urginea maritima* L. (Liliaceae) and *Euphorbia myrsinites* L. (Euphorbiaceae), 1: 25 ratio in distilled water controlled leaf miner *L. trifolii* as good as cyromazine (Civelek and Weintraub, 2004).

15.2.8 South American Tomato Pin Worm (*Scrobipalpuloides absoluta/Tuta absoluta*)

The South American tomato pinworm, *Tuta absoluta* (Meyrick) (Lepidoptera: Gelechidae), was first reported India in 2014 (Shashank *et al.*, 2015), but it was also reported in the north-eastern states (Meghalaya) in 2017 (Sankarganesh *et al.*, 2017).

It is effectively controlled by triflumuron and teflubenzuron as reported by Imenes *et al.* (1990). Other chemicals like cartap, permethrin and growth regulator thiocyclam-hydrogen oxalate were also found to give good control over tomato pin worm. Cartap and abamectin applications were effective in reducing the damage to both leaves and stem apices (Picanco *et al.*, 1995). Insecticide abamectin, abamectin + natural oil and abamectin + triona were found to reduce the pest population of *T. absoluta* below the economic threshold level (Branco *et al.*, 1996). Effective control of *S. absoluta* was obtained by spraying of β-cyfluthrin, cyfluthrin, cartap and propiofos (Lima and Machado, 1996). Reis and Souza (1998) reported that insecticides cartap, permethrin, teflubenzuron, triflumuron and abamectin were the most efficient chemicals with the efficiency of the latter being enhanced by the addition of emulsifiable oil. However, in recent times, failure of chemical control has been observed due to resistance of the pest to different insecticides like bifenthrin, permethrin, diflubenzuron, teflubenzuron, triflumuron owing to weather conditions or spatial dependence (Silva *et al.*, 2011). Hamdy *et al.* (2013) reported Pyridalyl, Chlorantraniliprole and Indoxacarb to be effective chemical controls for the insect while biochemical insecticides Spinetoram, Spinosad and Emamectin were also found very efficient.

Cultural control measures (*e.g.* crop rotation, selective removal and destruction of infested plant material) (Korycinska *et al.*, 2009), the use of natural enemies (parasitoids, predators, entomopathogens and nematodes) (Desneux *et al.*, 2010; Urbaneja *et al.*, 2012; Öztemiz, 2013) and resistant varieties of tomato (de Oliveira, 2012) are the viable and environment friendly options for controlling the insect.

Navarro (1988) recommended biological control of *S. absoluta* by *T. exiguum*. Along with it Thuricide HP (*B. thuringiensis* sub sp. *kurstaki*) applied at 500 g/ha gave good control over the pest. Hence, these biological agents were included in integrated control programmes. Leite *et al.* (1995) suggested integrated pest management strategies using deltamethrin plus *B. thuringiensis* + abamectin for controlling *S. absoluta*. Commercial formulates based on *Bacillus thuringiensis* could be a good alternative in control of *T. absoluta* and the first instar larvae were the most susceptible Integrated pest management strategies for *S.absoluta* included a combination of cultural, microbiological, biological and chemical method, the biological control being the main component. Parasitism by *T. pretiosum* was found to be very effective (Haji *et al.*, 1995).

The lethal and sublethal activity of citrus peel essential oils as emulsions and included in polyethylene glycol (PEG) nanoparticles (EO-NPs) was determined against the invasive tomato pest *Tuta absoluta* by Campolo *et al.* (2017). The results showed an overall good insecticidal activity of the compounds tested, with a higher mortality through contact on eggs and larvae by EO emulsions and through ingestion on larvae by EO-NPs. The nanoformulation also significantly reduced the visible toxic effects on the plants. Therefore, such natural compounds, especially when nanoformulated, could be successfully used in integrated pest management programs for *T. absoluta*.

15.2.9 Fruit Borer (*Neoleucinodes elegantalis*)

Flucythrinate, cypermethrin and quinalphos were found to be effective against *N. elegantalis* infesting tomato as observed by Reis *et al.* (1989). Abamectin gave best control of *N. elegantalis* followed by permethrin and triflumuronin staked tomato plants grown in the field (Reis and Souza, 1996).

Silva *et al.* (2017) studied the effects of different chemicals on different stages of the insect. Cartap hydrochloride was found to affect oviposition behavior. Moreover, methomyl and cartap hydrochloride exhibited high toxicity against the eggs and larvae, with higher than 80 per cent of mortality. These insecticides interrupted larval hatching and caused alterations in the chorion layer. Flubendiamide and deltamethrin demonstrated toxicity on *N. elegantalis* larvae; however, lufenuron, indoxacarb, methoxyfenozide, and chlorantraniliprole demonstrated low toxicity on both eggs and larvae, with lower than 70 per cent of mortality. Fruits treated with cartap hydrochloride had a deterrent effect. The ovicidal activity revealed by methomyl and cartap hydrochloride might provide new approaches regarding insecticide effects on eggs. Methomyl, cartap hydrochloride, flubendiamide, and deltamethrin demonstrated toxicity on larvae. The evaluation of the chorion of the eggshell in their study has clarified the toxic effect of methomyl and cartap hydrochloride on eggs.

Gonzalez (1996) suggested possible control of fruit borer, *N. elegantalis* using plastic covering. Miras *et al.* (1997) described a new trap, water pan trap (BM-94) which used water for capturing the fruit borer.

SPLAT® formulation (Specialized Pheromone Technology and Lure Aplication) developed and patented by ISCA Technologies (Riverside, California, USA), consisting of an amorphous and pasty emulsion composed of wax, oil, and water, which controls the release of semiochemicals and insecticides, containing an identified component of the *N. elegantalis* sex pheromone, was used for mating disruption attract and kill approach and two SPLAT applications controlled this pest effectively (Franca *et al.*, 2013).

15.2.10 Colorado Beetle (*Leptinotarsa decemlineata*)

Doroschenko *et al.* (1988) have recommended permethrin for treating outdoor tomatoes against Colorado beetle. Kulik (1990) got 100 per cent protection from Colorado potato beetle (*L. decemlineata*) using webs made of a textured raid of polyster fibres (an intermediate product in the textile mills) by spreading the webs on the ground in rings, around the transplanted plants. Field trials conducted with oxamyl (Vydate L) added to a 1.5 per cent Laponite organic colloid gel and pre-germinated tomato seed for protection against *L. decemlineata* showed that the treatment resulted in significantly fewer larvae and significantly less percentage of plant defoliation (Ghidiu and Storlie, 1995). Hunt and Whitefield (1996) suggested potato as a trap crop for *L. decemlineata* (Colorado potato beetle) in processing tomato.

Multiple releases of *Edovum puttleri* Grissell (Hymenoptera: Eulophidae) and *Podisus maculiventris* (Say) (Heteroptera: Pentatomidae) resulted in increased parasitism and predation rates of eggs of Colorado potato beetle (CPB), in small plots of staked tomatoes but doubt remains on effective control in larger areas and its economic feasibility (Tipping *et al.*, 1999).

Tomato seedlings infected with a recombinant potato virus X (PVX), which produces poplar chitinase, WIN6, having a role in defense against insect pests was found to exhibit insecticidal properties on Colorado potato beetle [CPB; *Leptinotarsa decemlineata* (Say)], and upon insect bioassay, only 47 per cent CPB neonates feeding on leaves containing >0.3 per cent w/w WIN6 developed to 2nd instar while 93 per cent of controls reached 2nd instar and thus transgenics with WIN6 may have potential to control the insect (Lawrence and Novak, 2006).

15.2.11 Aphids (*Myzus persicae*)

Imenes *et al.* (1990) found that methamidophos and acephate were most effective against *Myzus persicae*. Pymetrozine was found to give good control of aphids (Gollas and Bland, 1994). Fischer and Leger (1996) advocated biological control of *M. persicae* by predatory mite, *Macrophylus caliginosus* and of *Macrosiphum euphorbiae* by *M. caliginosus* as well as *Aphelinus abdominalis* (Trottin-Caudal*et al.*, 1996). Quark (a natural pyrethrum) and ripcord (cypermethrin) resulted in lowering of aphid population whereas quark performed better than ripcord (Anonymous, 1998). Tomato plants treated with soil drenches of 5 per cent, 10 per cent, or 20 per cent vermicompost extracts, at germination and at weekly intervals thereafter were effective in suppression of the pest (Edwards *et al.*, 2010)

15.2.12 Tomato Pinworm (*Keiferia lycopersicella*)

Methomyl was found to be effective in preventing damage by *Keiferia lycopersicella* when applied weekly starting from the day 0.5 larvae/plant (approx.) was observed (Wiesendorn *et al.*, 1990). Williamson and Murray (1993) found fenvalerate, trichlorfon and profenos to be highly effective against *K. lycopersicella*. Jansson *et al.* (1996) found solid formulations of emamectin benzoate to be effective against *K. lycopersicella*. An integrated pest management (IPM) programme based on intensive sampling, release of the parasitoid, *T. pretiosum*, use of a mating disruption technique and applications of *B. thuringiensis* along with abamectin was found to be very effective against *K. lycopersicella*. Biological control agents, parasitoids (*Apanteles dignus, Horogenes* [*Diadegma*] *blackburni, Parahormius pallidipes, Trichogramma pretiosum* and *Trichogramma brassicae*) have been used to date for its management in Canada (Shipp *et al.*, 2001). Mating disruption of *K. lycopersicella* was a component of an IPM program but it could not control tomato pinworm (Walgenbach, 2018).

15.2.13 Mites (*Tetranychus urticae, Aculops lycopersici*)

Baradaran-Anaraki and Daneshvar (1992) recommended acaricides, bromoprophylate, sulfur and karthane for control of *A. lycopersici*. Dicofol, volk oil or dusting of sulfur were found to control *A. lycopersici* (Monkman, 1992). Atanasov (1995) enumerated integrated control of mites which included application of *Phytoseiulus persimilis* race resistant to organophosphorus pesticides and treatment with selective acaricides. After the primary detection of damage caused by mites on the plant leaves, a corresponding quantity of *P. persimilis* was released. Later in the spring and at the beginning of summer, a treatment with propargite, fenpyroximate, dicofol or diafenthiuron was carried out. If *A. lycopersici* appeared on the plants, acaricides safe to *P. persimilis* were used. Drukker *et al.* (1997) and Gragnes and Leger (1998) recommended biocontrol of *T. urticae* by *P. persimilis*.

Van Leewen *et al.* (2005) observed that when spinosad was applied to the roots of tomato plants in rock wool, excellent control of spider mites was obtained, and spinosad possessing systemic properties, quantities as low as 1 mg/plant could protect tomato plants from mite infestation.

Rosemary essential oil (pure) and a rosemary oil-based pesticide (EcoTrol) were found to check the population of two-spotted spider mite (*T. urticae*) and also affected their oviposition behaviour, without having any harmful effect on predatory mite *P. persimilis* (Mirasmailli and Isman, 2006).

15.2.14 Fruit Fly (*Drosophila busckii*)

Flower extracts of *Lantana camara* were found to reduce the degree of rotting following damage by *Drosophila busckii* (Sinha and Saxena, 1990). It was also effectively controlled by application of leaf extracts of onion (Sinha and Saxena, 1999).

Sticky traps containing 0.25 ml of methyl salicylate captured (814 ± 55) *D. busckii* flies/trap on average compared to 12 ± 4 flies/trap in control traps and may be used for mass trapping of these fruit flies (Buda *et al.*, 2009).

15.2.15 Others

Species of *Trichogramma* namely *T. pretiosum* and *T. brevicapillum* were found to parasitise *Manduca* sp. Hence, these proved effective as biological control agents in processing tomatoes (Hoffmann *et al.*, 1990). Jasminoid acid was found to induce resistance in tomato plants against *Manduca sexta* at all stages of plant growth and even when plants were exposed mechanical stress due to wind (Cipollini and Redman, 1999).

Marris and Edwards (1994) observed that introduction of the parasitoid *Eulophus pennicornis* reduced the extent of leaflet damage and significantly decreased the pest population of tomato moth (*Laconobia oleracea*).

Sood and Nath (1998) reported an effective trap for fruit fly *Bactrocera tau* control. The trap comprised of attractant solution containing 0.1 per cent dichlorvos and of yellow colour (as it was most attractive to *B. tau*). Jaggery trap attracted the greatest number of *B. tau* followed by ethyl-methyl ketone + ammonium acetate + sugar, *etc*.

Zenner (1995) reported a new pest of tomato *Melanagromyza socolena*. The infestation caused bud drop and this increased with the advancement of the flowering period. The pteromalid parasitoids, *Heteroschema rugosopunctata* and *Syntomopus* spp. and a species of *Zaeucoila* were found to parasitise on this pest.

Prodiplosis longifolia was found to attack leaf, shoots, leaves and flower buds throughout the entire crop cycle. Fully formed fruits were also attacked. Biomel (a mixture of entomopathogenic fungi) and conventional chemical control offered significantly better control than other treatments (Delgado *et al*., 1999).

A new scale insect on tomatoes, *Pseudococcus viburni* was reported by Schoen and Martin (1999) which was predominant in green houses. Rigorous preventive chemical treatments were required to control *P. viburni*. Two potential biocontrol agents *i.e.*, *Cryptolaemus montrouzieri* and parasitoid *Leptomastix epona* were identified.

15.2.16 Nematodes

15.2.16.1 Root-knot Nematodes (Meloidogyne spp., M. incognita, M. arenaria, M. javanica)

These are very destructive pests. Above ground symptoms consist of foliage discolouration to pale than normal, unthrifty development, dwarfing, wilting in hot dry weather and in extreme cases plants die. Jovani (1994) listed important nematodes affecting soil and root of tomato crop. The nematodes were *Meloidogyne, Ditylenchus, Tylenchulus, Globodera, Heterodera, Xiphinema, Pratylenchus, Helicotylenchus, Paratylenchus, Longidorus and Rotylenchus*. The common species were *M. incognita, X. pachtaicum, D. dispsaci, D. destructor, T semipenetrans, G. rostochimensis, Heterodera schachitet*, and *Pratylenchus penetrans*.

Large and small roots developed swellings, which varied from spheroid galls to elongated spindles. The nematodes lived from one crop season to the next in the soil. The larvae could travel 30 cm per month in upper 60 cm of soil. They are killed at temperatures between 40 and 50°C. A young free-living larva developed into an egg-laying female in about 80 days at 14°C.

The nematode damage was minimized by use of resistant cultivars. A number of tomato cultivars, resistant to root-knot nematodes were released in the USA. Bailey (1941) was first to point out resistance in *L. peruvianum*. Taylor and Chitwood (1951) and Sasser (1954) reported *L. peruvianum* as highly resistant to *M. incognito* but susceptible to *M. arenaria and M. javanica*. Resistant cultivars available with Southern Tomato Exchange Programme were Anahu, Florida Hawaii Cross, Hawaii

Pests of Tomato

Tomato Fruit Worm (Left), and Epilachna Beetle (Right) Infestation

Jassid (Left), and Leaf Miner (Right) Infestation

South American Tomato Pinworm

Aphid (Left) and Pinworm (Right) Infestation

Pests of Tomato

Fruit Borer (Left), and Colorado Beetle (Right) Infestation

Mite Infestation

Fruit fly infestation

Pests of Tomato

White Fly Infestation

Nematode Infestation

55, Nemared, Atkinson and VNF-8. Singh and Choudhury (1972) reported Nematax as immune, Atkinson, VNF-8, 65N-215-1 and 65N-255-1 as resistant and S-120 as highly tolerant to root-knot nematode. Resistance was reported to be controlled by single dominant gene (Thompson and Smith, 1957). Tzortzakakis *et al.* (1999) reported that more than one cropping cycle with resistant plants was required with subsequent nematicide applications in the following susceptible crop, to reduce population densities below damage thresholds. Use of resistant cultivar as the previous crop in the autumn reduced population densities to undetectable levels but did not eliminate the nematode, which subsequently increased on the following tomato crop.

The extent of damage caused by three species of *Meloidogyne* was reported by Singh and Choudhury (1972). *M. arenaria* was found to be the most damaging as it produced maximum number of galls and caused maximum reduction in root and shoot length. However, Swarup and Sharma (1965) reported reduction in shoot growth with increasing population of *M. javanica* and *M. incognita*.

Several methods were suggested for control of root-knot nematode. These include use of chemicals, soil solarization, use of various amendments, biological control, modification of cultural practices and an integration of various control measures. Reduction of root gall index by carbofuran treatment was reported by Sharma and Khan (1994) and Enokpa *et al.* (1996). Root gall formation on tomato was significantly reduced when seedlings were treated with both carbofuran and fenamiphos (Patel *et al.*, 1989; Siddiqui *et al.*, 1998). Chemical control of *M. incognita* was also achieved by spraying of carbofuran and quinalphos, all of which were very effective (Reddy, 1988). Soil application of Abamectin (10 litres/ha) and Cadusafos (8 kg/ha) controlled root knot nematode effectively (Qiao *et al.*, 2011).

Soil solarization and fumigation were found to be effective in controlling the root-knot nematodes inhabiting in soil. Overman and Jones (1987) found that soil solarization in off season by sealing plots with clear polyethylene film followed by soil fumigation in autumn effectively controlled *M. incognita*. Soil solarization by covering green house with polythene film thereby raising the soil temperature by 7-14°C virtually eliminated natural populations of *M. incognita, M. acrita, M. javanica* and *M. arenaria* in the root zone of tomatoes (Khurramov, 1990). Solarization for 11 weeks with a single treatment of solar-heated water (75-90°C) applied at the beginning or end of the solarization period reduced root galling of tomatoes by *M. javanica* (Abu-Gharbieh *et al.*, 1991). Chellemi *et al.* (1994b) observed that soil solarization using a photo selective polyethylene mulch effectively reduced the nematodes populations of *Paratrichodorus minor, Rotylenchulus reniformis* and *Criconemella* spp. Soil solarization in summer effectively controlled the root-knot infestation by *M. javanica* in the field. Solarization followed by growing of a resistant variety was also very effective (Eddaoudi and Ammati, 1995). Solarization together with dazometor with calcium cyanamide gave the best result in controlling root-knot nematode (Fiume and Parisi,b1995). Groundnut and neem oil-cakes and mahua-cakes were reported to be effective in the suppression of nematode population.

Other amendments like sesame oil-cake, castor oil-cake, wheat straw and saw dust were also effective. These amendments were applied along with NPK (Gui *et al.*, 1990). Mukhtar *et al.* (1994) found that addition of organic soil amendments like castor oil-cake + neem leaf cake + saw dust gave maximum control of nematodes. Mustard oil-cake was found to be the most effective organic amendment for reducing *Meloidogyne* spp. in soil and galls on the roots (Khan *et al.*, 1995b). Curative application of neem formulations (*e.g.* 'aza' a refined neem based product) significantly reduced the number of egg masses and eggs per egg mass as compared with the control (Javed *et al.*, 2006).

Different cultural practices led to reduction of the nematode population in the field. Jain and Bhatti (1987) found that 3 summer ploughings resulted in severe reduction in *M. javanica* population. Colombo *et al.* (1994) reported application of a new organic fertilizer of a sea crustacean (commercial product 'ClandoSan') on *Meliodogyne* spp. and observed significant reduction along with methyl bromide.

Biological control of nematode was recommended by various workers as an alternative to chemical application. Osman *et al.* (1988) found that Dipel and SAN 415 (two strains of *Bacillus thuringiensis*) suppressed population of nematodes *M. javanica* and *Tylenchulus semipenetrans* and reduced egg hatchability. An isolate of *B. thuringiensis*, designated as CR-371 was shown to have nematicidal properties and significantly reduced galls caused by *M. incognita* both under green house and field condition. *Bacillus pumilus* strain ZHA90 increased plant height, root fresh weight, root dry weight, shoot fresh weight, and shoot dry weight and reduced root-galling numbers, but did not affect egg-mass numbers, whereas *Paenibacillus castaneae* strain ZHA296 significantly affected egg-mass and root-galling indices and shoot dry weight (Cetintas *et al.*, 2018). *Bacillus megaterium* reduced egg hatching and the number of second-stage juvenile (J2) of *Meloidogyne incognita* (Huang *et al.*, 2010), which resulted from nematotoxic compounds or extracellular hydrolytic enzymes of bacteria destroying nematode eggshell and juvenile cuticle. *Paenibacillus* spp. reduced the rate of egg hatching and J2 number of *M. javanica, M. hapla, M. incognita, M. enterolobii, M. chitwoodi* and *M. fallax in vitro* (Bakengesa, 2016). Son *et al.* (2009) showed that among 40 strains of *Paenibacillus* spp. screened, *P. polymyxa* GBR-462 and GBR-508 and *P. lentimorbus* GBR-158 showed the strongest nematicidal activities and prevented *M. incognita* eggs from hatching.

Population of *Rotylenchulus reniformis* was also reduced considerably by application of CR-371(Zuckerman *et al.*, 1993). Ekanayake and Jayasundara (1994) reported the efficacy of *Paccilomyces lilacimus* and *Beauveria bassiana* as biocontrol agents for control of *M. incognita*. Effective biological control of nematode has also been reported by Walters and Barker (1994), Khan and Saxena (1996), Step *et al.* (1996) and Saikia (1998). *T. harzianum* + *A. strictum* through their talc based formulation @ 1kg/ha with spore load of 2 x 10^8 cfu/g to which 5 per cent carboxy methyl cellulose is added as a sticker has been proved successful in management of root knot nematodes under field conditions (Goswami *et al.*, 2008).

Bacillus thuringiensis (Bt) crystal (Cry) proteins *Cry 6* intoxicate free-living nematodes like *M. incognita* and thus transgenic plants transformed with *Cry 6A* can confer resistance to plant parasitic nematodes and this paved way for further research on transgenics for nematode resistance (Li *et al.*, 2007).

15.3.0 Disorders

15.3.1 Blossom End Rot

This is a very common and destructive disorder. Lesions appear at blossom end of the fruit while it is green. Water-soaked spots appear at the point of attachment of the senescent petals. It enlarges rapidly to 1cm or more in diameter. The affected portion of the fruit becomes sunken, leathery and dark coloured. If plants are grown in adequate soil moisture to produce rapid succulent growth and to promote high rate of transpiration, they are more subject to this type of disorder. This occurs due to sudden change in the rate of transpiration especially in reduced moisture. Increasing amount of nitrogen is conducive to the occurrence of this disorder. The accumulation of calcium and magnesium are important factors. If calcium content falls below 0.2 per cent, the disorder occurs. Compared with normal fruits, the ones affected by blossom end rot were seen to have possessed considerably lower Ca concentrations while K and N contents increased with incidence of this defect, especially at the distal end of the fruit (Wada *et al.*, 1996). Drews and Frank (1987) from Germany reported that few tomato plants were affected by blossom-end-rot when supplied with 100 mg Ca/l at pH 5.8, whereas 40 per cent of plants were affected when supplied with 50 mg Ca/litre at pH 5.8 or with 200mg Ca/litre at pH 4.8. According to Vielemeyer and Weissert (1990) however, blossom end rot of fruits was not related to total Ca content of leaves or to Ca content in sap. Franco *et al.* (1994) stated that gypsum lowered the incidence of blossom-end rot and this effect was further enhanced by the protein hydrolysate. Applying the protein hydrolysate increased the contents of histidine (by 17 per cent), proline (by 15 per cent) and alanine (by 11 per cent) in ripe fruits. Low root pressure, an important factor was involved in the incidence of blossom end rot. Dekock *et al.* (1982) found the influence of truss size on blossom end rot; thinning increased the incidence of blossom end rot. Single foliar spray of 0.5 per cent calcium chloride solution at the time of fruit development was effective in controlling blossom end rot (Bhandari, 1963). According to Raychoudhury and Lele (1966) blossom end rot caused by imbalance in moisture content was controlled by balanced irrigation and staking. Application of poultry manure + limestone reduced blossom end rot (Prezotti *et al.*, 1987). To avoid blossom end rot, it was recommended that the growers maintain a phosphorus concentration in the root environment of at least 1. 7 mM and Ca concentration of 8-9 mM at high air humidity (Kreij, 1996).

15.3.2 Catface

The fruits with 'catface' were characterized by the distortion of the blossom end of the fruit. Such fruits had ridges, furrows, indentation sand blotches. It resembled blossom end rot but was distinctly difficult. Abnormal growing conditions during formation of the blossom appeared to cause distortion of growth of the cells of the pistil. As a result, the cells in the blossom end of the ovary died and turned dark to form a leathery blotch at the end of the fruit without the symptoms characteristic of blossom end rot.

15.3.3 Puffiness

As the fruit reached about two-third normal size, the outer wall continued to develop normally but remaining internal tissue growth was retarded. As a result, tomato fruits were light in weight, they lacked firmness and were partially filled. This was due to non-fertilization of ovule, embryo abortion after normal fertilization and necrosis of vascular and placental tissue after the fruit was well developed. Causal factors were high or low temperature and low soil moisture. High temperature and high soil moisture were predisposing factors.

Fruit puffiness was reduced by single application of 4-CPA or CPPU at concentrations of 20 and 20-25 mg/l, respectively (Kim and Jeong, 1996).

15.3.4 Sunscald

Exposed fruits of tomato either green or nearing ripeness scalded readily during extreme heat. The tissues had blistered water-soaked appearance. Rapid desiccation lead to sunken area which usually had white or grey colour in green or yellowish in red fruits. Adegoroye *et al.* (1989) from Canada reported various sunscald symptoms such as the inhibition of fruit softening, production of tissue and differential ripening of unripe parts that determine the force deformation characteristics of the affected fruits. The cultivars with heavy foliage which provided greater protection from sunrays usually had the least damage.

15.3.5 Cracking

Cracking of the surface of the fruit at the stem end was a common occurrence and often resulted in large losses. The cracks were of two kinds, one which radiated from the stem end and other developed concentrically around the shoulder of the fruit. Radial cracking was more common and caused greater losses than concentric cracking. Besides these, cuticular cracking was also often found on the skin of fruits.

Several environmental factors seemed to be involved in the cracking. It was common during rainy season when temperature was high, especially when rain followed long dry spells. Radial cracking was more likely to develop in full ripe than in fruits at mature green or turning stage. On the other hand, concentric cracking was relatively low on ripe fruits than mature green. Fruits exposed to sun developed more concentric crackings than those covered with foliage. Choi *et al.* (1998) reported that high night temperature and relative humidity and later harvesting dates increased the cracking rate. The contents of soluble solid, acidity and citric acid in fruits were also found to be higher in cracked than in normal fruits. The percentage of cracked fruits, however, decreased with increasing soil water tension *i.e.*, by decreasing irrigation rate.

Use of resistant cultivars like Sioux, Manalucie and Crackproof were recommended. Polygenic inheritance was reported in respect of cracking.

15.3.6 Gold Fleck

Appearance of gold colour flecks on fruits was the main feature of this disorder and chlorophyll was not properly converted to carotenoids (Kalloo, 1986). According to Janse (1988), in goldfleck, a fine spotting affected the calyx end of the fruits and

Physiological Disorders in Tomato

Blossom End Rot

Catface

Sunscalding

Puffiness

Fruit Cracking

Gold Flecking

Zippering

Blotchy Ripening

Seed sprouting

Yellow Shouldering in Fruit

sometimes extended over the whole skin. The affected fruits had a shorter shelf-life than unaffected ones. This disorder was most commonly found in late crops in glass houses with little or no heating. A high incidence of the disorder was associated with large differences in temperature and humidity during the day and night, a low K: Ca ratio, low Mg content in the medium and a low EC level. The K, Ca and Mg levels were about 7.0, 7.0 and 3.5 mmol/l, respectively with an EC level of 3.0m S/cm^2 for reducing the disorder (Janse, 1988). Gold flecks on the surface of tomato fruits were caused by deposits of calcium oxalate in the cells (Janse, 1988; Janse and Kreij, 1989). High Mg and low P concentrations in the nutrient solution with high EC value reduced the severity of the disorder. In rockwool the disorder was less than in soil (Janse and Kreij, 1989).

Researches conducted confirmed that the incidence of the two physiological disorders- gold specks and blossom-end rot were inversely correlated. It was reported that the cumulative Ca uptake by the tomato cultivars tended to decrease with increasing NH_4-N concentrations in the nutrients solution so as to minimize the incidence of gold specks whereby the occurrence of blossom-end rot was observed (Nukaya *et al.*, 1995).

15.3.7 Crease Stem

This disorder in tomato was characterized by thick stems, short internodes and internal necrosis of the stem with more severe symptoms, the crease became deeper and usually formed a slit right through the stem. The disorder sometimes induced the abortion of stem growth and flower truss development and poor fruit set. Terabayashi *et al.* (1995) suggested that the high sulphate concentration in the nutrient solution caused small reductions in the P, K, Ca and Mg contents of the stem and a large increase in the B content. Further, it was concluded that B prevented crease stem, though the effect of high sulphate remained unclear.

15.3.8 Internal Blackening

Internal blackening syndrome in tomato shared some features with blossom end rot although some differences between the two disorders were apparent. A study conducted by Reid *et al.* (1996) suggested that internal blackness could have resulted from increased root competition for photosynthate, leading to abnormal seed development. Root turnover was appreciable *i.e.*, 30-40 per cent of the roots survived for $\leq$ 28 days.

15.3.9 Lower Stem Swelling

Lower stem swelling was a disorder seen with increasing frequency in tomato plantlets. The symptoms included limited root growth, especially of the lateral or secondary roots and a typical swelling of the lower part the stem which sometimes resulted in stem breakage. Results from trials conducted by Ortiz and Gonzalez (1999) in Spain showed that common dinitroaniline herbicides contributed to stem swelling and lack of rooting, mainly when they were incorporated superficially. Damage decreased with deep incorporation of the herbicides and to some extent, when fertilizers were used.

15.3.10 Low Temperature Injury

Tomato was very sensitive to frost. At the maturity of the crop when it was exposed to near freezing temperature, the accumulation of sugars might lower the freezing point slightly. As the vines were frozen they darkened, withered and desiccated. The fruits became soft, water-soaked and dull coloured. Marangoni *et al.* (1989) stated that the primary site of damage appeared to be the chloroplast and that the ultrastructural damage observed could be explained in terms of membrane disruption. According to Nouyen and Mazliak (1991) the induction of chilling injury in tomato was accompanied by degradation of galactolipids of the pericarp.

As regards control of the low temperature injury, many workers reported differently though their basic ideas which revealed that heat treatment was ideal. According to Lurie and Klein (1991) the heat treated tomatoes ripened normally with normal colour development, chlorophyll disappeared, lycopene content increased, CO_2 and ethylene evolution increased to a climacteric peak and K^+ leakage increased with time. They stated that there appeared to be a relationship between the heat shock response and the protection of fruits from low temperature injury.

Lurie and Sabehat (1997) found that holding the fruits of tomato for 48 h at 38°C before storage prevented the chilling injury development during storage at 2°C upto 4 weeks. Pre-storage at temperature higher than the said value, however, caused considerable damage. Kluge *et al.* (1998) from Brazil reported no chilling injury in fruits warmed for 96 h at 25°C for 10 days followed by storage at 5°C. Nagetey and Mao (1999) suggested that hot water immersion of tomato fruits at 40°C for 15 minutes was the most effective treatment for reducing chilling injury without adversely affecting the fruit quality.

16.0 Seed Production

Seed production in tomato can be taken up in any part of India. The agronomical practices for raising the crop for seed production are the same as for growing tomato for vegetable purpose. Tomato is a self-pollinated crop. Natural cross-pollination varies with the cultivar, weather condition, insect pollinators and isolation. In a study at the Indian Agricultural Research Institute, New Delhi, the natural cross-pollination in autumn-winter crop was 3. 84 per cent in alternate rows of 60 cm, 1.74 per cent at 240 cm and interspaced with three crops. The cross-pollination was nil in spring-summer crop (Choudhury and Anothai-Choomsai, 1970). The isolation distance is recommended as 50 m for foundation crop. This will avoid mechanical mixture. The crop should be examined at various stages of plant growth for detecting off-types. Production of nucleus seed is made either by raising progenies of true-to-type single plant or by selecting elite plants from seed crop. The selected plants are tagged and well developed ripe fruits from these plants are collected for nucleus seed production.

16.1.0 Seed Extraction

Choudhury (1972) suggested the following methods for seed extraction from slimy mass of tomato.

16.1.1 Fermentation Method

In this method, the fruits selected for seed are completely lost. The selected ripe fruits are harvested and kept for a day in wooden containers. They are crushed by hand. No fruit juice should be allowed to drain out. Entire mass is kept for 24 to 72 hours depending upon high and low temperature. Flesh will float at the top and seed will settle down at the bottom. The fermented mass is removed and the seeds are cleaned 8-10 times with clean water and later on, they are dried. Longer fermentation may damage the seed.

16.1.2 Alkali Treatment

Selected fruits are cut into halves and the slimy mass containing seed are scooped out in a vessel with wooden stick, the mass is treated with 300 g of washing soda in 4 litres of boiling water in equal volume. When the mixture is cooled down, it is allowed to stand overnight. Next day, all the seed will settle down at the bottom. Seeds are washed thoroughly.

16.1.3 Acid Treatment

Slimy mass is separated and treated with commercial hydrochloric acid at the rate of 75 ml per 12 kg of material. Seed separation is due to rapid dispersion of colloidal mass from seed by acid. Instead of microbiological action of fermentation, the seed is separated from sliminess in 15 to 30 minutes. The acidified liquid is decanted off and seeds are washed well with water.

Besides, many workers studied on the seed extraction procedures in tomato. Pandita *et al.* (1996) observed that seed quality such as germination percentage, vigour index and field emergence decreased with increasing fermentation period and these were highest for seeds extracted at the pink stage. It was also noted that fruits harvested at turning stages and fermented for 24- 48 h yield the highest quality seeds. According to Das *et al.* (1997) on comparing the manual extraction with fermentation or extraction using HCl or NaOH suggested that the highest percentage of germination was observed for seeds extracted using the fermentation technique and after soaking in 2.50 per cent HCl for 30 minutes (79.05 and 81.87 per cent, respectively).

17.0 Crop Improvement

With tomato breeding being widely performed, after 1930, the goal of tomato breeding has been primarily yield and productivity followed by disease resistance to maintain production in challenging environment. Prior to 1920, tomato improvement depended largely upon selection of chance variants that originated as a result of mutation, spontaneous out-crossing or recombination of pre-existing genetic variation. Such selections succeeded in developing cultivars adapted to such widely different environments in different parts of the world. During the last century many hundreds of new cultivars have been developed to meet the diverse needs. Identification of important phenotypic trait in wild or related germplasm, their genetic basis, and introgression of those traits into advanced modern lines have been keys to the development of new cultivars in tomato. Besides, hybridization followed by selection, backcrossing for introgression of a desirable trait from one parent to another has led to varieties with higher yield and better quality. Exploitation of exotic resources and introgression of valuable genes into the gene pool has been the backbone of tomato improvement. Conventional breeding has resulted in development of varieties with resistance to biotic stresses with monogenic and oligogenic control and dominant resistance in F_1s and also enabled development of genotypes with desired agronomic traits, high yield, earliness, uniformity, better keeping quality, *etc.* (Passam *et al.*, 2007).

17.1.0 Genetics of Important Characters

Attempts for genetic mapping of morphological characters in tomato dated from the beginning of 20th century and in 1968 resulted in the development of the first "classical" map containing 153 morphological and physiological markers on all twelve linkage groups (Butler, 1968). By 1986 the classical map assigned to their respective chromosomes more than 300 morphological, physiological and disease resistance genes (Stevens and Rick, 1986; Mutschler *et al.*, 1987). Menda *et al.* (2004) developed an isogenic tomato "mutation library" in the genetic background of the inbred variety M82 as a basic resource for exploring gene function. A total of 13000 M_2 families, were visually phenotyped in the field and categorized into a morphological catalogue that includes 15 primary and 48 secondary categories (Menda *et al.*, 2004).

The number of gene loci mapped has increased to more than 1500, using classical genetics, isozyme and different DNA based markers *viz.*, Restriction Fragment Length Polymorphism (RFLP), Random Amplified Polymorphic DNA (RAPD), Amplified Fragment Length Polymorphism (AFLP), Simple Sequence Repeat (SSR), *etc.* A few important genes of great interest to plant breeders and their locations on different chromosomes have been documented in Tables 7 and 8.

Table 7: Some Important Qualitative Characters of Tomato*

Gene Symbol	*Description of Plant Characters*
a	Anthocyaninless, absence of anthocyanin in all plant parts
aa	Anthocyanin absent, absence of anthocyanin in all plant parts
ae	Entirely anthocyaninless
aw	Anthocyanin without, absence of anthocyanin in all plant parts
Aft	Anthocyanin fruit
alc	Alcobaca, delay in fruit ripening
Ad	Resistance to *Alternaria solani*
Asc	Resistance to *Alternaria alternata* f. sp. *lycopersici*
B	High β carotene pigment in fruit
bls	Baby lea syndrome, absence of anthocyanin in all plant parts
Br	Resistance to buckeye rot
br	Brachytic, internode shortened
bk	Beaked, pointed styler end of the fruit
bs	Brown seed, brown endosperm colour

Gene Symbol	*Description of Plant Characters*
bs-2	Brown seed-2, Brown endosperm colour
c	Potato leaf, reduced leaf segments
Cf, Cf-2, Cf-3,Cf-4, Cf-11	Resistance to specific race of *Cladosporium falvum*
cm	Leaves curly mottled, sensitive to environment
cl-2	Cleistogamous flower, Functioal male sterility, failure of the petals to open
Cor-1/2	Resistance to bacterial canker, *Clavibacter michiganensis* ssp. *michiganensis*
d	Dwarf, all plant parts shortened, leaf dark green and rugose
dg (*hp 2* ^{dg})	Dark green fruit and exaggerated photo-responsiveness
dl	Functional male sterility, growth of epidermal hairs on the anther surface is prevented
ex	Exserted stigma, viable pollens but long style that protrudes from theanther cone
f	Fasciated fruit with many locules
Fri	Resistance to Fusarium crown and root rot, *Fusarium oxysporum* f.sp. *radicis lycopersici*
gf	Green flesh fruit, persistent chlorophyll
Hr	Hirsute, long hair on upper leaf surface
h	Hair absent, incompletely dominant
hp, hp-1, hp-2, hp-3, hp-4,hp-5	High pigment, chlorophyll, carotenoids and ascorbic acid contents in fruit
I	Immunity, resistance to *Fusarium oxysporum* f. *lycopersici*
I-2	Immunity-2, resistance to *Fusarium oxysporum* f. *lycopersici*, race-2
I-3	Immunity-3, resistance to *Fusarium oxysporum* f. *lycopersici*, race-3
j	Jointless pedicel
j-2	Jointless pedicel-2
lg-5	Foliage uniformly light green
Lv	Resistance to powdery mildew, *Leveillula taurica*
Meu-1	Resistance to potato aphid, *Macrosiphum euphobiae*
Mi	Resistance to root knot nematode, *Meloidogyne* spp.
ms-1 to ms-55	Male sterility, pollen aborted or no pollen
nor	Non ripening, very retarded fruit ripening process
Nr	Never ripe, fruits ripen to dirty orange in colour
o	Ovate, fruit ovate or pyriform
og ^{c}	Old gold crimson, high lycopene and β carotene contents in fruit
pat and pat 2	Parthenocarpic fruit
Ph, Ph-2	Resistance to race '2' of *Phytophthora infestans*
ph1	Resistance to *Pyrenochaeta lycopersici*
ps	Positional sterile mutant, connate petals, normal viable pollens but non-dehiscent anther due to persistent stomium
ps-2	Positional sterile-2 mutant, normal flower with viable pollens but non-dehiscent anther due to structural alterations in the zone of anther dehiscence

Gene Symbol	Description of Plant Characters
pst	Persistent style resulting beak in fruit
r	Yellow flesh of the fruit
rin	Ripening inhibitor, delayed ripening and bright yellow fruit at maturity
s	Compound inflorescence
Se	Resistance to *Septoria lycopersici*
sh	Sherry
sl	Stamenless, male sterility
sp	Self pruning, determinate plant habit
Sm	Resistance to *Stemphylium solani*
Sw-5	Resistance to Tomato spotted wilt virus
sucr	Sugar (sucrose) accumulator gene
t	Tangerine fruit colour
Tm	Resistance to Tomato mosaic virus
Tm-2, Tm-22	High level of resistance to many strains of Tomato mosaic virus
Tv-1/2	Resistance to green house white fly, *Trialeurodes vaporariorum*
Ty-1, Ty-3, Ty-4, and *Ty-6*	Resistance to tomato yellow leaf curl virus (introgressed from *Solanum chilense*)
Ty-2	Resistance to tomato yellow leaf curl virus (introgressed from *Solanum habrochaites*)
ty-5	Resistance to tomato yellow leaf curl virus (introgressed from *Solanum peruvianum*)
u	Uniform ripening of fruit
Ve	Resistance to *Verticillium albo-atrum, V. dahliae*
y	Un-pigmented fruit epidermis

* Stevens and Rick (1986); Mutschler *et al.* (1987), Laterrot (1997), Atanassova and Hazra (2012).

Table 8: Some Important Genes Located on different Chromosomes*

Chromosome No.	Genes present
1	*br, bs, y, ms-32, Cf, Cf-4, ms-6, Nr-2*
2	*s, bk, ms-10, aa, o, ms-15, ms-2, ms-5, ms-10, ms-26, ms-35*
3	*ms-9, r, sucr*
4	*cm, sl*
5	*Rin*
6	*Mi, ms-16, ms-33, Cf-2, c, sp, B, ogc, cl-2, Ty-1*
7	*pst, lg-5, bs-2, Ph, I-3*
8	*ms-8, gf, Hr, ae, ms-4, ms-17*
9	*TM-2, Nr*
10	*nor, u, h, t, ms-31*
11	*j-2, Cf-3, j, ms-3, ms-7, ms-12, ms-14, ms-42, a, I-2, f,l*
12	*Cf-11*

* Stevens and Rick (1986), Mutschler *et al.* (1987), Laterrot (1997), Gorman and McCormick (1997).

17.2.0 Breeding Goals

The recent trend has been towards development of cultivars to meet the specific uses rather than multipurpose cultivars to meet several needs. Tigchelaar (1986) outlined the specific breeding objectives of tomato *viz.* (i) earliness, (ii) growth habit, (iii) machine harvestability, (iv) diseases and pest resistance, (v) nematode resistance, (vi) fruit quality for fresh market, and (vii) processing quality.

With fresh-market tomato, some of the important traits that have been of main focus for producer are yield or size, disease and insect resistance, heat and other abiotic stress tolerance, uniform/synchronous ripening, and harvesting with focus in mechanization, while for marketing are longer shelf life, minimum handling, and shipping damage; and for consumer are flavour and taste, appearance/colour, and forms of use like salad and cooking (Rodruguez *et al.*, 2011; Bauchet and Causse, 2012; Gascuel *et al.*, 2017).

The breeding goals for different categories of tomato as per Foolad (2007a) are as follows:

Universal Goals for Tomato Breeding

- ✰ Fruit yield
- ✰ Disease resistance
- ✰ Wide adaptability
- ✰ Early maturity
- ✰ Fruitset in extreme temperatures
- ✰ Resistance to rain induced cracking
- ✰ Tolerance to major ripe fruit rots
- ✰ Sufficient vine cover

Breeding Objectives for Fresh Market Tomatoes

- ✰ Large, round fruits
- ✰ Firmness and shelf life
- ✰ Uniformity in size
- ✰ Fruit shape and colour
- ✰ Appearance
- ✰ Freedom from external blemishes
- ✰ Texture
- ✰ Taste and flavour

Breeding Goals for Processing Tomatoes

- ✰ Compact and determinate plant habit
- ✰ Concentrated flowering
- ✰ Suited for mechanical harvest

- ☆ Ease of fruit separation from vine
- ☆ Fruit colour
- ☆ Fruit pH and total acidity
- ☆ Total soluble solids
- ☆ Fruit viscocity

Tomato for Fresh Market Purpose

For fresh market tomatoes besides nutritive quality, good flavour and/or texture, aroma, firmness and beautiful colouration are traits of primary importance. Tomato flavour is determined by a number of taste components, aromatic volatiles and a complex interaction between them (reviewed by Yilmaz, 2001), the ratio of reducing sugars and organic acids being considered as one of the most important among them (Bucheli *et al.*, 1999). Tomato contains two main types of sugars, sucrose and fructose. According to Stevens *et al.* (1979) sweetness is more affected by the content of fructose than glucose while citric acid is producing most of the sour taste in tomatoes as it is present in much greater quantities than malic acid in mature fruit. Over 400 volatiles were determined in tomato fruit, but only 30 of them were found to significantly impact tomato flavour (Petro-Turza, 1987). The flavour of tomato fruit is influenced by alcohols, aldehydes, carbonyls and sulfur compounds. The typical aroma of field-ripened tomato is due to carbonyls (32 per cent), short chain alcohols (10 per cent) and hydrocarbons, long chain alcohols and esters (58 per cent). The hexanal, trans-2-hexanal, 2-isobutyl thiazole, 2-methyl-2-hepten-6-one, geranyl acetone and formyl acetone are the important volatile compounds in fresh tomato. Sugars, acids and their interactions are important for sweetness, sourness and overall flavour intensity in tomatoes. Fructose and citric acid are more important to sweetness and sourness of tomato. The lipid fraction of tomato fruit is composed of triglycerides, sterols, sterol esters, free fatty acids and hydrocarbons. The ripe tomatoes contain 20 amino acids in addition to small amounts of tryptamine, S-hydroxy tryptamine and tyramine. Glutamic, aspartic, γ-amino butyric acid and glutamine comprise about 80 per cent of the total free amino acids and contribute to the taste. Some of the amino acids also serve as precursors of volatile components. Tomatine is a steroidal glycoalkaloid found in all genotypes, which is enzymatically changed to non-toxic form as the fruit ripens, therefore, tomatine does not pose any safety hazard in ripe tomato uses. Texture is an important component for the consumer's perception of tomato fruit quality (Causse *et al.*, 2003). The nature of the tomato fruit texture is complex and includes many traits, mainly sensory attributes such as juiciness, flesh firmness and mealiness (Szczesniak, 2002). Higher organoleptic value of tomato is becoming a subject of a number of studies aiming at the mapping of quantitative trait loci (QTLs) affecting fruit colour, firmness, elasticity, sugars and organic acids content (Causse *et al.*, 2004), identification of loci affecting flavour volatile emissions in tomato fruits (Tieman *et al.*, 2006). Investigations on the texture of tomato fruits at the rheological, sensory, morphological and genetic levels were carried out and attempts to correlate several parameters were made. Analyses were performed on tomato fruits from introgressed lines carrying quantitative trait loci (QTLs) associated with texture traits localized on different chromosomes, in two genetic backgrounds (Chaib *et al.*, 2007).

Tomato for Processing Purpose

Although tomatoes are commonly consumed fresh, over 80 per cent of the tomato consumption comes from processed products such as juice, paste, puree, ketchup and sauce. Processing tomatoes have to possess a number of rheological characteristics making them suitable for various processing applications (juice, ketchup, peeled tomato, *etc.*). Gould (1992) prepared a comprehensive general guideline enumerating seventeen traits that processing tomato varieties should possess such as: uniform fruit setting and ripening; adaptability to mechanical harvesting and bulk handling; uniform fruit size with no fruits smaller than 50 g and none larger than 90g; total solids content in excess of 5.5 per cent and preferably upward 8.5 per cent ; soluble solids content (Brix value) in excess of 4.5 per cent and preferably upward 7.5 per cent ; high acid (citric) content (minimum of 0.35 per cent and up to 0.55 per cent); low pH value (maximum of 4.4), tolerance to biotic and abiotic stresses, *etc.* However, it is difficult to cumulate in one variety so many traits (Barrett *et al.,* 1998). One of the most important characteristics in processing tomato breeding is the total soluble solids content (TSS or Brix value). Its improvement impacts taste and quality of fresh and processed products and genetic factors involved in tomato quality were studied and mapped with processing tomato cultivars crossed to related wild species (Lecomte *et al.,* 2004). Several studies revealed that isogenic lines developed with the wild tomato species *viz., S. pimpinellifolium, S. pennellii,* contained higher TSS content in fruits (Eshed and Zamir, 1995; Grandillo *et al.,* 1996; Hazra *et al.,* 2001). During the last decades, innovation has been one of the most important activities in tomato processing in order to be up to date with the technological changes, consumer demands and international competition which might result in the requirement of additional quality traits of processing tomato varieties.

Tomato for Distant Transporting Purpose

For long distance transport, fresh market tomatoes are picked up at the mature-green or breaker stage. Processing tomatoes are harvested when fully mature to optimize various quality parameters such as colour, flavour, texture, acidity and total solids in the fruits. Hence, to withstand the rigors of machine or hand harvesting and the transport to factory, the fruits need to be firm enough. Fruit longevity is correlated with fruit firmness (Philouze, 1995). Atanassova *et al.* (2008) found that genes with predominantly additive action controlled fruit firmness and longevity. Development of hybrids possessing fruits characterized by prolonged shelf life could be realized by incorporating in one of the parent lines, one of the genes that controled delayed fruit ripening: *alc* (*alcobaca*), *rin* (*ripening inhibitor*), *nor* (*non-ripening*). The development of long shelf life hybrids was welcomed by growers and processors. Further studies provided evidence however, that *rin* hybrids were more firm than normal cultivars but showed lower level of important flavour volatiles and scored lower in sensory tests even if they were harvested at a later stage of maturity (Baldwin, 1998). Now-a-days, the longevity of fruit, once considered as a trait of primary importance, is not appreciated if not combined with good flavour and/or texture.

Phenotypic characteristics have been used to study genetic differences and for genetic diversity analysis and cultivar development (Khadivi-Khud *et al.*, 2008; Terzopolous and Bebeli, 2010; Nikoumanesh *et al.*, 2011; Fufa *et al.*, 2005). Assessment of diversity in commercial cultivars provides the status of a crop in relation to its wild and domesticated relatives and assists in creating a breeding scheme to bring in the lost alleles due to extreme selection and bottlenecks during domestication or introduction (Ranc *et al.*, 2007). In the absence of genetic variation, breeding efforts in crops remain without any gain (Tanksley and Mccouch, 1997). This may lead to severe crop failure with the onset of new biotic or abiotic stress. Most of the disease resistant genes in modern varieties are not only the result of selection from these varieties but also originated from the wild species (Rick and Chetelat, 1995).

The jointless (*j2*) allele (Szymkowiak and Irish, 2006) was introgressed from *S. cheesmanii*. Compact fruit set by determinate growth, *sp* for self-pruning, was discovered from a natural mutation in the early 20th century. Four genes are known to control fruit shape: *sun* [van der Knaap *et al.*, 2004] and *ovate* (Ku *et al.*, 1999) control elongated shape like in Roma type whereas *fasciated* (FAS) (Cong *et al.*, 2007) and *locule number* (LC) (Munos *et al.*, 2011) control fruit locule and flat shape.

No single cultivar has been developed that can resist all races of *Xanthomonas* causing bacterial spot making this disease a very difficult problem to overcome (Bhattrai *et al.*, 2017). This explains the low correlation of bacterial spot with any other traits observed in this study as seen in a previous study (Bhattrai *et al.*, 2016), whereas resistance to fungal diseases like early blight, late blight, *Fusarium*, and *Verticillium* wilts and nematode resistance have been identified and are already being introgressed in new varieties (Scott and Gardener, 2006). Pyramiding the genes for resistance to diseases and pests has been one of the major goals of tomato breeding programs. Modern varieties developed have been reported to be introgressed with up to 12 different diseases and pest resistant genes (Bhattrai *et al.*, 2018).

17.3.0 Breeding Methods

Tomato is a self-pollinated inbred crop and until the early sixties of 20th century it was cultivated using open-pollinated varieties. Different breeding methods followed to develop line-bred varieties are mentioned below.

17.3.1 Selection

Selection has also played an important role in tomato improvement in India. Improved Meeruti was evolved at the Indian Agricultural Research Institute by selection from an Indian cultivar at Meerut in UP (Singh and Sikka, 1955). Pusa 120 was selected from segregating material from Hawaii. This cultivar has been found to be promising for nematode resistance, heavy yield, large round attractive fruits (Choudhury, 1967). Kamlanathan *et al.* (1969) and Rajgopal and Rao (1974) have recommended CO-1 tomato (a pure line selection from Kalianpur cultivar) for Tamil Nadu. Selection for resistance to root-knot nematode of cvs. Namura, Dar

Surkhana and Surkhan have been reported by Aramov and Dzhuraeva (1990). For green house cultivation in Italy, cvs. Restino, Boulba, Turquesa and Portanto were selected for resistance to root-knot nematode (Calabretta and Privitera, 1985). Some varieties developed through single plant selection from indigenous or introduced stock are Improved Meeruti, HS-110, Sonali, Pant Bahar, Pant T-1, Arka Vikas, Arka Alok, Arka Ashish, Arka Abha, Arka Saurabh, PKM-1, VC 48-1, NDT-5, NDT-10, *etc.* (Hazra and Som, 2006).

From a study on the improvement of hybrid tomatoes, Ghosh *et al.* (1995) opined that the highest variation occurred for yield/plants, plant height, fruit weight, locules/fruit, seeds/fruit and TSS (Total soluble solids). Values for heritability and genetic advance (as a percentage of mean) were high for fruit dry matter, seeds/ fruit, plant height, fruit weight, locules/fruit and yield/plant. Yield/plant was significantly correlated with fruit weight.

Study on genetic variability in exotic lines of tomato in off season revealed that the differences between the genotypes were highly significant for all characters studied (Barman *et al.,* 1998). Heritability estimates were very high for all the characters except for the days to fruiting. Furthermore, plant height, seeds per fruit and fruits per plant had high heritability estimates coupled with a high genetic advance, thus offering a scope for genetic improvement.

Dar and Sharma (2011) at SKUAST, Jammu noted high variability for yield attributes like yieldha^{-1}, fruits/plant, average fruit weight, and quality attributes like ascorbic acid, lycopene and beta-carotene in their study on 60 tomato genotypes. They also reported that the characters ascorbic acid, β-carotene and lycopene have high heritability and could be used for selection at early generation for improvement of these traits in breeding lines.

Prema *et al.* (2011) worked on cherry tomato and observed ample variation was present in all traits except plant height at 60 and 90 DAT. They estimated high PCV and GCV for average fruit weight, pericarp thickness, fruit yield/plant, lycopene content, fruit length, TSS and fruit width. They noted high heritability accompanying high genetic advance for average fruit weight, days to 50 per cent flowering. They also noted high heritability along with moderate genetic advance for plant height at 90 DAT, days to first flowering, and ascorbic acid content indicating governance of additive gene action for these characters, thus suggesting the scope of direct selection for improvement of these traits. Mukherjee *et al.* (2020) observed high, genotypic coefficient of variation, heritability coupled with genetic advance, for pericarp thickness, fruit firmness, ascorbic acid content, titratable acidity content, total antioxidant activity of fruit, PDI of ToLCV, and fruit yield per plant, which suggests that the characters concerned are conditioned by additive gene action and therefore, these characters would be more reliable for effective selection in cherry tomato.

Ahirwar *et al.* (2013) performed an experiment with 19 genotypes of tomato and observed that the PCV was of higher value than GCV for all the traits. They also noted that genetic advance as per cent of mean at five per cent was high for all considered traits except days to flower initiation and first fruit harvest.

Shankar *et al.* (2013) worked with 24 hybrids along with 11 parental lines of tomato and reported that PCV and GCV were high for number of average fruit weight, fruit number/cluster, yield/plant, plant height, lycopene, ascorbic acid and titrable acidity inferring that huge variability for these traits existed. They also observed high heritability besides high genetic advance as per cent of mean for primary branch number per plant, plant height, number of fruits per cluster, average fruit weight, fruit length, fruit width, pericarp thickness, locule number per fruit, lycopene, titrable acidity, ascorbic acid and shelf life. Thus simple selection based on phenotypic performance of the said traits would efficiently improve the genotypes.

Nwosu *et al.* (2014) performed an experiment at National Centre for Genetic Resources and Biotechnology, Nigeria on 19 accessions of tomato and confirmed that variability was present among the genotypes as depicted by the values of GCV and PCV. Very high genetic advance and heritability for days to flowering, fruit number per plant, fruit diameter, fruit length and fruit diameter was found and suggested that these traits may be improved in tomato by simple selection at early generations.

Kumar (2014) worked on fifty genotypes of tomato and observed high PCV and GCV for plant height, number of primary branches/plant, average fruit weight, number of locules/fruit, number of fruits/plant and fruit yield/plant. Moderate values were noted for pericarp thickness, total soluble solids and diameter of fruits, while they observed low GCV and PCV for days to 50 per cent flowering. They noticed high heritability along with high genetic advance in per cent of mean for most of traits except days to 50 per cent flowering.

Sherpa *et al.* (2014a) observed high heritability coupled with high genetic advance for number of flower clusters per plant, number of fruits per cluster, plant height, polar diameter, number of fruits per plant, fruit weight, pericarp thickness, total soluble solids, titratable acidity, ascorbic acid content, lycopene content and fruit yield per plant which were controlled by additive gene effects, indicating good response to selection for these characters in exotic tomatoes.

Prajapati *et al.* (2015) conducted an experiment during Rabi season of 2011 using 39 diverse germplasm of tomato and showed that number of fruits/plant possessed highest genotypic and phenotypic variance and test weight the lowest. Maximum genotypic coefficient of variation (GCV) and phenotypic coefficient of variation (PCV) were recorded for average fruit weight, number of seeds per fruit while the minimum were noted for days to 50 per cent fruit setting. They also observed greatest heritability (broad sense) for average fruit weight followed by number of secondary branches and lowest for test weight. On the other hand, they estimated maximum genetic advance as per cent of mean for average fruit weight and minimum for days to 50 per cent fruitset. They reported that presence of variability in the germplasm is effective for selection of elite genotypes.

Pujer *et al.* (2015) studied genetic variability and association of different traits in cherry tomato genotypes and reported high GCV and PCV for average fruit weight (67.76 and 67.9 per cent, respectively) and number of fruits per plant (56.45 and 57.28 per cent, respectively). On the other hand, heritability, genetic advance and

GCV were high for average fruit weight and number of fruits per plant indicating governance of additive gene action which suggested that selection will be effective in improving these traits. They recorded high heritability with low genetic advance for fruit yield per plant indicating that simple selection would not be effective in improvement of high yield and hence hybridization may be carried out.

Saleem *et al.* (2013) used twenty-five F_1 hybrids derived from 5 x 5 diallel mating scheme during 2009-10 and noticed that heritability was of high value for plant height, fruit length, fruit width, fruit weight, number of fruits per plant and fruit yield per plant. They recorded high genetic advance for plant height, fruit number/ plant, fruit length, fruit width, fruit weight, fruit number/plant and fruit yield/plant.

The work of Singh *et al.* (2017) on 35 determinate genotypes of tomato revealed the presence of high PCV and GCV for fruit yield/plant, average fruit weight, locules number/fruit, fruits/plant, plant height and number of primary branches/plant. They noted high heritability along with high genetic advance as per cent of mean for the different characters under study except for days to 50 per cent flowering. According to them, fruit yield per plant followed by average fruit weight, number of locules per fruit, number of fruits per plant and plant height were the top five traits which showed high level of genetic advance that indicated selection based on these traits for successful breeding programme.

Ullah *et al.* (2015) performed an experiment with 20 parental lines of tomato and estimated high heritability for fruit per plant (52.30) and fruit weight (46.32). Besides, they observed high heritability along with high genetic advance for fruits per plant (52.30), fruit weight (46.32) and flower per cluster (33.50). So, the selection based on such traits would result in improving fruit yield of tomato.

Kumar *et al.* (2016) worked on 60 genotypes of tomato to identify the variability present among the genotypes of tomato. They showed that high GCV and PCV were present for average fruit weight, yield/plant, number of branches at 60 DAT, polar diameter, pericarp thickness, number of locules/fruit, number of fruits/ plant, lycopene content, β-carotene content, vitamin C content, TSS: Acid ratio, thus inferring high variation for these traits and chances for improvement via selection. They also observed high heritability with high genetic advance as percent of mean for polar diameter, average fruit weight, number of fruits/plant, lycopene content, equatorial diameter, number of branches at 60 and 90 DAT and plant height at 60 and 90 DAT. Hence, they suggested that additive component was predominant for these traits, which was indicative of potential for improvement through direct selection.

Rai *et al.* (2016) worked on the genetic variability studies in tomato for yield and quality parameters at Solan on 56 tomato genotypes and revealed that PCV was slightly higher than GCV. They recorded high heritability and genetic gain for number of fruits/plant, fruit weight, fruit yield/plant, average fruit weight and lycopene content. From the experiment it was concluded that selection based on these traits help in identification of superior genotypes of tomato.

Ambresh *et al.* (2017) carried out experiment on tomato and reported high values of PCV and GCV for fruit yield/plant, moderate PCV and low GCV for pH of fruit and days to first flowering.

Bhandari *et al.* (2017) performed experiment on 66 genotypes of tomato and found high estimates of PCV and GCV for fruit number/cluster, fruit number/plant, average fruit weight, seed number/fruit, locules/fruit and fruit yield and moderate values for days to 50 per cent flowering, primary branches/plant and fruits/cluster. They also observed high heritability for all the traits, whereas high genetic gain for all traits except days to 50 per cent flowering.

Ligade *et al.* (2017) worked on twenty tomato genotypes and observed high values of GCV and PCV for characters *viz.*, number of fruits/plant, average fruit weight, number of locules/fruit, fruit yield/plant, plant height, number of fruits/cluster and TSS; moderate for traits *viz.*, pericarp thickness, flowers/cluster, β-carotene, vitamin C; low for traits days to first flowering, lycopene content and days to 50 per cent flowering.

Patel *et al.* (2017) found that the values of PCV were higher than the same for GCV for all the characters. This showed that environmental factors may be influencing expression of these traits and their betterment could be possible by different phenotypic selection, namely, disruptive, directional and stabilized. PCV and GCV was low for number of days to fruit initiation (8.471 and 6.193, respectively) On the other hand, they obtained high values of PCV and GCV values for average fruit weight (48.639 and 48.004, respectively), fruit number per plant (47.556 and 46.839, respectively) and primary branch number per plant (27.575 and 26.525, respectively) suggesting that these characters showed highest degree of variation.

Genetic variability among fifteen tomato genotypes was studied by Meena *et al.* (2018) for different traits. They observed that all the characters under study exhibited higher PCV than GCV. High PCV was estimated for average fruit weight and fruit yield/plant; moderate for branch number/plant, fruit number/cluster, number of clusters/plant, fruits/plant, fruit length, pericarp thickness and locules/fruit; low for days to 50 per cent flowering. They observed high GCV for fruit yield per plant; moderate for flowers/cluster, fruits/cluster, branches/plant, clusters/plant, fruits/plant, locules/fruit, pericarp thickness; low for TSS, fruit length and fruit width. High heritability was estimated for plant height, branches per plant, flowers per cluster, fruits per cluster, clusters per plant fruits per plant, pericarp thickness, fruit yield per plant and TSS, while low values were recorded for vitamin C. They noted high genetic advance for fruit yield per plant, average fruit weight, flowers per cluster, fruits per cluster, branches per plant, moderate for locules per fruit, fruits per plan, clusters per plant, pericarp thickness, TSS, and low for days to 50 per cent flowering.

Pandey *et al.* (2018) performed an experiment comprising of thirty genotypes of tomato and analysed data for 15 quantitative traits. Estimates of PCV and GCV were high for average fruit weight, locule number/fruit and pericarp thickness; moderate for branches/plant, flowers/inflorescence, fruits/plant, fruit diameter, plant height, days to maturity and fruit yield/plant; low for days to first and 50 per cent flowering, flower number and fruit number per cluster. On the other hand, they observed high heritability coupled with high expected genetic advance in per cent of mean for fruit yield/plant, fruit number/plant, fruit diameter, pericarp thickness, average fruit weight, days to maturity and plant height.

Singh *et al.* (2017) carried out an experiment on 20 genotypes of tomato and observed high PCV and GCV for locule number per fruit, pericarp thickness, fruit number per plant and average yield per plant.

Das *et al.* (2019) observed high heritability with high genetic advance for flower cluster per plant, flower per cluster, fruit per plant, number of locules per fruit, total chlorophyll content of leaf, total chlorophyll content of immature fruits, chlorophyll a content of immature fruits, chlorophyll b content of immature fruits, lycopene content of mature fruits and ascorbic acid content in induced mutants and their hybrids of tomato, suggesting selection would be rewarding for improvement of these traits.

17.3.2 Hybridization followed by Selection

The normal mode of reproduction and ease of manipulation of flower for controlled mating make tomato ideal for hybridization programme. In India, hybridization utilizing exotic cultivars as one of the parents was also attempted in evolving cultivars like Pusa Ruby and Pusa Early Dwarf. The following cultivars grown in India were developed by hybridization followed by selection have been listed in Table 9.

Table 9: Varieties Breed through Hybridization followed by Selection*

Variety	*Parentage*
Pusa Ruby	Sioux x Improved Meeruti
Pusa Early Dwarf	Improved Meeruti x Red Cloud
Marglobe	Marvel x Globe
Pusa Red Plum	*L. esculentum* x *L. pimpinellifolium*
Punjab Chhuhara	Punjab Tropic x EC 55055
Punjab Kesri	Punjab Tropic x EC 55055
HS-101	Sel. 2-3 x Exotic cultivar
HS-102	Sel. 12 x Pusa Early Dwarf
Hisar Arun (Sel. 7)	Pusa Early Dwarf x K2
Hisar Lalima (Sel. 18)	Pusa Early Dwarf x HS-101
Hisar Lalit	Bangalore x HS-101
Hisar Anmol (H-24)	Hisar Arun x *L. hirsutum* f. *glabratum*
Pusa Sheetal	Balkan (Bulgaria) x Jemnorrosnej (Russia)
Pusa Gaurav	Glamour (exotic) x Watch (exotic)
Sweet-72	Pusa Red Plum x Sioux
Arka Meghali	Arka Vikas x IIHR 554
Utkal Kumari	Selection 22 x BT-2
Paiyur-1	Pusa Ruby x CO-3
Azad T-2 (KS-2)	HB-5 x Kuber

* Hazra and Som (2006).

The cultivar Moryama, bred from cross between Rubin and M. Florida is suitable for canning purpose, resistant to blossom end rot and fruit necrosis and suited for once-over mechanical harvest (Turchenkov *et al.*, 1988).

Berry and Gould (1988) reported from USA that the immediate parentage of cv. Ohio 8243 is 01383 x 871-3 and it is a sixth generation selection. The pink-fruited cultivar Ozark Pink, also from U.S.A., has been derived from Traveler 76 x 75-18-3 cross (Mc-Ferran *et al.*, 1989). Bulgarian cultivar Venera, suitable for canning has been developed from a cross between Zlatno Vreteno (a *L. esculentum* x *L. pimpinellifolium* hybrid) and UC 82A from the USA (Georgiev *et al.*, 1986).

Mutation breeding was also utilized in developing cultivars like S-12 and CO-3.

17.3.3 Heterosis Breeding

Tomato hybrids began to take over the market at the end of sixties and the beginning of seventies (Atanassova and Georgiev, 2007). Their use increased throughout the following decades and in 1997, Duvick (1997) reported that 100 per cent of fresh market and 80 per cent of processing tomatoes in USA were F_1 hybrids. At present, there is a similar situation in several countries in Europe, Asia and Australia. The development of hybrids was probably the most important contribution to the genetic improvement of tomato during the last few decades. This phenomenon was widely used in developing early and high yielding commercial hybrids. Hybrids are preferred by growers and processors. They cumulate a larger number of economically important traits (multiple resistance to diseases, fruits with prolonged longevity, more efficacious plant coping with abiotic stress, *etc.*) than the open-pollinated ones and combine beneficial characters from both parents.

Tomato is a self pollinated crop where degree of heterosis is theoretically considered less (Gallias, 1988). However, the unusual high heterosis observed in tomato crop has been attributed to the fact that tomato was basically a highly out crossing genus which was later evolved into a self pollinated crop (Rick, 1965). Hybrids have replaced inbred and open pollinated varieties in tomato which has remarkably increased productivity (Grandillo *et al.*, 1999). Heterosis manifestation in tomato is in the form of the greater vigour, faster growth and development, earliness in maturity, increased productivity, better quality attributes, and higher levels of resistance to biotic stresses (Mahendrakar *et al.*, 2005; Seeja *et al.*, 2007; Hannan *et al.*, 2007; Gul *et al.*, 2010; Chattopadhyay *et al.*, 2011; Sherpa *et al.*, 2014b; Pandiarana *et al.*, 2015 a, b; Acharya *et al.*, 2018a). Likewise, considerable heterosis has been developed to produce greater vigor, faster growth and development, earliness in maturity, increased productivity, better quality and leaf curl disease tolerance in cherry tomato (Marbhal *et al.*, 2016; Pujer *et al.*, 2017; Renuka and Sadashiva, 2016; Mukherjee *et al.*, 2020).

Certain dominant characters can also be combined and heterosis has been exploited in F_1 hybrids in tomato to a great extent in the last 50 years in developed countries like U.S.A., Japan, *etc.* In India, heterosis in tomato resulted in an increased yield of 20 to 50 per cent (Pal and Singh, 1943; Choudhury *et al.*, 1965; Som and Islam, 1970; Chattopadhyay *et al.*, 2011; Shende *et al.*, 2012; Pandiarana *et al.*, 2015 a, b; Acharya *et al.*, 2018a).

Shende *et al.* (2012) estimated the range of heterobeltiosis to be up to 52.59 per cent for fruit number per plant and up to 38.78 per cent for fruit yield per plant. Tamta and Singh (2018) observed 149.24, 36.43, and 28.92 per cent higher heterosis compared to the better parent for fruit yield/plant in hybrids from crosses 'CLN2237 x Sweet-72', 'CLN2237 x Punjab Chhuhara', and 'PT2007-09 x Punjab Chhuhara', respectively. Promising trait specific heterotic crosses 'CLN2777-G x BCT-59' and 'CLN2777-A x BCT-82P (Chattopadhyay *et al.*, 2011), 'CLN2777-G x FEB-2 (Sherpa *et al.*, 2014b), 'CLN 2777 F x CLN 2777 E' and 'CLN 2777E x Ailsa Craig fulgens' (Pandiarana *et al.*, 2015 b), 'CLN-2777C x Utkal Urbashi' and 'CLN-2498D x Utkal Kumari' (Acharya *et al.*, 2018a) have been identified over the years.

Two very promising combinations were Red Jacket x Meeruti using closed anther mutant (Mittal *et al.*, 1962) and Pusa Ruby x Best of All (Choudhury *et al.*, 1965). Som and Islam (1970) reported that superior hybrid yield was obtained in a cross between Pusa Early Dwarf and Pusa 120, when Pusa 120 was used as pollen parent. Apart from high total yield the F_1 hybrids have specific advantage of higher early yield, number of fruit, size of fruit, improved quality, uniformity, adaptation to adverse condition and resistance to diseases (Esipova and Veselovsky, 1963; Stoner and Thompson, 1966; Tesi *et al.*, 1970; Boos and Kochneva, 1971; Phills and Hernandez, 1975; Rao and Choudhury, 1976; Kim *et al.*, 1984; Pilowsky *et al.*, 1989; Gardner, 1990; Chattopadhyay *et al.*, 2011; Sherpa *et al.*, 2014b; Pandiarana *et al.*, 2015 a,b; Acharya *et al.*, 2018a).

The cost of hybrid seed production can be minimized by the use of male sterile lines. Male sterility was reported by Rick in 1944. Larson and Paur (1948) obtained functional male sterile mutant in John Baer cultivar. Closed anther character was successfully incorporated in other cultivars (Singh *et al.*, 1966). In Bulgaria, the cultivar Standard 69 was developed by heterosis breeding using a male sterile line with resistance to tobacco mosaic virus, *Verticillium* and *Fusarium* wilts and nematodes as the maternal parent and an early line tolerant to unfavourable condition as paternal parent (Georgiev *et al.*, 1986). A good number of artificially induced and spontaneously isolated mutants have been reported by several workers. Masuda *et al.* (1998) described three induced male sterile mutants, but only few of them have the potentiality to produce hybrid seeds at commercial scale. Georgiev (1991) reported one such potential male sterile line (ms-10^{35}). The *ms* gene in this line is linked with a recessive marker gene (a) responsible for absence of anthocyanin. Hence, ms-10^{35} sterile plant can be identified at seedling stage and fertile plant can be rogued out in the nursery itself. Two stamenless mutants, *viz.*, stamenless-1 (*sl-1*) and stamenless-2 (*sl-2*) produce flowers without stamen when grown at higher temperature, while those grown at low temperature produce flowers with abnormal stamens and often with viable pollen. Similarly, in *ms-15* and *ms-33* mutants, low temperature (<30°C) is reported to be associated with fertility restoration (Sawhney, 1997). Earlier, Rick and Boynton (1967) were of the view that the variable male sterile (*vms*) plants express male sterility at a temperature of 30°C or above.

Hormonal imbalance seems to be associated with the expression of male sterility. In vegetative and floral parts (except pistil) of *sl-2* mutant, abscisic acid (ABA) content was observed to be more than the normal, especially in stamens.

The increase in ABA content in stamens was found to be coinciding with first sign of abnormality in anther. There was a drop in ABA level in leaves and stamens at low temperature when fertility was restored. Thus in the opinion of Singh and Sawhney (1998), male sterility in *sl-2* may be the manifestation of hormonal imbalance (high ABA) and low temperature regulation of male sterility is mediated through reduction in ABA content.

Cytoplasmic male sterility (CMS) is the most commonly utilized male sterility to produce commercial hybrid seeds. However, without restorer gene CMS can not be utilized in fruit vegetables like tomato, chilli, melon, *etc.* (Kumar *et al.*, 2000). Through protoplast fusion of *L. esculentum* with *Solanum acaule* and *S. tuberosum*; CMS hybrid plants have been isolated with different flower morphology than tomato. Restriction analysis of mtDNA of these sterile plants revealed that these plants originated due to the recombination between the parental mt-genome (Melchers *et al.*, 1992). The sterile cytoplasm from *L. peruvianum* has been transferred into *L. pennellii*. Subsequently, CMS *pennellii* has been successfully crossed with *esculentum*. This hybrid provides basis for the development of CMS system in tomato derived from sterile cytoplasm of *peruvianum* (Petrova *et al.*, 1999). However, the practical utility of these CMS would depend upon the identification of restorer gene in tomato.

Six indeterminate tomato hyrbrids were grown under a plastic tunnel during the winter season of 1988-89 to study their comparative performance in respect of time taken to fruit maturity, number of fruits per plant, total fruit per plant, average fruit weight and yield/m^2. Cv. Carmello gave the highest yield of 18.07 kg/m^2, followed by Sarras and Samar yielding 13.57 and 13.26 kg/m^2 respectively, while the lowest yield of 6.96 kg/m^2 was recorded in Polifemo, which was late in maturity taking 153.33 days for fruits to ripen. The highest average fruit weight of 163.33 g was recorded in Carmello and the lowest 975.67 g/fruit in Polifemo (Chaudhary *et al.*, 1995).

The result of a study (Rai *et al.*, 1998; Padianrana *et al.*, 2015 a, b; Acharya *et al.*, 2018a) on heterosis and inbreeding depression clearly suggests that, due to preponderance of non-additive genetic variances, exploitation of heterosis breeding would be of importance for genetic improvement of tomato. Mukherjee *et al.* (2020) also observed the preponderance of non-additive gene action in controlling the expression of fruit yield per plant and most yield components, and suggested heterosis breeding is the best strategy for improvement of traits in cherry tomato.

17.4.0 Breeding for Biotic Stresses

Tomato is highly prone to biotic stresses especially diseases, insects and nematodes. These factors contribute significantly on yield reduction and production of inferior quality fruits and in severe cases total crop failure may occur. Usually pesticides are applied for their control, but this is very expensive and some of them have high residual effects which cause health hazards. Development of resistant varieties, thus, is the most economical method for the control of these stresses. In the absence of a resistant variety, tolerant one can be utilized in the integrated pest management programme.

17.4.1 Diseases

Tomato is susceptible to over 200 diseases caused by pathogenic fungi, bacteria, viruses, or nematodes (Lukyanenko, 1991). Breeding for disease resistance remains a major goal of most public and private tomato breeding programs as new diseases achieve significance or new races of existing pathogens become established and the ultimate goal is to eliminate or significantly reduce the use of pesticides in tomato production by the use of host resistance. The greatest contribution of modern plant breeding to tomato improvement has been through development of cultivars with improved disease resistance. Resistance has been identified, and in many cases characterized, for more than 30 of the major tomato diseases (Foolad, 2007a). These mainly include diseases for which major resistance genes have been identified, including *Fusarium* wilt, *Verticillium* wilt, root-knot nematode, *Alternaria* stem canker, gray leaf spot, and some bacterial and viral diseases. However, horizontal (*i.e.*, field or polygenic) resistance has also been reported for several tomato diseases, where major genes for resistance to a particular pathogen or race are not found, such as early blight, powdery mildew, bacterial canker, and bacterial wilt. In most cases, tomato wild species have been utilized as the source of resistance for all tomato diseases.

Resistance to tomato-infecting begomoviruses has been successfully introgressed from *Solanum pimpinellifolium, Solanum peruvianum, Solanum chilense* and *Solanum habrochaites* (Ji *et al.*, 2007b). From these sources, a few resistance genes have been well characterized and mapped using molecular markers. A partially dominant major resistance gene, *Ty-1*, was introgressed from *S. chilense* accession LA1969 and mapped to the short arm of chromosome 6 (Zamir *et al.*, 1994). A major resistance QTL derived from *S. pimpinellifolium* (Hirsute- INRA) was mapped to a different position on chromosome 6 (TG153-CT83; Chague *et al.*, 1997). Hanson *et al.* (2000) mapped a dominant resistance gene, *Ty-2*, in *S. habrochaites*-derived line H-24, to the short arm of chromosome 11. A partially dominant major gene, *Ty-3*, derived from *S. chilense* (LA 2779 and LA 1932), was mapped to chromosome 6 (Ji *et al.*, 2007a). The *Ty-3* introgression derived from LA 2779 was found to be longer and linked to *Ty-1*. However, recent studies on fine mapping and characterization demonstrated that *Ty-1* and *Ty-3* are allelic and code for an RNA-dependent RNA polymerase (Verlaan *et al.*, 2013). An additional gene, *Ty-4*, was mapped to the long arm of chromosome 3. While *Ty-3* has a major effect that accounts for 60 per cent of the variation in symptom severity, *Ty-4* accounts for only 16 per cent of the variation (Ji *et al.*, 2009). Recently, a recessive resistance gene (*ty-5*) was identified on chromosome 4 in the lines derived from cultivar Tyking (Hutton *et al.*, 2012), which is suspected to be similar to the *Ty-5* locus that accounts for more than 40 per cent of the variation (Anbinder *et al.*, 2009). Most of these resistance sources are known to support virus replication. However, the level of virus accumulation is lower than the levels in susceptible cultivars. It is well established that the virus level in tomato lines carrying *Ty-1*/*Ty-3* is <10 per cent of the level found in susceptible cultivars (reviewed in Verlaan *et al.*, 2013). Similarly, a low level of virus accumulation and a positive correlation between virus level and disease severity were found in *Ty-2*-carrying lines (Barbieri *et al.*, 2010). Verlaan *et al.* (2013)

indicated the presence of additional resistance allele in Fla. 8680. It is likely that the putative resistance allele reported by Hutton *et al.* (2012) in Fla. 8383 and in Fla. 8638B is the same as that in Fla. 8680, since Fla. 8680 is the resistant parent in a cross from which Fla. 8383 is derived. This resistance allele is presumably inherited from *S. chilense* and was recently designated as *Ty-6* (Scott *et al.*, 2015). *Ty-6* is a major resistance gene located on chromosome 10 of tomato that provides resistance against monopartite and bipartite begomoviruses and complements resistance conferred by the known *Ty-3* and *ty-5* genes (Gill *et al.*, 2019). Resistant lines and hybrids with various combinations of these known *Ty* genes present a range of resistance levels, and differential responses to isolates, strains and species, reflecting the genetic diversity of tomato-infecting begomoviruses. For example *Ty-1* lines showed a low level of resistance against some Brazilian isolates under high inoculum pressure, and against bipartite Tomato mottle virus (ToMoV) in Florida. Also, *Ty-2* lines showed differential responses to TYLCV strains (Ji *et al.*, 2007b). In such a scenario, pyramiding multiple *Ty* resistance genes may improve the spectrum, durability and level of resistance, thereby providing resistance to diverse tomato-infecting Begomoviruses (Prasanna *et al.*, 2015).

For some tomato diseases, such as late blight (caused by oomycete *Phytophthora infestans*) and powdery mildew (caused by fungus *Oidium lycopersicum*), both vertical and horizontal resistances have been identified. All original characterization, disease evaluation, and incorporation of resistance genes were through phenotypic selection and traditional breeding protocols, and still today much of the disease resistance breeding in tomato is through the use of similar protocols.

The crop is highly susceptible to a number of fungal, bacterial and viral diseases. Due to the incidence of the diseases, crop losses to the extent of 95 per cent have been reported. Therefore, breeding for disease resistance has assumed greater importance than improvement of yield and quality. The basic requirement of any breeding programme for disease resistance is to identify useable source(s) of resistance. Sources of resistance to several diseases have been identified (Table 10).

Resistance to most diseases is governed by single dominant gene while resistance to leaf curl virus, bacterial wilt and bacterial spot is polygenically controlled and to bacterial canker by three dominant and one recessive gene (Lukyanenko, 1991; Hanson *et al.*, 1998; Ji *et al.*, 2007b). Several other studies suggested resistance to bacterial wilt under the genetic control of multiple recessive genes (Villareal and Lai, 1978), two independent genes (Tikoo *et al.*, 1983), single dominant gene (Scott *et al.*, 1988; Mahir *et al.*, 1993; Mane, 2013; Acharya *et al.*, 2018b), single recessive gene (Thakur *et al.*, 2004) and partially recessive gene (Monma *et al.*, 1997). Oliveira *et al.* (1999) suggested that genetic control of BW resistance depends on the interaction between genotype and bacterial isolates. The nature of resistance to different biotic and abiotic stresses is presented in Table 11.

Recent research has revealed resistance against nearly all serious pests and diseases are available amongst wild *Lycopersicon* species. The genus *Lycopersicon* can be divided into 2 groups: the *esculentum*-complex and the *peruvianum*-complex. The *esculentum*-complex includes 6 species, 3 of which (*L. esculentum, L. cheesmanii* and *L. pimpinellifolium*) are able to synthesize colour carotenoids in their fruits and

Table 10: Sources of Resistance to Important Diseases of Tomato*

Diseases	*Sources*
Fungal Disease	
Anthracnose	PI 272636
Brown root rot	*L. pimpinellifolium*
Collar rot	*L. pimpinellifolium* (Manapal)
Corky root	*L. peruvianum*, PI 262906, PI 198601
Foot and stem rot	*L. hirsutum, L. hirsutum* 66087 (IVT61292)
Early blight	*L. hirsutum* f. *glabratum* CI 943, *L. peruvianum* var. *dentatum*, *L. pimpinellifolium* 'Pan American', *L. peruvianum* B 6002177, NCEBR-1 and NCEBR-2
Fusarium wilt	
Race 1	*L. pimpinellifolium, L. hirsutum,* Walter, Columbia, Roma, HS 110, Homestead, Flora-Dade, Sel. 22 and Sel. 30.
Race 2	PI126915, Rustrel.
Race 3	*L. pimpinellifolium,* CR6
Grey leaf spot	*L. pimpinellifolium*, Manale, Manalucie,Campbell 28, Rotella, Caraiba, Ont. 7620 and Flora-Dade
Late blight	*L. pimpinellifolium, L. esculentum* var. *cerasiforme,* PT 251305, PI 126445, LA 1255
Leaf mould	*L. pimpinellifolium,* *L. hirsutum* f. *glabratum, L. peruvianum,* *L. pennellii*, Vegabond, Vinequeen, V548, Vantage, Wetham.
Leaf spot	*L. hirsutum, L. pimpinellifolium, l. esculentum* var. *cerasiforme,* Targinnie Red
Bacterial Disease	
Bacterial wilt	Louisiana Pink, PI 127805A Venus, Saturn, *L. pimpinellifolium*
Bacterial canker	*L. hirsutum* PI251305, *L. pimpinellifolium* lines Utah 737, Uttah 20
Bacterial spot	*L. esculentum* var. *cerasiforme*, Hawaii 7998
Bacterial speck	*L. pimpinellifolium* PI 112215, *L. hirsutum* f. *glabratum* PI 129157
Viral Disease	
Leaf curl	*L. pimpinellifolium* A 1921, *L. hirsutum* f. *glabratum* B 6013, *L. hirsutum* f. *typicum* A 1904
Tomato mosaic	*L. peruvianum* PI 128650, *L. hirsutum*
Tomato yellow leaf curl	*L. chilense, L. peruvianum,*
Spotted wilt	*L. pimpinellifolium*, Pearl Harbor

* Kalloo (1986).

3 others (*L. hirsutum, L. parviflorum* and *L. pennellii*) which are green-fruited. All representatives of this group are easily crossable with recently produced tomato cultivars. The *peruvianum*-complex includes 2 green-fruited species (*L. peruvianum* and *L. chilense*). These 2 species, which carry a number of economically important

traits, are sexually incompatible with *L. esculentum*. This sexual incompatibility can be overcome using bridge lines or by some tissue culture methods such as embryo-rescue and somatic hybridization (Lebeda and Mieslerova, 1998).

Table 11: Nature of Resistance to Disease, Insect Pests, and Environmental Stresses*

Nature of Resistance	*Diseases/Insect Pests/Environmental Stress*
A. Genetic factors	
(i) Monogenic dominant	Fusarium wilt, leaf mould, Rhizoctonia root rot, leaf spot, Verticillium wilt, spotted wilt virus, tobacco mosaic virus and root knot nematodes
(ii) Monogenic recessive	Fruit worm, root knot nematodes and fruit set at low temperature
(iii) Oligogenic factors	Bacterial canker, early bight, rhizoctonia root rot and spotted wilt virus
(iv) Polygenic factors	Leaf curl virus, bacterial wilt, bacterial canker, late blight and fruitset at low and high temperatures
B. Plant biochemical factors	
(i) Antibiosis	Fruit worm
(ii) Antibiosis and plant morphological factors	Aphids, White flies and spider mites

* Kalloo (1986).

Backcross method is commonly utilized for the transfer of resistant genes from donor parents to the desired variety. However, in certain cases selection, pedigree method, combination of backcross and pedigree methods, and population improvement methods have been employed. Intra specific hybridization has been attempted and the backcross method has been followed for transfer of resistant genes from the wild species to the cultivated variety. Now the resistant genes from the wild species have been mostly combined in commercial cultivars which can more easily be utilized as donor parent (Kalloo, 1986).

The goal of plant breeders who were interested in disease resistance is to get multiple disease resistance in commercially acceptable cultivars. A few among the multiple disease resistant cultivars/hybrids developed are as follows:

1. **Manalucie:** resistant to fusarium wilt, grey leaf spot, early blight, collar rot, leaf blight and root-knot nematode
2. **Manale:** resistant to *Fusarium* wilt, grey leaf spot, early blight and leaf mould
3. **New Hampshire Sure crop:** resistant to late blight and early blight (moderate)
4. **Tropi Red and Tropi Orange:** resistant to Fusarium wilt, grey leaf spot, blossom end rot, cracking and grey wall
5. **Southland:** resistant to collar rot, Fusarium wilt (highly resistant), moderately resistant to early blight and one or two races of late blight disease.

6. **Arka Rakshak**: resistant to leaf curl virus, bacterial wilt and early blight.
7. **Arka Samrat**: resistant to leaf curl virus, bacterial wilt and early blight.
8. **Arka Abhed**: resistant to leaf curl virus, bacterial wilt, early blight and late blight.
9. **Arka Apeksha**: resistant to leaf curl virus, bacterial wilt and early blight.

Fairly large number of multiple-resistant cultivars have been developed. Tropic, Walter, Royal Red Cherry and Verlioka are multiple disease resistant cultivars of tomato (Strobel and Walter, 1969; Beal, 1972; Jones *et al.*, 1980; Gavrish and Sysina, 1990).

The cultivars Ohio 8243 and Ozark Pink are resistant to race 1 of *Fusarium oxysporum* f. sp. *lycopersici* and race 1 of *Verticillium dahliae* as well as radial and concentric fruit cracking (Berry and Gould, 1988; McFerran *et al.*, 1989). From India, a tomato cultivar Punjab NR-7, derived from cross S-2 x NMRI was found to be highly resistant to *F. oxysporum* f. *lycopersici* (Nandpuri *et al.*, 1989). In Queensland, Australia, cvs. Scorpio, Redlands Summertaste and Redlander were developed for bacterial wilt resistance (Moffett, 1986).

The toxin responsible for causing wilt disease in tomato was isolated from *Pseudomonas solanacearum* race 1 biovar 3 by alcohol precipitation and ion exchange chromatography. Characterization of toxin revealed the presence of glucose, mannose and polypeptides. Cell lines of 5 wilt susceptible tomato varieties, growing in modified MS medium, were treated with purified toxin (EC50). Cells showing toxin insensitivity were isolated and regenerated to plants. One regenerant showed a high level of resistance to *P. solanacearum*, while 2 others had moderate resistance. Field tests of the regenerants for 3 years showed segregation for the characters. Resistant plants had a significant increased in rishitin production following toxin and pathogen treatments. The toxin induced physical damage of the membrane system in susceptible cell (Baruah *et al.*, 1995). Variability and interrelationship estimate studies for early blight disease severity index and other morphological characters in three populations of tomato suggested that determinate plants with densely spaced leaves are more favourable for disease spread even at later growth stages (Lohithaswa *et al.*, 1998).

A good source of resistance to tomato spotted wilt tospovirus (TSWV, a major limiting factor of tomato production in Brazil and other tropical areas) and to insect pests has been identified among some *Lycopersicon peruvianum* individuals. Due to incompatibility and early embryo abortion, it has not been possible by conventional methods to obtain hybrids between these sources and the commercial tomato (*L. esculentum*). Following an earlier report of the recovery of several interspecific hybrid plants by embryo rescue, this characterizes the hybrid nature of such plants and their progenies, as well as the screening tests for TSWV and leaf miner [*Scrobipalpuloides absoluta* (*Scrobipalpula absoluta*)] resistance. Seeds of F_1 plants were variable in size and colour; small green seeds had to be recultured again and were then able to develop directly into plants, without a callus phase. Large brown seeds were sown directly in soil. The large seeds gave rise to plants with more *L. esculentum* characteristics, whereas plants from small seeds resembled the *L. peruvianum* phenotype. F_2 and

F_3 plants varied in relation to leaf hairs, leaf shape and size, and fruit colour and size. Hybrid fruits varied in colour from green with stripes to yellow or orange at maturity, with a slightly larger size than those of *L. peruvianum*. All F_1 plants were self-incompatible. A few F_2 and F_3 plants obtained by sibcrosses were self-compatible. Allelic segregation for the acid phosphatase (*Apsl*) locus was detected among F_2 and F_3 individuals. Isoenzyme analysis of *Apsl* demonstrated the presence of 3 phenotypes corresponding to genotypes 1/1, +/1 and +/+. Screening tests among F_2 and F_3 hybrid populations were performed under controlled conditions and showed segregation for resistance to TSWV and to leaf miner. The recovery of interspecific hybrids through embryo culture was demonstrated to be an effective technique for overcoming crossing barriers within the genus *Lycopersicon* (Segeren *et al.*, 1993).

The green stemmed lines Ludan and Lunan were bred from crosses involving the purple-stemmed, large-fruited, determinate line 848. By crossing this line with 82-70 (a male-sterile line), Ludan was produced. It gives large fruit (mean weight 155-175 g) with firm flesh and good flavour. By crossing 848 with Huangfen and pollinating 82-70 with pollen from the F_1 hybrid and another variety (Aifen), Lunan was bred after 5 generations of systematic selection. It has a mean fruit weight of 133 g and out yields Aifen by 33 per cent. These *aw* lines were crossed with 5 pollen parents and the F_1 hybrids were numbered 901 (Ludan/Huangfen), 902 (Lunan/Danhuang), 903 (Ludan/Danma), 904 (Lunan/Huangfen) and 905 (Lunan/Danhuang). In early and total yields, 901 exceeded the control, Sufen, by 39.4-45.4 per cent and 21.64 per cent, respectively, while 902 and 903 gave a 13.5-14.44 per cent higher total yield than the control. The hybrids, which carry the resistance gene *Tm2nv*, all proved highly resistant to tobacco mosaic tobamo virus. In particular, 901 performed well, with determinate habit, vigorous growth and mean fruit weight of 141g (maximum up to 600 g). It is suitable for sheltered spring cultivation in the Jiangsu, Zhejiang, Anhui and Sichuan provinces of China (Xue, 1994).

Manapal, a non-processing tomato cultivar with the gene *Tm2nv*, which conditions yellow seedling and resistance to TMV (Tobacco Mosaic Tobamo virus), was crossed with Red Agate 140 and foreign cv. Macelo. From the segregating progenies, 3 lines showed stable inheritance of *Tm2nv*. Crossing with these lines, a number of combinations were obtained. 8746 x Red Agate 140 and 9076 x Red Agate140 were highly productive, resistant to diseases and suitable for processing (Zhou *et al.*, 1995).

Domestic and exotic virus resistant cultivars and wild relatives were compared for resistance to tobacco mosaic tobamo virus (TMV) and cucumber mosaic cucumovirus (CMV) by artificial inoculation during the seedling stage. Continuous screening by this method led to the selection of virus-resistant cultivars Zhongshu 4, Zhongshu 5, Zhongshu 6 and Zhongza 4, which have become important cultivars in China (Gao *et al.*, 1995).

Pilowsky *et al.* (1989) from Israel reported that F_1 hybrid TY-20 is the first tomato cultivar with tolerance of the white fly-borne tomato yellow leaf curl geminivirus. The tolerance trait was derived from *L. peruvianum*, recessive and controlled polygenically.

17.4.2 Insect Pests

Tomato is susceptible to number of insects which cause huge loss to the crop in most parts of the world. Various species of white flies, aphids, mites, Lepidopteran insects (*e.g.*, tomato fruit worm, *Heliothis armigera, H. zea*), Coleopteran insects (*e.g.*, Colorado potato beetle and tobacco flea beetle), Dipteran insects (*e.g.*, leaf miners and fruit fly), thrips, sinkbugs, and cutworms, attack tomatoes and many of them capable of causing devastating losses. Though chemical control is very common practice, use of resistant varieties is the most economical measure to check such pests. But resistance breeding is important for reducing the use of pesticides.

Resistance to major insect pests of tomato has been identified within the related wild species, in particular, *L. hirsutum* and *L. pennellii* (Tiglechar, 1986; Farrar *et al.*, 1994; Juvik *et al.*, 1982; Muigai *et al.*, 2003; Mutschler *et al.*, 1996). *L. hirsutum*, the most notable source of arthropod resistance, occurs in two distinct forms, *L. hirsutum* f. *typicum* and *L. hirsutum* f. *glabratum* CM Mull (Weston *et al.*, 1989), each showing resistance to at least 16 pest species (Farrar and Kennedy, 1991). Resistance to at least nine insect species has been reported in *L. pennellii*, including green house white fly, carmine and two spotted spider mites, and the potato aphid (Muigai *et al.*, 2003). Some insect resistance has also been reported in *L. esculentum* var. *cerasiforme*, *L. pimpinellifolium*, *L. cheesmanii* and *L. chmielewskii*, *L. peruvianum* and *L. chilense* (Farrar and Kennedy, 1991). Host resistance to insects and mites in this crop has been summarised in Table 12.

Table 12: Sources of Resistance to Insects and Mites in *Lycopersicon* spp.*

Common Name	*Source of Resistance*
Fruit borer	*L.hirsutum, L.hirsutum* f. *glabratum*
Colorado potato beetle	*L.hirsutum*
White fly	*L.hirsutum* f. *glabratum*, Red cloud, Strain B.
Leaf miner	*L. hirsutum, L.hirsutum* f. *glabratum*, Pearson VF6L
Tobacco horn worm	*L. hirsutum* f. *glabratum* PI 134417, PI LA 407
Potato aphid	*L. peruvianum, L. pennellii*
Tobacco flea beetle	*L.hirsutum*f. *glabratum*, PI 126449, *L.hirsutum*
Pinworm	*L.cheesmanii* f. *minor. L. glandulosum, L.hirsutum*
Green house white fly	*L.hirsutum, L.hirsutum* f. *glabratum, L. pennellii*
Carmine spider mite	*L. hirsutum*, PI 126449
Two-spotted spider mite	*L. hirsutum* f. *glabratum, L. pennellii*

* Kalloo (1988).

Several factors are involved in the expression of resistance attributes in tomato. Resistance in wild species is primarily due to glandular trichomes. The glandular trichome exudate from the leaflets of *L. hirsutum* f. *glabratum* PI 134417 has a non-alkaloid insecticide 2-tridecanone with ethanol which is toxic to the larvae of *H. zea* (Dimock and Kennedy, 1983). Similar effects of 2-tridecanone has been found on other insects like tobacco horn worm, pinworm, Colorado potato beetle and carmine spider mite (Kalloo, 1988). He also mentioned other chemicals like sesquiterpenes (curcumene and zingiberene) in *L. hirsutum* f. *glabratum*, sugar esters and tomatine in *L. pennellii* also associated with insect resistance in tomato.

Several morphological factors like fruits, stems or plant growth habit have been related to insect resistance. Environmental factors such as temperature, light duration and intensity, nutrition, insect population, stage of the crops, and plant density also influence the insect resistance as reported by number of workers. With the change of the factors, the chemical composition of plant also changes leading to variation in degree of resistance.

17.4.3 Root Knot Nematode

Root knot nematodes are responsible for marked reduction in tomato yields and sometimes for complete failures in many parts of the world. A number of *Meloidogyne* spp. are responsible for affecting the crop. Among these, *M. incognita* is the most common followed by *M. javanica*, *M. hapla* and *M. arevaria*. Though chemical control is most common, it is too expensive and with adverse effects on environment. This necessitates exploitation of non-chemical methods like crop rotation, fallow, trap crops, heat, flooding, biological control and resistant cultivars.

The nature of resistance to this pest has been studied by many workers and in their opinion, a single dominant gene controls the resistance. The first report on high level of resistance to *Meloidgyne* spp. in *L. peruvianum* was made by Bailey in 1941 which was later supported by other workers. With the use of embryo culture technique in interspecific hybridization, it has been possible to combine the resistance gene of *L. peruvianum* in cultivated cultivar by the backcross method after 1944 (Kalloo, 1988). After that a fairly large number of varieties resistant to *Meloidogyne* spp. was reported (Table 13).

Table 13: Host Resistance to Root Knot Nematodes in Tomato*

Root Knot Nematode	*Resistant Genotype*
M. incognita	Anahu, Nemared, Ronita, Piernita, Nemacross BBF_1, Nemared F_1, VFN 14, Monita, Rossol, Santa Cruz, Venus, Saturno, VFN 8, Atkinson, Monalucie, 66 N1, 569 N -10.
M. hapla	Rutgers, Ace, PI 270435.
M. javanica	Ronita, Rossol, Santa Cruz, Venus, Saturno,V14, V 275 NC 802, Healani, Anahu, Nematex, Bonus VFN, Atkinson, Manalucie 66 N1, 569N-10.
M. arenaria	Ronita, Rossol, Santa Cruz, Venus, Saturno, Atkinson, Manalucie 66 N1, 569 N-10
Meloidogyne spp.	*L. peruvianum, L. peruvianum* var.*dentatum*

* Kalloo (1988).

The root knot nematode resistance in tomato is controlled by *Mi* gene which is a single dominant or partially dominant. This gene not only exhibits resistance against *M. incognita* but to *M. javanica* and *M. arenaria*, as well. According to Sidhu and Webster (1975), the resistance to *M. incognita* was controlled by two dominant genes *LMiR1* and *LMiR2* which are closely linked and that *LMiR1* and *Mi* are either allelic or identical.

A number of factors may contribute to the resistance. Kalloo (1988) mentioned that ethylene has a direct relationship in increasing gall formation in roots of tomato by *M. javanica* when seedlings are treated with ethrel. Similarly, cytokinin application can block the hypersensitive reaction of tomato resistance to *M. incognita*. He also reported that phenol accumulation in tomato is associated with the resistance of *M. incognita* and a higher concentration of phenol checks the nematode incidence. Tomatine content of roots also has a negative correlation with the incidence of *M. incognita*. The threshold content of resistance ranged from 3.8 to 6.1 mg/g dry weight and highly resistant cultivars contained 11 to 12.4 mg/g dry weight. At lower temperature (10 to 15°C), root knot nematode escape may occur.

Development of resistant cultivar is the only alternative to the nematode affected areas of tropical and subtropical parts of the world since the most effective nematicides have been banned.

17.5.0 Breeding for Abiotic Stresses

Tomato is highly sensitive to different abiotic stresses namely, climate, salinity and moisture, *etc.* which markedly impair the yield and quality. Sources of resistance to various stresses have been identified and presented in Table 14.

Table 14: Source of Resistance to Various Abiotic Stresses*

Traits	*Sources of Resistance*
Cold tolerance (germination)	PI 120256, PI 174263, PI 342988 *L. hirsutum* f. *glabratum* B 6013
Fruits set	Precoce, Apedice, Montfavet 63-4 Super marmande, Severianin
Chilling tolerance	*L. hirsutum, L. hirsutum* f . *glabratum.*
Heat tolerance (germination al 40°C)	Nema 1200, P 28693, UC 82 L.
Fruits set	Saladette, PS Mini Rose (84-58), PI 205009, *L. pimpinellifolium* (Pan American), EC 130042, *L. cheesmanii*
Drought resistance	*L. pennellii*, EC 130042, Sel.28, PI 205009, *L. pimpinellifolium, L. cheesmanii.*
Salt tolerance	*L.cheesmanii, L. pennellii.*
Ozone tolerance	Heinz 1439, New Yorker, Tatiner, Pierette, Kharkov
Flood tolerance	PI 406966 (*L. esculentum*)

* Kalloo (1988).

17.5.1 Temperature Stress

Role of temperature needs no emphasis in tomato production. Night temperature is more important for fruit setting than day temperature and normal fruit set takes place at night temperature range of 15-20°C. When the temperature goes beyond these two extremes, the fruitset will drastically be reduced in open.

Wild relatives like *L. hirsutum* and *L. chilense* have the ability to resist temperature fluctuation of both low and high (Tewari, 1999). Very little information is available on the genetics of fruitset under low temperature. The gene for low fruitset is simply inherited and completely dominant over high fruit set (Kemp, 1965). Kalloo and Dudi (1982) were of the view that the number of fruitset at low temperature is mostly controlled by additive and additive x dominance components. Further, in segregating progenies of *L. esculentum* x *L. pimpinellifolium*, the fruitset at low temperature was also controlled by recessive factors (Kalloo and Banerjee, 1990). Among the factors responsible for low fruitset at low temperatures is poor anther dehiscence. Walker *et al.* (1990) suggested a chlorophyll fluorescence screening test for evaluation of chilling tolerance in tomato and *Solanum lycopersicoides* and *L. hirsutum* were found to be tolerant to low temperatures.

At higher temperature (>29°C), fruitset is seriously affected because of style elongation, lack of pollination, reduced pollen germination and slow pollen tube growth. This has been noticed by Abdalla and Verkerk (1967). A great variation in tomato accessions regarding fruitset at high temperature has been observed by several workers and summarized in Table 15.

Table 15: Genotypes Setting Fruits at High Temperature in *Lycopersicon* spp.*

Genotypes
HS-102, Sparton Red 8, OK-7-2, OK-7-3, Saladette, TAMU Saladette, HS-101, PI 262934, BL 6807, S 6916, CIA 5161, VF 36, Punjab Tropic, Marzano, P-4, Avalanche, Chikol, Merit, C 28, Red Rock, Narcarlang, Breeding Line 165, Red Cherry, KS-1, KS-2, Walter, Sweet 72, Sel.58, Sel.84, CL 9-0-0-1, Palican, Tropic, Caraibe, Severianin, L 569, L567, L 573, L 571, L 125, L 226, L 232, L 2972, L 3690, L 246, L 2991, Burgess, *L. pimpinellifolium*

* Kalloo (1988).

Though the genetics of fruitset at higher temperature is not clearly understood, but according to Schaible (1962), it is controlled by quantitative factors.

Genetic research on heat tolerance has revealed that this character is largely under the control of recessive genes. Contrary to previous reports, genetic analysis of heat tolerance in a 7-parent diallel revealed that this character appears to be a quasi-quantitative trait, *i.e.*, it is controlled by relatively a few major genes and an undefined number of modifiers (Opena *et al.*, 1993).

17.5.2 Moisture Stress

Tomato production under excess moisture conditions will be restricted. Even short periods of flooding due to rain may cause wilting. Kuo *et al.* (1982) conducted an experiment on screening for flood tolerance in the genus *Lycopersicon* in which a total of 4630 accessions were included. They observed that seven *L. esculentum* accessions (L-123, L-125, L-973, L-3072, L-3091, L-4313 and L-4360) and one accession of *L. pimpinellifolium* (L-4422) exhibited some degree of tolerance to flood for a short period at high temperature. According to Aloni and Rosenshtein (1982), proline accumulation is an indicator of cultivar sensitivity to the dehydration associated with flooding stress that proline may be involved in post stress recovery.

Twenty-eight F_2 progenies of tomato, derived from crosses between 8 parents differing in moisture stress tolerance, were evaluated for yield-related and morphological traits under stimulated moisture stress conditions at Rajendranagar, Hyderabad by Sreenivasulu and Rao (1996) during the post-rainy period of 1993-94. Genetic variability was large for number of branches per plant, leaf area, root dry weight (RDW), shoot dry weight (SDW), RDW/SDW, proline content, stomatal diffusive resistance (SDR), number of fruits per plant, average fruit weight and fruit yield per plant, indicating the possibility for improvement of these characters under moisture stress conditions. High heritability coupled with high genetic advance as a percentage of mean was observed for SDR and number of fruits per plant, indicating the presence of additive gene action and the possibility of improving these traits through selection.

According to Springer and Phills (1982), growth rate and leaf area were the indicators for evaluation of drought tolerance in tomato and cv. Saladette was found drought tolerant on the basis of these indicators. This variety also tolerated heat indicating a correlation between heat resistance and drought tolerance.

L. pennellii has been reported to be drought resistance on the basis of plant growth (Rick, 1982).

17.5.3 Salt Stress

Soil salinity poses a very serious problem especially in arid and semi arid parts of the world and thus limits the production of tomato. Development of salt tolerant types is the only economic approach to grow a profitable crop under such conditions. Almost all the cultivated varieties are sensitive to salinity while wild species like *L. peruvianum, L. cheesmani* spp. *minor* and *L. pennellii* exhibit some degrees of tolerance to sodium chloride (Tewari, 1999). According to Rush and Epstein (1981), salt tolerance is a heritable character.

The usefulness of marker-assisted selection (MAS) to develop salt-tolerant breeding lines from a F_2 derived from *L. esculentum* x *L. pimpinellifolium* was studied by Monforte *et al.* (1996). In the study interval mapping methodology of quantitative trait locus (QTL) analysis was used to locate more precisely previously detected salt tolerance QTLs. A new QTL for total fruit weight under salinity (TW) near TG24 was detected. Most of the detected QTLs [(3for TW, 5 for fruit number (FN) and 4 for fruit weight (FW)] had low R2 values, except the FW QTL in the TG180-TG48 interval, which explains 36.6 per cent of the total variance. Dominant and over dominant effects were detected at the QTLs for TW, whereas gene effects at the QTLs for FN and FW ranged from additive to partial dominance. Phenotypic selection of F_2 families and marker-assisted selection of F_3 families were carried out. Yield under salinity decreased in the F_2 generation. F_3 means were similar to those of the F_1 as a consequence of phenotypic selection. The most important selection response for every trait was obtained from the F_3 to F_4 where MAS was applied. While F_3 variation was mainly due to the within-family component, in the F_4, the FN and FW between-family component was larger than the within-family one, indicating an efficient compartmentalization and fixation of QTLs into the

F_4 families. Comparison of the yield of these families under control versus saline conditions showed that fruit weight is a key trait to success in tomato salt-tolerance improvement using wild *Lycopersicon* germplasm. The QTLs detected under salinity seem to be also working under control conditions, although the interaction family x treatment was significant for TW, thereby explaining the fact that the selected families responded differently to salinity.

17.6.0 Breeding for Quality

Quality is a component that adds value to crops and their products. There is a recent development of quality awareness among the people. It refers to many aspects like colour, size, nutrient contents, shelf-life and suitability for processing and conversion into value added products. In addition to raw tomatoes which are eaten directly as a raw vegetable or added to other food items such as salads and other forms of fast food, a variety of processed products are increasingly gaining acceptance. To cater the needs of processing industry, hybrids that are high yielding, early or late, uniform in maturity, rich in colour, vitamin C, TSS and acidity are to be bred. Besides, there is an urgent need to develop hybrids suitable for long distance transportation for internal market as well as for export.

Tewari (1997) suggested that a hybrid/variety should have the following characters:

a) **Pericarp thickness**: It should be above 0.5 cm which keeps the fruit firm, evaporation of water through surface will be less and can tolerate the jerks in transportation.

b) **Locules:** Minimum number of locules from 2-3 to avoid the fruit to become misshaped.

c) **Viscosity:** More viscosity will produce more product.

d) **TSS:** Hybrids/varieties having more than 5.5 per cent TSS are more suitable for processing.

e) **Ascorbic acid/acidity:** Both are important to make a better product.

f) **Fruit shape:** Oblong types are suitable for processing as they produce more pulp.

g) **Fruit size:** Fruits above 50 g are suitable for market as well as for processing industries.

Quality and flavor of processed products depend on chemical compositions of TSS, sugar, lycopene, acidity, and pH which vary with cultivar/hybrid (Balasubramanian, 1984; Chattopadhyay *et al.*, 2013). TSS is related to yield of concentrated tomato products in more concentrated form and yield of certain processed products are determined by fruit sugar content (Hewitt and Garvey, 1987). The major component of total solids is sugars, which also influence flavor. A primary determinant of tomato flavor is the ratio of sugars to acids. High soluble solids are correlated with tomato-like flavor (Stevens *et al.*, 1977). Generally, fructose and citric acid are more important to sweetness and sourness than glucose and malic acid, respectively (Stevens *et al.*, 1977). The pH makes for more tart taste than titratable acidity. Acidity influences storability of processed tomatoes and tomato products by inhibiting spore germination of thermophilic organisms.

17.6.1 Colour

Development of proper colour is an important quality parameter both for table purpose and for processing varieties from the point of view of consumer preference. Red and yellow are the two major colours of tomato which are controlled by the carotene content of the fruits. Red fruits have mostly lycopene and only a small quantity of carotene. In normal red fruit the ratio of lycopene and carotene is generally 12 to 15: 1. Besides, small amounts of phytoene, phytofluene, β-carotene, α-carotene, neorosporine, and lutein are also present. Lycopene is the pigment principally responsible for the characteristic deep-red color of ripe tomato fruit and tomato products. It has attracted attention due to its biological and physico-chemical properties, especially related to its effects as a natural antioxidant. Although it has no provitamin A activity, lycopene does exhibit a physical quenching rate constant with singlet oxygen almost twice as high as that of beta-carotene. Lycopene in fresh tomato fruit essentially occurs in the all-trans configuration. The main causes of tomato lycopene degradation during processing are isomerization and oxidation. Isomerization converts all-trans isomers to cis-isomers due to additional energy input and results in an unstable, energy-rich station. Determination of the degree of lycopene isomerization during processing would provide a measure of the potential health benefits of tomato-based foods. Thermal processing (bleaching, retorting, and freezing processes) generally cause some loss of lycopene in tomato-based foods. Heat induces isomerization of the all-trans to cis forms. The cis-isomers increase with temperature and processing time. The bioavailability of cis-isomers in food is higher than that of all-trans isomers. Lycopene bioavailability in processed tomato products is higher than in unprocessed fresh tomatoes (Chakraborty *et al.*, 2007; Acharya *et al.*, 2018a). The composition and structure of the food also have an impact on the bioavailability of lycopene and may affect release of lycopene from the tomato tissue matrix. Food processing may improve lycopene bioavailability by breaking down cell walls, which weakens bonding forces between lycopene and the tissue matrix making lycopene more accessible and enhancing the cis-isomerization.

Several natural mutant genes (*hp-1*, *dg* or *hp-2*^{dg}, og^{c}) alter normal phytochrome regulation and enhance carotenoid content in tomato fruit (Sacks and Francis, 2001). The *dg* (*hp-2*^{dg}) gene, located in chromosome 1 (Levin *et al.*, 2003) affects photo-morphogenesis regulatory gene *TDETI*, and also affects light signal-transduction machinary (Mustilli *et al.*, 1999). In the cultivated tomato species, anthocyanins are not produced in the fruit due to lack of expression of the chalcone isomerase gene (*CH1*) in the flavonoid biosynthetic pathway in the peel of tomato fruit (Mes *et al.*, 2008). Three mutant genes have been identified that can lead to production of anthocyanin content in the peel of the fruit are Anthocyanin fruit (*Aft*), Aubergine (*abg*) and atroviolacea (*atv*). The anthocyanin fruit, *Aft* gene was identified in crosses with *Solanum chilense* Dunal (Jones *et al.*, 2003). This gene is located in chromosome 10 and its presence in tomato leads to production of anthocyanin pigment mainly, delphinidin, malvidin and petunidin, as well as higher levels of the flavonols

Mutants of Tomato

High Pigment Tomato Mutants.

A. *hp-2^dg^/hp-2^dg^* mature-green fresh-market tomatoes on the vine. B. Mature-green tomatoes in cv. Ailsa Craig background: regular tomatoes (*left panel*), and *hp-1/hp-1* tomatoes (*right panel*). C. Mature-green and ripe-red tomatoes in cv. Moneymaker background: *hp-2^j^/hp-2^j^* (*left panel*), regular tomatoes (*right panel*). D. Mature-green tomatoes in cv. Manapal background: regular tomato (*Left*), *hp-2^dg^/hp-2^dg^* (*Right*). E. Ripe-red tomatoes in cv. Ailsa Craig background: *hp-1/hp-1* (*left panel*), normal tomatoes (*right panel*). F. Puree of tomato fruits in cv. Manapal background: *hp-2^dg^/hp-2^dg^* (*Left*), regular tomatoes (*Right*). G. Lycopride: *hp-2^dg^/hp-2^dg^* processing cultivar (Levin *et al.,* 2006).

Mutants of Ailsa Craig Background

Mutants of Tomato

Dark Green Fruit Mutant (*dg*)

High Pigment Mutant (*hp-1*)

Old Gold Crimsom (og^c) Mutant

Anthocyanin Fruit (*Aft*) Mutant

High β Carotene (*B*) Mutant

Chlorophyll B Deficient Fulgens (*ful*) Mutant

Parthenocarpic Fruit (*pat-2*) Mutant

Self-Pruning (*sp*) Mutant

quercetin, 3.6-fold, and kaempferol, 2.7-fold in tomato fruit (Jones *et al.*, 2003; Sapir *et al.*, 2008).

The genetics of colour in tomato is well documented. Fruit colour is controlled by three major independent genes, *B, R* and *T. B* controls β-carotene synthesis in the presence of *R* and *T*, whereas *R* and *T* control the synthesis of lycopene. Two genes namely, *hp* and *og* decide the colour. While the gene *hp* increases α-carotene thereby producing yellow shade while *og* gene increases the lycopene which increases reddish shade. A proper balance between *hp* and *og* genes can give a good fruit colour without sacrificing the nutritive value namely pro-vitamin A. As a whole, colour is controlled by og^c, *at, B, moB, r, t, v* and *hp* independent genes (Kalloo, 1988).

Two recessive mutant genes, *hp1* (*high pigment 1*; Yen *et al.*, 1997) and *hp2* (van Tuinen *et al.*, 1997), were identified a few decades ago and introgressed into several tomato cultivars and increased lycopene content. However, the adverse pleiotropic effects of these genes, such as slow germination and seedling growth, seedling mortality, inferior leaf coverage, brittle stems, low yield, reduced total acidity and soluble solid contents, high sensitivity to various pathogens and premature defoliation, have prohibited widespread commercial use of these genes (Jarret *et al.*, 1984; Wann, 1995). The old gold crimson and crimson gene (og^c, *cr*), increases fruit lycopene content at the expense of β-carotene (Ronen *et al.*, 2000; Thomson *et al.*, 1967), and has been incorporated in many recent tomato genotypes for increasing lycopene and cultivars containing og^c on average contain 25 per cent more lycopene than normal cultivars (Foolad, 2007a).

According to Thamburaj (1998) non-additive gene action is reported to be more important than additive gene action for both lycopene and carotenoids and both have moderate heritability estimates. Wild species like *L. hirsutum* and *L. hirsutum* f. *glabratum* may be incorporated in the hybridization programme to develop an acceptable red quality tomato as they contain high β-carotene (Kalloo, 1988).

From a study on combining ability based on morphological characteristics and attributes of fruit quality, Amaral *et al.* (1997) suggested to use the combinations Santa Clara x Angela I.5100, Santa Clara x Floradade and Angela I.5100 x Floradade for improvement of total carotenoids, β-carotene and vitamin C.

Three tomato genotypes high in lycopene and total soluble solids (TSS) (DT10, Pusa 120 and Pusa Sheetal) were crossed with 6 testers (Kumar *et al.*, 1997). The nine parents and their 18 F_1 hybrids were grown during 1993-94 and evaluated for a number of yield and quality related traits. The results indicated that heterosis breeding should be effective for the improvement of more yield and fruit number in tomato. For average fruit weight, however, selection will be more rewarding due to the prevalence of additive gene action. For inheritance of processing characters, *i.e.*, TSS, ascorbic acid, lycopene and acidity, non-additive gene action was predominant. Thus, heterosis breeding is recommended for the improvement of these characters. SCA effects for most crosses were related to the GCA effects of their parents, and the best cross combinations for all the characters studied involved at least one parent with high GCA effects. Hence high GCA effects can be used as criteria in selection of desirable parents for heterosis breeding for processing characters (Kumar *et al.*, 1997; Pandiarana *et al.*, 2015b).

Increasing both contents of β-carotene and intensity of red colour in tomato fruits may be achieved by coupling crimson og^c and *hp* pigment characters in one genotype. It was shown that a rise of β-carotene from 5-10 per cent up to 20 per cent in breeding lines did not cause any significant drop of the redness of fruits in relation to the standard and parental line. Further increase of the contents of β-carotene above 20 per cent of the total pigmentation (above 2 mg per 100 g) should be obtained by a rise of the lycopene level. It was observed that environmental factors exert a higher influence on redness and lycopene level than on carotene contents in tomato fruits (Potaczek and Michalik, 1998). Recently a conventional breeding approach was framed to combine lycopene enhancing mutant gene, *dg* or $hp\text{-}2^{dg}$ and anthocyanin synthesizing gene, *Aft* in one genotype in maximizing the accumulation of both lycopene and anthocyanin pigment in tomato fruit to develop a new variety "Purple tomato" with unique combination of high lycopene (6.13 mg/100 g fresh weight), anthocyanin (20.73 mg/100 g fresh weight) and ascorbic acid (31.56 mg/100 g fresh weight) contents (Hazra *et al.*, 2018).

17.6.2 Ascorbic Acid

In the quality breeding programme of tomato, *L. pimpinellifolium* and *L. peruvianum* have been reported to contribute genes for high ascorbic acid content. According to Tigchelaar (1986), there existed a relationship between ascorbic acid content and fruit size, shape and extent of fruit exposed to light. Small fruited genotypes are richer in ascorbic acid content and firm fruited cultivars exhibit lower content. According to Thamburaj (1998), the interspecific hybrid progenies exhibited substantial increase in ascorbic acid content. He also reported that as the ascorbic acid content in the fruit increases, the plant's resistance to certain diseases like fruit rot caused by *Phytophthora parasitica* is also increased. High ascorbic acid in addition to improve the nutrition, also aids in the better retention of natural colour and flavour of the products. The preponderance of the gene action for ascorbic acid content is additive in cherry tomato (Mukherjee *et al.*, 2020) and non-additive in tomato (Pandiarana *et al.*, 2015b).

17.6.3 Acidity

It is yet another important quality parameter which not only contributes to sourness but is also major factor in flavour intensity of tomato. If a fruit sample has a pH above 4.5 and below 0.35 g citric acid/100g of fresh weight, it is commercially undesirable as it would require increased processing time and temperature to avoid spoilage (Kalloo, 1991). There was an overwhelming response of non-additive gene action for control of titratable acidity in tomato (Pandiarana *et al.*, 2015b; Acharya *et al.*, 2018) and cherry tomato (Mukherjee *et al.*, 2020); hence, selection will bring no, or little, genetic improvement for the trait. Successful breeding methods will accumulate genes to form superior gene constellations interacting in a favorable manner. Hybrid breeding could be the best possible option for improving titratable acid content of tomato fruit. However additive and non-additive gene actions have been reported for pH of the juice (Thamburaj, 1998). Generally, small fruited varieties have more acids than large fruited ones. There is negative correlation between size/

weight of fruit and acidity. Locular tissues have more acidity than the pericarp tissues. So for high acidity, a variety with more locular area should be preferred.

17.6.4 Sugars and Solids

The total solids of the cultivated tomato comprise 4–7.5 per cent of its fresh weight, though this percentage can be much higher in some wild species (Fridman *et al.*, 2000). The total solids are composed of all fruit components except water and volatiles. In the cultivated tomato, the soluble and insoluble solids account for about 75 per cent and 25 per cent, respectively, of the total solids (Stevens, 1986). Reducing sugars glucose and fructose are the major components of the soluble solids (Azanza *et al.*, 1994). Sucrose is also present but in very small quantities, although some wild species of tomato, including *L. chmielewskii* and *L. hirsutum*, have higher concentration of sucrose (Chetelat *et al.*, 1995). The remaining soluble solids are composed of organic acids, lipids, minerals, and pigments. The insoluble solids include proteins, cellulose, hemicellulose, pectins, and polysaccharides, which determine fruit viscosity (Tiglechar, 1986; Hewitt and Garvey, 1987), which determines the quality of the processed products. The increase of total solids of tomatoes has often been found to enhance the yield of processed products that are based on the solids (Berry and Uddin, 1991; Wood, 1992). The taste and flavour are also improved by higher solids (Foolad, 2007a). Soluble solids in tomato have been found to be up to 4.6 in fresh market tomatoes and up to 6.3 per cent in most processing cultivars (Chen and Foolad, 1999). However, wild relatives *L. pimpinellifolium*, *L. chmielewskii* and *L. cheesmanii* have much higher concentrations (≤ 9–15 per cent) of soluble solids (Hewitt and Garvey, 1987; Rick, 1974). However, the complex, quantitative nature of the trait and the negative relationship between yield and TSS content in tomato (Georgelis *et al.*, 2004; Foolad, 2007a) and cherry tomato (Mukherjee *et al.*, 2020) have limited success of getting high yielding genotypes with high soluble solids. Both additive and non-additive gene actions for TSS content in tomato (Kalloo, 1998; Shende *et al.*, 2012; Acharya *et al.*, 2018) and cherry tomato (Mukherjee *et al.*, 2020) have been reported. The use of diallel selective mating (Jensen, 1970), or mass selection with concurrent random mating (Redden and Jensen, 1974), or restricted recurrent selection by intermating the most desirable segregants followed by selection (Shende *et al.*, 2012), might be useful breeding strategies for improvement of TSS content.

Sugar content is another important qualitative attribute which deserves due attention in the quality breeding programme. Both reducing and non-reducing sugars are found in tomato. Sugars contribute to sweetness and flavour may lead to increase in flavour. It is reported that at the pink stage, there is a higher content of sugars than at other stages. A single incomplete-dominant gene (*Fgr*) has been identified in *L. hirsutum* that increases the proportion of fructose over glucose, thus contributing to fruit sweetness (Levin *et al.*, 2000). Total soluble solids (TSS) which is correlated with sugar and ascorbic acid content is a quality component which can be improved by simple selection criterion commonly used for quality breeding. Beckles *et al.* (2012) reported some wild tomato species having high total soluble solid contents and depicted in Table 16.

Table 16: Sources of High Total Soluble Solids Content in Tomatoes*

Species	*Fruit Colour*	*Locus*	*Primary Storage Sugar(s)*	*Mechanism Reported for Increased Content*	*Reference*
S. cheesmanii	Yellow	—	Fructose, glucose	Increased sugar import during ripening	Balibrea *et al.*, 2006
S. pimpinellifolium	Red	—	Fructose, glucose	Increased invertase activity	Husain *et al.*, 2001a, 2001b
S. pennelli	Green	*Lin5*	Fructose, glucose	Apoplastic invertase with altered activity	Fridman *et al.*, 2004
S. chmielewskii	Green	*sucr*	Sucrose	Reduced fruit acid invertase activity	Chetelat *et al.*, 1995
S. habrochaites	Green	*Agp-L1*	Sucrose	*Agp2* and increased fruit starch	Petreikov *et al.*, 2009
S. habrochaites	Green	*fgr, frk*	Higher fructose, glucose	Epistatic interaction between *frk*3 and *fgr*	Levin *et al.*, 2004
S. neorickii	Green	—	Sucrose	Unknown	Schauer *et al.*, 2005
S. peruvianum	Green	—	Sucrose	Higher sucrose import during ripening	Stommel, 1992

* Beckles *et al.* (2012).

The inheritance of fruit firmness and its relationship with other traits was studied by Wang *et al.* (1995) in 6 tomato cultivars and the F_1, F_2 and back crossed generations from their 15 hybrid combinations. Fruit firmness was a quantitative character and its inheritance conformed to the additive and dominance models. Additive effects were predominant, and high firmness values had incomplete dominance. Fruit firmness of the hybrids tended to be intermediate between the parents, and in some combination heterosis was observed. Fruit firmness was correlated with fruit weight, compression tolerance, soluble solids content, fruit shape index and cracking resistance. Cultivar 8711-209 had the highest fruit firmness and was recommended for breeding improved fruit firmness lines.

Gene effects for fruit firmness, pericarp thickness and total soluble solids (TSS) were studied in the parents, F_1, F_2 and backcross generations of three crosses of tomato, involving four parents (FT5, FT12, Vl6 and EC174041) grown at Solan during 1998 by Dobhal *et al.* (1999). The character means over six generations were subjected to scaling tests. Fruit firmness and pericarp thickness were primarily governed by additive gene action, although the cross FT5 x V 16 also exhibited significant dominance effects for both traits. For TSS, the crosses showed different types of gene action, and the results indicated the equal importance of both additive and non-additive gene action with duplicate epistasis. Pedigree selection combined with biparental mating was suggested as a suitable method for improvement of fruit firmness in tomato.

From studies on diallel analysis of 7 x 7 diallel population without reciprocals and six generations of two promising crosses, Akhtar and Hazra (2013) reported that TSS was governed by both additive and dominance effects, but dominance effect was more important, whereas, for lycopene and beta-carotene there were governance of both additive and non-additive gene action, and duplicate epistasis was also observed. Presence of duplicate epistasis for lycopene and β-carotene content will decrease variation in the F_2 and subsequent generations and will hinder the pace of progress through selection. Single seed descent method with progeny row testing and deferred selection would be the best breeding method to develop line bred varieties with good fruit quality character.

Non-additive gene action and duplicate epistasis were identified as the most predominant types of gene effects controlling fruit yield and quality traits in crosses involving cultivars and induced mutants of tomato (Das *et al.*, 2020). They suggested that selection should be practiced in later segregating generations and intermating among the selected segregates followed by one or two generations of selfing could be advised to break the undesirable linkage and allow the accumulation of favourable alleles. The pattern of segregation of dark green fruit (*dg1*) in cross between cultivar and mutant genotype stated that a single recessive gene governing the 'dark green fruit' character and such fruits were formed when the mutant gene governing this character was in homozygous recessive condition (Das *et al.*, 2020).

17.6.5 Post-harvest Shelf-life

Besides nutritive and processing qualities, certain postharvest qualitative traits like slow ripening which leads to long shelf-life, thick skin facilitating long distance transport are also equally important in the quality improvement programme. Tomato, being a climacteric fruit, shows a dramatic increase in respiration at the onset of ripening accompanied by increased synthesis of phytohormone ethylene. In tomato, three single gene mutants namely *rin* (ripening inhibitor), *nor* (non ripening) and *Nr* (never ripe) have been identified located on the chromosomes 5, 10 and 9, respectively (Foolad, 2007b) that alter the fruit ripening process. The usefulness of these mutants for breeding tomato genotypes for better storage life has well been recognized (Thamburaj, 1998).

In the experiments of Barkai-Golan and Kopeliowitch (1985) fruits of the non-ripening tomato mutants *nor* and *rin* and those of the slow ripening mutants Alcobaca and Long Keeper were highly resistant to postharvest decay development during 30 days at 12°Cor 20°C as compared to the fruits of common cultivars. Incidence of decay in the *nor* hybrids was about 50 per cent and 30 per cent of that found in normal fruit at 12°Cand 20°C, respectively. The fruits of the mutant and of the *nor* hybrid were markedly resistant to *Rhizopus stolonifer* infection during cold storage. Mutschler (1984) has proposed that cv. Alcobaca possesses a new fruit-ripening gene *alc*. This mutant is controlled by a single gene (*alc*) located on the short arm of chromosome 10, about 20 cM apart from *u*, a gene conferring *uniform ripening* in tomato (Kinzer *et al.*, 1990).

Studies by Tigchelaar and Barman (1985) led to the conclusion that the cvs. Alcobaca, Long Keeper, Spanish Winter and Italian Winter all carry a similar allele at the *nor* locus which influences both fruit ripening and shelf-life. Kalloo (1991) opined that the firm-fruited processing cultivars with an extended shelf-life have become important in the fresh market industry too. The *rin, nor* and *alc* genes are likely to continue to have increasingly prominent significance in the breeding of fresh market cultivars, especially large-fruited ones with a longer shelf-life and with little loss of fruit quality during and after ripening.

Three ripening mutants, containing mutant genes *rin, nor* and *alc*, were crossed with Da Hong, a cultivar with normal ripening. The ripening mutants, their F_1 hybrids and ripening mutant and normal control cultivars were compared for ethylene and CO_2 evolution, firmness and shelf-life. Ripening mutants differed significantly from Da Hong normal tomatoes. Mutant fruits were typically non-climacteric. They lacked an obvious respiratory peak and concomitant rise in ethylene at the time when normal fruit undergo ripening. Colour development occurred slowly during this period. Fruit from *alc* mutants gave a climacteric peak of respiration but produced <35 per cent of the ethylene peak produced by Da Hong fruit. Fruit of normal cultivars softened rapidly during the ripening period, the greatest decrease in firmness occurring 60-68 days after anthesis. Fruit of the *rin* and *nor* genotypes exhibited only a slight softening during this same period of time, and retained the appearance of mature green fruit. Fruits with the *alc* gene exhibited intermediate values between *nor* and normal fruit. Fruit of heterozygous plants softened more slowly and changed colour several days after normal fruit. Fruit homozygous for *rin, nor* or *alc* had extremely long storage life regardless of the harvesting stage and their fruits could be stored for 90-100 days when harvested mature-green (Lu *et al.*, 1995).

F_1 hybrids incorporating the non-ripening genes *nor, rin* and *Nr* were evaluated for yield, fruit quality and shelf-life. The studies were started in 1984 and in the period 1985-92 a series of comparative experiments with F_1 hybrids was carried out. Slawa Nadrenii was the standard cultivar. Total and commercial yields were analysed. The storage life and percentage of sound fruits picked at mature green, breaker and ripe stages were estimated. F_1 hybrids of *nor* had the best fruit colour, firmness and shelf-life among the 3 groups of hybrids and also gave yields comparable to those of the standard, while the best (C36 x *nor* and S15 x *nor*) exceeded it significantly in some years. In the period 1985-89, the last harvesting took place in mid or late September and storage duration lasted 77 (1988) to 99 days (1986). When plants were grown in a non-heated plastic tunnel rather than the field, there was a 2-4 week delay in the last harvest and a shift in the storage period as well as fruit quality improvement. Ripe fruits of the best hybrids could be stored on average 2-4 weeks longer than those of Slawa Nadrenii, and breaker and green fruits 2-3 weeks longer. *nor* F_1 hybrids were also superior to Slawa Nadrenii with respect to fruit firmness. The stage of maturity at harvest affected final fruit colour, flavour and shelf-life in the hybrids. Breaker fruits combined good quality with longer shelf-life. Hybrid S15 x *nor* was registered as cultivar Rafal F_1 in 1996 (Seroczynska *et al.*, 1998).

On the basis of studies carried out in the Moscow region of Russia in 1997, Galitskii and Novikova (1999) were of the view that tomato cultivars Mytishchinskii and Sukromskii and the hybrids Natus, Semko Sindbad, Lelya, Energo and Gamma were suitable for growing in film-coated green houses. These were the fastest-ripening and/or gave the highest yields.

To slow down the fruit ripening, antisense technology as a means for genetic engineering have been used by some organisations. One of the first transgenic products released for commercial cultivation in USA is Flavr-Savr tomato with increased shelf-life.

For processing industry, synchronous and uniform ripening are the basic requirements of tomato. Heritability estimates suggested that selection for this character with the aim of once-over harvesting would be possible from segregating population. Induced mutagenesis was estimated to be a better tool for developing processing tomato including uniform ripening as in the case of Pusa Lal Meeruti. At IIHR, Bangalore specific varieties like Arka Saurabh, Arka Ahuti and Arka Ashish have been developed mainly to suit processing industries. Pusa Gaurav is another processing tomato variety developed at IARI, New Delhi whose ripe fruits are firm elliptical in shape have thick flesh and two well filled locules rendering themselves for easy transportation besides having high TSS and good storage quality (Thamburaj, 1998).

The most effective method of breeding synchronous cultivars is the combination of high synchrony of fruit maturity with their slower ripening. Besides, various visible parameters like absence of fruit cracking and blotchy ripening, *etc.* need to be upgraded. Correlation analysis of 13 lines compared for different fruit traits showed that strong shape was correlated with firmness, thickness of flesh and insoluble pectin content, while fruit cracking rate was strongly correlated with weight, number locules, fruit shape index, flesh thickness and soluble pectin. To develop high firmness and low cracking rate fruit lines, selection should be for elongated and proper sized fruit with thick flesh, and fewer locules and high insoluble and soluble pectin contents (Gao *et al.*, 1995).

17.7.0 Interspecific Hybridization

Studies on interspecific hybridization in tomato have been made by many research workers (Afify, 1933; Lindstrom and Humphrey, 1933; Humphrey, 1937; Powers, 1939, 1941; Powers and Lyon, 1941; Porte and Wellman, 1941; Pal and Singh, 1943; Porte and Walker, 1945) in respect of hybridization among the species of *Eulycopersicon* sub-genus. Initial interspecific hybridization mainly targeted major qualitative traits like disease and pest resistance governed by major genes, thus often neglecting the agronomic traits (Passam *et al.*, 2007).

In India, Pusa Red Plum cultivar (early, rich in vitamin C and sugar) was evolved by Pal and Singh (1943) from the interspecific hybridization between *L. esculentum* and *L. pimpinellifolium*. Similarly in the U.S.A., Porte and Walker (1945) evolved Pan America cultivar (resistant to *Fusarium* wilt) from this interspecific hybridization. Experiences of these research workers revealed that there is no barrier in making interspecific hybridization under *Eulycopersicon* subgenus.

While studying interspecific hybridization between the species of *Eriopersicon* subgenus and species of *Eulycopersicon* sub-genus, there are certain barriers in getting fertile hybrids. The crosses succeed only in one direction when the species of *Eriopersicon* is used as the male parent and not in the reciprocal cross (MacArthur and Chiasson, 1947; McGuire and Rick, 1954; Choudhury, 1959), the phenomenon being termed unilateral incompatiility. F_1 hybrids of the crosses are to be used as male parent only while making back-cross. Another common feature is that most of the species in *Eriopersicon* are self-incompatible (Lamn, 1950; McGuire and Rick, 1954) and this condition is transmitted to all hybrids with *L. esculentum* as if it is a dominant trait. The reproductive behaviour and breeding behaviour of the cultivated and major wild species of tomato are enlisted in Tables 17 and 18.

Table 17: Reproductive Features of different Wild Relatives of Tomato

Species	*Sub-genra/Common Name*	*Reproductive Features*
Solanum lycopersicum L.	Eulycopersicon (Common tomato)	Self pollinated
Solanum lycopersicum var. *cerasiforme*	Eulycopersicon (Cherry tomato)	Self pollinated
S.pimpinellifolium (Jusl.)	Eulycopersicon (Currant tomato)	Self pollinated and Cross pollinated
S. cheesmaniae	Eriopersicon (Wild)	Self pollinated
S. chmielewskii	Eriopersicon (Wild)	Cross pollinated
S. chilense	Eriopersicon (Wild)	Self-incompatible
S. *parviflorum*	Eriopersicon (Wild)	Self pollinated
S. peruvianum	Eriopersicon(Wild)	Self-incompatible
S. *habrochaites*	Eriopersicon (Wild)	Self-fertile and Self-incompatible
S. *pennellii*	Eriopersicon (Wild)	Self-incompatible

Table 18: Breeding Barriers in Major Tomato Species

Female / *Male*	*S. lycoper-sicum*	*S. pimpinelli-folium*	*S. habro-chaites*	*S. chilense*	*S. peruvi-anum*
S. lycopersicum	No barrier	No barrier	No barrier	Embryo abortion	Embryo abortion
S. pimpinellif-olium	No barrier	No barrier	No barrier	Embryo abortion	Embryo abortion
S. *habrochaites*	Unilateral incompatability	Unilateral incompatability	Self incompatibiliy	Embryo abortion	Embryo abortion
S. chilense	Embryo abortion	Embryo abortion	Embryo abortion	Self incompatibiliy	Embryo abortion
S. peruvianum	Unilateral incompatability	Unilateral incompatability	Unilateral incompatability	Embryo abortion	Self incompatibiliy

Many research workers have tried to overcome these difficulties. The techniques used by Choudhury (1959) to raise hyrbids of *L. esculentum* and *L. peruvianum* were (i) embryo culture, (ii) application of growth substances, (iii) grafting, (iv) mixed pollination and (v) crossing tetraploid *esculentum* with diploid *peruvianum*. He obtained some success by employing these techniques. However, useful genes, especially for disease resistance present in the species of *Eriopersicon* sub-genus could not be effectively incorporated to the cultivated species due to difficulties encountered in interspecific hybridization with wild relative. More attention is required in this field of research, especially for evolving cultivars resistant to tobacco mosaic and leaf curl virus diseases.

17.8.0 Other Aspects of Improvement

In addition to breeding new cultivars, there have been innumerable studies by many research workers looking into the inheritance of desired characters, linkage relationships and cytogenetics of interspecific hybrids. All these informations help the tomato breeder for developing a new cultivar. Such extensive studies have been possible in tomato because of its (i) short life cycle, (ii) ecological versatility, (iii) ease of culture, (iv) long flowering period, (v) very high level of self pollination under most conditions, (vi) ease of making controlled cross-pollination with relatively large yield of seeds per pollinated flower, (vii) large number of known genes, and (viii) favourable chromosome cytology, particularly in the pachytene stage. About 250 genes have been allocated to their chromosomes and chromosome arm contents have been well established by cytogenetic identification of centromere positions (Rick, 1974). These research findings have been effectively utilized by the geneticists and breeders in improving this crop and this has been further strengthened by the Tomato Genetics Co-operative.

Large resources of germplasm are available and yet unexploited opportunities (like the high water economy of *Solanum pennellii*) await investigation (Rick, 1973).

Single cell culture and cell hybridization also promote prospects for improvement of tomato (Rick, 1976). Several aspects of improvement have also been discussed in biotechnology.

18.0 Biotechnology

The genome size of tomato has been determined to be 950 Mb (2.0 pg/2c = 9.5 x 10^5 Kb/lc) comprising of 77 per cent heterochromatin and 23 per cent euchromatin (Peterson *et al.*, 1996).

Extensive investigations on different aspects of biotechnology have been carried out in tomato to overcome problems associated with this important vegetable crop.

18.1.0 Tissue Culture

18.1.1 Micropropagation

For the micropropagation of tomato direct regeneration from different explants like shoot-tips, cotyledon, hypocotyl, leaves and seeds as well as regeneration via callogenesis of the explants has been reported.

In tomato, adventitious shoot regeneration can be achieved either directly (Dwivedi *et al.*, 1990) or indirectly through an intermediate callus phase (Geetha *et al.*, 1998). Bhatia (2003) reported that both callus and shoots may be produced together. Shoot apex, nodal segments and root segments were successfully used for callus induction and regeneration (Jatoi *et al.*, 2001).

Polevaya *et al.* (1988) obtained shoot from apical and axillary meristems cultured on MS medium with added gibberellins, kinetin and glucose. Direct shoot bud differentiation, without intervening callusing, was induced in tomato leaf segments first by culturing them in 0.5mg/l BA and 0.25 mg/l NAA (morphogenetic medium) for 7 days followed by their transfer to 0.01mg/l NAA (Dwivedi *et al.*, 1990), Leaflets of 5 cultivars (Platense INTA, Planeuco INTA, Platauco INTA, Marmande and Rossol) were cultured on MS medium supplemented with BA and IAA or only BA. Shoot regeneration was best in medium containing BA and IAA and was higher in explants taken from autumn cultivated plants than from spring cultivated ones (Zorzoli *et al.*, 1988).

Costa *et al.* (2000) excised cotyledon explants of tomato cvs. Santa Clara, Firme mutant IPA-5 and IPA-6 from 8 to 10 day old seedlings. Four different shoot induction media supplemented with timentin (300 mg/l) were tested. When cotyledon explants were cultured in MS based medium with 1mg/l zeatin plus 0.1mg/l IAA along with timentin, there were higher regeneration frequencies and a greater number of elongated shoots.

Godishala *et al.* (2012) and Ting-ting *et al.* (2005) reported that 1.0 mg l^{-1} zeatin enhanced adventitious shoot regeneration in tomatoes. In addition, Jozef *et al.* (2004) reported that the addition of growth regulators in media influenced the number of shoots regenerated from tomato cotyledons.

Pino *et al.* (2010) found that 8 to 10 day-old cotyledons of tomato were superior to other source of explants, including hypocotyls, stems and leaves for promoting shoot organogenesis of tomato, while Cortina *et al.* (2004) and Ajenifujah-Solebo *et al.* (2012) reported the effective use of 10 - 13 day-old cotyledons. Ajenifujah-Solebo *et al.* (2012) also observed that in a culture medium of MS with 0.1 mg/litre IAA and 1.0 mg/litre zeatin 64-97 per cent of the cotyledonary explants produced shoots in three different cultivars, while rooting medium was MS + 0.1 ml/litre NAA.

Chaudhry *et al.* (2010) obtained maximum callogenesis from hypocotyls on MS medium supplemented with IAA (2 mg/l), NAA (2 mg/l), BAP (5 mg/l) and Kinetin (4 mg/l). It was 65.2 per cent for hypocotyls. For leaf discs maximum callogenesis was achieved on MS medium supplemented with IAA (2 mg/l), NAA (2 mg/l), BAP (2 mg/l) and kinetin (4 mg/l). It was 81.3 per cent for leaf discs. Calli were cultured on MS medium having concentrations of Zeatin (1 mg/l) and IAA (1 mg/l) for regeneration. They showed maximum regeneration of 69.2 per cent from hypocotyls.

18.1.2 Embryo Culture

Embryo culture has been used by tomato breeders to obtain interspecific crosses of incompatible species, which are of economical importance. Gradziel (1988) used a range of techniques for (i) highly self-incompatible *L. peruvianum*, (ii) cross between *L. peruvianum* as female parent and *L. esculentum* x *L. peruvianum* hybrid, and (iii) introgress genes from *Solanum lycopersicoides* x *L. esculentum*. The techniques used included bud pollinations where pollen germination on immature stigmas was enhanced by applying a germination medium, multiple pollination, auxin application, embryo culture and environmental conditioning.

Tolerance of low temperatures, cucumber mosaic virus and *Botrytis cinerea* was transferred from *S. lycopersicoides* to tomato with varying degrees of success. Saccardo *et al.* (1981) transferred useful characters from *L. peruvianum* to *L. esculentum*. They obtained lines which had resistance to race 1 of *Verticillium dahliae*, races 1 and 2 of *Fusarium oxysporum* or tobacco mosaic virus, and with agronomic characters such as synchronous ripening, large firm fruits, dwarf habit, numerous flowers per inflorescence, a high soluble solids content or male sterility.

Demirel and Seniz (1997) reported that the 24-day-old embryos isolated and cultured on modified MS medium showed good germination and growth. No fruits were obtained when *L. peruvianum* was the female parent, whilst aborted embryos were produced from crosses in which *L. peruvianum* was the male parent. Hybrid plantlets were obtained from *L. peruvianum* x *L. esculentum* crosses via callus formation from seeds with aborted embryos as an alternative to embryo culture.

For the transfer of valuable traits from wild species into the cultivated tomato, Lanzhuang and Adachi (1996) excised globular-stage embryos 13 and 15 days after pollination (DAP) and were cultured *in vitro*. Plants were regenerated from interspecific crosses of *Lycopersicon esculentum* cv. 'Early pink' x *L. peruvianum* PI 270435, and backcrosses of *L. esculentum* 'Giban (JF) No. 1' x ('Early pink' x PI 270435). Somatic embryos and single cotyledons emerged on hypocotyl sections of the embryos. Five to nine plantlets per embryo were obtained by clonal propagation. The hybrid nature of the plants was confirmed by comparing hybrids and parents in their ability to regenerate shoots from leaf segments *in vitro*, by comparing plant morphology and characteristics and by chromosome number.

Pico *et al.* (2002) obtained interspecific hybrids with the two *L. chilense* accessions (LA 1932 and LA 1963) that exhibited post-zygotic barriers and embryo rescue was successful in them for introgression of TSWV, TYLCV and PepMV.

Solanum lycopersicum cultivars Kashi Amrit and Megha Tomato-3 were crossed with *S. peruvianum* (WIR 3957) and *Solanum pimpinellofolium* (EC 521030) by Kharkongar *et al.* (2013) and embryo rescue was carried out. 25 days after pollination was found to be the optimum time for rescuing the embryos. MS medium supplemented with 1 mg/l GA_3, 0.1 mg/l NAA and 0.5 mg/l BAP was found to be the most effective for germination of the immature putative hybrid embryos. The confirmation of hybridity of the embryo rescued plants from the interspecific crosses of both the cultivated tomato varieties with *S. peruvianum* (WIR-3957) was done using RAPD markers.

Pollen from 'EELM-802', a genotype showing strong resistance to *Tomato chlorosis virus* (ToLCV), derived by selfing and selection from 'IAC-CN-RT' (a population obtained by embryo rescue in Brazil from a cross between the cultivated tomato species (*Solanum lycopersicum* L.) and the wild relative *S. peruvianum* L., exhibiting unilateral incompatibility and thus showing early embryo abortion, was used to pollinate the susceptible cultivar 'Moneymaker' by Encina *et al.* (2012), and proembryos were collected from developing fruits 17-24 days after pollination. Rescue of immature embryos was obtained from proembryos 21 days after pollination. One of the regenerated plantlets showed intermediate morphological characters between 'Moneymaker' and 'EELM-802'. True hybrid nature of that plant was further confirmed by using a codominant molecular marker showing polymorphism between the parents.

18.1.3 Anther Culture

Haploid callus cultures induced from anthers were successfully obtained in selected tomato races. Callus differentiation could be controlled both in the dark and in the light by adding hormones (NAA and kinetin at various levels) to defined minimal media, but development into plantlets was achieved only in the light. Callus cells could be induced to develop into seedless pseudo fruits (Krzysko and Rogozinska, 1975). Haploids could be maintained in culture on defined minimal media for at least one year (Gresshoff and Doy, 1972). Haploidy has been successfully used in tomato for producing pure line. To obtain pure male sterile mutants anther cuture has been used (Zamit *et al.*, 1980; Shchereva *et al.*, 1990).

In vitro cultures of anthers of eight self-incompatible diploids and two self-incompatible autotetraploid clones of *L. peruvianum* for callus induction and organ regeneration showed diploid, triploid, tetraploid, hexaploid and octoploid plants and cyto-chimeras. The occurrence of self-incompatibility in regenerated *S*-heteroallelic triploid revealed that, besides *S*-heteroallelic pollen, a balanced genome or a certain level of heterozygosity is necessary for the manifestation of self-compatibility (Ramulu and Devereux, 1979).

Oankh *et al.* (1986) cultured anthers of cultivar Roma and its male-sterile counterpart carrying *ms* 10^{35} and obtained haploid, diploid and mixoploid plants. The 477 regenerates included plants resembling the source (42 per cent of male-sterile and 34 per cent of normal plants) and also differing appreciably from it. Plants of the former group were mainly mixoploid.

Segu-Simarro and Nuez (2005, 2007), Shtereva *et al.* (1998), Zagorska *et al.* (1998, 2004) obtained regeneration of entire plants with a clear haploid origin but with a high percentage of mixoploid calli and at a low general efficiency of the process.

Corral-Martinez *et al.* (2011) analyzed the process of callus formation in anthers of tomato lines carrying the *ms* 10^{35} gene for male-sterility, using light and electron microscopy, flow cytometry and genetic analysis with morphological and molecular markers and reported production of haploid, double haploid and diploid calli, although at different frequencies. Diploid calli derived either from somatic cells or from the fusion of two genetically different haploid nuclei account for more than 90 per cent of the total of calli produced. Somatic calli are derived from the stubs of connective tissue.

18.1.4 Protoplast Culture and Fusion

Protoplasts have been isolated from different tissues such as leaves, cotyledons, fruits, callus and cell suspensions and used for somatic hybridization. A large number of intergeneric and interspecific hybrids using protoplast fusion has been produced. *L. esculentum* protoplasts have been successfully fused with *L. pennelli* (Bonnema *et al.*, 1991) with *L. chilense* (Bonnema and O'Connell, 1991),with *L.*

hirsutum (Jourdan *et al.*, 1993), with *L. peruvianum* (Sakata *et al.*, 1991) with *Solanum nigrum* (Jain *et al.*, 1988), with *S. ochranthum* (Kobayashi *et al.*, 1996), with *S. melongena* (Samoylov and Sink, 1996), with *S. bulbocastanum* (Brown *et al.*, 1996), with *S. lycopersicoides* (Hossain *et al.*, 1994) and with *S. tuberosum* (Smille *et al.*, 1979).

Leaf mesophyll protoplasts have been found very efficient for protoplast culture and regeneration of plants. Calli rapidly regenerated shoots (as early as 11 days after subculture) at high frequencies on basal medium + GA_3, BA and 2iP. Regenerated shoots were easily rooted on the basal medium used plus the appropriate auxin (dependent on cultivar) and plantlets were planted in soil within 2 months of isolation of protoplasts (Shahin, 1985). Tan *et al.* (1987) compared shoot regeneration from leaf discs and leaf mesophyll protoplasts in 11genotypes (hybrids, parental lines and true breeding varieties). There was no heterosis for regenerative capacity. A correlation was detected between regeneration from leaf discs and from protoplasts, suggesting that regeneration from leaf discs could be used as a preliminary screen for protoplast regenerative ability.

Preconditioning donor plants with temperature and light treatment has been found to enhance the yield of protoplasts. Tabaeizadeh *et al.* (1984) reported that leaf-mesophyll protoplasts obtained from plants grown in16-h light and then for 5 days in 9-h light produced significantly more cell colonies than similar protoplasts obtained from plants kept in16-h light. Bellini *et al.* (1990) found that low light intensity, a long photoperiod (16h) and 25/17°C temperature(day/night) were best for obtaining viable protoplasts.

Shahin and Spivey (1986) isolated protoplasts from cotyledons of UC 82, which is susceptible to *F. oxysporum* f. *lycopersici* race 2 and cultured on medium containing the non specific toxin fusaric acid. Plants regenerated from toxin-resistant callus (R_0 generation) were screened for resistance using the Fusarium slurry inoculation technique. The ratio of resistant to susceptible plants among the R_1 progenies was 3: 1. Analysis of the R_2 confirmed that resistant plants were either homozygous or heterozygous for the gene conferring resistance, indicating that the gene is dominant.

Wachocki *et al.* (1991) compared the organization of the mitochondrial genome in somatic hybrids and cybrids regenerated following fusion of protoplasts from *L. esculentum* and *L. pennellii* to assess the role of the nuclear genotype on the inheritance of organellar genomes. No organellar encoded traits were required for the recovery of either somatic hybrids or cybrids.

Very few reports are available on fusion experiments involving *L. chilense*. Bonnema and O'Connell (1991) recovered somatic hybrid plants following fusion of leaf mesophyll protoplasts isolated from tomato cv. UC 82 with protoplasts isolated from suspension cutured cells of *L. chilense* LA 1959. Indole Acetic Acid was used to select against the growth of unfused tomato protoplasts. Two somatic hybrids were recovered in a population of 16 regenerants.

Parokonny *et al.* (1997) used genomic *in situ* hybridization (GISH) to examine genome interactions in two allohexaploid (2n=6x=72), *L. esculentum* (+), *L. peruvianum* somatic hybrids and their seed progenies originated from subsequent backcrosses to *L. esculentum*. The ability of GISH to distinguish between chromatin derived from two closely related species *L. esculentum* and *L. peruvianum* (both 2n = 2x = 24), allowed the precise chromosomal constitution of somatic hybrids and their backcross progenies to be unequivocally established.

Kochevenko *et al.* (2000) regenerated fertile cybrid plants of three subclones, B1A, B3A, B4A from the single colony obtained after fusion of mesophyll protoplasts of plastome chlorophyll deficient mutant *L. peruvianum* var. *dentatum* (line 3767 and γ-irradiated mesophyll protoplasts of *L. esculentum* (cv. Quedlinburger Fruhe Liebe).

Samoylov *et al.* (1996) studied morphology, the extent of elimination of donor chromosomes and the organelle composition of highly asymmetric somatic hybrid plants between a interspecific tomato hybrid *L. esculentum* x *L. pennellii* (EP) as donor and a *S. melongena* (E) recipient. Morphologically the somatic hybrids most resembled eggplant but due to polyploidy, growth was slower relative to both fusion parents. The somatic hybrids produced flowers that were characterized by abnormal styles, stigmas and anthers which did not produce pollen.

Hossain *et al.* (1994) electro fused mesophyll protoplasts of tomato cv. Ohgata Zhiko with *S. lycopersicoides* and produced fertile tetraploid somatic hybrids.

Matsumoto *et al.* (1997) obtained somatic hybrid plants between the F_1 of the cross *L. esculentum* x *L. peruvianum* and *S. lycopersicoides* by symmetric electrofusion. Twenty putative somatic hybrid plants were randomly selected and analyzed. The hybrid nature of these plants was evaluated according to morphological characteristics and then confirmed by RAPD analysis.

Ramulu *et al.* (1996) reported microprotoplast-mediated chromosome transfer (MMCT) through fusion of small (subdiploid) microprotoplasts of a transgenic triploid potato (*Solanum tuberosum*) cell line with leaf protoplasts of tobacco (*Nicotiana tabacum*) and the wild tomato species *L. peruvianum.*

Schoenmakers *et al.* (1993) selected some 25 allotriploids from 450 somatic hybrids obtained from fusion of protoplasts of diploid tomato (genotypes C3 1-244 and ALRC x M8-7) and monohaploid potato (line 7322). The F_1 line C31-244 was derived from the back cross of the nitrate reductase-deficient (nia) mutant C31 and its wild type GT. A cytoplasmic albino mutant of tomato cv. Large Red Cherry (ALRC) grafted onto Moneymaker rootstock was crossed with MsK8 which has good regenerative ability. ALRC x M8-7 was selected from the totally albino F_1 population. The fusion of nitrate reductase-deficient tomato C31-244 with potato only gave rise to hybrid calli if selection was performed on media lacking ammonium. Parental microcalli were rarely obtained and did not regenerate. The fusion of cytoplasmic albino tomato ALRC x M8-7 with potato gave rise to 5 albino and 275 green hybrid calli and plants, respectively.

18.1.5 Somatic Embryogenesis

Various explants have been used for inducing somatic embryogenesis such as cotyledons, leaves, hypocotyl, *etc.* El-Farash *et al.* (1993) found that genotype, explant age and explant type influenced the callus formation and shoot regeneration.

From the hypocotyl explants direct regeneration of plantlets is possible. Newman *et al.* (1996) excised 10 mm hypocotyl section from 2-week-old seedlings grown on MS medium supplemented with or without BA. Regeneration through somatic embryogenesis and shoot organogenesis occurred in explants of all treatments, even on hypocotyls from seedlings grown on basal medium and sub cultured to medium without growth regulators.

18.2.0 Biochemical and Molecular Markers

18.2.1 Genetic Linkage and Identification of Genotypes

In tomato, there are over 1300 morphological, physiological (*e.g.*, male sterility, fruit ripening, fruit abscission), and disease resistance genes (Kalloo, 1991) of which only less than 400 have been mapped (Chetelat, 2002; Mutschler *et al.*, 1987; Tanksley *et al.*, 1993). The second generation of genetic markers, isozymes, became popular during 1970s and early 1980s. In tomato, 41 isozymic genes corresponding to 15 unique enzymatic reactions have been characterized, of which 36 have been mapped onto the 12 tomato chromosomes (Tanksley *et al.*, 1993). Despite their great advantages, isozyme markers are very limited in number and often are not polymorphic among closely-related genotypes (Foolad *et al.*, 1993).

The use of isoenzymes as markers in genetic mapping, introgression and other breeding work was examined by Kudryakova and Kalloo (1991). A review of the work on isoenzyme analysis in the genus *Lycopersicon* and applications in breeding practices is summarized hereunder.

Rick (1983) presented a comprehensive review with data concerning allozyme variability (including data on inheritance and linkage) in *L. esculentum*. Isozymes were used for genetic linkage studies and identification of genotypes by various workers (Kudryakova and Kalloo, 1991, Mori *et al.*, 1992).

Huang *et al.* (1985) analysed peroxidase isozymes of 103 cultivars and 38 hybrids at 15 different growth stages by polyacrylamide gel electrophoresis. Although different cultivars showed different patterns at any one growth stage, the combined patterns for all the growth stages were similar to each other and to the overall peroxidase isozyme pattern for tomato.

Tanksley and Kuehn (1985) crossed *L. hirsutum* LA 1777 as male parent to *L. esculentum* cv. 82T47-1 and the F_1 hybrid was backcrossed as male parent to *L. esculentum* cv. UC 82. Backcross individuals were assayed electrophoretically for isoenzymes of 6-phosphogluconate dehydrogenase (6-PGDH) and 13 other enzyme loci. LA 1777 was also crossed as male parent to primary trisomics (triplo 4, 5, 7, 11 and 12) of *L. esculentum*. From the results, 3 independent genes for 6-PGDH activity were identified.

Taxonomists have long argued the taxonomic origin and division of plants. Knowledge of the relative genetic distance among individual tomato genotypes or populations is useful in a breeding programme because it permits organization of germplasm and provides more efficient sampling. In the ongoing quest to classify material better, molecular markers are now playing an important role. Miller and Tanksley (1990) studied 156 accessions representing eight species of *Lycopersicon* using 40 single copy RFLP markers. They calculated genetic distances between accessions and species and obtained a dendrogram illustrating relationship among these accessions. Accessions belonging to the same species clustered together.

Klein-Lankhorst *et al.* (1991) used RAPD markers to develop genome-specific fingerprint of *L. esculentum*. By comparing fingerprints of *L. esculentum, L. pennellii* and the *L. esculentum* chromosome 6 substitution line LA 1641 which carries chromosome 6 from *L. pennellii*, three chromosome 6-specific RAPD markers were directly identified.

Williams and St Clair (1993) used RAPD and RFLP markers to assess the variability in tomato germplasm (*L. esculentum*) which included 46 accessions from vintage cultivars, modern cultivars, South American regional cultivars and wild *L. esculentum* var. *cerasiforme*. Two *L. cheesmanii* accessions served as an outgroup. 48 accessions were monomorphic at 135 of the 215 RAPD loci and 31 of the 48 RFLP loci that were assayed.

By 2008, more than 2500 markers, including RFLP, EST (Expressed Sequence Tag polymorphism), SSR (Simple Sequence Repeat) and COS (Conserved Orthologue Set) were mapped on 12 tomato chromosomes (Frary *et al.*, 2005; Barone *et al.*, 2009). They are intensively used for the identification and mapping of QTL (Quantitative Trait Loci) associated with important biological and agronomical traits, fingerprinting and Marker-Assisted Selection (MAS). The use of MAS in tomato breeding has been limited by number of polymorphic marker in the species and for economical reason. The work is continuing on the development of high efficiency and resolution markers, such as SNP (Single Nucleotide Polymorphism). Development of tomato EST database (currently 214,000 sequences), DNA microarray and genomes sequencing will contribute to emergence of new and more practical markers. Tomato euchromatin sequencing paves the way for a comparison of important genes and QTL (Foolad, 2007a). The DArT (Diversity Array Technology) is another technique with carrying high hopes for search of useful markers. It is rapid analysis of multiple loci in the genome. This technique allows discovery and mapping hundreds of polymorphic markers in a single experiment (Stevens and Robbins, 2007).

18.2.2 Molecular Maps

To facilitate the use of molecular markers in tomato genetics and breeding research, some efforts have been made to develop linkage maps based on mainly PCR-based markers. Mapping populations based on diverse traits have been developed from intra-specific and more from inter-specific crosses involving different accessions of *L. pimpinellifolium, L. pennellii, L. peruvianum, L. chmielewskii, L. cheesmanii, L. parviflorum, L. hirsutum with that of L. esculentum* (Foolad, 2007a). The different genetic linkage maps of tomato developed have been reviewed by Foolad (2007a) and presented in Table 19.

Table 19: Genetic Linkage Maps of Tomato (*Lycopersicon* spp.) Developed Based on different Intra- and Interspecific Crosses

Linkage Map	*Population Type(a)*	*Population Size*	*Number of Markers*	*Type of Markers*	*Reference*
L. esculentum x L. esculentum var. *cerasiforme*					
1. Cervil *x* Levovil	F_7-RIL	153	377	RFLP, RAPD, AFLP	Saliba-Colombani *et al.*, 2000
L. esculentum x L. pimpinellifolium					
1. M82-1-7 *x* LA1589	BC_1	257	120	RFLP, RAPD, Morphological	Grandillo and Tanksley 1996
2. NC84173 *x* LA722	BC_1	119	151	RFLP	Chen and Foolad 1999
3. Giant Heirloom *x* LA1589	F_2	200	90	RFLP, CAPS	Lippman and Tanksley 2001
4. E6203 *x* LA1589	C2F_6-BIL	196	127	RFLP	Doganlar *et al.*, 2002
5. NC84173 *x* LA722	F_{10} RIL	119	191	RFLP, RGA	Foolad, 2007a
6. NCEBR1 *x* PSLP125	F_2	172	256	RFLP, EST, RGA	Foolad, 2007a
7. NCEBR1 *x* PSLP125	F_8-RIL	172	255	TFLP, EST	Foolad, 2007a
L. esculentum x L. cheesmanii					
1. UC204B *x* LA483	F_2	350	71	RFLP	Paterson *et al.*, 1991
2. UC204B *x* LA483	F_7-RIL	97	132	RFLP	Paran *et al.*, 1995
L. esculentum x L. parviflorum					
1. E6203 *x* LA2133	BC_2	170	133	RFLP, SCAR, morphological	Fulton *et al.*, 2000
L. esculentum x L. chmielewskii					
1. UC82B *x* LA1028	BC_1	237	70	RFLP, isozyme	Paterson *et al.*, 1988
L. esculentum x L. hirsutum					
1. E6203 *x* LA1777	BC_1	149	135	RFLP	Bernacch1 and Tanksley NIL, B1997
2. E6203 *x* LA1777	NIL, BIL	111	95	RFLP	Monforte and Tanksley 2000
3. NC84173 *x* PI126445	BC_1	145	171	RFLP, RGA	217. Zhang *et al.*, 2002
L. esculentum x L. pennellii					
1. VF36 *Tm2*(a) *x* LA716 (high-density map of tomato)	F_2	67	1050	Isozyme, morphological, RFLP	Tanksley *et al.*, 1992; 229. Pillen *et al.*, 1996

Linkage Map	*Population Type(a)*	*Population Size*	*Number of Markers*	*Type of Markers*	*Reference*
2. Vendor *Tm2*(a) *x* LA716	F_2	432	98		de Vicente and Tanksley 1993
3. M82 *x* LA716	IL	50	375	RFLP	Eshed and Zamir 1995
4. VF36 *Tm2*(a) *x* LA716	F_2	67	1242	AFLP, RFLP	Haanstra *et al.*, 1999
5. E6203 (LA925) *x* LA716	F_2	83	1500	COS	Fulton *et al.*, 2002
6. E6203 *x* LA1657	BC_2	175	110	RFLP	Frary *et al.*, 2004
7. E6203 *x* LA716	F_2	83	152	SSRs, CAPs	Frary *et al.*, 2005
L. esculentum x L. peruvianum					
1. E6203 *x* LA1706	BC_3	241	177	RFLP, SCAR	Fulton *et al.*, 1997
L. esculentum var. *cerasiforme x L. pimpinellifolium*					
1. E9 *x* L5	F_6-RIL	142	132	SSR, SCAR	Villalta *et al.*, 2005
L. esculentum var. *cerasiforme x L. cheesmanii*					
1. E9 *x* L3	F_6-RIL	115	114	SSR, SCAR	Villalta *et al.*, 2005
L. peruvianum x L. peruvianum					
1. LA2157 *x* LA2172	BC_1	152	73	RFLP	van Ooijen *et al.*, 1994

18.2.3 Identification of Resistance to Biotic Stresses

Most sources of resistance to disease and insects are wild relatives and hence difficulties have been encountered while transferring resistance from wild species to the cultivated tomato *via* traditional breeding method which has restricted transfer of resistance to many tomato elite lines. This is in addition to the lack of suitable screening facility or expertise in many tomato breeding programs to develop cultivars with multiple disease and insect resistances. Thus, breeders have consistently sought more effective approaches for resistance breeding. Insect resistance in tomato has received even lesser attention than disease resistance, and few commercial cultivars have been developed with specific insect resistance, linkage drag being a major impediment in the process (Hartman and Clair, 1998; Kohler and Clair, 2005). During the past three decades, the use of molecular markers and MAS techniques have facilitated identification, mapping, and transferring of many disease and insect resistance genes and quantitative trait loci (QTLs) in tomato (Foolad, 2007a). The use of marker technology in disease resistance breeding in tomato is becoming a routine procedure.

Sequential development of acid phosphatase activity from the root and shoot of resistant and susceptible tomato cultivars following inoculation with *Meloidogyne incognita* in pot experiments indicated that enzyme activity was lower in the susceptible inoculated cultivar (Pusa Ruby) than its control and higher in the resistant inoculated cultivar (Pusa Nemamukt) (Sreedhar *et al.*, 1996).

Bashan *et al.* (1987) suggested that the oxidative phenol system contributes to resistance to *Pseudomonas syringae*. A direct correlation was found between disease index, poly phenol oxidase activity and phenol content in the tissue. Phenols extracted from inoculated resistant plants inhibited pathogen growth. Li *et al.* (1983) found the zymogram patterns of F_2 progeny of crosses between Manapal *Tm-2nv* (female), which carries *Tm-2nv* for resistance to tobacco mosaic virus and five other varieties fell into four groups corresponding with four groups distinguished by leaf shape and colour, similar to female parent, similar to male parent and two intermediate groups.

Van der Biezen *et al.* (1995) and Mesbah *et al.* (1999) reported that *Asc* gene was linked to RFLP markers TG422 and TG134 on the long arm of chromosome 3 for conferring resistance to *Alternaria alternata* f. sp. *lycopersici* in tomato.

Robert *et al.* (2001) used RFLP and CAPS markers and mapped QTLs responsible for resistance to *A. alternata* located on chromosomes 2, 3, 9, 12, and the source of resistance used was a wild species *L. cheesmanii*.

Zhang *et al.* (2003) using RFLP markers and resistance gene analogs (RGA) found 7 QTL associated with resistance to *A. solani*, one in each of chromosomes 3, 4, 5, 6, 8, 10, and 11. Location of some RGA overlaps with location of known genes or QTLs of resistance to the EB, which may suggest link between RGA and resistance to *Alternaria solani*.

Sarfatti *et al.* (1989) showed that a dominant *I2* gene confers resistance to *Fusarium oxysporum* f. sp. *lycopersici* race 2 in tomato, and is linked with the RFLP marker TG105B on the chromosome 11. They mapped the *I1* gene using RFLP markers on chromosome 7 between TF_20 and TG128 markers, which were responsible for resistance to race 1 (Sarfatti *et al.*, 1991). Staniaszek *et al.* (2007) developed CAPS marker TAO1902 that is useful to identify tomato genotypes with *I2* gene. Lim *et al.* (2006) reported the use of CAPS CT226 marker for detection of the *I3* gene responsible for resistance to race 3. The SCAR marker $P7\text{-}43DF_3/R1$ was found to be closely linked to the *I3* gene in tomato. This marker can be used for selection of inbreds with *Fol* race 3 resistance in tomato breeding programs (Barillas *et al.*, 2008).

Fazio *et al.* (1999) identified three RAPD (Random Amplified Pol-ymorphic DNA) markers - UBC 116, UBC 194 and UBC 655, linked to *Frl* gene, located on chromosome 9, responsible for resistance to *Fusarium* crown and root rot of tomato caused by *Fusarium oxysporum* f. sp. *radicis-lycopersici*. UBC 194 marker would be the most useful for selecting *Frl* because it is the most tightly linked.

Behare *et al.* (1991) mapped *Sm* gene, responsible for resistance to gray leaf spot caused by *Stemphylium* spp. on chromosome 11, linked with RFLP T10 marker at a distance of 6.8 CM.

Moreau *et al.* (1998) using RFLP and AFLP (Amplified Fragment Length Polymorphism) markers have located *Ph-2* gene contributing to the partial resistance to late blight. This gene is located on the end of long arm of chromosome 10 between RFLP markers CP105 and TG233. YiPeng *et al.* (2009) identified RAPD marker for resistance to late blight in tomato, designated as the CCPB272-03740 and tightly linked to resistant *Ph-3* gene. The AFLP marker L87 was successfully converted to

SCAR marker which is linked to *Ph-3* gene, and can be used for selection of tomato plant resistant to *Phytophthora infestans* (Mont.) deBary (PueHee *et al.*, 2010).

Stommel and Zhang (1998) using RAPD markers found QTLs responses for resistance to *Colleotrichum coccodes* causing anthracnose ripe rot.

Kawchuk *et al.* (1998) developed co-dominant, allele specific SCAR marker useful in the selection of tomato for resistance to *Verticillium dahliae*. Linkage analysis showed that genetic distance between marker and locus *Vr* is less than 0.67 CM ± 0.49 CM. Diwan *et al.* (1999) mapped this gene on the short arm of chromosome 9. It is strongly linked with RFLP marker, GP39.

The *Ol-1* gene governing resistance to powdery mildew pathogen *Oidium lycopersici* was mapped on chromosome 6. RAPD marker RAPD17 was amplified in the susceptible plants but failed to amplify in resistant ones. Localization of this marker is close to the *Asp-1* locus. Further analysis suggested that *Ol-1* locus is close to RFLP marker TG135 (Van der Beek *et al.*, 1994). Another gene responsible for resistance to *O. lycopersici*, *Ol-3* was found on chromosome 6, and several RFLP and SCAR markers, TG25/$SCAF_1$0 and H9A11 could be associated with it (Huang *et al.*, 2000). Bai *et al.* (2003) reported that resistance to *O. lycopersici* in *L. parviflorum* accession G1.1601 was controlled by three QTLs: *Ol-qtl1* on chromosome 6 in the same region as the *Ol-1*, *O1-qtl2* and *Ol-qtl3* on chromosome 12. Two SCAR markers, B432U, tg111 and seven CAPS markers, Aps1, dc21, tg25, dct136, ct129, ct99, Y258 were developed that could be used to monitor the target QTL genomic regions in marker-assisted selection.

Chunwongse *et al.* (1994) localized *Lv* gene, responsible for resistance to the pathogen *Leveillula taurica*, between RFLP markers CT211 and CT219 on chromosome 12.

The *Cf-2* gene, conferring to resistance to *Cladosporium fulvum* causing leaf mold in tomato, was mapped on chromosome 6 between RFLP markers TG232 and H2D1 (Dickinson *et al.*, 1993).

Bacterial speck caused by *Pseudomonas syringae* pv. *tomato* (*Pst*) can be present in all stages of development and in all parts of above-ground plant. Three RAPD markers S47, R110, RS120 (Martin *et al.*, 1991) and seven RAPD markers R-S7, R-Q14, R-X6, R-150n, R-151n, R174n, R-DD5 (Carland and Staskawicz, 1993) were found to be linked to *Pto* locus conferring resistance to bacterial speck caused by *Pseudomonas syringae* pv. *tomato*. Martin *et al.* (1993) developed RFLP marker TG538 that co-segregated with *Pto* on chromosome 5.

Van Heusden *et al.* (1999) using RFLP and SCAR markers mapped three QTLs *Rcm 5.0, Rcm 7.1, Rcm 9.0* located on chromosome 5, 7 and 9 specific to *Clavibacter michiganensis* ssp. *michiganensis* causing bacterial canker on tomato, while Kabelka *et al.* (2002) using RFLP markers, found two QTLs *Rcm 2.0* on chromosome 2 and *Rcm 5.1* on chromosome 5.

Two loci, *Rx1*, *Rx2* on chromosome 1 and one *Rx3* on chromosome 5, were described based on linkage to RFLP markers for *Xanthomonas* spp. causing bacterial spot (Yu *et al.*, 1995, Yang and Francis, 2007). Ballvora *et al.* (2001) converted two

AFLP markers, P11M6, P15M1 and four RFLP markers, CT242, TG432, TG441, TG598 linked to *Bs4* locus, into locus-specific PCR markers. The CAPS P11M6 marker and STS (Sequence Tagged Site) TG432 marker could be used in large scale screening.

RFLP and isozymes analysis showed the presence of *Cmr* gene conferring resistance to CMV located on chromosome 12 and it is linked to TG68 and CT79 markers (Stamova and Chetelat, 2000).

Stevens *et al.* (1995) using a near isogenic line developed a RAPD marker of the size of 2200 bp (base pair) linked to *Sw-5* gene for tomato spotted wilt virus. They determined the location of the gene for resistance to TSWV to be between the RFLP CT71 and CT220 markers, near the telomere on the short arm of chromosome 9. Stevens *et al.* (1996) also developed SCAR421 marker linked to the *Sw-5* gene conferring to resistance to TSWV.

Tanksley *et al.* (1992) using RFLP markers, located the *Tm-1* gene on the short arm of chromosome 2 near the rDNA region. Ohmori *et al.* (1996) developed SCAR markers, SCA151000, SCB101300, SCG12900, SCI101100, SCN091100 and SCN201400 based on the sequence of RAPD markers. Young *et al.* (1988) created a RFLP markers linked with *Tm-2a* gene.

Zamir *et al.* (1994) mapped the dominant *Ty-1* gene responsible for resistance to TYLCV on chromo-some 6, near the RFLP marker TG36. Another gene *Ty-2* responsible for resistance to this virus, originated from *S. habrochaites*, was located on chromosome 11 between RFLP markers TG393 and TG36 (Hanson *et al.*, 2000; Hanson *et al.*, 2006). Ji and Scott (2006) mapped *Ty-3* gene on the long arm of chromosome 6, between cLED-31-P16 and T1079 markers. Castro *et al.* (2007) developed a CAPS marker JB-1 associated with the *Ty-1* gene and this marker is useful in MAS to identify resistant plant.

A PCR-based codominant marker REX-1 was developed which is tightly linked to *Mi*, a dominant genetic locus in tomato that confers resistance to several species of root-knot nematode and homozygous and heterozygous individuals can be distinguished and the procedure can be used for rapid, routine screening (Williamson *et al.*, 1994).

Tomato DNA spanning the *Mi* locus that confers resistance to root knot nematodes locus was isolated as bacterial artificial chromosome clones, and 52 kb of contiguous DNA was sequenced by Milligan *et al.* (1998). Three open reading frames were identified with similarity to cloned plant disease resistance genes. Two of them, *Mi-1.1* and *Mi-1.2*, appear to be intact genes; the third is a pseudogene. A 4-kb mRNA hybridizing with these genes is present in tomato roots. Complementation studies using cloned copies of *Mi-1.1* and *Mi-1.2* indicated that *Mi-1.2*, but not *Mi-1.1*, is sufficient to confer resistance to a susceptible tomato line with the progeny of transformants segregating for resistance. The cloned gene most similar to *Mi-1.2* is *Prf*, a tomato gene required for resistance to *Pseudomonas syringae*. *Prf* and *Mi-1.2* share several structural motifs, including a nucleotide binding site and a leucine-rich repeat region, that are characteristic of a family of plant proteins, including several that are required for resistance against viruses, bacteria, fungi, and nematodes.

Mi-1, originated from the *Lycopersicon peruvianum* gene conferring resistance to the root-knot nematode *Meloidogyne incognita* was also capable of conferring resistance to an unrelated pathogen, the potato aphid *Macrosiphum euphorbiae* and this resistance was stable (Vos *et al.*, 1998). Kaloshian *et al.* (1995) also reported that *Mi* gene is tightly linked with *Meu-1* gene responsible for resistance to potato aphid (*Macrosiphum euphorbiae*) in tomato. Garcia *et al.* (2007) developed co-dominant SCAR (Sequence Characterized Amplified Region) marker *Mi23* linked with *Mi-1.2* gene (Godzina *et al.*, 2010).

Globodera rostochiens is another nematode on tomato. Resistance gene *Hero* has been introgressed from wild species *Lycopersicon pimpinellifolium* into the cultivated tomato. Ganal *et al.* (1995) mapped *Hero* gene between the RFLP markers H293 and CT229. The RFLP marker HR1 is closely linked to *Hero* gene.

The genes/QTLs for major diseases and insect pests of tomato and markers specific to them are depicted in Table 20.

Table 20: Genes/QTLs for Major Diseases and Insect Pests of Tomato and DNA and Isozyme Markers Specific to them

Disease	*Pathogen*	*QTL/ Gene*	*Marker*	*Chromo-some*	*Reference*
Root-Knot nematode	*Meloidogyne* spp.	*Mi, Mi-1.2*	Isozyme, CAPS, SCAR	6	Rick *et al.*, 1974; Gilbert *et al.*, 1955; Williamson *et al.*, 1994; Garcia *et al.*, 2007
Nematode (potato cyst)	*Globodera rosto-chiensis*	*Hero*	RFLP	4	Ganal *et al.*, 1995
Alternaria stem canker	*Alternaria alter-nata* f. sp. *lyco-persici*	*Asc*	RFLP	3	Grogan *et al.*, 1975; Van der Biezen *et al.*, 1995; Mesbah *et al.*, 1999
Black mold	*Alternaria alter-nata*	QTL	RFLP, CAPS	2, 3, 9, 12	Robert *et al.*, 2001
Early Blight	*Alternaria solani*	7 QTL	RFLP	3, 4, 5, 6, 8, 10, 11	Zhang *et al.*, 2003
Fusarium wilt	*Fusarium ox-ysporum* f. sp. *lycopersici* race 1, race 2	*I1, I2, I3*	RFLP, CAPS, SCAR	7, 11	Sarfatti *et al.*, 1991; Staniaszek *et al.*, 2007; Sarfatti *et al.*, 1989; Lim *et al.*, 2006; Barillas *et al.*, 2008
Fusarium crown and root rot	*Fusarium ox-ysporum* f. sp. *radicis-lycopersici*	*Frl*	RAPD	9	Fazio *et al.*, 1999
Grey leaf spot	*Stemphylium* spp.	*Sm*	RFLP	11	Behare *et al.*, 1991
Late blight	*Phytophthora infestans* (Mont.) deBary	*Ph-2Ph-3*	RFLP, RAPD, SCAR	10, 7	Moreau *et al.*, 1998,YiPeng *et al.*, 2009,PueHee *et al.*, 2010
Anthracnose ripe rot	*Colleotrichum coccodes*	QTL	RAPD	several chromo-somes	Stommel *et al.*, 1998

Disease	*Pathogen*	*QTL/ Gene*	*Marker*	*Chromo-some*	*Reference*
Verticillium wilt	*Verticillium dahliae*	*Ve*	SCAR, RFLP	9	Kawchuk *et al.*, 1998; Diwan *et al.*, 1999
Powdery mildew	*Oidium lycopersici*	*Ol-1, Ol-2*	RAPD, RFLP, SCAR, CAPS	6	Van der Beek *et al.*, 1994; Huang *et al.*, 2000; Bai *et al.*, 2003
	Leveillula taurica	*Lv*	RFLP, RAPD	12	Chunwongse *et al.*, 1994
Leaf mold in tomato	*Cladosporium fluvum*	*Cf-2*	RFLP	6	Dickinson *et al.*, 1993
Bacterial spot	*Xanthomonas euvesicatoria, X. vesicatora, X. peronars, X. gardneri*	QTL: Rx1, Rx2, Rx3, *Bs4*	RFLP, CAPS, STS	1, 5	Yang *et al.* 2007; Yu *et al.*, 1995
Bacterial speck	*Pseudomonas syringae* pv. *Tomato* (Pst)	*Pto*	RFLP, RAPD	5	Martin *et al.*, 1991;Martin *et al.*, 1993
Bacterial canker	*Clavibacter michiganensis* ssp. *michiganensis*	QTL (Rcm5.0, Rcm7.1, Rcm9.0, Rcm2.0, Rcm5.1)	RFLP, SCAR	2, 5, 7, 9	Van Heusden *et al.*, 1999; Kabelka *et al.*, 2002
Cucumber mosaic viruse	CMV	*Cmr*	RFLP	12	Stamova *et al.*, 2000
Tomato spoted wilt virus	TSWV	*Sw-5*	RAPD, RFLP, SCAR	9	Stevens *et al.*, 1995;Stevens *et al.*, 1996
Tabaco mosaic virus	ToMV	*Tm-1, Tm-2a*	RFLP, SCAR	2, 9	Ohmori *et al.*, 1996; Young *et al.*1988; Schroeder *et al.*, 1967; Tanksley *et al.*, 1992
Tomato yellow leaf curl virus	TYLCV	*Ty-1, Ty-2, Ty-3*	RFLP, CAPS	6, 11	Zamir *et al.*, 1994; Hanson *et al.*, 2000; Hanson *et al.*, 2006; Castro *et al.*, 2007

18.2.4 Identification of Resistance to Abiotic Stresses

Vallejos and Tanksley (1983) identified three QTLs responsible for growth at low temperatures. They were located on chromosome 6, 7, 12. Truco *et al.* (2000), working with the BC_1 generation and using RFLP markers mapped ten QTL associated with tolerance to low temperature.

Researchers have undertaken the identification of QTLs responsible for tolerance to salinity stress at different stages of plant development, germinating seeds, vegetative growth, mature plant (Foolad, 2007a). Foolad *et al.* (2001), using RFLP markers, located five QTLs on chromosomes 1, 3, 5, 6, 11, associated with tolerance to salinity stress. The source of tolerance was *L. pimpinellifolium* (Jusl.)

Mill. LA722. Salt tolerant species (*L. pennellii, L. peruvianum, L. cheesmanii*) are able to accumulate more potassium and sodium ions than sensitive species (Rush and Epstein, 1981; Tal and Shannon, 1983). Zamir and Tal (1987) using isoenzyme markers identified four loci associated with the uptake of Na^+ and Cl^- (enzymes *Prx-1, Prx-2, Tpi-2, Pgm-2, Aco-1*) and two loci responsible for uptake of K^+ (enzymes *Asp-1, Got-2*).

Drought is another important factor limiting the yield potential of tomatoes in the world. Using RFLP markers, four QTL responsible for this trait have been mapped, two of them, located on chromosome 1 and 9 (derived from *L. pimpinellifolium*) showed a stronger effect than the other two QTL (chromosome 8 and 12) derived from *L. esculentum* (Foolad *et al.*, 2003).

The genes/QTLs for major abiotic stresses of tomato and the markers specific to them are presented in Table 21.

Table 21: Genes/QTLs for Major Abiotic Stresses of Tomato and DNA and Isozyme Markers Specific to them

Stress	*QTL/Gene*	*Markers*	*Chromosome*	*Reference*
Cold tolerance	QTL	RFLP	1, 3, 5, 6, 7, 9, 11, 12	Truco *et al.*, 2000; Vallejos and Tanksley, 1983
Salt	QTL	RFLP, Isozyme	1, 3, 5, 6, 11	Foolad, 2007a; Foolad *et al.*, 2001; Zamir and Tal, 1987; Rush and Epstein, 1981; Tal and Shannon, 1983
Drought	QTL	RFLP	1, 9, 8, 12	Foolad *et al.*, 2003

18.2.5 Improvement of Fruit Quality

Five QTLs, *ao-6-2, ao7-2, ao10-1, ao6-3, ao-3-3* have been mapped for the makeup of total antioxidant capacity of the water-soluble fraction in tomato and they are located on chromosome 3, 6, 7 and 10, while for the synthesis of ascorbic acid, six corresponding QTLs *aa3-3, aa3-4, aa10-1, aa5-3, aa12-3, aa12-4* were mapped on chromosome 3, 5, 10, 12, while two QTLs *bc6-2, bc6-3* for β-carotene were located on chromosome 6, and the QTL donor was a wild species *L. pennellii* (Rousseaux *et al.*, 2005).

In order to locate the QTLs responsible for the main sugars and organic acids in tomato the candidate gene approach with RFLP markers was used. Seven QTL *ca-4F, ca-5E, ca-7B, ca-8E, ca-9F, ca-10B, ca-10E* were reported to be responsible for the synthesis of citric acid, for malic acid five QTLs, *ma-3FG, ma-4G, ma-4I, ma-8E, ma-12D*, for fructose four QTLs, *fru-4I, fru-5G, fru-7F, fru-9D* and for glucose four QTLs, *glu-4I, glu-5G, glu-9D, glu-12D* (Causse *et al.*, 2004).

Using an AB-QTL mapping strategy (Backcross Advanced QTL mapping) and RFLP markers the quantitative trait loci responsible for quality of tomato fruit (size, weight, shape and color) have been identified (Szezechura *et al.*, 2011). Tanksley and Nelson (1996) were the first to apply AB-QTL to tomato. They found three QTLs, *fw3.1, fw10.1, fw12.1* to be responsible for the fruit weight, four QTLs, *fsz2b.1*, *fsz3.1, fsz10.1, fsz12.1* for fruit size and three of these QTLs were closely linked to the loci of fruit weight. Four QTLs, *fs2b.1, fs8.1, fs10.1, fs12.1* were associated with fruit shape and two QTLs, *ec5.1, ec12.1* with external fruit color (Frary *et al.*, 2004).

In tomato two loci have been identified that suppress flower formation and fruit pedicel abscission zone and these genes are *j* and *j-2* (Zhang *et al.*, 1994; Butler 1936, Rick, 1957). A jointless locus (*j*) was mapped between two RFLP markers TG523 and TG194 in a F_2 population of crosses between *L. esculentum* and *L. pennellii*. Additionally, five RAPD markers were mapped on chromosome 11. One of these, RPD158, was mapped to the opposite side of a *jointless* relative to TG523 (Wing *et al.*, 1994). A RAPD marker RPD140 linked to *j-2* gene was identified. For a chromosomal location of this locus, researchers converted RPD140 marker to RFLP marker, then mapped it on chromosome 12 (Zhang *et al.*, 2000). The genes/QTLs for tomato fruit quality and the different markers specific to them are presented in Table 22.

Table 22: Genes/QTLs for Tomato Fruit Quality and DNA and Isozyme Markers Specific to them

Traits	*QTL/Gene*	*Markers*	*Chromosome*	*Reference*
Antioxidant capacity	5 QTL	RFLP	3, 6, 7, 10	Rousseaux *et al.*, 2005
Ascorbic acid	6 QTL	RFLP	3, 5, 10, 12	Rousseaux *et al.*, 2005
β-carotene	2 QTL	RFLP	6	Rousseaux *et al.*, 2005
Citric acid	7 QTL	RFLP	4, 5, 7, 8, 9, 10	Causse *et al.*, 2004
Malic acid	5 QTL	RFLP	3, 4, 8, 12	Causse *et al.*, 2004
Fructose content	4 QTL	RFLP	4, 5, 7, 9	Causse *et al.*, 2004
Glucose content	4 QTL	RFLP	4, 5, 9, 12	Causse *et al.*, 2004
Fruit weight	3 QTL	RFLP	3, 10, 12	Frary *et al.*, 2004
Fruit size	4 QTL	RFLP	2, 3, 10, 12	Frary *et al.*, 2004
Fruit shape	4 QTL	RFLP	2, 8, 10, 12	Frary *et al.*, 2004
Fruit color (external)	2 QTL	RFLP	5, 12	Frary *et al.* 2004
Jointless	*j, j-2*	RFLP, RAPD	11, 12	Zhang *et al.*, 1994; Wing *et al.*, 1994; Zhang *et al.*, 2000

18.2.6 Ripening and Fruit Development

Heidrich *et al.* (1983) separated eight peroxidase isozyme bands following partial purification of the enzyme from ripe red fruits (cvs. unspecified). Research elsewhere indicated that the number of isozymes present depends on the stage of fruit maturity.

The activity of polygalacturonase (PG), a fruit ripening-specific cell wall pectinase, has been strongly implicated as a key determinant of softening during tomato fruit ripening and is hypothesized to have a more general role in the regulation of maturation of the fruit as a whole (Giovannoni *et al.*, 1993).

Della Penna *et al.* (1990) described the construction and expression of a chimeric gene that allowed developmentally regulated the expression of polygalacturonase in ripening-impaired mutant (*rin*) tomato fruit. It was shown that expression of the chimaeric polygalacturonase gene in *rin* tomato fruit resulted in the accumulation of all 3 polygalacturonase isoenzymes (PG1, PG2A and PG2B). Polyuronide solubilization and polyuronide depolymerization both reached their maximal levels in transgenic *rin* fruit prior to the appearance of PG2 isoenzymes.

18.2.7 Mechanical Harvesting

For mechanized production of tomato, compact and highly determinate cultivars suitable for once-over machine harvest are necessary. Most of the fresh market tomato cultivars for field production are determinate, although with larger vine than processing types.

The determinate growth habit in tomato was first reported by Yeager (1927) and the gene *self-pruning* controlling it (*spsp* = determinate) was first characterized by McArthur (1932). The *sp* gene was originally mapped onto the short arm of tomato chromosome 6 using a classical linkage map of tomato and later it was mapped to the same location on the tomato molecular linkage map (Grandillo and Tanksley, 1996; Patterson *et al.*, 1988). The *sp* gene was fine mapped, cloned, and physically characterized by Carmel-Goren *et al.* (2003) and Pnueli *et al.* (1998). The introduction of the *sp* allele into processing tomato cultivars transformed the industry by creating a major modification in plant architecture.

18.3.0 Genetic Transformation

18.3.1 Methods of Genetic Transformation

Various methods of genetic transformation have been achieved in tomato. They are (i) *Agrobacterium*-mediated transformation, which is most widely used method of gene transfer (Fillatti *et al.*, 1987; Agharbaoui *et al.*, 1995), particularly *A. rhizogenes* (Shahin *et al.*, 1986; Lipp-Joao and Brown, 1994), (ii) pollen-mediated transformation (Chesnokov *et al.*, 1989), (iii) protoplast transformation (Jongsma *et al.*, 1987), (iv) microinjection (Ilyubaev *et al.*, 1988), (v) direct transformation through seeds (Glavinich *et al.*, 1988), (vi) microprojectile (Baum *et al.*, 1997), (vii) electroporation (Nakata *et al.*, 1992) and (viii) liposome-mediated transformation (Rosenberg *et al.*, 1990).

As with other dicotyledonary crops, *Agrobacterium*-mediated transformation is most widely used method of gene transfer in tomato also. A wide range of explants is successfully transformed. Although transformation through cotyledon has been the most common, other explants like leaf, hypocotyls and stem are also used with high transformation frequency.

Root transformation was reported by using *A. rhizogenes* (Shahin *et al.*, 1986). They infected punctured tomato cotyledons with *A. rhizogenes* strain A4 containing the binary vector pARC8, which carried a chimaeric *nptII* gene conferring kanamycin resistance. nptII activity was detected in transformed calli and in transgenic plants, all of which were phenotypically normal and fertile and set seeds.

Enhanced transformation of tomato root clones cocultivated with *A. tumefaciens* in the presence of 20 μM acetosyringone was observed by Lipp-Joao and Brown

Genetic Transformation in Tomato

Different Stages of *Agrobacterium*-Mediated Transformation of Tomato.

In vitro-germinated seedlings on GM2. (a) Pre- cultured cotyledon explants on PCM. (b) Shoot initiation on selection medium after 5 wk. (c) Non-transformed cotyledons on selection medium after 4 wk. (d) Shoot development from transformed callus on selection medium after 12 wk of culture. (e) Multiple shoots from regenerating callus on selection medium. (f) Well-rooted plantlets on selection medium. (g) Acclimatization of a transgenic plant. (h) Mature transgenic tomato T0 plants under containment conditions bearing normal fruits (Rai ***et al.,*** 2012).

(1993). For cotyledon explant, 50 mg/l kanamycinin selection medium and 200 mg/l cefotaxime to remove *Agrobacterium* after three days of co-cultivation was suitable for improved transformation efficiency (Ryu *et al.*, 1998).

Chaithra *et al.* (2015) transformed tomato cultivar Pusa Ruby by biolistic gun method to introgress *Cry2Ax* gene that was carried by binary vector pCAMBIA2300 with the help of EnCAMV35S promoter assembled on to disarmed *Agrobacterium tumefaciens* strain LBA4404. Leaf explants from 8-day old seedling were used and callus from them were transformed with GUS reporter gene by biolistic gun method. The putative transformants has similar morphology as their non-transgenic counterparts, but showed slow growth and finally regeneration of two plants was possible.

Different tomato cultivars have been transformed to express *cry1Ab, cry1Ac, cry2A, cry2Ab* genes giving effective control of tomato fruit borer, but the final outcome of transformation frequency varied as reported as 6 per cent in Pusa Ruby (Vidya *et al.*, 2000), 40 per cent in Micro Tom (Sun *et al.*, 2006), 18 per cent in Arka Vikas (Hanur *et al.*, 2015) through *Agrobacterium*-mediated transformation. The success of transformation is determined by the gene integration and its optimum expression in the plant and its successive generations. Biolistic gun method has been tried by some workers and up to 75 per cent transformation has been reported when bombarded twice at a distance of 9 cm at 1100 psi (Chaithra *et al.*, 2015), although low regeneration of transformed plants may be the result of injury following the technique (Sidhu *et al.*, 2014) which has not made this technique much popular.

18.3.2 Disease and Insect Resistance

Disease resistance: Two strategies *viz.*, coat protein and satellite RNA have been used to obtain transgenic plants resistant to viral diseases.

Nelson *et al.* (1987) produced transgenic tomato plant expressing the coat protein (CP) of the common (UI) strain of tobacco mosaic virus (TMV). Kunik *et al.* (1994) cloned the tomato yellow leaf curl (bigemini) virus (*tylcv*) gene that encodes the capsid protein (*vi*) into an *Agrobacterium Ti*-derived plasmid. This plasmid was used to transform plants from an interspecific tomato hybrid *L. esculentum* x *L. pennellii* (F_1) which is sensitive to TYLCV disease.

Yang *et al.* (1995) regenerated 42 transgenic tomatoes from leaf discs following *A. tumefaciens*-mediated transformation with a chimeric gene containing a cloned cDNA encoding the cucumber mosaic cucumovirus (CMV) coat protein (*cmv-cp*). The *cmv-cp* gene was stably integrated into the genome of the transgenic plants and expressed in plant cells. SreeVidya *et al.* (2000) introduced the coat protein gene of *Physalis* mottle tymovirus (PhMV) through *Agrobacterium* transformation of cotyledonary leaves. The T_1 progenies segregated in monogenic ratio showing Mendelian inheritance.

Insect resistance: Fischhoff *et al.* (1987) determined the structure of an insect control protein gene from *Bacillus thuringiensis* subsp. *kurstaki* strain HD1 and generated truncated forms of the gene. Two of these truncated genes were incorporated into vectors and *via Agrobacterium*-mediated transformation were introduced singly into tomato plants. The chimeric genes were expressed in the transgenic plants, conferring tolerance to lepidopteran larvae. Larvae of *Manduca sexta, Heliothis zea* and *Heliothis virescens* exhibited up to 100 per cent mortality when fed on leaves of the transformed plants. The introduced gene was stably transmitted to the plants, progeny segregating as a single dominant Mendelian factor.

Salm *et al.* (1994) produced transgenic tomato plants exhibiting insect resistance due to the introduction of modified *cry1A(b)* and *cry1C* genes of *B. thuringiensis*. Limited modifications at selected regions of the coding sequences of both genes were sufficient to obtain resistance against *Spodoptera exigua, Heliothis virescens* and *Manduca sexta*.

Insecticidal crystal proteins from *Bacillus thuringiensis* (*viz., Cry1Ac, Cry1Ab, Cry2A, Cry2Ax1*) have been successfully engineered into tomato for resistance against tomato fruit borer and *Cry1Ac* has been found to be highly efficient against tomato fruit borer due to its binding ability to different receptors (Bravo *et al.*, 2004; Purcell *et al.*, 2004), though the rate of pore formation is lesser as compared to *Cry1Ab* (Kato *et al.*, 2006).

Much work has been done to develop modified tomato with *Cry1Ac* and Mandaokar *et al.* (2000) used a synthetic *cry1Ac* gene modifed for plant codon usage and possessing G + C content of 47.7 per cent and introgressed into tomato variety Pusa Ruby by *Agrobacterium tumefaciens* strain carrying pBinBt3 to transform the cotyledonary explants. High level of lepidopteran insect resistance was obtained in the putative transformants.

Koul *et al.* (2014) used a modified truncated version of *cry1Ab* gene for the development of transgenic plants over expressing *Bt Cry1Ab* toxin in the popular commercial tomato variety Pusa Early Dwarf through *Agrobacterium tumefaciens* carrying binary vector pBIN200 harbouring modified and truncated 1845 bp *cry1Ab* gene followed by extensive screening and selection of progenies initiating at T_0 transgenic plants. Stable transgenic tomato lines with high levels of expression were obtained in T_4 generation with high resistance to lepidopteran especially *Helicoverpa* insects and in T_6 there was no retardation in plant growth, development and fruit yield in comparison to non-transgenic plants.

Hanur *et al.* (2015) transformed Arka Vikas, a popular tomato cultivar of India, using hypocotyl explant through *Agrobacterium* strain EHA105 mediated transformation. Bt *Cry2A* gene was introgressed through *Agrobacterium* along with 35S CaMV promoter, OCS terminator and the selectable *nptII* marker. The laboratory assays revealed that moderate to high lepidopteran insect resistance in the transgenic *Bt* plants.

18.3.3 Herbicide Resistance

Fillatti *et al.* (1987) regenerated over 100 transgenic plants containing 2 *nptII* genes and mutant *aroA* gene which conferred tolerance to glyphosate and 80 per cent of these both exhibited NPTII enzyme activity and produced the mutant AROA protein. Progeny of self-pollinated AROA positive plants were tolerant to glyphosate at concentrations of 0.84 kg/ha. The tolerance phenotype segregated in a manner indicating Mendelian inheritance.

18.3.4 Fruit Ripening

Major progress has been made over the last few years in the identification and regulation of tomato ripening genes. At least 25 genes showing elevated expression during ripening have been cloned and several including polygalacturonase which modifies fruit texture have been found to be ripening-specific (Grierson and Fray, 1994). Inhibition of expression of polygalacturonase, pectinesterase, ACC synthase, ACC oxidase and phytoene synthase has been achieved in transgenic plants using antisense technology. The Flavr Savr™ tomato was developed using antisense RNA to regulate the expression of polygalacturonase (PG) in the ripening fruit. This enzyme is one of the most abundant proteins in ripe tomato fruit and has long been thought responsible for softening in ripe tomatoes. The Flavr Savr™ tomato is the first genetically engineered wholefood to be sold in commerce (Kramer and Redenbaugh, 1994). Following is the details of some of the approaches used for altering ripening in tomato.

Nagata *et al.* (1995) examined the ripening characteristics of tomatoes transformed with the ACC synthase gene in sense or antisense orientation. Tomato plants (cv. Alisa Craig) were transformed by *A. tumefaciens*. Forty-three independent antisense and 26 sense transformants were obtained. A marked retardation in red colour development was observed in one antisense transformant and was reversed by treatment of fruits with ethylene.

Redenbaugh *et al.* (1992) considered the different aspects of food safety of transgenic tomato Flavr Savr and concluded that the tomatoes are food and no unintended changes were detected in Flavr Savr tomatoes to affect their status as tomatoes. The Flavr Savr tomato only differs in terms of characters related to pectin and the presence of the novel *kanr* gene and APH (3′)II.

Phytoene synthase from the bacterium *Erwinia uredovora* (*crt*B) has been overexpressed in tomato (*Lycopersicon esculentum* Mill. cv. Alisa Craig) as reported by Fraser *et al.* (2002). Fruit-specific expression was achieved by using the tomato polygalacturonase promoter, and the CRTB protein was targeted to the chromoplast by the tomato phytoene synthase-1 transit sequence. Total fruit carotenoids of primary transformants (T_0) were 2–4-fold higher than the controls, whereas phytoene, lycopene, β-carotene, and lutein levels were increased 2.4-, 1.8- and 2.2-fold, respectively. The biosynthetically related isoprenoids, tocopherols plastoquinone and ubiquinone, were unaffected by changes in carotenoid levels. The progeny (T_1 and T_2 generations) inherited both the transgene and phenotype. Determination of enzyme activity and Western blot analysis revealed that the CRTB protein was plastid-located and catalytically active, with 5–10-fold elevations in total phytoene synthase activity. Metabolic control analysis suggested that the presence of an additional phytoene synthase reduced the regulatory effect of this step over the carotenoid pathway. The activities of other enzymes in the pathway (isopentenyl diphosphate isomerase, geranylgeranyl diphosphate synthase, and incorporation of isopentenyl diphosphate into phytoene) were not significantly altered by the presence of the bacterial phytoene synthase.

The antisense-mediated decrease in either of two tomato ethylene receptors, *LeETR4* or *LeETR6*, results in premature ripening (Tieman *et al.*, 2000; Kevany *et al.*, 2007).

A construct was generated consisting of an *LeETR4* RNAi inverted repeat sequence fused to the transcriptional promoter of *Tfm7*, a gene that is expressed specifically in immature fruits (Santino *et al.*, 1997), to specifically reduce *LeETR4* expression throughout fruit development and transgenic plants were generated by *Agrobacterium*-mediated transformation into the tomato cultivar Flora-Dade, a large-fruited fresh market variety, by Kevany *et al.* (2008). Six T_0 lines with significantly earlier ripening were identified in initial screening. Of these, three T_1 lines that exhibited both a decrease in time from anthesis to breaker stage and a reduction in *LeETR4* transcript throughout fruit development were chosen for further characterization. Transgenic lines began ripening between 5 and 7 days earlier than controls. No significant effects were observed on the time from breaker stage to fully ripe, nor were there differences in the colour of ripe fruits. *LeETR4* transcript levels were reduced by as much as 73 per cent in immature fruit and 95 per cent in ripening fruit. Although *Tfm7* expression has been reported to be specific to immature fruit, the RNAi effect persisted into ripening fruit. This gene specific decrease in expression was not seen in non-target tissues, such as leaves. Expression analysis of other family members showed no decrease in transcript levels in transgenic plants. Protein blot analysis confirmed that LeETR4 protein levels were correspondingly decreased at all stages of fruit development relative to non-transgenic control fruit.

19.0 References

Abad, M.K.R., Fathi, S.A.A., Nouri-Ganbalani, G. *et al.* (2020) *Int. J. Trop. Insect Sci.*, **40:** 39–48.

Abak, K., Bascetincclik, A., Baytorun, N., Altuntas, O. and Ozturk, H. H. (1994) *Acta Hortic.*, **366**: 149-54.

Abd El-Gaid, M.A., Al-Dokeshy, M.H. and Nassef, D.M.T. (2014) *Asian J. Crop Sci.*, **6**(3): 254-261

AAbdalbagi, A.H., Salih, O.S. , Abdu, E. A. and Ahmed, M. E. N. (2010) *Res. J. Agric. Biol. Sci.*, **6**(5): 665-669.

Abdalla, A. A. and Verkerk, K. (1967) *Netherlands J. Agric. Sci.*;**16**: 71-76.

Abdalla, M. Y., Al Mihanna, A. A., Al-Rokibah, A. A. and Ibrahim, G. H. (1999) *Ann. Agril. Sci.*, **44**: 511-21.

Abdalla. A. A. and Verkerk, K. (1970) *Netherlands J. Agric. Sci.*, **18**: 105-110.

Abdel-Hafeez, A. T. (1971) *Netherland J. Agric. Sci.*, **19**: 67-75.

Abdel-Hafeez, A. T. and Verkerk, K. (1969) *Netherlands J. Agric. Sci.*, **17**: 50-59.

Abo-Elyousr, K.A.M. and Asran, M.R. (2009) *Arch. Phytopath. Plant Prot.*, **42**: 573-578.

Abou-Hadid, A. F., El- Beltagy, A. S., Gaofer, S. A. and Youssef, S. M. (1993) *Acta Hortic.*, **323**: 3 15-321.

Abou-Hadid, A. F., Zayed, A. M., El-Behairy, U. A. and El-Behairy, A. S. (1989) *Egyptian J. Hort.*, **16:** 111-118.

Abreu, P. E. and Meneses. J. F. (1994) *Acta Hortic.*, **366**: 183-194.

Abreu, P. E., Monteiro, A. A. and Meneses, J. F. (1994) *Acta Hortic.*, **366**: 195-200.

Abu-Gharbieh, W. I., Saleh, W. I. H. and Al-Banna, L. (1991) *FAO Plant Production and Protection Paper*, **109**: 69-77.

Abuye, F. and Achamo, B. (2016) *J. Biol., Agric. Healthcare*, **6:** 54-62

Acharya, B., Dutta, S., Dutta, S., and Chattopadhyay, A.(2018a) *Int. J. Veg. Sci.*, **24** (5): 407-423.

Acharya, B., Ghorai, A.K., Dutta, S., Maurya, P.K., Dutta, S., Mandal, A.K., Chattopadhyay, A. and Hazra, P. (2018b) *Australas. Plant Pathol.*, 47 (6): 591-600.

Acosta, M. T. (1981) *Herbicides Hortofructicultura*, p. 17.

Adams, P. (1986) In: *Mineral Nutreition, The tomato Crop* (Eds. J.G. Atherton and J. Rudich), Chapman and Hall, London.

Adams, P., Graves, C. J. and Winsor, G. W. (1978a) *J. Hort. Sci.*, **53**: 115-122.

Adams, P., Graves, C.J. and Winsor, G. W. (1978b) *Sci. Hort.*, **9:** 199-205.

Adams, S.R., Cockshull, K.E. and Cave, C.R.J. (2001) *Ann. Bot.*, **88:** 869-877.

Adamson, R. M. and Mass, F. F. (1971) *Can. Agric.*, **16:** 30-31.

Adamson, R. M. and Mass, F. F. (1976) *HortScience*, **11:** 212-213.

Adegoroye, A. S., Jolliffe, P. A. and Tung, M. *Food* A. (1989) *J. Sci. Agric.*, **49:** 95-102.

Adekiya, A. O. and Agbede, T. M. (2009) *Emir. J. Food Agric.*, **21**: 10-20

Adhikari, T. B. and Basnyat, R. C. (1998) *Canadian J. Plant Path.*, **20**: 283-7.

Adriaenssens, W. (1997) *Proeftuinnieuws*, **7:** 44-46.

Adriaenssens, W. (1998) *Proeftuinnieuws*, **8:** 22-23.

Adsule, P.G., Dan, A. and Tikko, S.K. (1980) *Indian Food Packer*, **34**: 18-20.

Afify, A. (1933) *Genetica*, **15:** 225-240.

Agar, L.T., Abak, K. and Yarri, G. (1994) *Acta Hortic.*, **368:** 742-753.

Agharbaoui, Z., Greer, A. F. and Tabaeizadeh, Z. (1995) *Plant Cell Rep.*, **15:** 102-105.

Ahammed, G.J., Zhang, S., Shi, K., Zhou, Y. H. and Yu, J. Q. (2012) *Plant Growth Regul.*, **68:** 87–96.

Ahirwar, C.S., Vijay, B. and Vinay. P. (2013) *Int. J. Agric. Sci.*, **9**(1): 172-176.

Ahluwalia, M. S. and Singh, B. (1966) *Indian J. Ecol.*, **23:** 93-98.

Ahmed, F.A., Sipes, B.A. and Alvarez, A.M. (2017) *Afric. J. Agric. Res.*, **12**: 684-691.

Ajenifujah-Solebo, S.O.A., Isu, N.A., Olorode, O., Ingelbrecht, I. and Abiade, O. O. (2012) *Afric. J. Plant Sci.*, **6,** doi: 10.5897/AJPS12.054.

Akbar, A., Din, S., Ahmad, M., Khan, G. and Alam, S. (2014) *J. Nat. Sci. Res.*, **4:** 2225-09921.

Akhtar, K. P., Matin, K., Mirza, J. 1-1., Shakir, A. S. and Rafique, M. (1994) *Pak. J. Phytopath.*, **6:** 125-129.

Akhtar, K. P., Shakir, A.S. and Sahi, S.T. (1999) *Pak. J. Phytopath.*, **11:** 25-29.

Akhtar, S. and Hazra, P. (2013) *Afr. J. Biotechnol.*, **12**: 2869-2875.

Akintokun, A. K., Ojesola, C. O., Akintokun, P. O. and Oloyede, A.R.(2019) *Nig. J. Biotech.*, **36**: 94-102.

Al-Amri, A. M. S. (1997) *J. King Saud. Univ., Agric. Sci.*, **9:** 21-38.

Alcantar, G., G., Villarreal R. M., and Aguilar S. A. (1999) *Acta Hortic.*, **366:** 385-391.

Alexander, L. J. and Hoover. M. M. (1955) *Ohio Agri. Exp. Sta. Res. Bull.*, **752:** 76.

Ali, A. A. and El-Said, N. M. H. (1987) *Arab J. Plant Protect.*, **5**: 24-30.

Ali, M.R., Mehraj, H. and Jamal Uddin, A. F. M. (2015) *J. Biosci. Agric. Res.*, **6**: 512-517.

Almasoum, A. A. (1998) *South Indian Hort.*, **46:** 35-8.

Almunqedhi, B.M., Kassem, H.A., Al-Harbi, A.R. (2017) *J. Scientific Engg. Res.*, **4**: 114-122

Aloni, I. J. and Rosenshtein, G. (1982) *Physio. Plant.*, **56:** 513-17.

Altuhaish, A., Altuhaish, A. and Tjahjoleksono, A. (2014) *Emirates J. Food Agric.*, **26:** 716-722.

Alvarenga M.A.R. (2004) Tomate: produção em campo, em casa-de-vegetação e em hidroponia. Lavras: Editora UFLA, p. 400.

Alvino, A. D., and Ria, R. and Zerbi, G. (1986) *Acta Hortic.*, **366**: 117-124.

Amaral, A. T. J. Jr. do., Casali, V. W. D., Finger, F. L., Cruz, C. D. and Amaral, A. T. A. Jr. do (1997) *Ciencia-e-Agrotechnologica*, **21:** 313-23.

Amin, H.P., Czank, C., Raheem, S., Zhang, Q.Z., Botting, N.P., Cassidy, A. and Kay, C.D. (2015) *Mol. Nutr. Food Res.*, **59**: 1095–1106.

Amini, J. and Sidovich, D.F. (2010) *J. Plant Prot. Res.*, **50**: 173-178

Amoako, J. and Yeboah-Gyan, K. (1991) *Acta Hortic.*, **366:** 225-259

Anbinder, I., Reuveni, M., Azari, R., Paran, I., Nahon, S., Shlomo, H., Chen, L., Lapidot, M. and Levin, I. (2009). *Theor. Appl. Genet.*, doi: 10.1007/s0012-009-1060-Z.

Andrus, C. F. (1946) *Pl. Dis. Reptr.*, **30:** 269-270.

Andrus, C. F. and Ruehle, G. D. (1953) *Markel Grower J.*, **82:** 14-15.

Andrus, C. F., Reynard, C. B., Jorgensen, H. and Eades, J. (1942) *J. Agric. Res.*, **65:** 339-346.

Andrus, C.F., Reynard, G.B. and Wade, B.L. (1942) *US Dep. Agric Circ.*, p. 652.

Angadi, V., Rai, P.K. and Bara, B.M. (2017) *J. Pharmacognosy Phytochem.*, **6**: 807-810.

Anjaneyulu, K., Murthy, S. V. K. and Iyengar, B. R. V. (1998) *J. Nuclear Agric. Biol.*, **27:** 122-125.

Anju Kumari, Bhardwaj, M. L. and Guleria, S. P. S. (1993) *Hort. J.*, **6**: 89-92.

Ankara, S. and Arin, L. (1998) *Bache*, **27:** 35-46.

Anonymous (1998) *Pyrethrum Post*, **20:** 29-3 l.

Anonymous (2018) In: *Horticultural statistics at a glance.* Department of Agriculture and Cooperation, Govt. of India, New Delhi.

Anonymous (2020) https: //njaes.rutgers.edu/tomato-varieties/#I accessed on 15th July, 2020.

Antignus, Y., Cohen, S., Mor, N., Masika, Y. and Lapidot, M. (1996) *Plasticulture*, **112:** 15-20.

Antoniou, P. P., Tjamos, E. C. and Panagopoulos, C. G. (1995) *Plant Pathol.*, **44:** 438-47.

Apahidean, S., Indrea, D., Apahidean, M. and Paven, I. (1991) *Buletinal Institutului Agronomic Cluj-Napoca. Seria Agricultura si Horticultura*, **45:** 53-57.

Apahidean, S., Indrea, D., Apahidean, M., Paven, I. and Step, I. (1989) *Bulletin Institutului Agro-nomic Cluj-Napoca Seria Agricultura*, **43:** 49-54.

Apodaca-Sanchez, M. A, Zavaleta-Mejia, E. and Garcia Espinoza, R.O. (1994) *Fitopotologia*, **29**: 86-93.

Ara, N., Bashar, M.K., Begum, S. and Kakon, S.S. (2007) *Int. J. Sustain. Crop Prod.*, **2**: 35-39

Araki, Y. (1994) *J. Japanese Soc. Hort. Sci.*, **63:** 91-97.

Aramov, M. and Dzhuraeva, L. (1990) *Sel'skoe Khozyaistvo Uzbekistana*, **1:** 35-36.

Argerich, C. A., Bradford, K.J. and Tarquis, A. M. (1989) *J. Exptl. Bot.*, **40:** 593-598.

Arora, S. K., Pandita, M. L. and Pratap, P. S. (1981) *Haryana J. Hort.* Sci., **10:** 102-106.

Artes, F. and Escriche, A. J. (1994) *J. Food Sci.*, **59:** 1053-1056.

Artes, F., Garcia, F., Marquino, J., Cano, A. and Fernandez-Trujillo, J. P. (1998b) *Postharvest Biol. Technol.*, **14**: 283-296.

Artes, F., Sanchez, E. and Tijskens, L. M. M. (1998a) *Lebensmittel Wissen Schaft and Technologic*, **31:** 427-431.

Arushi, Malannavar, A.B. and Banyal, D.K. (2018) *Int. J. Curr. Microbiol. App. Sci.* **7**(07): 21-31.

Asada, S. and Ono, M. (1996) *Appl. Entomol. Zool.*, **31**: 581-586.

Asada, S. and Ono, M. (1997) *Acta Hortic.*, **437:** 289-292.

Asahira, T. (1974) In: *The Plant Root and its Environment* (Ed. E. W. Garson), Charlottesville, U.S.A., pp. 29-61.

Asahira, T. and Kelly, W. C. (1966) *Proc. Amer. Soc. Hort. Sci.*, **89:** 563-570.

Asahira, T., Bryan, H.H. and Byrnc, J. M. (1975) *J. Amer. Soc. Hort. Sci.*, **100**: 19-22.

Asamizu, E. and Ezura, H. (2009) *J. Jap. Soc. Hort. Sci.,* **78** (1)**:** 3–5

Atanasov, N. D. (1995) *Acta Hortic.*, **412:** 546-550.

Atanassova, B. and Georgiev, H. (2007) In: *Genetic improvement of Solanaceous crops* (Eds. M.Razdan and A.K.Mattoo), Vol. 2. Science Publishers, Enfield, USA, pp. 113–151.

Atanassova, B. and Hazra, P. (2012) In: *Hand book of Vegetables* (Eds. K. V. Peter and P. Hazra), Studium Press LLC, Houston, Texas, USA, pp. 533-593.

Atanassova, B., Balacheva, E., Molle, E. and Georgiev, Hr. (2008) *Acta Hortic.*, **789:** 49–54.

Atta-Aly, M. A., Riad, G. S., El-Lacheene, Z. S. and El-Beltagy, A. S. (1999) *J. Plant Growth Regul.*, **18:** 15-24.

Attitalla, I. H., Quintanilla, P. and Brishammar, S. (1998) *Acta Phytopathologia et Entomologica Hungarica*, **33:** 89-95.

Aung, L. N. (1978) *1st International Symposium on Tropical Tomato*, pp.79-93.

Averswald, I. I., Drews, M. and Krumbein, A. (1996) *Gartenbau Wssenschaft*, **61:** 77-83.

Awas, G., Abdisa, T., Tolosa, K. and Chali, A. (2010) *Afr. J. Agric. Res.*, **6**: 2978-2981.

Aykroyd, W. R. (1963) *ICMR Special Series*, 42.

Ayub, M., Khan, M. Khan, and Amin, M. (1998) *Sarhad J. Agric.*, **14:** 475-481.

Ayuso, M.C., Nuez, F., Costa, J. and Cuartero, J. (1987) *Investigacion Agraria, Production Protecion Vegetales*, **2:** 251-259.

Ayyar, T.V.R. (1963) In: *Handbook of Economic Entomology for South India*, Govt. of Madras, pp.263.

Azanza, F., Young, T. E., Kim, D., Tanksley, S. D. and Juvik, J. A. (1994) *Theor. Appl. Genet.*, **87**: 965–972.

Baboy, L. and Sabiti, K. (1994) *Tropicultura*, **12:** 95-97.

Badekar, H. (1977) *Veg. Sci.*, **4**: 1-5.

Baevre, O. A. (1990) *Norwegian J. Agric. Sci.*,**4:** 225-32.

Bai, Y., Huang, C.H., Van der Huist, R., Meijer-Dekens, F., Bonnema, G. and Lindhout, P. (2003) *MPMI*, **16**: 169-176.

Bailey, L. I. I. (1949) In: *Manual of Cultivated Plants* (2nd edn.), The Macmillan Co., New York.

Bailey, R. M. (1941) *Proc. Amer. Soc. Hori. Sci.*, **38:** 573-575.

Bains, P. S., Bennypaul, I.I.A. and Mina, M. (1999) *Plant Dis.*, **83:** 488.

Bajpai, P. N. and Bajpai, V. K. (1963) *Allahabad Farmer*, **37:** 1-4.

Bakengesa, J.A. (2016) *Thesis in Master of Science in Nematology*, Univ. of Gent, p.31.

Bakker, J. C. (1989) *Groenten en Fruit*, **45:** 49.

Baldwin, E.A. (1998) *Proc. Florida State Hort. Soc.*, **111:** 103.

Balibrea, M.E., Martinez-Andujar, C., Cuartero, J., Bolarin, M.C. and Perez-Alfocea, F. (2006) *Funct. Plant Biol.*, **33**: 279–288.

Ballvora, A., Schornack, S., Baker, B. J., Ganal, M., Bonas, U. and Lahaye, B.J. (2001) *Mol. Genet. Genomics* **266**: 639-645.

Bambhaniya, V.S., Khanpara, A.V. and Patel, H.N. (2018) *J. Pharmacognosy Phytochem.*, **7**: 1471-1479

Banda, H. J. and Paxton, R. J. (1991) *Acta Hortic.*, **288:** 194-198.

Bandopadhyay, C.R., Bid, N.N. and Das, R.K. (1972) *Proc. 3rd Int. Symp. Subtrop. and Trop. Hort.*, Bangalore, pp.148-149.

Bano, F., Mahmud, T., Shah, S. M. and Awan, M. R. (1987) *Pak. J. Agric. Res.*, **8:** 195-198.

Baradaran-Anaraki, P. and Daneshvar, I.I. (1992) *Appl. Entom. Phytopath.*, **59:** 25-27.

Baranov, M. I. (1982) *Referativny i Zhurnal*, **9:** 55.347.

Baranwal, V. K. and Nasim, A. (1997) *Indian Phytopath.;* **50:** 297-299.

Barbolina, L. and Arkhipchenko, I. (1999) *Ekologiczne aspekty mechanizacji -nawozenia, Ochromy roslin upgrawy gleby I zbioru roslin uprawnych. Racenzowane Materialy VI Miedzynarodowego Symp.*, Warszawa, Polska, 23-24wrzesien1999r (Eds. E. Bien), pp. 76-82.

Barillas, A.C., Mejia, L., Sanchez-Perez, A. and Maxwell, D. (2008) *TGC Reports*, **58**: 11-17.

Barkai-Golan, R. and Kopeliowitch, E. (1986) *Hasso Eh.*, **66:** 2514-2516.

Barman, D., Shanna, C. K., Singh, I. P., Sardana, S. and De, L. C.(1998) *Int. J. Trop. Agric.*, **13**: 265-268.

Barone, A., Di Matteo, A., Carputo, D. andFruscinate, L. (2009) *Current Genomics*, 10: 1-9.

Barrantes, J. L. F. (1989) *Investigacion Agricola*, **3:** 15-17.

Barroso, M. R., Meneses, J. F. and Mexia, J. T. (1999) *Acta Hortic.*, **491:** 125-130.

Bartkowski, K. and Nowosielski, O. (1988) In: *Proc. Int. Cong. On Soilless Culture, Int. Soc. For Soilless Culture*, pp. 83-91.

Baruah, S. J. N., Deka, P. C., Nishio, T. and Dore, C. (1995) *Acta Hortic.*, **392:** 115-121.

Bar-Yosef, B. (1988) In: *Optimal Yield Management*, pp. 175-184.

Bascetincelik, A., Abak, K., Baytorun, N., Ozturk, H. O. (1994) *Acta Hortic.*, **66:** 141-148.

Bashan, Y., Okon, Y. and Henis, Y. (1987) *Canadian J. Bot.*, **65:** 366 - 372.

Batu, A. and Thompson, A. K. (1998) *Turk. J. Agri. Forestry*, **22:** 405-410.

Bauchet, G. and Causse, M. (2012) In: *Genetic diversity in plants*, (Ed. M. Caliskan), InTech Open, Rijeka, Croatia, pp. 133–162.

Baum, K., Groning, B. and Meier, I. (1997) *Plant J.* **12:** 463-469.

Baytorun, A. N., Topcu, S., Abak, K. and Dasgan, Y. (1999) *Gartenbau Wissenschaft*, **64:** 33-39.

Baytorun, N., Abak, K., Tokgoz, I.I. and Altuntas, O. (1994) *Acta Hortic.*, **366:** 125-132.

Beal, P. R. (1972) *Qd. Agric. J.*, **98**: 190.

Beckles, D.M., Hong, N., Stamova, L. and Luengwilai, K. (2012) *Fruits*, **67**: 49-64

Behare, J., Laterrot, H., Sarfatti, M. and Zamir, D. (1991). *Mol. Plant Microbe Interactions*, **4:** 489-492.

Behboudian, M. H. and Lai, R. (1994) *HortScience*, **29:** 1456-59.

Behboudian, M. H. and Tod, C (1995) *HortScience*, **30:** 490-491.

Belichki, I. (1983) *Gradmarka-i Lozarska Nauka*, **20:** 57-63.

Bellini, C., Chupeau, M. C., Gervais, M. Vastra, G. and Chupeau, Y. (1990) *Plant Cell Tissue Organ Cult.*, **23**: 27-37.

Benoit, F and Ccustermans, N (1998) *Proeftuinnieuwis*, **8**: 14

Benoit, F and Ceustermans, N (1986) *Revue de l' Agricultura*, **39:** 963-968.

Benoit, F and Ceustermans, N (1987) *Soilless Culture*. **3**: 3-7.

Bernacchi, D. and Tanksley, S. D. (1997) *Genetics*, **147**: 861–877.

Berry, S. Z. and Gould, W. A. (1988) *HortScience*, **23:** 930.

Berry, S. Z. and Uddin, M. R. (1991) In: *Genetic Improvement of Tomato*, (Ed. G. Kalloo), vol. 14, Springer, Berlin, Germany, pp. 197–206.

Bhandari, H.R., Srivastava, K. and Reddy, G.E. (2017) *Int. J. Curr. Microbiol. Appl. Sci.*, **6**(7): 4131-4138.

Bhandari, K. R. (1963) *Lal Bagh*, **8:** 5-6.

Bhardwaj, C. L., Thakur, D. R. and Jamwal, R. S. (1995) *Indian J. Agric. Sci.*, **65:** 148-511.

Bhattarai, K., Louws, F.J., Williamson, J.D. and Panthee, D. R. (2017) *Plant Path.*, **66** (7): 1103–1109.

Bhattarai, K., Louws, F.J., Williamson, J.D. and Panthee, D. R. (2016) *Horticulture Res.*, **3:** Article ID 16035.

Bhattarai, K., Sharma, S. and Panthee, D.R. (2018) *Int. Journal Agron.*, **2018**: Article ID 4170432, https: //doi.org/10.1155/2018/4170432.

Bhore, S. J., Nadgauda, R. S. and Gad re, R. V. (1999) *Indian J. Expt Biol.*, **37:** 102-3.

Bickers, U., Oerke, E.C. and Dehne, H.W. (1999) *12th International Reinhardsebrunn Symp. Germany*, pp.131-136.

Biró, B., Dudás, A., Wass-Matics, H., Kocsis, T., Pabar, S., Tóth, E., Szalai, M.Z. and Kotroczó, Z (2018) *Acta Agraria Debreceniensis*, (150), 93-105.

Bisaria, A.K. and Rastogi, G. R. (1988) *Madras Agric. J.*, **75**: 271-275.

Blaeser, P. and Steiner, U. (1999) *12th International Reinhardsebrunn Symp., Germany*, pp. 491-499.

Blain, J. (1998) *Agriculture*, Canada, **45:** 24-26.

Blanca, J., Montero-Pau, J., Sauvage, C., Bauchet, G., Illa, E., Díez, M.J., *et al.* (2015) *BMC Genomics*, **16**: 257, doi: 10.1186/s12864-015-1444-1.

Blando, F., Calabriso, N., Berland, H., Maiorano, G., Gerardi, C., Carluccio, M.A. and Andersen O.M. (2018) *Int. J. Mol. Sci.*, **19**: 169.

Blazquez, C.H. (1977) *Plant Dis. Rep.*, **61**: 1002–1006.

Bleyaert, P. (1987) *Boer en de Tuinder*, **93:** 18-19.

Blum, L.E.B. (2000) *Ciência Rural, Santa Maria*, **30**: 761-765

Boe, A. A., Do, J. Y. and Salunke, D. K. (1967) *Indian J. Hort.*, **24:** 160-165.

Bohn, G. W. and Tucker, C. M. (1940) *Mo. Agri; Exp. Sta. Res. Bull.*, pp. 311.

Boivin, C., Gosselin, A. and Trudel, M. J. (1987) *HortScience*, **22:** 1266-1268.

Bonde, R. and Murphy, E.M. (1952) *Maine Agri. Exp. Sta. Bull.*, pp.497.

Bonnema, A.B. and O'Connell, M. A. (1991) *Plant Cell Rep.*, **10**: 629-632.

Bonnema, A.B., Melzer, J. M. and O' Connell, M. A. (1991) *Theor. Appl. Genet.*, **81:** 339-348.

Boos, G. Vand Kochneva, V. N. (1971) *Tr. Prikl. Bot. Genet. Sel.*, **45:** 100.

Bora, L. C. (1995) *Plant Health*, **1:** 74-6.

Borna, Z. (1969) *Roc. Wyz. Szk. Roi. Poz.*, **46:** 3-8.

Borowski, E. (1995) *Annales Universitatis Mariae Curie-SklodowskaSectio EEE Horticultura*, **3**: 111-118.

Boulard, T., Baille, A and Gall, F. Le (1991) *Agronomie*, **11:** 543-553.

Bourbos, V. A. and Skoudridakis, M. T. (1996) *Phytoparasitica*, **24:** 277-280.

Bourbos, V. A., Skoudridakis, M. T. and Barbopoulou, E. (1999) *Acta Hortic.*, **487:** 275-278.

Branco, M. C., Franca, F. H. and Fontes, R. R. (1996) *Hortic. Brasil.*, **14:** 36-38.

Branthome, X. and Ple, Y. (1987) *Acta Hortic.*, **200:** 165-712.

Bravo, A., Gómez, I., Conde, J., Muñoz-Garay, C., Sánchez, J., Miranda, R., Zhuang, M., Gill, S.S. and Soberón, M. (2004) *Biochem. Biophys. Acta*, **1667**: 38–46.

Brooks, W. M. (1969) *Ohio State Univ. Exp. Bull.*, pp. 513-519.

Brown, C. R., Yang, C. P., Mojtahedi, H., Santo, G. S. and Masuclli, R. (1996) *Theor. Appl. Genet.*, **92:** 572-576.

Brown, M. M., Hall, J. L. and Ho, L.C. (1997) *Physiologia Plantarum*, **101:** 533-539.

Brundell, D. J., Ng, C. and Steenstra, D. R. (1982) *New Zealand Commercial Grower*, **37:** 25.

Bubici, G. and Cirulli, M. (2008) In: *Integrated Management of Diseases Caused by Fungi, Phytoplasma and Bacteria* (Eds. A. Ciancio and K.Mukerji), Vol. 3. Springer, Dordrecht. https: //doi.org/10.1007/978-1-4020-8571-0_12.

Bucheli, P., Voirol, E., Torre, R., Lopez, J., Rytz, A., Tanksley, S.D. and Petiard, V. (1999) *J. Agric. Food Chem.*, **47**: 659–664.

Buda, V., Radžiute, S. and Lutovinovas, E. (2009) *Verlag der Zeitschrift fur Naturforschung*, **64** c: 267 – 270

Buddenhagen, I. W., Sequera, C. and Kelmen, A. (1962) *Phytopath.*, **52:** 726.

Buitelaar, K. (1987) *Groenten en Fruit*, **43:** 36-37.

Buitelaar, K. and Janse, J. (1987) *Groenten en Fruit*, **43:** 48-49.

Buitelaar, K. and Janse, J. (1990) *Groenten en Fruit*, **46:** 38-39.

Burton-Freeman, B.and Sesso, H.D. (2014) *Adv. Nutr.*, **5**: 457–485.

Butler, L. (1936) *J. Hered.*, **27**: 25-26.

Butler, L. (1968) *Rep. Tomato Genet. Coop.*, **18:** 4–6.

Buysens, S., Hofte, M. and Poppe, J. (1993) *Mededelingen van de Faculteit Landbouwwetenschappen*, **58:** 1279-1286.

Byrne, J.M., Hausbeck, M. K. and Latin, R. X. (1997) *Plant Dis.*, **81:** 1167-1172.

Caamal-Maldonado, J.A., Jiménez-Osornio, J.J., Torres-Barragán, A. and Anaya, A.L. (2001) *Agron. J.*, **93:** 27–36.

Cabibel, M. and Ferry, P. (1980) *Annales de Technologie Agricole*, **29:** 27-45.

Cal, A., de Pascual, S. and Melgarejo, P. (1994) In: *Brighton Crop Protection Conf*, Pests and Diseases 1994, Farnham. U. K., British Crop Protection Council, **3:** 1127-1132.

Calabretta, C. and Privitera. S. (1985) *Tecnica Agricola*, **37:** 243-257.

Calabretta, C., Nucifora, M. T., Ferro, Band and Natale, V. Di (1994) *Acta Hortic.*, 361: 530-544.

Calvert, T. (1959) *J. Hort. Sci.*, **34:** 154-162.

Cameron, P. J., Walker, G. P., Henman, T. J. B. and Fletcher, J. D. (1995) *Proc. 48th. New Zealand Plant Protection Conf.*, Angus Inn, pp. 289-293.

Campiglia, E., Caporali, F., Radicetti, E. and Mancinelli, R. (2010a) *Eur. J. Agron.*, **33:** 94–102.

Campiglia, E., Mancinelli, R., Radicetti, E. and Caporali, F. (2010b) *Crop Prot.* **29:** 354–363.

Campolo, O., Cherif, A., Ricupero, M., Siscaro, G., Grissa-Lebdi, K., Russo, A., Cucci, L.M., Di Pietro, P., Satriano, C., Desneux N., Biondi, A., Zappalà, L. and Palmeri, V. (2017) *Scientific Rep.*, **7:** 13036 DOI: 10.1038/s41598-017-13413-0

Cantliffe, D. J. and Abebe, Y. (1994) *Proc. Florida State Hort. Soc.*, **106:** 177-183.

Capoor, S. P. (1958) *Mycological Res. Workers Conf.*, ICAR, Simla.

Carland, F.M. and Staskawicz, B.J. (1993) *Mol. General Genet.* **239**: 17-27.

Carmel-Goren, L., Liu, Y. S., Lifschitz, E. and Zamir, D. (2003) *Plant Mol. Biol.*, **52:** 1215–1222,

Caruso, G. and Postiglione, L. (1993) *Revista di Agronomia*, **27:** 211-219.

Castellane, P. O., Araujo, J. A. C., deMelo, W. J. and Gabarra, P. A. (1994) *Ente Antonomo Fiere di Verona*, pp.13.

Castro, A.P., Blanca, J.M., Diez, M.J. and Vinals F.N. (2007) *Eur. J. Plant. Pathol.*, **117**: 347-356.

Causse, M., Buret, M., Robini, K. and Verschave, P. (2003) *J.Food Sci.*, **68:** 2342–2350.

Causse, M., Duffe, P., Gomez, M.C., Buret, M., Damiadux, R., Zamir, D., Gur, A., Chevalier, C., Lemaire-Chamley, M. and Rothan C. (2004) *Journal of Experimental Botany*, **55**(403): 1671-1685.

Cerne, M., Vrhovnik, I., Stopar, M. and Heblin, D. (1994) *Acta Hortic.*, **366**: 119-124.

Cetintas, R., Kusek, M. and Fateh, S.A. (2018) *Egyptian J. Biol. Pest Control,* **28:** 7, doi: 10.1186/s41938-017-0008-x.

Cevri, H., Ozer, M., Serve, R., Deviren, A. and Durceylan, E. (1999) *Acta Hortic.*, **491:** 61-66.

Chaib, J., Devaux, M.F., Grotte, M.G., Robini, K., Causse, M., Lahaye, M. and Marty, I. (2007) *J.Expt. Bot.*, **58:** 1915–1925.

Chaithra, N., Gowda, P.H.R. and Guleria, N. (2015) *Int. J. Agri. Env. Biotech.* **8:** 795-803.

Chakraborty, I., Vanlalliani, Chattopadhyay, A. and Hazra, P. (2007) *Veg. Sci.*, **34**(1):26-31

Chakraborty, R. C. and Sadhu, M. K. (1994) *Indian J. Agric. Sci.*, **64:** 608-612.

Chand, J. N. and Thakur, D. P. (1969) *Indian J. Hort.*, **26:** 87-88.

Chand, T. and Prasad, V.M. (2017) *Bull. Env. Pharmacol. Life Sci.*, **7:** 80-87.

Chand, T., Prasad, V.M. and Lal, B. (2018) *Plant Arch.*, **18:** 177-189.

Chandravanshi, S. S., Singh, B. P. and Thakur, P. (1994) *Indian Phytopath.*, **47:** 241-244.

Charles, W. B. and Harris. R. E. (1972) *Canadian J. Plant Sci.*, **52:** 497-506.

Chatterjee, M.L. and Mondal, S. (2012) *Acta Hortic.*, **958**: 147-153.

Chattopadhyay, A., Chakraborty, I. and Siddique, W. (2013) *J. Food Process. Technol.*, **4**: 222. doi:10.4172/2157-7110.1000222.

Chattopadhyay, A., Dutta. S., Dutta. P. and Hazra, P. (2011) *Int. J. Plant Breed.*, **5:** 88-93.

Chaudhry, Z., Abbas, S., Yasmin A., Rashid, H., Ahmed, H. and Anjum, M.A. (2010) *Pak. J. Bot.*, **42**: 155-163.

Che, C.T. and Hsieh F.K. (1996) *Chinese J. Entomol.*, **16:** 167-175.

Chellemi, D. O., Olson, S. M. and Mitchell, D. J. (1994a) *Plant Dis.*, **78:** 1167-1172.

Chellemi, D. O., Olson, S. M., Scott, J. W., Mitchell, D. J. and Me Sorley, R. (1994b) *J. Nematol.*, **25:** 800-805.

Chen, F. Q. and Foolad, M. R. (1999) *Genome*, **42**: 94–103.

Chen, G., Chen, Y., Wu, D.H. and Wu, X.Y. (1998) *J. South China Agri. Univ.*, **20:** 36-40.

Chen, J.J., Sun, Y.W. and Sheen, T.F. (1999) *HortScience*, **34:** 852-854.

Chen, X. L. and Su, X. H. (1991) *Acta Agriculturae Nucleatae Sinica*, **5:** 25-30.

Chermiti, B., Onillon, J. C. and Aicha, B. B. (1993) *Mededelingen van de Faculteit Landbowwe-tenscliappen*, Universiteit Gent, **58:** 1057-1066.

Chesnokov, Y.V., Sedov, G.I. and Vikonskaya, N. A. (1989) *Izvestiya Akademii Nauk Moldavskoi SSR Biologicheskie i Khimicheskie Nauki*, **6:** 61-62.

Chetelat, R. T. (2002) *Report TGC*, **52**: 41–62.

Chetelat, R.T., Deverna, J.W. and Bennett, A.B. (1995) *Theor. Appl. Genet.*, **91**: 334–339.

Chiaranda, F. Q., Tedeschi, P. and Zcrbi, G. (1982) *Rivista di Agronomia*, **16:** 47-50.

Chiesa, A., Moceia, S., Frezza, D. and Filippinide Delfino, S. (1997) *Agricultura Tropicaet Subtropica*, **31:** 71-81.

Child, A. (1990) *Feddes Repert*, **101:** 209–235.

Chinaswamy, G. N. (1967) *Madras Agric. J.*, **54**: 144-146.

Chirinos, D. T. and Geraud-Pouey, F. (1996) *Interciencia*, **21:** 31-36.

Cho, I.H., Jo, S.J.C. and Kwon, Y.S. (1997) *RDA J. Horticulture. Sci.*, **39:** 140-144.

Cho, J.Y., Seo, D.S. and Chung, S. (1996) *J. Korean Soci. Hort. Sci.*, **37:** 633-635.

Choi, Y.H., Rhee, H.C., Kweon, G.B. and Lee, J.I. (1998) *RDA J. Agro-Environment Sci.*, **40**: 111-117.

Cholakov, D. (1987) *Rasteniev dni Nauki*, 24: 94-97.

Choudhury, B. (1959) *Indian J. Hort.*, **16:** 102-106.

Choudhury, B. (1967) *Proc. Res. Found. Bihar Agri. Coll., Sabour*, **19**: 67-68.

Choudhury, B. (1972) In: *Vegetables, National Book Trust,* New Delhi.

Choudhury, B. and Anothai-Choomsai, M. L. (1970) *Indian J. Agric. Sci.*, **40**: 805-812.

Choudhury, B. and Singh, S. N. (1960) *Indian Hort.*, **17:** 148-511.

Choudhury, B. B. and De, R. (1972) *Proc. 3rd bit. Symp. Subtrop. and Trop. Hort;* Bangalore, pp.143.

Choudhury, B. B. and De, R. (1975) *Soil Sci. and Plant Nutr.*, **22:** 57-62.

Choudhury, B., Punia, R. S. and Sangha, K. S. (1965) *Indian J. Hort.*, **22:** 59-69.

Choudhury, B., Tewari, R., Mathai, J., Joshi, G. C., Kesvan, V., Chakrabarty, A. K., Kalda, T. S. and Sivakami, N. (1973) *Proc. 2nd General Congress, SABRAO,* New Delhi.

Choudhury, S., Islam, N., Sarkar, M.D. and Ali, M.A. (2013) *Int. J. Sustainable Agric.*, **5**: 25-28.

Christova, D. and Schuster, G. (1991) *J. Phytopath.*, **132:** 168-171.

Chunwongse, J., Bunn, T.B., Crossman, C., Jiang, J. and Tanksley, S.D. (1994) *Theor. Appl. Genet*. **89**: 76-79.

Churata-Masca, M. G. C., Fabiani, M. A. Castelane, P. D. (1987) *Hortic. Brasil.*, **5:** 375.

Cipollini, D. F. Jr. and Redman, A. M. (1999) *J. Chem. Eco.*, **25:** 271-281.

Civelek, H.S. and Weintraub, P.G. (2004) *J. Econ. Entomol.*, **97**: 1581–1586.

Clayton, E. E. (1923) *Amer. J. Bot.*, **10:** 71-78.

Clinton, G. P. (1909) *Cnn. (State) Agri. Exp. Sta. Ann. Rpt.*, **1907-1908:** 657-658.

Cobelli, L., Scannavini, M., Antoniacci, L. and Brunelli, A. (1995) *Informatore Fitopotologico*, **45:** 42-54.

Cockshull, K. E., Graves, C. J. and Cave, C. R. J. (1992) *J. Hori. Sci.*, **67:** 16-24.

Cohen, S. and Nitzany, E. E. (1966) *Phytopath.*, **56:** 1127-31.

Colombo, A., Cosentino, S., Nucifora, A. and Sortino, O. (1994) *Acta Hortic.*, **361:** 406-413.

Conde, B. and Connelly, M. (1995) *Agonte* (*Darwin*), No. 624, pp. 2.

Cong, B., Barrero, L. S. and Tanksley, S. D. (2008) *Nature Genetics*, **40**: 800–804.

Contangelo, P., Camele, I., Malinconico, P. and Rana,G. L. (1994) *Informatore Agrario*, **50:** 49-55.

Cools, M. I. and Jansen, A. B. (1986) *Groenten en Fruit*, **41:** 36-37.

Cooman, A., Ubaque, I.I., Schrevcrs, E. and de Rijck, G. (1999) *Acta Hortic.*, **507:** 85-90.

Corral-Martinez, P. and Nue, F. Segui-Simarro (2011) *Euphytica* **178**: 215–228.

Correll, J. C., Gordon, T. R. and Elliott, V. J. (1988) *Phytopath.*, **78:** 512-519.

Cortina, C. and Culianez-Macia, F. (2004) *Plant Cell Tiss. Org. Cult.*, **76**: 269-275.

Costa, M. G. C., Nogueira, F. T. S., Figueira, M. L., Otooni, W. C., Brommonschenkel, S. H. and Cecon, P. R. (2000) *Plant Cell Rep.*, **19:** 327-332.

Coyne, D. P. (1968) *HortScience*, **3**: 39.

Curme, J. H. (1962) *Pre. Plant Sci. Symp., Campben Soup. Co.*, New Jersey, pp. 99-108.

Czapski, J. and Saniewski, M. (1995) *Acta Agrobotanica*, **48:** 27-35.

D'Amore, R., Oliva, A., Iovieno, P., Lahoz, E., Piro, E. and Magnifico, V. (1999) *Acta Hortic.*, **487**: 199-205.

D'Arcy, W.G. (1972) *Ann. Missouri Bot. Gard.*, **59:** 262–278.

Dar, R.A. and Sharma, J.P. (2011) *Int. J. Plant Breed. Genet.*, **5**: 168- 174.

Daramola, O.S., Adigun, J.A., Olorunmaiye, P.M. and Adeyemi, O.R. (2020) *Int. J. Veg. Sci.*, doi: 10.1080/19315260.2020.1795961.

Darwin, S., Knapp S. and Peralta, I. E. (2003) *Systematics and Biodiversity*, **1**: 29- 53.

Das I., Hazra, P., Longjam M., Bhattacharjee, T., Maurya, P.K., Banerjee, S. and Chattopadhyay, A. (2020) *J. Genet.*, **99**:56, doi.org/10.1007/s12041-020-01209-7.

Das, R. K., Baruah, G. K. S. and Paul, S. R. (1997) *Adv. Plant Sci.*, **10:** 165-167.

Dastane, N. G., Kulkarm, G. N. and Charian, E. C. (1963) *Indian J. Agron.*, **8:** 405-408.

Datnoff, L. E., Nemec, S. and Pernezny, K. (1995) *Biological Control*, **5**: 427-431.

Datta, A. K. (1961) *Indian Farming*, **11:** 1-7.

Daubney, I. I. A. (1961) *Proc. Amer. Soc. Hort. Sci.*, **78:** 445-449.

Davis, J. F. and Webb, R. E. (1965) *Proc. Amer. Soc. Hort. Sci.*, **88:** 557-563.

Davis, J. M. and Gardner, R. G. (1994) *HortScience*, **29:** 613-615.

Davis, R. M. and Lungle, J. C. (1961) *Plant Physiol.*, **36:** 153-162.

Davis, R. M., Miyao, E. M., Mullen, R. J., Valencia, J., May, D. M. and Gwynne, B. J. (1997) *Plant Dis.*, **81:** 601-603.

Ddamulira, G., Idd, R., Namazzi, S., Kalali, F., Mundingotto, J. and Maphosa, M. (2019) *J. Agric. Sci.*, **11**: 48-55.

de Aguiar, R.A., da Cunha, M.G. and Lobo Junior, M. (2014) *Biological Control*, **74**: 1–5.

de Azevedo, V.F., Abboud, A.C.S. and Carmo, M.G.F. (2010) *Hortic. Brasil.*, **28:** 389-394.

De Candolle, A. P. (1884) *Origin of Cultivated Plants;* NewYork.

de Oliveira, C.M. (2012) *Ciência. Agrotecnol.* **36**(1): 45- 52.

De Souza, A., García, D., Sueiro, L., Licea, L. and Porras, E. (2005) *Spanish J. Agric. Res.*, **3**: 113-122.

de Vicente, M. C. and Tanksley, S. D. (1993) *Genetics*, **134**: 585–596.

Decoteau, D. R., Kasperbauer, M. J. and Hunt, P. G. (1989) *J. Amer. Soc. Hort. Sci.*, **114:** 216-219.

Dekock, P. C., Inkson, R. H. E. and Han, A. (1982) *J. Plant Nutri.*, **5:** 57-62.

Delgado, S. A., Mesa, N. C., Estrada, E. I. and Zuluaga, J. I. (1999) *Revista Colombiana de Entomologia*, **25**: 137-142.

Dell' Amico, J., Rodriguez, P., Alfonso, J. L. and Morales, S. (1996) *Cultivos Tropicales*, **17:** 39-41.

Della Penna, D. Lashbrook, C. C. Toenjes, K., Giovannoni, J. J. Fischer, R. L. and Bennett, A. B. (1990) *Plant Physiol*, **94:** 1882-1886.

Demers, D. A., Dorais, M., Wcim, C. H. and Goseslin, A. (1998) *Sci. Hort.*, **74:** 295-306.

Demirel, F. and Seniz, V. (1997) *Acta Hortic.*, **447:** 237-238.

Deng, D., McGrath, P.F., Robinson, D.J. and Harrison, B.D. (1994) *Ann. Appl. Biol.*, **125** : 327-36.

Deng, L. P., Guo, Y.I.I., Jia, Y. K. and Sheng, J.I.I. (1988) *Hereditas, China*, **10:** 9-10.

Desneux, N., Wajnberg, E., Wyckhuys, K.A.G., Burgio, G., Arpaia, S., Narváez-Vasquez, C.A., González-Cabrera, J., Ruescas, D.C., Tabone, E. and Frandon, J. (2010) *J. Pest Sci.*, **83**: 197-215.

Dessalegne, L., Anchala, C. and Mariam, S. G. (1999) *Agritopia*, **14: 4**.

Devanathan, M. and Ramanujam, K. (1995) *Madras Agric. J.*, **82:** 228-229.

Dhaliwal, M. S., Jindal, S. K., Dhaliwal, L. K., Gaikwad, A. K. and Sharma S. P. (2017) *Int. J. Veg. Sci.*, **23**: 4 –17.

Dhanvantari, D. N. and Brown, R. J. (1993) *Can. J. Plant Path.*, **15:** 201-205.

Dhuria, H.J and Shukla, V. (1972) *Proc. 3rd Int. Symp. Subtrop. and Trop. Hort., Bangalore*, pp.145.

Dibyantoro, A.L.H. and Siswojo, S. (1988) *Bull. Penelitian llortikultura*, **16**: 67-72.

Dickinson, M.J., Jones, D.A. and Jones, J.D.G. 1993. *Mol Plant-Microbe Interact*, **6**: 341–347.

Dillard, H.R., Johnston, S. A., Cobb, A. C. and Hamilton, G. H. (1997) *Plant Dis.*, **81:** 677-681.

Dimock, M. B. and Kennedy, G. G. (1983) *Entomol. Exp. Appl.* **33**: 263-68.

Dinakaran, D. and Rarnakrishnan, G. (1996) *Plant Dis. Res.*, **11:** 148-150.

Ding, J., Chen, B., Xia, X., Mao, W., Shi, K. *et al.* (2013) *PLoS ONE*, **8**: e70080. doi: 10.1371/journal.pone.0070080.

Dinu, M., Dumitru, M.G. and Soare, R. (2015) *Bulgarian J. Agric. Sci.*, **21:** 998-1004.

Diwan, N., Fluhr, R., Eshed, Y., Zamir, D. and Tanksley S.D. (1999) *Theor. Appl. Genet.* **98**: 315-319.

Dixit, A., Sharma, D., Sharma, T.K. and Bairwa, P.L. (2018) *Int. J. Current Microbiol. Appl. Sci.*, **6**: 197-203.

Dobhal, V. K., Kohli, U. K. and Mehta, D. (1999) *J. Hill. Res.*, **12**: 31-33.

Dobreanu, M. and Indachescu, O. (1980) *Lucrari Stiintifice, Institutul de Cercetari Si Proiectari Pentru Valorificarea Si lndustrializarea Legumelor Si Fructelor*, **11:** 37-42.

Dodan, D.S., Shyam, K.R. and Gupta, S.K. (1994) *Pl. Dis. Rep.*, **9**: 141-144.

Dodds, G. T., Trenholm, L., Rajabipour, A, Madramootoo, C. Aand Norris, E. R. (1997) *J. Amer. Soc. Hort. Sci.*, **122:** 491-498.

Doganlar, S., Frary, A., Ku, H. M. and Tanksley, S. D. (2002) *Genome*, **45**: 1189–1202.

Dogterom, M.H., Mattconi, J. A. and Plowright, R. C. R. (1998) *J. Eco. Entomol.*, **91:** 71-75.

Doolittle, S. P., Porte, W. S. and Beechan, A. (1946) *Phytopath.*, **36**: 685.

Dorais, M., Gosselin, A. and Trudel, M. J. (1988) *Acta Hortic.*, **230:** 405-412.

Dorais, M., Gosselin, A. and Trudel, M. J. (1991) *Sci. Hort.*, **45:** 225-234.

Doroschenko, N. D., Gushchin, B. E., Voelkova, L. and Mal'tscva, N.I. (1988) *Agrokhimiya*, **3**: 95-97.

Dostal, H.C. and Wilcox, G. E. (1971) *J. Amer. Soc. Hori. Sci.*, **96:** 5.

Drews, M. and Frank, R. (1987) *Arcliiv fur Gartenbau*, **35:** 417-424.

Drews, M. and Heissner, A (1982) *Archiv fur Gartenbau*, **30**: 387-404.

Drozdov, S. N., Shabalina, L. P. and Klykova, V. V. (1988) *In Vliyanie Faktorov Soedy, i Fiziologiya Akti-vnykh Veshchestv na Pruduktivnost; i Ustoichchivost Rastenii. Petrozavodsa*, USSR, 48-54.

Drukker, B., Janssen, A., Ravensbrg, W. and Sabelis, M. W. (1997) *Exp. and Appl. Acarology*, **21:** 507-518.

Duden, R. and Schlarzus, H. (1982) *Gordian*, **82:** 172-175.

Dufour, L. and Taillens, J. (1994) *Revue Suisse de Viticulture, d ' Arboriculture et d ' Horticulture*, **26:** 85-88.

Dunger, R. P. (1962) *Indian Fmg*., 12: 11.

Duvick, D.N. (1997) In: *Book of abstracts of the International Symposium "The Genetics and Exploitation of Heterosis in Crops"* pp. 17–22.

Dwivedi, K., Srivastava, P., Verma, H. N. and Chaturvcdi, H. C. (1990) *Indian J. Exp. Biol.*, **28:** 32-35.

Dyakunchak, S. A., Bolakhonenkova, V. D. and Varvarina, N. A. (1997) *Kartofe/'i Ovoshchi*, **2:** 26-27.

Eddaoudi, M. and Ammati, M. (1995) *Afro- Asian J. Nematol.*, **5:** 28-33.

Ekanayake, H. M. R. K. and Jayasundara, N. J. (1994) *Nematologia Mediterranea*, **22:** 87-88.

El-Abd, S. O., El-Beltagy, A. S. and Han, M. A. (1986) *Acta Hortic.*, **190:** 389-386.

Elad, Y., Gucllino, M. L., Shtienberg, D. and Aloi, C. (1995) *Crop Prot.*, **14:** 105-109.

El-Beltagy, M. S., Sawan, O. N.,Mohamcdien, S. A., El-Deltagy, A. S. and Maksoud, M. A. (1986) *Acta Hortic.*, **190:** 481-496.

El-Farash, E. M., Abdalla, H. I., Taghian, A. S. and Ahmad, M. H. (1993) *Aust. J. Agric. Sci.*, **24**: 3-14.

El-Gizawy, A. M., Abdallah, M. M. F., Gomma, H. M. and Mohamed, S. S. (1993) *Acta Hortic.*, **190:** 349-350.

El-Hassan, H. S. (1986) *Acta Hortic.*, **190:** 425-428.

Ellis, G. H. and Hamner, K. C. (1943) *J. Nutrition*, **25:** 539-553.

El-Sayed, S. F. and El-Fadaly, K. A. (1991) *Bull. Fae. Agri. Univ., Cairo*, **42:** 359-376.

El-Sayed, T.R., Jacquiod, S., Nour, E.H., Sbattu2rensen, S.J. and Smalla, K. (2020) *Front. Microbiol.*, **10:** 2835. doi: 10.3389/fmicb.2019.02835.

El-Sherif, A. F., Shala, S. M. and Youssef, R. A. (1993) *Egyptian J. Hort.*, **17:** 123-129.

Encina, C.L., Carmona-Martin, E. and Fernández-Muñoz, R. (2012) *Acta Hortic.*, **935**: 91-93.

Enokpa, E. N., Okwujiako, I. A. and Madunagu, B. E. (1996) *Global J. Pure Appl. Sci.*, **2**: 131-136.

Erlandsson, G. (1987) *Rapport Institution en for Tradgards vetenskap, Sveriges Lantbruks universitel*, **50:** 37.

Eshed, Y. and Zamir, D. (1995) *Genetics*, **141**: 1147–1162.

Esipova, Z. I. and Veselovskij, I. A. (1963) *Zap. Leningr. Skh. Inst.*, **92:** 41.

Esquerra, E. B. and Bautista, O. K. (1990) *ASEAN Food J.*, **5:** 27-33.

Etebu, E., Nwauzoma, A. and Bawo, D. (2013) *J. Biol. Agric. Healthcare*, **3:** 51-61.

FAO (2020) FAO website http: //www.fao.org/faostat/en/#data/QC accessed on 15th July, 2020.

Farkas, G. L. and Krialy, Z. (1962) *Phytopath. J.*, **44:** 105-150.

Farneti, B., Zhang, W., Witkowska, I. and Woltering, E.J. (2010). *Acta Hort.*, **877:** 1191-1196.

Farrar, R. R. J. and Kennedy, G. G. (1991) *In: Genetic Improvement of Tomato*, (Ed. G. Kalloo), vol. 14, Springer, Berlin, Germany, pp. 122–141.

Farrar, R.R.J., Barbour, J.D. and Kennedy, G.G. (1994) *Entomologia Experimentalis et Applicata*, **71**: 211–226,.

Fazio, G., Stevens, M.R. and Scott, J.W. (1999) *Euphytica*, **150**: 205-210.

Fernandez, J. E. and Dailey, B. J. (1994) *J. Agric. Engg. Res.*, **58:** 201-210.

Feuilloley, P., Mckikdjian, C. and Lagicr, J. (1994) *Plasticulture*, **104**: 33-46.

Filho, A.B.C., Rezende, B.L.A., Barbosa, J.A. and Grangeiro, L.C. (2011) *Anais da Academia Brasileira de Ciências* (*Annals of the Brazilian Academy of Sciences*), **83**: 1109-1119.

Fillatti, J. J., Kiser, J., Rose, B., Comai, L. and Jones, R. A. (1987) *Tomato biotechnology: Procedings of a symposium held at the University of California, Davis, California*, August 20-22, 1986, pp.199-210.

Fillinger, S. and Elad, Y. (2015) *Botrytis-the Fungus, the Pathogen and Its Management in Agricultural Systems* Springer, p.486.

Fischer, S. and Leger, A. (1996) *Revue Suissede Viticultured' Arboricultureet d'Horticulture*, **28**: 111-112.

Fischhoff, D. A., Bowdish, K. S., Perlak, F. G., Marrone, P. O., McCornmick, S. M., Niedermeyer, J. G., Dean, D. A., Kusano Kretzrner, K., Mayer, E. G., Rochester, D. E., Rogers, S. G., Fraley, R. T. and Kretzmer, K. T. K. (1987) *Biotechnology*, **5:** 807-813.

Fiume, F. and Fiume, G. (2006) *Comm. Appl. Biol. Sci, Ghent University*, **71**: 897 – 908

Fiume, F. and Parisi, B. (1995) *Informatore Fitopatologico*, **45:** 44-50.

Fiume, G and Fiume F. (2008) Biological control of corky root in tomato. *Commun. Agric. Appl. Biol. Sci.* **73**(2): 233-248.

Fletcher, J. T., Smcwin, B. J. and Cook, R. T. A. (1988) *Pl. Pathology*, **37:** 594-598.

Flett, S. P., Ashcroft, W. J., Jeric, P. H. and Taylor, P. A. (1991) *Aust. J. Exp. Agric.*, **31:** 279-283.

Foolad, M. R., Jones, R. A. and Rodriguez, R. L. (1993) *Plant Cell Rep.*, **12** (5): 293–297.

Foolad, M.R. (2007a) *Int. J. Plant Genomics*, Article ID 64358, **10**: 1-52.

Foolad, M.R. (2007b) In: *Genetic improvement of Solanaceous crops*, Vol. 2 Tomato, *Science Publisher*, **2**: 521-592.

Foolad, M.R., Zhang, L.P. and Lin, G.Y. (2001) *Genome*, **44**: 444-454.

Foolad, M.R., Zhang, L.P. and Subbiah, P. (2003) *Genome*, **46**: 536-545.

Forsberg, A. S. (1989) *Vaxtsky dasnotiser*, **53:** 144-148.

Franca, S.M., Oliveira, J.V., Badji, C.A., Guedes, C.A., Duarte1, B.L. R., Oliveira, C.M. and Breda, M.O. (2017) *J. Agric. Sci. Tech.*, **19:** 133-144.

Francescangeli, N. Ferraho, J. Bussilaechi, H. and Lara, M.A. (1994) *Horticultura Argentina*, **13**: 58-64.

Francisco, Y., Borroto, C. G., Ortega, J. R. and Legan, C. (1987) *Centro Agricola*, **14:** 3-8.

Franco, J. A., Banon, S. and Madrid, R. (1994) *Sci. Hort.*, **57:** 283-292.

Frary, A., Fulton, T. M., Zamir, D. and Tanksley, S. D. (2004) *Theor. Appl. Genet.*, **108**: 485–496,.

Frary, A., Fulton, T.M., Zamir, D. and Tanksley S.D. (2004). *Theor. Appl. Genet.*, **108**: 485-496.

Frary, A., Xu, Y., Liu, M. J., Tedeschi, S. and Tanksley, S.D. (2005) *Theor. Appl. Genet.*, **111**: 291-312.

Fraser, P.D. and Bramley, P.M. (2004) *Prog. Lipid Res.*, **43**: 228–265.

Fraser, P.D., Romer, S., Shipton, C.A., Mills, P.B., Kiano, J.W., Misawa, N., Drake, R.D., Schuch, W. and Bramley, P.M. (2002) *Proc. Natl. Acad. Sci., USA*, **99**: 1092-1097.

Frey, P., Prior, P., Marie, C., Kotoujansky, A., Demery, D. T. and Trigalet, A. (1994) *Appl. Environ. Microbiol.*, **60:** 3175–3181.

Fridman, E., Carrari, F., Liu, Y.S., Fernie, A.R. and Zamir, D. (2004) *Science*, **305**: 1786–1789.

Fridman, E., Pleban, T. and Zamir, D. (2000) *Proc. Nat. Acad. Sci.* USA, **97**: 4718–4723.

Fryer, H.C., Ascham, L., Cardwell, A. B., Frazier, J.C. and Willis, W. W. (1954) *Proc. Amer. Soc. Hort. Sci.*, **64:** 465-371.

Fufa, H., Baenziger, P. S., Beecher, B. S., Dweikat, I., Graybosch, R. A. and Eskridge, K. M. (2005) *Euphytica*, 145: 133–146.

Fulling, B. A., Tigehelaar, E. C. and Latin, R. (1995) *Plant Dis.*, **79**: 223-233.

Fulton, T. M., Beck-Bunn, T., Emmatty, D., Eshed, Y., Lopez, J., Petiard, V., Uhlig, J., Zamir, D. and Tanksley, S.D. (1997) *Theor. Appl. Genet.*, **95**: 881–894.

Fulton, T. M., Grandillo, S., Beck-Bunn, T., Fridman, E., Frampton, A., Lopez, J., Petiard, V., Uhlig, J., Zamir, D. and Tanksley, S.D. (2000) *Theor. Appl. Genet.*, **100**: 1025–1042.

Fulton, T. M., van der Hoeven, R., Eannetta, N. T. and Tanksley, S.D. (2002) *The Plant Cell*, **14**: 1457–1467.

Furui, H. Nakuma, T. Ishiguro, Y. and Kiso, M. (1997) *Nippon Nogeikagaku Kaishi*, **71**: 777-782.

Galitskii, VI. and Novikova, N.P. (1999) *Kartofel-i-Ovoshchi*, **1**: 10.

Gallagher, P. A. (1972) *Proc. Par. Glass house Conf. Dublin*, pp.13-18.

Gallegly, M. E. (1952) *Phytopath.*, **42**: 461-462.

Gallegly, M. E. (1962) *Phytopath.*, **52**: 12-18.

Gallegly, M. E. and Marval, M. E. (1955) *Phytopath.*, **45**: 103-109.

Gallegly, M. E. and Walker, J. C. (1949) *Amer. J. Bot.*, **36**: 613-623.

Gallias, A. (1988) *Euphytica*, **39**: 95-284.

Gamer, L. C. and Bjorkman, T. (1997) *HortScience*, **32:** 227-229.

Gamer, L. C. and Bjorkman, T. (1999) *HortScience*, **34:** 848-851.

Ganal, M.W., Simon, R., Brommonschenkel, S., Arndt, M., Phillips, M.S., Tanksley, S.D., Kumar, A. (1995) *MPMI*, **8**: 886-891.

Gao, Y., Lu, Y., Lin, W., Tian, J. and Cai, K. (2019) *Microorganisms*, **7**: 676; doi: 10.3390/microorganisms7120676.

Gao, Z. H., Li, S. D., Dai, S. S., Zhu, D. W., Zhang, W. S., Chen, X. W., Wang, B., Feng, L. X., Cai, S. H., Yang, C. R.,Tian, R.Y. and Zhu, D. W. (1995) *Acta Hort.*, **402**: 321-28.

Garcia, B.E., Mejía, L., Salus, M.S., Martin, C.T., Seah, S., Williamson, V.M. and Maxwell, D.P. (2007) www.plantpath.wisc.edu

Garcia-Gonzalez, J. and Sommerfeld, M. (2016) *J. Appl. Phycol.*, **28:** 1051–1061 DOI 10.1007/s10811-015-0625-2.

Gardner, R. G. (1990) *HortScience*, **25**: 989-990.

Gardner, R. G. (1993) *HortScience*, **28**: 348-349.

Gardner, R. G. (1999a) *HortScience*, **34**: 940.

Gardner, R. G. (1999b) *HortScience*, **34**: 941-942.

Garedal, L. and Lundegardh, B. (1997) *Biol. Agric. Hort.*, **14:** 291-301.

Garita, V. S. Bustamante, E. and Shattock, R. (1999) *Manejo Integrado de Plagas*, No.51, pp. 47-58.

Gascuel, Q., Diretto, G., Monforte, A. J., Fortes, A. M. and Granell, A. (2017) *Front. Plant Sci.*, **8:** article no. 652.

Gautam, R., Dhankar, R. B. and Kaloo (1981) *Haryana J. Hort. Sci.*, **10:** 81-85.

Gavrish, S. F.and Sysina, E. A. (1990) *Kartofel ' i Ovoshchi*, **3**: 32-34.

Genanew, T. (2013) *World J. Agril. Sci.*, 9: 29–37.

Gent, M. P. N. and Malerba,V. (1994) *Hort. Tech.*, **4:** 277-281.

Georgelis, N., Scott, J.W. and Baldwin, E. A. (2004) *J. Amer. Soc. Hort. Sci.*, **129**: 839–845.

Georgiev, I. (1991) In: *Genetic Improvement of tomato, Springer-Verlag, Heidelberg,* Germany, pp. 93-98.

Georgiev, Kh. Vladimirov, 13. Danailov, Zh. and Popova, D. (1986) *Rasteniev' dni Nauki*, **23:** 67-69.

Gerardi, C., Albano, C., Calabriso, N., Carluccio, M., Durante, M., Mita, G., Renna, M., Serio, F. and Blando, F. (2018) *Food Chem.*, **240**: 1184–1192.

Gerlach, W. W. P. Matalavea, S. and Tuionoula, M. (1986) *Alafua Agril. Bull.*, **11**: 33-36.

Ghanbari, F. and Sayyari, M. (2018) *Sci. Hort.*, **229:** 167-174.

Gharezi, M., Joshi, N. and Sadeghian, E. (2012) *Nutr. Food. Sci.*, **2:** 8 DOI: 10.4172/2155-9600.1000157.

Ghidiu, G. M. and Storlie, CA (1995) *J. Agric. Entomol.*, **12:** 77-84.

Ghini, R. Bettiol, W. Spadotto, C. A. Moraes, G. J. de, Praiba, L. C. and Mineiro, J. L. de C. (1993) *Summa Phytopathologica*, **19**: 183-189.

Ghosh, P. K., Syamal, M. M., Rai, N. and Joshi, A. K. (1995) *Ad. Plant Sci.*, **8:** 207-213.

Giametta, G. (1989) *Proc. Amer. Soc. Hort. Sci.*, **68:** 437-442.

Gilbert, J. C. and Mc Gurie, D. C. (1956) *Proc. Amer. Soc. Hort. Sci.*, **68:** 408-432, 437.

Gilbert, J.C. and McGuire, D.C. (1955) *Report Tomato Genet Coop.*, **5:** 15.

Giovannoni, J. J. Della-Penna, D. and Bennett-Fischer, L. (1993) *Hort. Rev.*, **13:** 67-103.

Girish, N. and Umesha, S. (2005) *Arch. Phytopath. Plant Prot.*, **38:** 235-243.

Glavinich, R. D., Toropkina, M. N. and Zhidkova, V. A. (1988) *Selektsiya i Semenovodstvo Moskva*, **6:** 20-21.

Gleason, M.L., MacNab, A.A. Pitblado, R.E. Ricker, M.D. East, D.A. and Latin, R.X. (1995) *Plant Dis.*,**79:** 113-121.

Godishala, V., Kairamkonda, M., Kagithoju, S., Mangamoori, L. and Nanna, R.S. (2012) *Aust. J. Crop Sci.*, **6**: 31-35.

Godzina, M., Staniaszek, M. and Kie³kiewicz, M. (2010) *Veget. Crops Res. Bull.*, **72**: 25-33.

Gollas, G. and Bland, S. (1994) *Proc. 47th. New Zealand Plant Protection Conf.*, pp. 125-127.

Gomaa, N.A., Mahdy, A. M. M., Fawzy, R.N. and Ahmed, G.A. (2016) *Middle East J.Agril. Res.*, **5**: 479-486

Gonzalez, J. M. (1996) *Boletin de Entomolgia Venezolana*, **11:** 59-60.

Gonzalez, P.R., Zancada, C., Verdugo, M. and Salas, M.L. (1996) *J. Hort. Sci.*,**71:** 453-460.

Goode, M. J. and Sasser, M. (1980) *Pl. Dis. Rep.*, **64:** 831-834.

Gopalakrishnan, C. and Asokan, R. (1998) *Proc. Ist, Natl. Symp. 011 Pest Management in Hort. Crops: environmental implications and thrusts*, pp.215-721.

Gopalan, C., RamaShastri, B.V. and Balasubramanian, S.C. (1989) In: *Nutritive Value of Indian Foods*, National Institute of Nutrition, ICMR, Hyderabad, India.

Gorman, S.W. and McCormick, S. (1997) *Crit. Rev. Plant Sci.*, **16:** 31–53.

Gosselin, A. and Trudel, M. J. (1982) *Canadian J. Plant Sci.*, **62**: 751-757.

Goswami, J., Pandey, R.K., Tewari, J.P. and Goswami, B.K. (2008) *J. Environ. Sci. Health Part B*, **43:** 237-240.

Gould, W.A. (1992) In: *Tomato production, processing and technology*. 3rd ed. CTI Publications Inc., Baltimore, MD, U.S.A.

Gouli, R.C. (2008) *Regional Agricultural Research Station (RARS), Lumle.*

Gradziel, T. M. (1988) *Diss. Abst. Intern. B. Sci. Engin.*, **48:** 147.

Gragnes, and Leger, A. (1998) *Revue Suisse de Viticulture d' Arboriculture et d' Horticulture*, **30**: 65-71.

Graifenberg, A., Giustiniani, L., Petsas, S. and Jardin, P. D. (1986) *Acta Hortic.*, **176:** 83-89.

Grandillo, S. and Tanksley, S. D. (1996) *Theor. Appl. Genet.*, **92** (8): 957–965.

Grandillo, S., Ku, H.M. and Tanksley, S.D. (1996) *Mol. Breed.*, **2:** 251–260.

Grandillo, S., Ku, H.M. and Tanksley, S.D. (1999) *Theor. Appl. Genet.*, **99**: 978–987.

Green, J. C. (1980) *Acta Hortic.*, **100:** 129-135.

Gresshoff, P. M. and Doy, C. H. (1972) *Planta*, **107:** 161-170.

Grierson, D. and Fray, R. (1994) *Euphytica*, **79:** 251-263.

Grimault, V., Anais, G. and Prior, P. (1994) *Plant Pathol.*, **43:** 663–668.

Gromovikh, T.I. Gukasian, V.M. Golovanova, T.I. and Shmarlovskaya, S.V. (1998) *Mikologia iFitopatologiya*, **32**: 73-78.

Grunzweig, J.M., Rabinowiteh, H. D., Katan, J., Wodner, M. and Ben-Tai, Y. (2000) *Plant Growth Regulation*, **30**: 233-239.

Gudero, G., Hussien, T., Dejene, M. and Biazin, B. (2018) *J. Biol. Agric. Healthcare*, **8**: 94-109

Gui, A., Saifullah and Shah, S. F. A. (1990) *Sarliad J. Agric.*, **6**: 95-97.

Guimaraes, A. M., Pavan, M. A. and Kurozawa, C. (1997) *Fitopatologia Brasileira*, **22:** 142-147.

Guistiniani, L., Graifenberg, A., Botrini, and Barsanti, L. (1997) *Colture Protette*, **26**: 81-85.

Gul, R., Rahman, H., Khalil, I.H., Shah, S.M.A. and Ghafoor, A. (2010) *Afr. J. Biotech.*, **9**: 4144-4151.

Gullino, M. L., Aloi, C. and Garibaldi, A. (1991) *Bulletin SROP*, **14:** 211-215.

Gunawan, O. S. (1988) *Bull. Penelitian Hortikultura*, **16:** 114-116.

Gunawan, O. S. (1989) *Bull. Penelitian Ilortikultura*, **17:** 41-44.

Gupta, D. K. (1988) *Indian J. Mycology and Plant Pathology*, **17:** 62-63.

Gupta, P. R. and Babu, B. R. R. M. (1998) *Proc. 1st. Natl. Symp.on Pest Management Hort. Crops: environmental implications and thrusts*; Bangalore, pp. 75-80.

Gupta, S. Arora, D. K. and Srivastava, A. K. (1995) *Soil Biol. Biochem.*, **27**: 1051-1058.

Gupta, S.K. and Bharat, N.K. (2008) *Pestology*, 32: 17-19

Gustafsson, G. and Weich, R. (1991) *Rapport- Institutionen Lantbrukets Buggnadsteknik, Sveriges Lantbruksunivese*, **72**: 64.

Gyori, L. (1986) *Kerteszet es Szoleszet*, **35**: 4.

Haanstra, J.P.W., Wye, C., Verbakel, H. *et al.* (1999) *Theor. Appl. Genet.*, **99**: 254–271.

Haji, F.N.P. Freire, L.C.L. Roa, F.G. Silva, C.N. da, Souza Junior, M. M. and Silva, M. I. V. da (1995) *Anais da Sociedade Entomologica do Brasil*, **24**: 587-591.

Hakim, A., Kaukovirta, E., Pchu, E., Voipio, I. and Purvis, A. C. (1997) *Adv. Hort. Sci.*, **11:** 142-146.

Hall, C. B. and Augustine, J. J. (1981) *HortSci.*, **6:** 780-781.

Hamar, N. and Olabi, M. (1985) *Kertgasdasag*, **17**: 1-6.

Hamdy, E.M., Hanafy and El-Sayed, W. (2013) *Aus. J. Basic Appl. Sci.*, **7**: 943-948

Hameedunnisa, Begum (1998) *South Indian Hort.*, **46**: 88-90.

Hamner, K. C., Bernstein, L. and Maynard, L. A. (1945) *J. Nutr.*, **21:** 85-97.

Han, H. S. and Lee, K. D. (2005) *J. Agric. Biol. Sci.*, **1:** 176-180.

Hanafy, A. and Gehan, A.E.E. (2018) *Asian J.Res.Biochem.*, **3**: 1-8.

Hand, D. and Slack, G. (1988) *Grower*, **109**: 27-34.

Hankawa, Y. (1988) *Bull. Hiroshima Perfectual Agril. Expt. Sta.*, **5l:** 65-77.

Hankawa, Y., Nakazawa, K. and Hayashi., H. (1988) *Bull.Iliroshima Perfectual Agril. Expt. Station*, **51:** 79-84.

Hanna, H. Y., Millhollon, E. P., Herrick, J. K. and Fletcher, C. L. (1997) *HortScience*, **32**: 224-226.

Hannan, M.M., Ahmed, M.B., Razvy, M.A., Karim, R., Khatun, M., Haydar, A., Hossain, M. and Roy, U.K. (2007) *Amer. Eurasian J. Scientific Res.*, **2**: 146-150.

Hanson, P., Green, S. and Kuo, G. (2006) *Rep.*, **56**: 17–18.

Hanson, P., Green, S.K., Kuo, G. (2006) *Report Tomato Genet Coop.*,**56**: 22-25.

Hanson, P.M., Bernacchi, D., Green, S., Tanksley, S.D., Muniyappa, V., Padmaja, A.S. *et al.* (2000) *J. Amer. Soc. Hort. Sci.* **125**: 15-20.

Hanssen, I., Lapidot, M. and Thomma, B. (2010) *Molecular plant-microbe Interactions*, **23**: 539-548.

Hanur, V.S., Reddy, B., Arya, V.V. and RamiReddy, P.V. (2015) *J. Agric. Sci. Tech.*, **17**: 1805-1814

Harbaoni, Y. and Yellodt, I. (1980) *Mededelingen van de Faculteit Landbo11-Wwetensclwppen Rejksuniversiteit Gent.*, **45:** 1315-1324.

Harris, A. R. and Nelson, S. (1999) *Microbiol. Res.*, **154:** 123-130.

Hartman, J. B. and St. Clair, D. A. (1998) *Crop Sci.*, **38**: 1501–1508,.

Hartz, T. K., Miyao, G. Mullen, R. J. Chaim, M. D. Valencia, J. and Brittan, K. L. (1999) *J. Amer. Soc. Hori. Sci.*, **124:** 199-204.

Hasna, M.K., Orgen, E., Persson, P., Martensson, A. and Ramert, B. (2009) *Crop Prot.*, **28**(2): 155-161

Hassan, I. A. Bender, J. and Weigel, I. J. (1999) *Gartenbauwissenschaft*, **64:** 152-157.

Hayashi, M., Sugahara. T. and Nakajima, I. (1998) *Environ. Cont. Biol.*, **36:** 97-104.

Hayati, J. and Varma, J. P. (1985) *Indian J. Virology*, **1:** 152-156.

Hayslip, N. C., Walker, J. M. and Gilbert, D. G. A. (1958) *Dire. Fla. Agri. Exp. Sta.*, S-11, p.11.

Hayslip, N. C.and Iley, J. R. (1967) *Proc. Fla. Sta. Hori. Sci.*,**79**: 132-139.

Hazra, P. and Som, M.G. (2006) In: *Vegetable Science*. Kalyani Publishers, pp.37-108

Hazra, P., Das, R.K., Pandit, M.K. and Sahu, P.K. (2001) *Indian J.Genet.*, **61:** 143–146.

Hazra, P., Longjam, M. and Chattopadhyay, A. (2018) *Sci. Hort.*, **239**:253-258.

Heidrich, E. Lorenz, G. and Schreier, P. (1983) *Food Chem.*, **10:** 285-296.

Heinze, P. H. and Craft, C. C. (1953) *Proc. Amer. Soc. Hort. Sci.*, **62:** 397-407.

Hermus, R. (1986) *Iffigation Farmer*, **13:** 16-17.

Heuvelink, E. (1995) *Phys. Plantarum*, **94**: 447-452.

Hewitt, J. D. and Garvey, T. C. (1987) In: *Tomato Biotechnology*, (Eds. D. J. Nevins and R. A. Jones), vol. 4, Alan R Liss, New York, NY, USA, pp. 45–54.

Hibar, K., Daami-Remadi, M., Ayed, F. and El Mahjoub, M. (2007) *Int. J. Agric. Res.*, **2:** 687-695.

Hobson, G. E. and Harman, J. E.(1986) *Acta Hortic.*, **190:** 167-173.

Hoffmann, M. P., Wilson, L. T., Zalom, F., Hilton, R. J. and Weakley, C. V. (1990) *Californian Agricultutre*, **44**: 20-23.

Holsteijn, G. P. A. van (1989) *Groeten en Fruit*, **46**: 26-27.

Holsteijn, G. P. A. van and Janse, J (1990) *Groenten on Fruit*, **45:** 34-35.

Holsteijn, G.van (1988) *Groeten on Fruit*, **45:** 30-33.

Hong, J.H. and Lee, S. K. (1999) *J. Korean Soc. Hort. Sci.*, **40**: 525-528.

Hossain, M. M. and Mohanty, B. K. (1999) *Orissa J. Hori.*, **27:** 44-47.

Hossain, M., Ara, N., Islam, M., Hossain, J. and Akhter, B. (2014) *Int. J. Agril Res. Innovation Technol.*, doi: 4.10.3329/ijarit.v4i1.21090.

Hossain, M., Imanishi, S. and Matsumoto, A. (1994) *Breed. Sci.* **44**: 405-412, 435-436.

Houghtaling, H. B. (1935) *Bull. Torrey. Bot. Club*, **62**: 242-253.

Howlett, F. S. (1962) *Proc. Plant Sci. Symp. Campbell Soup*.Co., New Jersey, pp. 59-64.

Huang, C.C., Van De Putte, P.M., Haanstra Van Der Merr J.G., MeijerDekens, F. and Lindhout, P. (2000) *WIiley Heredity*, **85**(6), doi: 10.1046/j.1365-2540.2000. 00770.x

Huang, S., Xu, B.S., Tan, B.S., Zhao, Y.P., Wang, J.Q. and Qiao, J.L. (1999) *Acta Agric. Universitatis Henanensis*, **33**: 138-142.

Huang, Y. F., Wang, Q.Y. and Wang, H.T. (1985) *Acta Gen. Sinica*, **12**: 193-199.

Huang, Y., Xu, C., Ma, L., Zhang, K., Duan, C. and Mo, M. (2010) *Eur. J. Plant Pathol.* **126**: 417–422

Humphrey, L. M. (1937) *Cytologia (Tokyo)*, **8**: 306-318.

Hunt, D.W. and Whitefield, G. (1996) *Canadian Entomologist*, **128:** 407-412.

Hurd, R.G. and Graves, C.J. (1985) *J. Hort. Sci.*, **60:** 359-371.

Husain, S.E., James, C., Shields, R. and Foyer, C.H. (2001a) *New Phytol.*, **150**: 65–72.

Husain, S.E., Thomas, B.J., Kingston-Smith, A.H. and Foyer, C.H. (2001b) *New Phytol.*, **150**: 73–81.

Hutton, S.F., Scott, J.W. and Schuster, D.J. (2012) *HortScience*, **47**: 324–327.

Ibrahim, A., El-Gindy, A., Abou-HadidA. and Abdel-Har. A. (1989) *Egyptian J. Hort.*, **16**: 192-198.

Ihsan, S. A. J., Voican, V. and Atanasiu, N. (1988) *Lucrari Stiintifice, Iustitutul Agronomic Nicolae Balcescu', Bucuresti, Seria B, Horticultura*, **31:** 31-8.

Iizuka, H., Tataki, H., Kurihara, K., Suda, T. and Hoshino, F. (1988) *Gunma J. Agric. Res.*, **5:** 55-58.

Iloba, C. and Achor, C. C. (1989) *Trop. Ecol.*, **30**: 233-243.

Ilyubaev, S.U., Karabacv, M. K., Kadyrzhanova, D. K., Aitkhozhin, M. A. and Butenko, R. G. (1988) *Biologiya kul'tiviruemykh kletok i biotekhnologiya*,**1:** J 93-194.

Imenes, S.D.L., Campos, T.D. De, Takernatsu, A. P., Myasto, A. and Silva, MAD. da (1990) *Anais da Sociedade Entomologica do Brasil*, **19:** 281-289.

Inaba, M. and Crandall, P. G. (1987) *Proc. Florida State Hort*. Soc., **99:** 143-145.

Iqbal, S.M., Ghafoor, A. and Haqqani, A.M. (2003). *Int. J. Agric. Bicl.*, **4:** 618-620.

Islam, K., Islam, M.S. and Ferdousi, Z. (2011) *J. Life Earth Sci.*, **6:** 75-80.

Ito, I.I. and Kawai, S. (1994) *Research Bull. Aichi-ken Agril. Research Center*, **26:** 191-199.

Iwahori, S. (1965) *J. Jap. Soc. Hori. Sci.*, **34:** 33-41.

Iwahori, S. and Lyon, J. M. (1970) *J. Amer. Soc. Hort. Sci.*, **95:** 88-91.

Iwahori, S. and Takahashi, K. (1963) *J. Jap. Soc. Hort. Sci.*, **32:** 299-302.

Jain, R. K. and Bhatti, D. S. (1987) *Tropical Pest Management*, **33:** 122-124.

Jain, S. M., Shahin, E. A. and Sun, S. (1988) *Plant Cell Tissue Organ Cult.*, **12:** 189-192.

Jaksch, T. and Kell, K. (1997) *Gemuse (Munchen)*,**33:** 345-346.

Jamwal, R. S., Sant Prakash and Thakur, D. R. (1995) *Himachal J. Agril. Res.*, **21:** 27-31.

Janes, H. W. and McAvoy, R. J. (1991) *Hort. Tech.;* **1:** 110-114.

Janse, J. (1988) *Groenten en Fruit*,**43:** 30-31.

Janse, J. (1990) *Groeten en Fruit*, **45:** 30-33.

Janse, J. and Kreij, C.De (1989) *Groenten en Fruit*, **45:** 34-35.

Jansson, R. K., Peterson, A. R., Halliday, W.R., Mookerjee, P.K. and Dybas, R.A. (1996) *Florida Entomologist*, **79**: 434-449.

Jarret, R. L., Sayama, H. and Tigchelaar, E. C. (1984) *J. Amer. Soc. Hort. Sci.*, **109**: 873–878.

Jarvis, W. R. (1988) *Phytoprotection*, **69:** 49-64.

Jatoi, S.K., Sajid, G.M., Sappal, H., Baloch, M.S., Qureshi, A. and Anwar, R. (2001) *Online J. Biol. Sci.*, **1**: 1141-1143.

Javanmardi, J. and Moradiani, M. (2017) *Int. J. Veg. Sci.*, **23:** 31-41.

Javed, N., Gowen, S.R., Inam-ul-Haq, M, and Anwar, S.A. (2007) *Crop Prot.*, **26** (4): 530-534

Jemakinde, O. A. and Onuveme, I. C. (1980) *Qua. J. Int. Agri.*, **19:** 25-28.

Jenkins, J. A. (1948) *Econ. Bot.*, **2:** 379-392.

Jensen, E. (1989) *Gartner Tidende*, **105:** 114-115.

Ji, P., Momol, M.T., Olson, S.M., Pradhanang, P.M. and Jones, J.B. (2005) *Plant dis.*, **89**: 497-500.

Ji, X., Lu, G., Gai, Y., Zheng, C. and Mu, Z. (2008). *FEMS Micro. Eco.*, **65:** 565–573.

Ji, Y., Scott, J.W., Hanson, P., Graham, E., Maxwell, D. P. (2007a) In: Tomato Yellow Leaf Curl Virus Disease (Ed. H. Czosnek), *Springer, The Netherlands.* pp. 343-362.

Ji, Y., Scott, J.W., Schuster, D.J., Maxwell, D.P. (2009) *J. Amer. Soc. Hort. Sci.*, **134**: 281–288.

Jimenez F. M. (1991) *Plasticulture*, **92**: 11-18.

Johnson, A. (2015) *University of Minnesota Extension* https: //extension.umn.edu/ diseases/leaf-mold-tomato accessed on 10th August, 2020.

Joi, M. B. and Khade, B. M. (1981) *J. Maharashtra Agri. Univ*: **6:** 69-71.

Jolliet, O. Bailey, B. J. Hand, D. J. and Cockshull, K. (1993) *Acta Hortic.*, **328:** 115-24.

Jomova, K. and Valko, M. (2013) *Eur. J. Med. Chem.*, **70**: 102–110.

Jones J.P. and Jones J.B. (1984) *Proc. Fla. State Hortic. Soc.*, **97**: 216–218.

Jones, C.M., Mes, P., Myers, J.R. (2003) *J. Hered.*, **94:**449–456.

Jones, J.B., Jones, J.P., Stall, R.E., and Zitter, T.A. (1991) *Compendium of tomato diseases.* The American Phytopathological Society. APS Press, St. Paul, MN.

Jones, R. A. Smith, P. G. Millett, A. H. and Kimble, K. A. (1980) *HortScience*, **15:** 98.

Joshi, G.C. (1970) *Ph.D. Thesis*, Division of Horticulture, IARI, New Delhi, India.

Joshi, S. Tittal, G. L. Singh, R. P. and Mehta, J. S. (1980) *Haryana J. Hort.* Sci., **9:** 192-194.

Jourdan, P., Montage, T. and Berry, S. (1993) *Plant Sci. Limer.*, **91:** 55-65.

Jovani, V. (1994) *Bull. OEPP*, **24:** 423-427.

Jozef, G., Zuzana, L. and Zuzana, J. (2004) *Biologia Bratislava* **59**: 405-408.

Juvik, J. A., Berlinger, M. J., Ben-David, T. and Rudich, J. (1982) *Phytoparasitica*, **10**: 145–156.

Kabelka, E., Franchino, B.M and Francis, D. (2002) *Phytopathology* **92**: 504-510.

Kadam, D. D. and Deore, B. P. (1990) *J. Maharashtra Agril. Universities*, **15:** 264-265.

Kakar, K. L., Bhala, O. P. and Dhaliwal, M. S. (1990) *J. Insect Sci.*, **3:** 57-61.

Kalia, P., Rattan, R. S. and Saini, S. S. (1980) *South Indian Hort.*, **28:** 144-147.

Kalloo, G. (1986) *Tomato, Allied Publishers* Pvt. Ltd., NewDelhi.

Kalloo, G. (1988) *Vegetable Breeding*, CRC Press, Inc. Boca Ralon, Florida.

Kalloo, G. (1991) *Genetic Improvement of Tomato*, Springer-Verlag, Heidelberg, Germany.

Kalloo, G. and Banerjee, M. K. (1990) In: *11th Meet Eucarpia Tomato Working Group, 6-8 March*, Malaga,Spain, pp. 99-103.

Kalloo, G. and Dudi, B. S. (1982) In: *21st Int. Hori. Cong.* Aug.29-Sept. 4, Hamburg, pp. 1453 (Abst.).

Kaloshian, I., Lange, W.H. and Williamson, V.M. (1995). *Proc. Natl. Acad. Sci. USA*, **92**: 622-625.

Kamaev, I. N. (1996) *Russian Agril. Sci.*, **11**: 28-31.

Kamlanathan, S. and Thamburaj, S. (1970) *Madras Agric. J.*, **57:** 9-10.

Kamlanathan, S. Sunderarajan, S. Thamburaj, S. and Vcnkatram, K. (1969) *Madras Agric. J.*, **56:** 69-573.

Kanski, L., Naumann, M. and Pawelzik, E (2020) *Front. Plant Sci.*, 11: 472, doi: 10.3389/fpls.2020.00472.

Karpenko, V. N., Chernyaeva, I. I. and Sobeshchanski, I. N. V. (1989) *Vcstnik Sel'skokhozyaistvennoi Nauki* (*Moskva*), **9**: 96-99.

Kasrawi, M. A., Khraishi, N. and Tabaza, Y. (1994) *Hort Tech.*, **4:** 394-398.

Kasselaki, A.M., Goumasa, D., Tammb, L., Fuchsb, J., Cooperc, J. and Leifert C. (2011) *NJAS - Wageningen J. Life Sci.*, **58**: 145– 147

Kato, T., Higuchi, M., Endo, R., Maruyama, T., Haginoya, K., Shitomi, Y., Hayakawa, T., Mitsui, T., Sato, R. and Hori, H. (2006) *Pesticide Biochem. Physiol*, **84:** 1–9

Kaur, C., George, B., Deepa, N., Singh, B. and Kapoor, H.C. (2004) *J. Food Sci. Technol.*, **41**: 479-486.

Kaushik, M. P. I., Sharma, J. K. and Singh, I. (1974) *Plant Sci.*, **6**: 51-53.

Kawamura, N., Namikawa, K., Fujiura, J., Ura, M. and Ogawa, Y. (1987b) *Res. Rep. Agri. Machi, Kyoto Univ.*, **17**: 1-7.

Kawamura, N., Namikawa, K., Fujiura, T. and Ura, M. (1987a) *In: Proceedings of the international symposium on agricullural mecltanizallon and international cooperation in high technology era, Univ. of Tokyo*, pp.132-138.

Kawchuk, L.M., Hachey, J. and Lynch, D.R. (1998) *Genome*, **41**: 91-95.

Kayanas, K. and Surmeli, N. (1995) *Turkish J. Agric. Forestry*, **19:** 277-285.

Kegler, H. (1994) *Arch. Phytopath. Plant Prot.*, **29:** 119-132.

Keinath, A. P. DuBose, V. B. and Rathwell, P. J. (1996) *Plant Dis.*, **80:** 1277-1282.

Keithly, J. H., Yokoyama, H. and Gausman, H. (1990) *Plant Growth Regulation*, **9:** 127-316.

Kelbert, D. G. A. and Walter, J. M. (1954) *Circ., Fla. Agric. Exp. Sta.*, **S-72**: 8.

Kemp, G. A. (1965) *Proc. Amer. Soc. Hort.* Sci., **86:** 565-568.

Kemp, G. A. (1968) *Canaian. J. Plant Sci.*, **48**: 281-286.

Keularts, J. L. W. and Lindquist, R. K. (1989) *Environment. Entomol.*, **18**: 499-503.

Kevany, B.M., Taylor, M.G. and Klee, H.J. (2008) *Plant Biotechnol. J.*, **6:** 295–300.

Kevany, B.M., Tieman, D.M., Taylor, M.G., Dal Cin, V. and Klee, H.J. (2007) *Plant J.*, **51:** 458– 467.

Khachik, F., Carvalho, L., Bernstein, P.S., Muir, G.J., Zhao, D.Y. and Katz, N.B. (2002) *Exp. Biol. Med.*, **227**: 845–851.

Khadivi-Khub, A., Zamani, Z. and Bouzari, N. (2008) *Hort. Environ. Biotechnol.*, **49:** 188–196.

Khalid, M., Shaukat, A., Aslam, M. and Ahmad, S. (1995) *Pakistan J. Phytopath.*, **7:** 128-131.

Khan, M. A., Ahmad, M. and Saeed, M. A. (1995a) *Pakistan. J. Phytopath.*, **7**: 166-168.

Khan, M., Ayub, M. and Mohibullah, J. (1995b) *Pakistan J. Nematol.*, **13:** 25-28.

Khan, T. A. and Saxena, S. K. (1996) *Pakistan J. Nematol.*, **14:** 1 I 1-116.

Kharkongar, H.P., Khanna, V.K., Tyagi, W., Rai, M. and Meetei, N.T. (2013) *Agrotechnol* **S11**: 004. doi: 10.4172/2168-9881.S11-004

Khavari, N. R. A. (1996) *Photosynthetico*, **32:** 471-474.

Khoshkam, S., Seyed, Z. and Aeen, A. (2014) *Int. J. Farming Allied Sci.*, **3**: 659-663.

Khudairi, A. K. (1972) *Amer. Scientist*, **60**: 596- 707.

Khurramov, Sh. Kh. (1990) *Zashchita Rastenii*, **4:** 12-13.

Kikuta, K. and Frazier, W. A. (1947) *Proc. Amer. Soc. Hort. Sci.*, **49**: 256-262.

Kim, B., Seo, Y., Jang, K.S. and Cho, K.Y. (1997) *Korean J. Plant Path.*, **13:** 257-261.

Kim, I. and Jeong, C. (1996) *J. Korean Soc. Hort. Sci.*, **37:** 187-192.

Kim, K. Y. (1988) *Research Reports of the Rural Dev. Administration. Horticulture, Korea Republic*, **30:** -35.

Kim, K. Y., Takahashi, K. and Nagaoko, M. (1982) *J. Korean Soc. Hort. Sci.*, **33**: 93-108.

Kim, S. J., Kwack, S. N. and Lee, C. H. (1984) *Res. Rep. Off. Rural Dev.* (Korea), **26:** 30.

Kinet, J. M. (1977) *HortScience*, **6**: 15-26.

Kinzer, S. M., Schwager, S. J. and Mutschler, M. A. (1990) *Theor. Appl. Genet.*, **79**: 489–496.

Kirimi J.K., Itulya, F.M. and Mwaja, V.N. (2011) *Afr. J. Hort. Sci.*, **5:** 50-60

Klein-Lankhorst, R. M., Vermunt, A., Weide, R., Liharska, T. and Zabel, P. (1991) *Theor: Appl. Genet.*, **83:** 108-114.

Kluge, R. A., Rodrigues, D. S. and Minami, K. (1998) *Horticulture Brasileira*, **16:** 4-6.

Kluge, R.A. Rodrigues D.S. Kalil, G. P. C. Russo, R., Lucas, M. B. and Minami, K. (1996) *Scientia Agricola*, **53:** 6-13.

Kobatake, H., Osaki, T. and Inovyl, T. (1981) *Proc. Kansas Plant Prod. Sci.*, **19:** 238-244.

Kobayashi, R. S., Stommel, J. R. and Sinden, S. L. (1996) *Plant Cell Tissue Organ Cult.*, **45:** 73-78.

Kochevenko, A., Ratushnyak, Y. Kornyeyev, D. Stasik, O. Porublyova, L. Kochubey, S. Suprunova, T. and Glcba, Y. (2000) *Plant Cell Rep.*, **19:** 599-597.

Kodandaram, M.H., Halder, J. and Rai, A.B. (2014) *Indian J. Plant Prot.*, **42**: 333-337

Kohler, G. R. and St. Clair, D. A. (2005) *J. Econ. Entomol.*, **98**: 988–995.

Koning, A. De (1988) *Groenten en Fruit*, **43:** 32-33.

Koning, A. N. M. De (1989) *Acla Hortic.*, **260:** 267-273.

Korycinska, A. and Moran, H. (2009) *Plant Pest Factsheet*. Sand Hutton, York, UK: FERA.

Koul, B., Srivastava, S., Sanyal, S., Tripathi, B., Sharma, V. and Amla, D.V. (2014) *SpringerPlus*, **3:** 84.

Kramer, M. G. and Redenbaugh, K. (1994) *Euphytica*,**79:** 293-297.

Kreij, C. De (1996) *J. Plant Nutr.*, **19:** 361-377.

Krishnamoorty, A. S. and Bhaskaran, R. (1990) *J. Biol. Cont.*, **4:** 52-54.

Krzysko, K. and Rogozinska, J. H. (1975) *Ilodowla Roslin Aklimatyzacja I Nasiennictwo*, **19**: 23-27.

Ku, H.M., Doganlar, S., Chen, K.Y. and Tanksley, S.D. (1999) *Theor. Appl. Genet.*, **99:** 844–850.

Kudryakova, N. V. and Kalloo, O. (1991) *Theor. Appl. Genet.*, **14:** 277-281.

Kuksal, R. P., Singh, R. D. and Yadav, J. P. (1977) *Prog. Hort.*, **9:** 13-20.

Kulik, A. V. (1990) *Coshcliita Rastenii*, **7:** 40.

Kumar, D., Kumar, R., Kumar, S., Bhardwaj, M.L., Thakur, M.C., Kumar, R., Thakur, K.S., Dogra, B.S., Vikram, A., Thakur, A. and Kumar, P. (2013). *Int. J. Veg. Sci.*, **19**(4): 313-323

Kumar, J., Thareja, R. K., Banarjee, M. K.and Arora, S. K.(1998) *Res. Dev Rep.*, **5:** 51-54.

Kumar, M., Meena, M.L., Kumar, S., Maji, S. and Kumar, D. (2013) *Asian J. Hort.*, **8:** 616-619

Kumar, R., Kumar, R., Singh, R.K., Rai, V.P., Singh, M. and Singh, P.K. (2017) *Curr. Hort.*, **5**(1): 30-39.

Kumar, R., Ram, C.N., Yadav, G.C., Deo, C., Vimal, S.C. and Bhartiya, H.D. (2014) *Plant Arch.*, **14**(1): 367-371.

Kumar, S. and Sugha, S.K. (2000) *Indian Phytopath.*, **53:** 105-106.

Kumar, Sanjeet, Banerjee, M. K. and Kalloo, G. (2000) *Veg. Sci.*, **27:** 1-24.

Kumar, T. P., Tewari, R. N. and Pachauri, D. C. (1997) *Veg. Sci.*, **24:** 34-38.

Kunavin, G. A. (1982) *Vernshk. SH Nauki Kazakslttana*, **8:** 4647.

Kundu, P., Adhikary, N.K., Saha, M., Ghosal, A. and Sahu, N.C. (2019) *Curr. J. Appl. Sci. Technol.*, **35**(4): 1-8.

Kunik, T., Salomon, R., Zamir, D., Navot, N., Zeidan, M., Michelson, I., Gafni, Y. and Czosnck, H. (1994) *Biotechnol.*, **12**: 500-504.

Kuo, G. G., Tsay, S. S., Chen, B. W. and Lin, P. Y. (1982) *HortScience*, **17:** 76-78.

Kusainova, G. S. (1985) In: *Progress. Priemy v Selektsii Plodoovoshch. Kul'tur.*, pp.43-48.

Kuwata, I., Saitoh, H. and Shimada, K. (1994) *Bull. Aomori Agril. Expt. Station*, **34:** 1-10.

Kyzlink, V. Micova, K. and Jelink, R. (1981) *Sbornik iysoke Skoly Chemicko Technologicke V Praze, Potraviny E*, **51:** 69-83.

Lamn, R. (1950) *Hereditas*, **36:** 509-511.

LanZhuang, C. and Adachi, T. (1996) *Plant Breed.*, **115**: 251-256.

Larkin, R. P. and Fravel, D. R. (1998) *Plant Dis.*, **82:** 1022-1028.

Larson, R. E. and Paur, S. (1948) *Proc. Amer. Soc. Hort. Sci.*, **52:** 354-364.

Laterrot, H. (1997) *Proc. EUCARPIA Tomato, Meeting of XIII EUCARPIA Toamto group*, 19–23 January, Jerusalem, Israel, p.18.

Latunde-Dada, A. O. (1993) *Plant Pathol.*, **42:** 522-529.

Lawrence, S.D. and Novak, N.G. (2006) *Biotechnol. Lett.* **28:** 593–599.

Laykin, S., Edan, Y., Alchanatis, V., Regev, R., Gross, F., Grinshpun, J., Barlev, E., Fallik, E. and Alkalai, S. (1999) In*: ASAEN Annual International Meet, CSAE-SCGR,Toronto, Ontario, Canada*, p.16.

Lebeda, A. and Micslcrova, B. (1998) *Zahradnictvi*, **25:** 53-65.

Lecomte, L., Saliba-Colombani, V., Gautier, A., Gomez-Jimenez, M.C., Duffé, P., Buret, M. and Causse, M. (2004) *Mol. Breed.*, **13:** 1–14.

Lee, J.M. (2003) *Chronica Hort.*, **43:** 13–19.

Lee, J.W. and Kim, K.Y. (1999) *J. Korean Soc. Hori. Sci.*, **40:** 412-415.

Lee, S.J., Kim, S., Kim, J.C., Lee, M.R., Hossain, M.S., Shin, T.S., Kim, T.H., Kim, J.S. (2017) *Bi°Control.* **62:** 639–648.

Lee, Y. B. and Lee, B. Y. (1994a) *J. Korean Soc. Hort. Sci.*, **35:** 421428.

Lee, Y. B. and Lee, B. Y.(1994b) *J. Korean Soc. Hort. Sci.*, **35:** 309-317.

Leela, D., Dhuria, H. S. and Shukla, V. (1972) *Proc. 3rd Int. Symp. Subtrop. and Trap. Hort., Bangalore*, p.77.

Leeper, P. W., Cox, E. L., Scully, B. T., Oerther, G. F., Burns, E. E. and Lineberger, R. D. (1993) *HortScience*, **28:** 346-347.

Lehmann, C. O. and Schwanitz, F. (1965) *Kulturpflanze*, **8**: 545-585.

Leite, D., Bresciani, A.F., Groppo, A.G., Pazini, W.C. and Gravena, S. (1995) *Anais da Sociedade Entomologica do Brasil*, **24:** 27-32.

Leonardi, C., Guichard, S. and Bertin, N. (2000) *Sci. Hort.*, **84:** 285-296.

Levin, I., Ric de Vos, C.H., Tadmor, Y., Bovy , A., Lieberman, M., Oren-Shamir, M., Segev, O., Kolotilin, I., Keller, M., Ovadia, R., Meir, A. and Binob, R. J. (2006) *Isr. J. Plant Sci.*, **54** : 179–190.

Levin, I., Frankel, P., Gilboa, N., Tanny, S., Lalazar, A. (2003) *Theor. Appl. Genet.*, **106**: 454–460.

Levin, I., Gilboa, N., Yeselson, E., Shen, S. and Schaffer, A. A. (2000) *Theor. Appl. Genet.*, **100**: 256–262.

Levin, I., Lalazar, A., Bar, M. and Schaffer, A.A. (2004) *Ind. Crop. Prod.*, **20**: 29–36.

Lewis, D. (1953) *J. Hort. Sci.*, **28:** 207-220.

Lewis, M.L.I. and Miller, S.A. (2005). *Acta Hortic.*, **695:** 197-204.

Li, B.J., Zhu, G.R., Zhao, K.H. and Qiao, Y. (1999) *Acta Phytopathol. Sinica*, **29:** 63-67.

Li, M., Schmidt, J.E., LaHue, D.G., Lazicki, P., Kent, A., Machmuller, M.B., Scow, K.M. and Gaudin, A.C.M. (2020) *Front. Plant Sci.*, **11:** 360, doi: 10.3389/fpls.2020.00360.

Li, X.Q., Wei, J.Z., Tan, A. and Raffi, V. (2007) *Plant Biotechnol. J.*, 5: 455–464.

Li, Y. X. Geng, Y. X. Gao, W. Q. and Li, J. G. (1983) *Acta Hort. Sinica*, **10:** 115-118.

Li, Y.H. (2013) The Connecticut agricultural experimental Station, http: //www.ct.gov/caes

Ligade, P.P., Bahadur, V. and Gudadinni (2017) *Int. J. Curr. Microbiol. Appl. Sci.*, **6**: 1775-1783.

Lim, G.T.T., Wang, G.P., Hemming, M.N., Basuki, S., McGrath, D.J., Carroll, B.J. and Jones D.A. (2006) *Australasian Plant Path.*, **35** 671-680.

Lima, J. O. G. De and Machado, W. A. (1996) *Revista Ceres.*, **48:** 337-345.

Lin, B.W., Gong, C.C., Song, H.F. and Cui, Y.Y. (2017) *Br. J. Pharmacol.*, **174**: 1226–1243.

Lin, T. and Lan, G.H. (1997) *J. Agric. Forestry*, **46:** 2946.

Lindstrom, E. W. and Humfrey, L. M. (1933) *Genetics*, **18:** 193-209.

Lipari, V. (1982) *Rivista della Ortoflorofrutticoltura Italiana*, **66:** 59-73.

Lipp-Joao, K. H. and Brown, T. A. (1993) *Plant Cell Rep.*, **12**: 422-425.

Lipp-Joao, K. H. and Brown, T. J. (1994) *J. Exp. Bot.*, **45:** 641-047.

Lippman, Z. and Tanksley, S. D. (2001) *Genetics*, **158**: 413–422.

Liptay, A. and Hoffman, K. (1988) *Acta Hortic.*, **230:** 533-538.

Liptay, A. and Moore, L.A. (1989) *Acta Hortic.*, **253:** 225-23l.

Liptay, A., Tan, C.S., Ramsey, R., Drury, C.F., Hunt, D.W.A., Jewett, T., Weaver, S. and Woodrow, L. (1998) *Canadian J. Plant Sci.*, **78:** 473-475.

Liu, J. and Wang, X. (2020) *ÿþPlant Methods* **16:** 83, doi.org/10.1186/s13007-020-00624-2

Liu, J., Tian, S., Meng, X. and Xy, Y. (2007) *Postharvest Biol. Technol.*, **44**(3): 300-306.

Liu, T., Cheng Z., Meng, H., Ahmad, I. and Zhao, H. (2014) *Sci. Hort.*, **170**: 159-168.

Locascio, S.J., Hochmulli, G.J., Rhoads, F.M., Olson, S.M., Smajstrla, A.G. and Hanlon, E.A. (1997) *HortScience*, **32:** 230-235.

Locke, T. and Fletcher, J. T. (1988) *Plant Pathol.*, **37**: 381-384.

Lohithaswa, H. C., Bhushana, H. O. and Basavarajaiah, D. (1998) *Crop Improve.*, **25:** 135-38.

Lozano, M. G., Escobar, I., Lorenzo, P., Guzmun, M. and Urrestarazu, M. (1998) *Acta Hortic.*, **458:** 363-367.

Lu, C. G., Xu, H. L., Yian, R. C., Yu, W. G. and Zhu, D.W. (1995) *Acta Hortic.*, **402:** 141- 150.

Luckwill, L. C. (1943) *Royal Hort. Soc. J.*, **68:** 19-25.

Luckwill, L.C. (1943) *Aberdeen University Studies*, **120:** 1–44.

Lugo, M. L., Liu, L. C. and Almodovar. L. (1988) *J. Agric., Univ. Puerto Rico*, **72:** 291-296.

Lukyan, A. M. (1988) In: *Soversh Tekhnol. Priemov Vozdel. Ovoshcheii Vyssh. S'edob. Gribov*, 26-29.

Lukyanenko, A. N. (1991) *In: Genetic Improvement of Tomato* (Ed. G. Kalloo), vol. 14, Springer, Berlin, Germany, pp. 99–119.

Lukyanenko, A.N. (1991) In: *Genetic Improvement of Tomato, Springer-Verlag, Heidelberg, Germany*, pp. 99-119.

Lurie, S. and Klein, J. D. (1991) *J. Amer. Soc. Hori. Sci.*, **116:** 1007-1012.

Lurie, S. and Sabehat, A. (1997) *Postharvest Biol. Technol.*, **11:** 57-62.

Lurie, S., Klein, J. D., Watkins, C., Ross, G., Boss, P. and Fcruson, I. F. (1993) *Acta Hortic.*, **343:** 283-285.

Lutwama, J. J. and Matanni, B. A. (1988) *Bull. Entomol. Res.*,**78:** 173-179.

Lutz, J. M. (1944) *U.S. Dept. Agri. Circ.*, p. 695.

Maboko, M.M. and Du Plooy, C.P. (2014) *HortScience*, **49**: 438–440. 2014.

MacArthur, J. W. and Chiasson, L. P. (1947) *Genetics*, **32:** 165-177.

Macedo, M.A., Albuquerque, L., Maliano, M. R., Souza, J. O., Rojas, M., Inoue-Nagata, A. and Gilbertson, R. (2017) *Archives of Virology*, DOI: 10.1007/s00705-017-3662-0.

Madhavi, D.L. and Salunkhe, D.K. (1998) In: *Handbook of Vegetable Science and Technology: Production, Composition, Storage and Processing* (Eds. D.K.Sakunkhe, S.S. Kadam), Marcel Dekker Inc, New York, USA.

Mahadevan, N. R. (1998) *Insect Environmen*, **3**: 92.

Maharaj, R., Arul, J. and Nadeau, P. (1999) *Postharvest Biol. Technol.*, **15**: 13-23.

Maharana, T., Pati, P. N. and Sahu, G. S. (1995) *Adv. Hort. Forestry*, **4**: 91-94.

Mahendrakar, P., Mulge, R. and Madalageri, M.B. (2005) *Karnataka J. Hort.*, **1:** 1-6.

Maisonneve, B. and Philouze, J. (1982) *Agronomic*, **2:** 453-458.

Majidi, H., Minaei, S., Almassi, M. and Mostofi, Y. (2014) *J. Food Sci. Technol.*, **51**(9): 2155–2161.

Malfa, G. La (1993) *Agricoltura Mediterranea*, **123:** 267-272.

Malfa, G. La, Leonardi, C. and Romano, D. (1995) *Agricoltura Mediterranea*, **125:** 404-412.

Malik, G., Masoodi, L., Nabi, S.U., Sharma, A. and Singh, D.B. (2017) ICAR-Central Institute of Temperate Horticulture, J&K, India.

Mallick, M. F. R. and Muthukrishnan, C. R. (1980) *Veg Sci.*, **7:** 6-13.

Mallik, S.C. and Bhattacharya, B. (1996) *Environ. Ecol.*, **14**: 301-303.

Mametsuka, S., Yamamoto, Y. and Sibato, Y. (1991) *Agric. Hort.*, **66:** 937-942.

Mandaokar, A.D., Goyal, R.K., Shukla, A., Bisaria, S., Bhalla, R., Reddy, V.S., Chaurasia, A., Sharma, R.P., Altosaar, I., and Kumar, P.A. (2000) *Crop Prot.*, **19:** 307-312

Mane, R.S. (2013) *Ph.D. Thesis*, University of Agricultural Sciences, Dharwad, India.

Mao, W., Lewis, J. A., Lumsden, R. and Hebbar, K. P. (1998) *Crop Prot.*, **6**: 535-542.

Marajan, W.A., Hadad, M.A., Gafer, M.O., Khalifa, K.M., Hatim, S.A. and Abdelrhman, M.A. (2017) *IJRDO-J. Agriculture Res.*, **3:** 61-73.

Marangoni, A. G. Smith, A. K., Yada, R. Y. and Stanley, D. W. (1989) *J. Amer. Soc. Hort. Sci.*, **114**: 958-962.

Marchesi, G. (1986) *Colture Protette*, **15:** 39-42.

Maresca, A (1998) *Macchine e Motor Agricoli*, **56:** 65-67.

Marre, E. and Murneek, A. E. (1953) *Plant Physiol.*, **28:** 255-266.

Marris, G. C. and Edwards, J. P. (1994) *Proc. Brighton Crop Prot. Cong, Pests and Dis.*, **3:** 1139-1144.

Martin, C. (1988) *Revue Horticole*, 283: 26-27.

Martin, G.B., Vicente, M.C. and Tanksley, S.D. (1993) *Mol. Plant Microbiol. Interactions* **6**: 26-34.

Martin, G.B., Williams, J.G.K. and Tanksley, S.D. (1991) *Proc. Natl. Acad. Sci.*, **88:** 2336-2340.

Martinez, E. R. (1987) *Centro Agricola*, **14**: 42-49.

Martinez, E. R. and Careaga, C. L. (1987) *Centro Agricola*,U: 68-76.

Martinez, G. S. B., Garbi, L. M., Arturi, F. M. J. and Asborno, L. M. D. (1998) *Agro-Giencia*, **14:** 201-206.

Martínez-Damián, M.T., Cano-Hernández, R., Moreno-Pérez, E. C., Sánchez-del Castillo, F., and Cruz-Álvarez, O. (2019) *Revista Chapingo Serie Horticultura*, **25**(1): 29-43.

Masarirambi, M.T., Brecht, J.K. Sargent, S.A. and Sims, C.A. (1996) *Proc. Foridia State Hort. Soc.*, **108**: 268-272.

Maschio, L. M. de and Sousa, G. F. (1982) *Pesquisa Agropecuaria Brasileira, Agronomia*, **17:** 1309-1315.

Masuda, M., Furuichi, T., Ma, Y. and Kato, K. (1998) *J. Jap. Soc. Hort. Sci.*, **67:** 583-588.

Mathur, K. and Bhatnagar, G. C. (1993) *Indian J. Mycol. Plant Pathol.*, **23:** 141-144.

Mathur, K. and Shekhawat, K. S. (1986) *Indian J. Mycol. Plant Pathol.*, **16:** 235-236.

Matsumoto, A., Imanishi, S., Hossain, M., Escalante, A. and Egashira, H. (1997) *Breed Sci.*, **47:** 327-333.

Mauromicale, G., Davino, M., Nuciforo, S., Sortino, O., Raccuia, S. and D'Urso, F. (1996) *Colture Protette*, **25:** 73-79.

Mayno, A. (1998) *Gemuse (Muncher)*, **34:** 668-670.

Mazumder, N. (1998) *J. Mycol. Plant Pathol.*, **28:** 189-192.

McAvoy, R. J. and Giacomelli, G. A. (1986) *Acta Hortic.*, **187:** 107-113.

McAvoy, R. J. and Janes, H. W. (1988) *Sci. Hort.*, **35:** 161-166.

McClean, A. P. D. (1940) *Dept. Agri. For. S. Africa Sci. Bull.*, pp. 225.

McCue, G. A (1952) *Ann. Missouri Bot. Gdn.*, **39:** 289-348.

Mcferran, J. Goode, M. J. Scott, S. J. and Montgomery, F. W. (1989) *HortScience*, **24:** 712-713.

Mcferran, J., Goode, M. J. and Sistrunk, W. A. (1982) *Arkansas Farm Res.*, **31**: 1.

McGlasson, W. B. and Adata, I. (1977) *Aust. J. Pl. Physiol.*, **3**: 809-817.

McGovern, R. J. Vavrina, C. S. Noling, I. W. Datnolf, L. A. and Yonce, H. D. (1998) *Plant Dis.*, **82:** 919-923.

McGuire, D. C. and Rick, C. M.(1954) *Hillgardia*, **23:** 101-124.

McLaughlin, R. J. and Sequeira, L. (1988) *American Potato J.*, **65**: 255-268.

McMohan, R. W., Stewart, C. R. and Gladon, R. J. (1990) *J. Amer. Soc. Hori. Sci.*, **115:** 292-301.

Mehta, A. K. and Mathai, P. J. (1975) *Haryana J. Hori. Sci.*, **4:** 167-176.

Mehta, P. M. and Bhavarnaryana, K. (1981) *Comp. Physiol. Ecol.*, **6**: 177-179.

Mekuria, T., Blaeser, P., Steiner, U. and Dehne, I-I. W. (1999) *In: 12th International Reinhardsbrunn Symp.* Germany, pp.483-490.

Melchers, G., Mohri, Y., Watanabe, K., Wakabayashi, S. and Harada, K. (1992) *Proc. Nat. Acad. Sci.* (USA), **89**: 6832-6836.

Menda, N., Semel, Y., Peled, D., Eshed, Y. and Zamir, D. (2004) *Plant J.*, **38:** 861–872.

Meneses, J. F., Monteiro, A. Aand Abreu, P. E. (1994) *Plasticulture,* **101**: 3-12.

Mennan, H., Ngouajio, M., Isýk, D. and Kaya, E. (2009) *Phytoparasitica*, **37:** 385–396.

Menzies, J. G. and Ehret, D. L. (1997) *Acta Hortic.*, **450:** 466-457.

Mes, P.J., Boches, P. and Myers, J.R. (2008) *J. Amer. Soc. Hortic. Sci.*, **133**: 262–269.

Mesbah, L.A., Kneppers, R.J.A., Takken, F.L.W., Laurent, P., Hille, J. and Nijkamp, H.J.J. (1999) *Mol. Gen. Genet.* **261**: 50-57.

Messiha, N., van Diepeningen, A., Farag, N., Abdallah, S., Janse, J., van Bruggen, A. (2007) *Eur. J. Plant Pathol.*, **118:** 211–225.

Messina, V., Dom´ýnguez, P.G., Sancho, A.M., de Reca, N.W., Carrari, F. and Grigioni, G.(2012) *Int. J. Electrochemistry,* Volume 2012, Article ID 687429, 7 pages doi: 10.1155/2012/687429

Miah, A. J., Rahman, M. Z., Adhikary, S. K. and Rahman, M. M. (1995) *Bangladesh J. Plant Pathol.*, **11:** 17-18.

Michalik, H.B., Akowski, J. and Potaczck, H. (1986) *Biuletyn Warzywniczy*, **26:** 291-308.

Michel, V. V., Wang, J. F.,Midrnore, D. J. and Hartman, G. L. (1997) *Plant Pathol.*, **46:** 600-610.

Midan, A. A., Malash, N. M. El-sayed, M. M. (1985) *Ann. Agric. Sci. Ai11 Shams Univ*; **30**: 1413-1431.

Mikhov, M. and V'lchev, S. (1995) *Rasteniev "dni Nauki*, **32:** 9-12.

Miller, J. C. and Tanksley, S. D. (1990) *Theor Appl. Genet.*, **80**: 437-448.

Miller, J. C. and Tanksley, S. D. (1990) *Theor. Appl. Genet.*, **80:** 437–448.

Milligan, S.B., Bodeau, J., Yaghoobi, J., Kaloshian, I., Zabel, P. and Williamson, V.E. (1998) *The Plant Cell*, **10**: 1307–1319.

Minuto, A., Spadaro, D., Garibaldi, A., and Gullino, M.L. (2006) *Crop Prot.*, **25**: 468-475

Miras, B., Issa, S. and Jaffe, K.(1997) *Agronomia Tropical* (*Maracay*), **47**: 315-130.

Miresmailli, S., Isman, M.B. (2006) *J. Econ. Entomol.*, **99** (6): 2015–2023.

Mishra, A. (1996) *J. Res., Birsa Agril. Univ.*, **8**: 25-31.

Misra H.P. (2012) *SAARC J. Agri.*, **10**: 63-70.

Mittal, S.P., Thomas, T. A. and Singh, H. B. (1962) *Indian J. Hort.*, **9**: 50-60.

Mizahi, Y. (1982) *Plant Physiol.*, **69**: 966-970.

Mladenov, M., Ilieva, E. and Ali, A.Z. (1989) *Rasteniev" dni Nauki*, **26**: 86-91.

Moccia, S. and Katcherian, F.(1999) *Horticulture Argentina*,**16:** 5-10.

Moffett, M. L. (1986) *ACIAR Bacterial Wilt News Letter*, **1**: 1-2.

Mohammad, S. E. (1988) *Mesopotamia J. Agric.*, **20**: 359-366.

Mohan, K. S., Asokan, R. and Gopalakrishnan, C. (1996) *Pest Manage. Hort. Ecosyst.*, **2**: 1-8.

Mohri, K., Umeda, S. and Tsurumi, G. (1987) *Bull. Univi Osaka Perfect*, 13.**39:** 85-90.

Mohseni-Moghdam, M. and Doohan, D. (2017) *Weed Technol.*, **31**: 441-446

Monaco, C., Perello, A., Alippi, I. E. and Pasquare, A. O. (1991) *Adv. Hort. Sci.*, **5:** 92-95.

Moneruzzaman, K.M., Hossain, A.B.M.S., Sani, W., Saifuddin, M. and Alenazi, M. (2009) *Aus. J. Crop Sci.*, **3**: 113-121

Monforte, A. J. and Tanksley, S. D. (2000) *Genome*, **43**: 803–813.

Monforte, A. J., Asins, M. J. and Carbonell, E.A. (1996) *Theor. Appl. Genet.*, **93:** 765-72.

Monkman, K. D. (1992) *Monthly Bull. Dept. of Agriculture, Fisheries and Parks. Bermuda*, **63**: 19-21.

Monma, S., Sakata, Y. and Matsunaga, H. (1997) *Japan Agric. Res. Quart.*, **31**:195–204.

Moore, E. L. and Thomas, W. O. (1952) *Proc. Amer. Soc. Hort. Sci.*, **60:** 289-294.

Moore, W. D. and Reynard, G. B. (1945) *Phytopath.*, **35:** 933-935.

MooSeong, I., Choi, E.J. and Choi, J.E. (1996) *RDA J. Agric. Sci. Crop Prot.*, **38:** 473-476.

More, N. B. and Shinde, P. M. (1991) *J. Maharashtra Agril. Univ*,**16:** 169-712.

More, R. (1978) *Kali Kenkyu* **15** Kali Kenyu, Tokyo.

Moreau, P., Thoquet, P., Olivier, J., Laterrot, H. and Grimsley, N. (1998) *Molecular Plant Microbe Interactions*, **11**: 259-269.

Mori, H., Nakagawa, N., Ono, T., Yarnangishi, N. and Imaseki, H. (1992) *Proc. Internatl. Symp. Cellular and Molecular Aspects of Biosynthesis and Action of the Plant Hormone Ethylene, Agen*, France, August 31-September 4, 1991, pp.1-6.

Mortley, D.G., Smith, C. D. and Demchak, K.T. (1991) *J. Amer. Soc. Hort. Sci.*, **116**: 659-662.

Moss, G. (1982) *Proc. 21st Int. Hort. Cong.*, Hamburg, GFR, Abst. No. 1931.

Moussa, A. G. and Doss, M. (1986) *Acta Hortic.*, **190:** 451-454.

Mozarkar, D. R., Paradkar, V. K., Upadhyay, P. C. and Sharma, T. R. (1991) *Orissa J. Hort.*, **19**: 27-29.

Mu˜nos, S., Ranc, N., Botton, E. *et al.* (2011) *Plant Physiol.*, **156**: 2244–2254.

Muigai, S. G., Bassett, M. J., Schuster, D. J. and Scott, J. W. (2003) *Phytoparasitica*, **31**: 27–38.

Mukherjee, D., Maurya, P.K., Banerjee, S., Bhattacharjee, T., Chatterjee, S., Chatterjee, S., Mandal, A.K., Maji, A. and Chattopadhyay A. (2020) *Int. J. Veg. Sci.*, **26**: 211-248.

Mukhopadhyay, S. and Som, M. G. (1969) *Indian J. Mycol. Res.*, **7**: 23-28.

Mukhtar, K., Ahmad, R., Javed, N. and Khan, S. H. (1994) *Pakistan J. Phytopath.*, **6:** 152-154.

Muller, C. H. (1940a). *USDA Misc. Pub.*, **382:** 1–28.

Muller, C. H. (1940b) *Nat. Hort. Mag*, **19:** 157-160.

Murai, T. (1988) Bull. *Shimane Agril. Expt. Sta.*, **23**: 1-73.

Mursanov, V. P. and Otreper, V. G. (1977) *Agrokhimiya*, **8:** 83-86.

Musard, M. (1988) *Iufos, Centre Technique Interprofessionnel des Fruits el Legumes*, France, **46**: 19-24.

Mutschler, M. A (1984) *J. Amer. Soc. Hort. Sci.*, **109**: 500-503.

Mutschler, M. A., Tanksley, S. D. and Rick, C. M. (1987) *Rep. Tomato Genet. Coop.*, **37**: 5–34.

Mutschler, M., Guttieri, M., Kinzer, S., Grierson, D. and Tucker, G. (1988) *Theor. Appl.Genet.*, **76:** 285–292.

Mutschler, M.A., Doerge, R.W., Liu, S.C., Kuai, J.P., Liedl, B.E.and Shapiro, J.A. (1996) *Theor. Appl. Genet*, **92**: 709–718.

Nagal, G., Verma, K.S. and Rathore, L. (2016) *Adv. Life Sci.* **5**(3): 1081-1084.

Nagata, M., Mori, H., Tabei, Y., Sato, T., Hirai, M., Imascki, H., Nito, N., Looney, N. E., Nevins, D. J. and Ha levy, A. H. (1995) *Acta Hortic.*, **394:** 213-218.

Nagetey, F. K. and Mao, L.C. (1999) *Acta Agriculturae Zhejiangeusis*, **11**: 67-72.

Nainwal, N. C., Jaiswal, R. C. and Sanjiv Kumar (1992) *Prog. Hort.*, **24:** 70-73.

Nakaho, K., Inoue, H., Takayama, T. and Miyagawa, H. (2004) *J. Gen. Plant Path.*, **70:** 115–119.

Nakata, K. Tanaka, H.,Yano, K.andTakagi,M. (1992) *Japanese J. Breed.*, **42:** 487-495.

Nandpuri, K. S., Kunwar, J. C. and Singh, H. (1971) *Punjab Hort. J.*, **11**: 87-89.

Nandpuri, K. S., Singh, S. and Mahajan, R. (1989) *Indian J. Nematology*, **18**: 121-122.

Nargis, S., Gunasckaran, M., Lakshmi, S. and Selvakumar, P. (1998) *Orissa J. Hort.*, **26:** 47-49.

Nariani, T. K. (1968) *Plant Dis. Rep.*,**52:** 595-596.

Nasrin, T.A.A., Molla, M.M. and Hossain, M.A. (2008) *Bangladesh J. Agric. Res.*, **33**: 579-585

Naumann, K. (1980) *Nachrichtenblatt fur den Pflanzenschutz in der DDR*, **34:** 157-159.

Navarro, M. A. (1988) *Colloques de l'INRA*,No. **43**, pp.453-458.

Nederhoff, E. and Rijsdijk,T. (1990) *Groenten en Fruit*, **45:** 28-29.

Neelamathi, D. and Karivaratharaju, T. V. (1998) *Seed Res.*, **26:** 92-93.

Nelson, R.S., McCormick, S., Dube, P., Niedermeyer, J., Anderson, E. J., Fraley, R.T. and Beachy, R.N. (1987) *Plant Physiol.*, **63** (4 (Supplement)): 139.

Newman, P. O., Krishnaraj, S. and Saxena, P. K. (1996) *Int. J. Plant Sci.*, **157:** 554-560.

Nightingale, G. T., Schermerhorn, L. G. and Robbins, W. R. (1928) *N. J. Bull.*, p. 461.

Nikoumanesh, K., Ebadi, A., Zeinalabedini, M. and Gogorcena, Y. (2011) *Sci. Hort.*, **129**: 108–118.

Nilsen, S., Hovland, K., Dons, C. and Sletten, S. P. (1983) *Sci. Hort.*, **20:** 1-14.

Niranjana, R., Gayathri, R., Mol, S.N., Sugawara, T., Hirata, T., Miyashita, K., Ganesan, P. (2015) *J. Funct. Foods*, **18**: 968–985.

Nishina, H., Ontsuka, M., Umakoshi, K. and Hashimota, Y. (1996) *Acta Hortic.*, **440:** 245-250.

Nobuoka, T., Oda, M. and Sasaki, H. (1996) *J. Jap. Soc. Hort. Sci.*, **64:** 859-865.

Norrie, J., Graham, M. E. D. and Gosselin, A. (1994) *J. Amer. Soc. Hort. Sci.*, **119:** 163-168.

Nouyen, V.X. and Mazliak, P. (1991) *C11FLORA*, pp. 311-318.

Nowaczyk, P. and Nowaczyk, L. (2000) *Acta Physiol. Plant.*, **22:** 309–311.

Nukaya, A., Goto, K., Jang. H., Kano, A. and Ohkawa, K. (1995) *Acta Hortic.*, **401:** 381-388.

Nunez, M.F. and Magnuson, B.A. (2014) In: *Anthocyanins in Health and Disease* (Eds. T.C.Wallace and Giusti), CRC Press, Boka Raton, pp. 1-11.

Oankh, K., Zagorska, N. and Georgiev, H. (1986) *Genetika I Selektsiya*,**19:** 425-427.

Oda, M. (1995) *Jpn. Agr. Res. Qrtly*, **29:** 187–194.

Oda, M. (1999) http: //www.agnet.org/library/eb/480/.

Oda, M., Nagaoka, M. Mori, T. and Sci, M. (1994) *Sci. Hort.*, **58:** 259-264.

Odell, G. B. and Cantliffe, D. J. (1987) *Proc. Fla. State Hort. Soc.*, **99:** 303-306.

Odland, T. E., Bell, R. S. and Smith, J. B. (1950) *U/U Bull.*, p.309.

Ohmori, T., Murata, M. and Motoyoshi, F. (1996) *Theor. Appl. Genet.*, **92**: 151-156.

Ohta, K., Hasegawa, R. and Ilosoki, T. (1995) *Environ. Cont. Biol.*, **33:** 297-298.

Okasha, A. M., Sharif, F. M., Al-Hussein, F. A. and Jameel, W. N. (1989) *Arab J. Plant Protect*, **2:** 126-132.

Olejnik, A., Kowalska, K., Kidon, M., Czapski, J., Rychlik, J., Olkowicz, M. and Dembczyñski, R. (2016) *Food Funct.*, **7**: 557–564.

Oliveira, W.F., Giodano, L.B. and Lopes, C.A. (1999) *Fitopatol Bras.*, **24**:49–53.

Olmstead, R. G. and Palmer, J. D. (1997) *Syst. Bot.,* **22:** 19–29.

Olympios, C., Kcrkidcs, P., Diamantopoulos, V., Grillas, E. and Karipidis, C. (1994) In: *International Conference 011 land and water resources management in the medeterranean region,*Vacnzano, Bari, Italy, 4-8 Sept, 1994. **5:** 181-196.

Omotude, C. T. and Adeniyi, O. R. (1995) *Turrialba,* **45**: 133-136.

Opena, R. T., Chen, J. T., Kuo, C. G., Chen, H. M. and Kuo, C. G. (1993) In: *Proc. Int. Symp, Adaptation of food crops to temperature and water stress,* pp. 321-334.

Ortas, I. (2013). *Academic J.,* **8:** 1048-1055.

Ortiz, R. and Gonzalez, C. (1999) *Acta Hortic.,* **487**: 243-5.

Os, E. A., Van Runia, W. T. and Buuren, J. V. (1998) *Acta Hortic.* **458:** 377-382.

Osman, G. Y., Salem, F. M. and Ghattas, A. (1988) *Anzeiger fur Schadlingskunde,* **61:** 35-37.

Oudard, E. (1999) *PHA1 Revue Ilorlicole,* **407**: 17-20.

Overman, A. J. and Jones, J. P. (1987) *Proc. Fla. State Hort. Soc.,* **99:** 315-318.

Ozawa, A., Saito, T. and Ota, M. (1999) *Japanese J. Appl. Entom.* Zool., **43:** 161-168.

Ozbun, J. L., Bontomel, C.E., Adik, S. S. and Minges, D. A. (1967) *Proc. Amer. Soc. Hort. Sci.,* **91:** 566-572.

Ozcelik, N and Akilli, M (1999) *Acta Hort.,* **491:** 155-160.

Ozguven, A.I., Paksoy, M. and Abak, K. (1997) *Acta Hortic.,* **463:** 243-249.

Ozguven, A.I., Paksoy, M., Yilmaz, C. and Tatli, H. (1998) *Bache,* **27:** 73-84.

Öztemiz, S. (2013) *Afr. J. Biotechnol.* **12**(15): 1882-1887.

Pal, B. P. and Tandon, R. N. (1937) *Indian J. Agric. Sci.,* **7**: 363-393.

Pal, D. P. and Singh, H. B. (1943) *Indian J. Genet.,* **3:** 115-116.

Palmer, J. D. and Zamir, D (1982) *Proc. Natl. Acad. Sci. U.S.A.,* **79:** 5006–5010.

Palozza, P., Simone, R.E., Catalano, A. and Mele, M.C. (2011) *Cancers,* **3**: 2333–2357.

Palumbo, A. D., Morra, L., Sanaja, V. O. and Picascia, S. (1996) *Colture Protette,* **25:** 61-65.

Pandey, C. S. and Padhiar, R. G. (1998) *Indian J. Environ. Toxicol.,* **8**: 44-46.

Pandey, D. K. (1995) *Sci. Hort.,* **62:** 57-62.

Pandey, R. P. Solanki, P. N. Saraf, R. K. and Parihar, M. S. (1996) *Punjab Veg. Grower,* **31:** 1-5.

Pandey, R.P., Kumar, N. and Mishra, S.P. (2018) *J. Pharmacognosy Phytochem.,* **7**: 3387-3389.

Pandiarana, N., Chattopadhyay, A., Seth, T., Shende, V. D., Dutta, S. and Hazra, P. (2015a) *N. Z. J. Crop Hortic. Sci.,* **43** (4): 282-293.

Pandiarana, N., Shende, Varun Durwas, Seth, Tania, Chatterjee, Soumitra, Dutta, Subrata and Chattopadhyay, Arup (2015b) *Journal of Applied and Natural Science*, **7** (2): 606-615.

Pandita, M. L. and Bhatnagar, D. K. (1981) *Haryana Agric. Univ. J. Res.*, **11**: 8-11.

Pandita, M. L., Arora, S. K. and Singh, Kirti (1976) *Haryana J. Hort. Sci.*, **5:** 565-572.

Pandita, V. K., Randhawa, K. S. and Modi, B. S. (1996) *Gartenbauwissenschaft*, **61:** 33-36.

Paneru, R.B. and Aryal, S. (2004) *Advances of Horticulture Research in Nepal*, pp. 391-395

Pansare, P. D., Desai, B. B. and Chavan, U. D. (1994) *J. Maharashtra Agric. Univ.*, **19:** 462-463.

Papadopoulos, A. P. and Hao, X. (1997) *Sci. Hort.*,**70:** 165-178.

Papadopoulos, A. P. and Ormord, D. P. (1990) *Canadian J. Plant Sci.*, **70:** 565-573.

Papadopoulos, A. P. and Ormrod, D.P. (1988) *Canadian J. Plant Sci.*, **68:** 1206-1218.

Papadopoulos, A. P. and Pararajasingham, S. (1998) *Hort. Technol.*, **8:** 193-198.

Paran, I., Goldman, I., Tanksley, S. D. and Zamir, D. (1995) *Theor. Appl. Genet.*, **90**: 542–548.

Pardossi, A., Tognoni, F. and Lovemore, S. S. (1988) *Acta Hortic.*, **229:** 371-378.

Pardossi, A., Tognoni, F. and Serra, G. (1989) *Colture Protette*, **18:** 81-84.

Park, H. J., Chinnan, M. S. and Shewfelt, R. L. (1994) *J. Food Sci.*, **59:** 568-570.

Park, H.B., Kim, J.C., Kwon, S.W., Kong, J.S., Kong, S.W. and Wang, K.H. (1999) *J. Korean Soc. Hort. Sci.*, **40:** 200-204.

Park, Y.S. and Yu, S. (1998) *J. Crop Prot.*, **40:** 44-49.

Parokonny, A. S., Marshall, J. A., Bennett, M. D., Cocking, E. C., Davey, M. R. and Power, J. B. (1997) *Theor. Appl. Genet.*, **94:** 713-723.

Pascual, B., Maroto, J. V., Bardisi, A., Lopez-Galarza, S., Alagarda, J. and San Bautista, A. (1999) *Acta Hortic.*, **487:** 117-120.

Passam, H.C., Karapanos, I.C., Bebeli, P.J., Savvas, D. (2007) *Eurpoean J.Plant Sci. Biotechnol.*, **1**: 1-21

Pasture, E. (1971) *Tuinbouwberichten*, Belgium, **35:** 328-330.

Patel, H. V., Patel, S. K. and Patel, D. J. (1989) *Indian J. Plant Protect.*, **17** L19-22.

Patel, J. J., Patel, N. C., Jyani, D. B. and Patel, J. R. (1998) *Indian J. Agric. Sci.*, **68:** 782-783.

Patel, K., Patel, A., Patel, D. and Vashi, J. (2017) *Int. J. Chem. Stud.*, **5**: 573-576.

Patemotte, S. J. (1988) *Mededelingen van de Faculteit Landbouwwetenschapprn Rijksuniversiteit Gent*, **53:** 657-661.

Paterson, A. H., Damon, S., Hewitt, J. D. *et al.* (1991) *Genetics*, **127** (1): 181–197.

Paterson, A. H., Lander, E. S., Hewitt, J. D., Peterson, S., Lincoln, S.E. and Tanksley, S. D. (1988) *Nature*, **335**: 721–726.

Pavlov, P., Gyurov, S. and Parrnakov, D. (1983) *Fiziologiya na Rasteniyata*, **9:** 65-70.

Pecaut, P. and Laterot, H. (1962) *Proc. 16th, Int. Hori. Congress*, **2:** 119-123.

Peet, M. M. and Willits, D. H. (1995) *HortScience*, **30:** 65-68.

Pek, H.H., Tan, H.L., Lam, W. and Yong, T.A. (1996) *Singapore J. Primary Indust.*, **24:** 19-24.

Pentangelo, A., Carboni, A., Grassi, G., Giordano, I. and Ragozzino, A. (1999) *Acta Hortic.*, **487**: 171-178.

Peralta, I.E. (2007) 3rd Global Botanic Gardens Congress, Wuhan, China, April 15-20, 2007.

Peralta, I.E. and Spooner, D.M. (2001) *Amer. J. Bot.*, **88:** 1888–1902.

Peralta, I.E. and Spooner, D.M. (2005) *Monogr. Syst. Bot. Missouri Bot. Gard.*, **104:** 227-257.

Peralta, I.E., Knapp, S., Spooner, D.M. and Lammers, T.G. (2005) *Syst. Bot.* **30:** 424–434.

Peralta, I.E., Spooner, D.M. and Knapp, S. (2008) *Syst. Bot. Monogr.*, **84:** 1–186.

Peregrine, W.T.H. and Binahmad, K. (1982). *Trop. Pest Mgt.*, **28:** 71–76.

Pereira, J. F. M. and Reisser Junior, C. (1998) *Agropecuaria Clima Temperado*, **1:** 73-77.

Perez, I., Dell' Amico, J. M., Rodriguez, P., Echevarria, I., Reynaldo, I. and Camejo, D. (1997) *Cullivos Tropicales*,**18**: 30-35.

Pernezny, K., Datnoff, L.E., Mueller, T. and Collins, J. (1996) *Plant Dis.*, **80**: 559–563.

Perveen, K. and Bokhari, N.A. (2020) *Plant Protect. Sci.*, **56:** 191-196

Petreikov, M., Yeselson, L., Shen, S., Levin, I., Schaffer, A.A., Efrati, A. and Bar, M. (2009) *J. Am. Soc. Hort. Sci.*, **134**: 134–140.

Petrikova, K. (1991) *Acta Univ. Agril. Fac. Hort.*, **6**: 37-42.

Petro-Turza, M. (1987) *Food Rev. Intern.*, **2:** 309–351.

Petrova, M., Yulkova, Z., Gorinova, N., Izhar, S., Firon, N., Jacquemin, J. M., Atanassov, A. and Stoeva, P.(1999) *Theor. Appl. Genet.*, **98**: 560-566

Petruchuk, T. E. (1987) In: *Sistemy lntensivnogo Kul.tivirovaniya Rastenii*, p.41-45.

Phatak, S.C., Wittwer, S.H. and Teubner, *E.g.* (1966) *Proc. Amer. Soc. Hort. Sci.*, **88**: 527-531.

Philips, S., Honma, S. and Murakeshi, H. H. (1964) *Euphytica*, **13:** 231-236.

Phills, D.R. and Hernandez, T.I. (1975) *HortSci.*, **10**: 149.

Philouze, J. (1995). *PHM Revue Horticole*, **364:** 50–54.

Picanco, M., Guedes, R.N.C., Leite, G.L.D., Fontes, P.C.R. and Silva, E.A. da (1995) *Horticultura Brasileira*, **13**: 180-183.

Picha, D. H. and Hall, C.B. (1982) *HortScience*, **17**: 634-635.

Picó, B., Herraiz, J., Ruiz, J.J. and Nuez, F. (2002) *Sci. Hort.*, **94**: 73-89

Pilatti, R.A., Gariglio, N.F. and Gutierrez, H.F. (1994) *Horticultura Argentina*, **13**: 82-86.

Pillen, K., Pineda, O., Lewis, C. and Tanksley, S. D. (1996) In: *Genome Mapping in Plants* (Eds. A. H. Paterson and R. G. Landes), Austin, Tex, USA, pp. 281–308.

Pilowsky, M., Cohen, S., Oen Joseph, R., Shlorno. A., Chen, L., Nahon, S. and Krikun, J. (1989) *Hassadeh*, **69**: 1212-1215.

Pimpini, F., Berti, A. and Zanin, G. (1986) In: *Compt es Rendus de la I 3e Conference du Columa*, 165-175.

Pimpini, F., Gianquinto, G., Babbo, G. and Xodo, E. (1987) *Cotlure Protette*, **16**: 63-73.

Pino, L.E., Lombardi-Crestana, S., Azevedo, M.S., Scotton, D.C., Borgo, L., Quecini, V., Figueira, A. and Peres, L.E.P. (2010) *Plant methods* **6:** 23.

Pino, M. A., Domini, M.E., Terry, E., Bertoli, M. and Espinosa, R. (1994) *Cultivos Tropicales*, **15**: 60-63.

Pivot, D., Reist, A. and Gilloiz, J.M. (1999) *Revue Suisse de Viticulture d 'Arboriculture et Horticulture*, **31**: 265-269.

Plant, Z. (1997) *Acta Hortic.*, **449**: 591-597.

Pnueli, L., Carmel-Goren, L., Hareven D., Gutfinger, T., Alvarez, J., Ganal, M. Zamir, D. and Lifschitz, E. (1998) *Development*, **125**: 1979–1989.

Pokharkar, D.S., Chaudhary, S.D. and Verma, S.K. (1999) *Indian J. Agric. Sci.*,**69**: 185-188.

Polevaya, V.S., Yashina, S.G. and Olcinikova, T.A. (1988) *Fiziologiya Rastenii.*,**35**: 612-617.

Polizzi, G. and Asero, C. (1994) *Acta Hortic.*, **366**: 345-52.

Poncu, J. (1987) *Productia Vegetaba. Horticulturae*, **31**: 9-11.

Porte, W.S. and Walker, H.B. (1945) *Phytopath.*, **35**: 931-933.

Porte, W.S. and Wellman, F.L. (1941) *USDA Circ.*, **584**: 1-19.

Potaczek, H. and Michalik, H. (1998) *Veg. Crop. Res.*Bull., **49**: 13-20.

Powell, C.A. and Stoffella, P.J. (1998) *HortScience*, **33**: 283-284.

Powers, L. (1939) *J. Genet.*, **39**: 139-170.

Powers, L. (1941) *J. Agric. Res.*,**63**: 171-174.

Powers, L. and Lyon.A. (1941) *J. Agric. Res.*,**63**: 129-148.

Prajapati, S., Tiwari, A., Kadwey, S., and Jamkar, T. (2015) *Int. J. Agric., Env. Biotechnol.*, **8**: 245-251.

Prasanna, H.C., Sinha, D.P., Rai, G.K., Krishna, R., Kashyap, S.P., Singh, N.K. Singh, M. and Malathi, V.G. (2015) *Plant Pathol.*, **64**: 256–264.

Prema, G., Indiresh, K.M. and Santhosha, H.M. (2011) *Asian J. Hort.*, **6**: 207-209.

Pressman, E., Shaked, R., Rosenfeld, K. and Hefetz, A. (1999) *J. Hort. Sci. Biotech.*,**74**: 101-104.

Prezotti, L. C., Balbino, J. M., De. S., Stock, L. A. and Ferreira, L. R. (1987) *Pesquisa em Andamento, Empresa Capixaba de Pesquisa Agropecuaria EMCAPA*, **45:** 9.

Pritchard, F. J. (1922) *U. S. Dept. Agri. Bull.*, **101**-5.

Pruthi, H.S. and Samuel, C.K. (1939) *Indian J. Agric. Sci.*, **9**: 223-275.

Przybylska, S. (2019) *Int. J. Food Sci. Tech.*, **55**: 11–32.

PueHee, P., Young, Ch., HyunRan, K., Kyeong, Ho Ch., DaeGeun, O., KiTaek, K. (2010) *Korean J. Breed. Sci.*, **42**: 139-143.

Pujer, A., Jagdeesha, R.C. and Badiger, M. (2017) *Int. J. Adv. Biol. Res.*, **7**:543–548.

Pujer, P., Jagadeesha, R.C. and Suresh (2015) *Green Farming*, **6**(1):30-32.

Pulupol, L. U., Bchboudian, M. H. and Fisher, K. J. (1996) *HortScience*, **31**: 926-929.

Purcell, J.P., Oppenhuizen, M., Wofford, T., Reed, A.J. and Perlak, F.J. (2004) In: *Handbook of Plant Biotechnology* (Eds. P., Christou and H. Klee), John Wiley and Sons, pp. 1147–1163.

Qiao, K., Liu, X., Wang, H., Xia, X., Ji, X. and Wang, K. (2012) *Pest Manag. Sci.*, **68:** 853–857

Rahman, A. K. M. M., Haque, Md. M. and Hossain, S. M. M. (1994) *Punjab Veg. Grower*, **29**: 13-14.

Rahman, M., Nahar, M.A., Sahariar, M.S. and Karim, M.R. (2015) *Prog. Agric.*, **26:** 32-37

Rai, A.K., Vikram, A. and Pandav, A. (2016). *Int. J. Agric. Env. Biotechnol.*, **9**(5): 739-744.

Rai, G.K., Rai, N.P., Kumar, S., Yadav, A., Rathaur, S. and Singh, M. (2012) *In Vitro Cell.Dev.Biol.–Plant*, doi: 10.1007/s11627-012-9442-3.

Rai, N., Syamal, M. M., Joshi, A. K. and Kumar, V. (1998) *Indian J. Agric. Res.*, **32**: 21-27.

Raj, H. and Kapoor, I. J (1997) *Indian Phytopath.*, **50**: 387-395.

Raj, H. and Kapoor, I. J. (1993) *Zeitschuftfur Pflanzenkraukheitenund Pflanzenschutz*; **100**: 652-661.

Raj, H. and Kapoor, I. J. (1996) *Indian Phytopath.*, **49**: 355-361.

Rajagopal, D. and Trivedi, T.P. (1989) *Tropical Pest Management*, **35:** 410-413.

Rajasekharan, L. R. and Shanmuzadin, K. G. (1981) *South Indian Hort.*, **29**: 95-99.

Rajgopal, V. and Rao, I. M. (1974) *Aust. J. Bot.*, **22**: 429-435.

Ramesh, R., Joshi, A., Ghanekar, M. (2009) *World J. Microbiol. Biotechnol.*, **25:** 47–55.

Ramin, A. A. (1998) *Iranian J. Agric. Sci.*, **29**: 139-146.

Ramirez, R., Santos, R., Bracho, F., Sandovall. A. and Castrode Rincon, C. (1998) *Revista de la Facultad de Agronomia, Universidad del Zulia*, **15**: 534-544.

Ramulu, K. and Devreux, M. (1979) *Incompatibility Newsl.*, **11**: 68-71.

Ramulu, K. S., Dijkhuis, P., Rutgers, E., Blaas. J., Krens, F. A., Dons, J. J. M., Colijn Hooymans, C. M. and Verhoeven, H. A. (1996) *Genome*, **39**: 921-933.

Rana, H. C. and Barevadia, T. N. (1995) *Indian J. Environ. Toxicol.*, **5:** 75-76.

Ranc, N., M˜uos, S., Santoni, S. and Causse, M. (2008) *BMC Plant Biology*, **8:** article no. 130.

Randhawa, K. S. and Nandpuri, K. S. (1977) *Veg Sci.*, **4**: 61-65.

Ranieri, A., Giuntini, D., Lercari, B., Soldatini, G.F. (2004) *Prog. Nutr.*, 6: 44-49.

Rao, M. S., Reddy, P.P., Mittal, A., Chandravadana, M.V., and Nagesh, M. (1996) *Nematologia Mediterranea*, **24**: 59-62.

Rao, R. M. and Choudhury, O. (1976) *Indian J. Agric. Sci.*, **46**: 597.

Rappaport, L. (1960) *Naturwissenschaften*, 12: 1-2.

Rattan, R. S. and Saini, S. S. (1969) *Veg. S*ci.,**6**: 21-26.

Rattink, H. (1993) *Mededelingen van de Faculteit Landbouwwetenschappen, Universiteit Genet*,**58**: 1329-1336.

Ravestijn, W. van (1990) *Groentenen Fruit*,**46**: 36-37.

Ravestijn, W. van and Sande, J. van der (1991) *Acta Hort.*, **288**: 204-212.

Ravinder Kumar and Srivastava, B. K. (1997) *Crop Res.* (Hisar), **14**: 137-141.

Ravinder Kumar and Srivastava, B. K. (1998b) *Crop Res.* (Hisar), **15**: 279-280.

Ravinder Kumar and Srivastava, B. K. (1 998a) *Indian J. Agric. Res.*, **32**: 239-242.

Rawat, U. S. and Pawar, A. D. (1993) *Plant Protect. Bull.*, **45**: 34.

Raychoudhury, S. P. and Lele. V. C. (1966) *Indian Hort.*, **10**: 39-43.

Raymundo, L. C., Chichester, C. O. and Simpson, K. N. (1976) *Agric. Food. Chem.*, **24**: 59-64.

Read, P. E. and Fieldhouse, D. J. (1970) *J. Amer. Soc. Hort. Sci.*, **95:** 73-78.

Reda, F., Mousa, O. M., Sejny, M. J. and Nawar L. S. (1986) *Egyptian J. Phytopath.*, **17**: 37-33.

Reddy, D. V. R. and Wightman, J. A. (1988) *Adv. Dis. Vector Res.*, **5:** 203-220.

Reddy, P.P.(1988) *Indian J. Hort.*,**45**: 166-168.

Redenbaugh, K., Hatt, W., Martineau, D., Kramer, M., Sheehy, R., Sanders, R., Houck, C. and Emlay, D. (1992) *Safety assessment of genetically engineered fruits and vegetables: a case study of the Flavr Savr™ tomato*, pp. xvii+ 267.

Regmi, R., Poudel, S., Regmi, R.C. and Poudel, S. (2018) *J. Agric. Forestry Univ.*, **2**: 127-131.

Reid, J. B., Winfield, D., Sorensen, I. and Kale, A. J. (1996) *Ann. of Appl. Biol.*, **129**: 137-149.

Reid, M. E. (1926) *Bot. Gaz.*, **130**: 180-186.

Reis, P. R. and Souza, J. C. de (1996) *Anais da Sociedade Entomologicado Brasil*, **25**: 65-69.

Reis, P. R. and Souza, J. C. De (1998) *Cienciae Agrotechnologia*, **22**: 13-21.

Reis, P. R., Souza, J. C. de and Malta, A. W. O. (1989) *Anais da Sociedade Entomologica do Brasil*, **18**: 131-144.

Reist, A., Gautrat, M. P., Gillioz, J. M., Leger, A., Pivot, D. and Corbaz, R. (1995) *Revue Suisse de Viticulture, d'Arboriculture el d' Horticulture*, **27**: 209-213.

Renuka, D.M., and Sadashiva, A.T. (2016) *Plant Archives*, **16**:654–658.

Resende, R. de O. and Cupertino, F. P. (1996) *Informe Agropecuario (Belo Horizonte)*, **18**: 19-27.

Reynard, N. G. and Andrus, C. F. (1945) *Phytopath.*, **35**: 25-36.

Rich, A. E., Richards, M. C. and Meader, E. M. (1962) *Pl. Dis. Reptr*, **46**: 136.

Rick, C. M. (1944) *Science*, **99**: 543.

Rick, C. M. (1950) *Evolution*, **4**: 110-122.

Rick, C. M. (1958) *Econ. Bot.*, **12**: 346-357.

Rick, C. M. (1960) *Proc. Natl. Acad. Sci. U.S.A.*, **46:** 78–82.

Rick, C. M. (1973) *Genes. Enzymes and Populations* (Ed. A. M. Srb), New York, pp. 225-269.

Rick, C. M. (1974) *Handbook of Genetics* (Ed. R. C. Kind), New York, pp. 247-280.

Rick, C. M. (1974) *Hilgardia*, **42**: 493–510.

Rick, C. M. (1976) *Evolution of Crop Plants*, (Ed. N. W. Simmonds) Longman, London and New York.

Rick, C. M. (1982) *In: Plant improvement and somatic cell genetics, Academic Press*, New York, pp. 1-28.

Rick, C. M. (1983) In: Isozymes in Plant Genetics and Breeding (Eds. S. D. Tanksley and T. J. Orton), Elsevier, Amsterdam, pp. 147-167.

Rick, C. M. and Boynton, J. E. (1967) *Amer. J. Bot.*, **54**: 601-611.

Rick, C. M. and Chetelat, R. T. (1995) *Acta Hortic.*, **412:** 21–38.

Rick, C. M. and Dempsey, W. H. (1969) *Bot. Gaz.*, **13**: 180-186.

Rick, C. M. and Lamm, G. (1955) *Amer. J. Bot.*, **42**: 663-675.

Rick, C.M. (1957) *Am. J. Bot.*, **43**: 687-696.

Rick, C.M. (1963). *Evolution*, **17:** 216–232.

Rick, C.M. (1965) *Adv. Genet.*, **8**: 267-382.

Rick, C.M. (1973) In: *Genes, Enzymes, and Populations* (Ed. A. M. Srb), Plenum, New York. pp. 255–269.

Rick, C.M. (1979) In: *The biology and taxonomy of Solanaceae* (Eds. J. G. Hawkes, R. N. Lester and A. D. Skelding), Academic Press, New York. pp. 667-677.

Rick, C.M. (1979) In: *The Biology and Taxonomy of Solanaceae.* (Eds. J. G. Hawkes, R. N. Lester and A. D. Skelding), *Linn. Soc. Symp. Ser. 7.* Academic Press, New York. pp. 667–677.

Rick, C.M. (1982) In: *Plant Improvement and Somatic Cell Genetics* (Eds. I.K. Vasil, W.R Scowcroft and K.J. Frey), Academy Press, New York, pp. 1-28.

Rick, C.M. (1982a) In: *Plant Improvement and Somatic Cell Genetics* (Eds. I.K. Vasil, W.R Scowcroft and K.J. Frey),. Academic Press, New York. pp. 1–28.

Rick, C.M. (1982b) *Biol. Zentralbl.,* **101:** 185–198.

Rick, C.M. (1986a). *Acta Hortic.,* **190:** 39–47.

Rick, C.M. (1986b) *In: Solanaceae, Biology and Systematics* (Ed. W. G. D'Arcy), Columbia Univ. Press, New York. pp. 477–495.

Rick, C.M. (1995) In: *Evolution of crop plants* [Eds. J. Smartt and N.W. Simmonds), Second edition, Longman Scientific and Technical, Essex, UK. pp. 452-457.

Rigney, C. J. and Wills, R. B. H. (1981) *HortScience,* **16**: 550-555.

Rijal, S. and Dahal, B.R. (2019) *Acta Scientific Agriculture,* **3**: 41-46.

Rijsdijk, A., Koning, A. N. M. De and Nederhoff, E. (1991) *Groenteneh Fruit, Glasgroventen;* **1:** 42-43.

Risse, L. A., Moffitt, T. and Dow, A. (1980) *Proc. Fla. Sta. Hort. Soc.,* **93**: 310-331.

Ristaino, J. B., Perry, K. B. and Lumsden, R.D. (1991) *Phytopath.,* **81**: 1117-1124.

Rivard, C. L. and Louws, F.J. (2008) *HortScience,* **43**: 2104–2111. 2008.

Rivard, C.L. and Louws, F.J. (2006) *North Carolina Coop. Ext. Serv. Bul. Ag-675.*

Rivas G. G. P., Lastra, R. and Hilje, L. (1994) *Manejo Integrado de Plagas,* **31**: 12-16.

Robert, V.J.M., West, M.A. and Inai, S. (2001) *Mol. Breed.,* **8**: 217-233.

Robertson, G. H., Mahoney, N. E., Goodman, N. and Pavlath, A. E. (1995) *J. Exp. Bot.,* **46:** 667-673.

Rodr´ýguez, G. R., Mu˜nos, S., Anderson, C.*et al.* (2011) *Plant Physiol.,* **156:** 275–285.

Rondon, A., Flores, Y., Soto, E. and Mujica, Y. (1995) *Revistade la Facultad de Agronomia. Universidad del Zulia.* **12**: 1-13.

Ronen, G., Carmel-Goren, L., Zamir, D. and Hirschberg, J. (2000) *Proc. Natl. Acad. Sci. U.S.A.,* **97**: 11102–11107.

Roorda Van Eysenga, J. P. N. L. (1971) *Fertilization of tomatoes with phosphate, Centre for Agricultural Publishing and Documentation,* Wageningen No. 775.

Roque, A. (1936) *Ann. Rept. Agri. Exp. Sta., Rio Piedras, Puerto Rico,* 1935-36, pp. 47-48.

Rosenberg, N., Dekel Reichenbach, M., Navot, N., Gad, A. E., Altman, A. and Czosnek, H. (1990) *Acta Hortic.,* **280**: 509-516.

Rotem, J. and Bashi, E. (1977) *Phytoparasitica,* **5:** 45–58.

Rousseaux, M.C., Jones, C.M., Adams, D., Chetelat, R., Bennett, A. and Powell, A. (2005) *Theor. Appl. Genet.* 111: 1396-1408.

Roy, R. S., Singh, R. K. and Bari, H. P. (1954) *Indian J. Hort.*, **11**: 131-137.

Roy, V. S. (1966) *Andhra Agric. J.* **13**: 177-180.

Rudich, J. L., Almer, D. K., Genzenberg, C. and Hare, S. (1977) *J. Hort. Sci.*, **52**: 91-399.

Ruiter, A. de, Eijnde, J. van den and Steen, J. ver (1988) *Groenten Fruit*, **43**: 46-47.

Runia, W. T., Michielsen, J. M. G. P., Kuik, A. J. Van and O. S. E. A. van. O. (1997) *In: Proc. 9th International Cong., on Soilless Culture, Netherlands*, pp. 395-407.

Rush, D. W. and Epstein, E. (1981) *J Amer. Soc. Hort. Sci.*, **106**: 699-704.

Rush, D.W. and Epstein, E. (1981). *Plant Physiol.* **68**: 1308-1313.

Russo, L., Dostal, H. C. and Leopold, A. C. (1968) *Biol. Sci.*, **18**: 109.

Ryloski, I. (1979) *J. Amer. Soc. Hort. Sci.*, **104**: 835-838.

Ryu, S. K., Park, Y.D., Sul, J. W. and Choi, G. W. (1998) *Korean J. Hort. Sci. Technol.*, **16**: 213-214.

Sabine, J. (1819) *Royal Hort. Soc. Trans.*, **3**: 342-354.

Saccardo, F., Ancora, G., Ramulu, K. and Philouze, J. (1981) *Genetics and Breeding of Tomato*, Proceedings of the meeting of the Eucarpia tomato working group, Avignon, France, May1981, pp. 235-242.

Sachs, R. M., Sisto, I., Jenkins, B. M. and Forister, G. W. (1992) *Sci. Hort.*, **49**: 135-146.

Sahin, U., Ozdeniz, A., Zulkadir, A. and Alan, R. (1998) *Turkish J. Agric. For.*, **22**: 71-79.

Saikia, A. K. and Muniyappa, V (1989) *Trop. Agric.*, **66**: 350-354.

Saikia, I., Puzari, N. N. and Shadeque, A. (1990) *Veg. Sci.*, **17**: 23-25.

Saikia, M. K. (1998) *Ann. Agri. Bio. Res.*, **3**: 91-95.

Saimbhi, M. S., Kanwar, J. S. and Nandpuri, K. S. (1976) *J. Res., Punjab Agri. Univ, Ludhiana*, **13**: 285-289.

Sajad, A.M., Jamaluddin and Abid, H.Q. (2017) *Int. J. Curr. Res. Rev.*, | **9**(5): 12-16

Sakata, Y., Nishio, T., Narikawa, T. and Monma, S. (1991) *J. Jap. Soc. Hori. Sci.*, **60**: 329-335.

Saklani, VD. and Mathai, P. J. (1977) *Indian J. Hort.*, **34**: 64-68.

Sakthivel, T. and Thamburaj, S. (1998) *South Indian Hort.*, **46**: 198-199.

Saleem, M.Y., Iqbal, Q. and Asghar, M. (2013) *Pak. J. Agric. Sci.*, **50**: 649-653.

Salguero, V and Morales, J. (1994) *Manejo Integrado de Plagas*, **31**: 25-28.

Saliba-Colombani, V., Causse, M., Gervais, L. and Philouze, J. (2000) *Genome*, **43**: 29–40.

Salinas, M., Capel, C., Alba, J.M., Mora, B., Cuartero, J., Fernandez-Munoz, R., Lozano, R. and Capel, J. (2013) *Theor. Appl. Genet.* **126**: 83–92.

Sallam, N.M.A. (2011) *Pl. Path. J.*, **10**: 187-191.

Salm, T. Vander, Bosch, D. Honee, G., Feng, L. Munsterman, E. Bakker, P. Stiekema, W. J. Visser, B. and Van der Salm, T. (1994) *Plant Mol. Biol.*, **26**: 51-59.

Salminen, M. S. (1997) *In: Proc. 9th. International Congress on Soilless Culture*, pp. 409-415.

Samen, F.A.A., Goussous, S.J., Jendi, A.A. and Makhadmeh, I.M. (2015) *Int. J. Pest Manag.*, **61:** 320-328, DOI: 10.1080/09670874.2015.1059969

Samimy, C. (1993) *HortScience*, **28**: 883-885.

Samnotra, R. K. Gupta, AK. and Jindal, K. (1998) *Environ. Ecol.*, **16**: 740-741.

Samoylov, V. M. and Sink, K. C. (1996) *Theor. Appl. Genet*; **92**: 850-857.

Samoylov, V. M., Izhar, S. and Sink, K. C. (1996) *Theor. Appl. Genet.*, **93**: 268-274.

Samson, R. W. *et al.* (1942) *Phytopath.*, **32**: 16.

Samunder, B. S. (1964) *Indian. J. Hort.*, 21: 65-76.

Sandhu, K. S., Singh, D., Singh, J. and Saimbhi, M. S. (1993) *Proc. Indian Soc. Weed Sci. International Symp., Hissar*, **3:** 211-212.

Sandyan, J. S. and Banerjee, M. K. (1981) *Indian J. Agric. Res.*, **15**: 92-94.

Sanok, W. J., Sieczka, J. B. and Creigton, J. P. (1981) *Proc. North eastern Weed Soc.*, **35**: 160-162.

Sapir M., Oren-Shamir, M., Ovadia, R., Reuveni, M., Evenor, D., Tadmor, Y., Nahon, S., Shlomo, H., Chen, L., Meir, A. and Levin, I. (2008) *Journal of Heredity*, **9**(3):292–303.

Sarfatti, M., Abu-Abied, M., Katan, J. and Zamir. D. (1991) *Theor. Appl. Genet.*, **82**: 22-26.

Sarfatti, M., Katan, J., Fluhr, R. and Zamir, D. (1989). *Theor. Appl. Genet.*, **78**: 755-759.

Sarkar, N. R. (1985) *Diss. Abs. Inter.*, **13**. 46: 902.

Sasai, K. (1991) *Scientific Reports Miyagi Agril. College*, No. **39**, pp. 1-9.

Sasser, J. N. (1954) *Maryland Agri. Exp. Sta. Tech. Bull.* A-77, p. 11.

Sastry, S. M. and Singh, S. J. (1974) *Indian J. Hort.* **31**: 178-181.

Satpathy, S., Rai. Samarjit and Chattopadhyay, M. (1999) *Insect Environ.*, **5**: 117-118.

Sattar, M. H. A. and Kassem, N. A. (1991) *Arab. I. Plant Protect.*, **9**: 9-T3.

Satti, S. M. E. (1986) *Acta Hortic.*, **190**: 455-457.

Satti, S. M. E. and Oebker, N. F (1986) *Acta Hortic.*, **190**: 347-354.

Sauser, B. J., Giacomelli, G. A. and Janes, H. W. (1998) *Acta Hortic.*, **456**: 87-92.

Sawhney, V. K. (1997) In: *Pollen Biotechnology for Crop Production and Improvement* (Eds. K. R. Shivanna and V. K. Sawhney), Cambridge Univ. Press, pp. 183-198.

Sayre, C. B., Robinson, W. B. and Wishnetsky, T. (1953) *Proc. Amer. Soc. Hort. Sci.*, **61**: 381-387.

Schaible, L. W. (1962) *Proc. Plant Sci. Symp. Campbell soup. Co.*, New Jersey, pp. 89-98.

Schalk, J. M. and Robbin, M. L. R. (1987) *HortScience.*, **22**: 30-32.

Schauer, N., Zamir, D. and Fernie, A.R. (2005) *J. Exp. Bot.*, **56**: 297–307.

Schelt, J. Van and Altena, K. (1997) *Proc. of the Section Exptl. Appl. Entom. of the Netherlands Entomological Soc.*, **8**: 151-159.

Schoen, L. and Martin, C (1999) *Phytoma*, **514**: 39-40.

Schoenmakers, H. C. H., Wolters, A. M. A., Nobel, E. M., Koornneef, M. and De Klein, C. M. J. (1993) *Theor. Appl. Genet.*, **87**: 328-336.

Scholz-Dobelin, H. (1997) *Gemuse (Munchen)*, **33**: 246-248.

Schrevens, E., Neyens, K., Verreydt, J. and Baerdemaeker J. De (1995) In: *Proc. of a workshop*, Leuven, Belgium, 14-15 Sept, 1993 (Eds. J. De Baerdemaeker, B.McKenna, M. Janssens, A. Thompson, F. Artes Calero, E.Hohn. and Z. Somogyi), Luxembourg, Dg XLA, Commission of the European Communities, pp. 155-173.

Schwabe, W. W. (1986) *Acta Hortic.*, **190**: 309-314.

Scott, J. W. and Farley, J. D. (1983) *HortScience.*, **18**: 114-115.

Scott, J.W. and Gardner, R.G. (2006). *Genetic improvement of Solanaceous crops.* Vol. 2: Tomato, Science Publisher. pp. 421-456.

Scott, J.W., Somodi, G.C. and Jones, J.B. (1988) *Proc. Fl. St. Hort. Soc.*, **101**:390–392. Mahir, A.M., Ismail, A., Diong, K.S. (1993) *ACIAR Proc.*, **45**:154–157.

Seddon, B. and Schmitt, A. (1999) *12th Inter: Reinharsbrunn Symp; Germany*, pp. 423-428.

Seeja, G., Chandramony, D. and Saraswathi, P. (2007) *Res. Crops*, **8**: 219-221.

Segarra, G., Reis, M., Casanova, E. and Trillas, M.I. (2009) *J. Plant Pathol.*, **91**: 683-689

Segeren, M. I., Siqueira, W. G., Sondahl, M. R., Lourencao, A. L., Medina-Filho, H. P. and Nagai, H. (1993) *Revista Brasileira de Genetica*, **16**: 773-83.

Seguý´-Simarro, J.M. and Nuez, F. (2005) *Acta Physiol. Plant.*, **27**: 675–685

Seguý´-Simarro, J.M. and Nuez, F. (2007) *J. Exp. Bot.*, **58**: 1119–1132

Seresinhe, P. S. J. W. (1998) *Trap. Agric. Res. and Extn.*, **1**: 87-93.

Seroczynska, A., Niemirowiez, S. K. and Korzeniewska, A. (1998) *Folia Horticulturae*, **10**: 2-14.

Seth, J. N. and Dhandar, D. G. (1970) *Prog. Hort.*, **2**: 5-13.

Seth, M. L. and Dhanraj, K. S. (1972) *Phytopath, Z.* **73**: 365-370.

Sethi, J. N. and Kuksal, R. P. (1977) *Prog. Hort.*, **9**: 43-50.

Shahin, E. A. (1985) *Theor. Appl. Genet.*, **69**: 235-240.

Shahin, E. A. and Spivey, R. (1986) *Theor. Appl. Genet.*, **73**: 164-169.

Shahin, E. A. Sukhapinda, K. Simpson, R. B. and Spivey, R. (1986) *Theor. Appl. Genet.*, **72**: 770-777.

Shankar, A., Reddy, R.V., Sujatha, M., Pratap, M. (2013) *IOSR Journal of Agriculture and Veterinary Science*, **4**: 31-34.

Sharma, C. B. (1971) *Indian, J. Hort.*, **28**: 228-233.

Sharma, C. B. and Mann, H. S. (1971) *Indian J. Hort.*, **29**: 228-233.

Sharma, C. B. and Shukla, V. (1972) *Proc. 3rd Int. Symp. Subtrop. and Trop. Hort.*, Bangalore.

Sharma, C. D. and Mann, H. S. (1972) *Indian J. Hort.*, **30**: 322.

Sharma, G. C. and Khan, M. L. (1994) *Orissa J. Agric.* Res., **7**: 92.

Sharma, J. N. and Singh, M. (1963) *Indian J. Agron.*, **7**: 279-284.

Sharma, K. C. Singh, A. K. and Sharma, S. K. (1999) *Ann. Agric. Res.*, **20**: 399-402.

Sharma, N. K. and Tiwari, R. S. (1995) *Recent Hort.*, **2**: 96-99.

Sharma, N. K. and Tiwari, R. S. (1996) *Recent Hort.*, **3**: 82-85.

Sharma, P. D. and Dashora, L. K. (1999) *Scientific Hort.*, **6**: 99-103.

Sharma, R. K.; Pandey, M. D. and Pandey, D. S. (1978) *Plant Sci.*, **10**: 163-165.

Sharma, S. L., Sohi, H. S., Chowla, S. C. and Reskinn, S. N. (1994) *Indian J. Agric. Sci.*, **44**: 323-324.

Shashank, P.R., Chandrashekar, K., Meshram, N.M., Sreedevi K.(2015) *Indian J. Entomol.*, **77**: 323 – 329.

Shchereva, L., Kruleva, M. and Zagorska, N. (1990) *Genetika I Selektsiya*, **23**: 185-193.

Shende, V.D., Seth, T., Mukherjee, S. and Chattopadhyay, A. (2012) *SABRAO J. Breeding Genet.*, **44** (2) 302-321.

Sherpa, P., Seth, T., Shende, V. D., Pandiarana, N., Mukherjee,S. and Chattopadhyay, A. (2014a) *Electron. J. Plant Breed.*, **5**(3):552-562.

Sherpa, P., Seth, T., Shende, V. D., Pandiarana, N., Mukherjee, S. and Chattopadhyay, A. (2014b) *J. Appl. Nat. Sci.*, **6** (2): 625-632.

Shi, C.H. (1999) *Zhejiang Nongye Kexue.*, **4**: 182-183.

Shipp, J.L., Ferguson, G.M. and Hunt, D.W.A. (2000) In: *Biological control programmes in Canada*, 1981-2000 (Eds. P. G.Mason and J. T.Huber), CABI Publishing p. 139

Shirakawa, T., Sasaki, T. and Ozaki, K. (1991) *Japan Agric. Res. Quarterly*, **25**: 27-32.

Shiskoff, N. and Campbell, R. N. (1990) *Plant Dis.*, **74**: 889-894.

Shittu, GA and Adeleke, J. A. (1999) *Global J. Pure Appl. Sci.*, **5**: 27-30.

Shivapuri, A., Bhargava, A.K. and Chippa, H.P. (2001) In: *Proceeding of Sclerotinia 2001, the XI International Sclerotinia Workshop* (Eds.C.S. Young and K.J.D. Hughes), 8-12, July, 2001, Central Science Laboratory, York, England., pp: 177-178.

Shoda, M. (1999) *Proc. 13th. Forum for appl. biotech., Universiteit Gent*, **64**: 275-280.

Shorey, H. H. and Gerber, R. G. (1996) *Environ. Entomol.*, **25**: 1401-1405.

Shrestha, K.K. and Ashley, R. (2007) *Nepal Agric. Res. J.*, **8**: 67-76

Shtereva, L.A., Zagorska, N.A., Dimitrov, B.D., Kruleva, M.M. and Oanh, H.K. (1998) *Plant Cell Rep.*, **18**: 312–317.

Shukla, A and Yadav, H. S. (1998) *Adv. Plant Sci.*, **11**: 265-270.

Shukla, Y. R., Jagmohan Kumar and Arya, P. S. (1999) *Scientific Hort.*, **6**: 105~109.

Shulgina, L. M. and Andrienko, Z. (1980) *Referativnyi Zhurnal*, **8**: 55-90.

Siddiqui, M. A., Azam, M. F. and Saxena, S. K. (1998) *Tests Agrochem. Cult.*, **19:** 8-9.

Sidhu, G. and Webster, J. M. (1975) *Canadian J. Genet. Cytol.*, **17**: 323-28.

Sidhu, M.K., Dhatt, A.S., Sandhu, J.S. and Gosal, S.S. (2014) *Int. J. Agri. Env. Biotech.*, **7:** 679-687.

Silva Junior, A. A. and Giorgi, E. (1992) *Boletim Tecnico-Empresa de Pesquisa Agropecuaria e Difusao de Tecnologia de Santa Catarina S. A.* **59:** 23.

Silva, G.A., Picanco, M.C., Bacci, L., Crespo, A.L., Rosado, J.F. and Guedes, R.N.C. (2011) *Pest Manage. Sci.*, **67:** 91–920

Silva, R.S., Arcanjo, L.P., Soares, J.R.S., Ferreira, D.O., Serrão, J.E., Martins, J.C., Costa, Á.H. and Picanço, M.C. (2018) *Neotrop. Entomol.* **47:** 318–325.

Simidchiev, Kh. and Kanazirska, V. (1995) *Rasteniev dni Nauki*, **32**: 32-34.

Sims, W. L. and Kashmere, R. F. (1972) *Calif Agric.*, **26**: 3-4.

Singer, S. M. (1990) *Egyptian J. Hort.*, **17**: 77-84.

Singh, B. and Choudhury, B. (1972) *Haryana J. Hort. Sci.*, **1**: 63-68.

Singh, D. and Chakal, B. S. (1978) *Haryana J. Hort. Sci.*, **7**: 82-86.

Singh, D. K., GulshanLal, Tewari, R. P. and Jain, S. K. (1995a) *Ann. Agric. Res.*, **16**: 125-127.

Singh, D. Sandhu, K. S. Saimbhi, M. S. and JaswinderSingh (l995b) *J Res., Punjab Agril. Univ.*, **32**: 151-153.

Singh, H. B. and Sikka, S. M. (1955) *Indian Farming*, **5**: 18-25.

Singh, J. P. (1994) *Recent J. Hort.*, **1**: 80-83.

Singh, K. and Upadhyay, S. K. (1967) *Horticulture, Srinagar*, **2**: 3-9.

Singh, N., Dotasara, S.K., Kherwa, B.S. and Singh, S. (2017) *J. Entomol. Zool. Stud.*, **5**: 1403-1408

Singh, S. J., Sastry, K. S. M. and Sastry, K. S. (1973) *Indian J. Agric. Sci.*, **43**: 669-672.

Singh, S. N. and Choudhury, B. (1966) *Indian J. Hort.*, **23**: 156-167.

Singh, S. N. and Joon, B. S. (1950) *Curr. Sci.*, **19**: 182-184.

Singh, S., Thakur, D. P. and Singh, J. P. (1986) *Indian Phytopath.*, **36**: 723-724.

Singh, Santokh and Sawhney, V. K. (1998) *J. Exptl. Bot.*, **49**: 199-203.

Sinha, P. and Saxena, S. K. (1990) *Acta Botanica Indica*, **18**: 101-103.

Sinha, P. and Saxena, S. K. (1999) *Indian J. Agric. Sci.*, **69**: 651-653.

Sivaiah, K.N. (2012) *Agrotechnol*, 1: 2 http: //dx.doi.org/10.4172/2168-9881.S1.004

Sivaprakasham, N. (1998) *Indian J. Agric. Sci.*, **68**: 801-802.

Sivaprakasham, N. Balasubramanian, G. Jayaraman, V. Narayanan, A. and Venkatesan, S. (1986) *Madras Agric. J.*, **73**: 614-617.

Siviero, P., Sandei, L. and Cacchioli, C. (2000) *Informatore Agroria*, **56**: 45-47.

Smid, E. J., Hendriks, L., Boerrigter, H. A. M. and Gorris, M. (1996) *Postharvest Biol. Technol.*, **9**: 343-350.

Smille, R. M., Melchers, G. and Wettstein, D. von (1979) *Carlsberg Res. Commun.*, **44**: 127-132.

Snapp, S. S. (1994) *Agronomy J.*, **86**: 754-758.

Snobar, B. A., Denis, Parid and Suwwan, M. A. (1988) *Dirat.*, **15**: 35-45.

Sobolewski, J. and Robak, J. (1994) *Ochrana Roslin*, **38**: 4-7.

Som, M. G. and Choudhury, B. (1975) *Veg. Sci.*, **2**: 95-99.

Som, M. G. and Choudhury, B. (1976) *J. Soc. Exp. Agric.*, **1**: 53-54.

Som, M. G. and Islam, M. A. (1970) *Hortic. Sci.*, **2**: 111-115.

Son, S.H., Khan, Z., Kim, S.G. and Kim, Y.H. (2009) *J. Appl. Microbiol.*, **107**: 524–532

with Song, W., Zhou, L., Yang, C., Cao, X., Zhang, L. and Liu,X. (2004) *Crop Prot.*, **23**(3):243-247.

Sood, P. and Nath, A. (1998) *J. Hill Res.*, **11**: 171-177.

Souri, M.K., Sooraki, F.Y. and Moghadamyar, M. (2017) *Hortic. Environ. Biotechnol.* **58:** 530–536.

Souza Machado, V., Pitblado, R., Ali, A. and May, P. (1999) *Acta Hortic.*, **487**: 139-143.

Splittstoesser, W. B. and Vandemnark, J. S. (1971) *J. Amer. Soc. Hort. Sci.*, **96**: 546-567.

Splittstoesser, W. E. (1988) *In: Proc. of the plant growth regul. soc. of Amer.* (Ed. A. R.Cooke), Ithaca, New York, pp. 206-221.

Spooner, D.M., Anderson, G.J. and Jansen, R.K. (1993) *Amer. J. Bot.*, **80:** 676-688.

Springer, F. L. and Phills, D. R. (1982) *HortScience*, **17**: 296.

Sreedhar, M., Ganguly, A. K. and Pankaj, A. (1996) *Nematologia Mediterranea*, **24**: 273-277.

Sreenivasulu, B. and Rao, P. V. (1996) *J. Res. APAU.* **24**: 71-72.

Sreenivasulu, R., Reddy, M.S.P., Tomar, D.S., Sanjay, M.S.S. and Reddy, B.B. (2019) *Int.J.Curr.Microbiol.App.Sci.*, **8**: 1442-1452.

SreeVidya, C.S. Manoharan, C. RanjitKumar, C.T. Savithri, H.S. and LakshmiSita, G. (2000) *J. Plant Physiol.*, **156**: 106-110.

Srinivasan, K., Moorthy, P. N. K. and Raviprasad, T. N. (1994) *Int. J. Pest Management*, **40:** 56-63.

Srinivasan, P. M. and Basheer, M. (1961) *Indian Farming.*, **11**: IS.

Srinivasan, R. (2010) *AVRDC Publication*, No.10- 740. p.97.

Srivastava, R. P. and Srivastava, K. K. (1966) *Punjab Hort. J.*, **4**: 110-116.

Stajcic, S., Cetkovic, G., Canadanovic-Brunet, J., Djilas, S., Mandic, A. and Cetojevic-Simin, D. (2015) *Food Chem.*, **172**: 225–232.

Stamova, B.S. and Chetelat, R.T. (2000). *Theor. Appl. Genet.*, **101**: 527-537.

Stancheva, I. (1997) *Rasteniev "dni Nauki*, **34**: 50-51.

Staniaszek, M., Kozik, E.U., Marczewski, W. (2007) *Plant Breeding* **126**: 331-333.

Stansly, P. A., Liu, T. and Vavrina, C. S. (1998) *J. Eco. Entomol.*, **91**: 686-692.

Starck, J. R. and Wojcik, A. (1995) *Annals of Warsaw Agricultura University S G G W. Horticulture*, **17**: 45-47.

Step, H.Z.A., El-Behadli, A. H., Al-Zahroon, H. H., Antoon, B. G. and Georgees, M. S. (1996) *Dirasat, Series 13, Pure and Appl. Sciences*, **23**: 13-16.

Stevens, C., Khan, V. A., Igwegbe, E. C. K., Plopper, L. D., Backman, P., Rodriguez-Kabana, R., Collins, D. J., BrownJ. E. and Wilson, M. A. (1998) In: *4th Int. Cong. on plastics in Agri., Israel, March, 1997* (Eds. S. Ben Yehoshua), Jerusalem, Israel, Laser pages publishing, pp. 402-408.

Stevens, M. A. (1986) *Plant Breeding Rev.*, **4:** 273–311.

Stevens, M.A. and Rick, C.M. (1986) In: *The Tomato Crop: a scientific basis for improvement* (Eds. J.G. Atherton and J. Rudich), Chapman and Hall, London. pp. 35-109

Stevens, M.A., Kader, A.A. and Albright, M. (1979) *J. Amer.Soc.Hort.Sci.*, **102:** 724–731.

Stevens, M.R. and Robbins, M.D. (2007) In: *Genetic improvement of Solanaceous crops, Vol. 2 Tomato*, Science Publisher. pp. 239-260.

Stevens, M.R., Lamb, E.M. and Rhoads D.D. (1995) *Theor. Appl. Genet.*, **90**: 451-456.

Sticher, L., Mauch-Mani, B. and Metraux, J.P. (1997) *Ann. Rev. Phytopathol.*, **35**: 235–270.

Stommel, J.R. (1992) *Plant Physiol.*, **99**: 324–328.

Stommel, J.R. and Zhang, Y.P. (1998) *HortScience*, **33**: 514.

Stoner, A.K. and Thompson, A. E. (1966) *Euphytica*, **15**: 377.

Stoppani, M. I., Rodriguez, J. P. and Uviedo, R. H. (1994) *Acta Hortic.*, **357**: 103-121.

Storey, H. H. (1931) *Nature*, **128**: 187-188.

Stravato, V. M. (1993) *InformatoreAgrario*, **49**: 36-37.

Streck, N. A., Buriol, G. A. and Schneider, F. M. (1996) *Pesquisa Agropecuaria Brasileira*, **31:** 105-112.

Streck, N. A., Buriol, G. A., Andriolo, J. S. and Sandri, I. (1998) *Pesquisa Agropecuaria Brasileira*, **33**: 1105-1112.

Strobel, J. W. and Walter, J. M. (1969) *Fla. Agric. Exp. Stn. Circ.* S-198, p. 9.

Su, N. R. (1974) *Agron. Coop. WorkshopMuscle School, Alabama*, p. 68.

Subedi, N., Testen, A. L., Baysal-Gurel, F. and. Miller, S. A (2015) *Plant Dis.*, **99:** 285-285

Subhan (1989) *Bull. Penelitian Hortikultura*, **17**: 97-103.

Sugiyama, T. (1962) *J. Jap. Soc. Hort. Sci.*, **31**: 141-145.

Sultana, S., Naser, H.M., Akhter, S. and Begum, R.A. (2016) *Bangladesh J. Agric. Res.*, **41**: 411-418.

Summers, W. L. (1996) *HortScience*, **31**: 291.

Sun, D., Fang, W., Zhang, B.Z. and Li, S.W. (1999) *Acta Hort. Boreali* Sinica, **14**: 75-78.

Sun, H.J., Uchii, Watanabe, S. and Ezura, H. (2006) *Plant Cell Physiol.* **47:** 426-431.

Surlekov, P., Toskov, K. and Dimov, I. (1995) *Rasteniev "dni Nauki*, **32**: 13-15.

Suszkiw, J. (1997) *Agricultural Research (Washington)*, **45**: 20-21.

Suwwan, M. A. (1986) *Dirasat*, **13**: 77-91.

Swarup, G. and Sharma, R. D. (1965) *Indian J. Exptl. Bio.*, **3**: 117-118.

Szczechura, W., Staniaszek, M. and Habdas, H. (2011) *Tomato Molecular Markers Vegetable Crops Research Bulletin* **74**: 5-23.

Szczesniak, A.S. (2002) *Food Quality and Preference*, **13:** 215–225.

Szekely, I. Baicu, T. and Alexandri, A. A. (1990) *Academia de Stiinte Agricole Si Silvice*, **23**: 245-524.

Szmidt, R. A.K., Hall, D. A. and Hitchon, G. M. (1988) *Acta Hortic.*, **221**: 371-378.

Szymkowiak, E. J. and Irish, E. E. (2006) *Planta*, **223**: 646–658.

Tabaeizadeh, Z., Bunisset-Bergounioux, C. and Perennes, C. (1984) *Physiol. Veg.*, **22**: 223-229.

Tajul, M.I. and Toyota, K. (2004) *Microbes Environ.*, **19**: 53-60.

Takacs, A. and Dolej, S. (1998) *Novenyvedelem*, **3**: 257-259.

Takahashi, I.I., Koshio, K. and Ota. Y. (1993) *J. Japanese Soc. Hort. Sci.*, **62**: 389-937.

Takeuchi, T. (1991) *Bull. Prefectural Agric. Expt. Station*, **32**: 12-14.

Tal, M. and Shannon, M.C. (1983) *Aus. J. Plant Physiol.*, **10**: 109-117.

Talekar, N.S., Opena, R.T. and Hanson, P. (2005) *Crop Prot.*, **25:** 461–467.

Tamta, S. and Singh, J.P. (2018) *Int. J. Veg. Sci.*, **24:** 2, 169-179.

Tan, H., Cao, L., He, Z., Su, G., Lin, B., Zhou, S. (2006) *World J. Microbiol. and Biotechnol.*, **22:** 1275–1280.

Tan, M. M. C., Colijn-Hooymans, C. M., Lindhout, W.I.I. and Kool, A.J. (1987) *Theor. Appl. Genet*. **75**: 105-108.

Tan, X.W., Ikeda, I. and Oda, M. (1999) *Soil Sci. Pl. Nutri.*, **45**: 609-616.

Tanaka, K. and Yasui, H. (1986a) *Bull. Veg. and Ornament. Crops Res. Sta.*, A (Ano, Mie), **14**: 169-187.

Tanaka, K. and Yasui, H. (1986b) *Bull. Veg. and Ornament. Crops Res. Sta.*, A (Ano, Mie), **14**: 159-168.

Tanaka, M. and Komochi, S. (1982) *Res. Bull. Hokaido Nat. Agri. Exp. Sta.*, **35**: 83-99.

Tanis, C. (1990) *Groeten en Fruit*, **46**: 52-60.

Tanksley, S. D. (1993) In: *Genetic Maps: Locus Maps of Complex Genomes,* (Ed. S. J. O'Brian), Cold Spring Harbor Laboratory Press, Cold Spring Harbor, NY, USA, pp. 6.3–6.15,

Tanksley, S. D. (2005) *Theor. Appl. Genet.*, **111**: 291–312.

Tanksley, S. D. and Bernatzky, R. (1987) In: *Plant Biology, Vol. 4, Tomato Biotechnology* (Eds. D. J. Nevins and R. A. Jones, R. Liss Alan), New York, NY, USA, pp. 37–44,

Tanksley, S. D. and Orton, T. J. (1983) *Isozymes in Plant Genetics and Breeding*, Elsevier, Amsterdam, The Netherlands.

Tanksley, S. D., Ganal, M.W., Prince, J. P. *et al.* (1992) *Genetics*, **132**: 1141–1160.

Tanksley, S. O. and Kuehn, G. D. (1985) *Biochemical Genetics*, **23**: 441-454.

Tanksley, S.D. and Loaiza-Figueroa, F. (1985) *Proc. Natl .Acad. Sci.*, USA **82**:5093–5096. Yang, W.C., Chen, J., Zhang, X.M. and Francis, D.M. (2007) *Sci. Agric. Sin.*, **40**(2):283–290.

Tanksley, S.D. and McCouch, S.R. (1997) *Science*, **277:** 1063–1066.

Tanksley, S.D. and Nelson J.C. (1996) *Theor. Appl. Genet.*, **92**: 191-203.

Tanksley, S.D., Bernachi, D., Beck-Bunn, T. *et al.* (1998) *Euphytica*, **99**: 77–83.

Tarakanov, G. I. Gavrish, S. f. Shaumyan, I. K. and Morev, VY. (1987) *Selektsiya i Semenovodstvo, USSR*, **4:** 40-42.

Tasdelen, O. and Bayindirli, L. (1998) *J. Food Process. Preserv.*, **22**: 303-320.

Taylor, A. L. and Chitwood, B. G. (1951) *Pl. Dis. Reptr.*, 35: 97.

Teasdale, J.R. and Abdul-Baki, A.A. (1998) *HortScience*, **33:** 1163–1165.

Terabayashi, S., Sugimoto, A., Ohshita, K. and Namkiti, T. (1995) *Acta Hortic.*, **396:** 131-136.

Terry, E. Pino, M. de los A. and Medina, N. (1995) *Cultivos Tropicales*, **16**: 69-71.

Terzopoulos P.J. and Bebeli, P.J. (2010) *Sci. Hort.*, **126** (2): 138–144.

Tesi, R. (1983) *Colture Protette*, **12**: 17-22.

Tesi, R. and Tosi, D. (1986) *Rivista della Ortoflorofruitticoltura Italiana*, **70**: 165-170.

Tesi, R., Graifenberg, A. and Greatini, M. (1970) *Riv. Ortoflorofrutticital*, **54**: 269-292.

Tewari, R. N. (1997) *In: Compendium of Lectures of Summer School on Hybrids and Hybrid Seed Production Technology of Vegetable Crops,* ICAR, New Delhi, 7th to 26th April, 1999, pp. 105-109.

Tewari, R. N. (1999) In: *Compendium of lectures of Summer School on Advances in Breeding of Temperate and Tropical Vegetables for Biotic and Abiotic Stresses, ICAR,* New Delhi, 17th May- 6th June, 1999, pp. 96-98.

Thakur, A.K. and Pandey, M. (1999) *Indian J. Plant Physiol.,* **4**: 221-223.

Thamburaj, S. (1998) In: *Souvenir of Silver Jubilee National Symposium 011 Emerging Scenario in Vegetable Research and Development, Varanasi, India, 12-14 December,* 1998, pp. 53-59.

Thoke, N. G. (1966) *Diamond Jublee Special Number Nagpur Agric. College,* 1966-1967.

Thomas, J. H., Drew, R. L. L. and Goodenough, P. W. (1982) *Ann. Appl. Biol.,* **100**: 197-202.

Thompson, A. E., Tomes, M. L., Erickson H. T., Wann, E. V. and Armstrong, R. J. (1967) *Proc. Amer. Soc. Hort. Sci.,* **91**: 495–504.

Thompson, B. D. Gill, D. D. and Halsey, I. I. (1960) *Proc. Fla. Sta. Hort. Soc.,* **73**: 160-166.

Thompson, H. C. (1927) *Cornell Univ. Agri. Exp. Sta. Mem.,* **107**: 1-73.

Thompson, H. C. and Kelly, W. C. (1957) In: *Vegetable Crops (5th cdn.),* McGraw-Rill, Book Company Inc., New York.

Thompson, I. J. and Smith, P. G. (1957) *Pl. Dis. Reptr.,* **41**: 180-181.

Thompson, P. A. (1974) *HortScience,* **9**: 35-54.

Tieman, D.M., Taylor, M.G., Ciardi, J.A. and Klee, H.J. (2000) *Proc. Natl. Acad. Sci. USA,* **97**: 5663– 5668.

Tieman, D.M., Zeigler, M., Schmelz, E.A., Taylor, M.G., Bliss, P., Kirs, M. and Klee, H.J. (2006) *J. Expt. Bot.,* **57:** 887–896.

Tigchelaar, E. C. (1986) In: *Breeding for Vegetable Crops,* M. J. Bassett, Ed., AVI, Westport, Conn, USA, pp. 135–171.

Tigchelaar, E. C. (1986) *In: Breeding Vegetable Crops, AVI, Westport, Conn,* pp. 135-I 71.

Tigchelaar, E. C. and Barman, R. J. (1985) *Rep. Tomato Genet. Coop.* **35**: 20-21.

Tikoo, S.K., Anand, N. and Kirishna, R. (1983) In: *15th International Congress of Genetics,* 22-23 December, 1983, New Delhi, India

Tikoo, S.K., Mathai, P.J. and Kishan, R. (1979) *Curr. Sci.,* **48:** 259–260.

Tillmann, M. A. A, Piana, Z. Cavarianl, C. and Minami, K. (1994) *Scientia Agricola,* **51**: 260-263.

Tingting, C., Zhou, S., Sauve, R., Bhatti, S. and Powell, W. (2005) *SNA Res. Conf.,* **50**: 628-631.

Tipping, P.W., Holko, C.A., Abdul-Baki, A.A. and Aldrich, J. (1999) *Biological Control,* **16**(1): 35-42.

Tipu, M.M.H., Amin, M., Dhar, M. and Alam, M. A. (2014) *J. Agric. Sci. Technol.,* **3:** 12-14

Tiwari, A.K. and Singh, D.K. (2014) *Indian J. Hill Farming,* **27**(2): 38-40

Tiwari, K.N., Mal, P.K., Singh, R.M. and Chattopadhyay, A. (1998) *J. Agric. Engin.*, **35** (2): 41-49.

Tiwari, R. N., Choudhury, B. and Pachauri, D. C. (1990) *Indian Hort.*, **35**: 4-5.

Traka-Mavrona, E. Bletsos, P. Grahadellis, M. Spanomitsios, G. (1997) *Acta Hortic.*, **379**: 171-178.

Traka-Mavrona, E., Bletsos, F., Grafiadellis, M. and Spanomitsios, G. (1995) *Acta Hortic.*, **379**: 289-296.

Tran, D.T., Hertog, M.L.A.T.M., Tran, T.L.H., Quyen, N.T., Van de Poel, B., Mata, C.I. and Nicolaï, B.M. (2017) *Front. Plant Sci.*, **8**: 608. doi: 10.3389/fpls.2017.00608.

Tresky, S. and Walz, E. (1997) *Organic Fmg. Res. Foundation* Bul. 4.

Tripathi, N. N. and Grover, R. K. (1975) *Zeitschrift fur Pflanzenkrankheiten and Pflanzenschutz*, **82**: 531-539.

Trottin-Caudal, Y., Fournier, C. and Millot, P. (1996) *Infos* (Paris), **123**: 30-35.

Truco, M.J., Randall, L.B., Bloom,A.J. and St Clair D.A. (2000) *Theor. Appl. Genet.*, **101**: 1082-1092.

Trudel, M. J. and Ozbun, J. L. (1970) *J. Exptl. Bot.*, **69**: 881-886.

Tsambanakis, J. (1984) *Proc. 3rd Conference on Protected Vegetables and Flowers*, pp. 26-27.

Tsuda, T. (2012) *Mol. Nutr. Food Res.*, **56**: 159–170.

Tu, C. C., Cheng, Y. Y. and Hwang, S. C. (1987) *Res. Bull. Taiwan District Agril. Improvement Station*, Taiwan, **21**: 33-41.

Tu, C. C., Cheny, Y. H. and Hwang, S. C. (1991) *Plant Protect. Bull.* (Taipei), **33**: 95-102.

Tu, J. C. and Zheng, J. M. (1994) *Mededelingen-Faculteit Landbouwkundige en Toegepaste Biologische Wetenschappen*, Univsiteit Gent, **59**: 951-958.

Tucker, D. T. (1976) *Ann. Bot.*, **40**: 1033-1042.

Turchenkov, G. I., Avdeev, Yu, I. Ivanova, L. M. Boeva, T. V. and Panfilova, T. A. (1988) *Kartofel'I Oveshchi*, **1**: 33.

Tuzel, I. H., Uc, M. A. and Tuzel, Y. (1994) *Acta Hortic.*, **366**: 381-8.

Tuzel, Y. (1994) *Acta Hortic.*, **366**: 175-82.

Tzortzakakis, E. A., Verdejo Lucas, S., Ornat, C., Sorribas, F. J. and Goumas, D. E. (1999) *Crop Protection*, **18**: 159-162.

Tzouramani, I., Mattas, K. and Grabiadellis, M. (1995) *Acta Hortic.*, **412**: 245-249.

Ullah, M.Z., Hassan, L., Shahid, S.B. and Patwary, A.K. (2015) *J. Bangladesh Agric. Univ.*, **13**: 65–69.

Umesha, S. (2006) *Crop Prot.* **25**: 375-381.

Upadhyay, N.K., Ratan, V., Yadav, V.K., Kumar, A., Awasthi, D., Chandra, S. and Rai, J.P. (2019) *Int. J. Curr. Microbiol. App. Sci.*, **8**: 2733-2743.

Urbaneja, A., González-Cabrera, J., Arnó, J. and Gabarra, R. (2012) *Pest Manage. Sci.*, **68**: 1215-1222.

Ushchekov, A. T. (1994) *Zaschita Rastenii* (Moskva), No. 3, pp. 56-57.

V'lchev, N. (1987) *Rasteniev. dni Nauki*, **24**: 19-23.

Vaishampayan, J. and Singh, H.N. (1995) *Indian J. Entomol.*, **57**: 224-232.

Vakalounakis, D. J. (1991) *Plant Dis.*, **75**: 795-797.

Vakalounakis, D. J. (1992) *Plant Dis.*, **76**: 43-46.

Vakis, N. J. and Photiades, I. (1989) *Technical Bulletin - Agricultural Research Institute*, Cyprus **108**: 10.

Vallejos, C.E. and Tanksley, S.D. (1983) *Theor. Appl. Genet.*, **66**: 241-247

Van der Beek, J.G., Pet, G. and Lindgout, P. (1994) *Theor. Appl. Genet.*, **89**: 467-473.

Van der Biezen, E.A., Glagotskaya, T., Overduin, B., Nijkamp, H.J.J. nd Hille, J. (1995) *Mol. Gen. Genet.*, **247**: 453-461.

Van der Knaap, E., Sanyal, A., Jackson, S. A. and Tanksley, S. D. (2004) *Genetics*, **168**: 2127–2140.

van Heusden, A.W., Koornneef, M., Voorrips, R.E. (1999) *Theor. Appl. Genet.* **99**: 1068-1074.

van Leeuwen, T., Dermauw, W., Van De Veire, M. *et al.* (2005) *Exp. Appl. Acarol.*, **37**: 93–105.

van Ooijen, J. W., Sandbrink, J. M., Vrielink, M., Verkerk, R., Zabel, P. and Lindhout, P. (1994) *Theor. Appl. Genet.*, **89**: 1007–1013.

Vanitha, S. C., Niranjana, S., Mortensen, C. and Umesha, S. (2009) *Biocontrol*, **54**: 685-695.

Vargues, A. C., Campo, J. L. and Monteiro, A. A. (1994) *Acta Hortic.*, **357**: 317-324.

Varis, S. and George, R. A. T. (1985) *HortScience.*, **60**: 376-386.

Varis, S. and Ozuyaman, T. (1994) *Acta Hortic.*, **366**: 417-422.

Varshney, C. K. and Rout, C. (1998) *Bull. Environmental Contamination and Toxicology*, **61**: 188-193.

Vasudeva, R. S. and Lal, T. B. (1948) *Indian Fmg.*, **8**: 183-186.

Vasudeva, R. S. and Samraj, J. (1948) *Phytopath.*, **38**: 364-369.

Veken, L. vander, Michels, P. Feyen, J. and Benoit, F. (1982) *Sci. Hort.*, **18**: 9-23.

Verkerk, K. (1964) *Neth. J. Agri. Sci.*, **12**: 57-68.

Verlaan, M.G., Hutton, S.F., Ibrahem, R.M. *et al.* (2013) *PLoS Genetics*, **9**: e1003399.

Verma, A. N., Srivastava, D. C. and Shanna, R. K. (1970) *Mysore Agri. Sci.*, **4**: 107-109.

Verma, T. S. and Sharma, R. K. (1981) *Haryana J. Hort. Sci.*, **10**: 212-215.

Verwer, F. L. J. A. W. (1975) *Bedrij Bontwikkeling*, **6**: 767-769.

Vezina, F., Trudel, M. J. and Gosselin, A. (1991) *Can. J. Pl. Sci.*, **71**: 923-932.

Vielemeyer, H. P. and Weissert, P. (1990) *Gartenbauwissenschaft*, **55**: 209-212.

Vikmane, Y. A., Zervekya, D., Maurinya, J. H. and Rudzite, L. T. (1981) *Referativny Zhurnal*, **7**: 55. 321.

Villalta, I., Reina-Sanchez, A., Cuartero, J., Carbonell, E. A. and Asins, M. J. (2005) *Theor. Appl. Genet.*, **110**: 881–894.

Vittum, M. T. and Flocker, W. J. (1967) *Proc. Amer. Soc. Agron.*, pp. 675-685.

Voican, V., Davidescu, V. and Atanoasiu, N. (1989) *Lucrari Stiintifice, Institutul Agronomic 'Nicolae Balcescu', Bucuresti, Seria B, Horticultura*, **32**: 15-20.

Volin, R.B., Pohronezny, K. and Simone, G.W. (1989) *Plant Dis.*, **73**: 1018–1019.

Vorayos, T. Fujieda, K. Okuda, H. and Ichiki, Y. (1992) *Bull. Inst. Tropical Agriculture, Kyushu Univ.*, **15**: 1-47.

Vos, P., Simons, G., Jesse, T., Wijbrandi, J., Heinen, L., Hogers, R., Frijters, A., Groenendijk, J., Diergaarde, P., Reijans, M., Fierens-Onstenk, J , de Both, M., Peleman, J., Liharska, S., Hontelez, J. and Zabeau, M. (1998) *Nature Biotechnol.*, **16**: 1365-1369

Wachocki, S. E., Bonnema, A.B. and O'Connell, M. A. (1991) *Theor. Appl. Genet.*, **81**: 420-427.

Wada, T., Ikeda, H., Ikeda, M. and Furukawa, I.I. (1996) *J. Jap. Soci. Hort. Sci.*, **65**: 553-558.

Wadi, J. A. (1999) *Egyptian J. Hort.*, **26**: 167-176.

Walgenbach, J. F., Leidy, R. B. and Sheets, T. J. (1991) *J. Eco. Eutomol.*, **84**: 978-986.

Walgenbach, J.F. (2018) In: *Sustainable Management of Arthropod Pests of Tomato*, Eds. Wakil, W., Brust, G.E., Perring, T.M. Academic Press, pp. 323-339.

Walker, J. C. and Foster, R. E. (1946) *Amer. J. Bot.*, **33**: 259-264.

Walker, M. A., Smith, D. M., Pauls, K. P. and Mc Kersie, B. D. (1990) *HortScence*, **25**: 334.

Wallace, T.C., Slavin, M. and Frankenfeld, C.L. (2016) *Nutrients*, **8**: 32.

Walter, J. M. and Conover, R. A. (1952) *Phytopath.*, **42**: 197-199.

Walter, J. M. and Hayslip, N. C. (1961) *Circ., Fla. Agri. Exp. Sta.*, **S-131**: 8.

Walter, J. M. and Kelbert, D. G. A (1953) *Circ. Fla. Agri. Exp. Sta.*, **59**: 10.

Walters, S. A. and Barker, K. R. (1994) *J. Nematol.*, **26**: 600-605.

Wang, F., Li, J. F., Li, G. Y and Zhu, D. W. (1995) *Acta Hortic.*, **402**: 253-258.

Wang, T. C., Black, L. L. and Hsieh, W. H. (1994) *Plant Pathology Bull.* **3**: 181-185.

Wann, E. V. (1995) *HortScience*, **30**: 186–193.

Ward, G. M. (1963) *Proc. Amer. Soc. Hort. Sci.*, **83**: 695-699.

Warnock, S. J. (1988) *HortScience*, **23**: 669-673.

Watenbe, K., Takahashi, B. and Inove, H. (1977) *Bull. Coll. Agri, Vet. Medicine, Nihon Univ.*, **34**: 26-35.

Watson, R. D., Heinrich, E. C. and Harvey, W. R. (1954) *Idaho. Agri. Exp. Sta. Res. Bull*, **27**: 20.

Watt, VM. (1931) *Ark. Bull.*, p. 267.

Weitang, S., Ligang, Z., Chengzong, Y., Xiaodong, C., Li-Qun, Z. and Xili, L. (2004) *Crop Prot.*, **23:** 243-247.

Went, F. W. (1944) *Amer. J. Bot.*, **31**: 135-150.

Went, F. W. (1945) *Amer. J. Bot.*, **32**: 469-479.

Went, F. W. and Cosper, L. (1945) *Amer. J. Bot.*, **32**: 543-554.

Weston, P. A., Johnson, D. A., Burton, H. T. and Snyder, J. C. (1989) *J. Amer. Soc. Hort. Sci.*, **114**: 492–498

Whitner, B. F., Walter, J. M. Kelbert, D. G. A. and Hayslip, N. C. (1962) *Circ., Fla. Agri. Exp. Sta.*, **S-137:** 8.

Wibowo, S. T. (2007) *Ph.D. Thesis, Bogor Graduate School*, Bogor Agricultural University.

Widder, I. E. and Lorenz, O. A. (1982) *J. Amer. Soc. Hort. Sci.*, **107**: 960-964.

Wien, H. C. and Minotti, P. L. (1988) *J. Amer. Soc. Hort. Sci.*, **113**: 342-347.

Wiesendorn, W. D., Trumble, J. T. and Oatman, E. R. (1990) *J. Eco. Entomol.*, **83**: 212-216.

Williams, C. E. and St Clair, D. A. (1993) *Genome*, **36**: 619-630.

Williamson, T. M. and Murray, R. C. (1993) *Bull. Res. and Development Divn.*, Ministry of Agriculture, Jamaica, **68**: 45-51.

Williamson, V.M., Ho, J.Y., Wu, F.F., Miller, N. and Kaloshian, I. (1994) *Theor. Appl. Genet.*, **87**: 757-763.

Willits, D. H. and Peet, M. M. (1994) *North Carolina Flower Growers' Bulletin*, **39**: 1-5.

Wing, R.A., Zhang, H.B. and Tanksley, S.D. (1994) *Mol. Gen. Genet.* **242**: 681-688.

Winsor, G. M., Messing, J. H. L. and Long, M. I. E. (1965) *J. Hort. Sci.*, **40**: 118-132.

Wisler, G. C., Duffus, J. E., Liu, Hsing Yeh, Li,Ruffui and Falk, B.W. (1997) *Calif. Agric.*, **51**: 24-26.

Wittwer, S. H. (1960) *Mich State. Agri. Exp. Sta. Bull.*, **228**: 5-30.

Wittwer, S. H. and Teubner, F. G. (1956) *Quart. Bull., Mich. Exp. Sta.;* **38**: 588-594.

Wolf, S. and Rudich, J. (1988) *Sci. Hort.*, 34: 1-11.

Wood, M. (1992) *Agricultural Research*, USDA, **40**: 4–5.

Workneh, T.S., Osthoff, G. and Steyn, M. (2012) *J. Food Sci. Technol.*, **49**: 685–694.

Wu, X., Wu, F., Zhou, X., Fu, X., Tao, Y., Xu, W., Pan, K. and Liu, S. (2016) *Front. Plant Sci.* **7:** 846. Wyenandt, C.A., Rhodes, L.H., Riedel, R.M. and Bennett, M.A. (2008) *HortScience*, **43**: 807–810.

Xing, J. M. Li, X. Y. Feng, S. L. Gan, X. I.I. Fu, Y.Q. and Zhang, M. Y. (1988) *Chinese J Biol. Coll.*, **4**: 101.

Xing, Y., Wang, X.F. and Papadopoulos, A. P. (1997) *Hort Technol.*, **7**: 169-176.

Xu, H. J., Zhou, Y. J. and Li, J. M. (1999) *China Vegetables*, **5:** 24-26.

Xu, H. J., Zhou, YJ. and Li, J. M. (1997) *China Vegetables*, **4:** 20-22.

Xu, H. L., Gauthier, L. and Gosselin, A. (1994) *J. Hort. Sci.*, **69**: 829-832.

Xue, B. Y. (1994) *China Vegetables*, **4**: 32-33, 50.

Xue, Q. Y., Chen, Y., Li, S. M., Chen, L. F., Ding, G. C., Guo, D. W. and Guo, J. H. (2009) *Biol. Control.*, **48:** 252–258.

Yaffa, S., Sainju, U. M., Singh, R.P. and Reddy, K. C. (2000) *HortScience*, **35**: 1258-1262.

Yamaguchi, K., Sano, T., Arita, M. and Takahashi, M. (1992) *Ann. Phytopath. Soc. Japan*, **58**: 188-194.

Yang, C. C. and Chinnan, M. S. (1988) *Transactions of the ASAE, American Society of Agricultural Engineers*, **31**: 920-925.

Yang, R. C., Xu, I. L. L., Long, M. S., Yu. W. G., Lu, C. G., Wu, G., Pan, N. S. and Chen, Z. L. (1995) *Jiangsu J. Agric. Sci.*, **11**: 40-44.

Yang, W. and Francis, D.M. (2007). *Genetic improvement of Solanaceous crops*. Vol. 2: Tomato, Science Publisher. pp. 379-420.

Yang, W.C., Chen, J., Zhang, X.M. and Francis, D.M. (2007) *Sci. Agric. Sin.*, **40**(2):283–290.

Yang, W. M., Chung, S. J. and Yang, S. Y. (1989) *Abstracts of Communicated Papers. Korean Soc. Hort. Sci.*, **7**: 54-55.

Yang, X.C. and Mu S.S. (1998) *J. Southwest Agric. Univ*, **20**: 165-169.

Yanuriati, A., Savage, G. P. and Rowe, R. N. (1996) *Proceedings of the nutrition Society of New Zealand*, **21**: 137-141.

Yao, G., Zhang, F. and Li, Z. (1994) *Chinese J. Biol. Cont.*, **10**: 106-109.

Yassin, A. M. and Nour, M. A. (1965) *Ann. Appl. Bioi.*, **56**: 207-217.

Yasui, M., Nishimatu, T., Fukada, M. and Maekawa, S. (1991) *Appl. Entomol. Zool.*, **26:** 271-274.

Yawalkar, K. S., Agarwal, J. P. and Bokde, S. (1962) In: *Manures and Fertilizer, Agri. Hort. Publishing House*, Nagpur, India.

Yelle, S., Gosseline, A. and Trudel, M. J. (1987) *Canndian J. Plant Sci.*, **67**: 899-907.

Yen, H.C., Shelton, B.A., Howard, L.R., Lee, S., Vrebalov, J. and Giovannoni, J.J. (1997) *Theor Appl Genet.*, **94**: 1069–1079.

Yi, Y. K. and Sul, K. C. (1998) *Korean J. Plant Pathol.*, **14**: 744-746.

Yildiz, F., Yildiz, M., Delen, N., Cofikuntuna, A., Kinay, P. and Türküsay, H. (2007) *Turk. J. Agric. For.*, **31:** 319-325.

Yilmaz, E. (2001) *Turkish J.Agric. For.*, **25:** 149–155.

YiPeng, Q., HaiTao, L., ZiJun, Z., QingDao, Z. (2009) *Acta Hort. Sinica.*, **36:** 1227-1232.

Yogev, A., Raviv, M., Kritzman, G., Hadar, Y., Cohen, R, Kirshner, B. and Katan, J. (2009) *Crop Prot.*, **28**: 97-103.

Yooboon, T., Pengsook, A., Ratwatthananon, A., Pluempanupat, A. and Bullangpoti, V. (2019) *Chem. Biol. Technol. Agric.*, **6:** 5 https: //doi.org/10.1186/s40538-019-0143-6,

Young, N.D., Zamir, D., Ganal, M.W. and Tanksley, S.D. (1988) *Genetics*, **120**: 579-585.

Yu, J. Q., Komada, I.T., Yokogama, H., Yamamoto, M., Terada, T. and Matsui, Y. (1997) *J. Hort. Sci.*, **72**: 989-996.

Yu, Z.H., Wang, J.F., Stall, R.E. and Vallejos, C.E. (1995) *Genetics*, **141**: 675-682.

Yýldýz, H. and Baysal, T. (2006) *J. Food Engg.*, **75**: 327-332

Zagorska, N.A., Shtereva, A., Dimitrov. B.D. and Kruleva, M.M. (1998) *Plant Cell Rep.*, **17**: 968–973.

Zagorska, N.A., Shtereva, L.A., Kruleva, M.M., Sotirova, V.G., Baralieva, D.L. and Dimitrov, B.D. (2004) *Plant Cell Rep.*, **22**: 449–456.

Zamir, D., Ekstein-Michelson, I., Zakay, Y., Navot, N., Zeidan, M., Sarfatti, M., EshedHarel, E., Pleban, T., Oss, H., Kedar, N., Rabinowitch, D. and Czosnek, H. (1994) *Theor. Appl. Genet.*, **88**: 141-146.

Zamir, D., Jones. RA and Kedar, N. (1980) *Plant Sci. Lett.*, **17**: 353-361.

Zappala, L., Biondi, A., Alma, A., Al-Jboory, I.J., Arno, J. Bayram, A., Chaillex, A., El-Arnaouty, Gerling, D., Guenaoui, Shaltiel-Harpaz, L., Siscaro, G., Stavrinides, M., Tavella, L., Aznar, R.V., Urbaneja, A., Desneux, N. (2013) *J. Pest Sci.* **86:** 635–647.

Zavaleta-Mejia, E. and Gomez, R. O. (1995) *Fitopatologia*, **30**: 35-46.

Zehnder, G.W., Sikora, E.J. and Goodman, W.R. (1995) *Crop Prot.*, **14:** 683-687.

Zenner, De P. T. (1995) *Revista Colombiana de Entomologia*, **21**: 9-14.

Zhai, Y. M., Shao, X. H., Xing, W. G., Wang, Y., Hung, T. T. and Xu, H. L. (2010) *J. Food Agric. Environ.*, **8:** 709 - 713.

Zhang, H.B., Budiman, M.A. and Wing, R.A. (2000) *Theor. Appl. Genet.* **100**: 1183-1189.

Zhang, H.B., Martin G.B., Tanksley, S.D. and Wing, R.A. (1994) *Mol. Gen. Genet.*, **244**: 613-621.

Zhang, L.P., Khan, A., Ni˜no-Liu, D. and Foolad, M.R. (2002) *Genome*, **45**: 133–146.

Zhou, Y.J., Xu, H.J. and Zhu, D.W. (1995) *Acta Hort.*, **402**: 227-32.

Zhukova, P. S., Zabara, Y. M. and Anikhovskaya, T. (1987) *Ovoshchevodstvo Minsk*, **7:** 71-77.

Ziauddin, A. T. M. and Rukshana, F (1999) *In: ASAE/CSAE-SCGR An1111al International Meeting*; Toronto, Ontario. Canada, pp. 14.

Zidan, Z. H., Afifi, F. A., Sobeiha, A.K., Moawad, A. G. and El-Malky. K. (1994) *Arab Univ, J. Agril. Sciences*, **2**: 179-185.

Zipori, I., Berlinger, M.J., Dayan, E., Dahan, R., Shmuel, D., Mordechai, S. and Aharon, Y. (1988) *Hassadeh*, **68**: 1711-1713.

Zipori, Y., Dayan, E. and Shmuel, D. (1988) *Hassadeh*, **68**: 1508-1509.

Zorzoli, R., Coiutry, E. T., Prado, E.A., Mroginski, L. A. and Picardi, L. A. (1988) *Turrialba*, **38:** 332-336.

Zuckerman, B.M., Dicklow, M.B. and Acosta, N. (1993) *Biocont. Sci. Tech.*, **3**: 41-46.

Zwart, H. F. De (1998) *Acta Hortic.*, **456**: 347-353.

BRINJAL

Y.A. Lyngdoh, S. Hazra, M.G. Som and T.K. Maity

1.0 Introduction

Brinjal or eggplant or aubergine is one of the most common, popular and principal vegetable crops grown in India and other parts of the world. It can be grown in almost all parts of India except in higher altitudes, all the year round. A number of cultivars are grown throughout the country depending upon the yield, consumer's preference about the colour, size and shape of the various cultivars. Brinjal is of much importance in the warm areas of fareast, being grown extensively in India, Bangladesh, Pakistan, China and Philippines. It is also popular in France, Italy and the United States. In Asia and the Mediterranean region, eggplant ranks among the top five most important vegetable crops (Frary *et al.*, 2007). It is highly productive and usually finds its place as the poor man's crop. In India, it is being consumed as a cooked vegetable in various ways.

India is one of the largest brinjal producing countries in the world covering an area of 0.73 million hectares with a production of 12.800 million tonnes having the productivity of 17.53 t/ha (Anonymous, 2018). However, the maximum production of brinjal in India is from West Bengal with a production of 3027.75 tonnes with a percentage share of 23.69 per cent. The major destinations of brinjal exports from India are Sri Lanka, Maldives, Singapore, Sudan, and Mali. The specification of brinjal for export to international markets particularly in South-East and South-West countries are glossy purple, round shaped fruit having green calyx and devoid of spines in any plant parts (Singh and Pandey, 1993). The crop is extremely variable in India owing to its wide range of fruit shapes and colours, ranging from oval or egg-shaped to long club-shaped; and from white, yellow, green through degrees of purple pigmentation to almost black. The early European name eggplant suggests that the first introductions were small fruited and gradually various sizes and shapes were evolved due to selection and other breeding methods.

2.0 Composition and Uses

2.1.0 Composition

Brinjal has been a staple vegetable in many tropical countries since ancient times. It is liked by both poor and rich. Contrary to the common belief, it is quite high in nutritive value and can well be compared with tomato (Choudhury, 1976a). The food value of brinjal is given in Table 1.

Bajaj *et al.* (1979) reported that on an average the oblong fruited brinjal cultivars are rich in total water soluble sugars, whereas the long fruited cultivars contain large amounts of free reducing sugars, anthocyanin, phenols, glycoalkaloids (as solasodine), dry matter and amide proteins. High anthocyanin content and low glycoalkaloid content are considered essential, regardless of how the fruit is to be used. For processing purpose, the fruit should have a high dry matter content and low level of phenolics. Bajaj *et al.* (1981) further observed that on an average the round types of brinjal had higher polyphenol oxidase activity and glycoalkaloid content (cv. PBR-91-1) than the long types. Bitterness in brinjal is due to the presence of glycoalkaloids and saponins (Mariola *et al.*, 2013) which are of wide occurrence in plants belonging to Solanaceae family. Generally, high amount of glycoalkaloids (20 mg/100 g fresh weight) produces a bitter taste and off flavour. Usually glycoalkaloid contents vary from 0.37-4.83 mg/100 g fresh weight in most of the commercial cultivars of India (Bajaj *et al.*, 1981). The discolouration in brinjal fruits is attributed to high polyphenol oxidase activity. The cultivars which are least susceptible to discolouration are considered suitable for processing purpose.

Sidhu *et al.* (1982) in their studies concluded that there is much variation in the chemical constituents in fruits of different cultivars, and chlorophyll, true protein and total phenols are influenced by other constituents of the fruits besides the effect of dry matter, especially on anthocyanin and ortho dihydroxy phenols.

Eggplant is one of the vegetables having the highest antioxidant capacity (Cao *et al.*, 1996), owing to its high content in phenolics (Singh *et al.*, 2009). It also has a very low caloric value and is considered to be one of the healthiest vegetables for its high content of vitamins, minerals and bioactive compounds for human health (Raigón *et al.*, 2008; Plazas *et al.*, 2014; Docimo *et al.*, 2016). Eggplant is also ranked as one of the top 10 vegetables in terms of oxygen radical absorbance capacity ORAC (Cao *et al.*, 1996). The bioactive properties of eggplant are mostly associated with high content in phenolic compounds which mainly includes hydroxycinnamic acid conjugates (Whitaker and Stommel, 2003; Plazas *et al.*, 2013), which are mostly phenolic acids, particularly chlorogenic acid (5-O-caffeoylquinic acid and its isomers) accounting for 70 per cent to 95 per cent of total phenolics in the fruit flesh (Stommel and Whitaker, 2003; Stommel *et al.*, 2015) and anthocyanins in the fruit skin (Mennella *et al.*, 2012). Both phenolic acids and anthocyanins have multiple properties beneficial for human health (Plazas *et al.*, 2013; Braga *et al.*, 2016).

Table 1: Composition of Brinjal (per 100 g of edible portion)*

Moisture	92.7 g	Sodium	3.0 mg
Protein	1.4 g	Potassium	2.0mg
Fat	0.3 g	Copper	0.17 mg
Minerals	0.3 g	Sulphur	44.0 mg
Fibre	1.3 g	Chlorine	52.0 mg
Carbohydrates	4.0 g	Vitamin A	124 I.U.
Calcium	18 mg	Thiamine	0.04 mg
Magnesium	16mg	Riboflavin	0.11 mg
Oxalicacid	18 mg	Nicotinic acid	0.09 mg
Phosphorus	47mg	Vitamin C	12.0 mg
Iron	0.9mg		

* Aykroyd (1963).

Studies on organic and mineral element analysis of mature brinjal fruit as conducted by Flick *et al.* (1977) revealed that both Cu content and polyphenol oxidase activity were highest in the purple colour fruit and lowest in white cultivar. Similarly, Fe and catalase were highest in the green and lowest in the white cultivar. The findings suggested that the green cultivar had the better processing properties than the more popular purple cultivar. Flick *et al.* (1978) further reported that white cultivar contained twice as much crude fibre as the purple and green cultivars. Amino acid contents were higher in the purple cultivars and lowest in the white ones. Bujdoso and Videki (1976) on the other hand, reported that largest brinjal fruits produced the highest crude protein content in Romanian cultivars.

Brinjal is also an important source of plant-derived nutrients which is valued for its composition in phytochemicals and especially minerals such as P, K, Ca, and Mg (Rodriguez *et al.*, 2018; Arivalagan *et al.*, 2013). Raigon *et al.*, 2008 and Arivalagan *et al.*, 2012 studied the mineral composition and the nutritional value of eggplant landraces and germplasm accessions from different regions to identify the mineral rich germplasm for breeding purposes and found considerable variation among the varieties and germplasm accessions, revealing the existence of ample variation in brinjal, which can be exploited for the selection of germplasm for nutritionally improved characteristics. It has also been reported that organic management and fertilization have a positive effect on the accumulation of certain beneficial minerals (K, Ca, Mg) and phenolic compounds in eggplant (Raigon *et al.*, 2010)

Irrigation has a variable effect on the mineral composition and vitamin (ascorbic acid, nicotinic acid and P-active substances) contents of brinjal fruits (Dikii *et al.*, 1980). They reported that in most cases the content of dry matter, total sugars, proto pectins, cellulose, acids and total fat were reduced but water soluble pectin content was increased by the application of irrigation. Spacing also influences the ascorbic acid content of brinjal fruit. Gutierrez *et al.* (1976) observed that highest (10.6 mg/100 g) and lowest (5.9 mg/100 g) fruits ascorbic acid contents were obtained in plant density of 31746 and 20408 plants/ha, respectively. Spacing, however, had no significant effect on the content of soluble carbohydrates.

2.2.0 Uses

The unripe fruit is primarily used as a cooked vegetable for the preparation of various dishes in different regions of the world. It has got much potential as raw material in pickle making and dehydration industries (Singh *et al.*, 1963). It is supposed to contain certain medicinal properties and white brinjal is said to be good for diabetic patients (Choudhury, 1976a). It can also cure toothache if fried brinjal fruit in sesame oil is taken and acts as an excellent remedy for those suffering from liver complaints (Chauhan, 1981). Eggplant is versatile in nature which can be used in different culinary purposes and therefore, brinjal is often described as the 'King of vegetables' (Choudhary and Gaur, 2009). In addition, it is also used for the treatment of asthma, bronchitis, diabetes, dysuria and dysentery (Daunay *et al.*, 2000; Subramanyam *et al.*, 2013). Eggplant is well known for its medicinal properties and has also been recommended as an excellent remedy for liver complaints and diabetic patients (Sabolu *et al.*, 2014). Seeds of brinjal are used as a stimulant (Nadkarni, 1972).

Uses of Brinjal

Different Recipes of Brinjal in Indian Cuisine

3.0 Origin, Taxonomy, Botany and Cultivars

3.1.0 Origin

The brinjal, eggplant or aubergine (French name), a normally self-fertilized annual is of uncertain origin. A wild form with many small fruits sometimes called as var. *insanum* is found on the Bengal plains of India (Martin and Rhodes, 1979). India is a centre of variation of brinjal, one of many names probably derived from the Arabic. The cultivated brinjal is undoubtedly of Indian origin and has been in cultivation for long time (Thompson and Kelly, 1957). Vavilov (1928) was of the opinion that its centre of origin was in the Indo-Burma region. According to Purewal (1957), it was found growing wild in India. Eggplant is an old age crop and the earliest known written record of cultivated eggplant in Chinese literature dates back to 59 BC (Wang *et al.*, 2008). It is known by various common names in Sanskrit documents, some of which dates back to 300 BC (Daunay and Janick, 2007).

Various forms, colours and shapes of brinjal are found throughout South-East Asia, suggesting that this area is an important centre of variation and possibly of origin. Burkill (1935) reported that wild yellow-fruited types after the name *Solanum ferox* are found in Malaysia and it is not certain whether it is a separate species. According to Zeven and Zhukovsky (1975), it originated in India but has a secondary centre of variation in China. In China, it has been known for the last 1500 years. Wild brinjal is an armed perennial shrub or herb with bitter fruits and it occurs in India (Bhaduri, 1951). Sampson (1936) suggested the African origin of this crop, but there is no evidence that *Solanum melongena* is native of the area though there are spiny African brinjal plants.

Solanum incanum is believed to be the closest relative of *S. melongena* as revealed by phylogenetic studies (Lester and Hasan, 1991) which encompasses up to 27 species and 81 taxa (Bitter, 1923) endemic to eastern Africa and the Middle East. On the basis of their distribution and morphological characters these taxa have been divided into four separate and diverse groups (A-D) of wild eggplant relatives (Lester and Hasan, 1991). *S. incanum* which belongs to group C have been originated from the plants of groups A and B which spread from Africa to the Middle East and became adapted to local conditions. *S. incanum* then reached Asia towards the east and diverged to produce *S. melongena*. Group D represents the types of plants that were moved to South Africa and adapted to the arid conditions there (Daunay *et al.*, 2001). Like its ancestor *S. melongena* is also a complex species consisting of four distinct groups (E-H) (Lester and Hasan, 1991; Mace *et al.*, 1999). It can be said that *S. melongena* belongs to Group E (*S. insanum*) which arose as revertants from group G forms (Lester and Hasan, 1991; Mace *et al.*, 1999) and this group includes the weedy plants with wild attributes. Group F also known as *S. cumingii* is the likely ancestor of domesticated eggplant and includes weedy forms found widely in Indo-China and Indonesia. Group G *S. ovigerum* represents the domesticated primitive cultivars and landraces with small round or oblong fruits which grows throughout the Southeast Asia. Group H *S. melongena* includes the large-fruited advanced cultivars that are grown in agricultural fields and green houses worldwide.

3.2.0 Taxonomy

S. melongena is a member of the Solanae tribe within the Solanaceae family. Eggplant is classified as one of the "spiny solanums", subgenus *Leptostemonum* including 350–450 species of New and Old World plants having leaf and stem prickles (Levin *et al.*, 2006) and comprising of almost one third of the genus being distributed worldwide. The genomics in genus *Solanum* are evolving at a moderate pace compared to other plant species (Doganlar *et al.*, 2002). The common name 'eggplant' comprises of three closely related cultivated species, endemic to the Old World, belonging to the genus *Solanum* L., subgenus *Leptostemonum* (Dunal) Bitter and two sections. Section one i.e Melongena consists of two species a) *S. melongena* L. (brinjal eggplant, aubergine) – synonyms: *Solanum cumingii* Dunal, *Solanum pressum* Dunal, *Solanum undatum* Poiret sensu Ochse and b) *S. macrocarpon* (Gboma eggplant) – synonyms: *Solanum integrifolium* Poiret var. *macrocarpum*, *Solanum melongena* L. var. *depressum* Bail. The other section is *Oliganthes* comprising of *S. aethiopicum* (scarlet eggplant) synonyms: *Solanum integrifolium* Poiret, *Solanum integrifolium* Poiret var. *microcarpum*, *Solanum gilo* Raddi, *Solanum naumannii* Engl., *Solanum pierreanum* Pailleux and Bois, *Solanum zuccagnianum* Dunal.

There are 3 main botanical varieties under the species *melongena* (Choudhury, 1976a). The round or egg-shaped cultivars are grouped under var. *esculentum*. The long, slender types are included under var. *serpentinum* and the dwarf brinjal plants are put under var. *depressum*. *Solanum* is a very large genus. Among the 22 Indian species, there is a group of 5 related ones, all prickly and diploids with 2n = 2x = 24, namely *melongena, coagulans, xanthocarpum, indicum* and *maccanii* (Choudhury, 1976b). Out of these species, *Solanum incanum* (often considered synonymous with *S. coagulans*) produces fertile hybrids in crosses with *S. melongena*. It appears that *S. melongena* is more closely related to *S. incanum* than to any other species. Occurrence of natural hybridization between *S. incanum*, and *S. melongena* has also been reported by Viswanathan (1975). Two other species, *S. khasianum* and *S. aviculare*, having the same chromosome number, have gained importance as sources of solasodine used for synthesis of steroid hormones (Choudhury, 1976b). The cross between *S. khasianum* and *S. melongena* is only successful when *S. melongena* is used as a female parent. Two other species are also commonly referred to as eggplants, *S. macrocarpon* (gboma eggplant) and *S. aethiopicum* (scarlet eggplant) and are both cultivated in Africa as fruit and vegetable crops. Wild and weedy forms of *S. incanum* cluster separately from *S. melongena*, and scarlet and gboma eggplants fall outside of the *incanum-melongena* complex (Weese and Bohs, 2010).

There is some relationship between the different species of *Solanum*. Studies on interspecific relation and hybridization are of great significance because sometimes interspecific hybridization may be necessary to incorporate the desirable genes to cultivated species. Anjaneyulu and Ramakrishnan (1968), after an elaborate study on the reaction of a large number of *Solanum* spp. and the cultivars of *S. melongena* to little leaf disease, indicated that all the 36 cultivars and species except *S. auriculatum* which was immune to this disease, proved susceptible to the little leaf disease. Among those species important ones are: *S. pubescens, S. seaforthianum,*

S. xanthocarpum, S. trilobatum, S. tuberosum, S. melongena var.*insanum, S. incanum, S. torvum, S. sisymbrifolium, S. ciliatum, S. robustum, S. wightii, S. capsicoides, S. khasianum, S. aviculare, S. gilo, S. integrifolium, S. testiculatum*. They also reported that out of 173 cultivars of brinjal tested so far, none was found to be resistant to the little leaf disease. Rajasekharan (1969) studied the inter-relationship of some common *Solanum* species occurring in South India. Eight *Solanum* spp.were crossed in all possible combinations out of which *S. indicum x S. melongena; S. xanthocarpum x S. trilobatum* crosses were successful. Schaff *et al.* (1980) produced interspecific hybrids in reciprocal crosses by using 11 *S. melongena* types and an African accession of *S. macrocarpon. S. indicum* is a highly variable species and further studies regarding the breeding behaviour of different forms of *S. indicum* with other species are necessary to determine affinities between the species (Choudhury, 1976b).

Taxonomic relationship study between the tetraploid wild forms (n = 24) with and without thorns and diploid (n = 12) cultivated type of *S. incanum* from West Africa revealed that wild tetraploids are not related to cultivated diploid *S. incanum* and that the thorny form should be designated *S. spinosum* and the thornless derivative should be treated as a botanical variety of the latter (Anaso and Ozo, 1990). Yamazuki and Takushi (1991) designated a new species *S. miyakojimense* which was previously treated as synonymous with *S. indicum, S. anguivi* or *S. trilobatum. S. miyakojimense* is similar to *S anguivi* but has subterete prickles, smaller and thicker leaves, short calyx-lobes and a smaller, white corolla. Study on segregation of isozymes in selfed progenies of a synthetic amphidiploid between *S. integrifolium* and *S. melongena* suggested that it was not caused by variation of chromosome number but by genetic segregation of isozyme genes. The genome of the synthetic amphidiploid was indicated to be unstable (Isshiki *et al.*, 2000).

Karihaloo *et al.* (2002) revealed that all members of the eggplant complex had high similarity to each other based on the studies of the seed protein profiles by SDS-PAGE. On the other hand, Furini and Wunder (2004) reported that the eggplant complex showed a very large morphological variation, based on the AFLP markers which proved to be more informative than RAPD markers and allozymes. The results of Karihaloo *et al.* (2002) based on the seed protein study, and Furini and Wunder (2004) indicate that *S. macrocarpon* is more related to *S. aethiopicum* (section *Oliganthes*) than to *S. melongena*. The fact that *S. macrocarpon* and *S. aethiopicum* were domesticated and are cultivated mainly in Africa supports their relatively similar topology in the dendrogram (Furini and Wunder, 2003). Singh *et al.* (2006) confirmed a close genetic relationship between *S. melongena* and *S. incanum* by using RAPD technique.

3.3.0 Botany

Brinjal is an annual herbaceous plant having a tap root system but on transplanting the seedlings, it changes the type of root system to a rather fibrous one as is true with tomatoes and peppers. The above ground portion of the plant is erect, compact and well branched. The leaves are large, simple, lobed and the underside of most cultivars is covered with dense wool-like hairs. Inflorescence is often solitary but sometimes it constitutes a cluster of 2-5 flowers. Solitary or clustering of flowers is the characteristics of a cultivar. The flowers are large and showy with the corolla purple in colour. Islam *et al.* (2018) found 53 per cent light violet, 37 per cent bluish violet and 10 per cent pale violet flowers among all brinjal genotypes. Dash *et al.* (2019) reported three flowering patterns in brinjal *i.e.* solitary, cyme and mixed. The flowers are hermaphrodites and the stamens dehisce at the same time the stigma is receptive so that self-pollination is the rule, although there is some cross-pollination by insects. Davidar *et al.* (2015) reported that the wild and cultivated brinjal have gamopetalous flowers i.e fused petals which are produced singly or in small clusters of 2–4 flowers at the apex of stems, are flat and usually five-lobed. The fruit is pendant and is a fleshy berry borne singly or in clusters. The colour of the mature fruits varies from purple, purple black, yellowish, white, green and striped, depending upon the cultivars. The seeds are borne on the fleshy placentae and the placentae with the seeds completely fill the locular cavity.

Variability in Shape, Size and Colour of Brinjal.

Krishnamurthi and Subramaniam (1954a) reported 4 types of flowers, depending on the length of styles, *viz.* (i) long-styled with big size ovary, (ii) medium-styled with medium size ovary, (iii) pseudoshort-styled with rudimentary ovary, and (iv) true short-styled with very rudimentary ovary.

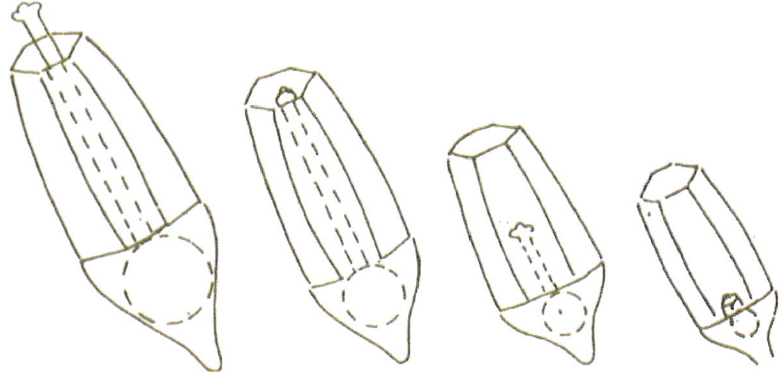

Types of Brinjal Flowers on the Basis of Length of Styles.

a. long-styled (big ovary), b. medium-styled (medium ovary), c. pseudoshort-styled (rudimentary ovary), d. true short-styled (very rudimentary ovary).

They reported that long and medium-styled flowers produced fruits, whereas pseudo-short and true short-styled flowers did not set any fruit. However, chances of cross pollination are more in long-styled flowers. Popova (1961) reported that the position of the stigma in relation to the stamens varies with cultivars and can also vary in different flowers of the same cultivar. Stigmas are found either above, on the same level as or below the stamens and the highest percentage of fruit set is found where the stigma is above the stamens. He also observed that at a temperature of 20-22°C and 50-55 per cent relative humidity, pollen viability is retained for 8-10 days. Repeated pollination with pollen from different plants increased both fruit and seed set. Chen (2001), also reported that the pollens remain viable for 8-10 days at a temperature of 20-22°C with a relative humidity of 50-55 per cent. Nagasawa *et al.* (2001) observed that the ovaries of unpollinated flowers began to senescence three days after anthesis. Study on problems associated with lack of fruit set in short-styled flowers of brinjal revealed that it may not be due to poor viability or germinability of pollen grains or due to lack of pollen-stigma interaction. Short-styled length cannot be considered as a constraint as hand pollination did not encourage fruit set. Lack of fruit set is due to some problems associated with the maternal organ (Vijayalalitha, 1999).

Long (Left), Medium (Middle) and Pseudoshort (Right) Styled Flowers of Brinjal.

Anthesis in brinjal flower, starts from 7.30 a.m. and continues upto 11 a.m. Peak time for anthesis is 8.30 a.m. to 10.30 a.m. The pollen dehiscence starts from 9-30 to 10 a.m. It has been observed that anthesis and dehiscence are mainly influenced by daylight, temperature and humidity (Sidhu *et al.*, 1980). Vijay *et al.* (1977) reported that anthesis in Pusa Purple Long and Supreme cultivars commenced at 5.35 a.m. which continued upto 7.35 a.m. with peak at 6.05 a.m. Despande *et al.* (1978), however, described that maximum anthesis in brinjal was between 6 and 8 a.m. and the stigma was most receptive at the time of opening of flower. Korat *et al.* (2018)

observed that among the different pollination time, pollination at 9.00 to 11.00 am recorded significantly higher fruit set per cent with higher number of fruit set per plant, fruit weight, fruit length, fruit girth, mature fruit yield per plant, seed yield per plant, number of seeds per fruit, seed weight per fruit and 1000 seed weight as compared to other pollination times.

Fruit setting in brinjal cultivars with long-styled flowers varied from 70 to 86 per cent, while short-styled flowers did not set fruit (Prasad and Prakash, 1968). Chadha and Saimbhi (1977), on the other hand, in their studies on flower types in 29 cultivars, showed that all the cultivars bear flower clusters along with a solitary flower and in most of the cultivars, the flowers in cluster are either short-styled and medium or all medium-styled. However, the fruiting habit in a cultivar was not directly related to the occurrence of different flower types in cluster. Fruit set rates are higher for long-styled flowers as compared to short-styled flowers, therefore, short styled flowers are not totally sterile (Chen, 2001). Pollination efficiency of the long and medium-styled flowers was the greatest in the range of 38.8-50.8 per cent (Sekara and Bieniasz, 2008).

3.4.0 Cultivars

There is a large variation available in brinjal with respect to plant type, fruit colour, shape and size. Two main fruit types, namely, round and long, are cultivated throughout India. The following important cultivars are grown in India.

The following cultivars/hybrids have evolved at Indian Agricultural Research Institute (IARI), Pusa, New Delhi.

Pusa Purple Long

It is a selection from the mixed Batia cultivar grown in Punjab, Delhi and Western Uttar Pradesh. The plants are semi-erect in habit. It is an early maturing cultivar becoming ready for picking in 100 to 110 days. Fruits are glossy, light purple in colour, 25-30 cm long, smooth and tender. It is suitable for spring and autumn plantings and gives an average yield of 275 quintals per hectare. Singh and Sikka (1955) observed that this cultivar is moderately resistant to shoot borer and to the little leaf disease.

Pusa Purple Cluster

It is a very early maturing cultivar becoming ready for picking in 75 days after transplanting. Plants are erect and tall, leaves and stems are purplish in colour. Fruits are small, dark purple in colour and borne in clusters. Relatively higher yield has been obtained in some regions of India. On an average, each fruit weighs about 21 g with bearing of 50 fruits per plant. Gill *et al.* (1978) reported that this cultivar is resistant to bacterial wilt (*Pseudomonas solanacearum*) under green house and field conditions and to little leaf disease under natural conditions.

Pusa Kranti

This is a cultivar with a dwarf and spreading growth habit. Fruits are oblong and stocky than slender with attractive dark purple colour. Leaves are long, narrow with highly serrated and light green in colour. Stems are medium thick and half green, half purple in colour. This cultivar is good for both spring and autumn plantings under North Indian conditions. On an average,each fruit weighs about 68 g with a bearing of 22 fruits per plant.

Pusa Anmol

It is a hybrid cultivar evolved from a cross between Pusa Purple Long and Hyderpur. It is an early maturing cultivar with an increased yield of about 80 per cent over Pusa Purple Long.

Pusa Purple Round

The plants are very tall with a thick stem of greenish purple in colour. The leaves are highly serrated with deep green in colour. On an average, each fruit weighs about 137 g with a bearing of only 6 fruits per plant. This cultivar is now being replaced by other high yielding cultivars. This cultivar was found to be resistant to shoot borer and little leaf disease (Singh and Sikka,1955).

Pusa Bhairav

This is an early, non-spiny, resistant to phomopsis fruit rot, dark purple, glossy, attractive,12-15 cm long fruits.

Pusa Ankur

Plants are semi-upright, medium tall with 6-8 branches. The fruits are small, round and purple in colour. The average yield is 560-580 q/ha.

Pusa Shyamla

The fruits are long, glossy, dark purple, weighing 80-90 g. It matures in 50-55 days. The average yield is 390 q/ha with an increase of 13 per cent over national checks KS-331.

Pusa Kaushal

Plants are non-spiny with erect branches, having light purple pigmentation partially on the younger leaves, mid-rib and veins. Its fruits are long, violet-purple with round distal end, each fruit weighing 80-90 g. It matures 55 days after transplanting. Average yield is 382 q/ha, which is 26.9 per cent higher than national check Punjab Sadabahar 301 q/ha.

Pusa Upkar

Plants are semi-upright, medium tall 70 cm with 5-6 branches and greenish purple leaves bearing 15 fruits. Fruits are round, dark purple with an average weight of 295 g. This cultivar has a yield potential of 650-670 q/ha.

Pusa Bindu

Fruits are borne in clusters of 4-5 having an average weight of 125 g each. This cultivar has a yield potential of 250-300 q/ha.

Pusa Uttam

Fruits are pendent, oval, medium, large sized and glossy with dark purple skin and green peduncle. The bearing habit is solitary. Single fruit weight is 250-300 g. It has yield potential of 425-450 q/ha.

Pusa Anupam

Fruits are cylindrical, tender, purple, borne in clusters with 3-5 fruits. Resistant to bacterial wilt and first picking is done 65-70 days after transplanting. Average yield is 370 q/ha.

Pusa Safed Baingan 1

It is the first white coloured oval fruited variety suitable for kharif season, weighing 50-60 g, maturing between 55-60 days with a yield potential of 40-45 t/ha.

Pusa Hara Baingan 1

It is the first green coloured oval fruited cultivar suitable for kharif season, weighing 210-220 g, maturing between 50-55 days with a yield potential of 35 t/ha.

Pusa Hybrid -5

Fruits are long, glossy, attractive, dark purple with partially pigmented peduncle with an average weight of 100 g taking 80-85 days from sowing to first picking. Average yield is 520-540 q/ha.

Pusa Hybrid-6

Fruits are round, glossy, attractive, violet purple with partially pigmented peduncle with an average weight of 200-250 g taking 85-90 days from sowing to first picking. It has an average yield potential of 450-470 q/ha.

Pusa Hybrid-9

Fruits are oval round, glossy, attractive, dark purple, partially pigmented peduncle with an average weight of 300 g giving a yield of 67 t/ha.

Pusa Hybrid-20

The fruits are long, dark purple, glossy, weighing 90-100 g and it takes about 55 days from transplanting to first harvest. Its average yield is 524.0 q/ha.

The following cultivars/hybrids evolved at the Indian Institute of Horticulture Research, Bangalore, India (Vijay *et al.*, 1971).

Arka Sheel

This improved purple long cultivar of brinjal has been developed from a collection from Coorg in Karnataka State. The fruits are medium-long, tender and possess an attractive deep purple skin colour which is retained much beyond the commercial stage. The fruit contains more edible flesh and less number of seeds. The plant is well branched and erect unlike the spreading type. It is a very high yielding and produces, on an average 394 quintals per hectare in 110 to 120 days after transplanting.

Arka Shirish

It is the extra long, green brinjal cultivar and is an improvement over the Irangeri brinjal of Karnataka State. The fruits are very tender, thick with attractive light green skin colour, with good edible flesh and half length of the fruit towards the stem end has no seed. The flesh texture is good with better cooking quality. It is very high yielding, each bears 25 to 30 fruits and produces,on an average, 384 quintals per hectare in about 115 days after transplanting.

Arka Kusumakar

This cultivar is an improvement over a local collection in Karnataka State. The fruits are small, finger-shaped and light green in colour. The fruits are borne in clusters of 5 to 7. They contain less seeds and have good flesh texture and cooking quality. It is a very high yielding and the dwarf plants bear about 70 to 75 fruits each. On an average, it produces about 397 quintals of fruits per hectare in about 110 to 120 days after transplanting.

Akra Navneet

This is a very high yielding round cultivar with shining deep purple colour with green calyx. The fruit has an excellent cooking quality with more edible flesh and less seeds. It yields about 392 quintals per hectare. Each fruit weighs about 398 g (Anon., 1981-82).

Arka Neelanchal Shyama

It is an early cultivar developed through selection, suitable for rabi season. Fruits are round and green with light purple shade, weighing upto 190-200 g per fruit. It is moderately tolerant to Phomopsis blight and yields 34 t/ha.

Arka Unnathi

High yielding bacterial wilt resistant cultivar derived from cross between Arka Kusumakar x IIHR-3. Fruits are green long with fleshy green calyx. Yield is 34-36 t/ha in 90-110 days.

Arka Harshitha

High yielding bacterial wilt resistant cultivar derived from a cross between Arka Kusumakar x IIHR-3. Fruits are green long with fleshy green calyx. The average yield is 36-38 t/ha in 120 days.

Arka Avinash

High yielding bacterial wilt resistant cultivar derived from a cross between Arka Kusumakar x IIHR-3. Plants are tall and spreading. The fruits are green long with fleshy green calyx. The yield is 40-42 t/ha in 95-115 days.

Arka Neelkanth

It is a derivative of a cross between Dingras Multiple Purple x Arka Sheel. Fruits are tender with slow seed maturity with no bitter principles. It is resistant to bacterial wilt. It gives a yield of 43 t/ha in 150 days.

Arka Nidhi

It is a derivative of cross between Dingras Multiple Purple x Arka Sheel. Fruits are purple, tender with slow seed maturity with no bitterness. It is resistant to bacterial wilt. It gives an average yield of 48 t/ha in 150 days.

Arka Anand

It is a high yielding F_1 hybrid which has been developed by crossing IIHR-3 x IIHR-322 and having resistance against bacterial wilt. Fruits are green, medium long with excellent cooking quality. Yield is 55-60 t/ha in 145-150 days

The following cultivars/hybrids have been released from Indian Institute of Vegetable Research (IIVR), Varanasi, Uttar Pradesh.

Kashi Komal

This is another hybrid having semi-upright plant growth habit, height 90-100 cm, with light green stem and leaves; first flowering takes place in 35-40 days after transplanting. Fruits are light purple, long, soft texture, average length 13 cm and diameter 3 cm. The picking starts in 65-70 days after transplanting and gives an average yield of 800 q/ha.

Kashi Prakash

Plants of this cultivar are semi upright, stems and leaves are green. Fruits are attractive with light green spots, calyx spiny, average weight 190 g. The picking starts in 80-82 after transplanting and gives an yield of 650-700 q/ha

Kashi Taru

Plants of this cultivar are tall and erect, height 120-130 cm, leaves and stem dark green; first flowering starts 45-50 days after transplanting. Fruits are long, purple, length 31 cm and diameter 5 cm. The picking starts 75-80 days after transplanting and gives a yield of 700-750 q/ha.

Kashi Sandesh

This is a hybrid having semi-upright (height 71.0 cm) plant habit with green stems, purplish green leaves and purple flowers appear in 45 days after transplanting. Fruits are purple, size medium, shape round, fruit length 12.4 cm, diameter 10.2 cm and weight 225.3 g. The picking starts in 76 days after transplanting. Gives an average yield of 780 q/ha.

Kashi Himani (IVBL-26)

Medium long, shiny white fruits with less seeds suitable for kharif purpose. Have medicinal value for diabetic patients. Cultivar is tolerant to fruit and shoot borer, moisture deficit and lodging. Yield 400- 430 q/h. Recommended for cultivation in Uttar Pradesh and notified vide gazette notification number S.O. 692(E), dated 05.02.2019.

The following cultivars/hybrids are reported from Punjab Agricultural University, Ludhiana, India.

Punjab Barsati

An early maturing brinjal cultivar developed from the cross between PPC and H4 and was released in 1987 for cultivation in Punjab. It is a dwarf, erect, thornless cultivar; fruits are medium, long and purple. The average yield is 35.5 tonnes/ha (Chadha and Sidhu, 1987).

Sada Bahar Baingan

It is an early maturing cultivar (Japanese Long x R 34). The dwarf, erect, thornless cultivar has long, thin and purple fruits. It is suitable for summer or autumn planting or as a ratoon crop and gives an average yield of 31.3 tonnes/ha (Chadha and Sidhu, 1987).

PBH 3

It is a small round brinjal, early maturing, fruits are shining purple, borne in clusters with an average yield of 257 q/acre.

PBH 4

It is an early maturing F_1 hybrid. The plants are medium in height, compact, thornless with green foliage. Its flowers are purple, borne in clusters and its fruits are long, medium sized, shining and purple black with green calyx. Its average yield is 267q/acre.

Punjab Raunak

It is an early maturing long fruited suitable for planting throughout the year and gives an average yield of 242 q/acre.

Punjab Bharpoor

Plants are medium-tall, compact, thorn-less with green foliage. Flowers are purple and borne in heavy clusters. Fruits are small-oblong, shining, deep purple with green calyx and borne in cluster of six-nine. It is resistant to bacterial wilt disease. Its average yield is 224 quintals per acre.

The following cultivars have evolved at Haryana Agricultural University (HAU), Hisar, Haryana.

PH-4

Plant are bushy, fruit are medium long to long and thin, deep purple in colour, flesh light green. Average yield 250-300 q/ha.

BR-112

Plants are bushy, fruits are round with bright purple colour and fleshy. Average yield is 250 q/ha.

The cultivars developed from Hisar Agricultural University, Haryana, India, are as follows:

Hisar Shyamal

Fruits are round, bright, dark purple in colour with an average yield of 300-325 q/ha.

Hisar Pragati

This variety is suitable for summer season, fruits are long bright dark purple in colour with an average yield of 325-350 q/ha.

Hisar Jamuni

Fruits are oblong, dark bright purple in colour and high yielding of 260-280 q/ha.

HLB-12

It is tolerant to high temperature and shoot and fruit borer giving an average yield of 200 q/ha.

The following cultivars/hybrids of brinjal have been evolved at Chandra Shekhar Azad University of Agriculture and Technology, Kanpur, Uttar Pradesh.

Azad Kranti

It is an early maturing cultivar becoming ready for picking in 100 to 105 days. Fruits are long to medium-long having purple colour with green calyx. On an average, each fruit weighs about 62 g and produces about 300 quintals per hectare.

Vijay Hybrid

It is a high yielding hybrid cultivar having oval to oblong shaped fruit of purple colour with green calyx. It takes about 102 days to harvest edible mature fruits. Each fruit weighs about 170 g and on an average, it yields about 400 quintals per hectare.

Azad B-2

Fruits are round, bright purple, resistance to phomopsis fruit rot. Average fruit yield is 350-400 q/ha.

Azad B-3

Fruits are long cylindrical, shining purple in colour with an average fruit yield of 400-450 q/ha.

Azad B-4

Fruits are round, light purple in colour and green calyx. Average fruit yield is 300-350 q/ha

Besides these, other promising cultivars are K-202-9, K-202-14 and T-3 (round cultivars). All these cultivars have been found promising in Bihar, Maharashtra, Gujarat and Uttar Pradesh under All India Coordinated Vegetable Improvement Project (Anon., 1981-82).

At Govinda Ballav Pant University of Agriculture and Technology, Pantnagar, Uttar Pradesh, some promising brinjal cultivars/hybrids are as follows.

Pant Samrat

It is a selection from local germplasms. Tall vigorous plants, medium long fruits in clusters, young leaves purplish green, resistant to bacterial wilt under field conditions, tolerant to shoot and fruit borer.

Pant Rituraj

It evolved from a cross between T3 and Pusa Purple Cluster. Semi erect plant with dark green foliage but light purple colour on new leaves, fruits almost round with slight tapering towards the bottom, dark purple in colour suitable for planting both in autumn and summer seasons.

Pant Brinjal Hybrid 1

This F_1 hybrid cultivar developed from a cross between PB 129 x PB 225. Medium tall plants, dark green leaves with purple tinge on young leaves, fruits long in clusters, soft flesh, resistant to bacterial wilt under field conditions.

The following cultivars of brinjal have been evolved at Assam Agricultural University (AAU), Jorhat, Assam

JC-1

Plants are tall, pedicel non spiny, fruit are long, medium in size, purple in colour with pointed apex. Tolerant to bacterial wilt, phomopsis blight, little leaf virus and borer. Maturity in 130 days with an average yield of 336 q/ha.

JC-2

Plants are tall, pedicel is spiny, fruit large, oblong, purple in colour with blunt apex. Tolerant to bacterial wilt, phomopsis blight, little leaf virus and borer. It matures in 130 days giving an average yield of 350 q/ha.

The following cultivars of brinjal have been evolved at Anand Agricultural University (AAU), Anand, Gujarat

GBH-1

The fruits are dark purple with single green styler spot at apex. This hybrid is tolerant to shoot and fruit borer. Average yield is 578 q/ha.

Oblong Brinjal-1

Fruits are attractive due to better shape, size, black in colour and shining. It is moderately tolerant to leaf disease. Average yield is 486 q/ha.

Anand Brinjal Small Round-2

The fruits are small round, light pinkish, shiny with non spiny calyx and an average yield of 408 q/ha.

The following cultivars of brinjal have been evolved at Mahatma Phule Krishi Viswavidyalaya (MPKV), Rahuri, Maharashtra.

Krishna

Plants are tall and hardy, resistant to lodging. Fruits are egg shaped, spiny purple coloured with white stripes, suitable for kharif and summer seasons. Average yield is 480 q/ha.

Phule Harit

Fruits are green in colour with white patch at apical end. It is suitable for kharif and summer seasons. Average yield is 200 q/ha.

Phule Arjun

Fruits are green with purple and white stripes, oval in shape and spiny. Suitable for kharif and summers seasons, average yield 450 q/ha.

The following cultivars/hybrids of brinjal have been evolved at Tamil Nadu Agricultural University (TNAU), Coimbatore, Tamil Nadu.

CO-1

Fruits are light green with white base and oblong shape. Medium sized fruits (50-60 g) and good keeping quality. Average yield is 24 t/ha.

MDU-1

Fruits are round, bright purple and average weight is 280 g with a yield of 34 t/ha.

PKM-1

Fruits are medium in size, drought tolerant and can withstand long distance transport. Average yield is 34 t/ha.

CO-2

Fruits are oblong with brown stripes. Suitable for Periyar and Coimbatore districts. Average yield is 35 t/ha.

COBH-2

F_1 hybrid developed by crossing EP 65 x Pusa Uttam. Fruits are medium in size, oblong in shape, dark violet, weighing 55 to 60 g. Average yield is 60 t/ha.

VRM-1

It has been developed by pureline selection. Fruits are oval in shape, glossy pink with green tinge at the distal end. Average yield is 45 t/ha.

The following cultivars/hybrids have been evolved at ICAR-Research Complex for Eastern Region, Ranchi, Jharkhand.

Swarna Pratibha

Fruits are medium sized and shiny purple in colour. Plants are resistant to bacterial wilt. First picking is done 75-80 days after planting. Average yield is 470-500 q/ha.

Swarna Shree

Fruits are creamy white, round to oval, soft and preferred for bharta preparation. Plants are moderately resistant to bacterial wilt. First harvest can be taken 75-80 days after planting. Yield potential of 55-60 t/ha.

Swarna Mani

Fruit are attractive, shiny purple in colour, round in shape and moderately resistant to bacterial wilt. First harvest 65-70 days after planting with an average yield of 60-65 t/ha.

Swarna Shyamli

Fruit are medium in size (250 g), round, attractive green colour with white stripes. Resistant to bacterial wilt. First harvest 35-40 days after planting and an average yield of 60-65 t/ha.

Swarna Shobha

Fruit are round, medium size, weight (300-350 g) and attractive milky white with purple tinge. Tolerant to bacterial wilt disease under field condition. First harvest 55-65 days after planting with a yield of 50-60 t/ha.

Swarna Abhilamb

Fruits are long (30-35 cm), weight (80-100g) and attractive purple blue colour. Resistant to bacterial wilt and phomopsis blight disease under field condition. First harvest 55-60 days after planting. Average yield is 60-70 t/ha.

Swarna Ajay (F_1)

The fruits are oblong and of medium length weighing 100-120 g and attractive light purple in colour. First harvest can be taken 55-65 days after planting. This hybrid has a yield potential of 70-75 t/ha.

Swarna Shakti (F_1)

Fruits are oblong, medium length (15-17 cm), weight (250-300 g) and attractive shiny light purple in colour. Resistant to phomopsis blight and moderately resistant to bacterial wilt under field condition. First harvest 55-65 days after planting and gives an average yield of 70-75 t/ha.

Swarna Mohit (F_1)

Fruits are oblong, medium length (10-15 cm), weight (100-150 g) and attractive light purple colour. Moderately resistant to bacterial wilt and phomopsis blight under field condition. First harvest 55-60 days after planting with an average yield of 65-75 t/ha.

Swarna Neelima (F_1)

Fruit are round, medium length (10-12 cm), weight (200-250 g) and attractive light purple colour. Resistant to bacterial wilt and phomopsis blight. First harvest 55-65 days after planting. Average yield is 70-75 t/ha.

HABR-21

Fruits are attractive with black colour and calyx is dark purple. Fruit length (18-20 cm), fruit breadth (7-8 cm), fruit weight (300-350 g), and yield per plant (2-3 kg). The mean yield potential ranges from 55-60 t/ha.

The following cultivars have been evolved recently at Bidhan Chandra Krishi Viswavidyalaya, Mohanpur, West Bengal, India.

Bidhan Supreme

This cultivar matures early (within 70 days after transplanting) and gives 29.52 per cent and more than 57 per cent yield than local check and national check varieties, respectively. It possesses good cooking quality and moderately field resistant against bacterial wilt disease and can easily be fitted well in intensive cropping system. The average yield of this cultivar is 289.27 q/ha.

Bidhan Super

This cultivar matures early (within 75 days after transplanting) and gives 28.00 per cent and more than 21 per cent yield than local check and national check varieties, respectively. It possesses round fruit with glossy blackish purple colour which is the most appealing trait for the export trade and can easily be fitted well in intensive cropping system. The average yield of this variety is 285.18 q/ha.

Similarly, at Punjab Agricultural University, Ludhiana, some promising cultivars, namely,**P-8, Punjab Bahar** (round types), **Punjab Chamkila, Sel. 4** (long types) and **Punjab Hybrid** (hybrid cultivar) have been evolved. In Tamil Nadu at Annamalai University, **Annamalai** (long type) cultivar has been found very promising. At Almora in Uttar Pradesh, long types brinjal cultivars, namely, **ARU-1**and **ARU-2C** have also been found very promising (Anon., 1981-82). In Orissa University of Agriculture and Technology, **Utkal Madhuri, Utkal Keshari, Utkal Jyoti** (Long types) and **Utkal Tarini** were promising cultivars. At Kerala Agricultural University (KAU), **Surya, Shweta** and **Haritha** have been developed.

Besides the above cultivars, **Makra** (spiny), **Elokeshi, Krishnanagar Purple Round, Krishnanagar Green Long** and **Muktakeshi** cultivars are popular inWest Bengal; **Cluster White, ICJ 885** (resistant to shoot and fruit borer) and **Wynad Giant** are popular in Tamil Nadu; **Nurki Long, Nurki Round** and **Banjari** are popular in Madhya Pradesh; **Banaras Giant** (very big size), **Muktakeshi** and **Ramnagar Giant** are popular in Uttar Pradesh; **Improved Muktakeshi** is popular in Bihar.

F_1 hybrids commercially marketed by Indo-American Hybrid Seeds in India are Long 13, Suphal and Round-14.

Long13

Tall, thorny, upright plant and early maturing. Fruits solitary, dark purple, 18 cm long. Widely adopted hybrid.

Cultivars and Hybrids of Brinjal

Pusa Shyamla

Pusa Uttam

Pusa Hybrid 5

Pusa Hybrid 6

Pusa Hybrid 9

Pusa Safed Baingan 1

Pusa Kaushal

Pusa Hara Baingan 1

Cultivars and Hybrids of Brinjal

Arka Neelanchal Shyama

Arka Harshitha

Arka Avinash

Arka Anand

Arka Unnathi

Kashi Taru

Kashi Sandesh

Kashi Komal

Cultivars and Hybrids of Brinjal

Swarna Shree   Swarna Mani   Swarna Shyamli

Swarna Prativha   Swarna Shobha   Swarna Ahilambh

Swarna Shakti   Swarna Mohit   Swarna Ajay   Swarna Neelima

Cultivars of Brinjal

Bidhan Supreme

Bidhan Super

Utkal Anusree

Utkal Madhuri

Utkal Tarini

Suphal

An early maturing vigorous hybrid which produces long, oval, glossy, deep purple fruits.

Round 14

Tall, thornless plant; fruit solitary, round, light purple, medium-sized with white prominent scar at blossom end.

Astrakom

It is a mid-early cultivar from Russia with compact growth habit; fruits cylindrical, weighing 150-160 g. The cultivar is suitable for processing and gives high yield (Turchenkov *et al.*, 1990).

Artica

An Italian F_1 hybrid with white fruit, obtained from the cross L.764-84B x L.519-81B. This cultivar is suitable for both outdoor and green house cultivation and is recommended for cooking, industrial processing and freezing (Acciarri *et al.*, 1989).

4.0 Soil and Climate

4.1.0 Soil

Brinjal can be grown practically on all soils from light sandy to heavy clay. Light soils are good for an early crop, while clay-loam and silt-loam are well-suited for high yield. Generally, silt-loam and clay-loam soils are preferred for brinjal cultivation. The soil should be deep, fertile and well-drained. The soil pH should not be more than 5.5 to 6.0 for its better growth and development.

4.2.0 Climate

Brinjal is a warm season crop and susceptible to severe frost. Climatic conditions, especially low temperature during the cool season causes abnormal development of the ovary (splitting) in flower buds which then differentiate and develop into deformed fruits during that season (Nothmann and Koller, 1973). A long and warm growing season is desirable for successful brinjal production. The late cultivars can tolerate mild frost to some extent than the early ones. Cool nights and short summers are unfavourable to its satisfactory yield. A daily mean temperature of 13° to 21°C is most favourable for better growth and yield (Choudhury, 1976a). Nothmann *et al.* (1979) observed that when brinjal is grown in warm season, the growth and flowering are normal and almost all fruits develop from the basal flowers. In cool season, growth is slow and fruit quality is poor both in size and colour. They also pointed out that the colour development on fruit seems to be related to vigour of the plant growth. Lombardi and Restaino (1981) reported that anthesis of flowers in different cultivars is influenced by both temperature and genotype, and flower set is affected more by genotype. They also observed that the genotype and temperature interact in their effect on leaf size also. The optimum temperature for growth and fruit set is 20 °C to 30 °C. However, high night and day temperature condition of 22 °C to 24 °C and 33 °C to 35 °C markedly reduce the fruit set and yield (Kalloo *et al.*, 1990; Kumar *et al.*, 2000; Mohanty and Prusti, 2002). *Ascophylum nodosum* extract could be used as an efficient stimulator of flowering and fruit setting for eggplant hybrids in field conditions in a temperate climatic zone (Pohl *et al.*, 2019).

Nikolov *et al.* (1987) observed that the yield of brinjal often does not reach the full biological potential due to flower drop or to impede flowering and fruit growth at the end of the season, all caused by adverse light, temperature and watering effects. A growth regulator formulation code-named '25-15', applied once or several times at flowering to green house-grown plants, was able to reduce these environmental effects, and increased fruit yields by 20-72 per cent. Sun *et al.* (1990) reported that fruit setting was affected neither by the natural light duration nor by the relative humidity but mainly be the average maximum temperature and precipitation during the first 5 days after the flowers opened. Ootake (1987) in Japan opined that cold winter condition was responsible for reported flesh browning without skin damage. Leonardi and Giuffrida (1999) were of the view that light, temperature and nutritive factors had significant effects on chlorophyll content and stomatal density and length of green house grown eggplant leaves. Specific leaf area (SLA) varied significantly only in relation to light level and N availability. Leaf expansion rate was most affected by temperature. Effect of mulch (0.5 mm thick clear polythene) and low tunnels on yield, earliness and root development of green house eggplants and on soil temperature were studied by Abak *et al.* (1992) and revealed that effect of mulch and low tunnels was to increase soil temperature by 2.30 and by 2.8, 2.4 and 1.60 at depths of 5, 15 and 30 cm, respectively. Root development was not influenced while the early and total yields were significantly increased by this treatment.

Different genotypes of eggplant were affected by low temperatures in different ways from each other in plant height, main stem diameter and leaf number. They concluded that all foreign F_1 hybrids were found more tolerant of cold than the Turkish originated open pollinated genotypes that showed inward curled symptoms on the leaves. Low light intensity under shading with black netting in the green house shortened stems and reduced stem diameter more than with white netting, and increased leaf area, lengthened the flowering period, significantly reduced yield and decreased the leaf chlorophyll -a: b ratio (Yi and Chen, 1999).

Watanabe *et al.* (1978) conducted an experiment with brinjal grown at 20° or 30°C and 2 light regimes (natural light or shading) which received either low or high NPK rates. At 20°C, the effect of both fertilizer rates was less pronounced than at 30°C. At 30°C, the adverse effect of shading at the time of bud differentiation was less pronounced in plants receiving high NPK but at 20°C it was highly pronounced at both fertilizer levels. Raising the temperature to 30°-33°Cand slightly decreasing soil aeration during the flowering to bearing period had no adverse effect on plant growth and fruit development (Ivanov and Nikolaeva, 1974). The effect of high temperature on fruit set of eggplant when grown with maximum temperature in the range of 36.7-39.8°C was investigated by Sanwal *et al.* (1997). Result revealed reduced fruit set at higher temperature associated with increased tip burning, reduced pollen germination and reduced pollen tube lengths. Takahashi *et al.* (1977) reported that at 24, 12 and 8 per cent natural light level, brinjal plants receiving high NPK grew faster, produced flower buds earlier and a large number of flowers than those under low NPK. Brinjal seeds germinate well at 25°C (Chauhan, 1981).

5.0 Cultivation

5.1.0 Seed Treatment and Raising of Seedlings

5.1.1 Seed Treatment

Halogen vapour treatment of fresh brinjal seeds (employing chlorine, bromine or iodine) for 8-48 h at 25°C, significantly slowed down the deterioration of seeds under accelerated as well as natural aging conditions (Rudrapal and Nakamura, 1988a). They (1988b) also suggested that the hydration-dehydration treatments extend the viability of the seeds by reducing free radical damage to cellular components.

Pre-sowing treatment of eggplant seeds with gibberellins improved seed vigour and germination rate, and resulted in faster and more uniform seedling emergence in the field (Si *et al.*, 1996).

Seed treatment of eggplant cv. Pusa Kranti with growth regulators and subsequent application of insecticides at intervals after, transplanting had a beneficial effect on yield and control of fruit and shoot borer. IAA (30 ppm) and application of phorate+ endisulfar resulted in the highest yield. Net profit was highest with IAA seed treatment (Choudhury and Saraf, 1996).

Rao and Bhatt (1990) reported that critical water potential for germination lay between -0.2 and -0.4M Pa. When seeds were primed at -0.6 M Pa for 10 days and then transferred to filter paper with distilled water, seeds germinated within 48 h compared with 72 h for unprimed controls.

Seed treatment with GA_3 (100 ppm) resulted in highest germination percentage, increased shoot and root length (Islam, 2006). Pre-sowing seed treatment with GA_3 (100 ppm) significantly improved field emergence, days to 50 per cent flowering, growth and yield parameters (Seervi *et al.*, 2018).

Seed treatment of Brinjal hybrid cv. Arka Navneet with GA_3 (200 ppm) significantly improved fruit yield, seed quality, growth and yield components (Satishkumar *et al.*, 2011).

Gonzales (2015) concluded that the used of vinegar as priming material is effective in improving germination of brinjal seeds. Application of 0.001 per cent concentration showed significant effects on percent germination and germination rate. He also reported that the higher the dose of vinegar the poorer is the germination. At 10 per cent concentration of vinegar it showed detrimental effect on seed germination.

5.1.2 Raising of Seedling

The time of seed sowing and transplanting the seedling in various agroclimatic situations primarily depend upon the temperature, rainfall and availability of irrigation facilities. In Northern India, there are generally two sowings, *viz.*, June to July for the autumn crop and in November for the spring-summer crop. A third rainy season crop can be raised by sowing the seeds in March. Seedlings of 4-6 weeks are generally transplanted in each case. In the East and South Indian conditions, brinjal can be grown all the year round, the main sowing season being during July to August. In the hills, the seed is sown in March to April and the seedlings are transplanted in May.

Seeds are sown in well prepared raised nursery beds 6 to 12 mm deep in row 5 cm apart. The beds should be well drained to prevent the attack of damping off in the early stage of crop growth. The seed beds should be fertile and compost, farm yard manure or leaf mould may be mixed with the soil. A small quantity of superphosphate may be used. Addition of humic acid to the growing medium and foliar fertilizer had positive effect on seedling growth of eggplant seedlings (Padem *et al.*, 1999). Humic acid at 2500 ml/1000 sq.m. in both ways of application influenced significantly seedling height and diameter, number of leaves, shoot fresh weight, shoot and root dry matter. Straw or other mulch materials may be used to cover the top soil till germination. According to Purewal (1957) brinjal seeds are light in weight having an average germinability of 75 to 80 per cent. To transplant one hectare of land, 500 to 750 g of seeds are sufficient for sowing. However, Choudhury (1976a) reported that about 375 to 500 g seeds are required for transplanting one hectare of land. Generally, about 250 seeds make one gram.

The use of cocopeat and vermicompost as soil less growing media in the nursery along with organic amendments *i.e. Trichoderma*, crucifer residues and AM fungus individually as well as in combination produced superior seedlings in respect of all plant growth attributes in comparison to the conventional soil media (Bhardwaj *et al.*, 2019).

5.2.0 Land Preparation and Transplanting

The preparation of land depends upon the type of soil and the previous crop of the land. As the crop remains in the field for a number of months, soil should, therefore, be thoroughly prepared by being ploughed 4 to 5 times before transplanting the seedlings. Bulky organic manures like well rotten cowdung manure or compost should be applied and incorporated well into the soil.

The seedlings are ready for transplanting when they attain a height of 15 cm with 4 leaves in 4 to 6 weeks. Under-aged (30 days) seedlings produced markedly lower yield compared to older ones (40-60 days) as reported by Chowdhury *et al.* (1991). The effect of different dates of planting on yield of three brinjal cultivars, Pusa Kranti, Pusa Purple Cluster and Malapur Thornless was studied by Reddy and Gowda (1987) and they reported that Pusa Kranti out-yielded the other cultivars. Duranti and Cuocolo (1988) transplanted brinjal seedlings on 21 May or at 10-day intervals between 21 June and 10 August. Yield declines from 74 tonnes/ha for the earliest transplanting date to 11 tonnes/ha for the late transplanting date. Fruit number/plant and average fruit weight also declined with later transplanting, and the mean length of the growing season shortened progressively from 167 to 86 days. The planting distance depends on the fertility status of the soil, growing season and cultivar. Usually, the spacing for the bushy, non spreading type should be 50 to 60 cm in both ways and for spreading cultivars row to row distance should be 75 to 90 cm and that for plant to plant 60 to 70 cm either in flat beds or ridges. According to Chauhan (1981) brinjal plants are transplanted at 45 x 60 cm in case

Raising of Seedlings and Field Growing of Brinjal

of long cultivars, at 60 x 75 cm and 90 x 90 cm in case of round and high yielding cultivars, respectively. Abutiate (1988) found that spacing of 90 x 60 cm were best with respect of marketable yield of fruits.

According to Sollapur and Hiremath (2016), significantly higher marketable yield of brinjal was found in wider spacing of 90 x 90 cm compared to closer spacing *viz.*, 90 x 75 cm, 90 x 60 cm, 75 x 75 cm and 75 x 60 cm. The increased yield obtained in the wider spacing could be attributed to higher number of fruits per plant, average fruit weight, fruit diameter and fruit yield recorded in wider spacing. Wider spacing offers lesser competition among the plants and larger amount of available nutrients and soil moisture for vigorous growth of fruit parameters in widely spaced plants as compared to closer spacing.

Sonavane *et al.* (2012) found that the highest growth characters were obtained with wider spacing of 90 x 120 cm. The yield contributing characters such as number of fruits plant, weight of fruits per plant and yield per ha were significantly highest with 90 x 120 cm spacing over rest of treatments. The incidence of shoot and fruit borer with this treatment was less as compared to the other treatments.

5.3.0 Irrigation

Irrigation is essential for brinjal cultivation in regions where there is little or no rain during the growing season. Brinjal being a shallow rooted crop needs irrigation at frequent intervals. Hence, timely irrigation is necessary for high yield. When there is a danger of frost, irrigation should be more frequent to keep the soil moist. About 100 to 110 cm irrigation are required for a successful brinjal crop (Choudhury, 1976a). Yawalkar (1969) was of the opinion that irrigation should be given according to local needs. In plains of India irrigation is to be given every 4th day during the summer season and at an interval of 10 to 15 days during winter. Abdelfattah and Abdel-Salam (1972) indicated that when brinjal plants were watered every 1, 2 and 3 weeks, the depth of root penetration increased as the interval between irrigation was lengthened, while the extension of the main stem, the number of branches per plant, the average fruit weight, the number and the weight of marketable fruits were reduced. Sadovski *et al.* (1978) observed that irrigation with waste water caused a 38 fold increase in microbial contamination but this could be reduced by the manipulation of the drip method of sub-irrigation or covering the soil and drip lines with plastic sheets. Sivanappan (1979) also pointed out that with drip irrigation the amount of water used was much less, weed growth was reduced and yield was usually higher. Irrigation requirement in brinjal was 29 per cent lower and the yield was 18 per cent higher by covering the rows with white transparent polythene film of 0.10cm thickness. Water use efficiency was 66 per cent higher than that in the open plot (Bangaland Gutal, 1986). Jaime *et al.* (1987) reported that with trickle irrigation the yield was highest by treatment with 180 kg N/ha but with furrow irrigation, 360 kg N/ha produced the maximum yield and the total volumes of water used during the whole experiment were 11000 and 6000 cm^3/ha with furrow and trickle irrigation, respectively.

Balakrishnan *et al.* (1993) studied the effects of two seed treatments and five foliar spray treatments on water stress in brinjal during summer season at

Periyakulam, Tamil Nadu. They observed that KCl sprays (1 per cent) resulted in significantly higher yields than the other treatments and also resulted in the lowest transpiration rate. Dry matter production was positively correlated with plant soluble protein content and yield was positively correlated with proline content. In an artificially developed water stress conditions in potted eggplants through either withholding water for 7 days or spraying with anti transpirant like cycocel (5mM), lime wash (2 per cent) or potassium chloride (1 per cent) at the flowering stage resulted in less plant height with reduced number of leaves per plant. Yield was also drastically reduced by these treatments while anti-transpirants have a slight edge over withholding of water for 7 days (Prakash *et al.*, 1993). Buzescu and Cenusa (1996) opined that reduced soil moisture content had an irreversible detrimental effect on the eggplant crop, causing massive abortion of flowers and spoiling fruit quality. They also suggested that eggplant prefers a balanced soil moisture regime, not lower than 60 per cent available water capacity but not exceeding field capacity. According to Jeevarani *et al.* (1998), nitrogen, potassium and calcium content was found to be significantly reduced in plants subjected to flooding stress. They also opined that anoxic conditions caused damage to the root primordial and affected the root metabolism which in turn affected the uptake of mineral nutrients. Rani *et al.* (1999) were of the view that on flooding anaerobiosis increases the ethylene content in brinjal plants which in turn causes epinasty of leaves and stem hypertrophy. Low levels of oxygen in the rootzone, due to flooding, reduced the uptake of nitrogen, phosphorus and calcium in the plants.

In trials in the Ofanto valley, Italy in 1988-89 daily maximum crop evapotranspiration (ETc) was determined by weighing lysimeter in crops of aubergine cv. Black Beauty. *etc.* ranged from 1.5 mm/day at transplanting to about 6 mm/day when growth was at a maximum and then fell to 2.0 mm/day by the end of the season. Crop coefficients calculated from class A pan evaporation were 0.25 at transplanting, 0.50 at flowering, 0.95 at the first harvest and 0.60at the last harvest. In a comparison of irrigation rates (no irrigation, and 25, 50, 75 or 100 per cent *etc.*), irrigating at 75 per cent *etc.* gave the highest marketable yields (Perniola *et al.*, 1992). In Italy, Barbieri and Deveronico (1989) obtained best results with a plant density of 4.6 ± 1.5 plants/m′ with an irrigation rate of 100 per cent evapotranspiration, corresponding to a rate of 6000 m^2/ha between June and Semptember.

Saline water has a profound effect on germination of eggplant seeds and enhanced rate of germination was noted when seeds were kept at 24°C in the dark in petridishes containing 0.8 per cent (w/v) agar and 1200 ppm sea salt. Hypocotyl and root growth rates on the contrary decreased with increasing salinity (Hashem, 1993). However, saline water had an adverse effect on the growth and development of eggplant. More salt concentration (≥ 1.0 per cent commercial sea salt) in irrigation water prevented root growth in the top 30 cm of soil while low concentrations delayed root growth only. Not only that higher salinity level (≥ 1.0 per cent) also reduced leaf area by 5 0 per cent and shoot growth and water uptake of the plant compared to control (no salt added). Marketable yields also reduced considerably Ruggiero *et al.* (1994).

From a study on analysis of quality parameters in eggplant grown under saline water irrigation, Sifola *et al.* (1995) reported that marketable fruit yield was reduced with 1 per cent NaCl. Saline water produced fruits of reduced length. Mesocarp firmness was highest in fruits with 1 per cent NaCl associated reduced water content. Mesocarp firmness and dry matter increased with decreasing irrigation frequent. Titrable acidity, reducing sugar and mineral ash contents increased with increasing salinity, while pH, total polyphenols and ascorbic acid contents decreased. Irrigation with 1 per cent NaCl shortened shelf life of fruits under ambient conditions, due to earlier and more pronounced tissue browning.

Mahata and Sarkar (2019) concluded that the use of pot under drip irrigation system in different treatments of brinjal cultivation gave higher yield with less application of water significantly as compared to conventional irrigation. In drip irrigation system usually less water is available to weeds, high efficiency in fertilizers use, less weed growth so low labour and relatively low operation cost and less evaporation losses of water as compared to conventional irrigation. The amount of irrigation water required under drip irrigation at 0.5 E_{pan} treatments with drip irrigation has been found lowest *i.e.*, 12 cm and highest value was 30 cm at 1.1 E_{pan} treatments with conventional irrigation it was due to application of different treatments, so it can be said that the water saving in 0.5 E_{pan} has found almost 60 per cent in compare to conventional brinjal cultivation in pot culture. Similarly yield was found highest at 0.5 E_{pan} treatments *i.e.*, 81.28 t/ha and water use efficiency was found highest at 0.5 E_{pan} treatments *i.e.*, 6.77 t/ha-cm. Drip irrigation in brinjal in pots is found much water saving practice. Drip irrigation itself is most water saving technique and cultivating the crops in pots with mulch further add to water saving since there is possibility of using entire rain water and less evaporation loss. The pot practice does not require the cultivable land and management of entire field. In consideration of all these aspects drip irrigation in pots and brinjal in particular may be a good proposition for cultivation towards saving of water and economic return.

In a study conducted by Adawadkar *et al.* (2019) to compare brinjal production under drip fertigation and traditional method of fertilizer application, at Dr. PDKV, Akola, the study indicated higher plant growth, more number of branches, higher number of fruits per plant and enhancement in the yield under all drip fertigation levels. Highest irrigation water use efficiency was found in treatment of drip fertigation at 150 per cent RDF followed by treatment of drip fertigation at 125 per cent RDF.

5.4.0 Manuring and Fertilization

One of the major aspects for high yield of brinjal is manuring. Therefore, balanced application of manures and fertilizers is very important for successful crop production. Brinjal being a long duration crop, required a good amount of manures and fertilizers.

Mehrotra *et al.* (1973) studied the deficiency symptoms of N, P, K, Ca, Mg, S, Fe and B in brinjal by growing them in water and sand culture. All the deficiencies had adverse effect on plant height, leaf and branch production. However, characteristic deficiency symptoms obtained by them are summarized below:

- ☆ N deficiency: general paling and stunted plants.
- ☆ P deficiency: dirty greyish green leaves and premature shedding.
- ☆ K and Fe deficiency: interveinal chlorosis of young leaves followed by yellowing and premature shedding.
- ☆ Ca deficiency: light green colour of young leaves along with necrotic spotting.
- ☆ Mg deficiency: inverted 'V' shaped interveinal chlorosis.
- ☆ B deficiency: death of apical bud coupled with browning of roots.

As regards methods of application, well-rotten farm yard manure should be thoroughly incorporated at the time of land preparation. In general, half of the total nitrogen, total phosphorus and potash should be given as basal dose, while remaining half of nitrogen in 3 equal doses: the first dose is to be given one and half month after transplanting, the second dose one month after the first application and the final at three and half months after transplanting. Diagnostic norms and identification of yield-limiting nutrient in brinjal using diagnosis and recommendation integrated system (ORIS) was studied by Raghupathi and Bhargava (1999) around Bangalore, India and the study revealed that the best nutrient ranges were 2.85-3.99 per cent for N, 0.33-0.8 per cent for P, 2.16-3.15 per cent for K, 0.64-0.96 per cent for Ca, 0.05-0.14 per cent for Mg and 0.54-0.82 per cent for S. The best micronutrient ranges were 69-274, 11-83,11-33, and13-26 ppm for Fe, Mn, Zn and Cu, respectively.

5.4.1 Nitrogen

The beneficial effects of nitrogen in increasing the production of brinjal have been reported by many workers. Iwata *et al.* (1968) were of the view that withholding of nitrogen at any stage of plant growth increases the percentage of flowers with short styles resulting in reduction of yield. Pusa Purple Long cultivar responded well to N fertilizers at 70kg/ha in respect of both fruit and seed yield (Seth and Dhandar, 1970). Similarly, Mertia and Chauhan (1970) reported that 89 kg N/ha was the most effective treatment for increasing leaf number and yield. Application of 125 kg N/ha significantly increased flower production (Chandrasekharan and George, 1973). Verma *et al.* (1975), on the other hand, observed that fruit yield rose linearly with increasing nitrogen although the variation due to the treatment with 100 and 150 kg N/ha was not pronounced. They also found that weight, length, diameter, volume and specific gravity of the fruit responded positively to increased nitrogen. Nitrogen application as top dressing is usually better utilized by plants than as a basal dressing (Nikolaeva, 1975).

In trials with brinjal cultivars Black Beauty, Florida Market, Local Long Black and Local Long White, El-Shal *et al.* (1986) reported that yields increased with N rates up to a maximum at 90 kg N/feddan (0.42 ha). Jose *et al.* (1988) used five different sources of N and obtained highest yield in cv. MDU-l with 50 kg N/ha as poultry manure + 50 kg N/ha as urea. Inorganic form of N showed early flowering when compared to organic sources or their combination.

Ramu and Muthuswamy (1964) obtained the maximum yield of brinjal with foliar spray of 4 per cent ammonium sulphate in the monsoon of 1962 and 2 per cent urea in summer of 1963. Mitra and De (1969) also found an increased yield in cv. Pusa Purple Long with half of the total nitrogen of 60 kg/ha as foliar spray. In Punjab, Singh and Sandhu (1970) observed that growth and yield significantly increased with increasing doses of N upto 75 kg/ha and by the foliar spray of urea at 1 per cent concentration. Higher concentrations of foliar spray are not beneficial. Similarly, in Andhra Pradesh, the highest fruit set and the largest plants were obtained in cv. Egg White with N at 50 kg as a basal dressing + 25 kg N/ha as foliar spray of 1 per cent urea after transplanting and a further 25 kg N/ha as foliar spray of 1 per cent urea at time of flowering (Rajeevan and Rao, 1980). According to Rizk *et al.* (1997), nitrogen fertilization in the form of ammonium sulphate produced the best plants in terms of growth and yield, urea was the worst.

In a trial at the Banaras Hindu University, fertilization with ammonium sulphate at the rate of 68 or 100 kg/ha or with lower doses of ammonium sulphate coupled with weekly sprays of NAA at 10 or 20 ppm or of cow's urine diluted to 1 in 500 litres, the highest yield resulted from ammonium sulphate at 100 kg/ha. The yield obtained from ammonium sulphate at 50 kg/ha + cow's urine was almost as high as that obtained from 68 kg/ha of ammonium sulphate (Pillai, 1967). Maurya and Singh (1978) noted that fruit size and yields were enhanced by treatment with urea at 1 or 2 per cent and IAA at 50 or 100 ppm either alone or in various combinations applied as foliar sprays at 15 days after transplanting. The best results were obtained with both substances applied together at higher concentrations.

5.4.2 NPK Fertilizers

Choudhury (1962) reported that a good crop of brinjal required about 84 to 112 kg N, 84 to 112 kg P and 56 kg K per hectare. Patnaik and Farooqui (1964) found that application of 90 kg nitrogen per hectare along with 45 kg each of P and K increased the fruit set and yield. Dhesi and Nandpuri (1965) suggested a dose of 112 kg nitrogen, 56 kg P and 28 kg K/ha for good yield. According to Singh (1969), the crop should be supplied with150 to 200 kg N/ha, 100 to 150 kg P/ha and 60 to100 kg K/ha depending upon the type of soil from medium to heavy fertility. Honda *et al.* (1970) studied the effect of slow acting nitrogenous fertilizers along with NPK mixtures and found that yield was highest when the slow acting nitrogenous fertilizers were used alone. Bandopadhyay *et al.* (1972) reported the highest yield was with 120 kg N + 80 kg each of P and K/ha in Bihar. Singh *et al.* (1988) obtained highest yield with cultivar PH-4 transplanted in early July and by application of 90 kg N and 40 kg P/ha.

From a trial on NPK fertilizers in combination with Spartin (secondary and trace elements) Chandrashekharan and George (1973) obtained increased yield in response to N and P but no response was recorded from Spartin or potash. Malyshkina (1976) recorded maximum yield at 120: 180: 60 kg/ha of N, P and K and N, P at 120: 180 kg/ha produced the best results. From a 3 year trial in Bulgaria, Doikova (1976) reported that Pat 120 kg + K at 240 kg/ha as basal treatment along with a single dressing at the onset of flowering with 240 kg N/ha was the most economic combination. It was further observed that dry matter content increased with higher fertilizer rate and FYM alone proved less effective than the mineral fertilizers (Doikova, 1977).

Som *et al.* (1992) suggested application of neem cake @ 50q/ha along with NPK fertilizer (80: 60: 60) for maximum benefits in relation to different growth and yield attributes as well as yield of brinjal.

Opara and Asiegbu (1996) conducted an experiment in a weathered tropical Ultisol in Nigeria to determine the optimum rate for eggplant and observed poultry manure at the rate of 15 t/ha appeared satisfactory for production, although no quadratic response of fruit yield was observed at the highest rate (20 t/ha). Yield responses to manure application were best predicted by a linear function within the range of 0 to 20 t/ha. Fruit yield increased with increased manure application were essentially due to more fruits harvested per plant and from higher average fruit weights. Leaf and fruit elemental concentrations did not increase in response to increasing manure rates.

By amending the sandy soil with liquid sewage and sludge, P, K, Ca and Mg contents were increased in the soil and a higher yield was obtained as compared to that treated with NPK in the form of fertilizers only (Gerber *et al.*, 1981).

From a study on the effects of salinity on eggplant, Richter *et al.* (1999) reported that the crop seemed to be includer plants as it can accumulate the salt which can be useful for osmotic adjustment against water stress due to salinity. Eggplants when grown in sand culture in a green house and supplied with a nutrient solution containing 0, 20, 40 or 60 meq NaCl or Na_2SO_4/l, showed moderately tolerant to salt. Increased salinity also hindered the absorption of K, Ca and Mg by plants. The ratio of the decrease in K absorption to the increase in Na absorption was found to be a good-indicator of salt tolerance in these species (Shimose *et al.*, 1991). The effects of NaCl salinity on germination, growth, gas exchange and yield of green house grown eggplant cv. Delica were studied by Chartzoulakis and Loupassaki (1997) in Greece. Plants were grown in a sand: perlite mixture (1: 3) and irrigated with half-strength Hoagland nutrient solution 0.10, 25, 50, 100 or 150 mM NaCl. Salinities upto 50 mM delayed germination, but did not reduce final germination percentage; it was reduced significantly at 100 and 150 mM NaCl. Plant height and leaf area were reduced significantly at salinities of 25, 50, 100 and 150 mM. Cl^- in leaves was always present at higher tissue concentrations than Na^+. Leaf growth was most sensitive parameter to NaCl salinity. Photosynthetic rate was inversely related to the concentration of either Na+ or Cl^- in the older leaf laminas, while no reduction was noticed in newly expanding leaves even of those grown at 150 mM NaCl. Total yield was reduced by 23, 41, 69 and 88 per cent at salinities of 25, 50, 100 and150 mM NaCl, respectively. Both fruit number per plant and fruit size were reduced by salinity.

In an experiment, Rastogi *et al.* (1979) obtained the highest average yield by treatment with 45 kg N/ha at spacing of 45 x 30 cm or 60 x 30 cm in the mid-hill region of Himachal Pradesh with Pusa Purple Long cultivar. In another experiment on fertilizer application and spacing, the highest yield was obtained with NPK at 80: 60: 40 kg/ha and a planting distance of 75 x 45 cm (Sulikeri *et al.*, 1977). Mangual

(1981) found that the maximum yield could be obtained at 90 x 60 cm and 60 x 60 cm spacing with 298 kg N, 200 kg P and 253 k per ha. In West Bengal, the application of 100 kg N/ha gave the highest yield with a spacing of 75 cm in-between rows and 60 cm in-between plants. The application of phosphorus, however, had no influence on fruit yield (Ghanti *et al.*, 1982).

The vitamin A, vitamin C, sugar and dry matter content of brinjal fruits increased at higher doses of N, P and K per hectare (Gnanakumari and Satyanarayana, 1971; Aliev, 1971). Dunyamalyev (1975) found that the best fruit quality (dry matter, total sugar and vitamin C) was obtained from plots receiving N, P and K at 150: 180: 90 kg/ha. Nitrogen fertilization did not affect the fruit size but significantly influenced the TSS, titrable acidity and the vitamin C content of the fruits (Addae-Kagya and Norman, 1977).

Though fertilizer application has pronounced effects on the vegetative growth, development, yield and quality of brinjal, there are several reports on the influence of fertilizers on pest and disease incidence. Symptoms of *Verticillium dahliae* were less severe by treatment upto 100 kg N as compared to 200 and 400 kg N/ha (Sivaprakasam and Rajagopalan, 1974). They also observed in 1971 that application of potash at the rate of 150, 300 and 600 kg/ha prolonged the incubation period of the *verticillium* wilt and reduced its intensity. The highest rate, however, had adverse effects on plant growth.

Chakraborty *et al.* (1974) observed that leaf hopper population on the crop increased significantly with fertilizer applications to a maximum at 120 kg N/ha. Fruit borer damage was however, not related to N application rates. The highest marketable yield was obtained in response to 140 kg N/ha. Sinha *et al.* (1976) reported that leaf hopper and fruit borer population were less under P and K treatment, while these increased with higher levels of nitrogen.

Venkatakrishnan *et al.* (2018) evaluated the effect of vermi-compost as well as the recommended dose of fertilizer (100: 75: 75 kg N, P_2O_5 and K_2O per hectare) in brinjal in a pot experiment with different doses and the treatment T_5 - 100 per cent RDF + Vermicompost @ 5 t/ha significantly recorded high fruit yield of 934.2 g per plant and brinjal stover yield of 403.7 g per plant.

Ali *et al.* (2019) carried out studies on the application of organic and in-organic fertilizers affecting the yield and vegetative growth of eggplant in Turkey, and found that there was no significant difference between the developed organic fertilizer (DOF), organic (OF) and inorganic fertilizers (CHF) being incorporated in the experimental plots suggesting that the organic fertilizers are competitive and can be used as an alternative to chemical fertilizer. The fruit yield obtained with the use of DOF, OF and CHF are 4593 kg/ha, 4375 kg/ha and 4579 kg/ha and plant fresh weight was 4018 kg/ha, 3818 kg/ha and 3882 kg/ha, respectively.

Shahid *et al.* (2019) carried out an experiment to understand the performance of eggplant cultivars *viz.*, Namli (F_1), Black Boy (F_1), BSS-513 and Meiyinqienquan in different organic and inorganic growing condition. The organic regime comprising of farm yard manure, spent mushroom compost and poultry manure at 25: 10: 5 tons/ha and inorganic regime consisted of the recommended dose of N, P, K at

100: 50: 50 kg/ha. It was observed that under the organic regime maximum plant height, number of fruits, fruit weight/plant, total yield and total soluble solids were produced whereas, maximum number of leaves/plant and chlorophyll content were produced under the inorganic regime. From this study, they further concluded that cultivar Namli should be grown for fruit yield while Black boy and BSS-513 for their better quality under the organic regime.

Fouda and Elhamied (2017) investigated the effect of foliar application with yeast and ascorbic acid under different fertilizer treatments and it was observed that the eggplants supplied with 100 per cent NPK fertilization and foliar application of yeast 5 g/litre gave the highest significant values for growth and yield components and quality of eggplant.

Thingujam *et al.* (2019) assessed the role of selected biofertilizer (*Azospirillum*, phosphate solubilizing bacteria or PSB) and micronutrient fertilizer (iron sulfate, zinc sulfate and borax) application in augmenting the yield and quality of eggplant cv. VNR-60 (F_1-hybrid). It was observed that along with regular chemical fertilizers, the application of bacterial fertilizers, alone as well as in combination with micronutrients, significantly improved the yield (13.10–15.35 t ha^{-1}) over the control (9.83–12.39 t ha^{-1}). It was further found that the eggplant yield and quality models developed through the artificial neutral network showed that fruit phosphorus (P) concentration was most important factor towards yield prediction followed by zinc (Zn) and soil available iron (Fe) concentration, respectively. Fruit nitrogen (N) concentration showed the highest sensitivity followed by soil available boron (B) and fruit potassium (K) concentration toward protein content. Therefore, the application of B emerged to be most beneficial along with *Azospirillum* and PSB in attaining best yield, soil nutrient availabilities and plant nutritional recoveries.

Kumar (2016) investigated the effect of integrated nutrient management practices on soil fertility and crop yield of hybrid cv. Hybrid Purple Long under field conditions by incorporating ten treatments. Results revealed that maximum plant height, root length, dry weight, chlorophyll content, LAI, number of flowers/plant, fruits/plant, crop yield/plant, crude protein, dietary fiber, total carbohydrates and total sugar of *S. melongena* was recorded in the treatment T_7 50 per cent RDF + vermi compost @ 5 t/ha.

Ramadan *et al.* (2019) studied the effect of planting date and foliar feeding on the productivity and fruit quality of brinjal and summer transplanting in April along with foliar spray of Potassium @ 3 ml/L or Raizante @ 2 ml/L significantly increased the plant height, branches and leaves numbers/plant, leaf area, total chlorophyll content, fruit yield and its quality.

5.4.3 Micronutrients

Brinjal do not have a high requirement for most of the minor elements, and deficiencies seldom a problem. However, continuous use of high analysis fertilizers often results to the deficiency of a number of micronutrients and brinjal is no way different from other crops. Application of micronutrients alone or in combination with major nutrients not only increases the yield but improves the quality of the produce also. Dobroljubskii (1955) reported that when the plants received manganese, zinc and cobalt alone or in combination, the flowering, fruiting and earliness were improved as compared to control. The best result was obtained when Mn was applied alone, followed by Zn + Co and Zn + Mn + Co. In another trial in glass house, the plants have been sprayed with Cu, Mn and Zn at 10, 16 and 20 ppm until the end of flowering. Although the nutrients had no effect on the vegetative growth, Cu increased the number of flower and fruit, while Zn improved the fruit weight and Mn showed a similar but lesser effect both on flowering and fruiting (Nagarajan and Shanmugasundaram, 1961). Jyotishi and Kashyap (1968) found maximum protein content in the fruit of cv. Black Beauty when 134.02 kg N and 22.34 kg P/ha were applied along with a foliar spray of Zn and also when 84.34 kg N and 44.67 kg P per hectare were applied along with a foliar spray of boron. Earlier, Jyotishi and Kashyap (1967) also determined the composition of brinjal leaves as affected by application of Mn, Zn, Fe and B. It was found that all foliar sprays increased the N content of leaves. B and Fe sprays reduced the P and Zn content of the leaves and Zn and B sprays reduced the Mn and Cu content, although not significantly. Treatment with B, Mn, Cu or Zn stimulated photosynthesis and increased yield by 11 to 23 per cent, particularly with B and Mn.

The highest yield was obtained by treatment with Cu and all the micronutrients used increased the sugar, dry matter and vitamin C content of the fruits (Aliev, 1971). Vaganov *et al.* (1975) observed that B, Mn, Zn, or Mo at 0.01 per cent increased the leaf water retaining capacity during almost all growth phases except flowering when sprayed 3 times. Shatrova (1977) indicated that foliar application of Mn or B, each at 0.05 to 0.15 per cent during bud formation and flowering decreased the incidence of wilt by 2-3 times. Mn was slightly more effective than B. Soil application of Co-sulphate (7 kg Co/ha) along with NP increased the Co content of the crop (Andryushchenko and Vyrodova, 1981). Masui *et al.* (1979) studied that soil application with methyl bromide had no effect on the growth of brinjal in glass house. However, the B content of the fruit rose with the increasing levels of methyl bromide. Basile and Lamberti (1981) also reported higher B content in fruits when the soil was treated with methyl bromide (60 g/m^2).

Bid *et al.* (1993) reported that in calcareous soils on young alluvium in Saran district of Bihar, India, zinc sulphate and copper sulphate at 10 and 20 kg/ha decreased Zn and Cu deficiency symptoms in eggplant and significantly increased yields. Regarding micronutrient application, the findings of Ravichandran *et al.* (1995) clearly indicated that application of Zn and Cu, both in soil and foliar as well, had beneficial effects on the number of fruits per plant yield and vitamin C content of the fruits. The recommendations were soil application of $ZnSO_4$ (25 kg/ha) along with 2 foliar sprays of 0.5 per cent Zn and $CuSO_4$ (12.5 kg/ha) with 2 foliar sprays of 0.15 per cent Cu both at 30 and 60 DAT for a silty clay loam soil deficient in Zn and Cu.

Sometimes the distal ends of young, fully developed eggplant leaves frequently become yellow. Midribs, veins and mesophyll show the symptoms, in contrast to the Mg deficiency known to occur in this crop which causes interveinal chlorosis over the entire leaf. To determine whether the symptoms are due to B deficiency, a green house experiment was conducted by Kreij and Basar (1997) in Netherlands in which eggplants were supplied with nutrient solutions containing different concentrations of B as Borax. At 200 mM B, leaf tip yellowing did not occur. The conclusion of the study is that leaf tip yellowing is due to B deficiency.

Ghazi (2018) studied the effect of foliar application of iron in the form of Fe-EDTA @ 300 mg/L and selenium as Na_2SeO_3 (sodium selenite) @ 5 mg/L on plant growth, yield, quality and chemical composition of eggplant with three different recommended dose of fertilizers *i.e*, NPK at 50 per cent, 75 per cent and 100 per cent and concluded that the application of NPK @ 75 per cent RD with foliar application of iron and selenium in combination together gave the highest significant values of all eggplant growth characters (plant height, number of leaves per plant, leaves fresh and dry weight and total chlorophyll content), yield (fruit length, fruit diameter and fruit yield) and quality (crude protein, total carbohydrates, dietary fiber and vitamin C) under the study conditions in Egypt.

5.5.0 Weed Control

Most vegetables are initially slow growing crops, incapable of offering any competition to the aggressive weeds. Infestation of weeds also increases insect pests and diseases of vegetables. Therefore, the weeds should be controlled as soon as they are seen, either by the traditional method of hand weeding and hoeing or by the application of herbicides. Frequent shallow cultivation should be done at regular intervals as to keep the field free from weeds and to facilitate soil aeration and proper root development. Deep cultivation is injurious because of the damage of roots and exposure of moist soil to the surface. The most serious weed in brinjal and other solanaceous crops is *Orbanchae* sp. It is a root parasite and it should be controlled very carefully (Chauhan, 1981).

Besides other beneficial effects, mulching has an influence on controlling the weeds. Mulching brinjal plants with black polythene film reduced weed growth, increased crop growth, improved earliness and increased yield (Raman, 1964; Petrov and Petkov, 1980). Several herbicides application under transparent polythene reduced the weed growth and increased yield (Gorske, 1979, 1981). But Petrov and Petkov (1980) reported that transparent polythene gave less satisfactory result than black polythene film and straw mulch was worse than no mulch. Sowinska *et al.* (2016) also reported that mulching the soil with polyethylene black foil or transparent foil after spraying with a herbicide was a beneficial agro-technical treatment that significantly increased the fruit yield of eggplant.

DCPA is the only herbicide still recommended for use in brinjal plants, although many other herbicides are not injurious to the crop when applied immediately after planting (Tanaka *et al.*, 1973). Americanos (1976) suggested that the best weed control without crop injury can be made by the pre-emergence application of chlorthal dimethyl, diphenamidor napropamide. *Digitaria sunguinalis*, another weed in brinjal has been reported in New Jersey, U.S.A. by Frye *et al.* (1978). This weed can be effectively controlled by acetanilides and napropamide. But stunted crop growth and yield reduction were observed by application of acetanilides, whereas the crop safety was maintained by the later. Brathwaite (1978) evaluated the effect of 6 herbicides on weed control, yield and quality of transplanted Early

Long Purple cultivar during wet season (June to December) in trifluralin at 1 kg/ ha gave the best yield with a good weed control for 8 weeks. All the herbicides tested controlled weeds for 4 weeks and no crop injury was observed. Filippov and Inshakova (1979) obtained the good weed control with pre-planting applications of treflan (trifluralin) at 0.9 to 1.5 kg/ha or devrinol (Napropamide) at 5- 7 kg/ha in brinjal plants. Doikova *et al.* (1980) reported effective control of weeds when reglone (diquat) and gramoxone (paraquat) were applied separately and in mixtures. Residues of reglone and gramoxone decreased in the soil on the 60th day after treatment and residues in fruits were below the norms set by the WHO.

In Sri Lanka, Ranamukaarachchi *et al.* (1989) observed on reddish brown soil that one hoeing 6 weeks after transplanting along with application of alachlor and linuron and 3 and 1.5 kg/ha, respectively gave the best weed control, highest yield and was most cost effective. Use of detergent or gasoline to control dodder weeds (*Cuscuta campestris*) in eggplant field was reported from Iraq by Al-juboory and Shati (1996). The results of their study clearly showed that detergent applied at 10 l/ha caused a 50 per cent growth reduction of *C. campestris* without damaging the crop plants. Gasoline at 20 l/ha gave a good level of control of *C. campestris*.

The weed growth interferes with the crop growth and yield in eggplant. In order to control the weed it is necessary to identify the main weed species so that it can be managed. A study was conducted by Marques *et al.* (2017) to identify the major weed species in the eggplant cv. Napoli in Brazil. The major weeds found were *Eleusine indica*, *Portulaca oleracea*, and *Cyperus rotundus* and coexistence between the weed community and the eggplant throughout the entire crop production cycle reduced eggplant fruit yield by 78 per cent.

5.6.0 Training

In trials conducted by Miyamoto and Hara (1988), brinjals were grown in single rows on 2 meter wide beds with 2 parallel shoots trained vertically, one on each side of the row, or in double rows on 3 m wide beds with 3 shoots trained vertically in the plane of the row. With the latter system, fruiting began earlier and the proportion of good quality fruit and total fruit yield were higher. Maaswinkel and Buitelar (1989) obtained highest returns/m^2 with a combination of 2 stems/ plant and 6.0 stems/m^2.

Moncada *et al.* (2013) studied the effect of grafting four eggplant cultivars (Longo, Black Moon, Birgah and Black Bell) on to *S. torvum* to evaluate the yield and quality. Significant differences were recorded for growth and yield among the grafted cultivars which may be due to different growth characteristics, graft affinity and compatibility. Longo and Black Moon were not influenced by grafting, whereas grafted plants of Birgah and Black Bell had a higher unmarketable production. Grafting onto *Solanum torvum* also changed the colour of fruits that were darker and with a less vivid color. Further, they also noticed that the browning of inner tissue was not influenced by grafting but total phenolic content was greater in the ungrafted plants.

It is widely known that *S. torvum* is a potential rootstock for grafting however, Sabatino *et al.* (2018) wanted to exploit the potentiality of new rootstock genotypes both wild and allied species of eggplant *S. torvum*, *S. macrocarpon*, *S. aethiopicum* (accession 1), *S. aethiopicum* (accession 2), *S. paniculatum* and *S. indicum*, and Msa 2/2 E7 and 460 CAL. eggplant hybrids on plant vigor, yield and fruit characteristics of eggplant F_1 hybrid Birgah. A high percentage of success was observed for *S. paniculatum*, Msa 2/2 E7 and 460 CAL giving a good vigour and yield with no undesirable effect in the quality as well as fruit composition.

5.7.0 Effect of Growth Substances

Investigations have revealed the beneficial effect of growth substances on fruit set in brinjal. Among the chemicals, para-chlorophenoxyacetic acid (4-CPA), 2, 4-dichlorophenoxyacetic acid (2, 4-D) and naphthalene acetic acid (NAA) are the most effective. Muthukrishnan and Srinivasan (1963) reported a significant increase in yield of 50 per cent by whole plant spray of 2,4-D at 2 ppm at intervals of 1week over a period of 60 to 70 days from the commencement of flowering.

Induced parthenocarpy, increased fruit set, advanced fruit maturation and significant increase in total yield were obtained by the application of 2,4-D at 2 ppm at flowering stage. Though short-styled and pseudo short-styled flowers do not normally set any fruit, the application of 2,4-D on pseudo short-styled, medium-styled, and long-styled flowers produced a higher percentage of fruit set compared to plants under control (Krishnamurthi and Subramanian, 1954b). In 1955, however, they found that fruit size was reduced by this treatment and 2,4-D at very low concentrations applied to seed before sowing inhibited germination.

Muthukrishnan and Srinivasan (1964) again noted that treatment with 2,4-D (at 0.0025 per cent in lanolin paste to cut style of brinjal flower) produced seedless fruit smaller in size and weight than seeded fruits. Treatment with 2,4-D at 2.5 ppm and 4-CPA at 20 ppm at the flowering stage caused earlier and more uniform ripened fruits (Even, 1970). Moriwaki and Ito (1980), on the other hand, found the best fruit set in brinjal by spraying with 2,4-D at a higher concentration. Nanda *et al.* (1973) recorded decreased flower abscission, increased fruit set and least number of small stone-like fruits by spraying with growth substances at flowering stage. Spraying of 2, 4-D (2.5 ppm), 4-CPA (20 ppm) and n-metatolylphthalamic acid (0.1, 0.5 per cent) promoted fruit set in brinjal (Northmann *et al.*, 1974). A mixture of 20 mg/l a-chlorophenoxyacetic acid and 500 mg/l Iprodione sprayed at weekly intervals on flower buds or flowers proved effective to improve fruit growth and control fruit rot (Ravestijn, 1988).

Olympios (1976) observed that NAA at 60 ppm alone or in combination with BA at 30 ppm, applied to open flowers during the winter and early spring months improved fruit set and fruit development. They also noted that spraying of plant with n-metatolyphthalamic acid at 250 and 500 ppm considerably increased the early yield. Though spraying of 4-CPA at 1 percent on brinjal flowers in the glass house and plastic tunnels increased fruit set and yield, but in field, the yield was not increased and it only improved the shape and colour of the fruits (Martinoni *et al.*, 1981). Balanced vegetative and reproductive growth and controlled premature flower drop were obtained by spraying with N as leaf, *i.e.*, 0.25 per cent 2,4-DNP (dinitrophenol) and 0.3 per cent 5 NG (p-nitro-o-methoxyphenol) to the brinjal plants (Yu, 1981). Mukherjee and Datta (1962) observed increased plant height, more N content, decreased total sugar content and small number of parthenocarpic fruits by the application of GA_3. Root-dipping of 1 month old seedling in ascorbic acid, GA_3, IAA or Thiourea advanced the flowering by 4 to 5 days and the highest yield was obtained from plants whose roots were dipped in solution containing GA_3+ ascorbic acid each at 250 ppm (Patil *et al.*, 1981).

Sharma (2006) investigated the effect of plant growth regulators on growth and yield of brinjal cultivars Pusa Purple Long and Pusa Purple Cluster by incorporating treatments *viz.*, fresh water (control), 40 ppm NAA, 10 ppm GA_3, 2 ppm 2,4-D, 300 ppm ethephon, 30 ppm BAP and 5 ppm triacontanol. The first spray was given at flowering stage and then at 20 days interval. The results indicated that the PGR's had no significant difference on growth parameters but the cultivars differed significantly. The interaction effect showed that PPL did not produced statistically different fruit number per plant with respect to growth regulators, while it had significantly higher fruit yield (17.76 t/ha) at 40 ppm NAA. PPC produced significantly higher fruit number per plant and higher fruit yield (19.25 t/ha) at 30 ppm BAP.

Patil *et al.* (2008) studied the effect of growth regulators and fruit load on seed yield and quality in brinjal hybrid seed production. The results revealed that GA_3 50 ppm recorded significantly more seed yield (33.96 g) per plant, germination, root length, shoot length, seedling vigour index and seedling dry weight compared to NAA 40 ppm. Four fruits could be retained in a plant which gave highest germination, root length, shoot length, seedling vigour index and seedling dry weight compared to retention of all fruits per plant.

A study on the relationship between flower development and gibberellic acid (GA_3) levels in parthenocarpic and non-parthenocarpic eggplant (*Solanum melongena* L.) genotypes was done and results showed that no relationship existed between the flower development and GA_3 levels in parthenocarpic and nonparthenocarpic eggplant genotypes (Boyaci *et al.*, 2011).

Patel *et al.* (2012) studied the effect of plant growth regulators on growth, yield and quality of brinjal (*Solanum melongena* L.) cv. Surati Ravaiya with nine treatments *viz.*, 2, 4- D @ 2, 4, 6 and 8 ppm as well as NAA @ 25, 50, 75 and 100 ppm along with control (water spray). Among the treatments the foliar sprays of 2, 4-D @ 4 ppm gave the highest yield of brinjal (64.35 t/ha), while plant height was maximum with 50 ppm NAA. Total soluble solids and acidity were found maximum with foliar spray of 100 ppm NAA in brinjal.

Netam and Sharma (2014) studied the efficacy of different plant growth regulators and their combinations of NAA, GA_3 and 2,4-D spray on fruit yield and quality in brinjal cv. Brinjal 3112. The plants were sprayed three times at 30, 60 and 90 days after transplanting. The highest number of branches, number of fruits, fruit weight, total soluble solid and nitrate reductase activity was recorded by the treatment combination of GA_3 @ 10 ppm and NAA @ 20 ppm.

Intercultural Operations and Spraying in Brinjal

Mahmood *et al.* (2016) investigated the effect of PGR's on fruit set, growth and development of brinjal cv Kajla. Plant growth regulators namely β-NOA (60 ppm), NOA (60 ppm) + BA (30 ppm), and BA (30 ppm), were applied to the emasculated flowers at anthesis to set parthenocarpic fruit while fruit set was achieved by natural pollination in control treatment. Results revealed that application of NOA alone and in combination with BA effectively promoted parthenocarpic brinjal with a high mean weight due to increasd fruit length and diameter but BA alone failed to set parthenocarpic fruit.

Application of the fruit-setting compound β-naphthoxyacetic acid (60 mg/litre used as an aqueous spray) resulted in the setting of seedless fruit (parthenocarpic fruit) that are similar in size and appearance to those set by natural pollination (Makrogianni *et al.*, 2018).

6.0 Harvesting and Yield

The fruit is harvested before it fully ripens and when it attains a good size and colour. The surface of the fruit should not lose its bright and glossy appearance. According to Choudhury (1976a) the fruits are edible from the time they are quarter grown, until they are near ripe. When fully ripe, the fruits become greenish yellow or bronze, their flesh turns dry and tough. At harvesting, the calyx and the stem-end are left attached to the fruit. The stalk of the fruits should be cut with a knife or other sharp instrument. The fruits, particularly large round should be handled with care. Narayanaswamy (1978) in a study with 5 brinjal cultivars obtained the highest yield from 19-day old fruits and with most cultivars this corresponded to the optimum maturity stage. Ascorbic acid content is also increased upto 19 days from flowering and thereafter declined.

The yield varies with the cultivars, the region and management practices, *etc.* However, the average yield varies from 250 to 300 quintal per hectare.

Usually, the fruits are packed in baskets for the markets. The Indian Standards Institution has recommended three grades of brinjal, *viz.*, Super, Fancy and Commercial (Choudhury, 1976a).

7.0 Storage

Storage extends the consumption period of vegetables. Brinjal fruits can be stored for 1-2 days in summer and 3-4 days in winter season provided they are kept in shade under ordinary conditions (Chauhan, 1981). Cold storage of brinjals was studied by Singh *et al.* (1952) and obtained good results on cooking even after 28 days when brinjal cv. Irangeri weighing 45 to 50 g stored at 47 to 50°F (8.3 to 10°C) with 87 to 90 per cent relative humidity. Small fruits weighing14-18 g did not store so well as big fruits. Smaller fruits showed greater loss in weight than big fruits and being tender were subjected to low temperature injury.

Valida *et al.* (2018) evaluated the fruit quality attributes of commercially mature eggplant cv Morena from harvest to end of shelf life during storage at 7-10°C and 21-25°C and results revealed that loss in fruit quality was due to shriveling (water loss) and *Phomopsis* rot at a temperature of 21-25°C resulting in much shorter shelf life than at 7-10°C in which the fruit deteriorated in quality mainly due to shrivelling.

Harvesting, Sorting, Grading, Packaging and Selling of Brinjal

Green brinjal had the longest shelf-life of 4 weeks on the basis of physiological loss in weight, discolouration, spoilage and CO_2 injury during storage at 8-10°C under controlled atmosphere with initial concentration of 5 per cent CO_2. Room temperature of 20-22°C was not, however, suitable for storage under modified atmosphere (Nagaraja and Mathew, 1964). From a study on controlled and modified atmosphere storage of eggplant in Yalova, Turkey, Kaynas *et al.* (1995) reported that the rates of increases insoluble solids, and losses in acidity, and starch and ascorbic acid contents were greater in air stored fruits than in MAP or CA stored fruits. Starch, total and reducing sugar contents were slightly higher in MAP and CA stored fruits. Weight loss was significantly greater in MAP compared with CA after 2 weeks but this difference disappeared with after 4 and 6 weeks. There were signs of severe CO_2 injury with storage in 10 per cent CO_2 (surface discolouration and calyx yellowing) for >2 weeks, but this was less at 5 per cent CO_2. The best external fruit quality was obtained with MAP, or CA storage at 3 per cent O_2+ 3 per cent CO_2. Under these conditions, fruits could be stored successfully for upto 6 weeks.

In storage, titratable acidity of fruits increased after 10 days, the maximum increase occurring at 20°C. Reducing and total sugars increased from 4th day to 14th day at 10 or 20°C, but decreased steadily with time at 5°C. Ascorbic acid content after 18 days at 10 or 20°C decreased by about 75 per cent.

Total polyphenol content initially increased, followed by a gradual decrease. The optimum storage temperature ranged from 5-10°C, the storage period being extended upto 15 days (Esteban *et al.*, 1989).

Wax emulsion can extend the storage life of brinjal (Agnihotri *et al.*, 1963). Waxing in 3 per cent concentration helped to reduce weight loss and retard change in acidity and starch, sugars and ascorbic acid contents. The storage life at room temperature (25-45°C) was extended by 30 to 40 per cent with wax emulsion. Waxing also reduced microbiological spoilage. The fruits treated with Waxol had the best appearance and the least physiological loss of weight (Singh *et al.*, 1989).

Singh *et al.* (2016) studied the formulation of improved and stable carnauba wax coating on eggplant in both packaged and unpackaged 35μ polypropylene pouches and revealed that packed eggplant treated with polyethylene glycol (PEG) and 0.5 per cent sodium alginate in CW emulsion remained acceptable upto 12 days since eggplant had a shelf life of only three days under ambient storage conditions.

Abe *et al.* (1974) studied the chilling injury and its effect on the pitting development in brinjal in Japan. Brinjal fruits, cv. Senryo were stored at 1, 6, 10 or 20°C in polythene bags. Fruits harvested in July showed severe pitting when stored at 1 or 6°C and this was increased by subsequently transferring them to storage at 20°C. With storage at 10 or 20°C no injury was found, but fruits stored at 10°C when transferred to 20°C decayed more rapidly than fruits stored continually at 20°C. In fruits harvested in October, no pitting was found in those store dat 6°C and after transferring them to 20°C. Storage at 10°C is, therefore, recommended for fruits picked in the cool season but storage above 10°C for those picked in the warm season.

A relationship between storage temperature and enzymatic activities during storage of eggplant fruits was studied by Fuster and Prestamo (1994) in Madrid,

Spain when fruits of the cultivar Moroda were stored at 4 or 12°C and 85 per cent R.H., or at 20°C and 50 per cent or at -24°C. Result showed that activities of peroxide and polyphenol oxidase (catecholoxidase) were always higher in peel than in pulp. With increasing storage time, the activities of both the enzymes in peel and pulp were higher at 20°C than at the lower temperatures, especially-24°C. The study indicated that this cultivar could be stored successfully for 8-10 days at 20°, for 16-21 days at 12°, for 21 days at 4°, and 56 days at -24°C.

Nakamura *et al.* (1986) reported that pitting and loss of quality in brinjal in cold storage was prevented to some extent by allowing the fruit to lose 0.5 per cent of the weight by wilting before storage. They (1987) suggested that brinjal grown under wet conditions were more sensitive to chilling injury than those grown under drier conditions, and watering 3 days before harvest led to severe chilling injury during storage. Fruits from plants grown with low levels of N or P were also sensitive to postharvest chilling but no effect was noticed from different applications of K. West Coast Green Round cultivar had the longest shelf-life (about 7 days). Pre-treatment with fungicidal wax emulsion and packaging in perforated polythene enhanced shelf-life (Narayanaswamy, 1978).

Eggplants for export from Israel are currently packed individually in PVC stretch film. However, these wraps are disliked by the wholesalers as these must be removed prior to marketing. Alternative bulk packaging for storage of fruits was assessed by Fallik *et al.* (1994). The best storage method involved packing of several fruits (10-15) pre-treated with a solution of prochloraz and NAA, inside an unperforated polythelene lining bag, together with 10 layers of tissue paper to prevent water condensation. This method prevented weight loss and maintained fruit quality during 14 days of storage at 12°C and an additional 3 days at 20°C. From a study Fallik *et al.* (1994) noted a relationship between chilling injury of eggplant fruits and growing conditions and time of harvesting. The study revealed that fruits exposed to high temperature (30°C) during their growth and development were very sensitive to chilling injury while in low temperatures (15°C) exposure, fruits were less sensitive. Packing disinfected fruits inside unperforated polythelene bags enabled them to be stored for 20 days at 8°C and an additional 3 days at 17°C while maintaining quality and preventing chilling injury whereas disinfected but unbagged control fruits showed severe injury. Polythene bags also showed the appearance of external and internal chilling injury symptoms in fruits harvested during both the high and low temperatures and stored at 6°C, compared with unbagged control fruits. Fruit transpiration in eggplant was measured by Diaz-Perez (1998) and opined that calyx was the main root for fruit water loss, accounting for at least 60 per cent of fruit transpiration. Fruit transpiration rate declined as the fruits increased in size. This decline in fruit transpiration was due to, both a reduction in the portion of fruit covered by the calyx and a reduction in the surface area: mass ratio of the fruits as they increased in size. The calyx covered 95 and 10 per cent of fruit surface area in a 5 and 550 g fruit, respectively. The high rate of water loss through calyx indicated that postharvest treatments intended to reduce water loss from the calyx may be beneficial in extending the shelf life of eggplant fruits.

In the Netherlands, external symptoms of cold damage to brinjal fruits appeared after 3 days at 3-4°C and after 6 days at 4-6°C, and internal flesh browning was evident after 7 and 11 days, respectively when the fruits were kept at 3-20°C and 70-75 per cent relative humidity for 3 days (Stork and Schouten, 1976). It was also observed that autumn harvested fruits were the most susceptible to the damage. The fruits remained unblemished even at optimum temperature (10-15°C) for a few days. Brinjal fruits should not be kept in the same store with other ethylene producing fruits. According to Tataru and Hristea (1977) in Romania, brinjal cv. Mammouth showed best in maintaining quality after storage at 10°C for 7 days among the cultivars studied. Uncini *et al.* (1977) also reported that brinjal fruits are better stored at 10°C than at 6°C and in perforated polythene bag than in open boxes.

Accelerated senescence of the calyx caused rapid deterioration of eggplant fruits during prolonged storage. Dipping the calyx in the solution containing 22 ppm NAA and 900 ppm prochloraz retarded senescence and controlled decay (primarily due to *Alternaria alternata and Botrytis cinerea*). Whole fruit dipping in the same solution gave better results than calyx dipping only. The calyx appeared fresh and green and the fruit was firm after 14 days of storage at 12°C and 3 days at 20°C and the decay incidence was less than 5 per cent only. The residual level of prochloraz in fruits that were wholly dipped was <0.34 mg/l (Temkin-Gorodeiski *et al.*, 1993).

Henz and Silva (1995) opined conservation of eggplant fruits by refrigeration and film-wrapping. The best results were obtained with fruits wrapped in perforated or non-perforated PVC film and held at 12°C for 16 days. The loss of fresh weight was minimum and there was no deterioration and no symptoms of chilling injury were observed.

Kumar *et al.* (2008) conducted an experiment to study the influence of storage containers and seed pelleting on seed quality in brinjal (*Solanum melongena* L.) during storage using brinjal hybrid seeds cv. Arka Navneet. Seeds were pelleted with Bavistin, $ZnSO_4$, $MnSO_4$, DAP and Arappu leaf powder and stored in paper and polyethylene bags under ambient conditions for 12 months. They reported that among the seed pelleting treatments, seeds pelleted with Bavistin (0.1 per cent) followed by *Albezia amara* leaf powder (250 g/kg) resulted in minimum quantitative losses with better seed quality parameters. The seeds stored in polyethylene (700 gauges) bags maintained better seed quality parameters with less quantitative losses in comparison with seeds stored in paper bags throughout the storage period.

8.0 Diseases and Pests

8.1.0 Diseases

Fungal Diseases

Brinjal is subjected to the attack of many diseases affecting roots, leaves, stems and fruits. The severity in any particular disease depends on the season and the region in which the crop is grown. Many of the diseases cause damage only in exceptional years, but a few are prevalent in many areas each year and cause varying levels of damage. Only the most common diseases are mentioned with their possible control measures.

8.1.1 Damping Off (*Pythium* spp., *Phytophthora* spp., *Rhizoctonia* spp.)

The fungus attack usually starts on the germinating seed spreading to the hypocotyl, basal stem and developing tap root. The affected seedlings are pale green and a brownish lesion is found at the basal portion of the stem. The lesion girdles the stem, later extending upwards and downwards. The affected tissue rots and the seedling collapses. The disease is very severe in the nursery. The fungus is soil borne.

The disease may be controlled if the nursery soil is sterilized before sowing and the seeds are treated with ceresan prior to the sowing. Hot water treatment (at 51.7°C for 30 minutes) of seeds has also been effective in controlling the disease (Nath, 1976; Choudhury, 1976a).

Alvarez-Garcia (1946) suggested soil-sterilization by steam and formaldehyde in the nursery soil but to minimize the chance of reinfestation, a soil application of Bordeaux mixture (4: 4: 50) at 0.5 gallon per sq foot was effective during the first week after emergence. Among a large number of fungicides tested as seed and soil treatments, several mercuric and copper compounds gave effective control.

A fungus, *Phytophthora parasitica var. macrospora* was identified by Sarejanni (1952) in Greece which caused serious damage to brinjal plants. The fungus infects either lower leaves, the fruit or the neck of the plants, and the last type of infection is the most serious as it causes the whole plants to wilt. There are three possible sources of infection - the seed bed, the irrigation water or the soil in the field. The author also recommended- (i) not to use any plants from an infected seed bed, (ii) to treat the seed bed before sowing with 2 application of 0.05 per cent copper sulphate when there is infestation in the locality, (iii) to remove infected plants from the field as soon as symptoms appear, (iv) to disinfect the irrigation water by putting a bag containing copper sulphate in the main channel (two disinfections at an interval of 8-10 days are usually sufficient), and (v) in infected fields, solanaceous crops should not be grown more often. In an investigation on the effect of some antagonistic fungi against the damping off diseases of brinjal plant, Bora (1977) reported that *Aspergillus niger* is found to be most effective. Chhabra and Sharma (1981) studied the combined effect of *Meloidogyne incognita* and *Rhizoctonia bataticola* on pre-emergence damping off of brinjal in pots and found that *M. incognita* and *R. bataticola* both were present in the substrate. They also suggested soil sterilization before sowing for better germination and seedling growth. According to Kaushal and Sugha (1995), seed treatment with carbendazim alone or in combination with thiram provided better control of pre- and post-emergence damping off than other fungicides.

Nahar *et al.* (2018) carried out effective treatment studies on the soil and seed to reduce the damping off disease on nurseries of Bangladesh. Soil treatment with *T. harzianum* combined with seed treatment with hot water increased the seedling emergence and produced 25–64 per cent more healthy seedlings. This combined treatment also improved the seedling performance (height, root length, lateral root development), and reduced stem girdling, a symptom associated with disease infection of transplants.

8.1.2 Phomopsis Blight (*Phomopsis vexans*)

The disease was first reported from Gujarat in 1914 and since then from many parts of India, its infestation as seedling blight, leaf spot and fruit rot was reported (Rangaswami, 1979). This is a serious disease of brinjal. In seedling infection, it causes damping off symptoms. The foliage is attacked any time during the season. When the leaves are infected, small circular spots appears which become grey to brown and have a light coloured centre with irregular blackish margins. The affected leaves may tum yellow and die. Lesions may also develop on petiole and stem, causing blighting of affected portions of the plant. The fungus perpetuates in the soil and lives on affected host tissues. It is specific to brinjal and once the seeds are sown or the seedlings are transplanted in the infested soil, the organism becomes active.

Panwar *et al.* (1970) reported that the causal organism of the disease remains viable for about 14 months in soil debris and the seeds from infected fruits are poor in germination.

Kalda *et al.* (1976) studied the resistance to *Phomopsis vexans* from 300 entries, including *Solanum spp.*, brinjal cultivars and interspecific F_1 hybrids. *Solanum xanthocarpum, S indicum, S gilo, S.khasianum, S sisymbrifolium* were highly resistant, *S. melongena* lines 11a and 264 were resistant. The reactions of the F_1 hybrids were variable. It was further observed that resistance to phomopsis blight was found to be recessive and governed by polygenes. Dominance gene effects were more pronounced than the additive in most of the crosses (Kalda *et al.*, 1977). Decker (1951) released two cultivars by hybridization, namely Florida Market and Florida Beauty resistant to this blight. Ramnagar Giant and KS-233 are the best cultivars against *Phomopsis* blight and showed resistant to moderately resistant response at different stages of disease development (Pandey *et al.*, 2002). Kumar *et al.* (2020) screened 31 lines of brinjal against phomopsis blight and observed the minimum phomopsis incidence on shoot in the genotype Pant Samrat, *Solanum gilo* and *Solanum aethiocum*. Both the wild species also recorded the minimum phomopsis incidence on the fruit.

The control measures suggested by different workers are: (i) use of seeds obtained from disease free plants, (ii) seed treatment by 0.1 per cent mercuric chloride or with other organomercurial fungicides, (iii) adaptation of suitable sanitary measures, (iv) growing of resistant cultivars, (v) hot water treatments of seeds at 50°C for 30 minutes,(vi) spraying at regular intervals in the nursery and in the main field with 1 per cent Bordeaux mixture or other chemicals, and (vii) a suitable crop rotation may also help in reducing the incidence of the disease. Jacqua and Gerion (1988) from France recorded that *Phomopsis vexans* in brinjal could be controlled with chlorothalonil (as Daconil) and mixtures of captafol plus carbendazim (ceretal) and maneb plus thiophanate methyl (peltar), when sprayed every 10-12 days.

Berua *et al.* (2008) studied the effectiveness of different fungicides for control of phomopsis twig blight and fruit rot in brinjal cv Utkal Tarini and results indicated that four sprays of carbendazim @ 0.1 per cent at 10 days interval during the initiation stage of the disease significantly recorded less incidence of twig blight (PDI – 7.86 per cent) and fruit rot (PDI – 6.42 per cent). Das *et al.* (2014) tested the efficacy of different fungicides Carbendazim (Bavistin 50 per cent WP), Captan (Captan-50 WP), Copperoxychloride (Blitox -50 WP), Mancozeb (Dithane-75) and Ridomil MZ -72 and two species of *Trichoderma* inhibiting the pathogen in vitro. Carbendazim 0.1 per cent was found most effective at the lower concentration and caused complete reduction of mycelial growth (100 per cent). *T. viride* was found to be most effective with 84 per cent inhibition followed by *T.harzianum* 78.22 per cent inhibition respectively over control after 7th days of incubation.

Anwar *et al.* (2018) evaluated the performance of different fungitoxicants and observed that propiconazole 25 EC @ 0.15 per cent proved to be most effective in reducing the disease by 81.88 and 81.42 per cent compared to check (no chemical used only water spray).

Jakatimath *et al.* (2017) evaluated the systemic and non-systemic fungicides against *Phomopsis vexans* and among the systemic, carbendazim, tebuconazole and hexaconazole @ 0.5 per cent recorded 100 per cent inhibition and in non systemic fungicides, chlorothalonil was most effective. Furthher, they also studied the effect of botanicals against the disease and the highest inhibition of mycelial growth was observed in 5 and 10 per cent garlic extract, kokum extract and onion extract.

Rohini *et al.* (2016) studied the biological management of phomopsis leaf blight by incorporating the beneficial phylloplane colonizing bacteria (PCB) *Eacillus subtilis* and rhizosphere colonizing bacteria (RCB) *Pseudomonas fluorescens* for proper plant growth promotion and control of the disease. Results indicated that a combination treatment of *P. fluorescens* Has-1/c as seed treatment followed by a foliar application of *B. subtilis* Br/ph-11 significantly reduced the disease incidence (18.0 per cent) and severity (0.54) in comparison with distilled water treated control (91 per cent and 6.0) suggesting that combined application of biocontrol agents is more efficient in improving the plant growth and controlling the disease over their individual application.

8.1.3 Leaf Spots (*Alternaria* spp., *Cercospara* spp.)

Rangaswami (1979) reported four different types of leaf spots in brinjal caused by the *Alternaria melongenae, Alternaria solani, Cercospora solani-melongenae* and *Cercospora solani*. Both the species of *Alternaria* produce the characteristic leaf spots with concentric rings. The spots are mostly irregular, 4-8 mm in diameter and may increase covering a large area of the leaf blade. The leaves may drop off due to severe infection. *Alternaria melongenae* infects the fruits also causing large deep-seated spots. The infected fruits tum yellow and drop prematurely.

The *Cercospora* leaf spots are characteristically chlorotic lesions, angular to irregular in shape, later turning greyish brown with profuse sporulation at the centre of the spot. Severely infected leaves drop off prematurely resulting in reduction of fruit yield.

The disease may be primarily controlled by maintaining proper field sanitation and by burning of diseased leaves. However, this can be controlled by spraying dithane Z-78, fytolan, blitox, *etc.* Dithane Z-78 (0.18 per cent) and bavistin (0.2 per cent) were the most effective fungicides showing the lowest incidence of leaf spot disease caused by *Cercospora melongenae* with the highest yields resulting in more net profits (Patel *et al.*, 1999).

Salih *et al.* (2019) observed that the fungicide Ortiva (Azoxystrobin) significantly reduced the disease severity of eggplant leaf spot disease up to 11.6 per cent as compared to control treatment which was 25 per cent indicating that it gave the best protection to the crop which was infected with leaf spot disease compared to the other fungicides.

According to Sharma and Sohi (1980), the moderately resistant cultivars to *Cercospora egenulae* are Manjari Gota, Black Round, Junagadh Sel.II (long), P-8, Pusa Purple Cluster and H-4.

Rajput and Choudhari (2018) evaluated various botanicals *in-vitro* by poisoned food technique to know their inhibitory effect on the growth of *A. alternata* of brinjal. The rhizome extract of turmeric (54.42 per cent) was found significantly superior in inhibiting mycelial growth which was followed by Garlic (50.67 per cent).

Biocontrol agents such as *Trichoderma asperellum, Trichoderma harzianum* and *Gliocladium virens* can be used for preparation of bioformulation for management of *Alternaria melongenae* (Upamanya and Dutta, 2019).

8.1.4 Wilt

This disease is becoming a limiting factor in successful brinjal cultivation. A number of pathogens associated with this disease include *Verticillium dahliae, Fusarium solani, Sclerotium rolfsii* and *Macrophomina phaseolina*.

Verticillium Wilt

This disease has been reported from several countries since long both in cultivated and wild plants. In India, it was first reported from Pune by Patel *et al.* (1949). Since then, the disease is prevalent in almost all the brinjal growing regions of the country (Rangaswami, 1979).

The most characteristic symptoms of infection by *Verticillium* spp. are found in the stem and roots. The infected plants become stunted in growth and generally, do not flower and set fruit. If the infection takes place after the flowering or fruit setting, the flowers and fruits become deformed, flaccid and finally drop off. The length-wise cut of the affected stem shows dark colour inside the vascular tissue. The first symptom is noticed on the lower leaves. The affected leaves turn yellow and then brown between the veins followed by wilting. Wilting proceeds from one margin to the other, till the leaves completely drop off. The pathogen is soil-borne and the primary inoculum usually comes from the soil. Since the pathogen is a soil inhabitant it can survive in the soil for 10 years or more in the form of its resting structures, that is, the microsclerotia (Lan *et al.*, 2017).

By increasing the soil temperature with plastic mulch prior to sowing, Katan *et al.* (1975) reported that mulching controlled weeds, enhanced plant development, decreased wilt (*Verticillium dahliae*) incidence by 60 to 90 percent and increased crop yield by 300 per cent.

Elmer (2000) studied the comparative effect of black plastic mulch on reducing the damage from *Verticillium* wilt on eggplants and concluded that the plots which were mulched increased the plant canopy three weeks after planting thereby reducing the percentage of symptomatic foliage eight weeks after planting as compared to the bare ground.

Soil fumigation is also effective in controlling this disease. Rosa (1954) obtained a significant increased yield per plant, weight of individual fruit and a lower wilt rate by fumigating the soil with Chloropicrin. Kodama *et al.* (1979) reported that solar soil sterilization inhibited disease in the vinyl house, while 29 percent disease infested plants were found in untreated plots.

Pirinc *et al.* (2017) studied the effect of VAM (*Glomus intradices*) on *Verticillium* wilt and yield of eggplant by application of fresh chicken manure followed by solarization. They concluded that there was a reduction in wilt by 17 per cent when treated with VAM and that the chicken manure should be used when fermented and applied during the non cultivation period of the crop to reduce the burning effect. In this experiment also there was an increase in the yield of eggplant by decreasing the disease factors.

Grafting brinjal plants on suitable rootstocks also provides effective method to minimize the disease infestation. Grafting wilt susceptible commercial brinjal plants of desirable type with high resistance to *Verticillium alboatrum* in small-scale field tests. Terminal grafts gave sturdier plants than approach grafts (Lockwood *et al.*, 1970). Gindrat *et al.* (1976) found that hybrid rootstock KNVF gave 92 to 440 percent increase in the yield of first quality fruit and harvesting was not delayed. Canevascini and Caccia (1976) also recommended the hybrid rootstock KNVF for grafting. Luxuriant vegetative growth and increased fruit yield were obtained when brinjal plants grafted on tomato hybrid KNVF and Roma VF compared to no-grafted plants (D 'ercole, 1978). Plants of cv. Rodo grafted on KNVF tomato rootstocks grew better and healthier and produced higher early, marketable and total yields than plants on their own roots. Pruning to leave 3 stems gave higher early and marketable yields than pruning to 2 stems (Ulinski and Glaps, 1989). In Italy, as there are no brinjal cultivar resistant to the main pathogens, a tomato rootstock (cv. KNVF) resistant to corky root, *Verticillium* spp. and nematodes is used (Trentini and Maioli, 1989).

Grafting of eggplant cv Tsakoniki on *Verticillium dahliae* resistant wild species *S. torvum* and *S. sisymbrifolium* had positive effects on growth, production and *Verticillium* wilt control (Bletsos, 2003).

Foliar application of Mn or Beach at 0.05 to 0.15 per cent during bud formation and flowering decreased the incidence of wilt by 2-3 times. Mn was slightly more effective than B (Shatrova, 1977). Potassium fertilizer application has a beneficial role on the control of the disease. Application of potash at the rates equivalent to 150, 300 and 600 kg/ha in a pot trial lengthened the incubation period of *Verticillium dahlia* and reduced its intensity. However, the highest level had adverse effects on plant growth (Sivaprakasam and Rajagopalan, 1971). Sivaprakasam *et al.* (1973) further observed that when K fertilizers were applied at the rate of 150, 300 and 600 kg/

ha, the alanine, methionine and proline content of the leaves decreased and their resistance to *Verticillium dahliae* increased. The artificial injection of those substances with the pathogen into a 45-day-old plants increased their susceptibility to the disease. Contrary to the K fertilizer application, N fertilizer had an adverse effect on the disease. Less severe symptoms of the disease were found when the plants received no nitrogen and 100 kg N per hectare as compared to those treated with 200 and 400 kg N/ha. With the high doses of N application, the symptoms appeared earlier than lower levels of nitrogen (Sivaprakasam and Rajagopalan, 1974).

Seed soaking in 0.05 per cent hydroquinine, 0.5 per cent methylene blue, 0.01 per cent para-nitrophenolor, 0.01 per cent ortho nitrophenol prevented the occurrence of both *Verticillium* and *Fusarium* wilt (Machuzhenko, 1972). As the pathogen is soil-borne, soil treatment with different chemicals is found to be very effective. In Romania, Costache (1977) obtained good control when the soil was treated with benlate (benomyl), topsinmethyl (thiophanatemethyl) and bavistin each at 0.05 or 0.1 per cent at an interval of 25 to 30 days. Rangaswami (1979) reported that foliar application as well as soil application of 0.1 per cent benlate was effective in controlling the disease in India.

Marois *et al.* (1982) tried to find out the biological control of *Verticillium* wilt under field condition in the USA. They reported that six isolates from soil-borne fungi reduced the disease in the glass house. When these were tested in two field conditions, *Talaromyces flavus* isolate reduced disease by 76 and 67 per cent and increased yields by 18 and 54 per cent in fruit weight and 22 and 71 per cent in fruit number. The findings of Matsubara *et al.* (1995) indicated that growth of eggplants in terms of plant height, number of leaves, leaf size and proximal diameter of the main stem after eight weeks of transplanting was more in VAM fungus infected (*Glomus etunicatum* and *G igasporamargarita*) plants than in non-inoculated plants. However, flowering was delayed. The appearance of *Verticillium* wilt was delayed and reduced by VAM fungus infection and the effects were more apparent in *G. etunicatum* than *G. margarita* and the incidence of fruits malformation caused by this disease was smaller, especially in *G. etunicatum*. VAM fungus infected plants yielded more fruits than the non-inoculated plants. From a study on increasing resistance to *Verticillium* wilt of eggplant by grafting, Zhao *et al.* (1997) concluded that there was a high affinity between the resistant stock *Solanum sisymbrifolium* and the grafting. No disease was found on the grafted plants in a plastic covered field. The survivality percentage of grafted plants was about 90 per cent which also grew normally. The strong root system of the grafted plants promoted the growth of the organs above soil and as a result the yield was increased.

Karagiannidis *et al.* (2002) studied the effect of arbuscular mycorrhizal fungus (AMF) *Glomus mosseae* and the soil borne pathogen *Verticillium dahlia* which revealed that there was higher root colonization as well as spore formation, also increasing the fresh and dry weight in eggplant. Further, they concluded that the AMF supercedes the pathogenic effect of *V. dahliae*.

Barchi *et al.* (2018) identified the first major QTL '*Vd40E05.01*' which showed resistance to *Verticillium wilt* on chromosome E5 explaining about 20 per cent of the PV.

A broad spectrum fungal strain QLP12 *Purpureocillium lilacinum* having anti-fungal property was isolated in China and evaluated to study the efficacy against *Verticillium* wilt both in green house and in the open field. In the green house, the incidence of *Verticillium* wilt decreased by 83.82 per cent with the soil pretreated with QLP12 fermentation broth and in the field there was a decline in the disease incidence by 40.1 per cent. They concluded that the strain QLP12 identified as *P. lilacinum* reduced the DI of eggplant *Verticillium* wilt and promoted plant growth via physiological and biochemical routes (Lan *et al.*, 2017).

Management of *Verticillium* wilt of eggplant cultivar Black Beauty was carried out by Alkolaly *et al.* (2018) by using fungicides, biocides and salicylic acid. Results indicated that fungicides Vincint (Flutriafol) 25 per cent SC and Meta Z (Metalaxyl + Thiram) 38 per cent WP were the most effective ones in reducing Verticillium wilt resulting disease control efficacy by 77 per cent and 76.6 per cent, respectively, both of them are systemic fungicides.

Use of resistant cultivars, however, controls the disease permanently. Dikii and Neklyudova (1975) reported that brinjal cvs. Black Beauty (USA), Pusa Purple Long (India) and K-2282(USA) were less affected by Verticillium species.

The disease symptoms due to other pathogens of the wilt complex are more or less similar. The critical period for *Fusarium solani* development in seedlings is about 20 days after inoculation (Govindaswamy, 1974). *Trichoderma hamatum* (isolate1) when multiplied in FYM and applied to soil recorded the least incidence of Fusarium wilt of eggplant followed by *T. viride* (isolate1) compared with control. When the antagonists were multiplied in coconut coir pith compost, *T hamatum* (isolateI) recorded the least incidence of wilt also (Sheela *et al.*, 1995).

The biocontrol agents, *Trichoderma viride and T. harzianum* (both 2 g each) and VAM fungi, *Glomus mosseae* and *G. fasciculatum* (*20 g*) brought significant reduction in the pathogenic effect of *F. oxysporum* on brinjal improving the plant growth and chlorophyll content (Wani, 2005).

Trichoderma viride and *T. hamatum* were the best biocontrol agents as manifested by the significant reduction in both disease severity and increase plant growth parameters and fruit yield components (Abdel-Monaim *et al.*, 2014).

In vitro antagonistic behavior of biocontrol agents against *F.solani* revealed that *T. harzianum* inhibited the radial growth of *F. solani* by 70.68 per cent followed by *T. asperellum* (68.71 per cent) and *G. virens* (61.41 per cent) (Upamanya and Dutta, 2019).

Joseph *et al.* (2008) evaluated the *in-vitro* efficacy of different extracts in Allahabad to control *Fusarium solani* f. sp. *melongenae* and from their studies it was observed that all the plant extracts showed significant reduction in the growth of pathogen but the most effective one was the use of 20 per cent *Azardiachta indica* followed by *Rheum emodi, Eucalyptus globules* and *Artemisia annua* against the disease which gave better results as they are biologically and environmental safe alternatives. Bashar *et al.* (2015) also reported the efficacy of the aqueous plant extract *Azardiachta indica* (20 per cent) against *F. oxysporum* and *F. solani.*

Bio-efficacy of the soluble powder formulation of *Aspergillus niger* was tested in a pot trial through seed treatment which revealed that the treatment "T5", a dextrose base formulation of *A. niger*, @ 10g per kg of seed showed maximum wilt disease control, (70.96 per cent) with a seed germination of 87.33 per cent (Patil *et al.*, 2017).

Bavistin 50 WP (50 per cent Carbendazim) at 400 ppm concentration inhibited the complete growth of *F. oxysporum* and *F. solani in-vitro* which was followed by Tall 25 EC (25 per cent Propiconazole) 400 ppm (Bashar *et al.*, 2015).

Mancozeb at 0.20 per cent concentration was effective in inhibiting the pathogen growth, increased plant growth, providing protection against fusarium wilt both *in-vitro* and green house studies (Naziya and Sharada, 2018).

The rhamnolipid biosurfactant from *Serratia rubidaea* SNAU02 was the most effective antagonist against *Fusarium oxysporum f. sp. melongenae* used for pot culture studies and the treatment including *Fusarium oxysporum f. sp. melongenae* (10^6 spores/ml) + 50 ml of 250 mg biosurfactant/ml added to soil + foliar spraying of biosurfactant (250 mg/ml) was most effective as a synthetic fungicide (Nalini and Parthsarathi, 2018).

Toppino *et al.* (2008) demonstrated that the *Fom* resistant trait was controlled by a single dominant locus (named Rfo-sa1) in the line '305E40' which is a double haploid derived from *Solanum aethiopicum* and this line also offers partial resistance to verticillium wilt.

Barchi *et al.* (2018) identified two QTL's affecting the resistance to *Fusarium* and *Verticillium* in an F_2 intraspecific population of 156 individuals bred from the cross '305E40' x '67/3'. The male parent was found to have partial resistance to fusarium wilt. The first major QTL *FomE02.01* explains 68 per cent of the phenotypic variation (PV) which was located at just 1.2 cM away from the locus Rfo-sa1 and was derived from the female parental line '305E40' lies on the end of chromosome 2 (E2) and another major QTL FomE011.01, which explains about 11 per cent of the PV, was derived from the male parental line '67/3'and located on E11.

Disease development with *Sclerotinia sclerotiorum* is much less in a green house covered with U.V. absorbing vinyl film than in a house with common agricultural vinyl film (Honda and Yunoki, 1980).

According to Marras and Corda (1970), PCNB and dicloran are effective against *Sclerotinia sclerotiorum*. An increasing incidence of collar rot caused by *Selerotium* (Corticium) *rolfsii* on commercial cultivars of eggplant in northern India was first reported by Singh and Dhancholia (1991). The best biocontrol agent against *S. sclerotiorum* was *T. asperellum,* followed by *T. harzianum* and G.virens inhibiting the growth of pathogen by (69.99 per cent),(69.15 per cent) and (63.63 per cent) (Upamanya and Dutta, 2019).

Wilting of brinjal plants may also arise due to the attack of bacteria and root-knot nematodes.

Bacterial Disease

8.1.5 Bacterial Wilt (*Ralstonia solanacearum*)

Bacterial wilt disease causes severe problem in brinjal cultivation. The characteristic symptoms of the disease are wilting, stunting and yellowing of the foliage followed by collapse of the entire plant.

According to Sohi *et al.* (1981), proper crop rotations reduce the disease infestation. In a 2 year trial with brinjal cv. Pusa Purple Long susceptible to *Pseudomonas solanacearum*, the following rotations were found to reduce the pathogen incidence: maize-okra-radish; maize-*Vigna* sp.-maize; okra-*Vigna* sp.-maize; or *Eleusine coracana*-brinjal (wilt resistant cv. Pusa Purple Cluster) – *Phaseolus vulgaris.*

Resistant genes may be found in brinjal cultivars or wild species, of which former is preferable as there is a less chance of unusual combination, occurrence of infertility or any undesirable wild character. The wild brinjal *S. melongena* var. *insanum* is reported to be resistant to bacterial wilt (*P. solanacearum*) and highly compatible with certain cultivars of brinjal (Srinivasan and Gopimony, 1969). Gopimony and George (1979) also found *S. melongena* var. *insanum* (small-fruited) is completely resistant to the disease. Dingras Multiple Purple, Sinampiro and Pusa Purple Cluster cultivars were the most resistant to *P. solanacearum* when several indigenous and exotic selections and cultivars were tested for resistance to this pathogen (Rao *et al.,* 1976). Similarly, at Kalyani in West Bengal, Mukherjee and Mukhopadhyay (1982) reported that Pusa Purple Long cultivar was the most susceptible and Pusa Purple Cluster the most resistant, while Black Beauty as susceptible, Pusa Kranti as moderately resistant and Improved Muktakeshi as resistant to bacterial wilt. They also observed that the virulence of the bacterium is related to the age of culture of bacteria after its isolation from the host plant. The root exudates of the susceptible cultivar enhanced the population of the bacterium, whereas that of the resistant cultivar Pusa Purple Cluster showed very little increase of the population. Akiba *et al.* (1972) observed that only Nihon Nassu, a cultivar of Japanese origin is resistant to *Pseudomonas solanacearum* when several brinjal cultivars have been inoculated with a mixture of the pathogen. Crossing it with the susceptible Florida Market indicated that resistance is due to one pair of dominant genes. Though its fruit is not of marketable quality, it appears to be valuable as a source of germplasm in breeding for resistance to this disease.

Yield performance of the brinjal lines/cultivars under disease free and diseased condition of bacterial wilt indicated that the yield loss ranged from 4.24 to 86.14 per cent. Among the lines/cutivars evaluated for nutrient contents, the cultivars JC-I and JC-1 x CH-157-16-4-1 were found to be of better quality as compared to the popular cultivar Pusa Purple Long. Cultivar JC-1 showed higher contents of crude fibre, protein, total sugar, TSS, calcium, phosphorus and potassium over Pusa Purple Long. The study concluded that line/cultivars AB-2, BB-3-1, JC-1, JC-1 x BB-13 and JC-I x CH-157-16-4-1were promising in respect of disease incidence and yield (Sabita *et al.,* 2000).

Among the parental lines and hybrids evaluated during the study, the crosses VNR-218 x BCB-11 and Arka Nidhi x KS-331 were found to be promising because of increase in the number of fruits, fruit weight, high fruit yield with low PDI values for bacterial wilt disease (Chattopadhyay *et al.*, 2012).

From the study on the management of bacterial wilt of brinjal, Pun and Dass (1997) suggested soil drenching of the crop with an organic formulation of asafoetida and turmeric powder in water which caused a substantial reduction in mortality of eggplants due to *Ralstonia solanacearum*. Three soil drenches at 20, 50 and 80 days after transplanting were the most effective for disease control.

In a rootstock trial, Mochizuki and Yamakawa (1979) found in Japan that *S. toxicarium* rootstock showed the highest resistance to this disease, followed by *S. torvum* and *S. integrifolium* (a standard Japanese rootstock). Scions on *S. torvum* rootstock were more vigorous and had greater yield than those on *S. integrifolium* but *S. toxicarium* rootstock resulted in rather poor fruit quality and yield. A study at Kerala Agricultural University, Kerala, India indicated that bacterial wilt susceptible but jassid resistant cultivars could be used for cultivations in summer if grafted on *S.torvum* rootstock (Singh and Gopalakrishnan, 1997).

Seeds of eggplant can be treated with talc based formulation of antagonistic *Pseudomonas fluorescens* (10 g/100 g of seeds) or soil application of P. *fluorescens* (50 g mixed with one kg of soil and incorporated in the nursery bed) for biological control of bacterial wilt. Also, before transplanting the seeds can also be dipped in either P. *fluorescens* or *B. subtilis* @ 259 talc formulation per litre of water solution 20-30 minutes (Ramesh, 2008).

Integration of poultry refuse (PR) @ 3 t/ha 3 weeks before transplanting with stable bleaching powder (SBP) @ 20 kg/ha given during final land preparation or $CaNO_3$ and Furadan 5G @ 20 kg/ha on the day of seedling transplanting reduced the bacterial wilt incidence as well as increased the plant growth and yield in brinjal (Faruk *et al.*, 2019).

Viral and Mycoplasmal Disease

8.1.6 Little Leaf

Little leaf disease of brinjal was first reported in India in 1939 by Thomas and Krishnaswamy. Earlier, this disease was supposed to be due to the virus but now it is confirmed that mycoplasma is the causal organism of the disease. This is a serious disease of brinjal throughout India. Little leaf affected plant is generally shorter in stature but possesses larger number of branches, roots and leaves than a healthy one. The leaves are malformed into tiny chlorotic structures. The petioles get shorter considerably, many buds appear in the axil of leaves, internodes get shortened and give the plants a bushy appearance. Flower parts are deformed leading the plant to be sterile. Infected plants do not bear any fruit and if fruiting occurs, the fruits are very hard. The mycoplasma is transmitted by leaf hopper *Cestitus* (*Hishimonus*) *phycitis* and *Amrasca biguttula biguttula*. *Datura stramonium* was reported as the natural weed host for brinjal little leaf phytoplasma in India (Singh *et al.*, 2012) and the leaf hopper *Hishimonas phycitis* was identified as putative vector (Azadvar and Baranwal, 2012). Choudhuri and Garg (1955) studied the effects of little leaf disease on the growth and physiology of brinjal and concluded that affected plants have stunted growth with numerous malformed leaves and the leaf chlorophyll content and total carbohydrate percentage decrease, while total nitrogen percentage increases. According to Anjaneyulu and Ramkrishnan (1972), the characteristic symptoms of the disease (attributed to a mycoplasma) were small leaves, stimulation of usually dormant axillary buds and proliferation of axillary branches, stunted growth, phyllody and virescence of floralparts. The growth of the fruits after infection was restricted and the fruits become hard and failed to mature. Sulochana and Solomon (1973) found the viral nature of the brinjal little leaf disease. Crystalline inclusions were observed in the cytoplasm of the sieve tubes and occasionally in other cells of the phloem when the stem and other parts of plant showed little leaf symptoms. These crystals are made up of spherical virus particles which are infective and form lesions of virus infection on plants of a test species.

Certain metabolic changes take place due to the infection of the disease. The respiration rate of plants of cultivars like Pusa Purple Long, Pusa Purple Round, Pusa Anmol, Pusa Kranti infected with little leaf disease was higher than that of healthy plants, and peroxidase and polyphenol oxidase activity were reduced in all cultivars (Mitra *et al.*, 1976). Mitra and Majumder (1977b) further reported that in cv. Pusa Purple Long, an increase in phenol content was noted during the first 40 days after infection and an increase in the polyphenol oxidase activity at 40 and 50 days after infection, while in other cultivars like Pusa Purple Round, T-2, Arka Navneet, both phenol content and enzyme activity were decreased at all stages of development.

Leaf N, Ca and Mg contents in cv. Pusa Purple Long were lower in diseased than in healthy plants and their P and K contents were higher (Sivaprakasam *et al.*, 1976).

Chakraborty and Choudhury (1974) observed that the disease increased dry matter content and decreased the moisture and total nitrogen content in the susceptible cv. Pusa Purple Long and *vice versa* in case of resistant species *Solanum integrifolium*. In both cases, ascorbic acid and phosphorus contents decreased but reducing non reducing sugars, starch and total carbohydrate contents increased initially but decreased subsequently. Chakraborty and Choudhury (1979) further worked on the effect of infection with little leaf disease at different stages of growth of brinjal plants and observed that earlier the plants were inoculated through insect vector, *Cestitus phycitis*, quicker was the appearance of symptoms and more adverse were the effects on plant growth and yield.

Roguing out of the diseased plants in the earlier stage of infestation and spraying with Ekatoxor Folidol till fruit set may help to check the spread of the disease (Choudhury, 1976a; Nath, 1976). Choudhury (1976a) also suggested the root dipping at the time of transplanting in 1000 ppm of tetracycline followed by three sprays at weekly interval for 4-5 weeks after transplanting check the disease. Temporary suppression of the disease has been recorded in Punjab by the application of tetracycline hydrochloride at the rate upto 500 ppm (Bindra *et al.*, 1972). Besides tetracycline, diseased plants also responded to the GA_3 application at 50 ppm (Verma *et al.*, 1975). Verma and Dubey (1976) found that among the different methods of tetracycline application, *i.e.*, spraying, wick feeding or girdling (via a strip of cotton tied round the stem) at various concentrations to brinjal seedlings graft inoculated with little leaf, the most effective method to control the disease was girdling followed by wick feeding. Among the cultivars and strains tested, the cultivar Brinjal Round showed most tolerant to this disease followed by Sel. 433-5 and Surati. In 1978, Verma and Dubey further observed that spraying and root immersion were the most effective method for tetracycline uptake among the spraying, girdling, root immersion or soil drenching of tetracycline at 50 to 1200 ppm to little leaf affected Pusa Purple Long cultivar.

According to Joi and Sonone (1977), the best disease control and highest yield were obtained when phorate at the rate of 1kg per hectare was applied to the seed bed followed by seedling dipping in aqueous solution of 0.05 per cent tetracycline along with 0.05per cent monocrotophos.

Gupta and Chakravarty (1975) studied the reaction of brinjal lines and cultivars to little leaf disease at Udaipur, Rajasthan under natural conditions and reported that cvs. Pusa Purple Cluster and Kartain were resistant to the disease. Under field conditions, Chakraborty and Choudhury (1975) obtained only 3 lines, namely, Sel. 212-1, Sel. 252-1-1 and Sel. 252-2-1 which were free from little leaf among 164 cultivars tested.

Early sown brinjal crops (3rd week of June) became less affected than those sown in July or August, due to differences in leaf hopper vector populations. Continuous cropping of the same cultivar increased disease incidence, as did the presence of weeds, tomato crops nearby. The cultivars Pusa Purple Cluster, Pusa Purple Round and Round local were more tolerant than others (Keshwal and Khare, 1986).

Witches broom and leaf yellowing symptoms were observed on two weed species, *Cannabis sativa subsp. sativa* (Noida, Uttar Pradesh) and *Portulaca oleracea* (IARI, New Delhi) with *Hishimonas phycitis* as the major leaf hopper species present in the brinjal fields. They also concluded that there was an association of two phytoplasma subgroups, 16Sr VI-D and II-D with BLL strains infecting brinjal plants in eight states of India on the basis of sequence comparison, phylogeny and RFLP analysis of 16S rDNA sequences (Kumar *et al.*, 2017).

Pandey and Chaturvedi (2019) carried out effective management strategy to control the little leaf disease and observed that seedling treatment for 20-30 minutes with (Streptomycin sulphate + Tetracyclin hydrochloride @ 150 ppm) + installation of yellow sticky traps @ 15/ha after 20 days of transplanting + destruction of the infected plants + spraying of need based foliar application of Streptomycin sulphate + Tetracyclin hydrochloride @ 150 ppm and imidacloprid @ 0.3 ml/l for the management of vector not only increased the yield from 46.29 to 69.83 per cent but also increased the net return from 57.86 to 89.35 per cent.

8.1.7 Mosaic

Plants infected with this virus are stunted in growth and show mosaic symptoms of leaves.

Sharma (1969) isolated 5 distinct viruses from brinjal plants infected by a wide spread sap-transmissible mosaic disease from Pune, Maharashtra. The isolates

Diseases of Brinjal

Cercospora melongenae **Infestation**

Sclerotinia **Rot Infestation**

Verticillium **Wilt Infestation**

Fusarium **Wilt Infestation**

Bacterial Wilt Infestation

Diseases of Brinjal

Little Leaf (Left) and Mosaic (Right) Infestation

A: Brinjal plant severely infected by leaf blight disease, B: Leaf showing the characteristic blight symptoms, C and D: Different varieties of brinjal fruits severely infected by *Phomopsis vexans*, E: *P. vexans* growing on Potato Dextrose Agar (PDA) medium, F: Conidia of *P. vexans* (20 μm) under light α microscope (45X).

are: isolate LL (a distinct strain of TMV) called as brinjal mild mosaic, isolate 4 (a distinct strain of CMV) known as common brinjal mosaic, isolate SS known as brinjal severe mosaic, isolate 6 known as brinjal ring mosaic and isolate 17 known as brinjal crinkle mosaic. With the exception of *Solanum indicum* IW 599, which was resistant to isolates SS and 4, the 12 *Solanum* species tested were susceptible to the above 5 viruses.

Singh *et al.* (1973) tried to measure the intensity of mosaic disease in the susceptible cultivar Kalianpur T-3 using 5 grades (A-healthy to E-severely affected). Losses in yield in terms of number of fruits per plant were 11 to 25 per cent in grade B, 55 to 62 per cent in grade C and 100 per cent in grades D and E.

Mayee *et al.* (1973) were of the opinion that mosaic disease recorded for the first time in Punjab is sap-transmissible. The incidence in long fruited cultivars varied from no infection in Pusa Purple Cluster to 27.12 per cent in Pusa Kranti and the round fruited cultivars from 3.83 per cent in T-3 to 10.72 per cent in Pusa Purple Round. In Trinidad, brinjal mosaic disease was first reported by Ferguson (1951). Rana and Vovlas (1971) were of the opinion that a strain of cucumber mosaic virus caused the mosaic in brinjal. A viral disease of brinjal caused by broad bean wilt vims appears to be the first report of its occurrence in Japan (Morita *et al.*, 1974). Ladipo (1976) reported that a virus transmissible by mechanical inoculation but not by *Myzus persicae*, was isolated from brinjal plant showing green mosaic symptoms.

Plants infected with tobacco ring spot virus are stunted and exhibit chlorotic rings as well as mosaic symptoms; the fruits are small, deformed and show a pattern of concentric rings. The virus is not transmitted by aphid species but there is some seed transmission. Naqvi and Mahmood (1976) observed that inoculation of young plants of Pusa Purple Long cultivar with brinjal mosaic virus reduced water uptake by 17 and 33 per cent at 10 and 30 days after inoculation respectively, as compared to non-inoculated plants. Shoot and root growths were also greatly reduced.

Joshi *et al.* (1979) reported that cucumber mosaic virus infection markedly increased soluble nitrogen and protein contents but reduced carbohydrate, polyphenol and ascorbic acid contents in brinjal fruits.

The removal of all the virus affected plants, as soon as they are identified, has proved advantageous. Proper sanitary measures may be beneficial in controlling the disease. Mayee (1977) found that viral activity of brinjal seed was entirely lost after 7 months storage at room temperature without any appreciable reduction in germination. According to Bharathi (1999), the extract of *Mirabilis jalapa* leaves completely inhibited cucumber mosaic virus (CMV) of eggplant while the plant extracts of *Prosopis chilensis, Bougainvillea spectabilis* and *Eucalyptus citriodora* showed 83, 75 and 58 per cent inhibition of CMV, respectively.

Use of resistant or tolerant cultivars is the only control measure to be adopted. From screening tests, Patil and Mali (1977) found that 6 cultivars were susceptible and only one, Pusa Cluster was tolerant to brinjal mosaic virus.

Prominent yellow mosaic symptoms were observed in Central India on eggplant which confirmed the association of begomovirus with the newly emerging disease of eggplant. Based on the high nucleotide sequence identity (97.6 per cent for DNA-A and 94.1 per cent for DNA-B) with the strains of ToLCNDV, phylogenetic analysis, the begomovirus isolated from diseased eggplant was considered as a variant of ToLCNDV and the ToLCNDV-IN was suggested [IN: Nag: Egg: 09] (Pratap *et al.*, 2011).

8.1.8 Other Minor Diseases

In addition to these diseases, fruit rot, root rot, collar rot (*Diplodina lycopersici*), ring rot (*Myrothecium roridum*), corky root (*Pyrenochaeta lycopersici*), *etc.* become frequently a serious threat to the crop in a particular region.

8.1.9 Market Diseases

Market surveys of the major eggplant growing areas in West Bengal, India were conducted by Roy (1997) during 1994 and 1995. The percentage loss of fruits and percentage of incidence of pathogen causing fruit rot diseases were recorded. *Phytophthora nicotianae* var. *nicotianae* was the pre-dominant pathogen and other fungi commonly isolated were *Curvularia lunata* (*Cochliobolus lunatus*), *Fusarium oxysporum, Phomopsis vexans, Pythium aphanidermatum, P. debaryanum* and *Rhizopus stolonifer*. Maximum incidence of fruit rot occurred during the monsoon months of June to October when temperatures and relative humidity usually remained high.

8.2.0 Pests

Insect pest infestation is one of the most limiting factors for accelerating yield potential of brinjal. The crop is prone to damage by various insects, although there is wide variability in their degree of infestation. Some of the important ones are described below.

8.2.1 Shoot and Fruit Borer (*Leucinodes orbonalis*)

It is an insect which acts as a limiting factor for the successful cultivation of this crop, particularly under conditions of high temperature and humidity. Seasonal incidence of *L. orbonalis* in relation to weather parameters and planting date revealed that moderate range of temperature coupled with high humidity was found favourable for this pest. Crop planted during March to September recorded higher level of shoot and fruit damage than the crops planted during rest of the months. Plants sustaining higher shoot damage were more vulnerable to attack at the age of 7-10 weeks from the date of planting (Tripathy and Senapati, 1998). Koundinya *et al.* (2017) studied the association of the fruit and shoot borer (*Leucinodes orbonalis* Guen) infestation in eggplant with morphological, yield, quality and weather variables during spring summer and autumn–winter crop and found out that the number of infested shoots per plant, per cent shoot infestation, number of infested fruits per plant, per cent fruit infestation and larvae per fruit were greater during the spring–summer months than the autumn–winter months. Further, they also reported that genotypes with purple colour flowers and fruits with long slender fruit shape and clustered fruiting habit had greater EFSB infestation.

Shoot and fruit borer alone is a serious menace and has been reported causing a loss of upto 70 per cent in yield (Srinivasan, 2009). It is a short pinkish larva of the pest that initially attacks the terminal shoots and bores inside, as a result of which the shoots wither and fry. In the later stage, it also bores into the young fruits by making holes and feeds inside which makes the fruits unfit for human consumption. In severe cases, it also causes rotting of the fruits.

As continuous cropping of brinjal is congenial for the infestation, crop rotation is suggested. As soon as the insect is detected, the affected parts should be picked and destroyed. As it is difficult to control the borer after its entry into the shoot or fruit, the control measure should be taken at the appropriate time just after hatching out of the larvae from the eggs.

Roy *et al.* (1973) on the other hand, tried three insecticides, namely, parathion, malathion and dimethoate in order to control this pest in West Bengal. They found that spraying of parathion 50 EC (0.04 per cent) 6 times at fortnightly interval starting one week after transplanting of the crop was comparatively more effective in reducing the infestation of fruits over control. They have, however, suggested that a thorough study on the bionomics of the pest is required for its effective control.

Good control of shoot and fruit borer was obtained with fortnightly application of sevin or 50 per cent W.P. at 28 or 56 g/4.54 litres (Thevasagayam and Canagasingham, 1961). From a 4 year trial, David (1964) recommended that spraying carbary 10.1 per cent (sevin 454 g of 85 per cent W.P. in 382 litres of water) and dylox 0.1 per cent (dipterex 454 g 80 per cent W.P. in 360 litres of water) were found effective in controlling this pest. The average fruit yield and fruit size were also increased. The results in respect of carbaryl were in conformity with those reported by Jotwani and Swarup (1963), David (1963) and Mote (1976). Mote (1976) observed that carbaryl at 0.2 per cent was found to be most promising and gave significantly higher yield of brinjal than other insecticides tested in Maharashtra. Three sprayings of carbaryl (0.2 per cent) were given at an interval of two weeks, the first spraying at 75 days after transplanting (at fruit setting stage).

Krishnaiah *et al.* (1976) studied the efficacy of 18 different insecticides (applied as sprays 20 days after planting and thereafter 2 weeks interval during the fruiting stage) on the control of shoot and fruit borer of brinjal and found that carbaryl at 1 kg a.i./ha was most effective insecticide among the other insecticides tested. Gahukar and Bagal (1976) also confirmed that carbaryl at 0.25 per cent was the best chemical in controlling this insect. Shoot and fruit borer was successfully controlled by the soil application of carbofuran at 6.0 kg/ha (Nath and Chakraborty, 1978).

Mohan *et al.* (1980), on the other hand, studied the effectiveness of insecticides during the rainy season of 1978 and obtained the highest marketable yield of 10313 and 10287 kg/ha by the application of fenvalerate (sumicidin 20 E.C.) and permethrin (permethrin 10 E.C.) each at 0.1kg/ha, respectively compared to 4047 kg/ha yield in the untreated control. All the chemicals were sprayed 5 times at 2 weeks interval till the fruiting stage commencing from 60 days after transplanting. In respect of chemical control of brinjal shoot and fruit borer, Deshmukh and Rattan Lal (1969) reported that carbaryl residue on brinjal fruits required 2 weeks (May-June), 3 weeks (September-November) and 4 weeks (December-January) to come down to the tolerance level of 10 ppm (which is actually 5 ppm). More *et al.* (1987) studied eleven insecticides for control of *Leucinode sorbonalis* and found that 0.008

per cent flucythrinate was the most effective pyrethroid, while 0.15 per cent monitor (methamidophos) was the most effective organophosphate and 0.02 per cent oncol (benfuracarb) was most effective carbamate. Agnihotri *et al.* (1990) found that among the synthetic pyrethroid insecticides, cypermethrin (0.01 per cent) and deltamethrin (0.00125 per cent) were the most effective in controlling *Leucinodes orbonalis* in two cultivars Pusa Kranti and Pusa Purple Long. Tripathy *et al.* (1997) from the study on bioefficacy of few insecticidal schedules against *L.orbonalis* in brinjal along with their cost effectiveness at Bhubaneswar, India opined that the pest can be controlled by application of triazophos and carbaryl sprayed alternately @ 0.5 kg a.i./ha at 15 days interval with high cost effectiveness.

Oxymatrine (1.2 EC @ 0.2 per cent) at an early vegetative stage and spinosad (45 SC @ 225 g a.i./ha) at fruiting stage were found to be effective against brinjal shoot and fruit borer, with maximum per cent reduction of fruit damage in spinosad which was on par with oxymatrine (Adiroubane and Raghuraman, 2008). The lowest mean shoot and fruit borer infestation was recorded when the crops were sprayed with spinosad 2.5 SC (50 g ai/ha) followed by indoxacarb 14.5 SC 50 g ai/ha, emamectin benzoate 5 SG 15 g ai/ha (Sandip *et al.*, 2009). Tracer-45 SC (Spinosad) @ 0.4 ml/liter could be the most effective insecticide in controlling brinjal shoot and fruit borer and also in getting highest yield of eggplant (Gowrish *et al.*, 2015; Akter *et al.*, 2017).

Latif *et al.* (2010) studied the efficacy of nine insecticides such as azadirachtin 0.03 EC, abamectin 1.8 EC, flubendiamide 24 WG, chlorpyriphos 20 EC, cartap 50 SP, carbosulfan 20 EC, thiodicarb 75 WP, cypermethrin 10 EC, and lambdacyhalothrin 2.5 EC belonging to different chemical groups against eggplant shoot and fruit borer in laboratory and field. In laboratory trial, carbosulfan and flubendiamide showed the highest toxicity against fourth instar larvae of *L. orbonalis* after 24 and 48h of exposure, respectively and in field trials, they reduced more than 80 per cent shoot and fruit infestation in winter, and 80 per cent shoot and 70 per cent fruit infestation in summer over control. Carbosulfan protected the highest amount of healthy fruit yield in both cropping seasons with flubendiamide showing similar efficacy. Thus, they concluded that carbosulfan and flubendiamide may be used for the control of *L. orbonalis* in eggplant. Lowest shoot and fruit infestation and highest fruit yield were obtained in plots treated with Flubendiamide 20 per cent WDG and Rynaxypyr (Chlorantraniliprole) 18.5 per cent SC followed by Emamectin benzoate 3.5 per cent SC (Biswas *et al.*, 2020).

The effect of five exposure periods to natural infestation of *L. orbonalis* and subsequently to six, five, four or three sprays with deltamethrin (20 g a.i./ha) as well as four and three sprays of neem oil 59 and 75 days after transplanting (DAT) were studied by Sharma and Chhibber (1999). The results indicated that exposure in the early crop growth stages resulted in more initial shoot damage and subsequent fruit damage. Six sprays from 30 DAT to 105 DAT with deltamethrin treated plots with minimum shoot damage also had the minimum fruit damage with 8.40 and 10.94 per cent, respectively, and were equally effective. Four or three sprays of neem oil after 59 and 75 DAT exposures were comparable with delta methrin and endosulfan under comparable exposure and number of sprays. However, six sprays with delta methrin were the most economical treatment. Neem oil was the least economical treatment. Yield of healthy fruits, however, was higher (264.46 per cent) than in endosulfan treated plots (233.59 per cent), but was statistically *at par*. Bhushan *et al.*(2011) evaluated the efficacy of pest management modules against EFSB and revealed that minimum shoot and fruit damage (9.32 and 14.83 per cent, respectively) was observed in the module having shoot clipping with alternate spraying of Multineem (1500 ppm) and Triazophos (35 per cent) plus deltamethrin (1 per cent) giving a maximum yield of 210.5 q/ha. Bio-intensive pest management (BIPM) utilizing the components of egg parasitoid *T. chilonis,* use of the bacterial bio-pesticide, *Bacillus thuringiensis* and use of pheromone trap with Lucinlure along with use of neemcake and removal of wilted twigs has been found successful in the reduction of fruit and shoot borer (Biswas *et al.,* 2020).

Now-a-days an increased emphasis is being given to develop resistant cultivars which seems to be the only solution to reduce the hazards caused by the use of toxic chemicals. Kashyap and Kalloo (1983) have reviewed the different works of insect resistance in vegetable crops. Several accessions of *Solanum melongena* have been screened against this insect pest and Pusa Purple Long (Singh and Sikka, 1955; Gill and Chadha, 1979); H-128 and H-129 (Srinivasan and Basheer, 1961); Aushey (Lal and Ahmed, 1965; Dhankhar *et al.,* 1977); Thorn Pendy, H-165 and H-407 (Padda *et al.,* 1971); SM202, Sel.519, Sci.520, Sel. 521 and Solan-ll (Lal*et al.,* 1976); PPC-2 (Dhankhar *et al.,* 1977); Dorley (Nawale and Sonone, 1977); Long Purple (Frempong and Buahin, 1977); PPC-17-4 and PVR-195 (Singh, 1981) have been reported to be tolerant or resistant. Singh and Sidhu (1988) reported that cultivars FBR 129-5 and SM17-4 were resistant to *Leucinodes orbonalis.*

At IARI, New Delhi, Subbaratnam and Butani (1981) reported that cultivars H-4, Pusa Kranti and A-61 were moderately tolerant to shoot and fruit borer, while Arka Kusumakar was tolerant to this insect pest. Hence they have suggested that these cultivars might be incorporated in the pest management programme. Similar results have been obtained by Raut and Sonone (1980) at Rahuri, Maharashtra. Out of 13 cultivars screened against this insect pest, they have reported that cultivars H-4, Pusa Purple Long, Pusa Kranti and SM-41 exhibited tolerance to fruit as well as shoot infestation. Rameshkumar *et al.* (2019) evaluated brinjal genotypes with resistance against BFSB based on biophysical and biochemical basis of resistance and observed that yield per plant, solasodine and phenol showed significantly negative correlation with fruit infestation and positive correlation with shoot infestation, polyphenol oxidase and total sugars. Therefore, the crosses Manaparai local x Spiny local, Sevathampatti local x Spiny local and Sevathampatti local x Seetipulam local recording maximum yield, less shoot and fruit infestation, lowest calyx length, low sugar, high polyphenol oxidase and high phenol could be used as resistant cultivar for further shoot and fruit borer resistance breeding programme. From a study on screening of eggplant cultivars in relation to biochemical characters for resistance to fruit and shoot borer, Darekar *et al.* (1991) reported that total sugars and free amino acids were positively correlated with fruits infestation, whereas polyphenol content was negatively correlated with attack. Dar *et al.* (2017) from their studies revealed that correlation coefficients of shoot and fruit infestation were strong and negative with total phenols and positive and significant with respect to total sugars. TSS and anthocyanin in peel were found as the phagostimulants, while the high total phenols content and 2,2-diphenyl-1-picrylhydrazyl (DPPH) free radical scavenging capacity were the feeding deterrents for EFSB (Koundinya *et al.*, 2017).

Alam *et al.* (1994) conducted a study to compare infestation by *Leucinodes orbonalis* of eggplant between non-grafted plants and plants grafted on wild *Solanum*. The lowest number of eggplant shoots and *L. orbonalis* infested fruits were recorded from plants grafted on wild *Solanum* as rootstock. Considering the positive performance, grafting technology can be accepted as a component of integrated pest management (IPM) package for eggplant.

Islam *et al.* (2019) measured the effects of the range of temperatures on the growth of EFSB and results indicate that the temperature significantly reduced the development and adult total lifespan of ESFB ($P < 0.05$). They also mentioned that the EFSB population increased with increase in temeperature. The per cent of infested fruits per plant exhibited a positive correlation with maximum, average and day temperatures and larvae per fruit had a positive association with maximum temperature, day temperature and helio-thermal units (Koundinya *et al.*, 2019). Javed *et al.* (2017) carried out experiments for management of EFSB by integration of non-chemical approaches to reduce infestation and found out that combination of practices (*T. chilonis* + hoeing + clipping) gave better results as compared to individual treatments in reducing pest infestations and increasing the yield followed by another combined treatment where hoeing, clipping and weeding were applied conjointly. Set up sex pheromone trap (Leucinlure) @ 12/ha to reduce the adult population of fruit borers (Bose *et al.*, 2020).

8.2.2 Epilachna Beetle (*Epilachna vigintioctopunctata*)

It is a polyphagous insect that feeds on the leaves of potato, brinjal, tomato, *etc.*, by scraping in a characteristic manner leaving the viens intact. The grubs are yellowish in colour and stout-bodied with stiff hairs on their bodies. The beetle is bronze to red, small spherical and mottled with black spots. In brinjal, the adult and grub feed voraciously on the leaves and tender parts of the plant and often cause serious damage when they appear in numbers. As a result of the feeding by both the grubs and adults on green chlorophyll, characteristic skeletonized patches which present lace-like appearance on leaves, are developed and later dry away.

Hand picking of grubs, adults and eggs along with infested leaves during the early stages of attack and destroying them are suggested to reduce the infestation.

Sandhu *et al.* (1973) evaluated the efficacy of 14 insecticides on the control of *E. vigintioctopunctata* in brinjal plants and reported that all of them gave over 90 per cent control of the pest and for both safety and effectiveness, carbaryl, and dichlorovos (all as 0.1 per cent sprays) can be recommended.

Treating susceptible eggplants with the steam distillate extract of wild *Solanum aethiopicum* decreased the setting of both adults and third-instar larvae of *Henosepilachna vigintioctopunctata* (*Epilachna vigintioctopunctata*). The steam distillate extracts reduced egg hatchability also (Rajendran, 1999). Among 42 brinjal selections and cultivars tested, Arka Shirish, Hissar Sel.1-4 and Shankar Vijar were found resistant to the insect (Raj and Kumaraswami, 1979).

Carbaryl 50 WP (0.10 per cent) proved to be the most effective against the hadda beetle followed by fipronil 5 SC and lamda cyhalothrin 5SC. *Beauveria bassiana* 1.25WP, *Steinernema carpocapsae* @ 2 billion IJs/ha and NSKE 5 per cent also significantly suppressed the pest population but had low cost benefit ratio (Jamwal, 2011). Quinalphos 1.5 D @ 25 kg/ha or chlorpyriphos 20 EC @ 2.5 ml/lit has been recommended for spraying (Bose *et al.*, 2020).

Natural enemies *viz.*, *Tetrastichus* sp. and *P. foveolatus* contribute to significant levels of immature stage mortality of Hadda beetle and are potentially important in pest management in commercial brinjal cultivation showing a large potential in IPM (Jamwal *et al.*, 2014).

Azadirachtin 1 per cent @ 2 ml/litre was found to be most effective in controlling leaf infestation by *E. vigintioctopunctata* (Gowris *et al., 2015*). Flower extract of *Eucalyptus globulus* at 0.5 per cent and seed extract of *Nerium indicum* at 0.2 per cent increased the total developmental period to 22.5 ± 1.5 and 22.7 ± 0.6 days respectively as compared to control in which larvae took 20.2 days to complete their development (Sharma and Saxena, 2016).

8.2.3 Jassids [*Amrasca biguttula biguttula, Cestius (Hishimonus) phycitis*]

Both nymphs and adults suck the sap from the lower surface of the leaves. The infested leaves curl upwards along the margins which may turn yellowish and show burnt up patches. They also transmit mycoplasma disease like little leaf and virus disease like mosaic. Fruit setting is adversely affected by the infestation.

Anand *et al.* (2013) studied the efficacy of six new insecticides against leaf hopper and found out that thiamethoxam 25 WG @ 35 g a.i. ha^{-1} recorded the lowest leaf hopper population which was followed by acetamiprid (20 SP @ 25 g a.i./ha), pymetrozine (50 WDG @ 120 g a.i./ha) and buprofezin (25 SC @ 25 g a.i./ha). The studies on bio-efficacy of leaf hopper population showed promising results, from the first spray it can be revealed that treatment with fenpropathrin 30 EC at 200 ml a.i./ha was found most effective while during the second spray the highest population reduction of leaf hopper was obtained in imidacloprid 70 WG at 35g a.i./ha treatment (Jadhav *et al.*, 2017).

Mitra and Majumdar (1977a) recommended the pre-transplant brinjal seedling dipping in 0.2 per cent carbofuran 75 per cent W.P. for 24 hours. They also observed that when the vectors were released onto the plants 15 days after dipping, they all died within 6 hours, when released 25 days after dipping they died within 48 hours and when released 30 days after dipping they died within 72 hours.

Leaves of brinjal infected with little leaf disease are preferred by *Cestius phycitis* rather than healthy leaves (Srinivasan and Caelliah, 1978).

Phorate granules at 1.0 kg/ha applied 15 days after transplanting, followed by 3 fortnightly sprays of carbaryl (0.2 per cent) or quinalphos (0.5 per cent) after fruit set gave effective control of jassids in Maharashtra, India (Mote, 1978a).

Mote (1978b) further recorded the lowest number of jassid nymph incidence per 30 leaves in the brinjal cultivars, namely, H-4, Round Green, Dorly, Aushey, Jumblimulayam and Long Purple indicating moderately resistant to jassids. Bindra and Mahal (1981) observed that Junagadh Sel.l, Aushey, R-34, H-4 and T-3 were resistant to *Amrasca biguttula biguttula*. They also studied that various morphological characteristics contributed to resistance, including long hairs and a high density of hairs on the midrib and lamina, more erect hairs on the midrib and a thin leaf lamina. Pusa Purple Long had the minimum population of nymphs and adults of jassids (Bilal *et al.*, 2017).

Varietal reaction of eggplant against jassid in field and morphological and biochemical resistance traits of each variety were also assessed in laboratory and green house by Gaikwad *et al.* (1991). The cultivars, Manjari Gota and Vaishali were resistant to jassid. Leaf thickness, midrib thickness and leaf area were positively correlated with infestation levels of *A. Biguttula biguttula*. Trichome length and density on leaves were negatively correlated with the infestation level. The total level of sugars, free amino acids and polyphenols in the leaves of eggplant was negatively correlated with the number of infesting *A. biguttula biguttula*. Egg parasites, namely, *Anagus empoascae, Aresconenocki, Stethynium empoascae, Erythmelus empoascae, Lymaenon empoascae* and *Oligosita* species have also been recorded in nature keeping the pest population under check (Butani and Varma, 1976).

8.2.4 Red Spider Mite (*Tetranychus neocaledonicus, Tetranychus cinnabarinus, Paratetranyehus indicus*)

These are small, polyphagous insects found in large colonies on underside of leaves covered with fine silkywebs. Protonymph is the most active feeding stage. As a result of their feeding, white specks appear on the leaves and these specks coalesce and appear as white patches. Ultimately, the entire affected leaf becomes discoloured and withers. The damage is more pronounced during warm and dry season. In severe cases, it may be found all over the leaf surface. Due to the removal of plant sap with chlorophyll and other plant pigments during feeding, the characteristic blotching of leaves is observed.

Carbophenothion, dioxathion, wettable sulphur, oxythioquinox and phosalone were equally effective in reducing the incidence of red spider mite over a 15 days period (Babu *et al.*, 1970).

Application of carbaryl + molasses, dicrotophos and monocrotophos effectively controlled mites and these were associated with lower total and reducing sugar contents and higher phenolics and K contents in the fruits (Uthamasamy *et al.*, 1976). Field efficacy of various insecticides against brinjal mite (*Tetranychus macfarlanei*) was tested and sevisulf (a mixture of carbaryl and sulphur) at 0.1 per cent was the most effective treatment giving 86.85 per cent mortality in 24 h after spraying and 63.98 per cent mortality after 15 days (Kumar *et al.*, 1990).

Seno *et al.*(1976) in Brazil reported that of the acaricides tested, AC85.258 (N-methyl carbamate) at 0.50 or 0.75 kg a.i./ha gave the best control of *Tetranychus uriticae* with a persistence of 15 days. In India, tarsonemid mite (*Hemitarson emuslatus*) was first noticed in brinjal in Punjab by Sandhu *et al.* (1974). High populations were observed on the leaves but no injury was reported except for barely visible brownish spots. When infested plants were sprayed with 9 pesticides at rates recommended for the control of other mites, the overall control achieved was only marginal.

Ghosh and Chakraborty (2014) studied the bio-efficacy of plant extracts against red spider mite by incorporating one botanical insecticide azadirachtin *i.e.* neem (neemactin 0.15 EC) @ 2.5 ml/L, and four botanical extracts, tulsi (*Ocimum tenuiflorum*) leaf extract @ 5.0 per cent, *Polygonum hydropiper* floral part extract @ 10.0 per cent, *Pongamia pinnata* fruit extracts @ 10.0 per cent, and garlic (*Allium sativum*) extract @ 5.0 per cent, and one treatment containing mixture of neem and dicofol (@ 2.5 ml + 1 ml/L and these were further evaluated and compared with the ability of Dicofol, a chemical insecticide (Kelthane 18.50 EC) @ 3ml/L to control the mite pest, *Tetranychus sp.* Dicofol treatment resulted in the best suppression of mite population (83.16 per cent suppression), closely followed by mixed formulation of botanical pesticide, azadirachtin and chemical pesticide, dicofol (71.41 per cent suppression). However, among the bio-pesticides including plant extracts *Polygonum* flower extract (at 10 per cent concentration) was the most effective providing moderate to higher control (51.04 per cent suppression).

A study was conducted by Masal *et al.* (2018) on effect of newer neonicotinoids on red spider mite infesting brinjal at Parbhani, Maharashtra and results indicated that dimethoate 30 EC (200 g ai/ha) which is the traditional organophosphate was found more effective against mites but the newer neonicotinoids including imidacloprid, thiamethoxam, acetamiprid, clothianidin, flonicamid, dinotefuran treatments increased mite population even more than untreated plots.

Bose *et al.* (2020) also recommended the spraying of either fenazaquin 10 EC @ 2 ml/lit., propargite 57 EC @ 2 ml/lit, spiromesifen 22.9 SC 0.8 ml/lit, dicofol 18.5 EC @ 2 ml/lit for the control of red spider mite.

8.2.5 Leaf Roller (*Eublemma olivacea*)

Carterpillars roll leaves and feed on chlorophyll while remaining inside the folds and thus lead a concealed life and as a result, the folded leaves wither and dry up. The caterpillars may also bore into the plant to wither.

Bt formulation @ 1 ml/litre was highly effective for early larval instars and for later instars, malathion 50 EC was found to be effective (Gowrish *et al.*, 2015).

8.2.6 Stem Borer (*Euzophera perticella*)

It is an oligophagous insect pest found mostly in the Indian subcontinent (Halder *et al.*, 2017). Pale white caterpillars attack stems and often kill the young plants. Due to the attack on the older plants, the infested plants become light yellow in colour, wither, growth is stunted and fruit yield is adversely affected. Halder *et al.* (2017) also assessed the damage caused by the borer and found out that it was more prominent in near maturing brinjal crop, ratoon brinjal as well as grafted brinjal. Occurrence of this pest was more during the summer months and less as the temperature decreases. Furthermore, they made a first comprehensive report of *P. euzopherae* as an endoparasitoid of brinjal stem borer, *E. perticella* which has become a serious pest in near maturing and ratoon brinjal in and around Varanasi.

Javed *et al.* (2017) studied correlation between the borer infeststion and the morphological characters and recorded positive significant correlation with stem diameter, number of branches per plant and plant height while negative significant correlation with number of hair/cm^2. It was observed that with an increase in plant height, stem diameter and number of branches per plant, there was a significant increase in infestation. On the other hand the cultivars with maximum number of hair/cm^2 showed a decrease in infestation and increased yield.

Removal and destruction of the affected plant or plant parts helps to reduce the infestation.

The crop may be sprayed with 0.03 per cent diazinon, 0.1 per cent lindane or 0.1 per cent carbaryl. The fruits should be picked before spraying and after spraying the fruits should not be harvested for the next 6-7 days.

8.2.7 Mealy Bug (*Centrococcus insolitus*)

The nymphs and adult females remain in large number on the undersurface of leaves and suck the sap from the plants making them pale and weak. As a result of their incidence, the plants wilt and dry. The insect is abundant during the rainy season. Mealy bugs attack the brinjal leaves only in the kharif crop and their abundance is more between 9 to 12 weeks old crop (Saha *et al.*, 2020).

Spraying of any systemic insecticide may afford some control. Banerjee and Basu (1956) in West Bengal reported that kerosene oil emulsion which was relatively more effective, burned the foliage. They also observed that in the field condition 0.1 per cent basudin gave 68.9 per cent control and in laboratory tests 0.2 per cent was slightly better. More than one application was needed to give adequate control. In West Bengal, Chatterjee (1954), on the other hand, worked on biological control of mealy bug and observed that *Hyperaspis maindroni* might provide satisfactory control of brinjal mealy bug if no insecticide was applied.

8.2.8 Cotton Aphid (*Aphis gossypii, Myzus persicae*)

It is a polyphagus species infesting a large number of plants throughout India. The aphid is small, soft, yellowish green or greenish brown and is found in colonies of hundreds on the tender shoot and the under surface of tender leaves sucking the cell sap and also found to colonize on flowers. Both nymphs and adults suck the sap of the leaves. In case of severe attack, the affected leaves curl, fade gradually and finally dry up. Black sooty mould develops on the honey dew excreted by the aphid which falls on the leaves and adversely affected the photosynthesis resulting the arrest of growth and reduction in fruit size. The occurrence of aphids is more during the peak flowering stage of the crop with 50 per cent population when brinjal is grown during the rabi season (Saha *et al.*, 2020).

Menezes *et al.* (1976) reported that among the numerous insecticides tested, methamidophosat 0.04 to 0.08 ml/l gave the best control (up to 100 per cent) of the green peach aphid on brinjal, followed by acephate at 0.075 to 1.0 g/l (upto 99.3 per cent) and methoate at 0.1 to 1.0 ml/l (upto 95.2 per cent). Spraying with 40 per cent nicotine sulphate (1: 600) gave a good check to this pest (Butani and Varma, 1976). Two sprays of biorational insecticides *viz.*, spinosad 45 SC @ 0.2 ml/l recorded significantly lower population (7.94 aphids/three leaves) followed by dimethoate 30 EC @ 1.6 ml/l (11.79 aphids/three leaves) in eggplant (Rajeshwari *et al.*, 2019).

Pests of Brinjal

Hadda Beetle Infestation (Left and Middle) and Adult Beetle (Right)

Jassid (Left) and Red Spider Mite (Right) Infestation

Shoot Borer Infestation

Fruit Borer Infestation

Pests of Brinjal

Stem Borer Infestation

Larvae (Left), Adult (Middle) and Symptom of Infestation (Right) of Leaf Roller

Mealybug (Left) and Aphid (Right) Infestation

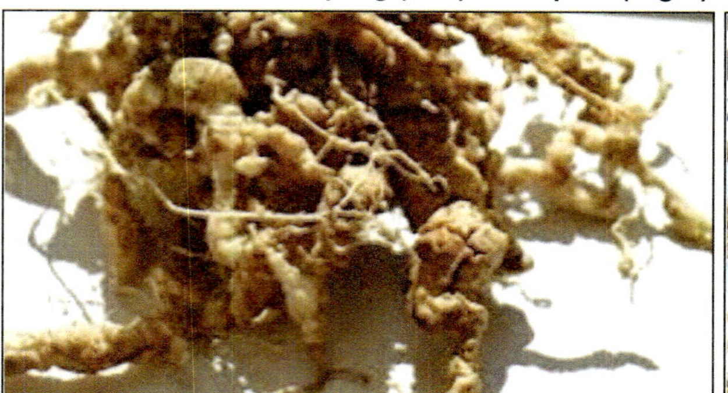

Nematode Infestation

Coccinella transversalis was considered as the major predator of aphids in brinjal (Borkakhati *et al.*, 2019). The ladybird beetles (*Menochilus sp.* and *Coccinella* sp.) and green lacewings are efficient predators of aphids and release of ladybird beetles @ 200 pairs per ha at fortnightly intervals can suppress the aphid population (Padwal *et al.*, 2016).

Ramanujam *et al.* (2018) studied the field efficacy of entomopathogenic fungi on brinjal and results revealed that the Ma-4 isolate of *Metarhizium anisopliae* suppressed the aphid population with a reduction of 90.79 per cent followed by Bb5a isolate of *Beauveria bassiana* with 83.69 per cent reduction over control in brinjal. They further concluded that four rounds of application of these isolates at 10 days interval, starting from 40 days after transplanting showed superior suppression of *A. gossypii* in brinjal.

Spraying betacyfluthrin 8.49 per cent + imidacloprid 19.81 per cent OD @ 70-80 ml in 200 l of water/acre or fenvalerate 20 per cent EC @ 150-200 ml in 240-320 l of water/acre or phorate 10 per cent CG @ 6000 g/acre or phosphamidon 40 per cent SL @ 250-300 ml in 300 l of water/acre or deltamethrin 1 per cent + triazophos 35 per cent EC @ 400-500 ml in 200 l of water/acre were found to be effective for control of aphids (Satyagopal *et al.*, 2014).

8.2.9 Bud Worm (*Pthorimoea blapsigona, Scorbipalpa blapsigona*)

A minor pest of brinjal is more common in South India. Pale white or pinkish small caterpillars of a small brown moth, bore into and feed on flowers and buds which may drop down in large number thus affecting the yield adversely.

If and when the pests appear, spraying with 0.05 per cent dichlorovos (DDVP) may control this insect. The braconid *Chelonus* (*Microchelonus*) sp. (*blackburni* sp.) was an ectoparasite of S. *blapsigona* larvae (Duodu, 1985). Bud worm can also be controlled by spraying of neem oil 2ml/litre (Satyagopal *et al.*, 2014).

8.2.10 Brinjal Lace Wing Bug (*Urentius echinus, Urentius sentis*)

This is a specific pest of brinjal. Nymphs and dark brown bugs with lace-like wings suck the sap from leaves which tum yellowish and are found covered with insect excreta. Affected leaves ultimately dry up. The summer crop suffers comparatively more. Bhandari and Sohi (1962) reported that in Punjab 50 percent plants were attacked by *U. sentis*.

Padwal *et al.* (2016) recommended the spraying of dimethoate 30 EC @ 1 lit/ha. Apply phorate 10 per cent CG @ 6000 g/acre (Satyagopal *et al.*, 2014).

8.2.11 Termites (*Trinervitermes biformis, Microtermes* spp.)

Termites or white ants are cosmopolitan pests having a very vast range of host plants and brinjal is no exception. These are abundant in sandy soil and do not thrive under conditions of heavy soil. Its infestations start soon after germination and may continue till harvest of the crop. The pest being subterranean, lives at the roots and stem below the ground level and tunnels upwards. The affected plants tum pale, wither and dry away.

Soil application with 5 per cent chlordane or heptachlorattherate of 22 kg per hectare is very effective in minimizing the attack of termites. In sandy or termite infested soils, the growing of crop should be avoided. Soil application of Aldrin is also very effective in controlling the termites.

Locating the termitarium, digging out the queen and destroying it are the only permanent remedy. The termite hill can also be fumigated with carbon disulphide or chloroform mixture. The crop residues should be destroyed as they are the source of infestation (Satyagopal, 2014).

8.2.12 Other Minor Insect Pests

Selepa docilis, a minor pest of brinjal was first observed from the semi arid region of Rajasthan (Pande and Yadava, 1979). Earlier, this was reported by Cherian and Pillai (1942) and Wishwanath *et al.* (1973) from other parts of the country. Azadirachtin 1 per cent @ 2 ml/litre was found to be highly effective against *S. docilis* during summer and kharif.

8.2.13 Impact of Insecticide Application on Non-Targets and Environment

From a study on the persistence kinetics of fenvalerate and deltamethrin in and on eggplant, Raha *et al.* (1993) opined that residues of fenvalerate in and on fruits were below the tolerance level of 2 ppm as specified by FAO/WHO immediately after application of the recommended rate, whereas residues of deltamethrin required 4.32 days to fall below the recommended tolerance level at same application rate. Waiting periods of 6.50 and 0.14 days were calculated for deltamethrin and fenvalerate, respectively, when double the recommended rate was applied. Approximately 50 to 70 per cent of the remaining insecticide on eggplant was removed during washing followed by cooking.

Based on produce efficacy and savings, farmers in the Philippines have switched to deep, point placement of carbofuran granules (44 days after transplanting) rather than their customary weekly foliar sprays for control of a number of insect pests like *Aphis gossypii, Amarasca biguttula biguttula, Bemisia tabaci, Ferrisia virgata, Luperomorpha serricornis* and *Epilachna vigintioctopunctata* in eggplant upto 129 days after placement of the insecticide (Litsinger and Apostol, 1994).

Eggplant in an experimental plot in Saitama, Japan in 1992 was continuously sprayed with the insecticides permethrin, milbemectin, phenthoate and imadacloprid to evaluate their side-effects. Effects on populations of pests and their natural enemies were assessed. The results revealed the importance of natural enemies such as anthocirids, *Orius* spp. Application of a pesticide may cause a resurgence of a pest population because of (i) the development of resistance by pest and (ii) non selective killing of predators. Imidacloprid, which was highly effective against *Hemiptera* and *Thrip palmi* caused a resurgence of *Tetranychus kanzawai*. Milbemectin, which had a minimal adverse effects on *Orius* spp. when used in combination with imidacloprid, maximized the latter's advantage while minimizing its disadvantage (Nemoto, 1995). The biological-integrated control strategy although was more expensive to administer than the chemical control strategy, there were advantages in that a decrease in the waiting period before harvest led to considerably reduced losses of production, and there were no problems for the environment and for the health of the workers or consumers. To maximize the use of biological pest control and minimize pesticide input, Bennison *et al.* (1996) conducted an experiment using IPM in glass house at Howard Davis Farm, Jersey, British Isles. They observed that *Trialeurodes vaporariorum* and *Thrips tabaci* were well controlled by *Encarsia formosa* and *Amblyseius cucumeris* (*Neoseiulus cucumeris*), respectively. *Tetranychu surticae* and *T. cinnabarinus* were well controlled, particularly by *Therodiplosis persicae* (*Feltiella acarigusa*), with a contribution by *Phytoseiulus persimilis* and one spot treatment with fenbutainoxide. Aphid species were well controlled by a combination of *Aphelinus obdominalis*, *Aphidius colemani* and *Aphidoletes aphidimyza*, with only one spot treatment of nicotine used to reduce damage by *Aulacorthum solani*. With the aim of finding management alternatives for *Thrips palmi*, Pinon *et al.* (1999) evaluated different commercial products in Cuba and the study revealed that yields were greatest with Melitox (natural insecticidal extract of *Melia azedarach* seeds) and sunflower oil. Furthermore, large number of natural enemies of the pest was observed in these plots. With the synthetic insecticides, the population were reduced after the first application but at the end of the experiment it was greater than in the other treatments.

Brinjal is one of the commercially grown crop in the country therefore, repeated occurrence of different major insect pests pose a serious threat to human kind as well as adversely affect the environment due to the tremendous application of chemical insecticides to combat the insect pests. Further, continuous use of chemical pesticides results in insecticide resistance, pest resurgence and pesticide residue. The frequent use of pesticides also increases the cost of production making eggplant expensive for the poor consumers. Therefore, an integrated pest management (IPM) technique could be adopted for pest management, to provide satisfactory control, keeping in mind that it should be simple and economical. An Integrated Pest Management strategy has been prepared by the Directorate of Plant Protection, Quarantine and storage for the management of various brinjal pests. This includes the use of resistant cultivars, sex pheromone trap for shoot and fruit borer, cultural, mechanical and biological control methods for manangement and control of insect pest of brinjal. Conserving the existing bio-control agents like Spiders, Coccinellids, Syrphid flies *etc.* in the field by avoiding, delaying and reducing the use of chemical pesticides, promoting the use of bio-pesticides including botanicals and microbial, and increasing the use of bio-control agents like egg parasitoids *Trichogramma* sp., *Telenomus sp., Encarsia spp.*; larval parasitoid *Bracon* sp., *Campcletis chlorideae, Chelonus blackburni*; predators like *Chrysopa* sp., *Coccinella* sp. are some of the management strategies to be considered. Spraying of the chemical pesticides can be strictly followed as per the recommendation of CIB and RC (www.cibrc.gov.in) as a last resort.

8.2.14 Nematodes

8.2.14.1 Root-knot Nematodes (Meloidogyne spp.)

These are the most common plant parasitic nematodes in India and their infestation is easily recognized by the characteristic root galls they cause. Brinjal is highly susceptible to the nematode. Due to the attack of the nematode, the plants become stunted and leaves show chlorotic symptoms. Fruiting is adversely affected.

Alam (1975) observed that when pot-grown brinjal plants are inoculated with 0-5000 *Meloidogyne incognita* larva and after 45 days they are uprooted, washed and placed in water containing flasks for 6 hours per day, the water uptake by the plants is reduced by the infection. The reduction in the water uptake is due to the decrease in root weight.

Hamada *et al.* (2000) reported soaking of seeds in ascorbic acid, pyridoxine and thiamine at both 50 and 100 ppm had a significantly negative effect on nematode reproduction. However, soaking seeds in thiamine treatment was highly effective in inhibiting nematode infection than the two other vitamins.

Masood and Husain (1975) at Aligarh University, studied the relationship between plant growth and root-knot infection due to the application of 22 water soluble amino acids as a soil drench or root dipping for 4 to 12 hours either before or after inoculation of *Meloidogyne incognita*. With soil drenches, plant growth is best with phenyl alanine and threonine when applied pre-inoculation. The intensity of infection is aggravated, however, by tryptophane and glutamic acid applied pre-inoculation and by serine, isoleucinc and tryptophane applied post-inoculation.

With root dip treatment prior to inoculation, development of root-knot infection is least with aspartic acid and valine, but it is increased by tryptophane and several other amino acids. Differences in treatment duration have little effect on plant growth, but nematode population usually declines with increasing duration.

Increased root total soluble phenols, free amino acids total proteins and *o*-dihydroxy phenols and unchanged total carbohydrates were noticed due to the inoculation with *M. incognito* in brinjal plants (Alam *et al.*, 1976).

Proper crop rotation with crop resistant to root-knot nematodes, like marigold *etc.* will help in the reduction of nematode population (Choudhury, 1976a). Different cropping sequences of brinjal with other crops, including capsicum, garlic, beet root and turnip have a relationship on the nematode population in the soil. *M. incognito* population rises with all cropping sequences, especially when preceded by capsicum. When brinjal is followed by beet root, they are further increased, but when followed by garlic they are greatly reduced (Khan *et al.*, 1976).

According to Verma *et al.* (1978), the yield of brinjal in a nematode infested soil in Rajasthan increased significantly over control when *Tagetes patula* or *Sesamum orientale* were grown within or between the brinjal rows. The yield also increased when the top soil to a depth of 15 cm was treated with aldicarb at 1.0 kg/ha. The root-knot index of *M. incognito, M. javanica* and *M. incognito var.acrita* was reduced by aldicarb or *Tagetes patula* or *Sesamum orientale* interplanting. Goyal *et al.* (1976) obtained the best result in relation to the control of *M. incognito* and *M. hapla* by the application of hexanema (dichlofenthion) at 45 litres/acre. The next best result obtained by them was the planting of 6-weeks-old *Tagetes erecta* seedlings in alternate rows of brinjal plants. Following and deep ploughing 2-3 times are also beneficial in minimizing the nematode population.

Patel and Dhillon (2017) evaluated sunhemp as a green manure amendment and its biomass content to RKN to reduce the root-knot nematode population in soil as well as in successive susceptible crop brinjal. They found that for reduction of nematode infestation in soil and roots of successive crop brinjal incorporation of sunhemp biomass @ 75 g/kg was effective. The reduction in nematode population may be due to production of pyrroloizidine alkaloids and monocrotaline by sunnhemp which possess high vertebrate toxicity and are toxic to nematodes.

Azotobacter chroococcum inhibited the hatching of egg masses of *M. incognito* and did not allow the larvae to penetrate into the roots of brinjal to form galls (Chahab and Chahal, 1988). Reduced number of galls and egg masses of *M. incognito* were noted when seeds and seedlings were treated with 50 and/or 100mg/ml water solutions of thiamine, cystin, tryptophan, tyrosin, aspartic acid and phenylalanine, or the plant growth regulators gibberellic acid, indol butyric acid and indol-3-acetic acid. The higher concentration of the tested substances was more effective in reducing the nematode infestation (Ibrahim *et al.*, 1999).

In experiments at 4 sites over 3 years at Cyprus Agricultural Research Institute, Nicosia conducted by Philis (1971), DD (dichloropropane-dichloropropene) gave better nematode control than DBCP (dibromochloropropane) which was slightly phytotoxic and increased yields though not significantly. The results were better when the fumigant was injected than when it was applied in furrows and the best rates were 45.2 and 52.2 l/donum. From pot experiments and field trials at Ludhiana, Punjab, Bindra and Soodan (1974) concluded that the most effective treatments against *M. incognita* in decreasing order of effectiveness were DD mixture at 280.82 litres, DBCP at 26.92 litres, phorate at 4.92 kg, DD mixture at 224.66 litres and DBCP at 20.19litres per hectare. Soil injection of 249 l/ha of ethylene dibromide one month before transplanting controlled root-knot nematode in cv. Black Beauty and increased yield by an average of 41 per cent in Italy (Vito and Lamberti, 1980). Other soil fumigants like DD and di-trapex (DD + methylisothiocyanate) and the granular nematicides aldicarb, carbofuran, DBCP and fenamiphos applied 14 to 60 days before transplanting were less effective. Rao *et al.* (1987) found that bare-root dip of brinjal seedlings with BPMC (fenobucarb), carbosulfan, chlorpyriphos, dimethoate and phosphamidon at 500 and 1000 ppm for 30 minutes significantly increased plant growth and reduced host infestation as well as multiplication of *M. incognita*. According to Stephan *et al.* (1989) fenamiphos 40 per cent EC at 5 ml/m^2, miral (isazofos) 10 per cent G at 25 g/m^2 and carborfuran 3 per cent G at 15g/m^2 were effective in controlling *M. javanica* on eggplant. The combined infection of *M. incognita* and *Fusarium solani* in brinjal was best controlled by carbofuran at 2 kg a.i./ha+ bavistin at 2000 ppm (Deol *et al.*, 1989).

Kaur and Kaur (2016) evaluated different chemicals for RKN management in brinjal nursery and results indicated that maximum reduction in RKN infestation was observed in treatment with application of Basamid @ 60 g/m^2 followed by Basamid @ 40 g/m^2.

Li *et al.* (2020) studied the efficacy of floupyram on controlling the RKN and the effects it had on the soil properties. A concentration of 60 g/ha of fluopyram applied with 200 L/ha irrigation water at 2 L/h flow velocity was found to be most effective. However, the functional diversity of the soil microbial community was significantly affected by fluopyram but the activities of soil urease and β- glucosidase decreased during the initial stages and recovered at later stages. Therefore, it can be that fluopyram has advantageous for efficient control of root-knot nematode with no deleterious effects on soil properties as well as chemigation is positive for application in Guangxi, China.

The water extract of garlic was found to be effective to control *M. javanica* on brinjal (Kashem *et al.,* 1994). No root gall was observed when the extract was incorporated in soil at 50 per cent dilution or higher concentrations. Satisfactory reduction of root galls was achieved even at the lowest concentration in which the extract was diluted 100 times. The material caused significant increase in root and shoot growth. Investigation on the management of nematodes associated with brinjal revealed that significant reduction of *Hoplolaimus indicus* (syn. *Basirolaimus indicus*), *Tylenchorhynchus vulgaris* and *M incognita* was recorded with the application of nematicides alone as well as in combination with neem leaves. Neem leaves with low doses of nematicides had a detrimental effect on nematode populations and a favourable effect on the plant growth which resulted in increased crop yield (Prasad *et al.,* 1994). According to Aziz *et al.* (1995), root dip treatment with monocrotophos, methamidophos and leaf extracts of neem (*Azadirachta indica*) and *Kanair* (*Nerium indicum* syn.*oleander*) was effective in controlling root-knot disease caused by *M. javonica* infecting eggplant. All the treatments enhanced plant growth and suppressed nematode development as the dip duration was increased (2 hrs) with their standard concentrations.

According to Burman *et al.* (1995) oil cakes in general proved more efficient in checking the nematode population and also alleviating the adverse effects on nematode on the growth of eggplants. Neem cake, castor cake (*Ricinus communis*) and mahua cake (*Madhuca longifolia*) each at 500 kg/ha were evaluated for the management of *M. incognit*a and *Rotylenchulus reniformis* infecting brinjal under field conditions and their effects were compared with carbofuran. Carbofuran with neem cake were effective in increasing plant growth parameters, while carbofuran, neem cake and castor cake reduce root-knot and reniform nematode population. All the treatments were effective inincreasing the fruit yield of the crop (Kumar and Vadivelu, 1996). Integration of *Paecilomyces lilacinus* with neem leaf aqueous suspension was attempted by Rao *et al.* (1997) for the management of root-knot nematode (*M. incognita*) on eggplant. Results revealed the suitability of neem leaf suspensions (5 and I 0 per cent) for mass multiplication of *P lilacinus*. Further, significant reduction in root-knot index and final population of *M. incognito* was observed in the seedlings which were given root-dip treatment in neem leaf suspension mixed with *P. lilacinus* spores. In addition, significant increases were observed on the colonization of *P. lilacinus* on the roots of eggplant and parasitisation of eggs of *M. incognita*, indicating the complementary interaction between these two components for the sustainable management of root-knot nematode on eggplant. Integrated management of plant parasitic nematodes with oilcakes, nematicides and ploughing revealed that nematicides found to be highly effective against commonly occurring plant parasitic nematodes on eggplant cv. Pusa Purple Long over oil cakes while application of compost showed poor results. The efficacy of these substances was enhanced in deep ploughed (40 cm) beds as compared to normal ploughed (20 cm). The beneficial effects of all the above treatments persisted even in the subsequent crop of tomato cv. Pusa Ruby (Siddiqui and Alam, 1999). Best control of *Meloidogyne incognita* was achieved when 10 per cent neem cake extract (at 20 ml/pot) was used with spores of *Paecilomyces lilacinus* (at 1 x 10^6 spores/ml). This gave maximum reduction in root galling (87.53 per cent) and egg masses (71.20 per cent) thereby increasing the yield (64.36 per cent) (Zareena and Das, 2014).

Prusty and Mahalik (2018) studied the efficacy of liquid formulation of biocontrol agents and results indicated that nursery treatment with *Trichoderma viride* @ 50 g/m^2 followed by application of vermi compost @ 1 t/ha enriched with *Purpureocillium lilacinum* @ 10 ml/kg recorded maximum decrease in the root galls per plant as well as final nematode population.

Verma and Choudhury (1974) reported that Sel.96-2, Sel. 419, Pol Baigan and the species *Solanum sisymbrifolium* are highly tolerant to *Meloidogyne* spp. under field and laboratory conditions. According to Yadav *et al.* (1975) Vijaya and Black Beauty cultivars were tolerant to *M. incognita*. Morra *et al.* (2000) reported that fruit yields were highest from plants grown on *S. torvum* and *M. incognita* populations were also lower on those plants. *Solanum torvum* was employed as a rootstock for cultivation of eggplant worldwide and was found to be resistant to *M. incognita, M. javanica, M. arenaria* and *M. luci* but susceptible to *M. hapla* (Serap *et al.*, 2018).

From another experiment at Udaipur, Rajasthan, Nandwana *et al.* (1980) reported that only Black Beauty, T-2, Vijaya and Banaras Giant cultivars showed relatively low numbers of nematode egg/root system. The number of eggs per root systems was considered more reliable criterion for determining the extent of infection than the number of galls per gram root.

The cultivars reported so far as tolerant to *Meloidogyne* spp. may be utilized in controlling the nematodes as well as in future breeding programme in order to evolve a resistant cultivar.

Pant and Pandey (2014) studied the effect of bio-control agents (*Trichoderma viride* and *Beauveria bassiana*) and organic matter (spent compost of oyster mushroom) in RKN of brinjal and results indicated that *T. viride* was the most effective among the treatments with maximum reduction in root knot nematodes and highest plant growth parameters. This is attributed to the presence of high concentration of toxic chemicals such as viridian, production of volatile and non-volatile antibiotics and release enzyme during the decomposition of brinjal plants.

El-Nagdi and El-Khair (2019) carried out an experiment in Egypt to determine the nematicidal activity of two *Bacillus sp viz., B. subtilis* and *B.pumilus* against *M. incognita* by conducting three experiments firstly when *Bacillus* spp. were applied as single or in combination, secondly by giving a single treatment at different doses and thirdly different times i.e three times of application. They concluded that the biocontrol agents used more than once or in combination were more effective than those used only once in controlling nematode parameters and thus increased the growth parameters compared to the controls. In another study,

B. subtilis and *Trichoderma harzianum*, alone or in combination, could suppress *M. incognita*, affecting the total final population of nematode, and enhanced the growth parameters of eggplant cv. Patra in the green house (Farfour and El-Ansary, 2013).

B. subtilis, *B. thuringiensis*, *Psudomonas fluorescens*, and *Serratia marcescens*, alone or as a mixture, inhibited the development of *Meloidogyne javanica* and highly reduced the numbers of galls and egg masses per root system and J2/250 soil as well as increased the dry weights of root and shoot in eggplants (Mokbel and Alharbi, 2014).

Faruk *et al.* (2019) studied the efficacy of integration of poultry refuse (PR) @ 3 t/ha 3 weeks before transplanting with stable bleaching powder (SBP) @ 20 kg/ha given during final land preparation or $CaNO_3$ and Furadan 5G @ 20 kg/ha on the day of seedlings transplanting which reduced the root-knot as well as increased the plant growth and yield in brinjal. The most effective treatment combination was PR + stable bleaching powder + Furadan 5G with early sowing.

Management of RKN is limited, so genetic engineering studies have been developed to disrupt the plant and nematode reaction. Papolu *et al.* (2016) developed a transgenic eggplant (from cultivar Pusa Purple Long) expressing a modified rice cystatin (OC-I1D86) gene under the control of the root-specific promoter, TUB-1 to evaluate the genetically modified nematode resistance. Five putative transformants were selected through PCR and genomic Southern blot analysis and expression of the cystatin transgene was confirmed in all the events using western blotting, ELISA and qPCR assay. Upon challenge inoculation, all the transgenic events exhibited a detrimental effect on RKN development and reproduction. The best transgenic line (a single copy event) showed 78.3 per cent inhibition in reproductive success of RKN. The results suggested that cystatins can play an important role for improving nematode resistance in eggplant and their deployment in gene pyramiding strategies with other proteinase inhibitors could ultimately enhance crop yield.

9.0 Seed Production

Cultivars of brinjal bear flowers in clusters as well as singly. The flower drop in single flower bearing cultivars is only 1 per cent, whereas 80 per cent is in the cluster. In Pusa Purple Long cultivar, the inflorescence may contain 4 to 5 flowers but there is fruit set in one or two flowers and rest of the flowers drop off. Generally, there are four types of flowers in brinjal as mentioned earlier.

Brinjal is a self-pollinated crop. The cone-like formation of the anthers favours self-pollination in the same way as in case of tomato (Kalda, 1979). Frydrych (1964) was of the opinion that self-fertility found in brinjal was partly due to the absence of pollinators. It has been mentioned by Das *et al.* (2009) that in India, only 17.8 per cent area of brinjal cultivation was under hybrid seed.

Though self-pollinated, it can out-cross to a considerable extent through insects, if more than one cultivar is grown in adjacent plots. Generally, out-crossing is more in case of long-styled flower. The extent of natural crossing depends upon insect activity and has been recorded from 0 to 48 per cent (Agrawal, 1980). Daskalov and Mutazov (1937) reported that 30 to 40 per cent cross-pollination occurred in some

Bulgarian types. Sambandam (1964a) found only 6.7 per cent natural crossing in brinjal. Under Delhi condition, the extent of natural crossing was 0.14 to 1.99 per cent (Choudhury and Anothai-Choomsai, 1970). Srinivas *et al.* (2016) indicated that in two eggplant hybrids having long, green and round, purple fruits, 80 per cent of fruits were set by long-styled flowers, whereas 20 per cent of the fruits were set by medium-styled flowers and no fruit set by short-styled flowers.

It is, therefore, essential to isolate cultivars to produce quality and pure seeds. Agrawal (1980) suggested to give isolation distances of at least 400, 200 and 100 metres for nucleus, foundation and certified seed production, respectively. But Choudhury (1976a) suggested 50 to 100 metres of isolation distance between cultivars for quality seed production of brinjal. The nucleus seed plots should also be isolated from commercial seed crop. Nucleus seed is raised by massing the selfed-seeds of selected plants.

Natural out-crossing coupled with hand emasculation can be exploited for hybrid seed production in brinjal with a reasonable success (Kanwar and Gill, 2000). Moreover, the isolation distance of 100 m presently recommended for certified seed can be reduced. They also opined a set of comprehensive multilocation tests are required to assess the minimum isolation distance for brinjal, so that new standards are prescribed that are based on the agroclimatic conditions of India.

Roguing in many cultivars may not be possible till the formation of the first fruit. Leaf and stem characters can easily be detected even in the nursery. The seed grower should be well acquainted with characteristics of the cultivar and should select true to type plants and rogue out the off-type and undesirable plants at different stages of crop growth.

The agronomic practices normally followed for a market crop are generally employed for seed production also. Eguchi *et al.* (1962) reported that in Japan, the most favourable conditions for growth and seed production occurred at 260 metres above sea level, owing to the wider range of day temperatures during the growing season. In Indian conditions, it can produce seeds both in plain and hills. Sowing of seed crop should be so adjusted that maturity does not coincide with heavy rains. The winter crop needs special protection from frost (Agrawal, 1980). Singh *et al.* (1964) reported that in the hills of Kulu Valley in Himachal Pradesh, the brinjal seeds should be sown in the nursery during the months of March and April for growing seed crop. Usually after 60-65 days, the seedlings are transplanted in the hills.

Brinjal responded well to nitrogen fertilization at 70 kg per hectare in respect of seed yield (Seth and Dhandar, 1970). There was, however, no significant response of phosphorus fertilization. They also reported that seed yield per plant and fruit quality were slightly better with a row to row spacing of 100 cm as compared to 75 cm.

Petrov *et al.* (1981) recommended that seeds should be collected from first or second tier fruits as those have a higher 100 seed weight, germination energy and germination rate than seeds collected from fruits beyond the second tier. They further suggested that fruit harvested at botanical maturity gave the highest quality seed and seed extracted from fruits kept for 5 days after harvest improved earliness and total yield in progeny. They also reported that for hybrid seed production ample pollen collected from many plants should be used and pollination may be repeated once or twice. Pollen should not be stored for more than 4 days and should be applied within 2-3 days of flower emasculation. Studies at Bangalore on the effect of the order of fruit formation on seed production in eggplant revealed that generally the first fruit was the largest in size and subsequently size decreased. Seed yield per fruit and 1000 seed weight followed a similar trend. The percentage of seed in the fruit (by weight) was lowest in the first fruit and increased in successive fruits to reach a maximum in the last fruit. There was no wide variation either in the germination percentage or vigour of the seeds obtained from different fruit (Naik *et al.*, 1996). According to Doijode (2000), position of seed in a fruit has a marked influence on viability and vigour during ambient storage and it is advantageous to select seeds from basal portion of fruit for getting the same. From an experiment on seed recovery, viability and vigour in brinjal as influenced by stage of fruit maturity and pickings Nehra *et al.*(1999) reported that ripened and over ripened fruits of first three pickings are suitable for seed extraction whereas data recorded on seed vigour suggest that only ripened fruits of first three pickings are suitable for quality seed production.

Crop protection measures should be taken in time. Sinha and Chakraborti (1982) evaluated the chemical control, schedules against important insect pests of brinjal seed crops and suggested effective pest control measures. The highest seed yield (3.6 to 7.1 q/ha) was obtained when the plants were sprayed 5 times, twice with dimethoate at 0.72 kg/ha followed by spraying of carbaryl at 2.4 kg/ha thrice.

For seed production, the fruits should be harvested when fully ripe, *i.e.* turning normal colour to yellow. The outer covering is peeled off and the flesh with seeds is cut into thin slices. In a hard skinned cultivar like Nurki (Purple Cluster Selection) both the basal and upper portion of the fruit which constitute the seed-free zones are chopped off and the remaining portion is used. The sliced materials are stored overnight so that the seed after washing with water and sieving in the morning can be dried during the day. The washing should always be done in the morning hours so that the seed is at least half dried till the evening or else there is a danger of its germination. Agrawal (1980) suggested that the seeds should be dried in partial shade to moisture content of 8 per cent or below.

Martins *et al.* (2012) determined the seed physiological maturity and the best moment for harvesting the fruits to extract the seeds so as to produce seeds with high quality. This work was being carried out in Brazil. The fruits were subjected to harvest at different periods of 49, 56, 63, 70, and 77 days after pollination (DAP) and harvested immediately or after a week. Results of germination and vigour of seeds have showed that the best period for harvesting the fruit is around 70 DAP and that seeds should be extracted immediately after harvest.

Acid extracted seeds of brinjal cv. Co. 1 were conditioned to 8 per cent moisture content and graded into sinkers and floaters by the water floatation method. The seeds that sank were of better quality than those floated, giving 83.5 per cent germination compared with 23.1 per cent (Selvaraj, 1988). Sachan *et al.* (2009) found out that acid fermentation @ 5 ml/kg pulp for 45 minutes was most effective for

Seed Production of Brinjal

Field View of Seed Production Plot

Matured Fruit Ready for Seed Extraction

Painting of Brinjal Fruit with Copper Oxychloride to Avoid Rotting

high seed recovery as well as attaining other seed quality attributes. Rahman *et al.* (2015) studied the physiological quality of eggplant seed by comparing the two extraction methods and results suggested that wet seed extraction without fermentation was comparatively better then dry method as the dry seed extraction method was labourious as well as difficult to accomplish and may only be used for small- scale seed extraction.

Long term storage of seeds following production under high ambient temperatures revealed that under Mediterranean conditions and in the absence of humidity control, long term (> 1-2years) storage of brinjal seeds should be carried out at ambient (25±5°C) rather than at low (5±2°C) temperature (Passam *et al.*, 1999).

The variation in seed yield is due to environmental factors, crop management practices and cultivars. The average seed yield is 100 to120 kg/ha (Choudhury,l 976a). In the region of Kulu Valley a seed yield of 590 to 880 kg per hectare in Pusa Purple Long, Pusa Purple Round and Pusa Clustered Selection cultivars were reported by Singh *et al.* (1964).

Flowering and fruit set are two most important factors determining the seed yield of cultivated brinjal (Korat *et al.,* 2018). The flower setting was improved by application of an improved plant growth regulators through seaweed extract Goemar BM 86 or by endogenous components, especially cytokinins, which enhance the nutrient partitioning in vegetative plant organs and increase in the transport of assimilates to the growing fruits (Pohl *et al.,* 2019). For hybrid seed production, crossing of four female flowers with one male flower (4: 1) was found to be significant for most of the seed yield parameters as compared to other crossing ratios in seed parent (Shivashankaragouda *et al.,* 2008).

10.0 Crop Improvement

10.1.0 Variability and Selection

As in many other crops, in brinjal too, the major efforts for the crop improvement have been directed towards stepping up the yield. Yield is very complex and polygenic in nature. The knowledge of the association among the yield components is the touch stone to a breeder to evolve potentially a better plant type through selection either from the existing genotypes or from the segregants of a cross. Hence the genetic potentialities of yield contributing characters and their inter-relationship should be properly assessed for improving the crop.

In brinjal, a strong association has been reported between the number of fruits and yield per plant (Singh and Nandpuri, 1974; Chadha and Paul, 1984; Kabir *et al.,* 1985; Chattopadhyay *et al.,* 2011; Shende *et al.,* 2017). High heritability and high genetic gain have also been observed for different characters like number of flowers per cluster, number of fruits per cluster, fruit length, fruit diameter, fruit weight, long and medium styled flowers per plant, number of fruits per plant and yield per plant by many workers (Kabir, 1981; Kalda *et al.,* 1988; Vadivel and Bapu, 1989a and 1990). These characters are suggested to be governed mostly by additive genes. Similarly, in segregating generations also high heritability and high genetic advance for number of fruits per plant, fruit weight and yield per plant have been noticed (Vadivel and

Bapu, 1989b; Nainar *et al.*, 1991). Path analysis in brinjal as conducted by Srivastava and Sachan (1973) and Kabir *et al.* (1985) revealed that number of fruits per plant exhibited maximum direct effects on yield. It is, therefore, considered useful to select the best genotype of brinjal per plant for effective improvement in this crop. In an experiment on variability and genetic parameters in brinjal, Mohanty (1999) recorded PCV was greater than GCV for all the traits. High heritability accompanied by high genetic gain and GCV were observed for average fruit weight, number of fruits and yield per plant which could be improved by selection.

Cultivars like Pusa Purple Long, Pusa Purple Round, Pusa Purple Cluster, Arka Sheel, Arka Shirish, Arka Kusumakar, Punjab Chamkila, Punjab Bahar, *etc.*, have been evolved through selection mostly from local genotypes. It is desirable to know the existing variability present for various vegetative, flower and fruit characters in different cultivars of brinjal. This type of information can be of immense utility for further crop improvement programme.

Maximum genotypic coefficient of variation (GCV) and phenotypic coefficient of variation (PCV) for number of flowers per inflorescence, number of fruits per picking and fruit girth, fruit weight, fruit yield per plant and shoot and fruit borer incidence, indicating that selection can be predicted to improve the brinjal genotypes for these characters (Ansari *et al.*, 2011; Samlindsujin *et al.*, 2017; Kumar *et al.*, 2020). However, Koundinya *et al.* (2017) not only observed higher phenotypic and genotypic co-efficient of variation for quatitative characters but also for anthocyanin in peel, total phenols and DPPH (2,2-diphenyl-l-picryl hydrazyl), free radical scavenging (FRS) capacity.

Highest estimates of heritability coupled with high genetic advance as percent of mean was estimated for the average fruit weight and number of fruits per plant, fruit weight, fruit yield per plant, number of secondary branches per plant and shoot and fruit borer incidence (Fatima *et al.*, 2011; Samlindsujin *et al.*, 2017), plant height, days to 1st flowering, days to 50 per cent flowering, number of fruits per plant, fruit weight, harvest index, fruit yield per plant, total sugar, anthocyanin in peel, total phenols and DPPH FRS capacity (Koundinya *et al.*, 2017), depicting that these traits were under the strong influence of additive gene action and hence simple selection based on phenotypic performance of these traits would be more effective. Karak *et al.* (2012) observed high heritability for fruit yield and other fruit characters except fruit weight.

Yield per plant has strong positive association with number of fruits per plant, plant height, leaves per plant and primary branches per plant, fruit weight, fruit length, fruit girth, harvest index, vitamin-A, total soluble solids and total phenols at both phenotypic and genotypic level (Karak *et al.*, 2012; Ahmed *et al.*, 2013; Koundinya *et al.*, 2017; Samlindsujin *et al.*, 2017; Sakriya *et al.*, 2020). Fruit yield per plant was positively and significantly correlated with number of pickings, fruit volume, number of fruits per plant and average fruit weight (Madhavi *et al.*, 2016).

Path analysis indicated that fruit weight was one of the major contributory factors to yield, fruit girth and leaves per plant (Karak *et al.*, 2012), number of fruits per plant imparted the highest positive direct effect on yield (Koundinya *et al.*, 2017) and number of long styled flowers per plant, number of short styled flowers per plant, days to first harvesting and shoot and fruit borer incidence showed positive direct effect in path co-efficient analysis (Samlindsujin *et al.*, 2017). Genotypic path coefficient revealed the highest positive direct effect on fruit yield followed by number of fruits per plant, plant spread, fruit width, fruit length (Ahmed *et al.*, 2013). These factors emerged as the most important fruit yield contributing characters which may be used as important selection parameters because of their probable conditioning of additive genes.

Jirankali *et al.* (2019) carried out studies on genetic variability for qualitative and quantitative characters in 100 brinjal genotypes and observed that high estimates of PCV, GCV, heritability coupled with high genetic advance for fruit weight, number of fruits per cluster and shoot borer infestation. Kumar *et al.* (2020) could also observed high estimates of genotypic co-efficient of variation, heritability (broad sense) and genetic advance as percentage of mean for the characters such as fruit yield per plant, number of fruits per plant, fruit weight, fruit girth and node at which first flowering appeared, suggesting additive gene action for expression of these characters indicating their possibility of improvement with simple selection.

Biological divergence in the landraces of eggplant was studied by Prasad and Singh (1998) under alfisols of Eastern India with 120 genotype, collected from all over the country. The results revealed that high and significant differences were observed among all the genotypes of yield per plant, fruit firmness, fruit length, fruit breadth, fruit weight, days to flowering, plant spread and plant height. It also revealed that biological diversity is not always related to genetic diversity and the clustering of germplasm lines suggested the dependence upon directional selection pressure applied for realizing yield in different regions. From a study on genetic divergence in brinjal, Sarma *et al.* (2000) were of the view that ecogeographic diversity of the genotypes was not related to genetic diversity. According to them, fruit circumference and average fruit weight were the main characters affecting grouping of genotypes. While studying the genetic variability in brinjal, Chandra *et al.* (1998) observed high PCV and GCV for number of seeds per fruit, seed weight, number of fruits per plant and plant height. However, the low estimate was recorded for length of fruiting period, days required to flowering and days required to first picking. High heritability with high genetic advance was observed for plant height, fruit set; number of fruits per plant, fruit length, fruit weight, seed weight and number of seeds per fruit emphasizing effectiveness of these characters for selection of high yielding genotypes as these characters are governed by additive gene effects. From another study, Chandra *et al.* (1998) recorded significant positive correlation between yield per plant, fruit weight, number of fruits per plant, fruit length and seed weight, while a negative correlation with number of days for first picking. Genetic divergence using D^2 statistic of 43 eggplant genotypes of different geographic origins revealed the existence of considerable diversity and the diversity among the genotypes as estimated by inter-cluster distance was adequate for improvement of the crop by hybridization and selection. The genotypes included in the diverse clusters can be used as promising parents for hybridization to obtain high heterotic response and thus better segregants in eggplants (Thirumurugan *et al.*, 1999).

Hazra *et al.* (2010) carried out heterosis studies in relation to multivariate genetic divergence in brinjal and observed significant differences among the 70 entries for all the characters taken into study but there was a lack of correspondence between geographical origin and genetic divergence of the entries indicating that either a common character constellation was manifested simultaneously in the genotypes or mutual balancing in character expression was operative in the genotypes of brinjal with no character contributing overwhelmingly towards divergence. They further concluded that while selecting parents for hybridization genetic distance should also be considered.

Bhatt *et al.* (2019) carried out genetic divergence studies in forty one genotypes of brinjal having two groups long purple and round purple including two checks Pant Rituraj and Pant Samrat and observed no direct relationship between the geographical distribution and genetic distance which was in compliance with the previous studies conducted by Doshi *et al.* (1998), Kumar *et al.* (1998), Mohanty and Prusti (2000).

A number of researches have carried out genetic divergence studies in brinjal in order to identify genetically diverse parents which will have a good combining ability resulting in high heterosis and release superior segregants in later generations. Though, the total numbers of genotypes vary from one study to another but through grouping of genotypes into different clusters it could be suggested that there is a presence of high degree of diversity in the material evaluated (Quamruzzman *et al.*, 2009, Muniappan *et al.*, 2010, Balaji *et al.*, 2013, Yadav *et al.*, 2017, Bhushan *et al.*, 2018, Arti *et al.*, 2018). They further concluded that the kind of genetic diversity found among the genotypes belonging to same geographic origin may be due to differences in adoption, selection criteria, and selection pressure in environmental conditions.

10.2.0 Inheritance Study

10.2.1 Qualitative Characters

The breeding method in any crop depends upon its breeding system, genetic architecture and the pattern of inheritance of characters. The inheritance of some qualitative characters in brinjal may be simple and governed by few genes only (Sambandam, 1964b; Sambandam, 1967; Baha-Eldin and Blackhurst, 1968; Sambandam, 1969; Thakur *et al.*, 1969; Swamy Rao, 1970; Rangaswamy and Kadambavanasundaram, 1973). In USA, Baha-Eldin and Blackhurst (1968) reported partial dominance of tall plants over short plants, early flowering over late flowering and round fruits over long fruits in a cross between Black Beauty and PI169651. On the contrary, Swamy Rao (1970) and Rangaswamy and Kadambavanasundaram (1973) observed dominance and partial dominance of long fruit over round fruit. Both of them reported dominance of clustered fruit over non-clustered fruit. Gopinath *et al.* (I987) reported from the segregation analysis of cross WCGR112-8 x Pusa Kranti that fruit shape was controlled by a single gene with oblong partially dominant over round.

A pair of duplicate genes appeared to control brinjal seedling colour (Rangaswamy and Kadambavanasundaram, 1973). In Punjab, Thakur *et al.* (1969) reported that fruit colour in brinjal is controlled by dominant complementary gene. They also observed that purple corolla is monogenically dominant over white and green flower and the presence of spike is controlled by a single dominant gene. The purple hypocotyl is dominant over green and controlled by a single gene (Sambandam, 1964a). In the year 1969, Sambandam reported that fruit skin and flesh colour is inherited independently of each other. He also concluded that purple pigmentation in different parts of the plant is controlled by several linked gene. The genes *Ph, Al, Pb, Pm, Pu, Pl, St* and *C* are concerned with the pigmentation of the hypocotyl, epicotyl, leaf midrib, leaf margin, upper and lower surface of calyx, anther suture and fruit skin, respectively. The genes *Ph, Pb, Pu, Pl, St* and *C* are linked. Segregation ratio of $F_2$and BC between cv. American (greenish/white fruit) and wild species *S. indicum* (green fruit) indicated that basic colour of young fruit is governed by the interaction of 3 non-allelic genes, previously tentatively designated *Gf, Wf* and *Grwf*. Purple skin is determined by double recessive *grwfgrwf*, while *Grwf* gives greenish white fruit. The modifier *Wf* changes the greenish white *Grwf* phenotype to white. *Gf* modifies expression of *Grwf* or *Wf* to give a green phenotype (Khapre *et al.*, 1989).

Kamani *et al.* (2007) observed that single dominant gene was responsible for the inheritance of spiny leaf, and possibly two complimentary genes for growth habit and three complimentary genes for fruit bearing habit. Patidar (2015) carried out inheritance studies on qualitative characters of brinjal on forty five hybrids and ten parents along with one standard hybrid ABH-1 which revealed dominance of purple flower over white; presence or absence of spines on leaves, calyx and stem, monogenic or pleiotropic in action; elongated (ellipsoid and cylindrical) fruit shape dominant over round (globular, ovoid and obovate) fruit shape and round shape dominant over oval (pear and club shaped) fruit shape; purple calyx colour dominant over green and fruit colour ranging from white purple to dark purple showing quantitative nature of gene expression.

Three independent complementary genes *D, P,* and ***Y*** affect the anthocyanin development in corolla, hypocotyl, and fruit. Four alleles were identified at *D* (in order of dominance), *d* inhibits anthocyanin colouration in the fruit only, *dt* inhibits fruit colouration and dilutes the flower color, and *dw* inhibits anthocyanin colouration in all plant parts. Of three alleles at *P* (in order of dominance), *p* inhibits anthocyanin colouration in the hypocotyl and fruit, *pw* inhibits anthocyanin colouration development in all plant parts. Of two alleles at *Y*, y inhibits anthocyanin development in all plant parts. Dominant *Ac* converts the anthocyanin delphinidin-3-rhamnoglucoside to nasunin and is accompanied by a change in fruit colour from vinaceous to purple. *Puc* (pigment under the calyx) permits the dark synthesis of anthocyanin in the fruit and *Sa* the formation of an anthocyanin stripe on the anthers. *Puc* and *Sa* appear tightly linked. At least two genes, Dil_1 and Dil_2 affect the color intensity and three alleles are indicated: dil^p represses plant color only and dil^f represses both plant and fruit color. A dominant gene *R* inhibits fruit pigmentation in certain genetic backgrounds. Partial linkage of Dil_1 with *Ac* and complete linkage of Dil_2 with *Puc* is indicated (Tigchelaar *et al.*, 1968).

Kumar (2017) carried out chi-square studies which revealed that calyx spininess, fruit pedicel prickles, leaf prickles on upper surface, and leaf pubescence segregated in 3: 1 ratio (single dominant gene), leaf blade colour and fruit shape trait segregated in 9: 3: 4 ratio and plant growth habit segregated in 9: 6: 1 in all the eight hybrid combinations taken for the study.

The qualitative characters of different plant parts like stem, leaf, flower, fruit and other characters have been studied by several workers. Genetics as well as dominance of these characters have been presented in Table 2.

Table 2: Genetics of Qualitative Characters in Eggplant*

Characters	*Genetics, Dominance and Symbols*
Stern colour/pigmentation	Single gene pair, *Pst pst*, Purple dominant over green
Purple pigmentation of stem, petiole and veins	All three controlled by duplicate genes designated *Pst1* and *Pst2*
Leaf vein colour, purple shade on leaves, corolla and calyx colour	Each governed by three or more non-allelic young genes.
Stem and leaf colour	Monogenic, Purple colour dominant over green leaf and stem
Spinisence	Monogenic dominant over spineless
Spinisence and purple pigmentation	Digenic dominant over spineless and green colour
Pigmentation on leaves and veins	Governed by two complementary genes, C_1a_1, C_1b_1, P_1v_1 and P_1v_2 respectively
Spines on stem, leaf and petiole	Single gene dominant
Spines of leaves and stem	Governed by three complementary genes of which any two must be dominant
Spines on calyx	Two complementary genes and an inhibitor
Erect versus spreading habit of plant	Controlled by two genes (Egr_1 and Egr_2) with erectness being dominant
Plant height	Monogenic, tall being dominant over dwarf
Fruit shape	Controlled by three partially dominant genes (*Ofa*, Ojb_1, Ojb_2, and Ojb_3)
Fruit colour	Purple dominant over green, variegated which is dominant over white and green.
	Green (*Gm*) is incompletely dominant over white (*gm*).
	Green streaked purple is dominant over greenish white
Fruit end shape	Narrow end dominant over round end
Bunching of fruit	Intermediate in F_1 hybrids
Male sterility	Controlled by single gene, dominant over solitary fruiting
	Controlled by two recessive nuclear genes, ms_1 and ms_2
	Under cytoplasmic control
	Controlled single recessive allele (*fms*) which is linked to purple colour of fruit

* Kalda (1998).

10.2.2 Quantitative Characters

Several quantitative characters in brinjal are polygenically inherited. These include metric traits as yield, size, shape, number of branches and number of fruits. Resistance to disease and pests is also sometimes polygenically controlled. Unlike the qualitative characters which have a simple inheritance, the polygenic characters are greatly influenced by environment and hence in these cases a phenotype is often not a true indicator of its genotype. In a quantitative character, the cumulative effects of the individual genes and their interactions, both allelic or dominance and non-allelic or epistasis, determine the genotypic value. The estimation of various types of gene effects would provide useful information for future crop improvement programme in brinjal (Scossirali *et al.*, 1972; Peter and Singh, 1973; Gill *et al.*, 1976; Gowda, 1977; Shende *et al.*, 2016; Dutta *et al.*, 2019).

Additive gene action in respect of number of fruits per plant, fruit length, number of flowers per inflorescence, number of long-styled flowers in brinjal have been observed by Peter and Singh (1973). They also observed that maturity and number of primary branches are controlled by over dominant genes; fruit weight per plant is controlled by dominant genes and the height of the plant is controlled by additive gene action with some overdominance. Additive effects for fruits per plant, fruit length and width have been reported by Vadivel and Bapu (1989a; 1990). Singh and Mital (1989), however, observed that days to flowering, plant height and yield per plant are controlled chiefly by low additive gene action. Gill *et al.* (1976) and Gowda (1977) observed additive gene effects to be more important than dominance in most of the economic characters in brinjal. While Shende *et al.* (2016) recorded non-additive gene action for the conditioning of days to first flowering, days to 50 per cent flowering, plant height, fruit length, phenol content, marketable fruit yield per plant, fruit borer infestation, total sugar, number of marketable fruits per plant and average fruit weight suitable breeding strategies for the condition of different characters under study by additive (total sugar content, fruit borer infestation), non-additive (days to first flowering, days to 50 per cent flowering, plant height, number of primary branches per plant, total fruit yield per plant, marketable fruit yield per plant), both additive and non-additive (fruit length, fruit diameter, fruit weight, number of fruits per plant, total fruit phenol content) effects were discussed by Dutta *et al.* (2019). Vaghasiya *et al.* (2000) raised six generations (P_1,P_2, F_1,F_2, $BC_1$and BC_2) of brinjal crosses KS233 x PLR1 and Green Round and reported that in KS233 x PLR 1, additive as well as non-additive gene effects were important for fruit yield/plant, fruits/plant, fruit girth and plant height. In Green Round x PLR1, all characters were under the control of both additive and non-additive gene effects. Selection in the segregating generations of the crosses is suggested for bringing out the desired improvement in this crop.

Deshmukh *et al.* (2014) carried out gene action studies on 9 parents, two commercial hybrids (Pusa Hybrid 6 and Pusa Hybrid 9) and 36 F_1 hybrids developed by using half diallel mating scheme which revealed the importance of non-additive systems operating in inheritance of yield and its economic components and that the degree of dominance was in range of over dominance range for all the characters and close to complete dominance for plant height.

Devmore *et al.* (2016) carried out an investigation to study the nature and magnitude of gene action among five crosses of brinjal involving six generation *viz.*, P_1, P_2, F_1,F_2, $BC_1$and BC_2 and found out that additive [d], dominance [h], additive x additive, dominance x dominance gene actions were equally important for primary branches per plant, fruit breadth, fruit weight, yield per plant and seeds per fruit in most of the crosses.

Kumar (2017) carried out gene inheritance studies and observed that days to first flowering, days to first fruit set and fruit length are controlled by additive gene action, whereas traits like plant height, yield per plant and average fruit weight are controlled by non additive gene action for four hybrids Rajendra Baigan-2 x BRBR-01, Rajendra Baigan-2 x BRBL-01, Rajendra Baigan-2 x Swarnamani, Rajendra Baigan-2 x Muktakeshi and over dominance in Rajendra Baigan-2 x BRBR-01 for days to first flowering, days to first fruit set, number of primary branches per plant, fruit length, fruit girth and yield per plant.

Mistry *et al.* (2018) carried out gene action studies for inheritance of twelve fruit yield traits which revealed the adequacy of simple additive dominance model for plant height, primary branches per plant, secondary branches per plant, fruit length, fruit girth and pedicel length indicating the absence of non-allelic interactions. Duplicate epistasis effects controlled all the traits in different cross-combinations suggesting that mild selection should be done in the earlier generations while intense in the later generations. However, in few cases average dominance ratio was more than unity indicating the importance of dominance gene effects. From these results, it can be concluded that for eggplant improvement, additive variation is of great importance and makes it possible to successfully select better individuals in segregating populations. For this reason, backcross, pedigree, single-seed descent or gametic selection methods are recommended for advancing the segregating populations.

10.3.0 Hybridization

A high yielding cultivar Pusa Kranti has been developed through hybridization at the Indian Agricultural Research Institute, New Delhi and has been recommended for both spring and autumn planting under North Indian conditions (Choudhury *et al.*, 1971). This cultivar has been bred by combining the good qualities of three parents, namely Pusa Purple Long, Hyderpur and Wynad Giant. The photo sensitive characters from Pusa Purple Long, upright habit, hardiness and glossiness from Hyderpur and bright purple colour of Wynad Giant have been incorporated into Pusa Kranti. Pusa Purple Long was first crossed with Hyderpur and the progeny was back-crossed to Wynad Giant. Pusa Kranti was thus evolved through selection from subsequent generations. Punjab Barsati, developed from a cross between PPC and H4 was released in 1987 for cultivation in Punjab. An early maturing cultivar Sada Bahar Baingan has also been evolved through hybridization between Japanese Long and R 34 (Chadha and Sidhu, 1987).

10.3.1 Hybrid Vigour

The occurrence of considerable hybrid vigour was recorded as early as 1892 by Munson in the USA and in 1926 in Japan by Nagai and Kida and hybrid brinjals are now commonly used in many countries, especially in Japan.

Experiments conducted mostly in Japan and India has shown distinct hybrid vigour in brinjal. Goto (1952) recorded marked yield increase in the F_1 generation of a series of crosses between Japanese cultivars. Daskaloff (1941) in Bulgaria reported higher yield ranging between 10 and 44 per cent in crosses with imported cultivars. Odland and Noll (1948) observed that F_1 hybrids exceeded the parental mean yields by 62 per cent and the high yield was due to more number of fruits produced rather than fruit size. In Illinois, Courter and Glover (1965) reported that Burpee hybrid and Black Magic hybrid gave the best result in brinjal. Popova *et al.* (1976) at Breeding and Genetics Institute in Bulgaria obtained increased yield in F_1hybrid to a great extent. Similarly, in Tennessee, Cordrey *et al.* (1979) observed that Black Bell hybrid was found superior to the recommended cultivar Florida Market.

In India, several reports are available on hybrid vigour in brinjal. Pal and Singh (1946, 1949) reported that hybrids in brinjal showed 48.8 to 56.6 per cent increased yield over the better parent. Mishra (1961) also observed increased yield in F_1hybrids in brinjal. Viswanathan (1973) reported that the most economic combination was Muktakeshi x Banaras Giant in which the increase in yield was over 100 per cent compared to the parental mean. Som and Mallik (1979) observed higher yield due to more number of branches, increase in fruit size, fruit number and fruit weight, and this increased yield was about 164 per cent and 134.10 per cent more than both the parents in a cross between Kalyani Green and Elokeshi. Pusa Anmol, a hybrid cultivar, has been evolved at the Indian Agricultural Research Institute from a cross between Pusa Purple Long and Hyderpur. It showed an early increased yield of about 80 to 100 per cent more over cv. Pusa Purple Long (Choudhury, 1966). The cross combination Pusa Purple Cluster x Sel. 5 exhibited 162.5 per cent heterosis for yield over better parent (Singh and Kumar, 1988). According to Som *et al.* (1989) 204 per cent heterosis for fruit yield per plant over mid-parent in hybrid of Muktakeshi x Pusa Purple Long was possible. Hybrid vigour in brinjal has also been reported by Thakur *et al.*(1968), Lal *et al.* (1974), Singh *et al.* (1978), Ram *et al.* (1981) and Salehuzzaman (1981). Some of the good general combiners for yield and most of the characters are APUl and Sel.5 (Singh and Kumar, 1988). H4 (Chadha and Hegde, 1989), Round White, Pusa Kranti, Bhubaneswar-4 and Keonjhar-1(Mishra and Mishra, 1990) which can be used in heterosis breeding programme. Heterosis for days to 50 per cent flowering, number of flowers per cluster, number of long-styled flowers per cluster, number of medium-styled flowers per cluster, number of fruits per cluster, length of fruit, diameter of fruit, number of marketable fruits per plant, fruit weight and total yield per plant was reported by Dubey *et al.* (1998).

The cost of hybrid seed production of brinjal is not high as compared to other vegetable and this can be further reduced by the use of male sterile lines. Fang *et al.* (1985) developed two male sterile lines by three back crosses to *S. melongena* after crossing *S. gilo* with *S. melongena* lines 9334A and 2578A and suggested that the male sterility is under cytoplasmic control. Khan and Isshiki (2016) reviewed the CMS systems in eggplant and classified it into two types namely the anther indehiscent type in which anther contains normal pollen but does not open to release and the pollen non-formation type in which anther of the male sterile lines is completely devoid of pollen. Furthermore, in pollen non-formation CMS systems, two independent dominant fertility restorer (*Rf*) genes were discovered.

Chauhan (1986) reported naturally occurring genie male sterile line with cultivar Pusa Purple Long. The male sterility is controlled by 2 recessive nuclear genes, ms_1 and ms_2. A homozygous male sterile line UGA 1-MS was derived from a male sterile spontaneous mutant of Florida Highbush (Rhatak and Jaworski, 1989). The functional male sterility of UGA1-MS is governed by a simple recessive allele, for which the symbol *fins* is proposed (Pathak *et al.*, 1991). Exploitation of hybrid vigour in brinjal is economical as each fruit contains large number of seeds as compared to other vegetables.

Several studies were conducted to assess the hybrid vigour of different hybrid combinations over better parent and standard check for yield and its components. The details are listed below in Tables 3 and 4.

Table 3: Studies Showing Promising Heterobeltiosis

F_1 Hybrid Combinations with Heterobeltiosis (per cent) for Fruit Yield	*Improved Characters*	*Source*
PR x PB-61 (58.80 per cent), PR x PB-60 (40.6 per cent)	Fruit yield/plant, fruit length, fruit diameter, total weight of fruits/plant	Panda *et al.* (2005)
EP 65 x Pusa Uttam (51.74 per cent)	Fruit number/plant, fruit yield/plant	Prabhu *et al.* (2005)
Arka Sheel x Green Round (153.01 per cent)	Number of flowers/inflorescence, number of fruits/cluster, average fruit weight, number of fruits/plant, fruit yield/plant	Shafeeq *et al.* (2007)
Arka Sheel x Green Round (162.89 per cent)	Number of flowers/inflorescence, number of fruits/cluster, average fruit weight, number of fruits/plant, fruit yield/plant	Shafeeq *et al.* (2007)
BCB 75 x BCB 15 (37.90 per cent)	Number of fruits/plant, fruit yield/plant	Das *et al.* (2009)
P-1 x P-20 (74.89 per cent)	Fruit yield	Chowdhury *et al.* (2010)
IGBL 67 x KS 327 (139.13 per cent)	Fruit yield and its contributing characters	Sao *et al.* (2010)
H 9 x S 16 (39.52 per cent)	Fruit yield/plant	Singh *et al.* (2012)
WBPF x PR (88.18 per cent)	Total fruit yield/plant	Biswas *et al.* (2013)
AB-07-08 x GP-180 (125.78 per cent), NDB-18 x AB-07-08 (86.25 per cent)	Plant height, fruit, yield/plant	Makani *et al.* (2013)
KS-314 x IC-90099 (94.72 per cent), KS-314 x PPC (85.10 per cent)	Fruit diameter, number of fruits/cluster, total yield/plant	Dubey *et al.* (2014)
BARI x PB-66 (83.19 per cent), BARI x PR (70.83 per cent), PB-71 x PS (42.64 per cent)	Average fruit weight, number of fruits/plant, total yield/plant	Ansari and Singh (2016)
AB-09-1 x AB- 12-10 (48.95 per cent), GJB-3 x JBL-08-8 (48.60 per cent), JBGR-1 x NSR-1 (35.10 per cent)	Fruit diameter, average fruit weight, number of fruits/plant, fruit yield/plant	Desai *et al.* (2016)

F_1 *Hybrid Combinations with Hetero-beltiosis (per cent) for Fruit Yield*	*Improved Characters*	*Source*
Pusa Purple Cluster x Pant Samrat (56.16 per cent)	Number of fruits/plant, fruit yield/plant	Sharma *et al.* (2016)
Gujarat Brinjal Round x BCB-10 (91.77 per cent), Gujarat Brinjal Round x 10 BRR Var-2 (62.77 per cent), BCB-10 x 10BRR Var-2 (52.27 per cent)	Marketable fruit yield/plant	Shende *et al.* (2016)
ABSR 2 x GPBRJ-31 (52.52 per cent), GOB 1 x AB 08-14 (45.31 per cent),GOB 1 x GP BRJ-204 (37.78 per cent),	Fruit yield/plant	Balwani *et al.* (2017)
Doli 5 x Kashi Taru (63.54 per cent)	Fruit yield/plant	Patel *et al.* (2017)
IBWL 2001-1 x Pant Rituraj (157.59 per cent)	Weight of marketable fruits/plant, fruit yield/plant	Pramila *et al.* (2017)
Bhavani Local x Palur 2 (77.90 per cent)	Fruit yield/plant	Kalaiyarasi *et al.* (2018)
Doli 5 x KS 331 (24.47 per cent)	Days to first picking, fruit volume, fruit weight, fruit yield/plant	Mistry *et al.* (2018)
IC 354611 x IC 310886 (162.39 per cent)	Number of fruits/cluster, fruit length, average fruit weight, number of fruits/ plant, fruit yield/plant	Rani *et al.* (2018)
GJLB-4 x JBR-15-08 (86.21 per cent)	Number of fruits/plant	Bhatt *et al.* (2019)
NSR 1 x Swarna Mani Black (46.00 per cent), NBR-14-01 x Swarna Mani Black (40.03 per cent), NSR-1 x Pant Rituraj (39.35 per cent)BCB 40 x KS-224 (50.22 per cent)	Fruit yield/plantMarketable fruit yield/ plant	Mohd Zeal *et al.* (2019)Dutta *et al.* (2019)

Table 4: Studies Showing Promising Standard Heterosis

F_1 *Hybrid Combinations with Standard Heterosis (per cent) for Fruit Yield*	*Improved Characters*	*Source*
PR x PB-61 (59.5 per cent), PR x PB-60 (41.3 per cent)	Fruit yield/plant, fruit length, fruit diameter, total weight of fruits/plant	Panda *et al.* (2005)
EP 65 x Pusa Uttam (51.74 per cent), EP 12 x MDU 1 (34.56 per cent)	Plant height, branches/plant, fruit number/plant, fruit yield/plant	Prabhu *et al.* (2005)
Arka Sheel x Green Round (75.87 per cent)	Number of flowers/inflorescence, number of fruits/cluster, average fruit weight, number of fruits/plant, fruit yield/plant	Shafeeq *et al.* (2007)
P-14 x P-5 (72.60 per cent)	Fruit yield	Chowdhury *et al.* (2010)
IGBL 70 x PPL (127.41 per cent)	Fruit yield and its contributing characters	Sao *et al.* (2010)
IBWL x Pusa Purple Cluster (46.86 per cent)	Total fruit yield/plant	Biswas *et al.* (2013)

F_1 Hybrid Combinations with Standard Heterosis (per cent) for Fruit Yield	Improved Characters	Source
GBL-1 x KS-331 (50.41 per cent), AB-07-08 x KS-331 (43.53 per cent)	Days to first picking, average fruit weight, fruit yield/plant	Makani *et al.* (2013)
BARI x PB-66 (64.36 per cent), BARI x PR (56.39 per cent),PB-71 x PS (40.45 per cent)	Average fruit weight, number of fruits/plant, total yield/plant	Ansari and Singh (2016)
AB-09-1 x AB-12-10 (46.29 per cent), AB-09-1 x AB-08-5 (42.02 per cent), AB-08-5 x JBL-08-8 (32.69 per cent)	Fruit diameter, average fruit weight, number of fruits/plant, fruit yield/plant	Desai *et al.* (2016)
Pusa Purple Cluster x Pant Samrat (31.03 per cent)	Number of fruits/plant, fruit yield/plant	Sharma *et al.* (2016)
GOB 1 x AB 08-14 (88.88 per cent), GOB 1 x GP BRJ-204 (71.42 per cent), ABSR 2 x AB 08-14 (70.23 per cent)	Fruit girth, fruit weight, total soluble solids, fruit yield/plant	Balwani *et al.* (2017)
GBL 1 x Doli 5 (48.28 per cent)	Fruit yield/plant, ascorbic acid and total anthocyanin content	Patel *et al.* (2017)
SMB115 x DBL02 (116.52 per cent), PB 66 x Pant Samrat (110.28 per cent), IBWL 2001-1 x Pant Rituraj (101.49 per cent)	Weight of marketable fruits/plant, fruit yield/plant	Pramila *et al.* (2017)
MDU-1 x Palur 2 (28.94 per cent)	Number of branches/plant, fruit girth, seedling dry weight	Kalaivarasi *et al.* (2018)
IC 261797 x IC 104101 (112.78 per cent)	Number of fruits/cluster, fruit length, number of fruits/plant, fruit yield/plant	Rani *et al.* (2018)
JBL-10-08-07 x Pant Rituraj (60.85 per cent)	Fruit girth, fruit yield/plant	Bhatt *et al.* (2019)
NSR x Swarna Mani Black (62.00 per cent), NBR-14-01 x Swarna Mani Black (55.39 per cent), AB-15-07 x Swarna Mani Black (50.09 per cent)	Number of fruits/plant, fruit yield/plant	Mohd Zeal *et al.* (2019)

10.3.2 Resistant Breeding

Most of the cultivars of brinjal are susceptible to a number of diseases and pests. The crop is also very susceptible to the attack of root-knot nematodes. Screening of cultivars and species of brinjal for resistance to diseases and pests is of great significance for evolving resistant cultivars. Efforts were made at the Indian Agricultural Research Institute, New Delhi, to find out sources of resistance against various brinjal diseases. Chakraborty and Choudhury (1975) from this Institute reported three brinjal lines, namely Sel.212-1, Sel. 252-1-1 and Sel. 252-1-2 as free from little leaf disease out of 164 germplasms tested. Similarly, Kalda *et al.* (1976) found that *S. xanthocarpum, S. indicum, S.gilo, S.nigram* and *S.sisymbrifolium* are highly resistant to phomopsis blight and *S.melongena* lines 11a and 264 are found resistant. Resistance to this disease is recessive and polygenically inherited (Kalda *et al.*, 1977). They further observed that dominance gene effects are more pronounced than additive in most of the crosses studied. Earlier, Decker (1951) released two brinjal cultivars, namely Florida Beauty as resistant to this blight. Quaiser (1987) reported that Pusa Dwarf and Annamalai are less affected by *Phomopsis fruit rot*, while 2 wild species, *S. gilo* and *S. integrifolium* are immune and may provide breeders with source of resistance.

Pradip and Singh (2017) screened two commercial cultivars of brinjal Pant Rituraj and Pant Samrat and wild relatives *S. gilo, S. aethiopicum, S. khasianum* and their interspecific hybrids against phomopsis blight. *S. gilo* and *S. khasianum* showed the highest value of resistance against leaf, stem and fruit infestation. Among the hybrids, Pant Rituraj x *S. gilo, S. gilo* x Pant Samrat were highly resistant. Therefore, crosses involving *S.gilo* as one of the parents showed high level of resistance against phomoposis blight and successful gene transfer from these wild relatives is possible through back cross except in *S.khasianum* which needs the support of biotechnological tools.

Srinivasan and Gopimony (1969) observed that wild variety *S.melongena* var. *insanum* is resistant to bacterial wilt (*Ralstonia solanacearum*). Later, Gopimony and George (1979) confirmed the findings of Srinivasan and Gopimony (1969). The cultivar Pusa Purple Cluster is also found to be resistant to bacterial wilt (Rao *et al.*, 1976; Mukherjee and Mukhopadhyay, 1982). *Solanum melongena* var. *insanum* carries a single dominant gene for resistance to bacterial wilt (Swaminathan and Srinivasan, 1971). Ushamani and Peter (1987) rated three lines SM6-1SP, SM6-1M and MS-7SP resistant to *Pseudomonas solanacearum*. Lalramhlimi *et al.* (2019) identified three elite brinjal genotypes 16/BRBW RES-2, 2012/BRBW RES-5 and 2012/BRBW RES-6, which had recorded high bacterial wilt resistance coupled with diverse horticultural traits, could be utilized as donor parents either for developing hybrids or to isolate promising lines, in segregating generation, resistant to this disease. Ali *et al.* (1990) from Japan reported that *S. ferox* and *S. toxicarium* were immune to 4 and 5 strains of *Pseudomonas solanacearum*, respectively. Moderate susceptibility was evident in *S. sisymbrifolium* against most of the strains; *S. integrifolium* and its hybrids with *S. melongena* were highly tolerant of 2-3 strains. They, however observed poorest level of resistance in *S. melongena* var. *insanum*, *S. gilo* and their hybrids with *S melongena*.

Namisy *et al.* (2019) screened eggplant germplasms including cultivated and wild against two strains of *R. solanacearum* Pss97 and Pss2016 and results indicated that one accession of *S. sisymbriifolium* (SIS1) and two accessions of *S. torvum* (TOR2 and TOR3) were resistant or moderately resistant to the strain Pss97. Three resistance QTLs (*EBWR2, EBWR14*, and *EBWR9*) control strains belonging to one or more phylotypes. By combining three resistance QTLs, breeders could able to develop varieties with a large spectrum of bacterial wilt resistance (Salgon *et al.*, 2017). Breeding for stable resistance is a challenge due to the fact that resistance in brinjal to bacterial wilt can be location-specific, and incorporation of resistance into large-fruited varieties has also been a major challenge to breeders for coming years, suggesting linkage between small fruit size and bacterial wilt resistance (Lalramhlimi *et al.*, 2019). Selection of cultivar with higher inner area of large vessel in both stem and roots and higher number of large over small vessel per sqmm of stem vascular area in breeding programme along with comprehensive investigation on anatomy of the vasculature to vascular bacterial wilt resistance in brinjal merits in development of durable resistant and/or tolerant cultivar against the disease (Chattopadhyay *et al.*, 2012; Ghosh *et al.*, 2005).

Accessions of *S. torvum* collected in Java Island represent a valuable source of resistance against both *Ralstonia* and *Fusarium* wilts and can be utilized for the improvement of eggplant (Gousset, 2005).

In France, several crosses were made between different *S.aethiopicum* genotypes and tropical varieties of *S. melongena* and despite sterility problems encountered in F_1, selection was successfully applied using an artificial inoculation test. Subsequently, a recurrent selection scheme involved 2 backcross to *S. melongena* was performed, yielding types with a high level of wilt resistance and a wide range of fruit shape and colour (Ano *et al.*, 1991).

Sakata *et al.* (1989) studied the potential sources of resistance of 43 non-tuberous *Solanum* sp.to *Pseudomonas solanacearum* and *Verticillium dahlia* for use in brinjal breeding. They found that high resistance to *V dahlia* was shown by BIRS0655 (*S.abutiloidea*), BIRS0839 (*S.torvum*), BIRS1034 (*S.caripense*), BIRS1171 (*S. persicum*), BIRS0246 (*S. scabrum*) and to *P solanacearum* by BIRS1816 (*S. carolinense*), and BIRS1489 (*S. violaceum*). According to Chaudhary and Sharma (1999), a single dominant gene controls the resistance to bacterial wilt which can incorporate resistance into the susceptible cultivars through' backcross pedigree method. Interspecific hybridization between *S.torvum* and *S.melongena* for resistance to *Verticillium* wilt clearly indicated that *S. torvum* is a good donor of resistant genes to eggplant cultivars (Bletsos *et al.*, 2000).

In respect of resistance breeding against pests, Dhankar *et al.* (1977) extensively screened brinjal cultivars and wild species against shoot and fruit borer (*Leucinodes orbonalis*). They found that *S. sisymbrifolium* and PPC-2 are the least susceptible lines to shoot and fruit borer. Genetic variability studies of eight eggplant genotypes and four related *Solanum* species, *S. gilo, S. anomalum, S. incanum* and *S. indicum* in relation to yield and shoot and fruit borer infestation indicated considerable genetic variability existed for all the characters studied. High genotypic and phenotypic coefficients of variation were observed for length and diameter of fruits, yield of fruits per plant, percentage of infested yield and percentage of infested fruits per plant. High heritability together with high genetic advance was observed for diameter and length of fruit and fruit yield (Behera *et al.*, 1999). Similarly, *S.torvum* is reported to be highly resistant to epilachna beetle. Singh and Sidhu (1988) concluded that the cultivars PBR-129-5 and SM17-4 can be used as a part of an integrated pest management programme. Dhankar (1988) reviewed the progress in resistance studies in brinjal against shoot and fruit borer and remarked the existence of low level of resistance in the cultivars screened. Screening of large numbers of cultivars and breeding lines with broad genetic base were emphasized. Mandal *et al.* (2019) identified some genotypes BCB-50, BCB-11, BCB-30, BCB-40, Garia, and Punjab Sadabahar as potential donors that could be passed on to breeders not only for FSB resistant breeding but also for improvement in yield and better fruit quality.

Solanum viarum, closely related and cross compatible to brinjal was found to be resistant to shoot and fruit borer. Therefore, hybridization programme was undertaken by Pugalendhi *et al.* (2010) to transfer the resistance traits of shoot and fruit borer and combine with a high yielding but susceptible variety EP 65. Selfing was done from F_2 onwards to obtain F_9 generation and ultimately, in the F_9 generation of the cross EP65 x *Solanum viarum*, recombinant inbred progenies namely HD 1 (34-55-25-10-08-07) and HD 2 (34-55-25-10-08-09) were identified based on their high marketable yield (4.07 kg/plant and 3.46 kg/plant, respectively) and the least infestation of shoot borer (7.69 per cent and 9.09 per cent, respectively) and fruit borer (6.67 per cent and 6.85 per cent, respectively).

Barik *et al.* (2020) screend 59 brinjal genotypes which includes germplasm accessions, high yielding varieties, F_1 hybrids and a wild species *Solanum sisymbrifolium* against epilachna beetle during the kharif season and results revealed that *S. sisymbrifolium* was found to be highly resistant to epilachna beetle with lowest infestation percentage and lowest number of grubs/adults per plant. Among the other genotypes BRINJAL No. 23, CHB-11 and CHB-35 were found to be moderately resistant with scores ranging from 1-2.

Kameya *et al.* (1990) successfully produced somatic hybrids between *S. melongena* and *S.integrifolium* by protoplast fusion with a view to transferring resistance to *Pseudomonas solanacearum* from *S. integrifolium* cv. Hiranasuto *S. melongena* cv. Shironasu and indicated the possibility of developing a resistant type. Interspecific somatic hybridization between *S. melongena* and *S.torvum* by mesophyll protoplast fusion has been successfully carried out in USA and *in-vitro* culture shoots from hybrids were resistant to *Verticillium dahlia* extract (Guri and Sink, 1988). In France, Sihachakr *et al.* (1989) also produced somatic hybrids plants by electrofusion of mesophyll protoplasts between the two species.

Vijaya and Black Beauty are tolerant to root-knot nematodes (*M. incognita*) as reported by Yadav *et al.* (1975).

Table 5: *Solanum* Species as Source of Resistance to some Insect-Pests and Diseases*

Species	*Insect-pests/Nematode*	*Diseases*
Solanum integrifolium	Red spider mite	Littleleaf, *Fusarium* wilt and *Verticillium* wilt
S. gilo	Shoot and fruit borer	Little leaf, *Phomopsis* fruit rot
S. indicum	-do-	*Fusarium* wilt, *Phomopsis* fruit rot
S. incanum	-do-	*Fusarium* wilt
S. sisymbrifolium	Root knot nematode	*Fusarium* wilt, *Phomopsis* fruit rot, *Verticillium* wilt
S. torvum	-do-	*Verticillium* wilt, *Phomopsis* fruit rot
S. aculeatissimum	-do-	*Verticillium* wilt
S. sodamaem	-do-	*Verticillium* wilt
S. xanthocarpum	-do-	*Phomopsis* fruit rot
S. khasianum	-do-	*Phomopsis* fruit rot
S. nigrum	-do-	*Phomopsis* fruit rot
S. testiculatum	-do-	*Phomopsis* fruit rot

* Kalda and Kumar (1999).

Resistance to different insect pests, nematode and diseases is available in a number of *Solanum* spp.are presented in Table 5.

10.3.3 Prospects

The use of F_1 hybrid cultivars is likely to increase as a result of appreciation of their attributes and efforts to economise in such seed production need proper attention. Breeding for disease and pest resistance will require constant surveillance and resistance is badly needed in respect of bacterial wilt (*Pseudomonas solanacearum*), phomopsis blight (*Phomopsis vexans*), little leaf, shoot and fruit borer (*Leucinodes arbonalis*), red spider mite and jassids. Breeding for resistance to root-knot nematode (*Meloidogyne* spp.) also requires special attention. Several related wild species are the possibilities for getting complete resistance sources and difficulties in such interspecific somatic hybridization by protoplast fusion will go a long way in developing resistant cultivars by exploiting related wild species possessing resistant characters.

11.0 Biotechnology

The biotechnological research on eggplant has been reviewed by Sen and Deka (2001). The ability of eggplant regeneration in tissue culture has allowed the application of biotechnology, particularly the exploitation of somaclonal variation, haploidy, somatic hybridization, and genetic transformation (Collonier *et al.*, 2001). Salient achievements in researches on various aspects of biotechnology in eggplants are summarized below:

11.1.0 Tissue Culture

11.1.1 Micropropagation

Polisetty *et al.* (1994) described a protocol for micropropagating eggplants from leaf explants and Bhansali and Ramawat (1993) from axillary buds. Low sugar concentration was important for shoot regeneration. Yamamoto and Matsumoto (1998) obtained differentiated adventitious bud on the hypocotyls derived callus. The regeneration of plantlets depended on explants type and genotype (Ionescu, 1998). Franklin *et al.* (2004) observed that eggplant could be efficiently regenerated from cultured roots of 15 days old seedlings on Murashige and Skoog (MS) medium containing 0.45 mM thidiazuron and 13.3 mM 6-benzyladenine with an 8-fold increase in the production of shoots per explants. They further concluded that the age of the explants and PGR's used in the medium play a crucial role for plant regeneration from root explants of eggplant.

Ferdausi *et al.* (2009) recorded the highest rate of callus induction in shoot tip (82.78 per cent) which was followed by mid-rib (78.71 per cent) and hypocotyl (74.88 per cent) with 2 mg/l NAA showing the best performance for callus proliferation and 0.5 mg/l BAP giving the highest percentage of root regeneration (57.13 per cent).

Mannade *et al.* (2013) carried out a study to standardize the factors affecting the *in–vitro* regeneration of brinjal. It was observed that the calluses of *S. melongena* could be successfully induced from *in vitro* germinated seedlings when explants from leaf and hypocotyls were cultured on MS media added with 2,4 D and BAP with significant differences between the explants and culture medium. Highest callusing was obtained in BAP (1.5mg/l) + 2, 4-D (2mg/l) for both hypocotyls and leaf explants with maximum shoot regeneration in MS media containing BAP (1.5mg/l) + Kinetin (1mg/l) at 14 days after callus inoculation.

In another study Satish *et al.* (2015) evaluated the effect of seaweed liquid extracts (SLEs) which were made from *Gracilaria salicornia, Padina gymnospora, Padina boergesenii, Gelidiella acerosa* and plant growth regulators (PGRs) for *in-vitro* mass propagation using hypocotyls and leaf disc explants of brinjal (*Solanum melongena* L.) cultivar Pusa purple long and reported that there was enhanced germination, high rate of shoot induction (96.2 per cent), proliferation (6 cm) and rooting (95.3 per cent) with 20 to 40 per cent of SLEs. Therefore, SLEs could serve as an alternative to phytohormones as they were easy to extract and gave quick and high-frequency mass propagation.

11.1.2 Anther Culture

Raina and Iyer (1973) were the first to report plant regeneration from anther culture in eggplant by regenerating homozygous diploids (double haploid) plants through the callus developed from anthers cultured at uninucleate pollen stage treated with colchicine. Production of haploid plant through anther culture of eggplant was reported by Isouard *et al.* (1979) and Dam usde Vaulx and Chambounet (1982). Eight day pretreatment at 35°C yielded an average of 12 plants per 100 anthers. With 2, 4-D 15 to 50 per cent regenerants were found to be diploid, while with NAA only haploid plants were regenerated. It was also reported that when anthers were incubated at high temperature (35 ± 2°C) under dark conditions for 7-8 days an improvement in the haploid plant formation was observed (Dumas de Vaulx and Chambonnet, 1982). Anther culture protocol could be improved by allowing a controlled environment condition for production of donor plants, a modified N_6 medium with zeatin for culture of anther callus and addition of $AgNO_3$ to the regeneration medium (Khatun *et al.*, 2006).

11.1.3 Somatic Hybrids

The first successful somatic hybrid in eggplant was developed between cultivars of *Solanum melongena* and *Solanum sisymbrifolium* by using polyethylene glycol (PEG) mediated protoplast fusion (Gleddie *et al.*, 1986). Somatic hybrids between *S.melongena* and one of its related species *S. aethiopicum* were obtained by Daunay *et al.* (1993) also with *S. torvum* (Guri and Singh, 1988) and *S. khasianum* (Sihachakr *et al.*, 1988). A variety of ploidy level in the hybrids ranging from tetraploid, hexaploid and mixoploid was reported. Among the hybrids two tetraploid somatic hybrid clones were found more productive. Production of interspecific somatic hybrids between eggplant and *Solanum torvum* was also reported by Guri and Sink (1988a). Li and Constabel (1988) reported fusion of *S. melongena* and *S.tuberosum*. Jarl *et al.* (1999) transferred tolerance to verticillium wilt in *S.torvum* in eggplant through protoplast fusion. A total of 12 hybrids were found to have a high degree of veticillium tolerance under field condition with the morphological character of eggplant including normal seed set. Asymetric somatic hybrids were obtained after fusion of protoplast of an interspecific tomato hybrid (*Lycopersicon esculantum* x *L.pennellii*) and *Solanum melongena*. Cytology revealed that 4 hybrids plants had chromosome number between 42 and 60 and all were sterile. Anther culture was also successfully employed to reduce the ploidy level of eggplant between somatic hybrids of *Solanum melongena* and the wild species i.e *S. integrifolium and S aethiopicum* to the diploid status so that the resulting plants can be easily crossed with the diploid cultivated eggplant for transfer of useful traits (Rotino *et al.*, 2001; Rizza *et al.*, 2001). Somatic hybrids have also been developed between *S. melongena, S aethiopicum* group *aculeatum* and *gilo* and were found to be highly resistant to bacterial wilt (Collonier *et al.*, 2001). These somatic hybrids were produced by electrical fusion of mesophyll protoplasts.

11.1.4 Somaclonal Variation

Somaclonal variations were observed among the plants regenerated through somatic embryogenesis induced by NAA or 2, 4-D (Hitomi *et al.*, 1998). It was observed that variations in leaf shape, plant height, fruit shape and pollen fertility were higher in the variants obtained by treatment with NAA than those with 2, 4-D. Plant species, the genotype as well as the explants used, the protocol being followed during *in-vitro* process, composition of the medium and the number of subcultures were some of the factors influencing the frequency of somaclonal variation (Cai *et al.*, 1990; Ducos *et al.*, 2003; Magioli and Mansur, 2005). Zayova *et al.* (2010) observed somaclonal variation in eggplants regenerated from cotyledon derived callus. The plants were propagated by node and shoot tip explants on Murashige and Skoog medium supplemented with 1.0 mg/l^{-1} BAP and rooted on the same medium but without growth hormones.

11.2.0 Biochemical and Molecular Markers

Isozymes have been used to study the genetic relationship in species of *Solanum* and hybrids and cutivars of egg plant.

11.2.1 Isozymes

Inheritance and linkage relationship of isoenzymes of aspartate amino transferase (AAT), alcohol dehydrogenase (ADH), phosphogluconate dehydrogenase (PGD), phosphoglucomutase (PGM) and shikimate dehydrogenase (SKDH) were studied in *Solanum melongena* and its wild relative *S.indicum* (Isshiki *et al.*, 1994). Evidence of Mendelian inheritance was obtained for loci: Aat-1, Adh-1,Adh-2.Pgd-1, Pgm-1, Pgm-2 and Skdh-1. Tests of 22 pairs of loci for independent assortment suggested 3 linked pairs: Aat-2 with Pgd-2, Adh-2 with Pgm-I and Pgd-Z with Pgm-2.

Karihallo and Gottlieb (1995) surveyed 29 isozyme loci to know the genetic relationship among 29 accessions of *S.melongena*, 33 accessions of weedy forms (referred to as *insanum*) and 2 accessions of wild forms (referred to as *incanum*). The results demonstrated that although the taxa included large morphological diversity but they had a very close genetic relationship.

Toppino *et al.* (2008) evaluated the extent of genetic transformation by employing biochemical and molecular markers to understand the gene exchanges between *S. melongena* and *S. aethiopicum* for introgression of useful traits. The isozyme markers which were studied in the 70 dihaploids segregated in a distorted manner which did not fit in any of the expected segregation ratios.

11.2.2 Molecular Markers

RFLP analysis of a polymerase chain reaction (PCR) amplified 3.2kb region of ctDNA bounded by the conserved sequences in rbcL and ORP 106 was performed in eggplant and its related *Solanum* species (lsshiki *et al.*, 1998). The results revealed that the restriction pattern divided the examined nine species into the following five clusters: (i) *S. melongena* and *S. incanum*, (ii) *S. virginianum* (=*S.surattense*), (iii) *S. torvum*, (iv) *S. aethiopicum* (=*S gilo*), *S aethiopicum* (=*S integrifolium*), *S violaceum* (=*S.indicum*) and *S.violaceum* (=*S.sanitwongsei*), and (v) *S.mammosum*.

The first sets of eggplant microsatellites was developed from the screening of small insert genomic libraries with di- and trinucleotide probes (Nunome *et al.*, 2003 a,b). A small set of SSR markers was developed from the genic DNA sequence lodged in public access databases (Stagel *et al.*, 2008), while Nunome *et al.* (2009) reported the identification of over 1,000 SSR markers from a screen of enriched gDNA and cDNA libraries. An extensive set of nearly 2,000 putative eggplant SSRs was later on isolated by Barchi *et al.* (2011) from RAD (restriction-site associated DNA) tags, of which a sub-set showed to be polymorphic among the parents of mapping populations. Among the molecular markers, SSRs have been revealed as the best available marker to study the relationships of different groups of cultivars of eggplant as they have wide genome and are highly variable, having a high repeatability, are easy to use, and are amenable to high throughput (Varshney *et al.*, 2005; Kalia *et al.*, 2011). Munoz-Falcon *et al.* (2008, 2009, 2011) have found that SSRs are much better than AFLPs at resolving the relationships between and within the cultivar groups as well as for assigning correct accessions to their cultivar groups. Also, Demir *et al.* (2010) compared SSRs with RAPDs and found that the former were more adequate than the latter to study the diversity and relationships of local landraces. In eggplant SSR markers have been applied in diversity, phylogenetic as well as linkage mapping studies and several hundred SSR markers are available in eggplant (Stagel *et al.*, 2008; Nunome *et al.*, 2009; Ge *et al.*, 2011; Tumbilen *et al.*, 2011; Barchi *et al.*, 2010, 2012; Vilanova *et al.*, 2012; Prohens *et al.*, 2012; Cericola *et al.*, 2013; Gramazio *et al.*, 2017) and among them genomic SSRs have proved to be more polymorphic than EST-SSRs (Munoz- Falcon *et al.*, 2011). Hurtado *et al.*, 2012 assessed the genetic diversity and relationships of 52 accessions of eggplant from three geographically distant secondary centers of diversity i.e China, Spain, and Sri Lanka using 28 morphological descriptors and 12 highly polymorphic genomic SSRs and found a wide variation for most of the morphological traits and also identified 110 alleles which were private for each origin. SSR markers were also developed from *S. melongena* genomic libraries enriched for AG/CT motif (Vilanova *et al.*, 2012) and, more recently, from transcriptome analysis of *S. incanum* and *S. aethiopicum*, two close relatives of the common eggplant (Gramazio *et al.*, 2016). Recently a high quality genome draft of eggplant, covering 1.06 Gb of gap free sequence with an N50 of 2.9 Mb, was produced by a combination of Illumina sequencing and optical mapping and was anchored to 12 pseudomolecules2 (The Eggplant Genome Consortium, 2017).

Portis *et al.* (2018) characterized the SSR markers of the eggplant (*Solanum melongena*) by using a recent high quality sequence of its whole genome and found nearly 133,000 perfect SSRs of which 15.6 per cent were complex, with two stretches of repeats separated by an intervening block of <100 nt. A total of 133k/178k eggplant-specific perfect/imperfect microsatellite loci have been reported, which were mined using the whole genome bioinformatics survey. The SSRs were classified according to their number of repeats and overall length, and were assigned to their linkage group.

An "Eggplant Microsatellite DataBase" (EgMiDB) was also developed from the work carried out by Portis *et al.*, 2018 which permits the identification of SSR markers in terms of their location on the genome, type of repeat (perfect vs. imperfect), motif type, sequence, repeat number and genomic/gene context. It is easy to use web interface and a highly customized primer designing tool to cater the needs of plant biologists and breeders. EgMiDB will thus contribute to uniform bioinformatics work flow strategy to achieve SSR genotyping tailored to eggplant, and represents a key tool for making progress in both basic genomics research and molecular breeding of the species. The database will be constantly updated as additional sequencing data for eggplant genotypes become available, allowing rapid and simple in silico identification of SSRs including those polymorphic

Tiwari *et al.* (2019) developed a mapping population of two elite cultivars of eggplant i.e Ramnagar Giant (*Solanum melongena*) and W-4 (*Solanum incanum*) to develop large scale SSR markers to be used in QTL mapping and in *S. melongena*, 11,262 SSR while in *S. incanum* 11,829 SSR containing regions were identified. Further, a total of 536 SSR markers were designed and screened for polymorphism of which, 157 markers produced polymorphism between the parental lines. The polymorphic SSRs will then be used for genotyping of RILs to map QTLs for various horticultural traits in eggplant and identification of candidate genes in response to biotic and abiotic stress.

11.3.0 Genetic Transformation

One of basic pre-requisite for the genetic engineering of a crop plant is the availability of a procedure for genetic transformation based on high frequency regeneration *in-vitro*. A number of researches have been done to establish the procedure to regenerate eggplants *in-vitro*. Genetic transformation of eggplant via *Agrobacterium* was first reported by Guri and Sink (1988b), using leaf explants and a cointegrate vector, although no success was achieved with a binary vector. Later, Fillipone and Lurquin (1989) reported the transformation of leaf and cotyledon explants using the wild supervirulent strain A281. Transgenic plants were obtained by Rotino and Gleddie (1990) and Fári *et al.* (1995), using organogenic regeneration systems. An optimization of factors that influence transformation efficiency, including length of pre and post-coculture periods, explant type, and genotype was performed using a TDZ-based organogenic system (Magioli *et al.*, 2000). The efficiency of transformation protocols based on organogenesis may be influenced by the antibiotic used to eliminate *A. tumefaciens*.

The production of transgenic eggplant through somatic embryogenesis either fails to occur (Fillipone and Lurquin, 1989) or is achieved with very low efficiency (Fári *et al.*, 1995). It has been demonstrated that both cocultivation with *Agrobacterium* and the presence of bactericidal antibiotics used in transformation protocols cause a reduction of 80-99 per cent in the number of embryo/explant. The inhibitory effect on somatic embryos development may result from the interaction between the physiological alterations caused by these treatments and the delicate processes of gene regulation that are induced in early culture stages (Magioli *et al.*, 2001). High frequency of somatic embryogenesis and plant regeneration in eggplant were reported by using leaf explants and cell suspension (Gleddie *et al.*, 1982). The somatic embryogenesis was highly responsive in the presence of α-naphthalene acetic acid (NAA) whereas in other auxins, only callus was formed. Cell suspensions were also embryogenic when grown under conditions which stimulated embryogenesis in leaf explants. Somatic embryos from leaf callus or cell suspensions cultured on hormone-free medium gave rise to whole plant. Plant regeneration was obtained from mesophyll protoplasts cultured in half-strength MS medium. Shoot regeneration took place in the presence of IAA and zeatin in MS medium. Whole plants were obtained after 3 months.

Following the establishment of basic protocols, successful introduction of agronomic traits into eggplant was achieved via *Agrobacterium* by many researchers (Rotino and Gleddie, 1990; Arpaia *et al.*, 1997; Kumar *et al.*, 1998; Magioli *et al.*, 2000; Prabhavathi *et al.*, 2002; Franklin and Sita, 2003). *Agrobacterium*-mediated transformation of eggplant has been accomplished using binary vector carrying a gene encoding neomycin phosphotransferase-II (nptII) conferring resistance to antibiotics such as kanamycin (Guri and Sink, I988b). Stable transformation to kanamycin resistance was also obtained by incubation of cotyledonary and young leaves with hypervirulent fully oncogenic strains of *A. tumefaciens*. As reported by Filipone (1990), the highest proportion of transformed explants and Largium was obtained from intact cotyledonary leaf pieces, while the highest *npt II* specific activity was detected in callus cells originating from superficially wounded cotyledonary leaf pieces. A novel time and labour saving method for transformation and regeneration of eggplants was described by Szasz *et al.* (1998). In this method co-cultivation with *Agrobacterium*, callus regeneration and organogenesis were carried out in one step via the decapitation of rooted hypocotyls. In most of the cases *nptII* gene was used as selectable marker and the data demonstrated Mendelian inheritance of this functional gene. Based on the procedures developed to transform eggplant using selectable marker gene, genes of economic importance could be introduced.

As discussed earlier brinjal is seriously affected by insect infestations, primarily the eggplant fruit and shoot borer (EFSB), *Leucinodes orbonalis* causing a yield between 30-60 per cent even when the crop has been frequently sprayed EFSB (Mondal and Akter, 2018). The larvae damages the eggplant shoots and flowers but the most serious damage is caused when they bore into the fruit making it unmarketable. Since the crop has to be continuously sprayed during its growing period there is high pesticide residue levels left on the fruit which is unfit for consumption, killing the beneficial insects, exposing the farmers to different chemical hazards and ultimately

Genetic Transformation in Brinjal

Agrobacterium-Mediated Genetic Transformation of Brinjal var. Arka Samhitha using Seed as Explant.

a. Seeds of brinjal var. Arka Samhitha used for the *A. tumefaciens* infection. b. Pre-cultured seeds were infected and co-cultivated with *A. tumefaciens* harbouring pCAMBIA 1301-bar plasmid. c. Brinjal seedlings after 6 days of inoculation on 5 mg l^{-1} BASTA®-containing MS medium. d-g. Different developmental stages of brinjal plantlets after 10, 15, 18, 23 days of seed inoculation on 5 mg l^{-1} BASTA®-containing MS medium. h. Hardened putatively transformed brinjal plantlets. i. Acclimatised putatively transformed brinjal plant showing resistance against 100 mg l^{-1} BASTA® under green house conditions. j. Acclimatised wild type brinjal plant after 100 mg l^{-1} BASTA® spray under green house conditions. k. Putatively transformed plants with brinjal showing resistant against 100 mg l^{-1} BASTA®. l. Canopy of putatively transformed plant with brinjal (Subramanyan *et al.,* 2013).

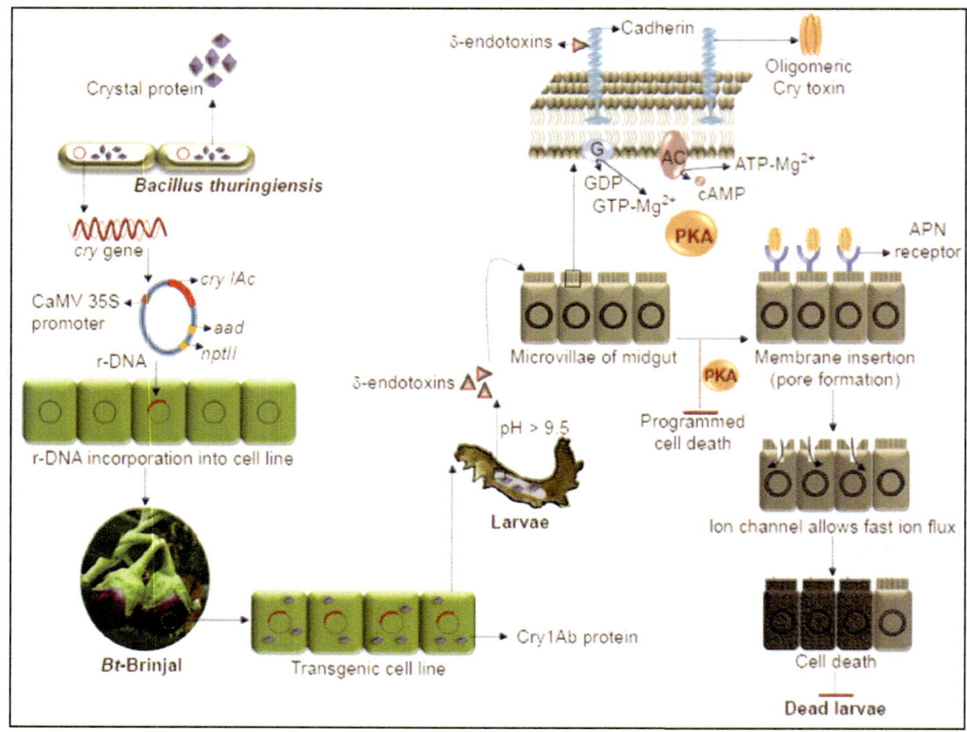

Flow Chart of the Transgenic Development in Brinjal (Biswas *et al.*, 2018).

adding to pollution exposes (Rahman, 2013). Therefore, with the background of genetic transformation, genetically engineered, insect resistant brinjal with the *cry1Ac* gene (Bt brinjal) was developed by the Indian based company Maharashtra Hybrid Seed Company (Mahyco) to provide an effective control of EFSB but till date this technology has not been approved by the Indian governement. Hence, an Agricultural Biotechnology Support Project II at Cornell University, supported by the United States Agency for International Development, facilitated the transfer of the Bt brinjal event ("EE-1") to the Bangladesh Agricultural Research Institute (BARI) and this event was introgressed into several local and commercially popular open-pollinated brinjal varieties (Shelton *et al.*, 2018). Nine Bt varieties underwent 7 years of green house and confined field trials by BARI in various geographic locations in Bangladesh to test their efficacy and environmental safety out of which four were subsequently approved for cultivation by the National Committee on Biosafety (NCB) of Bangladesh on October 2013. The released Bt varieties are BARI Bt Begun-1, BARI Bt Begun-2, BARI Bt Begun-3, and BARI Bt Begun-4 which are Bt isolines of Uttara, Nayantara, Kazla, and ISD006, respectively (Shelton *et al.*, 2018). A study was further conducted to assess the impact of Bt brinjal cultivation in market value chain in five districts in Bangladesh (Shelton *et al.*, 2020) and it was concluded from the study that there was an increased in the yield by 19.6 per cent with 21.7 per cent higher revenue, reducing the hours utilized by the labour for different field operations and reduction in the amount of insecticides being sprayed.

12.0 References

Abak, K. and Guler, H.Y. (1994) *Acta Hortic.*, **366:** 85-91.

Abak, K., Sari, N. and Pakyurek, A.Y. (1992) *DogaTurkTarimve Ormancilik Dergisi*, **16:** 212-221.

Abdelfattah, M.A. and Abdel-Salam, A.S. (1972) *BeitrageZurTropischen undsubtropischen Landwirtschaft und Tropenveterinarmedizin*, **10**: 239-246.

Abdel-Monaim, M.F., Abdel-Gaid, M.A., Zayan, S.A. and Nassef, D.M.T. (2014) *Inter. J. Phytopathol.*, **3**(1): 33-40.

Abe, K., Iwata, T. and Ogata, K. (1974) *J. Jap. Soc. Hortic. Sci.*, **42**: 402-407.

Abutiate, W.S. (1988) *Trop. Agric.*, **65**: 188-189.

Acciarri, N., Celani, G. and Porcelli, S. (1989) *Informatore Agrario*, **45**: 67-69.

Adawadkar, M.P., Deshmukh, M.M. and Wadatkar, S.B. (2019) *Int. J. Curr. Microbiol. App. Sci.*, **8**: 2793-2800.

Addae-Kagya, K.A. and Norman, J.C. (1977) *Acta Hortic.*, **53**: 397-401.

Adiroubane, D. and Raghuraman, K. (2008) *J. Biopestic.*, **1**(2): 124-129.

Agnihotri, B.N., Kapoor, K.L. and Shukla, B.L. (1963) *Indian Agric.*, **7**: 72-75.

Agnihotri, N.P., Sinha, S.N., Jain, J.J. and Chakrabarti, A.K. (1990) *Indian J. Entom.*, **52**: 373-378.

Agrawal, R.L. (1980) *Seed Technology. Oxfordand IBH Publishing Co.*, New Delhi, pp. 198-201.

Ahmed, N., Singh, S.R. and Lal, S. (2013) *Indian J. Agric. Sci.*, **83**(1): 93-95.

Akiba, F., Ribeiro, R. Del. D., Castro, L.A.B.De, Robbs, C.F., Kimura, O. and Sudo, S. (1972) *Arquivos daUniversidade Federal Ruraldo Riode Janeiro*, **2:** 17-21.

Akter, S., Alam, M. Z., Rahman, M. M., and Akanda, A. M. (2017). *The Agric.*, **15**: 49-57.

Alam, M.M. (1975) *Indian J. Mycol. Pl. Path.*, **5:** 15.

Alam, M.M., Ali, Q.G., Masood, A. and Khan, A.M. (1976) Indian *J. Expt. Biol.*, **14:** 517-518.

Alam, M.Z., Ali, M., Akanda, A.M., Choudhury, D.A.M., Haque, N.M.M., Hossain, N.M. and Ogata, K. (1994) *Bull. Inst.Trop. Agric.Kyushu Univ.*, **17:** 85-91.

Ali, M., Quadir, M.A., Okubo, H. and Fujieda, K. (1990) *Sci. Hort.*, **45:** 1-9.

Ali, M.A., Gencoglan, C. and Gencoglan, S. (2019) *Int. J. Plant Soil Sci.*, **29**(1): 1-9.

Aliev, D.A. (1971) *Trudy Azerbaidzhanskiz, Nauchno-Issledovatelskii Institute Zemledeliya*, **15:** 64-69.

Al-juboory, B.A. andbShati, R.K. (1996) *Arab J. Pl. Protection*, **14:** 36-40.

Alkolaly, A., Helal, M. R. and Mostafa, S. H. (2018) *Zagazig J. Agril. Res.*, **45**: 1675-1682.

Alvarez-Garcia, L.A. (1946) *J. Agric., Puerto Rico*, **30:** 69-96.

Americanos, P.G. (1976) *Technical Paper, Agric. Res. Inst.*, Cyprus, **14:** 11.

Anand, G.K.S., Sharma, R.K. and Shankarganesh, K. (2013) *Pestic. Res. J.*, **25**(1): 6-11.

Anaso, H.U. and Ozo, J.O. (1990) *Cytologia*, **55:** 1-14.

Andryushchenko, V.K. and Vyrodova, A.P. (1981) *Fizzologiya I Biokhimiya Kul'turnykh Rastenii*, **13:** 310-313.

Anjaneyulu, A. and Ramakrishnan, K. (1968) *Madras Agric. J.*, **55:** 142-143.

Anjaneyulu, A. and Ramakrishnan, K. (1972) *Mysore J. Agric. Sci.*, **6:** 147-154.

Ano, G., Hebert, Y., Prior, P. and Messiaen, C.M. (1991) *Agronomie*, **11:** 550-560.

Anonymous (1981-82) *Progress Report, All India Coordinated Vegetable Improvement Project*, Indian Council of Agricultural Research, NewDelhi.

Anonymous (2018) Horticultural statistics at a glance. Government of India, Ministry of Agriculture and Farmers' Welfare Department of Agriculture, Cooperation and Farmers' Welfare, Horticulture Statistics Division

Ansari, A.M. and Singh, Y.V. (2016) *Electron J. Plant Breed.*, **7**(2): 197-208.

Ansari, S.F., Mehta, N., Ansari, S. and Gavel, J.P. (2011) *Elec. J. Plant Breed.*, **2**(2): 275-281.

Anwar, A., Bhat, M., Ganaie, N., Ambardar, V.K. and Hassan, M.G. (2017) *The Pharma Inno. J.*, **6**(9): 318-323.

Arivalagan, M., Bhardwaj, R., Gangopadhyay, K.K., Prasad, T.V. and Sarkar, S.K. (2013) *J. Appl. Bot. Food Qual.*, **86**: 99-103.

Arivalagan, M., Gangopadhyay, K.K., Kumar, G., Bhardwaj, R., Prasad, T.V., Sarkar, S.K. and Roy, A. (2012) *J. Food Compos. Anal.*, **26**: 173-176.

Arpaia, S., Mennella, G., Onofaro, V. and Perri, E. (1997) *Theor. Appl. Genet.*, **95**: 329-334.

Arti, D., Sharma A.K. and Kumar, R. (2018) *Int. J. Curr. Microbiol. App. Sci.*, **7**(9): 2567-2572.

Awal, A., Rahman, M.M., Alam, M.Z. and Khan, M.M.H. (2017) *Jahangirnagar Univ. J. Biol. Sci.*, **6**(1): 35-43.

Aykroyd, W.R. (1963) *ICMR Special Rpt.*, 42.

Azadvar, M. and Baranwal, V.K. (2012) *Phytopatho. Mollicutes.*, **2:** 15-21.

Aziz, I., Ahmad, R. and Javed, N. (1995) *Pakistan J. Phytopath.*, **7**: 68-70.

Babu, P.C.S., Lakshmanan, P., Hanifa, A.M. and Gunasekharan, C.R. (1970) *South Indian Hort.*, **18**: 33-38.

Baha-Eldin, S.A. and Blackhurst, H.T. (1968) *Proc. Amer. Soc. Hort. Sci.*, **92:** 480-489.

Bajaj, K.L., Kaur, G. and Chadha, M.L. (1979) *J. Plant Foods*, **3:** 163-168.

Bajaj, K.L., Kaur, G., Chadha, M.L. and Singh, B.P. (1981) *Veg. Sci.*, **8:** 37-44.

Balaji, L., Reddy, P.S., Reddy, R.V.S.K. and Sivaraj, N. (2013) *J. Res. ANGRAU.*, **41**(1): 79-82.

Balakrishnan, K., Thandapani, V., Prakash, M. and Arumugam, R. (1993) *South Indian Hort.*, **14:** 148-151.

Balwani, A.K., Patel, J.N., Acharya, R.R., Gohil, D.P. and Dhruve, J.J. (2017) *J. Pharmacogn. Phytochem.*, **6**(5): 187-190.

Bandopadhyay, C.R., Bid, N.N. and Das, P.K. (1972) *Proc. 3rd Int. Sump. Subtrop. and Trop. Hort.*, Bangalore, pp. 148-149.

Banerjee, S.N. and Basu, A. (1955) *Sci. and Cult.*, **20:** 350-351.

Banerjee, S.N. and Basu, A.N. (1956) *Proc. Zool. Soc.*, Calcutta, **9:** 45-52.

Bangal, G.D. and Gutal, G.B. (1986) *Current Res. Reptr., Mahtma Pirnie Agril. Univ.*, **2**(Spl.No.): 33-36.

Barbieri, G. and Deveronico, L. (1989) *Irrigazione e Drenaggio*, **36:** 121-125.

Barchi, L., Lanteri, S., Portis, E., Acquadro, A., Valè, G., Toppino, L. and Rotino, G.L. (2011) *BMC Genom.*, **12**: 304.

Barchi, L., Lanteri, S., Portis, E., Stàgel, A., Valè, G., Toppino, L., and Rotino, G. L. (2010) *Genome*, **53**: 805-815.

Barchi, L., Lanteri, S., Portis, E., Valè, G., Volante, A., Pulcini, L., Ciriaci, T., Acciarri, N., Barbierato, V., Toppino, L. and Rotino, G.L. (2012) *PloS one*, **7**: e43740. https: / /doi.org/ 10.1371 /journal.pone.0043740

Barchi, L., Toppino, L., Valentino, D., Bassolino, L., Portis, E., Lanteri, S., Rotino, G.L. (2018) *Euphytica*, **214:** 20.

Barik, S., Dash, L., Das, M.P., Panigrahi, T., Naresh, P., Kumari, M., Sahum, G.S. and Acharya, G.C. (2020) *J. Entomol. Zool. Stud.*, **8**(2): 297-301.

Bashar, M.A., Akter, R. and Hossain, K.S. (2015) *Dhaka Univ. J. Biol. Sci.*, **24**(2): 209 213.

Basile, M. and Lamberti, F. (1981) *Rijksuniversiteit Gent.*, **15:** 175-178.

Behera, T.K., Singh, N. and Kalda, T.S. (1999) *Orissa J. Hort.*, **27:** 1-3.

Bennison, J.A., Sampson, C., Vautier, A. and Challinor, P.F. (1996) *Bull. OILB/SROP*, **19:** 7-10.

Berua, S.K., Mahanta, I.C. and Mohapatra, K.B. (2008) *J. Mycopathol. Res.*, **46**(1): 73-76.

Bhaduri, P.N. (1951) *Indian J. Genet.*, **2:** 75-86.

Bhandari, K.G. and Sohi, G.S. (1962) *Punjab Hort. J.*, **2:** 44-61.

Bhansali, R.R. and Jindal, S.K. (1997) *In: Recent Advances in Management of Ecosystem*, pp. 391-396.

Bhansali, R.R. and Ramawat, K.G. (1993) *J. Hort. Sci.*, **68:** 25-30.

Bharathi, M. (1999) *J. Mycol. Plant Path.*, **29:** 57-60.

Bhardwaj, A., Goswami, B.K., Bhardwaj, V. and Singh, N. (2019) *Plant Arch.*, **19**(2): 44-46.

Bhatt, P.K., Zapadiya, V.J., Sapovadiya, M.H. and Chetariya, C.P. (2019) *J. Pharmacog. Phytochem.*, **8**(4): 1384-1388.

Bhushan, A., Samnotra, R.K. and Kumar, S. (2018) *Int. J. Curr. Microbiol. App. Sci.*, **7**(6): 1647-1653.

Bhushan, S., Chaurasia, H.K. and Shanker, R. (2011) *The Bioscan*. **6**(4): 639-642.

Bid, N.N., De, A.K., Srivastava, B.K. and Kole, K.K. (1993) *In: Proc. Workshop on Micronutrients*, 22-23 January, 1992, Bhubaneswar,India.

Bilal, M., Javed, K. and Javed, H. (2017) *J. Entomol. Zool. Stud.*, **5**(2): 174-179.

Bindra, O.S. and Mahal, M.S.(1981) *Phytoparasitica*, **9:** 119-131.

Bindra, O.S. and Soodan, L.S. (1974) *Indian J. Hort.*, **31:** 286-290.

Bindra, O.S., Sohi, AS., Khatri, H.L. and Deol, G.S. (1972) *Curr. Sci.*, **41:** 819-820.

Biswas, L., Mehta, N. and Ansari, S.F. (2013) *Glob. J. Sci. Front. Res. Agric, Vet.*, **13**(9): 9-11.

Biswas, M., Mishra, I. and Mishra, B.K. (2020) *Int. J. Curr. Microbiol. App. Sci.*, **9**(2): 1442-1447.

Biswas, M., Mishra, I. and Mishra, B.K. (2020) *J. Entomol. Zool. Stud.*, **8**(2): 1562-1566.

Biswas, R., Banerjee, A., Halder, U. and Bandopadhyay, R. (2018) In: *Genetic Engineering in Horticultural Crops*, Elsevier Inc.,pp.155-167.

Bletsos, F., Thanassoulopoulos, C. and Roupakias, D. (2003) *Hort. Sci.*, **38**(2): 183-186.

Bletsos, F.A., Rounkias, D.G. and Thanassoulopoulos, C.C. (2000) *Acta Hortic.*, **522:** 71-78.

Bora, T. (1977) *J. Turkish Phytopath.*, **6:** 17-25.

Borkakati, R.N., Venkatesh, M.R. and Saikia, D.K. (2019) *J. Entomol. Zool. Stud.*, **7**(1): 932-937.

Bose, A.S.C., Rabeena, I. and Sathyan, T. (2020) *Biotica Res. Today.*, **2**(7): 673-676.

Boyaci, H.F., Oguz, A., Yazici, K.M. and Eren, A. (2011) *African J. Biotech.*, **10**(34): 6522-6528.

Braga, P.C., Lo Scalzo, R., dal Sasso, M., Lattuada, N., Greco, V. and Fibiani, M. (2016) *J. Funct. Foods.*, **20**: 411-421.

Brathwaite, R.A.I. (1978) *Weed Res.*, **18:** 143-146.

Bujdoso, G. and Videki, L. (1976) *Zoldsegtermesztesi Kutato'Intezet Bulletinje*, **11:** 29-37.

Burkill, I.H. (1935) *A Dictionary of the Economic Products of the Malay Peninsula*, Crown Agents for the Colonies, London, p. 2402.

Burman, Uday, Kaul, R.K. and Kumar, Praveen (1995) *Agro-Asian J. Nematology.*, **5:** 213-217.

Butani, K.D. and Varma, S. (1976) *Pesticides*, **10**: 32-35,38.

Buzescu, D. and Cenusa, M. (1996) *Anale Institutul de Cercetari Pentru Legumicultura Si Floricultura, Vidra*, **14:** 331-334.

Cai, T., Ejeta, G., Axtell, J.D. and Butller, L.G. (1990) *Theor. Appl. Genet.*, **79**(6): 737-747.

Canevascini, V. and Caccia, R. (1976) *Informatore di Ortoflorofrutticoltura*, **17:** 15-20.

Cao, G., Sofic, E. and Prior, R.L. (1996) *J. Agric. Food Chem.*, **44:** 3426-3431.

Cao, G.H., Sofic, E. and Prior, R.L. (1996) *J. Agric. Food Chem.*, **44:** 3426-3431.

Cericola, F., Portis, E., Toppino, L., Barchi, L., Acciarri, N., Ciriaci, T., Sala, T., Rotino, G.L. and Lanteri, S. (2013) *PLoS One*. **8**: e73702.

Chadha, M.L. and Hegde, R.K. (1989) *Indian J. Hort.*, **46**: 44-52.

Chadha, M.L. and Paul, B. (1984) *Indian J. Hort.*, **41**: 101.

Chadha, M.L. and Saimbhi, M.S. (1977) *Indian J. Hort.*, **34**: 426-429.

Chadha, M.L. and Sidhu, A.S. (1987) *J. Res., Punjab Agric. Univ.*, Ludhiana, **24:** 357.

Chahab, P.P.K. and Chahal, V.P.S. (1988) *In Advances in Plant Nematology* (M.A. Moqubool, A.M. Golden, A. Graffar, L.R. Krusberg, eds.), pp.257-263.

Chakraborty, A.K. and Choudhury, B. (1974) *Veg. Sci.*, **1:** 12-17.

Chakraborty, A.K. and Choudhury, B. (1975) *Proc. Indian Nat. Sci. Acad. B,*. **41**: 379-385.

Chakraborty, A.K. and Choudhury, B. (1979) *Veg. Sci.*, **6:** 66-69.

Chakraborty, D.P., Ghosh, G.C. and Dhua, S.P. (1974) *Fertl. Tech., India*, **11:** 192-194.

Chandra, A., Dhurvey, G.S. and Verma, B.K. (1998) *Abst. Silver Jubilee Nat. Symp. Emerging Scenario in Vegetable Research and Development*, PDVR, Varanasi, India, 12-14 December, p. 8.

Chandrasekharan, P. and George, C.M. (1973) *Agric. Res. J., Kerala*, **11:** 106-108.

Chartzoulakis, K.S. and Loupassaki, M.H. (1997) *Agric. Water Manage.*, **32:** 215-225.

Chatterjee, N.C. (1954) *Curr. Sci.*, **23**: 133.

Chattopadhyay, A., Dutta, S. and Hazra, P. (2011) *Veg. Crops Res. Bull.*, **74**: 39-49.

Chattopadhyay, A., Seth, T., Dutta, S., Ghosh, P.P., Chattopadhyay, S.B., Majumder, D. and Hazra, P. (2012) *Int. J. Veg. Sci.*, **18:** 376-392.

Chaudhary, D.R., Sharma, D.K. (1999) *Indian J. Hill Fmg.*, **12**: 94-96.

Chauhan, D.V.S. (1981) *Vegetable Production in India (3‴ edn.)*, Ram Prasad and Sons, Agra,India.

Chauhan, D.V.S. (1986) *J. Genet. Pl. Breed.*, **44:** 367-371.

Chen, N.C. (2001) Eggplant seed production. AVRDC, International Cooperators' Guide. Asian Vegetable Research and Development Center, Shanhua, Taiwan

Cherian, M.C. and Pillai, B.R. (1942) *Indian J. Ent.*, **4**: 236-237.

Chhabra, H.K. and Sharma, J.K. (1981) *Sci. and Cult.*, **47:** 256-257.

Choudhary, B. and Gaur, K. (2009) *ISAAA Brief No.*, **38**: 1-102.

Choudhury, B. (1962) *Indian Fmg.*, **12**: 34-40.

Choudhury, B. (1966) *Indian Hort.*, **10**: 56-58.

Choudhury, B. (1976a) *Vegetables (4th edn.), National Book Trust*, New Delhi, pp. 50-58.

Choudhury, B. (1976b) *In: Evolution of Crop Plants* (Ed. N.W. Simmonds), Longman Inc., London and New York, pp. 278-279.

Choudhury, B. and Anothai-Choomsai, M.L. (1970) *Indian J. Agric. Sci.*, **40:** 805-812.

Choudhury, B., Kalda, T.S. and Bhagchandani, P.M. (1971) *Indian Hort.*, **16:** 11-12.

Choudhury, P.K. and Saraf, R.K. (1996) *Orissa J. Hort.*, **20**: 42-47.

Choudhury, R.S. and Grag, O.K. (1955) *Indian J. Hort.*, **12**: 180-187.

Chowdhury, A.R., Amin, A.K.M., Uddin, K.S. and Haider, J. (1991) *Annals of Bangladesh Agric.*, **1:** 55-57.

Chowdhury, M.J., Ahmad, S., NazimUddin, M., Quamruzzaman, A.K.M. and Patwary, M.M.A. (2010) *The Agric.*, **8**(2): 8-13.

Collonier, C., Fock, I., Kashyap, V., Rotino, G. L., Dunay, M. C., Lian, Y., Mariska, I. K., Rajam, M. V., Servaes, A., Ducreux, G. and Sihachakr, D. (2001) *Plant Cell Tiss. Organ Cult.*, **65**: 91-107.

Collonier, C., Mulya, K., Fock, I., Mariska, I., Servaes, A., Vedel, F., Siljak-Yakovlev, Ducreux, G., Sihachakr, D. (2001) *Plant Sci.*, **160**: 301-313.

Cordrey, T.D., Coffey, D.L. and Brown, J.F. (1979) *Tennessee Farm and Home Science Progress Report.*, **112:** 41-42.

Costache, M. (1977) *Acta Horticulture*, **58:** 461-468.

Courter, J.W. and Glover, C.N. (1965) *Illinois Res.*, **7:** 17-18.

Damus de Vaulx, R. and Chambounet, D. (1982) *Agronomie*, **2**: 983-988.

Dar, S.A.,Wani, A.R., Rather, B.A. and Kandoo, A.A. (2017) *Chem Sci Rev Lett* **6**(23), 1931-1940.

Darekar, K.S., Gaikwad, B.P. and Chavan, U.D. (1991) *J. Maharashtra Agric. Univ.*, **16:** 366-369.

Das, S., Mandal, A.B. and Hazra, P. (2009) *J. Crop Weed.*, **5**(2): 25-30.

Das, S.N., Sarma, T.C. and Tapadar, S.A. (2014) *Inter. J. Sci. Res. Publications.*, **4**(9): 1-3.

Dash, S.P., Singh, J. and Sharma, D. (2019) *J. Pharma. Phytochem.*, **8**(2): 1574-1578.

Daskaloff, C.H. (1941) *Forchungsdienst*, **12:** 617-618.

Daskalov, H. and Murtazov, T. (1937) *News Inst. Pl. Indtr., Sofia*, **4:** 65-72.

Daunay, M.C., Chaput, M.H., Sihachakr, D., Allot, M., Vedel, F. and Ducreux, G. (1993) *Theor. App. Genet.*, **85:** 841-850.

Daunay, M.C., Lester, R.N., Hennart, J.W. and Duranton, C. (2000) *Capsicum Eggplant Newsletter*, **19**: 11-18.

David, A.L. (1963) *Madras Agric. J.*, **50**: 103.

David, A.L. (1964) *Indian Fmg.*, **14**: 34.

Davidar, P., Snow, A.A., Rajkumar, M., Pasquet, R., Marie-Christine, D. and Mutegi, E. (2015) *American J. Bot.*, **102**(1): 129-139.

Decker (1951) *Phytopath.*, **41:** 9.

Demir, K., Bakir, M., Sarikamis, G., and Acunalp, S. (2010) *Genet. Mol. Res.*, **9**, 1568-1576.

Deol, N.S., Chhabra, H.K. and Kaul, V.K. (1989) *Plant Dis. Res.*, **4:** 151-153.

D'ercole, N. (1978) *Informatore Agrario*, **34**: 3839-3841.

Desai, K.M., Saravaiya, S.N., Patel, A.I., Tank, R.V. and Patel, D.A. (2016) *Adv. Life Sci.*, **5**(13): 5426-5429.

Desgoabdem A.A., Bankapur, V.M. and Nalawadi, U.G. (1978) *Curr. Res.*, **7:** 174-175.

Deshmukh, S.B., Sawant, S.N., Narkhede, G.W. and Dod, V.N. (2014) *Middle-East J. Sci. Res.*, **21**(11): 2177-2181.

Deshmukh, S.N. and Rattan Lal (1969) *Indian J. Ent.*, **31:** 222,234.

Despande, A.A., Bankapur, V.M. and Nalawadi, U.G. (1978) *Curr. Res.*, **7:** 174-175.

Devmore, J.P., Bhave, S.G., Burondker, M.M., Dhekele, J.S. and Sawardekar, S.V. (2016) *Elec. J. Plant Breed.*, **7**(4): 1040-1045.

Dhankar, B.S., Gupta, V.P. and Singh, K. (1977) *Haryana J. Hori. Sci.*, **6**: 50-58.

Dhankar, D.S. (1988) *Tropical Pest Manage.*, **34:** 343-345,370.

Dhesi, N.S. and Nandpuri. K.S. (1965) *Vegetable Growing in Punjab, Punjab Agric. Univ.*, Ludhiana.

Diaz-Perez, J.C. (1998) *Postharvest Biol. Tech.*, **13:** 45-49.

Dikii, S.P. and Neklyudova, R.T. (1975) *Byulletin ' Vsesoyuznogo Ordena Lenina Institute Rastenievodstvaimeni NJ. Vavilova*, **50**: 65-69.

Dikii, S.P., Auikeenko, A.P. and Fonina, T.J. (1980) *Fiziologiya I Biokhimiya Kul'tumykh Rastenii*, **12**: 170-174.

Dobroljubskii, O.K. (1955) *Doklady Akad. Nauk.*, SSSR, **101:** 1135-1137.

Docimo, T., Francese, G., Ruggiero, A., Batelli, G., De Palma, M., Bassolino, L., Toppino, L., Rotino, G.L., Mennella, G. and Tucci, M. (2016) *Front. Plant Sci.*, **6:** 1233.

Doijode, S.D. (2000) *Veg. Sci.*, **27:** 199-200.

Doikova, M. (1976) *Gradinarstvo*, **57:** 28-31.

Doikova, M. (1977) *B Igarski Plodove, Zelenchutsi J Konservi*, **1:** 20-23.

Doikova, N., Petrov, K., Rankov, V. and Alandzhiski, D. (1980) *Gradinarstvo Ovoshcharstvo*; Lozarstvo, **25:** 61-67.

Doshi, K.M., Bhalala, M.K. and Kathiria, K.B. (1998) *Veg. Sci.*, **25**(1): 72-74.

Dubey, R., Das, A., Ojha, M.D., Saha, B., Ranjan, A. and Singh, P.K. (2014) *The Bioscan*, **9**(2): 889-894.

Dubey, R., Dixit, J. and Pandey, S. (1998) *Abst. Silver Jubilee Nat. Symp. Emerging Scenario in Vegetable Research and Development*, PDVR, Varanasi, India, 12-14 December, p. 36.

Ducos, J.P., Alenton, R., Reano, J.F., Kancnanomai, C., Deshayesl, A. and Petiard, V. (2003) *Euphytica*, **131**: 215-223.

Dunyamalyev, M.Z. (1975) *IzvestijaAN AzSSR, Biol. Nauk*, **5:** 67-70.

Duodu, Y.A. (1985) *Trop. Pest Manage.*, **31**(1): 73-75.

Duranti, A. and Cuocolo, L. (1988) *Advances Hori. Sci.*, **2:** 71-74.

Dutta, T., Banerjee, S., Bhattacharjee, T., Maurya, P.K., Dutta, S., Chattopadhyay, A. and Hazra, P. (2019) *Indian J. Hort.*, **76**: 438-445.

Eguchi, T., Eguchi, T., Oshika, Y., Yamada, H. and Watahara, T. (1962) *J. Japanese Soc. Hort. Sci.*, **31:** 33-38.

El-Nagdi, W.M.A. and Abd-El-Khair, H. (2019) *Bull. Natl. Res. Cent.*, **43**: 154.

El-Shal, M., Kamar, M.E. and El-Sharkaury, A.M. and Osman, E.A. (1986) *Alexandria J. Agric. Res. Egypt*, **31**: 213-224.

Esteban, R.M., Molla, E., Villarroya, M.B. and Lopez-Andreu, F.J. (1989) *Sci. Hort.*, **41:** 19-25.

Even, S. (1970) *Hassadeh*, **51:** 22-25.

Events in Solanaceae Evolution. Available at: https: //pag.confex.com/pag/xxv/

Fallik, E., Temkin-Gorodeiski, N., Grinberg, S. and Davidson, H. (1994) *Hassadeh*, **74:** 518-520.

Fallik, E., Temkin-Gorodeiski, N., Grinberg, S., Rosenberger, I., Shapiro, B. and Apelbaum, A. (1994) *J. Hort. Sci.*, **69:** 131-135.

Fang, M.R., Mao, R.C. and Xie, W.M. (1985) *Hortic. Sci.*, **12:** 261-266.

Farfour, S.A. and El-Ansary, M.S.M. (2013) *Egypt. J. Agronematol.*, **12**(1): 63-73.

Fári, M., Nagy, I., Csányi, M., Mitykó, J. and Andrásfalvy, A. (1995) *Plant Cell Reports*, **15**: 82-86.

Faruk, M.I., Islam, M.M., Khatun, F., Hossain, M.A. and Dey, T.K. (2019) *Bangladesh J. Agril. Res.*, **44**(3): 427-437.

Ferdausi, A., Nath, U.K., Das, B.L. and Alam, M.S. (2009) *J. Bangladesh Agril. Univ.*, **7**(2): 253-258.

Ferguson, I.A.C. (1951) *Plant Dis. Reptr.*, **35**: 102-105.

Filipone, E. and Lurquin, P.F. (1989) *Plant Cell Rep.*, **8**: 370-373.

Filippov, G.A. and Inshakova, K.P. (1979) *Khimiya v Sel'skom Khozyaistve*, **17:** 38-40.

Flick, F.L., Burnette, F.S., Aung, L.H., Ory, R.L. and St. Angelo, A.J. (1978) *J. Agric. Food Chem.*, **26**: 1000-1005.

Flick, G.L.Jr., Ory, R.L. and St. Angelo, A.J. (1977) *J. Agric. Food Chem.*, **25:** 117-120.

Fouda, K.F.and Abd-Elhamied, A.S. (2017) *J.Soil Sci. Agric. Eng., Mansoura Univ.*, **8**(11): 643-648.

Franklin, G. and Sita, G. L. (2003) *Plant Cell Reports*, **21**: 549-554.

Franklin, G., Sheeba, C.J. and Sita, G.L. (2004) *In Vitro Cell. Dev. Biol. Plant.*, **40**: 188-191.

Frary, A., Doganlar, S., and Daunay, M.C. (2007) *Genome Mapping and Molecular Breeding in Plants* (Ed.C. Kole) Berlin, Springer, pp. 287-313.

Frempong, E. and Buahin, G.K.A. (1977) *Bulletin de I'InstituteFondamental d' afrique Noire*, Serie A, **39**: 627-641.

Frydrych, J. (1964) Bull. *Vyzk, Ust. Zelin., Olomouc*, **8**: 27-37.

Frye, D.M., Ilnicki, R.D. and Meade, J.A. (1978) *Proc. Northeastern Weed Sci. Soc.*, **32**: 220-225.

Furini, A., and Wunder, J. (2004) *Theor. Appl. Genet.*, **108**: 197-208.

Fuster, C. and Prestamo, G. (1994) *Alimentaria*, **31**: 69-72.

Gahukar, R.T. and Bagal, S.R. (1976) *Indian J. Agric. Sci.*, **44**: 366-369.

Gaikwad, B.P., Darekar, K.S. and Chavan, U.D. (1991) *J. Maharashtra Agric. Univ.*, **16**: 354-356.

Ge, H., Li, H., Liu, Y., Li, X. and Chen, H. (2011) *Afr. J. Biotechnol.*, **10**: 9023-9031.

Gerber, J.M., Swiader, J.M. and Peck, T.R. (1981) *Illompos Res.*, **23**: 12-13.

Ghanti, P., Sounda, G., Som, M.G., Patra, A.P. and Jana, P.K. (1982) *Orissa J. Hort.*, **10**: 24-27.

Ghazi, D.A. (2018) *J. Soil Sci. Agric. Eng., Mansoura Univ.*, **9**(10): 525-532.

Ghosh, P.P., Dutta, S., Seth, T. and Chattopadhyay, A. (2015) *Indian Phytopath.*, **68** (4): 363-367.

Ghosh, S.K. and Chakraborty, K. (2014) *Res. J. Agr. Env. Sci.*, **1**(1): 26-31.

Gill, C. and Chadha, M.L. (1979) *Indian J. Hort.*, **36**: 67-71.

Gill, H.S., Arora, R.S. and Pachauri, D.C. (1976) *Indian J. Agric. Sci.*, **46**: 484-490.

Gill, H.S., Bhagchandani, P.M. and Thakur, M.R. (1978) *Indian Hort.*, **23**: 13-22.

Gindrat, D., Ducrot, V. and Caccia, R. (1976) *Revue Suisse de Viticulture, d ' Arboriculture et d ' Horticulture*, **8**: 71-76.

Gleddie, S., Keller, W. and Setterfield, G. (1986) *Theor. Appl. Genet.*, **71**: 613-621.

Gleddie, S., Keller, WA. and Setterfied, G. (1982) *In: Plant Tissue Culture* (Ed.A. Fujiwara), Maruzen, Tokyo, pp. 139-140.

Gnanakumari and Satyanarayana, G. (1971) *Indian J. Agric. Sci.*, **41**: 554-558.

Gonzales, L.M.R. (2015) *Int. J. Sci. Res. Pub.*, 5(3): 1-4.

Gopimony, R. and George, M.K. (1979) *Agric. Res. J., Kerala*, **17**: 7-10.

Gopinath, G., Madalageri, B.B. and Somasekar, C. (1987) *Mysore J. Agric. Sci.*, **21:** 329-331.

Gorske, S.F. (1979) *American Vegetable Growers*, **27:** 13-14.

Gorske, S.F. (1981) *Proc. Northeastern Weed Sci. Soc.*, **35:** 167-168.

Goto, K. (1952) *Japanese J. Breed.*, **1:** 196.

Gousseta, C., Collonniera, C., Mulyab, K., Mariskab, I., Rotinoc, G.L., Bessed, P., Servaesa, A. and Sihachakra, D. (2005) *Plant Sci.*, **168:** 319-327.

Govindaswamy, C.V. (1974) *J. Madras Univ.*, B,**39/40:** 1-23.

Gowda, D.M.V. (1977) *Mysore J. Agric. Sci.*, **11**: 426.

Gowrish, K.R., Ramesha, B. and Ushakumari, R. (2015) *Indian J. Entomol.*, **77**(1): 51-55.

Goyal, J.P., Sharma, H.C. and Pathak, V.N. (1976) *Udyanika*, **2:** 36-38.

Gramazio, P., Blanca, J., Ziarsolo, P., Herraiz, F. J., Plazas, M., Prohens, J. and Vilanova, S. (2016) *BMC Genom.*, **17**: 300.

Gramazio, P., Prohens, J., Borras, D., Plazas, M., Herraiz, F.J. and Vilanova, S. (2017) *Euphytica*, **213**: 264.

Gupta, D.K. and Chakravarty, B.P. (1975) *Indian J. Mycol. Pl. Path.*, **5:** 28.

Guri, A. and Sink, K.C. (1988a) *Theor. Appl. Genet.*, **76:** 490-496.

Guri, A. and Sink, K.C. (1988b) *J. Plant Physiol.*, **133:** 52-55.

Gutierrez, L.E., Minami, K., Camargo, T.P. and Mantovani, W. (1976) *Anais da Escola Superior de Agricultura 'Luiz de Queiroz'*, **33:** 259-265.

Halder, J., Kushwaha, D., Dey, D., Tiwari, S.K., Rai, A.B. and Singh, B. (2017) *Indian J. Agric. Sci.*, **87**(6): 801-803.

Hamada, A.M., El-Zawahry, A.M. and Al-Hakimi, A.M. (2000) *Assiut. J. Agril. Sci.*, **31**: 227-241.

Hashem, M.M. (1993) *Egyp. J. Hort.*, **18:** 123-130.

Hazra, P., Sahu, P.K., Roy, U., Dutta, R., Roy, T. and Chattopadhyay, A. (2010) *Indian J. Agric. Sci.*, **80** (2): 119-124.

Henz, G.P. and Silva, C. (1995) *Pesquisa Agropecuaria Brasileira*, **30:** 157-162.

Hitomi, A, Amgai, H. and Ejura, H.(1998) *Plant Breed.*, **117:** 379-383.

Honda, F., Amano, T. and Saito, H. (1970) *Bull. Horti. Res. Sta. D (Kurume)*, **6**: 57-73.

Honda, Y. and Yunoki, T. (1980) *J.A.R.Q.*, **14:** 78-83.

Hurtado, M., Vilanova, S., Plazas, M., Gramazio, P., Fonseka, H. H., Fonseka, R., and Prohens, J. (2012) *PLoS one*, **7**: e41748. https: //doi.org/10.1371/journal. pone.0041748

Ibrahim, J.K.A., Shahda, W.T. and Dawood, O.A.I. (1999) *Nematologia Mediterranea*, **27:** 31-33.

Ionescu, A. (1998) *Anale Iustitutul de Cercetari Pentru Legumicultura si Floricultura Vidra*, **15:** 17-27.

Islam, M. O., Islam, M. S., and Azad-Ud-Doula Prodhan, A. K. M. (2006) *J. Bangladesh Agril. Univ.*, **4**: 43-49.

Islam, M.O., Islam, M.S. and Azad-Ud-Doula Prodhan, A.K.M. (2006) *J. Bangladesh Agril. Univ.*, **4**(1): 43-49.

Islam, M.T., Chhanda, R.A., Pervin, N., Hossain, M.A. and Chowdhury, R.U. (2018) *Bangladesh J. Agril. Res.*, **43**(3): 499-512.

Islam, T., Roknuzzaman, A. H. M., Hassan, K., and Ullah, M. S. (2019). *Int. J. Trop. Insect Sci.*, **40**: 351–360.

Isouard, G., Raquin, C. and Demarly, Y. (1979) *C.R. Acad. Sci.*, **288:** 987-989.

Isshiki, S., Okubo, H. and Fujieda, K. (1994) *Euphytica*, **80:** 145-150.

Isshiki, S., Okubo, H. and Fujieda, K. (2000) *Euphytica*, **112:** 9-14.

Isshiki, S., Uchiyama, T., Tashiro, Y. and Miyazaki, S. (1998) *Euphytica*, **102:** 295-299.

Ivanov, S.M. and Nikolaeva, S.I. (1974) *Referativny I Zhurnal*, 5.55.484.

Iwata, M., Kozaki, I. and Narusawa, T. (1968) *J. Japanese Soc. Hort. Sci.*, **37:** 340-344.

Jacqua, G. and Gerion, A.L. (1988) *Bulletin Agronomique Antilles-Guyane*, No.7, pp. 39-43.

Jadhav, Y.T., Zanwar, P.R. and Shinde, D.S. (2017) *Int. J. Curr. Microbiol. App. Sci.*, **6**(3): 2520-2526.

Jaime, S., Casado, M. and Aguilar, A. (1987) *Anales de Edafologiay Agrobiologia*, **46:** 1375-1383.

Jakatimath, S.P., Mesta, R.K., Mushrif, S.K., Biradar, I.B. and Ajjappalavar, P.S. (2017) *J. Pure App. Microbiol.*, **11**(1): 229-235.

Jamwal, V.V.S. (2011) *Ph.D. Thesis*, Sher-e-Kashmir University of Agricultural Sciences and Technology,Jammu, India.

Jamwal, V.V.S., Ahmad, H., Sharma, D., Srivastava, K. and Ganai, S.A. (2014) *Indian J. Entomol.*, **76**(4): 356-358.

Jarl, C.I., Rietveld, E.M. and Haas, J.M.de (1999) *Plant Cell Rep.*, **18**: 791-796.

Javed, H., Hussain, S., Javed, K., Mukhtar, T., and Abbasi, N. (2017) *Pak. J. Agri. Sci.*, **54**(4): 763-768.

Javed, H., Mukhtar, T., Javed, K. and Ul Mohsin, A. (2017) *Pak. J. Agri. Sci.*, **54**(1): 65-70.

Jeevarani, N., Baskararajan, G. and Shakila, A. (1998) *Abst. Silver Jubilee Nat. Symp. Emerging Scenario in Vegetable Research and Development*, PDVR, Varanasi, India, 12-14 December, p. 168.

Jirankali, J.P. Reddy, N., Gangaprasad, S. and Manohara, S.N. (2019) *Int. J. Curr. Microbiol. App. Sci.*, **8**(3): 476-484.

Joi, M.B. and Sonone, H.N. (1977) *J. Maharashtra Agric. Univ.*, **2:** 74-76.

Jose, D., Shanmugavclu, K.G. and Thamburaj, S. (1988) Indian *J. Hort.*, **45**: 100-103.

Joseph, B., Dar, M.A. and Kumar, V. (2008) *Global J Biotech. Biochem.*, **3**(2): 56-59.

Joshi, R.D., Sarkar, K.R. and Kulshrcshtha, K. (1979) *Sci. and Cult.*, **45:** 112-114.

Jotwani, M.G. and Swarup, P. (1963) *Indian J. Entomol.*, **25:** 275-291.

Jyotishi, R.P. and Kashyap, R. (1967) *Jawahar Lal Nehru Krishi Vlswavidyalaya Res. J.*, **1:** 38-43.

Jyotishi, R.P. and Kashyap, R. (1968) *Indian J. Sci. Indust.*, **2**: 27-32.

Kabir, J. (1981) *M.Sc.(Ag.) Thesis, Bidhan Chandra Krishi Viswavidyalaya*,West Bengal,India.

Kabir, J., Som, M.G. and Das, N.D. (1985) *Indian Agric.*, **29**: 37-40.

Kalaiyarasi, G., Ram, S.R.R. and Saravanan, K.R. (2018) *Hortic. Biotechnol. Res.*, **4:** 35-38.

Kalda, Swarup, V. and Choudhary, B. (1976) *Veg. Sci.*, **3:** 90-101.

Kalda, T.S. (1979) *Mimeograph No. B-13, Training Section, National Seeds Corporation*, India.

Kalda, T.S. (1998) In: *Compendium of lectures of Summer School 011 Advanced Technology in Improvement of Vegetable Crops incuding Cole Crops*, ICAR, New Delhi, 24th May-6th June, 1998, pp. 132-140.

Kalda, T.S. and Kumar, R. (1999) In: *Compendium of lectures of Summer School 011 Advances in Breeding of Temperate and Tropical Vegetables for Biotic and A biotic Stresses*, ICAR, New Delhi,17th May-6th June, 1999, pp. 102-107.

Kalda, T.S., Suran, B.S. and Gupta, S.S. (1988) *South Indian Hort.*, **36**: 110-113.

Kalda, T.S., Swarup, V. and Choudhury, B. (1976) *Veg. Sci.*, **3:** 65-70.

Kalda, T.S., Swarup, V. and Choudhury, B. (1977) *Veg Sci.*, **4:** 90-101.

Kalia, R.K., Mai, M.K., Kalia, S., Singh, R. and Dhawan, A.K. (2011) *Euphytica*, **177:** 309-334.

Kalloo, G., Baswana, K. and Sharma, N.K. (1990) *XXIII Int. Hort. Cong. Florence, Italy. Aost.*, 1204.

Kamani, J.M., Monpara, B.A. and Dhameliya, H.R. (2007) *Natl. J. Plant Improv.*, **9**(1): 32-35.

Kameya, T., Miyazawa, N. and Toki, S. (1990) *Japanese J. Breed.*, **40**: 429-434.

Kanwar, J.S. and Gill, B.S. (2000) *Veg. Sci.*, **27**: 145-147.

Karagiannidis, N., Bletsos, F. and Stavropoulos, N. (2002) *Sci. Hort.*, **94:** 145-146.

Karak, C., Ray, U., Akhtar, S., Naik, A. and Hazra, P. (2012) *J. Crop Weed*, **8**(1): 86-89.

Karihaloo, J.L. and Gottlieb, L.D. (1995) *Theor. Appl. Genet.*, **90**: 578-583.

Karihaloo, J. L., Kaur, M., and Singh, S. (2002). *Genet. Resour. Crop Evol.*, **49**: 533-539.

Kashem, M.A., Ahmad, M.U., Hossain, I., Khan, A.A. and Aziz, A. (1994) *Bangladesh J. Pl. Path.*, **10:** 1-2.

Kashyap, R.K. and Kalloo (1983) *Haryana J. Hort. Sci.*, **12**: 101-108.

Katan, J., Greenberger, A., Alon, H. and Grinstein, A. (1975) *Phytoparasitica*, **3**: 69.

Kaur, D.N. and Kaur, S. (2016) *Plant Dis. Res.*, **31**(1): 99-105.

Kaushal, N. and Sugha, S.K. (1995) *Indian J. Mycol. and Pl. Path.*, **25**: 189-191.

Kaynas, K., Ozelko, S., Surrneli, N. and Abak, K. (1995) *Acta Hortic.*, **412:** 143-151.

Keshwal, R.L. and Khare, M.N. (1986) *Indian J. Mycol. Pl. Path.*, **16**: 237-240.

Khan, A.M., Siddiqui, Z.A., Alam, M.M. and Saxena, S.K. (1976) *Indian J. Agric. Sci.*, **46**: 439-441.

Khan, Md.M.R. and Isshiki, S. (2016) *The Hortic. J.*, **85**(1): 1-7.

Khan, R. (1979) In: *The Biology and Taxonomy of Solanaceae*. (Eds. J.Hawkes, R.N. Lester, and A.D. Skelding), Academic Press, London, pp. 629-636.

Khapre, P.R., Wanjari, K.B. and Deokar, A. (1989) *J. Maharashtra Agril. Univ.*, **13:** 88-97.

Khatun, F., Meah, M.B. and Nasiruddin, K.M. (2006) *Pak. J. Biol. Sci.*, **9**(1): 48-53.

Kodama, T., Fukui, T. and Nakanishi, Y. (1979) *Bull. Nara Agric. Exp. Sta.*, **10**: 89-92.

Korat, V., Patel, B.R., Kalyanrao and Gohil, D.P. (2018) *Veg. Sci.*, **45**(1): 22-26.

Koundinya, A. V. V., Pandit, M. K., and Saha, A. (2019) *Agric. Res.*, **8**: 309-319.

Koundinya, A.V.V., Das, A., Layek, S., Chowdhury, R. and Pandit, M.K. (2017) *J. App. Natural Sci.*, **9**(3): 1343-1349.

Kreij, C. de and Basar, H. (1997) *J. Plant Nutrn.*, **20**: 47-53.

Krishnaiah, K., Tandop, P.L. and Mohan, N.J. (1976) *Pesticides*, **10:** 41-42.

Krishnamurthi, S. and Subramanian, D. (1954a) *Indian J. Hort.*, **11:** 63-67.

Krishnamurthi, S. and Subramanian, D. (1954b) *Curr. Sci.*, **23**: 125-126.

Kumar, A., Dahiya, M.S. and Bhutani, R.D. (2000) *Haryana J. Hort.*, **11:** 63-67.

Kumar, A., Kumar, R., Ansar, M., Akhtar, S., Adarsh, A. and Kumar,V. (2020) *Int. J. Curr. Microbiol. App. Sci.*, **9**(3): 223-228.

Kumar, M., Madhupriya and Rao, G.P. (2017) *Biotech*, **7:** 7.

Kumar, M.K., Mishra, H.N., Patel, M. and Nayak, D.P. (2020) *The Pharma Inno. J.*, **9**(2): 416-419.

Kumar, P.A., Mandaokar, A., Sreenivasu, K., Chakrabarti, S.K., Bisaria, S., Sharma, S.R., Kaur, S. and Sharma, R.P. (1998) *Mol. Breed.*, **4**(1): 33-37.

Kumar, S. (2017) *M.Sc. Thesis*, BAU, Sabour, India.

Kumar, S. and Vadivelu, S. (1996) *Pest Manage. Hort. Ecosys.*, **2:** 71-74.

Kumar, S., Gowda, B. and Patil, S.S. (2008) *J. Hort. Sci.* **3**(2): 146-149.

Kumar, S., Jain, P.C. and Kumawat, S.R. (1990) *Indian J. Entom.*, **51:** 253-255.

Kumar, S.R., Verma, S.P. and Ganguli, D.K. (1998) *South Indian Hort.*, **46**(3-4): 251-255.

Kumar, V. (2016) *Adv. Plants Agric. Res.* **4**(2): 249-256.

Ladipo, J.L. (1976) *Plant Dis. Reptr.*, **60:** 1068-1072.

Lal, O.P., Sharma, R.K., Bhagchandani, P.M. and Chandra, J. (1976) *Veg. Sci.*, **3:** 111-116.

Lalramhlimi, B., Dutta, T., Jamir, I., Maurya, P.K., Bhattacharjee, T., Banerjee, S., Chatterjee, S., Mandal, A.K. and Chattopadhyay, A. (2019) *Int.J.Curr.Microbiol. App.Sci.*, **8**(6): 2370-2390.

Lal, S., Verma, G. and Pathak, M.M. (1974) *Indian J. Hort.*, **36**: 51-55.

Lan, X., Zhang, J., Zong, Z., Ma, Q. and Wang, Y. (2017) *BioMed Res. Inter.*, 2017: Article ID 4101357. DOI: 10.1155/2017/4101357

Latif, M.A., Rahman, M.M. and Islam, M.Z. (2010) *J. Pest Sci.*, **83**(4): 391-397.

Leonardi, C. and Giuffrida, F. (1999) *Acta Hortic.*, **491**: 215-220.

Li, G.G. and Constabel, F. (1988) *Curr. Plant Sci. Biotechnol. Agri.*, **7**: 203-204.

Li, J., Wang, C., Bangash, S.H., Lin, H., Zeng, D. and Tang, W. (2020) *PLoS One*, **15**(7): e0235423.

Litsinger, J.A. and Apostol, R.F. (1994) *Philippine Entomologist*, **9:** 286-301.

Lockwood, J.L., Yoder, D.L. and Beute, M.K. (1970) *Plant Dis. Reptr.*, **54**: 846-848.

Lombardi, D. and Rcstaino, F. (1981) *Colture Protette*, **10**: 31-36.

Maaswinkel, R. and Buitelaar, K. (1989) *Groenten en Fruit*, **45**: 46-47.

Machuzhcnko, V.I. (1972) *Trudy Khar'Kovskii Sel'Skokhozyaistvennyi lnstitut.*, **144**: 72-75.

Madhavi, N., Prasad, J.O., Bijalwan, P. and Sekhar, V. (2016) *Plant Arch.*, **16**(1): 426-430.

Magioli, C. and Mansur, E. (2005) *Acta Bot. Bras.*, **19**(1): 139-148.

Magioli, C., Rocha, A.P.M., Margis-Pinheiro, M., SachettoMartins, G. and Mansur, E. (2000) *J. Plant Biotech.*, **2**(1): 43-49.

Magioli, C., Rocha, A.P.M., Tarré, E., Santiago-Fernandes, L.D., Oliveira, D.E., Krul, W.R. and Mansur, E. (2001) *J. Plant Biotech.*, **3**(1): 19-25.

Mahata, C. and Sarkar, N. (2019) *Int. J. Pure App. Biosci.*, **7**(3): 632-636.

Mahmood, S., Uddin, M.K. and Rashid, A.K.M.B. (2016) *Int. J. Sustain. Crop Prod.*, **11**(2): 4-6.

Makani, A.Y., Patel, A.L., Bhatt, M.M. and Patel, P.C. (2013) *The Bioscan*, **8**(4): 1369-1371.

Makrogianni, D.I., Karapanos, I.C. and Passam, H.C. (2018) *J. Plant Growth Regul.* **37**: 937–946.

Malyshkina, M.S. (1976) *Referativny I Zhurnal*, **7**.55.568.

Mandal, R., Maurya, P.K., Bhattacharjee, T., Banerjee, S., Chattopadhyay, A. and Hazra, P. (2019) *J. Entomol. Zool. Stud.*, **7**(4): 1340-1347.

Mangual, Crespo, G. (1981) *J. Agric. Univ., Puerto Rico*, **65:** 250-257.

Mannade, A.K., Chopkar, R. and Srivastva, P. (2013) *J. Plant Dev. Sci.*, **5**(4): 369-373.

Mariola, P., Isabel, A., Santiago, V., Maria, H., Pietro, G., Francisco, J.H. and Jaime, P. (2013) *Not. Bot. Horti. Agrobo.*, **41**(1): 26-35.

Marois, J.J., Johnston, S.A., Dunn, M.T. and Papavizas, G.C. (1982) *Plant Dis. Reptr.*, **66:** 1166-1168.

Marques, L.J.P., Bianco, S., Bernardes, A., Filho, C., Bianco, M.S.and Lopes, G.S. (2017) *Rev. Caatinga, Mossoró*, **30**(4): 866-875.

Marras, F. and Corda, P. (1970) *Studi sassaresi*, **18:** 126-131.

Martin, F.W. and Rhodes, A.M. (1979) *Euphytica*, **28:** 367-383.

Martinoni, A., Cancvascini, V., Granges, A. and Hurter, J. (1981) *Revue Suisse de Viticulture, d'Arboriculture et d'Horticulture*, **13:** 109-112.

Martins, D.C., Vilela, F.K.J., Guimarães, R.M., Gomes, L.A.A. and Silva, P.A. (2012) *Revista Brasileira de Sementes*, **34**(4): 534-540.

Masal, M.S., Kadam, D.R. and Surwase, S.R. (2018) *Int. J. Curr. Micrcbiol. App. Sci.*, Special Issue-**6**: 579-584.

Masood, A. and Husain, S.I. (1975) *Indian J. Mycol. Pl. Path.*, **5:** 16.

Masui, M., Nukaya, A. and Ishida, A. (1979) *J. Japanese Soc. Hort. Sci.*, **48:** 55-60.

Matsubara, Y.I., Tamura, H. and Harada, T. (1995) *J. Japanese Soc. Hort. Sci.*, **64:** 555-561.

Maurya, A.N. and Singh, V.K. (1978) *Haryana J. Hort. Sci.*, **7:** 217-218.

Mayee, C.D. (1977) *Seed Sci. Tech.*, **5:** 555-558.

Mayee, C.D., Singh, J. and Khatri, H.L. (1973) *Punjab Hort. J.*, **13:** 253-256.

Mehrotra, O.N., Saxena, H.K. and Misra, P.H. (1973) *Progressive Hort.*, **4:** 93-100.

Menezes, F.B., Nakano, O. and Marechini, L.C. (1976) *Solo*, **68:** 21-24.

Mennella, G., Lo Scalzo, R., Fibiani, M., D'Alessandro, A., Francese, G., Toppino, L., Acciarri, N., de Almeida, A.E. and Rotino, G.L. (2012) *J. Agric. Food Chem.*, **60**(47): 11821-11831.

Mertia, N.S. and Chauhan, K.S. (1970) *Allahabad Fmr.*, **44:** 439-445.

Mishra, G.M. (1961) *Indian J. Hort.*, **19:** 305-317.

Mishra, S.N. and Mishra, R.S. (1990) *Indian J. Hort.*, **47:** 239-243.

Mistry, C.R., Kathiria, K.B., Sabolu, S. and Kumar, S. (2018) *Curr. Plant Biol.*, **16**: 27-31.

Mitra, D. K. and Majumdar, M. (1977a) *Pesticides*, **11:** 25-26.

Mitra, D. K., Majumdar, M. and Farkas, G.I. (1976) *Phytopathologische Zeitchrift*, **86:** 310-313.

Mitra, D.K. and Majumdar, M. (1977b) *Proc. Indian Nat. Sci. Acad.*, **43:** 228-234.

Mitra, P. C. and De, R. (1969) *Indian Fmg.*, **19:** 23-24.

Miyamoto, S.and Hara, M. (1988) *Agric. Hort.*, **63:** 1406-1408.

Mochizuki, H. and Yamakawa, K. (1979) *Bull. Veg. Ornamental Crops Res. Sta., Japan*, **6:** 11-18.

Mohan, N.I., Kumar, N.K.K. and Prasad, V.G. (1980) *Pesticides*, **14:** 19-21.

Mohanty, B.K. (1999) *Haryana J. Hort. Sci.*, **28:** 213-215.

Mohanty, B.K. and Prusti, A.M. (2000) *Orissa J. Hort.*, **28**(1): 14-20.

Mohanty, B.K. and Prusti, A.M. (2002) *Orissa J. Hort.*, **30**(1): 1-4.

Mohd Zeal, A., Jivani, L.L., Kachhadia, V.H., Valu, M.G. and Vadavia, A.T. (2019) *The Pharma Inno. J.*, **8**(9): 50-53.

Mokbel, A. A. and Alharbi, A. A. (2014). *Aust. J. Crop Sci.*, **8**: 1428-1434

Moncada, A., Miceli, A., Vetrano, F., Mineo, V., Planeta, D. and D'Anna, F. (2013) *Sci. Hort.*, **149:** 108-114.

Mondal, M.R.I. and Akter, N. (2018) Success story on Bt brinjal in Bangladesh. Available online at: http: //www.apaari.org/web/wpcontent/uploads/ downloads/2018/Bt per cent 20Brinjal per cent 20in per cent 20Bangladesh per cent 20final.pdf

More, G.D., Kadu, N. R.and Lokhande, S.P. (1987) *PKV Res. J.*, **11:** 178.

Morita, K., Arai, K., Doi, Y. and Yora, K. (1974) *Arm. Phytopath. Soc., Japan*, **40:** 288-294.

Moriwaki, K. and Ito, K. (1980) *Res. Bull. Aichi-Ken Agric. Res. Centre*, **12:** 83-90.

Morra, L., Troisi, A., Bilotto, M. and D' Amore, R. (2000) *Colture Protette*, **29:** 33-38.

Mote, U.N. (1976) *Veg. Sci.*, **3:** 128-130.

Mote, U. N. (1978b) *Veg. Sci.*, **5:** 107-110.

Mote, U.N. (1978a) *Pesticides*, **5:** 20-23.

Mukherjee, R. and Mukhopadhyay, S. (1982) *Veg. Sci.*, **9:** 122-125.

Mukherjee, R.K. and Datta, C. (1962) *Sci. and Cult.*, **28:** 476-478.

Muniappan, S., Saravanan, K. and Ramya, B. (2010) *Elect. J.Plant Breed.*, **1**(4): 462-465.

Munoz-Falcon, J.E., Prohens, J., Vilanova, S. and Nuez, F. (2008) *Euphytica*, **164:** 405-419.

Munoz-Falcon, J.E., Prohens, J., Vilanova, S., Ribas, F., Castro, A. and Prohens, J. (2009) *J. Sci. Food Agric.*, **89:** 320-328.

Munoz-Falcon, J.E., Vilanova, S., Plazas, M. and Prohens, J. (2011) *Sci. Hort.*, **129:** 238-246.

Muthukrishnan, C.R. and Srinivasan, K. M. (1963) *Indian J. Hort.*, **20:** 150-151.

Muthukrishnan, C.R. and Srinivasan, K.M. (1964) *Indian J. Hort.*, **21:** 170.

Nagaraja, N. and Mathew, A.G. (1964) *J. food Sci. Tech.*, **1:** 43-44.

Nagarajan, S.S. and Shanmugasundaram, S. (1961) *Madras Agric. J.*, **48:** 22-23.

Nagasawa, M., Sugijama, A., Mori, H., Shiratake, K. and Yamaki, S. (2001) *J. Plant Physiol.*, **158:** 235-240.

Nahar, N., Islam, M.R., Uddin, M.M., Jong, P., Struik, P.C. and Stomph, T.J. (2018) *Crop Protect.*, **112:** 177-186.

Naik, L.B., Prabhakar, M., Hebbar, S.S. and Doijode, S.D. (1996) *Seed Res.*, **23:** 71-74.

Nainar, P., Subbaiah, R. and Irrulappan (1991) *South Indian Hort.*, **39:** 32-36.

Nakamura, R., Fujii, S., Inaba, A. and Ito, T. (1986) *Scientific Rep. Fac. Agric., Okayama Univ.*, No. 68, pp. 19-25.

Nakamura, R., Fujii, S., Inaba, A. and Ito, T. (1987) *J. Japanese Soc. Hort. Sci.*, **55:** 490-497.

Nalini, S. and Parthasarathi, R. (2018) *Ann. Agrarian Sci.*, **16:** 108-115.

Namisy, A., Chen, J.R., Prohens, J., Metwally, E., Elmahrouk, M. and Rakha, M. (2019) *Agric.*, **9:** 157.

Nanba, B., Saito, Y. and Hasegawa, N. (1973) *J. Yamagata Agri. Forestry Soc.*, **30:** 37-46.

Nandwana, R.P., Yadav, B., Vanna, M.K. and Lal, A. (1980) *Indian J. Mycol. Pl. Path.*, **10:** 181-182.

Naqvi, Q.A. and Mahmood, K. (1976) *Comparative Physiol. Ecol.*, **1:** 51-52.

Narayanaswamy, A. (1978) *Mysore J. Agric. Sci.*, **12:** 186-187.

Nath, D.K. and Chakraborty, D.B. (1978) *Pesticides*, **12**: 27-28.

Nawale, R.N. and Sonone, H.N. (1977) *J. Maharashtra Agric. Univ.*, **2:** 184.

Naziya, B.and Sharada, M.S. (2018) *The Bioscan*, **13**(1): 285-290.

Nehra, B.K., Malik, Y.S. and Singh, N. (1999) *Haryana J. Hort. Sci.*, **28:** 91-93.

Nemato, H. (1995) *JARQ*, **29:** 25-29.

Netam, J.L. and Sharma, R. (2014) *IOSR J. Agric. Vet. Sci.*, **7:** 27-30.

Nikolaeva, S.I. (1975) *Referativny I Zhurnal*, **55**: 761.

Nikolov, M., Milcheva, K., Belcheva, S., Ivanova, I. and Nenkova, D. (1987) In: *Proceedings of the International Symposium of Plant Growth Regulators*, Part 2 (Eds. D. Lilov, G. Vassilev, C. Christov, T. Andonova), pp. 947-949.

Nothmann, J. and Koller, D. (1973) *Israel J. Botany*, **22:** 231-235.

Nothmann, J., Avielie, E. and Sachs, M. (1974) *Israel J. Agric. Res.*, **23:** 217-222.

Nothmann, J., Rylski, I. and Spigelman, M. (1979) *Sci. Hort.*, **11**: 217-222.

Nunome, T., Negoro, S., Kono, I., Kanamori, H., Miyatake, K., Yamaguchi, H., Ohyama, A. and Fukuoka, H. (2009) *Theor. Appl. Genet.*, **119:** 1143-1153.

Nunome, T., Suwabe, K., Iketani, H. and Hirai, M. (2003a) *Plant Breed.*, 122: 256-262.

Nunome, T., Suwabe, K., Ohyama, A. and Fukuoka, H. (2003b) *Breed. Sci.*, **53**: 77-83.

Odland, M.I. and Noll, C.J. (1948) *Proc. Amer. Soc. Hort. Sci.*, **51:** 417-420.

Olympios, C.M. (1976) *Hort. Res.*, **16**: 65-70.

Ootake, Y. (1987) *Res. Bull., Aichi-Ken Agri. Res. Cen., Japan*, No.19, pp. 328-335.

Opara, C.N. and Asiegbu, J.E. (1996) *Biol. Agric. Hortic.*, **13:** 341-350.

Padda, N., Mahapatra, A. and Sahoo, N. (1971) *Indian J. Agric. Sci.*, **47:** 597-601.

Padem, H., Ocal, A. and Alan, R. (1999) *Acta Hortic.*, **491:** 241-246.

Padwal, K.G., Singh, S.K. and Sharma, S.K. (2016) *Popular Kheti*, **4**(2): 48-53.

Pal, B.P. and Singh, H.B. (1946) *Indian J. Genet.*, **6:** 19-33.

Pal, B.P. and Singh, H.B. (1949) *Indian Fmg.*, **10:** 378-380.

Panda, B., Singh, Y.V. and Ram, H.H. (2005) *J. App. Horticulture*, **7**(2): 121-123.

Pande, Y.D. and Yadava, S.R.S. (1979) *Veg. Sci.*, **6:** 70-71.

Pandey, K.K., Pandey, P.K., Kalloo, G. and Chaurasia, S.N.S. (2002) *Indian Phytopath.*, **55**(4): 507-509.

Pandey, R.and Chaturvedi, A.K. (2019) *J. Exp. Biol. Agric. Sci.*, **7**(3): 295-300.

Pant, H. and Pandey, G. (2014) *J. Exp. Zool. India.*, **17**(2): 499-500.

Panwar, N.S., Chand, J.N., Singh, H. and Paracer, C.S. (1970) *J. Res., Punjab Agric. Univ.*, Ludhiana, **7:** 641-643.

Papolu, P. K., Dutta, T. K., Tyagi, N., Urwin, P. E., Lilley, C. J., and Rao, U. (2016). *Front. plant sci.*, *7*: 1122. https: //doi.org/10.3389/fpls.2016.01122

Passam, H.C., Akoumianakis, K. and Khah, E.M. (1999) *Seed Sci. Tech.*, **27:** 977-980.

Patel, A.A., Gohil, D.P., Dhruve, J.J. and Damor, H.I. (2017) *J. Pharmacogn. Phytochem.*, **6**(6): 975-978.

Patel, J.S., Sitapara, H.H. and Patel, K.A. (2012) *Int. J. Forestry Crop Improv.*, **3**(2): 116-118.

Patel, K.K., Yadav, F.D. and Shivani (1999) *J. Living World*, **6:** 16-18.

Patel, M.K., Qureshi, I.M. and Bhide, V.P. (1949) *Indian Phytopath.*, **2:** 245-246.

Pathak, S.C. and Jaworski, C.A. (1989) *HortScience*, **24:** 1050.

Pathak, S.C., Liu, J., Jaworski, C.A. and Sultanbawa, A.F. (1991) *J. Hered.*, **82:** 81-83.

Patidar, D. (2015) *Trends Biosci.*, **8**(1): 178-180.

Patil, B.N., Sontakey, P.Y. and Shirsat, S.P. (1981) *Indian J. Pl. Physiol.*, **24:** 145-149.

Patil, N.G. and Mali, V.R. (1977) *J. Maharashtra Agric. Univ.*, **2:** 82-84.

Patil, S.B., Merwade, M.N. and Vyakaranahal, B.S. (2008) *Indian J. Agric. Res.*, **42**(1): 25-30.

Patil, V.M., Patole, K.R., Paprikar, M.S. and Rajput, J.C. (2017) *Int. J. Adv. Res. Biol. Sci.*, **4**(11): 66-71.

Patnaik, B.P. and Farooqui, M.M. (1964) *Fertl. News*, **9**: 27.

Perniola, M., Rivelli, A.R. and Tarantino, E. (1992) *Irrigazione e Drenaggio*, **39:** 41-45.

Peter, K.V. and Singh, R.D. (1973) *Indian J. Agric. Sci.*, **43:** 452-455.

Petrov, K.H. and Pethov, L. (1980) *Nauchni Trudove, Vissh Selskostopanski Institut Vasil Kolarov*, **25**: 31-37.

Petrov, K.H., Doikova, M. and Popova, D. (1981) *Acta Hortic.*, **111**: 273-280.

Philis, J. (1971) *Phytopathologia Mediterranea*, **10:** 117-119.

Pillai, K.M.(1967) *Indian J. Agron.*, **12:** 1-6.

Pinon, M., Hernandez, L., Hernandez, A., Gomez, O., Casanova, A., Depestre, T. and Estrada, J. (1999) *Manejo Integrado de Plagas*, **53:** 84-86.

Pirinc, V., Cimen, I.and Ozaslan, C. (2017) *Environ. Eco.*, **35**(4): 2806-2810.

Plazas, M., Andújar, I., Vilanova, S., Hurtado, M., Gramazio, P., Herraiz, F.J. and Prohens, J. (2013) *Not. Bot. Horti. Agrobot.*, **41**(1): 26-35.

Plazas, M., Prohens, J., Cuñat, A.N., Vilanova, S., Gramazio, P., Herraiz, F.J. and Andújar, I. (2014) *Int. J. Mol. Sci.*, **15**(10): 17221-17241.

Pohl, A., Grabowska, A., Kalisz, A. and Sekara, A. (2019) *Agron.*, **9**: 432.

Polisetty, R., Khetarpal, S. and Patil, P. (1994) *Seed Res.*, **22:** 112-118.

Popova, D. (1961) *Izv. Centr: Nauc-izsled. Inst. Rasten.*, **12:** 187-207.

Popova, D., Murtazov, T., Petrov, K. and Daskalov, S. (1976) *Biologicheskii Zhurnal Armenii*, **29:** 67-72.

Portis, E., Lanteri, S., Barchi, L., Portis, F., Valente, L., Toppino, L., Rotino, G.L. and Acquadro, A. (2018). *Front. plant sci.*, **9**: 401. https: //doi.org/10.3389/ fpls.2018.00401

Prabhu, M., Natarajan, S. and Pugalendhi, L. (2005) *Veg. Sci.* **32**(1): 86-87.

Pradip, K. and Singh, Y.V. (2017) *Veg. Sci.*, **44**(1): 38-41.

Prakash, M., Ramachandran, K. and Nagarajan, M. (1993) *South Indian Hort;* **41:** 208-211.

Pramila, M.L., Kushwaha, and Singh, Y.P. (2017) *Int. J. Curr. Microbiol. App. Sci.*, **6**(11): 641-651.

Prasad, D., Nagia, D.K., Kumar, S. and Saini, M.L. (1994) *Plant Protection Bull.*, **46:** 31-33.

Prasad, D.N. and Prakash, R. (1968) *Indian J. Agric. Sci.*, **38:** 1053-1061.

Prasad, V.S.R.K. and Singh, D.P. (1998) *Abst. Silver Jubilee Nat. Symp. Emerging Scenario in Vegetable Research and Development*, PDVR, Varanasi, India, 12-14 December, p. 9.

Pratap, D., Kashikar, A.R. and Mukherjee, S.K. (2011) *Virol. J.*, **8:** 305.

Prabhavathi, V., Yadav, J.S., Kumar, P.A. and Rajam, M.V. (2002) *Mol. Breed.*, **9**(2): 137-147.

Prohens, J., Plazas, M., Raigon, M. D., Seguμ-Simarro, J., Stommel, J. and Vilanova, S. (2012) *Euphytica*, **186**: 517-538.

Prusty, A.and Mahalik, J.K. (2018) *Ann. Plant Protec. Sci.*, **26**(1): 202-206.

Pugalendhi, L., Veeraragavathatham, D., Natarjan, S. and Praneetha. (2010) *Electron. J. Plant Breed.* **1**(4): 643- 648.

Pun, K.B. and Dass, G.R. (1997) *Bacterial Wilt Newsletter*, **14**: 6.

Purewal, S.S. (1957) *Vegetable Cultivation in North India*, Farm Bull, Indian Council of Agricultural Research, New Delhi, No.36.

Quaiser, A. (1987) *Indian Phytopath.*, **40:** 98-99.

Quamruzzaman, A.K.M., Rashid, M.A., Ahmad, S. and Moniruzzaman, M. (2009) *Bang. J. Agric. Res.*, **34**(4): 705-712.

Raghupathi, H. B., and Bhargava, B. S. (1999). *Indian J. Agric. Sci.*, **69**: 490-493.

Raha, P., Banerjee, H., Das, A.K. and Aditya Chaudhury, N. (1993) *J. Agric. Food Chem.*, **41**: 923-928.

Rahman, S. (2013) *Sci. Total Environ.*, **445-446C:** 48-56.

Rahman,M.B., Hossain, M.M., Haque, M.M., Ivy, N.A. and Ahmad, S. (2015) *Adv. Crop Sci. Tech.*, **S1**: 005.

Raigón, M.D., Prohens, J., Muñoz-Falcón, J.E. and Nuez, F. (2008) *J. Food Comp. Anal.*, **21**: 370-376.

Raigón, M.D., Rodríguez-Burruezo, A. and Prohens, J. (2010) *J. Agric. Food Chem.*, **58**(11): 6833-6840.

Raina, S.K. and Iyer, R.D. (1973) *ZX. Pflanzenzuchtg*, **70**: 275-280.

Raj, K.G. and Kumaraswami, T. (1979) *Sci. Cult.*, **45:** 60-61.

Rajasekharan, S. (1969) *Annamalai Univ. Agric. Res. Ann.*, **1:** 49-60.

Rajeevan, P.K. and Rao, N.S. (1980) *Agric. Res. J.*, Kerala, **18**: 45-50.

Rajendran, B. (1999) *Pest Management in Hort. Ecosystem*, **5**: 11-16.

Rajeshwari, G., Sridhar, V., Chakravarthy, A.K. and Kumar, S.M. (2019) *Int. J. Chem. Stud.*, **7**(1): 584-588.

Rajput, R.B. and Chaudhari, S.R. (2018) *J. Pharmacog. Phytochem.*, **7**(4): 1306-1309.

Ram, D., Singh, S.N. Chauhan,Y.S. and Singh, N.D. (1981) *Haryana J. Hort. Sci.*, **10**: 201.

Ramadan, M.M.A., Fayza, M.A., Darwesh, and El-Zeiny, O.A. (2019) *J. Product. Dev.*, **24**(4): 831-850.

Raman, K.R. (1964) *Madras Agric. J.*, **51:** 79.

Ramanujam, B., Krishna, J. and Poornesha, B. (2018) *Pestic. Res. J.*, **30**(2): 159-162.

Ramesh, R. (2008) *Technical Bulletin No: 10,* ICAR Research Complex for Goa (Indian Council of Agricultural Research), Ela, Old Goa- 403 402, Goa, India.

Rameshkumar, D., Swarna Priya R., Muthukrishnan, N., Ravikesavan, R. and Savitha, B.K. (2019) *Int. J. Adv. Biol. Res.*, **9**(4): 270-274.

Ramu, N. and Muthuswamy, S. (1964) *Madras Agric. J.*, **51:** 80-81.

Rana, G.L. and Vovlas, C. (1971) *Phytopathologia Mediterranea*, **10:** 273-277.

Ranamukaarachchi, S.L., Nawalage, K., Surasena, J.M. and Godawita, K. (1989) *Sri Lankan J. Agric. Sci.*, **26:** 142-149.

Rangaswami, G. (1979) *Diseases of Crop Plants in India (2nd edn.)*, Prentice-Hall of India Pvt. Ltd., New Delhi, pp. 298-302.

Rangaswamy, P. and Kadambavanasundaram, M. (1973) *South Indian Hort.*, **21:** 1-6.

Rani, M., Kumar, S. and Kumar, M. (2018) *J. Environ. Biol.*, **39**: 710-713.

Rani, N.J., Shakita, A. and Bhaskarrajan, G. (1999) *South Indian Hort.*, **46:** 39-42.

Rao, C.V., Mani, A. and Rao, P.K. (1987) *Tropical Pest Management*, **33:** 137-138.

Rao, M.S., Reddy, P.P. and Nagesh, M. (1997) *Nematologia Mediterranea*, **25:** 249-252.

Rao, M.V.D., Sohi, H.S. and Vijay, O.P. (1976) *Veg. Sci.*, **3:** 61-64.

Rao, N.K.S. and Bhatt, R.M. (1990) *Gartenbauwissenschaft*, **55:** 41-44.

Rastogi, K.B., Korla, B.N., Peshin, S.N. and Saini, S.S. (1979) *Indian J. Agric. Sci.*, **49:** 680-682.

Raut, U.M. and Sonone, H.N. (1980) *Veg. Sci.*, **7:** 74-78.

Ravestijn, W.Van (1988) *Netherlands J. Agric. Sci.*, **36:** 283-290.

Ravichandran, M. Jerome, J.J. and Sriramachandrasekharan, M.V. (1995) *Annals Agric. Res.*, **16**: 282-285.

Reddy, B.S. and Gowda, P.M. (1987) *South Indian Hort.*, **35:** 386-389.

Richter, C., Heiligtag, D., Nagieb, M. and Mukhallalati, U. (1999) *Tropenlandwirt*, **100:** 3-16.

Rizk, F.A., Abdallah, A.M. and Shaheen, A.M. (1997) *Egyptian J. Hort.*, **24:** 239-259.

Rizza, F., Barchiesi, F., Mennella, G., Tacconi, M.G., Collonnier, C., Sihachakr, D., Fusari, F., Acciarrri, N. and Rotino, G.L. (2001) Proc. 11th Eucarpia Meeting on Genet. Breed. Capsicum and Eggplant. Antalya, Turkey, pp 117-120.

Rodriguez-Jimenez, J., Amaya-Guerra, C., Baez-Gonzalez, J., Aguilera-Gonzalez, C., Urias-Orona, V. and Nino-Medina, G. (2018) *Molecules*, **23:** 3210.

Rohini, H.G., Gowtham, P., Hariprasad, S., Singh, B. and Niranjana, S.R. (2016) *Biol. control.*, **101:** 123-129.

Rosa, M. (1954) *Ital. Agric.*, **91**: 395-400.

Rotino, G.L. and Gleddie, S. (1990) *Plant Cell Reports*, **9**: 26-29.

Rotino, G. L., Mennella, G., Fusari, F., Vitelli, G., Tocconi, M. G., D Alessandra, A., Acciarri, N. (2001) Proc. 11th Eucarpia Meeting on Genet. Breed. Capsicum and Eggplant. Antalya, Turkey, pp. 303-307.

Roy, H.S.G. (1997) *Environ. Ecol.*, **15:** 100-103.

Roy, P., Som, M.G. and Dasgupta, A. (1973) *Sci. Cult.*, **39:** 270-272.

Rudrapal, D. and Nakamura, S. (1988a) *Seed Sci. Tech*, **16:** 115-121.

Rudrapal, D. and Nakamura, S. (1988b) *Seed Sci. Tech*, **16:** 123-130.

Ruggiero, C., Pascale, S.de and Barbieri, G. (1994) *Rovosta do Agronomia*, **28:** 222-234.

Sabatino, L., Iapichino, G., D'Anna, F., Palazzolo, E., Mennella, G. and Rotino, G.L. (2018) *Sci. Hort.*, **228:** 81-90.

Sabita, J.N., Baruah, B.M. and Rachid, H.A. (2000) *Veg. Sci.*, **27:** 76-77.

Sabolu, S., Kathiria, K.B., Mistry, C.R. and Kumar, S. (2014) *Aus. J. Crop Sci.*, **8**(2): 243-250.

Sachan, C.P., Verma, R.K., Srivastava, A.K. and Singh S.K. (2009) *Ann. Hort.*, **2**(2): 217-218.

Sadovski, A.Y., Fattal, B. and Goldberg, D. (1978) *J. Fd. Prot.*, **41:** 336-340.

Saha, S., Adhikary, S. and Raychaudhuri, D. (2020) *World Sci. News.*, **146:** 255-273.

Sakata, Y., Mishio, T. and Mon'Ma, S. (1989) *In: EUCAR PIA VIIti, meeting on genetics and breeding on capsicum and eggplant*, Kragujevac, Yoguslavia, 27-30 June, 1989.

Sakriya, S.G., Vaddoria, M.A., Raval, L.J. and Balas, A.D. (2020) *Int. J. Chem. Stud.*, **SP-8**(2): 37-39.

Salehuzzaman, M. (1981) *SABRAO J. Breed. Genet.*, **13**: 25.

Salgon, S., Jourda, C., Sauvage, C., Dauna, M.C., Reynaud, B., Wicker, E. and Dintinger, J. (2017) *Front. Plant Sci.*, **8**: 828. doi: 10.3389/fpls.2017.00828.

Salih, Y. A. and Ridha, R. J. A. (2019) *Basrah J. Agric. Sci.*, **32:** 182-193.

Sambandam, C.N. (1964a) *Econ. Bot.*, **18:** 128-131.

Sambandam, C.N. (1967) *Econ. Bot.*, **21:** 309-311.

Sambandam, C.N. (1969) *Annamalai Univ. Agric. Res. Ann.*, **1:** 1-7.

Sambandam, C.N. (l964b) *Indian J. Genet.*, **24:** 175-176.

Samlindsujin, G., Karuppaiah, P., and Manivannan, K. (2017). *Intern. J Plant Sci.* 2017; **12**: 21-27.

Sampson, H.C. (1936) *Bull. Misc. Inf. Roy Bot. Gdn. Kew*, Add.Scr., **12:** 159.

Sandhu, G.S., Dhooria, M.S. and Singh, A. (1974) *Sci. Cult.*, **40:** 258-260.

Sandhu, G.S., Ramzan, M. and Bhalla, J.S. (1973) *Punjab Hort. J.*, **1**: 211-212.

Sandip, P., Chatterjee, M.L., Mondal, S. and Samanta, A. (2009) *Pestic. Res. J.*, **21**(1): 58-60.

Sanwal, S.K., Baswana, K.S. and Dhingra, H.R. (1997) *Annals Biol., Ludhiana*, **13:** 123-125.

Sao, A. and Mehta, N. (2010) *Dirasat, Agric. Sci.*, **37**(1): 36-45.

Sarejanni, J.A. (1952) *Ann. Inst. Phytopath., Benaki*, **6:** 14-18.

Sarma, S.K., Talukdar, P. and Barbora, M.C. (2000) *Annals Biol.* (Ludhiana), **16:** 67-70.

Satishkumar, Basegowda, Sharankumar and Shivakumar (2011) *Indian Hortic. J.*, **1**: 19-22.

Satish, L., Rameshkumar, R., Rathinapriya, P., Pandian, S., Rency, A.S., Sunitha, T. and Ramesh, M. (2015) *J. Appl. Phycol.*, **27:** 993-1002.

Satyagopal, K., Sushil, S.N., Jeyakumar, P., Shankar, G., Sharma, O.P., Boina, D., Varshney, R., Sain, S.K., Rao, S.N., Sunanda, B.S., Asre, R., Kapoor, K.S., Arya, S., Kumar, S., Patni, C.S., Krishnamurthy, A., Devi, U., Rao, K., Vijaya, M., Sireesha, K., Madhavilatha, Sreedharan, S., Chandel, R.P. and Kotikal, Y.S. (2014) AESA based IPM package for brinjal. p. 74.

Schaff, D.A., Boyer, C.D. and Pollack, B.I. (1980) *HortScience*, **15:** 419.

Scossiroli, R.E., Silnetti, E. and Brunelli, B. (1972) *Genetica Agraria*, **26:** 321-339.

Seervi, K.S., Bara, B.M., Dayal, A., Choyal, P. and Yadav, S.L. (2018) *Int. J. Curr. Microbiol. App. Sci.*, **7**(7): 1614-1618.

Sêkara, A. and Bieniasz, M. (2008) *Acta Agrobot.*, **61**(1): 107-113.

Selvaraj, J.A. (1988) *South Indian Hort.*, **36:** 32-35.

Sen, P. and Deka, P.C. (2001) In: *Biotechnology of horticultural crops*, (Eds.V.A. Parthasarathy, T.K. Bose, P.C. Deka), pp. 174-187.

Seno, S., Oliveira, C.A.L.De. and Velho, D. (1976) *Cientifica*, **4:** 236-241.

Serap, O., Ozalp, Z. and Devran, Z. (2018) *J. Plant Dis. Protec.*, **125:** 577-580.

Seth, J.N. and Dhandar, D.G. (1970) *Progressive Hort.*, **1**: 45-50.

Shafeeq, A., Madhusudan, K., Hanchinal, R.R., Vijayakumar, A.G. and Salimath, P.M. (2007) *Karnataka J. Agric. Sci.*, **20**(1): 33-40.

Shahid, U., Shah, K., Ahmed, I., Amin, N., Ara, G., Khan, S.I., Guttai, G. and Zainub, B. (2019) *Pure Appl. Biol.*, **8**(2): 1861-1875.

Sharma, A.K. and Chhibber, R.C. (1999) *Indian J. Entom.*, **61:** 242-251.

Sharma, A.K. and Saxena, R. (2016) *J. Biopest.*, **5**(1): 1-6.

Sharma, D.C. (1969) *Phytopath. Z.*, **65:** 341-345.

Sharma, M. D. (2006). *J. Inst. Agric. Animal Sci.*, **27**: 153-156.

Sharma, M.D. (2016) *J. Inst. Agric. Anim. Sci.*, **27:** 153-156.

Sharma, S.R. and Sohi, H.S. (1980) *Veg. Sci.*, **7:** 67-69.

Sharma, T.K., Pant, S.C., Kumar, K., Kurrey, V.K., Pandey, P.K. and Bairwa, P.L. (2016) *Int. J. Bio-resource Stress Manage.*, **7**(5): 964-969.

Shatrova, G.L. (1977) *Referativny I Zhurnal*, **55**: 611.

Sheela, J., Sivaprakasam, K. and Seetharaman, K. (1995) *Madras Agric. J.*, **82:** 199-201.

Shelton, A.M., Hossain, M.J., Paranjape, V., Azad, A.K., Raman, M.L., Khan, A.S.M.M., Prodhan, M.Z.H., Rashid, M.A., Majumder, R., Hossain, M.A.,

Hussain, S.S., Huesing, J.E. and McCandless, L. (2018) *Front. Bioengin Biotechnol.*, **6**: 106.

Shelton, A.M., Sarwer, S.H., Hossain, Md. J., Brookes, G. and Paranjape, V. (2020) *Front. Bioeng. Biotechnol.*, **8**: Article 498.

Shende, V.D., Chattopadhyay, A., Yadav, Y., Seth, T. and Mukherjee, S. (2016) *Agric. Res.*, **5**(3): 219-229.

Shende, V.D., Seth, T., Mukherjee, S. and Chattopadhyay, A. (2017) *J. Appl. Nat. Sci.*, **9** (1): 451 – 455.

Shimose, N., Sekiya, J., Kimura, O. and Suzuki, I. (1991) *Japanese J. Trop. Agric.*, **35:** 16-19.

Shivashankaragouda, B., Patil, M.N., Merwade, B.S., Vyakaranahal and Deshpande, V.K. (2008) *Indian J. Agric. Res.*, **42**: 7-12.

Si, Y.P., He, W.M. and Chen, D.K. (1996) *China Vegetables*, No. 2, pp. 22-23,27.

Siddiqui, M.A. and Alam, M.M. (1999) *Pakistan J. Nematology*, **17:** 129-136.

Sidhu, A.S., Kaur, G. and Bajaj, K.L. (1982) *Veg. Sci.*, **9:** 112-118.

Sidhu, A.S., Kolloand Pandita, M.L. (1980) *Haryana J. Hort. Sci.*, **9:** 207-217.

Sifola, M.I., Pascale, S.de and Romano, R. (1995) *Acta Hortic.*, No. **412:** 176-184.

Sihachakr, D., Daicour, R., Chaput, M.H., Barrientos, E., Ducreux, G. and Rossignol, L. (1989) *Theor. Appl. Genet.*, **77**: 1-6.

Sihachakr, D., Haicour, R., Chaput, M. H., Barrientos, E., Ducreux, G., Rossignol, L. Souvannavong, V. (1988) *Plant Sci.*, **57**: 215-223.

Silambarasan, V., Eswaran, R., Senthilkumar, N., Thangavel, P. and Thirugnanakumar, S. (2020) *Plant Arch.*, **20**(1): 9-15.

Singh, A.K., Singh, M., Singh, R., Kumar, S., and Kalloo, G. (2006) *Curr. Sci.* **90**: 357-364.

Singh, A.P., Luthria, D., Wilson, T., Vorsa, N., Singh, V., Banuelos, G.S. and Pasakdee, S. (2009) *Food Chem.*, **114:** 955-961.

Singh, B. and Kumar, N. (1988) *Veg. Sci.*, **15:** 72-78.

Singh, B., Joshi, S. and Kumar, N. (1978) *Haryana J. Hort. Sci.*, **7**: 95-99.

Singh, B.P., Singh, D.R. and Ram, S. (1973) *Indian J. Farm Sci.*, **1:** 111-112.

Singh, D. (1981) *Personal Communication; Punjab Agric. Univ*, Ludhiana.

Singh, D. and Dhancholia, S. (1991) *Himachal J. Agril. Res.*, **17:** 119-120.

Singh, D. and Sidhu, A.S. (1988) *Indian J. Entomol.*, **48:** 305-311.

Singh, D.P., Bhutani, R.D. and Pandita, M.L. (1989) *Haryana J. Hort. Sci.*, **18:** 142-145.

Singh, H. and Nandpuri, K.S. (1974) *J. Res., Punjab Agric. Univ.*, Ludhiana, **11:** 150-157.

Singh, H.B, Bhagachandani, P.M. and Thakur, M.R. (1964) *Indian J. Hort.*, **21:** 221-231.

Singh, H.B. and Sikka, S.M. (1955) *Indian Fmg.*, **5**: 18-21.

Singh, K. (1969) *Farm Bull. No. 24,* Punjab Agric. Univ., Ludhiana.

Singh, K. and Pandey, U.B. (1993) *Veg. Sci.,* **20**: 93-103.

Singh, K. and Sandhu, D.S. (1970) *Punjab Hort. J.,* **10**: 103-110.

Singh, K., Sidhu, A.S. and Kumar, A. (2012) *J. Hort. Sci.,***7**(2): 142-144.

Singh, K.K., Kapur, N.S. and Mathur, P.B. (1952) *Indian J. Hort.,* **9:** 16-19.

Singh, N., Rao, G.P., Madhupriya and Upadhayaya, P.P. (2012) *Phytopatho. Mollicutes.,* **2:** 69-71.

Singh, N.D. and Mital, R.K. (1989) *Indian J. Agric. Sci.,* **58:** 402-403.

Singh, P.K. and Gopalakrishnan, T.R. (1997) *Hort. J.,* **10**: 57-64.

Singh, S., Khemariya, P., Rai, A., Rai, A.C., Koley, T.K. and Singh, B. (2016) *LWT,* **74:** 420-426.

Singh, S., Krishnamurthi, S. and Katya I.S.L. (1963) *Fruit Culture in India, Indian Council of Agricultural Research,* NewDelhi, p. 412.

Singh, S., Singh, N. and Mangal, J.L. (1988) *Haryana J. Hort. Sci.,* **17:** 237-240.

Sinha, K.K., Chakraborty, D.P., Dasgupta, K.P. and Dhua, S.P. (1976) *Fertl. Tech. India,* **13:** 15-18.

Sinha, S.N. and Chakrabarti, A.K. (1982) *Seed Res.,* **10:** 58-65.

Sivanappan, R.K. (1979) *Punjab Hort. J.,* **19:** 83-85.

Sivaprakasam, K. and Rajagopalan, C.K.S. (1971) *Annamalai Univ. Agric. Res. Ann.,* **3**: 70-74.

Sivaprakasam, K. and Rajagopalan, C.K.S. (1974) *Plant Soil,* **40**: 217-220.

Sivaprakasam, K., Jagadeesan, M. and Kannapiran, C. (1976) *Annamalai Univ. Agric. Res. Ann.,* **8:** 146-150.

Sivaprakasam, K., Rajagopalan, C.K.S. and Vidhyasekharan, P. (1973) *Annamalai Univ. Agric. Res. Ann.,* **5:** 101-106.

Sohi, H.S., Rao, M.V.B., Rawal, R.D. and Kishun, R. (1981) *Indian J. Agric. Sci.,* **51:** 572-573.

Sollapur, D.L. and Hiremath, S.M. (2016) *J. Farm Sci.,* **29**(1): 137-139.

Som, M.G. and Malik, S.C. (1979) *Orissa J. Hort.,* **7**: 28-32.

Som, M.G., Hashim, H., Mandal, A.K. and Maity, T.K. (1992) *Crop Res.,* **5:** 80-84.

Som, M.G., Hazra, P., Mandal, A.R. and Maity, T.K. (1989) *In: Proc. Nat. Sem. Heterosis Breeding, Edited, Rakshit,* S.C., pp.172-178.

Sonavane, L.K., Patil, P.V. and Chavan, S.K. (2012) *J. Agric. Res. Technol.,* **37**(2): 196-200.

Sowinska, K.A., Krygier, M. and Turczuk, J. (2016) *Folia Hort.,* **28**(1): 19-24.

Srinivas, G., Jayappa, A.H. and Patel, A.I. (2016) *Adv. Life Sci.,* **5**: 1211-1215.

Srinivasan, K. and Caelliah, S. (1978) *Haryana J. Hori. Sci.,* **7**: 203-227.

Srinivasan, K. and Gopimony, R. (1969) *Agric. Res. J., Kerala.*, **7**: 39-40.

Srinivasan, P.M. and Basheer, M. (1961) *Indian Fmg.*, **11:** 19.

Srinivasan, R. (2009) AVRDC-The World Vegetable Center, Taiwan.

Srivastava, B.K. and Mathur, L.M.L. (1962) *Indian J. Ent.*, **24:** 229-235.

Srivastava, B.P. (1961) *Madras agric. J.*, **48**: 429-433.

Srivastava, L.S. and Sachan, S.C.P. (1973) *Indian J. Agric. Sci.*, **43:** 673-675.

Stàgel, A., Portis, E., Toppino, L., Rotino, G. L. and Lanteri, S. (2008) *BMC Genom.*, **9**: 357.

Stephan, Z.A., Michbass, A.H. and Shahir, C.W. (1989) *International Nematology Network Newsletter*, **6**: 25-26.

Stommel, J. R., Whitaker, B. D., Haynes, K. G. and Prohens, J. (2015) *Euphytica*, **205**: 823-836.

Stommel, J.R. and Whitaker, B.D. (2003) *J. Amer. Soc. Hort. Sci.*, **128:** 704-710.

Stork, H.W. and Schouten, S.P. (1976) *Groenten en Fruit*, **31**: 2237.

Subbaratnam, G.V. and Butani, D.K. (1981) *Veg. Sci.*, **8:** 149-153.

Subramanian, V., Chinnarajan, A.M., Rajendran, G., Chandrasekharan, J. and Dakshinamurthy, A. (1977) *Curr. Res.*, **6:** 192-193.

Subramanyam, K., Rajesh, M., Jaganath, B., Vasuki, A., Theboral, J., Elayaraja, D., Karthik, S, Manickavasagam, M. and Ganapathi, A. (2013) *Appl. Biochem. Biotechnol.*, **171**: 450-468.

Sulikeri, G.S., Bankapur, V.M. and Rao, M.M. (1977) *Curr. Res.*, **6:** 48-49.

Sulochana, C.B. and Solomon, J.J. (1973) *Phytopathologische Zeitschrift*, **77**: 339-347.

Sun, W.Q., Wang, D.B., Wu, Z.Z. and Zhi, J.R. (1990) *Sci. Hort.*, **44:** 11-13.

Swaminathan, M. and Srinivasan, K. (1971) *Agric. Res. J., Kerala*, **9:** 11-13.

SwamyRao, T. (1970) *Madras Agric. J.*, **57**: 508-509.

Szasz, A., Szilassy, D., Salanki, K., Fari, M. and Balazs, E. (1998) *Acta Agron. Hungaria*, **46:** 201-207.

Tanahashi, B., Watanabe, K., Suzuki, A., Inoue, H. and Furukawa, K. (1977) *Bull. Coll. Agric. Veg. Medicine, Nihon Univ.*, **34:** 14-25.

Tanaka, J.S., Romanowski, R.R.Jr. and Crozier, J.H.Jr. (1973) *Res. Rept., Hawaii Agric. Exp. Sta.*, **190:** 20.

Tataru, D. and Hristea, N. (1977) *Acta Hortic.*, **58:** 521-524.

Temkin-Gorodeiski, N., Shapiro, B., Grinberg, S., Rosenberger, I. and Fallik, E. (1993) *J. Hort. Sci.*, **68**: 689-693.

Thakur, M.R., Singh, K. and Singh, J. (1968) *J. Res. Punjab Agric. Univ.*, Ludhiana, **5:** 490-495.

Thakur, M.R., Singh, K. and Singh, J. (1969) *J. Res. Punjab Agric. Univ*, Ludhiana, **6:** 769-775.

The Eggplant Genome Consortium (2017) *The Eggplant Genome Reveals Key Events in Solanaceae Evolution*. Available at: https: //pag.confex.com/pag/xxv/meetingapp.cgi/Paper/23484.

Thevasagayam, E.S. and Canagasingham, L.S.C. (1961) *Trop. Agriculturist.,* **117:** 105-114.

Thingujam, U., Bhattacharya, K., Ray, K., Phonglosa, A., Pari, A. and Banerjee, H. (2019) *Commun. Soil Sci. Plant Anal.,* **51:** 70-85.

Thirumurugan, T., Babu, S., Sassikumar, D. and Ganesan, J. (1999) *Trop. Agril. Res.,* **11**: 143-148.

Thomas, K.M. and Krishnaswamy, C.S. (1939)*Proc. Indian Acad. Sci.,*B, **13:** 201-212.

Thompson, H.C. and Kelly, C.W. (1957) In: *Vegetable Crops, McGraw-Hall Book Co., Inc.,* NewYork, p. 611.

Tigchelaar, E.C., Janick, J. and Erickson, H.T. (1968) *Genetics*, **60**: 475-491.

Tiwari, S.K., Mishra, P., Singh, S., Singh, V.K., Kashyap, S.P., Singh, M., Tiwari, K.N., Naik, P.S. and Singh, B. (2019) *bioRxiv,* doi: 10.1101/560805.

Toppino, L., Mennella, G., Rizza, F., D'Alessandro, A., Sihachakr, D., and Rotino, G.L. (2008) *J. Heredity.,* **99**(3): 304-315.

Trentini, L. and Maioli, B. (1989) *Colture Protette*, **18**: 48-51.

Tripathy, M.K. and Senapati, B. (1998) *Orissa J. Hort.*, **26:** 37-41.

Tripathy, M.K., Das, B.C., Dash, S.K. and Patra, R. (1997) *Orissa J. Hort.*, **25**: 39-43.

Tu¨mbilen, Y., Frary, A., Mutlu, S. and Doganlar, S. (2011) *Int. Res. J. Biotechnol.*, **2:** 16-25.

Turchenkov, G.I., Avdeev, Yu, I., Alekseeva, V.S., Bazhmaeva, F.K. and Kigashpaeva, O.P. (1990) *Kartofel I Ovoshchi*No.3, p. 35.

Ulinski, Z. and Glaps, T. (1989) *Biuletyn Warzywniczy Supl. II*, pp. 179-182.

Uncini, L., Gorini, F. and Sozzi, A. (1977) *Acta Hortic.*, **58**: 383-388.

Upamanya, G.K. and Dutta, P. (2019) *Int. J. Curr. Microbiol. App. Sci.,* **8**(6): 1232-1240.

Ushamani, P. and Peter, K.V. (1987) *Agril. Res. J. Kerala*, **25**: 133-135.

Uthamasamy, S., Kumaresan, D., Menon, P.P.V. and Jayraj, S. (1976) *Indian J. Exp. Biol.,* **14**: 639-640.

Vadivel, E. and Bapu, J.R.K. (1989a) *South Indian Hort.,* **37**: 13-15.

Vadivel, E. and Bapu, J.R.K. (1989b) *Capsicum Newsletter*, No.7, pp. 86-87.

Vadivel, E. and Bapu, J.R.K. (1990) *South Indian Hort.*, **38**: 301-304.

Vaganov, A.P., Kulik, N.I. and Vinichenko, N.V. (1975) *Referativny I Zhurnal*, 2.55.762.

Vaghasiya, M.H., Kathiria, K.B., Bhalala, M.K. and Doshi, K.M. (2000) *Indian J. Genet.*, **60:** 127-130.

Valida, A., Rivera, F.R., Salabao, A., Benitez, M., Sudaria, E., Acedo, A. and Ekman, J. (2018) *Acta Hortic.*, **1213**: 297-302.

Vanna, M.K., Sharma, H.C. and Pathak, V.N. (1978) *Plant Dis. Reptr.*, **62:** 274-275.

Varshney, R.K., Graner, A. and Sorrells, M.E. (2005) *Trends Biotechnol.*, **23:** 48-55.

Vavilov, N.I. (1928) *Proc. 5th International Congress of Genetics*, New York, pp. 342-369.

Venkatakrishnan, G., Sathiyanarayanan, G. and Balaji, D.S. (2018) *Madras Agric. J.*, **105**(7-9): 437-441.

Verma, J.P., Rathore, S.V.S. and Dayal, R. (1975) *Progressive Hort.*, **6:** 25-30.

Verma, R.K. and Dubey, G.S. (1976) *Acta Botanica Indica*, **4:** 144-150.

Verma, R.K. and Dubey, G.S. (1978) *Veg. Sci.*, **5:** 36-39.

Verma, T.S. and Choudhury, B. (1974) *Veg. Sci.*, **1:** 55-61.

Vevai, E.J. (1970) *Pesticides*, **4:** 26-35.

Vijay, O.P., Dutta, O.P. and Nath, P. (1971) *Indian J. Hort.*, **29**: 67-68.

Vijay, O.P., Nath, P. and Jalikop, S.H. (1977) *Indian J. Agric. Sci.*, **47**: 288-291.

Vijayalalitha, S.J. (1999) *Adv. Plant Sci.*, **12:** 245-248.

Vilanova, S., Manzur, J.P. and Prohens, J. (2012) *Mol. Breed.*, **30**: 647-660.

Viswanathan, T.V. (1973) *Proc. Indian Acad. Sci.*, B, **77:** 176-180.

Viswanathan, T.V. (1975) *Curr. Sci.*, **44:** 134.

Vito, M.D. and Lamberti, F.(1980) *Informatore Fitopatologico*, **30:** 11-13.

Wani, A.H. (2005) *Indian Phytopathol.*, **58**(2): 228-231.

Watanabe, K., Takahashi, A. and Inoue,H. (1978) *Bull. Coll. Agri. Vet. Medicine*, Nihon Univ., **35**: 88-96.

Whitaker, B.D. and Stommel, J.R. (2003) *J. Agric. Food Chem.*, **51:** 3448-3454.

Wishwanath, B.N., Nagesh Chandra, B.K. and Balasubramanian, R. (1973) *Agric. Res. J., Kerala*, **11**: 160.

Yadav, B.S., Nandwana, R.P., Lal, A. and Verma, M.K. (1975) *Indian J. Mycol. Pl. Path.*, **5:** 17.

Yadav, N., Dhankhar, S. K., Chandanshive, A. V., and Kumar, V. (2017). *J. Appl. Nat. Sci.*, **9**: 1032-1035.

Yamamoto, Y. and Matsumoto, O. (1998) *J. Jap. Soc. Hort. Sci.*, **63:** 67-72.

Yamazuki, T. and Taskushi, A. (1991) *J. Jap. Bot.*, **66:** 46-48.

Yawalkar, K.S. (1969) *Vegetable Crops of India*, Agri-Horticultural Publishing House, Nagpur.

Yi, J.X. and Chen, J.II. (1999) *Jiangsu Agril. Sci.*, **6**: 62-65.

Yu, X. (1981) *Liaoning Agric.Sci.*, **4**: 32-34.

Zareena, S.K. and Das, V.V. (2014) *Global J. Bio-Sci. Biotech.*, **3**(1): 126-127.

Zayova, E., Vassilevska-Ivanova, R., Kraptchev, B. and Stoeva, D. (2010) *Biol. Divers. Conserv.*, **3/3**: 1-5.

Zeven, A.C. and Zhukovsky, P.M. (1975) *Dictionary of Cultivated Plants and their Centres of Diversity*, Wageningen, Netherlands, p. 219.

Zhao, Qing Chu, Zhao, Na and Zhang, Li (1997) *China Vegetables*, **6:** 7-9.

3

CHILLI

N.D. Bhutia, T. Bhattacharjee, C.R. Muthukrishnan, T. Thangaraj, R. Chatterjee and T.K. Maity

1.0 Introduction

Chillies are the green or dried ripe fruits of pungent forms of *Capsicum annuum* L. and sometimes *Capsicum frutescens*. Chilli is used in many forms, such as fresh or cooked vegetables, herbs or spices, and various kinds of processed products (Hazra *et al.*, 2011). It forms an indispensable adjunct in every house in the tropical world. Chilli fruit contains a broad variety of antioxidant vitamins especially vitamin A and C, capsaicin, which determine the great variability of the fruit's smell, flavour, taste and consequently consumer preference (Bhattacharya *et al.*, 2010). The colouring agents, capsanthin/capsorubin, have also been used in the feed of laying hens to colour the egg yolk and the skin of broilers apart from pharmaceutical and cosmetic industries (Tepic *et al.*, 2008). In India, *Piper nigrum* was commonly used in its place before the introduction of this valuable condiment.

2.0 Composition and Uses

2.1.0 Composition

The chief constituent of chilli pericarp is a crystalline colourless pungent principle known as capsaicin or capsicutin ($C_{18}H_{27}NO_3$), a condensation product of 3-hydroxy-4-methoxy benzylamine and decylenic acid and 7 closely related alkyl vanillyl amides, collectively referred to as capsaicinoids which produce a highly irritating vapour on heating. Capsaicin and dihydrocapsaicin are secreted by the outer walls of the fruit and they accounts for >80 per cent of capsaicinoids in the placenta of fruit (Yahara *et al.*, 1988; Bernal *et al.*, 1993; Kobata *et al.*, 1998). The degree of pungency in chili is measured by Scoville Heat Unit (Scoville, 1912). The content of capsaicin differs in cultivars of chillies (Ananthasamy *et al.*, 1960; Kamalan and Rajamani, 1963; Biswas *et al.*, 2011; Chattopadhyay *et al.*, 2011). Green chillies are rich in vitamin A and C and the seed contain traces of starch (Saimbhi *et al.*, 1977; Sayed and Bagavandas, 1980; Biswas *et al.*, 2011; Chattopadhyay *et al.*, 2011; Bhutia *et al.*, 2015). The food value of green chilli is given in Table 1.

Green chilli has antioxidants and antioxidative enzymes that provide protection against harmful reactive oxygen species and free radicals. The quality of green chilli depends on the content of antioxidants as well as the activity of antioxidative enzymes. The content of chain-breaking antioxidants in green pepper varies (Antonious *et al.*, 2006; Deepa *et al.*, 2007; Kim *et al.*, 1997; Panchal *et al.*, 2001; Biswas *et al.*, 2011). To a large extent, antioxidant composition of pepper depends on genetic factor (Bhattacharya *et al.*, 2010; Biswas *et al.*, 2011) and is affected by cultural practices and environmental factors (Mahendran and Bandara, 2000; Panchal *et al.*, 2001). Capsaicin and dihydrocapsaicin can directly scavenge free radicals (Bhattacharya *et al.*, 2010; Biswas *et al.*, 2011), and have wide application in food, medicine and pharmaceutical industries (Janaki *et al.*, 2015). The fruit contains the flavonoids quercetin, luteolin, apigenin, capsanthin, myristcin, hesperidin, scopoletin, and the phenolic compounds esters of ferulic and sinapic acids, which have antioxidant activities (Martínez-Flórez *et al.*, 2002).

Table 1: Composition of Green Chilli (per 100 g of edible portion)*

Moisture	85.7 g	Phosphorus	80 mg
Protein	2.9 g	Iron	1.2 mg
Fat	0.6 g	Sodium	6.5 mg
Minerals	1.0 g	Potassium	217 mg
Fibre	6.8 g	Copper	1.55 mg
Carbohydrates	3.0 g	Sulphur	34 mg
Calcium	30 mg	Chlorine	15 mg
Magnesium	24 mg	Thiamine	0.19 mg
Riboflavin	0.39 mg	Vitamin A	292 I.U.
Oxalic acid	67 mg	Vitamin C	111 mg
Nicotinic acid	0.9 mg	–	80 mg

* Aykroyd (1963).

The fruits also contain a fixed oil, red colouring matter which is non-pungent and yield 20-25 per cent alcoholic extract. Quantities of extractable color in chilli is estimated by a spectrophotometric process, and described by American Spice Trade association (ASTA) units. Generally, the higher the ASTA color value, the greater the effect on brightness or richness of the final product. The range of color content is classified as low (<70 ASTA units), medium (71-100 ASTA units) and high (>100 ASTA units) (Dhall, 2008). More than 120 ASTA color units are preferred in products for export (Tainter and Grenis, 1993). Indian chili varieties and hybrids contain ASTA color values from 50 to 160 (Babu *et al.*, 2014; Sreenivas *et al.*, 2019). The constituents of red chilli as dry matter 22.02 per cent, ascorbic acid 131.06 mg/100 g (fresh weight), oleoresin 66.53 ASTA units, colouring matter 67.38 ASTA units, capsaicin 0.34 per cent (dry weight), crude fibre 26.75 per cent and total ash 6.69 per cent have been reported (Bajaj *et al.*, 1980; Chattopadhyay *et al.*, 2011; Sreenivas *et al.*, 2019). Sun dried fruits of cv. Byadgi had highest capsaicin of 0.75 per cent (Nisar *et al.*, 1987).

Based on the most important quality aspects like capsaicin, capsanthin, oleoresins, ascorbic acid contents and antioxidant activities the cultivars/varieties LCA 235, LCA 248, LCA 206 and CA-586 (Reddy *et al.*, 1998); AC-588-1, BCC-62 and 'AC-465 (Bhattacharya *et al.*, 2010); AC-573, Bullet, AC-462, and BCC-49 (Biswas *et al.*, 2011); BCC-1, BCCH Sel-4 (Bidhan Chilli-4) and ACC-588 (Chattopadhyay *et al.*, 2011) were found better in quality than others. A survey was carried out by Mathur *et al.* (2000) to find a chilli pepper that contained high amount of capsaicin and dihydrocapsaicin which could be extracted for use as substitutes for orthochloro-ben-zalmalanitrile or chloroacetophenone (used for riot control). HPLC analysis of various Indian chilli cultivars identified the Tezpur cultivar popularly known as 'Bhut Jolokia' as containing the highest amounts of capsaicin and dihydrocapsaicin, contributing to a pungency rating of 1,001,304 SHU which is the hottest chilli known to date in India (Bosland and Baral, 2007). The taxonomic relationship of 'Bhut Jolokia' based on RAPD markers placed 'Bhut Jolokia' in a taxonomic position between *C. chinense* and *C. frutescens* with 'Bhut Jolokia' clustering more closely with the *C. chinense* group (Bosland and Baral, 2007). The Carolina Reaper, a cultivar of *Capsicum chinense*, with pungency of 22000000 SHU is the hottest chilli pepper known to date in the world surpassing the Trinidad Scorpion (1200000 SHU).

A continuous increase in the ascorbic acid content starting from matured green fruits up to fully matured red ripe stage is observed (Kumari *et al.*, 1998). The red ripe fruits thus contain higher ascorbic acid content than mature green fruits. Once the process of drying of red ripe fruit begins, there is a decline in the content. This decline continues as the process of drying progresses and further decline is observed in storage.

2.2.0 Uses

Chillies are widely used throughout the tropics and are major ingredients of curry powder in the culinary preparations. They are extensively used in Central America as constituents of dishes such as tamales and 'chile con curne'. In its powdered form, it constitutes red or cayenne pepper. Pepper sauces are made by pickling the pulp in strong vinegar or brine. Extracts of chillies are used in the production of ginger beer and other beverages. Cayenne pepper is incorporated in poultry feeds. *C. frutescens* is used in medicine as carminatives internally, besides being an external counter irritation. The green chillies are rich in rutin which is of immense pharmaceutical need (Purseglove, 1977).

Chilli is used in many forms, such as fresh or as cooked vegetables, as herbs or spices, and various kinds of processed products (Hazra *et al.*, 2011). Fresh green pungent fruits are generally used in salads, stuffing, and as a flavoring agent in cooked meals. The non-pungent varieties are cooked as vegetables or processed with other food items for flavor. They promote health benefits such as reducing obesity and diabetes (Vasconcelos, 2016). Pepper extracts are used in cosmetics and pharmaceuticals. Besides the use in feed as spice, peppers have also ornamental potential when grown in pots or gardens (Bosland and Votava, 2012). Several studies have found that capsaicin can be used as pain-relieving agent in chemotherapy or radiation therapy (Berger *et al.*, 1996), diabetic neuropathy and nerve pain (also called neuropathic pain). Capsaicin possesses the antioxidant, anti-mutagenic, anti-carcinogenic, and immunosuppressive activities having the ability to inhibit bacterial growth and platelet aggregation. Regular consumption of chilli fruit is helpful against hemorrhoids, varicose veins, anorexia, and liver congestion. Pure and processed form of chilli extracts is used externally as rubefacient and analgesic in case of back pain, rheumatism, articular and muscular pains and swollen feet and even as antidote in case of poisoning. The two non-foods and non-pharmacological application of capsaicin are in the preparation of "pepper sprays" as weapons of self-defense and in preparation of "natural and organic pesticide" (Welbaum, 2015).

3.0 Area and Production

Being an important crop of tropical and sub-tropical countries, cultivated in large areas of Asia, Africa, South, and Central America and southern Europe, the total area under chilli cultivation in the world is about 1.99 million hectares with

Uses of Chilli

Green and Dry Chilli Used in different Recipes of Indian Cuisine

production of 36.77 million tones for green chillies and 1.77 million hectares with production of 4.16 million tones for dry chillies (FAOSTAT, 2018). India is the major producer as well as exporter of the dried chilli in the world. The other important producers of chilli include China, Mexico, Peru, Turkey, Thailand, Indonesia followed by countries in tropical Africa mainly Ghana and Ethiopia (Table 2 and 3).

Table 2: Production Share of Chillies by Continents*

Continent	*Production Share (Green)*	*Production Share (Dry)*
Asia	68 per cent	72.5 per cent
Europe	8.8 per cent	2.8 per cent
Oceania	0.1 per cent	0 per cent
Africa	9.5 per cent	22.1 per cent
Americas	13.7 per cent	2.6 per cent

* FAOSTAT (2018).

Table 3: Top 10 Countries in Terms of Production of Green and Dry Chillies and Peppers*

Dry Chillies and Peppers			*Green Chillies and Peppers*		
Country	*Area (ha)*	*Production (tonnes)*	*Country*	*Area (ha)*	*Production (tonnes)*
India	781737	1808011	China, mainland	769,078	18184711
China, mainland	47753	321290	Mexico	156799	3379289
Ethiopia	152642	294299	Turkey	91973	2554974
Thailand	91675	247010	Indonesia	308547	2542358
Pakistan	65275	148114	Spain	20580	1275457
Bangladesh	101072	141177	Nigeria	97818	747367
Myanmar	107551	130335	Egypt	42132	713752
CÃ´te d'Ivoire	19804	117159	United States of America	21813	705790
Ghana	15299	115147	Algeria	22108	651045
Vietnam	70922	101548	Tunisia	19362	426503

* FAOSTAT (2018).

Chilli is a universal spice crop of India grown in almost all the states of the country. The total area under dry chilli cultivation in India is 7.8 lakh ha and total production is 18.08 lakh tonnes (FAOSTAT, 2018). The quality of the chilli varies from state to state. For instance, chilli of Karnataka is known for its oil content, Gujarat quality is majorly known for its sharp color while that of Rajasthan is well known for making pickles. The chilli cultivated in Assam is famous for its pungency and that grown in Andhra Pradesh is mostly used as vegetables and liked by non-vegetarians. Major chilli producing states include Andhra Pradesh, Maharashtra, Karnataka, and Madhya Pradesh. Apart from being a large consumer and producer of chilli, India is also the largest exporter of the crop.

According to FAO (2016), value generated from global production of peppers was 14.4 billion dollars, approximately forty times greater than that obtained in 1980. In 2013, nearly 200 million and 33 million tonnes of green and dry peppers, respectively, were produced worldwide.

4.0 History, Origin and Taxonomy

4.1.0 History and Origin

The archaeological evidence at sites located in southwestern Ecuador showed that chilli pepper was domesticated more than 6,000 years ago, and it is one of the first cultivated crops in the Central and South Americas (Bosland, 1996; Perry *et al.*, 2007). The genus *Capsicum* comprises about 38 species (USDA-ARS, 2011) of which only six species (namely *Capsicum annuum*, *C. baccatum*, *C. frutescence*, *C. chinense*, *C. pubescens* and *C. assamicum*) are cultivated. *C. assamicum* was recently identified as a distinct domesticated species from the North Eastern part of India (Purkayastha *et al.*, 2012). These domesticated species are derived from different ancestral stocks found in three distinct centers of origin: Mexico being the primary center of origin for *C. annuum*, the most widely cultivated species, with Guatemala as a secondary center; Amazonia for *C. chinense* and *C. frutescens*; and Peru and Bolivia for *C. baccatum* and *C. pubescens*. *C. annuum* and *C. frutescens* are widely distributed from Mexico through Central America and throughout the Caribbean region (Greenleaf, 1986).

The centre of diversity of the common cultivated pepper *Capsicum annuum* is probably Mexico with a secondary centre at Guatemala. *C. frutescens* is widely distributed throughout the tropical and subtropical America both in wild and cultivated forms which also have their origin in Central and South America and the genus is quite clearly South American in origin (Bukusov, 1930 and Shoemaker, 1953). *C. annuum* is not known in wild state and *C. frutescens* doubtfully so but it has now become naturalized in many parts of the tropics and is spread by birds. The long viability of seeds and the case with which they are transported assisted in the rapid spread in the tropics and subtropics (Purseglove, 1977).

Chilli and capsicum were unknown in Europe prior to the discovery of America. They were carried to the old world by the early explorers being introduced into Spain in 1493 (Boswell, 1951). The cultivation of chilli and capsicum spread rapidly from Spain to Europe. The Spaniards and Portuguese in their search for pepper were quick to seize this new and much more pungent form from the new world to distribute it widely.

5.0 Botany

C. annuum is a variable herb or sub-shrub sometimes woody at base, erect, much branches, 0. 5-1.5 m high, grown as an annual, storage taproot usually broken or arrested in growth on transplanting, and numerous profusely branched laterals develop, extending to one metre.

Variability in the Fruit Morphology of different *Capsicum* Species (Ramchiary *et al.,* 2014).

***Capsicum* Species Varieties and Pod Types.**

***Capsicum annuum* pod types are A) Asian, B) Cayenne, C) Chile de Arbol, D) Chiltepin, E) Hungarian Paprika, F) Jalape~no, G) New Mexican, and H) Poblano. A *Capsicum baccatum* pod type is I) Aji. *Capsicum chinense* pod types are J) Bhut Jolokia, K) Habanero, and L) Scorpion. A *Capsicum pubescens* pod type is M) Rocoto and a *Capsicum frutescens* pod type is N) Tabasco (Guzman and Bosland, 2017).**

Domesticated Species of *Capsicum*

Capsicum annuum ***C. frutescens***

Capsicum pubescens ***C. chinense***

Capsicum baccatum ***C. assamicum***

Main shoot is radial but lateral branches are cincinnate, one of the branches at each node remaining undeveloped and subtending bract or bracts are adnate and are carried up a lateral shoot to node above.

Leaves are variable in size, simple, petiole 0.5-2.5 cm long, lamina broadly lanceolate to ovate, entire, thin sub-glabrous, 1.5-12 x 0.5-7.5 cm. tip acuminate, base cuneate or acute.

Flowers usually borne singly and are terminal, but due to the form of branching, appear to be axillary, pedicels up to 1.5 cm long, calyx campanulate, shortly 5 dentate, 10-ribbed about 2 mm long enlarging and enclosing base of fruits, corolla rotate campanulate, deeply 5-6 partite, 8-15 mm in diameter, white or greenish, stamens 5-6 inserted near base of corolla, anthers bluish, deshiscing longitudinally, ovary 2-celled but often multiplying under domestication, style simple, white or purple, stigma capitate.

Fruit indehiscent, many seeded berry, pendulous or erect, borne singly at nodes, variable in size, shape, colour and degree of pungeny, linear conical or globose, 1-30 cm long, unripe fruit green or purplish, ripening to red or orange, yellow, brown, cream or purplish, seeds 3-5 mm long, pale yellow.

C. frutescens is perennial sub-shrub, living 2-3 years, similar in structure to *C. annuum*, but usually bears 2 or more erect berries at fruiting nodes, small and narrow, 0.7-2.5 x 0.3-1 cm, red or yellow, extremely pungent and var. *baccatum* has globose flowers.

5.1.0 Pollination

Bees, ants and thrips are the possible agents of pollination. Both self and cross-pollination occur, the latter being about 16 per cent (Odland and Porter, 1941; Murthy and Murthy, 1962; Free, 1975). From a study on foraging behaviour of wild bees at hot pepper flowers and its possible influence on cross pollination in South-Central Brazil, Raw (2000) concluded that the small native bees species pollinate flowers effectively and that their small foraging areas are important in keeping cultivars generally distinct where they are grown in close proximity of each other. Anthesis takes place sometimes after the flowers have opened. Flowers remain open for 2-3 days. The percentage of fruit setting is 40-50 (Purseglove, 1977). Studies on the minimum isolation distance in chilli revealed that the extent of cross pollination was 8.8 and 6.5 per cent when the distances were maintained 5 and 10 meters, respectively. The per cent of cross pollination decreased as the isolation distance increased and only 1.8 per cent cross- pollination was observed at the isolation distance of 150 meters from the pollinator. No cross-pollination was observed at the isolation distance of 200 meters and beyond (Kanwar, 1999).

Flowering begins 1-2 months after planting and it takes another month for fruiting. The flowers open in the morning between 2 and 10 a.m. (Gopalratnam, 1933). The anthers normally deshisce an hour after the flower opening. Flower opening and anther deshiscence to a large extent depend on the weather conditions. According to Sheela *et al.* (1998), the duration of flower opening at 8.00 a.m. and continued upto 1.00 p.m. with peak period from 9.00 a.m. to 10.00 a.m. in *C. frutescens*.

Anther dehiscence commenced at 8.00 a.m. and continued upto 11.00 a.m. with peak period of dehiscence between 9.00 to 10.00 a.m. Maximum receptivity of stigma was observed 24 hours prior and six hours after anthesis. During cold as well as cloudy days, the opening is delayed (Purseglove, 1977).

To address the deficit to the events associated with fruit production such as flowering and seed set in chilli, Aleemullah *et al.* (2000) examined flowering phenology in cv. Red Hot Glory, a representative of the Longum group of peppers. Measurements of bud length and width showed that these characteristics can provide an effective index of the number of days required by a bud to reach anthesis. Anthesis itself mainly occurred during the morning with a second, smaller peak of flower opening in the afternoon. Measurements made in spring and summer suggested that the daily start of anthesis is controlled by daylength. In flowers that opened before 17.00 h, anther dehiscence took place 1 h after the buds opened; in flowers that opened later, dehiscence was delayed until the following morning. This pattern of anther dehiscence suggests that this event is partially controlled by the flower's endogenous rhythms. Controlled pollinations showed that the period of female receptivity, as judged by seed set, lasted from 5 days before anthesis to 3 days after anthesis with maximum fertility occurring on the day of anthesis. Therefore, bud pollinations are possible, and for maximum seed set, should be made as close to the day of anthesis as possible. This study has provided baseline data that will aid the production of new, hybrid cultivars and will facilitate further studies on the factors affecting fertility in this respect.

6.0 Cultivars/Hybrids

Wide range of cultivars/hybrids with various pod and plant characteristics, capsaicin content, duration and yield are available in chilli. India is the largest producer of dry chilli. However, in northeastern India, several landraces of green chilli are traditionally grown which are basically preferred for its fresh flavor and high pungency levels (Bhutia *et al.*, 2019).

6.1.0 Cultivars

K-1

It is a selection from an Assam type chilli B72A. The plants are tall and compact. The fruits are 6.6 long and are shiny red with a capsaicin content of 0.35 mg/g. It yields 19 quintal dry pods/ha in a crop duration of 215 days which is 20 per cent higher than local cultivars.

K-2

This is a hybrid derivative of B70 and Sathur Samba. The plants are tall and compact. Fruits measure 7.3 cm long with a capsaicin content of 0.49 mg/g of fruit. The fruits are bright red and the crop matures in 210 days. It yields 15 quintal dry pod per hectare and is tolerant to thrips and fruit rot.

CO.1

It is a selection from a 'samba' type from Sathur of Ramanathapuram district in Tamil Nadu. The plants are erect, medium tall and compact with moderate branching habit. The fruits measure 7.3 cm long with a capsaicin content of 0.72 mg/g. The ripe pods are bright red and shiny. It yields 21 quintal dry chilli per hectare in duration of 210 days.

In a varietal evaluation trial, Elangovan *et al.* (1981) recognized CO.1 as the most superior cultivar with regard to yield and other yield attributes closely followed by K-2.

CO.2

A 'Gundu' type selection from CA (P) 63 is suited for both green and dry chilli. The plants are medium tall and erect and the fruits are thick red with more seeds. This cultivar is suited for raising in June-July, September-October or March-April seasons; this selection yields 43.50 quintal green chilli and 8.7 quintal dry chilli per hectare.

MDU-1

An induced mutant obtained by gamma irradiation of K-1 cultivar has been identified as a good yielder. Plants are tall and compact with large broad leaves and with clustered flowering habit (4-9 flower per node). Fruits are long, shiny with a capsaicin content of 0.70 mg/g. It yields 18.09 quintal dry chilli per hectare in crop duration of 210 days (Sethupathi and Ramalingam, 1978).

G-4

It is a 'Samba' type known as Bagyalakshmi in Andhra Pradesh and known for its high yielding potentiality. Plants are tall, dense and fruits are 8.8 cm long. The pedicel is attached with fruit firmly. The fruits are bright red and contain 0.52 mg capsaicin per gram of fruit.

G-5 (Andra Jyothi)

This is evolved from a cross between G-2 and 1331 which is tolerant to thrips. The fruits of this cultivar are short and 'Gundu' type. Plants are tall and dense, and the fruits are bright red measuring 5.1 cm in length and 6.3 cm in girth. The capsaicin content is 0.65 mg/g of fruit.

NP 46A

This variety is evolved at IARI, New Delhi, India. Plants are dwarf, dense and spreading and this is a 'Samba' type. The pods are red measuring 10.7 cm long. The cultivar is less seeded and contains 0.53 mg capsaicin per gram of fruit.

Pusa Jwala

This is an early cultivar of 'Samba' type. This is derived from a cross between NP 46A and Puri Red. The plants are dwarf and spreading in habit. The fruits are long, thin and usually curved. The pods are red. The dried fruits have shrunken skin which is not liked by the traders and hence more suited as green chilli. The fruits contain capsaicin of 0.48 mg/g of fruit

Pant C-1

It is an advanced generation selection from a cross between NP 46A and a local cultivar (Kandhari) resistant to leaf curl and mosaic virus. The plants are erect and 50-60 cm tall. First fruiting commences in two months and picking commence 100 days after transplanting. Fruits are green when immature and red when ripe. Fruit surface is smooth with a blunt apex. Fruits are erect, 6-7 cm long and are highly pungent. Leaf curl and mosaic disease incidences are very low.

Pant C-2

It is a selection from a cross involving the same parents as that of Pant C-1. Plants are tolerant to mosaic and leaf curl virus, fruits are pendant and the yield is 14 quintal dry chilli per hectare.

Sindhur

This is a dual purpose chilli selected from CA 960. The fruits have deep red thick pericarp. The fruits measure 8.2 cm in length with capsaicin content of 0.81 mg/g of fruit. It yields 9.50-13.40 quintal dry chilli per hectare in duration of 180 days (Murty and Bavaji, 1980).

JCA 154

This is a pickling cultivar from J. N. Krishi Viswavidyalaya, Jabalpur, Madhya Pradesh, India. Fruits are dark green when unripe and bright red when ripe. Fruits mature in 115-120 days.

Pusa Sadabahar

This is a multipurpose chilli cultivar of India. It can be consumed green, used as red powder and for the manufacture of oleoresin. It bears fruits in cluster of 6-14 that facilitate bunch harvesting. The average fresh fruit yield of Pusa Sadabahar is 75-100 q/ha and dry yield 15-20 q/ha (Tewari, 1991).

Phule Jyoti

Developed through selection by Mahatma Phule Krishi Vidyapeeth, Rahuri, Maharashtra (India). Fruits are dark green in colour and medium in fruit length. It can be used both as green and red dry chilli. It is tolerant to leaf curl and powdery mildew diseases and resistant to lodging and good for kharif and summer planting. It produces an average yield of 30 q/ha for red dry fruits as against the average yield of 19.68 q/ha by cv. Musalwadi and 17.57 q/ha by Pant C-1 (Lawande *et al.*, 1998).

Phule Suryamukhi

This cultivar is also developed through selection from Mahatma Phule Krishi Vidyapeeth, Rahuri, Maharashtra (India). The fruits of this cultivar are medium in length and green in colour which can be used both as green and red dry chilli fruits for powder. It is tolerant to leaf curl and powdery mildew diseases and good for kharif and summer planting. The cultivar produces highest yield of 28 q/ha for red dry fruits (Lawande *et al.*, 1998).

Arka Lohit

It is a selection from local collection IIHR-324. The Fruits are dark green turning deep red at maturity (Capsanthin 0.21 per cent) and are highly pungent (Capsaicin 0.71 per cent). Suitable for irrigated and rainfed areas. Average yield is 250 q/ha (Green Chilli) and 30 q/ha (Dry chilli).

Arka Suphal

The Fruits are green, smooth, medium long (6-7 cm x 1 cm), pendent, turning deep red at maturity. This variety is suitable for irrigated and rainfed areas. Tolerant to powdery mildew. Average yield is 250 q/ha (Green Chilli) and 30 q/ha (Dry chilli).

CO 3

It is a selection from a cultivar of Sri Lanka. The fuits are long, slender and dark green and suitable for export. Average yield is 40 q/ha.

CO 4

The fruits are dark green, less pungent and suitable for high density planting. Average yield of green chilli is 30 q/ha.

Phule Mukta

It is resistant to Powdery mildew and Fusariaum wilt. Fruits are small, dark green and smooth. This variety is sutaible for kharif season. Average yield of dry chilli is 15 q/ha.

Jawahar 218

It is developed from Kalipeeth x Pusa Jwala. The plant is dwarf with spreading habit. The foliage is light-green with erect fruits (8.5 x 0.92 cm). The thin wrinkled light green turns rosy red at maturity and is early and high yielding.

Punjab Lal

It is developed from Perennial x Long Red. The Plants are dwarf bushy with dark green foliage. The fruits are erect, medium size (4.25 x 0.79 cm) with dark-green colours that turns dark red at maturity. It is rich in capsacin (0.7 per cent), oleoresin, dry matter and red pigment. It is also resistant to TMV.

Kashi Anmol

This is an improved population derived from two cycles of simple recurrent selection from a Sri Lankan introduction. Plants are determinate, dwarf (60-70 cm) with nodal pigmentation on stem and bear green attractive pendant fruits. First picking starts from 55 days after transplanting and yield 200 q/ha in only 120 days of crop. Average green fruit yield is 250 q/ha.

Kashi Gaurav

The plants are bushy with dark green foliage and bears dark green fruits that turn dark red after 95-100 days of sowing. The fruits are 9-11 cm long with

1.1-1.2 cm thick, pendant and pungent. Average red ripe fruit yield is 150 q/ha. It is tolerant to anthracnose, thrips and mites. This variety is a very good general combiner for hybrid combinations and 50 per cent flowering occurs in 35-40 days after transplanting. It is a sturdy plant that can withstand low temperature (87.23 per cent pollen viability at 10°C). It is suitable for fresh green chilli with mild pungency.

Kashi Sinduri

The plants are spreading type and are resistant to anthracnose disease. The plant bears 50 per cent flowering in 30-35 days after transplanting. The fruits are green and turn dark red at maturity (95-100 days after sowing) with 10-12 cm long and 1.1-1.3 cm thick, pendant, non-pungent and high oleoresin content (15 per cent). Average yield of red ripe fruits is 140 q/ha.

Hisar Shakti

The fruits are upright in clusters, pointed, thin, medium to long, highly pungent, tolerant to leaf curl mosaic virus, fruit rot and powdery mildew. Average yield is 170-190 q/ha.

Hisar Vijay

It is an early fruit bearing similar to Hisar Shakti but with blackish tinge on fruit surface. It is tolerant to leaf curl mosaic virus, fruit rot and powdery mildew. Average yield is 150-170 q/ha.

PKM 1

Developed at TNAU, Coimbatore, Tamil Nadu, India. It is a cross between CO_2 x Ramanathapuram Gundu. This cultivar is suitable for semi-dry condition. Fruits contain high capsaicin 0.36 per cent and are gundu type. Average yield is 300 q/ha.

Arka Abhir

A pure line selection from Dyavanur Dubba (a local Byadagi variety grown in Kundgoltaluk, Dharwad District). Plants tall, fruits light green, wrinkled turning deep red on maturity and less pungent. Yields are 20 q/ha dry chilli in 160-180 days.

Arka Neelanchal Prabha

A mid- season cultivar developed through selection suitable for both green and red chilli. Fruits are medium long, first harvesting starts at 65-70 days of planting. It is moderately tolerant to anthracnose and yields 100-120 q/ha.

Bidhan Chilli-4 (BCCH Sel-4)

Selection from indigenous collection obtained from Coochbehar district of West Bengal. Growth habit is semi prostrate, average fruit length is 6.23 cm; average fruit diameter is 1.20 cm and average fruit weight is 1.63g. Fruit of this variety contains 131.32 mg/100g ascorbic acid; 0.12 per cent capsaicin; 31.00 per cent oleoresin, and ASTA colour value is 200. It can be used as dual purpose (green and dry chilli). It has got high tolerance against PepLCV disease and can easily be fitted well in intensive cropping system. First harvest can be done within 65 days from the date of transplanting. However, the periodic harvest can be continued up to 120 days. BCCH Sel-4 exhibited green fruit yield of 112 q/ha and dry fruit yield of 22 q/ha.

Swarna Praphulya

Swarna Praphulya was developed through pureline selection from local germplasm. Plant is vigorous with 90-100 cm plant height and resistant to bacterial wilt under natural field conditions. Fruits are long (6.0-6.5 cm), pungent, dark green and dark red when ripe. Average yield is 200-250 q/ha.

Samrudhi

Released by UAS, Bangalore, India suitable for growing under rainfed conditions.

Kashi Abha (VR-339)

Fruits are short, staut with blunt apex and highly pungent. Tolerant to biotic (anthracnose, CLCV, thrips and mites) and abiotic stress (low and high temperature), yield is 150 q/ha.

Jwala Sakhi

It is evolved by cross involving Vellanotchi x Pusa Jwala. Fruits sulphary green, long, succulent, yields 196 q/ha of green chilli in 125 days. Tolerant to little leaf and leaf spot.

Jwala Mukhi

It is developed at KAU, Vellayani, India by using same parents of Jwala Sakhi. Plants dwarf, fruits long (9.6 cm) succulent, dark green with light pungency yield 225 q/ha of green chilli in 137 days, tolerant to little leaf and leaf spot.

X-235

It is cross between G4 x Anther mutants. Early maturing plant spreading with short internodes. Leaves small dark green. Flowers with yellow anthers as markers. Fruits are 5-6 cm long and pointed tip.

Punjab Sindhuri

Plants are dark green, compact and medium tall. It is an early maturing variety and first picking (red fruits) is possible 75 days after transplanting. Fruits are long (7.14cm), thick skinned, dark green when immature and deep red when mature. Fruits are pungent (1 per cent capsaicin content) and rich in vitamin C content (155 mg/100g). Average yield of red ripe fruits is 187.72 q/ha.

Punjab Tej

Developed by Punjab Agricultural University, Ludhiana, Punjab, India. Plants are light green, spreading and medium tall. It is an early maturing cultivar and first picking (red fruits) is possible after 75 days of transplanting. Fruits are long (6.80 cm), thin skinned, light green when immature and deep red when mature. Fruits are highly pungent (1.32 per cent capsaicin content) and rich in vitamin C (115 mg/100g). Average yield of red ripe fruits is 138.32 q/ha.

Punjab Surkh

Developed by Punjab Agricultural University, Ludhiana, Punjab, India. Plants are medium tall, leaves dark green, fruits long (7cm), green when immature and dark red on maturity. It is a dual purpose cultivar which makes it suitable for salad and drying. It is tolerant to fruit rot and moderately resistant to mosaic virus. Fruits have capsaicin content of 0.80 per cent on dry weight basis. It is an early bearing variety which yields 197.60 q/ha of red ripe fruits.

Punjab Guchhedar

It is a selection from material introduced from Indonesia developed by Punjab Agricultural University, Ludhiana, Punjab, India. Plants are tall. Fruits are small (5cm), erect and borne in clusters of 5-16. Fruits have destalking habit which leaves the stalk while picking. It is tolerant to fruit rot, highly resistant to mosaic and tolerant to leaf curl virus. Fruits are rich in capsaicin (0.98 per cent) and deep red in colour. Fruits are late bearing and yield 148.20 q/ha.

6.2.0 Hybrids

Arka Harita

It is an F_1 hybrid, fruits are dark green in colour and turn to red at maturity light. The length of the fruit is 10 cm with 1cm width. It is resistant to powdery mildew and viruses. Average yield of green chilli is 382 q/ha and 58.60 q/ha dry chilli.

Arka Meghana

It is a CMS based high yielding F_1 hybrid with resistance to powdery mildew and viruses. Fruits are dark green and turn red at maturity with length of 10.6 cm and width of 1.2 cm. Average yield is 350 q/ha green chilli and 50 q/ha dry chilli.

Arka Sweta

It is a CMS based high yielding F_1 hybrid for fresh market. Suitable for kharif and Rabi seasons under irrigated conditions. Fruit is smooth and medium (11-12 cm x 1.2-1.5 cm) with light green colour and turn red at maturity. It is tolerant to viruses and the duration of hybrid is 160 days. It yields 330 q/ha fresh fruits and 50 q/ha dried fruits.

Arka Khyati

It is a CMS based high yielding F_1 hybrid for fresh market. The fruits are smooth surfaced having size of 12 x 1cm with light green colours that turn deep red on maturity. It is medium pungent and turn wrinkled after drying, tolerant to viruses. The yield of fresh chilli is 400-450 q/ha and dry chilli is 50-55 q/ha.

Kashi Surkh

It is a CMS based hybrid. Plants are semi determinate (1-1.2 m), erect with nodal pigmentation on stem. Fruits are light green, straight, length 11-12 cm, suitable for green as well as red fruit production. First harvest starts after 55 days

Cultivars of Chilli

Arka Suphal

Arka Abhir

Arka Lohit

Kashi Abha

Kashi Surkh

Pusa Jwala

Kashi Gaurav

Bidhan Chilli-4

Hybrids of Chilli

Arka Khyati

Arka Sweta

Arka Meghana

Arka Harita

Hybrids of Chilli

Kashi Tej (CCH-4)

Kashi Ratna

Hybrid-Kashi Early

CH-27

BSS-453

of transplanting. Green fruit yield is 240 q/ha whereas red fruit yield is about 140 q/ha. This hybrid is tolerant to thrips, mites and viruses.

Kashi Early

The plants are tall (100-110 cm height) without nodal pigmentation on dull green stems and bear pendant fruits. Fruits are long (8-9 x 1.0-1.2 cm), attractive, dark green and turn bright red at physiological maturity, pungent with smooth surface. First picking of the green fruits starts at about 45 days after transplanting. The average yield of this hybrid is 250 q/ha (red ripe).

Kashi Tej (CCH-4)

CMS based F_1 hybrid dual purpose (green and dry), early maturing hybrid variety, high pungency, tolerant to Anthracnose, fruit rot disease and thrips. Average yield is 142 q/ha.

Kashi Ratna (CCH-12)

CMS based F_1 suitable for green chilly purpose tolerant to anthracnose and thrips with a yield potential of 200-220 q t/ha. Semi erect, early maturing variety, fruit contains 0.62 per cent (93450 SHU) capsaicin and 175.6 mg/100g vitamin C (ascorbic acid).

CH-3

This hybrid has been developed by crossing MS 12 x S-2530 at Punjab Agricultural University, Ludhiana, Punjab, India. It is an early maturing hybrid with dark green foliage and pendent fruits. The fruits are long (8.2 cm). The colour of immature fruit is dark green and turns dark red at maturity. The fruits are mild in pungency with 0.51 per cent capsaicin content, high dry matter (22.5 per cent) and good in vitamin C (109.95 mg/100g) content. The deep red coloured fruits make it specially suitable for making chilli paste for export purposes. The average yield of red ripe fruits is 271.70 q/ha.

CH-1

It is a hybrid between MS12 x LLS, developed by Punjab Agricultural University, Ludhiana, Punjab, India. Its plants gain one metre height and more branching. Plants continue bearing fruits for a long time. Fruits are light green when immature and attain deep red colour at maturity. Fruits are of medium size (6.62 cm long) and weigh 2.7 g each. This hybrid is tolerant to viral and fungal diseases. Its fruits are highly suitable for drying and used as salad. Its fruits are used in processing industry. Fruits on an average have capsaicin content of 0.80 per cent. Yield of red ripe chilli is 247 q/ha.

CH-27

It has been developed by using nuclear male sterility at Punjab Agricultural University, Ludhiana, Punjab, India. The plants are spreading, tall and continue to bear fruits for a long time. Fruits are medium long (6.7 cm), thin skinned, light green when immature and deep red when mature with high seed content. Fruits are pungent (0.7 per cent capsaicin), high in dry matter (26 per cent) and rich in colouring matter (242 ASTA units). The hybrid is resistant to leaf curl virus, fruit rot and root knot nematode, and tolerant to sucking pests such as thrip and mite. Average fruit weight is 3.6 g and yield of red ripe fruits is 240 q/ha.

7.0 Soil and Climate

7.1.0 Soil

An ideal medium for growing chilli is a light loamy soil rich in lime. Still it can be grown in varieties of soils provided they are well-drained and rich in organic matter. In sandy soils, the crop can be grown with success provided adequate irrigation and fertilizer practices are adopted. As a rainfed crop well drained black soils are also suitable. Acidic and alkaline soils are not suitable for chilli growing. Chilli can be grown in saline soils but the germination and early vigour of plants are affected by the salinity in the soil. The germination of chilli crop is found to be satisfactory up to a salt level of 4000 ppm with a soil pH of 7.6 and E.C. 0.2 (Kaliappan and Rajagopal, 1970). Soil moisture did not affect the fruit set but the fruit drop was higher (71.2 per cent) at 10 per cent moisture level compared to 55.6 and 57.4 per cent at 20 and 30 per cent field capacity, respectively (Rhee and Park, 1975).

Hirota *et al.* (1999) were of the view that relative growth rate decreased with increasing salinity and was related to retardation of net assimilation rate (NAR). Leaf area ratio did not decrease with increasing salinity. Leaf expansion rate was inhibited by salinity. Salinity did not influence the growth balance between leaves and the other organs. Retardation of NAR was related to obstruction of stomata conductance and transpiration. According to Richter *et al.* (1999), yield of chilli was reduced by increased salinity from 0 to 8 dSm^{-1} in the saturation extract. However, they opined that the crop seems to include plant as they can accumulate the salt which can be useful for osmotic adjustment against water stress due to salinity.

7.2.0 Climate

Chilli is grown in both tropical and subtropical areas at altitudes ranging from sea level to 2000 metres above (Joshi and Singh, 1975). A temperature ranging from 20 to 25°C is ideal for chilli. It is killed by frost. A warm humid climate favours growth while dry weather enhances fruit maturity. As a rainfed crop, it is grown in areas receiving an annual precipitation of 25-50 inches. But too heavy a downpour is detrimental as it leads to poor fruit set, besides causing rotting of the fruit. Higher night temperature was found to be responsible for the higher capsaicin content (Ohta, 1962). The detrimental effects of climate in growth and development of chilli and its mitigation and adaptation strategies are discussed and reviewed by Bhutia *et al.* (2018).

7.2.1 Temperature

Changes in climatic factors more so in temperature bring about considerable variation on the vegetative characteristics of chilli. Informations on the effect of temperature on growth of plants, number of flowers produced and fruit set are available in detail. The interval between seedling emergence and flowering prolongs as air and soil temperatures fall. Growth rate of chilli plants increased as soil temperature rose and the total plant weight after 100 days of growth at various soil temperatures was greater with the rise in temperature. A soil temperature of 10°C retarded plant development, whereas 17°C caused normal development. The top continued to develop with higher temperature but root development was retarded at soil temperature of 30°C and above (Rylski, 1972). It was clear that lower temperatures were more harmful to chilli than supra-optimal temperatures. Temperature also influenced fruit set and size. Low night temperature (8-10°C and 15°C) increased the fruit set and parthenocarpic development of fruits. Similar low night temperatures at flower opening stage reduced the pollen viability leading to formation of parthenocarpic seedless or seed deficient fruits. High night temperature (20°C) during flower development was a prerequisite for well shaped elongated pods. The highest length/diameter ratio in both fertilized and non-fertilized fruits was obtained with a high night temperature up to anthesis and a low temperature (8-10°C) thereafter.

From a study on the effect of different day and night temperature regimes on the growth of hot pepper, Pak *et al.* (1996) reported that plant height was influenced by day temperature rather than night temperature and by interaction of day and night temperature. Stem length was influenced by the difference between day and night temperatures within the range 20-28°C. As average daily temperature increased, growth as determined by leaf expansion rate increased but node number at which the first flower was initiated decreased. They recommended that daily temperature should be maintained at 24°C to keep plants compact and night temperature should be >20°C to raise average daily temperature to promote early flower initiation.

7.2.2 Light

The effects of light intensity on photosynthetic characteristics, growth and development in hot pepper were investigated by Chen and Jiang (1998) and suggested that plant height, leaf area and chlorophyll and carotenoid contents increased and leaf thickness and specific leaf weight decreased with increasing shade (65 per cent). Net photosynthetic rate and yield were highest under 30 per cent shade. Fruits grown fewer than 30 per cent shade contained the highest content of soluble sugar. Vitamin C content of fruit decreased with increasing shade.

8.0 Cultivation

8.1.0 Raising of Seedlings

Chilli is grown from seeds which are very light and remain viable for 2-3 years. Nursery beds are made in areas with full sunshine. Area selected for nursery is ploughed to a fine tilth. Well decomposed farm yard manure or compost is applied at the rate of 4 kg/m^2 to the seed bed. Furadan is applied to the nursery bed 15 prior to sowing of seeds. Raised beds of one metre width are ideal for nursery. Seedlings from about 1 kg seeds are necessary for planting one hectare. The seeds are dressed with thiram or captan at 3 g/kg of seeds, 6 hours before sowing to prevent any seed-borne disease. The seeds are sown thinly in lines 5 cm apart, sand and well decomposed compost are sprinkled over the seeds. To avoid damage by ants and other soil insects, chlorpyriphos @ 0.25 per cent is applied along the borders as well as on the beds. A fine mulching with paddy straw can be given soon after sowing to prevent excessive moisture loss. The mulch is removed as soon as the seeds start germinating. Drenching of nursery beds with chlorothalonil (0.2 per cent) + carbendazim (0.1 per cent) at fortnightly interval is found effective against damping off disease. Nursery beds are covered with 200 μm ultraviolet (UV)-stabilized polyethylene film supported by bamboo poles with open sides to protect seedlings from rain and direct sunlight. To prevent the attack of white fly, the vector of chilli leaf curl virus, nursery beds are to be covered with 40-60 mesh nylon nets. Water is applied by rose can daily or alternate day depending on the soil moisture status. Hardening of seedlings is generally done by withholding irrigation water 4-6 days before transplanting of seedlings.

Now-a-days plug trays are being used to produce quality seedlings of chilli. Plug trays, after bleaching (1 per cent), are filled with a mix of loamy soil, sand, and cocopeat at 2: 1: 1 ratio by volume. Seeds, after treatment with Thiram (3 g′″kg^{-1} of seed), are sown at a shallow depth in the middle of the plug tray and covered with finely sieved well rotten deciduous tree leaves to improve, and prevent the soil from drying. After sowing, trays are covered with straw until germination which normally takes 8 to 10 days and are hand-watered daily. Trays are drenched with chlorothalonil @ 0.2 per cent + carbendazim @ 0.1 per cent to avoid damping-off disease. Plug trays are kept under cover with 200 μm ultraviolet (UV)-stabilized polyethylene film supported by bamboo poles with open sides to protect seedlings from rain and direct sunlight. Plug trays are also covered with insect proof net to prevent the spread of leaf curl virus.

The effects of dehydration temperature (20, 35 or 50°C) and period (l-72 h), and seed moisture content during storage, on the germination of primed red pepper seeds were investigated by Jeong and Cho (1996) in Korea Republic. Primed seeds were surface-dried at 20°C for 3h (17.2 per cent seed moisture content or dried at 35°C for 3 h to 7.8 per cent seed moisture content) and seeds were stored in air tight glass bottles at 4°C for upto 3 months before germination at 15 or 35°C. Final germination percentage was not influenced by dehydration temperature or period. The longer the dehydration period the more days required for germination, especially when germinated at l5°C. After storage for 3 months surface dried seeds germinated faster than dried back seeds when germinated at 15°C. Seed moisture content of surface dried seeds decreased from 17.2 to 8.3 per cent after 3 months of storage. Seed moisture of dried back seeds slightly decreased with storage. According to Radhe Shyam *et al.* (1996), germination percentage of different genotypes of chilli declined as duration of aging treatment increased, although the extent of the decrease depended on the genotype. Positive effects of low doses of γ-ray irradiation (0-2400 rad) on germination and growth of chilli were observed by Lee *et al.* (1998). Irradiated seeds, particularly the 1000 and 2000 rad treatments, germinated better than control seeds. Plant height was increased in the 100, 200 and 400 rad treatments, but decreased in the 2400 rad treatment, compared with control. Nutrient contents of leaves were significantly increased in the 800 and 1200 rad treatments. Electrical conductivity of the water used for seed germination was lower in irradiated seeds than in the control. In Maharashtra and Andhra Pradesh, the seeds are sown in nursery beds from the third week of May to the middle of July. In the Gangetic regions, sowing are done in June, in Punjab and Rajasthan where the harvesting is to be completed before the onset of frost, it is sown in April- May. Under Kanataka conditions, July planting of the cultivars NP 46A, Pusa Jwala and Pant C-1 responded favourably with an yield of 129 q/ha green chilli as against November (30.1 q/ha) and December plantings (31.6 q/ha), with NP 46 A giving the highest yield (Gowda and Gowda, 1982). In Southern districts where rainfed cultivation is in vogue, chilli best grown in the south-west and north-east monsoon seasons coinciding with May-June and September-October and March-April. In Andhra Pradesh, the highest yield was obtained in August planting in cultivars Pusa Jwala and NP 46A (Rao and Suryanarayana, 1983).

The land is ploughed 3-4 times to get a fine tilth. Farm yard manure is incorporated during the last ploughing. For irrigated crop, ridges and furrows are formed at the recommended spacing. Seedlings of 4-6 weeks old attaining a height of 6½ (15.24 cm) should be transplanted on one side of the ridges after initial irrigation.

8.2.0 Spacing

Trials with different spacings indicated the superiority of 30 x 30 cm yielding 1596 kg dry pods/ha which account for 11.4 and 35.83 per cent higher than 45 x 30 cm or 45 x 45 cm spacings, respectively (Selvaraj *et al.*, 1974). In MDU-1, a closer spacing of 30 x 20 cm is reported to yield 2358 kg dry pods/ha when applied with nitrogen at 120 kg/ha (Ramachandran and Subbiah, 1981). Chougule and Mahajan (1979) suggested 45 x 30 cm spacing for cultivar Pusa Jwala, while Boominathan *et al.* (1971) obtained maximum yield of K-1 chillies by planting at 30 x 30 cm spacing. Levy *et al.* (1983) reported that increase in plant density resulted in less lateral braching making the fruits easier to harvest.

According to Revanappa *et al.* (1998), plant height was greatest and plant spread and girth least with the highest plant density (60 x 30 cm). Yield was also more with highest plant density and decreased with decreasing plant density. Sharma and Kumar (2017) observed that chilli yield per plant was significantly influenced by spacing levels. The maximum yield (316.67 g/plant) was recorded from the widest spacing (60 x 60 cm) and differed significantly from that of the other spacing (60 x 30 cm, 60 x 45 cm).

8.3.0 Irrigation

A light irrigation is given during the third day of transplanting and thereafter at weekly interval. Gap filling is done during second irrigation after 10 days of transplanting.

In arid and semi-arid regions, irrigation water is very scarce and the excessive consumptive use of water owing to high rate of transpiration can be reduced by spraying anti-transpirants on the foliage. In Andhra Pradesh, Suryanarayana *et al.* (1983) reported that irrigation at 12 days interval along with spraying of 20 ppm Alachlor solution as anti-transpirant was most economical in Pusa Jwala cultivar. Weeding and hoeing are done before top dressing every time and thereafter depending upon the weed intensity. Earthing up is done on the 30th day after planting. Effect of supplemental irrigation and nitrogen fertilization on chilli studied by Prabhakar and Naik (1997) revealed that plants receiving 40 mm CPE exhibited the highest dry matter production, fruit yield, nitrogen uptake and water use efficiency. Imposition of water stress during the reproductive stage resulted in a lower reduction of yield compared with stress imposed during vegetative phase. Nitrogen uptake and water use efficiency were also improved in the increased nitrogen rates. According to Selvaraj *et al.* (1998), irrigation to chilli at 0.8 IW/CPE gave the highest yield with a water consumption of 542 mm. Scheduling irrigation at 40 mm depth and at a 0.75 IW/CPE ratio (compared with 0.25 or 0.50) produced the highest red chilli yield under semi- arid conditions of South India (Prabhakar and Hebbar, 1998). In experiments at Madurai, chilli was grown on a sandy clay loam of pH 7.l and irrigated through all furrows, alternate furrows and through the check basin at IW/CPE ratio of 0.45, 0.56, 0.75 or 0.90. Mean dry chilli yields were highest when plants were irrigated through all furrows and at on IW/CPE ratio of 0.90 or (Subramanian *et al.,* 1998). Pitcher method of irrigation has relatively low evaporation and leaching losses compared with other methods and also uses less water. Impact of saline water by pitcher method on chilli was conducted by Chauhan *et al.* (1999) at Kota, India and the method was concluded to be very effective for this crop. It was also recommended that when using saline water, salt resistant and shallow rooted crop such as chillies could be grown near the pitcher for effective utilization of saline and high residual sodium curbonate water in sandy loam soil of arid and semi-arid regions.

The effects of water stress on phenylpropanoid metabolism was studied in Padron pepper plants (*C. annuum var. annuum*) growing in a green house in La Coruna, Spain by Estrada *et al.* (1999). The soluble phenolic and lignin contents of fruits from control plants were higher than those of water-stressed plants. The amount of capsaicinoids (capsaicin and dihydrocapsaicin) in fruits of water-stressed plants was higher than that in control plants, especially under water stress.

According to Xie *et al.* (1999), drip irrigation increased chilli yield through providing either favourable soil moisture conditions or unfavourable conditions for *Phytophthora* root rot incidence. Chilli yield increased by 33 – 48 per cent and water consumption reduced by 34 - 50 percent due to the adoption of drip irrigation system. With drip irrigation, the application of mulch increased the chilli yield by 32 percent, and lowered the water consumption by 28 percent (Nijamudeen and Dharmasena, 2012). Ashwani *et al.* (2020) recorded that 43 per cent water could be saved in drip system as compared to surface irrigation in chilli. They also found that N-fertigation level supply beyond 100 per cent RDN had no significant effect on yield and WUE even reduced the NUE, whereas, with the reduction of RDN from 100 to 75 per cent, significance reduction was observed in yield and other parameters.

8.4.0 Manuring and Fertilization

Chilli has a long growing season and therefore, needs a judicious management in the application of manures and fertilizers. Fertile soils with sufficient humus are most desirable for growing chilli. Sub-optimal levels of major nutrients often lead to deficiency symptoms in chilli (Mehrotra *et al.*, 1968). Deficiency symptoms observed are as follows.

8.4.1 Nitrogen

Stunted growth, leaves pale green, older leaves are small and uniformly pale green, start getting bleached from margin inwards until finally the entire leaf is bleached to pale white.

8.4.2 Phosphorus

Leaves become small and bluish green in the beginning, later turning to dirty greyish green.

8.4.3 Potassium

Plant growth becomes retarded to a lesser extent in comparison with nitrogen and phosphorus. The leaves are normal green but smaller in size with crinkled surface. Foliar symptoms start with the appearance of very small whitish necrotic spots dispersed over the entire lamina in the older leaves.

The development of extractable colour in chilli fruits (cvs. Americana and Fibola) was compared by Boronat *et al.* (1999) between plants on sand *vs.* perlite substrates and between plants with control nutrient solution and those subjected to K deficiency from fruit set to complete maturation. The contents of chlorophyll –a and -b and total chlorophylls decreased quickly during fruit development in all cases, especially between 60 and 70 days after fruit set. The lowest levels were observed in K-deficient plants. Red, yellow and total carotenoids increased rapidly during fruit maturation, especially towards the end, in all cases. In both cultivars, the highest levels of total carotenoids occurred in plants on perlite substrate. The ratio of red to yellow carotenoids (R/Y) increased during fruit maturation, ultimately reaching a stable level, with higher values for Fibola initially, but higher values for Americano at the end of maturation (suggesting greater precocity in Fibola). The potassium deficient treatment had negative effects on precocity and commercial quality of decreasing the synthesis and accumulation of carotcnoid pigments (Boronat *et al.*, 1999).

Nutrients deficiency symptoms in chilli were studied by Balakrishnan (1999). According to him, nitrogen deficiency was manifested by pale yellow colour of older leaves, phosphorus by bluish-green colour on older leaves, potassium by marginal chlorosis and subsequently marginal scorching, magnesium by interveinal chlorosis of the matured leaves and zinc deficiency by producing of small leaves in shorter internodes. Plant growth and fruit yield were reduced considerably in nitrogen deficient treatment but was only slightly affected by calcium deficiency.

8.4.4 NPK

Considerable work on fertilization in chilli has been done in Tamil Nadu (Ramanathan, 1965; Iruthayaraj, 1970; Iruthayaraj and Kulandaivel, 1973; Rajagopal *et al.*, 1976).

In a fertilizer-cum-spacing trial with K-1 and a local chilli cultivar, highest yield (1596 kg dry pod/ha) was obtained when the plants were spaced at 30 x 30 cm and applied with a fertilizer level of 50: 25: 25 kg of N, P and K per hectare (Sethupathi, 1978). Application of phosphorus had no response, while nitrogen had a linear response. Potassium had no influence on yield, duration of flowering, plant height and fruit ripening. He noted that plant height, fruit length, fruit girth, fruit weight, dry fruit weight and seed weight were not influenced either by spacing or by fertilizer. The aphid population was reduced by 24.9 per cent when potassium level was increased from 60 kg/ha to 90 kg/ha and 120 kg/ha. Capsaicin content of ripe pods was significantly influenced by the application of N, P and K but was less when K was omitted. The highest capsaicin content of 17.2 mg/100 g was obtained when N, P and K were applied at the rate of 80: 0: 35 kg/ha. Plants without N, P and K had only13.4 mg/1 00 g and the lowest capsaicin (10.4 mg/100 g) was noted at 0: 35: 0 and 120: 0: 0 N, P and K/ha (Subbiah *et al.*, 1980).

Prabhakar *et al.* (1987) recorded highest yield (82.04 q/ha) of green chilli in cv. G-3 with a nitrogen level of 90 kg/ha. The yield increased from 45.26 to 59.45 q/ha by increasing the rates of phosphorus from 0 to 60 kg/ha. Singh and Srivastava (1988) recommended 120 kg N and 60 kg P/ha for high yield. Increase in yield and yield attributes by application of nitrogen and phosphorus had been reported by many workers (Ramachandran and Subbiah, 1981; Rao and Lall, 1986). Ahmed and Tanki (1991) worked out the optimum dose of nitrogen and phosphorus to be 120 kg/ha and 82.8 kg/ha, respectively. Nitrogen efficiency at optimum dose was found higher compared to phosphorus. Altaf *et al.* (2019) observed that the application of FYM @ 8 ton per hectare along with recommended dose of nitrogen, phosphorus and potassium fertilizers (100: 50: 50 kg/ha) proved best treatment for yield increase in chilli cv. Ghotki as well as improved nutrient uptake. Hence, organic manures along with inorganic fertilizers have positive effect on growth, nutrient uptake, yield and quality of chilli.

Based on many experiments conducted in various parts of the country, optimum fertilizer doses have been standardized for an economic productivity. Best responses were obtained to nitrogen and phosphorus each at 90 kg/ha at Kamal (Relwani, 1963). Under Punjab conditions, a dose of 56 kg each of N and P and 28 kg K/ha along with 20 cartloads of well rotten farm yard manure were advocated for better yield (Dhesi and Nandpuri, 1965). Phosphorus and Potassium were applied at the time of transplanting, while half of the nitrogen was applied as side dressing when the plants were established in the main field and the next half at the time of flowering. The application of still lower levels of N, P and K at 34 kg/ha each produced the best result in Karnataka (Dhulapanaver, 1965). In Maharashtra, a fertilizer schedule of 45 and 22.5 kg N and P/ha, respectively, was considered necessary for higher yield. Application of 350 kg ammonium sulphate, 175 kg single super phosphate and 100 kg potassium sulphate per hectare, besides 50 cartloads of farm yard manure was found optimum under Delhi condition (Choudhury, 1967). Naeem *et al.* (2002) suggested that chilli cv. sanam should be fertilized with 90-60-30 kg NPK ha $^{-1}$ under the agroclimatic conditions of Peshawar, Pakistan. In Andhra Pradesh, India Hari *et al.* (2009) observed that the treatment comprising of 340: 60: 80 kg/

ha of NPK recorded the highest fresh and dry pod yield (112.43 and 50.58 q/ha, respectively) and cost: benefit ratio (1: 5.2).

The use of a weak starter solution at the time of transplanting was reported to prove beneficial. In Pune, 50 kg nitrogen per hectare in a single dose at the time of commencement of flowering produced the best result for rainfed chilli. Any nitrogen dose exceeding 56 kg per hectare had adverse effects on yield and the profit thereof (Khare and Umrani, 1969). The light soils of Haryana required 57 kg each of N and P and 27 kg K per hectare besides 23 to 34 tonnes of organic manures. Considering the yield and quality of fruits, application of 100 kg N in 4 equal split doses to chilli cv. G-3 was observed to be more beneficial (Dod *et al.,* 1983). Added nitrogen significantly increased the yield of dry pods. But K and N x K interaction did not have any marked effect on yield indicating the poor response of chilli to K application. Though 105 kg N per hectare recorded maximum yield of dry chilli, it was at par with 70 kg N and hence 70 kg N/ha can be considered optimum for CO. 1 chilli (Subbiah, 1983a). Added nitrogen improved the plant height and number of fruits but did not influence the length and girth of fruits (Subbiah, 1983b). A dose of 100: 90: 50 kg/ha N, P and K was found to be optimum under semi-arid conditions and a relatively higher dose of 150: 120: 60 kg/ha N, P and K was recommended for adoption in irrigated green chilli (Nathu and Pundrick, 1971). Spraying 75 kg N as 1.5 per cent urea on the 30th day and subsequently at fortnightly interval accelerated flowering in G-4 cultivar and gave higher yield of pods when 50 kg N was applied through soil and 25 kg N as foliar sprays. Mallanagouda *et al.* (1995) studied the performance of chilli under different intercropping systems and fertility levels and recorded highest dry yield with intercropping with garlic and recommended NPK and FYM compared to control (no NPK or FYM). Mallangouda *et al.* (1995) further noted the performance of chilli as influenced by different companion crops and fertilizer rates. They reported that growth parameters as well as yield and yield components were markedly influenced in chilli + garlic cropping systems followed by chilli + onion cropping systems at the recommended dose of NPK and FYM.

Housain *et al.* (1989) obtained high yield with foliar sprays of Zn alone and of Zn + B + Fe and with soil application of Devimicroshakti (a commercial formulation of Zn, B, Mn, Fe, Cu and S). Effect of copper ore tailings on yield and quality of chilli was investigated by Basavaraja *et al.* (1998). Copper ore tailings (COT), a waste material of copper/gold mining, increased the dry chilli yield when applied @ 0.5 t/ha. Higher rate of COT (1.5 t/ha) increased the ascorbic acid content of green chilli and decreased the incidence of powdery mildew and anthracnose. From a solution culture experiment on *C. annuum* cv. MI 2 exposed to pH 5.5 without Al (control), or pH 3.8 with or without Al, Malkanthi *et al.* (1995) reported that shoot and root dry weight were the highest in the control and markedly decreased by low pH with or without Al. Concentrations of K, Ca, Mg, Mn and Zn in roots and shoots were significantly decreased at the low pH. P-concentration was highest at the lower pH with all, particularly in the roots.

Subbiah (1994) observed no significant effects on yield of chilli with bio-fertilizer application. However, it had some significant effect on nutrient uptake. Shrihari and Sreenivasa (1998) studied the optimum dose of arbuscular mycorrhizal (*Glomus macrocarpum*) inoculum for chilli in two soil types and opined that 50 and 60 g per 10 kg of black clay and red sandy loam, respectively for better P-nutrition, growth and yield of the crop. Nofiyanto *et al.* (2017) observed that the application of pellet MOSS (a mixture of duckweed (*Lemna minor* L.), cow manure, zeolite with a ratio of 1: 1: 1 and addition of 1: 2 tapioca starch composition ratio as pellet adhesive) can improve soil fertility, germination, growth and production of chili. Pellet MOSS has the potential to be developed as a qualitative bio-organomineral fertilizer with a dose of fertilizer recommendation of 25 tons per ha.

8.5.0 Mulching

Yoon *et al.* (1995) suggested mulching (black polyethylene film with rice husk) in chilli in relation to soil water management in drought areas of Korea Republic. Mulching increased soil water content and yields compared with controls. The effects of mulches were investigated within integrated crop management of hot pepper at tropical lowland sites in Indonesia by Vos and Sumari (1997). According to them, mulches affected soil temperature, light reflection and soil nutrient concentrations after the last harvest. Improved crop performance and production with increase in fertilizer efficiency, control of evaporation, leaching and soil erosion makes plastic mulches suitable for use in integrated crop management. Kumara *et al.* (2016) observed a significant increase of 38 per cent green chilli yield in mulched (black polythene mulch film of 25 micron thickness), compared to unmulched plots. Mulched treatment reduced labour cost up to 39 per cent compared to unmulched. The use of black polythene mulch in chilli brought the benefit cost ratio from 2.61 to 2.81. On the basis of benefit cost ratio it is economically viable for the farmers to adopt drip irrigation at 80 per cent ET with mulching for green chilli which shows BC ratio of 2.71 (Burghate, 2014). Narayan *et al.* (2017) recorded the highest soil moisture retention (16.74 per cent), lowest weed density (74.81 g/plot dry weight), highest number of fruits (140/plant), with maximum fruit weight (9.99 g), total fruit yield (463.08 q/ha) and B: C ratio (3.49) under black plastic mulch (double coated) in chilli under temperate conditions.

8.6.0 Weed Control

According to Husain *et al.* (1998), major weed flora recorded in chilli comprised of *Cyperus rotundus, Cynodon dactylon, Parthenium hysterophorus, Chenopodium album, Physalis minima, Euphorbia hirta and Lagasca mollis*. Herbicidal treatments (trifluralin, fluchloralin, pendimethalin, alachlor and butachlor) reduced the weed population production significantly compared to unweeded control. Reduction in dry matter production of weeds and higher fruit yield were recorded under weed free treatment. Among the herbicides, application of trifluralin followed by pendimethalin both applied at the time of transplanting at 1.0 kg/ha were effective in controlling weeds and increasing fruit yields of 55.16 and 52.95 q/ha, respectively.

Soil incorporation of EPTC 10 days prior to transplanting followed by the application of nitrofen and alachlor provided the best control of weeds without any adverse effects on chilli (Saimbhi and Randhawa, 1976). Spraying Tok-E.25 @ 2 litres a.i./ha with one hand weeding recorded an yield of 9468 kg of ripe pods/

Different Stages of Chilli Growing under Open Field

ha followed by hand weeding alone (9144 kg/ha). Likewise, lasso at the rate of 1.5 litres with one hand weeding and lasso 2 litres with one hand weeding ranked the next best with 7104 kg of ripe pods/ha, respectively. Tok-E at 2 litres followed by one hand weeding on the 30th day resulted in an increased income compared to hand weeding alone. Evaluation of herbicides for control of weeds in chilli was studied by Singh *et al.* (1995) in India. They reported that pendimethalin (1.0 kg/ha) was the most effective (87.3 per cent) on grassy weeds followed by metolachlor @ 1.0 kg/ha (80.0 per cent) and oxyfluorfen at 0.3 kg/ha (73.4 per cent). However, best control of broad leaved weeds was achieved by them with oxyfluorfen @ 0.3 kg/ha (92 per cent) followed by lower doses of oxyfluorfen (0.1-0.2 kg/ha) and pendimethalin @ 1.0 kg/ha (74.4 per cent). Oxyfluorfen @ 0.3 kg/ha resulted in improved crop growth and fruits/plants, leading to the highest dry chilli yield (2210 kg/ha). Rajput *et al.* (2003) recorded the highest net returns by application of fluchloralin 2 kg a.i./ha + hand-weeding at 45 days after transplanting. Singh and Verma (2015) observed that the application of propaquizafop (10 per cent EC) 100 g ai.i./ha resulted in significant reduction in dry weight of weeds and highest fruits/plant and fruit yield over propaquizafop (10 per cent EC) 50 g/ha and fenoxaprop-p-ethyl 9.3 per cent EC, respectively and were at par with propaquizafop (10 per cent EC) 62.5 g/ha. Integrated weed management with propaben® (metolachlor 20 per cent w/v + prometryn 20 per cent w/v) + HW (40 days after transplanting) or butachlor (50 per cent w/v) each at 2.0 kg ha^{-1} as pre-emergence + HW resulted higher chilli fruit yield and net return than multiple HWs or sole application of either herbicide through efficient weed control at a lower cost (Daramola *et al.*, 2020).

8.7.0 Effect of Growth Substances

Germination was hastened by soaking the seed in 30, 50 or 70 ppm NAA for 4-5 days and the resulting plants reached the flowering stage 5-7 days earlier with larger fruits and seeds but fruit maturation was delayed by 7 days. Seed number was markedly increased with 30 and 50 ppm compared with the untreated controls (Hariharan and Unnikrishnan, 1983).

A large number of growth substances have used for regulating growth and flowering improving the yield of chilli. Mote *et al.* (1975) reported that foliar application of 50 ppm NAA at full bloom stage effectively controlled flower drop with an increase in yield up to 41 per cent. Similar reduction in flower drop and increase in yield by using NAA were recorded by Chandra *et al.* (1976) and Sharma *et al.* (1980). MH and Ethrel @ 3000 and 200 ppm, respectively, suppressed flowering in chilli (Kim and Ho, 1978). Nagdy *et al.* (1979) reported that 400 ppm Ethrel improved fruit setting in both winter and summer crops and the fruit quality was also improved. Treatment with NAA @ 20 ppm at first flower opening followed by two sprays at an interval of 30 days was most effective in increasing yield, number of fruits/plant, fruit size, and contents of capsaicin, ascorbic acid, carbohydrate, protein and fat of chilli fruits (Patil *et al.*, 1985).

In chilli cv. K2, application of Triacontanol at 2 ppm, 30 days after transplanting and again at full bloom enhanced all growth and cropping characteristics, increased yield, improved fruit quality and increased nutrient uptake (Miniraj and

Shanmugavelu, 1987). Usha and Peter (1988) observed highest reduction in flower drop in the summer season with Triacontanol as 'Vipul' at 1 ml/2 I water and in the monsoon season with NAA @ 15 ppm. Application of Cycocel (ccc) with 1000 ppm at 30 days after planting, though plant height was reduced but it increased significantly the number of branches, leaves and fruits, fruit set and yield per plant (Deka and Shadeque, 1996). Significant varietal differences were observed in fruit length, width and volume, number of seeds per fruit, number of fruits and average weight of the fruit in trials with 4 cultivars of chilli when plants were sprayed 35 or 45 days after transplanting with 5 or 10 ppm cytokinin or 10 or 20 ppm NAA. Spraying with 20 ppm NAA had the greatest positive effect on yield in all cultivars. Growth regulators had no significant effect on fruit length or volume (Revanappa *et al.*, 1998).

Effect of growth regulators on the performance of chilli under coastal saline soils of West Bengal, India was studied by Munsi and Sadhukhan (1998) and reported that application of NAA (20 and 40 ppm), 2,4-D (2 and 4 ppm) and Ethrel (200 ppm) increased the yield of the crop. However, they obtained highest yield following application of NAA @ 40 ppm. However, growth regulator treatment increased yield with NAA at 20 ppm and also gave the highest growth parameters (Revanappa *et al.*, 1998). According to Joshi and Singh (1998), application of Paclobutrazol @ 300 ppm at flower bud initiation and second at 20 days later to chilli plants cv. Pant C-1 influenced significantly per cent fruit set, short styled flowers, pollen viability, number and weight of fruits per plant, seed weight per fruit and total fruit yield per hectare. They also reported that application of 2,4-D @ 2 ppm and NAA @ 40 ppm had beneficial effects on these characters. Chaudhary *et al.* (2006) obtained 28.75 per cent, 25.70 per cent, 13.61 per cent and 2.30 per cent higher fruit yield over control by application of 2 ppm 2,4-D, 5 ppm triacontanol, 40 ppm NAA and 10 ppm GA_3, respectively. The highest net profit and B : C ratio were recorded in case of 2 ppm 2,4-D. Chandini Raj *et al.* (2016) found the most effective treatment was NAA 75 ppm followed by NAA 50 ppm and GA_3 60 ppm under alluvial plains of West Bengal, India for production of chilli. Singh and Singh (2019) recorded that the application of GA_3 @ 150 PPM found significantly superior over other treatments in terms of plant height at 30 DAT, 60 DAT and at harvest (76.82 cm, 116.14 cm, 141.58 cm), number of branches/plant at 30 DAT and at harvest (28.37 and 50.73), fruit length (29.74 cm), fresh weight of 10 fruits (109.85 g), number of fruits/plant (127.91), fruit yield/plant (636.04 g), fruit yield/ha (25.70 t/ha) and vitamin C content (98.49 mg/100 g).

From the findings of several workers on the use of different growth regulators and its effects on chilli, Rajput and Parulckar (1996) made a comprehensive report in these aspects (Table 13).

9.0 Harvesting, Yield and Storage

9.1.0 Harvesting and Yield

In India, field-sown crops are mainly grown for ripe fruits, but in transplanted crop the first harvest is usually picked green to stimulate further flush of flowering and fruit set. In Tamil Nadu, the stage of picking either green or red depends upon the market needs. Flowering beings 1-2 months after transplanting and it takes another month for green fruits. Thereafter, ripe fruits are picked at interval of 1-2 weeks and harvesting continues over a period of about 3 months. The number of picking varies from 6-10 depending upon the season, cultivar and cultural practices. The ripe chillies are dried under sun for 8-15 days depending upon weather conditions, while commercially it is dried at about 54.4°C in 2 to 3 days. 100 kg of fresh ripe fruits yield 25-40 kg dry chilli depending upon the cultivars and thickness of the inner wall (Nath, 1976). Open pollinated varieties normally yield 7.5 to 12 tonnes green chilli and 1 to 2.5 tonnes dry chilli from a hectare.

Table 13: Response of growth regulators in chilli*

Growth Regulator	*Concentration (ppm)*	*Stages of Application*	*Responses*
NAA (Naphthelene acetic acid)	50	Full bloom stage	Control of fruit drop, and yield increase
NAA	10	First at flowering, second at 5 weeks later	Increased fruit set
NAA	60	Bloom stage	Increase in yield
MH (Maleic hydrazide)	3000	Preflowering	Suppression of flowering
Ethrel (Ethephon)	200	Preflowering	Improved flowering and fruit set
GA (Gibberellic acid)	50	Fruit setting	Decrease in flower shedding and increased fruit set
Ethrel	500	At maturity	Turning colour and increase in yield
Tricontanol	2	First spray at 30 days after transplanting and second at bloom stage	Reduction in flower drop and increased fruit yield

* Rajput and Parulekar (1996).

9.2.0 Storage

Green chilles can be stored for about 40 days at 0°C and 95 to 98 per cent relative humidity (Platenius *et al.*, 1934). Hasan and Soehardi (1978) reported that after 17 hours transport, weight loss and damage of fresh red chillies were least and quality remained best when they were packed in perforated plastic bags of 1 kg capacity and further packed into bamboo baskets, cartons and plastic bags. Influence of plant growth substances in enhancing the shelf-life of chilli fruits cv. Pusa Jwala was studied by Arora *et al.* (1998) when equal size healthy fruits, harvested at green mature stage were packed in card board boxes with newspaper lining as cushioning material after giving the treatments of dipping for about 10 minutes with GA_3 (50, 100 and 200 ppm), kinetin (25, 50 and 100 ppm) and control. Minimum physiological loss in weight (PLW) was observed from fruits treated with GA_3 200 ppm whereas maximum PLW in fruits was found in control. Maximum fruit ripening percentage was observed in control whereas minimum respiration,

Harvesting and different Drying Methods of Chilli

fruit ripening percentage and maximum retention of total chlorophyll was found in fruits treated with kinetin 100 ppm. Fruits treated with GA_3 200 ppm and kientin 100 ppm had the highest oleoresin content. Among different chemicals tried GA_3 200 ppm was found significantly superior in increasing shelf-life of chilli fruits followed by kinetin 100 ppm. Schweiggert *et al.* (2007) observed that blanching the whole fresh chilli pods for 5 min at 90 and 100°C, respectively, was found to be suitable for the production of chilli powders with low microbial loads and high carotenoid and capsaicinoid contents. In contrast to chilli, paprika pigments were more prone to process and storage degradation. Their losses might be minimised by the addition of antioxidants.

10.0 Diseases and Pests

10.1.0 Disease

Chilli is attacked by a number of diseases caused by fungi, bacteria and viruses.

Fungal Diseases

10.1.1 Damping Off

The disease caused by *Pythium aphanidermatum* and *Rhizoctonia solani* is of frequent occurrence in nurseries. The seed may rot or the seedlings may be killed before they emerge from the soil. The stem of young seedlings may also be attacked after emergences showing water soaking and shrivelling of stem which fall over and die. In a nursery, the disease may start in patches and in the course of 2-4 days, the entire lot of seedlings may be destroyed. The disease is most damaging on moist soils with poor drainage conditions.

The fungus is mainly soil-borne. When there is sufficient moisture it germinates and produces the mycelium which later forms the asexual state of reproduction. The fungus is capable of living for many years in soil. Partial sterilization of soil by burning trash in the surface helps in checking the disease. Steam sterilization of soil is commonly practised in Western countries. Providing better drainage by improving the soil texture and by forming raised beds with free drainage all round helps to avoid the disease. Sterilization of soil by drenching the soil (10.16 cm) deep with formaldehyde diluted 50 times with water or with some other effective chemical soil sterilant is also effective. Commencing from 10-15 days after the seed is sown, the seedlings may be protected by spraying with chlorothalonil (0.2 per cent) + carbendazim (0.1 per cent). Management of damping off of chilli caused by *Pythium aphanidermatum* with plant products and fungicides revealed that thiram gave the best control followed by garlic bulb extracts and leaf extracts of *Lawsonia inermis* (Kurucheve and Padmavathi, 1998).

Rhizosphere bacteria belonging to the fluorescent *Pseudomonas* spp. and *Bacillus subtillis* are receiving increasing attention for the protection of plants against soilborne fungal pathogens. Experiments were carried out with 2 strain of *P. fluorecens* and 4 strains of *B. subtillis* from various regions of Iran by Sharifi-Tehrani and Omati (1999) for the biocontrol of damping off (*Phytophthora capsici*) in sweet pepper. The results indicated that volatile metabolites and non-volatile metabolite caused maximum inhibition of the fungal growth. Two strains of *B. subtillis* were highly effective against *P. capsici* in soil microcosms. Strains from *P. fluroescens* proved some protection but less that from *B. subtillis*.

The efficacy of biocontrol agents *viz.*, *Trichoderma viride* and *Pseudomonas fluorescens* against damping off disease of chilli caused by *P. aphanidermatum* was evaluated by Manoranjitham *et al.* (2000) at Coimbatore, Tamil Nadu (India). Seed treatment with talc based formulations of biocontrol agents individually and in combination with various proportions of bacterial and fungal biocontrol agents revealed that population of *P. aphanidermatum* in soil was decreased. Pot culture studies showed reduction of disease incidence as well as increase in root length, shoot length and vigour of seedlings over control.

10.1.2 Anthracnose and Fruit Rot

This is one of the worst maladies in chilli found in India and it occurs in severe form in all the Southern States. The same disease is also reported from several other countries. The disease is caused by the fungus *Colletotrichum capsici*.

The disease symptoms appear mostly on ripened fruits and hence the disease is also called as ripe fruit rot. The spots are usually circular and sunken with black margins. The sunken spot is covered with a pinkish mass of fungal spores. As the disease advances the spots spread forming concentric markings with dark fructifications representing the fungal acervuli. The fruits with many spots drop off prematurely resulting in heavy loss of yield. The fungus may also attack the fruitstalk and spread along the stem causing die-back symptoms.

The fungus is externally seed borne and the secondary spread is through air borne inoculum. The disease spreads rapidly with wind rains during the rainy seasons. The fungus may not survive long in the soil but may survive on the dead twigs stored under dry conditions.

The seed borne infection is controlled by seed treatment with organomercuric compound. The secondary spread causing fruit infection may be checked by spraying the plants with copper fungicides at fortnightly interval beginning from the time the first bunch of fruits is almost mature. The die back was better controlled by 1 per cent Bordeaux mixture followed by 0.1 per cent cosan. But Bordeaux mixture caused a reduction in yield by 7. 9 per cent, while cosan 0.1 per cent recorded an yield increased of 13.5 per cent (Subramaniam *et al.*, 1971). Difolatan 80 per cent W.P. (captafol) sprays at 0.2 per cent a.i. monthly interval commencing from transplanting against anthracnose and dithane M-45, 75 per cent W.P. (mancozeb) against ripe fruit rot were advocated (Chauhan and Duhan, 1977). The best control of *Colletotrichum capsici* on chilli by foltaf (captafol) at 0.2 per cent followed by fytolan (copper oxychloride) @ 0.25 per cent and bavistin (carbendazim) @ 0.1 per cent. These compounds also gave good control of the die back phase of the disease and increased yields (Eswarmurthy *et al.*, 1988). According to Das and Mohanty (1988) the best control of *Colletotrichum capsici* was given by sprays of carbendazim, followed by benomyl and captafol. Seed treatment with benlate, bavistin and Agrosan GN @ 0.1 per cent is reported to reduce the incidence of *Colletotrichum capsici*. Benlate @ 0.1 per cent or zirum @ 0.2 per cent sprays once after fruit set followed by another spray 15 days later afforded the best control (Sindhan *et al.*, 1980). Bavistin 0.1 per cent spray was also reported to give good control of this disease (Sankaralingarn *et al.*, 1983). Captan was reported as an effective fungicide for *Colletotrichum* sp. and the possibility of its compatibility with pesticides like carbaryl, phosalone quinalphos and dimethoate was indicated (Raju and Rao, 1981). Kolte *et al.* (1994) suggested seedling dipping of chilli in a suspension of *Azospirillum* sp. for best control of shoot die back (*Colletotrichum capsici*). Yields of green and red chillies were highest for mixed dips of Azotobacter + *Azospirillum*. Biocontrol of fruit rot and die back of chilli using antagonistic microorganisms was investigated by Jeyalakshmi *et al.* (1998). Among the fungal antagonists, *Saccharomyces cerevisiae* exhibited the maximum reduction mycelia growth, followed by *Trichoderma viride*. Among the ten bacterial antagonists, *Bacillus subtilis* showed maximum growth reduction, followed by *Pseudomonas fluorescens* isolate 27. In pot culture experiments *S. cerevisiae* recorded the maximum reduction of fruit rot intensity and incidence, and die-back incidence, followed by *B. subtilis*, when sprayed 105 and 120 days after sowing. In an attempt to screen cultivars resistant to this disease, the cultivar Pant C-1 recorded only less than 5 per cent incidence of anthracnose and fruit rot and thereby exhibited some degree of resistance.

10.1.3 Frogeye Leaf Spot

This disease caused by *Cercospora capsici* may result serious defoliation in chilli. The fungus is seed borne but does not live in the soil. The *Cercospora* leaf spot is characterized by chlorotic lesions angular to irregular in shape and later turning greyish brown with profuse sporulation at the centre of the spot. Severely infected leaves drop off prematurely resulting in reduced yield.

The leaf spot disease is easily controlled by spraying 1 per cent bordeaux mixture or other effective fungicides. Spraying bavistin 0.1 per cent 5-6 times at fortnightly interval gave effective control of this disease (Sharma and Sohi, 1981). It was also reported that this disease can be controlled by foliar sprays of benlate 0.1 per cent, dithane M-45 (0.2 per cent), bayleton 0.1 per cent, baycor 0.1 per cent or difolatan 0.2 per cent (Venkata Rao and Shanrnugam, 1983).

5.1.4 Fusarium Wilt

It is caused by *Fusarium annuum* and characterized by wilting of the plant and upward and inward rolling of the leaves. The leaves turn yellow and die. The fungus can live for several years in the soil.

The control measure consists of soil treatment or use of wilt resistant cultivars. Drenching with 1 per cent bordeaux mixture or blue copper or fytolan may give protection. The disease can also be controlled by seed treatment with bavistin or vitavax C.

10.1.5 Powdery Mildew

The powdery mildew caused by *Oidiopsis taurica* syn. *Leveilluala taurica* has been reported from India, Mauritius, Hungary, Libya and elsewhere. It causes rapid defoliation of the plants.

Spraying of wettable sulphur, karathane (dinocap) or dusting of 10 per cent sulphur will control this disease. Bavistin (carbendazim) and calixin (tridemorph) were reported most effective (Jharia *et al.*, 1978). Later dinocap @ 0.05 per cent or thiovit (wettable sulphur) @ 0.3 per cent sprayed thrice at fortnightly interval was reported to give good control with an increased yield of 132-147 per cent (Gohakar and Peshney, 1981). Spraying of karathane 3-4 times @ 0.1 per cent at 10 day interval or aureofungin 50 ppm registered effective control of this disease (Raju and Srinivasa Rao, 1981).

Bacterial Diseases

10.1.6 Bacterial Leaf Spot

Xanthomonas vesicatoria can cause serious injury to capsicum and chilli on leaves as well as fruits. On young leaves the spots are yellowish green, on older leaves they are dark and water-soaked. They later appear with straw coloured centres and dark margins. On the fruits, small blister-like spots occur becoming warty in appearance later.

The bacteria are seed-borne and the seeds can be treated with corrosive sublimate. The bacteria are also soil-borne and hence following suitable crop rotation is essential. The nurseries and plants in the main field may be sprayed with a copper fungicide such as bordeaux mixture. Out of 19 chilli cultivars it was found that CA960, CAP(63), CO.I, G-5 and Xl97 were highly resistant, while G-4 and X206 were resistant (Dahivval and Kore, 1981).

From a study on the sources of resistance to bacterial wilt in Capsicum, Matsunaga and Monma (1999) were of the view that accessions of sweet and hot pepper which originated in Asia, *i.e.*, Japan, China *etc.* were high or moderate resistance to this disease while those from other areas such as North and South America, Europe *etc.*, had varying degrees of susceptibility.

Viral Diseases

As many as 18 viruses have been reported to occur naturally on chilli and in addition 16 others have artificially been transmitted (Ramchandran and Subbiah, 1981). Of these only the leaf' curl virus (Vasudeva, 1954) and chilli mosaic virus (McRae, 1924) were reported in India (Greenleaf, 1956).

Among the viral diseases leaf curl is most serious. As a result of the disease the leaves curl and become deformed. The leaves curl towards the midrib and the plants become stunted in growth. Fruiting is greatly affected. Flower buds abscise before attaining full size and anthers do not contain pollen grains.

There are yet no control measures known. The infected plants should be uprooted and burnt or buried to avoid further infection. However the incidence of the disease can be delayed and the intensity reduced by insecticidal sprays to kill the insect vectors. In field trials at Bangalore, one soil application of carbofuran or disulfoton @ 1.5 kg a.i./ha at 10 day interval reduced *Bemisia tabaci* populations, an insect vector of the disease and tobacco leaf curl incidence on *Capsicum annuum* to a great extent and an increased yield of 2 to 3 times over the control was recorded.

From a 2 year field trial for the evaluation of 8 different chemicals at Khandwa, Krishna *et al.* (1983) reported that thimet (phorate) @ 0.1 per cent applied after the seedlings became established and twice more at 15 day interval gave the best control of insect vector. Furadan (carbofuran) @ 0.03 per cent gave the next best result.

Field trials were also conducted at Khandwa (M.P.) to assess 64 cultivars for the resistance to leaf curl disease and it was found that only Karanja, Pant C-1, Sel. 46-1, IC-18253, IC-18885, JCA 196, Cross-218 and EC-121490 showed less than 30 per cent leaf curl incidence (Bhalla *et al.*, 1983). Field screening of chilli cultivars to viral diseases in Madhya Pradesh, India revealed that cv. JCA 248, JCA 218, Pant C-1, NP46-A, Pusa Jwala and JCA 196 were resistant to leaf curl vims (Sangar *et al.*, 1988). PSP-11 is resistant to cucumber mosaic cucumovirus, potato x potexvirus and tobacco mosaic tobamovirus (Tewari and Viswanath, 1988). A total of 11 *Capsicum* varieties were screened for resistance to CMV in India. Punjab Lal, Bengal Green - 1, Perennial were resistant, 6 genotypes were moderately resistant and 2 genotypes were highly susceptible (Singh and Singh, 1998).

Sources of curly top virus in *C. annuum* is presented in Table 15.

Table 15: Reactions of *Capsicum annuum* Genotypes to a Natural Epidemic of Curly Top Virus*

0 per cent to 25 per cent symptomatic plants within an accession
Burpee Chiltepin, NuMex Bailey Piquin, NuMex Twilight
26 per cent to 50 per cent symptomatic plants within an accession
Black Olive, Milky Way, Mississippi Sport, NuMex Pinata, Ordono, PI 439378 (Jamaica), Sujata, Tarahumara Chiltepin, Thai Hot
51 per cent to 75 per cent sympotomatic plants within an accession
Big Red, de Arbo!, Golden Giant II, Hot Portugal, Huasteco, Jigsaw, Long Red Cayenne, Mississippi Sport, NuMex Joe E. Parker, NuMex Sunflare, Nume Pinata, NuMex Sunglo, Peter Pepper, Pinguita de Mono, serrano, Valencia
76 per cent to 100 per cent symptomatic plants within an accession
Ancho, antohi Romanian, Apple, Big Bertha, Big Chile, Big Thai, Biscayne, Chilaca, Chile de Arbo), Chimayo, Cochiti, Como de Toro, Cubanelle, Flamingo, Goklen Summer, Guajilo, Jsleta, Jackpot, Jemez, Jingle Bells, Kishiner, Kori Sitaknc, LaMuyo, Large Red Cherry, Lavender apple Blander, Lemon Belle, Long Slim Cayenne, Mandarin Manconi, Messilla Vayenne. Mirasol, Mushroom, Negro, New Mexico 6-4, Numex Big Jim, NuMex Conquis- tador, NuMex Rio Hrande, NuMex Kishiner, NuMex Eclipse, NuMex Sunset, NuMex Sun- rise, NuMex Princes ole, Orange Belle, Pasilla, Perfecto,Pepperoncini, PI 599426 Numex Sunburst, Pimento Select, Pulla, Purple Beauty, Queen Bell, Sandian, San Fellipe, San Juna, Tsile, Santaka Santo Domingo, Suger Chille, Super Cayenne 11, Sweet Banana, Tejaswini, Thai Dragon, Yellow Francess Yellow Flame.

Symptomatic plants had visible symptoms of curly top disease.

* Bosland (2000).

Mosaic incidence in chilli can be reduced by spraying 0.05 per cent dimethoate or monocrotophos 6 times at fortnightly interval commencing from 21 days after transplanting or by spot application of carbofuran or phoratc granules @ 1.5 kg a.i. per hectare twice 15 and 60 days after transplanting.

The inhibitory effects of plant extracts against the infectivity of a disease were demonstrated by number of workers. Singh and Sharma (1999) studied the efficacy of 25 plant extracts against the infectivity of chilli mosaic virus on chilli cvs. Faizabadi and Surajmukhi at Ayodhya, Faizabad (India). According to them, *Datura stramonium* showed 100 per cent inhibition on both cultivars followed by *Solanum nigrum* (60 per cent), *Tamarindus indica* (50 per cent) and *Papaver somniferum* (40 per cent). Other leaf extracts showed different levels of inhibition for the different cultivars: *Argemone mexicana* (65-75 per cent), *Nicotiana tabacum* (65 - 75 per cent) and Arachis hypogaea (50 – 70 per cent). Only the leaf extract of *Peristrophe bicalyculata* showed no inhibitory activity against the virus. Dhawan and Rishi (1999) suggested growing sesame and wheat were the most effective non-host barrier trap crops in the management of viral diseases of chilli in summer and winter crops, respectively. The yield of chilli with the barrier crops was also enhanced.

In general, CMV resistant genotypes recorded higher contents of chlorophyll, phosphorus and magnesium and lower levels of zinc and manganese than susceptible genotypes. The study of Singh *et al.* (1998) revealed that virus infection showed a significant reduction in total chlorophyll, chlorophyll 'a' and 'b' contents and Zn, and an increase in P, K, Ca, Cu and Fe in susceptible genotypes compared with the resistant genotypes. The level of Mg was found relatively higher in resistant varieties but it showed very little increase after infection.

Chiemsombat *et al.* (1999) isolated virus (chilli vein banding mottle virus) genes from virus infected pepper cv. Bang Chang and cDNA of the viral coat protein gene was successfully cloned. Plant material was transformed with coat protein cDNA using *Agrobacterium tumefaciens*. Transgenic pepper plants possessing the viral coat protein gene were self-fertilized. Seedlings derived from transgenic peppers were resistant to virus infection after challenge inoculation.

Rahman *et al.* (1998) studied the integrated management of insect-borne viral diseases occurring on chilli for three years and found that use of nylon net and application of carbofuran granules in the nursery followed by insecticidal sprays in the main field check the insect borne viruses to lower levels. Soil application of carbofuran granules 3G @ 1 kg a.i./ha in the nursery followed by three foliar applications with monocrotophos and sulfex resulted in higher yield.

10.1.7 Other Diseases

Among the other diseases of chilli of minor importances are stem rot caused by *Macrophomina phaseoli*, dry rot caused by *Sclerotium rolfsii*, fruit rot caused by *Phomopsis* sp., fruit and stem rot caused by *Glomerella* sp.

Blossom-end rot, sun scald and salt-injury are non-parasitic troubles of chilli that sometimes cause considerable loss (Walker, 1952; Chowdhury, 1967).

The excessive development of lenticels, corkiness and necrosis at the base of the stalk, collar and main roots, decay of secondary roots and shoot wilt of *Capsicum annuum* L. are reported to be a physiological disorder of chilli due to waterlogging (Matta and Garibaldi, 1980).

Diseases of Chilli

Damping Off (Left), and Anthracnose (Middle, Right) Infection

***Cercospora* Leaf Spot (Left), and Dieback (Middle, Right) Infection**

Powdery Mildew (Left), and *Choanephora* Blight (Right) Infection

Bacterial Leaf Spot (Left) and Chilli Leaf Curl Virus (Right) Infection

Stored chilli fruits collected from different parts of India exhibited varying degrees of decay with respect to discolouration and breaking of the pericarp, detachment of the pedicel and spore dust formation within the fruit. Altogether 67 spp. of fungi were isolated from the stored chilli fruits, 3 spp. belonging to the Phycomycotina, 11 spp. to the Ascomycotina and 53 to the Deuteromycotina. *Aspergillus flavus, A. terreus, A. candidus, A. niger, A. sclerotiorum, Fusarium moniliforme* (*Gtbberellafujikuroi*), *F. sporotrichioides, Syncephalastrum racemosum, Paecilomyces varioti,* and *Penicillium corylophilum* were commonly associated with decaying chilli fruits stored in humid regions. Capsaicin content was decreased due to fungal infection (Prasad *et al.*, 2000).

10.2.0 Pests

Over 20 insect pests are known to infest chilli in India. Thrips and aphids are serious ones.

10.2.1 Thrips (*Scirtothrips dorsalis*)

The larva and adults infest tender leaves and feed on the sap causing curling of the leaf. Severe infestation causes heavy curling of leaves and stunting of the crop. Buds and flowers are also damaged. Yield loss due to thrips attack may range from 25-50 per cent in addition to the transmission of leaf curl virus (Butani, 1976). The insect reproduces sexually as well as parthenogenetically. The female thrips inserts the eggs into the veins of leaves and a female may lay 40-48 egg. The life cycle occupies from 10-20 days.

Thrips population can be minimized by spraying dimethoate @ 0.025 per cent at fortnightly interval commencing from three weeks after transplanting (Rangarajan *et al.*, 1974). Soil application of carbofuran at the 10th day after transplanting followed by azodrin @ 0.05 per cent spray at 60th, 75th and 90th day of transplanting controlled thrips effectively. Aldicarb at 1 kg/ha applied as granules to soil at sowing followed by 5 sprays of 0.05 per cent Monocrotophos at fortnightly interval afforded the best control of thrips (Mote, 1977; Reddy and Jagadish, 1980). CO. 1 chilli cultivar recorded a thrips incidence of 59.5 per cent per clipping per plant. Soil application of carbofuran @ 0.5 kg a.i./ha or phorate at 0.7 kg a.i./ha 15 days after transplanting, followed by 12 sprays of 0.04 per cent monocrotophos or 0.1 per cent triazophos or spray application of 0.04 per cent monocrotophos or 0.1 per cent triazophos alone at 10 days intervals, commencing from 40 days after transplanting, effectively checked the thrip population and gave the highest yield of green chillies. Granular application alone failed to check the pest population effectively (Patel *et al.*, 1997). According to Patnaik and Mahapatra (1997), application of methomyl @ 0.5 kg a.i./ha at 50 DAP and monocrotophos @ 0.5 kg a.i./ha at 75 DAP, irrespective of the fertilizer concentration used, gave a substantial reduction in the number of thrips (71.0 per cent) and the incidence of leaf curl (60.4 per cent).

According to Patel *et al.* (1999), soil application of carbofuran @ 0.5 kg a.i./ha 15 days after transplanting followed by continuous spray of triazophos 0.04 per cent at 10 days intervals 30 days after transplanting effectively controlled thrips and gave a higher green chilli fruit yield with a higher cost: benefit ratio.

10.2.2 Aphids (*Aphis gossypii, Myzus persicae*)

The aphid (*Aphis gossypii*) and green peach aphid (*Myzus persicae*) sometimes become serious on chilli. They suck the cell sap from leaves and petioles and cause considerable loss in yield.

Light paraffin, groundnut oil, heavy paraffin, BOC 20 and castor-oil spray reduced the *Aphis gossypii* population (Khatri and Sekhon, 1974). Spraying Methomyl @ 440 g/acre registered 100 per cent mortality of aphids within 72 hours of treatment. A combined spray application of Nuvacron at 0.05 per cent and Nuvan at 0.05 per cent sprayed 6 times at 15 days interval commencing from 15th day of transplanting controlled aphids effectively. The green peach aphid (*Myzus persicae*) can be controlled effectively by foliar application of ethiofencarb at 0.05 per cent and methamidophos 60 E.C. @ 0.05 or 0.06 per cent (Saivaraj *et al.*, 1979). They also reported that application of aldicarb, phorate and carbofuran, though effective, could leave toxic residues in the green fruits.

Reddy *et al.* (1981) obtained 100 per cent control of the aphid by the application of demeton methyl (0.05 to 0.2 per cent), fenvalcrate (0.01 to 0.02 per cent) or monocrotophos (0.05 to 0.01 per cent). The yield of dried fruits was highest (39.37 q/ha) when plants were treated with fenvalerate @ 0.02 per cent. Acephate application also showed satisfactory control of aphids @ Ohio, U.S.A. (Ladd *et al.*, 1982).

10.2.3 Mites (*Polyphagotarsonemus latus*)

Mites of different genera have been found feeding on leaves. The tiny spiderlike creature may be found in large numbers on the underside of leaves, covered with fine webs. Both nymphs and adults suck the cell-sap and devitalize the plants (Butani, 1976).

It is reported that spraying of phosalone (zolone) 35 E.C. at 3 ml/l controlled the mites on chillies. In 4 field trials for the control of *Hemitarsonemus* (*Polyphagotarsonemus latus*) the highest chilli yields were obtained in treatments with amitraz @ 0.02 per cent, propargite @ 0.03 per cent, Triazophos @ 0.04 per cent, Chlordimeform @ 0.04 per cent and cyanotox (cyanophos) @ 0.025 per cent. However, chlordimeform was mildly phytotoxic (Chou and Ng, 1982). Rao and Ahmed (1983) studied the efficacy of different insecticides against white mite of chilli and found that highest yield (1895 kg dry pods/ha) of chilli was obtained with a good control of mites when the plants were sprayed with plictran 50 W.P. (cyhcxatin) @ 0.5 kg a.i./ha applied first at the occurrence of mite and 3 more subsequent sprayings at 15 days interval. Leaf extracts of 9 plant species (*Adathoda vasica, Vitex negundo, Azadirachta indica, Aristolochia bracteata, Lippia nodiflora, Argemone mexicana, Sansevieria* sp., *Cissus quadrangularis* and *Aloe* sp.) reduced the yellow mite (*Polyphagotarsonemus latus*) population when applied as 5 per cent aqueous extract on chillies 30 days after transplanting in Tamil Nadu, India. Among the different extracts, *L. nodiflora* and *Aloe* sp. recorded the lowest mite population per leaf (Palaniswamy and Ragini, 2000).

10.2.4 Minor Insect Pests

In India, other pests of minor importance in chilli are pepper weevil, pepper maggot (Choudhury, 1967), *Opatroideefrater fabricius* (Puttaswamy and Gowda, 1976).

10.2.5 Nematodes (*Meloidogyne* spp., *Rotylenchus* spp.)

Root-knot nematode incidence has also been recorded in chilli like tomato, brinjal, *etc.* In India, the occurrence of root-knot nematode in chilli was reported in 1960 (Prasad, 1960). Of these three species *Meloidogyne incognita, M. javanica and M. arenaria*, the later one is most harmful to cultivar, Sathur Samba of *Capsicum annuum* (Rajagopalan *et al.*, 1971). It attacks the root zone, producing tiny galls on the roots. The attacked plants show wilting and withering of leaves which ultimately affect the growth and yield of the crop.

Soil fumigation with DD (dichloropropene-dichloropropane), EDB or nemagon is very effective in reducing the nematode population. Azam *et al.* (1979) reported that Rogor G (dimethoate) showed effective control of the nematode and best growth of chilli at Aligarh. Sivakurnar *et al.* (1979) were of the view that in glass house, red pepper was antagonistic to the nematode *Rotylenchus reniformis*. Application of carbofuran and aldicarb to red pepper at 100 mg/plant reduced the nematode population. From a field trial, they also reported that the population was reduced by 19.86 and 79.55 per cent after 1 and 4 months, respectively in red pepper and spot application of carbofuran, aldicarb, Phorate, disulfoton or CGA 12223 (azophos) @ 1 kg/ha and phorate with disulfoton each at 0.5 kg/ha at 15 days after transplanting brought about a decline in nematode population of 71.35 to 88.89 per cent and 98.13 to 99.77 per cent at 1 and 4 months after application, respectively. In Florida, soil fumigation by DD @ 75 1/ha gave excellent control of *Belonolaimus longicaudatus* and increased yields than control, provided there was 2 week gap between DD application and transplanting. Application of granular nematicides, aldicarb, phenamiphos (fenamiphos) or turbufos, @ 2.24 kg a.i./ha significantly reduced the nematode populations (Rhoades, 1981). Application of carbofuran, aldicarb, phorate or diazinon at a 0.40, 0.60 or 0.80 g/m^2 suppressed the gall and eggmass development of *Meloidogyne* sp. (Yadav and Mathur, 1990).

In screening for resistance with 19 chilli cultivars, none was free from nematodes but the cultivar G-2 was found to be tolerant with fewer incidences (Rajagopalan *et al.*, 1971). In Indonesia, cvs. Cpianas and Indragiri were reported to be resistant to *Meloidogyne* spp. (Marwoto, 1988).

C. annuum has resistance to root-knot nematodes (RKN) (*Meloidogyne* spp.), severe polyphagous pests that occur world-wide. Several single dominant genes confer this resistance. Some are highly specific, whereas others are effective against a wide range of species. The spectrum of resistance to eight clonal RKN populations of the major *Meloidogyne* sp., *M. arenaria* (2 populations), *M. incognita* (2 populations), *M. javanica* (1 population) and *M. hapla* (3 populations) was studied using eight lines of *C. annuum*. Host susceptibility was determined by counting the egg masses (EM) on the roots. Plants were classified into resistant (R; EM. ≤ 5) or susceptible (H; EM>5) classes. The French cultivar Doux Long des Landes was susccptible all nematodes tests. The other seven pepper lines were highly resistant to *M. arenaria*, *M. javanica* and one population of *M. hapla*. Variability in resistance was observed for the other two populations of *M. hapla*. One lines PM687, PM217, Criollo de Morelos 334 and Yolo NR were resistant to *M incognita*. To investigate the genetic basis of resistance in the highly resistant line PM687, the resistance of two progenies was tested with

Pests of Chilli

Thrips Infestation

Yellow Mite Infestation

Fruit Borer (Left), and Root Knot Nematode (Right) Infestation

Mealybug (Left) and Caterpillar (Right) Infestation

the two populations of *M. incognita*: 118 doubled-haploid (DH) lines obtained by androgenesis from F_1 hybrids of the cross between PM687 and the susceptible cultivar Yolo Wonder, and 163 F_2 progenies. For both nematodes populations, the segregation patterns 69 R/49 S for DH lines and 163 R/45 S for F_2 progenies were obtained at 22°C and at high temperatures (32°C and 42°C). The presence of a single dominant gene that totally prevented multiplication of *M. incognita* was thus confirmed and its stability at high temperature was demonstrated. This study confirmed the value of *C annuum* as a source of complete spectrum resistance to the major RKN (Djian-Caporalino *et al.*, 1999).

11.0 Seed Production

Chilli is a self-pollinated crop but cross-pollination also takes place to some extent. It is suggested that to obtain pure seeds, the two cultivars may be kept 250-400 m apart to avoid any cross pollination. Good, healthy and well developed fruits should be selected for producing seeds.

Seeds can be extracted from healthy and well development fruits. Seeds of Pant C-1 and Byadgeri Local harvested at the ripe fruit stage showed high germination capacity and seedling vigour. However, fruits could be harvested at the breaker stage for an ealry seed crop, without affecting its quality (Doijode, 1988). Seeds of cv. Co.1 and Co. 2 extracted from fruits collected 48 days after anthesis were best with regard to germination percentage, field emergence and vigour index. A simple mechanical seed extractor has been described by Muthukrishnan *et al.* (1983). Curry powder grinder after suitably adjusting the grinding plate can be advantageously put to use to break the pods into pieces while seeds are completely shed. A winnower or sieve shaker is used for separation of seeds fitted with BSS 6 wire mesh. This procedure brings down the extraction cost by 50 per cent and enhances germination percentage 0 to 90. The seeds treated with Captan or Ceresan at the rate of 2 g/kg of seeds and packed. The average seed yield per hectare varies from 50 to 80 kg (Choudhury, 1967). According to Dhanelappagol *et al.* (1994), germination was highest (92 per cent) in seeds from sundried fully ripe red coloured fruits dried at an air temperature of 40°C when compared the effect of stages of harvesting and methods of drying of chilli fruits on seed quality. Silva and Peiris (1997) were of the view that seeds packed in aluminium foil or in polythene (500 gauge), could be stored in cold room (20°C, 65 per cent R.H.) for 12 months (percentage germination of 78-81 and field emergency 40-52 per cent). Radheshyam *et al.* (1996) opined that seed moisture content and electrical conductivity of chilli seed were higher in seeds from fruits harvested at the turing stage but test weight, germination percentage and viability were lower. Germinability of chilli seeds declined during storage particularly after 19 months when seeds after sundrying to a moisture content of 6 per cent and following seed treatment with thiram, were placed in 700 gauge polyethelene bags or paper bags and stored in an insect free godown for 27 months. Seed moisture content remained about 6 per cent in poly- bags but in paper bags it responded to changes in R.H. and temperature in the godown (Sharma *et al.*, 1998).

Seed viability of chilli, in general, decreases with increasing seed storage time. The findings of Nagaraja *et al.* (1998) indicated that seeds of all cultivars of chilli (11

Method of Genetically Pure Seed Production in Chilli

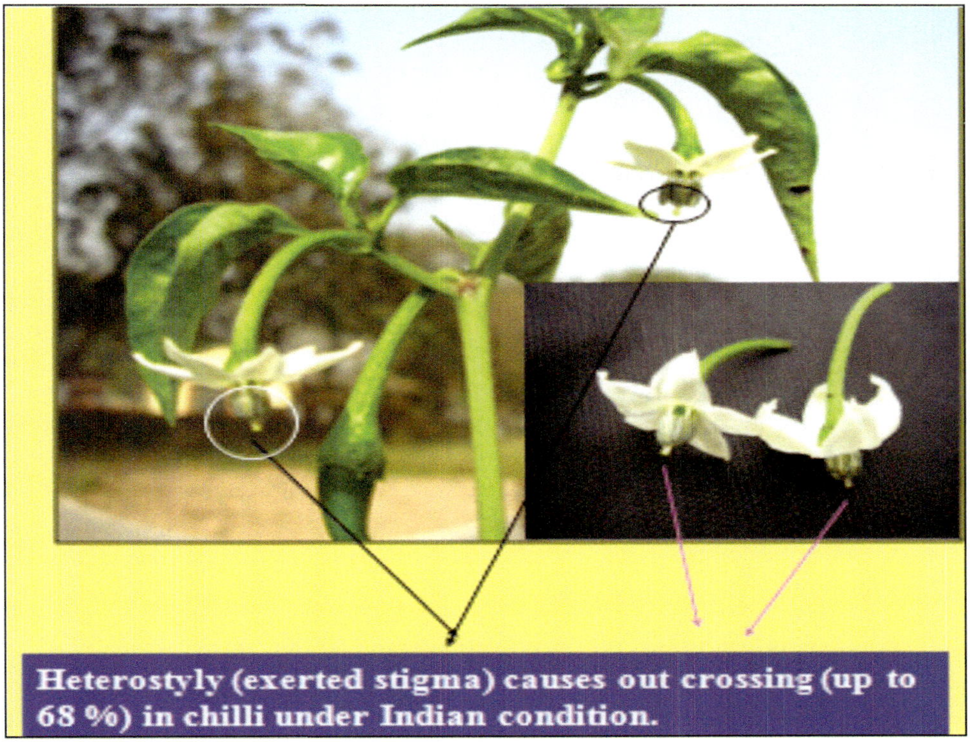

Heterostyly (exerted stigma) causes out crossing (up to 68 %) in chilli under Indian condition.

Cotton wading and covering with 16 mesh nylon net also prevent out crossing

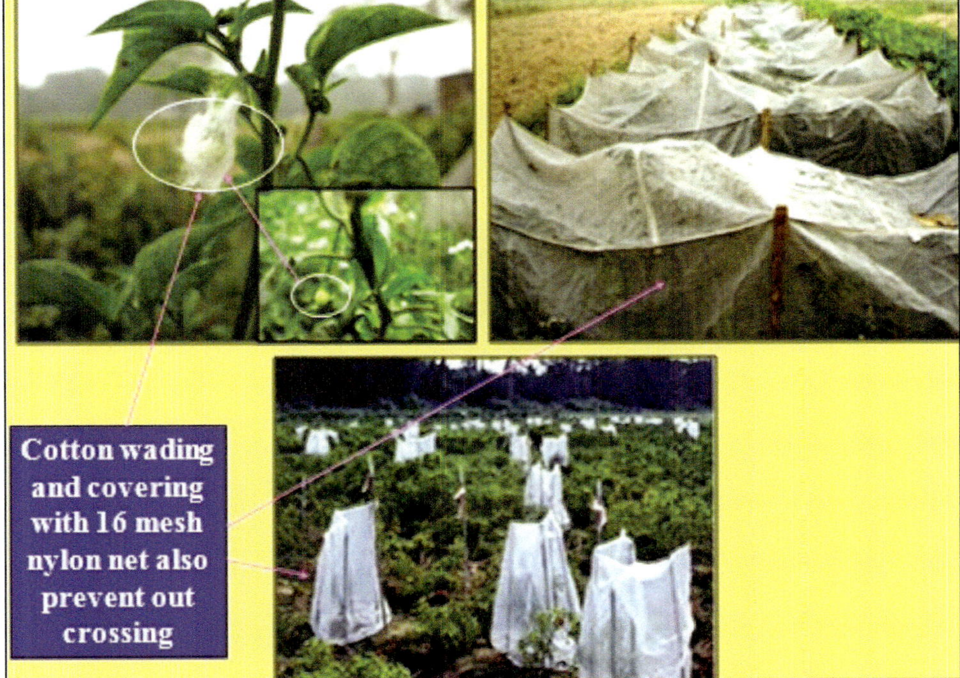

per cent seed moisture) tested produced certifiable germination (60 per cent) after 6 months' storage under ambient condition. Accelerate aging tests (seeds treated with captan 2 g/kg and subjected to aging at 42°C and 100 per cent RH for upto 8 days, finally stored for 6 months in polyethylene bags) indicated that seeds could be stored for upto 4 and 6 months in case of chilli cvs. Dyavanoor local and Hyadagi Kaddi, respectively with certifiable limits of germination (Patil and Nagaraja, 1999).

12.0 Crop Improvement

Development of chilli hybrids having high pungency, antioxidants, vitamins and appreciable tolerance against leaf curl virus is the present need in major part of the tropics (Bhutia *et al.*, 2017a). Extensive use of local landraces or non-selfed seeds of improved varieties, and tremendous biotic stresses have resulted in drastic reduction in optimum productivity of both fresh and dry chillies in India over the last couple of years (Chattopadhyay *et al.*, 2011). Moreover, most of the hybrids evaluated at national trials are less pungent and prone to attack of leaf curl disease which hinders the greater acceptability among the growers and consumers particularly in eastern part of India. Therefore, much concentrated efforts are necessary to improve its yield, quality and host plant resistance against viral diseases and pathogens. The proper choice of parents is a prerequisite in any sound breeding programme. To achieve these objectives classical methods such as mass selection, pedigree method, single seed descent, backcross, recurrent selection and hybridization are utilized in chili pepper breeding (Coon *et al.*, 2008; Kulkarni and Phalke, 2009; Nsabiyera *et al.*, 2013; Manzur *et al.*, 2014). The choice of the best method or combination of them depends mainly on the type of inheritance (monogenic, oligogenic, or polygenic) of traits to be improved (Lee *et al.*, 2013a). The Genes of major horticultural traits and disease resistance in capsicum are summarised in Table 16.

Table 16: Genes of Major Horticultural Traits and Disease Resistance in Capsicum

Plant Character	*Character Type and its Gene Symbol*	*Reference*
Anthocyanin coloration of leaves, stem, flower, and immature fruits	*A* (incomplete dominance) along with modifier gene *MoA*	Lippert *et al.* (1965), Odland (1960)
Plant height	Dwarf nature controlled by nine recessive genes *dw1* to *dw9*	Daskalov (1973a), Restaino (1989), Yazawa *et al.* (1991), Aniel *et al.* (2001)
Branchless	*Bl* (recessive gene)	Bergh and Lippert (1964)
Branching	*ct* (for plant habit), *fa* (clustered fruit habit) and *dt* (for determinate growth) along with modifiers Dominant genes Dt and Ct (in-determinate growth habit)	Mc Cammon and Honma (1984)
Leaf shape	Nl (narrow leaf), broad leaf (bl), small leaf (sl-1 and sl2), curved leaf (cl), folded leaf (fl), round tip leaf (rl-1, rl-2, rl-3)	Daskalov (1973b), Aniel *et al.* (2001)
Leaf pubescence	Two dominant genes *H* and *Sm* (HHSmSm: presence of pubescence, hhsmsm: glabrous leaves)	Shuh and Fontenot (1990)

Plant Character	*Character Type and its Gene Symbol*	*Reference*
Flowers	Multiple flowers by three dominant genes (*Mf-1, Mf-2* and *Mf-3*) *ef*: early flowering *lf*: late flowering *nf*: n o flowering	Shuh and Fontenot (1990), Pathak *et al.* (1985)
Fruit shapes	*P:* Pointed fruit shape *fb*: Non bulging fruit shape *ce*: Fruit base with enclosed calyx *O*: Round fruit shape with modifiers *up*-1 and *up-2*: Erect fruit *pf*: Parthenocarpy	Deshpande (1933), Daskalov and Poulos (1994), Peterson (1959), Gopalkrishnan (2007), Lippert *et al.* (1965), Pathak *et al.* (1983).
Immature fruit color	Three alleles of a recessive gene: *sw1*, *sw2* and *sw3* *sw1*: Sulphur white *sw2*: Yellowish green *sw3*: Cedar green *sw1* > *sw2* > *sw3*	Odland and Porter (1938)
Mature fruit color	y^+: red colour *y*: yellow colour *cl and* y^+ : brown colour *Ccs*: Capsanthin-capsorubin synthase enzyme that synthesizes red carotenoid pigment *Psy*: Locus responsible for development of fruit colour Psy/C2: Rate limting factor in carotenoid production	Boswell (1937), Smith (1950), Hurtado-Hernandez and Smith (1985), Lefebvre *et al.* (1995), Thorup *et al.* (2000), Huh *et al.* (2012a)
Pungency	*Pun*: Controls acyl transferase responsible for capsaicin synthesis *lov*: Non pungency due to loss of vesicles on the placental walls	Deshpande (1986), Greenleaf (1956), Daskalov (1986), Votava *et al.* (2000)
Beta carotene	*B, t* and *bc*: Confer high beta carotene contents	Chalukova *et al.* (1993), Daskalov *et al.* (1995)
Male sterility	Genetic male sterility: Total 20 genes have been identified, *ms-1* to *ms-20* Cytoplasmic male sterility: Major gene ms in interaction with *S* cytoplasm *Rf*: Restorer of fertility locus	Shifriss and Frankel (1969), Shifriss and Rylski (1972), Daskalov (1973a), Daskalov (1986), Shifriss (1973), Meshram and Narkhade (1982), Pathak *et al.* (1983), Peterson (1959), Novac *et al.* (1971) Daskalov (1973a)
Tobacco mosaic virus	L^3, L^2, L^1, L^+ series of multiple alleles L^2: localization of TMV L^1: imperfect localization of TMV L^+: Mottling $L^3 > L^2 > L^1 > L^+$	Homes (1937), Boukema (1980)
Cucumber mosaic virus	*cm*: Recessive gene and 4 QTLs	Singh and Thakur (1977), Ben *et al.* (2001)
Tomato spotted wilt virus	*Tsw*: Hypersensitive resistance to TSWV	Moury *et al.*, 1997a, b.

Plant Character	*Character Type and its Gene Symbol*	*Reference*
Bacterial leaf spot	*Bs1, Bs2, Bs3, Bs4*: Hypersensitive Resistance *bs5* and *bs6*: nonhypersensitive recessive resistance *gds*: General defense system	Cook and Guevara (1984), Csillery *et al.* (2004)
Phytophthora disease	*Psr*: Stem resistance to *Phytophthora* *Pfo*: Foliar resistance *Pfr*: Fruit rot resistance	Sy *et al.* (2005), Walker and Bosland (1999)
Anthracnose resistance	*Anr1*: Resistance to *Colletotrichum dematium* Anr2, Anr3, Anr4: Resistance to *Colletotrichum gleosporoides* *Anr5*: Resistance to *Colletotrichum capsici*	Park *et al.* (1990), Lin *et al.* (2013)
Bacterial Wilt	Two genes with incomplete dominance	Matsunaga *et al.* (1998)
Powdery Mildew	Three genes: *Imr-1*, *Imr-2*, *Imr-3*	Shifriss *et al.* (1992)
Root-knot nematodes	N: Resistance to *M. incognita acritaMe1, Me2, Me3, Me4, Me5:* Resistance to *Meloidogynae* spp. *Me6*: *M. arenaria* and *M. javanica* *Mech1* and *Mech2*: Suppresses nematode resistance	Pegard *et al.* (2005), Djian-Caporalino *et al.* (2004)
Bentazon herbicide resistance	*Bzt*: Tolerance	Fery and Harrison (1990)

Based on inheritance pattern of the gene(s) different breeding approaches are deployed in chilli improvement programme.

Mass Selection

Mass selections refers to a method of crop improvement in which individual plants are selected on the basis of phenotypes from a mixed population, their seeds are bulked and used to grow the next generation. Mass selections were successufully utilized by indigenous peoples of tropical America, whereby seeds of the best plants were saved for the next growing season. This method should be used for populations with genetic variability and selected in environments where the traits express themselves and for those of high heritability (Nsabiyera *et al.*, 2013).

Pedigree

Pedigree refers to record of the ancestry of an individual selected plant. Pedigree breeding is a method of genetic improvement of self pollinated species in which superior genotypes are selected from segregating generations and proper record of the ancestry of selected plants are maintained in each generation. 'NuMex Joe E. Parker' is an example of a cultivar developed by this method (Bosland *et al.*, 1991). It originated as a single plant selection from a field planted to an open-pollinated population of 'New Mexico 6-4'. A pedigree selection protocol was subsequently followed for three generations, under green house conditions. Each line produced was then evaluated for more than 25 horticultural traits.

Single Seed Descent

A breeding procedure used with segregating populations of self pollinated species in which plants are advanced by single seeds from one generation to the next is referred to as single seed descent method. This method is also utilized in the development of recombinant inbred lines (RILs). Moreira *et al.* (2009) utilized this method to obtain lines resistant to bacterial spot.

Recurrent Selection

Recurrent selection is a breeding method that involves the selection of individual plants from a population, followed by intercrossing to form a new population. The intent of this system is to produce transgressive segregants: individuals that contain greater resistance as a result of the genetic segregation resulting from crosses between parents that already have some degree of resistance. Current efforts that utilize this method for pepper breeding have focused on developing pepper cultivars resistant to *Verticillium dahliae* and *Phytophthora capsici*.

Heterosis

Heterosis is the superiority of F_1 hybrid over the parents in a given characteristic, assessed not by absolute value and appearance but by its usefulness for practical advantages in given environments. Heterosis in desirable direction (hybrid vigour) is the ultimate aim of a breeder. Reports on simultaneous improvement of green and dry chili yield, and quality traits through heterosis breeding have been documented (Bosland and Votava, 2000; Singh *et al.*, 2012; Chaudhary *et al.*, 2013; Bhutia *et al.*, 2015; Sreenivas *et al.*, 2020). Their study depicted that hybrid vigour is available for commercial production of chilli hybrid, and that isolation of pure lines from the segregating generation of heterotic F_1s is an alternative approach to improve fruit yield, quality and viral disease tolerance. It has been found that closely or distantly related parents exhibited low heterosis, but crosses between parents of intermediate divergence classes tended to show higher heterosis for fruit yield and other important traits (Geleta *et al.*, 2004; Bhutia *et al.*, 2015; Sreenivas *et al.*, 2020).

Backcross

Backcross refers to crossing of F_1 with either of its parents. When F_1 is crossed with homozygous recessive parent, it is known as test cross. Backcross is used particularly for traits controlled by one or few genes. The backcross method uses a successful cultivar as a recurrent parent, following an initial cross between that successful cultivar and a separate individual that serves as a donor parent for a given desired character. Following successive backcrosses to the recurrent parent, the breeder arrives at a cultivar that is almost identical to the recurrent parent but contains the additional desired trait from the donor parent. An example of a successful pepper cultivar developed via the backcross method is 'Greenleaf Tabasco'. In 1950, Walter Greenleaf hybridized the Tabasco pepper with an accession of *C. chinense* that was resistant to tobacco etch virus (TEV). At the third backcross, Greenleaf made hybridization to a second TEV resistant *C. chinense* donor parent. The recessive mode of inheritance for TEV resistance required that alternate generations be screened. In 1970, after several backcrosses, 'Greenleaf Tabasco' was introduced (Greenleaf, 1986).

Mutation Breeding

Mutation breeding is a means by which mutants are generated in peppers, either to improve economically important traits or to eliminate deleterious traits. Although not a major breeding method, it may be a means of producing novel mutants of interest. Mutations can be induced by ionizing radiation or chemically. Mutations of interest may then be incorporated into commercially acceptable lines by traditional breeding methods, such as the ones described above. Bhargava and Umalkar (1989) used both gamma radiation and chemical mutagens to produce an array of pericarp mutations. Alcantara *et al.* (1996) described the optimal conditions necessary for seed mutagenesis in *C. annuum* when ethyl methanesulfonate (EMS) was used as the mutagen.

The achievements made by using different breeding methods in chilli improvement are discussed in Table 17.

Table 17: Achievements Made by different Breeding Methods in Chilli Improvement

Breeding Approach	*Significant Achievements/ Varieties Released*	*References*
Introduction	✰ NuMex Centennial (Mexico) ✰ CO 4 (introduced from Sri Lanka in India) ✰ CO 3 (introduced from Sri Lanka in India)	TNAU Portal: http://agritech.tnau.ac.in/horticulture/horti_vegetables_ hilli.html Vidhi J: http://www.biologydiscussion com/vegetable-breeding/top-7-breeding-methods-ofcapsicum-india/68448
Mass Selection	✰ It is still used in Mexico to select seed for Poblano, guajillo and other traditional capsicum landraces. ✰ Heritage New Mexico 6-4: Selection made from NuMex Big Ji	DeWitt (2014)
Pureline Selection	✰ G 1, G 2, G 3, G 4, NP 46A, K 1, Co 1,CO.2, Musalwadi, Sindhur, Patna Red, Pant C 1, PLR1, KI	Gopalakrishnan (2007), Ramachandran (2013)
Single plant Selection	✰ New Mexico No. 6 ✰ NuMex Conquistador: single plant selection from 'New Mexico 6-4 ✰ NUMEX JOE E. PARKER: single plant selection from population of 'New Mexico 6-4 ✰ NuMex Sweet: single plant selection from population of 'New Mexico 6-4	DeWitt (2014)

Breeding Approach	*Significant Achievements/ Varieties Released*	*References*
Pedigree Method	✰ Andhra Jyoti, Pusa Jwala, Pusa Sadabahar, X 235, K2, Punjab Lal and Jawahar 218 (India) ✰ New Mexico No. 9 Sandia: Cross of New Mexico No. 9 and California Anaheim Rio ✰ Grande 21: Cross of New Mexico No. 6 and Anaheim ✰ Española Improved: hybridization between 'Sandia' and a Northern New Mexico line of chilli ✰ Numex Sunrise, Numex Sunset and Numex Eclipse: All three cultivars originated from a hybridization between 'Permagreen,' a green bell pepper, and 'New Mexico 6-4 ✰ Numex RNaky: cross of 'Rio Grande 21,' 'NewMexico 6-4,' Bulgarian paprika, and an early-maturing native type ✰ Numex Sunburst, Numex Sunflare and Numex Sunglo: derived by pedigree breeding from a seed source from India in Mexico ✰ Numex Primavera, NuMex Memorial Day' and 'NuMex Thanks, Numex Garnet, NuMex Primavera' ✰ AVPP0506, Berke's Joy, AVPP0105, AVPP0206, AVPP0303, AVPP0409, AVPP0411, AVPP0512, AVPP0514 (from World Vegetable Centre)	Ramachandran (2013), Dewitt 2014; http://www.biologydiscussion.com/vegetablebreeding/top-7-breeding-methods ofcapsicumindia/ 68448; AVRDC https://avrdc.org/seed/ improved-lines/chili-capsicum
Backcross method	✰ Pyramiding of genes conferring resistance to PMMoV, PVY and TSWV in sweet Charleston capsicum lines (BC: Backcrossing; L4: The gene conferring resistance to PMMoV pathotype 1, 2, 3. c Tsw: The gene conferring resistance to TSWV; dPVY: Potato Virus Y pathotype 1-2) ✰ Introgression of heat shock protein (Hsp70 and sHsp) genes into the Malaysian elite chilli variety Kulai (*capsicum annuum* L.) ✰ Resistance has been successfully introgressed into commercial capsicum cultivars, including resistance to tobamoviruses from *capsicum chacoense* and *C. chinense* ✰ Resistance to tomato spotted wilt virus (TSWV) from *C. chinense* and *C. baccatum* ✰ Resistance to anthracnose fruit rot from *C. chinense* ✰ Resistance to *Phytophthora capsici* from *C. annuum* cv. CM334 and resistance to bacterial leaf spot disease from *C. annuum* and *C. chacoense*. ✰ *p-AMT* and *Pun1* markers were used to develop a new fresh cultivar containing capsinoids, named 'Maru Salad'.	http://www.biologydiscussion.com/vegetable-breeding/top-7-breeding-methods-ofcapsicum-india/68448; Boukema (1980), Cook and Guevara (1984), Kim and Hartmann (1985), Hibberd *et al.* (1987), Boiteux and de Avila (1994), Berzal-Herranz *et al.* (1995), De la Cruz *et al.* (1997), Voorrips *et al.* (2004), Vallejos *et al.* (2010), Mallard *et al.* (2013), Hoang *et al.* (2013), Soler *et al.* (2017), Liu *et al.* (2014), Usman *et al.* (2008)

Breeding Approach	*Significant Achievements/ Varieties Released*	*References*
Heterosis breeding	☆ No. 1 Zao Fong: First F_1 hybrid of China ☆ Based on CGMS system: Kashi Surkh, Kashi Early, (IIVR, India) ☆ Arka Meghana, Arka Sweta, Arka Harita (IIHR, India) ☆ YU JING NO 2, (China) ☆ F1 Hybrid Coral and F_1 Hybrid Dara, Clover Seeds, Hong Kong China ☆ VNR38, VNR108, VNR174, VNR Seeds, India ☆ VNR200, VNR332 ☆ F_1 Forever Tropicasem, Senegal Many SSA countries ☆ Remington, F_1 Alpha Seeds, South Africa Many SSA countries ☆ F_1 TSS AVRDC No.4 Suntech Seeds, Taiwan ☆ F_1 TSS AVRDC No.2 Yung Shan Seeds,Taiwan ☆ F_1 Hsing AVRDC No.3 Suntech Seeds, Taiwan ☆ Yun High Pungency No.1 Horticulture Research Institute, YAAS, China China ☆ Ulka F_1, Masaya 315, East-West Seeds, India ☆ Yuvraj IN ☆ Super F_1, Muria F_1 East-West Seeds, Thailand ☆ Hybrid Indus Seeds, India ☆ Based on GMS system: ☆ CH-1, CH-3, CH-27 multiple disease resistant hybrid PAU, Ludhiana, India	Tong (1998), Gopalkrishnan (2007), Lin *et al.* (2013), Dhaliwal *et al.* (2014); IIHR Website: https://www.iihr.res.in/division-varities/786; IIVR website: https://www.iivr.org.in/iivrvarieties/ by-crop
Mutation Breeding	☆ Horgoskaslatka-X-3–resistant to CMV; Yugoslavia Karasz ☆ Albena: Attractive fruits with better flavor, Bulgaria S. Daskalov, Institute of Genetics,Sofia 113 ☆ Krischimski ran: Hybrid variety with high yield, earliness and improved fruit quality, Bulgaria S. Daskalov and L. Milkova Institute Institute of Genetics, Sofia 113 ☆ MDU 1: Compact plant type with higher yield and capsaicin content, Tamil Nadu Agriculture University, India ☆ Lyulin: Hybrid variety based on induced male sterility, very early and high yield, Institute of Genetics, Sofia 113	Daskalov (1986)

The breeding methods hitherto adopted in chilli varied from introduction, mass selection, reciprocal recurrent selection, *etc.* (Lippert, 1975). Back-crossing and mass pedigree method of selection might prove useful in chilli. Inter and intra-crossings among groups of chilli cultivars might enable the breeders to get a large number of recombinants for simultaneous improvement (Sohi *et al.*, 1977).

Muthukrishnan *et al.* (1983), however, have suggested the following aspects to be considered in future breeding programmes:

1. Evolution of cultivars to suit both internal and export markets, wider adaptibility and quality, *viz.* better pod colour and shape, high capsaicin, carotene and ascorbic acid contents.
2. Cultivars which are early, determinate and compact with shorter phase have to be developed.
3. Resistance breeding needs emphasis breeding lines resistance to viruses (TMV, leaf curl, potato mosaic, *etc.*), thrips, fruit rot and nematodes.
4. Exploitation of hybrid vigour.

The cultivars developed in India through introduction, selection, hybridization and possibility of exploiting hybrid vigour are discussed below:

12.1.0 Introduction and Selection

In India introduction and selection as methods of crop improvement has led to the release of many strains for direct cultivation. The 'G' series of Lam Research Station (Andhra Pradesh), 'L' series of Pant Nagar (Uttar Pradesh), 'Co' series of Coimbatore and 'K' lines of Kovilpatti (Tamil Nadu) are noteworthy. In the types CA 1068, Oosimilagai, Lavang mirche and Golgonda Mirche (of lam Research Station), the heavy yielding and high capsaicin characters have been combined. Kalianpur Red, Kalianpur Yellow, Kalianpur Selection-I and Kalianpur Chaman are valued both for Rajendra Agriculture University, Bihar. These fruits are long (9.1-9.8 cm), shiny red and with high seed content (55-58 per cent), besides they are claimed to be partially resistant to die-back and virus. The cultivar JCA 154 is a pickling type developed from J.N. Krishi Viswavidyalaya, Jabalpur, Madhya Preadcsh. Rajpura Long Red and Long Red are selections best suited for Punjab conditions. The type CA (P) 247 faired better in Tamil Nadu which was later released as CO.1. CA 960 cultivar was identified as a calry flowering type (Suthanthirapandian and Rangaswamy, 1983). CO. 1 is superior in dry pod weight with minimum loss on drying (Vadivel *et al.*, 1983). The line K-1 was isolated as a superior yielder in Tamil Nadu (Rajagopal and Muthukrishnan, 1980). Sindhur is a dual purpose chilli selected in Andhra Pradesh. Selection for capsaicin content in *C. frutescens* led to the development of Pusa Sadabahar with 12 per cent capsaicin (Tewari, 1989).

12.2.0 Breeding Behaviour and Genetics

The breeding behaviour of cultivated peppers was well understood besides recognition of 5 species of *Capsicum viz. C. annuum, C. frutescens, C. pubescens, C. pendulum* and *C. chinense* (*sinense*) that were known to have brought into cultivation (Shoemaker, 1953).

12.2.1 Crossing Behaviour

1. *C. annuum* x *C. f rutescens*

In general, about 2 per cent of the seeds were viable when *frutescens* was used as female parent. The F_1 plants ranging from completely pollen sterile to partially fertile were obtained. Back crosses to both parents had been obtained, but both F_2 and back cross plants showed much pollen sterility. The 'L' gene for tobacco mosaic resistance was successfully transferred from *frutescens* to *annuum* (Holmes, 1937). The transfer of resistance to the tobacco etch virus from *frutescens* had also been reported (Greenleaf, 1956).

2. *C. annuum* x *C. chinense*

Crosses can be made in either direction but are much easier when *C. annuum* is used as the female parent. Only an occasional successful cross was obtained in the other direction. The F_1 plants ranged from completely pollen sterile to moderately fertile and back cross to both the parents were obtained. Considerable sterility occurred in F_2 and back cross plants but gene interchange would be made with ease.

3. *C. annuum* x *C. pendulum*

No viable F_1 seed had ever been obtained using *C. pendulum* female parent but F_1 plants could be obtained by culture of partially developed embryos. With *C. annuum* as the female, no seed development had been observed. The F_1 plants were highly self-sterile but an occasional F_2 plant had been obtained. It appears probable that genes can be transferred from one species to the other but with difficulty. No successful back crosses to either parent had been obtained as yet.

4. *C. annuum* x *C. pubescens*

These species are completely cross-sterile. Crosses rarely resulted in fruits and those obtained were without seed.

5. *C. frutescens* x *C. chinense*

Crosses would be obtained in either direction although with considerable difficulty when *C. chinense* was used as the female parent. The F_1 plants were completely to moderately self-sterile. Backcrosses to both the parents were obtained.

6. *C. frutescens* x *C. pendulum*

Crosses were made in either direction, but more readily when pendulum was used as the female parent. The F_1 plants were highly sterile and viable F_2 seeds were very rarely formed. Back crosses to each parent resulted in only an occasional viable seed.

7. *C. frutescens* x *C. pubescens*

No fruit had ever been obtained from repeated pollinations in both directions.

8. *C. chinense* x *C. pendulum*

Crosses were made with some difficulty with *C. pendulum* as the female parent.

A few F_2 seeds were obtained but none was viable. A limited number of back-crosses were successful.

9. C. chinense x C. pubescens

Fruits were formed when *C. chinense* was used as the seed parent. The fruit at maturity contained many seeds with fully developed embryos but without endosperm. Such embryos could be grown only on artificial media. The F_1 plants were completely sterile and could not be back-crossed to either parent.

10. C. pendulum x C. pubescens

No successful crosses could be obtained.

The chromosome number for all the above species is 2n = 24.

11. C. annuum x C. chacoense

C. chacoense is a wild species of capsicum. Kumar *et al.* (1988) found that F_1 hybrids were partially fertile by using *C. chacoense* as female parent. They also opined that F_1 plants formed mostly bivalents at meiosis with a few multivalents and univalents suggesting substantial homology between the genomes of the two species.

Interspecific hybrid performance and meiotic chromosome behaviour of F_1 hybrids were studied by Tong and Bosland (1999) to elucidate genetic relationship between *C. tovarii* and other *Capsicum* sp. *C. tovarii* was hybridized, as a female and a male parent, to *C. annuum, C. chinense, C. frutescens, C. chacoense, C. galapogense, C. baccatum, C. praetermissum, C. cardenasii, C. eximium and C. pubescens*. When the hybridization of *C. baccatum* x *C. tovarii* was performed, F_1, F_2 and backcross progenies were successfully obtained. In addition, a successful hybridization of *C. praetermisum* x *C. tovarii* was also obtained. A cytological investigation of F_1 hybrids of *C. baccatum* x *C. tovarii* revealed that most meiotic chromosomes paired as bivalents. However, multivalents, chromosome bridges and chromosome lags were observed. These results suggest that *C. baccatum* differs from *C. tovarii* by at least a chromosomal reciprocal translocation. Crosses of *C. tovarii* to *C. chinense* and *C. frutescens* produced plump seeds, but none of the seeds germinated. Hybridization of *C. tovarii* to *C. pubescens, C. eximium and C. cardenasii* did not produce seed. These hybridization results indicate that *C. tovarii* is genetically more closely related to the *C. baccatum* complex than to the *C. annuum* complex or the *C. pubescens* complex.

12.2.2 Estimation of Genetic Parameters

The genetic co-efficient of variation measures the range of genetic variability shown by the plant character and help to compare the genetic variability present in various characters (Sanghi *et al.*, 1964). Lush (1949) reported that heritability (broad sense), specifies the proportion of the total variability that is due to genetic causes, or the ratio of the genetic variance to the total variance. Johnson *et al.* (1955) had suggested that heritability estimates along with genetic gain is usually more helpful than the heritability alone in predicting the resultant effect from selecting the best individuals.

In several studies green and ripe fruit yield/plant, dry fruit yield/plant, plant height, plant spread (E-W), plant spread (N-S), primary branches/plant, fruit length, fruit diameter, vitamic C, oleoresisn, capsaicin and beta carotene contents of fruit exhibited high phenotypic and genotypic variation in chilli (Devi and Arumugam, 1999b; Rosmaina *et al.*, 2016, Pandiyaraj *et al.*, 2017, Murmu *et al.*, 2017, Bhutia *et al.*, 2017b; Yogeshkumar *et al.*, 2018, Vaishnavi *et al.*, 2018; Sreenivas *et al.*, 2019). The period taken for flowering, girth of fruit, per cent of seeds to fruit weight and ascorbic acid content showed a high heritability but low genetic advance suggesting the possible role of non-additive gene effects (Jamal Hussain, 1977). High heritability for most of the yield component characters has been observed by many workers (Raju *et al.*, 1984; Shah and Pant, 1986; Ado and Samarrawira, 1986; Meshram, 1987; Depestre and Espisona, 1988; Chattopadhyay *et al.*, 2011; Hasanuzzaman *et al.*, 2012; Bhutia *et al.*, 2017b; Yogeshkumar *et al.*, 2018; Sreenivas *et al.*, 2019). High genetic advance has been reported as well for plant height, number of primary branches, fruit length, fruit width, days to first flower and number of fruits per plant showing the prospect of improving yield by selection for these characters (Shah and Pant, 1986; Meshram, 1987; Chattopadhyay *et al.*, 2011; Bhutia *et al.*, 2017b; Sreenivas *et al.*, 2019). Studies on genetic variability in hot pepper by Jabeen *et al.* (1999) revealed that fruit yield per plant, fruit number per plant, seed number per fruit, pericarp thickness and average fruit weight have fixable additive gene effects and should be considered as useful selection criteria.

12.2.3 Identification of Selection Indices

Correlation measures the mutual relationship between various plant characters and determines component characters on which selection can be based for improvement in yield. Estimation of genotypic and phenotype correlation among various characters may provide information necessary in a breeding programme when selection is based on two or more characters simultaneously. Selection of one trait results in correlated response in several other traits. Path analysis is simply standardized partials regression co-efficient which splits the correlation co-efficient into the measures of direct and indirect effects of a set of independent variables on the dependent variable.

The set of characters identified as selection indices by different workers based on genetic variability parameters, correlation and path coefficient analysis is presented in Table 18.

Table 18: Different Set of Characters as Selection Indices in Chilli

Characters to be Considered in Selection	*Reference*
Fruit length and fruit girth	Singh and Singh (1974)
Height of plant, number of primary branches, length of fruits, girth of fruit, pericarp thickness, seed content and number of fruits	Jamal and Hussain (1977)
Fruit numbers/plant, fruit weight and plant height	Korla and Rastogi (1977);
Fruit weight, fruits/plant and fruit width	Despestre *et al.* (1989)
Fruits/plant, Fruit weight and dry yield/plant.	Sahoo *et al.* (1990).

Characters to be Considered in Selection	Reference
Fruits/plant, fruit weight and fruit length	Kataria *et al.* (1997), Devi *et al.* (1999).
Total dry weight, leaf area index, leaf number and plant height	Aliyu *et al.* (2000)
Fruits/plant, days to first fruit harvest and fruit weight	Munshi *et al.* (2000), Jose *et al.* (2002)
Fruits/plant, days to first fruit harvest and fruit weight	Jose *et al.* (2002)
Fruits/plant	Nandadevi and Hosamani (2003)
Fruits/plant, green fruit weight and dry fruit weight	Choudhary and Samadia (2004)
Fruits/plant, Fruit weight, yield/plant and fruit length.	Sreelathakumary and Rajamony (2004)
Yield/plant and fruit/plant.	Singh *et al.* (2005)
Fruits per plant, fruit weight, primary branches/plant, fruit length, fruit diameter and plant height	Smitha and Basavaraja (2006)
Fruits/plant, fruit weight, fruit length and 1000 seed weight	Sarkar (2006)
Number of fruits/plant, green fruit length for green chilli, weight of dry fruit and the number of fruits per plant for dry chilli	Chattopadhyay *et al.* (2011)
Fruit weight, yield/plant and fruit length	Chaudhary *et al.* (2013)
Plant height, fruit length, number of fruits plant-1, fruit weight and fruit yield	Amit *et al.* (2014)
Number of fruits/plant	Bhutia *et al.* (2015)
Fruits/plant, fruit weight, number of branches, fruit length, fruit diameter and plant height	Bharadwaza *et al.* (2015)
Fruits per plant, coupled with large fruit length, high fruit girth and fruit weight	Rohini and Lakshmanan (2016)
Plant spread, fruit length and fruit diameter	Vidya *et al.* (2018)
Ripe fruit yield/plant and dry fruit weight	Sreenivas *et al.* (2019)

Among the intercorrelations, the association between number of branches and number of fruits per plant was positive and a similar trend was indicated between fruit girth and number of fruits per plant and between fruit girth and fruit length (Balakrishnan, 1980). The number of fruits was positively related to the number of flowers produced between 66 and 86 days of crop duration (Suthanthirapandian and Sivasubramanium, 1978). Capsaicin content is negatively correlated with fresh fruit weight (Jiang *et al.*, 1987), fruit diameter, length and thickness (Fuentes and Mora, 1989). The direct and indirect effects of the various yield components had been identified (Singh and Singh, 1974; Gill *et al.*, 1974; Korla and Rastogi, 1977).

12.2.4 Genetic Diversity, Combining Ability and Gene Action

Plant breeding is the creation of evolution at the will of human kind and is dependent on genetic variation (Frankel and Benett, 1970). In fact, the achievement in this direction depends as genetic diversity available for the concerned crop. Among the several statistical methods developed for measuring the divergence between populations, multivariate analysis has been a useful tool in the quantitative estimation of genetic diversity. The value of multivariate analysis (Mahalanobis's D^2 statistics, 1936) has well been demonstrated for the assessment of genetic diversity in biological populations and choosing the parents to be hybridized through reduction of large amount of data about the parents to manageable proportion. The genetic diversity of chilli through multivariate analysis has been done by many workers (Sarkar, 2006; Krishnamurthy *et al.*, 2013; Hasan *et al.*, 2014; Bhutia *et al.*, 2017b; Vanitha and Jansirani, 2017; Sreenivas *et al.*, 2020).

Genetic Diversity in Chili Pepper Fruits from *Capsicum* Gene Bank of Embrapa Temperate Agriculture (Photo: Henrique Padilha).

The combining ability concept proposed by Sprague and Tatum (1942) is a measure of gene action. According to them, general combining ability (GCA) is the average performance of a parent across all cross combinations and specific combining ability (SCA) is the deviation of specific cross from the average of the parents involved. Genetic diversity, together with combining ability, is a criterion for selection of parents in developing of a hybrid. Low magnitude of heterosis occurs when cross involves very closely related parents, but high heterosis is the consequence of crossing between parents of intermediate divergence (Geleta and Labuschagne, 2006; Bhutia *et al.*, 2015). Analysis of combining ability of sets of inbred lines plays an important role in chili breeding for inbred line selection, heterotic group classification, heterotic pattern identification, and hybrid development (Gaddagimath *et al.*, 1988; Do Rego *et al.*, 2009; Bhutia *et al.*, 2015; Sreenivas *et al.*, 2020).

Muthukrishnan *et al.* (1983) observed a narrow ratio between GCA and SCA for most of the characters, *viz.* plant height, number of branches, earliness, fruit length, fruit girth, number of fruits per plant and 1000 seed weight which suggested the possible involvement of both additive and non-additive gene action with a predominance of non-additive gene action. For the improvement of these characters one or two cycles of recurrent selection could be resorted to (Muthukrishnan *et al.*, 1983). Rao and Chhonkar (1984) reported that fruit weight was controlled by non-additive gene effects. Ribeiro and Costa (1990) also stated that capsaicin content showed predominance of additive gene effects and the magnitude of narrow sense heritability indicated that the capsaicin content can be effectively changed by simple selection. From a study on gene action of yield components in F_1 generation of chillies, Devi and Arumugam (1999a) reported that additive gene action was more important than non-additive gene action for all the yield components studied except fruit length. In green chilli Bhutia *et al.* (2015) observed overwhelming response of non-additive gene action for the control of number of primary branches per plant, days to 50 per cent flowering, fruit length, fruit width, number of seeds per fruit, test weight of seed, vitamin C, PDI of leaf curl virus and fruit yield per plant, hence selection will bring no or slow genetic improvement. The study of predictability ratio (Baker, 1978) in 6 x 6 diallel mating for dry chilli improvement (Sreenivas *et al.*, 2020) exhibited a desired response for non-additive gene action in controlling expression of most traits including dry fruit yield per plant, but not for fruit length, fruit diameter, and capsaicin content which were controlled by additive and non-additive gene action, and dry fruit weight, percent disease index (PDI) of leaf curl virus controlled by additive gene effect.

The successful breeding methods will be those that accumulate the genes to form superior gene constellations interacting in a favorable manner. The importance of non-additive gene action for most of yield components indicates heterosis breeding to be the best possible option for improving these traits in chilli. For the control of both additive and non-additive gene action for certain characters, some form of recurrent selection, such as diallel selective mating or bi-parental mating in early segregating generations might prove to be effective approaches. Restricted recurrent selection by intermating the most desirable segregants followed by selection might also be a useful breeding strategy for the exploitation of both additive and non-additive type of gene action. Due to the predominance of additive gene action in certain characters, selection of such traits should be done in later generation when the effects of non-additive gene action have been minimized and those of additive gene action effects have been fixed.

12.2.5 Inheritance Study

Pendant fruit character is monogenically dominant, while the erect fruit character is recessive (Sayed and Bagavandas, 1980). Crosses between the mutants Cr. 197-57-8-80-11 with marcescent (withered and ephemeral) calyx, and IHR 354-28-1 with erect habit, indicated that each character is controlled by a single independent recessive gene (Deshpande *et al.*, 1986). Shifriss (1987) reported independent assortment between male sterility gene (ms), erect fruit position gene (*up*) and fruit anthocyaninless (*ae*) gene. According to Loaiza-Figueroa and Tanksley (1988) absence of capsaicin was controlled by a single recessive gene (*c*). The umbrella phenotype is controlled by three major recessive genes, *et* and *dt* determining plant growth habit and *fb* determining fruit bearing habit (McCammon and Honma, 1984). Long pedicel was partially dominant over short and polygenic in inheritance (Subramanya and Ozaki, 1980). Probably three major dominant genes control double flowers but more genes are required to produce 3 to 4 flowers per node (Subramanya, 1983).

Inheritance of 10 quantitative traits related to plant and fruit development was studied in an intraspecific cross between a bell-type *C. annuum* (Grossum Group) cv. Maoro and a small fruited pungent chilli line (*C. annuum* var. *annuum* Longum group) cv. Perennial. Estimates of broad and narrow sense heritabilitics, coefficients of genetic variance and genotypic correlations were obtained from the segregation of 120 F_1 families grown at Qiryat Gat, Israel during summers of 1996 and 1997. Days to first ripened fruit, plant height and pedicel length exhibited heterosis and transgressive segregation. Days to first ripened fruit and total soluble solids had low narrow-sense heritabilities. The other traits studied had moderate to high narrow-sense heritability estimates. Most of the genetic variation associated with traits that affect the size of the fruit and its shape was additive. The highest genetic correlation coefficients among pairs of traits were found between fruit weight and each of the 3 width characters: fruit diameter, pericarp thickness and pedicel diameter. In contrast, fruit weight had a low correlation coefficient with fruit length, indicating that the size of the pepper fruit in this cross was determined primarily by its width (Ben-Chaim and Paran, 2001).

The inheritance of mature fruit colour in peppers (*Capsicum* spp.) is controlled by several genes. However, the inheritance of the transition of colours the fruit undergo during ripening has not been described extensively. The jalapeno cultivar NuMex Pinata (NP) is unique in the transition of colours and fruit undergo as they mature (light green, through yellow and orange, and finally red). The inheritance of a unique gene which affects foliage colour and fruit colour transition was studied in reciprocal crosses between Early Jalapeno (EJ) and NP. Foliage and fruit colour transition phenotypes of the entire F_1 population were that of EJ, regardless of whether EJ was the female or male parent, indicating that the NP phenotype is recessive and displays no maternal effects. Data from the segregating F_2 population indicated that the inheritance of foliage colour and fruit colour transition phenotype of NP is due to single homozygous recessive gene. The gene responsible is designated Ira (Votava *et al.*, 2000).

Genetics of resistance to different diseases and nematodes have been studied by several workers and summarized in Table 19.

Table 19: Genetics of Biotic Stress Resistance in Chilli*

Disease	*Mode of Genetic Control*
Anthracnose	Recessive, additive, dominant and epistatic effects are important
Cercospora leaf spot	Recessive, three complementary genes, additive and dominance being important
Damping off	Digenic
Bacterial leaf spot	Monogenic dominant
Soft rot	Oleigogenic
Mosaic (TEV)	Monogenic and dominant
Mosaic(TMV)	Monogcnic and dominant
Root knot nematode (*M. incognita*)	Monogenic and dominant

* Peter and Pradeepkumar (1999).

12.3.0 Hybridization

The commercial cultivars K-2, G-5, Pusa Jwala, Pant C-1 and Pant C-2 are examples of hybrid derivatives in commercial cutivation. Yellow anther in chilli is a rare character and this was transferred to G-4 as a marker gene by crossing with yellow anther mutant and selecting G-4 like plants with yellow anthers in F_2 and in subsequent generations. X 235 has been developed which is superior to G-4 in yield and fruit character (Srirarnachandramurthy *et al.*, 1983). The selection G-3 satisfies the trade requirement of Sri Lanka and yields heavily. X 200, a hybrid involving this line with Lavang Mirche and G-2 and X 197 involving G-3 and Assam chillies are superior to G-3 in yield, pungency and are also resistant to thrips which can replace G-3. To suit the needs of the United States market, chilli pods without calyx and with stalk of medium size with red pericarp have to be developed. The Japanese lines Huntaka and Santaka combine these attributes in them but are less preferred for their shy bearing. X 206, a hybrid developed by crossing G-3 and Huntaka, recorded yield similar to G-3, besides inheriting the red colour and shape of Huntaka. G-5 is a thrip resistant hybrid involving G-2 and B-31 (Murthy and Bavaji, 1980). The hybrid Pusa Jwala, product of NP 46A and Puri Red is claimed to be mosaic virus resistant. In Tamil Nadu, X 5, a hybrid from a cross between Sathur Samba and B 24 A2 yields 22.8 per cent higher than K-1 developed at Tamil Nadu from a cross between 1370 A and Sathur Samba is tolerant to thrips and fruit rot. The hybrid derivatives Chamakkar, Puri Red and Puri Orange are reported to be resistant to leaf curl and mosaic virus. A line designated LCA 206 with good colour retention in storage was developed from cross G-3 x Huntaka (Murthy *et al.*, 1987). Rising of hybrid seeds is not very difficult. The techniques for doing emasculation and hybridization are very simple and they are as follows.

12.3.1 Emasculation

Emasculation may be done either early in the morning or in the previous evening before the flowers open and calyx are still covering the anthers and stigma. With the help of a pair of forceps, the petals are easily parted and anthers are removed and bagged. If there is no intention of doing any emasculation on the following 1 or 2 days, all inoperated buds which are likely to open on these days must be removed and the whole plant covered with a muslin cloth bag.

12.3.2 Hybridization Technique

Late in the morning or early in the afternoon of the following day, fresh flowers should be plucked from intended male parent which have been previously bagged and pollen dusted on the stigmas of the emasculated flowers. The pollen is dry and powdery and easily dislogged from the anthers with a gentle shake. To facilitate pollination, the petals of the pollen flowers may be cut off. After pollination the bag should be replaced and labels denoting details of parents and date of pollination should be tied outside the bag. Crossing is generally found to be successful when the plant is in full bloom and natural setting is high. Crossing should be avoided as far as possible during cloudy weather as setting is very low in such conditions.

12.4.0 Hybrid Vigour

The experience on the possiblility of exploiting the hybrid vigour and heterosis in chilli has shown considerable promise and work in this respect involving intervarietal crosses need to be emphasized. The identification of male sterile lines like G-2 (Murthy and Lakshmi, 1981) indicated the feasibility of hybrid seed production. Earlier, a male sterile mutant in *C. annuum* has also been reported by Shifriss (1973). Pushpa and Shambhulingappa (1981) isolated spontaneous male sterile plants from a field population of Jwala. It exhibited a more vigorous growth than normal plants and the flowers were shrunken with a multilocular ovary. The present interest on exploitation of heterosis is an encouraging trend in chilli improvement even though hybrid sweet peppers have been common with growers. The genetic base of chilli is large and more number of unexploited cultivars still exist (Heiser Jr., 1976).

Balakrishnan *et al.* (1983) showed promise for F_1 hybrid seed production for increased yield. Earlier report by Lippert (1975) has shown that yield components are generally involved in increasing the yield of chilli hybrids but the number of fruits per plant tended to be the most important contributory character towards higher yield.

Mak (1987) observed significant heterosis for yield in hybrid Irian x CM 31 and CPK x 0110-1. Ashtankar and Jaipurkar (1988) reported that the hybrid G-4 x Pant C-1, X 235 x Jawahar 218 and Jwala x K-2 out yielded the improved standard variety Jawala quite significantly. Cao and Su (1988) observed heterosis in hot pepper for early and total yield whereas period and disease index exhibited negative heterosis. Heterosis was particularly high for dry fruit yield per plant, especially in crosses J 218 x CA586 (110.4 per cent heterosis over the better parent), Pusa Jwala x Sindhur (98.1 per cent) and BR Red x G-4 (89.8 per cent) (Mishra *et al.*, 1989).

In green chilli Bhutia *et al.* (2015) observed heterobeltiosis for fruit length, number of fruits per plant, number of seeds per fruit, vitamin C content of fruit, capsaicin content of fruit and PDI of leaf curl virus in the cross 'BCCH Sel-4 x AC-575'. Hybrids 'BCC-30 x G-4' and 'BCCH Sel-4 x Beldanga' were considered promising in dry chilli for simultaneous improvement in dry fruit yield, fruit quality and leaf curl virus disease tolerance (Sreenivas *et al.*, 2020).

12.5.0 Disease Resistance

Phytophthora capsici Leonian is one of the most destructive pathogen of peppers which attacks all parts of the plant, causing symptoms such as root rot, crown rot, stem lesion, fruit rot, damping-off of seedlings and so on (Ristaino, 1990; Quirin *et al.*, 2005). Resistant sources for this pathogen have been identified by many researchers, such as 'CM334' from Mexico, 'AC2258', 'PI201232' and 'PI201234' from Central America, among which 'CM334' has found to get the highest resistance level (Bosland and Lindsey, 1991; Ortega *et al.*, 1992; Quirin *et al.*, 2005). Thabuis *et al.* (2003) mapped QTL for *P. capsici* resistance in three *C. annuum* populations derived from independent resistance and susceptible parental lines. Of the 18 QTL detected, one major loci on chromosome 5 was common in all the three populations, while another locus was common to two populations and the remaining were cross-specific.

KAU Cluster, White Kandari and Pant C-1 was reported to be resistant to many of the races of *Pseudomonas solanacearum* (Hibberd *et al.,* 1984) whereas Pant C-1 was found to be resistant to four *P. solanacearum* isolates and moderate resistance to *Phytophthora capsici* (Peter *et al.,* 1984).

Boukema (1980) reported resistance to TMV in *C. chinense*. Virus resistant line PSP- 11 has been developed in India from *C. frutescens*. It is resistant to cucumber mosaic cucumovirus, potato X potexvirus and tobacco mosaic tobamovirus (Tewari and Viswanath, 1988). Field screening of chilli cultivars to viral diseases in India revealed that cvs. JCA 248, JCA 218, PC-1, NP 46-A, Pusa Jwala and JCA 196 were resistant to leaf curl virus (Sangar *et al.,* 1988). In Indonesia, cvs. Cipanas and Indragiri were reported to be resistant to *Meloidogyne* spp. (Marwoto, 1988). Field reaction of chilli germplasm to cucumber mosaic virus, potato X virus, potato Y virus, tobacco mosaic virus and chilli leaf curl virus showed that one or other type of disease occurred in all the genotypes except 76262 (Rishi and Dhawan, 1988).

Anthracnose, fungal disease caused by *Colletotricum* genus, originates losses in *Capsicum* peppers production, both before and after harvest, in countries around the world such as China, South Korea, United States and Brazil (Mahasuk *et al.,* 2013). Breeding programs have been developing methods that aimed to identify and incorporate resistance genes in peppers cultivars (Pereira *et al.,* 2011). However, understanding about resistance genes control and inheritance for anthracnose is not clearly yet. Sun *et al.* (2015) suggests different kinds of genes (QTLs) that control resistance to immature and mature fruits. Park *et al.* (2012) evaluating healthy and deseased fruits (*Capsicum annuum*) by chromatography, identified molecules acted as phytoalexin and helped in response against the fungus. Silva *et al.* (2014) found accessions of *C. baccatum* var. *pendulum* species promising to be introduced in breeding programs (UENF 1718 e UENF 1797). Two SNP maps were constructed from two chili populations including *Capsicum annuum* 'Bangchang' x *C. chinense* 'PBC932', and *C. baccatum* 'PBC80' x 'CA1316', aiming to identify QTLs for anthracnose resistance. The validated SNPs will be greatly beneficial to select for the anthracnose resistance in chilli breeding programs (Mahasuk *et al.,* 2016).

Number of species showing resistance against different diseases and nematodes are presented in Table 20.

13.0 Biotechnology

Capsicum (pepper and paprikas) are grown all over the world and are predominantly tropical while paprikas are grown in cooler climates. *Capsicum* has been well exploited biotechnologically and cultivars derived from anther culture were grown in an area of 38,400 ha in China (Li *et al.,* 1995). One of the promising varieties developed through anther culture was Haihua 3. Many diseases caused by fungus and viruses are likely to be controlled by the development of transgenies carrying genes for resistance. Parthasarathy and Niral (2001) presented a detailed review on biotechnology of *Capsicum*.

Table 20: Different Species of Capsicum Showing Resistant to Biotic and Abiotic Stresses*

Species	*Biotic and Abiotic Stresses*
C. chinense	TMV, poly virus, root rot, tomato spotted wilt virus, potato virus X, drought, aphids, thrips
C. chacoense	TMV, bacterial spot, poty virus, anthracnose, aphids,Thrips
C. baccatum var. pendulum	CMV, potato virus Y, powdery mildew, drought, highTemperature
C. ciliaum	Root rot, water logging.
C. eximium	Potato virus Y, leaf curl, drought, high temperature.
C. flexuosum	CMV, potato virus X, soil salinity.
C. cadenasii	Aphids, thrips, low temperature.
C. pubescens	Powdery mildew
C. microcarpum	Powdery mildew

* Dasmunshi (1999).

13.1.0 Micropropagation

Micropropagation in *Capsicum* has been reported with shoot buds, nodes, axillary buds and shoot-tips. This method offers scope for maintaining and propagating male sterile lines and for cloning *Capsicum* genotypes.

Swamy (1983) reported that generation of shoot buds and multiple shoots occurred from cotyledon and shoot-tip explants, respectively of *C. annuum* cultured on MS medium supplemented with various growth regulators of which BA was the most effective for regeneration. IAA, NAA and IBA induced roots from explants.

Sun and Wang (1989) developed a technique for rapid mass propagation by *in vitro* culture of stem meristems using *C. annuum* cv. Qiemen, *C. frutescens* form 310, *C. praetermissum* form 234 and *C. baccatum* form 85008 on supplemented MS medium. All the explants of different species cultured on optimum medium for 3-4 weeks developed into shoots with 3-4 leaves. The best explants were those excised from 3-week-old seedlings, Regeneration of plantlets through micropropagation has been reported using nodal segment hypocotyl leaf and cotyledon (Christopher, 1996), apical and axillary bud (Gupta *et al.*, 1998).

Bhutia *et al.* (2016) reported an efficient *in vitro* regeneration protocol for a chilli cultivar of Sikkim namely Dalle Khursani (*Capsicum annum* L.). MS medium containing 4 mg/l Thidiazuron (TDZ) showed the best result multiple shoot regeneration and among the explants cotyledons showed the best response. Regenerated shoots elongated and rooted well on MS medium containing 2 mg/l GA_3 + 1 mg/l IAA.

13.2.0 Anther Culture

Investigations have been done on anther culture of capsicum in order to produce dihaploids with many desirable traits.

13.2.1 Genetics of Anther-Cultured Plants

Genetic expression of major characters in sweet pepper lines derived by anther culture has been studied by many workers. Chen (1984) found in the study of 74 H_2 and 5 H_3 lines of *C. annuum* var. *grossum* derived by anther culture that there was considerable variation in the botanical and agronomic characters between lines derived from the same parent. Characters were uniform within lines and between generations of the same line. More than 30 per cent of the lines were superior to the original material induced by anther culture in total yield and 28 per cent had a lower disease index. Some 20 per cent were superior to the control Tongfeng 37.

The main fruit characters of lines derived by anther culture from a sweet x hot, *C. annuum* F_1 hybrid and grown on up to 5 generations (H_1-H_5) were inherited from the parent and plants in H_2-H_5 generations were uniform (Jiang and Li, 1984). A new sweet pepper (*C. frutescens* var. *grossum*) cultivar Haihua 3 was developed by *in vitro* culture from uninucleate pollen of the breeding line 78203. Li and Jiang (1990) reported that Haihua 3 has a wide adaptability in China and a high tolerance to diseases. Li *et al.* (1995) developed 6 cultivars directly from anther culture and a further 6 hybrids with parents developed from anther culture have been cultivated on over 38,400 ha throughout China.

Anther culture has been used to develop double haploid lines showing resistance to different diseases and pests. Abak *et al.* (1982) evaluated a total of 27 doubled haploid *C. annuum* plants from the cross PM 217 (resistant) x Yolo Wonder (susceptible) for both root and stem resistance to *Phytophthora capsici*. There was a continuous gradation between complete susceptibility and resistance and a good correlation was observed between the results of both tests.

Vaulx *et al.* (1982) used a total of 212 doubled haploid lines obtained from six heterozygous (L+Ll) *C. annuum* hybrids between plants susceptible (L+/L+) and resistant (Ll/Ll) to tobacco mosaic virus (Po pathotype) and found that the resistance: susceptibility segregation ratio was 1: 1. Selassie *et al.* (1986) produced 4 homozygous lines by anther culture, resistant to 3 strains of potato virus Y and one of pepper veinal mottle virus (PVMV) from the F_1 of Perennial x Florida.

Hwang *et al.* (1998) screened 33 breeding lines of peppers for resistance to bacterial spot (*Xanthomonas campestris* pv. *vesicatoria*). Plants derived from anther culture of resistant lines were screened against and analyzed for inheritance of resistance. Nine lines including PI163184 were resistant to the disease and PI183439 and PI244669 were moderately resistant.

13.2.2 Characterization of Anther-Cultured Plants

Anther culture is an efficient tool for producing viable haploids and generating completely homozygous plants in *C. annuum*. However, spontaneous doubling of chromosomes is also common. Wang *et al.* (1991) cultured anthers of 19 cultivars subjected to anther culture and only Squarehead and Lu-tai produced embryoids from which haploid plantlets developed. Diploidization which was induced by treatment with 0.2 per cent colchicine for 24 h, resulted in flowering and seed set. Barcaccia *et al.* (1999) obtained sixty-nine androgenic plants from 4 accessions by culturing anthers *in vitro* under carefully controlled conditions (on an average 4.7 plantlets per 100 anthers cultured). Thirty-three of these plants spontaneously underwent chromosome doubling and were cytologically proven to be doubled haploids.

Gyulai *et al.* (1999) used the spontaneous doubled haploid (DH-R_1) plants developed from F_1 hybrids of 'blocky type' pepper (2n = 24) in anther culture and advanced generation DH-R_2 in a green house for flow cytometric analysis, which revealed plants with haploid (n) and diploid (2n) genomes. DNA samples of DH-R_2 individuals were pooled in 9 groups and subjected to RAPD, SSR and ISSR-PCR analysis. All 9 pools of DH-R_2, plants were discriminated. In total, the 26 primers generated 54 scorable PCR bands. Of these 19 (35.2 per cent) were polymorphic.

13.2.3 Factors affecting Androgenesis

The success of androgenesis depends on genotype, pretreatment and supplementation of media. Using young mother plants and frequent subculturing was found to increase the androgenic yield in different *C. annuum* gentoypes (Kristiansen and Andersen, 1993). Mityko *et al.* (1995) used fifteen genotypes (four breeding lines, seven cultivars and four F_1 hybrids) and established a new system for qualifying the androgenic response. For use in practical breeding, a minimum of 5 per cent of plant regeneration was proposed as the criterion for a fair response. Accordingly, one excellent, one good and eight fair responses were identified among the genotypes investigated.

The importance of genotypic influences in tissue dedifferentiation and *in vitro* callus proliferation has been well understood (Novak, 1974; Ionescu *et al.*, 1989; Matsubara *et al.*, 1998). Mityko and Fari (1997) studied the anther culture response of more than 500 pepper cultivars, breeding lines and F_1 hybrids. A poor androgenic response was observed in waxy, dark green block-type, tomato-shaped pepper and Cayenne/chilli- type pepper. The Hungarian cv. Feherozon showed an excellent response (>30 per cent plant regeneration) with the highest frequency of responsive anthers (48.3 per cent) and the largest number of regenerated plants (75.8 per cent).

Prechilling of flower buds has been recommended for success in anther culture. Morrison *et al.* (1987) excised anthers from prechilled flower buds of *C. annuum* cv. Emerald Giant x *C. chinense*, cultured in the dark at 35°C for 8 days and 25°C for 4 days and then floated on liquid medium overlaying a solid medium supplemented with 2.0 per cent charcoal. Embryos developed from microspores below the surface of the liquid on the charcoal medium. These embryos, when transferred to fresh culture medium without charcoal, developed into haploid and diploid plantlets. 2,4-D alone or in combination with cytokinin is commonly used for induction of embryoids. The response varies between genotypes. Yoon *et al.* (1991) cultured anthers in the mid- uninucleate to early bi cellular stage, from 5 F_1 hybrids of hot pepper on Dumas de Vaulx basic medium supplemented with various growth regulators. Efficiency of anther culture was dependent on donor genotype and induction of embryoids from microspores and was best on medium containing 0.1 mg Kin and 0.1 mg 2,4-D/l or medium containing 0.1 mg 2,4-D and 0.1 mg

BAA/I. Thidiazuron (TDZ) has also been reported to be very effective in inducing androgenesis. The best results on callus induction were seen in the hybrid with *C. chinense* on MS medium with TDZ on morphogenesis from callus in the hybrid with *C. eximium*, and in direct embryoid induction in Kalinkov and Byala Kapiya on MS medium with carrot extract (Pandeva *et al.*, 1990).

13.3.0 Protoplast Culture

Protoplasts can be isolated from cotyledons, hypocotyls, leaf mesophylls and callus tissues. Viable protoplast was isolated from leaf and cotyledon explants of *Capsicum* (Saxena *et al.*, 1981; Diaz *et al.*, 1988; Murphy and Kyle, 1997). Saxena *et al.* (1981) isolated protoplasts from 15-day-old leaves in an MS solution containing cellulase, macerozyme and mannitol. Isolated protoplasts were cultured in TN media supplemented with 2,4-D, NAA, BA, sucrose and mannitol. About 5 per cent of the protoplasts regenerated cell wall and formed callus in 6 weeks. Subsequent transfer of the callus to the callus to shoot and root regeneration media gave rise to plants. Successful isolation of protoplasts and its subsequent regeneration depended on previous growth conditions of source plants and concentration of mannitol used. Along with this, light intensity also played an important role.

13.4.0 Somatic Embryogenesis

Somatic embryogenesis in *Capsicum* has been induced directly and indirectly *via* callus. Various explants such as cotyledons, leaves, embryos, *etc.* have been used for indirect embryogenesis. But for direct embryogenesis only embryos have been used.

A protocol was developed for the regeneration of pepper cv. Ace through somatic embryogenesis in liquid medium (Buyukalaca and Mavituna, 1996). For embryo-genie callus formation, sections of mature zygotic embryos as explants were placed on basal MS medium supplemented with 9.05 µM 2,4-D and 3 per cent sucrose. Embryogenic callus was transferred to liquid basal MS medium supplemented with 4.52 µM 2,4-D and 3 per cent sucrose. Embryos were matured in half-strength MS medium containing 1.89 µM ABA, and converted into plants both in vivo and in vitro at up to 97 per cent efficiency.

Direct somatic embryogenesis and plant regeneration in pepper has been reported by some workers. Harini and Sita (1993) used immature zygotic embryos for direct embryogenesis and developed a protocol for in vitro regeneration of plants. Binzel *et al.* (1996) reported that plants were regenerated from immature zygotic embryos *via* direct somatic embryogenesis. Somatic embryos were formed directly, without any intervening callus, on the zygotic embryo apex, embryo axis and cotyledons on MS medium containing 2,4-D (4-18 µM) and TDZ (10 µM) and a high concentration of sucrose (6-10 per cent).

13.5.0 Production of Secondary Metabolites

The main secondary metabolite in capsicum is capsaicinoid and pungency is due to the presence of capsaicin. It is the secondary metabolite and main capsaicinoid produced by capsicum (Bhutia *et al.*, 2018)

Plant cell culture has been employed in Capsicum for the production of secondary metabolite like capsaicin (Yeoman *et al.*, 1980; Chavez-Moctezuma and Lozoya-Gloria, 1996). Identification of cell lines that produce the required secondary metabolite in large quantities is crucial (Yamada and Fujita, 1983). Yeomen *et al.* (1980) manipulated culture media to increase the production of capsaicin from *C. frutescens* and reported an increase in the capsaicin by adding precursor-like vanillylamine and isocapric acid, eliminating sucrose from the cultural media and decreasing the level of N to 5 per cent in the MS solution.

Nunez-Palenius and Ochoa-Alejo, (2005) conducted an experiment to estimate the effect of phenylalanine, cinnamic or caffeic acids, vanillin, vanillylamine, p-coumaric acid and ferulic acid on the total production of capsaicinoids in cell culture of *Capsicum annuum* cv. Tampiqueno-74. MS liquid medium supplemented with 6.25 µM 2, 4-D, 0.66 µM BA and 3 per cent sucrose was used for cell suspension culture. 100 µM concentration of each of the treatment was added to estimate their effects on capsaicinoids production. The highest total capsaicinoid content was recorded in cultures grown in the presence of vanillin followed by cells treated with vanillylamine, p-coumaric acid and ferulic acid, while cell cultures treated with phenylalanine and cinnamic or caffeic acids did not influence capsaicinoids production

Somaclonal Variation in Chilli (Grozeva and Todorova, 2015).

SA (Salicylic acid) and MeJa (Methyl jasmonate) treatment to *Capsicum chinense* cell culture provoked an increase in capsaicinoid accumulation. The cultures were exposed to 5, 50 and 500 µM of either SA or MeJa for 24 hours. Both treatments lead to the accumulation of capsaicinoids. The response was better on culture with SA treatment. The response increases with dose, reaching its maximum in cultures exposed to 500 µM SA (Gutierrez-Carbajal *et al.*, 2010).

Kehie *et al.* (2013) reported that osmotic stress induced capsaicin production in cell cultures of *Capsicum chinense* Jacq. cv. Naga king chilli. Hypocotyl explant was cultured on MS media supplemented with 2 mg/l 2,4-D and 0.5 mg/l Kinetin for induction of callus which was then used for suspension cultures to estimate the effects of different osmoticum. Addition of sucrose, mannitol and NaCl enhanced the accumulation of capsaicin in suspension cultures of King chilli. Among the various concentrations of sucrose, mannitol and NaCl treatment, 87.6 mM sucrose, 164.68 mM Mannitol and 40mM NaCl showed significant production of capsaicin. However, addition of NaCl on media containing 87.6 mM sucrose showed decreased level of capsaicin accumulation.

Ancona-Escalante *et al.* (2013) exposed *in vitro* cultured placental tissues of *Capsicum chinense* jacq. to SA and MeJa treatment. Both treatments resulted in increased accumulation of capsaicinoids, and the induction by SA was more pronounced. Maximum response was observed in tissues exposed for 24 hrs to 500 µM SA, and no further increase was observed after a longer exposure.

The plants regenerated from the explants that are treated with elicitors like Sa and MeJa could be producing higher amount of capsaicinoids including capsaicin. The protocol to regenerate the plants from cells or tissues which has been treated with SA was standardized in Dalle Khurasani (*C. annuum* L.) (Bhutia *et al.*, 2018). However, the amount of capsaicin in fruits obtained from those plants were not determined.

13.6.0 Biochemical and Molecular Markers

Biochemical markers, which include the variation of enzymes called isozymes, are based on the principle that allelic variation exists for many different proteins. Isozymes are protein molecules that are separated electrophoretically based on their charge. Gels are stained for specific enzyme activity and, by doing so, allelic and non-allelic proteins can be identified. The use of isozymes in pepper research has focused predominantly on measuring genetic variability and clarifying systematic and phylogenetic relationships in the Capsicum genus.

The introduction of biochemical techniques like Sodium dodecyl sulphate polyacrylamide gel electrophoresis (SDS-PAGE), isozyme markers has been particularly helpful in deducing systematic relationships between groups where morphological and cytological data were not corollary. SDS-PAGE is an economical, simple and extensively used technique for describing the seed protein diversity of crop germplasm. Yatung *et al.* (2014) evaluated thirty chilli genotypes SDS-PAGE, isozyme markers and observed considerable variation in banding pattern of total protein.

A major shortcoming of isozyme analysis is the small number of isozyme loci available.

Wang and Ma (1987) undertook analytical eletrophoresis on Capsicum germplasm. Based on the electrophoretic pattern of peroxidase isozymes extracted from a functional leaf at the flowering stage, they assigned 8 species into 4 groups: group A containing *C. annuum, C. frutescens* and *C. chinense*; group B containing *C. chacoense* and *C. pubescens*; group C consisting of *C. praetermissum*; and group D containing *C. baccatum* and *C. eximium*. Eighty accessions of *C. annuum* var. *annuum*, most of them Chinese land races, could be grouped into 3 types: 73 were of type A, 5 of type B and 2 of type C.

Andrzejewski (1992) suggested the application of isozymes as genetic markers in the breeding of pepper based on the results of isoenzymatic analyses of superoxide dismutase, glutamate oxaloacetate transaminase (aspartate aminotransferase), phosphoglucomutase, and isocitrate dehydrogenase and shikimate dehydrogenase in *Capsicum annuum, C. baccatum* and *C. chacoense* and progeny of their crosses.

13.7.0 Molecular Markers

Molecular markers are DNA-based genetic markers that can detect polymorphism both at the level of specific loci and at the whole genome level (Bhutia *et al.*, 2018). Molecular markers have proven invaluable towards overcoming limitations of morphological and biochemical markers. The major advantages of molecular markers are it can distinguish individuals without relying on complete development of the plant (*i.e.* tissue from a plantlet may be analysed rather than waiting for the plant to exhibit some morphological feature) and the environment does not alter its expression. Molecular markers are used to characterize the genetic structure of individuals and populations. When markers are identified with a gene or genes of interest, the marker(s) can be used as a selection criterion by plant breeders. They are used in combination with other markers to construct genetic maps, and are used in linkage studies. Selection via molecular markers eliminates the need for costly and sometimes inefficient screenings, and speeds up the process of cultivar development.

Advances in the science of genomics has led to the identification of thousands of DNA markers in pepper which include DNA based Restriction Fragment Length Polymorphism (RFLP), Amplified Fragment Length Polymorphism (AFLP), Random Amplified Length Polymorphic DNA (RAPD), Cleaved Amplified Polymorphic Sequence (CAPS), Simple Sequence Repeats (SSR), Inter Simple Sequence Repeats (ISSR) and more recently, Single Nucleotide Polymorphisms (SNPs). These molecular markers are used for DNA fingerprinting, cultivar identification, genome mapping and finding the genetic loci/quantitative trait loci (QTL) for economically important traits in pepper.

In peppers, several inter and intra specific genetic maps covering the whole *Capsicum* genome have been developed using wide range of markers (Table 21). This genetic map helps to dissect capsicum genome into many parts and associate with the economically important traits by QTL mapping or using the Candidate gene approach.

Several researchers reported mapping and identification of many QTL/ candidate genes in peppers for fruit size, shape and color, male sterility system and

Table 21: Genetic Linkage Map Constructed in *Capsicum* by using DNA Based Molecular Markers

Population/Cross Type	*Mapping Population*	*Population size*	*Number and Type of Markers Tested*	*Number of Linkage Groups*	*Map Length (cM)*	*Average Marker Distance (cM)*	*Reference(s)*
Inter specific linkage map							
C. annuum cv. Doux des lands (CA50) x *C. chinense* cv. CA4	BC_1	46	85 RFLP	14	–	–	Tanksley *et al.* (1988)
C. annuum "CA133" x *C. chinense* "CA4"	F_2	46	192 RFLP	19	720	-	Prince *et al.* (1993)
C. annuum cv. NuMex RNaky x *C. chinense* "PI159234"	F_2	75	1,007 (RAPD, AFLP, RFLP)	13	1,245.7	9.0	Livingstone *et al.* (1999)
C. annuum cv. TF68 x *C. chinense* cv. Habanero	F_2	107	580 (150 RFLP, 430 AFLP)	16	1,320	7.5	Kang *et al.* (2001)
C. annuum "Maor" x *C. frutescens* wild genotype"BG 2,816"	BC_2	248	92 RFLP	12	1,100	–	Rao *et al.* (2003)
C. annuum cv. TF68 x *C. chinense* cv. Habanero	F_2	107	333 (46 SSR, 287 RFLP)	15	1,761.5	5.3	Lee *et al.* (2004a)
C. annuum "NuMex RNaky" x *C. frutescens* "BG2814 6"	F_2	396	728 (489 SSR, 195 AFLP, 8 specific PCR markers, 36 RFLP)	16	1,358.7	–	Ben Chaim *et al.* (2006)
C. annuum "TF68" x *C. chinense* "Habanero"	F_2	107	243(180 SSR, 63 RFLP)	14	2,201.5	9.1	Yi *et al.* (2006)
C. frutescens var. BG2814 6 x *C. annuum* cv. NuMex RNaky	F_2	94	587 COSII	12	1,613	6.9	Wu *et al.* (2009)
C. annuum "NB1" x *C. chinense* "Jolokia"	F_2	94	116 SNP (HRM)	12	1,167.9	–	Lee *et al.* (2013b)
C. annuum "NuMex RNaky" (RN) x *C. chinense* "PI159234" (CA4)	F_2	90	512 (214 IBP, 143 COSII, 48eSNP, 107 other	12	2,335.6	4.5	Park *et al.* (2014)
C. annuum "BA3" x *C. frutescens* "YNXML"	F_2	154 and 147	224 (129 InDel, 95 SSR)	13	1,249.7	5.6	Tan *et al.* (2015)

Population/Cross Type	*Mapping Population*	*Population size*	*Number and Type of Markers Tested*	*Number of Linkage Groups*	*Map Length (cM)*	*Average Marker Distance (cM)*	*Reference(s)*
C. annuum "Bangchang" x *C. chinense* "PBC932"	F_2	126	214 SNPs	12	824	3.85	Mahasuk *et al.* (2016)
C. frutescens "Tabasco" x *C. annuum* "P4"	F_2	90	1,361 bins (array)	12	1,392.3	–	Hulse Kemp *et al.* (2016)
C. annuum "BA3" x *C. frutescens* "YNXML"	F_2	297	3,826 bins	12	1,628.8	0.45	Cheng *et al.* (2016)
C. chinense "740" x *C. annuum* "CA1"	F_2	150	9,038 SLAF	12	1,586.8	0.18	Zhu *et al.* (2019)
Intra specific linkage map							
C. annuum "Maor" x *C. annuum* "Perennial"	F_3	180	177 (RAPD, RFLP, AFLP)	12	1,740	–	Ben Chaim *et al.* (2001)
C. annuum cv. H3 x *C. annuum* cv. Vania	Doubled haploid	101	543 (50 RFLP, 56 RAPD, 434AFLP, 3 Morphological)	20	1,513	13.0	Lefebvre *et al.* (2002)
C. annuum cv. Perennial x *C. annuum* cv. Yolo Wonder	Doubled haploid	114	630 (133 RFLP, 164 RAPD, 325AFLP, 1 Isozyme, 4 SpecificPCR, 3 Morphological)	26	1,668	12.5	Lefebvre *et al.* (2002)
C. annuum cv. Yolo Wonder x *C. annuum* cv.Criollo de Morelos334	F_2	151	208 (67 RFLP, 28 RAPD,109AFLP, 2 Specific PCR, 2Morphological)	18	685	13.7	Lefebvre *et al.* (2002)
C. annuum cv. K9-11 x *C. annuum* cv. AC2258	Doubled haploid	179	518 (382 AFLP, 122 RAPD, 7SCAR, 3 RFLP, 4 CAPS)	16	1,043.1	4.6	Sugita *et al.* (2005)
C. annuum cv. Manganji x *C. annuum* cv. Tongari	Doubled haploid	–	374 SSR, AFLP, CAPS, RAPD(106 newly developed SSR)	13	1,042.4	2.8	Minamiyama *et al.* (2006)

Population/Cross Type	*Mapping Population*	*Population size*	*Number and Type of Markers Tested*	*Number of Linkage Groups*	*Map Length (cM)*	*Average Marker Distance (cM)*	*Reference(s)*
Yolo Wonder (*C. annuum*) x Criollo de Morelos 334 (CM334) (*C. annuum*)	Recombi-nant in bred lines (RIL)	297	587 (507 AFLP, 40 SSR, 19RFLP, 17SSAP, 4 STS)	49	1,857	5.71	Barchi *et al.* (2007)
California Wonder (*C. annuum*) x LS2341 (JP187992) (*C. annuum*)	Doubled haploid	94	253 (151 SSRs, 90 AFLPs, 1°CAPSs, 2 STS)	12	1,336	<10	Mimura *et al.*(2012)
C. annuum var. K9-11 x *C. annuum* var.MZC–180	Doubled haploid	184	597 (69 cDNA derived SSRs,196 genomic SSRs, 107 previously reported SSRs, 34 SNP/InDel, 164 AFLP, 27 RAPD)	12	2,028	< 4.0	Sugita *et al.* (2013)
C. annuum "BA3" x *C. annuum* "B702"	F_2	178	251 InDel	12	1,178.01	5.01	Li *et al.* (2015)
C. baccatum "PBC80" x *C. baccatum* "CA1316"	F_2	146	403 SNPs	12	1,270	3.15	Mahasuk *et al.* (2016)
C. annuum "Perennial" x *C. annuum* "Dempsey"	RILs	120	2,578 bins (whole genome re-sequencing)	12	1,372.2	0.53	Han *et al.* (2016)
C. baccatum "Goldenaji" x *C. baccatum* "PI594137"	F_2	93	395 HRM	12	1,056.2	2.7	Lee *et al.* (2016)
C. annuum var. PM702 x *C. annuum* var. FS871	RILs	146	9,328 SLAF	12	2,009.7	0.22	Zhang *et al.* (2019)
Integrated map							
C. annuum var. Perennial x *C. annuum* var. Yolo Wonder (DH200); *C. annuum* var. Vat x *C. annuum* var. CM334 (DH591); *C. annuum* var. Yolo Wonder x *C. annuum* var.CM334 (DH702)	Doubled haploid	169 lines (94DH200 + 44DH591 + 31 DH702)	85 (RFLP, RAPD)	14	820	–	Lefebvre *et al.* (1995)

Population/Cross Type	*Mapping Population*	*Population size*	*Number and Type of Markers Tested*	*Number of Linkage Groups*	*Map Length (cM)*	*Average Marker Distance (cM)*	*Reference(s)*
C. annum cv. NuMex RNaky x *C. chinense* cv. PI 159,234 (EGOD); *C. annuum*cv. 100/63 x *C. chinense* cv. PI 152,225 (EGOE); *C. annuum* cv. Maor x *C. annuum*cv. Perennial (EGOT); *C. annuum* cv. H3 x *C.annuum* cv. Vania (HV); *C. annuum* cv.Perennial x *C. annuum* cv. Yolo Wonder (PY);*C. annuum* cv. Yolo Wonder x *C. annuum* cv.Criollo de Morelos 334 (CM334) (YC)	F_2 (EGOD); BC1 (EGOE); F_2 (EGOT); Doubled haploid (HV); Doubled haploid (PY); F_2 (YC)	75 (EGOD); 83 (EGOE); 180 (EGOT); 98 (HV); 94 (PY); 151 (YC)	2,262 (1,528 AFLP, 440 RFLP,288 RAPD, 3 Isozyme, 3Morphological)	81	1,832	0.8	Paran *et al.* (2004)
Two interspecific (*C. annuum* "TF68" x *C. chinense* "Habanero") (SNU2,3 and SNU6), and Two intra specific (*C. annuum* "CM334" x *C. annuum* "Chilsungcho")(SNU4 and SNU5)	F_2	107 (SNU2,3), 100 (SNU4), 100 (SNU5),93 (SNU6)	805 SSR, RFLP, STS, AFLP,WRKY, rRAMP (interspecific), 745(intraspecific)	–	1,858 (inter-specific) 1,892 (intra-specific)	2.2 (interspecific), 2.5 (intraspecific)	Lee *et al.* (2009)
C. baccatum var. UENF 1616 x *C. baccatum* var. UENF 1732	F_2	203	183 (42 SSR, 85 ISSR, 56 RAPD)	16	2,547.5	14.25	Moulin *et al.* (2015)

fertility restorer genes and metabolites such as pungency and disease resistances. Of the many reported, only few listed in Table 22 could be applied in pepper translational research as they were shown directly associating with the trait of interest.

Table 22: Identified QTL/Genes which Could be Used in Molecular Breeding of Capsicum

Cross/Species	*QTL/gene*	*Trait*	*References*
Capsicum sp.	*fs*10.1	Fruit elongation	Barchi *et al.* (2009)
C. annuum	*CaOvate*	Fruit shape	Tsaballa *et al.* (2011)
PI 152225 (*C. chinense*) x Kelvin, 4751 (*C. annuum*)	*cl, c2, y*	Fruit color	Popovsky and Paran (2000)
Capsicum sp.	Phytoene synthase (*PSy*), phytoenedesaturase and capsanthin, capsorubin synthase (*Ccs*) genes,*LcyB, CrtZ-2*	Fruit color	Ha *et al.* (2007), Rodriguez-Uribe *et al.* (2012)
Capsicum sp.	*Pal, Ca4h, Comt, 3-ketoacyl-ACP*synthase (*Kas*), putative aminotransferase (*pAmt*), *Acl, Fat,* SB2-149 and SB1-158 clones(similarity to *pAm*t and *Kas*). SB2-66, SB2-115, SB2-115	Capsaicinoid biosynthesis	Curry *et al.* (1999), Kim *et al.*(2001), Aluru *et al.* (2003)
NMCA 30036 (*C. chinense*)	*Pun 1* (locus C), *At3*	Capsaicinoid biosynthesis	Blum *et al.* (2002)
Capsicum sp.	*pun1*^{2}	Capsaicinoid biosynthesis	Stewart *et al.* (2007)
C. annuum	$CaMF_2$	Pollen development	Chen *et al.* (2011)
C. annuum	Methionine sulfoxide reductase (*B2*)	Plant defense	Oh *et al.* (2010)
C. annuum	*Ipcr*	Resistance to *P. capsici*	Reeves *et al.* (2013)
C. annuum cv. 'Bukang'	*Cmr1* (cucumber mosaic resistance 1)	Resistance to CMV	Kang *et al.* (2010)
C. annuum	*CaBtf3*	Defense related genes	Hun *et al.* (2012b)
C. annuum	*CaWRKYb*	Resistance gene	Lim *et al.* (2011)
C. annuum and *C. chinense*	*Phyto.5.2*	*P. capsici* resistance	Quirin *et al.* (2005)
C. annuum L.	*CaMi*	Nematode resistant gene	Chen *et al.* (2007)
Capsicum sp.	*Pcpme 1–9*	*Pme* genes	Li *et al.* (2011)
CM33 (*C. annuum*)	*CaRGA2*	*P. capsici* resistance	Zhang *et al.* (2013)
Capsicum sp.	*caWRKY40*	Resistance to *R. solanacearum*	Dang *et al.* (2013)
Capsicum sp.	*Bs1, Bs4, bs5, bs6, gdr*	Resistance to *Xanthomonas campestris* pv. *vesicatoria*	Jones *et al.* (2002, 2004)

Cross/Species	QTL/gene	Trait	References
C. annuum	*$CaPO_2$*	Defense against *Xanthomonas campestris* pv. *vesicatoria*	Choi *et al.* (2007)
C. annuum	*$CaMLO_2$*	Mildew resistance locus O gene	Kim and Hwang (2012)
Capsicum sp.	*pvr1, pvr21, pvr22, pvr3, pvr4, pvr5,pvr6, Pvr7*	Resistance genes	Caranta *et al.* (1996), Kyle and Palloix (1997), Grube *et al.* (2000)
C. annuum cv 'Bukang'	*Cmr1*	Resistance gene	Kang *et al.* (2010)
C. chinense	*Tsw.*	Resistance gene	Moury *et al.* (2000)
Capsicum sp.	*L1, L3, L4*	Resistance gene	Lefebvre *et al.* (1995), Ben-Chaim *et al.* (2001), Matsunaga *et al.* (2003), Tomita *et al.* (2008)
C. annuum	*CaWRKYb, CaPR-10, CaPR-1,CaPR-5*	Reduce resistance pathogens is related gene	Lim *et al.* (2011)
C. annuum	*Me4, Mech1, Mech2*	Root knot nematode resistance gene	Djian-Caporalino *et al.* (2001, 2007)
C. annuum	*CaSD1*	Senescence-delaying	Seo *et al.* (2013)
C. annuum	*CaMsrB2*	Methionine sulfoxide reductaseB2 gene	Oh *et al.* (2010)
C. annuum	*orf456*	Cytoplasmic male sterility	Kim *et al.* (2007)
C. annuum	*PmsM1-CAPS*	Genetic male sterility	Lee *et al.* (2010a)
C. annuum	*ms1, ms3*	Genetic male sterility	Lee *et al.* (2010b, c)

Lefebvre *et al.* (1995) studied nuclear RFLP between pepper cultivars of *C. annuum*. Forty-one nuclear probes, encompassing the different linkage groups of a previously published map, were used to examine RFLPs of cultivated peppers. Wang *et al.* (1996) surveyed fourteen diverse pepper (*Capsicum* spp.) accessions by RAPD analysis: A high degree of polymorphism was obtained from 4 random decamer primers which produced 11 reproducible and effective amplification fragments useful for identification between species. On the basis of cluster analysis the accessions were divided into 4 groups containing (i) 2 accessions of *C. chinense*, (ii) 4 C. *annuum* accessions, (iii) 5 C. *annuum* accessions, and (iv) 3 *C. baccatum* accesions. The results indicated that RAPD markers could be effectively and reliably used for the classification of *Capsicum* spp. Lefebvre *et al.* (1995) also constructed an intraspecific integrated molecular linkage map of pepper using RFLP and RAPD markers and segregating doubled haploid progenies.

Ninety accessions of pepper (*Capsicum annuum*) germplasm collected from 16 countries of Europe, Asia and Australia were investigated and compared using microsatellite DNA (inter simple sequence repeat, ISSR) and RAPD markers (Wang and Fan, 1998). There were 151 polymorphic bands produced by 5 primers using ISSR, while 130 polymorphic bands were produced by RAPD and all 90 accessions could be distinguished. ISSR primer SCRI1418 was most useful for fingerprinting.

The Bs2 resistance gene of pepper confers resistance against the bacterial pathogen *Xanthomonas campestris* pv. *vesicatoria* (*X. vesicatoria*). As a first step toward isolation of Bs2, Tai *et al.* (1999) identified molecular markers tightly linked to the gene by RAPD, DNA and AFLP analyses of near-isogenic lines.

13.8.0 Genetic Transformation

Genetic transformation is a process that involves the introduction and expression of foreign genes in a host organism. It has provided an alternative approach for capsicum improvement program. The major advantage which transgenics technology offers is that it overcomes interspecific or intergeneric barriers and enables transfer of useful genes or novel traits into capsicum.

Liu *et al.,* reported the first capsicum transformation work in 1990. Major capsicum transformation work has been done for disease resistance particularly against viruses, *viz,* tobacco mosaic virus (TMV), pepper mild mottle virus (PMMV) (Lee *et al.,* 2004b), tomato mosaic virus (ToMV) (Shin *et al.,* 2002a), and cucumber mosic virus (CMV) (Shin *et al.,* 2002b). Such transgenic virus resistance mechanism utilizing viral coat protein regions and satellite RNA is currently known as RNA silencing (Voinnet, 2001). Transformation and overexpression of TsiI, a tobacco pathogenesis related (PR) gene in capsicum displayed broad spectrum resistance against different pathogens like PMMV, CMV, bacterial pathogen *Xanthomonas campestris* pv. *vesicatoria,* and a fungal pathogen *P. capsici* (Shin *et al.,* 2002a).

RNA silencing approach has been used to identify a new gene ketoacyl-ACP reductase (CaKR1) in capsicum responsible for producing nonpungent fruits (Koeda *et al.,* 2019). Capsicum transformation studies has most commonly used GUS gene (b-glucuronidase) as reporter gene, CaMV 35S as promoter, and nopaline synthase (NOS) as terminator gene (Lee *et al.,* 2004b). Direct transformation using gene gun has been attempted recently in *C. frutescens* (Chee *et al.,* 2018). However, the exploitation of transgenic technology in capsicum is slow as capsicum is highly recalcitrant to transformation and regeneration process. Table 23 summarises the details on studies of genetic transformation in chilli pepper.

The shooting frequency was 100 per cent on MS medium supplemented with 4-6 mg/IBA + 0.5 mg/l IAA Transient expression of the *gus* gene was detected and kanamycin-resistant shoots were obtained from cotyledons. Wang *et al.* (1991) reported regeneration and transformation of sweet pepper (*C. frutescens*). Segments of cotyledon, hypocotyl and true leaf taken from 10 to 30 day old seedlings of 3 cultivars (Shuangfeng, Zhongjiao 2 and Zhongjiao 3) were cultured on MS medium supplemented with various phytohormones for induction of callus, shoots and roots. *Agrobacterium tumefaciens* transformed plantlets were obtained using strains C58 and GV3111 and Horch's leaf disc method.

Zhu *et al.* (1996) obtained transgenic sweet pepper plants from *Agrobacterium* mediated transformation Fertile transgenic sweet pepper (*C. annuum* var. *grossum*)

Table 23: Genetic Transformation Studies in Chilli Pepper

Species	*Transformation System*	*Tissue*	*Results*	*References*
C. annuum	Agrobacterium strains A282 and C58; pCV 3850 (*nptII*, gus)	Hypocotyl, cotyledon and leaf	GUS positive	Liu *et al.* (1990)
C. annuum	*A. tumefaciens* strain LBA 4404, p5T35 AD (acetolactate synthase (ALS) gene, gus, pSLJ 1911, (*nptII*, gus), pWTT2039 (hpt, gus)	Cotyledon	Transformed tissues,GUS positive	Engler *et al.* (1993)
C. annuum	*A. tumefaciens* strain LBA 4404, pROK1/105 (CMV I17 N-Satellite RNA)	Cotyledon	Transformed plants(4 per cent transformation frequency)	Lee *et al.* (1993)
C. annuum	*nptII A. tumefaciens* strain GV 3III-SE,(CMV-CP, *nptII*)	Young leaves	Transformed plants expressing (CMVCP) gene	Zhu *et al.* (1996)
C. annuum	*A. tumefaciens*	Young leaves	Transgenic lines	Kim *et al.* (1997)
C. annuum	pROK1/105 (cmv. Satellite RNA, *nptII*)	Cotyledon	Transgenic plants	Manoharan *et al.* (1998)
C. annuum	*A. tumefaciens* strain EHA 105,(pBI 121)(*nptII*, gus)	Wounded	Hairy roots	Jayashankar *et al.* (1997)
C. annuum	*A. rhizogenes* strain K 599 (gus)	HypocotylCotyledon	40.8 per cent transgenic plants	Li *et al.* (2003)
C. annuum	*A. tumefaciens* strain LBA 4404(*nptII*) *A. tumefaciens* strain EHA 105, LBA	CotyledonHypocotyl	Transformed plantsinvolving callus phase	Lee *et al.* (2004a)
C. annuum	4404,pCAMBIA 2300(TMV-CP,PPI1) *A. tumefaciens* strain	Hypocotyl	Transformed shoots	Delis *et al.* (2005)
C. annuum	LBA4404, pBI121(nptII) PEG mediated transformation, pCAMBIA:: Ac(hpt, gus)	Mesophyll protoplast	Efficient transient expression and transformation	Jeon *et al.* (2007)
C. frutescens	EHA 101 harbouring binary vector pTCL5	Hypocotyl segments	Transformation	Hasnat *et al.* (2008)

A T-DNA map of the pCAMBIA2301 vector; B Generation of *C. annuum* var. Pusa Sadabahar (a–g) and Pusa Jwala (h-n) transgenic plants. Hypocotyl (a, h) and cotyledonary (b, i) explants on selection regeneration medium (SRM); (c, d, j, k) Regenerated shoots on regeneration medium; (e, l) Elongated shoots on rooting medium; (f, m) Rooted plants on soilriteTM for hardening; (g, n) Potted plants in green house (Mahto *et al.*, 2018).

plants were regenerated at high rates from explants that were cocultivated with *A. tumefaciens* strain GV3111-SE harbouring a plasmid that contains the cucumber mosaic cucumovirus coat protein (*cmv-cp*) gene.

Manoharan *et al.* (1998) established a protocol for regeneration and Agrobacterium mediated genetic transformation in hot chilli (*C. annuum* cv. Pusa Jwala). High frequency regeneration of shoot buds from cotyledonary leaves was achieved using MS medium supplemented with 0.5 mg/I TDZ. The *A. tumefaciens* strain EHA 105, carrying a binary vector plasmid pBI 121, was used for the transformation. The transgenic nature of the regenerated plants was confirmed by the histochemical staining of GUS, PCR and Southern hybridization analyses of the *nptII* gene.

High frequency of plant regeneration was observed by Lim *et al.* (1999) when hot pepper cotyledon segments were cultured on MS medium supplemented with 0.1 mg NAA and 2.0 mg zcatin/1 or 10.0 mg IBA and 1.0 mg BNI. The bar (herbicide resistance) gene was also introduced into hot pepper and progenies of the transgenic plants were obtained. Stable inheritance (3: 1 ratio) of *bar* was confirmed by PCR analysis and treatments with of 5 g/1 bastar.

13.9.0 Gene Transfer

Tissue culture, in combination with molecular techniques, has been successfully used to incorporate specific traits through gene transfer.

Dong *et al.* (1992 and 1995) obtained transgenic pepper plants containing a CMV (cucumber mosaic cucumovirus) satellite RNA cDNA by *A. tumefaciens* mediated transformation. Transformants were regenerated on selection medium and some flowered and set fruit. Compared with non-transgenic plants, most transgenic plants had about 10 days' delay in disease symptoms, lower disease index and slow CMV development.

cmv-cp (cucumber mosaic cucumovirus coat protein) gene transformation into pepper and expression in the offspring of the transgenic plants has been reported by Zhang *et al.* (1994). *cmv-cp* genes were introduced into pepper by coculturing cotyledon explants of 3 cultivars with *A. tumefaciens*. Inoculation with the virus showed that the transgenic plants and their offspring were resistant to CMV.

13.10.0 *In vitro* Conservation

In vitro conservation studies in Capsicum have been restricted to seeds and pollen grains. Cooling (-196°C) and rewarming (30°C) up to 50 cycles over a period of a few months slightly reduced the viability. Belletti *et al.* (1990) also observed similar result on seed viability by cooling and rewarming.

Kristof and Barnabas (1983) stored pollen grains of *C. annuum* varieties Soroksari and Hajtato and Hatvaki Hajtato at -196°C for 3 and 10 months and their capacity for germination and fertilization was retained. Bezdickova (1988) germinated pollen of *C. annuum* lines CW in liquid medium following storage at 4°C for up to 16 days but declined thereafter. Germination and viability were maintained following storage at both below -zero temperatures for up to 66 days.

14.0 References

Abak, K., Pochard, E. and Dumas de Vaulx, R. (1982) *Capsicum Nwsl.*, **1**: 62-63.

Achal Shah, S.D. and Pant, C.C. (1986) *Prog. Hortic.*, **18**: 270-272.

Ado, S.G. and Samarrawira, I. (1986) 12th Annual Conference of the Genetics Society of Nigeria held at Forestry Research Institute of Nigeria (FRIN), Ibadan, 10-14 February, 1986, p. 5.

Ahmed, N. and Tanki, M.I. (1991) *Haryana J. Hort. Sci.*, **20:** 114-118.

Alcantara, T.P., Bosland, P.W. and Smith, D.W. (1996) *J. Hered.*, **87**: 239-241.

Aleemullah, M., Haigh, A. M. and Holford, P. (2000) *Aust. J. Exp Agric.*, **40:** 755-762.

Aliyu, L. (2000) *Biol. Agric. Hortic.*, **18**: 29-36.

Altaf, M.A., Shahid, R., Altaf, M.A., *et al.* (2019) *Hortic. Int. J.*, **3:** 217 222.

Aluru, M.R., Mazourek, M., Landry, L.G., Curry, J., Jahn, M. and O'Connell, M.A. (2003) *J. Exp. Bot.*, **54:** 1655-1664.

Amit, K., Ahad, I., Kumar, V. and Thakur, S. (2014) *J. Spices Arom. Crops*, **23**: 170-177.

Ananthasamy, T.S., Kamat, V.N. and Pandya, H.G. (1960) *Curr. Sci.*, **29:** 271.

Ancona-Escalante, W.D.R., Baas-Espinola, F.M., Castro-Concha, L.A., Vazquez-Flota, F.A., Zamudio-Maya, M., and Miranda-Ham, M.D.L. (2013) *Plant Cell Tis. Cul.*, **113:** 565-570.

Andrzejewski, P., Kohler, K.U. and Pluschkell, W. (1992) *Steel Res.*, **63**: 242.

Aniel, K.O, Anitha, V., Roseline, S.K. and Raja Rao, K.G. (2001) *Capsicum Eggplant Newsl.*, **20:** 72-75.

Anonymous (1998) *FAO Production Year Book*, **52:** 133-134.

Antonious, G.F., Kochhar, T.S., Jarret, R.L. and Snyder, J.C. (2006) *J. Environ. Sci. Health*, **41**: 1237-1243.

Arora, S.K., Brar, J. and Mehta, R. (1998) *Abst. Silver Jubilee Nat. Symp. Emerging Scenario in Vegetable Research and Development*, 12-14, December, 1998, Varanasi, India, p. 130.

Ashtankar, C.M. and Jaipurkar, M.A. (1988) *Ann. Plant Physiol.*, **2:** 193-203.

Ashwani, Kumar, S., Jhorar, R.K. and Makkar, R. (2020) *Int. J. Chem. Stud.*, **8**: 1557-1566.

Aykroyd, W.R. (1963) *TCMR Special Report Series*, No.42.

Azam, M.F., Khan, AM. and Sexena, S.K. (1979) *Indian J. Mycol. Plant Path.*, **8:** 65.

Babu, P.S., Guravaiah, M., Hatti, I., and Srikanth, K. (2014) *Int. J. Adv. Res. Biol. Sci.*, **1**: 184-191.

Bajaj, K.L., Kaur, G. and Sooch, B.S. (1980) *Veg. Sci.*, **7:** 48-57.

Balakrishnan, K. (1999) *Indian J. Plant Physio.*, **4:** 229-231.

Balakrishnan, P. (1980) M.Sc. (Ag.) *Thesis, Tamil Nadu Agric. Univ.*, Coimbatore.

Balakrishnan, P., Muthukrishnan, C.R. and Irulappan, I. (1983) *Proc. Nat. Seminar Tomato, Chillies, Tamil Nadu Agric. Univ.*, Coimbatore, pp. 142-146.

Barcaccia, G., Tomassini, G. and Falcinelli, M. (1999) *J. Genet. Breed.*, **53:** 251-254.

Barchi, L., Bonnet, J., Boudet, C., Signoret, P., Nagy, I., Lanteri, S., Lefebvre, V. (2007) *Genome*, **50:** 51-60.

Barchi, L., Lefebvre, V., Sage-Palloix, A.M., Lanteri, S. and Palloix, A. (2009) *Theor. Appl. Genet.*, **118:** 1157-1171.

Basavaraja, P.K., Shashidhara, G.B., Manjunathaiah, H.M., Ekbote, S.D., Jagadeesha, R.C. and Prabhu, S.T. (1998) In*: Proceedings of the National Seminar*, Madikeri, Karnataka, India, pp. 99-102.

Belletti, P., Lanteri, S., Lepori, G., Nassi, M.O. and Quagliotti, L. (1990) *Adv. Hori. Sci.*, **4:** 118-120.

Ben, C. and Paran, I. (2001) *Theor. Appl. Genet.*, **102:** 1016-1028.

Ben Chaim, A., Borovsky, Y., Falise, M., Mazourek, M., Kang, B. C., Paran, I. and Jahn, M. (2006) *Theor. Appl. Genet.*, **113:** 1481-1490.

Ben-Chaim, A., Grube, R.C., Lapidot, M., Jahn, M. and Paran, I. (2001) *Theor. Appl. Genet.*, **102:** 1213-1220.

Berger, A., Henderson, M., Nadoolman, W. and Duffy, V. (1996) *J. Pain Symptom. Manage.*, **10:** 243-248.

Bergh, B.O. and Lippert, L.F. (1964) *Am. Nat.*, **99:** 159-166.

Bernal, M.A., Calderon, A.A., Pedreño, M.A., Muñoz, R., Barcelo, A.R. and Merino de Cáceres, F. (1993) *J. Agric. Food Chem.*, **41:** 1041-1044.

Berzal-Herranz, A., de la Cruz, A., Tenllado, F., Diaz-Ruiz, J.R., Lopez, L., Sanz, A.I., Vaquero, C., Serra, M.T. and Garcia-Luque, I. (1995) *Virology*, **209:** 498-505.

Bezdickova, A. (1988) *Capsicum Newsl.*, Abst., pp. 287-295.

Bhalla, P.L., Bagchel, B.S. and Krishna, A. (1983) *Proc. Nat. Seminar Tomato, Chillies,* Tamil Nadu Agric. Univ., Coimbatore, pp. 147.

Bharadwaza, R.K., Prasad, V.M. and Srinivas, M. (2015) *Environ. Eco.*, **33**: 1626-1629.

Bhargava, Y.R. and Umalkar, G.V. (1989) *Acta Hortic.*, **253:** 233-238.

Bhattacharya, A., Chattopadhyay, A., Majumder, D., Chakraborty, A. and Pal, S. (2010) *Intl. J. Veg. Sci.*, **16:** 201-211.

Bhutia, K.L., Bhutia N.D. and Khanna, V.K. (2019) *J. Agric. Forest Meteorol. Res.*, **2**: 77-83.

Bhutia, K.L., Meetei, N.G.T. and Khanna, V.K. (2016) *Agrotechnology*, **5**: 1-4.

Bhutia, K.L., Meetei, N.T. and Khanna, V.K. (2018) *J. Pharma. Phyto.*, **7**: 1008-1012.

Bhutia, N.D., Kumar, P. and Chattopadhyay, A. (2017b) *Indian J. Hortic.*, **74**: 135-138.

Bhutia, N.D., Seth, T., Shende, V.D., Dutta, S. and Chattopadhyay, A. (2015) *Sci. Hort.*, **182:** 47-55.

Bhutia, N.D., Sureja, A.K., Arya L., Munshi, A.D. and Verma, M. (2017a) *Indian J. Hort.*, **74**: 615-617.

Binzel, M.L., Sankhla, N., Joshi, S. and Sankhla, D. (1996) *Plant Cell Rep.*, **15:** 536-540.

Biswas, A., Bhattacharya, A., Chattopadhyay, A., Chakravarty, A. and Pal, S. (2011) *Int. J. Veg. Sci.*, **17**: 224-232.

Blum, E., Liu, K., Mazourek, M., Yoo, E.Y., Jahn, M. and Paran, I. (2002) *Genome*, **45:** 702-705.

Boiteux, L.S. and de Avila, A.C. (1994) *Euphytica*, **75:** 139-142.

Boominathan, H., Mani, L.S., Rajamani, A. and Krishnamurthy, A. (1971) *Madras Agric. J.*, **58:** 520-522.

Boronat, M., Madrid, R., Egea, C. and Collados, I. (1999) *ITEA Production Vegetal*, **95:** 125-135.

Bosland, P.W. (1996) In: *Capsicums*: Innovative use of an ancient crop (Ed. J. Janick), Progress in New Crops, ASHS Press, Arlington, pp. 479-487.

Bosland, P.W. (2000) *HortSci.*, **35:** 1321-1322.

Bosland, P.W. and Baral, J.B. (2007) *HortScience*, **42**: 222-224.

Bosland, P.W. and Lindsey, D.L. (1991) *Plant Dis.*, **75:** 1048-1050.

Bosland, P.W. and Votava, E.J. (2012) Peppers: Vegetable and Spice *Capsicums* J. Atherton and A. Rees (eds), CABI Publishing, p. 199.

Bosland, P.W. and Votava, E.J. (2000) Centre for Agriculture and Bioscience International, Wallingford, UK.

Bosland, P.W., Iglesias, J. and Tanksley, S.D. (1991) *HortScience*, **26:** 215-216.

Boswell, V.R. (1937) Yearbook of Agriculture. U.S. Govt. Printing Office, Washington, pp. 176-206.

Boswell, V.R. (1951) *Natl. Geogr.*, **96:** 145-217.

Boukema, I.W. (1980) *Euphytica*, **29:** 433-439.

Bukusov, S.M. (1930) *Bull, Appl. Bot. Genet. Plant Breed. Suppl.*, **47:** 261-273.

Burghate, R. (2014) In: *3rd International Conference on Agriculture and Horticulture*, October 27-29, 2014, Hyderabad International Convention Centre, India.

Butani, D.K. (1976) *Pesticides*, **10:** 38-41.

Buyukalaca, S. and Mavituna, F. (1996) *Plant Cell Tiss. Cult.*, **46:** 227-235.

Cao, J.S. and Su, Z.Y. (1988) *Acta Hortic. Sinica*, **15:** 57-63.

Caranta, C., Palloix, A., Gebre-Selassie, K., Lefebvre, V., Moury, B. and Daube'ze, A.M. (1996) *Phytopathol.*, **86:** 739-743.

Chalukova, M., Daskalov, S., Lukarska, E. and Baralieva, D. (1993) *Capsicum Eggplant Newsl.*, **12:** 57-58.

Chandini Raj, A., Holebasappa, K., Hore, J.K. and Das, Kalyan Sundar (2016) *Research on Crops*, **17**: 320-323.

Chandra, R., Murthy, R.S. and Murthy, M.S. (1976) *Curr. Res.*, **5:** 196-197.

Chattopadhyay, A., Sharangi, A.B., Dai, N. and Dutta, S. (2011) *Chilean J. Agric. Res.*, **71**: 350-356.

Chaudhary, A., Rajesh, K. and Solankey, S.S. (2013) *African J. Biotechnol.*, **12**: 6605-6610.

Chaudhary, B.R., Sharma, M.D., Shakya, S.M. and Gautam, D.M. (2006) *J. Inst. Agric. Anim. Sci.*, **27:** 65-68.

Chauhan, M.S. and Duhan, J.C. (1977) *Pesticides*, **11:** 17-18.

Chauhan, V., Singhania, R.A., Singh, A.K. and Kumar, A. (1999) *Indian J. Agric. Res.*, **33:** 62-66.

Chavez-Moctezuma, M.P. and Lozoya-Gloria, E. (1996) *Plant Cell Rep.*, **15:** 360-366.

Chee, M., Lycett, G.W. and Foan, C. (2018) *Electron. J. Biotechnol.*, **34**: 51-58.

Chen, C., Chen, G., Hao, X., Cao, B., Chen, Q., Liu, S. and Lei, J. (2011) *Plant Sci.*, **181:** 439-448.

Chen, R.G., Li, H.X., Zhang, L.Y., Zhang, J.H., Xiao, J.H. and Ye, Z.B. (2007) *Plant Cell Rep.*, **26:** 895-905.

Chen, X.S. (1984) *Acta Hort. Sinica*, **11:** 113-118.

Chen, Y.H. and Jiang, J.Z. (1998) *Acta Agric. Sanghai*, **14:** 46-50.

Cheng, J., Qin, C., Tang, X., Zhou, H., Hu, Y., Zhao, Z. and Hu, K. (2016) *Scientific Reports*, **6:** 332-393.

Chiemsombat, P., Atathom, S., Paosang, T., Warin, N., Sukprakarn, S., Patarapuwadol, S., Eiemkaeng, S. and Bunchote, A. (1999) *In the 37th Kasetsart University Annual Conf.*, Bangkok, Thailand, 3-5 February, 1999.

Choi, H.W., Kim, Y.J., Lee, S.C., Hong, J.K. and Hwang B.K. (2007) *Plant Physiol.*, **145:** 890-904.

Choudhary, B.S. and Samadia, D.K. (2004) *Indian J. Hort.*, **61**: 132-136.

Choudhury, B. (1967) *Vegetables, National Book Trust*, New Delhi.

Chougule, A.B. and Mahajan, P.R. (1979) *Veg. Sci.*, **6:** 73-80.

Christopher, E.P. (1996) McGraw-Hill Book Company, Inc. NewYork.

Chuo, S.K. and Ng, B.B. (1982) *Singapore J. Primary Industries*, **10:** 64-70.

Cook, A.A. and Guevara, Y.G. (1984) *Plant Dis.*, **68:** 329-330.

Coon, D., Votava, E. and Bosland, P.W. (2008) New Mexico State University. *Res. Rep.*, **73**: 1-8.

Csillery, G., Szarka, E., Sardi, E., Mityko, J., Kapitany, J., Nagy, B. and Szarka, J. (2004) *In Proceedings of 12th Eucarpia meeting on genetics and breeding of capsicum and eggplant*, 17-19 May 2004, Noordwijkerhout, The Netherlands, pp. 147-153.

Curry, J., Aluru, M., Mendoza, M., Nevarez, J, Melendrez, M. and O'Connell, M.A. (1999) *Plant Sci.*, **148:** 47-57.

Dahivval, A.L. and Kore, S.S. (1981) *South Indian Hort.*, **29:** 226.

Dang, F.F., Wang, Y.N., Yu, L., Eulgem, T., Lai, Y., Liu, Z.Q., Wang, X., Qiu, A.L., Zhang, T.X., Lin, J., Chen, Y.S., Guan, D.Y., Cai, H.Y., Mou, S.L. and He, S.L. (2013) *Plant, Cell Environ.*, **36:** 757-774.

Daramola, O.S., Adigun, J.A. and Adeyemi, O.R. (2020) *J. Crop Improve.*, doi.org/10.1080/15427528.2020.1795770.

Das, S.R. and Mahanty, K.C. (1988) *Indian J. Plant Protect.*, **16:** 109-111.

Daskalov, S. and Poulos, J.M. (1994) *Capsicum Eggplant Newsl.*, **13:** 15-26.

Daskalov, S. (1973a) *Genet. Plant Breed.*, **6:** 401-408.

Daskalov, S. (1973b) *Genet. Plant Breed.*, **6:** 419-429.

Daskalov, S. (1986) In: Micke *et al.* (ed) Mutation breeding review. International Atomic Energy Agency/FAO, Vienna, No. 4, p. 25.

Daskalov, S., Chalukova, M., Baralieva, D. and Lukarska. E. (1995) *Proc. 9th Eucarpia Meeting on Genetics and Breeding of Capsicum and Eggplant*, 21-25 August, 1995, Budapest, Hungary. pp. 24-27.

Dasmunshi, A. (1999) In: *Compendium of Lectures of Summer School on Advances in Breeding of Temperate and Tropical Vegetables for Biotic and Abiotic Stresses*, ICAR, New Delhi, 17th May to 6th June, 1999, pp. 133-140.

De la Cruz, A, Lopez, L., Tenllado, F., Dýaz-Ruýz, J.R., Sanz, A.I., Vaquero, C., Serra, M.T. and Garcýa-Luque I (1997) *Mol. Plant-Microb. Interact.*, **10:** 107-113.

Deepa, N., Kaur, C., George, B., Singh, B. and Kapoor, H.C. (2007) *LWT Food Sci. Technol.*, **40**: 121-129.

Deka, P.C. and Shadeque, A. (1996) *Hort. J.*, **9:** 141-147.

Delis, M., Garbaczewska, G., Niemirowicz-Szczytt, K. (2005) *Acta Biol. Crac.*, **17:** 193-198.

Depestre, T. and Espinosa, J. (1988) *Agrotechniade Cuba*, **20:** 87-90.

Deshpande, A.A., Anand, N. and Pathak, C.S. (1986) *Serviciode Investigation Agraria*, 25-29.

Deshpande, R.B. (1933) *Indian J. Agric. Sci.*, **3:** 219-300.

Despestre, T. (1988) *Agrotecniade Cuba*, **20**: 115-18.

Devi, D.S. and Arumugam, R. (1999a) *Crop Res.*, **18:** 108-111.

Devi, D.S. and Arumugam, R. (1999b) *Crop Res.*, **18:** 112-114.

Devi, D.S., Arumugam, R. and Devi, D.S. (1999) *Crop Res.*, Hissar **17**: 90-93.

DeWitt, D. (2014) Catapult, p. 396.

Dhaliwal, M.S., Jindal, S.K. and Cheema, D.S. (2014) *Agric. Res. J.*, **52**: 127-129.

Dhall, R.K. (2008) *J. Res. Punjab Agric. Univ.*, **45**: 156-160.

Dhanelappagol, M.S., Shashidhara, S.D. and Kulkarni, G.N. (1994) *Karnataka J. Agric. Sci.*, **7:** 36-39.

Dhawan, P. and Rishi, N. (1999) *Capsicum and Eggplant Newsl.*, No.18, pp. 52-55.

Dhesi, N.S. and Nandpuri, K.S. (1965) *Vegetable Growing in Punjab Agric.* Univ., Ludhiana.

Dhulapanaver, C.V. (1965) *Indian J. Agron.*, **10:** 165-169.

Diaz, I., Powder, J.D. and Diaz-Ruiz, J.R. (1998) *Curr. Plant Sci. Biotechnol. Agric.*, **7:** 115-116.

Djian-Caporalino, C., Berthou, F., Fazari, A., Lefebvre, V. Palloix, A., Pegard, A. and Pijarowski, L. (2004) *Proc. 12th Eucarpia Meeting on Genetics and Breeding of Capsicum and Eggplant*, 17-19 May, 2004, Noordwijkerhout, the Netherlands, p. 180.

Djian-Caporalino, C., Fazari, A., Arguel, M.J., Vernie, T., VandeCasteele, C., Faure,.I, Brunoud, G., Pijarowski, L., Palloix, A., Lefebvre, V. and Abad, P. (2007) *Theor. Appl. Genet.*, **114:** 473-486.

Djian-Caporalino, C., Pijarowski, L., Fazari, A., Samson, M., Gaveau, L., O'Byrne, C., Lefebvre, V., Caranta, C., Palloix, A. and Abad, P. (2001) *Theor. Appl. Genet.*, **103:** 592-600.

Djian-Caporalino, C., Pijarowski, L., Januel, A, Lefebvre, V., Daubeze, A, Palloix, A, Dalmasso, A. and Abad, P. (1999) *Theor. Appl. Genet.*, **99**: 496-502.

Do Rego, E.R., Do Rego, M.M., Finger, F.L., Curz, C.D. and Casali, V.W.D. (2009) *Euphytica*, **168**: 275-287.

Dod, V.N., Joshi, A.T., Kale, P.B. and Kulwal, L.V. (1983) *Proc. Nat. Seminar Tomato, Chillies, Tamil Nadu Agric. Univ.*, Coimbatore, p. 152.

Doijode, S.D. (1988) *Haryana J. Hort. Sci.*, **17:** 94-96.

Dong, C.Z., Jiang, C.X., Feng, L.X. and Guo, J.Z. (1992) *Acta Hortic. Sinica*, **19:** 180-184.

Dong, C.Z., Jiang, C.X., Feng, L.X., Li, S.D., Gao, Z.H., Guo, J.Z. and Zhu, D.W. (1995) *Acta Hortic.*, **402**: 78-86.

Elangovan, M. and Suthanthirapandian, I. R. (1981) *South Indian Hort.*, **29:** 159-160.

Engler, D.E., Guri, A.Z., Lauritism J.A., Schloemer, L.M.P. (1993) *USA Patent*, **5:** 262-316.

Estrada, B., Pomar, F., Diaz, J., Merino, F. and Bernal, M.A. (1999) *Sci. Hort.*, **81:** 385-396.

Eswaramurthy, S., Pappiah, C.M., Muthusamy, M., Muthusamy, S., Mariappan, V., Jeyasekar, R., Natarajan, S., Devid, P.M.M. and Gomathinayagam, P. (1988) *Pesticides*, **22:** 38-40.

FAO, (2016) Available: http: //www.fao.org. Accessed 2016 Sept 10.

FAOSTAT (2018) Agricultural production. http: //www.faostat.fao.org.

Fery, R.L. and Harrison, H.F. Jr. (1990) *J. Am. Soc. Hort. Sci.*, **123:** 1008-1011.

Frankel, O.H. and Bennett, E. (1970) In: *Genetic resources in plants - their exploration and conservation* (Ed. OH. Frankel and E. Bennett). IBP Hand Book, Blackwell, Oxford, pp. 7-17.

Free, J.B. (1975) *Trap. Agric.*, **52:** 353-357.

Fuentes, R.G. and Mora, W.C. (1989) *Capsicum Newsl.*, **7:** 47-48.

Gaddagimath, N.B., Hiremath, K.G., Goud, J.V. and Patil, S.S. (1988) *J. Maharashtra Agri. Univ.*, **13:** 307-309.

Geleta, L.F., Labuschagne, M.T. and Viljoen, C.D. (2004) *Plant Breed.*, **123**: 467-473.

Geleta, L.F. and Labuschagne, M.T. (2006) *South African J. Plant Soil*, **23**: 73-77.

Gill, H.S., Asawar, B.M., Thakur, P.C. and Thakur, T.C. (1974) *Indian J. Agric. Sci.*, **47:** 408-410.

Gohokar, R.T. and Peshney, N.L. (1981) *Indian J. Agric. Sci.*, **51:** 663-665.

Gopalakrishnan, T.R. (2007) In: Peter KV (ed) Vegetable crops. Horticulture science series-4. New India Publishing, pp. 77-86.

Gopalratnam, P. (1933) *Madras Agric. J.*, **21**.

Gowda, K.K. and Gowda, P.M. (1982) *South Indian Hort.*, **30:** 4-8.

Greenleaf, W.H. (1956) *Phytopathol.*, **6:** 371-375.

Greenleaf, W.H. (1986) In: *Breeding vegetable crops*. AVI Publishing Co., Westport, pp. 67-134.

Grozeva, S. and Todorova V. (2015) *Elect. J. Biol.*, **11**(1): 17-22.

Grube, R.C., Blauth, J.R., Arnedo, M.A. and Jahn, M.K. (2000) *Appl. Genet.*, **101:** 852-859.

Gupta, C.G., Lakshmi, N. and Srivalli, T. (1998) *Capsicum and Eggplant Newsl.*, **17:** 42-45.

Gutierrez-Carbajal, M.B., Monforte-Gonzalez, M., Miranda-Ham, M.de L., Godoy-Hernandez, G. and Vazquez-Flota, F. (2010) *Biologia Plantarum*, **54**: 430-434.

Guzman, I. and Bosland, P.W. (2017) *Appetite*, **117**: 186-190.

Gyulai, G., Gemesne, J.A., Sagi, Z.S., Zatyko, L., Heszky, L. and Venczel, G. (1999) *Capsicum Eggplant Newsl.*, **18:** 40-43.

Ha, S.H., Kim, J.B., Jong-Sug, Park, J.S., Lee, S.W. and Cho, KJ (2007) *J. Exp. Bot.*, **58:** 3135-3144.

Han, K., Jeong, H.J., Yang, H.B., Kang, S.M., Kwon, J.K., Kim, S. and Kang, B.C. (2016) *DNA Res.*, **23:** 81-91.

Hari, G.S., Kumar, A.K. and Reddy, A.V. (2009) *Research on Crops*, **10**: 101-102.

Hariharan, M. and Unnikrishan. (1983) *Acta Bota. Ind.*, **11:** 161-163.

Harini, I. and Sita, G.L. (1993) *Plant Sci. Limer.*, **89:** 107-112.

Hasan, M.J., Kulsum, M.U., Ullah, M.Z., Hossain, M. and Mahmud, M.E. (2014) *Int. J. Agric. Res. Innov. Technol.*, **4:** 32-35.

Hasan, T.S.M. and Soehardi (1978) *Bull. Penelitian Hortikultura*, **6:** 21-27.

Hasanuzzaman, M., Hakim, M.A., Fersdous, J., Islam, M.M. and Rahman, L. (2012) *Plant Omics J.*, **5**: 337-344.

Hasnat, R., Abbasi, N.A., Hafiz, I.A., Ahmad, T. and Chudhary, Z. (2008) *Pak. J. Bot.*, **40:** 2655-2662.

Hazra, P., Chattopadhyay, A., Karmakar, K. and Dutta, S. (2011) In: *Modern Technology in Vegetable Production*. New India Publishing Agency, New Delhi, p. 413.

Heiser (Jr.), C.B. (1976) *Evolution of Crop Plants*, (Ed. N.W. Simmonds), Longman, London, pp. 265-268.

Hibberd, A. M., Stall, R.E. and Bassett, M.J. (1984) *Phytopathol.*, **74:** 791.

Hibberd, A.M., Bassett, M.J. and Stall, R.E. (1987) *Phytopathol.*, **77:** 1304-1307.

Hirota, O., Villavicencio, E., Chikushi, J., Takeuchi, S. and Nakano, Y. (1999) *J. Fac. Agric.*, Kyushu University, **44:** 39-47.

Hoang, N.H., Yang, H.B. and Kang, B.C. (2013) *Sci. Hort.*, **161:** 8-14.

Homes, F.O. (1937) *Phytopathol.*, **24:** 984-1002.

Housain, S.A., Mohammad, S. and Rao, B.V.R. (1989) *Indian J. Agron.*, **34:** 117-118.

Huh, J.H., Kang, B.C., Nahm, S.H., Kim, S., Ha, K.S., Lee, M.H. and Kim, B.D. (2001) *Theor. Appl. Genet.*, **102**: 524-530.

Huh, S.U., Choi, L.M., Lee, G.J., Kim, Y.J. and Paek, K.H. (2012a) *Plant Sci.*, **197:** 50-58.

Huh, S.U., Kim, K.J. and Paek, K.H. (2012b) *Biochem. Biophys. Res. Comm.*, **417:** 910-917.

Hulse Kemp, A.M., Ashrafi, H., Plieske, J., Lemm, J., Stoffel, K., Hill, T. and Deynze, A.V. (2016) *Hortic. Res.*, **3:** 16036.

Hurtado-Hernandez, H. and Smith, P.G. (1985). *J. Hered.*, **76:** 211-213.

Husain, LR., Ghode, P.B., Desmukh, D.T., Patil, B.M. and Sapate, B.G. (1998) *Abst. Silver Jubilee Nat. Sump. Emerging Scenario in Vegetable Research and Development*, 12-14, December, 1998, Varanasi, India, p.117.

Hwang, J.K., Paek, K.Y and Cho, C.H. (1998) *Acta Hortic.*, **461:** 301-307.

Ionescu, A., Tanasescu, M., Suta, E., Tanasescu, C., Ionescu, C. and Poncu, J. (1989) *Cereetari de Genetica Vegetala Animala.*, **1:** 237-245.

Iruthayaraj, M.R. (1970) *Pl. J. and Agriest.*, **47:** 257-261.

Iruthayaraj, M.R. and Kulandaivel (1973) *Indian J. Agron.*, **13**: 22-24.

Jabeen, N., Ahmad, N. and Tanki, M.I. (1999) *Appl. Biol. Res.*, **1:** 87-89.

Jamal Hussain, H.S. (1977) *M.Sc. (Ag.) Thesis Tamil Nadu Agricultural University*, Coimbatore.

Janaki, M.L., Naidu, N.L., Ramana, C.V. and Rao, M.P. (2015) *Plant Arch.*, **15**: 441-446.

Jayashankar, S., Bagga, S. and Phillips, G.C. (1997) *HortScience*, **32:** 454.

Jeon, J.M., Ahn, N.Y., Son, B.H., Kim, C.Y., Han, C.D. and Kim, G.D. (2007). *Plant Cell Tiss. Org.Cult.*, **88:** 225-232.

Jeong, Y.O. and Cho, J.L.(1996) *J. Korean Soc. Hort. Sci.*, **37**: 522-525.

Jeyalakshmi, C., Durairaj, P., Seetharaman, K. and Sivaprakasam, K. (1998) *Indian Phytopath.*, **51:** 180-183.

Jharia, H.K., Khare, M. N. and Chand, A. (1978) *C11rr. Res.*, **7:** 46-48.

Jiang, J.Z., Wang, D.H., Wang, Z.Y. and Han, Y. S. (1987) *Sci. Agric. Sinica*, **20:** 39-43.

Jiang, Z.R. and Li, C.L. (1984) *Acta Hort. Sinica*, **11:** 191-194.

Johnson, H.W., Rabinson, H.F. and Comstock, R.E. (1955) *Agron. J.*, **47**: 314-318.

Jones, J.B., Lacy, G.H., Bouzar, H., Stall, R.E. and Schaad, N.W. (2004) *Syst. Appl. Micro.*, **27:** 755-762.

Jones, J.B., Minsavage, G.V., Roberts, P.D., Johnson, R.R., Kousik, C.S., Subramanian. S. and Stall, R.E. (2002) *Phytopathol.*, **92:** 273-277.

Jose, L., Khader, K.M.A. and Jose, L. (2002) *Capsicum Eggplant Newsl.*, **21:** 56-59.

Joshi, M.C. and Singh, D.P. (1975) *Indian Hort.*, **20**: 19-21.

Kaliappan, R. and Rajagopal, A. (1970) *Madras Agric. J.*, **57:** 233.

Kamalan, N. and Rajamoni, T.S. (1963) *Indian J. Agric. Sci.*, **35:** 67-68.

Kang, B.C., Nahm, S.H., Huh, J. H., Yoo, H. S., Yu, J.W., Lee, M.H. and Kim, B.D. (2001) *Theor. Appl. Genet*, **102:** 531-539.

Kang, W.H., Hoang, N.H., Yang, H.B., Kwon, J.K., Jo, S.H., Seo, J.K., Kim, K.H., Choi, D. and Kang, B.C. (2010) *Theor. Appl. Genet.*, **120:** 1587-1596.

Kanwar, J.S. (1999) *Abst. 4th Agril. Sci. Cong.* Sustainable Agricultural Exportheldal Jaipur, India, 21-24 February, 1999, p. 177.

Kataria, G.J., Pandey, H.M. and Vaddaria, M.A. (1997) *Gujarat Agric. Uni. Res. J.*, **22**: 18-21.

Kehie, M., Kumaria, S. and Tandon, P. (2013) *Sci. Horti.*, **164:** 1-8.

Khare, R.D. and Umrani, K. (1969) *Indian J. Hort.*, **26:** 63-68.

Khatri, H.L. and Sekhon, I.S. (1974) *Indian J. Agric. Sci.*, **43:** 667-669.

Kim, B.S. and Hartmann, R.W. (1985) *Plant Dis.*, **69:** 233-235.

Kim, D.H., Kang, J.G. and Kim, B.D. (2007) *Plant Mol. Biol.*, **63:** 519-532.

Kim, D.S. and Hwang, B.K. (2012) *Plant J.*, **72**: 843-855.

Kim, H.K. and Ho, Q.S. (1978) *J. Korean Soc. Hori. Sci.*, **19:** 110-116.

Kim, M., Kim, S., Kim, S., Kim, B.D. (2001) *Mol. Cells*, **11:** 213-219.

Kim, S.H., Kim, Y.H., Lee, Z.W., Kim, B.D. and Ha, K.S. (1997) *J. Korean Soc. Hort. Sci.*, **38**: 384-390.

Kim, S.J., Lee, S.J., Kim, B.D. and Paek, K.H. (1997) *Plant Cell Rep.*, **16:** 825-830.

Kobata, K., Todo, T., Yazawa, S., Iwai, K. and Watanabe, T. (1998) *J. Agric. Food Chem.*, **46:** 1695-1697.

Koeda, S., Sato, K., Saito, H., Nagano, A.J., Yasugi, M., Kudoh, H. and Tanaka, Y. (2019) *Theor. Appl. Genet.* **132**: 65-80.

Kolte, S.O., Korekar, V. and Lokhande, N.M. (1994) *Orissa J. Agric. Res.*, **7:** 80-84.

Korla, and Rastogi, K. (1977) *Punjab Hort. J.*, **17:** 155-156.

Krishanmurthy, S.L., Reddy, M.K. and Mohanrao, A. (2013) *Veg. Sci.*, **40**: 210-213.

Krishna, A., Nema, A.G., Nadkami, P.G. and Bhalla, P.L. (1983) *Proc. Nat. Seminar Tomato,* Chillies, Tamil Nadu Agric. Univ., Coimbatore.

Kristiansen, K. and Andersen, S.B. (1993) *Euphytica*, **67:** 105-109.

Kristof, L. and Barnabas, B. (1983) *Kerteszeti Egyetem Kozlemenyei*, **47:** 59-65.

Kulkarni, M. and Phalke, S. (2009) *Sci. Horti.* **120:** 159-166.

Kumar, O.A., Panda, R.C. and Rao, K.G.R. (1988) *Euphytica*, **39:** 47-51.

Kumari, A.L., Reddy, G.K. and Savaji, J.N. (1998) *Abst. Silver Jubilee Nat. Sump. Emerging Scenario in Vegetable Research and Development*, 12-14 December, 1998, Varanasi, India, p. 12.

Kurucheve, V and Padmavathi, R. (1998) *Indian Phytopath.*, **51:** 379-381.

Kyle, M.M. and Palloix, A. (1997) *Euphytica*, **97:** 183-188.

Ladd, T.L.Jr., Reichard, D.L. and Krueger, H.R. (1982) *J. Econ. Ent.*, **75:** 879-881.

Lawande, K., Musmade, A.M., Memane, S.A., Pawar, D.B., Supe, V.S., Kulkarni, S.S. and Warade, S.D. (1998) *Abst. Silver Jubilee Nat. Sump. Emerging Scenario in Vegetable Research and Development*, 12-14 December, 1998, Varanasi, India, p. 64.

Lee, E.K., Kim, J.S., Lee, Y.K. and Lee, Y.B. (1998) *J. Korean Soc. Hort. Sci.*, **39:** 670-675.

Lee, H.R., Bae, I.H., Park, S.W., Kim, H.J., Min, W.K., Han, J.H., Kim, B.D. (2009) *Mol. Cells*, **27:** 21-37.

Lee, J., Han, J.H., An, C.G., Lee, W.P. and Yoon, J.B. (2010a) *Breed Sci.*, **60:** 93-98.

Lee, J., Park, S.J., Do, J.W., Han, J.H., Choi, D., and Yoon, J. B. (2013b) *Kor. J. Hort. Sci. Tech.*, **31:** 473-482.

Lee, J., Yoon, J.B., Han, J.H., Lee, W.P., Do, J.W., Ryu, H., Kim, S.H. and Park, H.G. (2010b) *Plant Breed.*, **129:** 35-38.

Lee, J., Yoon, J.B., Han, J.H., Lee, W.P., Kim, S.H. and Park, H.G. (2010c) *Euphytica*, **173:** 55-61.

Lee, J.M., Jahn, M.M. and Yeam, I. (2013a) *Euphytica*, **194**: 417-424.

Lee, J.M., Nahm, S.H., Kim, Y.M., and Kim, B.D. (2004b) *Theor. Appl. Genet.*, **108:** 619-627.

Lee, S.J., Kim, B.D. and Paek, K.H. (1993) *Korean J. Plant Tiss. Cult.*, **20:** 289-294.

Lee, Y.H., Kim, H.S., Kim, J.Y., Jung, M., Park, Y.S. and Lee, J.S. (2004a) *Plant Cell Rep.*, **23:** 50-58.

Lee, Y.R., Yoon, J. B. and Lee, J. (2016) *Mol. Breed.*, **36:** 61.

Lefebvre, V., Palloix, A., Caranta, C. and Pochard, E. (1995) *Genome*, **38:** 112-121.

Lefebvre, V., Pflieger, S., Thabuis, A., Caranta, C., Blattes, A., Chauvet, J.C. and Palloix, A. (2002) *Genome*, **45:** 839-854.

Levy, A., Palenitch, D., Mehagem, E., Admati, E., Shifriss, C. and Barzilai, M. (1983) *Hassadeh*, **63:** 1626-1628.

Li, C.L. and Jiang (1990) *Acta Hortic. Senica*, **17:** 39-44.

Li, C.L., Ye, B.J. and Zhu, D.W. (1995) *Acta Hortic.*, **402:** 442-444.

Li, D., Zhao, K., Xie, B., Zhang, B. and Luo, K. (2003) *Plant Cell Rep.*, **21:** 785-788.

Li, P., Feng, B., Wang, H., Tooley, P.W. and Zhang, X. (2011) *J. Basic Microbiol.*, **51:** 61-70.

Li, W., Cheng, J., Wu, Z., Qin, C., Tan, S., Tang, X. and Hu, K. (2015) *Mol. Breed.*, **35:** 32.

Lim, H.T., Lee, G.Y., You, Y.S., Park, E.J., Song, Y.N., Yang, D.C. and Choi, K.H. (1999) *Acta Hortic.*, **483:** 387-396.

Lim, J.H., Park, C.J., Huh, S.U., Choi, L.M., Lee, G.J., Kim, Y.J. and Paek, K.H. (2011) *Biochem. Biophys. Res. Comm.*, **411:** 613-619.

Lin, S., Chou, Y., Shieh, H., Ebert, A. and Kumar, S. (2013) *Chron. Hort.*, **53**: 21-27.

Lippert, F. (1975) *Crop Sci.*, **15:** 323-325.

Lippert, L.F., Bergh, B.O. and Smith, P.G. (1965). *J. Hered.*, **56**: 30-34.

Liu, W., Parrott, W.A., Hildebrand, D.F., Collins, G.B. and Williams, *E.g.* (1990) *Plant Cell Rep.*, **9:** 360-364.

Liu, W.Y., Kang, J.H., Jeong, H.S., Choi, H.J. and Yang, H.B. (2014) *Theor. Appl. Genet.*, **127**: 2503-2513.

Livingstone, K.D., Lackney, V. K., Blauth, J. R., van Wijk, R. and Jahn, M. K. (1999) *Genetics*, **152:** 1183-1202.

Loaiza-Figueroa, F. and Tanksley, S.D. (1988) *J. Hered.*, **79:** 314-315.

Lush, J.L. (1949) *Proceedings of the American Society of Animal Production*, **33:** 293-301.

Mahalanobis, P. (1936) *Proceedings of the National Institute of Sciences in India*, **12:** 49-55.

Mahasuk, P., Chinthaisong, J. and Mongkolporn, O. (2013) *Breed. Sci.*, **63**: 333-338.

Mahasuk, P., Struss, D. and Mongkolporn, O. (2016) *Mol. Breed.*, **36**: 1-10.

Mahendran, S. and Bandara, D.C. (2000) *Trop. Agric. Res.*, **12:** 95-106.

Mahto, B.K., Sharma, P., Rajam, M.V., Reddy, P.M. and Dhar-Ray Swatismita (2018) *Indian J. Plant Physiol.*, **23**(3): 573–581.

Mak, C. (1987) *Capsicum News L.*, No.6, pp. 47-48.

Malkanthi, D.R.R., Yokoyama, K., Yoshida, T., Moritsuga, M. and Matsushita, K. (1995) *Soil Sci. Plant Nutri.*, **41:** 161-165.

Mallanagouda, B., Sulikeri, G.S., Murthy, B.G. and Prathibha, N.C. (1995) *Indian J. Agron.*, **40:** 277-279.

Mallard, S., Cantet, M., Massire, A., Bachellez, A., Ewert, S. and Lefebvre, V. (2013) *Mol. Breed.*, **32:** 349-364.

Manoharan, M., Sree Vidya, C.S. and Lakshmi Sita, G. (1998) *Plant Sci.*, **131:** 77-83.

Manoranjitham, S.K., Prakasam, V., Rajappan, K. and Amutha, G. (2000) *J. Mycol. Pl. Pathol.*, **30:** 225-228.

Manzur, J.P., Oliva-Alarcón, M. and Rodríguez-Burruezo, A. (2014) *Sci. Hort.*, **170:** 203-210.

Martínez-Flórez, S., González-Gallegos, J., Culebras, J.M. and Tuñon, M.J. (2002) Nutr. *Hosp.*, **7:** 271-278.

Marwoto, B. (1988) *Buletin Penelitian Hortkultura*, **16:** 1-4.

Mathur, R., Dangi, R.S., Dass, S.C. and Malhotra, R.C. (2000) *Curr. Sci.*, **79:** 287-288.

Matsubara, S., Yamamoto, M., Man, H.J. and Murakami, K. (1998) *Scientific Rep. Fac. Agri. Okayama Univ.*, **87:** 117-122.

Matsunaga, H. and Monma, S. (1999) *J. Japanese Soc. Hort. Sci.*, **68:** 753-761.

Matsunaga, H., Saito, T., Hirai, M., Nunome, T. and Yoshida, T. (2003) *J. Jpn. Soc. Hort. Sci.*, **72:** 218-220.

Matsunaga, H., Sato, T. and Monma, S. (1998) *Proceedings of 10th Eucarpia meeting on genetics and breeding of capsicum and eggplant*, 7-11 Sept 1998, Avignon, France, pp. 172.

Matta, A. and Garibaldi, A. (1980) *Informatore Fitopatologico*, **30:** 17-20.

Mc Cammon, K.R. and Honma, S. (1984) *Theor. Appl. Genet.*, **68:** 541-545.

McRae, W. (1924) *Econ. Bot. Part III Mycology Aim. Rept. Board Scientific*, India.

Mehrotra, O.N., Saxena, H. K. and Misra, P.H. (1968) *Fertl. News*, **13:** 34-36.

Meshram, L.D. (1987) *PKV Res. J.*, **11:** 104-106.

Meshram, L.D. and Narkhade, M. N. (1982) *Euphytica*, **31:** 1003-1005.

Mimura, Y., Inoue, T., Minamiyama, Y. and Kubo, N. (2012) *Breeding Sci.*, **62:** 93-98.

Minamiyama, Y., Tsuro, M., and Hirai, M. (2006). *Mol. Breed.*, **18:** 157-169.

Miniraj, N. and Shanmugavelu, K.G. (1987) *South Indian Hort.*, **35:** 362-366.

Mishra, R.S., Lotha, R.E., Mishra, S.N., Paul, P.K. and Mishra, H.N. (1989) *Capsicum News Letter*, No. **7:** 49-50.

Mityko, J. and Fari, M. (1997) *Acta Hortic.*, **447:** 281-287.

Mityko, J., Andrasfalvy, A., Csillery, G. and Fari, M. (1995) *Plant Breed.*, **114:** 78-80.

Moreira, S.O., Rodrigues, R., Araújo, M.L. and Araújo, M.L. (2009) *Ciênc. Rural.*, **39:** 1387-1393.

Morrison, R.A. (1987) *Dissertation Abst. Internatl. B. Sci. Eng.*, **48:** 187.

Mote, U.N. (1977) *Veg. Sci.*, **4:** 148-150.

Mote, U.N., Patil, A.V. and Wavhal, K.N. (1975) *Res. J. Maharashtra Pirnie Agric. Univ.*, **6:** 57-60.

Moulin, M.M., Rodrigues, R., Ramos, H.C.C., Bento, C.S., Sudre, C.P., Goncalves, L.S.A. and Viana, A.P. (2015) *Gen. Mol. Res.*, **14:** 6683-6694.

Moury, B., Palloix, A., Selassie, K.G. and Marchoux, G. (1997a) *Euphytica*, **94:** 45-52.

Moury, B., Pflieger, S., Blattes, A., Lefebvre, V. and Palloix, A. (2000) *Genome*, **43:** 943-951.

Moury, B., Selassie-Gebre, K., Marchoux, G., Daubeze, A.M. and Palloix, A. (1997b) *European J. Plant Pathol.*, **104:** 489-498.

Munshi, A.D., Behra, T.K. and Singh, G. (2000) *Indian J. Hortic.*, **57**: 157-159.

Munsi, P.S. and Sadhukhan, S.K. (1998) *J. Andaman Sci. Assoc.*, **14:** 70-72.

Murmu, D.K., Das, B., Yanzone, R. and Barman, R. (2017) *Int. J. Curr. Microbiol. Appl. Sci.*, **6**: 2002-2012.

Murphy, J.P. and Kyle, M.M. (1994) *Plant Cell Rep.*, **13:** 397-400.

Murthy, N.S.C., Bavaji, J.N. and Rao, Y.R. (1987) *Capsicum Newsl.*, No.6, pp. 54-55.

Murthy, N.S.R and Murthy, B.S. (1962) *Andhra Agric. J.*, **59:** 161-165.

Murthy, N.S.R. and Lakshmi, N. (1981) *Indian J. Bot.*, **4:** 42-47.

Murty, N.S.C. and Bavaji, J.N. (1980) *Indian Cocoa Arecanut Spices. J.*, **4:** 36-38.

Muthukrishnan, C.R., Thangaraj, T. and Muthuswami, S. (1983) *Proc. Nat. Seminar Tomato*, Chillies, Tamil Nadu Agric. Univ., Coimbatore, p. 111.

Naeem, N., Muhammad, I., Khan, J., Nabi, G., Muhammad, N. and Badshah, N. (2002) *Asian J. Plant Sci.*, **1:** 599-601.

Nagaraja, A., Jagadeesha, R.C. and Yogeesha, H.S. (1998) *Karnataka J. Agric. Sci.*, **11:** 807-809.

Narayan, S., Makhdoomi, M.I., Malik, A., Nabi, A., Hussain, K. and Khan, F.A. (2017) *J. Pharmacog. Phytochem.*, **6**: 1733-1735.

Nagdy, G.A., Fouad, M.K. and Mahmoud, W.S. (1979) *Res. Bull. Fac. of Agric. Ainshams Univ.*, **1150**: 17.

Nandadevi and Hosamani, R.M. (2003) *Capsicum Eggplant Newsl.*, **22:** 43-46.

Nath, P. (1976) *Vegetables for the Tropical Region*, ICAR, New Delhi, p. 21.

Nathu, L. and Pundrick, K.C. (1971) *Punjab Hort. J.*, **11:** 82-86.

Nijamudeen, M.S. and Dharmasena, P.B. (2012) *Annals of the Sri Lanka Department of Agriculture*, **4:** 1-6.

Nisar, A., Krishnappa, K.M., Vpperi, S.N. and Khot, A. (1987) *Curr. Res.*, **16**: 161-162.

Nofiyanto, R.T., Wati, V.R., Setiawati, S.R., Noviandi, W.D., Kuscahyanti, A. and Fuskhah, E. (2017) *IOP Conf. Series: Earth and Environmental Science* **102**, doi: 10.1088/1755-1315/102/1/012070.

Novac, F., Betlach, J. and Dubovshi, J. (1971) *Z. Pflanzenzucht*, **65:** 129-140.

Novak, F.J. (1974) *Zeitschrift fur Pflanzenzuchtung*, **72:** 46-54.

Nsabiyera, V., Ochwo-ssemakula, M., Sseruwagi, P., Ojewo, C. and Gibson P. (2013) *Int. J. Plant Breed.*, **7**(1): 12-21.

Nunez-Palenius, H.G. and Ochoa-Alejo, N. (2005) *In Vitro Cellular Dev. Bio-Plt.*, **41:** 801-805.

Odland, M.L. (1960) *Proc. Am. Soc. Hort. Sci.*, **76:** 475-481.

Odland, M.L. and Porter, A.M. (1938) *Proc. Am. Soc. Hort. Sci.*, **36:** 647-651.

Odland, M.L. and Porter, P.M. (1941) *Proc. Amer. Soc. Hori. Sci.*, **38:** 385-388.

Oh, S.K., Baek, K.H., Seong, E.S., Joung, Y.H., Choi, G.J., Park, J.M., Cho, H.S., Kim, E.A., Lee, S., Choi D (2010) *Plant Physiol.*, **154:** 245-261.

Ohta, Y. (1962) *Seiken ziho*, **11:** 73-77.

Ortega, R.G., Espanol, C.P. and Zueco, J.C. (1992) *Plant Breed.*, **108**: 118-125.

Pak, H.Y., Son, K.C., Gu, *E.g.*, Lim, K.B. and Kim, B.H. (1996) *J. Korean Soc. Hort. Sci.*, **37:** 568-576.

Palaniswamy, S. and Ragini, J.C. (2000) *Insect Environ.*, **6:** 25-26.

Panchal, S.C., Bhatnagar R., Momin R.A. and Chauhan, N.P. (2001) *Capsicum Eggplant Newsl.*, **20:** 19-22.

Pandeva, R., Zagorska, N. and Simonova, N. (1990) *Genet. Breed.* (Bulg), **23**: 137-145.

Pandiyaraj, P., Saraladevi, D., Hepziba, S.J. and Das, A. (2017) *Int. J. Agric. Sci.*, **9**(14): 4081-4083.

Paran, I., van der Voort, J.R., Lefebvre, V., Jahn, M., Landry, L., van Schriek, M. and Peleman, J. (2004) *Mol. Breed.*, **13:** 251-261.

Park, H.K., Kim, B.S. and Lee, W.S. (1990) *J. Kor. Soc. Hort. Sci.*, **31:** 91-105.

Park, S.W., Jung, J.K., Choi, E.A., Kwon, J.K., Kang, J.H., Jahn, M. and Kang, B.C. (2014) *Mol. Breed.*, **34:** 963-975.

Parthasarathy, V.A. and Niral, V. (2001) *Biotechnology of horticultural crops*, **2**: 146-173.

Patel, J.J., Patel, N.C., Patel, B.H. and Patel, J.R.(1999) *Indian J. Entom.*, **61:** 345-350.

Patel, N.C., Patel, J.J., Jayani, D.B. and Patel, J.R. (1997) *Indian J. Agril. Sci.*, **67:** 75-76.

Pathak, C.S., Deshpande, A.A. and Singh, D.P. (1985) *Capsicum Newsl.*, **4:** 41-42.

Pathak, C.S., Singh, D.P. and Deshpande, A.A. (1983) *Capsicum Newsl.*, **2:** 99-100.

Patil, K.N. and Nagaraja, A. (1999) *Karnataka J. Agric. Sci.*, **12:** 189-190.

Patil, U.D., Sangale, P.B. and Desai, B.B. (1985) *Curr. Res. Reptr.*, **1**: 39-43.

Patnaik, H.P. and Mohapatra, S. (1997) *Mysore J. Agril. Sci.*, **31:** 159-163.

Pegard, A., Brizzard, G., Fazari, A., Soucaze, O., Abad, P. and Djian-Caporalino, C. (2005) *Phyt.*, **95**: 158-165.

Pereira, M.J.Z., Massola, J.N.S, Sussel, A.A.B., Sala, F.C., Costa, C.P. and Boiteux, L.S. (2011) *Hortic. Brasil.*, **29**: 569-576.

Perry, L., Dickau, R., Zarrillo, S., Holst, I., Pearsal, D.M., Piperno, D.R., Berman, M.J., Cooke, R.G., Rademaker, K., Ranere, A.J., Raymond, J.S., Sandweiss, D.H., Scaramelli, F., Tarble, K. and Zeidler, J.A. (2007) *Sci.*, **315:** 986-988.

Peter, K.V. and Pradeep kumar, T. (1999) *Abst. Silver Jubilee Nat. S11mp. Emerging Scenario in Vegetable Research and Development*, 12-14 December, 1998, Varanasi, India, pp.141-148.

Peter, K.V., Goth, R. W. and Webb, R.E. (1984) *HortScience*, **19:** 277.

Peterson, P.A. (1959) *Genetics*, **44:** 407-419.

Platenius, H. (1934) *Plant Physiol.*, **9**: 671-680.

Popovsky, S. and Paran, I. (2000) *Theor. Appl. Genet.*, **101:** 86-89.

Prabhakar, B.S., Srinivas, K. and Shukla, V. (1987) *Indian Cocoa, Arecanut and Species J.*, **11**.

Prabhakar, M. and Hebbar, S.S. (1998) *In Water and nutrient management for sustainable production and quality of spices: Proceedings of the National Seminar*, Madikeri, Karnataka, India, pp.131-135.

Prabhakar, M. and Naik, L.B. (1997) *Ann. Agric. Res.*, **18:** 34-39.

Prasad, B.K., Sahoo, D.R., Kumar, M. and Narayan, N. (2000) *Indian Phytopath.*, **53:** 42-44.

Prasad, S.K. (1960) *Indian J. Ent.*, **22:** 127-28.

Prince, J.P., Pochard, E. and Tanksley, S.D. (1993) *Genome*, **36:** 404-417.

Purkayastha, J., Alam, S.I., Gogoi, H.K. and Singh, L. (2012) *Ozean J. App. Sci.*, **5:** 1.

Purseglove, J.W. (1977) Tropical Crops Dicotyledons I and II. Longman, London, pp. 524-525.

Pushpa, G. and Shambhulingappa, K.G. (1981) *Sci. Cul.*, **47:** 61-63.

Puttaswamy and Gowda, B.L.V. (1976) *Curr. Res.*, **5**: 100.

Quirin, E.A., Ogundiwin, E.A., Prince, J.P., Mazourek, M., Briggs, M.O., Chlanda, T.S., Kim, K.T., Falise, M., Kang, B.C. and Jahn, M.M. (2005) *Theor. Appl. Genet.*, **110:** 605-612.

RadheShyam, Arora, S.K. and Singh, J. (1996) *Ann. Agric. Res.*, **17:** 311-312.

Rahman, M.A., Chiranjeevi, C., Reddy, T.B. and Bhattiprolu, S.L. (1998) *Abst. Silver Jubilee Nat.Sump. Emerging Scenario in Vegetable Research and Development*, 12-14 December, 1998, Varanasi, India, p.180.

Rajagopal, A. and Muthukrishnan, C.R. (1980) *South Indian Hort.*, **28:** 8-11.

Rajagopal, A., Khader, M., Subbiah, K.K. and Hussain, J. (1976) *South Indian Hort.*, **24: 73**.

Rajagopalan, P., Seshadri, A.R. and Muthukrishnan, T. S. (1971b) *Madras Agric. J.*, **58:** 843.

Rajagopalan, P., Seshadri, A.R. and Muthukrishnan, T.S. (197la) *Madras Agric. J.*, **58:** 604.

Rajput, H.D., Singh, K., and Kushwha, H.S. (2003) *Indian J. Agron.*, **48**: 136-138.

Rajput, J.C. and Parulekar, Y.R. (1996) In: *Handbook of Vegetable Science and Technology*, Marcel Dekker, Inc. New Work.

Raju, D.N.N., Singh, D.P. and Pathak, C.S. (1984) *Acta Agron. Academiae Scientiarum*, **33:** 173-177.

Raju, K.S. and Rao, G.S. (1981) *Indian Cocoa, Arecanut, Spices J.*, **5**: 27.

Raju, K.S. and Srinivasa, G. (1981) *Haryana J. Hort. Sci.*, **10:** 86-87.

Ramachandran, R.K. (2013) In: *Breeding of vegetable crops*. Narendra Publishing House, pp. 205.

Ramanathan, N. (1965) *Fert. News*, **10:** 7-12.

Ramchandran, S. and Subbiah, K.K. (1981) *South Indian Hort.*, **29:** 178-181.

Ramchiary, N., Kehie, M., Brahma, V., Kumaria, S. and Tandon, P. (2014) *Plant Biotech. Rep.*, **8:** 101–123.

Rangarajan, A.V., Janagarajan, A. and Sivagami, R. (1974) *Andhra Agric. J.*, **21:** 95.

Rao, C.H. and Lal, G. (1986) *Veg. Sci.*, **13:** 17-21.

Rao, D.M. and Ahmed, K. (1983) *Proc. Nat. Seminar Tomato, Chillies Tamil Nadu Agric. Univ.*, Coimbatore, pp. 160-162.

Rao, G.U., Ben Chaim, A., Borovsky, Y. and Paran, I. (2003) *Theor. App. Gene.*, **106:** 1457-1466.

Rao, P.V. and Chhonkar, V.S. (1984) *South Indian Hort.*, **22:** 26-32.

Rao, V.B. and Suryanarayana, V. (1983) *Indian Cocoa, Arecanut Spices J.*, **6:** 68-70.

Raw, A. (2000) *Ann. Bot.*, **85**: 487-492.

Reddy, A.S., Rao, G.S., Rao, B.H.K. and Narayana, K.L. (1981) *South Indian Hort.*, **29:** 192-195.

Reddy, D.N.R. and Jagadish, A. (1980) *Pesticides*, **4**: 22-23.

Reddy, G.K., Kumari, A.L. and Bavaji, J.N. (1998) In: *Silver Jubilee Nat. Symp. Emerging Scenario in Vegetable Research and Development*, 12-14 December, 1998, Varanasi, India, p.10.

Reeves, G., Monroy-Barbosa, A. and Bosland, P.W. (2013) *Phyto.*, **103:** 472-478.

Relwani, L.L. (1963) *Indian J. Agron.*, **7:** 291-299.

Restaino, F. (1989) In Proceedings of 7th EUCARPIA meeting on genetics and breeding of Capsicum and eggplant, 27-30 June 1989, Palanka, Yugoslavia, pp 55-59.

Revanappa, Nalawadi, U.G. and Chetti, M.B. (1998) *Karnataka J. Agric. Sci.*, **11:** 122-126.

Rhee, D.A. and Park, S.K. (1975) *J. Korean Soc. Hort. Sci.*, **16:** 99-105.

Rhoades, H.L. (1981) *Nematropica*, **11:** 121-127.

Ribeiro, A. and Costa, C.P. (1990) *Brazilian J. Genet.*, **13**: 815-823.

Richter, C., Heiligtag, B., Nagieb, M. and Mukhallalti, U. (1999) *Tropenlandwirt*, **100**: 3-16.

Rishi, N. and Dhawan, P. (1988) *Plant Dis. Res.*, **3**: 69-70.

Ristaino, J.B. (1990) *Phytopathol.*, **80**: 1253-1259.

Rodriguez-Uribe, L., Guzman, I., Rajapakse, W., Richins, R.D. and O'Connell, M.A. (2012) *J. Exp. Bot.*, **63:** 517-526.

Rohini, N. and Lakshmanan, V. (2016) *J. App. Nat. Sci.*, **1**: 25-32.

Rosmaina, Syafrudin, Hasrol, F., Yanti, Juliyanti and Zulfahmi. (2016) *Bulgarian J. Agric. Sci.*, **22**: 431-436.

Rylski, I. (1972) *J. Amer. Soc. Hort. Sci.*, **97**: 648-651.

Sahoo, S.C., Mishra, S.N. and Mishra, R.S. (1990) *Indian J. Agric. Sci.*, **60:** 834-835.

Saimbhi, M.S. and Randhawa, K.S. (1976) *Punjab Hort. J.*, **16:** 146-148.

Saimbhi, M.S., Kan, G. and Nanduri, K.S. (1977) *Qualitas Plantarum*, **27:** 171-175.

Saivaraj, K., Kumaraswamy, T. and Jayaraj, S. (1979) *Pesticides*, **13**: 20-21.

Sangar, R.B.S., Katwale, T. R., Saraf, R.K. and Parihar, H.S. (1988) *Farm Sci. J.*, **3:** 69-71.

Sanghi, A.K., Bhatnagar, M.P. and Sharma, K. (1964) *Indian J. Genet. Plant Breed.*, **24:** 164-167.

Sankaralingam, A., Suthanthirapandian, I.R. and Ravikumar, V. (1983) *Proc. Nat. Seminar Tomato, Chillies, Tamil Nadu Agric. Univ*, Coimbatore, pp.165.

Sarkar, S. (2006) *Ph.D. Thesis*, submitted to Bidhan Chandra Krishi Viswavidyalaya, West Bengal, India.

Saxena, P.K., Gill, R., Rashid, A. and Maheshwari, S.C. (1981) *Protoplasma*, **90:** 357-360.

Sayed, S. and Bagavandas, M. (1980) *South Indian Hort.*, **28:** 42-47.

Schweiggert, U., Kurz, C., Schieber, A. and Carle, R. (2007) *Eur. Food Res. Technol.*, **225**: 261–270.

Scoville, W.L. (1912) *J. Am. Pharmal. Asso.*, **1**: 453-454.

Selassie, G.K., Pochard, E., Marchoux, G. and Thouvenel, J.C. (1982) *VI Meeting on Genetics and Breeding on Capsicum and Eggplant*, 21-24 October, 1986, Zaragoza, Spain, pp. 189-191.

Selvaraj, K.V., Iqbal, I.M., Dwood, M.S., Krishnasamy, S.M., Muralidharan, V. and Hariharan, M.S. (1998) *Proceedings of the National Seminar*, Madikeri, Karnataka, India, pp.161-163.

Selvaraj, K.V., Subramaniam, S. and Rajasekharan, R. (1974) *Madras Agric. J.*, **61:** 889.

Seo, E.Y., Yeom, S.I., Jo, S.H., Jeong, H.J., Kang, B.C. and Choi, D.I. (2013) *Mol. Cells*, **33:** 415-422.

Sethupathi, Ramalingam (1978) *Madras Agric. J.*, **62:** 84-86.

Shah, Lal, S.D and Panth. C.C. (1986) *Prog. Hortic.*, **18:** 270-272.

Sharifi-Tehrani, A. and Omati, F. (1999) *Proc. 51st Int. Symp. Crop protection*, Gent. Belgium, 4 May, 1999, Part-II, pp. 419-423.

Sharma, J.P., Thikekar, D.G. and Singh, I.J. (1980) *Balwant Vidyapeth J. Agric. Sci. Res.*, **19:** 36-41.

Sharma, R. and Kumar, R. (2017) *Int. J. Pure App. Biosci.*, **5:** 524-527.

Sharma, S.N., Goyal, K.C. and Kakralya, B.L. (1998) *Seed Res.*, **26:** 83-86.

Sharma, S.R. and Sohi, H. S. (1981) *Indian J. Mycol. Plant Path.*, **11:** 30-34.

Sheela, K.B., George, T.E. and Peter, K.V. (1998) *Abst. Silver Jubilee Nat. Sump. Emerging Scenario in Vegetable Research and Development*, 12-14 December, 1998, Varanasi, India, p.11.

Shifriss, C. (1973) *Euphytica*, **22:** 527-529.

Shifriss, C. (1987) *HortSci.*, **22:** 944-945.

Shifriss, C. and Frankel, R. (1969) *J. Amer. Soc. Hort. Sci.*, **94:** 385- 387.

Shifriss, C. and Rylski, I. (1972) *Hort. Sci.*, **7:** 36.

Shifriss, C., Pilowsky, M. and Zacks, J.M. (1992) *Phytoparasitica*, **20:** 279-283.

Shin, R., Han, J.H., Lee, G.J., and Park, K.H. (2002b) *Transgenic Res.*, **11:** 215-219.

Shin, R., Park, J.M., An, J.M. and Park, K.H. (2002a) *Mol. Plant Microb. Interact.*, **15:** 983-989.

Shoemaker, J.S. (1953) *Vegetable Growing, Chapman and Hall Ltd.*, London.

Shrihari, P.C. and Sreenivasa, M.N. (1998) *Karnataka J. Agric. Sci.*, **11:** 696-700.

Shuh, D.M. and Fontenot, J.F. (1990) *J. Amer. Soc. Hortic. Sci.*, **115:** 499-502.

Silva, S.A.M., Rodrigues, R., Gonçalves, L.S.A., Sudré, C.P., Bento, C.S., Carmo, M.G.F. and Medeiros, A.M. (2014) *Trop. Plant Pathol.*, **39**: 335-341.

Silva, S.G.R.de and Peiris, C.N. (1997) *Trop. Agril. Res.*, **6:** 23-30.

Sindhan, G.S., Shah, A. and Roy, A.J. (1980) *Punjab Hort. J.*, **20:** 84-87.

Singh, D.K., Tewari, P. and Jain, S.K. (2012) *Pantnagar J. Res.*, **10**: 61-65.

Singh, J. and Thakur, M.R. (1977) In: *Proceedings of 3rd EUCARPIA meeting on Capsicum and working group*, Montfavet, Avignon, France, pp 119-126.

Singh, K. and Srivastava, K. (1988) *Indian J. Hort.*, **45:** 319-24.

Singh, M.D., Laisharam, J.M. and Bhagirath, T. (2005) *Indian J. Hortic.*, **62:** 203-205.

Singh, M.J. and Singh, J. (1998) *Plant Dis. Res.*, **13**: 184.

Singh, M.J., Singh, J. and Cheema, S.S. (1998) *Plant Dis. Res.*, **13:** 125-128.

Singh, N. and Singh, A. (1974) *Indian J. Farm Sci.*, **11:** 12-16.

Singh, R., Jangir, R.P. and Poonia, B.L. (1995) *Indian J. Agril. Sci.*, **65:** 723-726.

Singh, R.K. and Verma, S.K. (2015) In: *25th Asian-Pacific Weed Science Society Conference on "Weed Science for Sustainable Agriculture, Environment and Biodiversity"*, 13-16 October, 2015, Hyderabad, India.

Singh, S. and Singh, T. (2019) *J. Pharmacog. Phytochem.*, **8**: 2021-2023.

Singh, S.S. and Sharma, R.K. (1999) *J. Living World*, **6:** 18-21.

Sivakumar, C.V., Lakshmanan, P. and Balasubramaniam, M. (1979) *Nematologie*, **2:** 165-168.

Smith and Basavaraja (2006) *Karnataka J. Agric. Sci.*, **19**: 975-977.

Smith, P.G. (1950) *J. Hered.*, **41:** 138-140.

Sohi, A.C., Yap, T.C. and Graham, K.M. (1977) *SABRAO J.*, **9:** 127-134.

Soler, S., Debreczeni, D.E, Vidal, E., Aramburu, J., López, C., Galipienso, L. and Rubio, L (2017) *Ann. Appl. Biol.*, **170**: 286.

Sprague, G.F. and Tatum, L.A. (1942) *J. Amer. Soc. Agron.*, **3:** 923-942.

Sreelathakumary, I. and Rajamony, L. (2004) *J. Trop. Agric.*, **42**: 35-37.

Sreenivas, M., Bhattacharjee, T., Sharangi, A.B., Maurya, P.K., Banerjee, S., Chatterjee, S., Maji A., Mandal, A.K., Chakraborty, I. and Chattopadhyay, A. (2020) *Int. J. Veg. Sci.*, **26**: 457-486.

Sreenivas, M., Sharangi, A.B., Banerjee, S., Maurya, P.K., Bhattacharjee, T. and Chattopadhyay, A. (2019) *Int. J. Curr. Microbiol. App. Sci.*, **8**: 1881-1899.

Sriramachandramurthy, N., Venkateshwaralu, P. and Ramavao, Y. (1983) *Proc. Nat. Seminar Tamoto, Chillies, Tamil Nadu Agric. Univ*; Coimbatore, p. 119.

Stewart, C., Mazourek, M., Stellari, G.M. O'Connell, M. and Jahn, M. (2007) *J. Exp. Botany*, **58:** 979-991.

Subbiah, K. (1983a) *Proc. Nat. Seminar Tomato, Chillies, Tamil Nadu Agric. Univ.*, Coimbatore, p.156.

Subbiah, K. (1983b) *Proc. Nat. Seminar Tomato, Chillies, Tamil Nadu Agric. Univ.*, Coimbatore, p.154.

Subbiah, K. (1994) *Madras Agric. J.*, **81:** 277-279.

Subbiah, K., Helkiah, J., Ravikumar, V. and Rajagopal, C.K. (1980) *South Indian Hort.*, **28:** 103-104.

Subramaniam, K.S., Srinivasan, S. and Shanmugan, N. (1971) *Madras Agric. J.*, **58:** 548.

Subramanian, P., Krishnasamy, S. and Devasagayam, M.M. (1998) *South Indian Hort.*, **46:** 99-101.

Subramanya, R. (1983) *HortSci.*, **18:** 747-749.

Subramanya, R. and Ozaki, H.Y. (1980) *Natl. Pepper Conf. Univ. California, Davis.* Abs. No.4.

Sugita, T., Kinoshita, T., Kawano, T., Yuji, K., Yamaguchi, K., Nagata, R. and Todoroki, A. (2005) *Breed. Sci.*, **55:**

Sugita, T., Semi, Y., Sawada, H., Utoyama, Y., Hosomi, Y., Yoshimoto, E. and Ohyama, A. (2013). *Mol. Breed.*, **31:** 909-920.

Sun, C., Mao, S.L., Zhang, Z.H., Palloix, A., Wang, L.H. and Zhang, B.X. (2015) *Sci. Hortic.*, **181**: 81-88.

Sun, Z.J. and Wang, M. (1989) *Capsicum Newslett.*, **42:** 8-9.

Suryanarayana, V., Raju, K.T.R. and Rao, D.V.S. (1983) *Veg. Sci.*, **10:** 77-83.

Suthanthirapandian, I.R. and Sivasubramanium, V. (1978) *Madras Agric. J.*, **65:** 334-36.

Suthanthirapandian, I.R. and Rangaswamy, P. (1983) *Proc. Nat. Seminar Tomato, Chillies, Tamil Nadu Agric. Univ*ersity, Coimbatore, p.122.

Swamy, T.C.N. (1983) *Thesis Abst.*, **9:** 340-341.

Sy, O., Bosland, P.W. and Steiner, R. (2005) *J. Amer. Soc. Hort. Sci.*, **130:** 75-78.

Tai, T., Dahlbeek, D., Stall, R.E., Peleman, J. and Staskawicz, B.J. (1999) *Theor. Appl. Genet.*, **99:** 1201-1206.

Tainter, R.D. and Grenis, T.A. (1993) In: *Food Science and Technology*, VCH Publishers, New York.

Tan, S., Cheng, J.W., Zhang, L., Qin, C., Nong, D.G., Li, W.P., Hu, K. L. (2015). *PLoS ONE, 10*, e0119389. https: //doi.org/10.1371/journ al.pone.0119389.

Tanksley, S.D., Bernatzky, R., Lapitan, N. L. and Prince, J.P. (1988). *Proceedings of the National Academy of Sciences of the United States of America*, **85:** 6419-6423.

Tepic, A.N., Dimic, G.R., Vujicic, B.L., Kevresan, Z.S., Varga, M. and Sumic, Z.M. (2008) *APTEFF*, **39:** 77–83.

Tewari, V. P. (1991) *Indian Hort.*, **35:** 29-31.

Tewari, V.P. (1989) *Capsicum News Letter No.*7, 41.

Tewari, V.P. and Viswanath, S.M. (1988) *New Botanist*, **15:** 1985-1986.

Thabuis, A., Palloix, A., Pflieger, S., Daubeze, A.M., Caranta, C. and Lefebvre, V. (2003) *Theor. Appl. Genet.*, **106**: 1473-1485.

Thorup, T., Tanyolac, B., Livingstone, K., Popovsky, S., Paran, I. and Jahn, M. (2000) *Proc. Natl. Acad. Sci.* USA **97:** 11192-11197.

Tomita, R., Murai, J., Miura, Y., Ishihara, H., Liu, S., Kubotera, Y., Honda, A., Hatta, R., Kuroda, T., Hamada, H., Sakamoto, M., Munemura, I., Nunomura, O., Ishikawa, K., Genda, Y., Kawasaki, S., Suzuki, K., Meksem, K. and Kobayashi, K. (2008) *Theor. Appl. Genet.*, **117:** 1107-1118.

Tong, N. (1998) *Chile Pepper Inst. Newsl.*, **7**: 1-3.

Tong, N. and Bosland, PW. (1999) *Euphytica*, **109:** 71-77.

Tsaballa, A., Pasentsis, K., Darzentas, N. and Tsaftaris, A. (2011). *BMC Plant Biol.*, **11:** 46.

USDA-ARS (2011) Grin species records of Capsicum. National Germplasm Resources Laboratory, Beltsville, Maryland

Usha, P. and Peter, K.V. (1988) *Veg. Sci.*, **15:** 185-189.

Usman, M.G., Rafii, M.Y., Martini, M.Y., Yusuff, O.A., Ismail, M.R. and Miah, G. (2008) *Cell Stress Chaperones*, **23**: 223-234.

Vadivel, E., Arumugam, R. and Muthuswami, S. (1983) *Proc. Nat. Seminar Tomato, Chillies, Tamil Nadu Agric. Univ.*, Coimbatore, pp. 125.

Vaishnavi. B.A., Khanm, H. and Bhoomika, H.R. (2017) *Int. J. Agric. Sci. Res.*, **7:** 259-266.

Vallejos, C.E., Jones, V., Stall, R.E., Jones, J.B., Minsavage, G.V., Shultz, D.C., Rodrigues, R., Olsen, L.E. and Mazourek, M. (2010) *Theor. Appl. Genet.*, **121:** 37-46.

Vanitha and Jansiran (2017) *Int. J. Agric. Sci. Res.*, **7**: 15-20.

Vasconcelos, C.S. (2016) Ph.D.Thesis, Universidade Federal de Pelotas - RS, Brazil.

Vasudeva, R.S. (1954) *Sci. Rept. Indian Agricultural Research Institute*, New Delhi.

Vaulx, R.D.de, Pochard, E. and Chambonnet, D. (1982) *HortScience*, **28:** 14-18.

Venkata Rao, A. and Shanmugam, N. (1983) *Proc. Na1. Seminar Tomato, Chillies, Tamil Nadu Agric. Univ.*, Coimbatore, p. 103.

Vidya, C., Jagtap, V.S. and Santhosh, N. (2018) *Int. J. Curr. Microbiol. Appl. Sci.*, **7**: 3265-3268.

Voinnet, O. (2001) *Trends Genet.*, **17:** 449-459.

Voorrips, R.E., Finkers, R., Sanjaya, L. and Groenwold, R. (2004) *Theor. Appl. Genet.*, **109:** 1275-1282.

Vos, J.G.M. and Sumarni, N. (1997) *J. Hort. Sci.*, **72:** 415-424.

Votava, *E.g.*, Balok, C., Coon, D. and Bosland, P.W. (2000) *J. Heredity*, **91:** 60-61.

Walker, J.C. (1952) *Disease of Vegetable Crops, McGraw-Hill Book Company, Inc.* NewYork, London.

Walker, S.J. and Bosland, P.W. (1999) *J. Am. Soc. Hortic. Sci.*, **124:** 14-18.

Wang, J.J. and Fan, M.J. (1998) *J. Agril. Res. China*, **47:** 267-282.

Wang, J.J., Fan, M.J., Lo, S.F. and Liu, S.Y (1996) *J. Agri. Res. China*, **45:** 370-381.

Wang, M. and Ma, D.H. (1987) *Capsicum Newsl.*, **6:** 25.

Wang, Y.W., Yang. M.Z., Pan, N.S. and Chen, Z.L. (1991) *Acta Bot. Sinica*, **33:** 780-786.

Welbaum, G.E. (ed) (2015). *Family Solanaceae in Vegetable production and practices.* CAB International Publisher.

Wu, F., Eannetta, N.T., Xu, R., Durrett, R., Mazourek, M., Jahn, M. M. and Tanksley, S. D. (2009) *Theor. App. Gene.*, **118:** 1279-1293.

Xie, J., Cardenas, E.S., Sammis, T.W., Wall, M.M., Lindsey, D.L. and Murray, L.W. (1999) *Agric. Water Manag.*, **42**: 127-142.

Yadav, B.D. and Mathur, N. (1990) *Int. Nematol. Net. Newsl.*, **7:** 40.

Yahara, S., Izumitani, Y. and Nohara, T. (1988) *Tetrahedron Letter*, **29:** 1943-1946.

Yamada, Y. and Fujita, Y. (1983) In *Handbook of Plant Cell Culture*, Vol. 1 McMillan, NewYork, pp. 717-728.

Yatung, T., Dubey, R.K., Singh, Upadhyay, V.G. and Singh, S. (2014) *Aust. J. Crop Sci.*, **8**: 369-377.

Yazawa, S., Sao, T. and Namiki, T. (1991) *J. Jpn. Soc. Hort. Sci.*, **58:** 609-618.

Yeoman, M., Miedzybrodzka, M.B., Lindsey, K. and MeLauchlan, W.R. (1980) In *Plant Cell Culture: results and perspectives*, Elsevier, NorthHolland, Amsterdam, pp. 327-343.

Yi, G., Lee, J. M., Lee, S., Choi, D., and Kim, B. D. (2006) *Theor. Appl. Genet.*, **114:** 113-130.

Yogeshkumar, H.J., Ajjappalavara, P.S., Megharaj, K.C., Patil, H.B., Revanappa, Meenakshi, S. and Gollagi, S.G. (2018) *Int. J. Chem. Stud.*, **6**: 879-881.

Yoon, J.T., Kim, C.G., Park, N.K., Lim, J.H., Choi, D.J. and Choi, B.S. (1995) *RDA J Agril. Sci. Soil Fert.*, **37**: 251-254.

Yoon, Y.J., Lee, K.S. and Chang, S.K. (1991) *J. Korean Soc. Hori. Sci.*, **32:** 8-16.

Zhang, X.F., Wang, G.Y., Dong, T.T., Chen, B., Du, H.S., Li, C. B. and Geng, S.S. (2019) *BMC Plant Bio.*, **19:** 167.

Zhang, Y.L., Jia, Q.L., Li, D.W., Wang, J.E., Yin, Y.X. and Gong, Z.H. (2013) *Int. J. Mol. Sci.*, **14:** 8985-9004.

Zhang, Z.J., Zhou, Z.X., Liu, Y.J., Jiang, Q.Y., You, M., Liu, G.M. and Mi, J.J. (1994) *Acta Agriculturae Boreali Sinica*, **9**: 67-71.

SWEET PEPPER

T. Bhattacharjee, S. Banerjee, C.R. Muthukrishnan, T. Thangaraj, R. Chatterjee and T.K. Maity

1.0 Introduction

Sweet pepper is known by various names such as *Capsicum*, bell pepper, Simla Mirch, vegetable paprika, *etc.* Fruits are generally blocky, square, elongate thick fleshed, three to four lobed, and non-pungent. They are used for salads, stuffing, cooked as a vegetable, pickled, or processed and are appreciated worldwide for their flavour, aroma, color and as an important source of provitamin A (at the red stage) and vitamin C. Its fruits are important constituents of many recipes. Its consumption is increasing all over the world with the increase in fast food industries and many hidden uses. It has become a multibillion dollar industry, as well as a part time hobby for home gardeners. Fruits (actually inflated berries) are available in the market year round with different shapes, colours and sizes. The crop is mostly cultivated under polytunnels or glass houses in European countries where night temperature, light intensity, relative humidity, *etc.* are so controlled that quality and yield of crop is improved (Uffelen *et al.*, 1988; Bakker and Uffelen, 1988; Leonardi, 1996; Siwek and Libik, 1999).

2.0 Composition and Uses

2.1.0 Composition

Sweet peppers are very rich in vitamins, even more so than tomatoes especially in A and C. The red bell peppers have different types of pigments. Capsanthin accounted for about 36 per cent of the total carotenoid content, β-carotene and violaxanthin for about 10 per cent each, cryptoxanthin and capsorvbin for about 6 per cent each and cryptocapsin for about 4 per cent (Curl, 1962). The vitamin C content was found as high as 321 mg (Simon, 1960). The vitamin C content is, however, reduced by indoor cultivation, close spacing and irrigation (Szoke, 1959). Butkevic (1967) recorded 280 - 480 mg vitamins per 100 g fresh weight. Vitamin B and B_2 ranged from 0.02 to 0.1 mg per 100 g of fresh weight. Folic acid amounted to 1.3 - 2.9 mg/100 g dry weight and nicotinic acid 6-10 mg. Bell pepper contains 92.4 per cent water and the food value per 100 g of edible portion is energy 29 calorics, protein 1.2 g, calcium 11 mg, vitamin A 870 I.U., ascorbic acid 175 mg, thiamine 0.06 mg, riboflavin 0.03 mg and niacin 0.55 mg (Joshi and Singh, 1975). Large quantities of water supplied shortly before picking, either by rain or by irrigation and higher temperature during ripening reduced vitamin C content of fruit (Spaldon and Pevna, 1962). According to Jo *et al.* (1999), the ascorbic acid content of flesh was highest followed by placenta, and lowest in seeds. Furthermore, ascorbic acid content of red mature fruit was higher than that of immature fruits.

They are appreciated worldwide for their flavor, aroma, color and are a rich source of capsaicinoids, carotenoids (some of them with provitamin A activity), flavonoids, ascorbic acid (vitamin C), and tocopherols (vitamin E) (Howard and Wildman, 2007; Wahyuni *et al.*, 2011). The amount and composition of these metabolites varies among genotypes, and are affected by many conditions such as fruit maturity, cultivation systems and processing methods (Markus *et al.*, 1999; Howard *et al.*, 2000). The presence of some of these metabolites may function as general defense mechanism against various biotic and abiotic stresses (Aloni *et al.*, 2008; Schulze and Spiteller, 2009; Park *et al.*, 2012). Its consumption is increasing all over the world with the increase in the fast food industries and many hidden uses.

2.2.0 Uses

Sweet pepper, whatever may be the colours, may be eaten cooked or raw, sliced in salads. In stews a little sweet pepper imparts a novel flavour. Ripe pimentos baked with white fish make a delicious dish (Herklots, 1972). Mild or sweet peppers are also used for pickling in brine, baking and stuffing. Diced green or red sweet peppers are sometimes mixed with sweet corn or other vegetables (Shoemaker and Teskey, 1955). Based on fruit shape and size, more than 20 market types of sweet peppers are known (Poulos, 1994). These can be grouped into four broad market categories: (i) fresh market (green, red, multi-color whole fruits), (ii) fresh processing (sauce, paste, canning, pickling), (iii) industrial extracts (paprika/oleoresin, capsaicinoids and carotenoids), and (iv) ornamental (plants and/or fruits). These market types reflect that, besides conventional nutritional food uses (salads, stuffing, cooked as a vegetable, pickled, or processed), there are a number of alternative food (*e.g.* paprika oleoresin, medicines) and non-food (*e.g.* spiritual, ethno-botanical) uses for bell peppers.

3.0 Area and Production

It is widely cultivated throughout temperate and subtropical Europe (Spain, Bulgaria, Hungary, Romania, Italy, and Yugoslavia), USA, Central America (Mexico), South America (Argentina and Peru), Africa (Nigeria, Ghana, and Tunisia) and Asia (China, Indonesia, Sri Lanka, Pakistan, Turkey, Korea, and India) (Govindarajan, 1985; Hazra *et al.*, 2011). In India it is generally grown in the cooler weather particularly in the periphery of cities. Mixed stock of *Capsicum annuum* including both hot and sweet peppers was introduced in India by the Portuguese from Brazil in 1584. However, chillies or hot pepper spread much more widely and rapidly than sweet pepper in India. India has the largest area under sweet pepper with an average productivity of 13.2 tones/ha (APEDA, 2010). In India, it is cultivated over an area of about 46,000 ha with annual production of 288,000 MT (National Horticulture Board, 2016). This crop is extensively cultivated in hills of Himachal Pradesh, Uttar Pradesh, Jammu and Kashmir, Andhra Pradesh and Nilgiris during summer months and as an autumn crop in Karnataka, Maharashtra, Tamil Nadu, Bihar, West Bengal, and Madhya Pradesh. Major sweet pepper growing areas lay around Bangalore, Belgaun and Mysore of Karnataka, Pune and Thane in Maharashtra, Nilgiri hill areas of Tamil Nadu, Ranchi of Jharkhand, Darjeeling hill and some parts of 24 Parganas of West Bengal, hill regions of Uttar Pradesh and Uttaranchal, Himachal Pradesh and Jammu and Kashmir. In Himachal Pradesh, it is grown as an off season crop during the summer and rainy seasons and is economically important to small and marginal farmers (Sood *et al.*, 2009). India contributes 17.9 per cent of the total production (Panigrahi *et al.*, 2013). A recent estimate of sweet pepper production in major Indian states is given in Table 1.

Uses of Sweet Pepper

Different Recipes of Sweet Pepper in Indian Cuisine

Table 1: Sweet Pepper Production in India*

State	Production (000 tonnes)	Share (per cent)
Karnataka	65.27	20.04
Himachal Pradesh	57.76	17.74
Haryana	40.05	12.30
Jharkhand	33.03	10.14
Madhya Pradesh	23.91	7.34
Maharashtra	22.96	7.05
Tamil Nadu	16.69	5.13
Uttarakhand	15.70	4.82
Jammu and Kashmir	13.63	4.19
Odisha	6.99	2.15

* National Horticulture Board (2018).

4.0 History, Origin and Taxonomy

4.1.0 History and Origin

Historians agree on the new world origin of *Capsicum*. Dried pod's discovered from burial tombs in Peru are believed to be more than 1000 years old. On account of lack of references on *Capsicum* in ancient languages, it is generally viewed that no *Capsicum* is indigenous to the old world (De Candolle,1986). Capsicums are historically associated with the voyage of Columbus's first expedition (1492-93), when he observed that natives of the new world used a colourful red fruit called aji or axi with most of their foods (Boswell, 1951; Govindarajan, 1985). Columbus introduced it to Europe, and from there the Portuguese introduced it to Africa and Asia. He thought that he had reached India and called it red pepper to differentiate it from the black pepper already known in Europe. Tropical South America, especially Brazil, is thought to be the center of diversity of peppers (Shoemaker and Teskey, 1955). In India, it was introduced by the Britishers in the 19th century in Shimla hills. In India, the term capsicum is often used for the fruits of sweet peppers.

4.2.0 Taxonomy

The genus *Capsicum* is a member of the family Solanaceae. Early taxonomic treatments of the genus resulted in description of nearly 100 supposedly good species and botanical varieties. There are many cultivars differing from each other in shape and colour of the fruits, pungency and position of fruits. One species, namely, *C. frutescens* (syn. *C. annuum*) and 5 varieties, namely, *cerasiforme, conoides, fasciculatum, grossum* and *longum* were recognized later (Bailey, 1948). But subsequently *C. frutescens* and *C. annuum* were regarded as valid species (Heiser and Smith, 1953). Two species can be distinguished as follows:

(i) Plants annual, fruits borne singly: *C. annuum*

(ii) Plants perennial, fruits borne in groups: *C. frutescens*.

Table 2: Morphological Characters that Generally Differentiate the Domesticated Species of Capsicum

Species	*Flowers per Node*	*Calyx*	*Corolla Colour*	*Corolla Lobe Basal Spots*	*Anther Colour*	*Seed Colour*	*Remarks*
C. annuum var. *grossum annuum* (Sweet pepper and chilli)	1 (-5)	No ring; often teeth	White to dingy white (rarely purple)	None	Blue-purple	Straw (tan)	Cultivated across the world. Wild species belongs to *C. annuum* var. *minimum*
C. frutescens (Tabasco pepper)	Usually 2-4(1-6)	No ring;usually no teeth	Greenish white or greenish	None	Blue-purple	Straw (tan)	Cultivated across the world.
C. chinense	(1-) 2 (-5)	Annular ring; no teeth	Greenish white or white	None	Blue	Straw (tan)	Cultivated in Amazon and Africa. It is supposed to be a progenitor of *C. frutescens*
C. baccatum var. *pendulum*	1 (-2)	No ring; teeth	White (cream) or greenish-white	Yellow-green	White toYellowish	Straw (tan)	Cultivation is restricted to South America. Wild types belongs to *C. baccatum* var. *baccatum*
C. pubescens	1	No ring; teeth	Purple orpurple-white	None	Purple (purple white)	Black (brown/black)	This Andean species is presumed to be a progenitor of one of the three American wild species, *viz*., *C. eximium*, *C. cardenasii* and *C. tovari* since *C. pubescens* shows affinities with them

Other species of cultivated pepper, namely *C. annuum* L. var. *annuum*; *C. baccatum* L. var. *pendulum* (Willd.) Eshbaugh, syn. *C. pendulum* (Willd.); *C. chinense* Jacq.; *C. frutescens* L. and *C. pubescens* Ruiz and Pav. were also recognized.

Heiser and Pickersgill (1969) and Heiser Jr. (1976) have described different species of *Capsicum*. Even though 20 wild species have been reported, only 5 species are domesticated, the keys for which have been presented byHeiser (1985), Greenleaf (1986), Eshbaugh (1993) in Table 2.

Seven botanical varieties are also considered and it is quite doubtful whether it is justified, as they all intercross and intermediates do occur (Redgrove, 1933; Chittenden, 1956).

var. *abbreviatum* Fingerh.: Wrinkled peppers; fruits generally ovate, wrinkled, 5 cm long or less.

var. *accuminatum* Fingerh.: Chillies; fruits linear-oblong, over 9 cm long, usually pointed, pungent, widely grown in India.

var. *cerasiforme* (Miller) Irish: Cherry peppers; fruits globose with firm flesh, 1.2- 2.5 cm in diameter, red, yellow or purple, pungent.

var. *conoides* (Miller) Irish.: Cone peppers; tabasco type, fruit erect, conical, about 3 cm long, very pungent.

var. *fasciculatum* (Sturt.) Irish.: Cluster peppers; fruits clustered, erect, slender, about 7.5 cm long, very pungent. As the fruits are not borne singly, it is probable that these are forms of *C. frutescens*.

var. *grossum* (L.) Sendt.: Sweet pepper, bell pepper, paprika; fruits large with basal depression, inflated, green, red,yellow, orange, cream, white or purple coloured, flesh thick and mild.

var. *longum* (DC.) Sendt.: Long peppers; fruits mostly drooping, tapering at the apex, 20-30 cm long, red, yellow or ivory, often mild, calyx not embracing the fruit.

Presently, *C. pendulum* is included or designated as *C. baccatum* var. *pendulum*. A number of other species are also recognized, especially with different characters with reference to corolla colour, shape, anther colour, presence or absence of calyx (Table 3).

Table 3: Distinguishing Morphological Characters of *Capsicum* Species

Species	*Corolla Colour*	*Corolla Throat Spot*	*Corolla Shape*	*Another Colour*	*Calyx Teeth*	*Flowers' Node*
C. chinense	White to greenish white	None	Rotate	Blue	Present	3-5
C. galapogense	White	None	Rotate	Yellow	None	1
C. chacoense	White	None	Rotate	Yellow	Present	1
C. schottianum	White	Yellow	Rotate	Yellow	None	5-7

Species	Corolla Colour	Corolla Throat Spot	Corolla Shape	Another Colour	Calyx Teeth	Flowers' Node
C. microcarpun C. pendulum	White	Green to Yellow	Rotate	Yellow	Present	1-2
C. oraeternussum	White to lavender	Yellow	Rotate	Yellow	Present	1
C. eximium	-Do-	Yellow	Rotate	Yellow	Present	2-3
C. pubescens	Purple	None	Rotate	Purple	Present	1
C. cardenasii	Blue	Greenish yellow	Campa-nulate	Pale Blue	Present	1-2
C. scolnikianum	Yellow	-	-Do-	-	Present	-

Sweet pepper may have evolved from an ancestral form in the Bolivia/Peru area. Domestication, human selection and breeding have produced many varieties of sweet pepper from the great natural diversity of the *Capsicum annuum* var. *grossum* differing greatly in fruit size and shape, pungency and ranging in colours from orange, red, yellow, cream, white, purple to green. Different cultivar groups of sweet peppers can be classified primarily on the basis of fruit shape and size in the following way (Hazra *et al.*, 2011).

Bell	Large, blocky fruits with a blunt end of 3 to 4 lobes, thick walled, sweet, fruit colours, green, red, yellow, cream, white orange, violet.
Cheese	Mostly globose shaped fruits, small to medium in size and of various shape, mild, yellow to green maturing to red.
Cuban	Fruits of irregular blunt shape, mildly pungent, thin walled.
Long Wax	Fruits long, tapered but in some cases blunt, mildly pungent
Pimiento	Fruits large, cone or heart shaped, thick walled, not pungent.

5.0 Botany

Its plants are herbaceous, tender, woody shrubs. Leaves are spirally inserted with whole or slightly sinuate blades. The shape of the corolla is short tubiferous, and the carpels are located slantwise to the medium plant, because the peduncle frequently gets twisted. The stamens are aduate to the corolla base. The fruit is an inflated berry, sometimes with a few stone cells (sclerified inclusions in the fleshy portion) (Knapp, 2002). The fruits are divided by septa and have 3-5 locules. The diversification of fruit shapes is striking *e.g.*, blocky (or lantern- or bell-shaped), globose, oblong (sausage-shaped), ovoid, conical, cylindrical, banana-like (curved); and smooth, grooved, lumpy or wrinkled. The ovules are borne on the central receptacle and partly on the septa (Somos, 1984). Its leaves are long petiolated, leaf blades ovate to broad lanceolate, green to dark green in color. The corolla is mostly white, but may be purple. Calyx dentation is acuminate. The anthers are cylindrical with a slight incision on the lower part, greyish violet in color. Flowers are mostly borne singly. The seed is straw colored, flat and slightly curved with a narrow hilum and a pointed obtuse rostrum.

Different Groups of Sweet Pepper

Bell Pepper

Mini Sweet Pepper

Cubanelle Pepper

Banana Pepper

Fresno Pepper

Cherry Pepper

Poblano Pepper

Jalapeno

Serrano Pepper

Thai Pepper

Pickersgill (1977) stated that the basic chromosome number of bell pepper is 2n = 24 which is characteristic of the family. She also states that in *C. annuum* one of the karyotypes having two pairs of acrocentric chromosomes is confined to Mexico. It is thus assumed that domesticated *C. annuum* is from Mexico.

5.1.0 Pollination

Sweet pepper flowers are complete, that is they have calyx, corolla, anthers, and stigma. It can cross-pollinate because of its protogynous condition (Bosland and Votava, 2000) and exerted position of stigma in some cases. The odourless flowers are visited by insects both for sugary nectar, which is mostly hexoses and low in daily amount (greatest on the day of anthesis), and also for their pollen (Rabinowitch *et al.*, 1993; Vogel, 1998; Raw, 2000; Roldán-Serrano and Guerra-Sanz, 2004. Solitary bees, honeybees, bumblebees, aphids and thrips are likely to transfer the pollen grains, especially those that obtain pollen by buzz pollination, shaking the anthers (Andrews, 1995; de Ruijter *et al.*, 1991; Kristjansson and Rasmussen, 1991; Kubišová and Háslbachová, 1991; Shipp *et al.*, 1994; Raw, 2000; Dag and Kamer, 2001). Pickersgill (1977) has reported 2 to 90 per cent cross-pollination in open pollinated crops. Breeders and seed producers thus need to undertake precautionary measures to prevent uncontrolled cross-pollination (Bosland 1993). Anthesis commenced at 7.15 am which continued up to 11.15 am with the peak at 7.15 am, and anther dehisced after 30 minutes of anthesis. The stigma became receptive from the day of anthesis and remained receptive up to 2 days after anthesis (Vijay *et al.*, 1979). Crosses in bell pepper can be made any time during the daytime, but the best time is early morning or late afternoon (Greenleaf, 1986). In field research *Capsicum* should be considered facultative cross-pollinating species (Odland and Porter, 1941; Tanksley, 1984).

6.0 Soil and Climate

6.1.0 Soil

Although sweet pepper can grow in almost all types of soil, well drained clay loam soil is considered as ideal for cultivation of sweet pepper (Hawthorn and Pollard, 1954). It can stand acidity to a certain extent (Thompson and Kelly, 1957). The method of soil surface preparation did not affect either yield or quality but for commercial cultivation, level and raised beds were found more suitable than sunken beds (Velev, 1973). On sandy loam soil, crop can successfully be grown provided manuring is done heavily and the crop is irrigated properly and timely. It can tolerate soil pH around 6-6.5 (Joshi and Singh, 1975). Soil should also have good drainage facility. Sweet pepper has bi-directional root system and it is important that there is free unhampered root growth and root aeration for successful crop growth (Hazra *et al.*, 2011).

6.2.0 Climate

Sweet pepper performs well in mild climate hence, considered a cool season crop. For a high yield of good quality fruit, Bakker and van Uffelen (1988) found that mean air temperatures of 21-23°C were optimal during vegetative growth, followed by 21°C during fruit growth. The minimum temperature for growth and development is 18°C, below which growth is trivial, with plants in the 5-15°C range growing poorly (Sanders *et al.*, 1980). A temperature ranging from 25-27°C during daytime and 18-20°C at night is ideal. It can not withstand extreme temperatures, below 15°C and above 30°C significantly affect growth and fruit set (Dorland and Went, 1947; Bakker and van Uffelen, 1988). Maximum flower set occurs when day and night temperatures are between 21°C and 16°C. Flowers drop when the night temperature is above 24°C. Yields are high when the daily air temperature during fruit set ranges between 18-32°C. Fruits do not set when the mean daily temperature is above 32°C, or is below 16°C or when cooler, the fruits are malformed (Hawthorn and Pollard, 1954; Olareweju, 1988; Aloni *et al.*, 1999a). High temperature and low relative humidity at the time of flowering increased the transpiration pull resulting in abscission of buds, flowers and small fruits (Cochran, 1936). The crop is susceptible to frost. High relative humidity is good for growth and yield, fruit set and yield however, this condition is favourable for the development of foliar and fruit diseases. It is highly sensitive to environmental factors. Hence, its cultivation under poly-tunnels or poly-houses is preferred. Use of misters and foggers inside the poly house to maintain the slightly higher relative humidity and marginally reduced temperature to the extent of 30-40°C significantly increase fruit set and yield. Providing shade through poly or net house during summer and only shade net in open during other seasons is beneficial.

Sweet pepper can be grown under open field conditions and in protected structures *i.e.*, net house, poly house, walking tunnels, low tunnels, *etc.* (Singh and Sirohi, 2006) but it lacks adaptability to varied environmental conditions in open (Yoon *et al.*, 1989). Naik (2005) reported that among three growing conditions namely medium cost poly house, low cost poly house and net house, the medium cost poly house recorded higher yield. The favourable environmental conditions prevailing in medium cost poly house might have helped in better growth of roots and shoots, which directly helped in better vegetative growth and finally improving the yield attributing parameters. Based on the results of several glass house trials Uffelen *et al.* (1988) recommended that in sweet pepper relative humidity at night should be kept at 85-90 per cent and in the day at 80 per cent. The effects of fruit load, defoliation and night temperature on the development of fruits of sweet pepper cv. Mazurka was investigated in a climate controlled plastic house in Israel by Aloni *et al.* (1999a). Deformed flowers were observed following removal of all fruits and/ or after partial defoliation (50 per cent) and at a lower night temperatures (12°C). Fifteen days after treatment, the concentrations of reducing sugars and starch in flower buds of defruited plants were higher than in control plants. Defoliation decreased the concentrations of reducing sugars and starch in flower buds of plants which bore 2 developing fruits. Complete defruiting affected the shape of subsequent fruits developing from flower which were 3 days pre-anthesis at the time of defoliation. They suggested that flower morphology is partly controlled by source-sink relationships. Assimilates that are normally translocated to developing fruits may be transported following fruit removal, to flower buds which swell, leading to the development of deformed fruits. Investigations on modifying the microclimate in tunnels covered with polyethylene (PE) and polypropylene (PP) were conducted by Siwek and Libik (1999) near Krakow, Poland. The tunnels were covered with PE film and non- woven PP. Both materials had either a natural colour (transparent for PE and white for PP) or were blue, red or green. Under these covers plants of sweet pepper cv. Mira were planted in early May. The transmittance of radiation through PE film and non woven PP differed considerably. The film limited the quantity of radiation slightly but the non woven material reduced the radiation rate markedly, acting as an optical filter. As a result of shading and ventilation the temperature in the tunnels covered with porous non woven PP was lower, which had a strong influence on the growth and yield of fruits. The plants under PP developed slower, produced less biomass and the ripening of fruits was delayed. The yield from the tunnels covered with PP was lower compared to that from film tunnels. The highest yield was obtained under white non-woven PP, where the shading rate was lowest. Ilic *et al.* (2011) evaluated the influence of different colored shade nets (photoselective) on the plant development, yield and quality of bell pepper. Pepper was grown under four different coloured shade-nets (pearl, red, blue and black) with different relative shading (40 per cent and 50 per cent). Exposure to full sunlight was used as a control. Used colour-shade nets improved productivity by moderating climatic extremes. The total fruit yields (t/ha) under the coloured shade nets were higher by 113 to 131 per cent, relative to the open field condition. Ghosal and Das (2012) evaluated suitability of a low-tech naturally ventilated green house for cultivation of capsicum in warm and humid climate and reported that the yield of capsicum per square meter in the green house was found to be 2.34 times more than open field condition.

6.2.1 Temperature

Changes in climatic factors more so in temperature bring about considerable variation on the vegetative characteristics of capsicum.

Sweet pepper produces parthenocarpic fruits or fruits with few seeds when night temperature is relatively low. The effect of low temperature is greater before anthesis than afterwards (Rylski, 1973) (Table 3). Fruit weight, length, girth, pericarp thickness and number of seeds per fruit were high with 25°/18°C day and night temperatures followed by 36°C maintained constantly during day and night. At 18°C maintained constantly during day and night the fruits of sweet pepper were flattened and unsuitable for market (Ali and Kelly, 1982). High day temperature (20-24°C) along with low light intensity (30 per cent shade) or higher night temperature alone (24°C) promted flower drop, while high day temperature alone did not increase flower drop in sweet pepper (Rylski and Halevy, 1974; Rylski and Spigelman, 1982). Kristof and Bahadli (1986) reported that height of plants, number and surface area of leaves increased proportionally with increasing temperature up to 28°C. The light intensity modified the growth rate. Temperature condition above 25°C combined with low light levels is unfavourable for paprika growth. El-Bahadli (1988) also summarised from phytotron study that high temperatures (> 25°C) were not beneficial, especially when combined with the low light intensity, and at 31°C bud formation was inhibited even at high intensity. With respect to evaluation of hardiness capacity to low temperature of sweet pepper cv. Predicts F_1 in a non-heated green house in Argentina, Rodriguez-Rey *et al.* (2000) found that the crop exhibited adjustments in membrane permeability, which avoided excessive desiccation of tissues without increasing relative content of water, when thermal conditions improved or tissue injury index values increased. These changes were not associated with increasing contents of reducing sugars. From the study, it was evidenced that sweet pepper had a limited capacity of adaptation to low temperature.

Bakker and Uffelen from the Netherlands (1988) reported that for raising glass house sweet pepper the optimum 24 hour mean temperature for vegetative growth was between 21°C and 23°C. The effect of day/night temperature amplitude on vegetative growth and on yield was of minor importance compared with the effect of the 24 hour mean temperature. Uffelen *et al.* (1988) further observed that day and night temperature regime of 24°-25°C/15°-16°C should be maintained until fruit set, when the day temperature should be lowered to 23°C, followed by 21°C from the end of February onwards.

Table 4: Effect of Night Temperature before and after Anthesis on Seed Setting in Sweet Pepper*

Night Temperature before Anthesis (°C)	*Night Temperature after Anthesis*	*Number of Seeds per Fruit*
18-20	8-10	251
18-20	18-20	252
8-10	8-10	235
8-10	18-20	204

* Rylski (1973).

The effect of night temperature is very much crucial in the growth, development and yield of sweet pepper. In this relation, the study of EI-Saeid *et al.* (1996) is quite relevant. It revealed from their report that night heating improved productivity and fruit quality. The highest yield and lowest numbers of deformed fruits were obtained from the warmest treatment (16°C). The contents of pigments (chlorophylls a and b and carotenoids) and total soluble solids of fruits were non-significantly increased by heating. The content of vitamin C in fruit was significantly increased by heating. Heating also decreased the amount of ABA and increased the amount of auxins (IAA) and GA like substances in fruits. High temperature stress caused abnormal pollen and anther development, resulting in decrease in pollen viability, in vitro germination and pollen tube growth. These effects were greater in a thermosensitive cultivar than in a thermo-tolerant one (Han *et al.*, 1996). According to Aloni *et al.* (1999b), scanning-electron rnicrographs show that initiation of fruit cracking in bell pepper is by formation of mini-cracks on the cuticle layer; these then enlarge into cracks and traverse the epidermal cells. Pepper cultivars differ in their sensitive to cracking, partly because of differences in fruit pericarp thickness. Limitation of night transpiration by high humidity or low temperature increased the turgor potential of the fruit pericarp and enhanced cracking. Likewise, leaf pruning suppressed night transpiration and concomitantly increased cracking. Fruit cuticle transpiration was low in the early stages of fruit development and to increase as the fruit became susceptible to cracking. Environmental conditions which enhanced cracking also increased to cuticle transpiration in a sensitive pepper cultivar but not in a less sensitive one. They concluded from their findings that fruit cracking in bell pepper occurs because reduction in night transpiration causes high turgor pressure in the fruit. As the cuticle becomes permeable to water, the inner turgor pressure causes the cuticle to crack. Difference in cultivar sensitivity may be partly due to differences in pericarp thickness and cuticle properties, but additional factors may be involved.

6.2.2 Light

Short day conditions (9-10 hours light) stimulated plant growth and increased the productivity by 21-24 per cent, besides improving the quality of sweet pepper (Egorova, 1975). Shading the plants increased the leaf surface, cell division and cell expansion (Schoch, 1972). Reduction of solar radiation by 50 per cent increased the fresh weight of peduncle, whole fruit, pericarp, placenta and seeds but no effect on dry weight and dry matter content of the placenta, pericarp, seeds, the capsaicin content of placenta of the whole fruit and the ascorbic acid content of the fruit was evidenced (Bigotti, 1974). On the other hand, 30 per cent reduction is solar radiation almost doubled the yield of sweet pepper due to an increase in both number of fruits and fruit size (Quaglitto, 1976). Number of days from sowing to flowering and percentage of flower drop increased as the shade increased. But capsaicin content was not affected by different shade intensities (Jeon and Chung, 1982). Calcium and magnesium deficiency reduced the chlorophyll content and such plants were unable to make efficient use of light intensities over 31 lux (Stanev *et al.*, 1982). After 65 days of development, fruits held in the dark weighed 15 per cent less than those receiving light whereas the seed weight remained unaltered (Acchireddy *et al.*, 1982). Hou *et al.* (1987) reported that shading in capsicum with readily available used plastic film or with white plastic net reduced leaf, fruit and soil temperature, fruit yeilds were highest when the plants were pinched and shaded with the plastic film. El-Bahadli (1988) observed that higher light intensity increased the assimilating surface (in terms of leaf area) strengthened the root system, increased leaf number/ plant and stimulated bud formation in capsicum.

It has been indicated that NH_4 supplied at low concentration was more deleterious for capsicum plants grown under high light intensity, and that when the light was reduced, the plants experienced alleviation in the toxic effect of NH_4 (Zornoza *et al.*, 1989). Leonardi (1996) opined shading reduced vegetative and fruit growth, fruit precocity and yield and increased plant height on green house pepper in Spain. It also reduced chlorophyll content, stomatal density, transpiration rate, stomatal conductance and photosynthetic rate. Sweet pepper cv. Delphin was grown under natural or supplementary lighting that extended the photoperiods to 16, 20 or 24 hours. Increasing the photoperiod to 16 and 20 h increased fruit yields, but continuous light (24 h) decreased yields compared to the 20 hours photoperiod (Demers *et al.*, 1998). In another experiment by them, plants were exposed to a photoperiod of 14 or 24 hours and either pruned to 1 fruit every four nodes or not pruned. During the first week of treatments, plant grown under continuous light grew more slowly than plants exposed to a 14 hours photoperiod. The shoot mass and yields of plants grown under a 14 hours photoperiod were equal to or higher than plants under continuous light. The result also suggested providing continuous lighting for a few weeks to improve growth and yields. Limiting the number of fruits per plant increased shoot mass and decreased yield, but had no effect on the general response of the plants to photoperiod. Leaf mineral composition was not affected by photoperiod, indicating that reduced growth and yields under continuous light were not due to unbalanced mineral nutrition. Leaf starch and sugar contents were increased under continuous light. However, fruit pruning treatments did not modify the pattern of starch and sugar accumulation under the different photoperiod treatments. Reduced growth and yields measured under a 24 hours photoperiod are probably explained by starch and sugar accumulation in leaves as a result of leaf limitations rather than a sink limitation. Shading (30 per cent) had no significant effect on gas exchange parameters, photosynthesis, or ribulose-bisphosphate carboxylase (Rubisco) activity and amount. Stomatal conductance was slightly smaller in shaded plants, leading to improved water use efficiency. During periods of high light intensity and duration and high temperature, shaded plants exhibited better growth and higher yields than control plants. Fruits from shaded treatments were of better quality than those from control plants (Centritto *et al.*, 1998).

Fruit yield and net photosynthetic rate were the highest and night respiratory rate was the lowest under 70 per cent light intensity. As light intensity decreased, the light compensation point declined, the chlorophyll contents increased and the low temperature flourescence emission spectra exhibited variations, some of which were related to cultivar when four cultivars sweet of pepper were grown at 100, 70 or 35 per cent of natural light intensity (Sui *et al.*, 1999). Yosepha (2007) observed that all sweet pepper cultivars showed more productivity under photo-selective shading, as there was an increase in number of fruits per plant by 30 to 40 per cent compared to the use of common black net. Different coloured nets were used for photo-selective shading and it found that the red net have some productivity advantage over the pearl and yellow nets. The risk of disease infestation by small insects like aphids, thrips and the incidence of insect borne viral diseases were recorded significantly lower under yellow net compared to all other nets.

Sweet pepper is sensitive to environmental factors. Owing to its sensitivity, its yield is affected significantly and the crop is restricted to a particular area and season. Efforts are being made to grow sweet pepper in the tropical and subtropical areas during summer season and introducing it for off season cultivation. Though some workers found that a reduction in light intensity caused an increase in the fruit yield, very few studies were made on the physiological basis of yield in relation to irradiation variations in this crop during summer cultivation. The findings of Bhatt *et al.* (1999) on response of sweet pepper to irradiance-photosynthesis, reproductive attributes and yield is very much informative in this regard (Tables 5 and 6).

Table 5: Leaf Area (LA, $m^2 plant^{-1}$), Total Dry Matter (TDM, g $plant^{-1}$), Specific Leaf Weight (SLW, g m^{-2}), Photosynthesis (P_N, μmol $m^{-2}s^{-1}$), Stomatal Conductance (g_s, mol $m^{-2}s^{-1}$) and Transpiration Rate (E, mol $m^{-2}s^{-1}$) in Sweet Pepper*

Cultivar	*Irradiance*	*LA*	*TDM*	*SLW*	P_N	g_s	*E*
Arka Gaurav	100	0.064	20.18	52.0	11.3	0.70	11.2
	75	0.076	26.52	58.0	10.6	0.69	12.4
	50	0.136	35.02	40.3	8.8	0.40	909
	25	0.170	24.65	30.3	5.1	0.34	8.4
Arka Mohini	100	0.078	24.95	70.0	9.8	0.68	12.3
	75	0.110	25.04	45.0	8.9	0.53	10.3
	50	0.150	28.65	30.3	7.5	0.41	8.8
	25	0.180	26.68	32.6	4.6	0.35	9.4
Varieties	SEm	0.003	0.42	1.5	0.16	0.02	0.15
	CD (0.05)	0.008	1.12	4.3	0.46	0.06	0.44
Treatments	SEm	0.004	0.60	2.1	0.23	0.03	0.22
	CD (0.05)	0.012	1.72	6.1	0.65	0.09	0.62

* Bhatt *et al.* (1999).

The results indicated that during summer season a reduction of 50 per cent irradiation caused better plant growth and fruit yield in sweet pepper. This may be utilized for off-season cultivation of this crop in tropical and subtropical areas.

Table 6: Influence of Irradiance on Chlorophyll Contents and Yield Attributes in Sweet Pepper*

Cultivar	*Irradiance*	*Total Chlorophyll*	*Chlorophyll Ratio*	*Fruit Chlorophyll*	*Flower and Fruit Drop*	*Fruit Dry Weight*	*Yield*	*LA/ Fruit Ratio*
Arka Gaurav	100	0.96	2.5	0.10	58	1.83	184	0.011
	75	1.21	2.3	0.13	63	3.43	184	0.016
	50	1.36	2.1	0.14	58	4.25	236	0.035
	25	1.60	2.0	0.16	54	3.31	147	0.057
Arka Mohini	100	0.88	2.6	0.11	60	2.44	123	0.017
	75	1.20	2.3	0.13	58	3.05	220	0.031
	50	1.25	2.2	0.16	54	5.54	228	0.058
	25	1.63	2.0	0.16	50	5.50	174	0.090

* Bhatt *et al.* (1999).

7.0 Cultivars

There are several cultivars of bell-shaped, non-pungent, mild, thick fleshed sweet peppers or Simla Mirch. Many of the Capsicum varieties as reported by Somos (1984) were known in the 16th century. Wafer (1699) referred to bell peppers in his diary. Among older cultivars Campana (1774), California Wonder (1828), and Ruby King (1884) were listed by Bassett (1986). Joshi and Munshi (2000) have mentioned common cultivars as: Non-pungent, green, turning red at maturity–California Wonder (various strains), Yolo Wonder (various strains), Bull Nose, Ruby King, World Beater, North Star, Bell Boy, Keystone Resistant Giant, HC-201, Arka Mohini, Arka Gaurav, Pusa Deepti, Florida (various strains), Chinese Giant, Early Giant and Jingle Bells; Non-pungent, green turning orange at maturity: Arianne, Golden Bell, Romanian, and Golden California Wonder. Pungent bell peppers include Bull Nose Hot, Romanian Hot, and Mexibell. With its increasing demand worldwide emphasis is being given to the development of hybrids. Beside the above listed cultivars, there are many open pollinated varieties and hybrids developed and recommended for cultivation in different countries. Poncavage (1997) has given a brief description of different types within bell pepper: Hottest bell pepper (Mexibell), Sweetest bell pepper (Arianne), Earliest ripening sweet bell pepper (Ace), Biggest stuffing pepper (Big Bertha), Tiniest sweet bell pepper (Jingle Bells), Oldest bell pepper (California Wonder) and Most productive bell pepper (Jingle Bells, Jupiter, North Star).

The important cultivars are California Wonder, Yolo Wonder, King of North, Early Giant, World Beater, Neapolitan, Early Cal wonder, Windsor A, Harris Early Giant, Vincdale, Bullnose, Russian 449, Chinese Giant, Ruby King, *etc.* Of the Ferrby-Morse cultivars, the earliest are World Beater, California Wonder and Oakview Wonder, the late yielders are Pimiento and Harris Early Giant (Herklots, 1972). The cultivar Yolo Wonder should little evidence of sensitivity to drought (Lecompte, 1962). The cultivars World Beater and California Wonder produce large, well formed plants under glass, Lednicka and Kalnikov also cropped well in hot beds (Rudzinska, 1959). Joshi and Singh (1975) described some bell pepper cultivars which are as follows.

California Wonder

This cultivar was introduced from USA and released by Indian Agricultural Research Institute for cultivation in India. Used for market gardening, plants are vigorous, upright and prolific. Fruits are 3-4 lobbed, upright, and smooth with medium thick sweet flesh. First picking is possible within 90-100 days after transplanting. Average yield is 18-20 t/ha.

Chinese Giant

Plants are vigorous and prolific bearer. Fruits are 3-4 lobbed, smooth, thick, sweet flesh and the skin is dark green in colour.

World Beater

Plants are upright, productive, fruits 4-lobbed, flesh thick, mild and sweet.

Yolo Wonder

It is an introduction from USA and released by Indian Agricultural Research Institute for cultivation in India. Large fruit, 3-4 lobbed, medium thick flesh mild and sweet. Its takes 65-70 days (green) 75-80 days (red) to first harvest and maturity.

Bharat

This is the first F_1 hybrid released by private sector (Indo-American Hybrid Seeds, Bangalore) in India (1973). Plants are vigorous, the fruits are bell shaped, each side measuring 8-10 cm diameter, dark green, thick fleshed and mostly 4-lobbed. Average fruit weight is 150 g and yields 20-25 t/ha. It is resistant to tobacco mosaic virus (TMV).

Some newly developed varieties/hybrids released by public sectors are as follows.

Arka Mohini

It is an improvement over IIHR 312-1-2 (Titan variety) followed by mass selection developed and released from Indian Institute of Horticultural Research, Bangalore. Recommended for release in 1984 by SVRC and in 1986 by AICRP (VC) at national level. Notified for entire country in 1986 by CSC on CSN and RV. Determinate plant growth habit with dark green foliage and pendent fruit bearing habit. Fruits are thick fleshed, 3-4 lobed, dark green colour during immature and turn red on ripening, blocky fruits, average fruit weight varies 180-200 g. Average yield is 20 t/ha in 160 days.

Arka Gaurav

Improvement over IIHR-214-1 (Golden Calwonder) followed by mass selection developed and released from Indian Institute of Horticultural Research, Bangalore. The variety was released in 1984 by State Varietal Release Committee and in 1986 by AICRP (VC) workshop for national level. Notified in 1986 by CSC on CSN and RV. Plant growth habit is Indeterminate with green foliage and fruit bearing habit is erect. 3-4 lobed fruits with thick fleshed, fruit colour is green which turn orange yellow on ripening, blocky fruits, average fruit weight 130-150 g. Average yield is 16 t/ha in 150 days.

Arka Basant

Improvement over IHR 225-1-1 (Var. Soroksan) followed by mass selection developed and released by Indian Institute of Horticultural Research, Bangalore. Recommended for release in 1984 by SVRC and in 1986 by AICRP (VC) at national level. Notified for entire country in 1986 by CSC on CSN and RV. Plant growth habit is indeterminate with yellowish green foliage, and have erect fruit bearing habit. Fruits are thick fleshed, 2-3 lobed, conical, cream coloured during immature stage, which turn orange red on maturity and average fruit weight 50-80 g. Average yield is 15 t/ha in 180 days.

Arka Athulya (Hybrid)

High yielding F_1 hybrid developed and released by Indian Institute of Horticultural Research, Bangalore. Plants are continuous in growth habit having dark green foliage. Blocky shaped fruits are firm with 3-4 lobes, average fruit weight 100-120 g suitable for fresh green market. Suitable for both *kharif* and *rabi* season cultivation under open field conditions. Average yield varies 45-50 t/ha in 140-150 days. The hybrids are tolerant to powdery mildew disease.

Pusa Deepti (Hybrid)

F_1 hybrid released by Indian Agricultural Research Institute, substation Katrain for cultivation under temperate and subtropical regions of Northern India. Plants are medium, bushy and erect, conical fruits having yellowish green colour turns dark red at maturity. Fruits are ready for first picking 70-75 days after planting. Average yield varies 300-350 q/ha.

KT 1 (Hybrid)

Another F_1 hybrid released by Indian Agricultural Research Institute, substation Katrain. It is high yielding F_1 hybrid resistant to anthracnose and fruit rot disease.

Solan Bharpur

This cultivar was developed and released by Dr. Y. S. Parmar University of Horticulture and Forestry, Solan. Plants having more number of fruits. Fruits are deep green in colour, average fruit weight 50-60 g. Average fruit yield is 30 t/ha. This variety is tolerant to *Phytophthora* fungal disease.

VL Shimla Mirch 2

This cultivar was developed and released from Vivekananda Parvatiya Krishi Anusandhan Sansthan (VPKAS), Almora. Suitable for cultivation Irrigated/rainfed/organic/inorganic conditions of Uttarakhand hills. Good fruit quality with average yield of 162.7 q/ha under organic condition and 259 q/ha under inorganic condition. Plants are resistant to fruit rot, wilt and powdery mildew disease.

VL Shimla Mirch 3

This cultivar was developed and released from Vivekananda Parvatiya Krishi Anusandhan Sansthan (VPKAS), Almora suitable for poly house conditions. Fruits mature within 70-100 days after transplanting. Number of fruits per plant varies 8 to 12 with average fruit weight 80-120 g. Average fruit yield is 150 q/ha under organic condition and 300 q/ha under inorganic condition. Suitable for cultivation under the Hills and plains of Uttarakhand. Plant has moderate resistance to powdery mildew, fruit rot and seedling rot complex.

Some newly developed popular varieties/hybrids released by private sectors are as follows.

Indira (Hybrid)

F_1 hybrid released by Indo American Hybrid Seeds, Bangalore. The plants are vigorous and medium tall with dark green foliage. 3-4 lobed fruits with thick pericarp and fruits are glossy dark green colour. Fruits are ready for first picking 60-65 days after transplanting. Average fruit weight is 170 g and yield may reach up to 330 q/ha.

Orobelle (Hybrid)

High yielding F_1 hybrid released by Syngenta. Fruits are green at immature stage and turn deep yellow on maturity, pendent bearing habit, blocky in shape

Cultivars of Sweet Pepper

Yolo Wonder

Arka Mohini

Arka Basant

Arka Athulya

California Wonder

Arka Gaurav

Exotic Cultivars of different Groups of Sweet Pepper

Sweet Pepper Breeding Lines Maintained at Bidhan Chandra Krishi Viswavidyalaya, West Bengal, India

with 3-4 lobes. Average fruit weight is 95 g and each plant bears 15-16 fruits. Fruits mature 125 days after transplanting and ready for harvesting.

Bomby (Hybrid)

High yielding F_1 hybrid released by Syngenta. Fruit colour turned deep red from green during maturity blocky in shape with 3-4 lobes. Plants bear about 18-20 fruits each 100 g weight. Fruits mature 125 days after transplanting and ready for harvesting.

Indra (Hybrid)

Blocky, medium early, high yielding F_1 hybrid released by Syngenta. During maturity fruits become red in colour. Fruits are shiny, good quality, blocky fruits have a medium thick wall and hardly any shoulder with average fruit weight is 170 g. Fruits are ready for harvest within 70 days after transplanting.

Burgmans (1983) reported that **Superset 19** and **Early Bountiful** are two high yielding and early maturing sweet pepper cultivars as compared to **Bell Boy** in Hawkes Bay of New Zealand. **Supereset 19** was best suited to processing. In Brazil, **Ruby King, Early Calwonder** and **Ikeda** were reported with yields of 15.8-19.5, 15.5-19.5 and 16.5-18.4 tonnes/ha, respectively (Muniz and Almeid, 1988).

Liliana SC/81, a new cultivar from Cuba with an average yield of 17 tonnes/ha, growth period of 165 days and having adequate resistance of *Xanthomonas vesicatoria* and viruses have been reported by Espinosa and Depestre (1986). Fruits are sweet, aromatic and deep red when ripe.

Redland Sweet Sue, a cultivar from Australia is highly productive (20-35.3 tonnes/ha), non-pungent, smooth skinned and pale yellow-green fruits which turns orange-red at maturity. It is partially resistant to *Xanthomonas campestris* pv. *visicatoria* and is resistant to strain 1 of potato Y virus (Hibberd *et al.*, 1988).

The fruits of cv. **Rio Grande Gold** from USA are 60 mm long, 34 mm wide, conical and suitable for fresh consumption or for processing. It is resistant to tobacco mosaic virus, potato Y virus and pepper mottle virus (Villalon *et al.*, 1988).

NuMex Conquistador is a new non-pungent cultivar of USA with fruits having round shoulders, pointed tips, thick flesh and 2 locules. Mean fruit length is 15.7 cm and fruit weight is 86 g (Bosland *et al.*, 1991).

Haihua 3 is developed by anther culture from breeding line 782031 in China. It has plant height of 37 cm, canopy width of 29.6 cm, fruit weight of 28.5 g and fruit number of 7.2/plant and given high yield. This cultivar has wide adaptability and a high tolerance of diseases (Li and Jiang, 1990).

8.0 Cultivation

8.1.0 Raising of Seedlings and Transplanting

Seedlings are first raised in the nursery beds and then transplanted in the main fields. Normally, 5-6 nursery beds (300 x 60 x 15 cm each) are sufficient for one hectare cultivation. Seeds should be sown in rows at 8-10 cm apart to get healthy seedlings.

The seeds should be dressed with thiram or captan @ 3 g per kg of seeds before sowing to prevent any seed-borne diseases. Aljaro and Martinez (1988) reported that treatment of seeds of cv. California Wonder 300, with magnesium sulphate solution at 14 bars and 25°C for 9 days increased marketable fruit yields from 42.6 to 62.8 tonnes/ha. Desai *et al.* (1987) reported of highest laboratory (95.33 per cent) and field (84.67 per cent) germination with seeds treated with chlormequat @ 600 ppm; NAA @ 10 ppm also gave very good results. Soaking of seeds in aerated solutions of $NaNO_3$, KNO_3 or GA_3 and in distilled water for specified times, followed by drying, improved both earliness and uniformity of germination (Rogers and Pike, 1989). About 1 to 2 kg of seeds are required for 1 hectare cultivation depending on the cultivar (Choudhury, 1967). The seeds should be covered with a thin layer of soil manure mixture or any other media and irrigated with sprinkler to maintain optimum moisture till the seeds germinate (Arya, 1983). The highest germination rates for fresh (9.7 per cent) and dry (98 per cent) seeds were obtained when the fruits were harvested 52 days after flowering (Dharmatti and Kulkarni, 1987). Gonzalez *et al.* (1989) observed that the optimum temperature range for germination of capsicum seeds pre-imbibed in water in light or darkness was 25-30°C. The sweet pepper is generally sown in August for the autumn-winter crop and in November for the spring-summer crop (Choudhury, 1967). In the hills of North Bengal, sowing of seeds in the month of March-April (under cover) and September-October, showed maximum yield in cultivar Califomia Wonder. Kossowski and Hortynska (1984) obtained highest yields of best quality fruits from sowing on 15th March and transplanting 56 days later with spacing at 40 x 40 cm. Very good fruit yields of bell peppers were obtained by transplanting in the field on 1st August. Transplanting on 15 August also provided acceptable yields (Bracy and Constantin, 1988).

Stout and healthy seedlings can be produced by using plug trays. Plug trays, after bleaching, are filled with a mixture of loamy soil, sand and cocopeat at 2: 1: 1 ratio by volume. Seed, after treatment with thiram or captan (3 g/kg of seed), are sown at a shallow depth in the middle of the plug tray and covered with finely sieved well rotten deciduous tree leaves to improve, and prevent the soil from drying. After sowing, trays are covered with straw until germination which normally takes 7 to 8 days and are hand watered daily to maintain proper moisture. Trays are drenched with chlorothalonil @ 0.2 per cent + carbendazim @ 0.1 per cent to avoid damping off disease. Plug trays are covered with 200 μm ultraviolet (UV)-stabilized polyethylene film supported by bamboo poles with insect proof nets covered in the sides to protect seedlings from rain, direct sunlight and vector attack. Seedlings are generally hardened by withholding water 4 days before transplanting.

The seedlings having 4-5 leaves should be transplanted. The nursery beds or plug trays should be irrigated before lifting of seedlings. Transplanting should be done in the evening or during cloudy day followed by irrigation. Seedlings should be transplanted in rows. Stoicescu (1964) suggested using 60 day old seedling for transplanting.

Shishlakov and Mayatovich (1988) from USSR described the intensive method of sweet pepper cultivation. In this method, graded seeds treated with TMTD (Thiram) were sown at 8-10 g/m^2 on 7-10 February in heated plastic houses in rows 8-10 cm apart. The seedlings were raised without pricking out and gave a yield of 600-800 transplants/m^2. These transplants were hardened off and planted out at 70 x 15-16 cm in early May. The first selective harvesting by hand usually occurred on 5-15 July and continued at 5 to 6 day intervals until October. Preconditioning of seeds had a positive influence on germination potential and seedling performance as well of capsicum. It is apparent from the study of Thakur *et al.* (1998) who observed that preconditioned seeds by PEG 6000 or KNO_3 germinated more rapidly than those seeds presoaked in water and was also found to significantly improved both emergence and vigour index of capsicum cultivars.

8.2.0 Spacing

The seedlings are transplanted to the field in rows at a distance of 30 to 60 cm, depending upon the area where grown as well as the cultivar (Hawthorn and Pollard, 1954). Rows spaced at 90 cm and plants spaced at 50-60 cm are fairly common. Spaldon and Pevna (1961) concluded that in sweet pepper cv. Cecei Sweet Pendent, plant densities had no definite influence on either the yield or the quality of the crop. They suggested 60 x 25 cm spacing with provision of sprinkler irrigation and 60 x 35 cm spacing with furrow irrigation. Reduction in plant population was found to increase the total seed weight but lowered the seed production per plant and per fruit in cultivar Ikeda (Da Silva *et al.,* 1971). In red pepper, increase in plant density was found to reduce the pigment content of fruit (Dimitrov, 1960). Growing of 2-3 plants in groups at 25-30 cm apart in rows and 60 cm between the rows showed no significant difference in yields from those obtained at 35 cm apart in rows and 50 cm between the two rows (Spaldon and Gromova, 1958). Velev (1967) opined that spacing had little effect on plant growth but influenced yield. Planting in 4 row sunken beds with 30 cm between rows and 60 cm ridge between the beds showed yield of 53.36 tonnes compared to 35.91 tonnes per hectare in 2 row sunken beds with 50 cm between the rows and 70 cm ridge between the beds. Rodendurg and Smid (1973) reported 16.5 per cent fewer yields with 2 row planting compared to planting in 3 rows. The 45 x 45 cm spacing was found optimum for bell pepper (Rastogi *et al.,* 1980).

In trials with cultivars California Wonder, Selection 5 and Selection 7, Srinivas (1980) found that fruit yield increased with plant density from 77.1-97.5 q/ha at 50 x 50 cm to 97.6-133.9 q/ha at 50 x 30 cm. White capsicum cv. Albaregia should be grown under optimum light conditions, at 75000-80000 plants/ha in the open or at 6-8 plants/m^2 in protected cultivation where it has shown a potential yield capacity of 60-70 tonnes/ha (Moor *et al.,* 1989).

Hanice (1988) described vertical soilless system for production of pepper in Yugoslavia. Green house capsicum cv. Poplava was grown in vertical columns of high-density polymer, each column being 2 m high with 4 sections each containing 25 litres of substrate and holding 6 plants in pockets measuring 8 x 9 cm.

Uffelen and Tanis (1989) reported from Netherlands that for Spanish paprikas, two stems/plant and 3.5 plants/m^2 are best. Leaf area index, biomass production and yield/m^2 were significantly higher for treatment 1 (T_1) plants (8 plants/m^2 with one stem) than for treatment 2 (T_2) plants (4 plants/m^2 with 2 stems and no flowers

Cultivation Practices of Sweet Pepper

removed) or treatment 3 (T_3) plants (4 plants/m^2 with 2 stems and all flowers of one stem removed). T_1 and T_2 plants had the same pattern of dry matter distribution. In T_3 plants, total dry matter production was not affected by the removal of flowers from one of the stems, but more dry matter was distributed to vegetative parts. Yield/m^2 was the lowest in T_3 plants. The fruiting pattern was not affected by treatments (Zhang, 1999).

Yield component analysis from a study on the influence of plant density and covering indicated that fruit size was larger under black polypropylene covered treatments in comparison with uncovered treatments irrespective of plant densities. Total marketable yield was significantly higher under cover. In addition to increasing plant density enhanced total marketable yield. The interactions of cover and density had no significant influence for total marketable yield (Znidarcic and Osvald, 1999).

8.3.0 Irrigation

Sweet pepper requires a regular water supply throughout its cycle, since it is sensitive to water deficiency and to excessive water in its growth substrate (Carvalho *et al.*, 2011). The first irrigation should be given just after transplanting. Later, the field should be irrigated as and when required (Singh and Singh, 1970). Irrigation is essential in arid and semi-arid regions and is also important in humid regions (Thompson and Kelly, 1957). Sadykov and Mikhoet (1981) in an experiment with capsicum cv. Podorok Moldovy reported average yield of 228.3 centners per hectare from non-irrigated plots compared to 360.8 centners per hectare from plants irrigated at 70-75 per cent field capacity. More frequent irrigation, however, decreased the yield. Sprinkler irrigation in the field and mist irrigation in plastic green house were found to reduce the incidence of *Leveillula taurica* (Moens and Welvaert, 1981).

Leon and Montalvo (1986) in trials with cv. California Wonder found that maximum water consumption was during flowering and fruiting before maturation of the fruit began. The highest yield of 20 tonnes/ha was obtained by irrigation at 85 per cent of field capacity which involved 14 irrigation application of total of 3082 m^3 water/ha. The degree of crop sensitivity to water deficit, during the different development stages of a plant, is an important aspect to consider in tropical semiarid regions. The effect of water deficit of flower dynamics and fruit production of sweet pepper was evaluated under several water regimes, during three consecutive years by Jaimez *et al.* (2000) in Venezuela. The study revealed that low water availability prior to flowering reduces the number of flowers produced and retards the occurrence of maximum flowering while a water deficit during the period between flowering and fruit development reduced final fruit production. The study suggests that a water deficit during this last stage affects final yield to a great extent. Dysko and Kaniszewski (1989) observed that when Capsicum cv. Lamuyo was grown in plastic tunnels with trickle irrigation, highest yield was obtained when irrigation was applied at 15 kg Pa moisture potential.

Maintenance of high and stable moisture level is crucial when capsicum grown in small container (Pudelski, 1996). The findings of Rajic *et al.* (1997) on irrigation water requirements for sweet pepper indicated that the crop was deficient in water during the growing season by an average of 208 mm. According Orta (1997), the most suitable method tested for estimation of evapotranspiration of sweet pepper was the FAO modification of the Penman method. Growth and water relations in seventeen bell pepper cul tivars were tested when plants exposed to 4 and 8 days of water stress at the 43rd day after transplanting in a pot experiment at Solan, Himachal Pradesh (India) by Thakur *et al.* (1999a). Height was inhibited both at 4 and 8 days of stress, particularly with the longer stress period. Leaf diffusive resistance increased under water stress. Diurnal changes in transpiration revealed significant cultivar differences between control and stressed plants. Relative water content (RWC) decreased with increasing stress levels, and drought injury index increased with increasing stress but exhibited significant cultivar differences. Further, they (1999) reported that single [38 day after transplanting (DAT)] and double (25 and 38 DAT) application of bioextracts were effective in lessening of detrimental effects of water stress on growth and yield characters in bell pepper at critical stage (flower initiation). Bioextracts treatment maintained higher height, growth, number of branches, flowers and fruit yield in stressed plants compared with stress only. Double application of bioextracts was more effective compared with single application. Bioextracts improved fruit quality by maintaining higher ascorbic acid and capsaicin contents as well as higher acidity and TSS in fruits under stressed condition.

Under water stress conditions, foliar application of triacontanol (20, 50 ppm), extracts of *Lantana camara, Punica granatum, Berberis lycium* and potassium (0.5 and 1.0 per cent) promoted *Capsicum* yield (Thakur *et al.*, 2000).

Irrigation interval at 50±5 mm E_{pan} and K_{cp} 3 plant-pan coefficient could be recommended for green pepper irrigation to save labour cost and time (Ertek *et al.*, 2007).

Santos *et al.* (2017) observed the Piché evaporimeter-based irrigation management resulted in a lower water application per plant, with production (marketable and non-marketable) similar to lysimetry in green house grown sweet pepper. However, the drainage lysimeter-based management resulted in higher quality characteristics, especially when associated to production on coconut fiber substrate. Therefore, they advocated that additional research involving different environmental conditions is required to identify the optimal pot size, varieties, and irrigation frequency to select the irrigation management system and the most suitable substrate for the crop, which result in a higher yield and greater fruit quality.

8.4.0 Manuring and Fertilization

The nutrient removal of capsicum cultivar Yolo Wonder showed that for production of 419 q/ha, the N, P, K, Ca and Mg uptake were 132.9, 43.9, 147.8, 124.6 and 19.9 kg/ha, respectively (Graifenberg *et al.*, 1983) which indicated high nutrient uptake by the plants. Correct manuring practices with both organic and inorganic nutrients were found to increase the growth and yield. Hsieh and Hsu (1994) when compared organic manures with chemical fertilizer on sweet pepper found significantly higher plant height, fruit size, fruit number and yield with the organic manures. Poultry manure composed with a microbial supplement had the highest yield, and increase of 77 per cent over the chemical fertilizer. However, fruit Fe and Mn levels were lower in the organic treatments, as were Na and Ni. Fruit quality was better with the organic treatments. Addition of humic acid to the growing medium and foliar fertilizers had positive effect on seedling growth of pepper seedlings (Padem *et al.*, 1999). Humic acid at 2500 ml/1000 m^2 in both ways of application influenced significantly seedling height and diameter, number of leaves, shoot fresh weight, shoot and root dry matter.

In a green house experiment in Brazil, effect of organic fertilizer application on sweet pepper was studied by Ribeiro *et al.* (2000) and it revealed that above ground dry matter increased with increasing vermicompost rate up to 600 g and root dry matter upto 400 g/pot, then decreased. When it was studied in the field, yield was greater with organic than mineral fertilizer, but did not differ between organic sources. Addition of NPK had no significant effect in green house or the field.

Application of vermicompost (2.50 t/ha) was found better for getting higher yield in bell pepper at Hyderabad, India (Naik, 2018).

8.4.1 Nitrogen

Campbell and Swingle (1965) reported maximum yield of sweet pepper at 25 cm spacing with 40 or 80 kg nitrogen and 100 kg K per hectare and at 50 cm spacing with 80 kg nitrogen and 100 kg K per hectare. During the earlier part of the flowering, an increase in the nitrogen level increased flower production but did not affect the percentage of fruit set, while during the later part of the period an increase in the nitrogen level reduced flower production but increased the percentage of fruit set (Maynard *et al.*, 1962). The number of days required for flowering was found to increase with high dose of nitrogen application (Gill *et al.*, 1974). With cultivar Arka Gaurav, Singh and Naik (1990) obtained highest net returns at a spacing of 40 x 50 cm with 50 kg N and 150 kg P_2O_5/ha.

Hartz *et al.* (1993) observed yield and mean fruit size of drip-irrigated bell pepper peaked at 168 to 252 kg N/ha, with additional N having a suppressive effect on crop productivity. Petiole NO_3^--N concentration >5000 µg·/g through the early fruit bulking stage was associated with maximum yields. Analysis of fresh petiole sap by nitrate-selective electrode was highly correlated with dry petiole NO_3^--N and appears to be a viable, on-farm technique to supplement conventional laboratory tissue testing. Chlorophyll content, as estimated by the Minolta SPAD 502 meter, was weakly correlated with whole-leaf N; however, the ratio of SPAD values to an in-field reference of known N sufficiency may be a useful measure of relative crop N status.

The study of Garcia-Bañuelos *et al.* (2017) revealed differences between cultivar/ rootstock combinations for nitrogen use efficiency and yield in bell pepper. The greatest N assimilation was presented with the combination Fascinato/Robusto, which coincided with the maximum yield of all cultivar/roostock combinations evaluated. Hence grafting bell pepper plants can be used as a quick and effective method to improve NUE, which has beneficial implications for a reduction in environmental pollution and agricultural sustainability.

8.4.2 Phosphorus

Fruit yields were highest with 12 g N/m^2 when plants of *C. annuum* cv. Lamuyo were grown under controlled conditions in an experimental green house (Baghour *et al.*, 2000). Nitrate reductase activity and leaf Chlorophyll content were greatest in this treatment.

The results of the study on response of bell pepper grown in plastic houses to nitrogen fertigation by Walid *et al.* (1999) indicated that the yield and marketable number of fruits increased with N application. The highest yield was obtained with the addition of 150 kg N/ha. The maximum growth rate and the maximum rate of dry matter accumulation in the fruits occurred 90-150 days after planting. This may indicate that the peak of the pepper N requirement and utilization would be during the same period of the maximum growth rate. Increasing the rates of N applied increased the N uptake by the plants and at the same time stimulated the uptake of K and P through the synergistic effect of N on both nutrients. The role of nitrogen in enhancing the seed yield of bell pepper has been reported but to the high cost of synthetic fertilizers, there is great need to work on environmental friendly fertilizers like biofertilizers.

In an experiment on nitrogen sources (pre-plant sulphur coated urea, preplant polyolefin resin-coated urea and fertigated ammonium nitrate) and nitrogen rate (90, 135, 180 or 225 kg/ha) Guertal (2000) concluded that nitrogen sources rarely affected the partitioning of harvested peppers into grade groups. Adding nitrogen from any source decreased per cent of nonmarketable yield and percentage of marketable yield that was large (U.S. grade 1) peppers. There were few consistent differences in pepper yield or quality due to nitrogen sources. Given cost consideration of drip application and nitrogen sources, slow-release nitrogen materials may be a viable option for small- scale growers not using drip application systems.

Sharma and Gajbhiye (1998) observed that for achieving higher seed yield with maximum net returns per unit area in bell pepper under mid hill conditions of Himachal Pradesh (India), application of natrain (0.5 kg/ha seed treatment + 1 kg/ha seedling treatment + 5 kg/ha soil treatment) along with 100 kg nitrogen was found suitable for getting high and good quality seed of bell pepper.

Experiment showed that side dressing of phosphorus is beneficial to sweet peppers when heavy rains have leached fertilizer from the root zone of the plants during the early stages of growth (Hortenstine, 1963). Ozaki and Hortenstine (1963) reported that application of phosphorus increased early and total yields of peppers and both band placement and side dressing were found equally effective.

8.4.3 Potassium

Spaldon and Ivanic (1968) opined that higher rates of potassium increased the uptake of P and K. Application of potassium increased the total dry matter and yield of red pepper. Treatment with potassium chloride showed an increased yield compared to potassium sulphate (Ivanic, 1963). Effect of fertilizer levels and spacings on flowering, fruit set and yield of sweet pepper cv. Bharat was studied by Shrivastava (1996). The study revealed that first flowering and 50 per cent flowering were delayed by 4 and 4-6 days, respectively, in plants receiving the highest rate of fertilizers (N: P: K:: 300: 250: 250 kg/ha). First fruit set was also delayed in these plants. The highest number of fruits per plant (10.66), fresh weight per fruit (128 g), yield per plant (637.5 gm) and yield per hectare (92.95 q/ha) were observed in plants treated with 250 kg N + 200 kg P + 200 kg K/ha. These attributes also decreased with increased spacing (60 x 60 cm).

Sweet pepper plants produce much larger fruit with considerable carbon sink demands. Fruit set displays an undulating pattern, which might influence current K requirements (Marcelis *et al.*, 2004). Therefore, splitting the K dose may have positive effects on yield, but the timing or synchronization with fruit set should be considered. This may provide an explanation for the inconsistent response of sweet pepper to the split K dose.

8.4.4 Nitrogen, Phosphorus and Potassium

In soils containing abundant organic matter, application of P and K at 100 and 250 kg per hectare and where the soil is poor in organic matter, application of N, P and K at 100, 50 and 150 kg per hectare was found optimum (Bottini, 1967). In chernozem soil, application of 120 kg per hectare each of N, P and K was suggested by Ludilov and Ludilov (1977). Application of 200 kg N/ha, 112.5 kg P_2O_5/ha, and 75 kg K20/ha and an intra-row spacing of 45 cm gave the highest yields (97.26 kg/ha) of quality seeds cv. California Wonder (Dharmatti and Kulkarni, 1988). From a two years experiment Russo (1999) opined that rate of NPK can affect fruit colour of both pungent and non-pungent peppers and also influence green and total yields.

In sand culture with capsicum cv. California Wonder, Suwandi and Subur (1986) from Indonesia obtained greatest vegetative growth and highest fruit yields by supplying a 1387 ppm solution of a 15N: 15P: 15K fertilizer or a solution containing 988 ppm Ca $(NO_3)_2$. $4H_2O$ + 658 ppm KNO_3 + 497 ppm $MgSO_4$, $7H_2O$) + 272 ppm KH_2PO_4 + 88 ppm vegimax (a biochemical fertilizer). The volume of nutrient solution required increased from 250 ml/day for each plant during vegetative growth to 325 ml/day for each plant during the reproductive stages.

Marti and Mills (1991) observed that nitrate (NO_3) form of nitrogen is preferred by capsicum. Uptake of nutrients was highest during fruit development and immediately after fruit harvest indicating fruit removal promotes nutrient uptake. According to Golez (1995), increasing potassium rates reduced leaf Ca and Mg contents but had no marked effect on Na.

Ortas (2013) recorded the highest yield by the application of 200 N kg/ha. K addition had less effect on plant yield. Increasing the N level statistically increased the yield and N, P and K concentration. N, P and Zn content were affected by N addition. Also plant tissue P and Zn concentration was affected significantly by the N addition rather than K addition.

Shilpa *et al.* (2018) recommended that the conjoint use of 75 per cent recommended dose of NP (100: 75 kg/ha) + combined application of vermicompost and enriched compost @ 2.5 t/ha + PGPR (*Bacillus subtilis*) along with full recommended potash (55 kg/ha) and FYM (20 t/ha) as basal application resulted in significantly maximum vegetative growth and fruit yield with B: C ratio of 2.73: 1.

8.4.5 Lime and Magnesium

Application of lime increased yield of Florida Giant pepper when the acid soluble Ca level in the soil was below 300 lb/acre. On such soil an application of 2000 lb limestone resulted in increased yield of marketable peppers (Ozaki and Hortestinc, 1961). Magnesium sufficient soil produced significantly larger plants than magnesium deficient soil or magnesium treated deficient soil (Dempsey and Boswell, 1979). Treatment with Hidronit (30 per cent N) at 2 per cent and magnesium sulphate at 1 per cent was found to enhance the yield of Kalocsai 622 and Szegedi 20 cultivars (Lorincz- Csapo *et al.*, 1982).

8.4.6 Micronutrients

Spraying with EDTA, chelates of Cu, Zn, Mn and Fe five times at 15 day interval increased the yield of pepper cv. Vinedale. The colour of the fruits was found to improve by spraying with B, Cu and Zn or Cu and Zn (Navrot and Levin, 1976). Seeds when treated with zinc chloride or zinc sulphate (25, 50, 75 and 100 mg/1) for 20-24 hours before sowing and plants sprayed with similar solution at two-leaf stage, the treated plants showed 1.3 times more leaves and were taller than control plants (Shestunov and Zolotukhin, 1973). Among the different micronutrients, Cu and B had a beneficial effect on capsaicin content. Cu and Mn also had beneficial effect on capsaicin accumulation in fruit dry matter (Nowak, 1980). According to Salim *et al.* (1995), cadmium had a marked effect on the growth of plants. Foliar treated plants were affected more than root-treated plants. The least growth inhibition and Cd uptake was found in plants grown in soil with high organic matter content. Both growth inhibition and Cd uptake were increased in plants treated with acidic, saline and hot Cd solutions. Titanium as a beneficial element for *C. annuum* has been reported by Carvajal and Alcaraz (1998). It promoted biomass production and nutrient absorption, increased the activities of several enzymatic systems (catalase, peroxidase, lipoxygenase and nitrate reductase) and increased the concentrations of malic acid, leaf chlorophylls, fruit carotenoids and ascorbic acid. Titanium decreased the starch content in plants. It mainly accumulated in the cytoplasm and chloroplasts of leaf cells. Besides, its application also increased the concentration of Fe^{++} (active form of Fe) in leaves, fruits, leaf chloroplasts. High concentrations of the active form of Fe could be responsible of the promotive effects of titanium.

A pot experiment was conducted by Azam *et al.* (2016) in the green house to study the effect of foliar applied different concentrations (5, 10 and 15 mM) of calcium chloride and potassium chloride to alleviate drought stress in bell pepper plants. Calcium and potassium significantly improved all parameters, *i.e.*, photosynthesis rate, transpiration rate, leaf free proline, leaf osmotic potential, fruits weight, *etc.* Calcium chloride at 10 mM showed better results than other treatments under normal irrigation while calcium chloride at 15 mM showed better results under drought conditions. So, it may be concluded from present study that drought tolerance in bell pepper could be improved by foliar application of calcium and potassium chlorides.

In bell pepper under green house conditions shoot pruning is done to train plant to grow upright in order to facilitate light penetration all over the leaf canopy, improve fruit set and obtain early fruit ripening and high yield of large sized fruits. Awalin *et al.* (2017) found that the treatment combination, shoot pruning plus mixed micronutrients @ 100 ppm each: B, Zn, Cu and Mn as H_3BO_3, $ZnSO_4$, $CuSO_4$ and $MnSO_4$ showed best potentiality on the growth and yield of capsicum.

Singh *et al.* (2017a) observed that application of RDF (120: 80: 80:: NPK kg/ha) + humic acid 10 kg/ha as soil application + humic acid 0.1 per cent as foliar spray + micronutrient mixture (0.5 per cent Zn + 0.2 per cent Boron + 0.5 per cent Mn) as foliar spray was found best for sustaining the higher growth, fruit yield and quality of bell pepper under poly house condition.

One of the most frequently appearing disorders in sweet pepper feeding in covered cultivation as well as in open fields, which happens in the period of most intense fruit growth, is blossom-end rot (BER). The main cause of this disorder is limited water absorption by plants, which is closely related to Ca transport (Suzuki *et al.*, 2003; Cobanero *et al.*, 2004). In order to check this disorder Buczkowska *et al.* (2016) found that foliar feeding of calcium (1 per cent concentration) 3 to 5 times caused a limited number of fruits with BER symptoms as compared to control fruits. The use of $Ca(NO_3)_2$ as a source of calcium had a positive effect on the accumulation of vitamin C and carotenoids as compared with other fertilizers. Reduced Ca spraying proved to be beneficial in terms of fruit yield and concentrations of carotenoids.

8.4.7 Nutrient Content in Leaf Tissues

The nutrient content in plant tissue at different stages was found to vary among the cultivars. In sweet pepper cultivar Avelan, the nutrient uptake was higher than in cultivar Ikeda at 75 days after planting but this difference was found to reduce at the end of the crop cycle at 115 days after planting (Fernandes *et al.*, 1971).

It appears that nitrogen was translocated from leaf to fruit tissue. The phosphorus concentration in the leaf tissue was related to the nitrogen content and stage of plant development. Under conditions of nitrogen stress, the phosphorus content in leaf increased significantly as the fruit matured (Thomas and Heilman, 1964).

8.4.8 Time of Application

Matev (1966) in an experiment with sweet pepper cv. Sivrija 600 concluded that earlier top dressing was beneficial for higher yield. Application of 40 tonnes farmyard manure and two-third of P and K (in trials of 120 kg N, 240 kg P and 180 kg K/ha) and top dressing with the rest of NPK increased yield by 43 per cent compared to control (Petkov, 1972). Application of 150 kg nitorgen per hectare in three equal splits, at planting, 30 days and 60 days after planting gave consistently higher yield of sweet pepper cv. California Wonder under Hessarghata (Bangalore) condition (Srinivas and Prabhakar, 1982). According to Song (1987) applying N fertilizer 5 times from anthesis to harvest increased yields by 167-232 per cent over those of control plants.

8.5.0 Weed Control

Two weedings at 30 and 60 days after planting led to the highest yield of sweet pepper. In California, diphenamid @ 4.8 kg a.i. per hectare, dinitramine at 1 kg a.i.

per hectare and pendimethalin at 1.25 kg a.i. per hectare alone or in mixture with EPTc @ 3 kg a.i. per hectare incorporated prior to transplanting gave satisfactory weed control of bell pepper (Gluhaich, 1981). Basalin @ 0.48 kg a.i./ha applied 5 days after transplanting followed by one hoeing 25 days after transplanting proved to be best treatments in terms of weed control and crop yield (Lankroo *et al.*, 1990).

Ackley *et al.* (1998) conducted a field studies, in 1993, 1994, and 1995, with the chemical clomazone at 390 g ai/ha and rimsulfuron at 35 g ai/ha as pre-plant incorporated (PPI) and postemergence (POST), respectively, for weed control and crop response in transplanted bell pepper 'Keystone RG3'. Common lambsquarters and jimsonweed control was 77 to 95 per cent by clomazone treatments but was found variable by rimsulfuron treatments at 41 to 93 per cent common lambsquarters control and 42 to 68 per cent jimsonweed control; common ragweed control by both herbicides was unacceptable. They found that clomazone controlled common lambsquarters and jimsonweed without injury to bell pepper plants. From this study they concluded that the solanaceous weeds were not well controlled by rimsulfuron, and bell pepper was not sufficiently tolerant to rimsulfuron to permit its use in this crop.

In Ontario, field trials were conducted over a 3-year period to evaluate the effect of tank mixtures of sulfentrazone (100 or 200 g ha^{-1} a.i.) with either s-metolachlor (1200 or 2400 g ha^{-1} a.i.) or dimethenamid-p (750 or 1500 g ha^{-1} a.i.). The combination of sulfentrazone + dimethenamid-p provided good to excellent control of all weed species except velvetleaf. Based on this study, sulfentrazone and dimethenamid-p have potential for minor use registration in pepper (Robinson *et al.*, 2008).

Sweet pepper (*C. annuum*) plants weeded 3 weeks intervals producing highest values against weeding at 6 and 9 weeks intervals and recommended to farmer growing *Capsicum annuum* in Asaba agro-ecological area (Agbogidi and Okonmah, 2011).

Figueroa *et al.* (2016) found that clomazone, pendimethalin, and S-metolachlor were the most effective treatments applied after transplanting, whereas all herbicides tested were selective for bell pepper when applied and incorporated before transplanting. Their study on bell pepper in irrigated Chilean fields shows that timing of herbicide application is critical to prevent injury to this crop.

8.6.0 Mulching

In Korea, black polyethylene mulch was recommended for capsicum (Kwon *et al.*, 1988). VanDerwcrken and Wilcox-Lee (1988) observed that yields from plots that were mulched but not irrigated similar to those from plots that were sprinkler irrigated but not mulched.

In National Horticultural Research Institute, Ibadan, sweet pepper ('California wonder') was grown on plots with melon, cucumber and pumpkin as live mulches for the study of sweet pepper fruit yield and its components such as number of fruits per plant and fruit weight. The highest significant yield were obtained from plants with melon mulch in 2006 and 2007, respectively, while the use of pumpkin as a cover crop for the production of sweet pepper significantly controlled weeds in two years, but resulted reduction in fruit yield for the same respective years (Akintoyea and Adebayo, 2013).

Komla (2013) tried different organic mulches on growth and yield of sweet pepper at the University of Ghana Forest. From this study he found that application of organic mulch compared to the control resulted in increased plant height, canopy size and stem diameter and also significantly influenced yield and yield components of sweet pepper in the dry season. Rice husk mulch was found best in increasing the weight of fruits per plant, total fruit yield and mean fruit weight per plant.

Black polyethylene mulch showed to be an efficient technique to reduce irrigation number and water volume applied in sweet pepper and plants showed good osmotic adjustment consequently, tolerance to water stress (Klar and Jadoski, 2004). Canul-Tun *et al.* (2017) studied the effect of plastic mulch films of different colours (black, aluminum, silver, and white) on soil temperature, growth, nutrimental status of two bell pepper varieties grown under shade house conditions. From this study they found that an increased amount of total content of nitrogen (N), potassium (K) and sulfur (S) in leaf and total yield was higher in black plastic mulch followed by the aluminum, silver and white plastic mulch, bare soil recorded the lowest yield.

Singh *et al.* (2017b) did an experiment with different coloured plastic mulches applied in bell pepper which resulted maximum rise in soil temperature by using black/black plastic mulch followed by silver/black mulch as compared to no mulch application in April. Highest fruit yield was obtained by using silver/black polyethylene mulch as compared to no mulch plots followed by black/black mulch. Under black/black coloured plastic mulches, maximum reduction in weed dry matter and maximum ascorbic acid content in fruit was recorded.

8.7.0 Effect of Growth Substances

In pepper, growth substances showed various effects on plant or plant parts through their action on the growth processes. Spraying of gibberellic acid at 25 ppm 3 times during the vegetative growth caused flowering 18 days later than control (La Red and Cucchi, 1966). Wally *et al.* (1971) reported that treatment with 2, 4-D at 2 ppm at fortnightly interval during the flowering season of winter and summer grown sweet pepper (transplanting in December and April) increased flowering in the winter crop. Spraying with Aminozide (1250-5000 ppm) retarded growth by causing loss of apical dominance, while GA_3 (25-200 ppm) accelerated apical development and resulted in flower abortion on the latter shoots. NAA (50-150 ppm) caused degeneration of flower buds at the first flowering node and delayed the appearance of flowers at the second and third nodes. Ethephon (500-1500 ppm) delayed flower initiation, caused the abortion of flowers already formed, suppressed growth of the main axis and promoted development of the lateral buds (Rylski, 1972). Kim and Ho (1978) recorded highest yield of red pepper cv. Dae Jang Kak by the application of maleic hydrazide at 3500 ppm towards the end of the growing season. Foliar sprays of morphactin caused epinasty, weakening of apical dominance and increase in the number and length of the lateral branches. At higher concentrations (100 or 1000 ppm) treatment with morphactin suppressed stamen formation and reduction of floral parts (Jayakaran, 1972). Number of fruit/plant, fruit yield/plant and the ratio of fruit length to diameter increased with the application of 500 or 1000 ppm of Atonik (isomers of nitrophenol) during June to October cultivation (Djumaijah and Aliudin, 1986). Suwandi (1988) from Indonesia found that application of Mixtalol (containing triacontanol) at 3.0 ml/l to young plants (3 fully expanded leaves) and to flowering plants gave highest increase in plant growth and yield. Thakur *et al.* (1999b) reported that application of 5 ppm Mixtalol (triacontanol) gave the highest fruit yield in both irrigated and water stress conditions (5 days of water stress at the 43rd day after transplanting) while paclobutrazol at the same dose was ineffective.

The effectiveness of different bioregulators for enhancing fruit yield and quality was investigated in a green house in Spain by Bclakbir *et al.* (1998). The commercial bioregulators CCC (Chtoromcquat chloride), NAA, GA_3 and Biozyme ® (GA_3 + IAA + Zeatin + Micronutrients) were sprayed on to plants at flower initiation, followed by two additional applications at 30 day intervals. Biozyme induced increased yields, but almost 40 per cent of the fruits were not marketable. Application of NAA produced the highest yields of marketable fruits. Fruit fimmess were not affected by bioregulator applications as compared to control. Fruit ascorbic acid and citric acid concentrations were increased by GA_3, and CCC increased fruit soluble solid content. Biozyme increased fruit fractose, sucrose, carotenoid and lycopene concentrations. However, these bioregulator applications had not effects on fruit ca or pH. Foliar sprays of 10 ppm NAA when the flowers first appeared and 15 days after this or on both dates reduced the flower drop, improved fruit set and increased fruit size (Sharma *et al.,* 1999). According to them, the most efficacious treatment was that involving 2 NAA sprays.

Ethylene products in sweet pepper caused a high rate of flower abscission due to the resultant imblance in ethylene and auxin metabolism in the flower. Aguirre *et al.* (1995) measured ethylene production at different stages of flowering in order to determine the best time to apply exogenous ethylene inhibitors and concluded that petal opening stage was the most critical one when exogenous application of ehtylene inhibitors would be most effective in preventing abscission. Mojecka-Berova and Kerin (1995) studied regulation of green pepper vegetative growth and fruit bearing capacity with paclobutrazol and reported that it reduced plant height compared with untreated controls after 30 and 45 days particularly when applied to the soil (25 or 50 mg/l). Application of 25 mg paclobutrazol/l to the soil significantly increased early yields, individual fruit weight and pcricarp weight whereas foliar application (same concentrations) tended to reduce them. In capsicum cv. Jrini grown in hydroponic culture in the green house, triacontanol (5 mg/l) treatment increased cumulative yield compared with control by 50 per cent ; plant height and leaf number were also increased, but not significantly (Kapitsimadi, 1995). Meir *et al.* (1996) suggested that methyl jasmonate might mediate the plant's natural response to chilling stress, and by its application might provide a simple means to reduce chilling injuries in chilling susceptible commodities.

Use of ecofriendly by-products has emerged as one of the important strategics for sustainable agriculture. Biozyme, a biproduct obtained from a seaweed alga (*Ascophyllum nodosumy*) and known to be rich in cytokinin and auxin precursor, enzymes and protein has been reported to enhance the growth and yield of crop by affecting the better utilization of nutrients. In sweet pepper cv. California Wonder, a study has been made by Kumar *et al.* (2000) on photosynthetic response to biozyme in relation to fruit yield. Results indicated that the best treatment was normal fertilizer (100: 60: 60 kg/ha NPK) + 4 ml/1 biozyme sprayed at 50 DAT in terms of photosynthetic rate, canopy photosynthesis and fruit yield. According to Ruiz *et al.* (2000), application of biozyme (GA_3 at 32.2 mg/l + IAA at 32.2 mg/l + zeatin at 83.2 mg/l + chelated micronutrients) increased NO_3^- assimilation, due possibly to a greater availability of these bioregulators in the leaves and increased nitrate reductase and nitrite reductase activities. This treatment also produced the greatest foliar dry weight and fruit yield per plant.

Maboko and Du Plooy (2015) conducted 2-year trials to evaluate the effect of plant growth regulator's (PGRs) on plant growth, yield, and quality of sweet peppers grown hydroponically. Sweet pepper plants were subjected to two levels of gibberellic acid (GA_3) (10 and 15 $mg \cdot L^{-1}$), two levels of naphthalene acetic acid (NAA) (15 and 30 $mg \cdot L^{-1}$), and four combinations of NAA and GA_3 (10 $mg \cdot L^{-1}$ GA_3 + 15 $mg \cdot L^{-1}$ NAA, 10 $mg \cdot L^{-1}$ GA_3 + 30 $mg \cdot L^{-1}$ NAA, 15 $mg \cdot L^{-1}$ GA_3 + 15 $mg \cdot L^{-1}$ NAA, and 15 $mg \cdot L^{-1}$ GA_3 + 30 $mg \cdot L^{-1}$ NAA) applied to plants at flower initiation in a nontemperature-controlled tunnel. The experiment resulted application of GA_3 at 10 and 15 $mg \cdot L^{-1}$ alone or in combination with NAA increased plant fresh and dry mass.

In the Horticultural Farm of Sher-e-Bangla Agricultural University, Dhaka, Bangladesh an experiment was conducted on the application of Gibberellic Acid (GA_3) @ 30 ppm alone or with 4-Chloro Phenoxy Acetic Acid (4-CPA) of different dose and number of time on sweet pepper (*Capsicum annum* L.). The highest yield was obtained from three sprays of 4-Chloro Phenoxy Acetic Acid (4-CPA) @ 45 ppm + Gibberellic Acid (GA_3) @ 30 ppm treatment combination (Akhter *et al.*, 2018).

Zeist *et al.* (2018) evaluated the effect of isolated and combined foliar application of boron, calcium, and the plant growth regulator Stimulate on fruit yield and gas exchange in bell peppers. They reported that the foliar application of boron from the beginning of flowering improved the photosynthetic yield and the yield of marketable fruits cultivated in the field. Whereas, the foliar application of calcium and Stimulate did not improve gas exchange and fruit yield. The most common effects of boron were an increase in the number of marketable fruits and foliar spraying with calcium from the beginning of flowering increased the firmness of commercial fruits.

9.0 Ripening, Harvesting, Yield and Storage

9.1.0 Ripening

The bell-type peppers are usually harvested and sold when they are of suitable market size and are still green. There is but a limited demand for the mature red specimens. The pimiento (Georgia), and paprika are harvested when completely red (Christopher, 1958). Now-a-days different chemicals are used to accelerate ripening as well as inducing the fruit colouration. In hot pepper, when harvest red-ripe, the total carotenoid content of the pericarp increased by 120 per cent before reaching its maximum 5-6 weeks after harvest, whereas in sweet red peppers the increase was only 65 per cent 3-4 weeks after harvest. The hot cultivars showed a drop in sugar content during after ripening, losing 32 per cent in 7-8 weeks and the sweet cultivar lost 25 per cent sugar in 5-6 weeks (Benedek, 1959). In bell pepper fruits, the rate of CO_2 production and the pattern of colour changes during ripening were similar to those of tomato but unlike the tomato no critical curve of respiration was obtained (Pratt and Lyons, 1959). Sims *et al.* (1970) reported that spraying with ethrel at 500 ppm on red peppers increased the percentage of red fruits harvested 23 days later from 7 to 49 per cent and in sweet peppers harvested 22 days later increased the red fruit from 9 to 28 per cent. Treatment with 2-chloroethyl phosphonic acid (CEPA) enhanced, whereas CCC inhibited chlorophyll degradation and carotenoid formation in ripening red pepper fruits (Kamicnska and Chrominski, 1971). Haung (1973) reported that the yield and the percentage of turning half red and full red fruits of capsicum cvs. Yang-Chiao and Jalapeno were increased by treatment with ethephon at 200 or 400 ppm. The optimum colour appeared in 80 per cent of the fruit of Yang-Chiao and Jala at 11-20 days and 30-35 days, respectively, after ethephon application. Spraying with ethephon at 200-3200 ppm on full grown bell pepper plants were found to accelerate fruit colour development, fruit and leaf drop and leaf yellowing (Knavel and Kemp, 1973). Benoit and Ceustermans (1974) suggested spraying of ethephon at lower concentration when the fruits were green. Sims *et al.* (1974) also observed that ethephon (0.5-1.25 kg/ha) treatment resulted in less green fruit and there was a trend towards improved ripening.

Like chillies, sweet pepper fruits undergo deep morphological, physiological and metabolic transformations in terms of pigment composition and content during ripening. The genotype, maturity and growth conditions affect these changes in fruit composition (Marín *et al.*, 2004). The green colour of sweet pepper fruits is due to chlorophyll pigments, violet/purple pigments colour due to anthocyanins and the yellow-orange colors are provided by α- and β-carotene, zeaxanthin, lutein and β-cryptoxanthin and carotenoids such as capsanthin, capsorubin and capsanthin 5,6-epoxide confer the red colours (Sun *et al.*, 2007). Carotenoids are responsible for yellow to red colours in fully mature pepper fruits and they are stored in chromoplasts (Deli *et al.*, 2001). The carotenoid synthesis pathway begins with the synthesis of phytoene by phytoene synthase during development and ripening pepper fruits (Hirschberg, 2001). Several desaturation reactions convert phytoene to orange coloured β-carotene, and then β-carotene is oxygenated to form xanthophylls, such as β-cryptoxanthin, zeaxanthin, and antheraxanthin. In *Capsicum* species, capsanthin/capsorubin synthase (*Ccs*) synthesizes two red pigments, capsanthin and capsorubin (Shah *et al.*, 2014). During pepper fruit colour formation, phytoene synthase (*Psy*), lycopene-β-cyclase (*Lcyb*), β-carotene hydroxylase (*Crtz*), and capsanthin/capsorubin synthase (*Ccs*) genes are responsible in the carotenoid biosynthesis pathway (Guzman *et al.*, 2010).

Roy *et al.* (2019) studied the colour development and ripening of sweet pepper which expressed four distinct fruit colour at fully matured stage *viz.*, deep green (Royal Wonder), yellow (Arya, C/4 yellow), orange-red (8/4) and red (C/4, Ayesha and Baby Bell). During immature stage, colours of all the breeding lines were green in different shades (light green, green, and dark green). After 55 to 69 days after anthesis depending on the genotype distinct colour in the fruits (red, orange, yellow and purple) started developing. They also noticed that fruits of all the genotypes turned different shades of red colour depending on the genotype during ripe stage.

9.2.0 Harvesting

The large sweet peppers are usually picked while they are still green colour but full- grown when sold in the market. Pimiento and Paprika cultivars are harvested when fruits are dark red ripe (Thompson and Kelly, 1957). Cochran (1943) stated that the most favourable time to harvest fruits of Pimiento pepper for seed appeared to be between the 50 and 60 day old stages or when the fruits attained a bright to deep red colour. Sweet peppers are picked with an upward twist which leaves a piece of stem attached (Christopher, 1958). Young, immature peppers are rather soft and yield readily to mild pressure of the fingers. Green fruits ready for harvest are relatively firm and crisp (Shoemarkcr and Teskey, 1955).

Harvesting time had a great influence on yield and fruit characteristics. Green harvests increased yield because of the higher number of fruits per plant while harvesting at the yellow or red ripening stage increased mean weight, length, basal diameter and mesocarp thickness of marketable fruits (Miccolis *et al.*, 1999). Molinari *et al.* (2000) were of the view that red and yellow pepper varieties could be harvested commercially at the onset of colour change and ripened at 20°C and 90 per cent RH while maintaining high quality.

9.3.0 Yield

The yield of sweet pepper varies with cultivars/hybrids and cultivation methods. Average yield of 300-400 bushels of green peppers and 3-4 tonnes of processing pepper per acre (Christopher, 1958) and 35-50 pounds of pepper seed (Hawthorn and Pollard, 1954) can be expected. In Hongkong, California Wonder cultivar was found to be the best yielder with more than 1 pound of fruits per plant (Herklots, 1972). Joshi and Singh (1975) opined that a good crop of capsicum may yield 10-12 tonnes of quality fruits per hectare if proper care is taken during its growth. There has been a high yield difference between the green house and the open field grown varieties ranging from 50 per cent to 150 per cent (Singh *et al.*, 2011), and this may be due to its protective ability against major biotic and abiotic stresses, which reduces the effect of high disease incidence, excess rainfall, water logging, and provide controlled environment (3-4°C higher temperature than open field condition) to the crop (Singh *et al.*, 2003; Singh *et al.*, 2010). Nkansah *et al.* (2017) recorded increased fruit yield of same sweet pepper variety by 200 per cent in fully controlled green house as compared to open field condition. They obtained yield of sweet pepper in open yield and green house condition to the tune of 5.02 (Pepper 1) to 12.57 t/ha (California Wonder) and 9.23 (F_1 Nobili) to 21.34 t/ha (Kukulkan), respectively in Ghana.

Complete defoliation during either flower bud formation or fruit stage resulted in an average yield reduction of 61.7 per cent over control (Muro *et al.*, 1994). Canopy

Harvesting and Selling of Sweet Pepper in Supermarket

shape has positive effects on the yield of many crops. Complete defoliation during either flower bud formation or fruit stage resulted in an average yield reduction of 61.7 per cent over control (Muro *et al.*, 1994). Canopy shape has positive effects on the yield of many crops. Cebula *et al.* (1998) reported that pruning to a single shoot increased leaf blade thickness, light penetration, early and total yields and fruit ascorbic acid content compared with pruning to 2 shoots though average fruit weight did not differ markedly between canopy shapes.

9.4.0 Storage

Sweet pepper can be stored for moderately long periods at temperatures between 7°C and 13°C, depending upon the variety and the stage of maturity (Meir *et al.*, 1995). Main factors of quality degradation of sweet pepper during storage include decay development, susceptibility to chilling injury and shriveling associated to rapid water loss, which limits storage at temperatures above 7°C (Meir *et al.*, 1995). Water loss in commercial bell pepper has been identified as the main physiological factor limiting pepper fruit quality and prolonged storage (Maalekuu *et al.*, 2003). Furthermore, reduced temperatures do not completely inhibit decay development during prolonged storage and red pepper has been found as the most perishable (Mohammed *et al.*, 1991; Meir *et al.*, 1995).

Sweet peppers can be kept in good condition for at least 40 days at 0°C and at relative humidity of 95-98 per cent. Shrinkage of the fruits stored under these conditions was only 4 per cent in 40 days (Platenius *et al.*, 1934).

The content of total carotenoids increased and that of sugars decreased in red peppers during after-ripening. Respiration which continued with diminishing intensity for 30-40 days after harvest was shown to be dependent on organic acids. It is considered that organic acids are formed by the decomposition of sugars and are used for respiration and carotenoid synthesis. Respiration was found to be lower and carotenoid synthesis higher in darkness than in light indicated that the peppers should be placed for after ripening in darkened drying installations (Gracza and Bcndck, 1961).

Janse (1988), while comparing the effects of different day/night relative humidity regimes reported that fruits kept in low/high or high/low regimes were firmer and kept for about 2 days longer than those in constant high or low regimes. In a trial on the effects of day/night temperature regimes ranging between 16/15°C and 28/21°C, fruits from plants grown under the lowest regime were firmest and kept for 18.9 days, compared with 11.1 days for those in the highest regime.

Fruits of the capsicum cv. Feherozon were stored successfully, without chilling damage, for 4-5 days at 5°C in circulating air and 85-89 per cent RH. For long storage (10-12 days) the optimum temperature was 7 to 8°C (Saray, 1988). Gao *et al.* (1987) studied storage properties of 16 cultivars and found that Tingling Three Ribs, World Champion, Qiemen, Three Ribs and 12-2 showed the best storage quality. Ascorbic acid content increased during storage. Cultivars with thicker cuticle, denser epidermis cells and a more orderly cell arrangement stored well. Barros *et al.* (1994) studied the postharvest storage conditions for sweet pepper when fruits of cv. Agronomico 10 G were dipped in a 1 per cent hypochlorite solution, dried, graded and packed (1) in plastic bags or (2) on aluminium trays covered with PVC film (0.0165 mm). In some cases $KMnO_4$ was added as an ethylene absorbent, and the fruits were held either in ambient conditions or at 8°C for 30 days. The least storage weight loss (11.6 per cent) was obtained in the later treatment without $KMnO_4$ in cold storage.

Mature green fruit of bell peppers cv. Jupiter and Capistrano were stored at 5 and 10°C, in air or 3 per cent O_2 alone or in combination with 10 per cent CO_2 and the fruits were stored for 6, 12 and 18 days then transferred to 15°C for 3 days to assess shelf life by Mercado *et al.* (1995) in Spain. According to them, there were no visual symptoms of chilling injury of fruits stored in air at either temperature after 12 days storage. Low O_2 had no beneficial effect on fruit quality. High CO_2 (with air or 3 per cent O_2) produced signs of chilling injury on fruits stored at 5°C, even after 6 days of storage bud did not affect fruit quality at 10°C. Respiration and ethylene production rates, after transfer to ambient conditions, were higher in fruits stored at 5°C and under 10 per cent CO_2 than in those stored at 10°C. They further reported that green fruits cv. Galaxy showed chilling injury at 5°C after 14 days. Fruit quality was best when stored at 5 per cent CO_2. Red fruits were less cold sensitive than green fruits. While comparing the storage conditions on chemical composition and market acceptability of fruits of sweet pepper, Ahmed *et al.* (1996) opined that controlled storage (8°C, 85-90 per cent R.H.) was considered the best as fruits could be stored for 4 weeks with minimum loss in weight and ascorbic acid content together with good market acceptability. According to Fallik *et al.* (1997), dipping commercially harvested sweet red peppers for 2 minutes in 1 or 2 per cent potassium bicarbonate reduced decay development after storage and shelf-life stimulation to a commercially acceptable level of 5-8 per cent compared with untreated or waterdipped controls. Higher concentrations of potassium bicarbonate (3 per cent) significantly reduced fruit quality. Manzano and Zambrano (1995) opined that Prolong, a commercial wax coating at 0.5 per cent (w/v) prolonged the storage life of fruits of sweet pepper at a storage temperature of 0 or 5°C. Perforation in the packaging film greatly affected O_2 and CO_2 concentrations and water condensation but much less so the in-package relative humidity. Consequently, regulation of the extent of perforation could be used to control atmosphere composition and water condensation in the package without a significant increase in fruit weight loss. Sweet peppers packaged without a significant increase in fruit weight loss. Sweet peppers packaged in perforated film lost less weight during storage and maintained higher quality than fruits stored in open carton boxes, and at the same time demonstrated lower decay levels than fruits sealed in non-perforated packages. Perforation as well as the introduction of hygroscopic materials, such as NaCl, effectively reduced the percentage of *Botrytis* decay in fruits (Ben-Yehoshua *et al.*, 1996).

The findings of Wiendi *et al.* (1996) on storage of fruits of sweet pepper using gamma irradiation revealed that shelf life increased from an average of 10 days in controls (non-irradiated, non-wrapped fruits held at 25-27°C) to 41 days by refrigeration. Irradiation of the fruits combined with refrigeration extended shelf life to 49-58 days. Wrapping in PVC film was the least effective treatment for extending shelf life and was unsuitable for fruits held at 25-27°C. Weight loss was reduced most effectively by refrigeration, followed by irradiation and wrapping in thick film. Estrada *et al.* (1997) carried out an experiment to determine the changes in fruit weight, size and colour (chlorophyll and carotenoids) and capsaicinoid accumulation during the ripening process in *C. annuum* cv. Podron. First harvested fruits were small and undeveloped but fruit length and weight increased gradually from 21 to 28 days, reaching a peak at 35 days, when peppers were fully developed and mature green. At 42 days (when peppers were red), fruit weight began to decrease. Chlorophyll content of fruits increased until 28 days after flowering, whereas carotenoid was low until 35th day, after which it increased markedly at 42 days after flowering. Fruit capsaicinoid content was initially low, but increased slightly between 21 and 28 days after flowering and markedly at the end of fruit development. Capsaicin was the major component of total capsaicinoids throughout fruit development. Chlorinated water treatments and dipping time (0, 50, 100, 150 and 200 µg/ml chlorine solutions for 20, 30 and 45 min dipping) had no major effect on external colour or total sugar content (Nunes and Emond, 1999). However, total chlorophyll, soluble solids content and ascorbic acid content decreased with increasing period of dipping and chlorine concentration. A 20 min dipping time in 50-100 µg/ml chlorinated solutions can be used without reducing the overall quality of capsicum. Carbohydrate changes in ripening *C. annuum* (*grossum*) in rclatin to texture degradation was studied by Prabha *et al.* (1998) and they reported that fruits showed striking changes in texture when ripened at room temperature. Fruits exhibited significant weight loss and a change in textural firmness due to extensive cellular damage. The associated biochemical changes during ripening were: (i) an increase in free sugar levels, (ii) an increase in *in situ* hydrolysis of hemicellulose fractions and (iii) a general increase in the activities of cellulase, α-mannosidase, laminarinase, polygalacturonase, galactanase, mannanase and β-galactosidase and increased hemicellulase. Cold storage (8°C) prevented textural changes, resulting in a longer shelf-life for cold stored fruits compared with fruits stored at room temperature. From a study on optimizing the dehydration conditions of green bell peppers for quality criteria, Sigge *et al.* (1999) reported maximum colour retention and L-ascorbic acid content were attained in drying conditions of 55 and 60°C (15-40 per cent RH) as well as at 65 and 70°C (15 per cent RH). Developing fruits on lower nodes were major reproductive sinks for photosynthates in sweet pepper cv. Arka Gaurav. The removal of these fruits resulted in faster growth of other fruits on upper nodes. Sink-manipulated plants had a higher net photosynthetic rate at the later stages of plant development (Bhat and Rao, 1997). Hot water dipping alone or in combination with low density polythelene film packaging extended the shelf-life of sweet pepper (Gonzalez-Aguilar *et al.*, 1997). Treatment of *Capsicum* fruits with hot water at 45°C for 4 min piror to storage at 8°C markedly reduced the incidence of fungal infections (Gonzalez-Aguilar *et al.*, 1999). However, hot water treatment induced shrivelling during storage. When hot water-treated fruits were subsequently placed in low density polyethylene bags, storage quality improved (retention of firmness, reduction of water loss, retardation of colour change and alleviation of chilling injury). Total soluble solids, titratable acidity and pH values in fruits were generally not affected by these treatments. A 4 min dip at 53°C followed by packaging with 0.065 mm low density polyethylene film was very effective at maintaining quality. This treatment inhibited respiration rate, reduced decay, retained turgidity and green colour and maintained excellent overall quality after 28 days of storage at 8°C. Hot water treatment combined with polyethylene film packaging is a promising technique for improving the storage quality of capsicum. The effect of temperature and perforated polythylene packaging on quality and storability of breaker and red pepper fruits was studied by Kosson *et al.* (1998). The weight losses of breaker and red pepper fruits stored in polyethylene perforated packages were 5-20 times lower than in unpackaged ones. Fruits stored at 8°C for 4-5 weeks were of higher quality than those stored at 15°C for 2-3 weeks. Use of perforated polyethylene packaging retarded colouring in breaker fruits.

Pre-harvest sprays of $CaCl_2$ increased the weight of fruits and improved firmness retention in storage, decreased decay, increased pericarp wall thickness and insoluble pectic fractions but had no effect on water loss in storage. Calcium could improve firmness retention through enhancing the pectic composition of cell walls and pericarp wall thickness (Toivonen and Bowen, 1999).

From the study on permeance to oxygen of detached fruit, Bower *et al.* (2000) reported that cuticle is composed of a coherent film. The low rate of gas diffusion through the cuticle may allow recycling of respired CO_2 by photosynthesis in the fresh before harvest.

Modified atmosphere packaging (MAP) is a technology used to extend the shelf-life of fruits and vegetables of high commercial value. Packaging of pepper in perforated plastic films and storing them at optimum temperature could be one of the viable alternatives to extend shelf-life. Manolopoulou *et al.* (2010) have demonstrated that by using polyvinyl chloride (PVC), low density polyethylene (LDPE) and medium density polyethylene (MDPE) packaging of different O_2/CO_2 permeability, it is possible to generate a model associating mass loss, firmness of bell peppers and the packaging films. Packaging bell peppers with polyethylene bags at low temperature (7.5°C) reduced the water loss by 40-50 per cent and maintained the fruit quality (Meir *et al.*, 1995). The use of packaging films has been shown to increase the shelf-life of perishable produce establishing a beneficial in-package atmosphere containing low O_2 and high CO_2 and reduced water loss (Kader *et al.*, 1989). Xtend® films (StePac LA Ltd, Israel) have been developed to modify atmosphere and humidity for prolonged packaging. Use of Xtend® film and polyethylene film to modify the atmospheric conditions have shown to maintain the quality of bell pepper during storage at an optimum temperature of 7°C (Sharma *et al.*, 2018). Furthermore, Xtend® packaging proved to be ideal for the storage based on prevention of decay, minimizing weight loss, maintaining firmness, elasticity and scored well in the sensory evaluation.

Bayogan *et al.* (2017) studied the efficiency of the evaporative cooler (EC) in prolonging the shelf life of sweet pepper. As compared to ambient condition, the EC could able to provide a storage environment with lower temperature, lower vapor pressure deficit and a higher relative humidity. They found minimum weight loss in EC as well as the extent and rate of moisture loss. Results also showed that physical characteristics of sweet pepper in terms of visual quality, color, firmness, and shriveling were maintained better when stored under EC condition. The rate of the chemical changes for total soluble solids and titratable acidity for sweet peppers stored in EC condition were slowed down contributing to its extended shelf life. Furthermore, samples in EC condition exhibited a longer shelf life of a maximum of 9 more days than those stored in ambient condition

10.0 Growing Sweet Pepper under Protection

Being a high valued crop, the demand of capsicum is increasing day by day in world-wide but the prevailing environmental conditions in the temperate regions in general and in the European countries in particular restricts its cultivation in the open. Similarly in tropical and subtropical parts for round the year cultivation, the situations necessitate the crop to grow under protection. The trend of crop production under protected conditions is increasing due to high productivity and improved shelf life of fruits (Rai *et al.*, 2004). Few works relating to protected cultivation have been discussed earlier under different sub-heads. However, realising the importances of protected cultivation, a concise report on its different aspects has been outlined below:

10.1.0 Growing Structures

According to Rui *et al.* (1989) mean daily net photosynthetic rates of the plants grown in plastic film covered plots were 38.7 and 29.8 per cent higher than those of plants in the open plots at the peak fruiting and later stages, respectively. Chlorophyll and soluble protein contents were also higher for film covered plots.

In red pepper seedlings raised under plastic covers, plant height, leaf number, leaf area, leaf dry weight, stem dry weight and root dry weight were all higher with blue colour removing (BCR) film than with polyethylene film. Leaf fresh weight, leaf volume, percentage of leaf volume occupied by air spaces and volume of internal air spaces were all higher in leaves at the central position on the stem and higher with BCR than with polyethylene film regardless of leaf position. Leaf volume, leaf thickness decreased at higher leaf positions. Average leaf fresh weight, leaf volume, the volume of internal air spaces, the percentage of leaf volume occupied by air spaces, the volume of non-gaseous leaf contents were higher, and the leaf density and thickness were lower with BCR than with polyethylene film (Chung *et al.*, 1991).

According to Andriolo *et al.* (1991), plant growth and fruit production of sweet pepper cv. Cascadura Ikeda was better under cover compared to open when plants were grown especially during the winter months under cover in transparent polyethylene tunnels of 10 m wide, 25 m long and 3.5 m tall.

The influence of different cover materials consisting of plastic hoops supporting ethylene vinyl acetate (EVA), polyethylene (PE) or polypropylene (PP) of plastic green houses on growth and development of sweet pepper in Ljubljana, Slovenia was studied by Kacjan-Marsic and Osvald (1997). The study revealed that plants under plastic were taller and had more leaves than controls. The highest yields and the highest number of fruits/plant were obtained under EVA (800 g/plant and 14.9 fruits/plant, respectively) followed by PE (755 g/plant and 11.7 fruits/plant) which was due to highest increase in average daily high and low temperatures occurred in the EVA tunnel.

Yield of sweet pepper cv. Spartacus under Kool Lite Plus photo-selective film (PE film 150 µM thick with 7 per cent EVA and anti-condens, with reduced UVA and UVB absorption, increased blue and red light transmission and high short-wavelength IR reflection) were 6 per cent less than under Hytilux (PE film 180 µM thick with 11 per cent EVA and anti-condens) due to limited global light transmission. Fruits grown under Kool Lite Plus showed less incidence of blossom-end rot due to a better (cooler) climate.

Net house cultivation results in producing very early and also prolonged total yield (Nagalakshmi *et al.,* 2001). Capsicum produce high yield under naturally ventilated green house as compared to open field conditions. The protected cultivation offers several advantages to produce vegetables of high quality and productivity particularly during off -season when prices are higher thus using the land and other resources more efficiently (Sirohi, 2004).

In India, depending on the climatic condition of the region, different types of structures can be used. The two commonly used structures are shade net house or poly house. The shade net houses are built using granite stone pillars placed 2 ft deep into the soil. The pillars of about 12 ft high and 8 x 4 inch thick are used in the construction. The ends of the pillars are sealed with rubber to avoid tearing of the material. 50 per cent HDPE white shade net is fixed and it supports another shade net (35 per cent) either green or black in colour. This additional shade net is required during peak summer season. The shade houses are also protected on all sides with 40 mesh insect proof nylon net, which is UV stabilized. Takte *et al.* (2003) have used plastic films and shade nets in order to protect high value crops against various stresses like extreme atmosphere and insect-pests. Further, they added that ventilation played an important role in crop production under controlled conditions.

According to Singh and Sirohi (2006) the use of low cost protected structures like low plastic tunnels or walk-in-tunnels or shade net houses for offseason nursery production, small scale production of vegetables to fetch high price, virus free and insect free development of nursery in rainy season and good quality production of sweet pepper during rainy and post-rainy season. The significance of protected cultivation of capsicum due to its economic viability, high yield, better fruit quality, superior fruit appearance and ensured availability of capsicum throughout the year have confirmed by Murthy *et al.* (2009). They have further endorsed the promotion of this technology should be regulated by government and non-government institutions.

Medany *et al.* (2009) concluded that black net green house gave significantly the highest early crop yield, while white net green house significantly contribute to the highest plant height, number of leaves per plant, leaf area index and total fruit yield as compared to other green houses. An increase in yield to the tune of 113 to 131 per cent in bell pepper was observed when grown under pearl, red, blue or black coloured shade nets with shading of 40 per cent or 50 per cent in comparison to open cultivation (Ilic *et al.,* 2011). Ombodi *et al.* (2015) had reported 23-39 per cent of decrease in incoming radiation and 32- 46 per cent reduction in PAR under shade nets of white, green, yellow or red colour. However, yellow and green shade nets decreased yield while white shade nets in green house was reported best under unfavourable conditions.

Green house technology can be utilized for controlling the environmental parameters such as temperature, relative humidity, light intensity and duration, CO_2 level, Irrigation and nutrient supply, spacing, growing medium and root development (Baghel *et al.*, 2003; Wani *et al.*, 2011). Choudhary (2016) reported a manifold increase in the resource use efficiency through crop cultivation under protected structures as compared to the open-field conditions. Among the various protected structures, poly house production had been proven as more profitable protected technique for capsicum cultivation (Aruna and Sudagar, 2010).

In protective cultivation, production of vegetable is higher than the open field condition due to congenial inside microclimate and that provided better price. The protected cultivation comprises different devices and technologies namely windbreaks, irrigation soil mulches *etc.* and the structures which are green house, tunnel, rowcovers made the production throughout the year by modifying the natural environment (Trivedi and Singh, 2015). It will further prolong the harvest period, increase yield, quality improvement, and keep the availability of commodities frequently. Tunnel farming is one of the modified techniques of plastic-culture of modern farming. Tunnel farming was started first in 1948; it is the cheapest form of green house. Later on, more operational and economical forms of tunnel farming were introduced. The Plastic tunnels had diverse mechanism as tunnels are not heated-up artificially and are operative under minimum environment control as compared to green houses. Tunnels are more cost-effective than the green houses (Fatima *et al.*, 2017).

Poly house is more compatible for farming when compared to shade house. The roof of the poly house is covered with a transparent UV stabilized polyethylene film of 200 micron thickness. The sides of the poly house need to be covered with 200 micron polyethylene film up to a height of 3 ft. The remaining height on all sides is covered with a 40 micron white coloured insect proof net (Agri farming, 2019).

A study was carried out to determine the effect of different micro climate on the physiological parameter and yield of capsicum (*Capsicum annuum* L.) at Plasticulture farm, CTAE, Udaipur, Rajasthan, India (Shukla *et al.*, 2019). They were using four small structures (1, 2, 3 and 4) of 64 m^2 (16 m x 4 m) area each with varying cladding material *i.e.*, 75 per cent shade net, 40 mesh insect net, 200 micron LDPE UV stabilized sheet air vent on side and top provided with 75 per cent shade net, and 200 micron LDPE UV stabilized sheet and air vent on side and top provided with insect net, respectively. They observed that maximum plant height (174.7 cm), leaves per plant (64) and first harvesting (57.3 DAT) was found in structure-1. Early flower initiation (29 DAT) was found in Structure-2. Maximum number of flower per plant (23.7), cumulative number of flower per plant (18.20), highest individual fruit weight (94.4 g), highest fruit yield (1720 g/plant) were found in structure-4. The water use efficiency was found 14.62 kg/m^3, 13.52 kg/m^3, 15.17 kg/m^3 and 18.91 kg/m^3 under structure-1, structure-2, structure-3 and structure-4 respectively. Based on the results, structure-4 was found best in respect of yield, physiological yield, physiological parameters, maximum net income (Rs. 13813/- per structure *i.e.* Rs. 215.83/m^2 area) and B: C ratio (1.87) (Shukla *et al.*, 2019).

Six single-poly passively ventilated Quonset high tunnels were used for the experiment conducted by Lang *et al.* (2020). The shade treatments (no shade-cloth, 30 per cent light-reducing shadcloth, and 50 per cent light-reducing shade-cloth) were applied in June of each season. Within each shade treatment, seven colored bell pepper cultivars (Archimedes, Delirio, Flavorburst, Red Knight, Sirius, Summer Sweet, and Tequila) were grown. Both the 30 per cent and 50 per cent shade-cloth treatments reduced monthly average and maximum air temperatures within high tunnels, with the largest differences occurring in the months of July and August. The use of a shade-cloth reduced the incidence of sunscald by 59 per cent between no shade and 50 per cent shade-cloth treatments. While there was no difference between 30 per cent and 50 per cent shade treatments, the use of 50 per cent shade-cloth caused a decrease in both marketable number (32 per cent) and weight (29 per cent) of pepper fruit compared with the control. 'Tequila', 'Delirio', and 'Flavorburst' had more marketable fruit per plant. Shade treatments did not affect fruit soluble solids content (SSC), pH, or total titratable acidity (TTA). Shade treatments had no effect on Soil Plant Analysis Development (SPAD) readings, shoot biomass, the number of leaves per plant or the total leaf area per plant; however, plant height increased by an average 14.5 cm for plants under shadecloth treatments. Average leaf size was 11.2 cm^2 larger on plants grown under the 50 per cent shade-cloth, compared with the control. Several cultivar differences existed for each fruit quality and plant growth parameter. While differences in fruit quality and plant growth parameters were limited among shade treatments, decreasing marketable yield is concerned. This research suggested that Midwest growers should not exceed 30 per cent light-reducing shade-cloth on their high tunnels for colored bell pepper production.

10.2.0 Green House Environment

The effects of CO_2 concentration in the range 300-1100 µmol/mol on leaf conductance (g) and rate of crop transpiration (E) of sweet pepper cv. Mazurka were investigated in spring. In 2 green house compartments (154 m^2) that were simultaneously exposed to different CO_2 concentrations, leaf conductance of the upper leaves was measured with a steady state diffusion parameter and crop transpiration rates were measured with 3 weighing lysimeters per green house compartment. Multiple regression equations, describing the effects of photosynthetic active radiation (PAR), vapour pressure deficit (VPD) between the leaf and air, CO_2 and, optionally, leaf temperature on g, were fitted to the measured data. The fitted regression curves demonstrated that a 100 µmol/mol increase in CO_2 reduced leaf conductance by about 3 per cent, at any CO_2 concentration, VPD and PAR, if VPD and PAR remained constant. Measured rates of crop transpiration were highly correlated with radiation and were in reasonable accordance with the Penman-Monteith combination equation. With this equation it was estimated that a 10 per cent decrease in g would reduce rate of crop transpiration (E) by 1.5-3 per cent at high levels of leaf conductance (g) (high radiation) and by 4-7 per cent at low leaf conductance (g) (dull weather), at least if VPD remained constant. In a green house-crop system, however, owing to thermal and hydrologic feedbacks, an increase in

CO_2 concentration leads to a considerable increase in the UPD between the leaf and air. This would enforce the effect of CO_2 on leaf conductance (g) and counteract the effect of CO_2 on reduce rate of crop transpiration (E), because the driving force for transpiration would be enhanced. Thus, in general, the apparent response of leaf conductance (g) to changes in CO_2 concentration would be far greater than 3 per cent, whereas the apparent response of reduce rate of crop transpiration (E) would be relatively small (Nederhoff *et al.*, 1992).

Leonardi (1994) studied the respone of sweet pepper cv. Rino to minimum temperature maintaining just above the minimum required for plant growth (about 10°C until the end of vegetative growth and about 13°C during flowering) grown in a plastic green house. Increased yield and advanced harvesting were recorded at these temperatures which did not produce any deleterious effects on fruit quality. Spraying flowers with 10 or 20 ppm GA_3 had no effect on yield or fruit quality except for slightly reducing the pericarp firmness (20 ppm GA_3).

10.3.0 Cultivation in Green House

10.3.1 Cultivars Suited for Green House

Different agronomic and fruit morphological characteristics were studied by Scivittaro *et al.* (1999) in 11 sweet pepper hybrids grow under plastic covers in Campinas, Sao Paulo, Brazil, from November, 1995 to July 1996. Revealation of the results indicated that cvs. Fresco and Luis produced the greatest yields whereas Lilac and Orange Wonder produced the lowest yields. The greatest pulp thickness was observed in Ivory, Orange Wonder and Rupia which differed significantly from Juncal, Luis and Cardinal. Fresco and Rupia showed the best performance in the red and yellow fruit groups, respectively.

Gong *et al.* (1999) evaluated 42 lines of sweet pepper for maturing habit, yield level, fruit characteristics, growth habit, yield level, fruit characteristics, growth habit and virus resistance and identified two lines suitable for protected cultivation. P 83-Y-10 is a high yielding, virus resistant, middle maturing, thick flesh and lantern shaped sweet pepper line, suitable for cultivation in the piped shed and green house. P 97-02-2 is a high-yielding, virus resistant, middle-late-maturing, thick flesh and long lantern-shaped sweet pepper line, which can be used in breeding for long-term-culture varieties for cultivation in large multi-arch green house.

Characteristics and chemical composition of fruits of 5 cultivars grown in plastic polytunnels were studied by Kawecki *et al.* (1998). Cultivar Bryza produced the most fruits/plant and the lightest fruits. Ascorbic acid content was highest in cv. Ino and lowest in Bryza. Cvs. Monsun and Zefir fruits were judged to be of the highest quality.

Singh *et al.* (2004) conducted a study under net house with an objective to produce early, uniform large sized fruits with minimum use of insecticides. Out of seven capsicum hybrids, the hybrid 'Bharat' was significantly higher in yields having marketable yield (1.118 kg/plant) and total yield (1.170 kg/plant) as compared to other hybrids. Fruits were more uniform, larger in size, and mature one month earlier to conventional cultivation (Singh *et al.*, 2004).

Kurubetta and Patil (2009) studied the performance of capsicum hybrids, *i.e.*, Orobelle, Bomby and Indra under naturally ventilated poly house and naturally ventilated shadowhall. The quantitative parametres like earliest flower initiation (33.00 days), first fruit harvesting (86.00 days) and highest fruit set (49.81 per cent) and the quality parametres like fruit weight (160.00 g), fruit volume (320.00 cc), rind thickness (0.91 cm) and shelf life (8.62 days) were significantly maximum under naturally ventilated poly house than naturally ventilated shadowhall. Among the hybrids, Bomby recorded higher fruit weight (158.50 g), fruit volume (310.00 cc) while Indra recorded higher rind thickness (0.87 cm) and shelf life (8.60 days).

Kanwar and Sharma (2010) evaluated performance of capsicum hybrids and recommended hybrids Bharat, BSS-519 and Spinx as most promising under green house conditions in the cold arid region of Ladakh. There were no significant differences observed for the plant characters like plant height, number of fruits per plant and days to first harvest, however hybrid Bharat was earliest for marketable yield and had longest fruit span.

Farooq *et al.* (2015) investigated the growth and fruit yield of five capsicum hybrids namely Orobelle, Figaro, Green Beauty, Mighty, Capistrano with Yolo Wonder as control grown under plastic tunnel and the observations were made for number of fruits per plant, fruit weight per plant (kg), length of fruits (cm), diameter of fruits (cm), pericarp thickness (mm), number of locules per plant and fruit yield (t/ha). Among five hybrids Orobelle ranked first for number of fruit per plant (43.47), fruit weight per plant (1.96 kg) and fruit yield (51t/ha) followed by hybrids Figaro (32.84, 1.72 kg, 48.57 t/ha) and Capistrano (41.48, 1 76 kg, 45.90 t/ha), respectively.

Jakhar *et al.* (2017) identified Indra as the best suited cultivar for shade net cultivation in arid region of Rajasthan.

10.3.2 Growing Media

The suitability of bark substrates and of new organic-mineral fertilizer for the year-round green house cultivation of capsicum cv. Bruinsma Wonder and Bell Boy was studied by Pudelski (1988). According to him, higher commercial yields were generally obtained on the substrate consisting of 70 per cent beech (*Fagus sylvalica*) bark + 20 per cent pine (*Pinus sp.*) bark + 10 per cent chicken manure while Koncentrat Podlozowy, a new mineral organic fertilizer containing brown coal and brown coal ash had no appreciable effect.

Paschold and Zengerle (1996) from green house trials with cv. Bell Boy planted in February at 4 plants/m^2 in ridged beds with 7.5 or 20 litres substrate per meter of row or in thin-layer beds with 20 litres substrate per meter of row, reported that total yields of grade 1 fruits by early November were between 14 and almost 20 kg/m^2 on ridged beds compared with about 14 kg for thin-layer beds. They also opined that higher yield on the ridge system was mainly due to better water and nutrient distribution.

According to Salama and Mohammedien (1996), plant height and number of fruits per plant were increased significantly over controls (grown in clay) when plants were grown on rice straw and/or legume (pea and bean) wastes under a 10 cm layer of clay. All waste treatments led to significant increase in early and total yields compared with controls. A 50: 50 mix of rice straw plus legume wastes in layers resulted in the highest yields and best fruit characteristics. The use of rice straw as a growing media proved effective for better production of capsicum under vinyl house (Choe *et al.*, 1991).

While studying sweet pepper production in a closed system in mound culture with special consideration to irrigation scheduling, Paschold and Zengerle (1998) opined that amount of water required by plants depended on the amount of substrate used and its water holding capacity. According to them, yields were good from both substrate treatments (peat or wood fibres). Mound culture (mounds of 7.5 litres) was better than thin layer culture (20 litres) in terms of yield.

An agro-ecological evaluation of growing media made from rock containing zeolites with regard to green house cultivation of sweet pepper cv. Lastochka in Russia by Loboda and Chuprikova (1999) revealed that the powdered rocks were more effecttive than organic media, reducing liquid fertilizer requirements (by <50 per cent), and increasing crop yields (by 15-58 per cent) and crop quality (increased contents of sugars, vitamins, cellulose and dry matter). Lead, mercury, copper and zinc contents of sweet peppers did not exceeded Russian maximum residue levels, while contents of cadmium and arsenic decreased. Nitrate contents also decreased.

Lee and Liao (2007) reported that the SRC (sugarcane residue compost substrate) was used as substrate basket culture of sweet pepper to compare the training method of single stem, double stem and no training at the densities of four to six plants in each basket in the following experiment. The highest marketable fruit yield (49,952 kg/ha) was achieved by double-stem training at a density of six plants per basket resulted in higher proportion of large sized fruits.

Gungor and Yildirim (2013) used growing mixture as peat: perlite: sand (1: 1: 1 v: v: v) in polythene bags to study their effect on different qualitative and quantitative characters of capsicum grown under green house conditions. A significant improvement in fruit size and weight was reported when mixture was used as media while Vitamin C, Total Soluble Solids (TSS), fruit count and yield was highest when peat was used as media.

10.3.3 Spacing

Cebula (1989) studied the effect of spacing and pruning on some vegetative and generative processes in sweet pepper under glass house condition and reported that highest yield and good fruit quality were obtained by spacing the plants at 80 x 30 cm and by pruning to leave two leading shoots and leaving one reproductive bud with two accompanying leaves on each tier plus one leaf on the side shoot.

Pignon and Granges (1992) from results of trials with cultivars planted at different densities in Switzerland recommended that the optimum plant density was 5 plants/m^2 giving an yield of about 8 kg/m^2.

In green house studies, *C. annuum* plants cv. Bendigo F_1 were spaced at 1.5, 3.0 or 6.0 plants/m^2 and pruned to 4, 2, or 1 shoot plant, respectively, to give a constant 6 shoots/m^2. Similarly a shoot density of 8 shoots/m^2 was produced from 2, 4 and 8 plants/m^2 pruned to 4, 2 and 1 shoot/plant, respectively. The number of leaves/plant was positively correlated with the number of shoots/plants. Limiting shoot number/plant while proportionally increasing plant population resulted in more effective coverage of the soil by the canopy (increase in leaf area index). The transmittance of photosynthetically active radiation in the plant profile was more beneficial with plants at a wider spacing but with a higher number of shoots/plant. Early and total yields/unit area increased with plant density; plants with one shoot at a density of 8 plants/m^2 produced the highest yield without any change in quality (Cebula *et al.,* 1995c).

According to Lorenzo and Castilla (1995), the higher value of leaf area index achieved with the higher plant density (3.2 plants/m^2) produced more efficient radiation interception, resulting in significantly higher total yield, marketable yield and top quality than the lower plant density (2.0 plants/m^2).

Dasgan and Abak (2003) reported that a spacing of 80 x 30 cm between row to row and plant to plant, respectively with three shoots per plant was more economical for cultivation of capsicum under glass house. They also observed that the fruit quality parameters such as fruit weight, fruit length, fruit diameter, fruit volume, fruit dry matter, TSS and pH of flesh were not significantly influenced by plant density and number of shoots per plant.

Mantur *et al.* (2007) reported that capsicum fruit yield, numbers of fruits per plant with high mean fruit weight were significantly more when planted at a spacing of 45 x 30 cm compared to a relatively wider spacing of 45 x 45 cm during summer season.

Zende (2008) had studied response of Orobelle cultivation against different growing condition, spacing and training practices and had reported that close spacing (45 cm x 30 cm) and 4 shoots training practices was most economical in capsicum however spacing of 45 cm x 45 cm with 3 shoots training practices were provided highest yield. Further, capsicum grown under poly house condition (NVPH) have resulted highest marketable fruit yield with excellent and export grade fruit quality. He also confirmed highest benefit: cost ratio with spacing of 45 cm x 30 cm under NVPH and it was most profitable for farmers in summer season.

Similarly, Kumar and Chandra (2014) reported greater flower count, fruit count and yield of Indra hybrid under different spacing, however in comparison to 45 cm x 30 cm of spacing with 4 shoot per plant training levels, closer spacing of 45 cm x 30 cm was most profitable option.

10.3.4 Planting Time

Growth, yield and fruit quality of 14 imported hybrid pepper cultivars grown in a green house on 20th September and 12th November during 1988-89 and 1989-90 in green house at Cairo University, Egypt were evaluated and compared with those of cv. California Wonder by Shehata *et al.* (1990). In general, early yield was higher from planting in November and total yield from planting in September.

While evaluating the selected pepper F_1 hybrids grown under protected cultivation for growth and yield, Mohamed (1992) reported that the cv. GSK proved to be the most suitable for production in winter and spring time at El-Harm Station, Egypt. Also the Gedeon F_1 hybrids proved to be the main hybrids for production in late winter and spring. For summer time the Glaxi F_1 hybrids was found to be the most suitable for growing under high temperature and gave the highest total production.

Manna and Ray (2015) carried out a study to focus on determination of optimum time of transplanting of the capsicum seedlings to maximize the yield component and profit component and to minimize the insect-pest attack. Their study showed that early as well as delayed transplanting of capsicum seedlings also faced the problem of rain and associated surge of insect pest infestation in the months of September, October and February as well as increased temperature which reduced Yield. From the benefit: cost ratio analysis, it was recorded that the capsicum seedlings transplanted in the First Week of November fetched highest Profit.

Dhaliwal *et al.* (2017) conducted an experiment to investigate the comparative effect of two growing environments (open field *vs.* poly house), three planting dates (15 October, 30 October and 15 November), and four mulch treatments (black polythene, clear polythene, paddy straw, and no-mulch) on growth and yield of bell pepper. The highest fruit number, fruit weight, early yield, marketable yield and total yield were obtained when the crop was planted in the poly house on 30 October under black polythene mulch. Thus, the use of poly house and black plastic mulch, when combined with appropriate planting date, improves early and total yields of bell pepper in areas where the production is constrained by sub-optimal temperature conditions.

Thakur *et al.* (2018) conducted an experiment by taking three levels of spacing 45 cm x 30 cm (4.4 plants/m^2), 45 cm x 45 cm (2.94 plants/m^2) and 45 cm x 60 cm (2.22 plants/m^2) and three levels of numbers of shoots per plant *viz.* two shoots, three shoots and four shoots. Among the different spacing level, the spacing S_1 (45 cm x 30 cm) recorded maximum plant height (137.46 cm) and yield (82.13 t/ha) of capsicum

10.3.5 Manuring and Fertilization

In Almeria, bell pepper cv. Lamuyo was planted in September in a polyethylene-clad green house and grown with 3 rates of mineral N fertilizer (118, 280 or 438 kg/ha; 125, 308 or 494 kg/ha, and 160, 340 or 520 kg/ha). The mean above ground biomass crop growth rate was 5.6-5.9 g/m^2 daily for crops receiving medium or high rates of N, but there were considerable variations during the growing cycle due to differences in air temperature. Average fresh fruit yields were between 8.0 and 11.0 kg/m^2. Nutrient uptake evaluations indicated that mineral N fertilization over 340 kg/ha as is usual in the area was excessive (Martinez-Raya and Castilla, 1993).

Basoccu *et al.* (1993) studied the effect of nutrition and substrate water content on growth under protection of pepper seedlings and on fruit production in the field and confirmed that increasing rates of N fertilizer increased plant growth and dry matter (DM) accumulation and affected DM distribution among the leaves, stem and roots. Low substrate water content reduced the DM content of the leaves, shoots and total plant. The total yield in the field was not affected by any nursery treatment, but a positive relationship was shown between plant height at transplanting and early fruit production.

A preliminary study on the nutrient management and water supply of sweet pepper in modern green houses was conducted by Chen *et al.* (1998) in China. The study revealed that as crops grew, the quantity and concentration of the nutrient supply needed to be increased. The most important factor affecting water supply was light intensity and the irrigation volume can be determined from the lighting intensity and the electrical conductivity of drainage water from the substrate. Nutrient formulae could also be adjusted by analysing various nutrients in the drainage water.

The study on yield and quality of fruits of sweet pepper cv. Mayata grown in protected condition in relation to N and K application as side-dressing in Sao Paulo, Brazil by Silva *et al.* (1999) indicated that N application up to an estimated 27.0 g/m^2 increased shoot, leaf and root dry matter. N application had no effect on fruit yield, weight, length or diameter, however, the beneficial effects of N were only observed when K application was low (13.3 g/m^2).

Silber *et al.* (2005) also examined fertigation frequency on the occurrence of plant disease and reported that increased fertigation frequency reduced blossom end rot disease incidence in capsicum.

Kumar and Verma (2009) had explored suitability of low cost NVPH for year around production of capsicum. They have further reported application of 250 kg of water soluble fertilizers (NPK) per hectare as fertigation in combination with black polythene mulching was most suitable for off-season capsicum production under protected condition as it was reported with maximum plant height, flower count, fruit count, fruit set percent and yield per plant or per square metre.

Lal and Kanaujia (2013) conducted integrated nutrient management study on growth, fruit yield and quality of capsicum cv. California Wonder under low cost poly house. The maximum plant height (54.46 cm), number of leaves per plant (38.89), number of fruits per plant (10.48), fruit weight (85.06g), fruit length (8.56 cm), fruit diameter (5.63 cm), fruit yield (29.70 t/ha) were recorded with the application of 50 per cent NPK + 50 per cent FYM + biofertilizer and had higher net returns with cost: benefit ratio of 1: 8.16.

Nijamudeen *et al.* (2013) recorded that normal application of nitrogen (320 kg/ha) for sweet peppers in green house condition could not fulfill the development of yield potential under low head drip fertigation practice fixed into the naturally ventilated green house. They observed that the nitrogen rate increased from 320 to 400 kg/ha with the normal dose of 360 kg/ha potassium enhanced the average yield from 56.1 to 62.2 t/ha. This average yield was 11, 26 and 104 per cent higher than the fertilizer application of normal dose, dry liquid fertilizer and the control, respectively. However, it was also observed that an additional amount of potassium from 360 to 450 kg/ha with the N level of 400 kg/ha did not improve yield. However, the growth attributes indicated by plant height, dry matter content and LAI were superior with increase in N rate from 240 to 320 t ha^{-1}. As far as yield is concerned, the best results were obtained after enhancing the N fertilizer by 25 per cent while keeping the normal dose of K fertilizer.

Sharma (2016) tried to find out best combination of NPK as well as the effect of different NPK levels on growth, yield and yield contributing characters of capsicum under low cost poly house Results of the study revealed that application of 50 per cent more of the recommended dose (100: 75: 75 kg/ha) recorded significantly higher plant height, fruit length, fruit breadth, fruits per plant and fruit yield per plant than other levels of NPK. These effects culminated in increased fruit yield/ha with increasing level of NPK. On an average, 50 per cent more of the recommended dose of NPK recorded 96.10 per cent higher yield over 50 per cent less than recommended dose of NPK.

Ima-obong *et al.* (2017) studied the growth and yield responses of green pepper to varying levels of manures under high tunnels and open field conditions. The findings showed that the 5 t/ha of poultry manure and 200 kg of NPK produced the highest total fruit yield, plant height, more number of branches per plant. High tunnel produced 3.1 fruits per plant having 102.8 g average fruit weight, which was higher than the open field conditions in which 1.7 fruits per plants having 32.3g average fruit weight were obtained. Moreover, in the rainy season high tunnel enhanced the pepper production as compared to open field conditions which was near failure as the open field conditions did not protect the crop against heavy rainfall, hails, cold wind and high humidity but also promoted disease, *i.e.*, bacterial wilt in plant and fruits.

10.3.6 Irrigation

According to Teodoro *et al.* (1993), the best results with regard to plant growth, yield and fruit quality were obtained when the plants of sweet pepper cv. Agronomico 10G were irrigated (at 4 l/h from drippers spaced at 1 m apart) after depletion of 30 per cent of available soil water.

The water requirement of sweet pepper cv. Gedeon F_1 grown in plastic green houses at Kassasin, Egypt were determined by Abou-Hadid *et al.* (1994) using measurements of solar radiation or water loss from a Class A pan. Trials over 2 seasons revealed that for all aspects of plant growth and yield measured, the radiation method was the best way of scheduling irrigation in plastic green houses. Soil electrical conductivity and soil, air and leaf temperatures were higher with the Class A pan method. The pan evaporation method can, however, be used if there is a suitable correction made for water loss through drainage.

Chartzoulakis *et al.* (1997) studied the water requirements of sweet pepper cv. Sonar grown in an unheated green house using tensiometers. They opined that maximum evapotranspiration (ETm), when soil water potential was maintained at values higher than -20 kPa, ranged from 0.5 to 4.0 nun/day with a whole season irrigation requirement corresponding to 348 mm. Water supplied to restore 85 per cent of ETm had no effect on plant growth and fruit yield; with irrigation to restore 65 or 40 per cent of ETm plant growth declined and the total yield was reduced by

26 and 47 per cent, respectively. Both fruit number per plant and fruit size were affected by the amount of water applied.

Kumar *et al.* (2016) laid down an experiment with seven treatments having three irrigation depths, two intervals and a farmers practice. The perusal of the pooled data revealed that different levels and frequencies of irrigation produced significant differences with regards to different traits under study. The treatment with 0.50 cm of irrigation at alternate day resulted in maximum fruit breadth (6.53 cm), average fruit weight (95.20 g), number of fruits plant^{-1} (18.26), total biomass production (165.38 g plant^{-1}) and highest economic yield plot (36.17 kg plot^{-1}) and along with moderate water productivity (0.25 kg m^{-2} cm^{-1}) and may be reccmmended for commercial cultivation of capsicum in the hilly regions under naturally ventilated poly houses. However, the treatment with 0.25 cm of irrigation at alternate day recorded maximum water productivity (0.43 kg m^{-2} cm^{-1}) and total soluble solids (5.30 per cent) with moderate yield (30.58 kg plot^{-1}) and lowest root rot incidence (2.76 per cent), may be suggested for the high water scare areas for the successful cultivation of capsicum under protected conditions.

Salunke *et al.* (2017) conducted an experiment comprising of three growing environments *viz.*, naturally ventilated poly house, shadehouse and open field and three irrigation levels at 100 per cent *etc.*, 80 per cent *etc.* and 60 per cent *etc.* in capsicum. Higher yield with minimum crop water requirement gave maximum water use efficiency which was observed in treatment drip irrigation at 60 per cent *etc.* under poly house *i.e.* 30.29 q/ha^{-cm}. However, minimum water use efficiency was noticed in treatment drip irrigation at 100 per cent *etc.* under shadehouse *i.e.* 12.25 q/ha^{-cm} although it was having higher yield than open field but it required maximum consumptive use. The significantly superior yield was recorded under the growing condition *i.e.*poly house and irrigation level drip irrigation at 60 per cent *etc.* over rest of the combinations (Salunke *et al.*, 2017).

10.3.7 Mulching

Changes in the light intensity and air and soil temperatures in the screened plastic houses and their effects on plant photosynthesis, growth and development were observed by Sivek and Kolev (1991) when cv. Albena was grown in unheated houses with single-, double- or treble-layer of polyethylene protection, without soil muching or in identical houses with polyethylene mulch. The highest average yield was produced by plants provided with a treble-layer of polyethylene cover with or without mulching, but total productivity was maximum in double-layered houses with soil mulching. Early planting (late March) showed some advantage over later planting (early April).

The effect of mulching with transparent or black plastic film in green house production of sweet pepper on soil temperature, light conditions, vegetative growth and generative development of plants was investigated by Cebula (1995 a,b). The temperature of the soil was on average 2°C higher under transparent and black plastic mulch at depths of 5 and 12 cm compared with the unmulched control. The transparent film ensured higher soil temperatures during the day, while the loss of heat energy at night was prevented to a greater degree by the black mulch. The vegetative growth of plants measured by their height, number of leaves per plant, and the area and thickness of leaf blades, were distinctly more intensive in mulched stands, slightly better results being noted with the transparent film than with the black one. In relation to light conditions, the reflection of PAR at the level of the main branches was almost doubled when transparent mulch was used compared with the unmulched control and black plastic mulch. During the entire period of fruiting the highest percentage of fruitset was recorded in transparent plastic mulch and the lowest in the unmulched control. Compared with the unmulchcd control, yields were 38.6 and 19.0 per cent higher with transparent and black mulches, respectively. Mulching also favourably affected the marketable quality of the fruits and the earliness.

A study was conducted to determine the effects of plastic film mulches on root zone temperature (RZT), soil water status, incidence of thrips and Tomato spotted wilt (TSW), plant growth, gas exchange, accumulation of mineral nutrients, and fruit yield in bellpepper by Diaz-Perez, 2010. Plastic mulch color influenced the micro-environmental, physiological, and yield responses of bell pepper plants. Plastic film mulches differed in their soil-warming ability with RZTs in both spring and fall being highest in black mulches and lowest in silver mulches. The percentage of photosynthetically active radiation (PAR) reflected from the mulches was highest in silver mulches and lowest in black mulches. The mean RZT under the plastic mulch decreased with increasing percentages of reflected PAR. The number of thrips per flower and the incidence of TSW in mature plants were not significantly different among plastic mulch treatments. The number of thrips per flower had no relationship with the percentage of reflected PAR or with RZT. Plastic mulch treatments had no significant effect on soil water status. Neither soil waters content nor soil water potential had a relationship with RZT. In the fall season, during the first 28 days after transplanting, plant growth attributes were among the highest in silver mulches and the lowest in black mulches. Gas exchange and accumulation of mineral nutrients in the leaves and the fruit were not significantly affected by plastic mulches. Both marketable and total yields were higher on silver mulches and lowest on black mulch in the fall, whereas they were in general higher on silver with black strip mulch and lowest on white and silver mulches in the spring. The reduced plant growth and fruit yields in black mulches during the fall were probably the result of the increased RZTs, and thus higher heat accumulation, that resulted in higher plant heat stress conditions compared with silver and white mulches. Fruit yield decreased with mean seasonal RZTs above 27.5°C. The optimal range of RZT for bell pepper fruit yield was computed to be 25 to 27.5°C or less.

Lodhi *et al.* (2019) recorded that the fruiting and fruit characteristics of bell pepper were significantly influenced by nutrients and mulching. Highest number of fruits per plant (9.95), highest fruit weight (69.18 g), maximum fruit length (9.95 cm), diameter (7.37 cm) and volume (97.07 cm^3) were observed in the treatment with 200 kg N ha^{-1} + 80 kg P, O... ha-1 + paddy straw mulch @ 7t/ha^{-1}.

10.3.8 Pruning

According to Esiyok *et al.* (1994), pruning caused a successive decrease in the total number of fruits/plant and total yield with each reduction in number of stems/plant when cvs. Kandil and 11 B-14 were either not pruned or pruned to 2, 3 or 4 stems at the beginning of branching. Cebula (1995c) observed that defoliation of plants to different extent ranging from slight to severe had no adverse effect on marketable yield and there was no correlation between the extent of defoliation and fruit quality.

In trials with green house grown sweet pepper plants, leaf area index, biomass production and yield were significantly higher for treatment 8 plants/m^2 with one stem (1) than for treatment 4 plants/m^2 with 2 stems and no flowers removed (2) or treatment 4 plants/m^2 with 2 stems and all flowers of one stem removed (3) (Zhang, 1999). Treatment 1 and 2 had the same pattern of dry matter distribution. In treatment 3, total dry matter production was not affected by the removal of flowers from one of the stems, but more dry matter was distributed to vegetative parts. Yield was lowest in treatment 3. The fruiting pattern was not affected by the treatments.

According to Alsadon *et al.* (2013) two experiments were performed to study the effect of pruning systems on vegetative growth, yield and quality traits of three hybrid bell peppers: 'Pasodoble', 'Lirica' and 'Sondela'. Hybrids were grown under green house conditions in drip fertigated soil culture and plants were pruned leading to one main branch, two and four side branches. Vegetative growth, yield and quality traits were affected by cultivars or pruning systems and their interactions. 'Pasodoble' had the highest early and total yields due to the relative rapid fruit set and fruit number. 'Lirica' had large fruit size (heavier, longer, wider and thicker pericarp fruits). Meanwhile, 'Sondela' exhibited superior fruit quality (vitamin C, titratable acidity, total soluble solids and total sugars). Pepper plants pruned to one branch resulted in a significant increase in early yield, fruit size and internal fruit quality with a decrease in total fruit yield followed by plants pruned to two branches. However, plants pruned to four branches produced the highest yield, due to higher number of fruits plant^{-1}. The best fruit number and total yield were obtained by pruning 'Pasodoble' plants to 4 branches. On the other hand, pruning 'Lirica' plants to one branch improved fruit weight, fruit size and gave thicker flesh width. Regarding to fruit quality traits, 'Sondela', especially under one branch pruning system represented a vital source of vitamin C.

Singh and Kaur (2018) conducted a study with three cultivars Indra, Bachata and Inspiration pruned with three intensities *i.e.* two shoots, four shoots and un-pruned (control). The results showed that the two stem pruning treatment significantly affected the plant growth and flowering characteristics in terms of plant height (194.75 cm), number of days to first flower (20.25 days) in cv. Indra. Same cultivar when left un-pruned recorded more number of fruits per plant (23.64 fruits), but fruit yield per plant (4.14 kg) was found maximum in four stem pruning treatment in Indra than other cultivars.

10.4.0 Pollination

Pollination of sweet pepper cv. Delphin in green houses by honeybees revealed that total crop weight, individual fruit weight and seed number per fruit were higher in the presence of bees, but fruit number was higher without bees. The number of malformed fruits was lower with bees (Ruijter *et al.*, 1991).

A colony of *Bombus terrestris* was introduced to a plastic tunnel containing *Capsicum annuum* plants; some other tunnels were opened to allow access by honey bees (*Apis mellifera*) from nearby hives, and some plants were enclosed by net cages to prevent insect pollination (Porporato *et al.*, 1995). Bumble bees and honey bees both foraged on *C. annuum* flowers, but the former were better pollinators in the tunnels. The attractiveness of the 4 cultivars tested varied greatly - Quadrato di Cuneo, Quadrato d 'Asti and Pathos were visited more by bees than Como di bue. Fruits produced by the caged plants, especially on Quadrato di Cuneo and Quadrato d 'Asti, were fewer, smaller and contained less seed than fruits produce by bee-pollinated plants (Porporato *et al.*, 1995).

Experiments were carried out in southern Quebec, Canada, to assess the possibilities of using *Eristalis tenax* as pollinator of *C. annuum* (cv. Bell Boy) under glass house conditions (Jarlan *et al.*, 1997). To determine the influence of insects on seed set and fruit weight, each flower was tagged before flowering and submitted to one of the following, conditions: (a) insect visits were prevented; (b) a specific number of visits was allowed; and (c) insects had unlimited access to the flower during its receptive period. Fruits were allowed to grow for 30 days and harvested. Higher seed set was observed in (b) and (c) as well as a shift towards a greater percentage of heavier fruits when compared with (a). The duration of insect visits significantly increased both seed set and fruit weight. It was suggested that *E. tenax* may possess desirable attributes for the pollination of *C. annuum* (Jarlan *et al.*, 1997).

Al-abbadi (2009) conducted a study on performance of *Bombus terrestris* L. and *Apis mellifera* L. different nuclei as pollinators on quality and quantity of the egg plant, sweet and chili pepper crops grown in plastic house compared with control. Bees showed no problem in visiting flowers at a temperature range between 16 and 36°C during daytime. Two honeybee nuclei pollinated plants produced significantly higher yield than the other treatments. Each section demarcated 15 plants labeled inflorescence; fruit set percentages for eggplant were 66, 59, 52 and 30 per cent, for sweet pepper were 82, 69, 55 and 49 per cent and for chili pepper were 83, 75, 61 and 35 per cent, for the two honeybees nuclei, one honeybees nuclei, bumblebees and control treatments, respectively. The average weight and length of individual fruits collected randomly from the demarcated plant were higher for two honeybee's nuclei, one honeybee's nuclei, bumblebees and control treatments, respectively. Moreover, fruits average number related to the weight showed significant differences among the four treatments, lower fruits number and high weight for two honeybees' nuclei and vice versa for control treatment. In conclusion, treatment of Solanaceous plants with honeybee's nuclei increased fruits quality, quantity and exhibited better appearance.

A study was carried out by Sharshir *et al.* (2019) for evaluating the effect of bumble bee (Hymenoptera: Apidea) and compared to nonbumble bee pollinated sweet pepper plants (*Capsicum annum* L.) variety Gedeon F_1 under plastic house. Results indicated that the effect of four methods of pollinations were in three categories; the first one was first class of mean weight each one sweet pepper fruit > 90g. It was found that the pollinated by bumble bee that held most mean fruit numbers of sweet pepper in February and March in two seasons (6.36, 6.956, 4.60 and 7.2 fruits/m^2 respectively) and weight of these fruits were (578, 685.5, 519.5 and

Sweet Pepper Production under Low Cost Poly House

Sweet Peeper Production under Climate-Controlled Green House

663.1 g/m^2, respectively) while control in the same category was (2.2, 2.4, 2.2 and 3.1 fruits/m^2, respectively and 246, 242.3, 270.7 and 288.3 g/m^2, respectively). The second category was the second class of mean weight each fruit of sweet pepper was 50 to 90 g. In this category it was clear that the highest mean number of fruits of pollinated with hand (6.39, 6.58, 12.2 and 7.4 fruits/m^2, respectively) and weight of these fruits were (454.7, 458.9, 731.7 and 597.4 g/m, respectively) while, bumble bee came in the second order in this category and the third category was third class of mean weight each one fruit sweet pepper respectively and weight of these fruits were (507.2, 478.2, 774.9 and 700.9 g/m^2, respectively). While considering the effect of bumble bee pollination on yield, the bumble bee plants had increased total yields in February by 30.54 per cent and 34.46 per cent and in March by 22.58 per cent and 22.40 per cent of two seasons, respectively.

10.6.0 Postharvest Shelf-life

In a series of glass house experiments carried out during winter and spring, the effects of environmental conditions, concentration of the nutrient solution (EC), and place of origin on several quality characteristics of capsicums were studied by Janse (1989). The shelf-life of capsicums originating from different local commercial nurseries varied from 5 to 17 days. Increasing EC level of the nutrient solution reduced the shelf-life, while it improved flavour. There was a tendency for changing RH regimes during day and night, resulting in an average 24-h humidity of about 80 per cent, to give the best results for shelf-life, compared with constant high and low humidity levels of 70 and 90 per cent, respectively. Fruits obtained from 12 different temperature regimes were both firmer and had a longer shelf-life when the 24 h temperature level was lower and temperature differences between day and night were smaller. Flavour also appeared to be better when the 24 h temperature level was lower. In all experiments, a significant positive correlation was found between firmness at harvest and shelf-life.

In studies carried out by Medany *et al.* (1990), Capsicum cv. Gedeon was grown in a warm-air-heated green house, a solar-heated green house, and an unheated green-house in Egypt. Fruit weight, size and flesh thickness were higher and fruit DM content was lower for fruits produced in the warm-air-heated house compared with those produced in the other green houses. Fruits were tested for shelf-life at room temperature with or without the peduncle and after dropping from a height of 1 m. Dropped fruits showed the greatest weight loss after 10-12 days of storage at room temperature. The presence of the fruit peduncle had little effect on weight loss. Fruits harvested from the warm-air-heated green house showed more weight loss than those from the solar-heated or unheated houses in the first season, but not in the second season. There was an interaction between the green house type and fruit treatments, fruits without peduncles or those that were dropped which had been harvested from the warm-air-heated green houses showed the greatest weight loss.

A laboratory experiment was carried out to determine the effects of harvesting stages (0, 25, 50, 75 and 100 per cent fruit colourations) and storage durations (0, 1, 2, 3 and 4 weeks) on physicochemical quality and shelf life of sweet pepper varieties (Telmo-Red and Velez-Yellow) under passive refrigeration system (PRS) (Tsegay *et al.*, 2013). Thirty (30) fruits of sweet pepper were packed in card-board boxes for each treatment and stored under PRS optimum storage conditions. Fruits were assessed for weight loss percentage, fruit firmness, total soluble solids, titratable acidity, postharvest decay percentage and shelf life. Total soluble solids were increased; whereas fruit firmness decreased with increasing harvesting stages. Weight loss percentage, postharvest decay percentage and shelf life increased; while fruit firmness decreased with increasing storage periods. Telmo variety showed significantly better postharvest quality and storability potential than Velez variety.

Mahajan *et al.* (2016) harvested the cultivar Indra at bright green colour stage and packed in different packaging films, *viz.* heat shrinkable film (15μ), cling film (15μ) and low density polyethylene film (LDPE 25 μ). After packaging, the fruits were stored under two different conditions, *i.e.* super-market conditions (18-20°C; 90-95 per cent RH) and ordinary market conditions (28-30°C; 60-65 per cent RH). The shrink and cling film significantly checked the loss in weight, firmness and decay incidence and maintained various qualities attributes like ascorbic acid, chlorophyll content and overall sensory quality of the fruits during shelf-life better than unwrapped control fruits. The in-pack bell pepper in shrink and cling film created a suitable headspace environment with desired level of O_2 and CO_2 concentrations, which resulted in a better retention of freshness of the vegetable for its marketability. On the other hand, LDPE film accumulated very high level of CO_2, which led to formation of fermenting odour and decay of fruits. The data revealed that shrink and cling film proved quite effective in prolonging the shelf-life of bell pepper fruits for 10 and 7 days under super market conditions and ordinary market conditions, respectively, as against 5 and 2 days only in case of unpacked control fruits.

Jadhav and Gurav (2019) conducted an experiment to determine the effectiveness of cold storage on the postharvest storage-life (days) of capsicum cv. 'Bachata F_1' and 'Inspiration F_1' at 10°C and 93 per cent relative humidity. The results showed that keeping the fruits of capsicum cv. 'Bachata F_1' (15 Days) and 'Inspiration F_1' (15 days) inside cold storage, recorded better retention of fruit qualities. The chilling symptoms were not observed inside cold storage conditions. Average post-storage shelf-life (days) was observed in cv. 'Inspiration F_1' and cv. 'Bachata F_1' to be 3.00 days and 3.68 days, respectively at room temperature after taking out of the cold room. The average weight loss was recorded in red cv. Inspiration F_1 [2.72 per cent day^{-1} (27.23 kg 1000 kg^{-1} day^{-1})] and in yellow cv. Bachata F_1 [2.47 per cent day^{-1} (24.73 kg 1000 kg^{-1} day^{-1})] at normal room temperature and ambient humidity. Skin shriveling was observed at normal room conditions in 'Bachata F_1' and 'Inspiration F_1'as on days 4.5 and 3.81 after taking out of the cold room, respectively. Shelf-life was recorded on days under normal room temperature for capsicum red and yellow cultivars as 3.5 and 4.5 days, respectively (Jadhav and Gurav, 2019).

11.0 Diseases, Pests and Disorders

11.1.0 Diseases

The serious diseases which occur in sweet pepper, *viz.* *Cercospora* leaf spot (*Cercospora capsici*), anthracnose and fruit rot (*Colletotrichum capsici*), bacterial leaf spot (*Xanthomonas vesicatoria*) and virus diseases also cause severe damages to sweet pepper. Jarvis *et al.* (1994) reported the first incidence of *Nectria haematococca* as a stem and fruit pathogen in sweet pepper grown hydroponically in Ontario, Canada which caused an estimated 50 per cent yield loss of the crop.

The symptoms of the diseases are almost similar to that mentioned in chilli and the control measures suggested in case of chilli are also effective in sweet pepper. In a trial on screening of 298 genotypes of sweet pepper resistant to anthracnose, *Cercospora* leaf spot and powdery mildew disease, Ullasa *et al.* (1981) observed that 16 were resistant and 6 moderately resistant to *Colletotrichum capsici*, 6 were resistant and 13 moderately resistant to *Leveillula taurica* and 8 were resistant to *Cercospora capsici*. Among the species, *Capsicum microcarpum* was resistant to *L. taurica* and *Cercospora capsici*. Hungarian Wax (*C. annuum*) cultivar was moderately resistant to both *Colletotrichum capsici* and *Cerospora capsici*, and PI-288982 (*C. annuum*) to *L. taurica* and *Colletotrichum capsici*.

Tecto 60 PM (thiabendazole) at 1.8 kg a.i./ha gave the best control of *Cercospora capsici* in capsicum cv. Casca Dura Ieda (Muniz and Teixeira, 1987).

In Yugoslavia, Soroksari cultivar showed 41-73 per cent field infection by mosaic virus and the disease incidence was reduced by seed treatment with NaOH (Aleksic and Marinkovic, 1981).

The *Phytophthora* blight disease caused by *Phytophthora capsici* causes damping off of seedlings, root rot, stem canker, leaf blight and fruit rot in older plants of sweet pepper (Chupp and Sherf, 1960). Schub (1983) reported that splashing rain and run off water appeared to be the major means of spread of the disease. In a rapidly enlarging spot, the leaf becomes water soaked and bordered by the white layer of the pathogen. Well drained soil, free from weeds, crop rotation, seed treatment are the general treatments suggested for keeping the disease in check. Susuri *et al.* (1979) in Yugoslavia recorded marked reduction of the infection by using 0.2 per cent Mycodifil and 0.15 per cent Difolatan. Kim *et al.* (1975) in an investigation observed the cultivars Saegochu and Kimjanggochu as resistant to *Phytophthora* blight in South Korea. Pollen viability and fruit rot (*Phytophthora nicotianae* var *nicotianae*) incidence were reduced in all treatments compared to control when seeds of bell pepper cv. *California* Wonder were treated with 0.05, 0.10, 0.15 and 0.20 per cent ethyl methane sulfonate (EMS) concentrations. All the treatments induced a high level of variation for the two observed characters. Completely disease free plants were observed in 0.15 and 0.20 per cent EMS treatment (Yadav and Sharma, 1995). Soil solarization delayed the onset of *Phytophthora capsici* on sweet pepper by 4-9 weeks compared with control when grown in polyhousc in Calabria, Italia during 1991-92 (Polizzi *et al.*, 1994). The control effectiveness of an organic amendment (Chicken manure), a fungicide (fosetyl- aluminium) solarization (with transparent and black polyethelene) and their combinations was studied to compare the epidemiological progress of blight in pepper caused by *P capsici* and to determine the economic feasibility of the treatments by Chavez-Alfaro *et al.* (1995) in Mexico. According to their findings, the treatments fosetyl-aluminium + chicken manure + transparent polyethylene, chicken manure + transparent polyethylene and fosctyl-aluminium + transparent polyethylene showed the best disease control, the lowest disease progress and highest yield. The highest marginal rates of economic returns were shown by chicken manure, chicken manure + transparent polyethylene and fosetyl-aluminium + chicken manure + transparent polyethylene.

According to Ziv *et al.* (1994), the use of sodium and potassium bicarbonates, horticultural oil and surfactant solution reduced the incidence of *Leveillula taurica* mildew on plant foliage and *Alternaria alternata* on harvested *capsicum* fruits compared with untreated controls. Potassium and sodium bicarbonate solution (0.5 per cent) gave consistently better protection against pre- and postharvest disease of *Capsicum* than penconazole, horticultural oil or surfactant. They also suggested that the bicarbonates could be used as part of an integrated pest management strategy.

Jimenez *et al.* (1988) studied the response of different cultivars of sweet pepper to bacterial wilt (*Pseudomonas solanacearum*) in Costa Rica and found that cv. Cholo was the most resistant.

Chen *et al.* (1995) suggested interplanting pepper with wheat as a means of avoiding aphid damage and the transmission of virus. The incidence and index of virus disease were reduced by 76.5 and 93.1 per cent, and the income from pepper cultivation was enhanced when interplanting of pepper with wheat (3 row of wheat: 2 row of pepper) was made. Avilla *et al.* (1997) suggested polypropylene covers must be placed immediately before transplanting in area of high virus pressure and under high yield expectations, to obtain virus control and to enhance the growth of sweet pepper. Formulated neem oils may help to reduce or delay the spread of non-persistent plant viruses in addition to direct control of insect (Lowery *et al.*, 1997).

Fajinmi and Oladiran (2000) reported maize/pepper intercropping gave more agronomic advantages over the sole crop for fruit number, fruit thickness and fruit length of pepper leading to heavier yield of the crop over sole pepper (control plot). In the intercropping treatment, infection of pepper veinal mottle virus (PVMV) was less than 4 per cent.

A field experiments were conducted simultaneously inside the poly house and rain shelter for management of fungal diseases of capsicum with seven treatments and three replications by Power (2017). The treatments included two biocontrol agents (*Trichoderma viride* (*asperellum*) and *Pseudomonas fluorescens*), two systemic fungicides (tebuconazole and difenoconazole) and two contact fungicides (mancozeb and copper hydroxide). Among the treatments, soil solarisation + soil application of *Trichoderma asperellum* + seed treatment with carbendazim + mancozeb (2 g/kg seed) + foliar spray with mancozeb (0.2 per cent) was the most effective for management of *Cercospora* leaf spot in both poly house and rain shelter, followed by soil solarization + seed treatment and foliar spray with *Pseudomonas fluorescens* (20g/l) and soil solarisation + soil application of *Trichoderma viride* (*asperellum*) and these were statistically on par. Among the treatments, soil solarisation + soil application of *Trichoderma asperellum* + foliar spray with tebuconazole (0.1 per cent) was the most effective for management of powdery mildew in rain shelter and poly house, followed by soil solarisation + soil application of *Trichoderma* + foliar spray with difenoconazole (0.05 per cent) and these were on par. Correlation analysis was performed with the meteorological data recorded during the experiment and per cent disease severity (PDS) at periodic intervals and it was found that there is significant positive correlation with RH in rain shelter and temperature in poly house in case of *Cercospora* leaf spot. In powdery mildew, there was a significant negative correlation with PDS and temperature inside the protected structures.

11.2.0 Pests

There are several kind of insects that includes aphids, cutworms, European cornborer, flea beetles, hornworms, pepper weevil and the pepper maggot found to attack capsicum. Nature of damage due to the attack of aphids is similar to that of chilli and control measure suggested in case of chilli is also effective is capsicum. According to Fiume (1993), application of methomyl followed by inoculation with *Verticillium lecanii* and introduction of *Chrysoperla carena* as integrated control of *Myzus persicae* and *Aphis gossypi* was the most effective treatment.

Pepper weevil (*Anthonomus eugenii*) and pepper maggot (*Spilographa electa*) are also serious pests in capsicum. The adult of pepper weevil is a black beetle with grey or yellow markings (Choudhury, 1967). The larva hatching from eggs lay beneath the surface of the pod feeds within the pod. There may be 5 to 8 generations in a year because various peppers and brinjal plants are host of the weevil. Infested large fruits are likely to be mis-shapcn or discoloured. Control measure includes application of suitable insecticide at regular intervals beginning at the time of fruit setting (Hawthorn and Pollard, 1954).

In case of pepper maggot, the larva or maggot is about 1.5 cm long when full grown and the adult is two-winged, yellow striped fly, about 0.75 cm long. The larva infests the pods of peppers and sometimes causes serious losses in some areas (Thompson and Kelly, 1957). The injury in many ways is similar in its effects to that caused by the weevil. Repeated sprays with malathion can control the insect (Choudhury, 1967).

Cutworm (*Agrostis sp.*) is a problem to the capsicum cultivation in Bulgaria (Nikolov, 1981). Similarly, European cornborer (*Ostrinia nubilalis*) has become a capsicum pest in Georgia, U. S. A (Dupree *et al.*, 1979) and this has been reported from Canada also (McLeod, 1981).

Burgio and Maini (1994) explored the possibility of biological control of the main pests of sweet pepper grown in a tunnel in Italy and reported that different preparations of *Bacillus thuringiensis* subsp. *kurstaki* were equally effective in reducing numbers of 1st and 2nd instar larvae of *Ostrinia nubilalis*, the most important pest. The best time to apply *B. thuringiensis* subsp. *kurstaki* was when the first females of the pest were captured. Adults of other species of Lepidoptera were also trapped but subsequent fruit damage by their larvae was insignificant. *Tetranychus urticae* was effectively controlled by one release of the phytoseiid predator Phytosetulus *persimilis* in July at a rate of 7 predators/m^2 when the density of prey was 0.1-0.3/leaf.

Yosepha *et al.* (2007) observed that all sweet pepper cultivars showed more productivity under photo-selective shading, as there was an increase in number of fruits per plant by 30 to 40 per cent compared to the use of common black net. Different coloured nets were used for photoselective shading and it was found that the red net had some productivity advantage over the pale and yellow nets. The risk of disease infestation by small insects like aphids, thrips and the incidence of insect borne viral diseases were recorded significantly lower under yellow net as compared to all other nets.

Kaur *et al.* (2010) studied that thrips and mites were serious pests on sweet pepper. The root-knot nematode population was not recorded at the time of transplanting in most net houses, but gradually increased over the growing season in capsicum. They recommended that the farmers should apply prophylactic insecticide sprays to manage sucking insect-pests in nethouse.

Anwar *et al.* (2013) conducted a survey to determine the occurrence of plant-parasitic nematodes on vegetable crops, *i.e.*, cucumber, tomato, chili and bell peppers grown in plastic tunnels. The presence of nine species of plant-parasitic nematodes *viz.*, *Aphelenchus avenae*, *Helicotylenchus dihystera*, *Hoplolaimus columbus*, *Meloidogyne javanica*, *M. incognita*, *Pratylenchus penetrans*, *Radopholus smilis*, *Tylenchorhynchus claytoni* and *Xiphinema* sp. were recorded. Un-uniform foliage growth, stunted growth, chlorotic leaves and wilted plants in patches were observed under the tunnels where the vegetables were grown. The infestation of nematodes in tunnel grown vegetables needed significant control measures to tackle nematode infestation by the use of integrated pest management strategies.

11.3.0 Physiological Disorders

Bakker and Uffelen from the Netherlands (1987) reported problem of skin cractking in capsicum. The cracking around the shoulder of fruits is often associated with fluctuations of temperature and humidity. The highest incidence of skin cracks in the earliest harvested crop (14th April) occurred at high day temperature and average RH. Low night and low day RH combination resulted in a low incidence of the disorder, but total were also depressed by this regime. Average temperature of around 19°C resulted in low incidence of the disorder. Raising the constant EC level from 1.5 to 6.0 mS lowered the incidence of skin cracks from 45 to 25 per cent.

The incidence of blossom-end-rot increased markedly as EC levels rose, but that of skin cracking declined. Little or no blossom-end-rot occurred when the EC was lowered to 1.5 mS or was maintained at that level. Fruit firmness and shelf life were adversely affected by high EC levels (Uffelen and Bakker, 1987).

A physiological plant disorder not known to occur in soil-grown plants was observed in 2 hydroponic green house crops at Gainesville, Florida (USA). The stem became swollen below the cotyledon level, where longitudinal cracks developed on the epidermis. Rotting occurred in the wounds and reached the internal stem vessels. Adventitious roots abscised leading to a major reduction of water absorption. The affected plants wilted with no diagnosable pathogenic disease. In other countries (*i.e.* Spain and Canada), where green house sweet pepper has been grown in soilless media, the disorder has been termed 'elephant's foot', based on a description of the stem appearance. Irrigation emitters close to the stem may be directly related to the cause of the disorder. Moving the emitter gradually away from the stem base partially alleviated the problem. In Spain and Canada, young pepper seedlings are transplanted into a rockwool block up to the level of the cotyledons in an attempt to

Diseases of Sweet Pepper

Anthracnose (Left), and Cercospora Leaf Spot (Right) Infection

Damping Off (Left), and Phytopthora Blight (Right) Infection

Powdery Mildew (Left), and Southern Blight (Right) Infection

Diseases of Sweet Pepper

Bacterial Spot Infection

Bacterial Wilt Infection

Cucumber Mosaic Virus (Left), and Tobacco Mosaic Virus (Right) Infection

Tomato Spotted Wilt Virus Infection

Pests of Sweet Pepper

Aphid (Left), and Leaf Miner (Right) Infestation

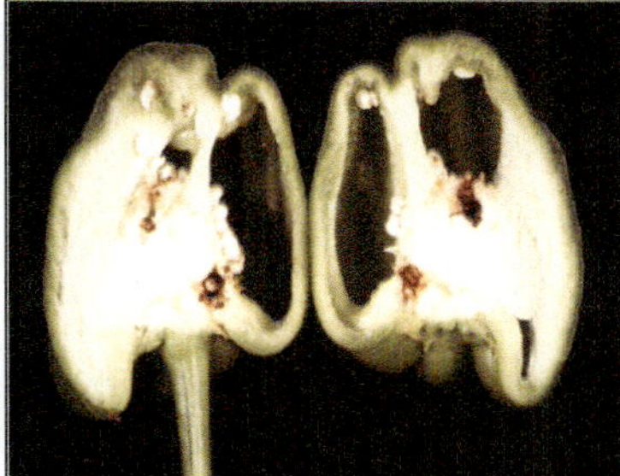

Sweet Pepper Weevil Infestation

Spider Mite Infestation

Disorders of Sweet Pepper

Sunscalding in Sweet Pepper

Blosson End Rot in Sweet Pepper

Different Stages of Fruit Cracking in Sweet Pepper

avoid the disorder. The block is later placed over a soilless media bag or rockwool slab. The irrigation emitter is moved from the top to the base of the block once the roots grow into the container media. Factors other than emitter location, such as irrigation and fertilization frequency, and/or the use of certain soilless media *per se*, also might lead to the development of 'elephant's foot' (Jovicich *et al.*, 2000).

12.0 Seed Production

In seed-to-seed method, transplanting should be followed for commercial seed production. Optimum spacing may be followed to obtain high quality seed.

The capsicum is generally cross-pollinated crop. It crosses easily with chillies and thus deteriorates fast in quality, if proper isolation distance is not maintained during seed production stage. The isolation distance between two cultivars of capsicum should be kept 200 meters for foundation seed and about 100 meters for certified seed production (Joshi and Singh, 1975). Off-type plants should be removed as soon as these are observed. The small-leaved plants can be detected from the plants with large-leaved and *vice versa* and should be rogued accordingly. The number of rouguing should be 3-4 depending on the purity of seed desired. Field inspections of seed crops should be done at least twice or thrice as it makes the seed crop eligible for certification.

The fruit should be picked when red ripe, and cut and crushed or macerated by machine. Seed is to be washed to make it free from pulp and skin. After washing it should be dried without delay in the sun. Picking of pods may be done according to climatic conditions. In case, there is no danger of rains at the maturity time, the pods may be picked in one lot but where there is some danger of rains picking may be done in 2 or 3 instalments. In Kulu Valley, Himachal Pradesh, India, the average seed yield of California Wonder cultivar was 105 kg/ha with maximum of 225 kg (Singh *et al.*, 1964). Naik *et al.* (1997) studied the effect of fruit maturity on seed quality in sweet pepper cv. Arka Gaurav and reported that seed dry weight increased until fruits reached the full yellow stage. Thousand-seed weight reached a maximum at the next and final stage when fruits had begun to wrinkle. Percentage germination was also highest at this stage.

Seed moisture contents had marked effect on the seed vigour of a crop. According to Song *et al.* (1999), as the duration of aging increased, seed vigour was reduced greatly when seed moisture content was above 6.0 per cent and was reduced slowly when seed moisture content was below 6.0 per cent. They also opined that the seed moisture content of 6.0 per cent was the upper limit for sweet pepper. The results of a study on maintaining and improving the vigour of pepper seeds by Wu *et al.* (1999) showed that lime had the same effect as silica gel when seeds were stored with different desiccants after ultradrying. They concluded that ultradrying raised the vigour of seeds, and greatly improved their storability. The vigour of the ultradried seeds could be further improved by polyethyleneglycol (PEG) osmoconditioning. Treating seeds with GA, especially those with reduced vigour, was effective and was a simple and practical method. The effects of dehydration temperature (20, 35 or 50°C) and period (1-72 h), and seed moisture content during storage, on the germination of primed red pepper seeds were investigated by Jeong and Cho (1996) in Korea Republic. Prime seeds were surface-dried at 20°C for 3 h (17.2 per cent seed moisture content) or dried at 35° for 3 h (7.8 per cent seed moisture content) and seeds were stored in airtight glass bottles at 4°C for up to 3 months before germination at 15 or 35°C. Final germination percentage was not influenced by dehydration temperature or period. The longer the dehydration period the more days required for germination, especially when germinated at 15°C. After storage for 3 months surface dried seeds germinated faster than dried-back seeds when germinated at 15°C. Seed moisture content of surface dried seeds decreased from 17.2 to 8.3 per cent after 3 months of storage. Seed moisture of dried-back seeds slightly decreased with storage.

13.0 Crop Improvement

Germplasm collection and conservation is the major step in improvement of sweet pepper and this material can be used in hybridization programme. Quality specifications such as size, shape, colour and non-pungency make the task of developing new hybrids/cultivars very challenging. Breeders must set objectives for breeding of sweet pepper in order to obtain varieties or hybrids that meet the requirements of growers and consumers. The main general objectives of breeding sweet pepper are follows (Negi *et al.*, 2018):

1. High yield
2. Earliness
3. Desirable fruit shape and size: oblate or round
4. Superior fruit qualities like high vitamin C and pigment content, pleasing flavour, *etc.*
5. Resistance to diseases like fruit rot, cercospora leaf spot, powdery mildew, common TMV, *etc.*
6. Resistance/tolerance to insects like thrips, mites, aphids, fruit borer.
7. Resistance/tolerance to abiotic stresses like heat, water stress, salinity.

The varieties of sweet pepper developed in India through introduction, selection, hybridization and possibility of exploiting hybrid vigour are discussed below:

13.1.0 Introduction and Selection

The first step is introduction of genetically diverse germplasm to enrich the base as a source of variability. Plant introduction involves exploration, collection, purification, conservation, evaluation, storage and utilization. Wild forms of cultivated species, wild species, land races, old and new cultivars, and other genetic materials are important. In bell pepper, International Board of Plant Genetic Resources (IBPGR), Rome, and Asian Vegetable Research and Development Centre (AVRDC), Taiwan, are important international sources for enriching the germplasm base beside personal collection from institutes and research stations working for the improvement of this crop. Open pollinated cultivars introduced from different countries are commonly used as a source of new variability. In India, heterozygous material obtained from different sources has been utilized to develop new varieties by following pure line breeding methods (Joshi and Berke, 2015). New germplasm should be observed under isolation for plant quarantine to avoid the possibilities of introducing new pathogens. The isolated germplasm found free from pathogens can be utilized in the improvement of bell pepper. If variability is not available in the germplasm as in the case of bell pepper it can be generated by hybridization.

There are many cultivars of sweet pepper throughout the world. American seedsmen recorded as much as 125-150 cultivars of sweet pepper in 1901 but most of these are thought to be synonymous (Boswell, 1937). Curtis (1935) developed cultivar Windsor A, which is a cross between California Wonder and Bountiful. Cochran (1934) described Truhart Perfection Pimiento which he developed in Georgia from single plant selection from Perfection Pimiento. The Pennsylvania Agricultural Experiment Station developed Pennwonder by crossing between Harris Earliest and California Wonder.

California Wonder, Chinese Giant, World Beauty and Yolo Wonder have been found promising as direct introduction in case of sweet pepper. Yolo Wonder is tolerant to tobacco mosaic virus. Russian introduction R 449 performed well under Delhi and Bihar conditions. Extremely pungent forms tabasco and the West African types also showed promise (Thomas *et al.,* 1983).

A large fruited capsicum Hongyhuan 773 was selected from a local variety in China (Deng, 1988). Hibberd *et al.* (1988) reported that cv. Redlands Sweet Sue had been derived from a single F_3 plant of Yolo Y x Hungarian Yellow Wax cross in Australia.

In USA cv. Tambel 2 was developed by series of crosses involving Grande Rio 66, Agronomico 8, TB 400 and Keystone Resistant Giant 3 (Villalon, 1986). Another cultivar from USA Rio Grande Gold has been developed from Caloro x TAES 74123-1 (Villalon *et al.,* 1988). It is nonpungent, small yellow wax pepper.

Dwarf ornamental Alppers Marbles and Riot were derived by pedigree breeding from Nosegay x Fiery Festival in USA. The new varieties are dwarf (150-250 mm tall) compact and bear prolific crops of attractive, highly coloured fruit (Baggett and Kean, 1988). Promising determinate, clustered bell pepper lines for once-over harvest was selected in F_6 and F_7 from crosses between MI 221 and Early Thickset hybrid (McCammon, 1989).

Daskaloff (1968) found some mutants from irradiated progeny of sweet pepper which were thought good for genetic studies as well as gene location. Male sterile mutant was obtained by similar means. Thakur *et al.* (1980) irradiated seeds of California Wonder with five doses of X-rays, *viz.,* 5, 10, 15, 20 and 20 Krad and found that the germination rate was divergently linear and the trends for mortality after germination in M_1 generation was reverse. Fruit shape was highly influenced in M_2 generation resulting in tomato shaped as well as pointed fruits. Flower buds without sepals and protruded stigma were found at 10 and 15 krad doses, respectively. Male sterile mutants with shrivelled anthers were detected both in 5 and 10 krad doses. Results of experiments carried out by Irulappan *et al.* (1981) showed that seeds of California Wonder when treated with gamma ray, the LD_{50} doses were between 10 and 20 krad for plant height, leaf area and dry weight of the seedlings and below 10 krad for fresh weight of the seedlings. A new white capsicum cultivar Alba Regia has been developed in Hungary. It is anthocyanin free mutant of cv. Sorokasari, ripens a few days earlier and produces 10-15 per cent higher yields than Soroksari (Moor, 1989).

Negi *et al.* (2018) reported that some introduced hybrids World Beater, Ruby King and King of North (IAHS, Bangalore) are performing very well under Indian condition.

13.1.1 Pureline Selection

Pureline is a progeny of single homozygous plant. Initial seed stock is space planted and superior plants are selected and harvested separately in the year. Next year, individual plant progenies are grown and progeny showing superior performance is bulk harvested and evaluated further with check cultivars in replicated trial.

Negi *et al.* (2018) reported that many varieties of sweet pepper in India have been developed through pureline selection. As for example Arka Mohini is a selection from Titan of USA by IIHR, Bangalore. Arka Gaurav is a selection from Golden Calwonder of USA by IIHR, Bangalore and it is an indeterminate cultivar tolerant to bacterial wilt. Arka Basant is a selection from Hungarian variety 'Soroksari' by IIHR, Bangalore and it has excellent keeping and cooking qualities. Nishat 1 is a selection from Capsicum Sel-2 (SKUAS and T, Srinagar).

13.1.2 Pedigree Method

It is a selection procedure of superior plants in the segregating generations by keeping pedigree records. Pedigree selection method in bell pepper is most widely used for developing varieties with high yield, earliness, quality, resistance to diseases and to abiotic stresses (Joshi and Berke, 2015). Hybridization is done between superior cultivars along with maintenance of pedigree record. Varieties developed through pedigree method are Spartan Garnet (California Wonder x Dwarf Pimiento Selection from variety Santanka), Spartan Emerald (Morgold x California Wonder) and Sonnette F_2 line (Morgold x California Wonder) x Keystone Resistant Giant (Negi *et al.*, 2018).

13.1.3 Backcross Method

Back crossing has been an essential method for introgressing genes from hot pepper into sweet pepper and from interspecific crosses. For fruit quality, disease resistance, or any other specific attribute a pure line can be improved by backcrossing to incorporate novel genes. In bell pepper generally 4-5 backcrosses are required to produce near-isogenic lines for the novel gene in the pure line. Backcrossing is used whenever disease resistance, improvement of male sterile line or any other simply inherited morphological trait is to be transferred because its application is convenient in bell pepper (Joshi and Berke, 2015). Incorporation of resistance (powdery mildew) from *C. baccatum* var. *pendulum* has been possible in sweet pepper through backcross breeding (Negi *et al.*, 2018).

13.1.4 Heterosis Breeding

The goal of increasing productivity in the quickest possible time can be achieved by utilizing heterosis breeding. An F_1 hybrid variety is the result of a cross between two homozygous (but genetically distinct) pure lines. The pre-requisite is that all the F_1 plants should resemble each other phenotypically. Capsicum has a tendency towards autogamy and even though it is hermaphroditic in nature, self-fertilization cannot be avoided to 100 percent. Heterosis in sweet pepper has been reported for early and total yield by Novak and Chemalo (1975). Expression in the F_1 depends upon the selected parents' combining ability. It also depends upon the genetic divergence; geneally the wider the divergence, the higher the heterosis. The economic feasibility of hybrids depends upon the cost of hybrid seeds in relation to the value of increased yield. Heterosis breeding is feasible in bell pepper due to its low seed rate per hectare and only 400-600 g of seed is sufficient to raise the seedlings for planting one hectare.

Heterosis averaging between 2 and 100 per cent has been reported for various economic traits including early maturity, plant height, certain yield components and fruit quality traits (Khalil *et al.*, 1987; Lee and Shin, 1989; Joshi *et al.*, 1995; Pandey *et al.*, 2002; Sood and Kaul, 2006; Sood and Kumar, 2010; Sharma *et al.*, 2013b; Rao *et al.*, 2016; Devi and Pathania, 2018; Chakraborty *et al.*, 2020; Aditika *et al.*, 2020; Banerjee, 2020). Leskovar *et al.* (1989) reported that hybrid bell pepper had more vigorous shoot and root growth than open pollinated varieties, however, gain in productivity were realized only under favorable environmental conditions. The first hybrid of sweet pepper 'Bharat' was developed by IAHS, Bangalore in India (Joshi and Berke, 2015).

13.1.5 Mutation Breeding

Breeding of modern sweet pepper cultivars focuses on improving economically important traits, including yield and nutrient and secondary metabolite content (Jeong *et al.*, 2012; Lee *et al.*, 2016; Solomon *et al.*, 2019). These programs require vast genetic diversity, which can be promoted by mutagenesis to alter nucleotide sequences, potentially creating novel alleles useful for crop improvement (Hwang *et al.*, 2014).

Physical mutagens have been used to produce mutant populations in sweet pepper (Daskalov, 1986; Jo *et al.*, 2016). EMS, a chemical agent, has been commonly applied to induce mutations in seeds. EMS induces random point mutations at a high frequency, some of which can create novel stop codons in genes of interest (Arisha *et al.*, 2014). In pepper, multiple studies have used mutant populations obtained via EMS mutagenesis (Hwang *et al.*, 2014; Jabeen and Mirza, 2002; Bosland, 2002; Paran *et al.*, 2007; Arisha *et al.*, 2015; Siddique *et al.*, 2020). In particular, an EMS mutant population of the sweet pepper cultivar "Maor" was created to investigate the genes that regulate flower and plant architecture (Chaim *et al.*, 2006; Cohen *et al.*, 2014; Jo and Kim, 2019). In addition, Hwang *et al.* (2014) developed a mutation population that consisted of 5,023 M_2 lines of a pepper (*C. annuum* L.) cultivar, 'Yuwol-cho'. Jeong *et al.* (2012) successfully isolated a line that is resistant to the tobacco *etc*.h virus (TEV) from this population by the screening of a novel resistance allele of a gene encoding a TEV host factor based on an Induced Local Lesions IN Genome (TILLING) approach. Recently Siddique *et al.* (2020) employed EMS to generate genetic variability in a dwarf and compact pepper germplasm, Micro-Pep. They observed and characterized M_1, M_2, and M_3 mutant phenotypes in plant growth and habit, leaf color and morphology, flower characteristics, and fruit color and morphology. Two of the phenotypes showed stable inheritance up to the M3 generation. With TILLING, they discovered nine putative point mutations in the eIF4E gene, four of which were confirmed by sequencing. Their Micro-Pep EMS mutant population showed great genetic diversity, making it a valuable resource for reverse genetic studies. Some promising mutated cultivars Albena, Lyulin, Horgoskaslatka-X-3 have also been reported (Negi *et al.*, 2018). According to the official data published by the FAO/IAEA program (http://mvd.iaea.org), 16 pepper cultivars have been bred through mutagenesis technology so far.

13.2.0 Breeding Behaviour and Genetics

Capsicum is basically a self-pollinated crop, but it may be referred to an often cross pollinated crop due to appreciable degree of cross pollination (on an average 16 per cent) and open nature of floral morphology. To maintain the purity of the varieties, isolation distance of 400-500 m is safe for seed production. Major pollinators are honey bees (*Apis* species), ants and wind. Anthesis generally starts from 5.30 am and majority of the flowers open by 9.00 a.m. however, it depends on the prevailing temperature and humidity condition. Anther dehiscence occurs after anthesis. The pollen grains in the anther become fertile a day before anthesis with maximum fertility on the day of anthesis. The stigma receptivity is the maximum at anthesis and the receptivity continues upto two days after. At comparatively high temperature condition, emasculation is advisable 2-3 days prior to anthesis and cross pollination is carried out on the day of anthesis. At high temperature condition, fair crossing success can be achieved if hybridization is taken up either early in the morning or late afternoon. Pollen viability is retained for a day in ambient condition. Pollens can be kept in refrigerator under airtight condition for 2-3 days. Dried pollens can be stored in sealed vial at -5.0°C for 10 days.

13.2.1 Crossing Behavior

Crossing behavior or interspecific hybridizations allow a transfer of genes of interest between different species, especially those involved in disease resistance (Bosland and Votava, 2000), allowing breeders to develop genetically superior genotypes. However, for the success of such a transfer, the species must be genetically close, minimizing problems of incompatibility and thus enabling hybridization. Moreover, it is imperative to know about the direction of the cross, since, in some species, the interspecific cross is possible in one direction only (Prestes and Goulart, 1995).

Incompatibility barriers in interspecific crosses may be expressed before fertilization in prezygotic barriers, and/or after fertilization, in postzygotic barriers. Among the former we can cite the non-germination of pollen grains on the stigma of the receptor plant and a slow pollen tube growth towards the ovule, and among the latter embryo death caused by endosperm degeneration, a lack of vigor and plant sterility of the hybrid plant (Pickersgill, 1997).

As a general rule, the closer the species involved in the cross are genetically the easier it will be to breed hybrids and the more fertile the progenies (Singh, 2002); thus, the existence of barriers in interspecific crosses indicates the genetic distance of species. The genus *Capsicum* comprises three gene complexes, formed on the basis of crossability between species: 1) Complex *C. annuum*, *C. annuum* (varieties *annuum* and *glabriusculum*): *C. frutescens*, C. chinense, C. chacoense and *C. galapagoense* (Zijlstra *et al.*, 1991); 2) Complex *C. baccatum*: *C. baccatum* (varieties *baccatum, pendulum* and *praetermissum*) (Zijlstra *et al.*, 1991) and *C. tovarii* (Tong and Bosland, 1999); and 3) Complex *C. pubescens*: *C. cardenasii*, *C. eximium* and *C. pubescens* (Pickersgill, 1991, Zijlstra *et al.*, 1991).

Costa *et al.* (2009) evaluated the compatibility of the cross between *C. annuum* and *C. chinense* (Casca Dura Ikeda) and seed vigor and germination of the interspecific hybrids. The authors reported that all interspecific crosses bore fruit, with fruiting rates ranging from 8.9 per cent to 40.0 per cent and seed germination from 0.0 per cent to 87.5 per cent.

The pollen viability of hybrids from the species *C. annuum* var. *annuum* and *C. chinense* was on average 11.35 per cent (*C. annuum* var. *annuum* x *C. chinense*) and 12.2 per cent (*C. chinense* x *C. annuum* var. *annuum*), suggesting that these plants are sterile or produce very few viable pollen grains. Monteiro *et al.* (2011) reported a pollen viability of the hybrids of these two species between 34 and 87 per cent. The low values found in this study were not expected, since both species belong to the same gene complex, however, spontaneous translocations were observed between *C. annuum* and *C. chinense*, which can generate unbalanced and therefore unviable gametes (Moscone, 1990).

According to Martins *et al.* (2015) species of *C. annuum*, *C. chinense*, *C. frutescens*, *C. baccatum* var. *pendulum*, and *C. baccatum* var. *baccatum* were crossable with each other and the number of hand-pollinations and of resulting fruits and plants was registered. The resulting hybrids were evaluated for their pollen viability and in the crosses that produced no hybrids; the *in vivo* germination of pollen grains was assessed. Some crosses generated hybrids, while others produced no fruit or, despite generating fruits, no plants grew from them. The crossability rate between species of the complex *C. annuum* and *C. baccatum* ranged from 2.2 per cent to 3.7 per cent, and was 14.6 per cent between species of the complex *C. annuum*. A pre-fertilization barrier was observed in the non-germination of pollen grains, as well as post-fertilization barriers consisting of embryo death, lack of vigor and hybrid sterility.

A diverse and representative collection of 18 accessions from four cultivated species of *Capsicum* were used for crossability studies, including *C. annuum* (12), *C. baccatum* (3), *C. chinense* (2), and *C. frutescens* (1). More than 5000 crosses were made and over 1000 embryos were rescued (Manzur *et al.*, 2015). *C. chinense* performed as a good bridge species between *C. annuum* and *C. baccatum*, with the best results being obtained with the cross combination [*C. baccatum* (♀) x *C. chinense* (♂)] (♀) x *C. annuum* (♂), while *C. frutescens* gave poor results as bridge species due to strong prezygotic and postzygotic barriers. Virus-like-syndrome or dwarfism was observed in F_1 hybrids when both *C. chinense* and *C. frutescens* were used as female parents. Regarding the ER strategy, the best response was found in *C. annuum* (♀) x *C. baccatum* (♂) crosses. First backcrosses to *C. annuum* (BC_1s) were obtained according to the crossing scheme [*C. annuum* (♀) x *C. baccatum* (♂)] (♀) x *C. annuum* (♂) using ER. These results provide breeders with useful practical information for the regular utilization of the *C. baccatum* gene pool in *C. annuum* breeding.

In a study of Wei *et al.* (2019), eight fertile F_1 hybrids were produced from pepper (*Capsicum annuum* L.) and the wild relative (*C. frutescens*), as female and male, respectively, by interspecific hybridization. Interspecific hybrids were identified using conventional morphological descriptors and SSR molecular markers. The results showed that significant differences in agronomic traits existed among cultivated pepper, wild relatives, and interspecific F_1 hybrids. Interspecific F_1 hybrids presented intermediate values, although they were closer to the wild species in most of the agronomic traits. Analysis of SSR markers clearly showed that interspecific F_1 hybrid had bands from the paternal and maternal accessions, which indicated that F_1 hybrid was heterozygous. These results could provide hybrid for breeder to transfer genes of interest from wild relative, *C. frutescens*, to cultivated pepper, which is an important step for introgression breeding.

13.2.2 Estimation of Genetic Parameters

Estimation of variability is a pre-requisite for success in any breeding programme. Assessment of genetic variability in the germplasm by estimating various parameters of variability helps to identify the potential genotypes for their use either directly as varieties or as parents in future crop improvement programmes. The programme in breeding solely depends on magnitude of variability for the characters which are under improvement. Therefore, the knowledge on the estimates of variability in respect of yield and its heritable components in the genetic material with which the breeder is working is essential for straightening selection strategies. It is also necessary to partition the total variability into heritable and non-heritable components *viz.*, phenotypic coefficient of variability (PCV), genotypic coefficient of variability (GCV) and further to compute heritability and genetic gain for various traits of interest to the breeder.

Sharma *et al.* (2010) recorded high phenotypic coefficient of variation (PCV) and genotypic coefficient of variation (GCV) for fruit yield per plant and ascorbic acid content indicating that these traits had wide genetic variability and would respond better to selection. High heritability and high genetic advance were recorded for average fruit weight, fruit yield per plant, fruit diameter, number of lobes per fruit, days to first harvest, leaf area and ascorbic acid content indicating the role of additive gene action for the inheritance of these traits.

Ahmed *et al.* (2012) observed high GCV and heritability accompanied by moderate to high genetic gain for number of fruits and fruit weight which could be improved by selection in early generations. Fruit length and fruit diameter exhibited high heritability coupled with moderate GCV and low genetic gain which needs selection over several successive generations following hybridization for their identification.

In an experiment with ten diverse lines of sweet pepper and forty five crosses generated from them Baseerat *et al.* (2013) recorded moderate values of PCV and GCV for vitamin C while total chlorophyll and total carotenoids exhibited high PCV and GCV. Heritability in broad sense was found to be high for all traits (greater than 91 per cent).

Pandey *et al.* (2013) observed high heritability along with high GCV and high genetic gain for fruit weight, fruit yield per plant, number of fruits per plant, total chlorophyll and fruit width.

Higher genotypic and phenotypic coefficients of variation were recorded by Kumari (2013) for number of fruits per plant (30.49 per cent and 30.63 per cent), average fruit weight (30.85 per cent and 30.03 per cent) and fruit yield per plant (32.12 per cent and 32.26 per cent) indicating that these traits had wide genetic variability and would respond better to selection. High heritability coupled with high genetic gain was observed for number of fruits per plant (57.85 per cent and 57.85 per cent), average fruit weight (60.62 per cent and 60.62 per cent) and fruit yield per plant (65.80 per cent and 65.80 per cent) indicating the role of additive gene action for the inheritance of these traits.

Rana *et al.* (2015) recorded that the phenotypic coefficient of variation (PCV) values was higher than genotypic coefficient of variation (GCV) demonstrating close association between phenotype and genotype for all the traits. High heritability estimates coupled with high genetic gain was observed for fruit yield per plant, which reveals that this character is under additive gene effects and thus more reliable for effective selection.

According to Sharma *et al.* (2019) thirty five genotypes including three check cultivars were evaluated for different horticultural traits. High heritability (92.97 per cent) coupled with high genetic gain (50.86 per cent) was obtained for fruit yield. Hence, selection can prove effective for improvement in fruit yield in bell pepper. High heritability coupled with moderate genetic gain was obtained for number of fruits per plant, average fruit weight and pericarp thickness indicating some scope for their improvement through selection.

Anuradha and Sood (2019) found in an experiment that the estimates of GCV were lower than the respective PCV, indicating the influence of environmental factors on the expression of the traits studied. The estimates of PCV and GCV were high for marketable fruits per plant, marketable fruit yield per plant and capsanthin content depicting the presence of substantial variability and would response better to selection. High heritability coupled with high genetic advance was observed for fruit length, marketable fruits per plant, marketable fruit yield per plant, capsanthin content and ascorbic acid revealing the importance of additive gene action and could be improved through direct selection.

A study was conducted with 46 genotypes by Thakur *et al.* (2019). Among all, genotypes DPCBWR-14-39 (644.29 g), DPCBWR-14-36 (610.17 g), DPCBWR-14-31 (601.96 g), DPCBWR-14-35 (517.88 g), DPCBWR-14-11 (510.34 g), DPCBWR-14-7

(510.14 g) and DPCBWR-14-29 (499.47 g) were the highest yielders and these were also among the top genotypes for economic traits *viz.*, marketable fruits per plant, fruit yield per plant, fruits per plant and average fruit weight. The PCV was invariably higher than their corresponding GCV for most of the characters, indicated close association between phenotype and genotype. High PCV as well as GCV was observed for fruit yield per plant, fruits per plant, marketable fruit yield per plant, marketable fruits per plant, capsanthin content and ascorbic acid depicting the presence of substantial variability and would respond better to selection. High heritability alongwith high genetic advance was observed for fruit yield per plant, fruits per plant, marketable fruit yield per plant, marketable fruits per plant, capsanthin content, TSS and ascorbic acid, respectively revealing the importance of additive gene action for the inheritance of these traits, and further improvement could be done through phenotypic selection.

13.2.3 Identification of Selection Indices

Yield is a complex character, determined by the interaction of several factors, including genetic, physiological and environmental factors (Zecevic *et al.*, 2011). In this way, the knowledge of the correlations or association of other agronomic characteristics with yield or even among them and the environmental influence in the expression of the studied characteristics are of fundamental importance in choosing the selection strategy. Although correlation studies are helpful in determining important yield contributing traits but when more number of variables in the correlation studies are included then the association becomes more complex. Path analysis is simply standardized partial regression co-efficient which splits the correlation coefficient into the measures of direct and indirect effects of a set of independent variables on the dependent variable.

The set of characters identified as selection indices by different workers based on genetic variability parameters, correlation and path coefficient analysis are presented in Table 7.

Table 7: Different Set of Characters Identified as Selection Indices in Sweet Pepper

Characters to be Considered in Selection	*Reference*
Leaf area and fruit breadth	Sharma *et al.* (1981)
Fruits per plant, harvest duration, marketable fruits per plant, and average fruit weight	Sood *et al.* (2009)
Average fruit weight and number of fruits per plant	Sharma *et al.* (2010)
Average fruit weight and number of fruits per plant	Kumari (2013)
Number of fruits per plant, fruit size, number of pickings, plant height and number of lateral branches	Thakur *et al.* (2013)
Days to first flowering, days to first fruit set, number of branches per plant, fruit length, fruit diameter, flesh thickness, average fruit weight, number of fruits per plant and average seed weight	Afroza *et al.* (2013)
Number of fruits per plant, average fruit weight and pericarp thickness	Sharma (2014)
Number of fruits per plant	Soares *et al.* (2017)

Characters to be Considered in Selection	*Reference*
Number of fruits per plant, fruit shape index, number of lobes per fruit, pericarp thickness and fruit yield per plant	Aditka *et al.* (2018)
Number of fruits per plant, fruit length, fruit diameter and pericarp thickness	Khan and Sridevi (2018)
Number of fruits per plant, fruit length, fruit breadth, average fruit weight, pericarp thickness and harvest duration	Sharma *et al.* (2019)
Fruit weight, number of fruits per plant, fruit breadth, thousand seed weight, fruit length and number of seeds per fruit	Thakur *et al.* (2019)
Fruit length, reducing sugar content, fruit weight, primary branches per plant, leaf number per plant, average leaf area, average number of seeds per fruit, pericarp thickness, number of fruits per plant and days to 1st flowering	Roy *et al.* (2019)
Number of branches per plant, number of fruit per plant	Esho (2019)
Fruits/plant and average fruit weight	Banerjee (2020)

According to Gazala *et al.* (2003), seed yield was significantly and positively correlated with days taken to first flower appearance, days to first picking, and pedicel length. Seed yield was found significantly and negatively correlated with fruit yield. Path analysis studies for seed yield revealed that average pedicel length contributed maximum direct effect towards the seed yield followed by days taken to first flower appearance and days to first picking, and these traits were the strongest forces influencing seed yield in bell pepper.

13.2.4 Genetic Divergence

The assessment of genetic divergence helps in identification of breeding lines with broad spectrum of genetic variability thus providing better scope to isolate superior recombinants.

The number of clusters and set of characters having maximum contribution towards divergence found by previous workers are presented in Table 8.

Table 8: Number of Clusters and Set of Characters having Maximum Contribution towards Divergence in Sweet Pepper

No. of Genotype Tested	*No. of Clusters Formed*	*Set of Characters with Maximum Contribution towards Divergence*	*Reference*
22	4	-	Mishra *et al.* (2004)
45	6	-	Farhad *et al.* (2010)
20	3	-	Buttow *et al.* (2010)
23	8	-	Monteiro *et al.* (2010)
24	3	Fruit length, fruit breadth, average fruit weight and fruit yield per plant	Rana *et al.* (2015)
25	4	-	Dabral *et al.* (2016)
20	3	-	Hussein and Mahdi (2017)
35	4	-	Sharma *et al.* (2017)
190	7	-	Pereira-Dias *et al.* (2019)

No. of Genotype Tested	*No. of Clusters Formed*	*Set of Characters with Maximum Contribution towards Divergence*	*Reference*
25	4	–	Negi *et al.* (2019)
78	-	Flavonoids, Total phenols, chlorophyll a and ascorbic acid	Araújo *et al.* (2019)
21	5	β-carotene content of fruit, fruits/plant, lycopene content of fruit and fruit weight.	Banerjee (2020)

13.2.5 Combining Ability and Gene Action

The concept of combining ability proposed by Sprague and Tatum (1942) is a measure of gene action. General combining ability is the average performance of a parent across all cross combinations and specific combining ability is the deviation of specific cross from the average of the parents involved. The *gca* variance is due to additive variance, whereas, sca variance is due to those of dominance and epistatic (additive x additive, additive x dominance and dominance x dominance) variance. Thus *gca* and *sca* variances act as diagnostic tools to detect the additive (linear) and non-additive (non-linear) gene action.

Table 9: Promising Donor Parents and Crosses Identified in Sweet Pepper

Donor Parents	*Crosses*	*Reference*
Yolo Wonder and Vinedale	Russian Yellow x Vinedale, Yolo Wonder x Russian Yellow, Cubanelle x All Big	Gill *et al.* (1973)
LPUNAL and AVELAR	LPUNAL x PIMAM, REDPE x L363 and CALIW x L363	Vallejo *et al.* (1997)
CW (T), Yolo Wonder, SPP RC-1, HC-201, Paprika and RC-1	CW (T) x SPP, CW (N) x RC-1, HC-201 x SPP, HC-201 x RC-1, Yolo Wonder x RC-1, Paprika x CW (N) and HC-201 x CW (N)	Johri *et al.* (2004)
EC 497637	California Wonder x Kandaghat Selection, PBC 631 x EC 497637	Sood and Kaul (2006)
KCP-04, KCP-11, KCP-13 and KCP-15	KCP-01 x BL	Kamble and Mulge (2008)
KCP-05 and KCP-13	KCP-01 x BL, KCP-12 x CW	Kamble *et al.* (2009)
EC-464107, Kandaghat Selection and Yolo Wonder	EC-464107 x Yolo Wonder, AC-48 x Solan Bharpoor and EC-464115 x California Wonder	Sood and Kumar (2010a)
Arka Mohini, IIHR-3341	Arka Gaurav x IIHR-4103, Arka Mohini x IIHR-4034 and Arka Mohini x IIHR-3341	Rao *et al.* (2016)
Kashmir Sel-1, DARL-10, UHFSP(Y)-11, Kashmir Sel-1, UHFSP(Y)-11 and DARL-01	ARCH-19 x Solan Bharpur, DARL-10 x California Wonder and UHFSP(Y)-11 x Yolo Wonder	Devi *et al.* (2018)
California Wonder, Solan Bharpur, Yolo Wonder, California Wonder, Nishath-1 and UHFBP-3	Solan Bharpur x Yolo Wonder, Yolo Wonder x Nishath-1, California Wonder x Solan Bharpur, California Wonder x UHFBP-3 and UHFBP-3 x KC-12	Aditika *et al.* (2020)
8/4, C/4, and Baby Bell	8/4 x Baby Bell, Arya x Baby Bell and C/4 x 8/4	Banerjee (2020)

From a series of studies promising parents and crosses were identified in sweet pepper over the years. A list of donor parents and crosses is presented in Table 9.

13.3.0 Hybridization

Germplasm collection and conservation is the major step in improvement of sweet pepper and this material can be used in hybridization programme. Quality specifications such as size, shape, colour and non-pungency make the task of developing new hybrids/cultivars very challenging. The goal of increasing productivity in the quickest possible time can be achieved by utilizing heterosis breeding. An F_1 hybrid variety is the result of a cross between two homozygous (but genetically distinct) pure lines. The pre-requisite is that all the F_1 plants should resemble each other phenotypically. The great success of hybrid cultivars is attributed to: i) Sufficiently large flowers, easy emasculation and pollen in abundance, ii) Heterosis for yield, iii) Large number of crossed seeds/fruit, iv) Easy deployment of dominant genes conferring resistance to diseases, v) Highly remunerative price of hybrid seed (Negi *et al.*, 2018).

A number of hybrids in sweet pepper have been developed over the years in India. Pusa Deepti (Kt-1) is a hybrid which has been developed by crossing Yolo Wonder with Runion Yellow, released by IARI Regional Station, Katrain (HP). This hybrid is tolerant to bacterial leaf spot and anthracnose. Solan Hybrid-1, Solan Hybrid-2 Solan Bharpur released by YSPUHF, Solan are resistant to fruit rot and virus. KTCPH- 3 and KTCPH- 5 are hybrids from a cross Yolo Wonder x HL-20 and Yolo Wonder x EC-143570, respectively, released by IARI Regional Station, Katrain. Punjab-27 and Bell Boy and No. 12 are F_1 hybrids developed by using GMS in bell pepper.

13.3.1 Hybridization Technique

13.3.1.1 Emasculation

Emasculation is done in the afternoon hours, a day prior to anthesis, because in some genotypes dehiscence of anthers takes place before anthesis. Emasculation is done with the forceps by gently parting the corolla and picking of anthers. Emasculated flowers are occasionally visited by honey bees and natural cross production ranges from 35-38 per cent. So, the emasculated flowers are either bagged or protected by thin cotton wad (Negi *et al.*, 2018).

13.3.1.2 Collection of Pollen

Fresh powdery pollen grains may be collected on the day of anthesis either by vibrator or by tapping the plucked flowers over a glass container. The pollens could be stored for a period of 1-2 months in a desiccators at 0°C temperature.

13.3.1.3 Pollination

Pollination is done in the morning hours, up to 10: 00 am, normally on the day of anthesis. Though stigma remains receptive up to 2 days after anthesis, yet pollination should be done on the day of anthesis to achieve maximum fruit set.

Stigma also becomes receptive a day prior to anthesis. Fruit sets in flower bud, pollinated a day prior to anthesis, has been found to be satisfactory (Negi *et al.*, 2018).

In this bud pollination method, emasculation and simultaneous pollination both are done in the morning hours. Flowering and fruit set requires bright sunlight, hence hybridization should be avoided in cloudy days. Spraying of 10 ppm NAA at flowering and seven days after, checks flower drop and increases fruit set by hybridization (Negi *et al.*, 2018).

13.3.1.4 Hand Pollination without Emasculation

Use of male sterility for economizing hybrid seed production was first reported by Shifriss and Rylski (1972). Breuits and Pochard (1975) developed the hybrid Lamuyo by using male sterile gene ms-509. Use of androcides has also been suggested by Kohli *et al.* (1981) for sterility induction. Spraying GA_3 @ 1000 ppm at 10 days interval from the onset of flowering for three times has caused complete male sterility which lasted throughout the cropping season. Chauhan (1980) used malic hydrazide and dalapon to develop male sterile plants in capsicum. These gametocides caused complete male sterility. In some cases hybrid seed yield can be doubled if the female line has conical shaped fruits, 15-20/plant, crossed with round/large blocky male parents. Some desirable combinations such as 19 x yw and 19 x 02, yw x 70, yw x 02 and 201 x 02 have been reported by Joshi *et al.* (1995) for commercial exploitation.

13.3.2 Hybrid Vigour

Depestre and Espinosa (1988) found high heterosis in F_1 over the mean parental value for number of fruits per plant (11.0-27.0 per cent) and yield per plant (16.6-29.5 per cent) when California Wonder was used as female parent.

In India, Joshi (1986) reported that Bullnosc x HC 20 I, HC 209 x Ruby King and Yolo Wonder x Bullnosc showed standard heterosis for initial yield over the best variety, HC 201. Kaul and Sharma (1988) reported that Sweet Canana x California Wonder, Osh Region x California Wonder and HC 20l x California Wonder recorded 34.0, 33.1 and 25.0 per cent heterosis, respectively over better parent for fruit yield per plant.

In a diallel cross, Miranda *et al.* (1988) from Brazil evaluated all possible F_1 hybrids and detected considerable heterosis for total yield per plant (varying from 5 to 54.7 per cent), fruits per plant (3.5 to 30.8 per cent), early fruit yield (9.1 to 109.2 per cent) and average fruit weight (9.3 to 52.9 per cent). BGH 18 and Amarclo Italiano were the most productive parents. A very early and cold tolerance sweet pepper F_1 hyrbrid Xingyan 1 has been reported from China (Li, 1989).

A genetic break-through was made when a dominant gene P_5 had been identified in one of the breeding lines of sweet pepper in the cultivar Vanderberg's Longe Rote (Vanderber's Long Red). This gene is considered to govern the secretion of a substance between fruit stalk and placenta aiding in easy separation of cork and stems. This gene manipulation will be of immense practical value in developing cultivars suited for mechanical harvest. The fruit without stalk is one of the desirable attributes in the USA markets. Crosses developed between this gene and other cultivars indicated that the expression of P_5 gene is modified by recessive gene affecting stem growth and fruit wrinkling. Through hybridization, the effect of this gene may be transferred to other types and a cultivar for destalking as well as mechanical harvesting may be evolved in near future.

In a Line x Tester mating, Sood and Kumar (2010b) from India studied 24 bell peper hybrids. On the basis of heterosis and mean performance, 'EC-464107' x 'Yolo Wonder' was the best cross-combination followed by 'Kandaghat Selection' x 'Solan Bharpoor', 'SKAU-SP-633-1' x 'Yolo Wonder', 'AC-48' x 'Solan Bharpoor', and 'EC-464115' x 'California Wonder' for fruit yield per plant.

Shrestha *et al.* (2011) evaluated twenty-three F_1 hybrids, their parents and three standard varieties (Special, Fiesta and President) based on their fruit and fruit quality traits at plastic house. Hybrids of 5AVS7 x SP32, SP12 x SP38, 5AVS7 x SP45, CO1234 x SP32, KNU1015 x SP32, 5AVS7 x SP34, 5AVS8 x SP51 and SP27 x SP25 expressed the highest positive heterosis on fruit number and yield per plant whereas highest positive heterobeltiosis was exhibited by the cross 5AVS7 x SP32 (87.2 per cent) and SP12 x SP38 (119.3 per cent). Hybrid of 5AVS7 x SP32 exhibited the highest heterosis for fruit number (104.0 per cent) and yield (141.2 per cent) per plant. Hybrids of 5AVS7 x SP45, 5AVS7 x SP32, and 5AVS8 x SP48 had highest positive standard heterosis on fruit yield per plant over Special, Fiesta and President. Highest positive standard heterosis had manifested on fruit yield per plant using

SP 32, SP 45, and SP 48 inbred lines, which can be used as the male parents with their counterpart female for commercial hybrid production.

In a diallel cross, Sharma *et al.* (2013a) observed wide range of heterosis over better parent and standard check in F_1 generation for marketable fruit yield and its attributing traits. The F_1 crosses PRC-1 x California Wonder and California Wonder x SSP had revealed the highest significant desirable heterosis both over better parent and standard check (California Wonder) for early maturity. The F_1 crosses (Rani Sel-1 x SSP, Rani Sel-1 x Sel-12-2-1, SSP x SP-316 and PRC-1 x California Wonder) showed appreciable heterosis over better parent and standard checks for marketable fruit yield per plant and quality traits in bell pepper. The high heterotic response as observed in these hybrids further supported by the predominant role of non-additive component in the inheritance of the characters studies. These cross combination could be exploited in heterosis breeding programme.

According to Bilashini and Pathania (2018), cross combinations *viz.*, DARL-10 x California Wonder, LC x Solan Bharpur, Kashmir Sel-1 x Solan Bharpur, Kashmir Sel-1 x Yolo Wonder and Kandaghat Sel-9 x Solan Bharpur exhibited the highest significant desirable heterosis over standard check (Indra) for early maturity. The F_1 crosses *viz.*, Kashmir Sel-1 x Solan Bharpur, ARCH-19 x Solan Bharpur, DARL-10 x California Wonder and UHFSP(Y) 11 x Yolo Wonder were the top heterotic cross combinations over standard check for marketable fruit yield per plant and number of marketable fruits per plant. These cross combinations could be exploited in future breeding programme for development of early and high yielding hybrids suitable for cultivation under protected structure in midhills of Himalayas.

A study was conducted by Nabeshima *et al.* (2019) for hybrid vigor on the fruit yield in Japanese sweet pepper (*Capsicum annuum* L.) to investigate dry matter production, canopy light-intercepting characteristics and canopy photosynthesis of F_1 cultivars to those of their parental lines. The hybrid vigor caused higher plant statures with richer foliage, which led to more effective vertical profiles of leaf area, light absorption, and photosynthesis within the respective canopies. These factors caused sufficiently higher canopy photosynthesis to produce greater dry matter production and maintain a higher dry matter partitioning rate into fruits, which resulted in higher yields. These results suggest that the vertical profiles of light-intercepting characteristics and photosynthesis within the canopies, which are evaluated by agro-meteorological methodology, are effective selection indices for breeding plants with higher yields.

13.4.0 Development and Evaluation of CMS Lines in Hybrid Breeding

The bell pepper genotypes are known to possess the fertility restorer (*rf*) gene (Yu, 1985, 1990; Shifriss, 1997; Kumar *et al.*, 2007; Zhang *et al.*, 2000), and a CMS line can be developed by transferring the maintainer gene from the bell pepper to newly identified source of S cytoplasm (Shifriss and Frankel, 1971). Shifriss and Frankel (1971) crossed 23 small fruited hot pepper lines with 3 bell pepper lines possessing the *rf* gene, for search the new sources of S cytoplasm. The segregation pattern in two populations, 1,005 and 1,006, indicated dominance of MF over MS and confirmed that sterility was controlled by a single recessive gene. Four cytoplasmic genetic male sterile lines, namely MS 1, MS 2, MS 3 and MS 4, and their counterpart CMS B lines were developed by Reddy *et al.* (2002) and found stable for sterility in various seasons', *viz. kharif and rabi*. Pákozdi *et al.* (2002) also reported that the CMS lines No. 202, 203, 204 and 205 did not form viable or functional pollen during three growing seasons, *viz.* May, June–July and August–September. Yazawa *et al.* (2002) succeeded in breeding of two stable CMS pepper, "P MS" and "Murasaki MS" with no fruit set. Liu and Gniffke (2004) bred three CMS lines, "CCA4759," "CCA4757" and "CCA5273" with little viable pollen and the lines set no fruits with seed under low temperature (<23.8°C/14.3°C day/night) at the World Vegetable Center, Taiwan. Five CMS sterile lines were evaluated by Gniffke, Lin, and Shieh (2009) at the World Vegetable Center. They found that 2 lines, "CCA7234" and "CCA7235," displayed stable sterility on all winter observation temperature dates. Recently, a genetically stable CMS-A line has been developed (Meena *et al.*, 2018) and being utilized in hybrid breeding programme (Meena, 2017).

13.4.0 Inheritance Study

The constrained of crop improvement is mainly due to lack of adequate information on the genetic control of different yield and yield related traits in sweet pepper. For the improvement of the yield, genetic information and efficient breeding methods are required (Hasanuzzaman and Golam, 2011) which depends mainly on the nature of gene action involved in the expression of quantitative traits (Sharmila *et al.*, 2007). Generation mean analysis is an efficient tool to understand the nature of gene effects involved in the expression of the character. The presence or absence of epistasis can be identified by the analysis of generation mean using the scaling test, that measures epistasis accurately whether complementary (additive x additive) or duplicate (additive x dominance and dominance x dominance) at digenic level.

13.4.1 Inheritance of Morphological Characters

A study was conducted by Mahmoud (2014) to evaluate six generations; *viz.*, P_1, P_2, F_1, F_2, BC_1 and BC_2 of sweet pepper cross (B10-24 x TS 6-33). They found additive gene effects (d) were significant in the inheritance of all characters, except fruit length, fruit diameter and pericarp thickness. On the other hand, dominance gene action (h) was more significant in the inheritance of all studied traits, except number of fruits/plant. Epistasis gene actions were found to be important in the inheritance of most traits. The minimum number of genes controlling the traits were one pair for plant height, fruit diameter, pericarp thickness, T.S.S, ascorbic acid and total yield/plant, while number of genes were estimated as two to three pairs for number of branches, two to four for fruit length, three to seven for number of fruits/plant and two pairs for average fruit weight.

Joshi and Nabi (2015) conducted a trial to study the genetic control of different characters. They found that selection will be effective in improvement of plant height, leaf length, leaf breadth, days to 50 per cent flowering and days to fruiting. Both selection and heterosis breeding will be effective in the improvement of number of branches at first node and yield per plant. Both the additive and dominance gene effects were found to contribute in the inheritance of yield per plant in all the crosses. There was dominance of male fertility over sterility and showed monogenic inheritance. On the basis of present findings, it appears that DKC-12A and DKC-12ms can be used to incorporate male sterility into bell pepper.

An experiment was carried out by Sharma *et al.* (2016) to study the genetics of yield and contributing traits with the help of six generations of six crosses evolved by utilizing four parents (PBC-631, IHR-546, California Wonder and Yolo Wonder). The presence of dominance components for yield per plant in all the crosses along with complementary type of interaction in IHR-546 x YW suggested the exploitation of heterosis breeding for improving yield per plant. Similarly, positive dominance components were recorded in most of the cross combinations for number of fruits per plant, which further indicate the importance of exploiting hybrid vigour for this trait. However, most of the crosses had negative additive component and positive additive x additive [i] gene interactions suggesting delaying the selection for improving average fruit weight. In the present study, the nature and magnitude of gene effect varied with different crosses for most of the traits.

A study was carried out by Devi and Sood (2018) to evaluate gene action for important horticultural traits in bell pepper. The results of the scaling tests revealed that additive-dominance model was inadequate in all the crosses for all the traits studied. For total fruits per plant, the magnitude of dominance [h] gene effects and additive x additive [i] gene interactions were positive and higher coupled with duplicate epistasis in the crosses C1, C2 and C4. For total fruit yield per plant, all the four cross combinations revealed positive and relatively higher magnitude of dominance [h] gene effects along with complementary gene action in cross C3 and duplicate type of epistasis (but low magnitude) in cross C4. Studies on the extent of heterosis suggested that the F_1 hybrid EC-464107 x SH-I (C4) was the most consistent for earliness, while the cross EC-464107 9 KS (C1) was more reliable for high fruit yield and can further be evaluated for yield and quality traits as well as resistance to bacterial wilt.

Very few attempts have been made in the past to study the genetics of fruit colour and shape of sweet pepper in India as well as in the World. Moreover, the genetic basis of fruit colour and shape has not so far been studied in detail. In this back drop, much concentrated efforts will be necessary to develop hybrids/lines having high yield, better fruit quality, attractive colour and shape. Some of few findings are presented below.

McArdle and Bouwkamp (1983) studied the genetics of different fruit characteristics and they found that fruit length and fruit apex configuration were simply Inherited and linked. A gene for fruit shape Index (length/width) could not be substantiated. Average locule number and fruit width were found to be quantitatively inherited. Fruit curvature and fruit base configuration probably are both heavily Influenced by environmental factors.

An experiment was conducted by Kumar *et al.* (2009a) at Regional Station, Katrain (HP), Indian Agricultural Research Institute, India to study the inheritance pattern of mature fruit colour in sweet pepper. Two inbreds each of 4 different mature fruit colours (red, yellow, orange and yellow orange) were crossed in complete diallele fashion to determine the dominance relationship. Red colour was found dominant over others whereas yellow-orange was recessive to all others. From analysis of a series of F_2 generations and test cross of all possible F_1 progeny it was established that when the c locus is silent (*i.e.*, homozygous recessive) the y locus controls the appearance of orange colour (Y_cc) vs. yellow orange (yycc). On the other hand when the y locus is homozygous recessive, the c locus controls the expression of yellow (yyC_) vs. yellow-orange (yycc) colour. Dominant genes at both the loci resulted in red colour while recessive yycc led to appearance of yellow-orange colour in the mature fruit of capsicum. No linkage between the loci was found and cytoplasmic inheritance of this trait was absent.

Vilarinho *et al.* (2015) studied the inheritance of fruit-related traits in a population generated by crossing two heirloom pepper (*Capsicum annuum*) cultivars, Round of Hungary and Bulgarian carrot. Inheritance of corrugated pericarp phenotype of 'Round of Hungary' behaved as a recessive trait controlled by two genes while round fruit shape behaved as a single gene.

Srivastava *et al.* (2019) used F_1 and F_2 generations developed from diverse parental stocks to report the inheritance pattern for different fruits traits. They found that pepper fruit bearing habit, its colour at unripe stage and fruit apex are simply inherited traits with or without epistatic interactions while fruit length is a quantitative character. Fruit orientation (drooping or erect) and fruit habit (solitary or cluster) showed segregation ratio of 3: 1 indicating monogenic inheritance of the trait where drooping character was dominant in nature over erect type and solitary fruit type was dominant over cluster type, fruit bearing habit (single pendant, single erect, cluster pendant and cluster erect) exhibited typical di-hybrid ratio of 9: 3: 3: 1, fruit colour at unripe stage (green, purple and mixture of green and purple) segregated in the ratio of 9: 3: 4 exhibiting recessive gene epistasis and fruit apex (acute or blunt) showed a ratio of 15: 1 indicating duplicate dominant gene epistasis. Fruit length exhibited quantitative inheritance with plants showing fruit length greater than the positive parent (parent showing higher fruit length among the two parents) and lesser than the negative parent (parent showing lesser fruit length among the two parents) from different crosses.

Roy *et al.* (2019) observed changes in fruit colour at ripe stage in F_1s developed through crossing between genotypes having contrasting fruit colour at harvestable maturity. Fruits of all 5 hybrids *viz.*, Royal Wonder x Arya (Green x Yellow), Royal Wonder x Ayesha (Green x Dark Red), 8/4 x Royal Wonder (Orange-red x Green), Baby Bell x Ayesha (Light Red x Dark Red), and Baby Bell x C/4 (Light Red x Dark Red) turned to red at ripe stage which clearly suggested that the single gene corresponding to the y locus which determine to be the capsanthin-capsorubin synthase (Ccs) gene for producing red fruit colour (Lefebvre *et al.*, 1998; Popovsky and Paran, 2000; Lang *et al.*, 2004) showed complete dominance over both green, yellow and orange-red fruit colour.

Recently, Banerjee (2020) conducted an inheritance study on fruit colour and shape of sweet pepper. Inheritance pattern of the cross involving red and yellow colour fruit controlled by 2 genes with duplicate recessive epistasis or complementary gene action, while the cross involving red and green colour fruit, and green and white colour fruit controlled by two genes with duplicate dominant epistasis. On the other hand, the inheritance pattern of the cross involving red and purple colour fruit and red and white colour fruit controlled by two genes with dominant and recessive epistasis or inhibitory gene interaction. He also studied the inheritance pattern of the cross involving blocky and elongate fruit shape which was controlled by two genes with recessive epistasis gene interaction or supplementary gene interaction.

13.4.2 Inheritance of Disease Resistance

According to Sharma *et al.* (1989) bell pepper suffers considerable losses from a strain of potato virus Y (PVYo-sbp). Crosses were attempted between two resistant lines Perennial and S41-1 and two highly susceptible bell pepper commercial cultivars California Wonder and Yolo Wonder. Studies of F_1's, F_2's, back crosses and F_3's indicated that Perennial and S41-1 carry a recessive gene imparting resistance to potato virus Y.

Bacterial spot caused by *Xanthomonas axonopodis* pv. *vesicatoria* is considered one of the most destructive diseases in sweet pepper. Inheritance of resistance to bacterial spot (BS) was investigated by Riva *et al.* (2004) with P_1, P_2, F_1, F_2, BC_1, and BC_2 generations obtained from the cross 'Hercules' (P_1) x UENF 1381 (P_2). All these generations (40 P_1 plants, 40 P_2 plants, 40 F_1 hybrids, 268 F_2 plants, 80 BC_1 and 80 BC_2 plants) were cultivated in a green house. These plants were inoculated with isolate ENA 4135 (10 3 cells mL $^{-1}$ per 1.0 cm 2 mesophile) to assess their reaction to bacterial spot. Broad-and narrow-sense heritabilities were 82.54 per cent and 50.17 per cent, respectively. The average dominance degree indicated presence of overdominance and at least three recessive genes control resistance inheritance.

In literature, variable reports on the genetics of bacterial wilt resistance in various solanaceous vegetable crops have been reported. Earlier, Tran and ByungSoo (2010) and Sharma *et al.* (2013b) in peppers have also reported inheritance of bacterial wilt by utilizing six generations (P_1, P_2, F_1, F_2, BC_1, and BC_2), whereas Lafortune *et al.* (2005) based their findings on F_1 and/or F_2 progenies only. Depending upon the source of resistance and progenies studied in capsicum, inheritance of bacterial wilt resistance was reported as digenic recessive (Thakur, 1990), two to five genes with additive effects (Lafortune *et al.*, 2005), duplicate epistasis (Tran and ByungSoo, 2010) and incomplete dominance to complete dominance (Sharma *et al.*, 2013b). The parental lines used in this study were different from those of earlier workers and this variation along with differences in the strains of the pathogen and different environmental conditions of study may perhaps be the reason for the discrepancy in results.

The genetic nature of bacterial wilt resistance was studied in bell pepper (*Capsicum annuum* L. var. *grossum* Sendt.) plants using the P_1, P_2, F_1, F_2 and backcross generations obtained from four crosses between four parental lines, *viz*, EC 464107, EC 464115, Kandaghat Selection and Sweet Happy I considered as standards for resistance and/or susceptibility by Devi *et al.* (2015). The evaluation trial was conducted in bacterial wilt sick plots along with two susceptible checks, namely Indira (Hybrid) and California Wonder (Open-pollinated variety) that were included after every 15th row to ensure uniform distribution of inoculum in the experimental

Inheritance Study of Fruit Colour in Sweet Pepper (Banerjee, 2020)

F_2 Segregates of the Cross Involving Green and White Colour Fruit

Inheritance Study of Fruit Shape in Sweet Pepper (Banerjee, 2020)

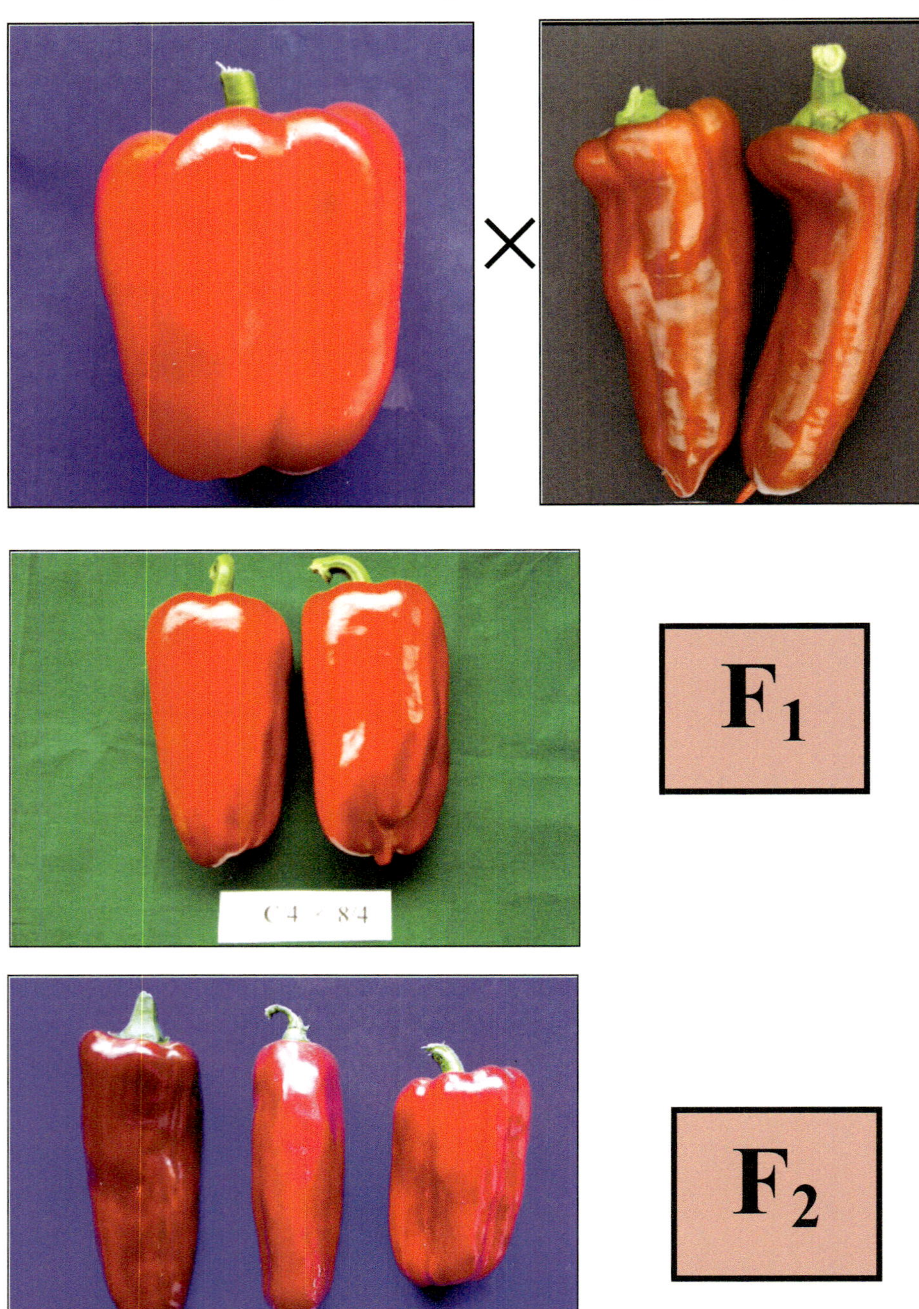

F_2 Segregates of the Cross Involving Blocky and Elongate Fruit Shape

fields. Simultaneously, in a separate block, the parents and their F_1's were also evaluated for twelve different horticultural traits. The findings revealed that bacterial wilt resistance was governed by a single dominant gene in the cross EC 464107 x Sweet Happy I, whereas two genes with dominant and recessive epistasis dictated resistance in the crosses EC 464107 x Kandaghat Selection and EC 464115 x Kandaghat Selection, respectively. Based upon the mean performance, the hybrid EC 464107 x Kandaghat Selection was the best in terms of total and marketable fruit yield, fruits per plant and fruit length, which underscores its importance to be exploited, especially during wet season. In addition, the resistant parents EC 464107 and EC 464115 can be used to develop bacterial wilt resistant F_1 hybrids as they expressed complete or near complete dominance for resistance in combination with the moderately resistant and susceptible varieties Kandaghat Selection and Sweet Happy I. Backcrossing with commercial cultivars and recombinant breeding accompanied by selection would also help to evolve bacterial wilt disease resistant, high yielding and horticulturally desirable pure line varieties.

13.5.0 Disease Resistance

Sweet pepper cultivars with improved genetic resistance to destructive diseases are among breeder's high priority. A resistant cultivar is cost effective and environmental friendly solution to control fungal, bacterial and viral diseases. Sweet pepper suffers from many diseases like *Phytophthora* leaf blight and fruit rot, leaf spot, powdery mildew, wilt, virus complex.

Phytophthora capsici Leonian is one of the most destructive pathogen of peppers which attacks all parts of the plant, causing symptoms such as root rot, crown rot, stem lesion, fruit rot, damping-off of seedlings and so on (Ristaino, 1990; Quirin *et al.*, 2005). Resistant sources for this pathogen have been identified by many researchers, such as 'CM334' from Mexico, 'AC2258', 'PI201232' and 'PI201234' from Central America, among which 'CM334' has found to get the highest resistance level (Bosland and Lindsey, 1991; Ortega *et al.*, 1992; Quirin *et al.*, 2005). Thabuis *et al.* (2003) mapped QTL for *P. capsici* resistance in three *C. annuum* populations derived from independent resistance and susceptible parental lines. Of the 18 QTL detected, one major loci on chromosome 5 was common in all the three populations, while another locus was common to two populations and the remaining were cross-specific. Transfer of the resistance factors from CM334 into a bell pepper genetic background was reported by Thabuis *et al.* (2003). A study was also conducted by Kumari and Sharma (2014) for screening of 15 genotypes of bell pepper against *Phytophthora* leaf blight and fruit rot disease under natural epiphytic conditions and revealed, genotype Feroz and LC-1 as moderately resistant to the disease reactions with lowest apparent infection rate and Kannaul Collection as highly susceptible to the disease reaction with highest apparent rate of infection.

Zhang *et al.* (2013) isolated and characterized a *P. capsici* resistance gene, *CaRGA2*, from a high-resistance pepper (*C. annuum*, CM334). Quantitative real-time PCR analysis indicated that *CaRGA2* was rapidly induced *by P. capsici* evidenced by showing very high and rapid expression in the resistant line compared to the susceptible cultivars, and when the gene was silenced, the resistance level was clearly suppressed, suggesting involvement of *CaRGA2* gene in conferring resistance against *P. capsici*. A total of 16 chromosomal intervals containing single or clusters of resistance QTL for root rot and (or) foliar blight were detected, thereby indicating a complex set of genetics involved in resistance to *P. capsici* (Ogundiwin *et al.*, 2005).

Xanthomonas campestris pv. *vesicatoria* (Xcv) causes the bacterial spot disease in sweet pepper. A total of six genes, four dominant (*Bs1* to *Bs4*) and three recessive genes (*bs5* and *bs6*, *gdr*) were reported conferring resistance to Xcv (Jones *et al.*, 2004; Csillery *et al.*, 2004). AFLP markers tightly linked to *Bs2* gene conferring resistance to *Xanthomonas campestris* pv. *vesicatoria* was detected and then subsequently cloned and expressed in tomato (Tai *et al.*, 1999a,b). Vallejos *et al.* (2010) identified *bs5* and *bs6* resistance genes in breeding line ECW12346 showing resistance to race P6 in *C. annuum*. The combined effect of these two genes gave full resistance to race P6 (indicating a positive epistatic interaction) and allele-specific PCR primers for *bs5* were developed to facilitate the genetic manipulation of this gene.

Lefebvre *et al.* (2003) mapped powdery mildew resistance QTL in a doubled-haploid population derived from crossing 'H3' (resistant) and 'Vania' (susceptible) by phenotyping in both natural field infection and artificial inoculation tests. A total of seven genomic regions including additive QTL and epistatic interactions were detected, of which two genomic regions were commonly detected by both the evaluation methods.

Sweet peppers are found to be infected by potyviruses such as potato virus Y (PVY), tobacco *etc.*h virus (TEV), pepper mottle virus, pepper veinal mottle virus and chili veinal mottle virus (Green and Kim, 1991). Although many resistance genes have been reported, Kyle and Palloix (1997) re-designated them into six resistance genes, namely *pvr1*, *pvr21*, *pvr22*, *pvr3*, *pvr4* and *pvr5*, based on their chronological order of identification. The first identified and cloned resistance gene against potyvirus was *pvr2* which was analyzed and mapped on chromosome 4 (Murphy *et al.*, 1998; Kang *et al.*, 2005). This recessive resistance gene against PVY was later found to be eukaryotic translation initiation factor 4E (eIF4E; Ruffel *et al.*, 2002). Another recessive resistance gene, *pvr6*, against potyviruses was mapped on chromosome 3 (Caranta *et al.*, 1996). *Pvr7*, a dominant resistant gene against the pepper mottle potyvirus (PepMoV) from C. chinense Jacq. 'PI159236' was reported to be tightly linked to the dominant potyvirus resistance gene Pvr4

(Grube *et al.*, 2000). This gene was mapped in pepper chromosome 10 using an interspecific pepper map. They reported the first identified dominant resistance gene cluster of *Pvr4*, *Pvr7* and *Tsw* [a gene conferring resistance to tomato spotted wilt virus (TSWV)] in *Capsicum* species. It has been reported that double and simultaneous mutations in translation initiation factors eIF4E (commonly known as *pvr2*) and eIF (iso) 4E9 (known as *pvr6*) are required to prevent pepper veinal mottle virus infection in pepper (Ruffel *et al.*, 2006; Hwang *et al.*, 2009; Rubio *et al.*, 2009).

Cucumber mosaic virus (CMV) is one of the most destructive viruses in the Solanaceae family. Caranta *et al.* (1997) first reported mapping three QTL for resistance to CMV using a DH lines derived from crossing between a partially CMV-resistant *C. annuum* perennial line and Yolo Wonder, a CMV-susceptible *C.*

annuum line. The three QTL were found explaining together a total of 57 per cent phenotypic variation. Kang *et al.* (2010) showed that the *C. annuum* cultivar 'Bukang' contains a single dominant resistance gene cucumber mosaic resistance 1 (*Cmr1*) against CMV (Korean) and CMV (FNY) strains.

Genetic analysis of resistance to Tobacco Mosaic virus (TMV) was reported to be governed by several alleles of the L gene. Subsequently *L1* gene showing TMV resistant was mapped in chromosome 11 of *C. annuum* genome (Lefebvre *et al.*, 1995; Ben-Chaim *et al.*, 2001).

Some *Capsicum* species and sweet pepper cultivars have been identified as good sources of resistance against different diseases and presented in Tables 10 and 11.

Table 10: Source of Disease Resistance in Sweet Pepper*

Species	*Disease Resistance*
C. chinense	Root rot, TMV, potato virus X, spotted wilt virus, poty virus, aphids, thrips, drought
C. chacoence	Bacterial leaf spot, anthracnose, TMV, poty virus, aphids, thrips.
C. baccatum var. *pendulum*	Powdery mildew, CMV, potato virus Y, drought, high temperature
C. ciliaum	Root rot, water logging
C. frutescens	Fruit rot, TMV,CMV, tobacco leaf virus, poty virus
C. eximium	Potato virus Y, tobacco leaf curl virus, drought, high temperature
C. flexuosum	CMV, potato virus X, soil salinity

* Negi *et al.* (2018).

Table 11: Disease Resistant Cultivars of Sweet Peppers*

Disease	*Cultivars*
Fruit rot	Hungarian Yellow Wax, Spartan, Chinese Giant, Yolo Y
Cercospora Leaf Spot	California Wonder, Hungarian Wax
Bacterial Leaf Spot	PI-163192, 260435, Aristotle, Revolution, King Arthur, Lafayette
Root Rot	Santaka, Charleston Belle, Charleston Hot, Carolina Wonder, Carolina Cayenne, and Mississippi Nemaheart
TMV	Sonnette, Keystone Resistant Giant, Yolo Y, Aladdin
Phytophthora Blight	Paladin
PYV	Aristotle, Lafayette

* Negi *et al.* (2018).

13.6.0 Nematode Resistance

Meloidogyne species, causing root knot diseases are important pathogens in many crop plants including *Capsicum* species. Many genes showing resistance to root-knot nematodes are reported mainly in *C. annuum*. Thies and Fery (2000, 2002) reported characterization of nematode resistance gene N. Genes that were stable at high temperature were reported in three distinct genotypes showing highly resistant phenotypes plants, namely PI 322719, PI 201234 and CM334. The genes *Me4, Mech1* and *Mech2* were shown to be specific to certain *Meloidogyne* species or populations, whereas *Me1, Me3* and *Me7* were non-specific and effective against wide range of root knot pathogen species (Djian-Caporalino *et al.*, 2001, 2007). Djian-Caporalino *et al.* (2001) reported mapping of Me3 and Me4 resistance genes in a segregating population derived from the cross of resistant parental line PM687 with a susceptible line and found both genes to be linked together. Fazari *et al.* (2012) found co-localization of *N*-gene in the *Me*-genes cluster on the P9 chromosome on pepper.

14.0 Biotechnology

The genome of *Capsicum annuum* represents one of the complex plant genomes assembled. The first whole-genome sequences of *C. annuum* ('CM334') and *C. chinense* (PI159236) were released by Kim *et al.* (2014). Genome sequence of the hot pepper provides insights into the evolution of pungency in *Capsicum* species. Subsequently, the genome sequences of *C. annuum* Zunla-1 and of Chiltepin (*C. annuum* var. *glabriusculum*) were reported by Qin *et al.* (2014). Whole-genome sequencing of cultivated and wild peppers provides insights into *Capsicum* domestication and specialization. A reference quality F_1 3.5 Gb genome of *C. annuum* from a single linked library was later assembled by Hulse-Kemp *et al.* (2018). These reports indicate that the capsicum genome contain very high degree of repetitive elements (80 per cent) and about 35k genes. Recently, whole genome resequencing of sweet pepper landraces was used to gain insights on sequence variation in genes of agronomic value (Acquadro *et al.*, 2020). Distinctive variations in miRNAs, resistence gene analogues (RGA) were detected. A detailed survey of genes affecting fruit size, shape and quality revealed highest frequency of variation in regulatory sequences. Many structural variations in the capsicum genome are presence/ absence variation (PAV), especially in RGAs and in the capsanthin/capsorubin genes. Resequencing have been used by Han *et al.* (2016) to construct a ultra high density bin map of *C. annuum* for high throughput QTL mapping of important horticultural traits. An ultra-high-density bin map facilitates high-throughput QTL mapping of horticultural traits in pepper (*Capsicum annuum*). These studies have provided important genomic resources for capsicum improvement including QTL mapping and candidate gene identification.

14.1.0 Micropropagation

Multiplication of F_1 hybrids through micropropagation if permitted is potentially cheaper for a crop. It is also potentially useful in maintenance of parental lines and production of artificial seeds through somatic embryogenesis. Regeneration of plant tissue explants from pepper has been achieved (Morrison *et al.*, 1986a; Sripichitt *et al.*, 1987). Success with anther culture for tissue regeneration and differentiation is genotype-specific and dependent on environmental conditions (Ochoa Alejo and Ireta-Moreno, 1990). Despite the varying degrees of out-crossing in bell pepper, populations are managed most commonly as for self-pollinated species, with the final goal of developing homozygous lines. Using doubled haploids as a tool for reducing generation time and fixing homozygosity has been a long time interest of many workers. Haploid-derived diploids may occur spontaneously or be induced by colchicine treatment in peppers. Haploids may arise from polyembryony, chemical mutagenesis; inter specific hybridization, or anther culture. Anther culture seems to be the most applicable for rapidly fixing new genotypes for the development of elite breeding lines (Li and Jiang, 1990; Morrison *et al.*, 1986a). Depending on the genotype, haploid regeneration from anther culture in pepper is generally between 1 and 20 per cent, although success up to 50 per cent has been reported (Dumas de Vaulx *et al.*, 1981; Morrison *et al.*, 1986b; Munyon *et al.*, 1989).

Induction and regeneration of haploid embryos from anthers or microspore culture facilitate recovery of recessive mutations and unique genetic recombinations. The doubling of the haploid genome results in fully homozygotic lines in shortened period which is important for creation of genetic diversity and a base of new varieties with higher quality. Production of doubled haploids (DHs) is a convenient tool to obtain pure lines for breeding purposes. Anther culture is still considered the method of choice for pepper double haploid production due to its simplicity (Germana, 2011). It is widely accepted that anther culture promotes the formation of microspore-derived embryos, and that the genotype has a remarkable influence in the percentage of microspores deviated towards embryogenesis and effectively transformed into embryos (Irikova *et al.*, 2011).

Parra-Vega *et al.* (2013) revealed that culture conditions have a significant influence on the presence of calli derived from anther walls, which could be minimized by reducing heat shock exposure and/or using a shed-microspore approach in sweet pepper. They concluded that culture conditions have a notable influence, even higher than the genotype, on the formation of callus from sweet pepper anther walls. This influence could be minimized in general by reducing heat shock exposure to a minimum when cultured in solid medium, or by using the shed-microspore approach in particularly sensitive genotypes.

14.1.1 Biochemical and Molecular Markers

Isozymes have been used to examine systematic relationships of *Capsicum cardenasii, C. eximium, C. pubescens, C. tovarii.* Analysis of the esterase isozymes in 15 accessions of *C. annuum* from United States, Mexico, Central America and South America showed three different isozyme patterns. Accessions from Peru, Central America and other accessions from the United States all shared a second pattern of isozyme banding. The third pattern of isozymes including of the hybrids produced when accessions from the first two pattern groups were crossed with one another. The isozyme data combined with the cytological data provide evidence for the theory that Mexico is the centre of domestication of *C.annuum* (Negi *et al.*, 2018).

Peroxidases are enzymes naturally found in plants. Gaspar *et al.* (1985) described peroxidases as isoenzymes associated to changes in physiological processes of plants under stress. According to Klar *et al.* (2006) successive water stress cycles may promote the development of characteristics responsible for improving plant tolerance to periods of low water availability. The peroxidase enzyme activity showed to be an efficient indicator of water stress in sweet pepper.

Schantz *et al.* (1995) studied ascorbate peroxidase (APX) system in *Capsicum annuum* during fruit ripening. A large increase in organelle APX activity was found during chloroplast-chromoplast transition whereas only a slight difference was detected in total fruit extracts. On native gels, four different isoforms were found in total fruit extracts but the patterns for red and green fruit were quite different. In isolated organelles, six isozymes were found and a comparison of the patterns showed significant differences. A cDNA encoding a cytosolic APX was cloned and sequenced. The corresponding transcript was shown to increase 3-4 folds during fruit ripening.

Following the publication of *Capsicum annuum* reference genome, a comprehensive characterization of simple sequence repeats in the capsicum genome was first reported by Cheng *et al.* (2016); thus paving the way for marker development. An average of 868,047.50, 45.50 and 30.00 SSR loci were identified in the nuclear, mitochondrial and chloroplast genomes of pepper, respectively. In recent years, various molecular markers including start codon tagged (SCoT), directed amplified minisattellite DNA (DAMD), single nucleotide polymorphism(SNP) markers have been employed in studies of genetic diversity and population structure of capsicum (Igwe *et al.*, 2019). Markers have also been used in genome wide association studies leading to the detection of association of SNP markers with various fruit stem and leaf related traits. Lee *et al.* (2016) have used SNP markers to construct a core collection from a large set of germplasm. RGA based markers closely linked to a major QTL on chromosome 5 conferring resistence to *P. capsici* have been developed by Korean researchers (Kim *et al.*, 2019). Apart from these developments, molecular markers linked to first flowering node, resistence to antracnose are identified in *C. annuum*. Markers linked to several male sterility genes are also identified in capsicum (Jindal *et al.*, 2019). In future, molecular markers are expected to find many novel applications in sweet pepper breeding including their use to elucidate the inheritance of many economically important characters and breeding of disease resistant cultivars.

14.2.0 Genetic Transformation

The first report on pepper genetic transformation using *Agrobacterium tumifacins* came in 1990 (Liu *et al.*, 1990). However, pepper is extremely recalcitrant to *in vitro* regeneration and genetic transformation (Li *et al.*, 2003). Since then several studies have demonstrated the potential of *Agrobacterium* mediated genetic transformation in sweet pepper (Wang *et al.*, 1991; Dong *et al.*, 1992; Zhu *et al.*, 1996; Harpster *et al.*, 2002; Kim *et al.*, 2002; Mihálka *et al.*, 2003; Li *et al.*, 2003; Heidmann *et al.*, 2011).

Wang *et al.* (1991) reported regeneration and transformation of sweet pepper (*C. frutescens*). Segments of cotyledon, hypocotyl and true leaf taken from 10 to 30 day old seedlings of 3 cultivars (Shuangfeng, Zhongjiao 2 and Zhongjiao 3) were cultured on MS medium supplemented with various phytohormones for induction of callus, shoots and roots. *Agrobacterium tumefaciens* transformed plantlets were obtained using strains C58 and GV3111 and Horch's leaf disc method.

Zhu *et al.* (1996) obtained transgenic sweet pepper plants from *Agrobacterium*-mediated transformation. Fertile transgenic sweet pepper (*C. annuum* var. *grossum*)

Genetic Transformation in Sweet Pepper

Regeneration Response of Sweet Pepper Explants.

Response of sweet pepper 'Fiesta' explants to cocultivation with agrobacterium carrying either the 35S: GUS construct (a–c) or the 35S:: BnBBM: GR construct (d–f). a Callus formation (arrow) and shoot-like structures (SLS; asterisk) after 3 weeks of culture; b leaf-like structures developing at the wounded edge of a cotyledon after 4 weeks of culture; c explant with callus (arrow) and SLS (asterisk) histochemically stained for GUS activity; d SLS, 4 weeks after treatment; e somatic embryo formation (arrows) on a newly emerged leaf, after 6 weeks of culture; f elongating SLS (asterisk) after 4 weeks on elongation medium (Heidmann *et al.,* 2011).

plants were generated at high rates from explants that were co-cultivated with *A. tumefaciens* strain GV3111-SE harbouring a plasmid that contains the cucumber mosaic cucumovirus coat protein (*cmv-cp*) gene.

Several explants including hypocotyls, cotyledon, and leaf have been used by Li *et al.* (2003). A highly efficient genetic transformation system using pepper cotyledons as explants was established by Li *et al.* (2003). The *nptII* and *gus* genes are generally used in genetic transformation of capsicum. Kim *et al.* (2002) have also explored the use of mannose as selecting agent instead of the antibiotic kanamycin and reported better transformation efficiency. A tissue culture independent *Agrobacterium* mediated in plant a transformation was later studied by Kumar *et al.* (2009b). Kumar *et al.* (2012) reported improved plant regeneration and *Agrobacterium* mediated transformation of red pepper in an auxin free medium. Genetic transformation in pepper is a novel tool to improve pathogen resistance. However, the progress in genetic transformation of sweet pepper has been slow as the crop is not very responsive to transformation.

15.0 References

Abou-Hadid, A.F., EI-Shinawy, M. Z., EI-Oksh, I., Gomaa, H. and EI-Beltagy, A.S. (1994) *Acta Hortic.*, **366**.

Acchireddy, N.R., Fletcher, J.S. and Beevers, L. (1982) *HortScience*, **17**: 635-637.

Ackley, J.A., Wilson, H.P. and Hines, T.E. (1998) *Weed Technol.*, **12:** 458-462.

Acquadro, A., Barchi, L., Ezio Portis1, Nourdine, M., Carli, C., Monge, S., Valentino, D. and Lanteri, S. (2020) *Sci. Rep.*, **10**: 9189.

Aditika, Kanwar, H.S., Kumar, R., Dogra, R.K., Sharma, R. and Singh, S. (2018) *Int.J.Curr.Microbiol.App.Sci.*, **7**(3): 2669-2673.

Aditika, Kanwar, H.S., Singh, P.S. and Singh, S. (2020) *Agric.Res.*, **9**:526-535.

Afroza, B., Khan, S.H., Chattoo, M.A., Mufti, S., Parveen, K. and Mukhdoomi, M.I.(2013) *The Asian J. Horti.*, **1** :358-360.

Agbogidi, O.M. and Okonmah, U.L. (2011) *I.J.A.B.R.*, **1:** 119-122.

Aguirre, I., Gutiérrez, M.C. and Cuartero, J. (1995) *Acta Hortic.*, 412.

Ahmed, N., Singh, S.R., Lal, S. and Mir, K.A. (2012). *Indian J. Plant Genet. Resour.*, **25**(3): 304-306.

Ahmed, N., Tanki, M.I., Mir, M. and Shah, G.A. (1996) *Capsicum Eggplant Newslett.*, **16**: 47-60.

Akhter, S., Mostarin, T., Khatun, K., Akhter, F. and Parvin, A. (2018) *The Agriculturists*, **16:** 58-64.

Akintoyea, H.A. and Adebayo, G.A. (2013) *Proc. 2nd All Africa Horticulture Congress*, Ibadan, Nigeria, pp. 543-548.

Al-abbadi, S.Y.A. (2009) *J. Biolog. Sci.*, **9**: 464-469.

Aleksic, Z. and marinkovic, N. (1981) *Plant Protect.*, **32**(1): 45-48.

Ali, A.M. and Kelley, W.C. (1982) *GFR Int. Soc. Hort. Sci.*, I. Abstr., No. 1562.

Aljaro, A.U. and Martinez, R.M. (1988) *Agric. Tecn.*, **48**(1): 33-38.

Aloni, B., Karni, L., Deventurero, G., Turhan, E. and Aktas, H. (2008) *J. Hortic. Sci. Biotechnol.*, **83**: 100-105.

Aloni, B., Karni, L., Rylski, I., Cohen, Y., Lee, Y., Fuchs, Moreshet, M.S. and Yao, C. (1999a) *J. Hortic. Sci. Biotechnol.*, **73**(6): 743-749.

Aloni, B., Pressman, E. and Karni, L. (1999b) *Ann. Bot.*, **83**: 529-534.

Alsadon, A., Wahb-Allah, M.A., Abdel-Razzak, H. and Abdullah, D. (2013) *Aust. J. Crop Sci.*, **7**(9): 1309-1316.

Andrews, J. (1995) The Domesticated Capsicums, new ed. University of Texas Press, Austin.

Andriolo, J.L., Buriol, G.A. Streck, N.A. and Fiorin, J. (1991) *Hortic. Bras.*, **21**: 191-204.

Anonymous. (2010) Agricultural and Processed Food Products Export Development Authority, Ministry of Commerce and Industry, Government of India.

Anonymous. (2016) National Horticulture Board Data Base, National Horticulture Board, Ministry of Agriculture and Farmers Welfare, Government of India.

Anonymous. (2018) National Horticulture Board Data Base, National Horticulture Board, Ministry of Agriculture and Farmers Welfare, Government of India.

Anuradha and Sood, S. (2019) *Int. J. Curr. Microbiol. App. Sci.*, **8**(8): 872-888.

Anwar, S.A., Mahdi, M.M., McKenry, M.V. and Qadir, A. (2013) *Pakistan J. Zool.*, **45**(3): 595-603.

Araújo, L.M., Neves, L.G., Sousa, D.A., Zeviani, W.M., Silva, L.R. and Marostega, T.N. (2019) *Horticultura Brasileira*, **37**: 210-214.

Arisha, M.H., Liang, B.K., Shah, S.N.M., Gong, Z.H. and Li, D.W. (2014) *Genet. Mol. Res.*, **13:** 10049-10061.

Arisha, M.H., Shah, S.N.M., Gong, Z.H., Jing, H., Li, C. and Zhang, H.X. (2015) *Front. Plant Sci.*, **6**: 399.

Aruna, P. and Sudagar, I.P. (2010) *The Asian J. Hortic.*, **4** (2): 336-37.

Arya, P. S. (1983) *Indian Fmg. Digest*, **16:** 13-16.

Avilla, C., Collar, J.L., Duque, M., Perez, P. and Fereres, A. (1997) *HortScience*, **32:** 882-883.

Awalin, S., Shahjahan, M., Roy, A.C., Akter, A. and Kabir, M.H. (2017) *J. Agric. Eco. Res. Int.*, **11**(3): 1-8.

Azam, M., Noman, M., Abbasi, N.A., Ramzan, A., Imran, M., Hayat, A., Abbas, H. and Akram, A. (2016) *Sci.Technol. Dev.*, **35**(2): 75-81.

Baggett, J.R. and Kean, D. (1988) *HortScience*, **23:** 1097.

Baghel, B.S., Gupta, M. and Tiwari, R. (2003) *Proc. All India Sem.* on Potential and Prospects for Protective Cultivation, organized by the Institute of Engineers Ahmednagar, December 12-13, 2003, Ahmednagar.

Baghour, M., Ruiz, J.M. and Romero, R. (2000) *J. Plant Nutrition*, **23:** 91-101.

Bailey, L.H. (1948) *Manual of Cultivated Plants*, (Rev.edn.), Macmillan Co., New York.

Bakker, J.C. and Uffelen, J.A.M.Van (1987) *Groenten en Fruit*, **43:** 42-43.

Bakker, J.C. and Uffelen, J.A.M.Van (1988) *Netherlands J. Agric. Sci.*, **36:** 201-08.

Banerjee, S. (2020) *Ph.D. thesis*, Department of Vegetable Science, Faculty of Horticulture, Bidhan Chandra Krishi Viswavidyalaya, Mohanpur, West Bengal, India.

Barros, J.C., daS. M.de, Goes, A.de, and Minami, K. (1994) *Scientia Agricola*, **51:** 363-368.

Baseerat, A., Khan, S.H., Chattoo, M.A., Mufti, S., Parveen, K. and Mukhdoomi, M.I. (2013) *Asian J. Hort.*, **8**(1): 358-360.

Basoccu, L., Nicola, S. and Smith, A.R. (1993) *Acta Hortic.*, **32:** 121-127.

Bassett, M.J. (1986) Avi Publishing, Westport.

Bayogan, R.R., Salvilla R., Majomot A.M.C. and Acosta J. (2017) *South Western J. Hort. Biol. Environ.*, **8**(1): 1-15.

Belakbir, A., Ruiz, J.M. and Romero, L. (1998) *HortScience*, **33:** 85-87.

Ben-Chaim, A., Grube, R.C., Lapidot, M., Jahn, M. and Paran, I. (2001) *Theor. Appl. Genet.*, **102**: 1213-1220.

Benedek, L. (1959) *Kiserl. Kozlem, Sect.*, **52:** 69-85.

Benoit, F. and Ceustermans, N. (1974) *Guinbouwberichten*, **38:** 250-254.

Ben-Yehoshua, S., Rodov, V., Fishman, S., Peretz, J., Asuncion, R. dela, Bums, P., Sornsrivichai, J. and Yantarasri, T. (1996) In: *Proc. of the Australasian postharvest horticulture conf. 'Sci. and Tech. for the fresh food revoltuion'* Melbourne, Australia, pp. 143-162.

Bhatt, R.M. and Rao, N.K.S. (1997) *Biologia Plantarum*, **39:** 437-439.

Bhatt, R.M., Srinivasa Rao, N.K. and Anand, N. (1999) *Indian J. Hort.*, **56:** 62-66.

Bigotti, P.G. (1974) *Revista della Orto flowfruithi Kultura*, Italiana, **58:** 187-93.

Bilashini, D.M. and Pathania, N.K. (2018) *Veg. Sci.*, **45**(1): 31-37.

Bosland, P.W. (1993) *Capsicum and Eggplant Newslett.*, **12:** 25-31.

Bosland, P.W. (2002) *J. Hered.*, **93:** 380-382.

Bosland, P.W. and Lindsey, D.L. (1991) *Plant Dis.*, **75**: 1048-1050.

Bosland, P.W. and Votava, E.J. (2000) CAB International, Wallingford, p. 204.

Bosland, P.W., Iglesias, J. and Tanksley, S.D. (1991) *HortScience*, **26:** 215-16.

Boswell, V.R. (1937) *Dept. Agri. Yearbook*, Washington DC.

Boswell, V.R. (1951) *Natl. Geogr.*, **96:** 145-217.

Bottini, E. (1967) *Pot. Symp.*, pp. 293-301.

Bower, J., Patterson, B.D. and Jobling, J.J. (2000) *Aust. J. Exp. Agric.*, **40**: 457-463.

Bracy, R.P. and Constantin, R.J. (1988) *Louisiana Agriculture*, **31:** 13,23.

Breutis, G. and Pochard, E. (1975) *Ann. Amehor. Plantes*, **25**: 399-409.

Buczkowska, H., Michalojc, Z. and Wierdak, R.N. (2016) *Turk. J. Agric. For.*, **40**: 222-228.

Burgio, G. and Maini, S. (1994) Boll.1st. Ent.G.Grandi University of Bologna, **49** :1-14.

Burgmans, J. (1983) *New Zealand J. Agric.*, **147**: 50-51.

Butkevic, S.T. (1967) *Inst. OnoseamZemledOvoseev*, **7**: 55-64.

Buttow, M.V., Barbieri, R.L., Neitzke, R.S., Heiden, G. and Carvalho, F.I.F. de (2010) *Ciencia Rural*, **40**(6) : 1264-1269.

Campbell, G.M. and Swingle, H.D. (1965) *Zenn. Home Sci.*, **49**: 30-32.

Canul-Tun, C.E., Ibarra-Jiménez, L., Valdez-Aguilar, L.A., Lozano-del Río, A.J., Cárdenas-Flores, A., Zermeño-González, A., Lozan°Cavazos, C.J., Valenzuela-Soto, J.H. and Torres-Olivar, V. (2017) *J. Plant Nutr.*, **40:** 1083-1090.

Caranta, C., Palloix, A., Gebre-Selassie, K., Lefebvre, V., Moury, B. and Daube'ze, A.M. (1996) *Phytopathology*, **86**: 739-743.

Caranta, C., Palloix, A., Lefebvre, V., Daube ze, A.M. (1997) *Theor. Appl. Genet.*, **94**: 431-438.

Carvajal, M. and Alcaraz, C.F. (1998) *Recent Res. Developments Phytochemistry*, **2**: 83-84.

Carvalho, J.A., Rezende, F.C., Aquino, R.F., Freitas, W.A. and Oliveira, E.C. (2011) *R. Bras. Eng. Agríc. Ambiental.*, **15**(6): 569-574.

Cebula, S. (1989) *Zeszyty NaukoweAkademii Rolniczej imHugonaKollataja w Krakowie, RozprawaHabilitacyjna.*, No. 130, p. 86.

Cebula, S. (1995a) *Folia Horticulturae*, **7**: 51-58.

Cebula,S. (1995b) *Folia Horticulturae,* **7**: 59-67.

Cebula, S. (1995c) *Zeszyty Naukowe Akademii Rolniczej imHugonaKollataja w Krakowie, Rozprawa Habilitacyjna.*, No.22, pp. 39-47.

Cebula, S. (1995a) *Folia Horticulturae*, **7**: 51-58.

Cebula, S., Salisz, A. and Kunicki, E. (1998) *Folia Hortic.* , **10**: 35-44.

Cebula,S. (1995b) *Folia Horticulturae*, **7**: 59-67.

Centritto, M., Delfine, S., Villani, M.C., Occhionero, A., Loreto, F., Alvino, A. and Marco, G. di (1998) In: *Photosynthesis: mechanisms and effects* (Ed.G. Garab), Vol. V, pp. 3773-3776.

Chaim, A.B., Borovsky, Y., Rao, G.U., Gur, A., Zamir, D. and Paran, I. (2006) *Isr. J. Plant Sci.*, **54**: 191-203.

Chakrabarty, S., Aminul Islam, A.K.M., Khaleque Mian, M.A. and Ahamed, T. (2020) *Open Agr. J.*, **14:** 34-43.

Chartzoulakis, K., Drosos, N. and Chartzoulakis, K.S. (1997) *Acta Hortic.*, **449**: 175-180.

Chauhan S.V.S. (1980) *J. Indian. Bot. Soc.*, **59**: 133-136.

Chavez-Alfaro, J.J., Zavaleta-Mejia, E. and Teliz-Ortiz, D. (1995) *Fitopatologia*, **30**: 47-55.

Chen, C.H., Yang, Z.J., Zhou, Q. and Cai, X.Y. (1998) *Acta Hortic. Sanghai*, **14**: 57-64 (Suppl.).

Chen, Z.J., Wu, G.J., Zhang, M.R. and Zhang, S.L. (1995) *Acta Phytophylacica Sinica*, **22:** 343-347.

Cheng, J., Zhao, Z., Li, B., Cheng, Q., Wu, Z., Diana, L., Saavedra, T., Luo, X., Cui, J., Rafael, F., Bustamante, R., Li, S. and Hua, K. (2016) *Sci. Rep.*, **6**: 18919.

Chittenden, F.J. (1956) *The Royal Hort. Dctionary of Gardening*; revised by P.G. Synha, Oxford Press, London.

Choe, J.S., Kang, K.H., Um, Y.C. and Choe, Y.H. (1991) *Proc. Int. Symp. Applied Tech. Green house, held in Beijing, China*, 7-10 October. 1991, pp. 185-189.

Choudhary, A.K. (2016) *Curr. Sci.*, **111**(2): 272-277.

Choudhury, B. (1967) *Vegetables, National Book Trust*, New Delhi.

Christopher, E.P. (1958) *Introductory Horticulture*, McGraw-Hill Book Company, Inc. NewYork.

Chung, S.J., Oh, B.R. and Seo, B.S. (1991) *J. Korean Soc. Hort. Sci.*, **32:** 286-291.

Chupp, C. and Sherf, A.F. (1960) *Vegetable Disease and their control*, The Ronald Press Company, New York.

Cobanero, F.J., Martinez, V. and Carvajal, M, (2004) *Plant Sci.*, **166**: 443-540.

Cochran, H.L. (1934) *J. Agric. Res.*, **48:** 737-748.

Cochran, H.L. (1936) *Cornell Memoir.*, pp. 190.

Cochran, H.L. (1943) *Proc. Amer. Soc. Hori. Sci.*, **43:** 229-234.

Cohen, O., Borovsky, Y., David-Schwartz, R. and Paran, I. (2014) *New Phytol.*, **202:** 1014-1023.

Costa, L.V., Lopes, R., Lopes, M.T.G., Figueiredo, A.F., Barros, W.S. and Alves, S.R.M. (2009) *Crop Breed. Appl. Biotechnol.*, **9:** 37-44.

Csillery, G., Szarka, E., Sardi, E., Mityko, J., Kapitany, J., Nagy, B. and Szarka, J. (2004) *Proceedings of the 12th Eucarpia Meeting on Genetics and Breeding of Capsicum and Egg-plant*, Noordwijkerhout, Netherlands, pp. 147-153.

Curl, A.L. (1962) *J Agric. Food Chem.*, **10:** 504-509.

Curtis, L.C. (1935) *Conn. Agric. Exp. Sta.Cire.*, 107.

Dabral, M., Kumar, S., Pant, S.C., Paliwal, A., Mishra, R.C., Bahuguna, P., Badhani, H.C. and Nautiyal, B.P. (2016) *Int. J.Agri. Sci.*, **8** (20): 1354-1356.

Dag, A. and Kamer, Y. (2001) *Am. Bee J.*, **141**(6): 447-448.

Dasgan, H.Y. and Kazim, A. (2003) *Turkish J. Agric. Fores.*, **27**: 29-35.

Daskaloff, S. (1968) *Theo. Appl. Genet.*, **38:** 370-372.

Daskalov, S. (1986) *IAEA*, **4:** 25.

De Candolle, AD.E. (1886) *Origin of Cultivated Plants*, Haner Publishing Company, pp. 288-90.

Deli, J., Molnar, P., Matus, Z. and Toth, G. (2001) *J. Agri. Food Chem.*, **49:** 1517-1523.

Demers, D.A., Gosselin, A. and Wien, H.C. (1998) *J. Amer. Soc. Hort. Sci.*, **123:** 202-207.

Dempsey, A. and Boswell, F.C. (1979) *HortScience*, **14**: 537-539.

Deng, Y.Y. (1988) *Zuowu Pinzhong Ziyuan, No.* 3, 45.

Depestre, T. and Espinosa, J. (1988) *Agrotechniade Cuba*, **20:** 87-90.

Desai, V. G.P., Patil, V.K. and Aniarkar, M.V. (1987) *SouthIndian Hort.*, **35:** 451-452.

Devi, J. and Sood, S. (2018) *Agric. Res.*, **7:** 112-119.

Devi, J., Sood, S., Vidyasagar and Singh, Y. (2015) *Indian J. Agric. Sci.*, **85**(10): 126-131.

Devi, M.B., Pathania, N.K. and Thakur, N. (2018) *J. App. N. Sci.*, **10** (1): 410-416.

Devi, M.B. and Pathania, N.K. (2018) *Veg. Sci.*, **45**(1): 31-37.

Dhaliwal, M.S., Sharma, S.P., Jindal, S.K., Dhaliwal, L.K. and Gaikwad A.K. (2017) *J. Crop Improv.*, **31**(6): 830-846.

Dharmatti, P.R. and Kulkarni, G. N. (1988) *Seed Res.*, **16:** 148-151.

Dharmatti, P.R. and Kulkarni, G.N. (1987) *South Indian Hort.*, **35:** 395-396.

Diaz-Perez, J.C. (2010) *HortScience*, **45**(8): 1196-1204.

Dimitrov, G. (1960) *Kiserl. Kozlem, Sect. C.*, **53:** 59-85.

Djian-Caporalino, C., Fazari, A., Arguel, M.J., Vernie, T., VandeCasteele, C., Faure, I., Brunoud, G., Pijarowski, L., Palloix, A., Lefebvre, V. and Abad, P. (2007) *Theor. Appl. Genet.*, **114:** 473-486.

Djian-Caporalino, C., Pijarowski, L., Fazari, A., Samson, M., Gaveau, L., O'Byrne, C., Lefebvre, V., Caranta, C., Palloix, A. and Abad, P. (2001) *Theor. Appl. Genet.*, **103**: 592-600.

Djumaijah and Aliudin (1986) *Buletin Penelitian Horticultura*, **13**: 43-47.

Dong, C., Jiang, C. H., Feng, L. and Guo, J. (1992) *Acta. Hortic. Sin.*, **19**: 184-186.

Dorland, R.E. and Went, F.W. (1947) *Amer. J. Bot.*, **34**: 393-401.

Dumas de Vaulx, R., Chambonnet, D. and Pochard, E. (1981) *French Agronomie*, **1**(10): 859-864.

Durpree, M., Dempsey, A.H., Beshear, R. and Tippins, H.H. (1979) *Res. Rept.Exp. Sta. Georgia Univ.*, 318, pp. 9.

Dysko, J. and Kaniszewski, S. (1989) *Biuletyn Warzyaniczy Suppl. II.*, 189-194.

Egorova, Z.M. (1975) *Referativnyi Zhurnal*, 10.55.641.

El-Bahadli, K. (1988) *J.Univ. Kuwait, Science*, **15:** 291-296.

El-Saced, H.M., Imam, R.M. and El-Halim, S.M.A. (1996) *Egyptian J. Hort.*, **23**: 145-165.

Ertek, A., Sensoy, S., Gedik, I. and Kucukyumuk, C. (2007) *Am. Eura. J. Agric. Environ. Sci.*, **2**(4): 349-358.

Eshbaugh, W.H. (1993) *New crops*. Wiley, New York, pp. 132-139.

Esho, K.B. (2019) *Plant Arch.*, **19**(2): 4316-4320.

Esiyok, D., Ozzambak, E. and Eser, B. (1994) *Acta Hortic.*, **366**: 293-300.

Espinosa, J. and Depestre, T. (1986) *Cienciay Tecnicaenla Agricultura. Hortalizas*, Papa, Granosy Fibras, **5**: 65-72.

Estrada, B., Pomar, F., Diaz, J., Merino, F. and Bernal, A. (1997) *Capsicum and Eggplant Newletter*, No. 16, pp. 60-63.

Fajinmi, A.A. and Oladiran, A.O. (2000) *Capsicum Eggplant Newsletter*, No.19, pp. 93-96.

Fallik, E., Grinberg, S. and Ziv, O. (1997) *J. Hort.Sci.*, **72**: 35-41.

Farhad, M., Hasanuzzaman, M., Biswas, B.K., Arifuzzaman, M., and Islam, M.M. (2010) *Bangladesh Res. Pub. J.*, **3**(3): 1045-1051.

Farooq, M., Ramzan, A., Riaz Chattha, M., Uzair Qasim, Nawab, N.N. and Hidyatulllah (2015) *Sci. Technol. Dev.*, **34**(3): 155-157.

Fatima, H., Almas, L. and Yasmin, B. (2017) *Southern Agricultural Economics Association 49th Annual Meeting*, Mobile, Alabama, February 4-7.

Fazari, A., Palloix, A., Wang, L., Hua, Y.M., Sage-Palloix, A.M., Zhang, B.X. and Djian-Caporalino, C. (2012) *Plant Breed.*, **131**: 665-673.

Fernandes, P.D., Oliveira, G.D. and Hoag, H.P. (1971) *Anaisda EscolaSuperior de Agricultura*, **28:** 145-151.

Figueroa, R., Pacheco, F., Echaiz, C., Cordovez, G. and Kuhn, N. (2016) *Weed Technol.*, **30:** 587-594.

Fiume, F. (1993) *Mededelingenvande FaculteitLandbouwwetenschappen, Universiteit Genet.*, **58:** 515-525.

Gao, R. X., Lin, G.R. and Wang, S.Q. (1987) *J. Shenyang Agricultural University*, China.

Garcia-Bañuelos, M. L., Sanchez, E., Gardea-Béjar, A.A., Muñoz-Marquez, E., Soto-Parra, J.M. and Ojeda-Barrios, D.L. (2017) *Emirates J. Food Agric.*, **29**(6): 420-428.

Gaspar, T., Pennel, C. and Castillo, F.J. (1985) *Physiol. Plant.*, **64**: 418-423.

Gazala, N., Ahmed N., Hussain, K., Qadir, R. and Dar, Z.A. (2003) *Asian J. Hortic.*, **2**(2): 181-183.

Germana, M.A. (2011) *Plant Cell Tissue Organ Cult.*, **104**: 283-300.

Ghosal, M.K. and Das, R.K. (2012) *Int. J. Agric. Sci.*, **8**(1): 220-223.

Gill, H.S., Thakur, P.C. and Thakur, T.C. (1973) *Indian J. Agric. Sci.*, **43:** 918-921.

Gill, H.S., Thakur, P.C. and Thakur, T.C. (1974) *Indian J. Hort.*, **31:** 74-78.

Gluhaich, P. (1981) *Proc.33'" Ann. Calif. Weed Conj.*, pp. 97-103.

Gniffke, P.A., Lin, S.W. and Shieh, S.C. (2009) AVRDC- The World Vegetable Center Publication No. 09-718. Shanhua, Taiwan. P.9.

Golez, A. (1995) *Roczniki Akademii Rolniczejw Poznaniu,Ogrodnictwo*, No. 23, pp. 25-31

Gong, F.R., Zhu, Y.Y., Wu, X.G. and Jin, H.J. (1999) *Acta Horticulturae Songhai*, **15:** 70-75.

Gonzalez, E., Iglesias, I. and Diaz, T. (1989) *Phyton.*, **49:** 83-87.

Gonzalez-Aguilar, G.A., Cruz, R., Baez, R.andWang, C.Y. (1999) *J.Fd.Quality*, **22:** 287-299.

Gonzalez-Aguilar, G.A., Cruz, R., Granados, M. and Baez, R. (1997) In: *Seventh International Controlled Atmosphere Research Conf. CA'97*. Proceeding Volume 4: Vegetables and Ornamentals, Davis, California, USA, No.18, pp. 66-72.

Govindarajan, V.S. (1985) *Crit Rev Food Sci Nutr.*, **22**(2): 109-176.

Gracza, L. and Benedek, L. (1961) *Kiserl. Kozlem, Sect. C.*, **54:** 87-96.

Graifenberg, A., Petsas, S. and Giustiniani, L. (1983) *Informatore diortiflorofrutticoltura*, **24:** 5-10.

Green, S.K. and Kim, J.S. (1991) Technical Bulletin no. 18, Asian Vegetable Research and Development Centre, Taipei, Taiwan

Greenleaf, W.H. (1986) In *Breeding Vegetable Crops*. AVI Publishing Co., Westport, Connecticut, USA, pp. 69-127.

Grube, R.C., Blauth, J.R., Arnedo, M.A. and Jahn, M.K. (2000) *Theor. Appl. Genet.*, **101**: 852-859.

Guertal, R.A. (2000) *Agron. J.*, **92:** 388-393.

Gungor, A. and Yildirim, E. (2013) *Pak. J. Bot.*, **45**(5): 1605-1608.

Guzman, I., Hamby, S., Romero, J., Bosland P.W. and O'Connell M.A. (2010) *Plant Sci.*, **179:** 49-59.

Han, K, Jeong, H.J., Yang, H.B., Kang, S.M., Kwon, J. K., Kim, S, Choi, D. and Kang, B.C. (2016) *DNA Res.*, **23:** 81-91.

Han, X.13., Li, R.Q., Wang, J.B. and Miao, C. (1996) *Acta Horticulturae Sinica*, **23**: 359-364.

Hanice, E. (1988) *Soiless Culture*, **4:** 23-26.

Harpster, M.H., Brummell, D.A. and Dunsmuir, P. (2002) *Plant Mol. Biol.*, **50**: 345-355.

Hartz, T.K., LeStrange, M. and May. D.M. (1993) *HortScience*, **28**: 1097-1099.

Hasanuzzaman, M. and Golam, F. (2011) *Aust. J. Crop Sci.*, **5**: 1868-1875.

Haung, H. (1973) *Mem.Coll.Agric.*, Taiwan, **14**: 1-11.

Hawthorn, L.R. and Pollard, H. (1954) *Vegetable and Flower Seed Production*, Blakiston Company

Hazra, P., Chattopadhyay, A., Karmakar, K. and Dutta, S. (2011) New India Publishing Agency, New Delhi, India. p. 478.

Heidmann, Iris, Lange, Brenda de, Lambalk, Joep, Angenent, Gerco C. and Boutilier, Kim (2011) *Plant Cell Rep.,* **30**: 1107–1115.

Heiser (Jr.), C.B. (1976) *Evolution of Crop Plants,* (Ed. N.W. Simmonds), Longman, London, pp. 265-268.

Heiser Jr., C.B. (1985) In: *Of Plants and People,* University of Oklahoma Press, Norman and London, pp. 142-154.

Heiser, C.B. and Pickersgill, B. (1969) *Taxon.,* **18:** 277-83.

Heiser, C.B. and Smith, P.G. (1953) *Econ Bot.,* **7**: 214-227.

Herklots, G. A.C. (1972) *Vegetables in South-East Asia,* George Allen and Unwin Ltd. London.

Hibberd, A. M., Gillespie, D., Nahrung, G.C., Persley, D.M. (1988) *HortScience,* **23:** 1095-96.

Hirschberg, J. (2001) *Curr. Opin. Plant Biol.,* **4:** 210-218.

Hortenstine, C.C. (1963) *Proc. Fla. Sta. Hort. Soc.,* **75:** 146-152.

Hou, G.Q., Chen, D.S. and Lui, B.Z. (1987) *Acta Hortic. Sinica,* **14**: 251-256.

Howard, L. and R. Wildman. (2007) *In Handbook of Nutraceuticals and Functional Foods.* CRC Press, Boca Raton, Florida, USA, pp. 165-191.

Howard, L.R., Talcott, S.T., Brenes, C.H. and Villalon, B. (2000) *J. Agric. Food Chem.,* **48:** 1713-1720.

Hsieh Ching Fang and HsuKuo Nan (1994) *Bull. Taichung Dist. Agril. Improvement Stn.,* No. 42, https: //www.agrifarming.in/ 2019.

Hulse-Kemp, A.M., Maheshwari, S., Stoffel, K., Hill, T.A., Jaffe, D., Williams, S. R., Weisenfeld, N., Ramakrishnan, S., Kumar, V., Shah, P., Michael, C.S., Church, D. M. and Deynze, A.V. (2018) *Hortic. Res.,* **5**: 4.

Hussein, I.A. and Mahdi, S.S. (2017) *8th International Conference on Agricultural, Environment, Biology and Medical Sciences,* 21-22 December, 2017, Dubai (UAE), p. 123.

Hwang, D., Jeong, H.J., Kwon, J.K., Kim, H., Kang, S.Y. and Kang, B.C. (2014) *Plant Genet. Resour.,* **12:** S141–S145.

Hwang, J.N., Li, J., Liu, W.Y., An, S.J., Cho, H., Her, N.H., Yeam, I., Kim, D. and Kang, B. (2009) *Mol. Cells,* **27:** 329-336.

Igwe, D.O., Afiukwal, C.A., Acquaah, G. and Ude, G.N. (2019) *Hereditas,* **156**: 32.

Ilic, Z., Milenkovic, L., Durovka, M. and Kapoulas, N. (2011) *Proceding 46th Croatian and 6th International Symposium on Agriculture,* Opatija, Croatia, pp. 529-532. Inc., NewYork.

Ima-obong, I. D., Ndueso, M.A. and Kayode, P.B. (2017) *Notulae Scientia Biologicae,* **9**(1): 137.

Irikova, T., Grozeva, S. and Rodeva, V. (2011) *Acta Physiol. Plant*, **33:** 1559-1570.

Irulappan, I., Ponnuswamy, V. andVadievel, E. (1981) *South Indian Hort.*, **29:** 223.

Ivanic, J. (1963) *Sobrn, vys. sk. polnhsp v. Nitre. Fall.*,**7:** 69-77.

Jabeen, N. and Mirza, B. (2002) *Asian J. Plant Sci.*, **1:** 425-428.

Jadhav, P.B. and Gurav, N.P. (2019) *Ann. Biol. Res.*, **10**(1): 1-9.

Jaimez, R.E., Vielma, O., Rada, F. and Garcia-Nunez, C.(2000) *J. Agron. Crop Sci.*, **195:** 113-119.

Jakhar, R.K., Singh, A.K. and Kumawat, N. (2017) *Environ. Eco.*, **35**(1A): 290-294.

Janse, J. (1988) *Groentenen Fruit*, **43**: 32- 35.

Janse, J. (1989) *Acta Hortic.*, **244:** 123-132.

Jarlan, A., Oliveira, D.de, Gingras, J., De Oliveira, D. and Richards, K.W. (1997) *Acta Hortic.*, **437:** 335-339.

Jarret, R.L. and P. Dang. (2004) *Georgia J. Sci.*, **62:** 117-133.

Jarvis, W.R., Barrie, S.D. and Khosla, S.K. (1994) *Plant Dis.*, **78:** 926.

Jayakaran, M. (1972) *Curr. Sci.*, **41:** 849-50.

Jeon, H.J. and Chung, H.D. (1982) *J. Korean Soc. Hort. Sci.*, **23**: 3-60.

Jeong, H.J., Kwon, J.K., Pandeya, D., Hwang, J., Hoang, N.H., Bae, J.H. and Kang, B.C. (2012) *Mol. Breed.*, **29:** 349-360.

Jeong, Y.O. and Cho, J.L. (1996) *J. Korean Soc. Hort. Sci.*, **37:** 522-525.

Jimenez, J.M., Bustamante, E., Bermudez, W.and Gamboa, A. (1988) *Manejo Integrado de plagas* No.7, 19-28.

Jindal, S.K., Dhaliwal, M.S. and Meena, O.P. (2019) *Plant Breed.*, **00:** 1-23.

Jo, M.H., Matsubara, S., Kwak, S.S., Lee, E.M., Song, N.H. and Woo, LS. (1999) *J. Korean Soc. Hort. Sci.*, **40:** 665-668.

Jo, Y.D. and Kim, J.B. (2019) *Quantum Beam Sci.*, **3:** 7

Jo, Y.D., Kim, S.H., Hwang, J.E., Kim, Y.S., Kang, H.S. and Kim, S.W. (2016) *Hortic. Environ. Biotechnol.*, **57**(6): 606–614.

Johri, S., Singh, R.V. and Mishra, A.C. (2004) *Capsicum Eggplant Newslett.*, **23**: 61-64

Jones, J.B., Lacy, G.H., Bouzar, H., Stall, R.E. and Schaad, N.W. (2004) *Syst. Appl. Micro.*, **27**: 755-762

Joshi, A.K. and Nabi, A. (2015) Green *Farming*, **6**(1): 21-25.

Joshi, M.C. and Singh, D.P. (1975) *Indian Hort.*, **20:** 19-21.

Joshi, S. (1986) *Capsicum Newsl.*, **5**: 33-34.

Joshi, S. and Berke, T. (2015) *J. New Seeds*, **6** (2-3): 51-74.

Joshi, S. and Munshi, A.B. (2000) Summer school and Advances in hybrid seed production technology, 24 April- 25 May, 2000, in Division of Seed Science and Technology, IARI, New Delhi, India, pp.133-196.

Joshi, S., Thakur, P.C. and Verma, T.S. (1995) *Veg. Sci.*, **22**(2): 105-108.

Jovicich, E., Cantliffe, D.J., Hochmuth, G.J. (2000) In: *Proc. Florida State Hort. Soc.*, No.112, pp. 310-312.

Kacjan-Marsic, N. and Osvald, J. (1997) *Zbornik Biotechniske Fakultete Univerze v Ljubljani, Kmetijstvo,* No.69, pp. 141-146.

Kader, A.A., Zagory, D. and Kerbel, E.L. (1989) *Critical Rev. Food Sci. Nutr.*, **28:** 1-30.

Kamble, C. and Mulge, R. (2008) *Crop Research Hisar*, **36**: 277-280.

Kamble, C., Mulge, R., Madalageri, M.B. and Jadeesha, R.C. (2009) *Karnataka J. Agric. Sci.*, **22:** 155-157.

Kamienska, A. and Chrominski, A. (1971) *J. Exp. Bot.*, **22:** 572-574.

Kang, B.C., Yeam, I., Frantz, J.D., Murphy, J.F. and Jahn, M.M. (2005) *Plant J.*, **42**: 392-405.

Kang, W.H., Hoang, N.H., Yang, H.B., Kwon, J.K, Jo, S.H., Seo, J.K., Kim, K.H., Choi, D. and Kang, B.C. (2010) *Theor. Appl. Genet.*, **120**: 1587-1596.

Kanwar, M.S. and Sharma, O.C. (2010) *J. Hill Agric.*, **1**: 88-89.

Kapitsimadi, C. (1995) *Acta Hortic*, **379:** 237-243.

Kaul, D.L. and Sharma, P.P. (1988) *Veg. Sci.*, **15:** 171-180.

Kaur, S., Kaur, S., Srinivasan, R., Cheema, D.S., Lal, T., Ghai, T.R. and Chadha, M.L. (2010) *Pest Manag. Hortic. Ecosys.*, **16**(2): 148-155.

Kawecki, Z., Michalik, L. and Wierzbicka, B. (1998) *Acta Academiae Agriculturaeac Technicae Olstenens is,* Agricultura, No.65, pp. 147-154.

Khalil, R.K., Malash, N.M. and El-Sayed, M.M. (1987) In: *Scientific International Technical Development Symposium on Hungarian Paprika* (red pepper), 17-19 September, Kalocsa Szeged, 1987, pp. 100-113.

Khan, I. and Sridevi, O. (2018) *Int. J. Chem. Stu.*, **6**(5): 1002-1006.

Kim, B.S., Chung, 13.J. and Lee, W.S. (1975) *J. Korean Soc. Hori. Sci.*, **16:** 214-219.

Kim, H.K. and Ho, Q.S. (1978) *J. Korean Soc. Hori. Sci.*, **19:** 110-116.

Kim, J.Y., Jung, M., Kim, H.S., Lee, Y.H., Choi, S.H., Lim, Y.P., Min, B.W., Yang, S.G. and Harn, C.H. (2002) *J. Plant Biotechnol.*, **4:** 129-134.

Kim, N., Kang, W.H., Lee, J. and Yeom, S.I. (2019) *BioMed Res. Int.*, 1-12.

Kim, S., Park, M., Yeom, S. I., Kim, Y. M., Lee, J. M. and Lee, H. A. (2014) *Nature Genet.*, **46**: 470-478.

Klar, A.E. and Jadoski, S.O. (2004) *Irriga. Botucatu.*, **9:** 217-224.

Klar, A.E., Jadoski, S.O. and Lima, G.P.P. (2006) *Irriga.*, **11**(4): 441-447.

Knapp, S. (2002) *J. Exp. Bot.* **53:** 2001-2022.

Knavel, D.E. and Kemp, T.R. (1973) *HortScience*, **8:** 403-404.

Kohli, V.K., Dua, I.S. and Saini, S.S. (1981) *Scientia Hort.*, **15:** 17-22.

Komla, A.N. (2013) M.Phil. Crop Science (Unpublished), Submitted to Crop Science Department, University of Ghana, Legon.

Kosson, R., Adamicki, F., Horbowicz, M. and Dobrzanska, J. (1998) *Veg. Crops.Res. Bull.*, **49:** 121-129.

Kossowski, M. and Hortynska, E. (1984) *Reczniki Akademii Rolniczejw Poznaniu, Ogrodnictwo*, **150:** 45-54.

Kristjansson, K. and Rasmussen, K. (1991) *Acta Hortic.*, **288**: 173-179.

Kristof, L. and Bahadii, K.E. (1986) *Kertgazdasag*, **18:** 17-22.

Kubišová, S. and Háslbachová, H. (1991) *Acta Hortic.*, **288:** 364-370.

Kumar, M. and Verma, Vedanta, V. (2009) *Acta Hortic.*, **807**(807): 389-394.

Kumar, M.A., Nanjareddy, K., Sreevathsa, R., Ganeshan, G. and Udayakumar, M. (2009b) *Sci. Hortic.*, **119**(4): 362-370.

Kumar, P., Dube, S.D. and Chauhan, V.S. (2000) *Veg. Sci.*, **27:** 54-56.

Kumar, P.R., Lal, S.K., Thakur, P.C. and Sharma, S.R. (2009a) *Indian J. Plant Genet. Resour.*, **22**(1): 36-40.

Kumar, R., Kumari, P. and Kumar, S. (2016) *Int. J. Bio-resour. Stress Manag.*, **7(6):** 1290-1296.

Kumar, R.V., Sharma, V.K., Chattopadhyay, B. and Chakraborty, S. (2012) *Physiol. Mol. Biol. Plants.*, **18**(4): 357-364.

Kumar, S., Singh, V., Singh, M., Rai, S., Kumar, S., Rai, S. K. and Rai, M. (2007) *Sci. Hort.*, **111**: 197-202.

Kumar, U. and Chandra, G. (2014) *J. Hill Agric.*, **5**(1): 9.

Kumari, S. (2013) *Asian J. Hort.*, **8**(1): 280-284.

Kumari, S. and Sharma, M. (2014) *Int. J. Farm Sci.*, **4:** 76-80.

Kurubetta, Y. and Patil, A.A. (2009) *Karnataka J. Agric. Sci.*, **22:** 1058-1061.

Kwon, Y.S., Lee, Y.B., Park, S.K. and Ko, K.D. (1988) *Res. Rep/.RuralDev. Admn., Hort.*, Korea.

Kyle, M.M. and Palloix, A. (1997) *Euphytica*, **97**: 183-188.

La Red, F.C. and Cucchi, NJ.A. (1966) *India*, **221:** 41-51.

Lafortune, D., Beramis, M., Daubeze, A. M., Boissot, N. and Palloix, A. (2005) *Plant Dis.*, **89**: 501-506.

Lal, S. and Kanaujia, S. P. (2013) *Ann. Hort.*, **6**: 170-177.

Lang K. M., Nair, A. and Moore, K.J. (2020) *HortScience*, **55**(4): 550-559.

Lang, Y.Q., Yanagawa, S., Sasanuma, T. and Sasakuma, T. (2004) *Breeding Sci.*, **54:** 33-39.

Lankroo, G.M., Sharma, P.P. and Singh, P. (1990) *Indian Cocoa, Arecanut and Spices J.*, **14:** 81-83.

Lecompte, S.B. (1962) *Proc. J6t1, Int. Hort. Cong.*, **1:** 419.

Lee, A.C. and Liao, F.S. (2007) *Acta Hortic.*, 761: 533-538.

Lee, H.Y., Ro, N.Y., Jeong, H.J., Kwon, J.K., Jo, J., Ha, Y., Jung, A., Han, J.W., Venkatesh, J. and Kang, B.C. (2016) *BMC Genetics*, **17:** 142.

Lee, Y.M. and Shin, D.Y. (1989) *Korean J. Breed.*, **21**(2): 138-142.

Lefebvre, V., Daubeze, A.M., Voort, R.J., Peleman, J., Bardin, M. and Palloix, A. (2003) *Theor. Appl. Genet.*, **107**: 661-666.

Lefebvre, V., Kuntz, M., Camara, B. and Pallois, A. (1998) *Plant Mol. Biol.*, **3**: 785-789.

Lefebvre,V., Palloix, A. Caranta, C. and Pochard, E. (1995) *Genome*, **38**:112-121.

Leon, M. and Montalvo, M. (1986) *Cienciay Tecnica en la Agricultura, Riego y Drenaje*, **9:** 7-18.

Leonardi, C. (1994) *Colture Protette*, **23:** 75-80.

Leonardi, C. (I 996) *Colture Protette*, **25:** 61-65.

Leskovar, D.L., Cantliffe, D.J. and Stoffella, P.J. (1989) *HortScience*, **64**(6): 711-716.

Li, C.L. and Jiang (1990) *Acta Hortic. Senica*, **17:** 39-44.

Li, D., Zhao, K., Xie, B., Zhang, B. and Luo, K. (2003). *Plant Cell Rep.*, **21:** 785-788.

Li, S. (1989) *Zuowu Pinzhog Ziyuan*, No. 2, p. 48.

Lippert, L.F., Smith, P.G. and Bergh, B.O. (1966) *Botanical Rev.*, **32**(1): 24-55.

Liu, W., Parrott, W.A., Hildebrand, D.F., Collins, G.B. and Williams, *E.g.* (1990) *Plant Cell Rep.*, **9:** 360: 364.

Liu, W.Y. and Gniffke, P.A. (2004) *Capsicum Eggplant Newslett.*, **23**: 85-88.

Loboda, B.P. and Chuprikova, O.A. (1999) *Agrokhimiya*, **2**: 67-72.

Lodhi, Y., Chakravorty, S., Prasad, B.V.G. and Chandrakar, S. (2019) *The Pharma Innov. J.*, **8**(6): 791-794.

Lorenzo, P. and Castilla, N. (1995) *Acta Horticulturae* (Eds. R.Fernandez-Munoz, J. Cuartero and M.L. Gomoz).

Lorincz-Csapo, E., Verzar-Petri, G. and Bugar-Meszaros, K. (1982) *Herba Hungarica*, **21**: 73-84.

Lowery, D.T., Eastwell, K.C. and Smirle, M.J. (1997) *Annals Appl. Bio.*, **130**: 217-225.

Ludilov, V.A. and Ludilova, M.I. (1977) *Khimiya Sel'skom Khozyaistve*; **15:** 50-53.

Maalekuu, K., Elkind, Z., Tuvia-Alkalai, S., Shalom, Y. and Fallik, E. (2003) *Adv. Hort. Sci.*, **17:** 187-191.

Maboko, M.M. and Du Plooy, C.P. (2015) *HortScience*, **50:** 383-386.

Mahajan, B.V.C., Dhillon, W.S., Sidhu, M.K., Jindal, S.K., Dhaliwal, M.S. and Singh, S. P. (2016) *Indian J. Agric. Sci.*, **86**(6): 738-742.

Mahmoud, I. (2014) *J. Plant Prod.*, **5**(11): 1931-1943.

Manna, D.C. and Ray, P. (2015) *Int. J. Bioresour. Sci.*, **2**(2): 71-80.

Manolopoulou, H., Xanthopoulos, G. and Douros, N. (2010) *Biosyst. Eng.*, **106**(4): 535-543.

Mantur, S.M., Patil, H.B., Biradar, S.R., Patil, S.R. and Patil, D.R. (2007) *Karnatka J. Agric. Sci.*, **20**(3): 578-580.

Manzano, J.E. and Zambrano, J. (1995) *Proc. Interamerican Soc. Trop. Hort.*, **39**: 39-45.

Manzur, J.P., Fita, A., Prohens, J., Rodríguez-Burruezo, A. (2015) *PLoS ONE*, **10**(12): e0144142.

Marcelis, L.F.M., Brajeul, E., Elings, A., Garate, A., Heuvelink, E. and de Visser, P.H.B. (2004) *Acta Hortic.*, 691.

Marín, A., Ferreres, F., Tomás-Barberán, F.A. and Gil, M.I. (2004) *J. Agr. Food Chem.*, **52:** 3861-3869.

Markus, F., Daood, H.G., Kapitany, J. and Biacs, P.A.J. (1999) *Agric. Food Chem.*, **47:** 100-107.

Marti, H.R. and Mills, HA (1991) *J. Plant Nutrition*, **14:** 1165-1175.

Martinez-Raya, A. and Castilla, N. (1993) *Agricoltura Mediterranea*, **123:** 43-46.

Martins, K.C., Pereira, T.N.S., Souza, S.A.M., Rodrigues, R. and Junior, A.T.A. (2015) *Crop Breed. Appl. Biotechnol.*, **15** (3).

Matev,T. (1966) *Grad. Iozar Nauka*, **3:** 715-722.

Maynard, D.N., Lachw, W.H., Check R.M., and Vernell, H.F. (1962) *Proc. Amer. Soc. Hort. Sci.*, **81**: 385-389.

McArdle, R.N. and Bouwkamp, J.C. (1983) *J. Hered.*, **74:** 125-127.

McCammon, K.R. (1989) *Capsicum News Letter* No. 7, p. 51.

McLeod, D.G.R.(1981) *Proc. Ent. Soc. Ontario*, **112**: 29-32.

Medany, M.A., Hadsanein, M.K. and Farag, A.A. (2009) *Acta Hortic.*, **807** :121-26.

Medany, M.A., Mahmoud, M.H., Gaafer, S.A. and Abou-Hadid, A.F. (1990) *Egyptian J. Hort.*, **17**: 57-67.

Meena, O.P. (2017) Ph.D. thesis, Punjab Agricultural University, Ludhiana.

Meena, O.P., Dhaliwal, M.S. and Jindal, S.K. (2018) *Breed. Sci.*, **68**: 404-412.

Meir, S., Kanner, J., Akiri, B. and Hadas, S.P. (1995) *J Agric Food Chem.*, **43**: 1813-1817.

Meir, S., Philosoph-Hadas, S., Lurie, S., Droby, S., Akerman, M., Zauberman, G., Shapiro, B., Cohen, E. and Fuchs, Y. (1996) *Canadian J. Bot*; **74**: 870-874.

Mercado, J.A., Quesada, M.A., Valpuesta, V., Reid, M. and Cantwell, M. (1995) *Acta Hortic.*, **412:** 134-142.

Miccolis, V., Candido, V. and Marano, V. (1999) *Acta Hortic.*, **491**: 205-208.

Mihálka, V., Balázs, E. and Nagy, I. (2003) *Plant Cell Rep.*, **21**: 778-784.

Miranda, J.E.C.D. and Costa, C.P.Da (1988) *Pesquisa Agropecuaria Brasileira*, **23:** 1269-1277.

Mishra, A.C., Singh, R.V., Ram, H.H. (2004) *Capsicum and Eggplant Newsl.*, **23** : 45-48.

Moens, M. and Welvaert, W. (1981) *Rijksuniversiteit Genl.*, **46:** 805-881.

Mohamed, M.H. (1992) *Egyptian J. Hort.*, **19**: 33-42.

Mohammed, M., Wilson, L.A. and Gomes, P.I. (1991) *Trop. Agric.Trinidad*, **68:** 81-87.

Mojecka-Berova, M. and Kerin, V. (1995) *Bulgarian J. Agril. Sci.*, **1:** 253-257.

Molinari, A.F., Castro, L.R.de, Antoniali, S., Pornehaloempong, P., Fox, A.J., Sargent, S.A. and Lamb, E.M. (2000) In: *Proc. Florida Slate Hort. Soc*; No. 112, pp. 143-146.

Monteiro, C.E.S., Pereira, T.N.P. and Campos, K.P. (2011) *Crop Breed. Appl. Biotechnol.*, **11:** 241-249.

Monteiro, E.R., Bastos, E.M., Lopes, ACde.A., Gomes, R.L.F. and Nunes, J.A.R. (2010) *Ciencia Rural.*, **40**(2): 258-263.

Moor, J. (1989) *Zoldsegtermesztezi Kutato Intezet Bulletinje*, **22:** 123-125.

Morrison, R.A., Koning, R.E. and Evans, D.A. (1986a) In: Hand Book of Plant Cell Culture, Vol. 4, Techniques and Application, New York, USA, Macmillan Publishing Co., pp. 552-573.

Morrison, R.A., Koning, R.E. and Evans, D.A. (1986b) *J. Plant Physiol.*, **126**: 1-9.

Moscone, E.A. (1990) *Brittonia*, **42:** 147-154.

Muniz, J.L.L. and Almeida, J.I.L.De (1988) *Horticultura Brasileira*, **6**: 18-19.

Muniz, J.O.L. and Teixeira, L.M. (1987) *Horticultura Brasileira*, **5:** 37.

Munyon, I.P., Hubstenberger, J.F. and Phillips, G.C. (1989) *Biol. J. Tissue Cult. Assoc.*, **25**(3): 293-296.

Muro, J., Lamsfus, C. and Fernandez, M.A. (1994) *Investigacion Agraria, Produccion y Proteccion Vegetates*, **9:** 53-62.

Murphy, J.F., Blauth, J.R., Livingstone, K.D., Lackney, V.K. and Jahn, M.K. (1998) *Mol. Plant Micro. Inter.*, **11**: 943-951.

Murthy, D.S., Prabhakar, B.S., Hebbar, S.S., Srinivas, V. and Prabhakar, M. (2009) *J. Horti. Sci.*, **4**(2): 148-152.

Nabeshima, S., Yasutake, D. and Kitano, M. (2019) *Environ. Con. Bio.*, **57**(2): 29-38.

Nagalakshmi, S., Nandakumar, N., Palanisamy, D. and Sreenarayanan, V.V. (2001). *South Indian Hort.* **49:** 345-346.

Naik, L.B., Hebbar, S.S. and Doijode, S.D. (1997) *Seed Res.*, **24:** 154-155.

Naik, R.K. (2005) Ph. D, Thesis, Univ. Agric. Sci., Dharwad, Karnataka, India.

Naik, R.M. (2018) *Int. J. Curr. Microbiol. App. Sci.*, **7**(5): 2166-2171.

Navrot, J. and Levin, I. (1976) *Exp. Agric.*, **12:** 129-133.

Nederhoff, N.M., Rijsdijk, A.A. and Graaf, R.de (1992) *Sci. Hort.*, **52**: 283-301.

Negi, R., Thakur, S. and Mehta, D.K. (2019) *Int. J. Agric. Sci.*, **11**(2): 7748-7750.

Negi, R., Thakur, S. and Sharma, P. (2018) *Int. J. Curr. Microbiol. App. Sci.*, **7**(4): 2272-2281.

Nijamudeen, M.S., Singh, M., Khanna, M., Singh, B., Singh, R., Pandey, R.N. and Hasan, M. (2013) *Indian J. Hort.*, **70**(1): 54-59.

Nikolov, N. (1981) *Gradinarska i Loszrska Nauka*, **18**: 55-60.

Nkansah, G.O., Norman, J.C. and Martey, A. (2017) *J. Hortic.*, **4:** 216.

Novak, F.J. and Chemalo, V. (1975) *Bull. Vyjk. Ins. Zeitudfsky*, **19:** 95.

Nowak, T. J. (1980) *Acta Agrobotanica*, **33:** 73-80, 81-92.

Nunes, M.C.N. and Emond, J.P. (1999) *J. Fd. Quality*, **22:** 353-361..

Ochoa-Alejo, N. and Ireta-Moreno, L. (1990) *Sci. Hort.*, **42**: 21-28.

Odland, M.L. and Porter, P. M. (1941) *Proc. Amer. Soc. Hori. Sci.*, **38**: 385-388.

Ogundiwin, E.A., Berke, M., Massoudi, T.F., Black, L.L., Huestis, G., Choi, D., Lee, S. and Prince, J.P. (2005) *Genome*, **48**: 698-711.

Olareweju, J.D. (1988) *Haryana J. Hortic. Sci.*, **18**: 285-288.

Ombodi, A., Pek, Z., Szuvandzsiev, P., Taskovics, Z.T., Kohazi-kis, A., Kovacs, A., Darazsi, H.L. and Helyes, L. (2015) *Nor. Bor. Horti. Agrobo.*, **43**(2): 398- 403.

Orta, A.H. (1997) *Turkish J. Agric. Forestry*, **21:** 513-517.

Ortas, I. (2013) *Sci. Res. Essays*, **8**(23): 1048-1055.

Ortega, R.G., Espanol, C.P. and Zueco, J.C. (1992) *Plant Breed.*, **108**: 118-125.

Ozaki, H.Y. and Hortenstine, C. C. (1961) *Proc. Soil. Soc. Fla.*, **21**: 50-55.

Ozaki, H.Y. and Hortenstine, C.C. (1963) *Proc. Soil. Soc. Fla.*, **22:** 89-92.

Padem, H., Ocal, A. and Alan, R. (1999) *Acta Hortic.*, **491:** 241-246

Pákozdi, K., Taller, J., Alföldi, Z., and Hirata, Y. (2002) *J. Cent. Eur. Agric.*, **3**: 150-158.

Pandey, V., Ahmed, Z. and Kumar N. (2002) *Veg.Sci.*, **29**(1): 66-67.

Pandey, V., Chura, A., Arya, M.C. and Ahmed, Z. (2013) *Veg. Sci.*, **40** (1): 37-39.

Panigrahi, T., Singh, Y.K. and Choudhary, S. (2013) *J. Agric.Vet. Sci.*, **4**(6): 15-18.

Paran, I., Borovsky, Y., Nahon, S. and Cohen, O. (2007) *Isr. J. Plant Sci.*, **55:** 125-131.

Park, J.H., Jeon, G.I, Kim, J.M. and Park, E. (2012) *Food Sci. Biotechnol.*, **21**(2): 543-550.

Parra-Vega, V., Begona, R. M., Alicia, S. and Seguý Simarro Jose,´M. (2013) *Plant Cell Tiss. Organ Cult.*, **112**: 353-360.

Paschold, P.J. and Zengerle, K.H. (1996) *TASPO Gartenbaumagazin*, **5:** 20-21.

Paschold, P.J. and Zengerle, K.H. (1998) *Acta Hortic.*, 458.

Pereira-Dias, L., Vilanova, S., Fita, A., Prohens, J. and Rodríguez-Burruezo, A. (2019) *Hortic. Res.*, **6**: 54.

Petkov, M. (1972) Gradinarska i Lozarska Nauka, **9**: 43-51.

Pickersgill, B. (1977) In: *Capsicum. Comptes Rendues 3'eme Congre's Eucarpia Piment. Montfavet, Avignon: INRA*, pp. 27-37.

Pickersgill, B. (1991) In: *Chromosome engineering in plants: genetics, breeding, evolution*, part B. Elsevier, Amsterdam, pp.139-160.

Pickersgill, B. (1997) *Euphytica*, **96**: 129-133.

Pignon, P. and Granges, A. (1992) *Revue Suisse de Viticulture, d' Arboriculture el d' Horticulture*, **24:** 353-356.

Platenius, H., Jamison, F.S.and Thompson, H.C. (1934) *Cornell Bull.*, pp. 602.

Polizzi, G., Agosteo, G.E. and Cartia, G. (1994) *Acta Hort.*, No. 366, pp. 331-338.

Poncavage, J. (1997) *Org. Gar.*, **44**(2): 30-36.

Popovsky, S. and I, Paran. (2000) *Theor. Appl. Genet.*, **101:** 80-89.

Porporato, M., Pinna, M., Manino, A. and Marletto, F. (1995) *Apicoltore Moderno*, **86:** 99-112.

Poulos, J.M. (1994) *Plant Breed Abstr.*, **64:** 143-155.

Power, D. (2017) Ph. D thesis, Department of Plant Pathology, College of Horticulture, Vellanikkara, p.82.

Prabha, T.N., Bhagyalakshmi, N. and Tharanathan, R.N. (1998) *Zeitschrift fur Lebensmittel- Untersuchung und – Forschung*, **206:** 121-125.

Pratt, I-I. K. and Lyons, J. M. (1959) *Proc. 9th Int. Bot. Congress*, **2:** 308-09.

Prestes, A.M. and Goulart, L.R. (1995) *Annu. Rev. Phytopathol.*, **3**: 315-363.

Pudelski, T. (1988) *Roczniki Akadenici Rolniczej w Poznaniu, Ogrodnictwo.*, **189:** 159-178.

Pudelski, T. (1996) *Zeszyty Problemowe Postepow Nauk Rolniczych*, **429:** 259-265.

Qin, C., Yub, C., Shena, Y., Fang, X., Chen, L. and Mind, J. (2014) *Proc. Natl. Acad. Sci.*, **111:** 5135-5140.

Quaglitto, L. (1976) Informatore Agrario, **32**: 22517-18.

Quirin, E.A., Ogundiwin, E.A., Prince, J.P., Mazourek, M., Briggs, M.O., Chlanda, T.S., Kim, K.T., Falise, M., Kang, B.C. and Jahn, M.M. (2005) *Theor. Appl. Genet.*, **110:** 605-612.

Rabinowitch, H.D., Fahn,' A., Meir, T. and Lensky, Y. (1993) *Ann. Appl. Bid.*, **123:** 221-232.

Rai, N., Nath, A., Yadav, D.S. and Patel, K.K. 2004. In *National Seminar on Diversification of Agriculture through Horticultural Crops*. 21-23, February, 2004, NDRI, Karnal, p.30.

Rajic, M., Skoric, M. and Stojiljkovic, D. (1997) *Acta Hortic.*, **499:** 193-198.

Rana, M., Sharma, R., Sharma, P., Bhardwaj, S.V., Kumar, S., Brar, N.S., Sharma, M. and Dogra, R. (2015) *Eco. Env. Cons.*, **21**(S): 429-435.

Rao, P.G., Reddy, K.M., Naresh, P., Chalapathi, V.V. and Indiresh, K.M. (2016) *Green Farming*, **7**(5): 1024-1028.

Rastogi, K.B., Korla, B. N.and Saini, S.S. (1980) *Punjab Hort. J.*, **20**: 88-90.

Raw, A. (2000) *Ann. Bot.*, **85**: 487-492.

Reddy, M.K., Sadashiva, A.T. and Deshpande, A.A. (2002) *Indian J. Genet.*, **62**: 363-364.

Redgrove, H. S. (1933) *Spices and Condiments*, Pitman, London. Republic, **30**: 9-17.

Ribeiro, L.G., Lopes, J.C., Martins Filho, S. and Ramalho, S.S. (2000) *Hortic. Bras.*, **18:** 134-137.

Ristaino, J.B. (1990) *Phytopathology*, **80:** 1253-1259.

Riva, E.M., Rodrigues, R., Pereira, M.G., Sudré, C.P. and Karasawa, M. (2004) *Crop Breed. Appl. Biotechnol.*, **4**: 490-494,

Robinson, D.E., McNaughton, K. and Soltani, N. (2008) *HortScience*. **43:** 1492-1494.

Rodendurg, D. and Smid, J. (1973) *Groenten en Fruit*, **29**: 841.

Rodriguez-Rey, J.A., Romero, E., Gianfrancisco, S., David, S. delC. and Amado, M.E. (2000) *Revista de la Facultad de Agronomia, Universidad del Zulia*, **17:** 10-19.

Rogers, B.A. and Pike, L.M. (1989) *J. Rio Grande Valley Hort. Soc.*, **42:** 25-31.

Roldán-Serrano, A.S. and Guerra-Sanz, J.M. (2004) *J. Hortic. Sci. Biotechnol.*, **79**(5): 717-722.

Roy, S., Chatterjee, S., Hossain, M.A., Basfore, S. and Karak, C. (2019) *Int. J. Chem. Stu.*, **7**(1): 1777-1784.

Rubio, M., Nicolaý¨, M., Caranta, C. and Palloix, A. (2009) *J. Gen. Virol.*, **90**: 2808-2814.

Rudzinska, M. (1959) *Biul Warzyw*, **4:** 207-222.

Ruffel, S., Dussault, M.H., Palloix, A., Moury, B., Bendahmane, A., Robaglia, C. and Caranta, C. (2002) *Plant J.*, **32**: 1067-1075.

Ruffel, S., Gallois, J.L., Moury, B., Robaglia, C., Palloix, A. and Caranta, C. (2006) *J. Gen. Virol.*, **87**: 2089-2098.

Rui, R.L., Nie, Y.Q., Tong, H.Y., Shen, X.T. and Zhu, Q.T. (1989) *Jiangsu Agric. Sci.*, **8**: 30-31.

Ruijter, A.de, Eijnde, J. van den and Steen, J. vander (1991) *Acta Hortic.*, **288**: 270-274.

Ruiz, J.M., Castilla, N. and Romero, L. (2000) *J. Agric. Food Chem.*, **48**: 2925-2929.

Russo, V.M. (1999) *J. Veg. Crop Prodn.*, **5**: 59-66.

Rylski, I. (1972) *Isreal J. Agric. Res.*, **22**: 3140.

Rylski, I. (1973) *J. Amer. Soc. Hort. Sci.*, **98**: 149-152.

Rylski, I. and Halevy, A. H. (1974) *Acta Hortic.*, **42**: 55-62.

Rylski, I. and Spigelman, M. (1982) *Sci. Hort.*, **17**: 101-106.

Sadykov, I. M. and Mikhoet, Zh. Yu. (1981) *Tn1dy Kuban S-Kh. Institut.*, **197**: 114-18.

Salama, G.M. and Mohammedien, S.A. (1996) *Egyptian J. Hort.*, **23:** 1-10.

Salim, R.,AI-Subu,M.M.andIsmail,YS.s.(1995) *J. Environmental Sc. Jealth Part A, Environmen- tal Sc. Eng. Toxic Hazardous Subs. Control*, **20**: 1659-1684.

Salunke, R., Wadatkar, S.B., Kale, M.U., Rao, K.V.R. and Deshmukh, M.M. (2017) *Agric.Update*, **12** (TECHSEAR-4): 995-1001

Sanders, D.C., Kirk, H.J. and Van der Brink, C. (1980) AG-236. North Carolina Agricultural Extension Service, Raleigh.

Santos, L.D.A., Taveira, G.B., Ribeiro, S.D.F.F., Silva Pereira, L., Carvalho A.O. and Rodrigues, R. (2017) *Protein. Exp. Purif.*, **132:** 97-107.

Saray, T. (1988) *Acta Hortic.*, **220**: 503-509.

Schantz, M.L., Schreiber, H., Guillemaut, P. and Schantz, R. (1995) *FEBS Letters*, **358**: 149-152.

Schoch,P. G.(1972) *J. Amer. Soc. Hort. Sci.*, **97**: 46-64.

Schub, R. L. (1983) *J. Agric. Sci., U.K.*, **100**: 7-11.

Schulze, B. and Spiteller, D. (2009) *Chembiochem.*, **10**(3): 428-429.

Scivittaro,W.B., Melo, A.M.T.A.de, Azevedo-Filho, J.A.de, Carvalho, C.R.L. and Ramos, M.T.B. (1999) *Hortic. Bras.*, **17:** 147-150.

Shah, S.N.M., Tian, S.L., Gong, Z.H. and Arisha, M.H. (2014) *Annu. Rev. Res. Biol.*, **4:** 1106-1120.

Sharma S., Singh Y., Sharma A. and Sharma K.C. (2016) *Indian J. Hortic.*, **73**(4): 526-531.

Sharma, A. (2014) M.Sc. Dr. Y.S. Parmar University of Horticulture and Forestry, Solan, India.

Sharma, A., Kumar, M. and Kumar, N. (2017) *J. Hill Agric.*, **8**(1): 31-34.

Sharma, A., Kumar, M., Kumar, N., Dogra, R.K. and Kumari, S. (2019). *J. Pharmacog. Phytochem.*, **8**(2): 646-649.

Sharma, K.D., Cardona, J.A., Sibomana, M.S., Herrera, N.G.S., Nampeera, E. and Fallik, E. (2018) *J. Nutr. Food Technol.*, **1**(2): 56-62.

Sharma, N., Kohli, U.K. and Sinha, B.N. (1999) *J. Hill Res.*, **12:** 74-76.

Sharma, O.P., Sharma, P.P. and Chowfla, S.C. (1989) *Euphytica*, **42:** 31-33.

Sharma, P.P., Saini, S.S. and Koria, 13.N. (1981) *Veg. Sci.*, **8:** 32-36.

Sharma, S., Singh, Y. and Sharma, A. (2013a) *Bioinfolet*, **10:** 795-799.

Sharma, S.K. and Gajbhiye, R.P. (1998) *Abst. Silver Jubilee Nat. Sump. Emerging Scenario in Vegetable Research and Development*, 12-14 December, 1998, Varanasi, India, p. 108.

Sharma, S.R. and Sohi, H. S. (1981) *Indian J. Mycol. Plant Path.*, **11:** 30-34.

Sharma, V. (2016) *Int. J. Sci. Environ. Technol.*, **5**(4): 2120-2125.

Sharma, V.K. Semwal, C.S. and Uniyal, S.P. (2010) *J. Hortic. Fores.*, **2**(3): 058-065.

Sharma, V.K., Punetha, S. and Sharma, B.B. (2013b) *Afr. J. Agric. Res.*, **8:** 4088-4098.

Sharmila, V., Ganesh, S.K. and Gunasekaran, M. (2007) *Genet. Mol. Biol.*, **30**: 80-84.

Sharshir, A.F. L., El-Aidy, F. and Mottaleb Gazzy, A. A. (2019) *Int. J. Adv. Sci. Res. Manage.*, **4**(4): 330-339.

Shehata, S.A., El-Hassan, E.A.A. and El-Sayed, S.F. (1990) *Bull. Fac. Agric. Uni11Cairo*, **l:** 809-824.

Shestunov, I.I. and Zolotukhin, A.A. (1973) *Izvestiya voronezhskogo Gosudarstvernogo Padagogicheshogo Institu.*, **144**: 65-69.

Shifriss, C. (1997) *Euphytica*, **93**: 83-88.

Shifriss, C. and Frankel, R. (1971) *J. Hered.*, **62**: 254-256.

Shifriss, C. and Rylsky, I. (1972) *HortScience*,**7:** 36.

Shilpa, Thakur, S., Sharma, M. and Sharma, A.K. (2018) *Int. J. Curr. Microbiol. App. Sci.*, **7**(3): 952- 961.

Shipp, J.L., Whitfield, G.H. and Papadopoulos, A.P. (1994) *Sci. Hort.*, **57**: 29-39.

Shislakov, V.A. andMyatovich, V.Ya (1988) *Kartofel'i Ovoshchi*, **5**: 29-31.

Shoemaker, J.S.and Teskey, B.J.E. (1955) *Practical Horticulture, Jol.11 Wiley and Sons, Inc.*New.

Shrestha, S.L., Luitel, B.P. and Kang, W.H. (2011) *Hortic. Environ. Biotechnol.*, **52**(3): 278-283.

Shrivastava, A.K. (1996) *Advances Plant Sci.*, **9:** 171-175.

Shukla, C., Acharya, S.K., Bhakar, S.R. and Jamrey, P.K. (2019) *Hort. Int. J.*, **3**(2): 61-65.

Siddique, M.I., Back, S., Lee, J., Jo, J., Jang, S., Han, K., Venkatesh, J., Kwon, J., Jo, Y.D. and Kang, B. (2020) *Plants*, **9:** 396.

Sigge, G.O., Hansmann, C.F. and Joubert, E. (1999) *J. Food Quality*, **22:** 439-452.

Silber, A., Bruner, M. and Kenig, E. (2005) *Plant Soil*, **270**: 135-146.

Silva, M.A.G.de, Boaretto, A.E., Melo, A.M.T.de, Fernandes, H.M.G. and Scivittaro, W.B. (1999) *Scientia Agricola*, **56:** 1199-1207.

Simon, J. (1960) *Bodenkultur*, **11:** 208-227.

Sims, W. L., Collins, H.B. and Gladhill, B.L. (1970) *Calif. Agric.*, **24:** 4-5.

Sims, W.L., Ririe, d., Brendler, R.A., Snyder, M.I., Wright, D.N., Schweers, VH.' and Oterli, P. P. (1974) *Calif. Agric.*, **28:** 3-4.

Singh, A., Kaushal, A., Garg, S. and Chawla, N. (2017b) *Indian J. Hort.*, **74**(2): 292-294.

Singh, A.K., Singh, A.K., Gupta, M.J. and Shrivastav, R. (2003) *Prog. Hort.*, **7:** 212-216.

Singh, A.K., Singh, B. and Gupta, R. (2011) *Indian J. Agric. Sci.*, **81** (10): 973-975.

Singh, B. and Sirohi, N.P.S. (2006) *Acta Hortic.*, **710**: 339-342.

Singh, B., Singh, A.K. and Tomar, B. (2010) *Indian Hort.*, **55** (4): 31-33.

Singh, D., Kaur, S., Dhillon, T.S., Singh, P., Hundal, J.S. and Singh, G.J. (2004) *Acta Hortic.*, **659**: 515-521.

Singh, I. and Kaur, A. (2018) *Indian J. Agric. Res.*, **52**(4): 414-418.

Singh, I.L., Bhagachandani, P.M. and Thakur, M.R. (1964) *Indian J. Hort.*, **21:** 221-231.

Singh, M., Ameta, K.D., Dubey, R.B., Pareek, S., Meena, N.L., Meena, S. and Lal, S. (2017a) *Chem. Sci. Rev. Lett.*, **6**(22): 1189-1193.

Singh, R.D. and Singh, N.P. (1970) *Indian Fmg. Digest*, **3:** 43-45.

Singh, R.J. (2002) CRC Press, London, p. 488.

Singh, R.V. and Naik, L.I. (1990) *Haryana J. Hort. Sci.*, **19:** 168-172.

Sirohi, N.P.S. (2004) In: International Conference on Vegetables 'Vegetables for Sustainable Food and Nutritional Security in the New Millenium. November 11-14, 2002 at Bangalore. p. 162.

Sivek, P. and Kolev, V. (1991) *Fiziologiya na Rasteniyata*, **17:** 18-25.

Siwek, P. and Libik, A. (1999) *International Agrophysics*, **13:** 119-124.

Soares, R. S., Silva, H.W., Candido, W.S. and Vale, L.S.R. (2017) *Com. Sci.*, **8**(2): 247-255.

Solomon, A.M., Han, K., Lee, J.H., Lee, H.Y., Jang, S. and Kang, B.C. (2019) *PLoS ONE*, **14:** e0216886.

Somos, A. (1984) Akademiai Kiado, Budapest, p. 302.

Song, S., Zheng, X. and Xing, B. (1999) *Acta Agriculturae Boreali-Sinica*, **14**: 129-132.

Song, S.J. (1987) *Acta Horticulturae*, **14**: 185-91.

Sood, S. and Kaul, S. (2006) *Veg. Sci.*, **33:** 178-179.

Sood, S. and Kumar, N. (2010a) *Adv. Hortic. Sci.*, **24** (2): 122-128.

Sood, S. and Kumar, N. (2010b) *Int. J. Veg. Sci.*, **16**(4): 361-373.

Sood, S., Sood, R., Sagar, V. and Sharma, K.C. (2009) *Int. J. Veg. Sci.*, **15**(3): 272-284.

Spaldon, E. and Gromova, V. (1958) *Agron. Cast.*, **5:** 43-55.

Spaldon, E. and Ivanic, J. (1968) *Potash Rev.*, **16**: 40.

Spaldon, E. and Pevna, V. (1961) *Agron. Fath.*, **3:** 25-38.

Spaldon, E. and Pevna, V. (1962) *S11bm. Vys. Sk. Polnohosp v Nitne, Agron Fall.*, **5:** 209-25.

Sprague, G.F. and Tatum, L.A. (1942) *J. Am. Soc. Agron.*, **3:** 923-942.

Srinivas, K. (1980) *Mysore J. Agric. Sci.*,**16** :396-399.

Srinivas, K. and Prabhakar, B.S. (1982) *Veg. Sci.*, **9:** 71-74.

Sripichitt, P., Nawatu, E. and Shigenaga, S. (1987) *Jap. J. Breed.*, **37**(2): 133-142.

Srivastava, A., Sharma, V. R., Dishri, M., Dikshit, A., Mangal, M. and Kalia, P. (2019) *Indian J. Hort.*, **76**(1): 86-93.

Stanev, V.P., Velichkov, D.K. and Pandev, S.D. (1982) *Comptes Rendus de Academic Bulgare des Science*, 35: 1271-73.

Stoicescu, M. (1964) *Grad. Via Liv.*, **13**: 16-19.

Sui, X., Jiang, J., Wang, Z. and Zhu, Y. (1999) *Acta Hortic. Sinica*, **26:** 314-318.

Sun, T., Xu, Z., Wu, C.T., Janes, M., Prinyawiwatkul, W. and No, H.K. (2007) *J. Food. Sci.*, **72:** 98-102.

Susuri, L., Pireva, I., Rogova, B. and Zogjari, A. (1979) *Zastita Bilja*, **30:** 353-56.

Suwandi (1988) *Buletin Penelitian Hortikultura*, **16:** 26-33.

Suwandi and Subur (1986)*Buletin Penelitian Hortikultura*, **14**: 8-14.

Suzuki, K., Kuroda,T., Miura, Y. and Murai, J. (2003) *Plant Dis.*, **87**(7): 779-783.

Szoke, T. (1959) *Bull. Hort. Licye.*, **14:** 318-25.

Tai, T., Dahlbeck, D., Clark, E.T., Gajiwala, P., Pasion, R., Whalen, M.C., Stall, R.E. and Staskawicz, B.J. (1999a) *Proc. Natl. Acad. Sci. USA*. **96**: 14153-14158.

Tai, T., Dahlbeck, D., Stall, R.E., Peleman, J. and Staskawicz, B.J. (1999b) *Theor. Appl. Genet.*, **99**: 1201-1206.

Takte, R.L., Ambad, S.N., Kadam, U.S. and Dhawale, B.C. (2003) *Proc of All India on Sem Potential and Prospects for Protective Cultivation*, Organized by the Institute of Engineers, Ahmednagar pp 117-119.

Tanksley, S.D. (1984) *HortSci.*, **4**: 580-582.

Teodoro, R.E.F., Oliveira, A.S. deandMinami, K. (1993) *Scientia Agricola*, **50**: 237-243.

Thabuis, A., Palloix, A., Pflieger, S., Daubeze, A.M., Caranta, C. and Lefebvre, V. (2003) *Theor. Appl. Genet.*, **106**: 1473-1485

Thakur, A., Thakur, P.S. and Bhardwaj, J. (1998) *Seed Res.*, **25:** 25-30.

Thakur, A., Thakur, P.S., Surekha, A. and Kanaujia, S. (1999b) *Annals Agric. Res.*, **20:** 313-317.

Thakur, B.R. (1990) Ph.D. thesis, Department of Plant Pathology, CSK Himachal Pradesh Krishi Vishvavidyalaya, Palampur.

Thakur, G., Singh, A.K., Maurya, P.K., Patel, P. and Kumar, U. (2018) *J. Pharmacog. Phytochem.*, **SP1**: 78-81.

Thakur, K.S., Kashyap, S. and Mehta, D.K. (2013) *Green Farm.*, **4**(5)**:** 634-636.

Thakur, M., Sood, S. and Gupta, N. (2019) *J. Pharmacog. Phytochem.*, SP1: 66-73.

Thakur, P. C., Gill, H.S. and Bhagchandani, P.M. (1980) *Veg. Sci.*, **7:** 118-21.

Thakur, P.S., Thakur, A. and Kanaujia, S.P. (1999) *Annals Agric. Res.*, **20:** 331-334.

Thakur, P.S., Thakur, A. and Kanaujia, S.P. (2000) *Indian J. Agril. Sc.*, **70**: 543-545.

Thies, J.A. and Fery, R.L. (2000) *J. Am. Soc. Hort. Sci.*, **125**: 71-75.

Thies, J.A. and Fery, R.L. (2002) *J. Am. Soc. Hort. Sci.*, **127**(3): 371-375.

Thomas, J. R. and Heilman, M.D. (1964) *Proc. Amer. Soc. Hort. Sci.*, **85:** 419-25.

Thomas, T.A., Singh, R. and Prasad, R. (1983) *South Indian Hort., 30th Commemoration Issue,*

Thompson, H.C. and Kelly, W.C. (1957) *Vegetable Crops, McGraw-Hill Book Company, New York.*

Toivonen, P.M.A. and Bowen, P.A. (1999) *Canadian J. Plant Sci.,* **79:** 411-416.

Tong, N. and Bosland, P.W. (1999) *Euphytica,* **109**: 71-77.

Tran, N. H. and Byungsoo, K. (2010) *Hortic. Environ. Biotechnol.* **51**: 431- 439.

Trivedi, A.K. and Singh, V.K. (2015) National workshop cum seminar on emerging prospects of protected cultivation in horticultrural crops under changing climate. Precision farming development center, Lucknow.

Tsegay, D., Tesfaye, B., Mohammed, A., Yirga, H. and Bayleyegn, A. (2013) *Int. J. Biotechnol. Mol Bio. Res.,* **4**(7): 98-104.

Uffelen, J.A.M.Van and Bakker, J.C. (1987) *Groenten en Fruit,* **43**: 30-31.

Uffelen, J.A. M.Van, Bakker, J.C. and Janse, J. (1988) *Groenten en Fruit,* **43**: 37.

Uffelen, J. A. M.Van and Tanis, K. (1989) *Groenten en en Fruit,* 44: 35.

Uffelen, J. A. M.Van and Tanis, K. (1989) *Groenten en Fn1il,* **44:** 35.

Uffelen, J.A. M.Van., Bakker, J.C. and Janse, J. (1988) *Groenten en Fruit,* **43:** 37.

Ullasa, B.A., Rawal, R.D., Sohi, H.S., Singh, D.P. and Joshi, M.C. (1981) *Plant Dis,* **65:** 600-601.

Vallejo, C.F.A., Ceballos, L.H. and Agudelo, A.E. (1997) *Acta Agronomica,* **47:** 25-36.

Vallejos, C.E., Jones, V., Stall, R.E., Jones, J.B., Minsavage, G.V., Schultz, D.C., Rodrigues, R., Olsen, L.E. and Mazourek, M. (2010) *Theor. Appl. Genet.,* **121**: 37-46

VanDerwerken, J.W. and Wilcox-Lee, D. (1988) *HortScience,* **23**: 985-88.

Velev, B. (1967) *Yrad Iozan Nauka,* **4**: 47-56.

Velev, B. (1973) *Lozarska Nauka,* **10:** 71-76.

Vijay, O.P., Singh, D.P. and Yadav, I.S. (1979) *Veg. Sci.,* **6**: 5-10.

Vilarinho, L.B.O., Silva, D.J.H., Greene, A., Salazar, K.D., Alves, C., Eveleth, M., Nichols, B., Tehseen, S., Jr. Khoury, J.K., Johnson, J.V., Sargent, S.A. and Rathinasababapathi, B (2015) *J. Am. Soc. Hortic. Sci.,* **140**(6): 597-604.

Villalon, B. (1986) *HortScience,* **21:** 328.

Villalon, B., Dainello, F. D. and Bender, D. (1988) *HortScience,* **23:** 1094-1095.

Vogel, S. (1998) *Flora,* **193**: 113-131.

Wafer, L. (1699) Franklin, NY, p. 107.

Wahyuni, Y., Ballester, A.R., Sudarmonowati, E., Bino, R.J. and Bovy, A.G. (2011) *Phytochem.,* **72**(11-12): 1358-1370.

Walid, Q., Mohammad, M.J., Husan, N. and Remon, Q. (1999) *Communication in Soil Science and Plant analysis,* **30**: 2499-2509.

Wally, Y.A., Gandoar, M.A. and El Fade, M.S. (1971) *Res. Bull. Ainshams Univ.*, 582, pp. 28.

Wang, Y., Yang, M., Pan, N. and Chen, Z.H. (1991) *Acta Bot. Sin.*, **33**: 780-786.

Wani, K.P., Singh, P.K., Amin, A., Mushtaq, F. amd Zahoor, A.D. (2011) *Asian J. Sci. Tech.*, **1** (4): 56-61.

Wei, J., Zheng, J., Yu, J., Zhao, D., Cheng, Y., Ruan, M., Ye, Q., Yao, Z., Wang, R., Zhou, G., Yang, Y., Li, Z., and Wan, H. (2019) *J. Agr. Sci. Tech.*, **21**(3): 761-769.

Wiendi, F.M., Arthur, V. and Wiendi, T.A. (1996) *Revista de Agriculture (Piracicaba)*, **71:** 21-32.

Wu, C., Fan, S. and Luo, G. (1999) *Acta Agriculturae Universitatis Jiangxiensis*, **21:** 505-508.

Yadav, A. and Sharma, P.P. (1995) *Annals. Agril. Res.*, **16:** 442-444.

Yazawa, S., Yoneda, H., and Hosokawa, M. (2002) *Capsicum Eggplant Newslett.*, **21**: 52-55.

Yoon, J.Y., Green, S.K., Tschanz, A. T., Tsou, S.C.S. and Changa, L.C. (1989) *Int. Symp. on Integrated Management Practices*, pp. 86-98.

Yosepha, S., Hanna, Y., Kira, R., Eli, M. and David, B. (2007) *HortScience*, **42** (4):

Yu, I. W. (1985) M.Sc. thesis, Kyung Hee University, Seoul.

Yu, I. W. (1990) *Ph.D. thesis*, Kyung Hee University, Seoul.

Zečević, B., Đorđević, R., Balkaya, A., Damnjanović, J., Đorđević, M. and Vujošević, A. (2011) *Genetika*, **43**(2) : 209-216.

Zeist, A.R., Zanin, D.S. Camargo, C.K. de Resende, J.T. Ono E.O. and Rodrigues, J.D. (2018) *Hortic. Bras.*, **36:** 498-503.

Zende, U.M. (2008) *M.Sc. thesis*, College of Agriculture, Dharwad, University of Agricultural Sci., Dharwad.

Zhang, B., Huang, S., Yang, G., and Guo, J. (2000) *Euphytica*, **113**: 155-161.

Zhang, Y.L., Jia, Q.L., Li, D.W., Wang, J.E., Yin, Y.X. and Gong, Z.H. (2013) *Int. J. Mol. Sci.*, **14**: 8985-9004.

Zhang, Z. (1999) *China Vegetables*, **6:** 14-18.

Zhu, Y.X., Ou Yang, W.J., Zhang, Y.F. and Chen, Z.L. (1996) *Plant Cell Rep.*, **16:** 71-75.

Zijlstra, S., Purimahua, C. and Lindhout, P. (1991) *HortSci.*, **26:** 585-586.

Ziv, O., Shifris, C., Grinberg, S., Fallik, E. andSadeh, A. (1994) *Hassadeh*, pp. 526-532.

Znidarcic, D. and Osvald, J. (1999) *Zbornik Biotehniske Fakultete Univerze v Lyubljani. Kmetijstvo*, **73**: 59-64.

Zornoza, P., Caselles, J. and Carpena, O. (1989) *J. Plant Nut.*, **12:** 306-317.

SALAD CROPS

LETTUCE

G. Kalloo, V.A. Parthasarathy, P. Karmakar, B.K. Singh and Tridip Bhattacharjee

1.0 Introduction

Lettuce (*Lactuca sativa* L.) is an important member of the Asteraceae or Compositae family, with very successful and diverse group of plants having global distribution. It is a major fresh vegetable and its leaves are commonly found in salad mixtures and sandwiches. In some eastern countries like China and Egypt, stems instead of leaves of lettuce are consumed, either cooked, raw, pickled, dried, or as a sauce. As a cool-season crop, lettuce is extensively grown on all the continents, particularly in temperate and subtropical regions. World lettuce production was at more than 26.50 million tons on about 1.2 million ha in 2017. About two-thirds of the total production area in the world is in Asia. China is the largest lettuce producer and accounts for about half of the world's total production and area, primarily for lettuce stems. Lettuce is also grown in large areas in India, Japan, Mexico, and Turkey.

Lettuce is one of the most important vegetable crops in temperate countries. It also occupies an important position in tropical and subtropical countries. It is being grown for over 2500 years and various forms of lettuce are known since 500 BC. In temperate countries, mostly it is grown under glass and plastic but sometimes it is grown as outdoor crop also. Lettuce occupies the largest area among salad crops and thus it is the most important salad crop of the world.

2.0 Composition and Uses

2.1.0 Composition

Lettuce is rich in vitamin A and minerals like calcium and iron. It also contains protein, carbohydrates and vitamin C. There is variation in nutritive values in different types of lettuces, cos or leaf types being the richest followed by butterheads and crispheads. The increased amounts are due to the increased number of green leaves exposed to light in the respective type (Ryder, 1979a). The nutritive composition of lettuce is shown in Table 1.

Table 1: Composition of Lettuce (per 100 g of edible portion)*

Moisture	93.4 g	Chlorine	23.0 mg
Fat	0.3 g	Calcium	50.0 mg
Fibre	0.5 g	Phosphorus	28.0 mg
Protein	2.1 g	Sodium	58.0 mg
Minerals	1.2 g	Sulphur	27.0 mg
Carbohydrates	2-5 g	Vitamin A	1650 IU
Magnesium	30.0 mg	Thiamine	0.09 mg
Iron	2.4 mg	Riboflavin	0.13 mg
Potassium	33.0 mg	Nicotinic acid	0.5 mg
Vitamin C	10.0 mg		

* Aykroyd (1963).

The nutritional value of lettuce varies greatly with type (USDA, 2005). Usually, Crisp-head types are reservoir of nutritional value whereas the Cos and Cutting types have appreciable amounts of ascorbic acid, vitamin A, vitamin K, folate, and the carotenoids β-carotene and lutein along with zeaxanthin. The Cutting types are especially enriched with vitamin A and β-carotene, having about 15-fold higher amounts than Crisphead types. Cos types are relatively rich in vitamin C and folate compared to the other lettuce types. Darker green and red leaves of certain lettuce varieties provide more overall nutrition than lighter green ones (Bunning and Kendall, 2012). A comparative nutritional value is presented in Table 2.

Table 2: A Comparative Nutritional Value of Various Types of Lettuce*

Nutrients	*Type of Lettuce*					
	Butter-head	*Cos*	*Crisp-head*	*Cutting (Green Leaf)*	*Cutting (Red leaf)*	*Stalk*
Protein (g)	1.35	1.23	0.90	1.36	1.33	0.85
Sugars, total (g)	0.94	1.19	1.76	0.78	0.48	–
Fatty acids, total saturated (mg)	29	39	18	20	–	–
Fatty acids, total mono-unsaturated (mg)	8	12	6	6	-	-
Fatty acids, total poly-unsaturated (mg)	117	167	74	82	–	–
Phytosterols (mg)	–	–	10	38	–	11
Calcium (mg)	35	33	18	36	33	39
Iron (mg)	1.24	0.97	0.41	0.86	1.20	0.55
Magnesium (mg)	13	14	7	13	12	28
Phosphorus (mg)	33	30	20	29	28	39
Potassium (mg)	238	247	141	194	187	330
Sodium (mg)	5	8	10	28	25	11
Zinc (mg)	0.20	0.23	0.15	0.18	0.20	0.27
Copper (mg)	0.016	0.048	0.025	0.029	0.028	0.040
Manganese (mg)	0.179	0.155	0.125	0.250	0.203	0.688
Selenium (mcg)	0.6	0.4	0.1	0.6	1.5	0.9
Carotene, beta (mcg)	1987	3484	299	4443	4495	–
Lutein + zeaxanthin (mcg)	1223	2312	277	1730	1724	–
Ascorbic Acid (mg)	3.7	24.0	2.8	18	3.7	19.5
Thiamin (mg)	0.057	0.072	0.041	0.070	0.064	0.055
Raiboflavin (mg)	0.062	0.067	0.025	0.080	0.077	0.070
Niacin (mg)	0.357	0.313	0.123	0.375	0.321	0.550
Pantothonic acid	0.150	0.142	0.091	0.134	0.144	0.183
Vitamin B-6 (mg)	0.082	0.074	0.042	0.090	0.100	0.050
Total folate (mcg)	73	136	29	38	36	46

Nutrients	Type of Lettuce					
	Butter-head	Cos	Crisp-head	Cutting (Green Leaf)	Cutting (Red leaf)	Stalk
Vitamin A (IU)	3312	5807	502	7405	7492	3500
Vitamin E (mg)	0.18	0.13	0.18	0.29	0.15	-
Tocopherol, gamma (mg)	0.27	0.36	0.09	0.37	0.24	-
Vitamin K (mg)	102.3	102.5	24.1	173.6	140.3	-

* USDA (2005).

2.2.0 Uses and Health Benefits

Its tender leaves and heads are chopped and used as salad with salt and vinegar. However, vitamin C is almost entirely lost if it is cooked. Lettuce is also known to be a sedative, diuretic and expectorant. Most lettuce is used in salads, either alone or with other greens, vegetables, meats and cheeses. Romaine lettuce is often used for Caesar salads. Lettuce leaves can also be found in soups, sandwiches and wraps, while the stems are eaten both raw and cooked. Lettuce has low calorie and zero fat and is considered as good source of vitamins and minerals. It is helpful in losing body weight, to reduce anxiety, to protect neuronal cell, maintain heart health due to the presence of omega-3-faty acid, to cure sleeping disorder and insomnia, can prevent cancer cell growth and also to lower cholesterol level in blood.

3.0 Origin, Domestication, Taxonomy and Genepool

Lettuce was first cultivated about 4500 BC. According to Ryder and Whitaker (1976), all the four species (*Lactuca sativa, L. serriola, L. varisa* and *L. saligna*) originated in the Mediterranean region. Lettuce was most likely introduced into India wither by the Portuguese or by the British during the 16th century.

Cultivated lettuce is believed to have been domesticated in the Mediterranean region from the wild species *Lactuca serriola* L. (Harlan, 1992) with the center of origin apparently southwest Asia (De Vries, 1997). Lettuce is known to have been cultivated at least 4,500 years B.C. as long-leaved Costype lettuce was depicted on the walls of Egyptian tombs (Lindqvist, 1960).

Before the domestication by humans, lettuce generally grew wild. Though it is still not quite clear about the progenitor of the modern day lettuce and which species were involved in the evolution that led to modern-day lettuce. But, it is assured to be that the *L. serriola* is one of or the only direct ancestor(s) (Kesseli *et al.*, 1991). The chromosome morphology of *L. sativa* and *L. serriola* are very similar, and they also hybridize each other easily and freely. Some taxonomist considered these two taxa as subspecies of the same species. It is almost certainly that changes in *L. serriola* caused by mutations led to the appearance of forms that were favoured by humans, particularly forms without spines on stems and leaves and with large seeds. They were then selected for use and further modified to fit human needs.

The existence of the primitive forms of the lettuce in the Middle East provides strong support for the idea that lettuce probably originated in the eastern

Uses of Lettuce

Different Recipes of Lettuce in Indian Cuisine

Mediterranean basin. Another piece of evidence is found in Egyptian tomb paintings dated from the Middle Kingdom, about 4,500 years ago. The cos lettuces are distributed close to the likely centre of origin and have considerable variation in leaf shape and length, flat and erect stature, open and closed heads, texture, and colour. It is likely that leaf or cutting lettuces, butterhead, Latin types, and Batavia-type crisphead were all selected from this rich source of variability (Ryder, 1999).

It belongs to the family Compositae and genus *Lactuca* having 8, 9 and 17 pairs of chromosomes. There are several wild species of *Lactuca* which are mostly used as source of resistance to insect pests and diseases.

L. sativa

It is the most diverse species having many forms that are nevertheless completely interfertile. It is quite likely that this species has been derived through selection from hybrids between *L. serriola* and a third species (Lindqvist, 1960).

L. serriola

This has usually small seeds; reflexed involucre; spines, shoot formation at the stem base; relatively rarrow leaves that many have entire or toothed margins and a rosetting tendency.

L. saligna

It is similar to *L. serriola* in several respects but differs in having very narrow, toothed leaves and bearing sessile flowers on a spike compared to pedicelled flowers on pani-cles as in *L. serriola*.

L. virosa

It is morphologically distinct. It has large flattened seeds; strongly rosetting; blue green leaves; includes biennial as well as annual forms.

Lactuca sativa, L. serriola, L. virosa and *L. saligna* belong to the group with 9 pairs of chromosomes and cross with varying degree of success with each other. The isolation of these species within the genus is possibly due to genetic barriers.

The primary genepool of *L. sativa* is constituted by its various cultivars, primitive landraces and by wild species with no crossing barriers that is the cosmopolitan species *L. serriola*, along with *L. aculeata, L. scarioloides, L. azerbaijanica, L. georgica, L. altaica* occurring in Asia and by *L. dregeana* from South Africa (Zohary, 1991). *Lactuca saligna* belongs to the secondary gene pool. The tertiary gene pool includes L. virosa and some other wild species which can be crossed with *L. sativa* only with difficulty (DeVries, 1997).

There are strong incompatibility barriers existing among the cultivated *Lactuca* species and the 17 chromosome species (Ryder and Whitaker, 1976). *L. sativa* and *L. serriola,* salad crops are most closely related and cross readily with each other. Both cross with difficulty with the other two species; *L. sativa* should be used as the male in a cross with *L. saligna* (Lindqvist, 1960). According to Thompson and Ryder (1961), the *sativa-virosa* cross yields sterile hybrid.

3.1.0 Classification and Morphological Types of *L. Sativa*

The species *L. sativa* is categorized by a very high genetic diversity resulting due to its polyphyletic origin and domestication syndrome (Kesseli *et al.*, 1991). A comprehensive survey of the cultivars of cultivated lettuce and its classification was provided by Rodenburg (1960). In recent time an inclusive overviews of taxonomic and phenotypic analyses of lettuce cultivars were presented by DeVries (1997) and Mou (2008). The crop comprises seven main groups of cultivars which also includes oilseed lettuce and phenetically different from each other. They are generally described as morphotypes. Different morphotypes of lettuce are as follows:

1. Butterhead lettuce (var. *capitata* L. *nidus tenerrima* Helm): A heading type with soft and tender leaves, eaten raw. It is most popular in England, France, the Netherlands and other western and central European countries. In last few decades many cultivars have been developed and grown in the USA.
2. Crisphead lettuce (var. *capitata* L. *nidus jäggeri* Helm) or Iceberg type: A heading type with thick crisp leaves and flabellate leaf venation, eaten raw. It is mainly cultivated in the USA. Nevertheless, it is also cultivated now in western and central European countries, including the Netherlands, the United Kingdom, France, Spain, Belgium, Germany, Poland and the Czech Republic, as well as in Japan, China, and Australia.
3. Cos lettuce (var. *longifolia* Lam., var. *romana* Hort.in Bailey): Plants with tall loose heads, which are sometimes tied up; oblong rigid leaves with a prominent midrib running almost to the apex, are eaten raw or cooked. The name of this morphotype is due to the Greek island Cos (Kos), from where it is evolved and the type has long been cultivated there. Cos lettuce is most common and popular in the Mediterranean countries of Europe, Western Asia and North Africa. Many landraces of this type supposed to have originated mainly from Egypt, Iran, Turkey and Syria.

***L. sativa, L. serriola* (1°) (Left), *L. saligna* (2°) (Middle), and *L. virosa* (3°) (Right).**

4. Cutting lettuce (var. *acephala* Alef., syn. var. *secalina* Alef., syn. var. *crispa* L.): Also known as gathering lettuce, loose-leaf, picking lettuce. These are non-heading type harvested as whole, open rosettes, occasionally as separate leaves, eaten raw. Cutting lettuces have been very popular

Morphotypes of Lettuce

Butterhead Lettuce

Crisphead Lettuce

Cos or Roman Lettuce

Cutting Lettuce or Loose Leaf Lettuce

Morphotypes of Lettuce

Stalk Lettuce or Asparagus Lettuce

Latin Lettuce

in the U.S., Italy, France, the Czech Republic and Slovakia. This type of lettuce is extremely heterogeneous. Cultivars of this segment may have entire, curled or fringed leaves, from non-lobed to deeply incised margins. The leaves are generally elongated or broad, having various shades of green, and various patterns and with different intensities of anthocyanin pigmentation.

5. Stalk (Asparagus) lettuce (var. *angustana* Irish ex Bremer, syn. var. *asparagina* Bailey, syn. *L. angustana* Hort. in Vilm.): It is also known as stem lettuce. Plants are with swollen stalks, which are eaten raw or cooked like asparagus. Leaves can be eaten raw in a very young stage or cooked like spinach. There are two types recognized within this group. The Chinese cultivars have light grey leaves resembling cos lettuce leaves; the second type has long lanceolate leaves with pointed apices. The stalk lettuce originated in Tibet, which would account for its extensive cultivation in China, in the Pamirs and India. But, the lettuce illustrated in Egyptian tombs is also stalk lettuce and dates back to about 2500 B.C.
6. Latin lettuce (without nomenclature): Plants have loose heads with thick leathery leaves, dark green color and are eaten raw. It is mainly cultivated in the Mediterranean countries, including North Africa, and in South America.
7. Oilseed lettuce: Because of the bitter taste of its leaves, this type is not eaten as a vegetable. Oilseed lettuce is characterized by a high percentage (35 per cent) of oil in the seeds, which is used for cooking. The oil contains Vitamin E, an essential nutrient. In Egypt, cultivation of oil-producing forms has continued to the present time.

4.0 Botany

It produces taproot which grows on the surface of the soil. Large number of leaves are produced of which outer leaves are usually green to dark green and inner leaves are light green to whitish which are tender and crisp. Inner tender leaves have better quality than outer leaves.

The inflorescence of lettuce (capitulum), contains approximately 24 florets. They are highly developed for self-pollination and the crop is therefore largely selffertilizing. However, some cross-pollination, up to 5 per cent, can be observed between lettuce cultivars. Inflorescence is head or capitulum consisting of a large number of flowers. Numerous small flowers remain closely together on a receptacle; floret with a single ligulate yellow petal, 5 stamens and 2 carpels. There is very high degree of self-pollination. Pollen and ovule are usually highly fertile. Anthesis starts at sunshine and flower fully opens by late morning. Pollination takes place from 9.0-9.15 a.m. (Pal, 1975). Fruit is a typical achene and seed exalbuminous.

5.0 Types and Cultivars

The variation in lettuce is limited primarily to vegetative characters like leaf length, shape, colour, texture and size, and heading type. The early cultivars were mostly pointed, narrow-leaves and non-heading types. Butterhead and crisphead types, developed later, are at present predominant in Europe and United States, respectively. It is not known when the hearting character first appeared, the first definite evidence for the existence of head lettuce being available in 1543. In Europe, butterhead cultivars developed from temperate and summer cultivars. Both butterhead and loose-leaf types are used for glass house production in France, Holland and UK. In the recent years, the popularity of crisphead types has increased in France, Spain, and North Africa. The crisphead lettuce is also imported to Europe from the United States (Ryder and Whitaker, 1976).

As early as 1904, Tracy classified the American types and recognized over 100 distinct varieties, many of which still find mention in the seed catalogues. The classes and sub-classes proposed by him are:

Class I: Butterhead varieties

Sub-class (i) Cabbage-heading varieties

(ii) Bunching varieties

Class II: Crisphead varieties

Sub-class (i) Cabbage-heading varieties

(ii) Bunching varieties

Class III: Cos varieties

Sub-class (i) Spatulate-leaved varieties

(ii) Lanceolate-leaved varieties

(iii) Lobed-leaved varieties.

Thompson (1951) and Rodenburg (1960) recognized six morphological types of lettuce: crisphead, butterhead, cos or romaine, leaf or bunching, stem and latin. A brief description of each type is described below:

5.1.0 Types

5.1.1 Crisphead

The cultivars are heading types with wrinkled non-wrapper leaves, round, large head (up to 1 kg), brittle-textured and tightly folded, with green outer and white or yellowish inner leaves, having excellent shipping and handling abilities. It includes most important commercial cultivars like New York 515, Imperial 44, 152, 456 (also called Cornell 456), 615 and 847. Great Lakes and Pennlake cultivars are commercially grown in the United States, Europe and many parts of the world including India.

5.1.2 Butterhead

The cultivars form relatively small, loose heads with broad, oily, crumpled, soft-textured leaves. They vary considerably in colour, size and appearance. They do not withstand shipping and handling as in crisphead. This class includes cultivars like Big Boston, White Boston, May King, Salamander and Waya-head. Bibb type

Cultivars of Lettuce

La Brillante (Left), Latin/Romaine (Middle), Merlot/Red Leaf (Right)

Green and Red Leafy Type

Stem Type

Cos Type

Iceberg Type

Butterhead Type

is small and dark green, with reddish margins and is also known as Limestone. It is considered by many as the highest quality lettuce grown in the United States. Butterhead types are most popular in Northern Europe and include summer and winter types. The English cultivars Borough Wonder, Cobham Green and Avondefiance; Dutch cultivars Reskia, Wonder Van Voorbing and May King and Belgian cultivar Hilde are adapted to warm weather.

5.1.3 Cos or Romaine

This group is not very popular and the cultivars have elongated leaves forming a loaf-shaped head; outer leaves are dark or light green with heavy ribs; inner leaves are finer and lighter in colour; coarse in appearance but eating quality is quite good. The principal cultivars are Eiffel Tower, Paris Island, Paris White and Dark Green Valmaine.

5.1.4 Leaf or Bunching

The cultivars in this class are non-heading or leaf type; colour, heat sensitivity and quality vary but all produce a rosette of leaves. The best known cultivars are Black-seeded Simpson, Prizehead, Australian and Salad Bowl for outdoor growing and Grand Rapids for green house production. This type is grown in the United States, Australia, Europe and also in India.

5.1.5 Stem

It is also called 'celery lettuce' and is grown for its thick stem which is peeled and eaten either raw or cooked as vegetable. The leaves are edible but inferior in quality to the leaves of the other types. The cultivar Celtuce is large-leaved, thick-stemmed and is grown in the United States and Europe. In general, stem lettuce is grown principally in the orient.

5.1.6 Latin

It resembles butterhead in forming loose head but the leaves are elongated and similar to cos. It is grown mainly in the Mediterranean countries and in South America. The cultivar Gallega is known as the first source of lettuce mosaic resistance (Van der Pahlen and Crnko, 1965). Other cultivars are Criolla, Verde, Criolla Blenca and Madrilene (Ryder, 1979a).

5.2.0 Cultivars

In India, cultivars Great Lakes (crisphead), Chinese Yellow and Slobolt (leaf type) are recommended (Choudhury, 1972). Different seed companies have also listed cultivars, Big Boston, Great Lakes, Black-seeded Simpson, Golden Ball, Little Gem, Iceberg, Imperial, Paris Cos and Wonderful, *etc.* (Chauhan, 1972).

In Netherlands, cultivars Norden and Plaza proved to be best for spring culture in glass house (Cools and Stijger, 1987) and cultivar Avanti has been reported to give good results both in open and under protected cultivation (Rijbrock, 1987). Some other reported lettuce cultivars are:

Tapaas, Salina and Ostinata (Stolk and Cools, 1981); Caton, 70-52, 70-53, INRN No. 4, Vilmorin 822 and Estella (Peron, 1975); Pascal, Columbus, Polo and Martine

(Dawson, 1982) and Salinas (Ryder, 1979b); Veronica (Vecchia *et al.*, 1989a); Gloria (Vecchia *et al.*, 1989b); Olympus (Jong and Dijk, 1990); Pacific (Ryder and Robinson, 1991) and Cortina (Jong and Dijk, 1990).

During the period between 1979 and 1995, 480 lettuce accessions from several countries, including the Netherlands, UK, Germany, Czech Republic, Russia, Italy, USA and Romania, were introduced into Bulgaria and studied along with 52 native forms and lines (Neikov, 1998). For spring production, the following cultivars (var. *capitata*) were good: Kitty from the Netherlands, Futura from Germany, Gordon Rouge from France and Bulgarian lines SN406, SN501 and SN504. Cultivars suitable for early spring harvest in non-heated green houses were also identified. Lines 139, 821 and 112 of the iceberg type were suitable for early spring cultivation. Line 139 showed high resistance to fungal diseases. Most of the Romaine-type cultivars were considered poorer. Wilhelm (1998) evaluated fifteen cultivars of lettuce in field and green house trials at Heidelberg in 1996 and 1997, respectively. Yields in field trials in 1996 were in the range 0.84-1.21 kg/m^2 and in 1997, 1.02-1.49 kg/m^2. Cultivars varied significantly in quality, with Cirilla and Medaillon ranking highest. In the green house, yields were in the range 0.85-1.0 and 0.7-0.92 kg/m^2, in 1996 and 1997, respectively. Arielle, Gala, Medaillon, Trophy and JW 1012 had the highest quality under green house conditions.

Behr (1993) studied the growing German demand for variegated types of lettuce, particularly Lollo Rossa. In 1992 the market volume (home production and net imports) was between 30-35000 t as compared to 180000 t of ordinary lettuce and 1000000 t of iceberg lettuce. Variegated lettuce has a rather longer growing season than ordinary green cultivars and seasonal variation in German prices and sales are analysed. Mainly from Italy, Spain and France the variegated lettuce was being exported.

The following are some of recently developed lettuce cultivars:

Cultivars 9547 and 9542

Lettuce cultivars 9547 and 9542 were identified for good field growth and vigour, bolting resistance and high yields under high temperature conditions in summer. These are ideal varieties for summer season leaf lettuce production in Shanghai and adjacent areas of the middle and lower reaches of the Yangtze River in China. They were significantly more resistant to heat and bolting than the leading domestic cultivar Guangdongruanwei and leading local variety Cuiye, and had quite high yields. The cultivars 9547 and 9542 are characterized by strong growth, late bolting and high yield under high temperature conditions. Compared with Guangdongruanwei and Cuiye, 9547 and 9542 bolted 12.6-18.7 and 9.8-15.9 days later, respectively. Yield per unit area under field cultivation in summer reached 931 and 922 kg/666.7 m^2 for 9547 and 9542, respectively, with an average yield 60.57 per cent and 92.28 per cent higher than Guangdongruanwei and Cuiye, respectively. Yield under substrate cultivation of 9547 and 9542 reached 1594 and 1174 kg/666.7 m^2, respectively, or 82.92 per cent and 34.73 per cent higher than Guangdongruanwei, respectively (Gong, 1998).

Salma (A new summer-autumn lettuce cultivar)

Salma is a mid-early cultivar with a potential yield of over 30 t/ha. It was obtained by hybridization and pedigree selection among lines identified among a collection of 150 local lines and varieties evaluated at Vidra during 1985-92. Heads are large or very large (420-500 g), heavily compacted and devoid of bitter taste. It is photoperiod insensitive, suitable for mechanized harvesting and tolerant of *Bremia lactucae* under natural infection (Iordan and Luca, 1998).

Svetlana (Nathasha) (Asparagus Lettuce)

Svetlana (Narasha) is a cultivar *L. sativa* var. *asparagina* selected from Chinese cultivar from the Ulan Bator area of Mongolia in 1979 and grown in the Moscow region of Russia. In terms of green leaf yield it is a mid-season type and in terms of stem yield a late type (93 days from emergence to stem maturity in the Moscow region). Typically its height is 93.5 cm, stem diameter 7.8 cm, leaf number 66, leaf length 38 cm, leaf width 7.5 cm, and green leaf yield 580 g per plant. Mean yield in the Moscow region with April sowing was 38 t/ha compared with 30 t/ha for the standard variety Duganskii. The leaves contain 94 per cent water, 1.1 per cent protein, 1.3 per cent carbohydrate and 0.4 per cent oil (Kononkov, 1996).

Impact

This cultivar was submitted for the protection of plant variety rights in Australia. It originated from the controlled pollination of Jackpot by Narromar, with 5 subsequent backcrosses to Jackpot of progenies selected for resistance to *Bremia lactucae*. Impact is resistant to this disease, possessing resistance genes *Dm1, Dm3* and *Dm5/8*. It is a Salinas-type, crisphead variety, of medium maturity (65 days from transplanting), producing a firm, round, well-covered head (Anon., 1992).

Magnum

This cultivar was also submitted for the protection of plant variety rights in Australia. It arose from the controlled pollination of Salinas by Capitan and is a Salinas-Vanguard type crisphead variety, producing firm, round, well covered heads in 47-54 days from transplanting. Magnum is similar to Target but produces larger (49.5 vs 48.7 cm diameter) and heavier (1112.5 vs 1031.5 g) heads. Magnum can be grown wherever Target is grown, maturing slightly after Target in autumn and before Target in spring (Anon., 1992)

Marksman (Iceberg Lettuce)

Submitted for the registration of plant variety rights in Australia, this medium-green Iceberg lettuce (*Lactuca sativa*) cultivar was derived by crossing E1 Toro (female) with Capitan (male), backcrossing 6 times to E1 Toro, and screening *in vitro* for resistance to an Australian isolate of downy mildew (*Bremia lactucae*) prior to each backcrossing. Marksman was evaluated for morphological and agronomic traits together with E1 Toro and Greenway in trials conducted at Narromine in 1993. Marksman produces a spherical firm head of 16 cm diameter, weighing 1191 g. It possesses the genes *Dm10* and *Dm11* for resistance to *B. lactucae* isolates SF_3 and TV, respectively, and is expected to be resistant to turnip mosaic potyvirus (Anon., 1994).

Diamond

Submitted for the registration of plant variety rights in Australia, this lettuce cultivar, derived from the controlled pollination of Van Sal 210 ??Salinas 105 by Alpha, is noted for its resistance to downy mildew (*Bremia lactucae*) Californian pathotypes I, IIA and III and its brown seed colour (Anon., 1994).

Elisa (Lettuce cultivar for summer cultivation)

Elisa (line S7M1AF426) was selected through the genealogical method from a cross between cv. Aurea and a plant from the germplasm which originated cv. Regina. It produces large, compact heads (35-40 cm diameter) with light-green, smooth leaves. It can be harvested within 50-70 days after sowing and has excellent bolting resistance in summer, flowering 10-15 days after Aurea. Elisa is resistant to lettuce mosaic potyvirus and showed very good performance in trials carried out in Braganca Paulista during November, 1990 and March, 1991 (Della Vecchia *et al.*, 1992).

Florida Buttercrisp (Corky root-resistant butterhead lettuce)

Florida Buttercrisp, a cross between Montello and Brazil 48, is a brown seeded lettuce with resistance to corky root (*Rhizomonas suberifaciens*) and premature bolting; in addition, it is tolerant of tip burn and lettuce mosaic potyvirus but susceptible to a *Bremia lactucae* strain resembling California Pathotype IV, mating type B. In southern Florida, Florida Buttercrisp can be grown during the autumn, winter and spring. It matures within 55-60 days after sowing, producing leaves of an attractive dull green colour similar to that of Dark Green Boston leaves, but more durable and somewhat thicker than leaves of other varieties. Heads weight about 300-360 g, and up to 450 g under favourable conditions. The head has a buttery-yellow centre and the basal end is pale green to white and attractive (Nagata *et al.*, 1992).

Glacier and Misty Dat and 16 Resistant Breeding Lines (Crisphead lettuce resistant to corky root)

Derived from a cross between the corky root (CR, *Rhizomonas suberifaciens*) resistant cv. Green Lake (conferred by a single recessive gene *cor cor*) and Salinas, the *Lactuca sativa* cultivars Misty Day and Glacier (released in 1991) have dull, dark green outer leaves that are slightly paler than those of Salinas. Heads are firm to hard at maturity, larger than those of Salinas under moderate to severe CR stress and the same size at lower stress. Leaf texture is relatively soft and the leaf surface is moderately crinkled or blistered. Both cultivars have a taller stem than Salinas and may be more susceptible to bolting induced by high temperatures. In CR-infested soil, root damage to Misty Day and Glacier was slightly to considerably less severe than that observed for Salinas and similar to or slightly more severe than that of Green Lake. Misty Day and Glacier are also resistant to pathotype I of *Bremia lactucae*. The 16 breeding lines related to Misty Day and Glacier are all descended from the same F_3 family (84-680) as the 2 cultivars and consist of $8F_5$, $5F_6$ and $3F_7$ populations. Morphologically similar to the cultivar, the lines vary primarily in leaf margin appearance, colour and texture and in extent to head cover (Ryder and Waycott, 1994).

Ice Cube, Blush and Mini-Green (Miniature crisphead lettuce cultivars)

Developed from a cross between 86-1024 [dwarf-1; a double mutant carrying early flowering (*Ef*) and dwarf (*dwf*) genes] and Salinas. These 3 lettuce cultivars have dull, dark-green outer leaves that are slightly darker than those of Salinas. The heads are about half the diameter (8-12 cm) of standard US cultivars, which makes them suitable for consumption by one person in one meal. Ice Cube and Blush are adapted for production in the western USA, especially California and Arizona, while Mini-Green tolerates higher temperatures during heading and may thus hold promise for eastern production areas and home garden use. Somewhat less susceptible to tip burn than Salinas, the 3 varieties are also slower to develop symptoms caused by *Bremia lactucae*. Ice Cube and Blush are virtually identical in appearance, except for a red tinge on the exterior leaves of the latter. Maturation times are slightly later than those of Salinas: up to 10 days in cool weather and about 5 days in warm weather. All 3 varieties withstood day-time high temperatures of 30-33°C in trials in New York and Pennsylvania. Field trials with Mini-Green showed it to have slightly higher nutritional quality than Salinas, albeit below that of Valmaine. The inbred gibberellic acid deficiency of the cultivars causes a delay in flowering and a reduction in seed set compared with Salinas, but can be eliminated by application of 2-3 sprays of 10-50 mM gibberellic acid about 3, 6 and 9 weeks after sowing (Waycott and Ryder, 1994).

Florida 202 and Everglades (Butterhead lettuce cultivars)

Florida 202 was derived from Montello Brazil 48, followed by 6 generations of pedigree selection and two mass seed multiplications. It has medium green foliage and matures in about 55 days after sowing, producing a large spherical head weighing on average 261 g with a buttery-yellow interior. Everglades was developed by crossing a single F_2 plant from [Line G (F_2) x Gallega) Dark Green Boston] with Troc, followed by 8 cycles of pedigree selection. It is a medium butterhead lettuce variety suitable for sowing from mid-October to February in southern Florida. The high quality head reached on average 312 g in 2 trials but is capable of producing up to 450 g or more. Both cultivars are resistant to lettuce mosaic potyvirus; Florida 202 is also resistant to corky root (*Rhizomonas suberifaciens*), bolting and tip burn. Everglades is tolerant of tipburn, resistant to a downy mildew resembling California Pathotype IV, mating type B, and spring yellow virus but moderately susceptible to corky root. Its seeds are thermotolerant, allowing 90 per cent or better germination of non-primed seed at a constant 35°C within 72 hours (Guzman *et al.*, 1992).

ID-BRI-Sulfonylurea Herbicide Resistant Lettuce

ID-BR1 was developed by the backcross procedure from *L. sativa* cv. Bibb and a sulfonylurea herbicide resistant prickly lettuce (*L. serriola*) biotype from northern Idaho, using Bibb as the recurrent parent for 5 generations. Throughout, the progeny of each backcross was screened at the 2-3 leaf stage with thifensulfuron and tribenuron at rates of 35.5 g a.i./ha. Experiments indicated that in the absence of herbicide, susceptible and resistant biotypes did not differ in dry weight, indicating that ID-BR1 is capable of yielding as much as Bibb. When averaged over 3 application rates (13.3, 326.6 or 53.2 g thifensulfuron and tribenuron/ha), however, treated plants showed a greater dry weight reduction than untreated plants, the reduction being greater for Bibb than for ID-BR1. The herbicide resistant germplasm line is recommended for incorporation of sulfonylurea herbicide resistance into commercial lettuce varieties (Mallory Smith *et al.*, 1993).

Onondaga

Onondaga is a cross between 'Saladcrisp' and 'Montello'. It is light green in colour, with a crisp, firm head. It resists physiological disorders like bolting and tip burn and is a mid-season cultivar, according to Robinson who is the only plant breeder who has ever been successful in breeding CMV resistance derived from *L. saligna* into domestic lettuce (McCandless, 1998).

5.2.1 New Cultivars of Lettuce

Cultivars of crisp head or iceberg lettuce: Summer Crisp, French Crisp, Ballade, Crispino, Ithaca, Legacy, Mission, Salinas, Summertime and Sun Devil.

Cultivars of butter head lettuce: Blushed Butter Oak, Buttercrunch, Carmona, Divina, Emerald Oak, Flashy Butter Oak, Kweik, Pirat, Sanguine Ameliore, Summer Bib, Tom Thumb, Victoria, and Yugoslavian red.

Cultivars of Romaine or Cos lettuce: Brown Golding, Chaos Mix II Black, Chaos Mix II White, Devil's Tongue, Dark Green Romaine, De Morges Braun, Hyper Red Rumple, Little Leprechaun, Mixed Chaos Black, Mixed Chaos White, Nova F_3, Nova F_4 Black, Nova F_4 White, Paris Island Cos, Valmaine, and Winter Density.

Cultivars of Looseleaf, Leaf, Cutting or Bunching: Austrian Greenleaf, Bijou, Black Seeded Simpson, Bronze Leaf, Brunia, Cracoviensis, Fine Frilled, Gold Rush, Green Ice, New Red Fire, Oakleaf, Perilla Green, Perilla Red, Merlot, Merveille De Mai,Red Sails, Ruby, Salad Bowl, and Simpson Elite.

6.0 Soil and Climate

6.1.0 Soil

Lettuce grows best in silt loams and sandy soils. These types of soils provide better drainage during cold weather and warm up more readily during the day, which is especially important in cooler times of the growing season. Prior to planting, the soil is typically amended to loosen clods and to improve overall quality of the soil for root development.

Well-drained sandy loam soil rich in organic matter is the best. Lettuce is highly sensitive to high acidic soil. In the neutral soil also it does not give good performance. Soil having 5.5-6.5 pH is the best.

6.2.0 Climate

Lettuce, in general, is a crop with distinct temperature requirements. Leaf lettuce is considered a cool-season crop. Optimal growing temperatures range between 23°C during the day and 8-9°C at night. Frost can damage the outer leaves of mature plants which can lead to plant decay. Temperature, light and CO_2 are

Different Types of Lettuce

1. Romaine, 2. Arugula, 3. Radicchio, 4. Watercress, 5. Butter, 6. Leaf lettuce, 7. Mizuna, 8. Endive, 9. Iceberg.

important climatic factors and germination, growth and development, and flowering are highly influenced by these factors.

6.2.1 Temperature

Lettuce requires cool climate and can successfully be grown at 13-16°C. In temperate countries, temperature becomes very low and at this temperature crop cannot be raised successfully and thus there is need to provide artificial heating in glass house. Maintenance of optimum temperature is economical for growing the crop. Both day and night temperatures are important and lettuce grown at 20/18°C exhibited better growth than those at 20/13°C or 20/17°C (Knaval, 1981). Furthermore, response of cultivar to temperature depends upon the light intensity and duration provided (Lorenz and Viebe, 1980). When atmospheric temperature is low, application of polythene film of 0.05 mm thickness with one cm perforation raises the temperature and produces early crop or advances the crop (Benoit and Ceustermans, 1979). High temperature (above 22°C) accelerates seed stalks and reduces the quality of heads and leaves. Drew and Brocklehurst (1990) observed that low temperatures (20/10°C, day/night) during production gave a low yield of large seeds and high temperatures (30/20°C) gave a higher yield of much smaller seeds; the highest yield came from medium temperatures (25/15°C) which gave medium sized seeds. Increased temperatures have been shown to decrease macronutrient content in red leaf lettuce while increasing flavonoids and phenolics (Sublett *et al.*, 2018). Reductions in soil temperature, air temperature and water loss from soil were proportional to the degree of shading for lettuce whereas minimum air temperature and relative humidity were similar inside and outside all types of tunnel (Buriol *et al.*, 1994). Shoot fresh and dry weights (g/plant) of butterhead lettuce increased significantly with temperature, while root dry weight followed the opposite trend and the average number of leaves per plant increased significantly with increasing solution temperature (Economakis and Chartzoulakis, 1997).

6.2.2 Light

Natural light is adequate for economical production of lettuce in subtropical countries. But in temperate countries, there is poor light condition, especially in winter, thus it is necessary to apply artificial light in glass houses. Usually incandescent and fluorescent lights are provided as supplementary light for production of lettuce in green houses. Application of supplementary light at early stage is more important than at later stage (Visser and Veeren, 1975). Low illumination of fluorescent light increased relative growth rate, net assimilation rate and leaf area. Furthermore, artificial illumination of fluorescent light 17 klux for 16 hours per day for 10 days produced extremely compact head (Dennis and Dullforce, 1974).

While comparing different types of light, it was observed that fluorescent tube produced highest chlorophyll and ascorbic acid. There was marked influence of photoperiod also and quick growth was noted at long photoperiod (Craker and Seibert, 1983). Photoperiod affects bolting of the plants and there is differential reaction of cultivars to photoperiod for bolting. Some cultivars bolted earlier, whereas other like Kagranes Sommer required very long photoperiod for bolting.

Long-day condition accelerated rate of leaf area, growth rate and head formation but did not influence the rate of leaf formation (Kalloo and Krug, 1980).

Kang *et al.* (2013) demonstrated the influence of light intensity and photoperiod on the growth and development of hydroponically grown leaf lettuce under close plant factory system. Four light intensity treatments, *viz.*, 200, 230, 260, and 290 $\mu mol \cdot m^{-2} \cdot s^{-1}$ PPFD provided LEDs with a combination of three different photoperiods 18/6 (1 cycle), 9/3 (2 cycles) or 6/2 (3 cycles) (light/dark) were used as treatments. The combination of 290-9/3 (light intensity-photoperiod) showed the highest plant height and fresh shoot weight, while plants grown at 290-18/6 exhibited the greatest root fresh weight, leaf dry weight, and longest root length. The greatest leaf width, maximum number of leaves, and greatest root dry weight were observed in the treatment combination of 290-6/2.

6.2.3 CO_2 Concentration

Generally, 300 ppm CO_2 is present in the atmosphere which is not adequate for the maximum production of crop. Thus, there is need to enrich CO_2 artificially up to 1000-1500 ppm in green house (Peet and Willits, 1982). Supplementary application of CO_2 is effective at low temperature and poor light conditions. Liquid CO_2 or combustion of organic fuel may be used as the source of CO_2 in green houses. Kawashima *et al.* (1989) studied the effects of supplying 600-750 ppm CO_2 to winter lettuces and found that weights of CO_2 treated crops were 1.5-2.0 times greater than those of untreated crops harvested at the same time.

7.0 Cultivation

7.1.0 Seeds and Sowing

About 400-500 g of seeds are required for one hectare of land. However, for green house crop much less quantity is required. Seeds are treated with 0.2 per cent ceresan before sowing.

Seed beds are sterilized with formalin to protect from soil-borne diseases or drenched with 0.2 per cent brassicol or captan. Seeds are sown in September-October either in line or broadcast. Transplanting date markedly influenced cultivar yield, quality and nitrate content of lettuce heads (Miccolis *et al.*, 2000; Kobryn, 2001 and Khalf *et al.*, 2003). Al-Harbi (2001) demonstrated that the greatest vegetative growth and yield were obtained in the first (23rd September) and second (14th October) planting dates. Delaying the planting date after mid-October resulted in less growth and early flowering. Sharma *et al.* (2001) observed that the early (1st August) and late (16th October) transplanting dates resulted in decreasing yield and other horticultural traits of lettuce cultivars. The maximum yield was obtained by the third (1st September) and fourth (16th September) transplanting. For transplanting one hectare, about 70-80 m^2 area of seed bed is sufficient. Soon after sowing, seed bed is irrigated with rose can. Beds are irrigated on alternate days and seedlings are sprayed with 0.2 per cent dithane M-45 or difolatan at 10-15 day interval. After 5-6 weeks of sowing, seedlings are ready for transplanting. Silva *et al.* (2017) obtained good result when the transplanting of seedlings was carried out 20 days after sowing with three definitive leaves, at a spacing of 0.30 x 0.30 m.

7.1.1 Germination

Sinska (1988) observed that seed treatment of cv. Grand Rapids with putrescine and spermidine, each at 1mM stimulated germination. The effect of spermidine was more pronounced. Ghazi and Matthees (1990) noticed opposing effects of strychnine and brucine on the germination and growth of lettuce. Strychnine nitrate inhibited whereas brucine sulphate stimulated radicle elongation at all concentrations from 1 x 10^{-4} to 8 x 0^{-4} M. Lettuce seed usually fails to germinate at temperature above 30°C. Several workers have shown that moist lettuce seed may go into dormancy when subjected to high temperature in the dark and its exposure to chilling at 4-6°C for 3-5 days results in breaking dormancy. It is a common practice in USA to chill lettuce seed to be planted in warm soil by placing it between folds of moistened cloth in a refrigerator or on blocks of ice (Thompson and Kelly, 1957).

Thermodormancy may result from imbibition of lettuce seed at temperature above 25°C (Hemphill Jr., 1982a). The factors responsible for poor stand establishment include inadequate soil moisture, soil crusting and soil temperature above 30°C. In spring planting, stand establishment problems may be aggravated by suboptimal soil temperatures (10-18°C) and poor uptake of plant nutrients, particularly phosphorus.

Stand establishment can be improved by use of anticrustants (Hemphill Jr., 1982b), fluid drilling of germinated seed (Currah *et al.*, 1974) and transplanting. According to the former author, phosphoric acid (75 per cent H_3PO_4) diluted with 2 volumes of water applied to plots at seedling significantly hastened the emergence of seedlings, increased the P content of seedings and also yields. While evaluating the different anticrustants, Hemphill Jr. (1982b) found that vermiculite was the most effective followed by H_3PO_4.

Seed dormancy may also be aggravated by the use of coatings, pellet shapes and methods of assembling the pellet. Fluid drilling is one of the methods to overcome seed dormancy. It includes a period of imbibition or pre-germination under optimum conditions followed by mixing of the seed in a gel material which is then extruded under pressure through the planting mechanism into the ground. According to Currah *et al.* (1974), the emergence period of lettuce seed planted in this manner was shorter than for dry seeds under the same conditions.

The dormancy in 'Romaine' lettuce was overcome by imbibing seed with osmotic solutions or with water before planting in warm soil. The combination of imbibition and gel showed more rapid emergence than did either treatment alone (Hemphill Jr., 1982a). It was also found that maximum germination of 'Minetto' lettuce seeds occurred after the seeds were imbibed for 16 hours. Most seeds primed in 1 per cent K_3PO_4 at 15°C for 9 hours then redried, germinated at 35°C (Guedes *et al.*, 1981). In the field, primed seeds emerged more rapidly and uniformly than non-primed ones, leading to advanced and uniform maturity.

Different kinds of light are also known to influence the seed germination in lettuce. According to Hsiao and Simpson (1971), far red and blue light inhibited germination of Grand Rapids seed imbibed for 24 hours at all water levels studied, red light was inhibitory at low levels and promoted germination at higher water levels in petridishes.

Different Stages of Lettuce Field Growing

7.2.0 Manuring and Fertilization

7.2.1 Organic Manuring

The use of organic fertilizers is gaining importance in view of the chemical residues that are feared to be left on the food crops. Various organic manures like Arbuscular mycorrhiza, vermicompost, farmyard manure, green manure, *etc.* are being widely used. They are considered to be environmental friendly. There is lot of studies reported on organic manures in lettuce. The response to manures varies between cultivars (Rodrigues and Casali, 2000). Organically grown lettuce had lower nitrite values ($P < 0.05$), 0.45 ± 0.08 mg/kg, than had conventionally grown lettuce, 0.83 ± 0.09 mg/kg and might be advantageous for groups such as growing children and strict vegetarians to take organically rather than conventionally grown vegetables (Perez Llamas *et al.*, 1996).

Mycorrhiza is very important for imparting drought and salt tolerance besides improving nutrient uptake. Under drought conditions, the uptake or metabolism of N forms is particularly affected in mycorrhizal fungi-colonized plants, depending on the mycorrhizal endophyte and the N source added. Thus the significance of arbuscular mycorrhizal fungus selection for plant growth in drought conditions is a consideration for management strategy (Azcon *et al.*, 1996). *Glomus deserticola* was the most beneficial endophyte for water use efficiency (Ruiz Lozano *et al.*, 2000).

Improved salt tolerance of mycorrhizal plants is commonly attributed to their better mineral nutrition, particularly phosphorus. However, the effect of arbuscular-mycorrhizal (AM) fungi on salt tolerance appeared to be based on physiological processes (increased CER, transpiration, stomatal conductance and WUE) rather than on nutrient uptake of N or P (Ruiz Lozano *et al.*, 1996). Arbuscular mycorrhizal (AM) fungi were inoculated into P-deficient soil treated with either organic or chemical fertilizer with cucumber as the first crop and lettuce as the second crop but without additional fertilizer application was AM inoculation indicated that AM could not serve as a substitute for P fertilizer. However, the other beneficial effects of AM on crop growth and yield could not be fulfilled with P fertilizer alone (Xu *et al.*, 2000). A combination of organic and mineral fertilizers gave higher yields of marketable heads, fewer rejects and a better average weight/head than mineral fertilizer alone (Tisselli, 1999).

VAM colonization of lettuce decreased with the use of chemical inputs. These included pesticides and P and N fertilizers. The addition of soil organic amendments, the occurrence of other host crops in the rotation, and soil C: P and C: N ratios were positively associated with VAM colonization of lettuce roots (Miller and Jackson, 1998).

Some of the *Rhizobium* species and other bacteria have also been tested for their use in lettuce. Chabot *et al.* (1996) examined *R. leguminosarum* bv. *phaseoli* strains P31 and R1, *Serratia* sp. strain 22b, *Pseudomonas* sp. strain 24 and *Rhizopus* sp. strain 68 were examined for their plant growth-promoting potential on lettuce. Lettuce inoculated with the strain R1 had a 6 per cent higher P concentration than the uninoculated control. In site 2 (infertile soil), the dry matter of lettuce shoots was significantly increased by seed inoculation with strain P31 and 24 plus 35 kg P/ha as superphosphate, or with strain 68 plus 70 kg P. Inoculation with PSM did not affect P uptake in lettuce in the less fertile soil in site 2. It is concluded that Rhizobia specifically selected for P solubilization function as plant growth promoting rhizobacteria with the non-legumes lettuce and maize. The P solubilization effect seems to be the most important mechanism of plant growth promotion in moderately fertile and very fertile soils when P uptake was increased with Rhizobia and other PSM. Inoculation of lettuce seedlings with *Azospirillum brasilense* strain Sp7 increased vitamin C content at all N rates. None of the treatments affected the leaf chlorophyll content (Agwah and Shahaby, 1993).

Vermicompost and FYM increased the yield of lettuces in comparison to the unamended control (Zarate *et al.*, 1997). Studies have shown that higher crop yields can be obtained after the use of compost. Mineralization of mushroom compost was quick (70 per cent during the 1st year), while that of green compost was slow. Nitrate levels in soil were higher after the use of mushroom compost in lettuces (Rooster, 1999). Application of earthworm compost (738 g/m^2 base application plus 150 g/m^2 applied 15 days after transplanting and 150 g/m^2 applied again 30 days after transplanting) gave similar results in terms of fresh weight, dry weight, head diameter and root length to application of NPK (3.3 g-7.14 g-10.32 g, plus 2.8 g $urea/m^2$ applied 15 days after transplanting and 2.8 g $urea/m^2$ applied again 30 days after transplanting) (Fernandes *et al.*, 1999). Vermicompost increases the content of sugars more than mineral ferti-lizers and lettuces grown in autumn contained more potassium and calcium (Nowak and Draszawka Bolzan, 2000).

Lettuce plant fresh weight and dry weight increased linearly with rising compost rates (Vidigal *et al.*, 1995; Lindner, 1996; Aguirre *et al.*, 1994). Rodrigues and Casali (1999) obtained highest estimated yields of 119.5, 119.4 and 153.9 g/plant with 37.7 t organic compost/ha with no mineral fertilizer application, 18.9 t organic compost/ha with half the recommended mineral fertilizer rate and 13 t organic compost/ha with the recommended mineral fertilizer rate, respectively. Organic compost application resulted in lower foliar N and Ca concentrations and higher foliar P, K and Na concentrations compared with mineral fertilizer application.

Under green house condition fresh weights of lettuce plants treated with sewage sludge or *L. leucocephala* were significantly higher than in other treatments. Leaf NO_3-N was higher in plants treated with sewage sludge than in other treatments, although all NO_3-N levels were below the threshold allowed by law (Castro and Ferraz Junior, 1998). Lettuce grown in a green house, when fertilized with 35 t/ha of urban waste compost with a Red-Yellow Latosol, clayey texture showed great variation within cultivars for fresh and dry matter yields, root dry matter and concentration of Zn, Mn, K and Na in leaf dry matter (Santos *et al.*, 1998; Misra and Mani, 1994). In all cultivars, Pb concentrations exceeded 0.50 µg/g of fresh matter, which is the tolerance limit for this metal in fresh leaves (Santos *et al.*, 1998). Lead content attained the maximum level of tolerance in lettuce fertilized with organic compost from urban waste (Costa *et al.*, 1994; Santos *et al.*, 1997).

Pumice medium fertilized with organic manure was found to be the best medium for soilless culture of lettuce (Turhan and Sevgican, 1999). Similarly, El Shinaway *et al.* (1999) and Maynard (1991) found chicken manure, with some modifications, could be used as an organic source under the nutrient film technique system. For environmental safety, the lower rate of layer hen litter (7 t/ha) is preferred to the higher rate (14 t/ha) because it is less than the critical 13 t/ha, considered to cause water contamination by the United States Environmental Protection Agency (Hamdar and Rubeiz, 2000). Rubeiz *et al.* (1998) recommended application of hen litter (3.4 per cent N, 0.86 per cent P, 15 per cent water) incorporated into the soil for lettuce cv. Amaranthus before planting 14 t/ha semi-decomposed poultry manure was incorporated; surface application of 7 t manure/ha gave the significantly highest yield of 20.0 t fresh matter/ha (Zarate *et al.*, 1997). The application of 27.5 t broiler manure or 18 t manure from laying flocks/ha (on a wet basis) gave similar lettuce yields as an application of 100 kg N/ha from NH_4NO_3 in 2 split applications (Rubeiz *et al.*, 1992). Application of dried pig manure was also reported to have given the highest yields 65 days after sowing (54.4 t/ha), an increase of 33.3 per cent above those supplied with NPK (Vidigal *et al.*, 1997). The best agronomic performance of the lettuce cv. 'Babá de Verão' was obtained via manuring with 15.6 t/ha of roostertree [*Calotropis procera* (Ait.) R. Br.)] applied at 10 days (spring) and 20 days (autumn-winter) before seedling transplanting (Souza *et al.*, 2017). Goat manure is potentially beneficial because of its lower carbon : nitrogen (C : N) ratio when compared to other manures (Souto *et al.*, 2005). The lower C : N ratio results in a higher mineralization speed of nutrients, an essential characteristic in the fertilization leafy vegetables, due to the rapid cycle of these crops. Studies that provide information on the management of organic fertilization by the application of goat manure in the lettuce crop are extremely important for the construction of technical subsidies that aim at raising productivity rates, reducing fertilizer costs, and greater profitability to the producer. However, there are few studies that have investigated the potentialilty of goat manure in lettuce production (Figueiredo *et al.*, 2012; Batista *et al.*, 2012; Peixoto Filho *et al.*, 2013; Sarmento *et al.*, 2019).

7.2.2 Nutrients and Inorganic Fertilizers

Lettuce plant has a small and shallow root system, therefore, surface soil should be rich in nutrients. Application of about 10-15 tonnes of farmyard manure supplemented with about 25 kg each of N and K, and 90 kg P/ha would give good results even on poor soils. On sandy and sandy loam soils without manure, 40-45 kg N and 75-100 kg each of P and K/ha should be applied for good growth. On silt loam and clay loam soils 25 kg each of N and K, and 50-75 kg P/ha may be used (Thompson and Kelly, 1957). Commercial lettuce production requires adequate levels of nitrogen (N), phosphorus (P), and potassium (K) to provide high-quality postharvest attributes needed for longer shelf life. The most economical treatment providing the highest yield and best postharvest quality was the combination of 225 kg/ha N and 112 kg/ha P (Hoque *et al.*, 2010).

Cadmium accumulation is a serious problem in many vegetables. The accumulation was reported to be reduced by higher NPK supply in the soil. The statuary EU acceptable levels of Cd exceeded on lettuce leaves when grown in polluted soil fertilized with ammonium nitrate (Gebski, 1998). Costa *et al.* (1997) suggested that the more expensive pure salts could be replaced by commercial fertilizers, particularly single superphosphate, for hydroponic cultivation of lettuce. Starter fertilizers injected below the seed of the drilled crop, or as spot applications around the planting module can contribute to the development of environmentally beneficial farming practices whilst maintaining the productivity and competitiveness of the horticultural industry (Stone *et al.*, 1999). Use of the NPK fertilizer (containing oxamide) at the lowest rate was recommended for lettuce production as it released N slowly enabling the plants to take it up completely thereby reducing environmental accumulation of nitrates (Belligno *et al.*, 1997).

7.2.2.1 Nitrogen

Sources of Nitrogen

El-Hassan (1990) reported that highest total and saleable yields and the highest total dry matter content were achieved with 40 kg N/0.42 ha and spacing at 10 cm. Lettuce absorbs and utilizes NO_3 and NH_4, forms of N. NH_4 induces better growth than NO_3 (Weir *et al.*, 1972). Thus source of N fertilizer is important. The response of crops to different sources of N depends on the leaching mobility or immobility, soil properties and effect of soil reaction, *etc.* Different sources of nitrogen as ammonium nitrate, ammonium sulphate, calcium nitrate and urea were compared and it was found that N sources did not affect yield, quality factor, head size and total N accumulation (Gardner and Pew, 1979). Regarding time of application of nitrogen fertilizer it was found that large quantity of nitrogenous fertilizer should be applied at the time of head formation (Zink and Yamaguchi, 1962).

There is leaching of nitrogen fertilizer and to prevent N leaching below the root zone after frequent irrigation, Nitrapyrin (a nitrification inhibitor) was tested. Application of N at 120 lb/acre along with Nitrapyrin gave significantly higher yield than N alone at 180 lb/acre (Welch *et al.*, 1979). The sulphur-coated urea is required to be applied only once as against 3 applications of uncoated urea. Experiments conducted have indicated that sulphur-coated urea (SCU) F9999 produced the most vigorous seedlings and early yield (Pew *et al.*, 1983). This economizes the cost of fertilizer application. Gastaldi and Sutton (1988) advocated use of constant fertilizer: irrigation water ratio of 2.86 m mols N/litre for good yields and high fertilizer uptake efficiency.

Adding hydrogels to the growing medium had a positive effect on the quality of leaves, which was manifested in a greater concentration of chlorophyll a+b and L-ascorbic acid in the biomass and in a decreased level of phenolic compounds, nitrates and nitrites together with higher nitrate reductase activity (Borowski and Michalek, 1998b). The presence of the hydrogels in the substrate resulted in a drop in calcium content and an increase in phosphorus and sodium in the case of Akrygel RP and Akrygel Na, and in potassium in the case of Akrygel K. Nitrogen in the ammonium form increased the content of chlorophyll a+b, L-ascorbic acid and phenols in the leaves, independent of the composition of the substrate, and decreased the content of nitrates and nitrites (Borowski and Michalek, 1998b). Among the hydrogels, the greatest absorptivity for water was shown by Akrygel K, and the least by Akrygel RP (Borowski and Michalek, 1998a).

Nitrogen is an important element and is recommended that N is added even if it does not appear to be necessary from the soil analysis (Bastelaere, 1998b). The incidence of dry tipburn in older leaves was clearly decreased by an increased N supply, especially at late planting (Sorensen *et al.*, 1994). A supraoptimal N supply not only increased the risk of tipburn but also the minimum N residue at harvest in the soil, and thus the risk of nitrate leaching (Brumm *et al.*, 1993). Restricting N supply affected growth and development independently, with the latter being less sensitive to the avai-lability of N (Burns, 1996). The effect of the form of applied N was more distinct during storage of the lettuce leaves both at room temperature (20°C) and in a cold chamber (5°C) (Rozek *et al.*, 1995). Foliar fertilization with Wuxal (composition 5 per cent N, 10 per cent CaO, 3 per cent MgO w/w + 0.3 per cent B, 0.1 per cent Co, 0.5 per cent Cu, 1.0 per cent Fe, 1.5 per cent Mn, 0.01 per cent Mo and 0.1 per cent Zn) (concentration of 0.5 per cent) twice produced leaves and stems of significantly higher fresh and dry weights (Abdel Razik, 1996). Cultan (controlled uptake long term ammonium nutrition) system involves injection of a concentrated solution of urea ammonium sulphate into the soil close to the rows. Nitrate residue in lettuce heads was lower in Cultan-treated plots (Rooster and Spiessens, 1998). It has also been suggested to apply ammonia by point injection (NH_4-depot method) from an environmentally friendly standpoint as well as furnishing good yields of high quality produce (Panak *et al.*, 1996). However, when used on soils low in N, starter fertilizer N applications are also necessary.

By changing the nutrient solution (NH_4: NO_3 ratio 20: 80) 2 days before harvest to water alone or 4 days before harvest to a solution with more ammoniacal N ($NH_{4:}$ $NO_{3:}$ 80: 20), leaf nitrate content was reduced only in *longifolia* cultivars (Santamaria *et al.*, 2000). Mean nitrate increased with nitric or uric N forms than with the ammoniacal one and delayed harvesting, lowered the mean nitrate content of the fresh product from 787 to 625 mg/kg in many cultivars (Miccolis *et al.*, 2000). NH_4: NO_3 ratio of 75: 25 and a salinity of 2.3 mS/cm was the best for promoting plant height, number of leaves/plant, root length, fresh and dry weights of top/plant, fresh weight of root/plant and total chlorophyll (Gabr, 1999). The lowest levels of nitrates and dry matter of iceberg-type lettuce were found in lettuce cultivated without N (Staugaitis and Viskelis, 2000). However, Gabr (1999) found that leaf nitrate content decreased with each increment of NH_4. Limiting the supply of nitrate to winter lettuce in green houses ensures low nitrate levels in the marketed produce (Seginer *et al.*, 1999). Effect of N form in the nutrient solution on DM yield depended on plant age and temperature of the nutrient solution; NH_4-N had a beneficial effect on 2-week-old plants growing at 7 and 14°C and on 4-week-old plants grown at 7°C (Borowski and Michalek, 1995).

The permissible nitrate concentration limit in lettuce in summer is 3500 mg NO_3 per kg fresh weight in many countries (Nadasy Iharosi, 1999). Nitrate content in heads at harvesting was not influenced by nitrogen fertilizer levels. However, fertilizer treatments can result in more leaf veins, leaf vein rot and yellow leaves (Bastelaere, 1998a). At a given N rate, the gradual release of N provided by controlled-release fertilizers inhibited lettuce nitrate accumulation to some extent (Tesi and Lenzi, 1998). The external leaves showed the highest nitrate concentrations, with significantly lower values in the internal ones (Pavlovic *et al.*, 1997). The concentrations of N-NO_{3-} at harvest ranged from 1639 to 4494 ppm in external leaves and from 651 to 1508 ppm in the internals for the different treatments (Rincon *et al.*, 1998). Omitting nitrogen from the nutrient solution four days before the second harvest, and then 7 days before the last harvest the nutrient solution was replaced with water lowered the nitrate content below the accepted nitrate level without any negative effects on yield (Kobryn, 1998). Reducing NO_3-N availability to 120 kg N/ha significantly reduced both nitrate content and fresh weight. Substituting 40 per cent of the available NO_3-N with NH_4-N significantly reduced nitrate content while fresh weight was unaffected (McCall and Willumsen, 1998). The effect of NH_4-N application on plant nitrate content was enhanced by applying the nitrification inhibitor dicyandiamide (DCD) (McCall and Willumsen, 1998). The application of calcium cyanamide 1 or 2 weeks before sowing or planting can reduce crop nitrate content and tipburn in lettuces besides reducing the incidence of several soilborne diseases such as Sclerotinia rot (Klasse, 1999). Application of reduced forms of N significantly improved the quality of the lettuce by reducing the accumulation of nitrates, especially in plants grown in peat, which has a slower rate of nitrification (Kowalska, 1997). Lettuces grown at pH 4.9 showed higher N and Mg contents when the N-source was urea and higher K and Ca contents when N was applied as ammonium nitrate (Stancheva *et al.*, 1997). Lettuce nitrate contents were lowest with ammonium sulphate and highest with ammonium nitrate (Kozik and Glen, 1995). Application of NH_4NO_3 alone resulted in significantly higher leaf NO_3 concentrations than applications of manure or those involving controlled release fertilizers (Arce *et al.*, 1996).

Effect of Nitrogen

The estimated rate of fertilizer-N to obtain maximum fresh and dry weights for butter-head lettuce and loose leaf lettuce, was about 155 kg N/ha for both the cultivars. At that rate, the estimated N uptake was about 145 kg/ha for loose head lettuce (Canasta) and 131 kg/ha for butterhead lettuce cv. Audran (Tei *et al.*, 2000). Considering that the N uptake with 0 kg/ha of fertilizer-N was 63 kg/ha for Audran and 58 kg/ha for Canasta, about 80 kg/ha of mineral nitrogen remained in the soil at harvest (Tei *et al.*, 2000). Osinga (1996) recommended 65 kg N/ha at planting and up to 150 kg N/ha at head filling was found to be very satisfactory. Yield and quality differences between treatments were negligible due to high mineralization of the soil. The dry and fresh matters of the aerial part were the most important characters to discriminate the cultivars as to its nutritional efficiency for N (Silva Ledo *et al.*, 2000). Highest yield and plant weight of iceberg-type lettuce were obtained by applying 135 kg N/ha with no effects on the contents of sugar, vitamin C and soluble solids in head lettuce, but they increased the amounts of nitrate and dry matter. N fertilizers increased N and reduced K content in lettuce (Staugaitis and Viskelis, 2000). Increasing then N rates up to 320 mg/kg gradually raised by N content of the lettuce leaves. Leaf N contents were highest when calcium nitrate was applied and the nitrate content averaged over the 80 mg/kg N (Nadasy Iharosi, 1999). The optimum yield of marketable heads of crisphead lettuce was obtained at a total nitrogen supply of 150 N kg/ha although only small differences were observed in the yield at 100 and 200 N kg/ha (Sorensen *et al.*, 1994).

Nitrogen-sulphur fertilizer combinations and growing season affects on sensory quality/acceptability and colour of hydroponically grown green house lettuce. Cuppett *et al.* (1999) found that plants receiving the highest treatment level (240 mg/l) of nitrogen were greener (sensory) and darker and duller (chromameter) than the other samples. Between seasons, it was found that lettuce grown in the spring was perceived as being greener and having a softer texture than lettuce grown in the winter and receiving 30 and 60 mg/l N were perceived as having firmer texture. N uptake and yield increased with increasing light intensity, but the content of soil N was maintained at a relatively low level (Zeng *et al.*, 1998; Lacertosa *et al.*, 1997). Fertilizer application at 12.5 kg/100 m^2 in the autumn on well-leached soils resulted in glassiness and rib blight. However, after soil disinfection with methyl bromide, an application of 25 kg/are was sometimes necessary. Greater application rates resulted in stagnated growth and lower crop weight (Bastelaere, 1999). Parente *et al.* (2006) observed that the application of nitrogen by fertigation increased yield compared with the unfertilized control, but with no difference between doses and rates of fertilization. On the other hand, nitrogen post-transplanting fertilization increased the nitrate content in leaves. The Lollo cultivars accumulated more nitrate than the other ones. The nitrate levels were greatly below the limit of 4.00 mg/kg fw imposed by EU Regulation No. 563/2002.

7.2.2.2 Phosphorus

Like other nutrients response of lettuce types to phosphorus vary between genotypes. P application has little effect on yields when the original soil P content was high, but increases yields when soil P was low (Wijk and Neuvel, 1998). With the increase in phosphorus level there was advance in the maturity (Wilson, 1976). The best rate of P fertilizer dressing was strongly linked to the P status of the soil and the cultivation method. Because of the strong phosphate requirement of lettuce, there is a big yield response to phosphate fertilization on soils with a low P level (Wijk *et al.*, 2000). Sanchez and El Hout (1995) found that all lettuce types showed large yield and quality responses to P fertilizer. Because environmental conditions affected yield potential, P rates required for optimal yield varied for lettuce types across seasons. They suggested that existing soil test based fertilizer recommendations for crisphead lettuce are adequate for other lettuce types currently grown. Yield of butterhead lettuces increased with P_2O_5 at 120 kg/ha rose with increasing soil pH levels up to about 70, and were slightly higher than those of unfertilized controls; at higher soil pH levels, there was no further response, and yields of the controls were higher (Neuvel and Wijk, 1998). Under optimal rainfall and soil moisture conditions at planting gross yields of crisphead and romaine lettuce ranged from 16.9 (control) to 25.5 t/acre at the highest P rates and plant survival was not influenced by P rate (Rice, 1999). Soil aggregate size and soil management practices affect phosphorus absorption from highly weathered soils and plant growth. Wang *et al.* (2001) showed in their pot experiments that total P in lettuce shoots, and the root dry weights of plants grown in the large aggregates (2-6 mm) were higher than for plants grown in the small aggregates (< 0.5 mm) after equal amounts of P were added to the separated aggregate fractions. Increased P uptake with increased aggregate size was attributed to increased P released from aggregates because of reduced P fixation. Their (Wang *et al.*, 2001) results suggest that soil management that favours soil aggregation may, in some cases, increase availability of applied P. Perhaps the distribution of soil aggregates should be considered in making P management decisions.

Leaf P content for optimum yields varied between 0.40 and 0.90 per cent while > 0.90 per cent P may cause leaf abnormalities. With high P supply, more Ca was necessary to overcome the detrimental effect of P. For maximum yields, a P supply rate of 540 mg P/kg soil is ideal (Claassens, 1994). The critical soil test P values required for either 95 or 99 per cent of maximum relative yield were 80 and 115 µg/g, respectively. The P concentrations of the wrapper leaf at early heading required for 95 or 99 per cent of maximum yield were 0.59 ± 0.03 per cent and 0.61 ± 0.03 per cent (spline regression, $R^2 = 0.80$), respectively (McPharlin *et al.*, 1996).

Band application of P fertilizer is better than broadcasting and it results in both higher net returns and large reductions in economically optimal phosphorus applications (Alvarez and Sanchez, 1995; McPharlin and Robertson, 1997). To enhance the uptake of Phosphorus, Gorelik and Pavlova (1997) used the nitrification inhibitor CMP which improved the uptake of phosphorus by plants, availability of phosphorus, and lettuce yield.

Among the various sources of phosphorus tested in green house, potassium dihydrogen phosphate improved the root systems with no effect on crop yield or quality (Vergote, 1998). Bull *et al.* (1997) reported that coarse thermophosphate (with micronutrients) was not effective in reducing soil acidity, while fine thermophosphate (with Mg) increased the soil pH. Due to lower solubility, coarse thermophosphate applications resulted in lower crop yields and fertilizer efficacy than those of fine thermophosphate.

7.2.2.3 Potassium

The deficiency symptom of K is chlorosis, followed by random necrotic lesions on the leaves (Berry and Carey, 1971). Application of K at high doses had no significant effect on yield but improved quality (Tafuri *et al.*, 1978). But according to Adams *et al.* (1978), both yield and quality were increased with the application of K fertilizer. Crop yields of butterhead lettuce (55 t/ha) did not vary significantly between treatments and dry matter production was not influenced by K fertilizer in sandy soils. Uptake of K increased with increasing fertilizer rate (Neuvel and Kanters, 1999). However, crop yield increased after the use of K fertilizer on clay soils. An application of 240 kg K/ha was required to compensate for K uptake. K fertilizer increased K levels in lettuce and soil (Neuvel *et al.*, 2000).

7.2.2.4 Calcium

Liming increases yields above the growth level of the unlimed untreated soil for lettuce. Soluble and exchangeable Cd, Ni and Zn were also reduced after liming indicating the benefit of pH adjustment in reducing relative solubility and plant uptake of metals as well as increasing crop yield in acid soils (Brallier *et al.*, 1996;

Molas and Szymanska, 1995). Laughlin *et al.* (1987) reported increase in lettuce yield with lime application from 4.96 t/acre at zero to 12.3 t/acre at 4 t lime/acre; it also rose with P from under 5 t/acre at 20 kg P/acre to 10-20 t/acre at 80 kg P/acre.

Head rotting of iceberg lettuce was mostly caused by tipburn, as a consequence of calcium deficiency in the heart of the heads. Misting, N fertilizer or Ca spray application can not prevent the occurrence of tipburn and differences in sensitivity to tipburn were found when the heads were stored at 30°C, 35 per cent RH and a light intensity of 1800 lux (Kraker *et al.*, 1996). Magnesium contained in dolomite meal had a positive effect on the synthesis of leaf pigments (chlorophyll, carotene) and also decreased the nitrate content in vegetables. The use of limestone and dolomite meal mixtures for neutralizing peat substrate increased the yields of lettuce (Jarvan and Teedumae, 1999; Jarvan, 1999).

7.2.2.5 Magnesium

Magnesium applied as Magniform (12 per cent Mg, 0.8 per cent Fe, 5 per cent N in amide form) under the rootball as 1 g tablet/plant increased the growth rate and Mg content compared to similar Mg rate applied as magnesium sulphate solution (Zeleny and Zelena, 1994). Increasing Mg rate from 50 to 100 mg/dm^3 and supplementary foliar application increased Mg and chlorophyll contents in leaves of lettuce both species (Biesiada *et al.*, 1998).

7.2.2.6 Micronutrients

It is recommended that for maximum economic return, fertilizer should be applied when highest nutrient requirements and assimilation rate takes place, namely, between 59 and 66 days after sowing as Hoagland solution in green house (Premuzic *et al.*, 1995). Selenium has been reported to influence nutrient uptake (enhanced the uptake of N and P, and reduced the uptake of K, Na, Ca, Mg, Fe, Mn, Cu and Zn), increased leaf and stem contents of sugars, chlorophyll and vitamin C [ascorbic acid], and decreased the contents of nitrite, fibre and the free amino acids, Cysteine, Methionine and that the effect of Se on protein metabolism is cultivar dependent (Shang *et al.*, 1998). In some soils there is zinc deficiency which can be noticed with the stunted growth, marginal purpling, rupture of margin of mature leaves and overall chlorosis to interveinal chlorosis (Berry, 1971). When these symptoms appear, zinc sulphate or other forms of zinc may be applied. Addition of Zn and Fe as either their basic salts or as commercial chelates is as effective as the chelates in providing Zn and Fe as well as being more environmentally safe (Arambarri *et al.*, 1994). According to Adams *et al.* (1978), molybdenum 0.5 g/m^2 can be applied in glass houses. At high nitrogen level there was better response of molybdenum. Plant molybdenum content increased as Mo content in nutrient solution increased. However, plant Mo uptake was decreased by high concentrations of tungsten (Gunes and Post, 1996). Agri-wave technology is where a special frequency sound wave and a micro-element fertilizer are applied to plants. Hou and Mooneyham (1999) treated in green house, 3 lettuce cultivars and found that the average FW of plants treated by Agri-wave technology has increased by 44.1 per cent over control. Compared with the control, FW was increased by 29.92 per cent in plants treated with sound alone, and by 16.19 per cent in plants treated with fertilizer alone.

7.2.3 Fertigation

The fertigation is better than the conventional application of fertilizers (McPharlin *et al.*, 1995; Goncalves *et al.*, 2000). The effect of irrigation effects on soil variables and lettuce yield, with and without fertilization using fertigation with urea and ammonium sulphate was reported (Klar *et al.*, 1995; Goncalves *et al.*, 2000). A total of 60 kg N/ha applied at 10, 20 and 30 days after transplanting increased the yields 22.6 per cent higher in fertigation treatments than conventional treatments (Klar *et al.*, 1995). Ammonium sulphate induces greater EC and lower pH values. The crop yield response to urea is better than to ammonium sulphate. However, Klar *et al.* (1995) found no difference in yield between ammonium sulphate and urea treatments. Such variations are due to the variation in genotype and soil type. Water and N management practices are very important in porous sandy soils. Lettuce yields have been reported to increase in response to water and N, and the highest yields were obtained in a range of 55 cm of water and 271 kg/ha N to 76 cm water and 270 kg/ha N (Sanchez, 2000). These water and N rates exceeded those typically required on finer-textured alluvial valley soils. Hence alternative production methods that enhance water and N use efficiencies, such as drip irrigation and/or the use of controlled-release fertilizers, should be considered for sandy soils. The fertilizer rates used in sprinkler or trickle irrigation were 30-60 per cent lower than the recommended rates on coastal sands (McPharlin *et al.*, 1995).

Sprinkler fertigation has also been reported to reduce the store of nitrate in the soil during the crop (limits pollution) and could reduce variations in the amounts of soil nitrate at harvest (Tourdonnet, 1999). Studies on nitrogen fertility requirement for iceberg lettuce grown on sand land with plastic mulch and drip irrigation indicated an improvement in yield by total N rates of 160-195 kg N/ha and thereafter decreased. Head firmness and head size were generally largest at this same range of N rates. Other quality factors such as core length and internal tip burn were minimized at the lowest N rate *i.e.*, 55 kg N/ha (Cantliffe *et al.*, 1998). Stepowska and Kowalczyk (2000) planted the seedlings of butterhead lettuce from plug trays in PE pots containing all the nutrient elements to bigger pots irrigated by a capillary system. The smaller ones, standing on a watering mat, were fertigated by a drip-line system. EC of nutrient solution was 1.8 mS/cm. The biggest lettuce heads (over 200 g), with the lowest level of nitrate (551 mg NO_3/kg FW), were obtained on the watering mat.

7.3.0 Irrigation

7.3.1 Methods

There are various systems of irrigation, *viz.*, furrow, surface, trickle and sprinkler practised for lettuce. Usually, for the field grown lettuce, furrow or surface irrigation is in commercial practice. Frequent and light irrigations have been found to be more effective for high yield and quality production. After planting, crop should be irrigated at an interval of 8-10 days. Although mostly furrow irrigation is applied, for good germination sprinkler irrigation is better. With furrow irrigation major part of nitrogen applied is leached and not utilized by the crop (Moore, 1970). However, furrow irrigation under proper management may affect distribution and movement of NO_3-N and total soluble salts in the soil, thus giving the same result as other methods (Pew and Gardner, 1978).

Trickle, subsurface and sprinkler irrigation have been attributed to higher yield than furrow method. Subsurface irrigation maintains uniform moisture in the soil (Sammis, 1980). Trickle irrigation increases yield by about 30 per cent over furrow irrigation (Rowse, 1974). Fertilizer can also be applied with trickle irrigation. High yield of good quality lettuce was produced from the plots irrigated during the main growing period with NPK in irrigation water (Sagiv *et al.*, 1981). Frequent application of low amounts of water (7 cbar regime) was more efficient in terms of water and energy use. Drip irrigation was favoured over microsprinkler irrigation as it reduces the likelihood of contaminating the foliage of crops destined for the fresh market (Slavik, 1996). Hanson *et al.* (1997) found that yields of lettuce obtained from the furrow and subsurface drip methods were similar and applied water of the drip methods were between 43 and 74 per cent of that of the furrow method. Subsurface irrigation uses 86 per cent less water than overhead irrigation and the irrigation type had a significant effect on some growth measurements by final harvest, 43 days after sowing with seedlings significantly taller (Ahmed *et al.*, 2000). Eberhard (2000) found that trickle irrigation reduced yield of lettuce due to superficial roots.

Buxton *et al.* (1999) recommended the controlled water table (CWT) irrigation system which consists of a capillary mat placed on a level bench with one end of the mat suspended in a small trough of nutrient solution. The CWT concept of hydroponic production does not require pumps nor large storage reservoirs. No runoff occurs; the only nutrient solution used is that required by plants and a minimum amount of evaporation from the ground cover surface. Excessive irrigation (5.3 kPa average SWT) results in low yields, head weight and head length besides marketable yields would show a pronounced water-N interaction (Thompson and Doerge, 1996). Chen *et al.* (2019) suggested that the combined plastic film mulching with micro-sprinkler irrigation was found to be the most suitable irrigation approach for increasing lettuce yield.

7.3.2 Effect of Water on Growth and Quality

Hamada and Testezlaf (1996) studied the growth analysis techniques on the effects of different water depths applied by trickle irrigation on the development of lettuce under field conditions. Five growth analysis *viz.*, leaf area index (IAF), crop growth rate (TCC), relative growth rate (TCR), net assimilation rate (TAL) and leaf area relation (RAF) were evaluated. They concluded that the best results for the leaf area (IAF) were obtained by the higher water depths applied (a difference of 35 per cent between the highest and lowest water depth treatments). It was possible to distinguish in the crop life cycle 2 different growth periods. At the beginning period, the TCC growth parameter had a rapid growth, and after reaching the peak it decreased until the final harvest. RAF showed that the crop development was affected by the different treatments, mainly in its final growth stage.

Rooting characteristics and root distribution of lettuce in sand beds is affected by substrate depth, the distance to a drain, drip lines and drip points, and the excess of nutrient solution applied. Schwarz *et al.* (1995) confirmed the hypothesis that a small excess and a large distance between drip points leads to local salt accumulations in the root environment and thus leads to a less homogeneous root distribution. Lettuce can be grown on sand beds with a recirculating nutrient solution provided that drip lines are well distributed in the bed and the daily nutrient solution excess is >30 per cent of demand. Cultivated crisphead lettuce has a shallower root system than its wild relative, *Lactuca serriola*. The wild species could provide germplasm for cultivated lettuces to extract more soil water from depth, which may improve irrigation efficiency (Gallardo *et al.*, 1996). FW was more sensitive than DW to reduction in water supply (Gallardo *et al.*, 1996).

Wood fibre substrates perform well for lettuce seedling production in cellular trays compared to standard commercially used press pots or peat-based growing media as wood fibres contain a high air volume to allow higher levels of water tension without danger of water saturation in the substrate (Gruda and Schnitzler, 2000). The response of the soil pF to the inner negative pressure of the ceramic pipe and its irrigation rate were reported by Liu *et al.* (2000). The soil pF was maintained at suitable values (1.3-1.5) for the growth of lettuce, while the pF in the usual soil cultivation fluctuated widely (0-1.9). This new cultivation system is useful and practical technique for lettuce vegetable production (Liu *et al.*, 2000).

7.3.3 Water Management

Plant fresh weight and yield were influenced by irrigation level and soil water potential and the highest yield (90.97 t/ha) and plant fresh weight were obtained in the 75 per cent treatment (Andrade Junior and Klar, 1996). A strategy based on tensiometer readings minimized the waste of water, requiring less irrigation and leading to less drainage loss with less stressed to the lettuce crop (Leenhardt *et al.*, 1998).

Different models and methods have been developed to schedule irrigation in lettuce. Crop evapotranspiration (ETc) by crisphead lettuce varies considerably with weather conditions and irrigation management. Accurate estimates of ETc enable irrigation applications to be matched more closely to crop requirements, which reduce excessive applications of irrigation water. Accurate assessments of ETc are difficult in crisphead lettuce because of high soil evaporation due to frequent irrigation and slow canopy development. Hence, Gallardo *et al.* (1996) developed a model to estimate *etc.* for crisphead lettuce using only daily reference evapotranspiration data as an input parameter. The model calculates crop transpiration and soil evaporation, and also ground cover development and daily radiation interception by the canopy. Oliveira *et al.* (1997) suggested that the Hargreaves equation can be used with satisfactory accuracy to estimate crop ET where all inputs for the combination methods are not available. Another method recommended is the Geisenheim method of irrigation timing which is based on the use of daily climatic data in the form of the modified Penman equation. The reference crop evapotranspiration (ETo) is multiplied by plant coefficients (kc) to determine when it is necessary to start irrigation. The use of the Geisenheim method made high yields possible without increasing the risk of nutrient leaching (Paschold *et al.*, 1997). Optimal use of weather forecasts to schedule irrigations was estimated to increase annual value by $1000/ha per year, much of which was derived from avoiding crop damage due to excessive soil moisture (Wilks and Wolfe, 1998). They analyzed a sequence of daily irrigation decisions utilizing weather forecast information was formulated for lettuces grown in USA, and solved using a stochastic dynamic programming algorithm. Crop response was represented using a simple but physiologically based model of lettuce growth modified to allow the stomatal conductance for CO_2 to depend on a simple soil moisture budget. In order to minimize the output of nutrients and pesticides to the environment, a plant dependent model for irrigating head lettuce in green houses was developed by Pollet *et al.* (2000) by estimating the lettuce evapotranspiration by means of the Penman-Monteith concept. The stomatal resistance and the amount of the transpiring surface needed for this model are estimated by the use of two submodels. The submodel for the transpiring surface was based on the Gompertz equation using the oudoor shortwave radiation as the only input parameter.

At least 4 mm of water should be given at each irrigation and the best depth for the tensiometer depends on the depth of rooting (Wilhelm and Mayer, 1996). The critical point at which watering was triggered lay between 120 and 400 hPa, extreme values being associated with the risk of under-watering or over-watering (Wilhelm and Mayer, 1996). Water use efficiency show a decreasing linear response with respect to irrigation levels. Yield and fresh weight show a quadratic response to the applied irrigation levels, with maximum values of 90.97 t/ha and 818.72 g/ plant for the 0.75 Class A pan evaporation level, respectively (Andrade Junior and Klar, 1997).

Yitayew and da Silva (1998) developed a Growing Degree Day (GDD) based crop coefficient for leaf lettuce. Marketable yield averaged 25 t/ha with crop water use efficiency of 9.8 kg of marketable yield per m^3 water. The risk of nitrogen leaching is higher for furrow irrigation systems than for drip. Oliveira *et al.* (2005) observed that the water use of subsurface drip-irrigated lettuce indicates an average demand of 117 mm. This amount basically reflects the water demand during the second half of the crop cycle that accumulated 1100°C-day, on an average. Crop coefficients based on Hargreaves, FAO Penman and FAO Pernman-Monteith peaked 0.88, 0.80 and 0.81, respectively in the range of 1000 to 1050°C-day. These values were lower than and occurred earlier than the maximum K_c predicted by the AZSCHED irrigation-scheduling model.

Lettuce is a shallow-rooted crop and is also susceptible to water stress (Allen *et al.*, 1998; Molina-Montenegro *et al.*, 2011). Kuslu *et al.* (2008) determined that the marketable yield, and macro and micro-element contents of lettuce irrigated with level-basin method under open field conditions significantly decreased in the lower irrigated conditions. Molina-Montenegro *et al.* (2011) obtained that the photosynthetic rate and biomass production in lettuce was significantly lower in the treatment receiving 50 per cent of irrigation water compared to the treatments receiving 100 and 75 per cent of irrigation water. Yazgan *et al.* (2008) found that lettuce yields in unheated green house condition significantly decreased with lower water applications, and the relationship between evapotranspiration and yield was linear. However, Capra *et al.* (2008) determined strong polynomial relationships between total water amount received and marketable yield of lettuce in two consecutive summer crop production conditions. Sahin *et al.* (2016) concluded that highly marketable yields in lettuce grown in a semi-arid region with a cool climate under high altitude conditions may be obtained with irrigation applications done considering water depths which equal 100 per cent of evaporated water from a Class A pan. However, more healthy lettuce production may be achieved with irrigation applications done considering the levels of 85 or 70 per cent of evaporated water from a Class A pan.

AlKhader and Rayyan (2013) recommended that application of Mono ammonium phosphate (MAP) fertilizer at the application rate of 375 kg P_2O_5/ha can be beneficial in terms of both plant fresh weight of lettuce and WUE, as this treatment of fertilizer can induce both relatively high crop yield and improve irrigation water productivity.

7.3.4 Effect of Salinity and Salts

The quality of water used for irrigation has great impact on yield and quality of lettuce. The consumption of water for the plants during the production cycle decreases with increasing levels of salinity and salinity causes a reduction in vegetative growth (Gervasio *et al.*, 2000). The highest commercial production of lettuces was obtained when the electrical conductivity of the saturated extract reached 0.2 dS/m; for each unit increase of conductivity, there was a reduction of 17 per cent in commercial production (Gervasio *et al.*, 2000). Beltrao *et al.* (1997) observed a steep yield reduction between 1 and 2 dS/m. A slow yield reduction was observed between 2 and 5 dS/m. Between 5 and 8 dS/m, yield was very low and near constant. Humidity also plays a role on the effect of salinity on lettuce. Bradbury and Ahmad (1996) irrigated lettuce with water of salinity ranging from 0 to 72 mM NaCl and observed greater leaf area, shoot: root ratio and dry weight when kept in a night/day RH regime of 92/100 per cent than in one of 62/82 per cent RH. There was no significant difference in the rate of assimilation or stomatal conductance between plants grown at high and low RH.

7.3.5 Pollutants

Lettuce, being a salad crop, is used for fresh consumption. Adequate care must be taken to see that the inputs used do not cause any pollution to the leaves. Barros *et al.* (1999) evaluated the sanitary and physico-chemical quality of water used in the irrigation of vegetable crops in Brazil. The irrigation water presented high faecal load (FC: 104 CFU per 100 ml) contaminating the vegetables. Lettuces irrigated with such water showed high levels of faecal contamination exposing the consumers of these vegetables to health hazards. Although the water contained considerable concentrations of salts, it did not present high risks of salinization or sodification of the soil. Another study by Guilherme *et al.* (1999) on the contamination of raw vegetables with intestinal parasites was investigated in samples collected from the market in Brazil. They found that 16.6 per cent of the samples were contaminated with intestinal parasites. Analysis of irrigation water samples showed that the water did not satisfy bacteriological standards of potability. They concluded that the contamination of vegetables occurs during the production phase and that a health education campaign directed at the producers is needed. Similarly, colonies and single cells of *Microcystis aeruginosa* and the hepatotoxin microcystin were retained by salad lettuce after growth with spray irrigation water containing the microcystin-producing cyanobacteria indicating the potential human exposure to cyanobacterial toxins via plant foods (Codd *et al.*, 1999). In field and glass house grown lettuces drip- and furrow-irrigated with waste stabilization pond effluent and with trickling filter effluent showed contamination levels under dry weather conditions in the order of 104 *E. coli* per 100 g (fresh weight) for lettuces (Bastos *et al.*, 1995).

Slugs also cause damage and contamination to lettuce. Their activity is highly dependent on sufficient soil moisture. Most of the feeding damage was caused by *Arion lusitanicus*. Morning irrigation gives a level of protection against slug damage as good as metaldehyde pellets combined with evening irrigation (Speiser and Hochstrasser, 1998). Maticic (1994) recommended proper water management of the soil (maintenance of water content in the soil within optimal range) to produce food with no toxic amounts of chemical substances. The drainage and irrigation should be considered as potential water management measures for environmentally sound and antitoxic food production.

The plant height and the number of leaves of *Lactuca sativa* L decreases under the influence of the gaseous pollutants compared to control inside rather than control outside, while the plant fresh and dry weight significantly decreases compared to control condition. According to the study conducted by several researchers,, there was reduction on number of chlorophyll and stomata index of *Lactuca sativa* L during the gaseous pollutant exposure period compared to control outside and control inside chamber. Reduction in physiological parameter was shown in number of chlorophyll of *Lactuca sativa* L.

7.4.0 Control of Disorders

Subsurface drip irrigation and associated mandatory minimum tillage practices signi-ficantly reduced the incidence of lettuce drop (*Sclerotium minor*) and the severity of corky root on lettuce compared with furrow irrigation and conventional tillage in California (Bell *et al.*, 1998). The suppression of lettuce drop under subsurface drip irrigation compared with furrow irrigation is attributed to differential moisture and temperature effects rather than to changes in the soil microflora or their inhibitory effects on *S. minor* (Bell *et al.*, 1998). The differential microclimates created by the irrigation methods do not affect *Bremia lactucae* infection, presumably when the mesoclimate is usually favourable for the pathogen. It is suggested that subsurface drip irrigation is a viable, long-term strategy for soil-borne disease management of lettuce (Subbarao *et al.*, 1997).

A high N dose at planting caused more internal scorch. Top-dressing with urea ammonium nitrate reduced the scorch risk compared with calcium nitrate. There was little difference between greater and daily irrigation intervals and the susceptibility depends on genotypes (Maync, 1997). Daily sprinkler irrigation applied in 3 short pulses (at 90-minute intervals) from sunset to midnight, between hearting until harvest, reduced tipburn incidence. This method of irrigation combined with Cultar application gave the best results (Obispo, 1997).

7.5.0 Weed Control

Weeds cause heavy loss in yield. Crop weed competition was studied by Robert *et al.* (1997). According to them, naturally occurring annual weeds at densities of 65-315/m^2 caused almost complete loss of marketable yield. After 2-3 weeks of crop emergence, removal of weeds is essential. For the control of weeds, hand weeding is a common practice and 3-4 hand weedings are adequate throughout the crop period.

Some herbicides were tested and found effective for the control of weeds. Incorporation of Benfluralin at 0.9 kg/ha in heavy soil before planting controlled most of the weeds. Also Propham at 2.5 and 3.75 kg/ha before planting gave satisfactory control of weeds in glass house (Lifshitz *et al.*, 1970). Several herbicides were tested in sandy loam soil and the pre-emergence herbicides which performed well were Propham + Diuron, Chlorpropham + Sulfallate and Propyzamide. Among these, Sulfallate was the best (Bradley and Hargreaves, 1977). Propyzamide at 1.5 kg/ha and Chlorpropham + Sulfallate at 1+4 kg/ha applied as preplanting were effective for the control of weeds (Cox, 1980). Post-emergence application of CDEC (Sulfallate) at 2 or 4 lb/acre enhanced weed suppression (Guzman and Dusky, 1981). Preplant irrigation by furrow or sprinkler followed by shallow tillage and lettuce planting 7 or 14 d later reduced in-crop weed density by up to 77 per cent and reduced hand-weeding and crop-thinning time by up to 50 per cent compared with the no preplant irrigation and no herbicide control. Preplant irrigation together with pronamide at either 0.67 or 1.34 kg ai/ha reduced weed density compared with the no preplant irrigation (Shem-Tov *et al.*, 2006). Pendimethalin at 1.1 kg/ha applied PRE or POST improved weed control by 57 and 42 per cent, respectively compared with pronamide. *S*-Metolachlor PRE at 0.7 kg/ha was less effective on weeds than pendimethalin, but maintained weed control level similar to pronamide. Thus Pendimethalin and *S*-metolachlor have potential for use in transplanted lettuce production and provide alternatives to current transplanted lettuce herbicides (Lati *et al.*, 2015).

Precision weeding can significantly reduce or even eliminate the use of herbicides in farming. To achieve high-precision, individual targeting of weeds, high-speed, low-cost plant identification is essential. Elstone *et al.* (2020) used the red, green, and near-infrared reflectance, combined with a size differentiation method, to identify crops and weeds in lettuce fields. Illumination was provided by LED arrays at 525, 650, and 850 nm, and images were captured in a single-shot using a modified RGB camera. A kinematic stereo method was utilised to compensate for parallax error in images and provide accurate location data of plants. The system was verified in field trials across three lettuce fields at varying growth stages from 0.5 to 10 km/h. In-field results showed weed and crop identification rates of 56 per cent and 69 per cent, respectively. Post-trial processing resulted in average weed and crop identifications of 81 per cent and 88 per cent, respectively (Elstone *et al.*, 2020).

7.6.0 Green House Cultivation

Lettuce is grown under glass or in plastic tunnel in temperate countries. Light, temperature and CO_2 are artificially provided as there is paucity of these factors in temperate countries. Seeds are sown in plastic or wooden tray or seed boxes in rows or broadcast. Soon after sowing, seeds are made firm in the medium. Seed boxes are irrigated with rose can or slightly dipped in water soon after sowing.

Adequate moisture is maintained for good germination and growth of the plants and seed boxes are kept at 12-15°C. After 12-15 days or when there is initiation of first true leaf, seedlings are pricked out in small pots or in small soil balls. Early pricking helps in quick growth and high early yield. White gravel or white synthetic material is kept in the balls for better reflection of light. After 4-5 weeks, these plants are shifted and planted in the green houses in bed at spacing of 45 x 45 cm or 45 x 30 cm or 30 x 30 cm depending upon cultivar. In green houses, 18-20°C/14-18°C day and night temperatures are maintained. Artificial light with fluorescent tube is provided and CO_2 at 1000 to 1500 ppm is maintained. Nutrients are also provided in higher quantity than field grown crop. In green houses, lettuce can be grown under hydroponics or nutrient film technique (NFT). Cresswell (1991) reported circulation of calcium nitrate solution (100 mg Ca/l) at night to be commercially acceptable means of reducing tipburn losses in lettuce crops grown using hydroponics. Zandvakili *et al.* (2019) assessed the effects of organic or chemical nutrient solutions on the growth and composition of lettuce in green house production. Hoagland and Arnon nutrient solution, an organic hydroponics solution, and no fertilization were factored with lettuce cultivars of different phenotypes, iceberg, romaine, loose head, and loose leaf. Lettuce growth was higher with Hoagland and Arnon solution than with the organic fertilizer or with no fertilization. The organic fertilizer increased growth above no fertilizer applied. Hoagland and Arnon solution gave the highest concentration of nitrogen, potassium, magnesium, and iron in the lettuce whereas lettuce grown with the organic fertilizer had the highest phosphorus. Plant nutrient accumulation differed slightly among the cultivars. Hoagland and Arnon solution produced high nitrate in leaves. Nitrate concentrations with organic or no fertilization were low. Loose head or loose leaf cultivars had higher nitrate than romaine or iceberg.

7.6.1 Soilless Cultivation

Growing leafy vegetables and salad crop in a floating system is the easiest and cheapest means of plant production, as this type of system provides high water and fertiliser efficiency and low environmental impact. Besides, soilless culture also allows direct control of the nutrient solution, making it possible to modify composition and concentration to attain expected results in relation to dry matter content, nitrate content or other organoleptic and structural (calcium) features of produce of salad crops including lettuce. Easy control of nutrient composition, lack of soil contamination, faster plant growth, shorter crop cycles, high product quality, and good consumer acceptance have made hydroponics an important alternative plant production technique (Gonnella *et al.*, 2002; Ohse, 2001; Petropoulos *et al.*, 2016). In tropical climates, a lettuce crop cycle is around 70 days in soil cultivation, whereas the duration is reduced to 30 days in hydroponic culture (Cometti *et al.*, 2013). Plants grown in hydroponics contain more minerals and chlorophyll compared to those produced in a conventional soil-based system (Coronel *et al.*, 2008). The macro- and micronutrient composition of a hydroponic solution determine plant growth, leaf number, leaf area, marketable yield, and crop quality including mineral and chlorophyll content (Fallovo *et al.*, 2009). Furthermore, Both *et al.* (1997) and Kang *et al.* (2013) reported that a light intensity of 200 to 300 $\mu mol{\cdot}m^{-2}\ s^{-1}$ and a CO_2 level of 400 to 600 $\mu mol{\cdot}mol^{-1}$ are suitable for the hydroponic culture of lettuce. Even though hydroponic culture can produce optimal plant growth (better yield and quality), its efficiency depends on many factors such as nutrient availability, crop genotype, growing method, and pest management (Spehia *et al.*, 2018). Although studies have been conducted on the hydroponic culture of lettuce, the results have been inconclusive about the influence of cultivar and nutrient solution composition on lettuce performance. For example, the fresh weight reported by Li *et al.* (2018) in a nutrient solution of 210 mg N/L was 88.8 g and 96.1 g for two lettuce cultivars. However, the fresh weight reported by Kowalczyk *et al.* (2014) in a nutrient solution of N, P, K of 140, 50, and 300 mg/L, respectively, was 245 g and 175 g for two other lettuce cultivars. Furthermore, the increasing N concentration on a hydroponic culture of lettuce from 0 to 60 mg/L increased shoots weight from 68.7 to 129.7 g (Soundy *et al.*, 2001). Sapkota *et al.* (2019) examined the effects of different nutrient solutions on the growth and weight of two lettuce cultivars grown in a floating hydroponic system. The lettuce cultivar Buttercrunch and the N3 nutrient composition (N = 250 ppm, K = 300 ppm, and Ca = 250 ppm) should be used for the hydroponic culture of lettuce to maximize lettuce growth and weight in similar environments.

Frezza *et al.* (2005) conducted an experiment with main aim to evaluate butter head lettuce quality in a soilless culture system. For this study they used butter head lettuce cultivar "Lores" and cultivated under two experimental growing condition. In the first experiment, floating system (FS) and soil culture (SC) were compared while in the second experiment, substrate culture (PC) with plastic gutter filled with perlite with three different levels of calcium in the nutrient solution were also compared. Growing cycles varied according to the culture system. In general, plants harvested from the soilless culture had a lower dry weight and leaf area, however significant differences were observed in productivity. Nitrate content was significantly affected by the production system. Ascorbic acid content showed no differences resulting from treatments. Floating system results showed some variation between the two years of the experiment. Colour intensity changes between production systems were non-significant. There was no variation in incidence of tip-burn relating to different calcium levels was observed in lettuce grown in perlite. The study conducted by Mahlangu *et al.* (2016) indicated that a solution containing N concentration of 100 and 120 mg/L may be sufficient to improve growth, yield and quality parameters of leafy lettuce grown in non-circulating hydroponic system.

7.7.0 Effect of Growth Substances

The effect of various growth substances on growth of lettuce has been studied extensively. The plant fresh weight increased with IBA and NAA at 50 or 100 ppm

Green House Cultivation of Lettuce

Soilless Lettuce Production under Protected Structure

(Singh and Saimbhi, 1967). The use of GA_3 at 10 mg/l stimulated respiration in lettuce seedlings and a peak was observed 12 hours after transfer to GA solution (Saxena *et al.*, 1978a). In another report, they observed that in dark germinated seedlings, 0, uptake attained a peak after about 9 hours compared with about 18 hours for light germinated seedlings (Saxena *et al.*, 1978b). Neocleous *et al.* (2007) observed that the application of GA_3 at 20 ppm and GA_3 + Kinetin at 200 + 20 ppm caused an increase in lettuce growth implying a shorter growing period with more crops per year and giving a high quality product. However, their role in lowering the tissue nitrate content was insignificant. The addition of 10^{-6} M GA_3 to the nutrient solution of a hydroponic floating system can promote growth and quality of lettuce (Miceli *et al.*, 2019).

8.0 Harvesting, Yield and Storage

8.1.0 Harvesting and Yield

Harvesting of lettuce depends on the type and the purpose for which it is grown. Head lettuce for market is allowed to develop a solid head, for home it is harvested before the head is well formed, and leaf lettuce for home use is harvested at any time but for market it is allowed to develop a full size. Of course, a higher yield will be obtained on a given area if plants are allowed to develop to full size than harvesting earlier, but the returns may be similar under both the practices (Thompson and Kelly, 1957). The standard size of crisphead for harvesting varies in different countries. The time to harvest is directly related to temperature and may range from 50 to 120 days. It is better not to harvest the lettuce immediately after a rain or while the dew is on it, because the leaves being crisp and brittle break easily in handling.

Lettuce should be properly graded and plants free from diseases and other injury should only be sent to the market. The average yield of lettuce per hectare varies from 10 to 14 tonnes (Sharma and Pandey, 1963).

8.2.0 Storage

Lettuce, if in good condition can be stored for a period of 3-4 weeks under refrigeration. For storage it is packed as for market and is placed in cold storage at about 32°F (0°C) and 90-95 per cent relative humidity. Storage helps to prevent market gluts and makes available the produce during the period of slack production (Thompson and Kelly, 1957).

Bessey (1960) found that pre-harvest field sprays of BA at 5-10 ppm helped lettuce to retain its fresh, green condition for an additional 3-5 days after packing. Postharvest treatments with BA also proved effective in delaying senescence (Zink, 1961).

The shelf life of the packed lettuce in high density polyethylene was 32 days at 4-5°C and 40 days at 0-1°C compared with 8-10 days for unpacked lettuce (Jeong *et al.*, 1990).

Fresh-cut lettuce is highly perishable with a shelf life of 2-3 days due to high water content and running metabolic activities. Shelf life and the visual quality of

Harvesting and Packaging of Lettuce

salad-cut lettuce can be affected by the production environment, vegetative maturity, storage temperature, choice of cultivar and its cut forms used (Couture *et al.*, 1993; Chiesa *et al.*, 2003). Cold storage and modified atmosphere (MA) packaging are used to extend its shelf life. Postharvest losses in relation to quality and quantity of food are a major problem all over the world, which can be controlled by optimal storage time and temperature (Khan *et al.*, 2012). Quality maintenance currently relies chiefly on the selection of suitable cultivars and the maintenance of low temperatures (optimum 1°C) during storage, distribution and retail (Ahvenainen, 2000). Postharvest storage time and temperature can have an influence on ascorbic acid content, antioxidant activity and total phenolic content (Lamikanra, 2002). The lettuce postharvest processing operation has several challenges due to large variations in respiration rate from different cultivars, large seasonal variation and postharvest storage duration prior to processing.

Results of the study conducted by Patil *et al.* (2017) revealed that whole leaf form of lettuce cultivars stored in refrigerated condition recorded minimum changes in physico-chemical parameters. With the advancement of storage period decrease in per cent moisture, ascorbic acid and polyphenol content was observed in all treatments irrespective of cultivars, forms and storage conditions. The maximum per cent browning and decay was observed in zero energy cool chamber (ZECC) storage which was minimum in refrigerated storage. The whole leaf form of lettuce cultivar Chinese Yellow recorded shelf life of 15 days when stored at refrigerated storage (5±1°C), whereas other cultivars (GKL-1, GKL-2 and GKL-3) recorded shelf life of 12 days. Whole leaf form of lettuce cultivars stored in ZECC condition recorded shelf life up to 6 days while freshcut lettuce recorded shelf life of 3 days in ZECC. Fresh-cut lettuce recorded accelerated physiological loss in weight, moisture loss with increase in browning and decay percentage. Decrease in per cent moisture, ascorbic acid and polyphenol content with increase in physiological loss in weight was recorded in all treatments over the period of storage irrespective of cultivars, forms and storage conditions.

9.0 Diseases, Pests and Disorders

9.1.0 Diseases

Lettuce is subjected to a number of diseases which are described below:

Fungal Diseases

9.1.1 Damping Off and Root Rot

This is caused by *Rhizoctonia solani* and *Pythium ultimum*. Damping off is of two types, pre-emergence and post-emergence. In the first case, seed germination is inhibited. Small seedlings are attacked at the soil surface in case of post-emergence. There is shrinking and brown rotting at the stem of the seedling at soil line and these seedlings thus fall down. In heavy wet soil at high temperature, there is more severity of the disease. Stem, leaves, petiole, midrib and complete head are also affected by the pathogens. *Pythium* causes severe stunting and rosetting followed by wilting and death of the plant. Lateral roots are severely rotted by this pathogen

(Coplin *et al.*, 1980). For control, proper aeration should be provided. Seed should be treated with 2 per cent ceresan and nursery soil should be drenched with 0.2 per cent captan or difolatan before sowing. Seedlings should be sprayed with 0.2 per cent difolatan at an interval of 10-12 days.

Phythium polymastum causes root rot. This disease is characterized by severe stunting and rosetting followed by wilting and death of the plant. Lateral roots and cortical tissues of the taproot are severely rotted. Control measures include removal of affected crop and crop rotation (Coplin *et al.*, 1980).

9.1.2 Downy Mildew

This is caused by *Bremia lactucae*. This is a serious disease of indoor and outdoor crops. Excess moisture enhances this disease. Light green to yellow areas develop on the leaves which later become brown. Cos and butterhead cultivars are more susceptible than crisphead and leaf types. Several fungicides were tested and CG 48988 at 8 oz/100 gal applied five times was highly effective followed by CG 38140 (Panlus *et al.*, 1997). Spraying of 0.2 per cent dithane Z-78 also controlled the disease. Resistant cultivars like Polo, Columbus and No. 83 (Cools and Stolk, 1980), and Soraya and Dolly (Engl *et al.*, 1982) can be recommended for cultivation.

Resistance to downy mildew has also been reported in cultivars Saffier and Mariska (Gustafsson, 1989).

9.1.3 Lettuce Drop

This is caused by *Sclerotinia sclerotiorum* and *S. minor*. Pathogen attacks stem near the ground and when there is serious attack it blocks translocation of solute. In the advanced stage, leaves dry and plants collapse. There is soft and watery rot on the stem and leaf bases. Fungi can persist for a long time in the soil. By continuous cropping of lettuce or other susceptible host in the same field, the chance of infection is increased. For controlling the disease soil sterilization with steam air mixture at 70-80°C is helpful (Sonneveld and Vooget, 1973). Significant control was found with DCNA 2-5 kg/ha applied at various times from preplant to early rosette stage but application immediately after thinning gave the best control. Mechanical soil application of DCNA at about 5 kg/ha provided better control than 6 kg/ha as post-thinning surface treatment (Marcum *et al.*, 1977). Single spray of benomyl at high concentration (3-4 kg/ha) at the rate of 4400 l/ha was more effective than standard multiple spray programme. Seed treatment or soil application did not control the disease (Hawthorne, 1979). Cultivars PI 250427, Taiwan, PI 251790, PI 255568, MSU 73-44 and Bibb has been reported to be more resistance to lettuce drop (Madjid *et al.*, 1983).

9.1.4 Botrytis Rot

This disease is caused by *Botrytis cinerea*. It is a serious disease in green houses. Small seedlings are usually attacked at nursery or early growing stage. Sometimes plants at late stage in the green houses are also attacked by the pathogen. Stem at soil line shows brown necrotic lesion and often the leaves touching the soil are also infected. Usually, there is vertical development of disease, thus showing one

Diseases, Pest and Physiological Disorders of Lettuce

Verticillium dahlia

Bacterial Leaf Spot

Downy Mildew

Dieback

Corky Root

Leaf miner

Tip Burn of Lettuce

sided development of infection. The disease spreads very fast at high temperature and wet conditions. Control of humidity and low temperature in the glass house are essential. Soil treatment and ample ventilation are also useful in controlling the disease in the glass house; in the field by maintaining the soil as dry as possible, and in transit by adequate refrigeration (Ryder, 1979a).

Bacterial Disease

9.1.5 Bacterial Rot (Slime)

This disease is caused by *Erwinia carotovora, Pseudomonas viridilivida* and *P. marginalis*. This is more a postharvest disease, particularly in transit and marketing. The disease spreads rapidly at high temperature. There is slimy decay of large internal leaves, decaying tissue looks like water-soaked which soon become brown. The conditions favourable for tip burn are conducive for this disease also. Weekly application of Cu-hydroxide or basic Cu-sulphate reduced the rot (Cho, 1977).

Viral Diseases

9.1.6 Lettuce Mosaic

There is much loss due to this disease, particularly in the dry season when there is high aphid population. This is a seed-borne disease and the leaves exhibit mottled yellow area. When the disease is severe plant fails to head. Sometimes, heads become irregular in shape. It causes internal tip-necrosis and stimulates rusty brown discolouration (Coakley *et al.*, 1973). Spraying of systemic insecticide to control the aphids is recommended. Mosaic free seed should also be used. Sometimes, there is incidence of sowthistle yellow vein virus which causes stunting and adversely affects the crop growth and yield (Duffuo *et al.*, 1970).

9.1.7 Big Vein

Yellow discolouration of veins is the primary symptom of big vein. Later, entire leaf becomes thickened. This is caused by seed-borne virus. Big vein disease can be controlled by soil application of methylbromide, dazomet, chloropicrin or 1,3-dichloropropene (White, 1980; Campbell *et al.*, 1980).

9.2.0 Pests

Aphid and cabbage looper are important insect pests of lettuce.

9.2.1 Aphid (*Myzus persicae, Myzus ascalonicus* and *Macrosiphum euphorbiae*)

Aphids attack the foliage of the plant and thus plant growth and quality are reduced. For controlling the pest, oxydemetonmethyl and monitor at 0.5 and at 0.75 lb/acre (McCalley and Wang, 1972) and fumigation with acetaldehyde 3-6 per cent or spray at 1.5-2 per cent (Stewart *et al.*, 1980) are recommended for *M. persicae*. Spraying of any systemic insecticide is effective for the control of aphid.

9.2.2 Cabbage Looper

It prefers lettuce as host for oviposition. Younger lettuce plants are more preferred for oviposition than older plants (Kishaba *et al.*, 1973, 1976). Any contact insecticide like malathion can be sprayed for the control of this insect.

9.3.0 Physiological Disorders

9.3.1 Tip Burn

Tip burn is visible on the lateral margins of the inner leaves of mature head. This is more common in glass house than in field crop. There are various reasons assigned for this disorder. These are high temperature (McGinty and Thompson, 1975), light, maturity and duration (Tibbitts and Rao, 1968), excess of nitrogen (Ashkar and Ries, 1971), calcium and boron deficiency and ontogenic age of plant (Crisp *et al.*, 1976), high Mn content (Honma and Bert, 1975), soil moisture content and high endogenous level of IAA. Mishagi and Grogan (1978b) reported an increase in head temperature of lettuce by enclosing mature intact plant with polyethylene in the field which increased the incidence of tip burn. The tip burn development is the manifestation of total calcium deficiency which is increased in plant during exposure to high temperature in growth chamber (Mishagi and Grogan, 1978a). This disorder has been found to be associated with the stage of rapid growth (Collier, 1983) and conditions favourable for rapid growth increase this malady. Close spacing resulted on lower incidence of tip burn because of slow growth (Cox *et al.*, 1976). Under wet condition where sprinkler irrigation is applied, the leaves have high Mn content and the disorder appears (Honma and Bert, 1975).

This disorder can be prevented or its appearance delayed, by raising the dark period and relative humidity and increasing Ca content in the leaves (Collier and Tibbitts, 1982). $CaCl_2$ can also be applied to reduce the incidence (Sonneveld and Ende, 1975).

Ozone at high concentration induced foliage injury to lettuce (Reinert *et al.*, 1972). Berahal and Kwolek (1975) have described non-parasitic troubles like crushing and bruising, russet spotting, and brown stain due to excessive CO_2 and pink and rib discolouration.

9.3.2 Other Disorders

Other than tip burn lettuce is also infected with a numbers of physiological disorders and is presented in Table 3.

10.0 Seed Production

Lettuce is a self-pollinated crop but some amount of cross-pollination has also so been reported. It is, therefore, suggested that two cultivars may be kept 50 m apart to produce pure seeds (Anon., 1971). The other agronomic practices for raising seed crop are more or less same as for the crop to be grown for leaf. There is no difficulty in the production of seeds of the leafy types. In the head types, a slight cut or tearing of upper leaves by hand helps in the seeds talk emergence. The harvesting of seed should be started when a substantial number of capitulums have burst and shown the pappus in the form of white hairs on the seeds, and good number of heads after opening have turned brown or dark brown but have not shown out the pappus (Nath, 1976). The leafy type produces about 500-600 kg seed

Table 3: Physiological Disorders of Lettuce*

Name of Physiological Disorder	*Plant Part Affected*	*Symptom*	*Cause*	*Management*
Bolting (seeders)	Head	Heads have elongated	Lettuce maturing	Production on moist muck land core inside during hot-dry weather and irrigation during hot-dry period
Brown rib	Leaves	The cells of the outside petioles turn pithy and brown. Leaf tissue also shows brown streaks, both inside and on the surface as this disorder develops.	High humidity and temperature, specially when rains shortly before harvest	Planting head lettuce on raised beds and avoiding irrigation during the last 2 weeks before harvest
Ribbiness/Pu-ffy heads	Head	The plants have bulky, prominent outer leaf ribs which do not fit tightly around the head, thus preventing norma head development	Excessively hot days with warm nights, shading by windbreaks, cloudy weather and low soil fertility increase the frequency of this physiological disorder	Maintaining a pH of 5.5-6.5 and irrigation during hot-dry periods
Rib- discoloration	Leaves	Light brown to black, oblong areas discolour inner surface of midribs the	High temperatures during growth	There is no control of rib-discoloration, but irrigating during hot weather can help reduce it
Russet spotting	Leaves	Numerous small tan to brown spots on the midribs of leaves which increases in number with storage periods	Induced by ethylene and temperatures above 3°C during storage	Maintain temperature near 0°C, holding in low Oxigen atmosphere andnot storing in same room with ethylene producing products reduces russet spotting development
Pink-rib	Head	Characterised by light to dark pink discoloration of larger midribs	Over-storage	Keep temperature near 0°C and avoid prolonged storage (>2 weeks)
Brown stain	Head	Characterised by lesions that are about 1 cm long. Their margins are darker than their centers, giving a halo effect	It is caused by CO_2 and aggravated by low oxygen and added CO	Do not hold intact lettuce heads where CO_2 concentrations might exceed 2 per cent. Shred and cut lettuce because it is less sensitive to CO_2 injury

* Masarirambi *et al.* (2018).

per hectare, whereas the head type (cv. Great Lakes) produces only about 100-125 kg/ha (Choudhury, 1972).

Seed production of lettuce may be started from a green house to produce breeder's seed. Breeder's seed is used to produce foundation seed that may be grown outdoors in progeny rows and checked for uniformity. Foundation seed is then used to produce stock seed in the field for sale to growers for market production. Lettuce seeds are produced in areas that provide isolation and a dry climate to minimize the impact of pathogens, minimal wind movement to prevent loss of seed, and sufficient warmth for seed maturation (Mou, 2008). Two major locations of commercial lettuce seed production are various part of Australia, Chile, China, southern France and Italy.

Different cultivars also differ in seed yield. The fields are inspected and rogued regularly to remove off-types and plants with diseases, particularly lettuce mosaic. Lettuce plants flower over a period of 50 - 70 days and seeds mature 12 - 17 days after flowering. Lettuce seeds may be mechanically harvested in two ways. One is the "shake method", in which seeds are usually harvested at two to three intervals by bending the seed head over and shaking the seeds into a container. The second, "cut method", is to combine heads at a given time, curing seeds on canvas sheets and then harvesting the seeds from seed heads. The "shake" method has the advantage of maximizing seed yield and allowing the separation of seed lots based on time of maturity, while the "cut" method is an once-over and therefore more rapid harvest operation (Mou, 2008).

11.0 Crop Improvement

Objectives of today's lettuce breeding programs can be divided into three general areas: resistance to diseases and insects, enhancement of production, and improvement in quality. These goals may overlap, *e.g.*, better disease or insect resistance may enhance production and quality, but they are discussed separately for clarity.

The breeding objectives during the early stages of lettuce improvement included development of heading, non-seed shattering, non-spiny, slow bolting (seedstalk formation) types, decrease in latex content, increase in seed size, uniform seed germination, increased leaf number and size. Later on, the emphasis continued on slow bolting, increased size and disease and pest resistance.

Hybridization is a technique most often used in a crop breeding program. The inbred nature of lettuce dictates the three principal breeding strategies used to improve the crop: pedigree method, backcross, and single-plant or mass selection. Attempts to produce F_1 hybrids commercially have not been successful. Lettuce pollen is heavy and sticky, and therefore is not easily transferred by wind. There is a dearth of pollinating insects for lettuce plants. Hand pollination cannot produce large numbers of seeds efficiently (Mou, 2008).

The breeding methods had been pure line selection, hybridization within and between the several forms, pedigree selection and back-crossing. Selection of occasional off-type variants for colour, size, and adaptability to different agroclimatic regions has been followed in many cases. Damarany (1994) carried mass selection for tipburn resistance in 4 cultivars and found significant differences in the 3 cycles of mass selection relative to the base population in both tipburn resistance and marketable heads percentage for all 4 cultivars. Tipburn was reduced from 50.9 to 28.2 per cent after 3 cycles of selection with an increased percentage of marketable head and with decreased bitterness score. The effect of additive and dominant genes was equally important in inheritance of rosette trait with dominant genes predominated in the inheritance of this trait (Zdravkovic, 1991). There was little evidence for heterosis and no attempt seems to have been made to produce F_1 hybrids, although male sterility genes have been reported (Ryder, 1979a). However, later Langton *et al.* (1990) recorded heterosis in crisphead lettuce hybrids by parents of dissimilar origin. Trials of F_2 segregants from one of these heterotic hybrids enabled predictions to be made for 10.2 to 24.1 per cent yield increase over the better parent by the retention and clonal multiplication of the 10 per cent highest yielding selections. Genetic coefficient of variability, heritability and genetic advance as a percentage of the mean were high for head size and weight, whereas these estimates were low for non-wrapper leaves (Thakur *et al.*, 1997).

Most cultivars in parts of the Mediterranean basin were cos and cos-like types, resembling the probable originals which continue to be very popular even at present. In Northern Europe, the emphasis was on butterhead lettuces which proliferated to Holland, England and France. In recent years, interest, of course, has been in evolving crisphead cultivars for both outdoor and green house production. In the United States and Australia, butterhead cultivars introduced from Europe became secondary in popularity to the crisphead type and more emphasis was laid on size, weight and ability to withstand the rigours of long distance shipment in cold and wet environment. The first cultivars to be evolved were New York types which were relatively large and could be shipped to long distances but were susceptible to brown blight.

Ryder (1985) reported that a partially dominant allele (*ef*) for early flowering in lettuce line USDA 56679 reduces days to flowering by about half. The backcross breeding procedure is useful to accelerate the transfer of useful alleles to a recurrent parent. In Egypt, Damarany (1989) noticed that cv. Dalady bolts readily with the rise in temperature in spring. With a view to improving bolting resistance via heat tolerance the cultivar Balady was crossed with 15 foreign cultivars which flowered later than Balady. The F_1 hybrids of Balady x Minetto 525 and Balady x Lobjoit cos had longer flowering time and higher fresh weight. Selection in F_3 lines for late flowering showed a 2.6 to 13.7 per cent delay in flowering compared to F_1 and decreased in fresh weight of 8.6 to 30.3 per cent. Gong (1998) identified lettuce cultivars 9547 and 9542 for good field growth and vigour, bolting resistance and high yields under high temperature conditions in summer besides they were significantly more resistant to heat and bolting than the leading domestic cultivar Guangdongruanwei.

Shannon *et al.* (1982) developed an improved screening technique to evaluate salt tolerance. He screened 85 US lettuce cultivars and breeding lines, and observed a considerable variability in this regard. The cultivars Climax, Climax 84, Shawnee,

Tom Thumb, Fulton and Wintergreen were reported to be the most tolerant. Better salt tolerance of Climax cultivar than that of Calmar had also been reported by Grant and Arthur (1982).

Robinson (1986) reported that the most effective ethyl methane sulfonate (EMS) treatment for inducing mutations was soaking the seed in 0.5 per cent EMS for 4 h at 20°C. It resulted segregation of 80 per cent of M_2 lines for seedling mutants. Chlorophyll mutant, leaf shape mutant and early flowering mutant have also been identified. A collection of different species was made by Whitaker and Provvidenti (1983) from Turkey and Greece for enriching the existing germplasms on which further improvement work depends.

Techniques of Hybridization and Use of Male Sterility

For ensuring the synchronization of flowering period, planting of parents with different flowering time generally planted staggeringly. As lettuce is a long-day plant, therefore artificial lighting may be provided in the green houses grown crop to fulfil photoperiodic requirement and enforce bolting, especially during winter season when the day length is short. The flower of lettuce consists of 10 to 20 florets that make up the flowering head or capitulum. Flowers generally open in the early morning during sunny and warm days, while on cloudy, cool days, they may not open until late morning. Few wild relatives may flower earlier than cultivated lettuce during the day; the early flowering parents can be kept in a cool dark room or growth chamber to "wait" for the late flowering ones. The floret opens its five (usually linked) petals and exposes the five anthers that fuse to form a tube through which the pistil emerges. The anthers shed pollen inside the stamen tube, and the elongating style with stigma sweeps the pollen along as it emerges from the top of the tube. The stigmatic lobes then begin to split, gradually form a "V" shape, and finally curl backwards, signalling the end of pollen reception (Mou, 2008).

Different type emasculation methods have been proposed over time, but most common method used by the lettuce breeders is a method advised by Oliver (1910), or a modification of it. The open flower generally washed with a stream of water using a wash bottle several times during the emergence of pistils to remove adhering pollen. Excess water may be removed by a stream of air or blown by mouth. The flower is left to air dry for a short time, and pollination is made by taking an open flower from the male parent and pressing it onto the emasculated flower before its stigmatic lobes begin to curl back. If it is desirable to remove pollen from a large number of flowers, a fine spray nozzle may be placed over the plants and the entire panicle can be washed intermittently or continuously during the pistil emergence period (Ryder and Johnson, 1974). To identify the seedling from crossed seed of crossed flowers *i.e.*, to distinguish hybrid and selfed seed, the male plants should carry dominant marker genes, preferably expressed in the seedling stage.

Genetic male sterility (GMS) is a useful system for the production of hybrid varieties in self-pollinated plants. Hayashi *et al.* (2011) obtained a GMS line developed from a spontaneous mutation in lettuce (*Lactuca sativa* L.). Genetic analysis revealed that the sterility was controlled by a recessive gene which was named *ms-S*.

11.1.0 Breeding for Low Nitrate and Cadmium Content

High nitrate content is considered a negative quality trait in lettuce and breeding efforts have been directed towards obtaining cultivars with reduced nitrate accumulation. Previous research has shown that with respect to the nitrate content in fresh matter, lettuce genotypes responded differently to changes in environmental conditions between experiments. Reinink (1992) estimated variance for nitrate content in the F_2 and F_3 generations of ten lettuce crosses. Estimates of wide-sense heritability of the F_2 from crosses between a high nitrate genotype and four low nitrate genotypes ranged from 0.44 to 0.74. Reinink (1993) also studied the inheritance and characteristics associated with genotype x environment (GE) interaction for nitrate content. He crossed two lettuce genotypes one parent showing nearly constant nitrate content and decreasing dry matter content contrary to the other parent. He observed a recombination of genes for low nitrate content from both parents, indicating the possibility of a further reduction of nitrate content (Reinink, 1993).

Cadmium accumulation is another negative factor in lettuce. Thomas and Harrison (1989) assessed cadmium concentration in the parents, F_1, F_2 and F_3 of PI278080 x PI140398 and PI169493 x PI278080. Broad-sense heritability estimates, calculated from parental and F_2 variances, were 0.38 and 0.29, respectively. Narrow-sense estimates, calculated by the standard-unit regression analysis of F_3 family means on F_2 parents, were 0.05 and 0.30, respectively. They found that selection for low concentration did not affect yield.

11.2.0 Resistance Breeding

Downy mildew, caused by the oomycete Bremia lactucae, is the most important disease in lettuce worldwide. *Bremia lactucae* is a biotrophic pathogen that primarily infects the foliar tissue, reducing yield and decreasing the quality of the marketable portion of the crop. The use of cultivars carrying dominant resistant genes (*Dm* genes) is the most efective strategy to control this disease. However, *B. lactucae* rapidly evolves to new virulence phenotypes that defeat individual *Dm* genes. For this reason, lettuce breeding programs continually seek new sources of resistance to downy mildew and new resistance genes (Crute, 1992; Lebeda *et al.*, 2014; Parra *et al.*, 2016).

For the last fifty years, breeding for resistance to *B. lactucae* has relied on the introgression of new genes from wild species (Crute, 1992; Beharav *et al.*, 2006; Parra *et al.*, 2016). Of the 100 wild Lactuca species described thus far, only 14 are known to be natural hosts of *B. lactucae* (Lebeda *et al.*, 2002). *L. serriola* and *L. saligna* have been used as sources of resistance to downy mildew in multiple breeding programs. *L. virosa* also possesses race-specifc resistance to *B. lactucae*, but its use in breeding programs has been restricted due to infertility barriers between *L. virosa* and *L. sativa*; however, interspecifc crosses followed by embryo rescue have occasionally allowed the transfer of resistance from *L. virosa* to *L. sativa* (Maisonneuve, 2003). Introgression of resistance from *L. serriola*, *L. saligna*, and *L. virosa* through repeated backcrosses to *L. sativa* has been carried out by several public and commercial breeding programs resulting in a large increase in the number of resistance genes being deployed (Michelmore *et al.*, 2013).

The genome of *L. sativa* cv. Salinas has been sequenced and assembled (Reyes-Chin-Wo *et al.*, 2017). The latest assembly covers 2.4 Gb of the total estimated 2.7 Gb lettuce genome; these sequences are ordered in nine chromosomal pseudomolecules (Reyes-Chin-Wo *et al.*, 2017). The 27 known *Dm* genes are located in major resistance clusters (MRC) that contain genes encoding nucleotide-binding leucine-rich repeats (NLRs) (Christopoulou *et al.*, 2015a, b). Currently, over 50 genes for resistance to lettuce downy mildew have been reported and genetically characterized to varying extents (Parra *et al.*, 2016). While 27 *Dm* genes have been mapped to individual clusters in the reference genome, another 23 resistance factors remain uncharacterized (Parra *et al.*, 2016). Recently, 11 new major resistance genes for lettuce downy mildew were introgressed from wild *Lactuca* species and mapped to small regions in the lettuce genome (Parra *et al.*, 2020).

In 1941, Jagger and others reported the release of three selections, Imperial 2, Imperial 3 and Imperial 7 from New York cultivars, resistant to brown blight. Crosses between New York cultivars and downy mildew resistant cos lettuce from France produced lines resistant to downy mildew. These lines were crossed with brown blight resistant strains and resulted in Imperial C and Imperial F, resistant to both the diseases and these strains became the principal cultivars in the United States. Later, Great Lakes cultivar was developed from the cross of an old cultivar, Brittle Ice with Imperial 152 followed by selection for 4 generations and a cross to imperial 615 (Barrons and Whitaker, 1943).

Great Lakes have large, dark green heads; slow bolting habit and excellent shipping quality. Further selection led to development of more strains of Great Lakes and eventual replacement of Imperial types in the United States. The Imperials and Great Lakes were susceptible to a downy mildew and cultivars Valverd and Calmar resistant to the new race were developed. Calmar was released in 1960 (Welch, 1965), and because of its greater size, uniformity and resistance to the new race of downy mildew became the dominant cultivars in California. Noorwood *et al.* (1985) observed that F_5 selection of cross Avoncrisp x Iceberg were more resistant to downy mildew than parent Iceberg. Over the last 30 years, six resistance alleles (*Dm2, Dm3, Dm6, Dm7, Dm11* and *Dm16*) located in two linkage groups have contributed to the control of downy mildew (*Bremia lactucae*) in lettuce crops grown under protection besides a genetically uncharacterized resistance factor, R18, has also begun to assume importance (Crute, 1992). It is predicted that variations of this integrated control strategy involving the use of appropriately selected *Dm* gene combinations may prove effective for some time (Crute, 1992).

The cultivar Vanguard derived from crosses including *L. sativa, L. serriola* and another wild species. *L. virosa*, was released in 1958 (Thompson and Ryder, 1961). Ryder (1979b, c) had reported the release of crisphead cvs. Salinas and Vanguard 75 resistant to lettuce mosaic. Merit, a tolerant cultivar to big vein disease was identified and tested, and resistant cultivars were developed (Ryder, 1979a). The cultivar Marquette resistant to corky root rot had been released by Sequeira (1970). Resistance to root aphid had been located in cultivars Avoncrisp and Avondefiance in the Imperial crisphead cultivar E-4 in California (Ryder, 1979a). Resistance to cabbage looper had been found in *L. serriola* and *L. saligna* (Kishaba *et al.*, 1973; Whitaker *et al.*, 1974). The interspecific cross between *L. sativa* lines 54671 and *L. seriola* has been found to hold good promise for developing material resistant to this insect (Kishaba *et al.*, 1980).

Simko *et al.* (2010) developed two romaine breeding lines of lettuce SM09A and SM09B resistant to the dieback disease and with good shelf life. SM09B was selected from a cross between Darkland and PI 491224. SM09A was developed from Green Towers (Darkland x PI 491224). In replicated field trials, these two breeding lines showed complete resistance to dieback. Testing of salad-cut lettuce in modified atmosphere packaging indicated slower decay in the two breeding lines compared with other dieback-resistant romaine varieties.

Aphid species infecting lettuce were *Nasonovia ribisnigri*, the most common aphid species, followed by *Macrosiphum euphorbiae, Pemphigus bursarius, Uroleucon sonchi* and *Myzus persicae.* Accessions with monogenic resistance to *N. ribisnigri* and *P. bursarius* showed complete resistance in all experiments (Reinink and Dieleman, 1993). Helden (1993) studied the monogenic absolute resistance of lettuce to aphid, *Nasonovia ribisnigri*. The only clear difference in behaviour with isogenic resistant and susceptible lettuce lines was apparent after a phloem sieve element was reached and feeding started and no chemical differences were found (Helden, 1993). Helden *et al.* (1996) found that the monogenic, dominant and absolute resistance factor of lettuce to the aphid *Nasonovia ribisnigri* is located in the phloem sap of the plant. Studies to isolate the allomones through bioassay-guided fractionation indicated that a fairly polar compound is involved (Crute, 1992).

11.3.0 Breeding for Longer Self-life

Lettuce is an extensively cultivated vegetable easily used as fresh-cut salads and is very popular to the consumers because of their convenience. Production and processing of fresh-cut lettuce is continually evolving, offering more products and becoming more efficient. Breeding new lettuce cultivars specialized for this market can offer further improvements, but has received attention only within the last decade. Investigations on improving the postharvest quality of lettuce thus far have largely focused on washing, packaging and storage (Gil, 2016), but there has been little research on breeding for improved postharvest quality, perhaps reflecting a relatively poor understanding of the genetic basis of these complex traits.

Many traits traditionally addressed in lettuce breeding (Ryder, 1986; Lebeda *et al.*, 2007; Simko *et al.*, 2014; Hayes and Simko, 2016) are also important for breeding cultivars specialized for fresh cut processing. Traits that are addressed specifically for fresh cut processing include novel plant architectures that improve production efficiency and yield, enhanced resistance to preharvest internal defects, reduced post processing discoloration, and extended shelf life.

Alterations in plant architecture may improve head weight and bulk harvest yield or the proportion of the plant that is useable for processing, or allow avoidance of some physiological defects. Advancements in alternate plant architectures have been primarily conducted by private seed companies, many of which result in cultivars protected by utility patents. As such, there is little peer reviewed literature reporting on the performance of cultivars possessing new plant architectures. Plants with short internodes increase the proportion of the plant that is leaves, increasing the yield of the useable portion of the plant. Short internodes also result in shorter cores, which are easier to remove during harvest. In romaine, a plant with short internodes in conjunction with narrow petioles (Ammerlaan, 2010) and/or longer petioles has been proposed as a model for mechanical harvesting. With this type of plant, harvesting could be conducted using equipment similar to what is used to harvest baby leaf lettuce. Leaves are harvested by cutting above the core, which is left attached to the roots. Currently, the vast majority of lettuce grown for fresh cut processing is hand harvested.

Plant weight is reciprocally related to plant density (Scaife and Jones, 1976), and plant stands have already been maximized (Simko *et al.*, 2014). Therefore, yield cannot be improved appreciably by increasing the number of plants per hectare or by increasing the diameter of the plants. Biomass yields may be improved by breeding for taller heads that arise from longer leaves. Cultivars with iceberg leaf characters, but with taller romaineshaped leaves, have been reported from iceberg x romaine crosses (Waycott, 2009).

Lettuce has extensive variation for leaf color, which includes a wide range of green colors and intensities. Dark green leaves may show more bruising, which appears as dark patches or black creases immediately after processing. Light green leaves may show less bruising. Newer romaine cultivars often have lighter green leaves compared with older cultivars, and therefore may be preferable for processors. In traditional romaine cultivars, blanched, yellow, or light colored tissue typically develops within the lettuce heart, and head closure is needed to form a heart. The closure of the top of the romaine plant increases the likelihood of developing tipburn, a physiological defect (Jenni and Hayes, 2010). Light green romaine plants can have more open heads that can reduce the risk of tipburn while still producing light colored leaves. Similar uses of light green leaf color in iceberg type cultivars are not known. Many stem lettuce accessions have solid midribs, whereas romaine cultivars often have a section of the midrib that is hollow. Increasingly, romaine cultivars are being bred with solid midribs (Avila and Mederos, 2013). Solid midribs add weight to each plant, which increases bulk harvest yields. Solid midribs also have reduced surface area after processing, which may reduce the amount of discoloration after processing.

The genetics for some of the traits needed to breed these types of cultivars are known to a degree. Leaf color is controlled by a number of major genes (Robinson *et al.*, 1983). Stem fasciation was quantitatively distributed in a single biparental cross (Eenink and Garretsen, 1980). Stem length, leaf length and heading characteristics appear to be controlled by multiple major and minor genes or QTLs (Hayes *et al.*, 2011b; Hayashi *et al.*, 2012; Jenni *et al.*, 2013).

Lettuce harvested for fresh cut processing must be free from any defects, damage, or disease. Rib discoloration is a physiological defect of iceberg lettuce that occurs under warm production conditions (Jenni, 2005). The defect is associated with heads in an advanced stage of maturity. The severity and incidence of rib discoloration are conditioned by several small effect and environmentdependent QTLs spread throughout the genome (Jenni *et al.*, 2013). Iceberg cultivars with elevated resistance have been developed for production in warm weather areas (Jenni and Yan, 2009). Tipburn is problematic in nearly all lettuce production regions, and is promoted by a diverse range of environmental conditions (Collier and Tibbitts, 1982). Plant architecture strongly affects the occurrence of tipburn; plants with open tops only rarely develop tipburn (Jenni and Hayes, 2010). Over mature heads are more likely to develop tipburn (Jenni and Yan, 2009). Cultural control methods are generally limited, and growing resistant cultivars, if available, is the most effective option. Several modern iceberg cultivars generally have adequate levels of resistance, while more work is needed in breed tipburn resistant romaine (Jenni and Hayes, 2010). Tipburn resistance is subject to extensive genotype x environment interaction, thus genotype performance varies from environment to environment. Breeding is conducted by continually discarding plants, lines, or families that have the defect. Extensive multilocation testing of the resulting inbred lines is needed to identify genotypes that consistently exhibit low tipburn. The genetics of tipburn resistance are only now being elucidated. Identifying QTLs that condition resistance independent of plant architecture is most valuable, since they can be bred into a diverse range of lettuce types. QTLs were identified in two intraspecific populations, two of which appeared to be independent of plant architecture (Simko *et al.*, 2014). The remaining QTLs co located with QTLs for core height, head closure, and head maturity. Five QTLs were identified in a cross between Emperor (an empire type iceberg) and El Dorado (a vanguard type iceberg) (Jenni *et al.*, 2013). One QTL for tipburn on LG 5, which co segregated with QTLs for head type, firmness, and stem length, controlled up to 70 per cent of the phenotypic variation. Inheritance of resistance QTLs from both parents in a single line resulted in genotypes with better tipburn resistance than either parent. Bolting is a normal development phase of lettuce that is required for flowering and seed production. However, plants that have begun to bolt are not useful for processing. Selection for late bolting cultivars is preferable for nearly every type of cultivar. Additionally, slow bolting may enable improved 'field holding' or 'holding ability' of the cultivar (Waycott and Ryder, 1992; Abbott, 2010). This is the ability to store an unharvested crop in the field without the plants becoming over mature. The trait could enable greater flexibility in harvesting and processing schedules and reduce waste. Several major genes and QTLs have been identified that control the time to initiation of bolting (Simko *et al.*, 2014). Premature bolting is when bolting occurs before the plant has reached a harvestable stage, and may occur in lettuce grown under higher than ideal temperatures (Simko *et al.*, 2014). Resistance to premature bolting is known, though the genetics of the trait have not been described. A group of romaine and iceberg cultivars, which are generally interrelated to each other, are typically grown in environments where premature bolting is a concern.

The cutting used in fresh cut processing of lettuce results in brown or pink discoloration on the cut surfaces via the phenylpropanoid pathway (Smyth *et al.*, 1998; Kim *et al.*, 2005). Symptoms typically occur within a few days after processing and are objectionable to consumers. Modified atmosphere packaging (MAP) is used in commercially prepared products to control the discoloration (Kim *et al.*, 2005). This is created by sealing packages with air and allowing the MAP to develop as a result of respiration and package film gas permeation (passive MAP) or by flushing the package with N_2 or a low O_2, high CO_2 atmosphere (active MAP) (Kim *et al.*, 2005). Breeding crops with reduced discoloration may reduce or eliminate the need for MAP. The physiology responsible for discoloration has been well studied (Saltveit, 1997). Phenylalanine ammonia lyase (PAL) controls the influx of metabolites into this pathway.

Polyphenol oxidase (PPO) and phenol peroxidase (POD) are primarily responsible for the oxidation of phenols and the resulting discoloration. In some crops, these genes may collocate with the region of a QTL that controls discoloration, thereby implicating these genes as the possible functional genes that control segregation for a trait in a breeding population. For example, a *PPO* gene in potato maps to the same region as a major QTL controlling browning (Werij *et al.*, 2007). The inheritance of pinking and browning resistance was investigated in an RIL population from the cross between the iceberg cultivar Saladin and the Batavia cultivar Iceberg (Atkinson *et al.*, 2013).

Fresh cut lettuce can exhibit short shelf life, becoming unsaleable as early as 8 days after harvest. Descriptions of the symptoms that characterize the end of shelf life vary. Zhang *et al.* (2007) described lettuce with breakdown, bruising, or damage. Hayes and Liu (2008) described salad pieces as waterlogged, limp, and slimy. In other reports, the appearance of browning or pinking defined the end of shelf life (Martýµnez Saµnchez *et al.*, 2012). Irrespective of the description, the symptoms are objectionable, and lettuce with even minimal defects is rejected, leading to waste at the distribution, retail, and consumer level. In the USA, about 40 per cent of all lettuce is wasted at the consumer level (USDA, ERS, 2014). The genetic relationship between shelf life and leaf developmental and biophysical traits was studied in a population of 60 RILs from a cross between the wild species *L.serriola* and Salinas (Zhang *et al.*, 2007). The relationship between shelf life and cell wall strength was pursued further through transgenic manipulation of xyloglucan endotransglucosylase/hydrolases (XTHs) (Wagstaff *et al.*, 2010). The inheritance of rapid decay was investigated in an RIL population derived from a cross between the iceberg cultivar Salinas 88 and La Brillante (Hayes *et al.*, 2014).

Improvements to plant morphology can increase yield, quality, and production efficiency. Crops must meet high stringencies for internal defects and cultivars specialized for fresh-cut often need enhanced levels of resistance to these problems. Development of cultivars that do not discolor could minimize or eliminate the need for MAP. New cultivars with longer shelf-life could reduce waste and increase the distribution system's efficiency. The nutritional content of fresh-cut lettuce can decline after processing. Research that addresses this issue may improve the popularity of fresh-cut lettuce. Lettuce and its wild relatives have genetic variation for plant morphology, quality, and resistance to abiotic stresses that can be used to breed improved lettuce cultivars. Quantitative trait loci (QTL) mapping has determined the inheritance of many traits useful for the fresh-cut industry. This knowledge should be used to devise molecular breeding methods that accelerate cultivar development (Hayes and Simko, 2016). Investigations on improving the postharvest quality of lettuce thus far have largely focused on washing, packaging and storage but there has been little research on breeding for improved postharvest quality, perhaps reflecting a relatively poor understanding of the genetic basis of these complex traits. Genetic variation for postharvest traits has been observed, demonstrating the potential for breeding lettuce for improved quality. Numerous investigations have identified significant differences in the nutritional quality of lettuce, particularly between red and green cultivars, with red varieties containing increased densities of phenolic compounds and anthocyanin (Damerum *et al.*, 2020).

12.0 Biotechnology

Biotechnological tools have been employed to produce transgenic for resistance to downy mildew and tomato spotted wilt virus. Results of some important investigations on biotechnology of lettuce are summarized below.

12.1.0 Tissue Culture

Tissue culture work on lettuce has been centered on micropropagation, embryo rescue, protoplast fusion and somatic embryogenesis. Lettuce has proved amenable to manipulation in tissue culture using a variety of media and growth regulators. Doer-shug and Miller (1967) were the first to regenerate shoots from explants of several tissues and several genotypes.

12.1.1 Micropropagation

Koevary *et al.* (1978) successfully induced shoot formation from leaf explants using IAA (28.5 µM) and kinetin (2.3 µM). Sasaki (1975a,b) employed two media to regenerate buds from etiolated hypocotyls. Maximum regeneration was induced by planting explants on an initial medium containing 2,4-D (0.45 µM) and kinetin (0.46 µM). Alconero (1983) induced plantlets to root by culturing with NAA (5.4-10.8 µM). Leaf petioles exhibited the best regeneration capacity (Sun *et al.*, 1998). Regeneration of plants from axillary buds of harvested and stored heads of field-grown Iceberg lettuce gave rise to viable plants (Bloksberg and Saltveit, 1986).

Lettuce is easy to culture *in vitro*; however, hyperhydricity has proved to be a serious problem after subcultures. The hyperhydric plantlets grow rapidly *in vitro*, but do not survive transfer to the green house. Teng and Liu (1993) described a method for reducing hyperhydricity of lettuce (cv. Great Lakes 118) shoots regenerated from culture on liquid and gelled media. In liquid medium, hyperhydricity was decreased by 30 per cent by using a filter paper raft or a polyester screen raft to prevent immersion of inoculum in the culture medium. In gelled medium, replacing agar with the water absorbent resin successfully prevented hyperhydricity without reducing organogenesis.

12.1.2 Embryo Culture

Various genes for disease resistance identified in wild *Lactuca* are difficult, even impossible to exploit in lettuce breeding due to sexual incompatibility between *L. sativa* and wild *Lactuca* spp. Maisonneuve (1987) obtained F_1 hybrids from crossing 2 accessions of *L. saligna* with 2 genetically male-sterile lines of *L. sativa* (sterility

Micropropagation of Transgenic Lettuce

The Course of Lettuce Micropropagation.

a. Plantlet multiplication *in vitro*; b. Early stage of formation of sequent plantlets from stimulated axillary buds of an initial plantlet; c. Fully developed multiplantlet before fission; d. Later stage of plantlet growth and rooting *in vitro*; e. Rooted plants to transfer to soil; f. *Ex vitro* growth of transgenic lettuce plant (on the right) in comparison to non-transgenic control constantly growing in green house (on the left) (Pniewski *et al.,* 2016).

controlled by *Ms7* and by 2 recessive genes respectively) and BC_1 hybrids were obtained from crossing 4 interspecific hybrids (*L. sativa* x *L. virosa*) with several commercial cultivars of *L. sativa*; both sets of hybrids were produced using embryo culture. From the 2 accessions of *L. saligna*, 17 and 83 hybrid plants respectively were obtained.

12.1.3 Protoplast Culture

Berry *et al.* (1982) observed only low plating efficiencies with leaf protoplasts of green house-grown plants but obtained plating efficiencies as high as 2-30 per cent for leaves, cotyledon and roots of aseptically grown seedlings and they were able to regenerate plants. Nishio *et al.* (1988) isolated protoplasts enzymatically from cotyledons and cultured on modified MS medium supplemented with BA and NAA. Colonies were obtained at a frequency of 300 per 7.5 x 10_4 protoplasts cultured and most of them regenerated plants following transfer to solid medium. Shoots were obtained from protoplasts of 7 lettuce cultivars within 2 months of culture. Sasaki (1975a) and Alconero (1983) regenerated shoots from calli grown from cell suspensions. Sasaki cultured cell suspension derived from hypocotyl explants on White's medium with 2,4-D (0.45 µM) and kinetine (0.23 µM).

12.1.4 Protoplast Fusion

Tsuzuki and Nakagawa (1991) used electrofusion to fuse mesophyll protoplasts of *Brassica campestris*, *L. sativa* and tri-hybrid cells were obtained. Matsumoto (1991) produced interspecific hybrids by protoplast electrofusion selected on the basis of the inactivation of *L. sativa* protoplasts in 20 µM iodoacetamide for 15 minutes and the inability of *L. virosa* protoplasts to divide in the culture conditions used. Of the 71 calli formed, 21 differentiated shoots on LS medium containing 0.1 mg NAA and 0.2 mg BA/l. Most regenerated plants exhibited intermediate morphology.

12.1.5 Somatic Embryogenesis

Xiaoli *et al.* (1992) induced somatic embryos in lettuce cotyledons cultured on MS medium containing either 2.0 mg BA and 0.2 mg/l NAA or 0.2 mg/l BA and 2.0 mg/l NAA. Both combinations induced a frequency of over 70 per cent. Explants cultured on medium with 2,4-D alone (2.0 mg/l) did not produce somatic embryos. Yu and Tang (1998) studied the best combination of hormone and explant during organogenesis of lettuce hypocotyl. MS medium supplemented with 0.88 mg/l NAA+ 0.1 mg/l BA proved to be the best.

12.2.0 Biochemical and Molecular Markers

12.2.1 Isozymes

Isozymes have potential for use as markers for resistance in practical plant breeding programmes. Isozyme variation was surveyed by Kasseli and Michelmore (1986) in cultivated lettuce and related wild species. Allozyme variation as detected by starch gel electrophoresis was scored among 31 collections of *L. sativa* and wild *Latuca* species and used to assess genetic diversities and phylogenies. Cole *et al.* (1991) used eight accessions each of *L. virosa, L. serriola, L. saligna* and *L. sativa* obtained from the Wellesbourne Vegetable Gene Bank. Using multivariate statistical techniques, 4 allozyme bands were identified which distinguished the 4 species together with an additional 2 bands which were related to the resistance of *Lactuca* species to *Pemphigus bursarius*.

12.2.2 Molecular Markers

12.2.2.1 Identification of Genotypes

Yamamoto *et al.* (1994) used PCR analysis with 21 originally designed RAPD primers and 8 sequence-specific primers for amplifying 4 specific DNA fragments of 12 lettuce varieties. Use of these primers revealed polymorphisms among varieties facilitating DNA fingerprinting.

Kesseli *et al.* (1991) examined the variation in RFLP loci in 67 accessions of *L. sativa* and 5 related species. Fifty-five clones were used to probe genomic DNA digested separately with 3 restriction endonucleases which defined 143 loci. Variation in *L. sativa* was sampled in detail and it was found to be closely related to *L. serriola* but not to any of the other species: *L. saligna*, *L. virosa*, *L. perennis* and *L. indica*. Kesseli *et al.* (1994) also constructed a detailed genetic map from the F_2 population of a single intraspecific cross of *L. sativa* (n = 9). It comprised 319 loci including 152 RFLP, 130 RAPD, 7 isozymes, 19 disease resistance and 11 morphological markers. Kesseli *et al.* (1993) mapped the recessive gene (*plr*) for resistance in lettuce to *Pseudomonas lactucae-radicis* to a cluster of genes for resistance to taxonomically diverse pathogens. Four RAPD loci linked to *plr* were identified by bulked segregant analysis of a previously unanalyzed F_2 popuation segregating for *plr*. RFLP and RAPD markers showed similar distributions throughout the genome and identified similar levels of polymorphism. RAPD loci were much quicker to identify.

Landry *et al.* (1987) also constructed a detailed linkage map using 53 genetic markers, including 41 RFLP loci, 5 *Bremia lactucae* resistance genes, 4 isoenzyme loci and 3 morphological markers. The markers fell into 9 linkage groups covering 404 cM, thought to represent 25-30 per cent of the genome. Most RFLP probes detected single segregating locus and when several locus were detected by a single probe the loci were generally linked, suggesting tandem duplications.

Selectively amplified microsatellite polymorphic locus (SAMPL) analysis is a method of amplifying microsatellite loci using generic polymerase chain reaction (PCR) primers. Forty-five cultivars of lettuce and five wild species of *Lactuca* were analyzed to determine the allelic diversity for codominant (SAMPL). From 3 to 11 putative alleles were found for each SAMPL; 2-6 alleles were found within *Lactuca sativa* and 1-3 alleles were found among the crisphead genotypes, the most genetically homogeneous plant type of *L. sativa*. This allelic diversity is greater than that found for RFLP markers (Witsenboer *et al.*, 1997).

12.2.2.2 Identification of Diseases and Genes of Resistance

At least 10 *dm* genes confer resistance to the oomycete downy mildew fungus *Bremia lactucae* map to the major resistance cluster in lettuce. Meyers *et al.* (1998) investigated the structure of this cluster in cv. Diana which contains Dm3 and concluded that the major cluster of disease resistance genes in lettuce contains at least nine downy mildew (*B. lactucae*) resistance genes (*dm*) spanning a genetic distance of 20 cM and a physical distance of at least 6 Mb. Anderson *et al.* (1996) used nine molecular markers that were genetically tightly linked to Dm3, to analyze nine independent deletion mutants and construct a map of the region surrounding Dm3. This analysis identified a linear order of deletion breakpoints and markers along the chromosome.

Montesclaros *et al.* (1997) investigated the interaction between two different potyviruses and resistant cultivars of lettuce. Turnip mosaic potyvirus (TuMV) and lettuce mosaic potyvirus (LMV) were used to inoculate several cultivars under different temperature regimes to characterize the resistance reaction. Resistance conferred by the recessive *mo* locus against LMV infection did not provide immunity. Virus accumulated in plant tissues to different levels depending on the genetic background of the cultivar suggesting that several genes were involved in the resistance phenotype.

12.2.2.3 QTL Mapping

Genes and QTL for disease resistance, plant growth and development, and seed germination traits have been mapped in lettuce. Using an $F_{2:3}$ *L. serriola x L. sativa* population, 13 QTLs with major effects on root architecture and water-use were mapped (Johnson *et al.*, 2000). *Lactuca serriola* develops a deep tap root and extracts water and nutrients from deep soil zones whereas cultivated lettuce (*L. sativa*) has a short taproot and numerous lateral branches largely restricted to the upper soil layers (Jackson, 1995). Because lettuce production requires relatively high inputs of fertilizers and water, the development of cultivars that are more deeply rooted would enhance agricultural sustainability and lower groundwater contamination by leaching (Still, 2007). Seventeen quantitative trait loci for seed germination characteristics and seedling quality were mapped in the interspecific F8 RILs derived from the cross between *L. sativa* cv. Salinas x *L. serriola* (UC96US23) and developed by Michelmore (Argyris *et al.*, 2005). Using the intraspecific F8 RIL resulting from the *L. sativa* cross between cv. Diplomat and cv. Margarita, Still and colleagues mapped 18 QTL associated with seed germination traits, including response to far-red light, high temperature germination, germination under osmotic stress, response to nitric oxides, and response to priming treatments(Still, 2007).

12.2.2.4 Development of Genomic Resources

Improvement of lettuce cultivars will be directed towards nutritional content, shelf life, and disease resistance through traditional and molecular-aided breeding methods. Some of these traits may be improved by genes from other sources far-removed from the *Lactuca* species. Gains in the characteristics will require the development of significant genomic resources as these traits are genetically complex or in some instances, genes coding for metabolic products are either missing or silent. To some extent wild germplasm may be used as a reservoir and efforts have been undertaken toward identifying and screening closely related species for resistance factors (Still, 2007). The development of virtually all horticultural traits will be facilitated by the development of genomic resources. To that end the Compositae Genome Project (CGP) has developed extensive EST data base that may allow lettuce more extensively to study like tomato.

The first phase of the CGP generated over 68,000 ESTs from *L. sativa* and *L. serriola*. The EST libraries were made from 10 pools of RNA from different tissues, developmental stages and environmental conditions of each of the two genotypes that had been used as parents for the core mapping populations for each species. These ESTs were assembled into ≤19,523 lettuce unigenes and may be accessed at the Compositae Genome Project web site (http://cgpdb.ucdavis.edu/). The EST sequence information has been released to Gen Bank, and incorporated by TIGR into their gene indices and by MIPS into their SPUTNIK EST database. A second phase of EST generation by the CGP is in the process of generating over 560,000 ESTs from 17 species within the Compositae. This will include an additional 70,000 ESTs for *L. sativa* and *L. serriola* and for the first time provide ca. 90,000 reads for *L.saligna, L.virosa* and *L.perennis* (Still, 2007)

12.3.0 Genetic Transformation

12.3.1 Explants and Protocols

Leaves and cotyledons have been used in transformation studies. Liu *et al.* (1996) developed an efficient trnsformation system. Young leaf sections (from either seminal or true leaves) of 2 lettuce cultivars were cultured on MS medium supplemented with 0.5 mg/l BA and 0.1 mg/l NAA. Duhu 366 produced the most adventitious buds and more than 1 shoot were regenerated per explant representing a regeneration rate of 87 per cent. After coculturing with *Agrobacterium* for 2 days, explants exhibited a 65 per cent regeneration rate on MS medium supplemented with 100 mg/l kanamycin. Regenerated plants showed a 50 per cent positive reaction frequency when leaf slices were incubated with X-Glu solution indicating that the *gus* (*uidA*) gene had been successfully transferred into the plant genome and was expressed at different levels.

Torres *et al.* (1993) used cotyledon explants to produce transgenic plants of cv. South Bay using *A. tumefaciens* vectors containing the *gus* reporter gene and the *nptII* gene for kanamycin resistance as a selectable marker. High frequency of transformation, based on kanamycin resistance and assays for GUS expression, was obtained with 24- to 72-h-old cotyledon explants cocultivated for 48 h with *A. tumefaciens* A208. Explants were then placed on selection medium containing 50 or 100 mg kanamycin, 100 mg cefotaxime and 500 mg carbenicillin/l for 10 days. Surviving explants were transferred every 14 days onto shoot elongation medium. Progenies of R_0 plants demonstrated linked monogenic segregation for kanamycin resistance and GUS activity.

Besides *A. tumefaciens*, other vectors like *A. rhizogenes* have also been used in transformation studies. Curtis *et al.* (1996) transferred *A. rhizogenes rolAB* genes into lettuce cv. Lake Nyah using a supervirulent strain of *A. tumefaciens*. Phenotypic characterization of transgenic populations indicated significant alterations in plant development, especially in the seed progeny of transgenic plants.

12.3.2 Bar Genes

Glyphosate resistant plants of lettuce cv. South Bay were produced by Torres *et al.* (1999) using *A. tumefaciens* containing plasmid carrying glyphosate oxidase and the *epsps* gene. An *in vitro* assay was performed to determine the sensitivity of South Bay leaf disc and seedling explants to varying glyphosphate concentrations. The I_{50} for glyphosate leaf discs was 53.8 μM and for glyphosate seedlings 7.6 μM. These findings will allow identification of glyphosate resistant transformants at an early stage.

12.3.3 Downy Mildew Resistance

Okubara *et al.* (1997) obtained one hundred and ninety-two independent primary transformants of lettuce cv. Diana by cocultivation with *A. tumefaciens* carrying constructs containing maize Ac transposase and Ds. R_2 families were screened for mutations of 4 genes (*dm*) for resistance to downy mildew caused by *Bremia lactucae*.

Dede (1998) introduced β-1,3-glucanase gene from *Arthrobacter* sp. driven by the 35S promoter singly into lettuce cultivars Cobham Green and Diana using the binary vector system of *A. tumefaciens*. The vector also contained NPTII selectable markers for kanamycin resistance. Southern blot analysis confirmed that the transformants containing the 2.37 kb HindIII fragment of the vector coding for β-1,3-glucanase. Transcription of the β-1,3-glucanase gene was confirmed by Northern blot analysis. Development of downy mildew pathogen, *Bremia lactucae* isolate CL9W was examined on leaf discs of transgenic lines and untransformed control plants. Susceptible controls showed 100 per cent sporulation 7 days after inoculation, while transgenic lines (with one exception) reached 95 per cent sporulation 10 days after inoculation. Transgenic line CG1-29 demonstrated a level of resistance similar to that of the resistant control cv. Diana.

12.3.4 Male Sterility

Male sterility is an economically trait useful in hybrid seed production. Curtis *et al.* (1996) produced male-sterile plants of cv. Lake Nyah using a genotype-independent transformation method employing *A. tumefaciens* with a binary vector. The latter carried the neomycin phosphotransferase II gene controlled by the nopaline synthase promoter and terminator sequences, enabling transformed shoots to be selected on medium containing kanamycin sulphate. All transformed plants were male-sterile. Plants regenerated from untransformed, excised cotyledons of Lake Nyah exhibited normal microspore development and were fertile. These results suggested that callose degradation may cause male sterility and provides the opportunity for producing F_1 hybrid seed in lettuce.

13.0 References

Abbott, A. (2010) The isolation of flowering time genes from lettuce to enable the manipulation of bolting time. PhD thesis (Warwick, UK: University of Warwick), p. 376.

Abdel Razik, A.H. (1996) *Alexandria J. Agril. Res.*, **41**: 359-368.

Adams, P., Graves, C.J. and Winsor, G.W. (1978) *J. Hort. Sci.*, **53**: 275-281.

Aguirre, Y., Marrero, S., Zapata, F., Chacin, F. and Albarracin, M. (1994) *Revista de la Facultad de Agronomia, Universidad Central de Venezuela*, **20**: 111-122.

Agwah, E.M.R. and Shahaby, A.F. (1993) *Annals Agril. Sci. Cairo*, **38**: 423-434.

Ahmed, A.K., Cresswell, G.C. and Haigh, A.M. (2000) *J. Hortic. Sci. Biotechnol.*, **75**: 350-354.

Ahvenainen, R. (2000) In: *Minimally Processed Fruits and Vegetables: Fundamental Aspects and Applications* (Eds. S.M. Alzamora, A. Lopez-Malo, M.S. Tapia), Aspen Publishers, Gaithersburg, MD. pp. 277-290.

Alconero, R. (1983) *HortScience*, **18**: 305-307.

Al-Harbi, A.R. (2001) *J. Veg. Crop Production*, **7**(1): 23-36.

AlKhader, Asad M.F, and Rayyan, Azmi M.Abu (2013) *Springerplus*, **2**: 563, doi: 10.1186/2193-1801-2-563.

Allen, R.G., Pereira, S., Raes, D. and Smith, M. (1998) FAO Irrigation and Drainage Paper No.56, Rome.

Alvarez, J. and Sanchez, C.A. (1995) *J. Sustainable Agril.*, **7**: 15-24.

Ammerlaan, A. (2010) Cos for mechanical harvest. United States patent US 7, 652, 196.

Anderson, P.A., Okubara, P.A., Arroyo-Garcia, R., Meyers, B.C. and Michelmore, R.W. (1996) *Mol. Gen. Genet.*, **251**: 316-325.

Andrade Junior, A.S. de and Klar, A.E. (1996) *Hortic. Bras.*, **14**: 27-31.

Anonymous (1971) Indian Minimum Seed Certification Standards, Central Seed Committee, Min. food Agric. Community Dev. Co-op., New Delhi.

Anonymous (1992) *Plant Varieties J.*, **5**: 23.

Anonymous (1992) *Plant Varieties J.*, **5**: 24-25.

Anonymous (1994) *Plant Varieties J.*, **7**: 28.

Anonymous (1994) *Plant Varieties J.*, **7**: 37-38.

Arambarri, P., Cabrera, F., Garcia, O. and Galvez, M. (1994) *Comm. Soil Sci. Plant Analysis*, **25**: 2697-2709.

Arce, A., Vallejo, A., Lopez Valdivia, L.M. and Hoyos, P. (1996) *Agricoltura Mediterranea*, **126**: 186-193.

Argyris, J., Truco, M.J., Ochoa, O., Knapp, S.J., Still, D.W., Lenssen, G.M., Schut, J.W., Michelmore, R.W., Bradford, K.J. (2005) *Theor. Appl. Genet.*, **111**: 1365–1376.

Ashkar, S.A. and Ries, S.K. (1971) *J. Amer. Soc. Hort. Sci.*, **96**: 448-452.

Atkinson, L.D., McHale, L.K., Truco, M.J., Hilton, H.W., Lynn, J., Schut, J.W., Michelmore, R.W., Hand, P. and Pink, D.A.C. (2013) *Theor. Appl. Genet.*, **126**: 2737–2752.

Avila, T. and Mederos, A. (2013) *Lactuca sativa* cultivar Solid King. United States patent US 8, 524, 981.

Aykroyd, W.R. (1963) *ICMR Special Rept. Series*, No. 42.

Azcon, R., Gomez, M. and Tobar, R. (1996) *Biol. Fertil. Soils*, **22**: 156-161.

Barrons, K.C. and Whitaker, T.W. (1943) *Michigan State Agric. Exp. Sta. Bull.*, **25**: 252-254.

Barros, M.J.M., Ceballos, B.S.O.de., Koning, A. and Gheyi, H.R. (1999) *Revista Brasileira de Engenharia Agricola e Ambiental*, **3**: 355-360.

Bastelaere, H. van. (1998a) *Proeftuinnieuws*, **8**: 23-25.

Bastelaere, H. van. (1998b) *Proeftuinnieuws*, **8**: 20-21.

Bastelaere, H. van. (1999) *Proeftuinnieuws*, **9**: 35-37.

Bastos, R.K.X. and Mara, D.D. (1995) *Water Sci. and Technol.*, **31**: 425-430.

Batista, M.A.V., Vieira, L.A., Souza, J.P., Freitas, J.D.B.F. and Bezerra Neto, F. (2012) *Revista Caatinga*, **25**: 8-11.

Beharav, A., Lewinsohn, D., Lebeda, A. and Nevo, E. (2006) *Genet. Resour. Crop Evol.*, **53**: 467–474.

Behr, H.C. (1993) *Gemuse Munchen*, **29**: 527-528.

Bell, A.A., Liu, L., Reidy, B., Davis, R.M. and Subbarao, K.V. (1998) *Phytopathol.*, **88**: 252-259.

Belligno, A., Muratore, G. and Izzo, R. (1997) *Agricoltura Mediterranea*, **127**: 126-133.

Beltrao, J., Trindade, D., Correia, P.J. and Chartzoulakis, K.S. (1997) *Acta Hortic.*, **449**: 623-627.

Benoit, F. and Ceusterrnans, N. (1979) *Phytotronic Newsletter*, No. 20, pp. 62-69.

Berahal, L. and Kwolek, W.F. (1975) *Plant Dis. Reptr.*, **59**: 1001-1004.

Berry, S.F., Lu, D.Y., Pental, D. and Cocking, E.C. (1982) *Z. Pflanzenphysiol.*, **108**: 31-38.

Berry, W.L. (1971) *Science*, **96**: 412-414.

Berry, W.L. and Carey, R. (1971) *J. Amer. Soc. Hort. Sci.*, **96**: 298-300.

Bessey, P.M. (1960) *Univ. Ariz. Exp. Sta. Rept.*, **189**: 5-8.

Biesiada, A., Kolota, E., Chohura, P. and Fiedorow, Z. (1998) *Roczniki Akademii Rolniczej w Poznaniu, Ogrodnictwo*, **27**: 19-25.

Borowski, E. and Michalek, S. (1998a) *Annales Universitatis Mariae Curie Sklodowska, Sectio EEE, Horticultura*, **6**: 103-116.

Borowski, E. and Michalek, S. (1998b) *Annales Universitatis Mariae Curie Sklodowska, Sectio EEE, Horticultura*, **6**: 117-128.

Borowski, E. and Michalek, W. (1995) *Annales Universitatis Mariae Curie Sklodowska, Sectio EEE, Horticultura*, **3**: 119-126.

Both, A.J., Albright, L.D. and Langhans, R.W. (1997) *Plant Growth Environ. Control Farm Manag. Prot. Cultiv.*, **456**: 45–52.

Bradbury, M. and Ahmad, R. (1996) *Pakistan J. Bot.*, **28**: 97-102.

Bradley, M.R. and Hargreaves, G. (1977) *Horticulture*, No. 29, pp. 35-44.

Brallier, S., Harrison, R.D., Henry, C.L. and Xue, D.S. (1996) *Water, Air and Soil Pollution*, **86**: 195-206.

Brumm, I., Schenk, M. and Gysi, C. (1993) *Acta Hortic.*, **339**: 125-136.

Bull, L.T., Lacerda, S. and Nakagawa, J. (1997) *Bragantia*, **56**: 169-179.

Bunning, M. and Kendall, P. (2012) Colorado State University Extension Fact Sheet No. 9.373.

Buriol, G.A., Streck, N.A., Gimenes, E.S. and Schneider, F.M. (1994) *Ciencia Rural, 1994*, **24**: 1, 1-6.

Burns, I.G. (1996) *Acta Hortic.*, **428**: 21-30.

Buxton, J.W., Jia, W.W. and Papadopoulos, A.P. (1999) *Acta Hortic.*, **481**: 281-287.

Campbell, R.N., Greathead, A.S. and Westerlund, F.V. (1980) *Phytopathol.*, **70**: 741-746.

Cantliffe, D.J., Hochmuth, G.J., Karchi, Z., Secker, I. and Ben Yehoshua, S. (1998) *14th Internatl. Cong., plastics in agriculture,* Tel Aviv, Israel, March 1997, 1998, pp. 421-427.

Capra, A., Consoli, S., Russo A. and Scicolone, B. (2008) *J. Irrig. Drain. Eng.,* **134**: 437-445.

Castro, S.R.P. de and Ferraz Junior, A.S.L. (1998) *Hortic. Bras.*, **16**: 65-68.

Chabot, R., Antoun, H. and Cescas, M.P. (1996) *Plant Soil*, **184**: 311-321.

Chauhan, D.V.S. (1972) *Vegetable Production in India* (3rd edn.), Ram Prasad and Sons, Agra.

Cheisa, A., Frezza, D., Fraschina, A., Trincheros, G., Moccia, S. and Leon, A. (2003) *Acta Hortic.*, **604**: 153-159.

Chen, Z., Han, Y., Ning, K., Luo, C., Sheng, W., Wang, S., Fan, S., Wang, Y. and Wang, Q. (2019) *PLoS ONE*, **14**(2): e0209329.

Cho, J.J. (1977) *Plant Dis. Reptr.*, **61**: 783-787.

Choudhury, B. (1972) *Vegetables*, National Book Trust, India, New Delhi.

Christopoulou. M., McHale. L.K., Kozik. A., Reyes-Chin-Wo. S., Wroblewski. T., Michelmore, R.W. (2015a) Mol. *Plant Microbe Interact.*, **28**: 751–765.

Christopoulou, M., Wo, S.R.C., Kozik, A., McHale, L.K., Truco, M.J., Wroblewski, T., Michelmore, R. (2015b) *Genes Genom. Genet.*, **5**(12): 2655–2669.

Claassens, A.S. (1994) *South African J. Plant Soil*, **11**: 145-146.

Coakley, S.M., Campbell, R.N. and Kimble, K.A. (1973) *Calif. Agric*, **27**: 6-7.

Codd, G.A., Metcalf, J.S. and Beattie, K.A. (1999) *Toxicon Oxford*, **37**: 1181-1185.

Cole, R.A., Sutherland, R.A. and Riggall, W.E. (1991) *Euphytica*, **56**: 237-242.

Collier, G.F. (1983) *J. Sci. Food and Agric.*, **34**: 264.

Collier, G.F. and Tibbitts, T.W. (1982). *Hortic. Rev.* (Am. Soc. Hortic. Sci.), **4:** 49–65.

Cometti, N.N., Bremenkamp, D.M., Galon, K., Hell, L.R. and Zanotelli, M.F. (2013) *Hortic. Bras.*, **31**: 287–292.

Cools, M.H. and Stijger, N.L.M. (1987) *Groenten en Fruit*, **43**: 28-29.

Cools, M.H. and Stolk, J.H. (1980) *Groenten en Fruit*, **36**: 32-33.

Coplin, D.L., Schmitthenner, A.F. and Bauerle, W.L. (1980) *Plant Dis.*, **64**: 63-66.

Coronel, G., Chang, M. and Rodríguez-Delfín, A. (2008) In: *International Symposium on Soilless Culture and Hydroponics; International Society for Horticultural Science*: Leuven, Belgium, Volume 843, pp. 137–144.

Costa, A.S.V.da., Silva, N.f.da., Martinez, H.E.P., Fontez, P.C.R. and Pereira, P.R.G. (1997) *Revista Ceres*, **44**: 169-179.

Costa, C.A.da., Casali, V.W.D., Loures, *E.g.*, Cecon, P.R. and Jordao, C.P. (1994) *Revista Ceres*, **41**: 238, 629-640.

Couture, R., Cantwell, M. I., Ke, D. and Saltveit, Jr. M. E. (1993) *HortScience*, **28**(7): 723-725.

Cox, E.F. (1980) *Hort. Res.*, **20**: 61-66.

Cox, E.F., Mc Kee, J.M.T. and Dearman, A.S. (1976) *J. Hort. Sci.*, **51**: 297-309.

Craker, L.E. and Seibert, M. (1983) *Canadian J. Pl. Sci.*, **63**: 277-281.

Cresswell, G.C. (1991) *J. Plant Nutr.*, **14**: 917-924.

Crisp, P., Collier, G.F. and Thomas, T.H. (1976) *Sci. Hort.*, **5**: 215-225.

Crute, I.R. (1992) *Euphytica*, **63**: 95-102.

Cuppett, S.L., McCluskey, M.M., Paparozzi, E.T. and Parkhurst, A. (1999) *J. Food Quality*, **22**: 363-373.

Currah, I.E., Gray, D. and Thomas, T.S. (1974) *Ann. Appl. Biol.*, **76**: 311-318.

Curtis, I.S., He, C.P., Powder, J.B., Mariotti, D., De Laat, A. and Davey, M.R. (1996) *Plant Sci. Limer.*, **113**: 113-119.

Damarany, A.M. (1989) *Assuit J. Agric Sci.*, **20**: 159-172.

Damarany, A.M. (1994) *Assiut J. Agril. Sci.*, **25**: 79-88.

Damerum, A., Chapman, M.A., Taylor, G. (2020) *Postharvest Biol. Technol.*, **168**: 111266.

Dawson, W.M. (1982) *Agril. in North Ireland*, **57**: 149-151.

De Vries, I.M. (1997) *Genet. Resour. Crop Evol.*, **44**: 165–174.

Dede, Y. (1998) *Turkish J. Agri. For.*, **22**: 313-321.

Della-Vecchia, P.T., Takazaki, P.E., Kikuchi, M. and Terencio Sobrinho (1992) *Hortic. Bras.*, **10**: 1, 44.

Dennis, D.J. and Dullforce, W.M. (1974) *Acta Hortic.*, **39**: 197-218.

Doershug, M.R. and Miller, C.O. (1967) *Amer. J. Bot.*, **54**: 410-413.

Drew, R.L.K. and Brochlehurst, P.A. (1990) *Ann. Bot.*, **66**: 63-71.

Duffuo, J.E., Zink, F.W. and Bardin, R. (1970) *Phytopathol.*, **60**: 1383-1384.

Eberhard, J. (2000) *Gemuse Munchen*, **36**: 8-10.

Economakis, C.E. and Chartzoulakis, K.S. (1997) *Acta Hortic.*, **449**: 1383-1384.

Eenink, A.H. and Garretsen, F. (1980) *Euphytica*, **29**: 653–660.

El Shinawy, M.Z., Abd Elmoniem, E.M. and Abou Hadid, A.F. (1999) *Acta Hortic.*, **491**: 315-318.

El-Hassan, E.A.A. (1990) *Bull. Fac. Agric., Univ. Cairo*, **41**: 407-417.

Elstone, Lydia, How, Kin Yau, Brodie, Samuel, Ghazali, Muhammad Zulfahmi, Heath, William P., and Grieve, Bruce (2020) *Sensors* **20**(2): 455, doi: 10.3390/s20020455.

Engl, G., Lorenz, H.P. and Schlaghecken (1982) *Gemuse*, **18**: 372-378.

Fallovo, C., Rouphael, Y., Cardarelli, M., Rea, E., Battistelli, A. and Colla, G. (2009) *J. Food Agric. Environ.*, 7: 456–462.

Fernandes, H.S., Morselli, T.B.G.A., Quijano, F.G., Cruz and L.E.C. da. (1999) *Revista Cientifica Rural.*, **4**: 54-57.

Figueiredo, C.C.D., Ramos, M.L.G., Pimentel, C.M.M. and Menezes, A.M.D. (2012) *Revista Horticultura Brasiliera*, **30**: 175-179.

Frezza, D., León, A., Logegaray, V., Chiesa, A., Desimone, M., and Diaz, L. (2005) *Acta Hortic.*, **697**(697): 43-48.

Gabr, S.M. (1999) *Alexandria J. Agril. Res.*, **44**: 251-262.

Gallardo, M., Jackson, L.E. and Thompson, R.B. (1996) *Plant Cell Environ.*, **19**: 1169-1178.

Gallardo, M., Jackson, L.E., Schulbach, K., Snyder, R.L., thompson, R.B. and Wyland, L.J. (1996) *Irrigation Sci.*, **16**: 125-137.

Gallardo, M., Snyder, R.L., Schulbach, K. and Jackson, L.E. (1996) *J. Irrigation and Drainage Engineering*, **122**: 354-359.

Gardner, B.R. and Pew, W.D. (1979) *J. Amer. Soc. Hort. Sci.*, **104**: 534-536.

Gastaldi, C.R. and Sutton, B.G. (1988) *Proc. 4th Internatl. Micro-Irrigation Cong.*, October 23-28, Albury Wodonga, Australia.

Gebski, M. (1998) In: *Ecological aspects of nutrition and alternatives for herbicides in horticulture* (Jakubczyk, H., Lata, B., Sadowski, A. and Whitehead, P. eds.), International Seminar, Warszawa, Poland, 10-15 June, 1997, pp. 23-24.

Gervasio, E.S., Carvalho, J. de A. and Santana, M.J.de (2000) *Revista Brasileira de Engenharia Agricola e ambiental*, **4:** 125-128.

Ghazi and Matthees, B.P. (1990) *J. Plant Growth Reg.*, **9**: 95-96.

Gil, M.I. (2016) *Acta Hortic.*, **1141**: 57-64

Goncalves, A.C.A., Rezende, R., Frizzone, J.A. and Folegatti, M.V. (2000) *Acta Hortic.*, **537**: 799-804.

Gong, F.R. (1998) *J. Shanghai Agril. College*, **16**: 227-231.

Gonnella, M.; Serio, F.; Conversa, G. and Santamaria, P. (2002) In: *VI International Symposium on Protected Cultivation in Mild Winter Climate: Product and Process Innovation*; International Society for Horticultural Science: Leuven, Belgium, 2002; Volume 614, pp. 687–692.

Grant, R.C. and Arthur, R.S. (1982) *HortScience*, **17**: 84.

Gruda, N. and Schnitzler, W.H. (2000) *J. Appl. Bot.*, **74**: 240-247.

Guedes, A.C., Cantiliffe, D.J. and Nell, T.A. (1981) *J. Amer. Soc. Hort. Sci.*, **106**: 121-126.

Guilherme, A.L.F., Araujo, S.M.de., Falavigna, D.L.M., Pupulim, A.R.T., Dias, M.L.G.S., Oliveira, H.S. de., Maroco, E. and Fukushigue, Y. (1999) *Tevista da Sociedade Brasileira de Medicina Tropica*, **32**: 405-411.

Gunez, A. and Post, W.H.K. (1996) *Agrochimica*, **40**: 166-172.

Gustafsson (1989) *Euphytica*, **40**: 227-232.

Guzman, V.L. and Dusky, J.A. (1981) *Proc. Fla. Sta. Hort. Soc.*, **93**: 271-273.

Guzman, V.L., Nagata, R.T., Datnoff, L.E. and Raid, R.N. (1992) *HortScience*, **27**: 852-853.

Hamada, E. and Testezlaf, R. (1996) *Ecossistema*, **21**: 62-65.

Hamdar, B.C. and Rubeiz, I.G. (2000) *Small Fruits Rev.*, **1**: 3-14.

Hanson, B.R., Schwankle, L.J., Schulbach, K.F. and Pettygrove, G S. (1997) *Agric. Water Manage.*, **33**: 139-157.

Harlan, J.R. (1992) Crops and Man. Am Soc Agron, Madison, WI

Hawthorne, B.T. (1979) *New Zealand J. Exp. Agric.*, **7**: 215-219.

Hayashi, E., You, Y., Lewis, R., Calderon, M.C., Wan, G. and Still, D.W. (2012) *Theor. Appl. Genet.*, **124**: 1487–1502

Hayashi, M., Ujiie, A., Serizawa, H., Sassa, H., Kakui, H., Oda, T. and Koba, T (2011) *Euphytica*, **180**: 429–436.

Hayes, R.J. and Liu, Y.B. (2008) *J. Amer. Soc. Hort. Sci.*, **133**(2): 228-233.

Hayes, R.J. and Simko, I. (2016) *Acta Hortic.*, **1141**: 65-76.

Hayes, R.J., Galeano, C.H., Luo, Y., Antonise, R. and Simko, I. (2014) *J. Am. Soc. Hortic. Sci.*, **139**: 388–398.

Hayes, R. J., Vallad, G. E., Qin, Q.M., Grube, R.C. and Subbarao, K.V. (2007) *Plant Dis.*, **91**: 439-445.

Hayes, R.J., Wu, B.M. and Subbarao, K.V. (2011) *Plant Breed.*, **130**: 388–393.

Helden, M. van (1993) *Bulletin OILB SROP.*, **16**: 69-19.

Helden, M. van., Wal, D. van der., Beek, T.A. van and Tjallingii, W.F. (1996) *Bulletin OILB SROP.*, **19**: 62-67.

Hemphill Jr., D.D. (1982a) *HortScience*, **17**: 256-257.

Hemphill Jr., D.D. (1982b) *J. Amer. Soc. Hort. Sci.*, **107**: 50-53.

Honma, S. and Bert, J.S. (1975) *Veg. Grower*, **23**: 24-26.

Hoque, M.M., Ajwa, H., Othman, M., Smith, R. and Cahn, M. (2010) *HortScience*, **45**: 1539-1544.

Hou, T.Z. and Mooneyham, R.E. (1999) *Amer. J. Chinese Medicine*, **27**: 131-141.

Hsiao, A.I. and Simpson, G.M. (1971) *Canadian J. Bot.*, **49**: 1359-1362.

Iordan, I. and Luca, L. (1998) *Anale-Institutul de Cercetari pentru Legumicultura si Floricultura Vidra*, **15**: 93-96.

Jackson, L.E. (1995) *Plant Cell Environ.*, **18**: 885–894.

Jarvan, M. (1999) *Sodininkyste ir Darzininkyste*, **18**: 113-121.

Jarvan, M. and Teedumae, A. (1999) *Proc. Estonian Acad. Sci., Biology, Ecology*, **48**: 236-245.

Jenni, S. (2005) *HortScience*, **40**: 2031–2035.

Jenni, S. and Hayes, R.J. (2010) *Euphytica*, **171**: 427–439.

Jenni, S. and Yan, W.S. (2009) *Plant Breed.*, **128**: 374–380

Jenni, S., and Hayes, R.J. (2010). *Euphytica*, **171**: 427–439.

Jenni, S., Truco, M.J. and Michelmore, R.W. (2013) *Theor. Appl. Genet.*, **126**: 3065–3079.

Jeong, J.C., Park, K.W. and Yang, Y.J. (1990) *J. Korean Soc. Hort. Sci.*, **31**: 219-225.

Johnson, W.C., Jackson, L.E., Ochoa, O., van Wijk, R., Peleman, J., St. Clair, D.A., Michelmore R.W. (2000) *Theor. Appl. Genet.*, **101**: 1066–1073.

Jong, M. De and Dijk, P. Van (1990) *Groenten en Fruit*, **46**: 33.

Jong, M. De and Dijk, P. Van (1991) *Groenten en Fruit, Glasgroenten*, **47**: 27.

Kalloo and Krung, H. (1980) *Gartenbauwissenschaft*, **45**: 116-120.

Kang, J.H., Kumar, S.K., Atulba, S.L.S., Jeong. B.R. and Hwang. S.J. (2013) *Hort. Environ. Biotechnol.*, **54(**6): 501-509.

Kawashima, N., Kurozumi, T. and Ohara, M. (1989) *Bull. Nara Agricultural Expt. Stn.*, **20**: 31-39.

Kesseli, R., Ochoa, O., and Michelmore, R. (1991) *Genome*, **34**: 430-436.

Kesseli, R.V. and Michelmore, R.W. (1986) *Heredity*, **77**: 324-331.

Kesseli, R.V., Paran, I. and Michelmore, R.W. (1994) *Genetics*, **136**: 1435-1446.

Kesseli, R.V., Witsenboer, H., Stanghellini, M., Vandermark, G. and Michelmore, R. (1993) *Mol. Plant Microbe Interact.*, **6**: 722-728.

Khalf, A.A.M., Feleafel, M.N., Badr, H.M. and Edress, A.A. (2003) *J. Agric. Res.*, **48:** 105-113.

Khan, A.S., Ahmad, N., Malik, A.U. and Amjad, M. (2012) Cold storage influences the postharvest pericarp browning and quality of litchi. *Int. J. Agric. Biol.*, **14**(3): 389-394.

Kim, J.G., Luo, Y., Tao, Y., Saftner, R.A. and Gross, K.C. (2005). J. Sci. Food Agric., **85**: 1622–1630.

Kishaba, A.N., McCrieght, J.d., Coudrief, D., Whitaker, T.W. and Pesh, G.R. (1980) *J. Amer. Soc. Hort. Sci.*, **105**: 890-892.

Kishaba, a.N., Whitaker, T.W., Berry, W. and Toba, H.H. (1976) *HortScience*, **11**: 216-217.

Kishaba, A.N., Whitaker, T.W., Vail, P.V. and Toba, H.H. (1973) *J. Amer. Soc. Hort. Sci.*, **98**: 367-370.

Klar, A.E., Fonseca, K.C.B., Banderalli, M. and Alves, D.R.B. (1995) *Dahlia Greidinger International Symposium on Fertigation: Technion* - Israel Institute of Technology, Haifa, Israel, 26 March-1st April, 1995, pp. 297-302.

Klasse, H.J. (1999) In: *Anac*, D. and Martin Prevel, P (eds.), *Improved crop quality by nutrient management, 1999*, pp. 233-235.

Knaval, D.E. (1981) *Hort. Res.*, **21**: 11-18.

Kobryn, J. (2001) *Annals Warsaw Agric. Univ.*, **22**: 43-48.

Kobryn, J. (1998) In: *Ecological aspects of nutrition and alternatives for herbicides in horticulture* (Eds. H.Jakubczyk,, B.Lata, A.Sadowski, and P.Whitehead,), *International Seminar*, Warszawa, Poland, 10-15 June, 1997, 1998, pp. 29-30.

Koevary, K., Rappaport, L. and Morris, L.L. (1978) *HortScience*, **13**: 24-28.

Kononkov, P.F. (1996) *Kartofel i Ovoshchi*, No. 3, p. 14.

Kowalczyk, K., Mirgos, M., Ba²czek, K., Niedzin´ ska, M. and Gajewski, M. (2014) *Acta Hortic.*, **1142**: 105–110.

Kowalska, I. (1997) *Folia Hortic.*, **9**: 31-40.

Kozik, E. and Glen, B. (1995) *Materialy ogolnopolskiej konferencji naukowej Nauka Praktyce Ogrodniczej z okazji XXV-lecia Wydzialu Ogrodniczego Akademii Rolniczej w Lublinie*, pp. 699-702.

Kraker, J. de., Osinga, K.J., Vlaswinkel, M.E.T. and Wijk, C.A.P. va. (1996) *Publicatie Proefstation voor de Akkerbouw en de Groenteteelt in de Vollegrond*, Lelystad, **81B:** 31-39.

Kuslu, Y., Dursun, A., Sahin, U., Kiziloglu, F.M. and Turan, M. (2008) *Spanish J. Agric. Res.*, **6:** 714-719.

Lacertosa, G., Montemurro, F., Capotorti, G. and Palazzo, D. (1997) *Rivista di Agronomia*, **31**: 72-77.

Lamikanra, O. (2002) In: *Fresh-cut fruits and vegetables Science, Technology and Market* (Ed. O. Lamikanra), CRC Press, Boca Raton, Florida. p. 99.

Landry, B.S., Kesseli, R.V., Farrara, B. and Michelmore, R.W. (1987) *Genetics*, **116**: 331-337.

Langton, F.A., Smith, J.W.M. and Edmondson, R.N. (1990) *Euphytica*, **49**: 15-23.

Lati, R.N., Mou, B., Rachuy, J.S., Smith, R.F., Dara, S.K., Daugovish, O. and Fennimore, S.A. (2015) *Weed Technology*, **29:** 827-834.

Laughlin, W.M., Smith, G.R. and Peters, M.A. (1987) *Agroborealis*, **19**: 34-38.

Lebeda, A., Køístkova, E., Kitner, M., Mieslerová, B., Jemelková, M. and Pink, D.A.C. (2014) *Eur. J. Plant Pathol.*, **138**: 597–640.

Lebeda, A.L., Ryder, E.J., Grube, R., Dolezalova, I., and Kristkova, E. (2007) In: Genetic Resources, Chromosome Engineering, and Crop Improvement (Boca Raton, FL, USA: CRC Press), pp. 377–472.

Lebeda, A., Pink, D.A.C., Astley, D. (2002) In: *Advances in downy mildew research* (Eds. P.T.N. SpencerPhillips, U. Gisi, A. Lebeda), Kluwer Academic Publishers, Dordrecht, pp 85–117.

Leenhardt, D., Lafolie, F. and Bruckler, L. (1998) *European J. Agron.*, **8**: 249-265.

Li, Q., Li, X., Tang, B. and Gu, M. (2018) *Horticulturae* **4**, 35.

Lifshitz, N., Elal, G. and Eshel, B. (1970) *Proc. 4th Israel Weed Control Conf.*, pp. 35-36.

Lindner, U. (1996) *Gemuse Munchen*, **32**: 94-98.

Lindqvist, K. (1960) *Hereditas*, **46**: 319-350.

Liu, F., Li, Y. and Cao, M.Q. (1996) *Acta Agric. Boreali Sinica*, **11**: 109-113.

Liu, M.C., Tanaka, M., Tanaka, A., Chen, H. and Kojima, T. (2000) *J. Soc. High Technology in Agril.*, **12**: 182-189.

Lorenz, H.P. and Viebe, H.J. (1980) *Sci. Hort.*, **13**: 115-123.

Madjid, A., Honna, S. and Lacy, M.L. (1983) *Sci. Hort.*, **18**: 201.

Mahlangu, R.I.S., Maboko, M.M., Sivakumar, D.,Soundy, P. and Jifon, J. (2016) *J. Plant Nutri.*, **39**: 1766-1775.

Maisonneuve, B. (1987) *Agronomie*, **7**: 313-319.

Maisonneuve, B. (2003) In: *EUCARPIA leafy vegetables* (Eds. T.J.L. Van Hintum, A. Lebeda, D.A. Pink, J.W. Schut), Noordwijkerhout, 19–21 Mar. 2003, pp. 61–67.

Mallory Smith, C.A., Thill, D.C. and Dial, M.J. (1993) *HortScience*, **28**: 63-64.

Marcum, D.B., Grogan, R.G. and Greathead, A.S. (1977) *Plant Dis. Reptr.*, **61**: 555-559.

Martýµnez Saµnchez, A., Luna, M.C., Selma, M.V., Tudela, J., Abad, J. and Gil, M.I. (2012) *Postharvest Biol. Technol.*, **63**: 1–10.

Masarirambi, M.T., Nxumalo, K.A., Musi, P.J. and Rugube, L.M. (2018) *American-Eurasian J. Agric. Environ. Sci.*, **18** (1): 50-56.

Maticic, B. (1994) In: *Proc. of the 2nd conference* (Eds. Campbell, K.L., Graham, W.D. and Bottcher, A.B.), Orlando, Florida, USA, 20-22 April, 1994, pp. 533-542.

Matsumoto, E. (1991) *Plant Cell Rep.*, **9**: 531–534.

Maynard, A.A. (1991) *Bull. Connecticut Agril. Expt. Stn.*, **894**: 13.

Maync, A. (1997) *Zeitz. Kultur. Landentwicklung*, Special Issue, X, XII.

McCall, D. and Willumsen, J. (1998) *J. Hort. Sci. Biotechnol.*, **73**: 698-703.

McCalley, N.F. and Wang, D.I. (1972) *J. Econ. Ent.*, **65**: 794-796.

McCandless, L. (1998) NYS Vegetable Conference on February 10, Cornell University.

McGinty, R.A. and Thompson, R.C. (1975) *Proc. Amer. Soc. Hort. Sci.*, **22**: 341-346.

McPharlin, I.R. and Robertson, W.J. (1997) *Australian J. Experimental Agril.*, **37**: 701-708.

McPharlin, I.R., Aylmore, P.M. and Jeffery, R.C. (1995) *J. Plant Nutr.*, **18**: 219-241.

McPharlin, I.R., Jeffery, R.C. and Pitman, D.H. (1996) *Australian J. Experimental Agril.*, **36**: 897-903.

Meyers, B.C., Chin, D.B., Shen, K.A., Sivaramakrishnan, S., Lavelle, D.O., Zhang, Z. and Michelmore, R.W. (1998) *Plant Cell*, **10**: 1817-1832.

Miccolis, V., Candido, V. and Posca, G. (2000) *Acta Horticulturae*, **533**: 533-541.

Miceli, A., Moncada, A., Sabatino, L. and Vetrano, F. (2019) *Agronomy*, **9**: 382; doi: 10.3390/agronomy9070382.

Michelmore, R.W., Truco, M.J. and Ochoa, O.E. (2013) https: //calgr eens.org/wp-conte nt/uploa ds/2013/07/Miche lmore –Lettuce-Breed ing.pdf.

Miller, R.L. and Jackson, L.E. (1998) *J. Agric. Sci.*, **130**: 173-182.

Misaghi, I.J. and Grogan, R.G. (1978a) *Phytopathol.*, **68**: 1744-1753.

Misaghi, I.J. and Grogan, R.G. (1978b) *Phytopathology*, **68**: 1738-1743.

Misra, S.G. and Mani, D. (1994) *Current Agril.*, **18**: 49-53.

Molas, J. and Szymanska, M. (1995) *Materialy ogolnopolskiej konferencji naukowej Nauka Praktyce Ogrodniczej z okazji XXV-lecia Wydzialu Ogrodniczego Akademii Rolniczej w Lublinie*, pp. 799-802.

Molina-Montenegro, M.A., Zurita-Silva, A.and Oses, R. (2011) *Cien. Inv. Agr.*, **38**: 65-74.

Montesclaros, L., Nicol, N., Ubalijoro, E., Leclerc-Potvin, C., Ganivet, L., Laliberte, J.F. and Fortin, M.G. (1997) *Theor. Appl. Genet.*, **94**: 941-946.

Moore, F.D. (1970) *J. Amer. Soc. Hort. Sci.*, **95**: 471-474.

Mou, B. (2008) In: *Handbook of Plant Breeding. Vegetables I. Asteraceae, Brassicaceae, Chenopodiaceae, and Cucurbitaceae* (Eds. J Prohens, F.Nuez), New York, Springer Science, pp.75–116.

Nadasy Iharosi, E. (1999) *Agrokemia es Talajtan*, **48**: 369-380.

Nagata, R.T., Guzman, V.L., Datnoff, L.E. and Raid, R.N. (1992) *HortScience*, **27**: 934-935.

Nath, P. (1976) *Vegetables for the Tropical Region*, ICAR, New Delhi, p. 94.

Neikov, S. (1998) *Rasteniev dni Nauki*, **35**: 877-880.

Neocleous, D., Papadopoulos, I. and Olympios, C. (2007) In: *Technical Bulletin 230* (Ed. C. Papachristoforou), Agricultural Research Institute, Ministry of Agriculture, Natural Resources and the Environment, Cyprus

Neuvel, J.J. and Kanters, F.M.L. (1999) *PAV Bulletin Vollegrondsgroenteteelt.*, No. Juni, pp. 20-21.

Neuvel, J.J. and Wijk, C.A.P. van (1998) *PAV Bulletin Akkerbouw*, May, 1998, pp. 15-17.

Neuvel, J.J., Weijers, G. and Titulaer, H.H.H. (2000) *PAV Bull. Vollegrondsgroenteteelt*, **4**: 12-16.

Nishio, T., Sato, T., Mori, K. and Takayanagi, K. (1988) *Japanese J. Breed.*, **38**: 165-171.

Noorwood, J.M., Johnson, A.G., O'brien, M. and Crute, I.R. (1985) *Zeitschrift fur Pflanzen Zuchtung*, **94**: 259-262.

Nowak, W. and Draszawka Bolzan, B. (2000) *Folia Universitatis Agric. Stetinensis, Agricultura*, **84**: 357-361.

Obispo, L.A. (1997) *Horticultura, Revista de Hortalizas, Flores y Plantas Ornamentales*, **118**: 13-15, 17-18, 20-22.

Ohse, S., Dourado-Neto, D., Manfron, P.A. and Santos, O.S.D. (2001) *Sci. Agric.*, **58**: 181–185.

Okubara, P. A., Arroyo-Garcia, R., Shen, K.A., Mazier, M., Meyers, B. C., Ochoa, O.E., Kim, S., Yang, C.H. and Michelmore , R.W.(1997) MPMI, **10**(8): 970-977.

Oliveira, A.S., Martin, E.C. Slack, D.C. and Pegelow, E.J. (1997) *Paper ASAE*, No. 972173, p. 12.

Oliveira, Aureo S. de, Martin, Edward C., Slack, Donald C., Pegelow, Edward J. and Folta, Allen D. (2005) *Rev. bras. eng. agríc. ambient.*, **9**: https: //doi.org/10.1590/ S1415-43662005000100006.

Oliver, G.W. (1910) New methods of plant breeding, Bulletin 167, U.S. Bureau of Plant Industry, Washington, DC, USA, pp. 12-13.

Osinga, K.J. (1996) *Publicatie Proefstation voor de Akkerbouw en de Groenteteelt in de Volleground, Lelystad*, **81B**: 40-45.

Pal, A.B. (1975) *Progressive Hort.*, **7**: 62-68.

Panak, H., Sienkiewicz, S. and Wojnowska, T. (1996) *Zeszyty Problemowe Postepow Nauk Rolniczych*, **429**: 249-254.

Parra, L., Maisonneuve, B., Lebeda, A., Schut, J., Christopoulou, M., Jeuken, M., McHale, L., Truco, M.J., Crute, I., Michelmore, R.W. (2016) *Euphytica*, 210: 309–326.

Parra, L., Nortman, K., Sah, A., Truco, M. J., Ochoa, O., Michelmore, R. (2020) *Theor. Appl. Genet.*, https: //doi.org/10.1007/s00122-020-03711-z.

Parente, A., Gonnella, M., Santamaria, P., Abbate, P. L and Conversa, G. (2006) *Acta Hortic.*, **700**: 137-140

Paschold, P.J., Zengerle, K.H. and Chartzoulakis, K.S. (1997) *Acta Horticulturae*, **449**: 289-295.

Patil, Supriya S., Dhumal, Sangram S., Patgaonkar, Deepak R., Garande, Vishnu K. and Kaur, Manpreet (2017) *Int. J. Environ. Agric. Biotech.*, **2** (3), doi.org/10.22161/ ijeab/2.3.29.

Pavlovic, R.M., stevanovic, D.R. and Kresovic, M.M. (1997) *Acta Horticulturae*, **462**: 885-889.

Peet, M.M. and Willits, D.H. (1982) *HortScience*, **17**: 948-849.

Peixoto Filho, J.U., Freire, M.B.D.S., Freire, F.J., Miranda, M.F., Pessoa, L.G. andKamimura, K.M. (2013) *Revista Brasileira de Engenharia Agrícola e Ambiental*, **17**: 419-424.

Perez Llamas, F., Navarro, I., Marin, J.F., Madrid, J.A. and Zamora, S. (1996) *Alimentaria*, **34**: 41-44.

Peron, J.Y. (1975) *Pepinieristes Horticulturs Maraichers*, No. 161, pp. 21-26.

Petropoulos, S.A., Chatzieustratiou, E., Constantopoulou, E. and Kapotis, G. (2016) *Not. Bot. Horti Agrobot. Cluj-Napoca*, **44**: 603–612.

Pew, W.D. and Gardner, B.R. (1978) *J. Amer. Soc. Hort. Sci.*, **103**: 324-327.

Pew, W.D., Gardner, B.R. and Bessey, P.M. (1983) *J. Amer. Soc. Hort. Sci.*, **108**: 448-453.

Pollet, S., Bleyaert, P. and Lemeur, R. (2000) *Acta Hortic.*, **519**: 151-161.

Pniewski, Tomasz, Czyz, Marcin, Wyrwa, Katarzyna, Bociag, Piotr, Krajewski, Pawe[3], Kapusta, Jo´zef (2016) Plant Cell Rep., doi: 10.1007/s00299-016-2056-1.

Premuzic, Z., De Los Rios, A., Clozza, M., Minino, H., Vilella, F. and Iorio, A.F. de (1995) *Horticultura Argentina*, **14**: 37, 68-73.

Reinert, R.A., Tingey, D.T. and Carterd, H.B. (1972) *J. Amer. Soc. Hort. Sci.*, **97**: 711-714.

Reinink, K. (1992) *Euphytica*, **60**: 61-74.

Reinink, K. (1993) *Sci. Hortic.*, **53**: 35-44.

Reinink, K. and Dieleman, F.L. (1993) *Bulletin OILB SROP*, **16**: 56-68.

Reyes-Chin-Wo, S., Wang, Z., Yang, X., Kozik, A., Arikit, S., Song, C., Xia, L., Froenicke, L., Lavelle, D.O., Truco, M.J., Xia, R., Zhu, S., Xu, C., Xu, H., Xu, X., Cox, K., Korf, I., Meyers, B.C. and Michelmore, R.W. (2017) *Nat. Commun.*, **8**: 14953.

Rice, R. (1999) *Hortscience*, **112**: 325-329.

Rijbroek, P. Van (1987) *Groenten en Fruit*, **43**: 64-65.

Rincon, L., Pellicer, C. and Saez, J. (1998) *Agrochimica*, **42**: 304-312.

Roberts, H.A., Hewson, R.T. and Ricketts, M.A. (1997) *Hortic. Res.*, 17: 39-45.

Robinson, R.W. (1986) *Mutation Breed. Newslett.*, **28**: 7.

Robinson, R.W., McCreight, J.D. and Ryder, E.J. (1983) *Breed. Rev.*, **1**: 267–293.

Rodenburg, C.M. (1960) *Varieties of Lettuce Inst. Verer. Tuin.*, Wageningen, Holland.

Rodrigues, E.T. and Casali, V.W.D. (1999) *Horticultura Brasileira*, **17**: 125-128.

Rodrigues, E.T. and Casali, V.W.D. (2000) *Revista Ceres*, **47**: 461-467.

Rooster, L. de (1999) *Proeftuinnieuws*, **9**: 29-32.

Rooster, L. de and Spiessens, K. (1998) *Proeftuinnieuws*, **8**: 25-28.

Rowse, H.R. (1974) *Plant and Soil*, **40**: 381-391.

Rozek, S., Leja, M., Myczkowski, J. and Mareczek, A. (1995) *Folia Hortic.*, **7**: 91-105.

Rubeiz, I.G., Farran, M.T., Khoury, R.Y. and Al Assir, I.A. (1992) *Communications in Soil Sci. Plant Analysis*, **23**: 725-731.

Rubeiz, I.G., Khansa, M. and Freiwat, M.M. (1998) *Communications in Soil Sci. Plant Analysis*, **29**: 161-167.

Ruiz Lozano, J.M., Azcon, R. and Gomez, M. (1996) *Physiologia Plantarum*, **98**: 767-772.

Ruiz Lozano, J.M., Gomez, M., Nunez, R. and Azcon, R. (2000) *Physiol. Plantarum*, **109**: 268-273.

Ryder, E.J. (1979a) *Leafy Salad Vegetables*, AVI Publishing Co., Inc., Westport, Connecticut.

Ryder, E.J. (1979b) *HortScience*, **14**: 283-284.

Ryder, E.J. (1979c) *HortScience*, **14**: 284-286.

Ryder, E.J. (1985) *J. Amer. Soc. Hort. Sci.*, **110**: 570-573.

Ryder, E.J. (1986) In: Breeding Vegetable Crops (Westport, CT, USA: AVI Publishing), pp. 433–474.

Ryder, E.J. (1999) Lettuce, Endive and Chicory, CABI Publishing, New York.

Ryder, E.J. and Robinson, B.J. (1991) *HortScience*, **26**: 437-438.

Ryder, E.J. and Waycott, W. (1994) *HortScience*, **29**: 335-336.

Ryder, E.J. and Whitaker, T.W. (1976) In: *Evolution of Crop Plants* (N.W. Simmonds ed.), Longmans, London and New York, p. 819.

Ryder, E.J., and Johnson, A.S. (1974) *HortScience*, **9**: 584.

Sagiv, B., Bar-Yosef, B. and Eliah, E. (1981) *Preliminary Rept., Division of Scientific Publication*, Bet Dagan No. 792, p. 42.

Sahin, Ustun, Kuslu, Yasemin, Kiziloglu, Fatih M. and Cakmakci, Talip (2016) *Journal of Applied Horticulture*, **18**(3): 195-202.

Saltveit, M.E. (1997) In: Phytochemistry of Fruit, Vegetables, Proceedings of the Phytochemical Society of Europe (New York, NY, USA: Oxford University Press), p. 205–220.

Sammis, T.W. (1980) *Agron. J.*, **72**: 701-704.

Sanchez, C.A. (2000) *HortScience*, **35**: 73-77.

Sanchez, C.A. and El Hout, N.M. (1995) *HortScience*, **30**: 528-531.

Santamaria, P., Generoso, C., Gonnella, M. and Elia, A. (2000) *Colture Protette*, **29**: 71-77.

Santos, I.C.dos., Casali, V.W.D. and Miranda, G.V. (1997) *Revista Ceres*, **44**: 53-62.

Santos, I.C.dos., Casali, V.W.D. and Miranda, G.V. (1998) *Pesquisa Agropecuaria Brasileira*, **33**: 157-161.

Sapkota, S., Sapkota, S. and Liu, Z. (2019) *Horticulturae* **5**, 72; doi: 10.3390/horticulturae5040072.

Sarmento, José Júnior A., Costa, Caciana C., Dantas, Maila V., Lopes, Kilson P., de Macedo, Ivando C., Bomfim, Silva Marinês P. and Barbosa, José Wilson da S. (2019) *J. Agric. Sci.*, **11:** 335-343.

Sasaki, H. (1975a) *J. Japanese Soc. Hort. Sci.*, **47**: 479-484.

Sasaki, H. (1975b) *J. Japanese Soc. Hort. Sci.*, **44**: 138-143.

Saxena, M.C., Rai, V.K. and Laloraya, M.M. (1978a) *Curr. Sci.*, **47**: 741-742.

Saxena, M.C., Rai, V.K. and Laloraya, M.M. (1978b) *Indian J. Plant Physiol.*, **21**: 106-112.

Scaife, M.A. and Jones, D. (1976) *J. Agric. Sci.*, **86**: 83–91.

Schwarz, D., Heinen, M. and Noordwijk, M. van (1995) *Plant Soil*, **176**: 205-217.

Seginer, I., Straten, G. van and Buwalda, F. (1999) *Acta Hortic.*, **507**: 141-148.

Sequeira, L. (1970) *Plant Dis. Reptr.*, **54**: 754-758.

Shang, Q.M., Gao, L.H. and Li, S.J. (1998) *J. China Agril. Univ.*, **3**: 67-71.

Shannon, M.C., McCreight, J.D. and Draper, J.M. (1982) *HortScience*, **17**: 75.

Sharma, D.K., Chaudhary, D.R. and Pandey, D.P. (2001) *Veg. Sci.*, **28**(1): 38-39.

Sharma, P.P. and Pandey, R.C. (1963) *Principles and Methods of Vegetable Cultivation* (Hindi), Asian Publ. House, Musaffarnagar, U.P., India.

Shem-Tov, S., Fennimore, S.A. and Lanini, W.T. (2006) *Weed Technclogy*. **20**: 1058–1065.

Sieard, D., Woo, S.S., Arroyo Garcia, R., Ochoa, O., Nguyen, D., Korol, A., Nevo, E. and Michelmore, R.W. (1999) *Theor. Appl. Genet.*, **99**: 405-418.

Silva Ledo, F.J.da., Casali, V.W.D., Melo Moura, E. de., Pereira, P.R.G. and Cruz, C.D. (2000) *Revista Ceres*, **47**: 271-285.

Silva, A.F.T., Avelino, R.C., Brito, L.P.S., Anjos, J.C.R., Júnior, J.V.S. and Cavalcante, M.Z.B. (2017) *Com. Sci.*, **8**(2): 265-274.

Simko, I., Hayes, R.J., Mou, B. and McCreight, J.D. (2014) In: Yield Gains in Major U.S. Field Crops (Madison, WI, USA: American Society for Agronomy, Crop Science Society of America, and Soil Science Society of America), pp. 53–86.

Simko, I., Hayes, R.J., Subbarao, K.V., and Sideman, R.G. (2010) *HortScience*, 45: 670-672.

Singh, K. and Saimbhi, M.S. (1967) *J. Res.*, Punjab agric. Univ., Ludhiana, **4**: 382-387.

Sinska, I. (1988) *Acta Physiol. Planta.*, **10**: 11-16.

Slavik, L. (1996) *Rostlinna Vyroba*, **42**: 381-384.

Smyth, A.B., Song, J. and Cameron, A. (1998) *J. Agric. Food Chem.*, **46**: 4556–4562.

Sonneveld, E. and Ende, J. Vabden (1975) *Neth. J. Agric. Sci.*, **23**: 191-201.

Sonneveld, E. and Vooget (1973) *Plant and Soil*, **38**: 415-423.

Sorensen, J.N., Johansen, A.S. and Poulsen, N. (1994) *Plant Foods for Human Nutrition*, **46**: 1-11.

Soundy, P., Cantliffe, D.J.; Hochmuth, G.J. and Stoffella, P.J. (2001) *HortScience*, **36**: 1066–1070.

Souto, P.C., Souto, J.S., Santos, R.V., Araújo, G.T. and Souto, L.S. (2005) *Revista Brasileira Ciência do Solo*, **29**: 125-130.

Souza, Enio, Lima, Edson, Júnior, Aurélio, Silveira, Lindomar, Bezerra Neto, Francisco and Abreu, Ellen. (2017) *Revista Caatinga*, 30: 391-400.

Spehia, R.S., Devi, M., Singh, J., Sharma, S., Negi, A., Singh, S. and Sharma, J.C. (2018) *Int. J. Veg. Sci.*, **24**: 557–566.

Speiser, B. and Hochstrasser, M. (1998) *Agril. Ecosys. Environ.*, **70**: 273-274

Stancheva, I., Arsova, A. and Dinev, N. (1997) *Bulgarian J. Agril. Sci.*, **3**: 143-152.

Staugaitis, G. and Viskelis, P. (2000) *Sodininkyster ir darzininkyste*, **19**: 82-97.

Stepowska, A. and Kowalczyk, W. (2000) *Annales Universitatis Mariae Curie Sklodowska-Sectio EEE, Horticultura*, **8**: 189-195.

Stewart, J.K., aharoni, Y., Hartsell, P.L. and Young, D.K. (1980) *HortScience*, **15**: 148-149.

Still, D.W. (2007) In: *Genome Mapping and Molecular Breeding in Plants*, Volume 5 Vegetables (Ed. C. Kole), Springer, Berlin, Heidelberg, New York.

Stolk, J.H. and Cools, M.H. (1981) *Groenten en Fruit*, **37**: 28-29.

Stone, D.a., Tyler, F.t. and Runham, S. (1999) *Soil Use and Management*, **15**: 128-132.

Subbarao, K.V., Hubbard, J.c. and Schulbach, K.f. (1997) *Phytopathol.*, **87**: 877-883.

Sublett, W., Barickman, T. and Sams, C. (2018) *Horticulturae.*, **4**: 48.

Sun, Y.F., Huang, J.H., Lu, R.J., Zhou, R.M. and Wang, Y.F. (1998) *Acta Agric. Shanghai*, **14**: 17-20.

Tafuri, F., Scarponi, L., Businelli, M. and Guisuiani, P.L. (1978) *Rivist di Agronomica*, **12**: 123-128.

Tei, F., Benincasa, P. and Guiducci, M. (2000) *Acta Horticulturae*, **533**: 285-392.

Teng, W.L. and Liu, Y.J. (1993) *Plant Cell Tissue Organ Cult.*, **34**: 311-314.

Tesi, R. and Lenzi, A. (1998) *Agricoltura Mediterranea*, **128**: 313-320.

Thakur, M.C., Joshi, A.K., singh, N.P. and Shukla, Y.R. (1997) *Horticultural J.*, **10**: 43-49.

Thomas, G.M. and Harrison, H.C. (1989) *J. Amer. Soc. Hort. Sci.*, **114**: 121-125.

Thompson, H.C. and Kelly, C.W. (1957) In: *Vegetable Crops*, McGraw-Hill Book Co., Inc., New York.

Thompson, R.C. (1951) *U.S. Dept. Agri. Farmers Bull.*, 1953.

Thompson, R.C. and Ryder, E.J. (1961) *USDA Tech. Bull.*, 1244.

Thompson, T.L. and Doerge, T.A. (1996) *Soil Sci. Soc. Amer. J.*, **60**: 163-168.

Tibbitts, T.W. and Rao, R.R. (1968) *Proc. Amer. Soc. Hort. Sci.*, **93**: 454-461.

Tisselli, V. (1999) *Informatore Agrario*, **55**: 45-47.

Torres, A.C., Cantliffe, D.J., Laughner, B., Bieniek, M., Nagata, R., Ashraf, M. and Ferl, R.J. (1993) *Plant Cell Tissue Organ Cult.*, **34**: 279-285.

Torres, A.C., Nagata, R.T., Ferl, R.J., Bewick, T.A. and Cantiliffe, D.J. (1999) *J. Amer. Soc. Hort. Sci.*, **124**: 86-89.

Tourdonnet, S. de (1999) *Acta Horticulturae*, **507**: 263-270.

Turhan, E. and Sevgican, A. (1999) *Acta Hortic.*, **491**: 405-408.

USDA (2005) USDA National Nutrient Database for Standard Reference, Release 18. Nutrient Data Laboratory Home Page, http: //www.nal.usda.gov/fnic/ foodcomp.

USDA, ERS. (2014) http: //www.ers.usda.gov/data products/food availability per cent 28per capita per cent 29 datasystem/.aspx.

Van der Pahlen, A. and Crnko, J. (1965) *Rev. Invest. Agro., B. Aiers. Ser.*, 5, 2, 25-31.

Vecchia, P.T. Della and Kikuchi, M. (1989a) *Hortic. Bras.*, **7**: 29-30.

Vecchia, P.T. Della and Kikuchi, M. (1989b) *Hortic. Bras.*, **7**: 29.

Vergote, N. (1998) *Proeftuinnieuws*, **8**: 19-20.

Vidigal, S.M., Ribeiro, A.C., Casali, V.W.D. and Fontes, L.E.F. (1995) *Revista Ceres*, **42**: 89-97.

Vidigal, S.M., Sediyama, M.A.N., Garcia, N.C.P. and Matos, A.T.de (1997) *Hortic. Bras.*, **15**: 35-39.

Visser, a.J.De and Veeren, J. Vaned (1975) *Neth. J. Agric. Sci.*, **23**: 219-223.

Wagstaff, C., Clarkson, G.J.J., Zhang, F., Rothwell, S.D., Fry, S.C., Taylor, G. and Dixon, M.S. (2010) *J. Exp. Bot.*, **61**: 1239–1248.

Wang, X., Yost, R.S. and Linquist, B.A. (2001) *Soil Sci. Soc. Amer. J.*, **65**: 139-146.

Waycott, W. (2009) Lettuce line P2428 50060900. United States Patent US 7, 601, 895.

Waycott, W. and Ryder, E.J. (1992) In: Adaptation of Food Crops to Temperature and Water Stress (Taipei, Taiwan: Asian Vegetable Research and Development Center), p. 531.

Waycott, W. and Ryder, E.J. (1994) *HortScience*, **29**: 333-334.

Weir, B.L., Paulson, K.N. and Lorenz, O.A. (1972) *Proc. Soil Soc. Amer. J.*, **36**: 462-465.

Welch, J.E. (1965) *Calif. Agric.*, **19**: 3-4.

Welch, N.C., Tyler, K.B. and Ririe, D. (1979) *Calif. Agric.*, **33**: 12-13.

Werij, J.S., Kloosterman, B., Celis Gamboa, C., de Vos, C.H., America, T., Visser, R.G.F. and Bachem, C.W.B. (2007) *Theor. Appl. Genet.*, **115**: 245–252

Wilhelm, E. (1998) *Gemuse Munchen*, **34**: 559.

Whitaker, T.W. and Prowidenti, R. (1983) *HortScience*, **18**: 285-288.

Whitaker, T.W., Kishaba, A.N. and Toba, H.H. (1974) *J. Amer. Soc. Hort. Sci.*, **99**: 74-78.

White, J.G. (1980) *Plant Pathol.*, **29**: 124-130.

Wijk, C. van, Veuvel, J., Alblas, J. and Titulaer, H. (2000) *Acta Hortic.*, **511**: 81-88.

Wijk, C.A.P. van and Neuvel, J.J. (1998) *PAV Bulletin Vollegrondsgroenteteelt*, February 21-23, 1998.

Wilhelm, E. and Mayer, N. (1996) *Gemuse Munchen*, **32**: 28-31.

Wilson, G.J. (1976) *New Zealand Commercial Grower*, **31**: 21.

Witsenboer, H., Kesseli, R.V., Fortin, M.G., Stanghellini, M. and Michelmore, R.W. (1995) *Theor. Appl. Genet.*, **91**: 178-188.

Xiaoli, Z., Yang, H., Wenjie, Y. and Ti, X. (1992) *Ann. Bot.*, **69**: 97-100.

Xu, H.L., Wang, R. and Mridha, M.A.U. (2000) *J. Crop Prodn.*, **3**: 313-324.

Yamamoto, T., Nishikawa, A. and Oeda, K. (1994) *Euphytica*, **78**: 132-148.

Yazgan, S., Ayas, S., Demirtas, C., Büyükcangaz, H. and Candogan, B.N. (2008) *J. Food Agric. Environ.*, **6**: 357-361.

Yitayew, M., Silva, E. da (1998) *ASAE Ann. Internatl. Meeting*, Orlando, Florida, USA, 12-16 July, 1998, ASAE Paper No. 982121, p. 41.

Yu, H.S. and Rang, W. (1998) *J. South China Agric. Univ.*, **19**: 21-23.

Zandvakili, Omid, Barker, Allen V., Hashemi, Masoud, Etemadi, Fatemeh, Autio and Wesley, R. (2019) *J. Plant Nutri.*, **42**(2): 1-11.

Zarate, N.A.H., Vieira, M. do. C. and Cabecas Junior, O. (1997) *Hortic. Bras.*, **15**: 65-67.

Zdravkovic, J. (1991) *Zbornik Radova Poljoprivrednog Fakultera, Univerziter u Beogradu*, **36**: 149-162.

Zeleny, F. and Zelena, E. (1994) *Agrochemia Bratislova*, **34**: 126-128.

Zeng, X.B., Qing, C.L. and Xie, D.T. (1998) *Chinese J. Appl. Ecol.*, **9**: 139-144.

Zhang, F.Z., Wagstaff, C., Rae, A.M., Sihota, A.K., Keevil, C.W., Rothwell, S.D., Clarkson, G.J.J., Michelmore, R.W., Truco, M.J., Dixon, M.S. and Taylor, G. (2007). *J. Exp. Bot.*, **58**: 1433–1449.

Zink, F.W. (1961) *J. Agric. Food Chem.*, **9**: 304-407.

Zink, F.W. and Yamaguchi, M. (1962) *Hilgardia*, **32**: 471-500.

Zohary, D. (1991) *Euphytica*, **53**: 31–35.

CELERY

P. Karmakar, V. Sagar, S. Chatterjee, S.C. Mallick and V.A. Parthasarathy

1.0 Introduction

Celery is considered as the third most important salad vegetable in the western countries including United States and European countries. Celery is an important salad crop primarily grown for its long fleshy leafstalks and looks somewhat like leafy onion. It ranks second in importance among salad crops. But it is not a commercially popular vegetable in India. In the high elevation of southern part of India, it is grown on limited scale for salad purpose (Yawalkar, 1969). It is also an important aromatic plant grown predominantly for its fresh herbs as salad crop in different parts of the world. The dried fruits are also used as spice. Celery is known as various vernacular name *viz.*, Celeri in French, Sellerie in German, Apio in Spanish, Salleri in Swedish, Karafs in Arabic, Selderiji in Dutch, Sedano in Italian, Aipo in Portugese, Syelderey in Russian, Serorjini in Japanese; Chin in Chinese; Karnauli or Ajmod in India (Malhotra, 2006). Celery was perhaps not under extensive cultivation up to the medieval period, though there is documentation available in the ancient literature that it was cultivated before 850 BC. It is believed to be that the celery production developed in the lowlands of Italy and gradually spread to England and France. The first document of its cultivation in France was reported during 1623. Though the present cultivated celery plants are a quite sweet, appetising and wholesome food but its wild ancestors were considered poisonous. In USA, France, and in other European countries, it is grown commercially and is available in the market throughout the year.

2.0 Composition and Uses

2.1.0 Composition

Celery is a rich source of water and antioxidants that are helpful in maintaining your health. Celery also contains vitamins A, C, and K, folate, essential minerals and over a dozen other antioxidants. It is much lower in calories than other green veggies. Celery leaves have the same earthy, herbaceous flavour as celery stalks. The flavour is savoury with a hint of saltiness since celery is naturally high in sodium. Celery leaves contain small amounts of many different minerals including potassium, calcium and manganese. Like the rest of the plant, celery leaves are good sources of both soluble and insoluble dietary fibre. Constituents of celery are given in Table 1.

Table 1: Composition of Celery (per 100 g edible portion)

Moisture	95 per cent	Na	88 mg
Energy	16 Kcal	K	284 mg
Protein	0.7 g	Vitamin A	127 IU
Fat	0.1 g	Thiamine	0.03 mg
Carbohydrate	3.6 g	Riboflavin	0.03 mg

Fiber	0.7 g	Niacin	0.30 mg
Ca	36 mg	Ascorbic acid	6.3 mg
P	26 mg	Vitamin B_6	0.03 mg
Fe	0.5 mg		

2.2.0 Uses and Medicinal Properties

The leafstalk and petioles of celery are eaten raw or used for the preparation of sauces, vegetable juices, stews, soups, *etc.* Celery is eaten primarily for its unique texture and crispness.

Celery seed is used as condiment in European countries and spice in India. Celery has important medicinal values and used as a flavouring material. Compared to many other vegetables, the nutritional quality of celery is rather poor. It contains 95 per cent water with some minerals. Celery is carminative stimulant, diuretic, tonic and nervine, useful in hysteria, promoting restfulness and sleep and diffusing through the system a mild sustaining influence. It is said to be very good for rheumatism.

Presence of compounds such as limonene, selinene, frocoumarin glycosides, flavonoids, and vitamins A and C are the reason that celery is the most widely used plant in traditional medicine. Some main constituents of celery with chemical structures are shown. Celery can prevent cardiovascular diseases, jaundice, liver and lien diseases, urinary tract obstruction, gout and rheumatic disorders. Research on rats shows that ethanol extracts of celery leaves increases spermatogenesis and also improves their fertility. Celery reduces glucose, blood lipids and blood pressure, which can strengthen the heart. Experimental studies show that celery has antifungal and anti-inflammatory properties. Moreover, its essential oils have antibacterial effects. Its seeds are useful in the treatment of bronchitis, asthenopia, asthma, chronic skin disorders, including psoriasis, vomiting, fever and tumours. The root of the celery is diuretic and it is used for the treatment of colic. Plants are an important source of natural active products, which vary, based on mechanism and biological properties. Various phytochemical compounds, especially polyphenols (such as flavonoids, phenolic acids and tansipropanoids) are responsible for collecting free radicals and antioxidant activities of plants (Kooti and Daraei, 2017).

3.0 Origin and History

Celery is believed to be originally from the Mediterranean basin. Ancient literature documents that celery or a similar plant form, was cultivated for medicinal purposes before 850 B.C. Celery was considered a holy plant in the classical period of Greece and was worn by the winners of the Nemean Games, similar to the use of bay leaves at the Olympic Games. The Nemean Games were conducted every second year, starting in 573, in the small city of Nemea in southern Greece in the Poloponnes peninsula.

Uses of Celery

Different Recipes of Celery in Indian Cuisine

The Romans valued celery more for cooking than for religion although much superstition was connected with it. The celery plant was thought to bring bad fortune under certain circumstances.

Celery was initially used as medicine hundreds of years before it was used for food. Celery is native to marshy low lands with a habitat that extends from Sweden, southward to Algeria, Egypt, Ethiopia and in Asia to the mountains of India. European colonists introduced celery to North America. Cultivation of the plant for food was first recorded in France in 1623. When production of celery for medicinal purposes was begun in Europe in the 16th century, it was still a primitive plant, although it had been known for centuries in the Mediterranean area where it probably originated. By the early part of the 18th century, there had been improvement of the wild type of celery previously transported to Italy, France, and England. As early as 1726, celery was being used in England to flavour soups and stews (http://vric.ucdavis.edu/selectnewcrop.celery.htm). It has been found growing wild in California and in New Zealand. Janes (1954) reported that it was a native British plant, cultivated there for centuries.

The Italians domesticated celery as a vegetable in the 17th century resulting in selections with solid stems. Early stalk celery had a tendency to produce hollow stalks. After years of domestication, selection eliminated this characteristic as well as bitterness and strong flavours. Early growers found that the naturally strong flavours could be diminished if grown in cooler conditions and also if blanched. Blanching is the practice of pushing dirt up around the base of the stalks to prevent sunlight from turning the stalks green.

4.0 Taxonomy, Botany and Cytogenetics

4.1.0 Taxonomy

Celery is one of three botanical varieties of the species *Apium graveolens* L. Celery is *A. graveolens* var. *dulce* (Miller) Pers.; celeriac is var. *rapaceum* (Miller) Gaud.; leafy types are var. *secalinum* Alef. Celery is closely related to parsley and is related also to carrot, parsnip, Florence fennel and the poisonous weeds wild parsnip and wild hemlock. All are members of the family Apiaceae (Ryder, 1979).

The taxonomic classification of celery documented by Malhotra (2006) is as follows: Division: Spermatophyta, Sub-division: Angiospermae, Class: Magnoliospida (Dicotyledoneae), Sub-class: Rosidae, Order: Apiales, Family: Apiaceae, Genus: *Apium and* Species: *graveolens.*

Besides on the basis of morphological characters, climate requirement, bolting behaviour, life cycle and growing season, celery can be also classified as presented below: Foliage colour: green or yellow/golden; Blanching habit: early or late; Bolting behaviour: slow or quick; Climate: temperate or sub-tropical; Life cycle: annual or biennial; Height: tall, intermediate or dwarf; Season: autumn or winter

Orton, 1984 provided a comprehensive classification of *Apium graveolens* L. According to the horticultural types, such as (a) *Apium graveolens* var. *dulce* which known as blanched celery, (b) *Apium graveolens* var. *rapaceum* or edible rooted celery and (c) *Apium graveolens* var. *secalinum* known as leafy type or smallage type.

Horticultural Types of Celery

Apium graveolens **var.** ***dulce***

Apium graveolens **var.** ***rapaceum***

Apium graveolens **var.** ***secalinum***

A. graveolens var. *secalinum* is the most popular celery in Asian and Mediterranean regions (Rubatzky and Yamaguchi, 1997). Out of the thee morphological types mentioned above, *Apium graveolens* var. *secalinum* (smallage type) has been reported to be commonly cultivated in India for seeds as spices and behaves annual in growth habit (Malhotra, 2006).

4.2.0 Botany

Celery (*Apium graveolens* L.) belongs to the family *Apiaceae* (Umbelliferae). It is a cool season biennial grown as an annual. *Apium graveolens* (2n = 2x = 22) occurs as a wild plant in wet places of temperate Eurasia. Cultivated celery had been derived from it by selection for size and succulence of the petioles. By the 17th century, garden celery had become distinct from the wild plant. Leafy types are referred to var. *seculinum*; blanched celery is var. *dulce* and celeriac (with swollen edible roots) is var. *rapaceum* (Smith, 1976). The celeriac is also known as turnip rooted celery (Pruthi, 1976). Intergeneric hybridization of celery and parsley (*Petroselinum crispum*) has been reported and yielded a range of new types in both crops (Smith, 1976). Celery is used almost exclusively fresh and only very small quantities are dried, canned or frozen usually in combination with other vegetables and in soups. Celery seed is also used for flavouring. Celery seed has a strong flavour like concentrated celery stalks. During the first year of growth, celery forms a short subconical stem bearing a spiral rosette of leaves on the end of an elongated petiole. Generally, the petiole is the most important edible portion. The petiole is crescentric in transection with prominent ribbing on the abaxial surface and a relatively smooth adaxial surface. The ribbing is a result of bundles of collenchyma on the abaxial side that run through the centre of the stalk and are responsible for the stringiness of celery.

Celery is generally a herbaceous annual or biennial erect herb, plants attain a height of 60–90 cm, having prominent branches producing well-developed leaves on long expanded petioles. Stems are branched, angular and clearly jointed. Leaves are radical, pinnate, deeply divided into three segments, once or twice divided and having toothed at apex. The leaflets are ovate to sub orbicular, 3-lobed, 2–4.5 cm long. The size of flowers is small, the inflorescence is a compound umbel as in Apiaceae family which bear white colour flower. Calyx teeth are botanically obsolete; having five petioles which are ovate in shape, acute with tip inflexed; carpels are semiterete, subpentagonal, having distinct primary ridges and are filiform. The fruit is a schizocarp containing two mericarps, which are sub-orbicular to ellipsoid, having a diameter of 1–2 mm. Schizocarp (fruits) splits and gives two mericarps and are much smaller than carrot seed.

Celery leafstalks are about 15-38 cm long, bearing 3 pairs and a terminal leaflet, coarsely serrated and alternately lobed or divided. Flower stalks are 60-100 cm high branched and leafy (Bailey, 1950). Flowers are small, white and borne in compound umbels. Fruit is rigid, 1 to 1.5 mm long and 1 mm in diameter, contains a small seed. Seeds are small brown, slow in germination (Knott, 1960), united or separated pericarp, and some with stalk ends. According to Choudhury (1970), 10 g seeds count about 26450. Seeds are somewhat bitter in taste.

4.3.0 Cytogenetic

Emsweller (1929) reported that most of the species of *Apium* are diploid and having somatic chromosome (2n) = 2x = 22. Hore (1977) investigated the structure and behavior of *A. graveolens* chromosomes and reported six major types, each with several subtypes. The main structural differences arose due to the length, location and number of constrictions of the chromosome. The cultivars of *A. graveolens* var. *dulce* reported to had either 4 long and 18 medium-to-short chromosomes or 8 long and 14 medium to-short chromosomes. A triploid type in celery that apparently resulted from a natural cross between celery and parsley also reported (Whitaker, 1941).

5.0 Cultivars

The cultivars of celery basically belong to the three botanical morphotypes of *A. graveolens* (http.//www.botany. com/apium.htm1).

Celery (*A. graveolens* var. *dulce*)

This is a popular vegetable type that grows wild in Europe. It is a bushy, mounded plant. Only the centre parts of the plant are used and the tough, outer stalks are trimmed off. There are three different colours, golden-green and a rare reddish-green. In its second year, this plant would produce small, yellowish flowers atop a 60-100 cm stem.

Leaf Celery (*A. graveolens* var. *secalinum*)

Also called French Celery or Chinese Celery, is grown for its pretty, fragrant leaves. It is a biennial that's grown as a long-season annual. This herb was used centuries ago in Europe and the Orient, where its ancestor was known as 'smallage'. It has thin, rounded stalks that grow up to 30 cm high. They have a strong flavour and usually become stringy as they age.

Celeriac (*A. graveolens* var. *rapaceum*)

This is a cultivar of celery. The edible, swollen root grows 5-10 cm across. It is not as popular as celery, but is used in stews and soups. The plants have open growth, with celery-like leaves growing on thin stem from the root. Celeriac root tastes somewhat like celery, but stronger.

Initially, blanched celery was grown by keeping soil and paper, around the developing petioles for developing white colour. This practice gives the stalks milder flavour and improves tenderness. Later, self-blanching cultivars were developed that were naturally pale yellow coloured without shading the stalks. The green celery, called as pascals, has poorer petiole quality and the American scientists developed long and straight varieties called the Utah types such as Tall Utah 52-70 which became the predominate cultivar grown in the US because of their exceptionally long straight petioles and mild flavour (http://vric.ucdavis. edu/selectnewcrop.celery.htm).

The following are some of the cultivars:

Green Types

Utah – Tall Utah 52-70 and selections (52-70R, Tender crisp, Florida 683, Fla. 2-13).

Pascal – Giant Pascal, Pascal 259-19 (Summer Pascal), Easy Blanching, Earligreen, Non-Bolting Green # 12 (or 96).

Self-blanching – Golden Self-Blanching (Golden Detroit).

Golden Plume – Cornell 19, Pink or Red (used for flavouring and decoration and not eaten fresh).

Sac Yuquin

Derived from a cross between an American variety and a Shanghai yellow-stemmed cultivar, Yuqin is a novel, high yielding celery cultivar with good quality characteristics. Mean yields over several years in experimental plots were 30 per cent higher than the yellow stemmed variety. Resistance to pests and extreme temperatures are superior to both parents, and it is suitable for cultivation in the south of Hainan Island and in the north of Jiangsu Province of China (Huang and Cao, 1997).

Wolf-249 and FBL 5-2M

FBL 5-2M was derived from 2 sequential single-plant selections extracted from Floribelle M9, with the progeny of the second selection being bulked and tested. Wolf-249 was developed by a combination of backcross and pedigree breeding methods. The breeding effort was begun in 1950 with the accession of the F4 breeding line identified as Cornell 50-256-2, which arose from a series of crosses involving elite New York germplasm and foreign introductions. Florida 2-14 was the recurrent parent in a number of crosses. Both Wolf-249 and FBL 5-2M have intermediate growth habits between the Crystal Jumbo and Ordinary Utah phenotypes. They are midseason genotypes, maturing 87-98 days after transplanting. Trials at 7 site-year environments revealed that Wolf-249 and FBL 5-2M out-yielded the standard cultivar Floribelle at every test site, giving average yield increases of 132 and 104 crates/ha, respectively. The superior yield of Wolf-249 was attributed to significantly higher yields of mature stalks in the 2.5-dozen-sized class (Scully *et al.*, 1995).

Shangnong Yuquin

Shangnong Yuqin is a cross between Chinese celery Huangxinqin and a European cultivar. It is intermediate between its parents for leaf and leaflet number, petiole cross sectional area and number of vascular bundles. Shangnong Yuqin is higher yielding than its parents, and crisper and more tender (Chao and Huang, 1994).

Florida Slobolt M68

This cultivar was developed from 2 crosses [Summer Pascal 259-19 x EES212 and June Belle x EES1740 (F_3)], followed by single plant selection. This mid-season,

Cultivars of Celery

Tall Uttah

Conquistador

Rivalry

Hudson

White Plume

Red Celery

Cultivars of Celery

Giant Pascal

Golden Self Blanching

medium-green, Utah-type celery cultivar is harvested in April to early May under Florida conditions. It matures 80-90 days after transplanting, somewhat earlier than June Belle and Florida 2-14. Plant height is 79-84 cm in a planting design with 60 cm row spacings and 17-20 cm between plants, making Florida Slobolt M68 5-8 cm taller than June Belle and 6-10 cm taller than Florida 683. In Florida's main production regions, Florida Slobolt M68 out-yielded June Belle and Florida 214. In the Everglades area, where bolting is a problem due to infrequent low temperatures, the 2 reference cultivars showed serious bolting, leading to yields of 1663-1836 crates/ha, compared with 2439 crates/ha for Florida Slobolt M68. This cultivar is susceptible to calcium and boron deficiencies (Wolf *et al.*, 1993).

Floribelle M9

Developed by pedigree breeding featuring single plant selection in the F_2-F5, and originating from a cross between breeding lines EES1624-2 and EES1625-2, this mid-season cultivar (92-99 days from transplanting to maturity) is intermediate in growth between Crystal Jumbo and Ordinary Utah phenotypes (on average 70 cm tall at a 60 cm row spacing). It is more robust in appearance than Earlibelle or Florida 2-14. It is more susceptible to boron related transverse cracking in the nodal region and longitudinal cracking than the other genotypes tested, and susceptible to blackheart induced by Ca deficiency, but more resistant to the complex of bacterial blight disorders than either Earlibelle or Florida 2-14. It is selected for resistance to seed thermo-dormancy and vigour and uniform seedling growth in summer (Wolf and Scully, 1992).

UC-T3 Somaclone (*Fusarium oxysporum* f. sp. *apii*, race 2)

US-T3 Somaclone was obtained by screening regenerated plantlets from a cell suspension culture of the cultivar Tall-Utah 527OR for resistance to *F. oxysporum* f. sp. *apii* race 2 by planting them in soil containing 10 g of barley straw inoculated with FOA2. In 7 field trials in California the S_1 progeny of UC-T3 Somaclone were consistently more resistant to FOA2 than Tall-Utah 527OR and Tall Utah 527OHK. UC-T3 Somaclone resembles Tall-Utah 527OR in colour, habit and resistance to pests and other diseases (Heath Pagliuso *et al.*, 1989).

RRL-85-1

This cultivar was obtained from a diverse germplasm collection consisting of over 50 accessions by mass selection and random mating of selected plants. It is recommended for cultivation for seed in Jammu. RRL85-1 has light green leaves, thick stem and small closely packed white flowers in compound umbels. It is medium tall (120-130 cm) and flowers and matures 110 and 180-190 days after sowing, respectively. Its seed yield is approximately 1.1 t/ha. The essential oil of RRL88-1 is commercially acceptable. RRL85-1 is moderately resistant to *Cercospora apii, Septoria petroselini, Phyllosticta apii, Puccinia bulata* and celery mosaic virus (Choudhury and Kaul, 1988).

Advantage, Pilgrim and Companion

These 3 cultivars are considered useful for both commercial production and breeding purposes. Advantage is an F6 selection from a cross between a strongly bolting-resistant selection of Spartan 162 and a line derived from Utah 52-70H. It is a midgreen Utah Pascal type that matures a week later than Utah 52-70H and is about 68 cm tall, 5 cm taller than Tall Utah 52-70HK. Petiole length from butt to first joint is about 33 cm and the shape is flaring to cylindrical with a compact to slightly open conformation and a full heart.

Pilgrim is an F_{10} selection from a cross between Emerson Pascal and strongly bolting-resistant selection of Spartan 62. It has good resistance to *Fusarium oxysporum* f. sp. *apii* race 2. Pilgrim has midgreen petioles, resembling those of Utah Pascal, and matures a week later than Utah 52-70HK. Petiole length from butt to first joint is about 36 cm and the shape is cylindrical with a slightly open conformation and medium heart.

Companion is an F_6 selection from a cross between June Belle and the breeding line MSU71-605 (celeriac x celery). It is moderately resistant to *Cercospora apii* and *F. oxysporum* f. sp. *apii* race 2. The petioles of Companion are green, resembling those of Florida 683, and it is slightly later maturing than Tall Utah 52-70HK. Plants are about 66 cm tall and petiole length from butt to first joint is about 30 cm. The shape is cylindrical with a compact formation and full heart (Honma *et al.*, 1986).

Sigfrido

This celery cultivar bred in Italy, is a self-blanching cultivar suitable for early cultivation under glass or plastic or for spring and summer cultivation in the open field. It is resistant to stem cracking (Anon., 1981).

Golden Spartan

Plants of this cultivar are erect, compact, uniform in size and height, and slightly taller than Cornell 19. Its harvesting date is similar to that of Golden Plume and Golden Self-Blanching. It is cold tolerant and does not bolt when grown with or without paper covers (Honma and Klompenberg, 1974).

Mission

Mission is a versatile celery cultivar that can be grown year-round and is the standard for bolting resistance. It is having compact plant with excellent girth, a uniform leaf and top appearance and beautiful dark green color.

Sonora

Ideal for summer and fall harvest, Sonora is a compact plant with smooth petioles that grow from a dark green stalk and has good bolting resistance.

Conquistador

It is versatile enough to be grown year-round and also boasts a high level of bolting resistance. It has become the cultivar of choice for winter-spring harvest periods.

Clean Cut

It is open pollinated cultivar; it took 125 days to mature from transplanting, excellent shipping quality, large heavy petioles of good length, relatively few side shoots.

Florigreen

It is an attractive, uniform, vigorous widely adopted green stalk cultivar developed to provide a distinct petiole type for fancy grade pack, tolerant to brown spot and having wide adaptability.

Transgreen

It is a thick petiole variety with high yield potential.

Other Cultivars

Par Cel, Safir and **Flora 55** are few cultivars of leaf celery; **Brilliant, Giant Prague, Mentor, President, Diamante** are few new celeriac cultivars those took about 100-120 to mature; while **Golden Boy,** with short stalks, **Tall Utah**, which has long stalks Monterey is an early cultivar.

Golden Plume, Golden Self Blanching, Michigan Improved Golden and Cornell 619 are the cultivars from golden segment. **Utah 52–70 R, Utah 52–70 HK, Utah 52–70, Florida 683, Tall Green Light** and **Tender crisp** are Utah type from green.

Vulsteke (1979) assessed the self-blanching celery cultivars and found best performance by **Avon Pearl, Golden Detroit Selection ZHZ. Golden Spartan, Asmer Jason** and **Lathom** Self-Blanching showed promise. **Loret** and **Golden Boy** (yellow) and **Claret and Utah 1130** (green) are good yielders (Vulsteke and Van Oost, 1989). Vulsteke *et al.* (1994) also found that the best performance for a combination of traits was given by **Golden Spartan** (yield 72.38 t/ha) and **Golden Fleece** (yielding 7 per cent less than Golden Spartan), followed by **Autumn Gold** (yielding 15 per cent less than Golden Spartan). The best dark-green cultivar was **Claret** (yield 75.38 t/ha).

The Indian Agricultural Research Institute, New Delhi, recommended **Standard Bearer** and **Wright Gieve Giant** vegetable type introductions of celery in India. **EC 99249-1 and RRL 85-1** are also considered good varieties for cultivation under Indian conditions for high essential oil content (Malhotra, 2006a).

6.0 Soil and Climate

6.1.0 Soil

Celery comes up well in a wide range of soils. However, specific cultivars perform well in a given soil type. Peat soil was found to be better for growing Tendercrisp cultivar. Recommended cultivars for cultivation on light sandy loam soils and supply to the vegetable processing industry are Spartacus, Victoria and Octavius (Vanparys, 2000).

Celery has a shallow root system and thus requires a highly fertile soil with good moisture-holding capacity. Though it is reported to be cultivated in a wide range of soils, peat and clay loam soils are usually well suited for production. Celery is moderately sensitive to salinity and grows best within a pH range of 6.0–6.6 in mineral soils and 5.5–6.0 in organic soils. The seed crop of celery under Indian conditions in Punjab state is grown on soils with an average pH of around 7.5 (Malhotra, 2006a).

6.2.0 Climate

Celery can be grown best in a climatic condition with a long, cool growing season, especially at night and where rainfall is well distributed or irrigation is assured. Optimum crop growth is reported, when mean temperatures range between 16°C and 21°C with wider adoptive cultivars capable of tolerating upper temperature ranges. Celery can be grown in subtropical regions of the India. Though celery is sensitive to sub zero temperatures but is well acclimatized to tolerate light frost for a short period of time. Leafy type celery are reported to be more heat tolerant than root celery or stalk celery type.

Celery responds to changes in temperature showing bolting. Use of bolting resistant cultivars and high temperatures during the early stages of cultivation should result in higher yields (Wiebe, 1999). Temperature also plays a crucial role in seed production. Seed yields were highest (18.8-19.5 kg/100 m^2) from stecklings sown on 1 June or 1 July. Sowing later than this period reduced seed quality (Dyduch, 1995). Light (intensity, quality and duration) also affects the bolting in celery. The least bolting (8-14 per cent) occurred in plants that had been held in unheated tunnels with the blue spectrum of light removed using cellophane paper filters. It is suggested that this treatment could be used during transplant production to obviate vernalization (Favaro and Bochinfuso, 1994). Eight-hour day lengths during chilling however markedly increased the number of plants forming sessile flowers. Short day lengths (8 h) after chilling decreased the proportion of young, but competent plants that bolted and flowered. Total darkness during chilling completely prevented any subsequent vernalization response either as bolting or as flowering (Ramin and Atherton, 1994). Supplementary light of 4400 lx shortened the period between pricking out and the time at which plant FW reached 5 g by 25-30 per cent, irrespectively of temperature regime (Folster, 1992).

7.0 Cultivation

7.1.0 Seeds and Sowing

Seed of celery germinates slowly and does not show uniformity. Seed quality of celery is affected by the branching habit of the seeding plant, plant spacing and harvest date. These factors also affect the subsequent variations in time of seedling emergence and seedling size, mainly due to difference in embryo size (Gray, 1987). Seed germination in celery could be easily hastened by treating the seeds with optimum temperature, light and osmoregulants. Brocklehurst *et al.* (1987) found the osmotic treatment of celery seeds with PEG (polyethylene glycol) to improve the germination and emergence, whereas KH_2PO_4 reduced the germination. Seed priming using fluid drilling technique was found to improve the rate, uniformity and percentage of germination at low and high temperatures (15 to 35°C) in celery (Globerson and Feder, 1987). The optimal treatment conditions were 300 g/l of polyethylene glycol (6000 MW) for 21 days at 10°C, which increased the average germination rate from 17.54 (control) to 87.9, and the total percentage germination at 21 days from 76 (control) to 81 per cent. However, treated seeds using this method should be sown immediately (Singh *et al.*, 1985). Furutani *et al.* (1985) observed that the removal of leachates improved the seed germination in light. They also advised that the treatment of seed at 10°C for 14 days prior to sowing helps to improve the emergence speed and uniformity. Celery seed germination was good at 15 or 20°C, poor at 25°C and nil at 30°C (Nakamura *et al.*, 1982). Specific gravity of seed separation technique developed by Finch Savage (1984) improved the percentage germination of seeds ready for sowing (with the longest radicles less than 2 mm) from 36.8 per cent to 91.7 per cent. Jardine *et al.* (1984) tried the hypobaric storage of celery transplants at 0°C for 2 weeks at 25 mm Hg or at atmospheric pressure and found no significant differences in bolting or yield compared with non-stored controls after transplanting to the field. Jeng and Sung (1989) suggested removal of lighter seeds from seed lots to get rapid and uniform germination. Zhang *et al.* (1998) recommended the use of low temperature (10°C) for 8-10 days to improve the germination vigour and index. At high temperature, priming the seeds with 3-4 ml of water or 1 per cent NaOCl solution, helped to improve the germination (Paerara *et al.*, 1993).

In India the celery is grown as seed spice covering a large area in the Punjab which is situated in the northern plains. It is grown during September-October and transplanted from mid-December to the first week of January (Randhawa and Kaur, 1995).

Celery crop may be raised from transplants or by direct seeding in the field. The time of sowing is determined on the basis of crop to be raised for fresh herb or seeds. In California, seed are sown in seed beds in July or sometimes in December–January. Eight to twelve weeks-old seedlings are transplanted in a well-prepared field. In other parts of the USA with more severe winters, crops are started in the early spring by sowing in a green house or hot beds and seedlings are dug in the autumn, rouged for off type and cold stored until planting time in spring. The tops are stored at 0°C and 90–95 per cent RH, maintaining moisture and good ventilation around the roots. The roots are placed in moist soil. The withered and decayed leaves should be removed when plants are transplanted at distance of about 90 cm between rows. The closer spacing results in higher seed yields. In some coastal areas of California, celery is seeded directly in the production beds (Malhotra, 2006a).

7.2.0 Spacing

Optimum spacing in celery is essential to obtain high yield from unit area. High density (10 x 15 cm) planting was helpful in celery to obtain high yield, lower crop weight and high quality (Vanparys, 2000). Plant spacing influences the plant height and petiole length more than number of leaves in celery (Starndberg, 1986). But, Balyan *et al.* (1990) noticed that the height is unaffected by row spacing. They found the number of branches and umbels to increase with increasing row spacing and concluded that 40 cm row spacing is optimum for seed and oil yield in celery. For Blanched celery 40 x 15 cm was found to be the optimum spacing (Vanparys, 1999). Stoffella *et al.* (1986) studied the effect of 10, 20 or 30 cm row spacing on fresh and processing yield of celery and found 10 cm spacing to give very low fresh market yield but very high yield for processing. Stoffella *et al.* (1988) also found 20 cm within-row spacing to be optimum for the fresh market yields. Malhotra (2006a) reported that row spacing of 40 cm gave maximum seed yield in this crop.

7.3.0 Irrigation

Irrigation significantly influences the yield and quality of celery crop. Irrigation increased dry matter and total sugar contents but decreased vitamin C contents. It also decreased α- and β-carotene and thiamine contents of celery (Evers *et al.*, 1997).

Relationship between water supply and quality of celery was studied by Ivanek and Soukup (1996) using two water tables (0.4 m and 0.8 m), unregulated and fallow control. Quantity of water required for irrigation was highest (3807 l/ha) at 0.4 m treatment. Irrigation at 50 mm CPE resulted in significantly higher seed yields compared with 74 and 100 mm CPE (Saini *et al* , 1989). Using lysimeters, water consumption of vegetables was worked out by Pfulb (1990). It was found that celery required more water than other vegetables. Shoot growth of celery was better with sprinkler irrigation than drip system (Tremblay and Senecal, 1990). But, drip irrigation was helpful in applying fertilizers and pesticides effectively. Drip-applied insecticide was found to be less expensive than a foliar application with air or ground equipment although it was limited to phloem-feeding insects, was slower in attainment, and at best was equal to or less than that obtained with a foliar spray (Royer *et al.*, 1988). The study of Breschini and Hartz (2002) suggested that commercial drip irrigation management in celery could be significantly improved by applying a few basic principles: i) more frequent irrigation; ii) lower volume per irrigation and iii) more closely matching irrigation volume to ETo and crop growth stage.

Use of diluted sea-water (1: 1) to increased yield of celery by 52 per cent compared with 35 per cent for tap water and 25 per cent for sea water (Rumasz *et al.*, 1999). Irrigation of celery at 0.7 and 0.5 bars increased the profit by 25.9 and 16.2 per cent which uses 204 litres and 2446 litres of water, respectively, to produce one kg of fresh matter (Nissen and Engell, 1996). Maticic *et al.* (1991) applied four levels of N (0-400 kg/ha) and six levels of irrigation (100-1022 mm water/growing period) in celery and recorded the highest yield with highest nitrogen rate and the highest irrigation level. Irrigating the celery plants with 100 per cent sewage water along with 50 kg N/ha produced the highest seed yield of 1390 kg and high net return/ha (Mahajan *et al.*, 1978).

Du *et al.* (2018) reported that plant yield, nitrogen agronomic efficiency, nitrogen physiological efficiency and nitrogen recovery efficiency were significantly increased under aerated irrigation in celery. They also suggested that air injection through subsurface drippers has a low cost and is simple to operate, so this technology can be widely used for cultivation of this crop.

Wang *et al.* (2020) found that individual plant weight of celery was higher under micro-sprinkler irrigation (MS), furrow irrigation under plastic film mulching (PF) and micro-sprinkler irrigation under the plastic film mulching (MSP) than under furrow irrigation (FI) in both autumn season crop (AC) and spring season crop (SC). In AC and SC, the economic yield of celery increases under MSP by 54.18 per cent and 49.55 per cent, whereas in case of PF, it is 30.37 per cent and 34.10 per cent. The irrigation amount of MSP was 151.69 and 179.91 m^3 667/m^2 in AC and SC, which was 23.13 per cent and 27.27 per cent lower than that of FI. The irrigation amount of PF was 151.69 and 196.78 m^3 667/m^2 in AC and SC, which was 23.13 per cent and 20.45 per cent lower than that of FI. So, it can be concluded that PF and MSP reduced the irrigation amount of celery cultivation in green house and soil evaporation content.

7.4.0 Manuring and Fertilization

Meng and Jiang (1993) found that nitrate content in petiole is positively correlated with that available in the whole plant. Fertigation had increased the marketable yield in celery (88.6 t/ha) as compared to the traditional application rate (70.5 t/ha) (Temperini *et al.*, 2000). Seedlings of celery when irrigated by drip or sprinkler systems with nutrient solutions supplying nitrogen as NO_3^-, NH_4^+ and urea in the proportions 2: 1: 0 or 2: 1: 3 (*i.e.*, 0 or 50 per cent as urea) improved the shoot growth and N and P concentrations (Tremblay and Senecal, 1990). On an average, celery accumulated 564 mg NO_3/kg fresh mass with the 100: 0 NH_4:

NO_3 ratio and showed that produce with a very low NO_3 content can be obtained by using NH_4. By increasing NO_3^- N percentage in the nutrient solution, NO_3 leaf content of celery was increased remarkably (7802 mg/kg fresh mass with the 0 : 100 NH_4: NO_3 ratio) (Santamaria *et al.*, 1999). Weiser and Scharpf (1989) analyzed the possibility of reducing 20 per cent of optimal level of nitrogen to the yield reduction in different vegetable crops and found yield reduction in celery was only slight. For a celery yield of 100 t/ha, the N requirement is 360 kg/ha; assuming a latent mineral N soil residue of 100 kg/ha not taken up by the plants, the total N requirement is thus 460 kg/ha (Demyttenaere *et al.*, 1988).

Ammoniacal form of nitrogen was more conducive to the disease, caused by *Fusarium oxysporum* f. sp. *apii* than nitrate irrespective of the accompanying ions. The incorporation of specific concentration of KCl with NO_3 [as Ca $(NO_3)_2$] resulted in complete control in the green house and field. Substantial disease control was achieved in the field with side applications of Ca $(NO_3)_2$ and KCl (Schneider, 1985). Phosphorus application at 48-80 kg/ha increased the celery yield by 150-200 per cent above the control (Paris *et al.*, 1999). P fertilization was studied with or without VA mycorrhizal treatment by Krikun *et al.* (1990). It was found that the VAM association was found to be essential in increasing P levels in tissues of celery. Biczak *et al.* (1998) applied N, P and K separately in nutrient solution to give soil contents of 88, 188 or 288 mg N, 225, 275 or 325 kg P_2O_5 and 280, 380 or 480 mg K_2O and found that the excessive fertilizer rates reduced the crop yield as well as the nutritive quality. Petiole levels of nutrients are helpful in reducing the fertilizer applications and improving the nutrient economy. Petiole P and K concentrations of 0.55 and 7.7 per cent, respectively, appeared to be sufficient for obtaining optimal yield (Sanchez *et al.*, 1990). Beverly and Anderson (1988) found maintenance of 2.5 g P per kg dry matter to be sufficient for celery production. Lee *et al.* (1999) investigated the effect of selenium on celery. Selenate treatments up to 4 mg/l did not adversely affect celery growth. Celery can tolerate salinity up to 10.5 dS/m without hampering plant dry matter production (Leonardi and Munoz Carpena, 1998). Addition of NaCl up to level of 10.0 mS/cm in celery grown in NFT enhanced quality by reducing the accumulation of $NO3^-N$ and the incidence of blackheart in young leaves. Blackheart was completely prevented by foliar application of calcium nitrate (Pardossi *et al.*, 1999).

Takahashi *et al.* (1987) studied the role of calcium on blackheart disorder and reported that high calcium levels reduced the Mn availability to the celery. Curly celery is a disorder in transplanted celery due to the high pH of the saturated medium extract (SME) and low B and Zn tissue concentrations (Tremblay *et al.*, 1989).

Celery can tolerate boron up to 10 mg/l (Francois, 1988). Celery plants harvested from treatments >10.0 mg of B/l were found to be immature, bitter tasting, and not of marketable quality.

Madrid *et al.* (2008) studied the influence of nitrate fertilizer on macronutrient contents of celery plants on soil-less culture. Two rates of nitrate (NO_3) supply were used in production of celery plants (*Apium graveolens* L., cv. 'Istar') grown on perlite or coir substrates. A standard nitrogen dose (CN) was compared to a concentration that was 25 per cent less (RN). Other nutrients were supplied at the same level

Field Growing and Harvesting of Celery

for both treatments. Plants were harvested after 85 days of growth, divided into leaves, petioles, internal stalks, basal disc and roots and the macronutrient contents [nitrogen (N), phosphorus (P), potassium (K), calcium (Ca), magnesium (Mg) and sulfur (S)] were analyzed. The ratios between elements (Na/K, Mg/Ca, and N/P) were also determined at the whole plant level. Commercial yield was statistically the same for CN and RN treatments. Nutrient content for plants cultivated on coir were slightly higher than those on perlite. Petioles and leaves showed the highest accumulation of K, Ca and Mg; the roots also had similar high levels. The basal disc and internal stalks tended to accumulate S. Nitrogen and P were uniformly distributed throughout the plant. Roots showed the highest accumulation of Na. On a whole plant basis, the nutrient contents for CN and RN were almost the same, with only a slightly higher N concentration for CN. Within the edible fraction (petioles + internal stalks), the RN treatment resulted in a higher concentrations for all the nutrients except N, compared to CN. Thus, celery plants cultivated with a nitrogen dose 25 per cent lower than that commonly used by producers had a balanced nutrient content and were safer for consumers.

Li *et al.* (2010) performed a green house pot experiment to study the effect of phosphorus (P) on celery (*Apium graveolens* L.) growth, quality and nutrient uptake under different calcium (Ca) and magnesium (Mg) levels in substrate culture. Results showed that there were significant interactions between P and the level of Ca and Mg in the growing media. Above ground fresh weight, total dry biomass, leaf area and P, Ca and Mg concentrations in celery leaves significantly increased from 0 to 124 mg/l P application. The opposite trends were observed for root/ shoot ratio, leaf chlorophyll, carotenoids, soluble protein, soluble sugar, vitamin C and nitrogen and potassium concentration in celery leaves. Medium Ca and Mg level (Ca 320 and Mg 192 mg/l) significantly increased celery above-ground fresh weight, total dry biomass and leaf area compared to low (Ca 160 and Mg 96 mg/l) and high (Ca 640 and Mg 384 mg/l) levels. This study suggested that 124–248 mg/l P applications under medium Ca and Mg level were appropriate for celery nutritional requirements.

Ahmed and Maprd (2017) observed that using biofertilizer potasiomag (*Basillus mucilaginosus*) + 50 per cent recommended dose of NPK (Ammonium sulphate (20.5 per cent N) at 300 kg/fed, calcium superphosphate (15.5 per cent P_2O_5) at 150 kg/ fed and potassium sulphate (48 per cent K_2O) at 150 kg/fed, recommended dose) have resulted in the greatest plant growth and oil compositions Apiol, Myristien, β-Pinene, Limonene and β- Phellandrene of celery.

7.5.0 Weed Management

Oxadiargyl is a good pre-emergent herbicide for transplanted celery crop. The crop tolerated dose rates ranging from 400 to 800 g a.i./ha (Tracchi *et al.*, 1997). Orabanche weed was successfully reduced by the application of 50 g glyphosate/ ha (Americanos, 1991). Runham *et al.* (2000) used paper mulch to reduce the soil temperature and weed growth in celery grown on sand and peat soils. Soil temperature beneath the paper mulch was lower than both the non-mulched and black polythene-covered plots.

Of the eight post-emergent herbicides tried in transplanted celery, prometryne at 1 lb/acre, oxyfluorfen at 0.0625 and 0.125 lb/acre, terbutryne at 1 and 2 lb/acre, bifenox at 1 and 2 lb/acre and chloroxuron gave a wide spectrum of weed control with good crop selectivity (Chism and Snyder, 1982). Combination of post-emergent applications of linuron, prometryn or thiobencarb plus 0.25 lb fluazifop-butyl/acre was more efficient than when used alone. Use of linuron generally reduced crop vigour and yield. Prometryn was sometimes phytotoxic to the crop (Dusky, 1983).

Charles *et al.* (2006) carried out field studies from 2002 to 2004 on Houghton muck soil to assess the impacts of cover crops and soil fertility regimes on weed populations and celery yield. The cover crops were oilseed radish, cereal rye, hairy vetch and a bare ground control. The fertility rates were full (180, 90, and 450 kg/ha nitrogen (N), phosphorus pentoxide (P_2O_5) and potassium oxide (K_2O), respectively), half (90, 45 and 225 kg/ha N, P_2O_5, and K_2O, respectively) and low (90 kg/ha N). Each cover crop treatment was combined with the low or half rate of fertilizer. An additional treatment with bare ground plus the full rate of fertilizer was added as standard practice. Treatments were maintained in the same location for the duration of the study. Major weed species were common chickweed, prostrate pigweed, shepherd's-purse, common purslane and yellow nutsedge. Each year, oilseed radish consistently produced the greatest biomass and provided over 98 per cent early season weed biomass suppression. Hairy vetch and cereal rye provided about 70 per cent weed suppression in early spring. Soil fertility level affected weed populations during the 2004 growing season. In 2004, weed biomass in treatments without cover crops or with vetch increased when greater amounts of fertilizer were applied. Within individual fertility levels, higher celery yields were recorded in the oilseed radish plots. For example, in the low fertility rate, celery yield was 34.8, 29.2, 23.9, and 24.4 ton/ha in the oilseed radish, cereal rye, hairy vetch and control plots, respectively in 2003. Overall, the results of this experiment indicate that when included in a system where hoeing and hand-weeding are the only weed control methods, cover crops can successfully improve weed management and celery yield on muck soils, allowing reduced fertilizer inputs.

Daugovish *et al.* (2007) recorded that pretransplant application of Flumioxazin at 0.1 and 0.2 kg ai/ha controlled broadleaf weeds near 100 per cent, suggesting that it can be an effective alternative to standard linuron and prometryn, which are applied post-transplant (POST) in celery.

Kaur and Gill (2012) carried out a study at Ludhiana, India to study the performance of different herbicides at varying doses, *viz.* pendimethalin 0.5, 0.75 and 1.0 kg/ha, trifluralin 0.75, 1.0 and 1.25 kg/ha and oxyfluorfen 0.15, 0.17 and 0.20 kg/ha applied after transplanting and combination of lowest dose of each herbicide followed by hand weeding 40 days after transplanting in comparison to hand weeding (at 20 and 40 days after transplanting) and unweeded/control in transplanted celery. All the herbicides and hand weeding treatments significantly reduced total weed population and their dry weight compared to unweeded/control. The lowest dose of oxyfluorfen with one hand weeding was most effective in reducing the population and dry weight of weeds and gave maximum weed control efficiency (91.9 per cent). The integration of oxyfluorfen @ 0.15 kg/ha with one hand weeding at 40 days after transplanting produced maximum number of branches and umbels per plant. The maximum seed yield of celery (13.0 q/ha) and net return (Rs. 32292 ha^{-1}) was recorded with oxyfluorfen 0.15 kg ha^1 followed by one hand weeding. The seed yield of oxyfluorfen 0.15 kg/ha integrated with one hand weeding was statistically similar to pendimethalin 1.0 kg/ha, pendimethalin 0.5 kg/ha supplemented with one hand weeding, oxyfluorfen 0.20 kg/ha as well as two hand weedings treatments.

7.6.0 Soilless Culture

Soilless culture is defined as any method for growing plants without the use of soil as a rooting medium, where soilless culture offers an ideal alternative crop production for traditional cultivation in soil when there is no soil available at all (Olympios, 2011). Many systems of soilless culture can be applied on rooftops gardens such as pot substrate, horizontal bag substrate bed substrate (Neveen Metwally, 2016) and modified NFT systems, such as A-shape NFT and flat NFT (Orsini *et al.*, 2012). Many crops and vegetables, such as celery (*Apium graveolens*), can successfully produce in soilless culture systems and green rooftops in urban cities.

Maharik and Mancy (2017) studied the effect of different soilless culture systems (beds, pots and bags), modified nutrient film techniques (A-shape NFT and flat NFT) as well as foliar spray with GA_3 (0, 10, 20, 30 and 40 mg/l), were tested on the growth and yield of celery plant (*Apium graveolens* var. *dulce* cv. Royal crown). A study was done on the rooftop of the Central Laboratory for Agricultural Climate (CLAC) at Dokki, Giza, during winter seasons of 2015 and 2016. Plant height, numbers of leaves, chlorophyll reading, fresh and dry weight/plant, yield per m^2, as well as NPK content in shoot of celery plant were determined. Results indicated that using substrate culture systems, during the two seasons, gave the highest significant values for plant height, number of leaves, fresh and dry weight per plant, nitrogen, phosphorus and potassium content as compared with modified nutrient film techniques. The highest values of chlorophyll reading were recorded using pots and A-shape NFT systems with no significant differences between them and with significant difference with other soilless culture systems. Results also indicated that increasing the concentrations of GA_3 up to 40 mg/l led to increase in plant height, fresh and dry weight per plant, nitrogen, phosphorus and potassium contents in comparison with control treatment (spray with distilled water). On the contrary, number of leaves and chlorophyll readings did not respond to increase of foliar spray of GA_3 concentrations as compared with control treatment. The highest total yield (Kg/m^2) was recorded using bed system followed by A-shape NFT system without a significant difference between them. Bed system gave the highest significant values of fresh weight per plant with the lowest number of plants per m^2 (16 plant per m^2), while A-shape NFT system gave the lowest values of fresh weight per plant with a greater number of plants per m^2 (24 plant per m^2) as compared with the others soilless culture systems.

7.7.0 Nutritional Deficiency Symptoms

Boron Deficiency

Due to boron deficiency brown checking or cracked stem observed in celery and these two are most obvious symptoms of the deficiency. On the inner or adaxial,

Marketing, Packaging and storage of Celery

surface, a brown necrosis occurs below the first leaflets and extends about halfway down the petiole. There may also be transverse cracks on this surface. On the outer, or abaxial surface, cracks occur along the ridges. This problem can be controlled by spraying a boron source such as sodium borate and borax directly on the foliage.

Magnesium Deficiency

Chlorosis or yellowing of older leaves particularly interveinally occurs in a number of celery production areas. The disorder is due to a deficiency of magnesium, can be alleviated in the field by weekly sprays of magnesium sulfate. Shortly after discontinuing the spray treatments, the yellow symptom returns. High calcium in the plant intensifies the disorder.

8.0 Harvesting, Yield, Postharvest Management and Storage

8.1.0 Harvesting and Yield

The environmental requirements of celery make it a potentially viable leafy vegetable crop for all-year-round production in many countries under controlled conditions.

Celery is harvested when the overall field reaches the desired marketable size of 35-41 cm stalk length, and before the outer petioles develop pithiness. Early harvests before the plants reach full size produce stalks with high market quality, and the price received may more than compensate for lower yield (Luo *et al.*, 2002). High quality celery consists of petioles that are well formed with thick, compact, and straight (Suslow and Cantwell, 1998). Petioles should be tender, light green and appear fresh.

Additional quality indices include the stalk and midrib length; freedom from defects such as blackheart, pithiness, seed stalks, cracks, and absence of insect damage and decay.

Sandwell and Wood (1972) found that mid-January sowing of celery cv. Lathom Blanching resulted in satisfactory crops by mid-July. Bolting is not a problem provided the crop is harvested as soon as it matures. Covering with plastic until early April produced the least bolting in celery (Maync and Riedel, 1986). The highest percentage of marketable self-blanching celery plants, cv. Long Dore Ameliore, was obtained from seedlings planted in green houses on 25 March or under perforated polyethylene from 4 April onward and harvested from the end of June (Benoit *et al.*, 1981). Dyduch (1994) developed a methodology using the cultivars Slow Bolting Green No. 96, Utah 52-70, Summer Pascal, Tendercrisp, Zefir and Helios to provide continuity of supply for fresh sales, processing and freezing. Heated and unheated green houses and plastic tunnels enabled harvesting between March and April with mean yields of 4.3-9.9 kg/m^2. Field covering with perforated plastic film enabled harvesting from May to July with mean yields of 2-7 kg/m^2. Open, unprotected cultivation provided harvests until the autumn frosts with mean yields of 5 t/ha on a mineral soil and 8 t/ha on a peat soil.

The machines used for harvesting include modified potato and onion harvesters for use in harvesting turnip-rooted celery (Meeldijk, 1984). A tractor-drawn or truck-mounted celery harvester is suitable for different row widths and soil conditions. The single-row unit, with 2 men, can be harvested up to 30 t/ha. Trimmed and cut stalks are carried on a conveyor belt to an accompanying truck. Other harvesters include a tractor-drawn and powered machine, able to cut celery below ground level at 5000 plants/ha (Candia, 1983). Combined disinfection with chlorine (100 ppm) and thiabendazole (500 ppm) immediately after cutting of petiole ends reduce storage decay in mechanically harvested celery (Aharoni *et al.*, 1978). The VCX harvester can lift and clean 100 ha of celery in 25-28 days (Hybl, 1977).

8.2.0 Postharvest Management

8.2.1 Precooling

Celery is cut 2.5 cm below the crown and coarse outer leaves trimmed. Prior to storage, the field heat should be removed as quickly as possible– ideally with hydro-cooling. Celery is typically hydro-cooled, hydro-vac-cooled, or vacuum-cooled with a chilled water spray application. Prompt pre-cooling to near 0°C is essential to maintain the freshness and crispness, as well as for extended storage (Luo *et al.*, 2002).

8.2.2 Grading and Packaging

Luo *et al.* (2002) presented a comprehensive review on the quality of celery. In California, fresh market celery is field packed in 27.2 kg (60 lb) cartons containing 48 stalks and 12.7 kg (28 lb) cartons containing 12 or 18 hearts. In Florida, it is packed in 7 size grades from 18 to 96 stalks per crate. Celery hearts are generally prepared from stalks that are smaller than regular size. They are trimmed to 20, 25 or 30 cm in length, and packed in 8 or 13 kg (18 or 28 lb) cartons (Peirce, 1987).

8.2.3 Storage

Celery lasts up well up to two months in storage with a minimum of loss if held at 0°C and a relative humidity of 95-100 per cent. Celery is not chilling sensitive and should be stored as cool as possible without freezing. The freezing point for celery is –0.5°C (Whiteman, 1957). At optimum conditions, it can be stored for up to 5 to 7 weeks with good quality (Hardenburg *et al.*, 1986). Storage-life is reduced to < 2 weeks at 5°C. Inner petioles may continue to grow during storage at > 0°C resulting in quality loss. Controlled atmosphere (CA) or modified atmosphere (MA) offers small to moderate benefits to celery in storage (Salveit, 1997). CA stored stalks maintain better texture and crispness than those stored in air (Gariepy *et al.*, 1984). Reduced O_2 (2 per cent to 4 per cent) and elevated CO_2 (3 per cent to 5 per cent) delayed senescence, leaf yellowing and decay (Leshuk and Saltveit, 1990). However, low O_2 or high CO_2 injuries may occur at < 2 per cent O_2 or >10 per cent CO_2, resulting in off-odours, off-flavours, and internal leaf yellowing. Shelf-life of celery stalks can benefit from 2 to 5 kPa O_2 and 5 to 10 kPa CO_2 atmospheres at 2–4°C (CSIRO, 1998; Suslow and Cantwell, 2000). It could be assumed that MAP of celery sticks should be designed to maintain both O_2 and CO_2 as close as possible to these levels. Gomez and Artes (2004) observed that MAP treatments improved the sensory quality, avoided the loss of green colour, decreased the development of pithiness and retarded the growth of microorganisms in minimally processed

Marketing and Packaging of Celery

Storage Disorder of Celery

celery sticks. In any treatment neither off-odours nor off flavours were detected. After 15 days at 4°C within the oriented polypropylene (OPP) bags a steady-state atmosphere of 6 kPa O_2 + 7 kPa CO_2 was reached and celery sticks stored under these bags showed the best quality.

Gomez *et al.* (2010) reported that green celery stalks could be stored at 0 to 4°C and about 95 per cent RH in air for 2 to 6 weeks. In those conditions undesirable quality changes like reducing shelf life occur and controlled atmosphere (CA) has been shown as useful for lowering those processes. Celery is an important source of mannitol, an uncommon sugar for vegetables and useful for pharmaceutical industries. Moreover, celery sugars, especially mannitol, may be used as a reduced-calorie sweetener. Horticulturally mature 'Trinova' celery stalks, free from defects and of uniform size were washed with chlorinated water, drained and stored in hermetic glass jars at 4°C during 35 days. Continuous humidified flows through air (as control), 5 kPa O_2 + 5 kPa CO_2 (CA5), 5 kPa O_2 + 15 kPa CO_2 (CA15) or 5 kPa O_2 + 25 kPa CO_2 (CA25) were applied. Throughout cold storage the respiration rate was monitored by using a closed system. At the beginning of their (Gomez *et al.*, 2010) experiment and at the end of cold storage sugar content was measured by liquid chromatography with a refractive index detector. The respiration rate was high at the beginning of storage, decreasing rapidly and being stable until the end. Compared to air, CA storage reduced respiratory activity. The reduction was mainly related to O_2 level and with almost no impact of varying CO_2, with no differences among CA treatments. Mannose was the predominant sugar followed by fructose and glucose. After storage, total sugar content decreased for air but not for CA5 and CA15, probably related to lower respiration rate which would retard loss of sugar. Fructose decreased for CA25 and non significant changes were detected for remaining sugars in any CA treatment. After 5 weeks of cold storage, a CO_2 level of 25 kPa was detrimental to sugar maintenance, and CA5 and CA15 provided the best results.

8.2.3.1 Storage Disorders

Crushing or cracking are signs of mechanical damage, and many lead to rapid browning and decay. Harvesting, packing and handling should be done with great care to prevent damage to the highly sensitive turgid petioles. Freezing injury starts at temperatures below –0.5°C. Mild freezing causes depressions in the tissues that subsequently turn brown. Severely frozen tissues develop wilted and water soaked appearance on thawing. Discolouration of vascular tissue, splitting of the cut ends, and bacterial decay are major problems limiting shelf-life of fresh-cut celery (Robbs *et al.*, 1996; Saltveit and Mangrich, 1996).

Gorini *et al.* (1973) carried out extensive studies on storage of celery. They found that reducing the leaves before storage at 0°C did not improve the stored product but was advisable for reasons of space. There was considerable less weight loss in CA storage (3 per cent CO_2 + 3 per cent O_2) rather than in a normal atmosphere. Vacuum precooling increases the weight loss but did not lessen rotting due to *Penicillium*, *Botrytis* and *Bacillus amilovorus*. Removing damaged leaves and stalks before storage was beneficial. Wrapping individual heads or boxes of heads in open polythene bags reduced weight losses but encouraged rooting. Pre-harvest treatments with thio-bendazole (TBZ) and benomyl solutions did not improve the stored product and postharvest dips increased weight loss in storage. Barkai Golan and Aharoni (1972) reported that TBZ at 0.1 per cent, applied to field-infected winter celery, inhibited the development of *Botrytis cinerea* and *Sclerotinia sclerotiorum* during storage. The incidence of petiole infection by *Septoria apii*, leaflet decay due to *S. apii* and *Cercospora apii* and leaflet yellowing.

9.0 Diseases, Pests and Disorders

9.1.0 Diseases

Serious damage by the pathogens (*Sclerotinia sclerotiorum, Fusarium oxysporum, Septoria apiicola, Pseudomonas* spp.), nematodes (*Meloidogyne* spp., *Ditylenchus dipsaci, Pratylenchus penetrans*) and insects (*Psila rosae, Philophylla heraclei* and *Liriomyza huidobrensis*) were reported in celery (Monnet and Thibault, 2000). Hot water treatment of celery seeds at 50°C for 30 minutes had good control of *Alternaria* spp., *Phoma* spp., *Septoria* spp., *Peronospora valerianellae, Xanthomonas* spp. Germination is not affected under these conditions. Increase of temperature up to 53°C had a similar effect but germination was affected (Nega *et al.*, 2001).

9.1.1 Septoria Leaf Spot/Blight

It is a fungal disease caused by *Septoria apiicola*. It is more prevalent when the crop residues are not properly eliminated. Crop rotation helps to check the disease to some extent. Daconil at 7.23 g a.i./l gave good control of the disease and produced the highest yield (Chinchilla and Mora, 1986). Timing of fungicide application is also an important in combating disease and preventing yield loss. Mudita and Kushalappa (1993) advised to initiate fungicide application as soon as the disease appears at transplanting stage. However, if the disease appears in later stage, fungicide spray could be delayed until a higher disease threshold is established. Use of triphenyl hydroxide was found to be most economic to control this disease (Velandia, 1992).

9.1.2 *Fusarium* Yellow

The disease is caused by the fungal pathogen *Fusarium oxysporum* f. sp. *apii*. Growing of resistant cultivars such as UC 8-1 and UC 10-1 helps to combat the disease incidence (Quiros *et al.*, 1993). Fumigation of soil with 98 per cent methyl bromide and 2 per cent chloropicrin (0.15 kg a.i./m^2) eradicates the pathogen from soil and roots (Awuah and Lorbeer, 1991). Bell *et al.* (1998) could reduce the disease incidence and severity by pre-plant chitin amendments to the soil.

9.1.3 Stem and Collar Rot

The disease is caused by the fungal pathogen *Sclerotinia sclerotiorum*. The disease causes dieback and finally kills the plants (Corato and Baeviello, 2000). Use of fungicides such as hymexazol and metalaxyl gave good control of the pathogen. *Coniothyrium cochlioides* is a fungal antagonist for this pathogen but it was effective in controlling the pathogen in lettuce and not in celery (Whipps *et al.*, 1993).

9.1.4 Bacterial Leaf Spot

The disease is caused by *Pseudomonas syringae* pv. *apii* (Koike *et al.*, 1994) or *Pseudomonas cichorii*. Hot-water treatment of seeds at 50°C for 25 min eliminates >99.9 per cent of seed contamination (Little *et al.*, 1997). High relative humidity was found to predispose the plants to this disease. Lesion size was correlated with mist duration (Su *et al.*, 1989). Severity of brown stem increased with higher temperatures, as measured by the length of characteristic brown streaks in the pith (Pernezny *et al.*, 2000).

9.1.5 Viruses

Celery mosaic virus was isolated from celery plants with vein clearing and leaf mottling symptoms. Cucumber mosaic virus was isolated from affected plants showing chlorotic mottle, rings and line patterns, leaf yellowing and reduction in leaf area with stunted growth (Gracia and Feldman, 1977) or necrosis of petioles (Crescenzi *et al.*, 1993). Virus affected plants were found to have low levels of furanocoumarins (0.6-9.9 μg/g fresh mass) compared to virus-free plants (0.9-48.6 μg/g fresh mass) (Ataga *et al.*, 1993). It was suggested that control of the disease may be achieved by introducing a celery-free period to break the cycle of virus transmission between successive crops (Alberts *et al.*, 1989).

9.2.0 Pests

Field scale vaccum unit was useful in celery crop to remove the insect pests such as agromyzid leaf miners, aleryoids, *Emboasca* sp. and aphids with the application of pneumatic removal. But the technique is not helpful in permanent solution to leaf miners due to highflying ability (Weintraub *et al.*, 1996). *Lygus hesperus* attacks young celery plants in green houses and mature plants in production fields, causing severe economic losses (Joseph, 2016).

The agromyzid (*Liriomyza huidobrensis*) is a serious pest in celery. It is polyphagous in nature, having fast multiplication rates and tolerance to medium-low temperature makes it difficult to combat the pest incidence. Use of biological control needs to be developed as the chemical control is ineffective (Bosio, 1994). Translaminar insecticides reduced the leaf miner population and increased the yield whereas yellow sticky traps were not found reliable (Weintraub and Horowitz, 1998). Abamectin at 0.17 kg/ha caused high mortality of all larval stages of leaf miner and prevented the emergence of adults. Thiocyclam (0.56 kg/ha) caused high mortality of all larval stages, but did not affect adult emergence from the surviving larvae. Cryomazine (0.84 kg/ha) acted most strongly against early-instar larvae before visible mines were present, caused little direct mortality of larger larvae, but prevented successful adult emergence (Grafius and Hayden, 1988).

9.2.1 Celery Looper

This insect pest (*Anagrapha falcifera*) damages the crop by defoliation. It is controlled by IPM strategies involving biocontrol either using polyhedral viruses (Tamez-Guerra *et al.*, 2000) or baculovirus (Farrar and Ridgway, 1997) or bacterial pathogen (*Bacillus thuringiensis* var. *tenebrionis* (Gelernter and Trumble, 1999).

Diseases of Celery

Cercospora Leaf Spot Infection

Septoria Late Blight Infection

Mosaic Infection

Powdery Mildew Infection

Sclerotinia Infection

Downy Mildew Infection

Pests of Celery

Root Knot Nematode (Left) and Aphid (Right) Infestation

Lygus Bug infestation

9.2.2 Mites

Broad mite (*Polyphagotarsonemus latus*) also found to affect celery. Use of acaricides like 50 per cent dienochlor WP, 40.8 per cent chlorpyrifos EC, 80 per cent sulphur DF, 25 per cent bromopropylate EC or 25 per cent azocyclotin WP was found to reduce the incidence if the pest control is taken up at early stage of infestation.

9.2.3 Nematodes

Stem nematode (*Ditylenchus dipsaci*) could be managed by IPM involving crop rotation, weed control and use of nematicides (Gentzsch, 1999). Other nematodes causing yield loss are *Pratylenchus bukowinensis*, *P. projectus* and *P. penetrans*. Crop rotation avoiding parsley and carrots (which are also hosts of these nematodes) but including cereals was recommended (Knuth and Schrameyer, 1991). Root knot nematode also affects the crop, which is controlled by use of nematicides aldicarb, phenamiphos and carbofuran (Rivera and Lopez, 1982). Bisessar *et al.* (1983) noticed that the roots of nematode-inoculated plants grown in heavy metal soil had significantly more nematode galls than did the roots of inoculated plants grown in non-metal soil and heavy metals, primarily nickel, predisposed celery to greater attack by *Meloidogyne hapla*.

9.3.0 Physiological Disorders

Pithiness is a major source of quality loss and decreased shelf-life in celery (Saltveit and Mangrich, 1996). It is characterized by the appearance of whitish regions and air spaces within the tissues and reduced tissue density and is caused by the breakdown of the internal pith parenchyma tissues of the petiole to produce aerenchyma. Pithiness may be induced by pre-harvest factors, including cold stress, water stress, pre-bolting (seed stalk induction) and root infection. Storage temperature has a major impact on development of pithiness after pre-harvest induction. Progressive development of pithiness is delayed by storage at 0°C.

Blackheart is a physiological disorder caused by cell death resulting from calcium deficiency and pre-harvest water stress. Internal leaves develop brown discolouration, which eventually become deep black. Brown check is a disorder related to boron deficiency. It appears as cracks on the inner petiole surface and is also referred to as crack stem. The exposed tissues become brown and susceptible to pathogen infection and decay.

10.0 Crop Improvement

The main objectives for improvement of celery are to evolve a plant type with non-pithy petioles and resistant cultivars to different diseases. Madzharova *et al.* (1973) developed new forms of leaf and root celery from a cross *Apium* x *Petroselinum* which had tender leaves rich in vitamin C, minerals, protein and sugar. The roots of some of the hybrids had doluble: insoluble pectin ratios which improved storage qua-lity. Inheritance study made by Pope and Munger (1953) indicated that a single gene controls the character for utilization of magnesium. A single dominant gene also governs pithiness in entire plant. The breeding goals of modern celery and celeriac are controlled and established by the local markets. The USA market is demanding resistance to biotic stress especially *Fusarium* yellow. Within Europe the common goals are high uniformity and yield, whereas bolting resistance, interesting petioles, and enlarged hypocotyl characteristics are regional breeding goals.

10.1.0 Genetics of Certain Characters

Quiros (2001) presented a comprehensive account of celery breeding in California. According to him, besides breeding for resistance, another important thrust of their programme is to develop the genetics of celery to expedite breeding by creating chromosome markers and a genetic map on the basis of isozymes, non-specific proteins, morphological traits (Quiros *et al.*, 1987; Huestis *et al.*, 1993; Yang and Quiros, 1995) and DNA based markers to tag useful genes such as those determining *Fusarium* resistance and male sterility. Inheritance patterns and linkage relationships of eight genes of celery using isozyme was reported by Arus and Orton (1984). They identified genes: *Adh1* and *Adh2* (alcohol dehydrogenase), *Pgm2* and *Pgm3* (phosphoglucomutase), *Pgi3* (phosphoglucose isomerase), *Mdh3* (malate dehy-drogenase) and *Sdh1* (shikimate dehydrogenase). There was also evidence for additional alcohol dehydrogenase and phosphoglucose isomerase loci. Segregation for petiole colour in one population identified a single locus, with dominant *A* (red) and recessive *a* (green) alleles. Linkages were detected between *Adh2* and *Pgm3* (1.04), *Pgi3* and *Mdh3* (9.07), and *Pgm2* and *Sdh1* (39.10). *Adh1* and *Adh2* segregated independently.

Outcrossing in celery ranged from 47 to 87 per cent, which a mean of 71.4 per cent. Outcrossing in the wild population was estimated to be 32.4 and 53.1 per cent (Orton and Arus, 1984). A procedure was described by Ochoa *et al.* (1986) for emasculating celery regardless of its degree of protandry. Umbels with 80-90 per cent open flowers but with undeveloped styles are selected and following removal of umbellules and flowers at the bud stage a water stream from a household cleaning fluid spray bottle is applied to force anthers and pollen from the flowers of the remaining umbellules. The flowers are then bagged. When the stigmas become receptive, usually 4-5 days later, they are pollinated. The procedure was verified using genetic markers. In another study, it was found that seed stalk branches cut and placed in water will root, extending the survival time sufficiently to produce viable seed. Male sterility is an important trait for the development of disease resistant F_1 hybrid (Quiros *et al.*, 1986). Finally, another research component is the adaptation of genetic engineering techniques that may be used as promising new tools by the breeders. This includes the development of procedures for gene transfer using *Agrobacterium tumefaciens* as a vector (Catlin *et al.*, 1988).

Interspecific hybridization is being widely attempted to transfer useful genes to cultivated celery. Madjarova and Bubarova (1974, 1978) developed a cultivar Festival 68 by crossing parsley with celery and the hybrid had morphological charac-teristics resembling those of the former. The leaves have a high content of ascorbic acid, sugars and essential oils. The leaf yields are 50-80 per cent higher than those of ordinary commercial varieties. New root forms from the same interspecific combi-nation, having intermediate characters in leaf rosette and higher contents of ascorbic acid, carotene, chlorophyll, essential oils and amino acids than either parents have been obtained, together with others which have larger roots than those of their parents and a long storage period, similar to that of parsley. Later, Honma and Lacy (1980) crossed celery cultivar Golden Spartan with parsley, which is immune to celery blight (*Septoria apiicola*).

Sethi and Sapra (1986) studied 7 quantitative characters of 25 local biotypes from Amritsar and Gurdaspur districts and 5 exotic cultivars and estimated genetic divergence by nonhierarchical Euclidien cluster analysis, with all 7 principal components taken into account. Members of highly divergent clusters were recommended as parental pairs in hybridization programmes. Choudhary and Kaul (1990) found that seed yield was positively correlated with all components. They suggested that selection for high number of umbels per plant can lead to genetic improvement.

10.2.0 Bolting Resistance

Low temperature during growing seasons causes yield losses from premature seed stem development (bolting). When bolting was greatest, losses in M68-27 and M68-29 S2, slow-bolting, early blight resistant breeding lines and their progenies ranged from zero to 49 per cent compared to 75 per cent loss in other varieties indicating the promise of these blight resistant varieties as bolt resistant (Wolf and White, 1978). A correlation between bolting-resistance and seed dormancy was established (Thomas, 1978). The cultivars which bolted early in response to a 4-week cold-treatment (7°C) given at the 5-leaf seedling stage had seeds which gave a high germination response to GA4/7 at 2 x 10-4M in the dark at 22°C. Conversely, bolting-resistant cultivars showed a low germination response (Thomas, 1978). Keleleer *et al.* (1987) observed marked differences in bolting resistance among cultivars following a 10-day treatment with 4°C at 60 days. The highest free ACC contents throughout the experiment and the highest capacity to turn free ACC into N-malonyl-ACC were found in the bolting resistant cultivar, which also retained the greatest capacity to respond to drought after a cold period. Wang *et al.* (2012) selected a slow-bolting line S80 and an excellent quality line S53 from numerous celery resources; after the cross of these two celery varieties and a single-plant selection, a slow-bolting celery cultivar was developed and released as "Juventus".

A deeper understanding of *Apium* bolting has been gained through transcriptomic data that identified orthologues of the flowering genes ELF7 and ELF8 (EARLY FLOWERING 7 and 8), FCA (FLOWERING TIME CONTROL), FLC (FLOWERING LOCUS C), RKF_3 (RECEPTOR-LIKE KINASE IN FLOWERS 3), TM6 (TOMATO MADS-BOX GENE 6), AP3 (APETALA 3), TFL2 (TERMINAL FLOWER 2), and PIE1 (PHOTOPERIOD INDEPENDENT EARLY FLOWERING 1) (Fu *et al.*, 2013). Allelic variations of these genes are potentially important for the late bolting trait and could be used in the future for molecular marker assisted selection and enable target development for the new emerging breeding techniques.

10.3.0 Heat Tolerance

Celery is semi-cold tolerant but not heat tolerant, thus making it difficult to cultivate in summer. Peng *et al.* (1998) identified four genotypes with significant heat tolerance. Li *et al.* (2015) studied the expression profile of the HEAT SHOCK FACTOR B2 (*AgHSFB2*) gene which confers protection against heat stress in celery cultivars Liuhe Huangxinqin, Jinnan Shiqin and Ventura. *AgHSFB2* upregulation differed between the cultivars, with Ventura accumulating faster higher transcript levels as a response to heat.

10.4.0 Resistance to Diseases

10.4.1 *Fusarium* Resistance

Fusarium yellows is a vascular disease caused by the pathogen *Fusarium oxysporum* f. sp. *apii*, race 2, resulting from the penetration of the fungus through the roots and invasion of the vascular tissue of the plants. These become yellow and stunted due to crown and root rot and eventually die. There is no practical means of controlling the disease, which is spreading in the celery fields of USA.

New stands of celery cannot be grown in infected fields. All commercially grown cultivars in USA are susceptible to *F. oxysporum* f. sp. *apii* in the field in California. Erfurt, Large Smooth Prague and the Hungarian introduction PI320912 were moderately resistant to the Californian isolate H171 of *F. oxysporum* f. sp. *apii* under green house conditions. Of many celeriac introductions tested accessions mainly from Turkey, were resistant (Opgenorth and Endo, 1979). Orton *et al.* (1984a) developed a celery breeding line UC1, derived from several accessions of celeriac (*A. graveolens* var. *rapaceum*) and 2 selections of celery (*A. graveolens* var. *dulce*). UC1 segregates for resistant to race 2 of *F. oxysporum* f. sp. *apii*. Resistance appears to be dominant and highly heritable under green house screening conditions. The number of genes involved is not known. Most plants of UC1 are vigorous, dark green and taller than those of 52-70R. During preliminary screening of 4-week-old seedlings of UC1, about two thirds of plants proved resistant to *F. oxysporum* f. sp. *apii*. UC1 was the first *Fusarium* resistant line bred. The genetics of resistance was also reported by Orton *et al.* (1984b) who found from their tests with race 2 of the pathogen, previously observed in the celeriac line PI 169001, F_2 populations derived from a cross of this line with a putatively tolerant selection from the susceptible cv. Tall Utah 52-70R showed segregation of resistance and susceptibililty. With one exception, F_2 segregation ratios were between 3: 1 and 15: 1 resistant/susceptible.

Results from progeny tests of 20 resistant F_2 plants were consistent with the conclusion that resistance in PI 169001 was conditioned by a dominant allele at a single locus and that segregation distortion was caused by segregating genes with quantitative effects on resistance contributed by the selected Tall Utah 52-70R parent or both parents. The F_2 segregation from a separate cross was also consistent with the hypothesis of a single dominant resistance gene and a quantitative resistance gene(s) in PI 169001. UC1 has served as the foundation material for further selection and release of the first commercial *Fusarium* resistant varieties developed by the seed industry. In an effort to improve UC1 for horticultural traits, a backcross programme was implemented to introgress the resistance to three popular celery varieties. Tall Utah 52-70R, TU 52-70HK and TU 52-75. Later research at California led to the release of eight advanced lines, including of which UC8, UC10, UC390S and UC862 have been the most popular for further improvement by seed companies (Quiros *et al.*, 1993)

Most commercial celery cultivars were rated highly or moderately susceptible to *Fusarium*. Research work carried out at Michigan also led to the release of certain resistant lines. Deacon and Tall Utah 52-70 HK were the most resistant commercially available lines tested and were rated moderately resistant while Summer Pascal and Golden Spartan also showed some resistance (Elmer *et al.*, 1986). Several experimental lines showed resistance, but many lacked desirable horticultural characteristics. Breeding lines 83-604 and 68-37, rated moderately-highly resistant, were horticulturally acceptable and released as the cultivars Companion and Pilgrim (Elmer *et al.*, 1986). Elmer and Lacy (1983) evaluated thirty cultivars and lines, including seven intergeneric crosses of celery x parsley for resistance to *F. oxysporum* f. sp. *apii* race 2 under field conditions in naturally infested soil and in the green house, using a colonized wheat straw soil infestation technique. Cultivars

Summer Pascal, Tall Utah 52-70 HK and Tendercrisp were rated as moderately resistant and were highest in fresh weights. Several lines of the intergeneric hybrid possessed resistance in segregating progeny.

Elmer and Lacy (1985) first reported the loss of resistance to race 1 in a green cultivar selected from Tall Utah 52-70, which is highly resistant to race 1. They compared isolates of race 2 from Michigan, California and New York and one isolated race 1 from France for virulence on 2 yellow and 9 green celery cultivars that varied in resistance to *Fusarium* yellows in Michigan. All race 2 isolates caused disease on yellow and green cultivars, but varied in virulence. Race 1 caused disease on Golden Detroit and on the green cultivar Bishop, but was avirulent on Golden Spartan and all other green cultivars. Ireland *et al.* (1987) found seasonal variation on the resistance pattern of the varieties. Pilgrim and breeding line MSU74-70 were highly resistant in both years, Deacon and Companion were moderately resistant in both years and Tall Utah 52-70 HK was moderately resistant in one year and moderately susceptible in the next year.

Tissue culture methods to employ somaclonal variants have been reported to be successful in selecting resistant lines. Ireland and Lacy (1987) found that somaclones regenerated from single cells were significantly more resistant to *F. oxysporum* f. sp. *apii* race 2 than the source cultivar, Tall Utah 52-70 HK (TU-HK). Heath Pagliuso *et al.* (1988) suggested two methods to screen putative resistant plantlets from cell culture *viz.*, placing plantlets on a mycelial mat for one month or planting them directly in *Fusarium*-infested soil. Analysis of progeny scores for root and crown decay of somaclonal variants showed that the new variation was heritable and appeared to be conditioned by more than one locus. Wright and Lacy (1988) initiated callus cultures from axillary buds of celery cultivars Florida 683 and Tall Utah 52-70 HK using shaken liquid cultures of single plant cells and cell clumps. Whole plants (somaclones) regenerated from embryoids formed in shaken cultures were screened in the green house for reaction to 3 fungal and 1 bacterial pathogen of celery: *Septoria apiicola* (late blight), *Cercospora apii* (early blight), *F. oxysporum* f. sp. *apii* (Fusarium yellows) and *Pseudomonas cichorii* (bacterial blight). Regenerated plants varied from highly susceptible to highly resistant to all 4 pathogens, whereas parent plants of Florida 683 were uniformly highly susceptible to all pathogens. Parent plants of Tall Utah 52-70 HK were moderately resistant to *F. oxysporum* f. sp. *apii* and highly susceptible to the remaining pathogens. Resistance to a given pathogen appeared to arise independently from resistance to other pathogens, since plants were rarely found with high resistance to 2 or more pathogens. They suggested that this technique may become an important way of introducing disease resistance into susceptible cultivars of celery.

In 2013, a new race, *F. oxysporum* f. sp. *apii* race 4, was identified in California (USA) and none of the commercial cultivars available display resistance to it. Research at the University of California, Davis, focused on searching resistance sources to this new race. 120 celery, 66 celeriac, 25 smallage, 5 celery x celeriac and 15 miscellaneous *Apium* species were screened. A celery from China and three potentially resistant celeriac accessions from Turkey were identified (Epstein, 2017).

10.4.2 Late Blight Resistance

Late blight, caused by the pathogen *Septoria apicola* is another important disease affecting celery. Although there are means to control late blight, the development of resistant varieties is expected to significantly reduce production costs and the environmental impact caused by the application of fungicides. Resistance to this disease has been found in the wild species from South America *A. chilense* (Ochoa and Quiros, 1989). *A. chilense* hybridizes readily with celery, but the F_1 hybrids have poor fertility due to the presence of chromosomal rearrangements. It has been possible, however, to obtain F_2 and backcross seed for the development of resistant lines. Pink *et al.* (1983, 1984) selected plants with partial resistance to *Septoria apiicola* and with a growth habit similar to celery were selected from a second backcross population of a cross between *A. graveolens* x *A. nodiflorum*.

Following self-pollination in cv. Praski, a line was produced which is tolerant of *Septoria apiicola*. This line produced progenies and some lines derived from these proved more resistant than available resistant varieties (Wasilewska and Masternak, 1987). Madjarova *et al.* (1978) obtained four types of plant differing in leaf form and a fifth similar to celery in leaf form but with resistance to *S. apii* from the cross celery female x parsley male. These forms are stated to have the celery caryotype.

Wiener Riesen (Viennese Giant) and Wiener Markt (Viennese Market) were the most resistant to late blight (Janyska, 1975). Lacy and Honma (1981) crossed the celery cultivar Golden Spartan as seed parent to *Petroselinum hortense*, with resistance to *S. apiicola*, using honey bees as pollinators in the green house and petiole colour (green being dominant over yellow) as a genetic marker. Of 1000 F_1 seedlings, three had green petioles indicating that these seedlings were hybrid. The F_2 from the selfed F_1 segregated 3 green: 1 yellow for petiole colour and showed a range of reactions from highly resistant to highly susceptible. Ochoa and Quiros (1989) found that wild species *A. chilense* and *A. panul* exhibited a few discrete yellow spots with very little, if any, pycnidial development. This type of resistance was not observed in any accession of celery screened. The degree of resistance was intermediate in the F_1 hybrids *A. graveolens* x *A. chilense* and *A. graveolens* x *A. panul*, indicating incomplete dominance.

10.4.3 Resistance to Virus Diseases

Commercial cultivars varied in reaction to celery mosaic virus (CeMV) inoculation. Symptoms were most severe on self-blanching cultivars but a high level of field resistance is some non-self-blanching ones such as Solid White and Fordhook was confirmed in green house tests (Walkey and Ward, 1984). Four local varieties showed fairly good resistance to the Yellows virus, such as those from the Georgian SSR (K637) and Turkey (K834), together with A Couper de Dinant (Dinant Cutting) from France and Sedano Bianca a Casta Picca from Italy (Kardaeva and Kletskova, 1978). D'Antonio *et al.* (2001) identified a single recessive locus for resistance to CeMV. Ruiz *et al.* (2001) found markers linked to this *cmv* gene. These markers can facilitate introgression and selection of the virus resistance in other celery varieties. McCormick (2006) attempted to produce celery and carrot plants resistant to CeMV and carrot virus Y (CarVY) using post transcriptional gene silencing technology. He was unsuccessful in obtaining celery resistant plants, presumably due to the recalcitrance of the used celery cultivars toward *Agrobacterium*-mediated transformation.

10.4.4 Leaf miner Resistance

Leaf miner (*Liriomyza trifolii* Burgess) is a major pest of celery, requiring control by constant insecticide spraying. Development of leaf miner resistant celery varieties will reduce the use of hazardous chemicals and environmental risks linked to their use. No resistance to this pest was observed in the cultivated collection, however an accession of *A. prostratum* from Australia was found to be practically immune in tests performed in the green house. No feeding or oviposition was observed in this species (Trumble and Quiros, 1988). *A. prostratum* was hybridized to celery to develop leaf miner resistant lines. Chemical analysis of the foliage indicated the resistance to this pest is not based on compounds noxious to humans (Trumble *et al.*, 1990). Trumble *et al.* (2000) examined celery cv. Conquistador, *A. chilense*, 23 backcross accessions of *A. graveolens* x *A. chilense* and two relatives for resistance to *L. trifolii*. They found that the resistance was not due to the undesirable compounds like carcinogenic and mutagenic linear furanocoumarins, or psoralen, bergapten and xanthotoxin (methoxsalen) present in both leaves and petioles of each accession.

10.5.0 Hybrid Breeding and Male Sterility

In Europe, hybrid breeding is one of the major breeding goals of celery and celeriac. Due to high uniformity, increased yield, vigour and disease resistance, hybrid cultivars attract growers and breeders equally. Genetic improvement of this crop accomplished mainly through open-pollination which results in high plant to plant variation and reducing the quality of the produces. Though hybrid cultivars are available in the market of Europe and USA, but the numbers of hybrids still remain quite less compare to open pollinated varieties. Due to lack of emasculation means for umbelliferous flowers of this crop, it is very difficult of obtaining F_1 hybrid seeds. Celery is highly self pollinated crop due to its tiny flowers and organized in inflorescences called umbels. The most efficient emasculation method is the use of genetic emasculation mechanism i,e., the use of totally male sterile lines. Genetic male sterility segregates and therefore these lines are vegetatively propagated in breeding programs. Conversely, cytoplasmic male sterility (CMS) and cytoplasmic genic male sterility (CGMS) are determined by mitochondrial encoded genes, hence are maternally inherited and are economically advantageous. The first celery male sterile line, the Iranian accession 'P1229526', described by Quiros *et al.* (1986), was determined by a single recessive gene, *ms*-1. Gao *et al.* (2006) identified a male sterile plant, '01-3A' in a commercial field trial of the inbred line '01-3'. They concluded that the male sterility was controlled by two nuclear recessive genes (Gao *et al.*, 2009) and they used this line to create the celery hybrid 'Jinqi 2' (Gao *et al.*, 2008). Reports on CMS sources in *Apium* are elusive.

10.6.0 Breeding for Phytoremediation

Celery is remarkably salt tolerant plant due to the enhance in mannitol biosynthesis, which is controlled by mannose-6- phosphate reductase (M6PR)

(Gao and Loescher, 2000). Experimental evidence in other plant species specify that metallothioneins (MTs) play a key role in heavy metal and salinity tolerance *via* ROS scavenging (Kumar *et al.*, 2012). These genes can be utilized as molecular markers for the development of heavy metal tolerant cultivars or can be engineered in order to obtain brand new tolerant genotypes in the gene pool. The pattern of expression of AgMT2 was investugated in celery cultivars Liuhe uangxinqin, Jinnan Shiqin, and Ventura (Chen *et al.*, 2015). There are numbers of studies available regarding the potential of different celery and celeriac cultivars to accumulate heavy metals and pollutants are available. Willaert *et al.* (1990) evaluated heavy metal accumulation in celeriac and white celery petioles and leaves grown on soils contaminated with heavy metals within the permissible levels. The accumulation of cadmium, manganese and zinc varied between the investigated cultivars. Twenty-seven Chinese *Apium* accessions were grown in cadmium and lead contaminated soil and the content of heavy metals in the shoots was evaluated; variations between the cultivars were observed, and Shuanggangkangbing accumulated the lowest amounts of cadmium and lead (Zhang *et al.*, 2013). The boron uptake in cultivars Emerson Pascal, an efficient boron user, and S48-54-1, which is susceptible to low boron, was investigated under low and high boron concentrations (Bellaloui and Brown, 1998). The boron uptake did not differ between the two cultivars but its distribution varied, with Emerson Pascal accumulating boron in its shoots and S48-54-1 in its roots. This indicates that selection of cultivars based on their capacity to accumulate boron in different organs is relevant for preventing the introduction of boron above the toxicity limit in the food chain.

10.7.0 Breeding for Nutraceuticals, Nutritional Quality and Food Safety

Celery and celeriac are generally considered as vegetable with low calorie and are well known medicinal plants. They are rich in fibres, various minerals like calcium, phosphorus, iron, potassium, and magnesium, numbers of vitamins *viz.*, B1, B2, B3, A, and C and flavonoids (Sowbhagya, 2014). There are very few reports available regarding the development and characterization of cultivars with increased nutritional quality. Huang *et al.* (2016) studied vitamin C levels and its metabolic pathway in the cultivars Liuhe Huangxinqin and Ventura and reported the last one containing increased levels of ascorbic acid in both leaves and petioles. The evidence is clear that variation in vitamin C content exists between celery cultivars and screening of gene expression can readily be used in breeding programs. Yao *et al.* (2010) analysed 11 celery cultivars and established a correlation between the phenols composition and antioxidant activity. Shad *et al.* (2011) also compared the nutraceutical composition in four genotypes with three celery cultivars and a wild accession. Variation between the content in different nutraceuticals was observed and the wild *Apium*, remain superior. Marongiu *et al.* (2013) analyzed the volatile fraction of aerial parts from two wild *Apium* accessions from Portugal and Italy and concluded that they possess strong antifungal activity against *Candida* and *Aspergillus*. The stronger antifungal activity of the Italian accession was attributed to the increase in neophytadiene content. Helay *et al.* (2015) studied the composition of different celery genotypes for limonene, tannins, phenolic acids, anthocyanins, chlorophyll, carotenoids and other essential oils. Though there was no variation reported for carotenoids level among the cultivars, but the other constituents significantly varied. Wild accessions and local cultivars can potentially improve the ongoing breeding programs by increasing the content of health promoting compounds.

11.0 Biotechnology

In celery, marked improvement in the regeneration of plants through tissue culture *in vitro* has been reported by various workers. The scope of somaclonal variation in *in vitro* derived plantlets has been established, particularly for disease resistance. Encapsulation technology has been standardised for *in vitro* conservation of genotypes or delaying plantlet regeneration.

11.1.0 Tissue Culture

11.1.1 Micropropagation

Cultures of celery were initiated from mature stem segments of field-grown plants and explants of aseptically grown seedlings. Micropropagation was achieved on MS medium through somatic embryogenesis (Pierik, 1991).

Silvertand *et al.* (1995) reported plantlet regeneration from flower stalk explants (5 mm discs) genotypically different plants of cv. Vrizo and cultured on MS medium supplemented with 1 mg BA and NAA/l. After 9 weeks the percentage of regenerating explants (PREX) and average number of regenerated shoots/explant (NSHO) were determined. Significant differences were found between the 29 plants; PREX ranged from 10 to 100 per cent and NSHO ranged from 0 to 40. A combination of 10 mg BA and NAA/l resulted in the highest total average PREX and NSHO for both genotypes. For all genotypes, PREX and NSNO decreased with increasing flower stalk length. Han *et al.* (2006) found that hypocotyl was most suitable material for establishing an efficient plant regeneration system in celery cultures.

11.2.0 Anther and Microspore Culture

Dohya *et al.* (1997) cultured anthers and microspores of *Apium graveolens* were *in vitro* to obtain haploid plants. Calluses developed from cv. Cornell 19 microspores at the tetrad stage on 1/2 MS medium supplemented with 2,4-D and B5 medium supplemented with 2,4-D. In Cornell 619, calluses developed from tetrad and early uninucleate microspores on B9 medium supplemented with 2,4-D. Adventitious embryos regenerated from calluses transferred to MS medium with or without NAA and BA and plantlets grew after 40 to 60 days.

11.3.0 Somatic Embryogenesis

Nadel *et al.* (1989) reported that embryogenic suspension cultures were established from petiole and leaf callus and routinely subcultured in a maintenance medium containing 2.3 µM 2,4-D and 0.88 µM BA. Somatic embryogenesis occurred in this medium, but was considerably improved in a regeneration medium containing 2.3 µM kinetin, without 2,4-D. Cultures thus maintained, contained embryogenic clumps, aggregated somatic embryos, and few free-floating single somatic embryos.

Addition of mannitol (3-4 per cent w/v) prevented cell lysis, greatly increased the number of single somatic embryos, improved their differentiation, and accelerated development of torpedo embryos. Fully-developed plantlets produced from the torpedo embryos were hardened off under mist for 10-14 days, resulting in plants ready for transplanting to the field. Kim and Janick (1989) reported that embryoids developed from callus clumps of leaf explants according to the standard pattern (through globular, heart and torpedo-shaped stages). Secondary somatic embryos occurred on the cotyledons and hypocotyls of primary embryoids.

Saranga and Janick (1991) reported improved protocol of somatic embryo production from leaf explants of celery cv. SB12 grown *in vitro* to form callus on a semisolid or liquid MS medium containing 2.3 µM 2,4-D, 2.3 µM kinetin, 0.5 g casein hydrolysate/l and 30 g sucrose/l. Callus was then transferred to liquid embryo production medium (EPM; MS medium containing 30 g sucrose/l). Proembryos were subcultured on fresh EPM for 6 days. Germinating embryos with cotyledons 1-2 mm in length and radicles 8-10 mm in length were subcultured in regeneration medium. The conversion of somatic embryos (artificial seeds) to seedlings was more successful when sucrose (1.5 or 3.0 per cent) was added to the regeneration medium. Calluses and somatic embryos were induced from seedlings of celery cv. Utah Improved 15. The frequency of diploids was highest in callus-derived R1 regenerants (162 cells) compared to R0 (115 cells) and R2 regenerants (94 cells). In somatic embryo-derived regenerants the highest frequency of diploids occurred in the R0 (Kim *et al.*, 1990). Highly embryogenic cell lines were recorded by Nadel *et al.* (1990) in *Apium graveolens* by inducing cell cultures from *in vitro* plants which were previously regenerated from somatic embryos. The early detection of embryogenic potential of new cell lines was found to be regulated by 2,4-D, mannitol and culture duration. Somatic embryos of 1400 µM in size provided the best growth rate and uniformity. Song *et al.* (2007) developed a protocol for rapid and efficient production of transgenic celery plants via somatic embryo regeneration from *Agrobacterium tumefaciens*- inoculated leaf sections, cotyledons and hypocotyls. These explants were excised from in vitro seedlings of the cvs. XP 166 and XP 85 and inoculated with *A. tumefaciens* strain EHA105 containing the binary vector pBISN1. PBISN1 has the neomycin phosphotransferase gene (nptII) and an intron interrupted b-glucuronidase (GUS) reporter gene (gusA). Co-cultivation was carried out for 4 d in the dark on callus induction medium (CIM): Gamborg B5 + 2.79 lM kinetin + 2.26 lM 2,4-dichlorophenoxyacetic acid (2,4-D) supplemented with 100 lM acetosyringone. Embryogenic calluses resistant to kanamycin (Km) were then recovered on CIM + 25 mg/l Km + 250 mg/l timentin after 12 weeks. Subsequently, a large number of Km-resistant and GUS-positive transformants, tens to hundreds per explant were regenerated via somatic embryogenesis on Gamborg B5 + 4.92 µM6 (d,d-dimethylallylamino)-purine (2iP) + 1.93 µM α-naphthaleneacetic acid (NAA) + 25 mg/l Km + 250 mg/l timentin after 8 weeks. Using this protocol, the transformation frequency was 5.0 per cent and 5.0 per cent for leaf sections, 17.8 per cent and 18.3 per cent for cotyledons, and 15.9 per cent and 16.7 per cent for hypocotyl explants of cvs. XP 85 and XP 166, respectively. Stable integration of the model transgenes with 1–3 copy numbers was confirmed in all ten randomly selected transgenic events by Southern blot analysis of gusA. Progeny analysis by histochemical GUS assay showed stable Mendelian inheritance of the transgenes. Thus, *A. tumefaciens*-mediated transformation of cotyledons or hypocotyls provides an effective and reproducible protocol for large-scale production of transgenic celery plants.

11.4.0 Somaclonal Variation

Donovan *et al.* (1994) analysed the potential in *in vitro* derived celery plants. The aim of this study was to related variation in the source of explant and the morphology of the newly initiated callus to the characteristics of the regenerant plantlet, of which the most important was resistance to *Septoria apiicola*. Greater variation was observed in plants regenerated from different seedlings than in plants derived from explants within seedlings or within callus. Evenor *et al.* (1994) selected somaclonal variants of celery (*Apium graveolens*) for resistance to *Septoria apiicola* by culturing in the presence of fungal culture filtrate or coculturing callus-derived single cells or small clusters with a hypovirulent or normal isolate of *S. apiicola* on solid medium. Resistant cells were developed that, unike their susceptible counterparts, were not killed by toxic compounds secreted by the fungus. Some of the plants regenerated from these cells showed different degrees of tolerance of *S. apiicola* in green house tests. Selfing the selected plants yielded tolerant progenies from which the most tolerant were further selected.

11.5.0 Biochemical and Molecular Markers

Yang and Quiros (1995) used F_2 population of two celery cultivated types (*Apium graveolens* vars. *rapaceum* and *secalinum*) to construct a linkage map consisting of 29 RFLP, 100 RAPD, four isozymes, one disease resistance and one growth habit markers. The map contains 11 major groups and 9 small groups and has a total length of 803 cM with an average distance of 6.4 cM between two adjacent loci. Some 10 per cent of the RAPDs segregated as codominant markers and their allelic homologies were tested by Southern hybridization.

Though molecular marker assisted selection has become a more and more important auxiliary means for conventional breeding. However, only a few molecular markers have been indentified in celery. After the development of isozyme markers and morphological markers (Arus and Ortan, 1984; Quiros *et al.*, 1987), Huestis *et al.* (1993) initially developed 21 restriction fragment-length polymorphism (RFLP) markers. Domblides (2008) used RAPD markers to discriminate celery cultivar. Then amplified fragment length polymorphism (AFLP) markers were used in celery for cultivar classification and genetic diversity analysis (Ju, 2007). In recent times simple sequence repeats (SSRs) as a marker system have been widely used. However, the lack of expressed sequence tags (ESTs) and genomic sequences prevents the development of SSR markers in celery. So far only 11 SSR markers have been reported (Acquadro *et al.*, 2006). Wang *et al.* (2011) utilized 11 SSR makers and 60 sequence-related amplified polymorphism (SRAP) primer combinations to study the genetic diversity of 68 celery accessions.

In case of *Apium graveolens*, till date only about 2,000 ESTs were deposited in Gen Bank database. Thus, extensive genomic or transcriptomic sequences are urgently required for *Apium graveolens*, which can be utilized for new genes discovery, molecular markers development, gene localization, and comparative genomics and so on. Given that the celery is a diploid with a large genome (3 x 109 bp) and a high degree of heterozygosity, it is infeasible to consider whole genome sequencing because of the high costs and time-consuming. Fortunately, the advent of transcriptome sequencing provides an alternative to the whole genome sequencing. Transcriptome sequences exclude non-coding DNA, therefore the sequences obtained must contain a high content of functional information and are beneficial to reveal the molecular mechanism of functional genes (Fu *et al.*, 2013).

11.6.0 Genetic Transformation

In an effort to improve the T-DNA-mediated transformation frequency in celery, the possible enhancement effect of multiple copies of *virG* genes in *Agrobacterium tumefaciens* strains upon the transient transformation was evaluated (Liu *et al.*, 1992). Four days after *A. tumefaciens* infection, histochemical beta-glucuronidase (GUS) assay were performed to determine the frequency of transient transformation of calli. Additional copies of octopine- and agropine-type *virG* genes in *A. tume-faciens* strains containing an agropine-type Ti-plasmid enhanced the frequency of transient transformation of celery. This enhancement ranged from 25 per cent to five-fold, depending upon the source of the *vir G* gene and the plant tissues inoculated. The results show that multiple copies of *vir G* in *A. tumefaciens* can greatly enhance the transient transformation frequency of celery and that this enhancement is influenced by both the type of Ti-plasmid harboured by *A. tumefaciens* and by the infected plant species.

11.7.0 *In vitro* Conservation

A method for the long-term storage of somatic celery embryos was devised by Nadel *et al.* (1990). Embryos survived well if stored slightly moistened at 4°C. Plants regenerated from somatic embryos, which were kept in cold storage for up to 24 weeks, developed well and were morphologically identical to those regenerated from non-stored embryos. Plant establishment was best if embryos and small plantlets were first grown in suspension culture with half-strength regeneration medium (RM) (half-strength MS supplemented with 1.15 µM kinetin, 1.5 per cent w/v sucrose and 4 per cent w/v mannitol) and then transferred to liquid half-strength RM with paper bridges or to a semi-solid medium on agar mixed with perlite.

Celery embryos and plantlets were found to be selectively released in a culture of immobilized Ca-alginate gel beads in which celery callus was entrapped under regeneration conditions. The feasibility of use of this process for celery embryogenesis in an artificial seed system was investigated by Suehara *et al.* (1995). The optimal concentrations of alginate gel for embryo and plantlet production was 2 per cent for the immobilized cell culture. Considering the maintenance of the gel bead structure and detrimental effect of $CaCl_2$ on plantlet development, 5 mM $CaCl_2$ supplementation gave the best result in terms of the number of heart and torpedo embryos and plantlets.

Production of encapsulatable units, scale-up of the production system and encapsulation technologies were developed by Onishi *et al.* (1994) as an integrated technology for transplant production using somatic embryos (SEs) of celery. Encapsulatable units, somatic embryos improved in their quality for encapsulation, showed 80 per cent conversion without encapsulation and any nutrient supply on a nursery medium in a green house.

12.0 References

Aharoni, N., Barkai Golan, R., Ramraz, Y. and Aviram, H. (1978) *Israel, Ministry of Agriculture, Inst. Technol. Storage Agric. Products: Scientific Activities, 1974-1977,* **184**: 34-35.

Acquadro, A., Magurno, F., Portis, E. and Lanteri, S. (2006) *Mol. Ecol. Notes.*, **6**: 1080–1082.

Ahmed, Z.A. and Maprd, A.R.C. (2017) *J. Plant Production, Mansoura Univ.*, **8** (2): 247 – 251.

Alberts, E., Francki, R.I.B. and Dietzgen, R.G. (1989) *Australian J. Agril. Res.*, **40**: 1027-1036.

Americanos, P.G. (1991) *Tech. Bull. Cyprus Agril. Res. Inst* , **137**: 5.

Anonymous (1981) *Sementi-Eletee*, **27**: 2, 44.

Arus, P. and Ortan T. (1984) *J. Hered.*, **75**: 11–14.

Ataga, A.E., Epton, H.A.S. and Frost, R.R. (1993) *Physiological and Molecular Plant Pathology*, **42**: 161-168.

Awuah, R.T. and Lorbeer, J.W. (1991) *Plant Disease*, **75**: 123-125.

Bailey, L.H. (1950) *The Standard Cyclopedia of Horticulture*, The Macmillan Co., New York.

Balyan, S.S., Chowdhary, D.K. and Kaul, B.L. (1990) *Indian Perfumer*, **34**: 168-170.

Barkai Golan, R. and Aharoni, N. (1972) *Israel J. Agric. Res.*, **22**: 157-159.

Bell, A.A., Hubbard, J.C., Liu, L., Davis, R.M. and Subbarao, K.V. (1998) *Plant Dis.*, **82**: 322-328.

Bellaloui, N. and Brown, P.H. (1998) *Plant and Soil.*, **198**: 153–158.

Benoit, F. and Ceustermans, N. (1988) *Plasticulture*, **79**: 37-43.

Benoit, F., Ceustermans, N. and Calus, A. (1981) *Acta Hortic.*, **122** (3): 121-131.

Beverly, R.B. and Anderson, D.L. (1988) *Proc. Soil and Crop Science Society of Florida*, **47**: 49-52.

Biczak, R., Gurgul, E. and Herman, B. (1998) *Folia Hortic.*, **10**: 23-34.

Bisessar, S., Rinne, R.J. and Potter, J.W. (1983) *Plant Dis.*, **67**: 11-14.

Bosio, G. (1994) *Informatore Agrario*, **50**: 71-74.

Breschini, S.J. and Hartz, T.K. (2002) *HortScience*, **37**(6): 894-897.

Brocklehurst, P.A., Dearman, J. and Drew, R.L.K. (1987) *Acta Hortic.*, **215**: 193-200.

Candia, E. Di. (1983) *Informatore di Ortoflorofrutticoltura*, **24**: 13-19.

Catlin, D., Ochoa, O., McCormick, S. and Quiros, C.F. (1988) *Plant Cell Rep.*, **7**: 100-103.

Chao, J.C. and Huang, M.J. (1994) *J. Shanghai Agril. Coll.*, **12**: 205-209.

Charles, K.S., Ngouajio, M., Warncke, D.D., Poff, K.L. and Hausbeck, M.K. (2006) *Weed Science*, **54**: 326–334.

Chen, Y.Y., Li, M.Y., Li,Y., Wang, F., Xu, Z.S. and Xiong, A.S. (2015) *Sci. Hort.*, **186**: 1–6. doi: 10.1016/j.scienta.2015.02.011.

Chinchilla, C.M. and Mora, D. (1986) *Agronomia Costarricense*, **10**: 51-55.

Chism, W.J. and Snyder, M.J. (1982) *Proc. Western Soc. Weed Sci.*, **35**: 60-62.

Choudhary, D.K. and Kaul, B.L. (1988) *Indian Farming*, **38**: 28-31.

Choudhary, D.K. and Kaul, B.L. (1990) *Indian Perfumer*, **34**: 176-177.

Choudhury, B. (1970) *Vegetables* (2nd edn.), National Book Trust, India, New Delhi, pp. 180-182.

Corato, U. de, and Baviello, G. (2000) *Informatore Fitopatologico*, **50**: 61-63.

Crescenzi, A., Ragozzino, A. and Alioto, D. (1993) *Petria*, **3**: 169-176.

CSIRO (1998) Fruit and vegetables storage and transport database, Australia: Sydney Postharvest Laboratory.

D'Antonio, V., Falk, B. and Quiros, C.F. (2001) *Plant Dis.*, **85**: 1276–1277.

Demyttenaere, P., Hofman, G., Vulsteke, G. and Ossemerct, C. (1988) *Mededelingen van de Faculteit Landbouwwetenschappen, Rijksuniversiteit Gent.*, **53**: 113-122.

Dohya, N., Matsubara, S. and Murakami, K. (1997) *J. Jap. Soc. Hort. Sci.*, **65**: 747-752.

Domblides, A., Domblides, H. and Kharchenko, V. (2008) *Proc. Latv. Acad. Sci., B Nat. Exact Appl. Sci.*, **62:** 219–222.

Donovan, A., Collin, H.A., Isaac, S. and Mortimer, A.M. (1994) *Allals Appl. Biol.*, **124**: 383-398.

Daugovish, O., Fennimore, S.A. and Smith, R. F. (2007) *Weed Technol.*, **21**: 719–723.

Du, Y., Niu, W., Zhang, Q., Cui, B., Gu, X., Guo, L. and Liang, B. (2018) *Soil Sci. Soc. Am. J.*, **82**: 606-613.

Dusky, J.A. (1983) *Proc. Fla. St. Hort. Soc.*, **96**: 93-96.

Dyduch, J. (1994) *Acta Hortic.*, **371**: 275-282.

Dyduch, J. (1995) *Zeszyty Problemowe Postepow Nauk Rolniczych*, **419**: 29-34.

Elmer, W.H. and Lacy, M.L. (1983) *Phytopathol.*, **73**: 786.

Elmer, W.H. and Lacy, M.L. (1985) *Phytopathol.*, **75**: 1361.

Elmer, W.H., Lacy, M.L. and Honma, S. (1986) *Plant Dis.*, **70**: 416-419.

Emsweller, S.L. (1929) *Proc. Amer. Soc. Hortic. Sci.*, **25**: 29-30.

Epstein, L. (2017) https: / /reeis.usda.gov /web/crisprojectpages/0230718-breeding-of-disease-resistantcelery.html.

Evenor, D., Pressman, E., Ben-Yephet, Y. and Rappaport, L. (1994) *Plant Cell Tissue Organ Cult.*, **39**: 203-210.

Evers, A.M., Ketoja, E., Hagg, M., Plaami, S., Hakkinen, U. and Pessala, R. (1997) *Plant Foods for Human Nutrition*, **51**: 173-186.

Farrar, R. Jr. and Ridgway, R.L. (1997) *Environ. Entomol.*, **26**: 1461-1469.

Favaro, J.C. and Bochinfuso, R. (1994) *Investigacion Agraria, Produccion y Proteccion Vegetables*, **9**: 63-71.

Finch-Savage, W.E. (1984) *J. Hort. Sci.*, **59**: 87-93.

Folster, E. (1992) *Gartenbau Magazin*, **1**: 51-53.

Francois, L.E. (1988) *J. Amer. Soc. Hort. Sci.*, **113**: 538-542.

Furutani, S.C., Zandstra, B.H. and Price, H.C. (1985) *J. Amer. Soc. Hort. Sci.*, **110**: 153-156.

Fu, N., Wang, Q. and Shen, H.L. (2013) *PloS One*, 8(2): e57686. https: //doi.org/10.1371/journal.pone.0057686.

Gariepy, Y., Raghavan, G.S.V., Plasse, R., Theriault, R. and Phan, C.T. (1984) *Acta Hortic.*, **157**: 193-201.

Gao, G.X., Jin, L.Z., Lu, Z.M. and Gu, Z.H. (2006) *Tianjin Agric. Sci.*, **12:** 9–11.

Gao, G. X., Jin, L. Z., Lu, Z. M., Lu, F. C. and Gu, Z. H. (2008) *Acta Hortic.*, **35**: 777.

Gao, G. X., Jin, L., Lu, F., Lu, Z., Ren, Z., and Yu, H. (2009) *J. Changjiang Veg.*, 14.

Gao, Z.G. and Loescher, W.H. (2000) *Plant Physiol.*, **124:** 321–330.

Gelernter, W.D. and Trumble, J.T. (1999) *Integrated Pest Management Rev.*, **4**: 301-306.

Gentzsch, D. (1999) *Gemuse Munchen*, **35**: 166.

Globerson, D. and Feder, Z. (1987) *Acta Hortic.*, **198**: 15-21.

Gomez, P.A. and Artes, F. (2004) *Lebensm.-Wiss. u.-Technol*, doi: 10.1016/j.lwt.2004.06.014.

Gómez, P., Artés-Hernández, F. and Artés, F. (2010) *Acta Hortic.*, **857**: 31-36.

Gorini, F.L. and Gasperetti, L. and Bentivegna, A. (1973) *Italia Agricola*, **110**: 1038-1050.

Gracia, O. and Feldman, J.M. (1977) *Plant Dis. Rep.*, **61**: 905-908.

Grafius, E. and Hayden, J. (1988) *Great Lakes Entomologist*, **21**: 49-54.

Gray, D. (1987) *Monograph, British Plant Growth Regulator Group*, **15**: 93-105.

Han, Q.X., Shen, H.L. and Zhu, X. (2006) *China Veg.*, **11**: 6–9.

Heath Pagliuso, S., Pullman, J. and Rappaport, L. (1988) *Theor. Appl. Genet.*, **75**: 446-451.

Heath Pagliuso, S., Pullman, J. and Rappaport, L. (1989) *HortScience*, **24**: 711-712.

Helay, A.A., Baek, J.P., Mady, E., Eldekashy, M.H.Z. and Craker, L. (2015) *J. Med. Active Plants.*, **4** (1): 1–7.

Hardenberg, R.E., Watada, A.E. and Wang, C.Y. (1986) U.S. Dept. of Agriculture, Agric. Hand Book, No. 66.

Honma, S. and Klompenberg, R.J. van (1974) *Res. Rep. Michigan Agril. Expt. Stn.*, **237**: 2.

Honma, S. and Lacy, M.L. (1980) *Euphytica*, **29**: 801-805.

Honma, S., Lacy, M.L. and Murakishi, H.H. (1986) *HortScience*, **21**: 1073-1074.

Hore, A. (1977) *Cytologia.*, **42**: 21-28.

Huang, M.J. and Cao, J.C. (1997) *J. Shanghai Agril. Coll.*, **15**: 144-146.

Huang, W., Wang, G.L., Li, H., Wang, F., Xu, Z.S. and Xiong, A.S. (2016) *Mol. Genet. Genomics.*, **291** (6): 2131–2143.

Huestis G, McGrath J.M. and Quiros C.F. (1993) *Theor. Appl. Genet.*, **85**: 889–896.

Hybl, O. (1977) *Gartenbau.*, **24**: 236-237.

Ireland, K.F. and Lacy, M.L. (1987) *Phytopathol.*, **77**: 1763.

Ireland, K.F., Elmer, W.H. and Lacy, M.L. (1987) *Phytopathol.*, **77**: 1712.

Ivanek, O. and Soukup, M. (1996) *Rostlinna Vyroba*, **42**: 430-432.

Janes, E.R. (1954) *The Vegetable Garden*, Penguin Hand Books, pp. 88-100.

Janyska, A. (1975) *Ochrana Rostlin*, **11**: 267-273.

Jardine, D.J., Dilley, D.R. and Price, H.C. (1984) *HortScience*, **19**: 869-870.

Jeng, T.L. and Sung, J.M. (1989) *J. Agril. For.*, **47**: 89-101.

Joseph, Shimat V. (2016) *Plant Health Progress*, **17**: 101-105.

Ju, J. (2007) *Chinese Agri. Sci. Bull.*, **23:** 120–123.

Kardaeva, N.S. and Kletskova, Z.I. (1978) *Byulleten Vsesoyuznogo Ordena Lenina i Ordena Druzhby Narodov Instituta Rastenievodstva Imeni N. I. Vavilova*, **80**: 76-77.

Kaur, R. and Gill, B.S. (2011) *Indian J. Ecol.*, **39**: 258-260.

Keleleer, A., Proft, M.P. de and Greef, J.A. de (1987) *Acta Botanica Neerlandica*, **36**: 195.

Kim, Y.H. and Janick, H. (1989) *HortScience*, **24**: 671-673.

Kim, Y.H., Yang, W.Y. and Cho, H.J. (1990) *Res. Reports Rural Dev. Administration, Biotechnology*, **32**: 18-22.

Knott, J.E. (1960) *Hand Book for Vegetable Growers*, John Wiley and Sons. Inc., New York.

Knuth, P. and Schrameyer, K. (1991) *Gemuse Munchen*, **27**: 136-140.

Koike, S.T., Little, E.L., Bishop, A.L. and Gilbertson, R.L. (1994) *California Agric.*, **48**: 32-34.

Kooti, W. and Daraei, N. (2017) *J. Evid. Compl. Altern. Medc*, **22**(4): 1029-1034

Krikun, J., Haas, J.H., Dodd, J. and Kinsbursky, R. (1990) *Plant Soil*, **122**: 213-217.

Kumar, G., Kushwaha, H.R., Panjabi-Sabharwal, V., Kumari, S., Joshi, R., Karan, R., Mittal, S., Pareek, S.L.S. and Pareek, A. (2012). *BMC Plant Biol.*, **12**: 1–15.

Lacy, M.L. and Honma, S. (1981) *Phytopathol.*, **71**: 234.

Lee, G.P., Park, K.W., Lee, J.M. (1999) *Acta Hortic.*, **483**: 185-192.

Leonardi, C. and Munoz-Carpena, R. (1998) *Acta Hortic.*, **458**: 257-261.

Leshuk, J.A. and Saltveit, M.E. (1990) In: *Controlled atmcsphere storage requirements and recommendations for vegetables*, CRC Press, Boca Raton DL, pp. 315-352.

Li, Y., Wang, T., Li, J. and Ao, Y. (2010) *Hort. Sci. (Prague)*, **37**: 99–108.

Li, Y., Xu, Z., Tan, G., Jia, X., Wang, F., and Xiong, A. (2015) *J. Nanjing Agric. Univ.*, **38:** 360–368.

Little, E.L., Koike, S.T. and Gilbertson, R.L. (1997) *Plant Dis.*, **81**: 892-896.

Liu, C.N., Li, X.Q. and Gelvin, S.B. (1992) *Plant Mol. Biol.*, **20**: 1071-1087.

Luo, Y., Suslow, T. and Cantwell, M. (2002) *Celery*, http: /www.ba.ars.usda.gov/ hb66/048celery.pdf.

Madjarova, D. and Bubarova, M. (1974) In: Antoszewski, *Proc. XIX Internatl. Hort. Cong., I. Section VII. Vegetables* (Eds. Harrison and C.C. Zych), Abstracts, p. 728.

Madjarova, D.J., Bubarova, M.G. and Madzharova, D.Zh. (1978) *Acta Hortic.*, **73**: 65-72.

Madrid, R., Lopez, M.D., Barba, E.M., Gomez, P. and Artes, F. (2008) *J. Plant Nutr.*, **31**: 55-67.

Madzharova, D., Benbasat, E., Baburova, M. and Chavdarov, I. (1973) *Grad. Lozar. Nauka*, **10**: 85-93.

Mahajan, V.P., Randhawa, G.S., Bains, D.S. and Sawhney, J.S. (1978) *Indian J. Agric. Sci.*, **48**: 270-273.

Maharik, Z.Y. and Mancy, A.G.A. (2017) *Middle East J. Agric. Res.*, **6**(4): 1189-1203.

Malhotra, S.K. (2006) In: *Spices, Medicinal and Aromatic Crops*, (Ed. J Singh), University Press, Hyderabad, India.

Malhotra, S.K. (2006) In: *Handbook of herbs and spices* (Ed. K.V. Peter), pp. 317-336.

Marongiu, B., Piras, A., Porcedda, S., Falconieri, D., Maxia, A., Frau, M.A., Goncalves, M.J., Cavaleiro, C. and Salgueiro, L. (2013) *Nat. Prod. Res.*, **27**: 1521–1527.

Maticic, B., Avbelj, L., Feges, M. and Lokar, V. (1991) *Roczniki Gleboznawcze*, **42**: 155-163.

Maync, A. and Riedel, W. (1986) *Deutscher Gartenbau*, **40**: 1756-1758.

McCormick, N. (2006). *Ph.D. thesis*, University of Melbourne, Melbourne, Australia.

Meeldijk, B.P. (1984) *Landbouwmechanisatie*, **35**: 783-785.

Meng, X.D. and Jiang, X.M. (1993) *J. Shandong Agril. Univ.*, **24**: 473-478.

Monnet, Y. and Thibault, J. (2000) *PHM Revue Horticole.*, **417**:66-67.

Mudita, I.W. and Kushalappa, A.C. (1993) *Plant Dis.*, **77**: 1081-1084.

Nadel, B.L., Altman, A. and Ziv, M. (1989) *Plant Cell Tissue Organ Cult.*, **18**: 181-189.

Nadel, B.L., Altman, A. and Ziv. M. (1990) *Sci. Hort.*, **44**: 9-16.

Nakamura, S., Teranishi, T. and Aoki, M. (1982) *J. Japanese Soc. Hort. Sci.*, **50**: 461-467.

Nega, E., Ulrich, R., Werner, S. and Jahn, M. (2001) *Gesunde Pflanzen.*, **53**: 177-184.

Nissen, M.J. and Engell, S.R. (1996) *Agro-Sur.*, **24**: 104-111.

Ochoa, O. and Quiros, C.F. (1989) *Plant Breed.*, **102**: 317-321.

Ochoa, O., D'-Antonio, V. and Quiros, C.F. (1986) *HortScience*, **21**: 1455-1456.

Olympios, C.M. (2011) *CIHEAM - Options Mediterraneennes.*, **31**: 307-324.

Onishi, N., Sakamoto, Y. and Hirosawa, T. (1994) *Plant Cell Tissue Organ Cult.*, **39**: 137-145.

Opgenorth, D.C. and Endo, R.M. (1979) *Plant Dis. Rep.*, **63**: 165-169.

Orsini, F., Marchetti, L., Michelon, N. and Gianquinto, G.P. (2012) n. 526476-LLP-1-2012-1-IT-Grundtvig-GMP (available at www.hortis-europe.net).

Orton, T.J. and Arus, P. (1984) *Euphytica*, **33**: 471-480.

Orton, T.J. (1984) *Plant Dis.*, **68**(7): 574–78.

Orton, T.J., Durgan, M.E. and Hulbert, S.D. (1984a) *Plant Dis.*, 68 (7): 574-578.

Orton, T.J., Hulbert, S.H., Durgan, M.E. and Quiros, C.F. (1984b) *HortScience*, **19**: 594.

Paerara, C.A., Qiao, P. and Cantliffe, D.J. (1993) *HortScience*, **28**: 20-22.

Pardossi, A., Bagnoli, G., Malorgio, F., Campiotti, C.A. and Tognoni, F. (1999) *Sci. Hort.*, **81**: 229-242.

Paris, P., Gavazzi, C. and Tabaglio, V. (1999) *Agricoltura Mediterranea*, **129**: 218-226.

Peirce, L.C. (1987) *Vegetables - Characteristics, Production and Marketing*, John Wiley and Sonds, pp. 240-245.

Peng, W.S., Mai, Z.S. and Yang, Z.C. (1998) *J. Shanghai Agril. Coll.*, **16**: 32-35.

Pernezny, K., Datnoff, L. and Collins, J. (2000) *Proc. Fla. Sta. Hort. Soc.*, **112**: 302-305.

Pfulb, E. (1990) *Gemuse Munchen*, **26**: 440-443.

Pierik, R.L.M. (1990) In: *Proc. of the International Seminar on New Frontiers in Horticulture*, Vol. 12.Organized by Indo-American Hybrid Seeds, Bangalore, India, November.pp. 25-28.

Pink, D.A.C., Walkey, D.G.A., Stanley, A.R.J., Carter, P.J., Smith, B.M., Mee, C., Miller, A. and Bolland, C.J. (1983, 1984) *Annual Report, NRS.*, Wellesbourne, UK, 1983-1984, pp. 58-59.

Pope, D.T. and Munger, H.M. (1953) *Proc. Amer. Soc. Hort. Sci.*, **61**: 472-480.

Pruthi, J.S. (1976) *Spices and Condiments*, National Book Trust, India, pp. 80-83.

Quiros, C.F., Rugama, A., Dong, Y.Y. and Orton, T.J. (1986) *Euphytica*, **35** (3): 867–875.

Quiros, C.F., D'-Antonio, V. and Greathead, A.S. (1993) *HortScience*, **28**: 351-352.

Quiros, C.F., Douches, D. and D'Antonio, V. (1987) *Theor. Appl. Genet.*, **74:** 203–208.

Quiros, F.C. (2001) *http: /vric.ucdavis.edu/veginfo/commodity/celery/fusariumcelery. html.*

Ramin, A.A. and Atherton, J.G. (1994) *J. Hort. Sci.*, **69**: 861-868.

Randhawa, G.S. and Kaur, S. (1995) In: *Advances in Horticulture* Vol. II – Medicinal and Aromatic Plants (Eds. K.L. Chadha and R. Gupta), Malhotra Publishing House, New Delhi, India, pp. 899–915.

Rivera, B.R. and Lopez, C.R. (1982) *Turrialba (IICA)*, **32**: 67-73.

Robbs, P.G., Bartz, J.A., McFie, G. and Hodge, N.C. (1996) *J. Food Sci.*, **61**: 444-448.

Rubatzky V.E. and Yamaguchi M. (1997) In: *World Vegetables*, Chapman and Hall, New York, USA, pp. 432–443.

Royer, T.A., Edelson, J.V., Bogle, C.R. and McCrate, S. (1988) *Pesticide Sci.*, **25**: 231-240.

Ruiz, J.J., Pico, B., Li, G.Y., D'Antonio, V., Falk, B. and Quiros, C.F. (2001) *J. Amer. Soc. Hortic. Sci.*, **126**: 432–435.

Rumasz, E., Koszanski, Z. and Wronkowska, H. (1999) *Folia Univ. Agric. Stetinensis Agric.*, **73**: 207-211.

Runham, S.R, Town, S.J., Fitzpatrick, J.C. and Fokkema, N.J. (2000) *Acta Hortic.*, **513**: 193-201.

Ryder E.J. (1979) Celery. In: *Leafy Salad Vegetables*, Springer, Dordrecht. https: // doi.org/10.1007/978-94-011-9699-4_3

Saini, S.S., Cheema, S.S., Randhawa, G.S. and Mahey, R.K. (1989) *Indian J. Agron.*, **34**: 341-345.

Saltveit, M.E. (1997) *Int. Contr. Atmos. Res. Conf.*, Vol. 4, Vegetables and Ornamentals. Univ. Calif., Davis CA, Postharv. Hort. Ser., **18**: 98-117.

Saltveit, M.E. and Mangrich, M.E. (1996) *J. Amer. Soc. Hort. Sci.*, **121**: 137-141.

Sandwell, I. and Wood, M.B. (1972) *Expt. Hort.*, **23**: 43-47.

Sanchez, C.A., Burdine, H.W. and Guzman, V.L. (1990) *Proc. Soil Crop Sci. Soc. Fla.*, **49**: 69-72.

Santamaria, P., Elia, A., Serio, F., Gonnella, M. and Parente, A. (1999) *J. Plant Nutrition*, **22**: 1091-1106.

Saranga, Y. and Janick, J. (1991) *HortScience*, **26**: 1335.

Schneider, R.W. (1985) *Phytopathol.*, **75**: 40-48.

Scully, B., Nuessly, G.S., Raid, R.N. and Stubblefield, R.E. (1995) *HorScience*, **30**: 1104-1105.

Sethi, K.L. and Sapra, R.L. (1986) *Internal J. Top. Agril.* **4**: 73-76.

Shad, A.A., Shah, H.U., Bakht, J., Choudhary, M.I. and Ullah, J. (2011) *J. Med. Plants Res.*, **5** (20): 5160–5166.

Silvertand, B., Jacobsen, E., Mazereeuw, J., Lavrijsen, P., Harten, A.van and Van-Harten, A. (1995) *Plant Cell Rep.*, **14**: 423-427.

Singh, H., Morss, S. and Orton, T.J. (1985) *Seed Sci. Technol.*, **13**: 551-558.

Smith, P.M. (1976) *Evolution of Crop Plants* (Ed. N.W. Simmonds), Longmans, London and New York, pp. 322-323.

Song, G,Q., Loskutov, A.V. and Sink, K.C. (2007) *Plant Cell Tiss. Organ Cult.*, **88**: 193-200.

Sowbhagya, H.B. (2014) *Crit. Rev. Food Sci.*, **54**: 389–398.

Stoffella, P.J., Bryan, H.H. and Shuler, K.D. (1988) *HortScience*, **23**: 988-991.

Stoffella, P.J., Williams, B.J., Bryan, H.H., Campbell, G.M. and Shuler, K.D. (1986) *Proc. Fla. Sta. Hort. Soc.*, **98**: 292-294.

Strandberg, J.O. (1986) *Proc. Fla. Sta. Hort. Soc.*, **98**: 294-299.

Su, C.C., Hsu, S.T. and Tzeng, K.C. (1989) *Plant Prot. Bull., Taiwan*, **31**: 346-357.

Suehara, K.I., Kohketsu, K., Uozumi, N. and Kobayashi, T. (1995) *J. Ferment.Bioengin.*, **79**: 585-588.

Suslow, T. and Cantwell, M. (1998) *Celery produce facts. Recommendations for maintaining postharvest quality.* http: //postharvest.ucdavis.edu/produce.

Suslow, T. and Cantwell, M. (2000) Recommendations for maintaining postharvest quality. Produce facts. California, USA: Department of Vegetable Crops.

Takahashi, H.W., Dechen, A.R., Carmello, Q.A. and Haag, H.P. (1987) *Anais da Escola Superior de Agricultura "Luiz de Queiroz" (Brazil)*, **44**: 1473-1501.

Tamez-Guerra, P., McGuire, M.R., Behle, R.W., Hamm, J.J., Sumner, H.R. and Shasha, B.S. (2000) *J. Econ. Entomol.*, **93**: 210-218.

Temperini, O., Colla, G., Saccardo, F. and Brancaleone, M. (2000) *Informatore Agrario.*, **56**: 95-98.

Thomas, T.H. (1978) *Sci. Hortic.*, **9**: 311-316.

Tracchi, G., Loubiere, P. and Montagnon, M. (1997) *Proc. Internatl. Conf.*, Brighton, UK, 17-20 November, 1997, **2**: 885-889.

Tremblay, N. and Senecal, M. (1990) *Agronomie*, **10**: 15-21.

Tremblay, N., Masson, J. and Gosselin, A. (1989) *Acta Hortic.*, **238**: 119-125.

Trumble, J.T. and Quiros, C.F. (1988) *J. Econ. Entomol.*, **81**: 602-607.

Trumble, J.T., Derks, W., Quiros, C.F. and Beier, R.C. (1990) *J. Econ. Entomol.*, **83**: 519-525.

Trumble, J.T., Diawara, M.M., Quiros, C.F. (2000) *Acta Hortic.*, **513**: 29-37.

Vanparys, L. (1999) *Proeftuinnieuws*, **9**: 13-14.

Vanparys, L. (2000) *Proeftuinnieuws*, **10**: 21-22.

Velandia, J. (1992) *Rev. ICA Instituto Colombiano Agropecuario (Colombia)*, **27**: 269-274.

Vulsteke, G. (1979) *Boer en de Tuindere*, **85**: 13, 27.

Vulsteke, G. and Van Oost, N. (1989) *Mededeling Provinciaal Onderzoek en Voorlichtingscentrum voor Land en Tuinbouw*, Beitem Roeselare, 1989, No. 297.

Vulsteke, G., Vanoost, N. and Seynnaeve, M. (1994) *Mededeling Provinciaal Onderzoek en Voorlichtingscentrum voor Land en Tuinbouw, Beitem Roeselar*, **343**: 4.

Walkey, D.G.A. and Ward, C.M. (1984) *J. Agric. Sci.*, UK, **103**: 415-419.

Wang, S., Yang, W. and Shen, H. (2011) *Sci. Hort.*, **129**: 1–8.

Wang, S., Luo, C., Xie, Y., Jiang, X., Wang, Y. and Wang, Q. (2020) *Research Square*, pp. 1-20.

Wang, W.T., Wu, F. and Gao, G.X. (2012) *Acta Hortic Sin.*, **39**: 1007-1008.

Wasilewska, I. and Masternak, H. (1987) *Biuletyn Warzywniczy*, **30**: 29-56.

Weintraub, P.G. and Horowitz, A.R. (1998) *J. Econ. Entomol.*, **91**: 1180-1185.

Weintraub, P.G., Arazi, Y. and Horowitz, A.R. (1996) *J. Crop Protec.*, **15**: 763-769.

Weiser, U. and Scharpf, H.C. (1989) *Gemuse*, **25**: 212-216.

Whipps, J.M., McQuilken, M.P. and Budge, S.P. (1993) *Pesticide Sci.*, **37**: 309-313.

Whitaker, T.W. (1941) *Proc. Am. Soc. Hortic. Sci.*, **39**: 346-348.

Whiteman, T. (1957) *USDA Mkt. Res. Rpt.*, **32**: 196.

Wiebe, H.J. (1999) *Gemuse Munchen*, **35**: 579-581.

Willaert, G., Verloo, M. and Vulsteke, G. (1990) *Rev. Agric.*, **43** (4): 599–615.

Wolf, E.A. and Scully, B. (1992) *HortScience*, **27**: 1235-1236.

Wolf, E.A. and White, J.M. (1978) *Proc. Fla. St. Hort. Soc.*, **91**: 324-327.

Wolf, E.A., White, J.M., Stubblefield, R.S. and Scully, B. (1993) *HortScience*, **28**: 754-755.

Wright, J.C. and Lacy, M.L. (1988) *Plant Dis.*, **72**: 256-259.

Yang, X.F. and Quiros, C.F. (1995) *Genome*, **38**: 36-44.

Yao, Y., Sang, W., Zhou, M.J. and Ren, G.X. (2010) *J. Food Sci.*, **75** (1): C9–C13.

Yawalkar, K.S. (1969) In: *Vegetable Crops of India*, Agri-Hort. Publishing House, Nagpur, pp. 295-298.

Zhang, Y.L., Sun, S.R., Wang, L.J., Qiao, B.J. and Lu, X.Y. (1998) *Acta Agric. Univ. Henanensis*, **32:** 73-75.

Zhang, K., Wang, J.B., Yang, Z.Y., Xin, G.R., Yuan, J.G., Xin, J.L. and Huang, C. (2013) *Front. Environ. Sci. Eng.*,**7**: 85–96.

PARSLEY

P. Karmakar, J. Devi, V.A. Parthasarathy and U.Parthasarathy

1.0 Introduction

Parsley (*Petroselinum crispum* syn. *Apium petroselinum*) belongs to the family Apiaceae. In America parsley is mostly used as a garnish, while in Europe and Middle East it is almost used as often as salt. It is mostly grown outdoors and is seasonally harvested (Navazio, 2012). Parsley is a leafy vegetable, rich in many biologically active compounds, and its name (*Petroselinum*) is derived from the Greek for 'rock celery' or the word Petroselinum is derived from the Greek word petros which means "stone," referring to the plant's habit of growing in rocky places. Selinon was the Greek word for parsley in ancient history. It can be distinguished from other leafy green herbs by its unique aroma.

Both the crowded, dense-leaved type and the broad open-growing type were described in the 4th century B.C. Parsley was introduced into England from Sardinia in 1548. European colonists brought parsley to the United States in the 17th century. It is also popular for container-grown plant for limited-space situations. Chopped parsley leaves are a popular decoration in Central Europe (similar to the use of coriander leaves in China, South East Asia and parts of India), mostly for soups and vegetables. The species name Parsley is known for over a thousand year ago in the Mediterranean (Katzer, 2002).

Parsley is a popular culinary herb, commercially cultivated as an annual plant in many parts of the world for its attractive and aromatic leaves. Parsley is the most widely cultivated herb in Europe, and the most used herb in the United States. It is used as a garnish rather than a component of the diet. The oil derived from the leaves can also be used in condiments and seasonings.

2.0 Composition and Uses

2.1.0 Composition

Parsley is considered as 'powerhouse' of nutrition, andis rich in B vitamins, vitamin C, β-carotene and zinc; it is an important dietary component for strengthening bone due to its high content of boron and fluorine, and also contains iron and calcium in an absorbable form.

The essential oils of leaves and roots have approximately the same composition. The main components (10-30 per cent) are myristicin, limonene and 1,3,8-*p*-menthatriene; minor components are mono- and sesquiterpenes. The curly varieties (var. *crispum*) tend to be richer in myristicin, but contain much less essential oil than var. *latifolium* (0.01 and 0.04 per cent, respectively). The leaves generally contain 0.04-0.4 per cent of volatile oil and the major constituents of this volatile oil are α-pinene, β-pinene, myrcene, β-phellandrene, 1,3,8-p-menthatriene and myristicin (Charles, 2004). The aroma of parsley is due to the presence of terpenes, which are toxic to many insects. Myristicin is a toxic phenylpropene/allylbenzene compound and has hallucinogenic properties, acting as a psychoactive at high intake levels (Hallström and Thuvander, 1997).

In contrast, the essential oil from the fruits (3-6 per cent) is either dominated by myristicin (60 to 80 per cent ; mostly in var. *tuberosum* and var. *crispum*) or by apiole (70 per cent in var. *latifolium*). A third chemical race shows allyl tetramethoxy benzene (55 to 75 per cent), which can also appear in apiol-dominated oils (up to 20 per cent).

The nutritional value of parsley is presented in Table 1.

Table 1: Nutritional Value of Parsley for One Serving (60 g) *

Nutrients	*Quantity*	*Nutrients*	*Quantity*
Total Calories	21.6	Calcium	82.8 mcg
Total carbohydrate	3.8 g	Iron	3.7 mg
Dietary fibre	2.0 g	Magnesium	30.0 mg
Total fat	0.5 g	Phosphorus	34.8 mg
Saturated fat	0.1 g	Potassium	332 mg
Monounsaturated fat	0.2 g	Sodium	33.6 mg
Polyunsaturated fat	0.1 g	Zinc	0.1 mg
Omega-3 fatty acids	4.8 mg	Copper	0.1 mg
Omega-6 fatty acids	69.0 mg	Manganese	0.1 mg
Protein	1.8 g	Selenium	0.1 mcg
Water	52.6 g	Pantothenic acid (B5)	0.2 mg
Ash	1.3 g	Folate (B9)	91.2 mcg
Vitamin A	5055 IU	Vitamin C	79.8 mg
Thiamine (B1)	0.1 mg	Vitamin E (α-tocopherol)	0.4 mg
Riboflavin (B2)	0.1 mg	Vitamin K	984 mcg
Niacin (B3)	0.8 mg	Choline	7.7 mg
Vitamin B6	0.1 mg		

* Daradkeh and Essa (2016).

2.2.0 Uses and Medicinal Properties

The finely-chopped leaves are used for flavouring sauces, soups, stuffing's, rissoles, minces, *etc.*, and also sprinkled over vegetables or salads. The leaves are also dried and powdered for culinary flavouring when fresh leaves are not available. In addition to the leaves, the stems are also dried and powdered, both as a culinary colouring and as a dye. The roots of the turnip-rooted variety are used as a vegetable and for flavouring. Medicinally, the two-year-old roots are used and the leaves dried for making parsley tea, and the seeds, for the extraction of an oil called apiol, which is of considerable curative value. The best kind of seed for medicinal purposes is that obtained from the Triple Moss curled variety. Parsley leaves are ready for use about 3 months after seeding. A few leaves at a time may be removed from each plant, or the entire bunch of leaves may be removed for use. Although parsley leaves are used most commonly in the fresh green condition as a garnish, their characteristic flavour and green colour can be retained if the leaves

Uses of Parsley

Different Recipes of Parsley in Indian Cuisine.

are dried rapidly. Dehydrated parsley flakes are produced from parsley grown in commercial fields.

Parsley has anti-inflammatory, antimicrobial, diuretic and hypoglycaemic properties due to its content of essential oil and phenolic compounds. Apiol, a phenylpropene is responsible for the abortifacient properties of parsley, and the herb may be used to treat menstrual disorders. Bruised leaves have been used to treat tumours, insect bites, lice, skin parasites, and contusions. At one time, parsley tea was used to treat dysentery and gallstones. Other reported traditional uses include treatment of diseases of the prostate, liver, and spleen. Historically, parsley also has been used in the treatment of anaemia, arthritis, and cancers, as an expectorant, antimicrobial, aphrodisiac, hypotensive, diuretic, and laxative. It also has been used as a scalp lotion to stimulate hair growth. However, there have been no clinical trials to confirm these uses.

3.0 Origin and History

The plant is of European (probably West Mediterranean) origin. Commonly, two different varieties are grown *viz.*, root parsley (var. *tuberosum*) which has a tender, edible root (used as aromatic vegetable), and leaf parsley, solely cultivated for its leaves for use as a garnish (Katzer, 2002). According to Linnaeus, wild habitat of parsley is Sardinia, from where it was brought to England and apparently first cultivated there in 1548 while Bentham considered it a native of the Eastern Mediterranean regions and De Candolle reported Turkey, Algeria and the Lebanon to be its home. Since its introduction into these islands in the sixteenth century it has been completely naturalized in various parts of England and Scotland. The Greeks held parsley in high esteem, crowning the victors with chaplets of parsley at the Isthmian games, and making with its wreaths for adorning the tombs of their dead.

Parsley is believed to be native to southern Europe, but it is now found throughout the world. It has been grown in Britain since at least the 16th century. Parsley has now become naturalized throughout Europe, North America, the West Indies, Algeria and Lebanon (Daradkeh and Essa, 2016). Being a native plant of Mediterranean region and is symbolic of victory and dedicated to Zeus. Parsley also was associated with the goddess Persephone, who was kidnapped by Hades but allowed leave the underworld to return to the earth's surface in spring and summer. Parsley was woven into crowns and worn at dinner by ancient Greeks to increase gaiety and to increase appetite. In antiquity, Greeks strewed their tombs with parsley, since it was considered by the ancients to be associated with oblivion and death. The Romans associated parsley with the god Mercury and also planted it on graves. Through the Medieval period, chewing parsley thought to relieve intestinal gas, cough, and effects of snakebite. Crushed parsley seeds were thought able to remove freckles, and to get rid of head and body lice. During the Dark Ages, Christians in Europe were forbidden to transplant parsley in the belief that doing so invited death and crop failure. Parsley is mentioned in Greek historical records as being used for cheering sportsmen by wearing crowns made of parsley; wreaths made from parsley were also used to adorn graves. Parsley was also used in Roman rituals. There are reports of it being sprinkled over dead bodies to remove the smell too. Parsley is used in the Jewish celebration of the Passover as well. It is mentioned as one of the plants in the gardens of Charlemagne and Catherine de Medici, and there is a rumour that parsley was popularized in France by Catherine de Medici (Daradkeh and Essa, 2016).

4.0 Botany, Cytogenetics, Types and Cultivars

4.1.0 Botany

The leaves are compound, alternately arranged, and are divided into two to three leaflets and the plant can grow to over 1 m tall, and as an annual (in tropical regions) or a biennial crop (in temperate areas). The typical flowering period is in the warmer months and the ideal temperature for pollination and seed production is 29–30°C (Teuscher, 2005). The roots are a faint yellow colour and carrot shaped. They can grow up to 20 cm in length and 5 cm in width. Hamburg root parsley has larger roots and is commonly used in European cuisine (Teuscher, 2005).

Parsley is mainly a biennial crop plant. In green house or warmer regions, it can be grown as semi-perennial. There are annual accessions also reported in the gene pool, which need not required vernalization for flowering. Accessions with biennial nature have vegetative growth first with rosette of leaves. The length of the tripinnate leaves with numerous leaflets varies in a wide range inside the species from around 8 to 25 cm. Generally, the leaves of var. crispum are shorter while var. *neapolitanum* has longer leaf.

After the exposure of first winter, the flowering stalk in biennial parsley will emerged immediately. The leaves of the flower stalk are smaller with simplified pinnation. Flowering plants may reach a height of 60–120 cm. At the end of the stalk umbels of different ranks are formed. Inside the umbels, a time gradient exists with the oldest partial inflorescences at the external circle and the younger unripe flowers in the internal circles. Such a gradient exists also inside the partial inflorescences, which group flowers arising from the same point. Each flower has five petals. The colour of petals varies between accessions from green to white. Protandry is common in the family Apiaceae and occurs in parsley as well. The five anthers start anthesis around 5 days before the two stigmata are receptive to pollen. At the time of receptiveness for pollination, the receptacle nectaries produce a sugar-rich liquid for rewarding pollinating insects. At this time, the flower gets shiny and the stigmatic papillae reach the length of flower radius. In a flower two carpels are fused to a bicarpellate pistil from which the schizocarpic fruits develop. The mature fruits can break into two mericarps of one single fruit (Marthe, 2020).

4.2.0 Cytogenetics

The somatic chromosome number of parsley is 2n = 2x = 22, while *P. segetum* reported to have somatic chromosome number 2n = 18 (Darlington and Wylie, 1955). Bennett and Leitch (1995) conducted flow cytometry analysis for analysing ploidy level of 50 parsley accessions obtained from the federal *ex situ* collection of agricultural and horticultural plants at the Leibniz Institute of Plant Genetics and Crop Plant Research at Gatersleben, Germany (IPK) and they detected diploid status

(2n = 2x = 22) for all the accessions and also measured the DNA content of diploid *Petroselinum* crispum as 2C = 4.0 pg.

4.3.0 Types and Cultivars

There are three cultivated varieties, namely, var. *latifolium* (broad-leaved), var. *crispum* (curly-leaved) both grown for their leaves, and var. *tuberosum* grown for its root (Katzer, 2002).

There are three botanical varieties of parsley grown in the United States which are as follows:

1. *P. crispum var crispum* or curly-leaf parsley: The leaves are bright green and finely cut with toothed-leaf edge. This is the most common variety that is used as a garnish.
2. *P. crispum var tuberosum*: It is known as hamburg or turnip-rooted parsley. This parsley has flat leaves, but is grown for its large, edible root which is prepared as a vegetable.
3. *P. crispum var neapolitanum* or Italian flat-leaved parsley: The leaves are dark green, flat, and less finely cut. This variety has a stronger flavour than curly-leaf parsley. This type of parsley is commonly used as a flavoring in sauces, soups and stews.

There are also two con-varieties reported in parsley *viz.*, con-var. *radicosum* is characterized with swollen taproots which are eatable and con-var *crispum* where the roots are not eaten. In con-var. *radicosum* leavesare not crispy; they belong to var. *tuberosum* (Bernh.). Non-crispy accessionsof var. *erfurtense* were available from gene banks. In parsley roots of con-var.*radicosum* are generally used as vegetable. On the other hand, in con-var. *crispum* includes various leaf types genotypes. The accessions with crispy leaves categorized in to var. *crispum* and with non-crispy leaves to var. *Vulgare* along with var. *neapolitanum* also known as Neapolitan or Italian parsley with fleshy and longer petioles (Marthe, 2020).

The cultivated cultivars are classified as the common plain-leaved, the curled-leaved, the Hamburg or broad-leaved and the celery-leaved. Of the variety *crispum*, or curled-leaved, there are no less than thirty-seven variations; the most valuable are those of a compact habit with close, perfectly curled leaves. The common variety bears close leaves, but is of a somewhat hardier nature than those of which the leaves are curled; the latter are, however, superior in every way. Hamburg parsley also known as German parsley, is available in specialty and ethnic markets. Hamburg parsley is used not for its leaves but for the turnip like root, which reaches lengths of 8 to 10 inches. It is roasted, fried or chopped up and incorporated into soup or stew.

The Hamburg, or turnip-rooted parsley, is grown only for the sake of its enlarged fleshy tap-root. This cultivar called the large rooted parsley, develops both a parsnip-like as well as a turnip-shaped form. Neapolitan, or celery-leaved parsley is grown for the use of its leaf stalks, which are blanched and eaten like those of celery. The plain-leaved parsley is not now much cultivated, the leaves being less attractive than those of the curled.

Botanical Varieties of Parley

Curly-Leaf (Left), Italian Flat-Leaf (Middle), and Turnip-Rooted (Right) Parsley.

Franz *et al.* (1982) evaluated parsley for industrial processing using smooth-leaved and 6 curly parsley cultivars grown on peaty soil and sandy loam. The smooth-leaved cultivars gave a higher average fresh weight yield than the curly cultivars. Vitamin C content of the smooth-leaved cultivars was higher than that of the curly cultivars, but the leaf essential oil content was lower. Avall *et al.* (1977) developed the cultivars Elegans and Kruse by selection from Extra Mosskrusig II [Extra Moss-curled II] and Paramount respectively. Both cultivars have a low spreading habit and are suitable both for forcing and for growing outdoors.

The cultivars Moss Curled Hilmar and Moss Curled Krausa are good for fresh market producing the highest yields (10.2 kg/m^2). Moss Curled No. 8 gave the highest early yield (2.3 kg/m^2) but growth was not so good and foliage not as dark as with some other cultivars (Muller, 1982). The most promising cultivars recommended for cultivation and as breeding material in Russia are Persil Commun [Common parsley], Listovaya Kudryavaya [Leafy Curly], Paramount and Stupicka Polodlouha [Stupice Half-long] (Archakova, 1978). Knyazeva (1975) found the best for yield and root quality in the cultivars K384 and K394 in Russia.

The most aromatic cultivar is Bardowicka followed by Halblange Berliner, Hanacka, Berlinska and Olomoucka Dlouha. Olomoucka Dlouha was noted for the best flavour followed by Halblange Berliner and Bardowicka. A correlation has been noted between the aromatic properties and root shape. Cultivars with long, thin and evenly tapering roots had the highest aromatic rating. No correlation was noted between the essential oil content and root aromatic rating (Ejsmond, 1979). Hamburger Schnitt [Hamburg Cutting] and Einfache Schnitt [Plain Cutting] have a relatively high percentage of apiol in the fruits and the fruit oil in the former contained 22 per cent 2,3,4,5-tetramethoxyallylbenzene. The turnip-rooted cultivar Halblange [Half-long] had the lowest ratios of myristicin to apiol (Franz and Glasl, 1976). Parsley cultivars belonging to *vulgare* group had the highest content of sugar, crude protein and carotene, and those of *radicosum* the highest content of ascorbic acid. Of var. *radicosum* the best were Kornevaya Sakharnaya [Sugar Root], Berlin and Hamburg, and of var. *vulgare*, Listovaya [Leafy], Obyknovennaya Dushistaya [Common Fragrant] and Aromatic Cutting (Kulikova, 1975).

Important Cultivars of Parsley

Forest Green Moss Curled

The highest content of ascorbic acid was found in Listovaya Kudryavaya, Grovera [Grover's Leafy Curled] and Turnip-rooted Early Short, Sugar. The highest content of trace elements was found in Listovaya Kudryavaya, Grovera, Kornevaya Sakharnaya [Sugar Root] and Lange Dicke Spate [Long Thick Late] (Zelenin, 1974).

Bubarova (1973) found that the essential oil of the celery-parsley hybrid, named Festival-68 was similar to that of parsley. The principal constituents in the hybrid essential oil were gamma-terpene and heptanol, in parsley myrcene and in celery myrcene and limonene. There are variations in essential oil content among the cultivars. Franz and Glasl (1976) found that Hamburger Schmitt [Hamburg Cutting] and Enface Schmitt [Plain Cutting] have a relatively high percentage of apiol in the fruits and the fruit oil in the former contained 22 per cent 2,3,4,5-tetramethoxyallylbenzene. The turnip-rooted cultivar Challenge [Half-long] had the lowest ratios of myristicin to apiol. Ejsmond (1979) reported that the most aromatic cv. was Bardowicka followed by Halblange Berliner, Hanacka, Berlinska and Olomoucka Dlouha. The cv. Olomoucka Dlouha was noted for the best flavour followed by Halblange Berliner and Bardo-wicka. A correlation was noted between the aromatic properties and root shape; cultivars with long, thin and evenly tapering roots had the highest aromatic rating. Simon and Quinn (1988) detected 2 per cent or less thymol content in 7 accessions, and this is claimed to be the first report of this compound in parsley leaf oil.

Banquet, Dark Moss Colored, Decorator, Deep Green, Forest Green, Improved Market Gardener, Moss Curled, and Sherwood are commonly grown parsley cultivars in USA.

5.0 Soil and Climate

5.1.0 Soil

Parsley requires an ordinary, good well-worked soil, but a moist one and partially-shaded position is best. Soil type did not affect yields but leaf colour was better on peat soil (Franz *et al.*, 1982). Highest overall yields and best soil physical condition were obtained by medium-deep ploughing. Soil surface compaction had a negative effect on yield (Kesik, 1994). In sunny areas with suitable environmental conditions in a humid soil with a pH of 5.3-7.3, parsley may grow up to 60-120 cm tall (Navazio, 2012).

5.2.0 Climate

It is a cool season crop that grows optimally at temperatures range of 7-16 °C, while plant growth is slows noticeably when temperature is below 7 °C. The germination of parsley seeds is slow and can reach up to 26 days under low temperature. Best germination obtained at 12-15°C (Marthe, 2020). Parsley leaves tolerate temperatures down to -6 to -8°C. Roots will survive the winter with temperatures down to -10°C. However, alternating frosts can lead to complete loss by rot. For seed production in the second year, it can be necessary to uproot the root ends of growing season and to replant them in spring.

Low temperature causes severe damage to parsley in outdoor condition (Brun *et al.*, 1985). Hence for parsley grown in the open to be overwintered in cold areas it is recommended that to prevent bolting sowing should not be too early, with late July and early August being suitable dates (Fu, 1991). A short exposure to root zone temperature (RZT) above 30°C retards growth. A relatively low daily average RZT did not compensate for the damage caused by a short daily high temperature exposure. Optimum temperature for curled parsley seemed to be about 21°C (Eidsten and Gislerod, 1986).

Growing conditions change when moving from South to North, being faster in northern latitudes than in Central Europe. The quality characteristics were influenced by changing climatic conditions during the years and by changes of local climate (Hardh, 1975). The quantity of vitamin C was increased by temperature changes, especially by low night temperature. Thus, the ascorbic acid contents were highest most frequently in the North (Hardh, 1975). However, Drazic and Jevdovic (1997) found that weather conditions had different and usually weak effects on the traits investigated.

Moore *et al.* (1997) examined the possible role of leaf cytosolic hexoses and the expression of mannitol metabolism as mechanisms that may affect the repression of photosynthetic capacity when plants are grown at 1000 versus 380 µl/l CO_2.

Parsley is sensitive to water stress, especially if it is planted in the summer and at the end of spring, and to increase production and improve quality, a permanent source of water should be provided. Both growth stage and parsley type determine the susceptibility of the plants to water stress (Najla *et al.*, 2012).

6.0 Cultivation

6.1.0 Seed and Sowing

6.1.1 Seed Treatment

Parsley seeds are slow to germinate and take up to a month. Hence seed treatment is usually followed. Seed priming in 0.5 MPa polyethylene glycol (PEG) enhances the early sown stands by 78 per cent (Rabin *et al.*, 1988). This priming effect on seedling establishment for the earliest planting led to a 67 per cent increase in early yield (96 days from sowing) and a 28 per cent yield increase over untreated seeds at a later harvest (Rabin *et al.*, 1988). Osmotic pre-treatment with PEG 6000 solution at –12 bars water potential for 3 weeks at 15°C under aerobic conditions reduced the radicle protrusion time to 1 or 2 days for seeds which had been surface-dried after treatment, to 2 or 3 days for seeds air-dried and stored for 1 week, and to 3 or 4 days for seeds given 10 weeks' dry storage (Ely and Heydecker, 1981). Hassell and Kretchman (1987) suggested soaking the seeds in aerated water for 72 h to remove an inhibitor in the seed coat and seed priming with PEG 6000 for increased yields. Pill (1986) recommended osmoconditioning in –12 MPa PEG for 3 weeks at 15°C for early emergence and higher seedling shoot fresh weight 24 days after planting than untreated seeds. Further improvement in earliness of emergence was achieved by fluid-drilling non-germinated imbibed seeds in hydroxyethyl cellulose gel.

Parsley seeds irradiated with UV light for 10 or 20 minutes/day following sowing began to germinate after 10 days, compared to non-irradiated seeds which germinated after 15 days. However, non-irradiated control plants were taller and healthier than irradiated plants (Barcelo *et al.*, 1981). Extraction of germination inhibitors with water markedly accelerated germination and improved germination percentage (Hovadik and Kratochvilova, 1977).

Six carriers possessing suitable physical properties as fluid carriers for pre-germinated seed were assessed for their effects on seedling emergence by Darby (1980). Synthetic clay or polyacrylate increased seedling emergence by 10 per cent compared with the sodium alginate control. Seedling emergence was least affected by moisture stress when synthetic clay was used as the carrier.

Parsley seeds are generally given a pre-treatment before sowing by soaking the seeds for 18 hours in 20 ± 2 °C warm water and subsequent re-drying at 30 °C maximum temperature is useful. Seed dormancy can be broken by stratifying in a refrigerator with around 5 °C or briefly freezing.

6.1.2 Sowing

The seeds are sown in drills, or broadcasted. For a continuous supply, three sowings can be done in early February, in April or early May, and in July and early August, the last being for the winter supply, in a protected cultivation. It can be sown in February for the summer crop and for drying purposes, but the seeds take several weeks to germinate, often up to a month. The main sowing is generally done in April as it germinates quickly and available for cutting throughout the summer. A mid-August sowing will furnish good plants for placing in the cold frames for winter use (www.botanical.com/botanical/mgmh/p/parsle 09.html).

An even broadcast sowing is preferable, if the ground is levelled. The seed should be drilled and slightly covered, not more than 1/2 inch deep and thinly distributed. If seeds are drilled, these should be 30 cm apart. It can also be raised as seedlings and when the seedlings are about 2-3 cm tall, they have to be thinned. About 20 cm from plant to plant is ideal spacing.

Different Stages of Field Growing and Harvesting of Parsley

Better field emergence and higher yields are determined by soil moisture supply at a given seed position (Pantielev, 1987). Covering seedbeds with transparent polyethylene sheeting advances germination by 2-4 days (Henriksen, 1985).

The most suitable time for sowing was late April to early May and the highest yield (14.4 t/ha) was obtained at a spacing of 30 cm (Vukasinovic and Bogdanovic, 1984). Depending on the cultivar cumulative crop yield was 2-5 times heavier from the August sown crop than the October sown ones. Yield increases proportionally with the density, the highest being from 15 x 15 cm (45/m^2) (Chrimes, 1984). Leaf surface formation in parsley is influenced by cultivar, cutting frequency and particularly sowing date. Best leaf surface formation was obtained with sowing in late autumn or early spring (Ovcharuk, 1997). For seed production, parsley should be grown in late April or early May. Sowing later than this often resulted in poor germination (because the surface soil dried out), poor overwintering and a reduced yield of seed (Nordest-gaard, 1984). The time from sowing to 50 per cent radicle protrusion of parsley seeds is about 1 month at 5°C and 1 week at 15°C (Ely and Heydecker, 1981).

Fu (1991) recommended sowing in late July and early August for parsley grown in the open to be overwintered in cold areas to prevent bolting. Jedras (1991) also recommended the optimum sowing date as 20 July. The number of plants per unit area affected the seed yield more than the date of sowing. Sowing date had no significant effects on 1000-seed weight or seed germination, although there was a tendency to reduce germination of seeds from the August sown crop (Jedras, 1989). The best storage of root parsley was obtained with roots of plants sown in April and of the 3 cultivars, Cukrowa stored best (Bakowski *et al.,* 1994).

The effect of soil preparation technique before sowing of seeds (classical method or in ridges using the Dutch method) on the growth and yield of parsley was investigated by Polak *et al.* (1999). Yield was 13.1 t/ha for the classical method of sowing and 23.3 t/ha for the ridge method. Blazewicz Wozniak (1998a) studied the effect of 3 methods of pre-sowing cultivation (ploughing, rotary tillage, cultivator tillage), 2 sowing dates, 2 sowing depths and 5 kinds of mulch (sand, peat, sawdust, bark, plastic film) on the growth and yield of Hamburg parsley cv. Berlinska. The longest parsley roots were obtained with spring ploughing. The roots had the greatest diameter and average weight but shortest after sowing at a depth of 2 cm. Date of sowing had no significant influence on the shape of roots or on the content of dry matter. Mulching with peat, saw dust, bark and plastic film decreased diameter and average root weight (Blazewicz Wozniak, 1998b).

The highest yield of leaves in root parsley was obtained with spring rotary cultivation, sowing on 23 April at a depth of 1 cm and mulching with peat, sawdust, bark or plastic film (Blazewicz Wozniak, 1997a; 1997b). Plant density was also highest in these treatments. Parsley leaf yield was positively correlated with plant density. There was a negative correlation between plant density and number, length and weight of leaves (Blazewicz Wozniak, 1997a).

6.2.0 Manuring and Fertilization

Zeolites and green house substrates have been used for the production of salad crops. Postnikov *et al.* (1996) tried two types of zeolite *viz.*, Tedzarni (Georgian

Republic) and Pegasski (Kemerovo, Siberia) for parsley. Yield of parsley was 10-30 per cent greater on zeolite than on organic soil. The nitrate content in the produce was 40-50 per cent lower in the crops grown on zeolite, and the contents of vitamin C [ascorbic acid], beta-carotene and Fe increased. Losses of nutrients from substrates and fertilizers were significantly reduced when zeolite was used.

6.2.1 Organic Fertilizers

Franken and Gnadinger (1994) studied the molecular aspects of the symbiosis between plants and arbuscular endomycorrhizal fungi in parsley cv. Hamburger Schnitt. Phosphate nutrition and low light conditions negatively influenced plant-fungal interactions in different ways. Palada *et al.* (2000) reported that without chemical fertilizers, legume green manure crops, particularly sunnhemp and hyacinth bean, increased the yield of culinary herbs like parsley in a crop rotation system.

Peyvast *et al.* (2008) applied vermicomposted cattle manure, in combinations with soil [0: 100, 10: 100, 20: 100, and 30: 100 (v/v)] and determined its effects on growth, yield and chemical characteristics of parsley in a green house. Results indicated that addition of vermicompost to soil increased plant height and numbers of leaves. Total yields and amounts of potassium, phosphorus, and total soluble solids were greater for parsley plants grown in the 10: 100 combinations.

Mylavarapu and Zinati (2009) observed that sequential applications of compost with and without fertilizer increased the fresh weight of parsley, nutrient uptake and soil nutrient concentrations, soil water retention and reduced soil bulk density.

Toaima-Wael *et al.* (2014) concluded that the parsley plants should be fertilized with compost manures before sowing during soil preparation at the rate of 15 m^3/feddan for producing highest herb yield with the highest quality specifications.

Pokhrel *et al.* (2017) observed that fertigation with chicken manure extract resulted in a higher biomass yield than with lupin sap. Plants receiving organic fertigation all had fewer and smaller leaves and lower fresh and dry weights aboveground than plants receiving inorganic fertigation. This was not due to reduced net photosynthetic rate, but all organic fertigation strategies resulted in high pH, high Na and Cl concentrations, high NH_4 to NO_3 ratios and nutrient imbalances in the root zone. Furthermore, lupin sap fertigation resulted in a high EC and very low P and micronutrient concentrations. These components may have been either individually or collectively responsible for the reduced plant growth and lower biomass yield compared with inorganic treatments.

Massoud *et al.* (2019) studied the effect of three FYM doses (2 t/feddan, 4 t/feddan and 6 t/feddan), four biofertilizers treatments (zero, yeast extract, biofertilizers mix, Super mix + bio + yeast) and Super mix foliar application compared to recommended NPK in parsley. The results indicated that vegetative growth, seed yield and essential oil per cent, total carbohydrates, vitamin C, total chlorophyll and NPK contents were increased with increasing FYM rate to reach the maximum values in 6t/feddan. On the other hand, biofertilizers treatments improved vegetative growth, seed yield and essential oil per cent with the superiority of Super mix + bio + yeast treatment. It could be recommended that 6 t/feddan FYM and Super mix + bio + yeast or yeast biofertilization treatments can significantly increase the growth rate, seed yield and essential oil percentage.

6.2.2 NPK

Being a salad crop, nitrogen plays an important role in parsley nutrition. The greatest increase in nitrate content of the harvested crop was recorded with 400 kg N/ha with little change in crop dry matter content due to N application (Hansen, 1978). A positive relationship was observed between the rate of ammonium nitrate application and nitrate accumulation, especially in parsley (Malakouti *et al.*, 1998). The largest nitrate accumulation in stems and leaves was observed with 400 kg N/ ha in samples collected at evening time (Malakouti *et al.*, 1998, 1999). Csizinszky (1999) evaluated the yield potential of parsley in multiple harvests using full-bed polyethylene mulch, seepage irrigation system with different N and K rates. Fresh weight yields in the first 2 harvests were highest with 49 or 98 kg N/ha and 40 or 80 kg N/ha. N should not be applied in the autumn as it is detrimental to over-wintering. Increased N in spring(up to 180 kg/ha) enhanced seed yield without reducing germination (Nordestgaard, 1984). The yields of marketable and undesirable forked roots of turnip rooted parsley were not affected by the N application rate. The NO_3-N content of fully grown leaves ranged from 195 to 580 mg/ka FW and that of roots from 154 to 290 mg/kg FW. Splitting the N application between pre-plant and side-dressing reduced root NO_3-N content compared with application of the same total amount as a side dressing only (Rumpel and Kaniszewski, 1994).

Irrigation generally increases the effectiveness of mineral fertilizers, even at high rates. Fertilization, irrigation and their interactions had greater effects on marketable yield than on total yield (Borna, 1976). Alexandrescu *et al.* (2000) found that nitrate content increased with: (a) increasing soil N content; (b) when plants were stressed (shaded); and (c) with application of foliar fertilizers. However, nitrate contents decreased in treatments with combined foliar and soil applications of fertilizers. Fertilizer application has been reported to influence the quality. The highest seed essential oil content (2.42 per cent) compared with 1.73 per cent (in control) was obtained with NPK + 2000 ppm Folifertile. The composition of the essential oil varied with treatment (Kandeel, 1991). Studies on the application of NPK @ 45 g N + 100 g P_2O_5 + 55 kg K_2O/m^3 or twice or three times this rate of NPK indicated that N and P rates, but not K rate, had a significant effect on total chlorophyll content, which increased as the rate increased. Total sugar and ascorbic acid contents did change in response to N, P or K application, but only in the case of P was there a clear increasing trend in both as application rate increased. Ascorbic acid content increased in response to increasing K rate. Applying N, P or K increased the activity of peroxidase and catalase particularly during the early phases of growth (Gurgul *et al.*, 1996).

Kolota (2011) found that single preplant nitrogen application was preferable for leaf growth and yield only in the first term of harvest which contained lower amounts of vitamin C and showed higher tendency for total N and nitrates accumulation compared with split doses. It was also reported that the most favourable method of leafy parsley nutrition was the application of 80 kg N/ha before seedlings transplanting and the same amount in two top dressing doses used directly after subsequent harvests.

Souri *et al.* (2018) reported that plant growth of parsley was increased by low compact urea pellet during 3-4 months growing period. It seems that the release rate of N in high compact urea pellet was not significant for plant nutrition; so under this treatment, despite application of 300 mg N/kg soil, plant growth did not improve. It is worth mentioning that pellet treatments resulted in the least nitrate content of leaves in this crop, whereas leaf N concentration in low compact urea pellet was highest among treatments. Therefore, for production of parsley in short periods, pellets with low compactness are preferred for optimum plant growth.

6.2.3 Micronutrients

Both copper and zinc ions have an inhibitory effect on the growth of parsley when applied in solution to the soil, copper being more toxic (Salim *et al.*, 1995a). The copper content was greatest in the leaves in comparison with the other parts of treated plants. Only very small portions of the added copper were translocated from soil into plants (Salim *et al.*, 1995b). Low levels of nickel fertilization, particularly 50 mg/kg clay soil, improved not only parsley leaf yield and quality, but also increased their safety for human consumption, as their nitrate and ammonium contents were significantly reduced (Atta Aly, 1999).

6.2.4 Contaminants

Many metallic contaminants like cadmium, heavy metals and radioactive compounds affect salad crops including parsley. Salim *et al.* (1995c) studied the effect of root-treatment of parsley plants with lead and cadmium in Palestine, where sewage water contaminated by heavy metals was used for irrigation. Both metal ions inhibited growth of plants treated with cadmium showing more symptoms of toxicity than those treated with lead. Cadmium was more concentrated in the edible parts of the treated plants whereas lead was more concentrated in the edible parts of cauliflower and spinach plants only. Metal ion concentration and total metal content of treated plants increased with the increase in concentration of cadmium or lead ions in solutions used for treatment. Assche and Mey (1977) studied the deleterious effects of heavy metals (Zn, Cu and Pb) on soil fungi and parsley. The addition cation exchangers consisting of polystyrene porous resins charged with Ca or Mn in granulated or dust forms to the soil restored normal plant growth and inhibited the uptake of heavy metal ions. Chibowski (2000) found potassium 40K isotope as the most prevalent element in the examined samples of parsley.

6.3.0 Irrigation

The best growth of parsley was obtained when the plants were under constant water flow from 3 weeks after sowing. In hot weather plant growth was best in pots in which the roots could penetrate through side apertures (Eidsten and Gislerod, 1986). As the plants are liable to become burnt up in very hot and dry summers, the rows should be liberally watered in dry weather and a sheltered position is preferred.

Ortmann mat irrigation is a simple system, in which a woven glass fibre mat on a table takes up water out of troughs. Water supply is controlled by the height difference between trough and table (Mayer and Wilhelm, 1999). Yields and efficiency of irrigation were determined mainly by the date of irrigation and the quality of irrigation water supplied (Varga *et al.*, 1996). Gyuro (1983) observed a significant negative correlation between root branching and rainfall during the period from sowing to 15 June in the semi-long rooted types and a highly significant negative correlation between branching and rainfall from sowing to 20 June in the long-rooted types. To prevent branching, selection and supplementary irrigation were recommended.

In trials with the parsley cv. Zaharat the maximum seed yield (1292.5 kg/ ha) was obtained from plants irrigated to maintain soil water capacity at 50 per cent before flowering and at 70 per cent after flowering (Gherman, 1979). Varietal differences in response to irrigation were also recorded by Tuza and Bogdan (1982).

Kaniszewski and Dysko (2008) carried out a research work on the influence of surface and subsurface drip irrigation on the yield and quality of roots of parsley grown on ridges and on flat ground. Irrigation water was supplied via drip lines, which in subsurface irrigation were placed at a depth of 50 mm below the surface of the ridges, along the centreline between two rows of plants. They found that both surface and subsurface irrigation used in the cultivation on ridges and on flat ground had a significant effect on the marketable yield of parsley roots. However, no significant differences in the yield between surface and sub-surface drip irrigation were found.

Gerami *et al.* (2016) reported that that increase of irrigation intervals had negative effect on morphological traits while there was no significant effect on essential oil content and yield.

6.3.1 Contamination of Water

Parsley is often eaten raw or with rich dressings which may threaten the public health, without clear outbreaks. Hence it is necessary to treat waste water effluents to a level where no residual contaminants are detected on irrigated crops (Armon *et al.*, 1994). Hence quality of irrigation is very important to prevent contamination and epidemics. The irrigation of vegetables with waste water has been practised in many countries. This water is contaminated with different serogroups of *Salmonella* B and C. These serogroups were detected on vegetables irrigated with the waste water effluents. However, in parsley *Salmonella* does not persist beyond 3 days after irrigation (Melloul *et al.*, 2001). Besides, such waste water contains high amounts of trace elements and heavy metals. Many of these are non-essential and are toxic to plants, animals and human beings (Kanwar and Sandha, 2000).

6.4.0 Interculture and Weed Control

Parsley grows easily and should be trimmed often otherwise the stalks will grow long with very few springs of leaves at the top. The beds should be renewed every two years, as the plant dies down at the end of the second season. When sowing parsley to stand the winter, a plain-leaved variety will often be found superior to the curled or mossy sorts, which succumb to frost.

Thermal weed control, which relies on heating plants until the cells burst (at 70-80°C), is particularly valuable for pre-emergence weed control because parsley is slow to germinate (Morelle and Thomas, 1994). Taupier Letage *et al.* (1994) suggested technique to encourage weed seed germination before sowing the crop and then one or two bare flame treatments before crop emergence for effective control of weeds. Timing of the thermal treatments was crucial.

Fenix (aclonifen), a novel herbicide was reported to be useful for parsley (Brendstrup and Kloster, 1998). Uptake takes place via cotyledons and seed leaves. On clay and sandy soil, Fenix can be applied pre- as well as post-emergence of weeds, while on organic soils only post-emergence applications are recommended. Fenix provides good control of *Sinapis arvensis, Polygonum lapathifolium, Galium aparine, Polygonum persicaria, Myosotis arvensis, Stellaria media, Chenopodium album, Capsella bursapastoris, Papaver* spp., *Thlaspi arvense, Silene noctiflora, Polygonum aviculare, Viola arvensis, Poa annua, Lapsana communis, Urtica urens, Brassica napus, Lamium purpureum, Polygonum convolvulus* and *Veronica* spp. Fenix has no effect on *Senecio vulgaris, Galeopsis* sp. and *Solanum nigrum* (Brendstrup and Kloster, 1998). Parsley showed almost complete tolerance to aclonifen up to 4.8 kg/ha. Overall weed control was very good at all rates, but *Aethusa cynapium,* some *Solanum* spp. and *Senecio vulgaris* were relatively tolerant (Roberts and Bond, 1984). Aclonifen at 2.4 and 3.6 kg/ha killed all common weeds except *Aethusa cynapium, Solanum nigrum* and *Senecio vulgaris.*

Preplant application of napropamide at 2.2 kg a.i./ha provided weed control for 2 weeks, but was subsequently ineffective against heavy infestations of *Portulaca oleracea*. Hand-hoed and glyphosate-treated plots (both with and without mulch) produced equivalent yields (Ricotta and Masiunas, 1991).

Linuron has been reported to give good control of both grassy and dicotyledonous weeds (Rapparini and Campagna, 1996; Van Vranken and Majek, 1985). Linuron at 1.0 kg a.i./ha was best and no residues were found in the leaves at harvest (Kameneva and Pen'Kov, 1973; Crosbic Cochrane, 1972).

7.0 Harvesting, Yield and Storage

7.1.0 Harvesting and Yield

Parsley leaves are ready for use about 3 months after seeding. A few leaves at a time may be removed from each plant, or the entire bunch of leaves may be removed for use. Although parsley leaves are used most commonly in the fresh green condition, their characteristic flavour and green colour can be retained if the leaves are dehydrated. Therefore, markets for parsley processed in this manner are growing (Anon., 2002).

Yield of parsley depends on cultivars, season and agro-climate. A yield of 14.4 t/ha was obtained at a spacing of 30 cm (Vukasinovic and Bogdanovic, 1984). Yield was 13.1 t/ha for the classical method of sowing and 23.3 t/ha for the ridge method (Polak *et al.*, 1999). Average yield/plant of parsley ranged from 14.9 to 22.1 oz from 4 pickings. Among the curled-leaf cultivars Afro, Forest Green, Sherwood and Unicurl had the highest yields while in plain-leaf type cv. Gigante generally had the highest yields (Hill, 1992). The earlier and more frequent the picking, the higher was the total number of leaves, picking 6 or 7 times from December/January to the end of April gave double the leaves/m^2 of picking 3 times from mid-March to the end of April. Yield/m^2 was lowest in December and increased month by month. The best leaves (based on a single leaf weight of 2.5 g minimum) were picked after mid-February. There were fewer yellow leaves in treatments with earlier picking dates and, the later the first picking, the sooner plants began to bolt (Hartmann and Waldhor, 1979). Irradiation with 1000-1500 rad has been reported to increase the yield of root parsley from 21.4 t/ha to 29.6 t/ha (root crop) and in case of leaf crop from 6.9 t/ha to 9.2 t/ha (Suciu *et al.*, 1981). Godzhonov (1982) found that the average seed yields from plots planted with parsley roots weighing 50, 100, 150 and 200 g were 800, 1060, 1036 and 900 kg/ha, respectively. Yield in the control, where ungraded roots were planted, was 831 kg/ha. Seed quality was also best in plants raised from 100 and 150 g roots.

Essential oil yield and other chemical parameters are influenced by the number of cuts. Chlorophyll contents increased gradually from the 1st to the 3rd cut (El Sherbeny and Hussein, 1993).

7.2.0 Storage and Postharvest Management

Least decay in storage at room temperature occurred when parsley was harvested at 70 days. Storage at 0°C and 84 per cent RH doubled the shelf life compared with storage at room temperature (Wally *et al.*, 1978). For the root parsley irrespective of cultivar, vitamin C content was highest in roots from the May sowing. The best storage was obtained with roots of plants sown in April and vitamin C, DM, sugar and nitrate contents in roots decreased during storage (Bakowski *et al.*, 1994). It was also reported that parsley should keep 2 to 2.5 months at 32oF (0°C) and 95 to 100 per cent relative humidity. High humidity is essential to prevent desiccation. Packaging in perforated poyethylene bags and using top ice are often beneficial. A controlled atmosphere of 10 per cent oxygen and 11 per cent CO_2 can help retain green colour and salability. Root of Hamburg parsley could be stored like carrots or parsnips. With the top removed, root parsely should keep several months at 32°F (0°C) (Anon., 2002).

For medicinal purposes the roots are collected in the second year, in autumn or late summer, when the plant has flowered. Parsley leaves can be dried in the oven on muslin trays till thoroughly dry and crisp, after which the leaves should be rubbed with hands or through a coarse wire sieve and the powder then stored in air and light tight containers to preserve the good colour. The oil is extracted from the 'seeds' or rather fruits, when fresh. Prior to processing, parsley leaves may be kept for >3 days in a non-cooled store and for up to 15 days in a cold store, assuming >50 per cent of the material should maintain its technological quality (Lisiewska *et al.*, 1997).

Food originated from fresh plant may either be minimally processed or not further processed at all prior to consumption. In a test of the storage of fresh parsley, leaves were sealed in polyethylene bags and stored at 4°C for 12 days. Quality characteristics colour, appearance (succulence and firmness) and aroma (odour, taste) were evaluated on a scale of 1to 5 on the 1st, 5th, 8th and 12th days. The scores were used to produce life curves, from which it was estimated that minimally processed parsley could stay fresh for a period of 23 days (Catunescu *et al.*, 2012).

8.0 Diseases and Pests

8.1.0 Diseases

Diseases which are reported to be affecting parsley are *Alternaria radicina, Septoria petroselini, Fusarium oxysporum, Erysiphe heraclei, Sclerotinia sclerotiorum,* alfalfa mosaic alfamovirus, celery mosaic potyvirus and *Enterobacter* sp. (Scholze *et al.*, 1996). Club root (*Plasmodiophora brassicae*) on parsley is a new record in Korea Republic (Hahm *et al.*, 1998). *Botrytis* spp., were associated with seeds of parsley (Moretto *et al.*, 1997). The *in vitro* antagonistic activity of some strains of bacteria and yeasts against 3 pathogenic fungi of seeds [*Fusarium oxysporum, Alternaria brassicicola* and *Stemphylium botryosum* (*Pleospora herbarum*)] indicated that two isolates, bacterium GOL2 and yeast $F_3$3, showing good effectiveness as biological seed dressings on parsley seeds infected by the same pathogenic fungi. Dry treatment with lyophilized preparations gave better results than liquid treatments in terms of effectiveness and persistence. Lyophilized preparations of strain GOL2 gave total protection for up to 6 weeks after treatment (Montuschi and Ricci, 1996). Seed samples of parsley showed the presence of *Alternaria* (the most frequent), *Rhizopus, Penicillium, Aspergillus, Fusarium* and *Trichothecium* (Dohroo and Gupta, 1994).

Fungal Diseases

8.1.1 Leaf Blight

The blight of parsley caused by *Septoria petroselini* is a very serious disease. The extent of damage by *S. petroselini* to parsley varies between varieties and ranged from 37 per cent in Decora to 100 per cent in Parus (Marthe and Scholze, 1997). The symptoms of the disease are small foliar lesions, somewhat angular and greyish brown, with a definite dark cinnamon-brown margin, and often contained dark pycnidia on the upper side of the foliage (Cerkauskas and Uyenaka, 1990). Small, oval, cinnamon-brown lesions on petioles were also occasionally observed (Cerkauskas and Uyenaka, 1990).

Hot water treatment and growing of tolerant cultivars like Grune Perle and the best methods of control (Marthe *et al.*, 1996). The method of hot water treatment is applied to the most important pathogens (*Alternaria* spp., *Phoma* spp., *Septoria* spp., *Peronospora valerianellae, Xanthomonas* spp. Hot water treatment proved effective against the investigated pathogens at 50°C for 30 minutes. Germination was not affected at 50°C but an increase up to 53°C resulted in a similar effect but germination could be reduced. The beneficial effects of the treatment were confirmed in field trials by reduced infestation with *Septoria* spp. (Nega *et al.*, 2001).

8.1.2 Damping Off

Damping off of seedlings results in extensively reduced parsley stands. The fungi most frequently isolated from field-grown seedlings with symptoms were

Fusarium oxysporum, F. solani, Pythium irregulare; P. ultimum and *Rhizoctonia solani* (AG-4, AG-1 and AG-2, type 2) (Hershman *et al.*, 1986). In tests, most isolates of *P. irregulare* and *P. ultimum* were highly pathogenic and caused extensive pre-emergence and post-emergence damping off. Similarly, most isolates of *R. solani* AG-4 and AG-2, type 2, were highly pathogenic. Isolates of AG-1, however, varied considerably in pathogenicity. No other fungi tested caused disease in parsley. In sand bed studies, a combination of *P. irregulare* and *P. ultimum* isolates was highly pathogenic at 15, 23 and 30°C. *R. solani*, however, was much less pathogenic at 15 than at either 23 or 30°C. Methyl bromide treatment of the soil followed by sowing after six days has been found to reduce incidence of diseases. Seed sown in soil treated at 125 g/m^2 emerged much faster than that sown in soil treated at 75 g/m^2. Both rates of treatment improved plant stands and increased yields compared with untreated plots in which 40 per cent of the crop had perished by mid-September. Treatment eliminated infection by *Verticillium dahliae* but 25 per cent of the plants were killed by *Phytophthora cactorum* infection. The treatments gave excellent nematode and weed control. No phytotoxicity was observed at rates up to 170 g/m^2 and plant losses were prevented by rates as low as 50 g/m^2 for 72 h, although with soil from the lower levels (60-80 cm) at least 85 g/m^2 was necessary (Cassini *et al.*, 1971).

Viral Diseases

8.1.3 Carrot Mottley Dwarf Disease

Stunting and rosetting of parsley plants, with the outer leaves turning from green to yellow to red or pink, were attributed to a combination of carrot mottle virus and carrot red leaf virus (Meunier and Verhoyen, 1987). The first virus was transmitted by mechanical inoculation and by the aphid *Cavariella aegopodii* in a persistent manner in the presence of the second virus. Measures to control the virus include the removal of umbelliferous weeds from the vicinity of carrot or parsley fields (especially in the autumn), crop rotation to prevent infected winter parsley or carrots from infecting the new spring seedlings of either crop, avoidance of planting parsley and carrot close together and foliar sprays of insecticide to eliminate aphid vectors.

8.1.4 Other Viral Diseases

Parsley latent virus, a new and prevalent seed-transmitted, but possibly harmless virus was reported by Bos *et al.* (1979). The virus was isolated from 38 of 54 parsley seed samples (17 of 24 cvs.) from 5 European countries. It could easily be detected in seedlings and in seeds germinated on moist filter paper. Strawberry latent virus was also detected in 5 samples. Parsley latent virus was symptomless in parsley and caused latent systemic infection of *Gomphrena globosa* and some spinach cultivars, and weak, often transient systemic symptoms on *Chenopodium amaranticolor, C. giganteum, C. glaucum* and *C. quinoa*. It was transmitted mechanically but not by aphids.

8.2.0 Pests

The most common pests of parsley are aphids, leafminers, thrips and whitefly.

Diseases of Parsley

***Septoria* Leaf Spot (Left), and *Alternaria* Blight (Right) Infection**

Bacterial Leaf Spot (Left), and Basal Soft Rot (Right) Infection

Crown and Root Rot (Left), and *Rhizoctonia* Crown and Collar Rot (Right) Infection

***Pythium* Rot Infection**

Pests of Parsley

Aphid (Left), and Carrot Weevil (Right) Infestation

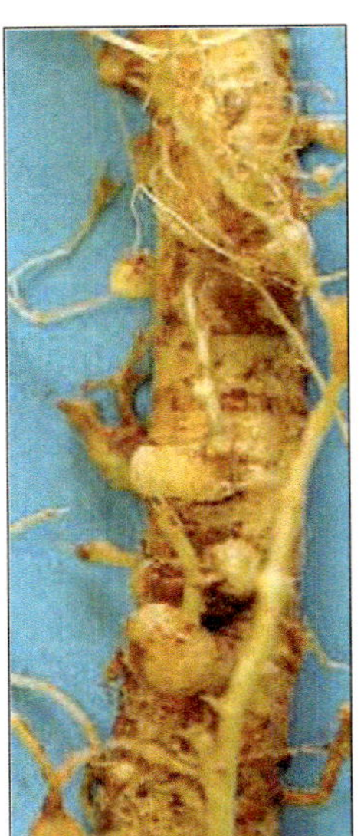

Nematode Infestion

8.2.1 Aphids

The species composition and intensity of aphid migration flights to a parsley fields in Belgium indicated *Myzus persicae, Aphis* sp., *Metopolophium dirhodum, Rhopalosiphum padi* and *Macrosiphum euphorbiae* were the most abundant species (each >5 per cent on the total number of aphids). The willow-carrot aphid (*Cava-riella aegopodii*) is the economically most important aphid species in Belgian parsley due to its ability to transmit a viral disease resembling carrot mottley dwarf, accounted for 2.18 per cent of the total number of aphids (Vercruysse *et al.*, 1999). It overwinters either in the egg stage on *Salix* spp (which do not harbour these viruses) or as apterae on wild Umbelliferae.

Studies on the effect of insect nets of various mesh sizes and treatments with pirimi-carb alone and with lambda-cyhalothrin, and with propamocarb on leaf reddening of parsley indicated that fine meshes (0.5 and 0.6 mm) prevented aphid intrusion and leaf reddening, but large meshes (1.6 mm) allowed more aphids to intrude, resulting in more leaf reddening. However, even large-mesh nets gave sufficient protection from leaf reddening. Compared to pirimicarb, the mixture of pirimicarb and lambda-cyhalothrin reduced the degree of leaf reddening to an economically acceptable level (Belyeart, 1994).

High incidence of mottley dwarf virus on carrot and parsley is closely linked with the numbers of aphids overwintering on carrots left in the soil during the winter; all overwintered crops of carrot and parsley should be removed, ploughed in or sprayed with an insecticide before the spring migration of aphids (Dunn, 1976).

The endoparasite *Aphidius gifuensis* has been used for the biological control of aphids on salad crops (Katyukha *et al.*, 1984). The optimum temperature for rearing the parasite is 20-25°C, at which the adult lifespan is 7-9 days, a generation is completed in 12-16 days and a single female lays 60-80 eggs. Its technical effectiveness is lowest in the winter at 8-12°C on parsley. The optimum parasite: host release rates is 1: 5-1: 10 at 8-12°C. Chemical insecticides are not recommended for use on salad crops because of their potential danger to consumers, and the tendency of recent years has been towards biological control of various sorts. One of the most successful methods of control has been by means of rearing and releasing *Chrysopa carnea* Steph. Tests have indicated that releases of second-instar larvae gave effective control of *Myzus persicae* (Sulz.) on parsley grown in green house conditions throughout the year (Beglyarov and Ushchekov, 1977).

8.2.2 Nematodes

A simulation model was developed to predict changes in the population densities of *Pratylenchus bukowinensis* and subsequent decreases in yield of parsley, the most sensitive host, under various crop rotations. By growing parsley once in four years economic damage was avoided and the nematode population declined. If parsley was grown alternatively once in 4 and then once in two years, the overall population decline was slower increasing the risk of damage to the crop (Viscardi and Brzeski, 1992). The nematode occurred mostly in the top 30 cm of soil and populations were highest 14 days after harvest (Coosemans, 1987). *Meloidogyne incognita* infestation can cause up to 73 per cent mortality under controlled inoculation with the nematodes (Radewald *et al.*, 1972). Numerous galls were found on the roots of inoculated plants, as well as infection by *Rhizoctonia solani*, *Pythium* and *Fusarium* spp. indicating that the fungi were introduced with the nematode inoculation from a field source and that death of some plants was probably due to a nematode/fungal interaction (Radewald *et al.*, 1972). Numeous galls were found on the roots of inoculated plants, as well as infection by *Rhizoctonia solani*, *Pythium* and *Fusarium* spp. indicating that the fungi were introduced with the nematode inoculation from a field source and that death of some plants was probably due to a nematode/fungal interaction (Radewald *et al.*, 1972). Crop rotations with a non-umbelliferous crop preferably cereals is an ideal method to control the nematodes.

9.0 Crop Improvement

Breeding of parsley is carried out with the objectives of improving the essential oil content and resistance to major pests and diseases besides improved yield. For many crops, yield and its contributing parameters are of main interest. This principle is also useful for parsley. Breeding objectives of this group for leaf production are size of yield, amount of marketable yield, essential oil content, fast re-growth after cutting, colour of leaves, crimping of the leaf blade tissue between the secondary veins, relation of leaves to stalk, height and density of plants. Breeding objectives for turnip-rooted parsley are yield of taproots, length and diameter of the taproot and a smooth taproot (Marthe, 2020).

9.1.0 Hybridization

Intergeneric hybridization between celery (*Apium graveolens* L.) and parsley has been reported by many workers (Bubarova and Boubarova, 1980a, b). The hybrids between celery and parsley have been reported to be resistant to *Septoria appicola* (Madzharova *et al.*, 1973; Madjarova *et al.*, 1978; Honma and Lacy, 1980).

Madjarova *et al.* (1979) presented information on essential oil content in F_4 hybrids between *A. graveolens* 'Pioner' [Pioneer] and *Petroselinum hortense* 'Listen' [Leafy] and in the celery parsley hybrid Festival 68. Seven terpenoid components were identified in Pioner, twelve in Listen and thirteen in Festival 68. The hybrids inherited all the terpenoids from celery and heptanol from parsley, while synthesizing new compounds of their own. The content of these new compounds was higher than that of the main celery components limonene and myrcene. Madzharova *et al.* (1978) found that the properties of 4 hybrid cultivars were intermediate between those of celery and parsley. Because of a reduction in the number of vascular bundles and essential oil canals in the stalks, the leaves had a finer structure and greater succulence than those of celery, and could be used as parsley. The total amino acid content of the hybrids exceeded that of the original parents.

A hybrid parsley, cv. Festival 68, was derived from a parsley-celery cross, producing 50-80 per cent more foliage than standard parsley cultivars (Madjarova and Bubarova, 1978). Festival 68 has the morphological characteristics resembling parsley. The leaves have a high content of ascorbic acid, sugars and essential oils. The leaf yields are 50-80 per cent higher than those of commercial cultivars (Madjarova *et al.*, 1974).

Crosses involving species of other genus such as *Daucus carota* has also been reported. Dudits *et al.* (1980) fused X-ray irradiated leaf protoplasts of parsley (2n = 22) with protoplasts isolated from cell culture of a nuclear albino mutant of *D. carota* (2n = 18). Introduction of genes from the irradiated parsley nuclei into the carrot genetic material was shown by the correction of the albino defect and by the appearance of parsley isozymes in selected green tissues and in regenerated plants. The high frequency of cells with 2n = 19 or 2n = 38 and the regeneration of plants with2n = 19 chromosomes indicated that the elimination of parsley chromosomes was incomplete.

9.2.0 Mutation

Very few reports are available on mutation in parsley even though there is a scope for improvement using mutagens. Turkov *et al.* (1973) reported a combination diploid-haploid in parsley. Huang *et al.* (1988) used genome subclones homologous to 6.6 kb of the simple-copy chalcone synthase gene in parsley to probe chromosome spreads. Autoradiographs showed a significant concentration of the labelled probe on the long arm of a sub-metacentric chromosome pair. This was the first report of chromosomal detection and localization of a unique endogenous gene in plants.

9.3.0 Disease Resistance

The production of phenylalanine ammonia lyase (PAL) and 4-coumarate: CoA ligase (4CL) mRNA in parsley is associated with *Phytophthora megasperma* f. sp. *glycinea* infection (Hahlbrock *et al.*, 1987). Somssich *et al.* (1988) characterized three nearly full length cDNAs complementary to mRNAs encoding 2 pathogenesis-related, class 1 (PR1) proteins in parsley by sequence analysis of 5′ and 3′ untranslated regions. Systemic acquired resistance (SAR) is an inducible plant defence state, the induction of which often requires the accumulation of salicylic acid (SA). Development of SAR is associated with both an accumulation of pathogenesis-related (PR) proteins and an enhanced capacity for activating local defence responses upon pathogen attack (Thulke *et al.*, 1999). It has been demonstrated that tissue priming and the resultant potentiation of local defence responses can be studied in a parsley cell culture/fungal elicitor (*Phytophthora megasperma*) model system. Pre-incubation with natural and synthetic inducers of SAR primed parsley culture cells for augmented elicitation of a variety of defence responses, indicating an important role for defence response potentiation in acquired plant disease resistance (Thulke *et al.*, 1999).

Parsley was known to respond to pathogen attack by the synthesis of furanocoumarins and to ultraviolet (UV) irradiation by the synthesis of flavone glycosides, whereas ozone treatment resulted in the induction of both pathways (Eckey Kaltenbach *et al.*, 1994). Ozone treatment (200 nl/l for 10 h) of parsley seedling resulted in an increased mRNA level of early genes within 3 h (pathogenesis related proteins PR1, PR2 and an elicitor-induced protein with unknown function (Eli 16), of intermediate induced genes within 6 h [phenylalanine ammonia-lyase (*pal*)], 4-coumaroyl-CoA ligase (4CL), [naringenin]-chalcone synthase (*chs*), and of late genes within 12 h [hydroxyproline-rich glycoprotein (*hrgp*)], peroxidase (*pod*) (Eckey Kaltenbach *et al.*, 1994). Wu and Hahlbrock (1992) studied the tissue-specific and light-dependent expression of four genes involved in phenylpropanoid metabolism in various parts of parsley seedlings using *in situ* RNA hybridization and immune-histochemical techniques. The genes encode phenylalanine ammonia-lyase (*pal*) and 4-coumarate: CoA ligase (*4cl*) (two enzymes of general phenylpropanoid metabolism), chalcone [naringenin-chalcone] synthase (*chs*) (the key enzyme of flavonoid biosynthesis), and bergaptol O-methyltransferase (*bmt*) (a late enzyme of the furanocoumarin pathway). In young leaves, *pal* and *4cl* genes are expressed predominantly and in a coordinated fashion in epidermal cells, young xylem cells of vascular bundles, and epithelial cells of oil ducts. The gene(s) encoding *bmt* was active exclusively in the epithelial cells of oil ducts, whereas *chs* gene expression was largely confined to the epidermis. In shoot and root apices, *pal*, *4cl*, and *chs* mRNAs are detectable at low levels without distinct patterns. The corresponding proteins, however, accumulate preferentially in the protoderm and pith meristem of the shoot apex and in root-cap cells. A gene encoding pathogenesis-related protein 1 (*pr1*) was analyzed in roots for comparison and was found to be expressed predominantly in the cortical cells of root tips. The expression of all genes investigated was, to a greater or lesser extent, dependent on light conditions and tissue age, with highest levels occurring in newly differentiated, light-exposed tissue. In the leaf epidermis of seedlings illuminated for 20 h, PAL and CHS mRNAs and proteins, as well as 4CL protein, were particularly abundant.

Kirsch *et al.* (2001) have identified the parsley *ELI17* gene as a particularly fast-responding gene possessing a new type of W box-containing, elicitor-responsive promoter element, E17. Highly selective E17-mediated reporter gene expression at pathogen (such as *Phytophthora sojae, Peronospora parasitica* and *Rhizoctonia solani*) infection sites in transgenic *Arabidopsis thaliana* plants demonstrated the potential of this promoter element for designing new strategies in resistance breeding.

9.3.1 Erwinia Resistance

Parsley leaves exhibit incomplete resistance following stomatal inoculation with *Erwinia chrysanthemi*, resulting in the development of local lesions in 58 per cent of inoculated leaves (Gijsegem *et al.*, 1995). Rapid activation of defense-related genes appears to be insufficient, although it may be necessary, to protect parsley from attack by *E. chrysanthemi* (Gijsegem *et al.*, 1995).

9.3.2 Powdery Mildew Resistance

Celery is a promising donor of resistance to powdery mildew (*Erysiphe umbelliferarum* f. sp. *anethi*) while varieties of parsley differed in resistance. Smooth-leaved forms could be differentiated in respect of resistance at the stage of 1-2 true leaves, while those with curly leaves showed no infection at this stage (Khlebnikov *et al.*, 1984).

9.3.3 Septoria Blight Resistance

Marthe *et al.* (1996) evaluated 7 parsley cultivars (Decora, Hilmar, Grune Perle, Clivi, Anja, Verta and Mooskrausse) for resistance to *Septoria petroselini* in the green house. Grune Perle gave the lowest percentage of leaf damage although none of the plants was totally resistant.

9.3.4 Carrot Mottley Virus Resistance

At National Vegetable Research Station, Wellesbourne, U.K., progenies of six selfed plants of the cultivar Claudia with apparent tolerance of carrot mottley dwarf virus were compared with the progeny of a selfed susceptible plant and unselected Claudia plants. In the seven progenies, 14-25 per cent of the plants were symptomless compared with 20 per cent in the Claudia control; 16 out of 40 symptomless progeny plants proved to be infected (Anon., 1976).

10.0 Biotechnology

Biotechnological research on parsley has been mostly on basic aspects using parsley as model system like carrot. Very few studies have been carried out on the use of this modern tool for improvement of parsley. Though basic studies have been carried out on transformation but their direct application to parsley improvement is limited.

10.1.0 Tissue Culture

Vandemoortele *et al.* (1996) developed a micropropagation protocol. They induced axillary shoots (rosettes) on surface sterilized axillary buds cultured on MS medium supplemented with various concentrations of benzyladenine (BA), kinetin or thidia-zuron. Kinetin resulted in a low proliferation rate. The concentrations of thidiazuron or BA that were optimal for shoot proliferation, resulted in shoots with a low rooting capacity. During the rooting treatment, these shoots showed signs of wilting. Exposure to 2.5 µM NAA for 2 or 4 weeks induced rooting. Two proliferation media (5 µM BA + 0.5 µM NAA acid or 5 µM kinetin + 2.5 µM NAA) resulted in only moderate proliferation but produced shoots that were easy to root. When these media were tested by repeated axillary proliferation on the same medium; the medium containing 5 µM BA and 0.5 µM NAA was optimal.

Sajina *et al.* (1997) reported formation of multiple shoots in MS medium supplemen-ted with BAP (1 mg/l) and IBA (0.5 mg/l) on nodal segments and shoot tips. They were also able to store the micro-shoots in 1/2 MS medium supplemented with 20 g sucrose/l by inducing minimal growth for about 4 months.

Vandemoortele *et al.* (1994) established callus cultures of curled-leaved types from petiole segments of plants harvested from the field. These calluses, subcultured on MS medium containing 2,4-D, NAA and BAP regenerated through two morphogenetic pathways. Somatic embryogenesis gave rise to plantlets with a radicle and cotyledons, while the more frequent caulogenesis led to microshoot formation from a single shoot meristem. Rooting occurred *ex vitro* during acclimatization of excised microshoots in perlite irrigated with three times diluted MS solution. Fully grown, fertile plants were obtained. Despite the curled leaved type of the donor plants, about 40 per cent of regenerated plants had plain leaves.

10.2.0 Biochemical and Molecular Markers

Very little work has been done on the use of molecular marker in parsley. Kawalleck *et al.* (1993) reported a group of isolated parsley cDNAs that are transcriptionally activated upon fungal infection or elicitor treatment to encode tyrosine decarboxylase (TyrDC). All four identified parsley *TyrDC* genes were cloned and encoded at least three TyrDC isozymes. Treatment of cultured parsley cells with fungal elicitor caused rapid and transient increases in TyrDC mRNA and enzyme activity levels. The nucleotide sequences are in the GenBank/EMBL data bank under accession numbers M96069, M96070, M96071 and M95685.

Domblides *et al.* (2010) used RAPD and ISSR markers to classify 32 samples of parsley (*Petroselinum crispum* (Mill.) Nym. ex A. W. Hill) including the root and leaf plants. In total, 51 RAPD and 66 ISSR markers were amplified by five RAPD and six ISSR primers. On average, the ISSR and RAPD spectra had 11 and 10.2 bands, respectively. The distance between root and leaf forms of parsley was reported 0.292.

The genetic variation and inter relationship among two different subspecies of parsley the plain leaf type (ssp. *neapolitanum*, Danert) and the curly leaf type (ssp. *crispum*), were analyzed using 5 RAPD and 5 ISSR markers. RAPD markers were more efficient than the ISSR assay regarding polymorphism detection in the examined parsley samples, where the highest polymorphism percentage detected by RAPD was 84.62 per cent, while the highest recorded for the ISSR marker was 58 per cent. The number of amplified fragments of genomic DNA of the 2 subspecies, using RAPD analysis, yielded 76 fragments, 49 of which were polymorphic. The size of specific amplicon ranged from 246 to 5685 bp. The 5 ISSR primers produced 60 bands, out of which 26 were polymorphic and the size of specific amplicons ranged from 157 to 807 bp. The polymorphic amplicons generated by RAPD and ISSR markers utilized could be useful for examining the genetic diversity and phylogenies of different Peroselinum crispum genotypes, which may be applied to identify elite lines and utilized in parsley improvement and genetic relationship assessment (Ibrahim *et al.*, 2017).

11.0 References

Alexandrescu, A., Gavriluta, I., Buzdugan, C., Beldiman, G. and Borlan, Z. (2000) *Proc. Romanian Acad. Series-B, Chemistry, Life Sciences and Geosciences*, **2**: 77-80.

Anonymous (1975) *26th Annual Report, National Vegetable Research Station*, U.K., p. 158.

Anonymous (2000) *Parsley, Commercial Vegetable Production Guides*, Oregon State University.

Archakova, L.I. (1978) *Byulleten' Vsesoyuznogo Ordena Lenina I Ordena Druzhby Narodov Instituta Rastenievodstva Imeni N. I. Vavilova*, **82**: 46-50.

Archakova, L.I. (1985) *Nauchno Tekhnicheskii Byulleten' Sibirskogo Otdeleniya VASKhNIL*, **50**: 59-65.

Armon, R., Dosoretz, C.G., Azov, Y. and Shelef, G. (1994) *Water Sci. Technol.*, **30**: 239-248.

Assche, C. van and Mey, I.W.de. (1977) *Acta Agrobotanica*, **30**: 167-179.

Atta Aly, M.A. (1999) *Scientia Hortic.*, **82**: 9-24.

Avall, H., Wredin, A. and Jonson, I. (1977) *Lantbrukshogskolans-Meddelanden A*, p. 27.

Bakowski, J., Michalik, H. and Umiecka, L. (1994) *Biuletyn Warzywniczy*, **41**: 157-167.

Barcelo, J., Torres, M. and Baztan, J. (1981) *Anales de Edafologia y Agrobiol.*, **40**: 337-345.

Beglyarov, G.A. and Ushchekov, A.T. (1977) *Zashchita Rastenii*, No. 2, pp. 25-27.

Belyeart, P. (1994) *Parasitica*, **50**: 67-73.

Bennett, M.D. and Leitch, I.J. (1995) *Ann. Bot.*, **76**: 113–176.

Beshanov, A.V. and Aspidova, Zh. V. (1979) *Byulleten' Vsesoyuznogo Nauchno Issledovatel'skogo Instituta Zashchity Rastenii*, **46**: 57-61.

Blazewicz Wozniak, M. (1998a) *Annales Universitatis Mariae Curie Sklodowska, Sectio EEE Horticultura*, **6**: 89-102.

Blazewicz Wozniak, M. (1998b) *Annales Universitatis Mariae Curie Sklodowska, Sectio EEE Horticultura*, **6**: 73-87.

Blazewicz Wozniak, M. (1997a) *Annales Universitatis Mariae Curie Sklodowska, Sectio EEE Horticultura*, **5:** 129-140.

Blazewicz Wozniak, M. (1997b) *Annales Universitatis Mariae Curie Sklodowska, Sectio EEE Horticultura*, **5:** 117-127.

Borna, Z. (1976) *Roczniki Akademii Rolniczej w Poznaniu, Ogrodnictwo*, **85**: 5-20.

Bos, L., Huttinga, H. and Maat, D.Z. (1979) *Netherlands J Plant Pathol.*, **85**: 125-136.

Brendstrup, I. and Kloster, I. (1998) *15th Danish plant protection conf. Side effect of pesticides, weeds. DJF-Rapport, Markbrug.*, **2**: 133-140.

Brun, R., Dauple, P. and Mario, R. (1985) *C.R. Seances de l'Academie d'Agriculture de France*, **71**: 267-272.

Bubarova, M. (1973) *Grandinarska I Lozarska Nauka*, **10**: 23-32.

Bubarova, M. and Boubarova, M. (1980a) *Genetika I Selektsiya*, **13**: 263-270.

Bubarova, M. and Boubarova, M. (1980b) *Genetika I Selektsiya*, **13**: 358-367.

Catunescu, G.M., Tofana, M., Muresan, C., David, A. and Stanila, S. (2012). *Bulletin UASVM Agriculture*, **69**: 205–212.

Cassini, R., Cassini, R., Alabouvette, C., Rouxel, F., Bonnel, L., Bremeersch, P. and Vasenave, J.G. (1971) *Phytiatrie Phytopharmacide*, **20**: 215-231.

Charles, D.J. (2004) In: *Handbook of Herbs and Spices*, Volume 2. (Ed. K.V. Peter), Woodhead Publishing, Abington, Cambridge, UK and CRC Press, Boca Raton Florida, pp. 231–234.

Cerkauskas, R.F. and Uyenaka, J. (1990) *Plant Dis.*, **74**: 1037.

Chibowski, S. (2000) *Polish J. Env. Studies*, **9**: 249-253.

Chrimes, J. (1984) *Grower*, **102**: 21-22.

Coosemans, J. (1987) *Mededelingen van de Faculteit Landbouwwetenschappen, Rijksuniversiteit Gent*, **52**: 625-629.

Crosbie Cochrane, H.S. (1972) *Agril. Northern Ireland*, **47**: 250-252.

Csizinszky, A.A. (1999) *J. Herbs, Spices and Medicinal Plants*, **6**: 11-22.

Darby, R.J. (1980) *Expt. Agril.*, **16**: 153-160.

Daradkeh, G. and Essa, M.M. (2016) In: *Leafy Medicinal Herbs: Botany, Chemistry, Postharvest Technology and Uses* (Eds D.C.P. Ambrose)

Darlington, C.D. and Wylie, A.P. (1955) Chromosome atlas of flowering plants, 2nd edn. George Allen and Unwin LTD, London, p. 205.

Dobrzanski, A., Anyszka, Z. and Palczynski, J. (1987) *Materialy Sesji Instytutu Ochrony Roslin*, **27**: 195-199.

Dohroo, N.P. and Gupta, S.K. (1994) *Plant Dis. Res.*, **9**: 175-177.

Domblides, A.S., Domblides, E., Kharchenko, V.A., Potekhin, G.A. (2010) *Moscow University Biological Sciences Bulletin* **65**(4): 152-154.

Drazic, S. and Jevdovic, R. (1997) *Selekcija I Semenarstvo*, **4**: 175-178.

Dudits, D., Fejer, O., Hadlaczky, G., Koncz, C., Lazar, G.B. and Horvath, G. (1980) *Mol. General Genet.*, **179**: 283-288.

Dunn, J.A. (1976) *Advisory Leaflet, Min. Agril., Fisheries and Food*, U.K., No. 603, p. 6.

Eckey Kaltenbach, H., Grosskopf, E., sandermann, H.Jr. and Ernst, D. (1994) *Proc. Royal Soc. Edinburgh, Section B, Biol. Sci.*, **102**: 63-74.

Eidsten, I.M. and Gislerod, H.R. (1986) *Plant and Soil*, **92**: 23-28.

Eidsten, I.M. and Gislerod, H.R. (1986) *Acta Hortic.*, **178**: 223-226.

Ejsmond, J. (1979) *Biuletyn Warzywniczy*, **23**: 405-417.

El Sherbeny, S.E. and Hussein, M.S. (1993) *Egyptian J. Physiol. Sci.*, **17**: 95-102.

Ely, P.R. and Heydecker, W. (1981) *Sci. Hort.*, **15**: 127-136.

Franceschetti, U. and Nassi, M.O. (1985) *Notiziario di Ortoflorofrutticoltura*, **11**: 169-174.

Franken, P. and Gnadinger, F. (1994) *Mol. Plant Microbe Interactions*, **7**: 612-620.

Franz, C., Fritz, D. and Eisenmann, J. (1982) *Gemuse*, **18**: 380-382.

Franz, C. and Glasl, H. (1976) *Qualitas Plantarum*, **25**: 231-246.

Fu, P.L. (1991) *Modernizing Agril.*,**8**: 28.

Gadzhonov, I. and Kr"stev, G. (1982) *Gradinarska i Lozarska Nauka*, **19**: 70-74.

Gerami, F., Moghaddam, P.R., Ghorbani, R. and Hassani, A. (2016) *Anais da Academia Brasileira de Ciências*, **88**: 2375-2385.

Gherman, N. (1979) *Analele Institutului de Cercetari pentru Legumicultura si Floricult.*,**5**: 47-50.

Gijsegem, F. van., Somssich, I.E. and Scheel, D. (1995) *European J. Plant Pathol.*, **101**: 549-559.

Godzhonov, I. (1982) *Gradinarska i Lozarska Nauka*, **19**: 59-64.

Gurgul, E., Herman, B. and Biczak, R. (1996) *Folia Hortic.*,**8**: 39-51.

Gyuro, G. (1983) *Mosonmagyarovari Mezogazdasagtudomanyi Kar Kozlemenyei*, **25**: 139-158.

Hahlbrock, K., Cretin, C., Cuypers, B., Frizemeier, K.H., Hauffe, K.D., Jahnen, W., Kombrink, E., Rohwer, F., Scheel, D., Schmelzer, E., Schroder, M. and Taylor, J. (1987) In: *Plant Mol. Biol. Proc. NATO Advanced Study Institute* (Eds. D. Wettstein von and N.H. Chua), 10-19 June, 1987, Carlsberg Lab., Copenhagen, denmark, pp. 399-406.

Hahm, Y.I., Kwon, M., Kim, J.S., Seo, H.W. and Ahn, J.H. (1998) *Korean J. Plant Pathol.*, **14**: 668-675.

Hallström, H. and Thuvander, A. (1997) *Natural Toxins*, **5**: 186–192.

Hansen, J. (1978) *Qualitas Plantarum*, **28**: 45-63.

Hardh, J.E. (1975) *Qualitas Plantarum*, **25**: 43-56.

Hartmann, H.D. and Waldhor, O. (1979) *Gemuse*, **15**: 316-318.

Hassell, R.L. and Kretchman, D.W. (1987) *Acta Hortic.*, **198**: 59-63.

Henriksen, K. (1985) *Meddedelse, Statens Planteavlsforsog*, **87**: 1827.

Hershman, D.E., Varney, E.H. and Johnston, S.A. (1986) *Plant Dis.*, **70**: 927-930.

Hill, D.E. (1992) *Bull. Connecticut Agril. Expt. Sta.*, No. 904, p. 8.

Honma, S. and Lacy, M.L. (1980) *Euphutica*, **29**: 801-805.

Hovadik, A. and Kratochvilova, H. (1977) *Bulletin, Vyzkumny a Slechtitelsky Ustav Zelinarsky Olomouc.*, **21-22**: 29-41.

Huang, P.L., Hahlbrock, K. and Somssich, I.E. (1988) *Mol. Gen. Genet.*, **211**: 143-147.

Ibrahim, H.M.M., Abou El-Leel, O.F., and Emam, K.A. (2017) *Middle East Journal of Agriculture Research*, **6** (1): 67-75

Jedras, L. (1991) *Biuletyn Warzywniczy*, **37**: 109-118.

Jedras, L. (1989) *Biuletyn Warzywniczy*, Suppl. II: 67-70.

Kameneva, E.A. and Pen'kov, L.A. (1973) *Khimiya v Sel'skom Khozyaistve*, **11**: 46-49.

Kandeel, A.M. (1991) *Ann. Agril. sci., Cairo*, **36**: 155-162.

Kaniszewski, S. and Dysko, J. (2008) *J. Elementol.*, **13**: 235-244.

Kanwar, J.S. and Sandha, M.S. (2000) *Agricultural Rev.*, **21**: 133-136.

Katzer, G. (2002) *http: II www-ang. Kfunigraz.ac.at/~ Katzer/engl/generic_frame.html? Petr_cri.html.*

Katyukha, S.A., Kulikov, N.S. and Kas'kova, L.A. (1984) *Zashchita Rastenii*, 7: 19.

Kawalleck, P., Keller, H., Hahlbrock, K., Scheel, D. and Somssich, I.E. (1993) *J. Biol. Chem.*, **268**: 2189-2194.

Kesik, T. (1994) *Proc. III Congr. European Soc. Agron., Padova Univ.*, Abano-Padova, Italy, 18-22 September, 1994, pp. 714-715.

Khlebnikov, V.F., Rudenko, N.M. and Zhuchenko, A.A. (1984) *Geneticheskie-osnovy selektsii sel'skokhozyaistvennykh rastenii I zhivotnykh*, **194:** 110-111.

Kirsch, C., Logemann, E., Lippok, B., Schmelzer, E. and Hahlbrock, K. (2001) *Plant J.*, **26**: 217-227.

Knyazeva, A.S. (1975) *Nauch. tr. Voronezh. s. kh. Int.*, **74**: 99-103.

Kolota, E. (2011) *Acta Sci. Pol.*, **10**: 145-154.

Kulikova, N.T. (1975) *Byulleten' Vsesoyuznogo Ordena Lenina I Ordena Druzhby Narodov Instituta Rastenievodstva Imeni N.I. Vavilova*, **52**: 58-64.

Lisiewska, Z., Kmiecik, W. and Budnik, A. (1997) *Folia Hortic.*,**9**: 21-29.

Madjarova, D., Bubarova, M., Hristova, I., Madzharova, D. and Khristova, I. (1979) *Herba Hungarica*, **18**: 185-195.

Madjarova, D. and Bubarova, M. (1974) *Proc. XIX Internatl. Hort. Song., I. Section VII, Vegetables*, pp. 721-764.

Madjarova, D.J. and Bubarova, M.G. (1978) *Acta Hortic.*, **73**: 65-72.

Madzharova, D., Benbasat, E., Bubarova, M. and Chavdarov, I. (1973) *Gradinarska I Lozarska Nauka*, **10**: 85-93.

Madzharova, D., Bubarova, M. and Genchev, S. (1978) *Gradinarska i Lozarska Nauka*, **15**: 59-65.

Malakouti, M.J., Navabzadeh, M. and Hashemi, S.H.R. (1999) *Improved crop quality by nutrient management*, Kluwer Academic Publishers, Dordrecht, Netherlands, pp. 43-45.

Malakouti, M.J., Navabzadeh, M. and Hashemi, S.H.R. (1998) *Soil Water J.*, **11**: 14-21.

Marlow, H. and Zschau, K. (1986) *Nachrichtenblatt fur den Pflanzenschutz in der DDR*, **40**: 237-239.

Marthe, F. and Scholze, P. (1997) *Gemuse Munchen*, **33**: 169-170.

Marthe, F., Scholze, P. and Pank, F. (1996) *Proc. Internatl. Symp. Breeding Res. on Medicinal and Aromatic Plants*, Quedlinburg, Germany, 30 June to 4 July, 1996, **2**: 250-253.

Marthe, F. (2020) In: Medicinal, Aromatic and Stimulant Plants, Handbook of Plant Breeding 12. (Eds. J. Novak, W.D. Blüthner).

Massoud, H.Y.A., Dawa, K.K., El-Gamal, S.M.A and Karkash, S.H.A. (2019) *J. of Plant Production*, **10**: 194 -1161.

Mayer, N. and Wilhelm, E. (1999) *Gemuse Munchen*, **35**: 119-120.

Melloul, A.A., Hassani, L. and Rafouk, L. (2001) *World J. Microbiol. Biotechnol.*, **17**: 207-209.

Meunier, S. and Verhoyen, M. (1987) *Mededelingen van de Faculteit Landbouwwetenschappen, Rijksuniversiteit Gent*, **52**: 1019-1025.

Montuschi, C. and Ricci, G. (1996) *Sementi Elette*, **42**: 25-28.

Moore, B.D., Palmquist, D.E. and Seemann, J.R. (1997) *Plant Physiol.*, **115**: 241-248.

Morelle, B. and Thomas, J.M. (1994) *Maitrise des adventices par voie non chimique, Communications de la quatrieme conference internationale IFOAM, Dijon*, France, 5-9 July, 1993, pp. 111-117.

Moretto, K.C.K., Barreto, M. and Inoue, R.Y. (1997) *Hortic. Brasil.*, **15**: 7-10.

Muller, T. (1982) *Gemuse*, **18**: 16-21.

Mylavarapu, R.S. and Zinati, G.M. (2009) *Sci. Hortic.*, **120**: 426–430.

Nega, E., Ulrich, R., Wermer, S. and Jahn, M. (2001) *Gesunde Pflanzen.*, **53**: 177-184.

Najla, S., Sanoubar, R. and Murshed, R. (2012) *Physiology and Molecular Biology of Plants*, **18**: 133–139.

Navazio, J. (2012) In: The Organic Seed Grower: A Farmer's Guide to Vegetable Seed Production (ed. J.Navazio), Chelsea Green Publishing, White River Junction, Vermont, pp. 108–111.

Nordestgaard, A. (1984) *Meddelelse, Statens Planteavlsforsog*, **86**: 1755.

Ovcharuk, V.I. (1997) *Visnik Agrarnoi Nauki*, **3**: 85, 87.

Palada, M.C., Crossman, S.M.A., Davis, A.M. and Colon, W. (2000) *Proc. 35th Annual Meeting, Caribbean Food Crops Soc.*, Castries, St. Lucia, 25-31 July, 1999, pp. 155-162.

Pantielev, N.Kh. (1987) *Vestnik Sel'skokhozyaistvennoi Nauki*, Moscow, **8**: 60-62.

Peyvast, G., Olfati, J.A., Madeni, S., Forghani, A. and Samizadeh, H. (2008) *International Journal of Vegetable Science*, **14:** 82-92.

Pill, W.G. (1986) *HortScience*, **21**: 1134-1136.

Pokhrel, B., Sorensen, J.N., Kristensen, H.L. and Petersen, K.K. (2017) *Can. J. Plant Sci.*, **97**: 1004–1013.

Polak, M., Ponican, J., Jech, J. and Angelovic, M. (1999) *Zemedelska Technika*, **45**: 1-8.

Postnikov, A.V., Romanov, G.A., Loboda, B.P. and Zekunov, A.V. (1996) *Kartofel'I Ovoshchi*, **5**: 21-22.

Rabin, J., Berkowitz, A.B. and Akers, S.W. (1988) *HortScience*, **23**: 554-555.

Radewald, J.D., Shibuya, F., Nelson, J., Brendler, R.A. and Vilchez, M. (1972) *Calif. Agril.*, **26**: 6-8.

Rapparini, G. and Campagna, G. (1996) *Informatore Agrario*, **52**: 55-61.

Renner, K. (1975) *Entomologische Blatter*, **75**: 76-78.

Ricotta, J.A. and Masiunas, J.B. (1991) *HortScience*, **126**: 539-541.

Roberts, H.A. and Bond, W. (1984) *Annals Applied Biol.*, **94**: Suppl. 1, pp. 48-49.

Rumpel, J. and Kaniszewski, S. (1994) *Acta Hortic.*, **371**: 413-419.

Sajina, A., Geetha, S.P., Minoo, D., Rema, J., Babu, K.N., Sadanandan, A.K. and Ravindran, P.N. (1997) *Biotechnology of Spices, Medicinal and Aromatic Plants. Proc. Natl. Sem. Biotechnol. of Spices and Aromatic Plants*, Calicut, India, 24-25 April, 1996, pp. 79-86.

Salim, R., Al Subu, M.M. and Isa, M. (1995a) *J. Env. Sci. Health. Part-A Environmental Science and Engineering and Toxic and Hazardous Substance Control*, **30**: 2123-2132.

Salim, R., Al Subu, M.M. and Isa, M. (1995b) *J. Env. Sci. Health. Part-A Environmental Science and Engineering and Toxic and Hazardous Substance Control*, **30**: 2057-2069.

Salim, R., Al Subu, M.M. and Isa, M. (1995c) *J. Env. Sci. Health. Part-A Environmental Science and Engineering and Toxic and Hazardous Substance Control*, **30**: 831-849.

Scholze, P., Marthe, F., Kramer, R., Proll, E., Rudi, Z. and Pank, F. (1996) *Proc. Internatl. Symp. Breeding Research on Medicinal and Aromatic Plants*, Quedlinburg, Germany, 30 June - 4 July, 1996, **2**: 247-249.

Simon, J.E. and Quinn, J. (1988) *J. Agric. Food Chem.*, **36**: 467-472.

Somssich, I.E., Schmelzer, E., Kawalleck, P. and Hahlbrock, K. (1988) *Mol. Gen. Genet.*, **213**: 93-98.

Souri, M. K., Rashidi, M. and Kianmehr, M. H. (2018) *J. Plant Nutr.*,**41**: 1405-1413.

Suciu, Z., Draganescu, I., Tabuia, M., Stela, S. and Bolcas, F. (1981) *Agronomie*, **18**: 207-211.

Taupier Letage, B., Meneuvrier, M., Vaute, G., Marquis Verjux, N. and Thomas, J.M. (1994) *Maitrise des adventices par voie non chimique, Communications de la quatrieme conference internationale I.F.O.A.M.*, Dijon, france, 5-9 July, 1993, pp. 359-364.

Teuscher, E. (2005) Medicinal Spices: A Handbook of Culinary Herbs, Spices, Spice Mixtures and their Essential Oils. (1st edn). Medpharm Scientific, Stuttgart, Germany and CRC Press, Boca Raton, Florida.

Thulke, O.U., Katz, V.A. and Conrath, U. (1999) *Modern fungicides and antifungal compounds II. 12th Internatl. Reinhardsbrunn Symp.*, Friedrichroda, Thuringia, Germany, 24th-29th May, 1999, pp. 373-379.

Toaima-Wael, I.M., Hamed, E.S. and Abd El-Wahab, M.A. (2014) *Egyptian J. Desert Res.*, **64**: 99-116.

Turkov, V.D., Nushikyan, V.A. and Drozdova, N.S. (1973) *Doklady Vsesoyuznoi Ordena Lenina Akademii Sel'skokhozyaistvennykh Nauk Imeni V.I. Lenina*, **12**: 15-16.

Tuza, S. and Bogdan, I. (1982) *Kertgazdasag*, **14**: 79-84.

Vandemoortele, J.L., Billard, J.P., Boucaud, J. and Gaspar, T. (1994) *46th Internatl. Symp. on Crop Protection: part IV*, Gent, Belgium, 3 May, 1994, **54**: 1455-1459.

Vandemoortele, J.L., Billard, J.P., Boucaud, J. and Gaspar, T. (1996) *Plant Cell Tissue Cult.*, **44**: 25-30.

Van Vranken, R.W. and Majek, B.A. (1985) *Proc. 39th annual meeting of the Northeastern Weed Sci. Soc.*, 1985, pp. 129-130.

Varga, G., Helyes, L., Dimeny, J., Somos, A., Csaky, A. and Nyujto, S. (1996) *Bull. Univ. agril. Sci., Godollo*, **1**: 173-182.

Vercruysse, P., Tirry, L. and Hofte, M. (1999) *Parasitica*, **55**: 185-193.

Viscardi, T. and Brzeski, M.W. (1992) *Agric. Ecosys. Env.*, **38**: 153-157.

Vukasinovic, S. and Bogdanvic, M. (1984) *Radovi Poljoprivrednog Fakulteta Univerziteta u Sarajevu*, **32**: 31-35.

Wally, Y.A., Sharaf, M.A. and Maksoud, M.A. (1978) *Egyptian J. Hort.*, **5**: 65-74.

Wu, S.C. and Hahlbrock, K. (1992) *Zeits. Naturforschung. Section C Biosciences*, **47**: 591-600.

Zelenin, V.M. (1974) *Tr. Perm. s kh. Int.*, **106**: 92-100.

LEAFY VEGETABLES

AMARANTHUS

V. Sagar, J. Devi, C.R. Muthukrishnan, I. Irulappan and T.K. Maity

1.0 Introduction

Amaranthus, also known as pigweed, is C_4 leafy vegetables of great diversity and plasticity that is grown during summer and rainy seasons in tropical Asia and Africa. Its popularity is continuously increasing in the Asian continent and elsewhere because of its attractive leaf colour, taste and adequate nutritional value. It is very cheap and has adequate protein with essential amino acids, such as methionine and lysine, dietary fibre, minerals, phytopigments and bioactive compounds, such as β-cyanin, chlorophyll, β-xanthin, carotenoids, β-carotene, vitamin C, phenolic compounds and flavonoids (Sarker and Oba, 2020). It fits well in a crop rotation because of its very short duration nature and large yield of edible matter per unit area. The estimation of the extent of its cultivation is not available. However, leafy types are usually grown in kitchen and market gardens. The grain amaranthus species are popular in the high lands of tropical and subtropical America. Recently, the grain types have gained some importance in the Indian plains also, particularly in Gujarat and Maharashtra. The grain types are usually grown as mixed crops along with cereals, pulses and vegetables.

2.0 Composition and Uses

2.1.0 Composition

The leaves and tender stems of amaranthus are rich in protein, minerals, vitamin A, C, K, riboflavin (B_2); niacin (B_3) and vitamin B_6. Phytochemical analysis of aerial parts of various *Amaranthus* spp. has established presence of active constituents like alkaloids, flavonoids, glycosides, phenolic acids, steroids, saponins, amino acids, vitamins, minerals, terpenoids, lipids, betaine, catechuic tannins and carotenoids (Peter and Gandhi, 2017). The nutritive values of raw and cooked amaranth compared with raw spinach leaves have been shown in Table 1.

Table 1: Nutritional Value of Raw and Cooked Amaranth Leaves, in Comparison with Spinach*

Nutrients	*Spinach Raw (Value/100 g)*	*Amaranth Raw (Value/100 g)*	*Amaranth Cooked (Value/100 g)*
Proteins (g)	1.20	2.46	2.11
Minerals			
Calcium(mg)	77	215	209
Iron (mg)	2.71	2.32	2.26
Magnesium (mg)	79.0	55.0	55.0
Phosphorus (mg)	49.0	50.0	72.0
Potassium (mg)	558.0	611.0	641.0
Sodium (mg)	79.0	20.0	21.0
Zinc (Zn)	0.53	0.90	0.88
Copper (mg)	0.136	0.162	0.158
Manganese (mg)	0.897	0.885	0.861

Nutrients	*Spinach Raw (Value/100 g)*	*Amaranth Raw (Value/100 g)*	*Amaranth Cooked (Value/100 g)*
	Vitamins		
Vitamin C (mg)	28.1	43.3	41.1
Riboflavin (mg)	0.189	0.158	0.134
Niacin (mg)	0.724	0.658	0.559
Vitamin B6	0.195	0.192	0.177
Folic acid (mcg)	194.0	85.0	57.0
Vitamin A RAE 1 (mcg)	469.0	146.0	139.0
Vitamin K 1140.0 –	482.9	1140.0	-
	Lipids		
Total saturated fatty acids (g)	0.063	0.091	0.050

* USDA National Nutrient Database for standard Reference.

Rattan *et al.* (2005) opined that long term intake of vegetables (cabbage, spinach and amaranthus) grown on sludge constitutes a possible health hazard to consumers as those contained high levels of lead, cadmium and chromium. Amaranth is known to accumulate high levels of oxalate, an anti-nutritional compound reported as potential risk factor for kidney stone development and its presence in food lowers calcium and magnesium availability. Two, dominant fractions of oxalate in amaranth were isolated: a fraction soluble in boiling water which was pre-dominantly in the form of K- and Mg-oxalate, and an associated insoluble residue which was predominantly Ca-oxalate. Most of the Ca in amaranth is unlikely to be available to the body. Oxalate concentration, calcium and magnesium concentrations were studied in 30 amaranth genotypes by Gelinas and Seguin (2007). They reported mean total oxalate concentration of 229 mg/100 g, with values ranged between 178 and 278 mg/100 g; mean Calcium concentration of 186 mg/100 g with values ranged between 134 and 370 mg/100 g; and average magnesium concentration of 280 mg/100 g with value ranging between 230 and 387 mg/100 g. Thus, amaranth can be considered a high source of oxalate, however, as most is in insoluble form and due to its high calcium and magnesium concentrations, oxalate absorbability might be low. Despite these facts, research on reducing the oxalate concentrate in leafy amaranthus is top priority for the breeders. Other toxic substances like cyanide (223 to 435 mg/kg dw) and nitrate (7.62 to 23.41 g/kg, dry weight basis) were also reported in *A. cruentus* vegetables and it was shown that the concentrations of cyanide and total oxalates increased during plant heading, whereas the amount of nitrates decreased (Musa *et al.*, 2011).

Seeds and plants of amaranthus have been a source of food for many cultures in the world. However, some species which grow wild or as weeds under severe climatic and soil conditions have not yet been examined for their potential as food sources. Contribution to the botany and nutritional value of some wild *Amaranthus* spp. of Nuevo Leon, Mexico was studied by Wesche-Ebeling *et al.* (1995). The wild species *A. retroflexus, A. viridis, A. palmeri* and *A. blitoides* were collected and examined as potential sources of vegetable green. Histochemical studies showed higher levels of starch in leaves of *A. viridis* and *A. blitoides* than in the other species, moderate amounts of tannins in all leaves, high protein concentrations in stems and leaves and moderate amounts of alkaloids in all tissues of *A. viridis* and *A. blitoides.* Anti-nutritional factors (nitrates, oxalates, cyanogenic glycosides, tannins and phytates) were quantified in plants at the pre flowering stage, but only nitrates (0.34-2 per cent DW) were found at concentrations above those generally considered as safe, but these were like those found in spinach. No cyanogenic glycosides were detected in any species. Bromatological analysis of whole or different plant parts before flowering and at seed maturity showed that mature whole plants or individual parts can be recommended as animal feed since they contain high levels of protein (20.6-24.7 per cent DW whole plant, 25.3-32.9 per cent DW leaves) and soluble carbohydrate (> 40 per cent). The composition of amaranthus plant was very similar to that of other food vegetables. These four species showed similar chemical compositions and had no detrimental chemicals which would deter their use as vegetable foods. Organoleptic taste preference studies would best indicate the stage of which plants should be harvested for human consumption.

Unsaturated fatty acids are important in the dietary management of heart diseases. The seed oil composition was determined for *A. hypochondriacus*, *A. cruentus* and *A. hybridus* (seeds obtained from Hungary). There were small quantitative differences in composition but in all species the major unsaturated fatty acid was linoleic acid (43-50 per cent). Histo-chemical evidence showed that the seed oil was located mainly in the germ layer just beneath the seed coat. This is easily separated mechanically, providing an oil-rich seed fraction. Therefore, Amaranth species could become useful medicinal plants (Mihalik *et al.*, 1993).

Two cultivars of amaranthus (*A. tricolor*, red and green type), were grown during the rainy and summer seasons at Vellayani, Trivandrum, India to assess the effect of maturity (10, 20 and 30 days after planting), leaf position (apex, middle and base) and postharvest storage (24, 48 or 72 h in the refrigerator, covered with a wet cloth, placed in a polythene bag, or the roots put in water) on the nutrient composition and organoleptic quality of leaves (Jijiamma and Prema, 1993). Ca, Fe and ascorbic acid content was unaffected by any of the treatments. In both seasons, the green cultivars were more nutritious than the red one. The red cultivar, however, when grown in the summer season, had the better organoleptic quality. Leaves were best stored in the refrigerator for no longer than 24 h, otherwise the ascorbic acid content was significantly reduced.

The different species of leafy amaranth has different nutrient content, it is evident from studies that β-carotene (1281.66 $\mu g\ g^{-1}$) and vitamin C (1848.15 $\mu g\ g^{-1}$) in *A. blitum*, which was relatively higher than *A. tricolor* (Sarker and Oba, 2020). Total phenol content is also higher in case of *A. blitum* than the *A. tricolor* (Khanam *et al.*, 2012; Sarker and Oba, 2020). The protein content of *A. blitum* (3.51 g 100 g^{-1}) is also reported to be higher than *A. tricolor* (1.26 per cent). Jimenez-Aguilar and Grusak (2017) reported a good amount of Mg, K and Ca in different species of amaranth. The concentration ranged from 1.5–3.5 mg/g for Ca, 5.5-8.8 mg/g for K, 1.8–4.5 mg/g for Mg, 0.5–0.9 mg/g for P, 3.2–5.5 mg gallic acid equivalents/g for phenols, and 38–90 µmol Trolox equivalents/g for antioxidant activities (FW), respectively. Further, Mg, Ca and K contents of different species of amaranth were much greater than kale, black nightshade, spider flower and spinach. Significant amount of variation for pigments like β-carotene, betalain chlorophyll is available across the different species *viz* β-cyanin (53.63 µg 100 g^{-1}), β-xanthin (58.38 µg 100 g^{-1}), betalain (112.01 µg 100 g^{-1}) and total carotenoids (1675.38 µg g^{-1}) in *A. blitum*, and almost similar amount in the *A. tricolor* (Khanam *et al.*, 2012; Sarker and Oba, 2020). Among the *A. tricolor*, the abundance of betalain, β-cyanin, and β-xanthin was 2-5-fold higher in the red leaf sectors than in the green leaf sectors. Amaranthin, which is the main β-cyanin produced in the genus *Amaranthus* first isolated from this plant.

2.2.0 Uses

Amaranth is a multipurpose crop supplying high nutritional quality grains and leafy vegetables for food and animal feed. With its attractive inflorescence coloration, it is also cultivated as ornamental plant. The fresh tender leaves and stem of amaranthus are delicious when cooked like other fresh leafy vegetables called as 'Sag'. The tiny seeds of grain amaranthus are also edible that are popped or parched and milled for flour or gruel. The flour or flaked forms are combined with wheat or other flours to make cereals, cookies, bread and other baked goods. In taste, nutritional value and yield, the grain compares favourably with maize and other true cereals. Its other useful potential includes anthocyanin (red) pigments to produce nontoxic natural dyes, microcrystalline starch for food and industry and squalene, specialized oil used in skin cosmetics, computer and pharmaceutical industries. The betalain extracted from the leaves of this plant can be used as natural colouring agent to develop different product including the anti-oxidant rich multi-functional packing material for food (Hu *et al.*, 2020). *A. tricolor* has been reported to contain some antimicrobial compounds, such as alkaloids, polyphenols, and terpenoids. Anti-microbial activity was reported against the *staphylococcus aureus*. The mechanism of action of *A. tricolor* crude extract, against *S. aureus* was associated with cell membrane depolarization, reduction of pH in, decrease of bacterial protein content, cleavage of cell DNA, and leakage of cytoplasm (Guo *et al.*, 2020).

3.0 Origin, Distribution and Taxonomy

3.1.0 Origin and Distribution

The word Amaranthus is basically derived from the Greek word *"anthos"* (flower) which means everlasting or unwilting. The leafy amaranthus is said to be the native of India (Nath, 1976). The centres of diversity for amaranthus are Central and South America, India, South-East Asia and the secondary diversity is the West Africa and East Africa (Wolosik and Agnieszka Markowska, 2019). A wide variation is reported to exist within each species in growth habit, disease resistance, taste and quality, thus offering considerable scope for future breeding programmes.

Among the leafy types, *A. tricolor* occupies a predominant position in India with different morphological forms in colour (red, scarlet, maroon, purple, yellow, cream, and green) and shape of leaves. The crop is mainly cultivated in mid and high hills of the Himalayan region as a pure as well as mixed crop. The crop is gaining

Uses of Amaranthus

Different Recipes of Amaranthus in Indian Cuisine.

popularity in other parts of the country including North western plains, North Eastern region and south Indian hills. *A. blitum* and *A. tristis* are the other species which are under cultivation. A few other are important weeds, *viz., A. spinosus, A. viridis* and *A. graecizans,* used as pot-herbs, and are widely distributed in road sides, waste places, in uncultivated and fallow lands. Similarly, grain amaranthus is grown along the whole length of the Himalayas from Kashmir to Bhutan, on the South Indian hills and also in the plains of India. *A. cruentus* is cultivated in Guatemala and *A. edulis* is restricted to Argentina. The species *A. caudatus* is chiefly cultivated in the Andes of America and is also distributed in Asia and probably in Africa.

3.2.0 Taxonomy

Genus *Amaranthus* belongs to order Caryophyllales, family Amaranthaceae and sub-family Amaranthoideae. It includes nearly 74 species, among which approximately55 species native to the Americas, and remaining 19 species natives to Eurasia, South Africa, and Australia/Oceania (Waselkov *et al.,* 2018). Out of these, three species, *A. caudatus* L., *A. cruentus* L., and *A. hypochondriacus* (L.) are mainly grown as pseudocereals in South America and South Asia, while two other *A. blitum* L. and *A. tricolor* L. are mainly grown as vegetable crops in Asia. Mild spinach-like flavour, high yields, ability to grow in hot weather and high nutritive value are few reasons for their popularity. However, few other species *viz., A. dubius, A. lividus, A. tristis, A. viridis, A. graecizans, A. spinosus, A. reftroflexus* and *A. hybridus* are also used for vegetable purpose (Singh and Singh, 2011; Gerrano *et al.,* 2017). *A. spinosus and A. tristis* has been identified for its medicinal value, while eight species are problematic agricultural weeds worldwide (Murugan *et al.,* 2013; Mosyakin and Robertson, 2003). Interspecific and intervarietal hybridization in nature have caused a wide variation differing in pigmentation of plant parts as well as in the inflorescence. Widespread nature of many species and variability has resulted in considerable synonymity (Das, 2016). Time to time attempts has been made by the researchers to classify the genus intragenerically. The genus has been classified traditionally into three subgenera, *viz.* subgenus *Acnida* (comprising dioecious members), subgenus *Amaranthus* (included the monoecious species) and subgenus *Albersia*. The subgenus *Acnida* includes nine dioecious species that are native to North America; subgenus *Amaranthus* consists of 20 monoecious species mostly native to the Americas and subgenus *Albersia* includes all the species traditionally included by many earlier authors in Amaranthus sect. *Blitopsis sensu lat.* Grain amaranths are included in subgenus *Amaranthus,* while vegetable and weedy amaranthus except *A. hybridus* are included in subgenus *Albersia* (Das, 2016).

4.0 Botany

It is an annual herb, erect or trailing, mostly unarmed and rarely armed, scarce to profuse branched; shallow to deep taproot system; stem green to purple or with mixed shades of these two colours, glabrous and succulent; leaf simple, alternate or opposite, exstipulate, entire, obovate to lanceolate, with acute tip, green to purple or with different shades of these two colours, purple colouration prominent in young leaves and fades away at maturity; inflorescence terminal and axillary,

Species of Amaranthus

A. hypochondriacus **(Left),** ***A. cruentus*** **(Middle),** ***A. caudatus*** **(Right)**

Inflorescences of Selected *Amaranthus* Species.

a. Apical and axillary inflorescences of ***A. blitum***; b. Hanging inflorescence of ***A. caudatus***; c. Terminal inflorescence of ***A. cruentus***; d. Terminal inflorescence of ***A. dubius***; e. Inflorescence of ***A. hypochondriacus***; f. Apical and axillary inflorescences of ***A. spinosus***; g. Axillary clusters of ***A. tricolor***; h. Terminal inflorescence of ***A. viridis.***

Leaf Colour Variation among *A. tricolor* Genotypes.

branched spikes differently coloured from green to deep purple and orange and with different shades of these colour, the basic unit of the inflorescence is dichasial cymes called glomerule; flower small, regular, mostly unisexual, monoecious, pentamerous, bracteate and bracteolate which are scaly, perianth parts 4-5, imbricate, membraneous, often persistent; stamens 2-5, placed opposite to perianth parts, often some reduced to staminodes, filaments either free or united below, anthers one or two celled; carpels 2-3, syncarpous, ovary superior with commonly one campylotropus ovule, style single to 2 or 3 branched; fruit a utricle, indehiscent, enclosed by the persistent perianth parts; seed varying in colour, black, brown or white, compressed, smooth, shiny with floury endosperm.

In general, most of the cultivated species are monoecious. The proportion of male and female flowers varies in the inflorescence. Each glomerule consists of an initial staminate flower and an indefinite number of pistillate flowers. The extent and degree of cross-pollination is governed by the proportion of the staminate and pistillate flowers in an inflorescence and also the position of the inflorescence in the plant. In monoecious species, including grain amaranthus two types of arrangement of the staminate and pistillate flowers in the glomerule of the inflorescence occur. In all species except *A. spinosus*, the first flower of each glomerule is staminate and the rest are pistillate. In *A. spinosus*, each glomerule bears flowers of the same sex with the pistillate flowers in the axils of the branches and at the base of the terminal inflorescences, while the staminate flowers are terminal on the main axis and lateral branches.

The cultivated grain types amaranth favour cross-pollination as most of them are monoecious, while vegetable types Blitopsis are self-pollinated due to the number of staminate flowers per glomerule and small axillary inflorescence. Dominant green perianth colour could be employed as marker gene to detect the extent of cross-pollination in grain types which ranges from 16 to 35 per cent (Walton, 1968).

The percentage of male flowers per glomerule is 0.5 to 1 in grain types and 10.0 to 25 in leafy types. From the breeding point of view, the small closely grouped flowers in glomerules make emasculation extremely difficult. The most satisfactory method of effecting crosses is to pollinate heavily as soon as the stigma is receptive and to remove the staminate flowers by hand. The stigma of the pistillate flowers is receptive several days prior to opening of the staminate flowers. *Amaranthus* species are characteristically wind pollinated, but the grain species favours insect pollination through their colourful inflorescences that are occasionally visited by bees (Das, 2016). The grain species and their close relatives are monoecious and abundant pollen production in these species leads to a high degree of self-pollination. Arrangement and sequence of anthesis of the unisexual flowers favour a combination of self and cross-pollinations (Sauer, 1976, 1980).

4.1.0 Cytology

Cytological studies revealed that with the exception of one tetraploid species, *A. dubius* (x = 16, n = 32, 2n = 64) belonging to Amaranthus section, all the other species examined so far have diploid number of either 32 (n = 16) or 34 (n = 37) chromosomes (Grant, 1959; Seth, 1968; Behara and Patnaik, 1974). Xu (1987) studied chromosome number of 8 amaranthus including *A. paniculatus* and *A. caudatus* and also reported 2n = 32 or 2n = 34. One form of *A. caudatus* however, have 2n = 30. The only report of x = n = 8 is by Malik and Ahmed (1963) in *A. blitum*. Thus, the genus is dibasic with x = 16 and 17, almost equally distributed in sections Amaranthus and Blitopsis. There is preponderance of x = 17 in section Blitopsis and therefore, in the genus as a whole. *A. graecizans*, a member of section Blitopsis is reported by Khoshoo and Pal (1970) to be dibasic.

The later chromosome counts indicated *A. tricolor* to be 2n = 2x = 34, *A. cruentus* to be 2n = 32 and *A. dubius* to be 2n = 64. The possibilities of crossing with related wild species or with grain amaranths have been reported. Many interspecific hybrids, both spontaneous and artificial, have been reported by Murray (1940) and Grant (1959) among the grain species and wild relatives. Pal and Khoshoo (1972, 1974) reported heterosis and nearly normal meiosis but with partial sterility in some experimental hybrids.

5.0 Cultivars

The popular cultivars of amaranthus developed in India and abroad are given below.

Co. 1 (*A. dubius*)

This cultivar has been developed by selection from a local type at Tamil Nadu Agri-cultural University, Coimbatore, India in 1968. It is mainly for tender green as well as matured stems. It yields 2.150 kg of greens per 3 sq m area (7165 kg/ha) at

25th day after sowing; the leaf-stem ratio is 2.0; lacks initial vigour but makes rapid growth after 30th day, thus making it specially suitable for the harvest; leaves dark green with ridged appearance; stem dark green, round and succulent; inflorescence terminal and axillary, takes 50 days for flowering and 90 days for seed maturity; plants tall 150 cm; seed yield per plant is 13 g with an yield of 1.5 t/ha. Seeds are small and black; tolerant to pests and diseases. This variety is recommended for cultivation in Tamil Nadu.

Co. 2 (*A. tricolor*)

It is a selection from local type at Tamil Nadu Agricultural University, Coimbatore and released in the year 1976. This cultivar is particularly suited for early harvest. It yields 3.230 kg of greens per 3 sq m area (10780 kg/ha) at 25th day after sowing; the leaf/stem ratio is 1.8; early germination and vigorous growth make it more suitable for early harvest as tender greens; leaves green, lanceolate, slightly elongated; stem green, succulent; spikes terminal and axillary, takes 42 days for flowering and 80 days for seed maturity; plants medium tall with 125 cm; seed yield per plant is 48.4 g and 2.80 t/ha; seeds bolder and black; tolerant to pests and diseases. This variety is recommended for cultivation in Tamil Nadu.

Co. 3 (*A. tristis*)

This cultivar has also been evolved by selection from local type at Tamil Nadu Agricultural University in 1981. This is the only type suitable for clipping of tender greens. This is also locally named as *Araikeerai* in Tamil Nadu. This cultivar is suitable for the entire Tamil Nadu State. It records an average yield of 30716 kg of greens per hectare as compared to the local type with 19431 kg/ha. The increased yield over local is 58.0 per cent. This cultivar lends itself for ten clippings commencing from 20 days after sowing and provides a continuous supply of luscious tender greens for a period of 3 months. The leaf-stem ratio is high and this enhances the palatability of the cooked greens to a great extent. The greens of this cultivar is nutritious with 35.9 mg of ascorbic acid per 100 g of fresh matter. As a seed crop, this cultivar flowers in 35-40 days after sowing and matures at 85-90 days. This variety is recommended for cultivation in Tamil Nadu.

CO.5 (*A. tricolor*)

This cultivar has developed by single plant selection from A-166-1. Double coloured type (green and pink leaves) and grown for dual purpose. It is also suitable for container cultivation. It gives a rosette growth in early stages and first harvest is possible in 25 days; yield 40 t/ha in 55 days. This variety is recommended for cultivation in Tamil Nadu.

Chhoti Chaulai (*A. blitum*)

This cultivar has been developed through selection at Indian Agricultural Research Institute, New Delhi. This is suitable for its leafy shoots. Plants erect with thin stem, slightly dwarf; leaves small, green in colour; responds well to cutting; flowers are borne in clusters in leafy axils. It is best suited for early summer and can be grown in rainy season also.

Badi Chaulai (*A. tricolor*)

It is also developed at Indian Agricultural Research Institute, New Delhi through selection. The economic product of this cultivar is leafy shoots. Stem thick, green; leaves large and green; responds to cutting and distinguishable by its much longer growing period. It is best suited for summer season.

Pusa Kirti (*A. tricolor*)

This cultivar developed at Indian Agricultural Research Institute, New Delhi is suitable for summer cultivation. The economic part is green and soft stem. The ratio of stem to leaf is 1: 3.4. The crop is ready in 30-35 days for first cutting which may continue up to 78-84 days.

Pusa Kiran (*A. tricolor*)

It is developed through hybridization between *A. tricolor* x *A. tristis* at Indian Agricultural Research Institute, New Delhi. This cultivar is most suitable for rainy season. Leaves are soft, green and wide. The ratio of stem to leaf is 1: 4.6. It provides first cutting in 21-25 days after sowing and continue up to 70-75 days.

Pusa Lal Chaulai (*A. tricolor*)

A high yielding, red pigmented cultivar developed at Indian Agricultural Research Institute, New Delhi is suitable for growing in both summer and rainy seasons. It has a stem-leaf ratio of 1: 5. First cutting of leaves can be taken 33 and 24 days after sowing in summer and rainy seasons, respectively. It yields on an average 490 and 450 q/ha in summer and rainy seasons, respectively.

Arka Suguna (*A. tricolor*)

A pure line selection from Taiwanese introduction (IIHR 13560) made at Indian Institute of Horticultural Research, Bangalore is ideal for multicut without becoming fibrous which also regenerates at a faster rate. It provides highly succulent stems along with attractive broad and green leaves. The first cutting is ready in about 25 days after sowing and subsequent cuttings can be taken at 10-12 days interval (up to 90 days). The cultivar is moderately resistant to white rust. It can be sown during March to September with the yield potential of fresh green of 270 q/ha.

Arka Arunima (*A. tricolor*)

It is a multicut, brown and dark purple leaves cultivar, developed at ICAR-Indian Institute of Horticultural Research, Bangalore. The first cutting becomes ready at 30 days after sowing. Leaves contain low antinutrient factors like oxalate and nitrates. It grows well in rainy and summer seasons and yields about 240 q/ha of greens.

Arka Varna

It is a pulling type amaranth cultivar with pink stem, petiole and veins, having ovate shape pinkish green leaves. Yield potential is around 10.58 t/ha in 30-35 days duration. The antioxidant activity is 417 mg/100 g FW. (AEAC units), nitrate and oxalate content, 37.6 mg/100 g FW and 1.42 g/100 g FW, respectively. This variety

Cultivars of Amaranthus

Arka Varna

Arka Samraksha

CO-2

Kashi Suhavaani

Arka Arunima

is suitable for growing throughout the year. Released for cultivation during 2019 and recommended for state Karnataka.

Arka Samraksha

This is also a pulling type variety of amaranth, suitable for growing throughout the year. The stem is green, leaves medium large, lanceolate shaped. Yield potential 10.91 t/ha in 30-35 days duration. The antioxidant activity 499 mg/100 g FW (AEAC units), nitrate and oxalate content 27.3 mg/100 g FW and 1.34 g/100 g FW respectively.

Sirukeerai (*A. polygonoides*)

It is a traditional cultivar in Tamil Nadu, suited for uprooting at 25 days after sowing. Leaves are small, ovate with blunt bifurcated tip and have long petiole. Collar region is dark pink and at leaf axil a miniature branch initiate.

Arun

This cultivar is having maroon red leaves with high yield and photo insensitive in nature. This variety is suitable for multiple cutting and released for cultivation in Kerala during 1992.

Mohini

This cultivar is having green colour attractive green foliage. This variety is photo-insensitive, high yielding and suitable for summer season. This is also released for cultivation in Kerala.

Krishnasree

This is a red amaranth cultivar, evolved through selection. It is high yielding (14.8t/ha) with high nutritive value and low anti-nutritional factors. It is having reddish brown leaves, fleshy stem and suitable for multi-cutting. This variety is released for cultivation during 2006 in Kerala.

Renusree

It is a variety of green amaranth evolved through selection. High yielding variety (15.5 t/ha) having green leaves and purple stem with low anti-nutritional factors. This variety is also released for cultivation during 2006 in Kerala.

Kashi Suhavaani

This cultivar is developed by ICAR-IIVR, Varanasi in 2019. It is having luxuriant plant growth with lusture green canopy, soft succulent green leaves, delayed flowering and high yield potential (30-33 t/ha). This one is suitable for growing during summer and rainy season. It is rich in protein content 15.50 per cent on dry weight basis and recommended for cultivation in Uttar Pradesh.

6.0 Soil and Climate

6.1.0 Soil

Amaranthus can be grown in a wide range of soils; however, sandy loam slightly acidic in nature is preferred. Heavy clay soils with very poor drainage or sandy soils with very poor water holding capacity are unsuitable for its cultivation. It requires a very good tilth and thorough field preparation for getting the best growth. Effect of soil pH on the growth and yield of amaranthus was studied by Ohshiro *et al.* (2016). They opined that growth and yield of amaranth lines were higher in grey soil (pH 8.4) than in dark red soil (pH 6.6) and red soil (pH 5.4). In one of the studies, Qin *et al.* (2013) reported that amaranth has high salt tolerance than the lettuce. At NaCl concentration below 5000 ppm, salt stress effect is insignificant on biomass output, leaf photosynthesis rate, Fv/Fm, photosynthesis pigment contents, activities of antioxidant enzymes, and inducing lipid peroxidation.

6.2.0 Climate

Amaranthus is widely distributed in both temperate and tropical regions of the world. *Amaranthus* species which grow under varying climatic conditions differ in their day length requirements and respond differently to changes in photo- and thermo-periodism. The optimum germination temperature for amaranth has been reported to be between 20 and 30 °C (ISTA 2010). *A. caudatus, A. cruentus* and *A. edulis* are short-day species, while the commonly grown grain amaranthus species, *viz., A. hypochondriacus* is reported to be day neutral. All the species, however, have early and late cultivars. Grain amaranthus type is highly resistant to drought, but lodges as a result of high wind velocity or heavy rainfall resulting in loss of yield. Amaranthus has a relatively low resistance to unfavourable edaphic factors and C4-species have no signs of high adaptivity and resistance to plant stress factors (Breus, 1999). Studies on the cold protection for *A. tricolor* using non-woven fabric during winter season revealed that non-woven cloth (30 g/m^2) gave the best plant yields, heights and lengths of internodes of the crop by increasing the air temperature to 17°C (control 16°C) and soil temperature to 18°C (control 17°C) around the plants (Chen and Lin, 1997).

The effects of 3 light levels (1403, 806 and 462 µmol/sec per sq m) on the severity of damping off caused by *Pythium myriotylum* in *A. hybridus* were tested by Sealy *et al.* (1990). The observed mortality (33 per cent, 69 per cent and 81 per cent, respectively) decreased as light intensity increased. The reduction in plant growth and maturity in a shaded location is related to the observed increase in susceptibility to damping off in such an environment. The effects of daylength on the growth of *A. caudatus* growing in pots are reported by Chandwade *et al.* (1994). Results indicated that daylengths of < 9 h were insufficient for adequate vegetative growth and that a day length of 12 h gave the highest yield of leaf material.

The influence of the environment on incidence of dieback, stem and leaf blights in *A. cruentus* cvs. NHAc 33, NHAc 30 and NHAc 100 were investigated for 6 seasons during 1987-88 in Nigeria by Abedanjo (1994). The lowest incidence of dieback (0 per cent) and stem blight (8 per cent) were recorded during the 1st season for NHAc 33 and NHAc 30, while NHAc 100 had 4 per cent leaf blight in the 3rd season (December-February). The lowest mean percentage of leaf blight was recorded for all cultivars by the 3rd season. Conversely, the highest incidence of symptoms caused by *Choanephora cucurbitarum* in cultivars occurred in the second season (August-October). The health of seeds was influenced both by the seasons of harvest, environmental conditions and seed variety. Precipitation, high temperature and humidity that accompanied harvests of the 1st and 2nd seasons greatly reduced seed quality. High quality seeds were obtained from NHAc 100 and NHAc 33.

7.0 Cultivation

It is grown throughout the year in tropics and in autumn, spring and summer seasons in temperate regions. Grain amaranths are grown in wide geographic areas ranging from tropical lowlands to 3500 m in the Himalayas. Altitudes above 1000 m in the tropics are considered best for grain amaranth cultivation though they are tolerant to drought condition and low soil fertility. Seed treatment with 0.1 M CCC (chlormequat) or 3 x 10^{-4} M ancymidol inhibited germination in *A. caudatus* seeds. Gibberellic acid application (3 x 10^{-4} M) however, tended to counteract these inhibitory effects (Kepezynski and Knypl, 1989). Essential oils extracted from a number of aromatic plants mixed with the top 0.5 cm of soil inhibited germination of *Amaranthus* sp. seeds. This effect depended on soil type and the most active essential oil was the aldehyde from *Cymbopogon citratus* (Dudai *et al.*, 1999). Seed treatment with 50-100 mg succinic acid/l stimulated growth and increased yields without affecting quality (Andrianova *et al.*, 1996). The duration for green crop is about 30 days, while it is 85-90 days for a seed crop.

The field is prepared to a fine tilth and beds of 2.0 m x 1.5 m size are usually made for growing this crop. The seed rate is 2 kg/ha for direct sowing and 1 kg/ ha for the transplanted crop. Olufolaji and Tayo (1989a) reported that highest total and edible shoot yields of *A. cruentus* at a sowing rate of 5 kg/ha. In the prepared beds, the seeds are sown in rows, spaced at 20 cm and later (10-15 days after sowing) thinned to a spacing of 20 cm between plants, leaving one or two plants per hole. Norman and Shongwe (1993) suggested closest spacing (45 x 45 cm) for *A. hybridus* which produced the minimum total yield/plant, gave the highest total yield/ha whereas the converse was observed with the widest spacing (90 x 60 cm). As per the recommendation, 1.2–3.5 kg seed/ha is to be planted to an average depth of 1.3 cm for *A. tricolor* varieties. Field studies were conducted at Emmaus, Pennsylvania, in the summer of 1997 and 1998 to determine the influence of three within-row plant spacings (0.3, 0.6 and 0.9 m) on the growth, yield and mineral composition of vegetable amaranth (*A. cruentus*). In 1997, root and shoot fresh weights per plant, stem diameter, leaf dry matter, shoot dry weight and marketable yield per plant increased as plant spacing increased, whereas, plant residue cover and marketable yield per hectare increased as plant spacing decreased.

Leaf Ca increased as plant spacing increased before seed head formation, whereas leaf P, K and Ca similarly increased after seed head formation. In 1998 root fresh weight, stem diameter, root dry weight, and marketable yield per plant increased as plant spacing increased, whereas, plant residue cover and marketable yield per hectare increased as plant spacing decreased. Although plant spacing did not affect leaf N, P, K, Ca and Mg content before seed head formation, both P and Ca increased as plant spacing increased after seed head formation (Igbokwe and Hollins, 2000).

Fasheun (1989) viewed that flat beds were better than ridges or raised beds since they conserve water and had a high infiltration rate. Kader Mohideen and

Field Growing of different Species of Amaranthus

Rajagopal (1975) studied the effect of transplanting in cultivars CO.1 and A. 62 (*A. leucocarpus*) and observed very little differences in the establishment between direct sowing and transplanting in both the cultivars. However, they observed that the mean heights at maturity were 147.6 (CO. 1) and 228.2 (A. 62) cm in the case of direct sown crop, whereas in the transplanted crop, the respective plant heights were 115.2 and 215.1 cm. To utilize the soil space between seed coconuts (sown at 30 ?30 cm) during the first 3 months after sowing, amaranthus was sown in 4 different ways (dibbling, flow line, broad cast and criss-cross). The flow line method at 1.5 kg seeds/ha gave the highest leaf yield of 26.9 t/ha; the broadcast method at 2 kg seeds/ha gave the lowest yield of 20.1 t/ha (Kumar and Pillai, 1990). Intercropping study of brinjal, radish and red amaranthus revealed that brinjal + red amaranth provided the highest land equivalent ratio. However, radish + red amaranthus had the highest benefit cost ratio due to its low cost of cultivation (Zaman *et al.*, 1999).

First watering is to be done soon after broadcasting and further irrigation are to be given once a week with a gentle flow of water. The effect of stressing plants at different physiological stages (vegetative, flowering and seed formation) on growth and seed yield of two cultivars of *A. cruentus* and *A. hypochondriacus* was investigated by Ayodele *et al.* (2000) in a green house where soil moisture stress of 0.60 MPa was imposed at those stages and the control (unstressed plants) maintained at field capacity. The two cultivars differed significantly in their response to soil moisture stress. Imposition of water stress at the vegetative stage reduced leaf area in both cultivars. Leaf area was reduced by 20 per cent in *A. hypochondriacus* and 18 per cent in *A. cruentus*, compared to their respective controls. Soil water stress at flowering and seed formation led to stem lodging in *A. hypochondriacus*. In *A. cruentus* however, the flowering stage appeared to be most critical. Imposition of soil moisture stress at flowering stage reduced seed yield by 19 per cent compared to 9 per cent reduction observed in plants stressed at the seed formation stage. According to Chukwuma (1988), irrigation at 50 per cent field capacity is recommended for optimum use of limited water supplies in the dry season. The effects of irrigation rate (rainfed, 2, 4 or 8 litres/ha) on the growth of green spinach (*A. tricolor*) were studied by Said and Lamm (1995) in Malaysia. Better crop yields and higher water use efficiencies were obtained with the highest irrigation rate treatments. In a second experiment, the water balance of point source emitters was compared with that of the spray-jet type. Water loss due to percolation from the point-source emitter system was more than twice that of the spray type. The field should be kept weed free by hoeing and weeding as and when necessary. According to Leela (1993), pendimethalin and fluazifopbutyl proved effective in short duration vegetable crops like amaranthus, bean, pea, *etc.* and yields of those crops were also increased.

7.1.0 Manuring and Fertilization

The manurial recommendation consists of incorporating 25 tonnes of farmyard manure at the time of last ploughing and a basal application of 50 kg N, 50 kg P and 20 kg K per hectare. However for clipping types of amaranthus (CO. 3), a higher fertilizer dose of 75 kg N, 25 kg each of P and K/ha is recommended. Singh *et al.* (1985) reported that N rate of 60 kg/ha applied in a split dose of one half as a basal dressing and the other half as a top dressing after the 5th clipping (8 weeks from sowing) of cv. CO. 3, gave the maximum plant height and optimum leaf: stem ratio and the highest leaf yield. Nath (1976) has recommended 25 to 30 tonnes of farmyard manure at the time of field preparation and 27 kg each of N and P and 54 kg K per hectare. An investigation conducted at Bidhan Chandra Krishi Viswavidyalaya, West Bengal indicated that the economic green yield in CO. 1 cultivar was obtained with the application of 30 kg N and 25 kg each of P and K per hectare (Ghanti *et al.*, 1983). According to Singh (1976), amaranthus should be manured at least with 15 tonnes of farmyard manure or compost per hectare followed by top dressing of 150 kg/ha of ammonium sulphate or calcium ammonium nitrate in four splits depending upon the pH of the soil. The first dressing with nitrogenous fertilizers should be given four weeks after sowing and the next three dressings after every cutting. It has been observed that *A. tricolor* plants grown in pot with 69 kg N/ha without shade produced highest number of leaves, largest leaf area and greatest dry matter on 49th day (Nawawi *et al.*, 1986). Hilman and Abidin (1987) opined that plant growth and yield increased with N rate when in screenhouse trials, N at 0, 45, 90 or 180 mg/kg soil was applied to the cultivars Giti Hijau, A 401, A 405 and A 406. The cv. Giti Hijau showed the best response to treatment with regard to plant growth and yield.

In Indonesia, Subhan (1989) obtained highest yields of *A. tricolor* with a split application of 110 kg N/ha. Leaf number and stem diameter was not affected by N application but plant height, leaf area and fresh weight increased with increasing N from 0 to 110 kg/ha. Higher yield in amaranthus with the application of nitrogen was also reported by Keshar *et al.* (1981), Purushothaman (1978) and Mathai (1978). According to Singh *et al.* (1996), nitrogen application at 90 kg/ha produced the highest amaranth (genotype RRC 41) yields in 1992-94 (12.0 Mg/ha green yield and 1.57 Mg/ha dry matter yield). Growth rate, yield and protein content of *A. cruentus* were studied by Auwalu and Tenebe (1997) under rainfed conditions in the Guinea Savanna ecological zone of Nigeria. They opined that at the highest N rate (180 kg N/ha), the crop had a very high growth rate coefficient (0.47 t/ha/day) before cutting. The growth rate coefficient was markedly reduced (0.21 t/ha/day) after cutting.

Total shoot and leaf dry matter yields increased significantly up to 45 and 60 kg N/ha. The crude protein content of leaves in the first harvest was high and averaged about 30 per cent on a dry weight basis.

According to Guo and Lu (1991), *A. tricolor* grown in low or high fertility calcareous soils showed an intermediate K-uptake ability but lower resistance to K starvation. According to Amelin (1999), effect of potassium fertilizers on nitrate accumulation depended on the biological and genotypic properties of the plants. The greatest decrease in nitrate content from potassium fertilizer occurred in cloudy and wet years and with increased levels of soil fertility the effect of potassium decreased.

Singh *et al.* (1986b) reported that plant Ca content was not influenced appreciably by N application; Mg content was negatively influenced while N had an inverse relationship with plant Fe content. Content and uptake of sulphur by *A. cruentus* cv. MT-3 in relation to macro and micronutrient fertilizer application was studied by Bobrzecka *et al.* (2000) and it was found that amaranth had a total sulphur content higher than cereals (0.23-0.28 per cent S in seeds, 0.26-0.35 per cent S in glumes, 0.55-0.66 per cent S in leaves and 0.25-0.34 per cent S in stems after harvesting). An increase in NPK rate from 290 to 435 kg/ha and an application of 5.0 kg Zn/ha limited S content and widened N: S ratio in each analysed plant organ. Introducing 2.0 kg B/ha (in the presence of 435 kg NPK) increased S uptake to 66.6 kg S/ha and gave the highest yield. Results indicate that amaranth might be considered as a species with high S demand.

Subbiah (1979) studied the influence of N and K in *A. dubius* and *A. tricolor*, nitrogen increased the plant crude protein, carotene and chlorophyll contents but K had no such marked effects. The best results with regard to crude protein and carotene content were obtained with 60 kg N/ha applied in 2 split doses, the first as a basal dressing and the second 8 weeks after sowing of cv. CO. 3 (Singh *et al.*, 1986a).

Panda *et al.* (1991) recorded highest fresh yield (12.7 t/ha) and protein content (4.9 g/100 g) of *A. tricolor* with 60 kg N + 20 kg P_2O_5 + 20 kg K_2O/ha with N applied as a full basal dose. The same treatment but with N applied as 50 per cent basal + 50 per cent foliar spray resulted in the next highest yield and protein content (12.1 t/ha and 4.7 g/100 g, respectively). In a field experiment in 1985-87 at Brahmvar, in coastal Karnataka, India, 6 local amaranthus cultivars were given 50: 25: 25 kg NPK/ha or 2 or 3 times this NPK rate. Leaf yield increased with increasing NPK rate. The best yielding cultivar was CO.2 (Lingaiah *et al.*, 1997). From a study on the fertility requirements of vegetable amaranth (*A. tricolor*) in an unimproved mineral soil containing 59 kg N, 14 kg P and 84 kg K/ha, Makus (1992) concluded that P was the most critical nutrient needed by the crop. It was shown that in the conditions of the Central Volga region of Russia, where the soils have low content of nitrogen, phosphorus and particularly of potassium, moderate dose of all 3 fertilizers (0.2 g/ka soil of each) met the physiological demands of *A. cruentus* (Breus, 1999). Huelsman *et al.* (2000) observed significant positive correlations between the total pest populations and nitrate levels in the soil and significant negative correlations between pest predator populations and nitrate levels in amaranthus.

The behaviour of amaranth to trace metal additions through application of high rates of sewage sludge to soils were studied by Haroon and Ramulu (1990) at Madurai, Tamil Nadu when two crops were raised on the same plot in succession and treatments were only imposed on the first crop. The study revealed that the total trace elements (Fe, Mn, Zn, Cu, Ni, Cr, Co and Pb) content of the soil after the first harvest increased with increasing sludge application (from 40 to 80 t/ha). Soil trace element contents decreased after the second crop except for Zn which increased. Compared with untreated controls, sewage irrigation increased yields of the first crop and sewage sludge application increased yields of the first crop. The yield of amaranthus increased as sludge application rate was increased. The residual effects of sludge application were marked with higher yield from the second than from the first crop. There was considerable variation in trace element content, however sludge application tended to increase Cu and Zn contents above accepted toxic levels. In an experiment on a sandy clay soil of pH 7.6, amaranthus was given 5, 10, 15 or 20 t urban sewage sludge/ha and the second crop was also grown in the same plot without further organic manuring (Paulraj and Ramulu, 1994). The results were compared with the application of 20 kg N + 22 kg P + 25 kg K/ha. Yield increased as sewage sludge rate increased in both the first and second crops. Crop Fe, Mn, Zn, Cu and Pb uptake increased with increasing sludge rate.

In green house experiments conducted in a dernopodzolic soil poorly supplied with nutrients, amaranth (*A. cruentus*) plants were highly responsive to seed inoculation with *Azotobacter* isolated from the rhizospheres of different crops. Bacterial inoculation was most effective when mineral nitrogen fertilizers were applied in moderate doses (Ozhiganova *et al.*, 1999). Sreeramulu *et al.* (1996) suggested that VAM inoculation of amaranthus reduces the P input required to obtain maximum leaf yields.

8.0 Harvesting and Yield

The optimum stage of harvest in most of the types of amaranthus studies could be fixed between 25 and 30 days after sowing to get the highest yield as well as nutritious and palatable greens. Saha *et al.* (2003) reported that yield of greens and dry matter content of stem and leaf was found to be superior at 30 DAS over the other stages of harvest, indicating that red amaranth should be harvested 25 and 30 DAS in December and January sowing respectively. However, according to Materechera and Medupe (2006), frequent cutting of amaranthus is necessary for both winter and summer crop. They recommended harvesting of amaranthus at 2-week intervals to get a greater number of leaves, high fresh weight and leaf dry matter per plant. In *A. tristis*, a total of 10 clippings can be obtained with the first clipping starting on the 25th day and thereafter at weekly interval. In *A. blitum* (Sirukeerai), 6 clippings are made in a duration of 60 days. Olufolaji and Tayo (1989b) from Nigeria suggested that planting of the indeterminate, late-flowering cultivars of *A. cruentus* at the start of the rains and continuously cutting back is a more profitable method of harvesting than uprooting at the optimum commercial stage. Maboko and Du Plooy (2012) conducted one study on the effect of plant densities and harvesting methods on hydroponically grown amaranth and concluded that plant population of either 16 or 25 plants/m^2, harvested by cutting, increased production of amaranth, as a result of a reduced number of inflorescences and increased leaf area.

9.0 Diseases and Pests

9.1.0 Diseases

Diseases are sometimes a severe problem in amaranth. Though, leaf spot and white rust are two diseases which affect the amaranth plants to a great extent, other diseases frequently pose problem in some cases.

In case of leaf spot, *Cercospora* sp. is the causal organism affecting this crop. This disease is characterized by the presence of numerous small brown circular spots on the leaves. In the beginning, the spots are small, roundish with concentric rings but later on these spots increase in size and sometimes they coalesce. It may be controlled by spraying Bordeaux mixture (5: 5: 50) or 0.3 per cent Blitox three times at an interval of 15 days (Nath, 1976).

Harvesting of Amaranthus

Diseases of Amaranthus

Phomopsis amaranthicola Infection on Leaves

White Rust Infection

Damping Off (Left) and Anthracnose (Right) Infection

In case of white rust, *Albugo bliti* is the causal organism. It is characterized by white, blister like circular or irregular pustules on the lower surface of the leaf and opposite each pustule on the upper surface a yellow patch develops. Severe infection causes leaves to die and turn brown, giving the field a blighted look. The disease can be controlled by following crop rotation and by spraying Dithane M-45 or any other copper fungicide at the rate of 1 kg per 500 litres of water. Severe losses incurred in *A. hybridus* crops near Bloemfontein, South Africa during 1999 were attributed to a root rot of the tap and secondary roots, which affected more than 90 per cent of the crop (Chen and Swart, 2000). Samples of pathogens from diseased roots were isolated and the most commonly identified species were *Fusarium oxysporum* and *F. sambucinum* (*Gibberella pulicaris*). The first report of a foliar blight by *Rhizoctonia solani* on *A. tricolor* in India was made by Nayar *et al.* (1996). The pathogen severely infected the crop during the monsoon period (August-September) and produced cream coloured spots on leaves which spread rapidly resulting extensive damage and economic losses. *Phomopsis amaranthophila* sp. *nov.*, a new coelomycete pathogenic to *A. tricolor* in Brazil has been described and illustrated by Inacio *et al.* (1999) which caused severe blight on leaves and stems. In *A. hybridus*, it was also found to be the causal agent of the observed stem and leaf blight (Rosskopf *et al.*, 2000). Based on the morphological characteristics and DNA sequence information they described the species as a new one of *Phomopsis*. A severe disease of amaranthus characterized by extensive blighting of the inflorescence, often accompanied by inflorescence dieback and breaking of the floral head was observed by Adebanjo (1990) in Nigeria. The disease is severely affected the seed production and susceptibility of the disease decreased with increasing inflorescence age. On pathogenicity test, the causal organism was identified as *Choanephora cucurbitarum*. *C. cucurbitacearum* (*C. cucurbitarum*) was first detected on amaranthus causing necrosis of stalks and petioles in the Congo. Sexual spores left on amaranthus debris after harvest are the main sources of inoculum (Makambila and Goma, 1993).

The first record of bacterial leaf spot caused by *Burkholderia* (*Pseudomonas*) *andropogonis* in *A. cruentus* in Japan was described by Murai and Goto (1996). Early symptoms include the presence of dark green water soaked spots on leaf blades, usually beginning on the lower leaves. These enlarge to form circular spots several mm in diameter, light brown to reddish brown in colour. Amaranthus rugose mosaic, a disease caused by cucumber mosaic virus (CMV) showing mosaic, rugose and stunting symptoms in *A. mangostanus* was reported by Chen *et al.* (1995).

Ikediugwe *et al.* (1994) reported that isolates of *Bacillus subtilis* from soil and Ogili (a local food condiment prepared from watermelon seeds) controlled *Choanephora* shoot disease of *A. hybridus*, in the green house. Application of either bacterial strain to plants 1 day before challenging with *C. cucurbitarum* prevented disease development. A single application of either strain prevented disease development from a subsequent challenge with the pathogen at any time over a 30-day period. Viable counts of the microflora on the shoot tips of the treated plants indicated that the inoculated bacteria multiplied and colonized the extending shoot tip whether or not the shoots were covered with polythene bags. Both strains of *B. subtilis* inhibited mycelial extension growth and conidial germination *in vitro*.

Root-knot nematodes have also been recognised in Amaranthus. Of the 13 Amaran-thus cultivars screened for resistance against *M. incognita* by Mohanty and Das (1989), Bhubaneswar Local and Kantei Khada exhibited a resistant reaction with gall index below 2.0.

9.2.0 Pests

Amaranthus Stem Weevil [*Hypolixus truncatulus, H. nubilosus* (Coleoptera; Curculionidae)]

This is the major pest of amaranthus, larvae tunnel the stems and adults feed on the tender leaves. The pest is found to be distributed in India, Mexico, Nigeria, South Africa and Kenya Percentage infestation varies from 35 per cent to 81 per cent at different regions in India. They completed at least three generations from April to November and total life cycle takes almost 58-64 days (Tara *et al.*, 2009). For the management of pest, remove and destroy promptly all the affected plants with grubs inside. Spraying of Neem seed kernel extract 5 per cent (w/v) is helpful to reduce the pest infestation (Rajeshkanna *et al.*, 2017).

Beet Webworm [*Spoladea recurvalis* (Lepidoptera; Pyralidae)]

The pest is also known as Hawaiian beet webworm that attacks on beet, beans spinach and amaranthus. This pest is distributed tropical and sub-tropical regions of Asia, Africa and Australia. The young caterpillars feed on epidermis and voraciously feed on the green matter. The older ones web the leaves and feed within. Severe attack of this pest results in complete skeletonization and drying up of the leaves within a short time. Deep ploughing of the field helps in exposing and killing of the pupae. Spray Azadirachtin 0.03 per cent (300 ppm) @ 1000-2000 ml in 200-400 l of water/acre or Azadirachtin 5 per cent w/w neem extract concentrate @ 80 ml in 160 l of water/acre (Seni, 2018).

Leaf miner [*Liriomyza huidobrensis* (Diptera; Agromyzidae)]

Known as serpentine leaf miner, this pest is highly polyphagous in nature feeding more than 15 families of plants including the amaranthus. The maggots produce the serpentine leaf mines which are white with dampened black and dried brown areas,that usually associated with the midrib and lateral leaf veins. Several larvae feeding on a single leaf may produce a secondary blotch like mine and leaf wilt may occur. Picking and destruction of infested leaves may be effective in reducing the population at early stage of infestation. Spray of Azadirachtin 0.03 per cent (300 ppm) @ 1000-2000 ml in 200-400 l of water/acre or Azadirachtin 5 per cent w/w neem extract concentrate @ 80 ml in 160 l of water/acre could be helped in management of the pest (Seni, 2018).

Aphid [*Myzus persicae* (Aphididae: Hemiptera)]

This pest is also polyphagous with a host range of more than 400 species in 40 different plant families (Bass *et al.*, 2014), and distributed throughout the world except in areas with extreme temperatures or moisture. Both adults and nymphs cause damage by sucking plant sap those results in yellowing and drying of leaves. If the damage is more severe, plant may show stunted growth, and curling of leaves

Pests of Amaranthus

Adult Weevil Infestation in Stem

Larvae of Weevil in Stem

Stem Canker Due to Weevil Attack

Leaf Miner (Left) and Flea Beetles (Right) Infestation

may occur followed by gradual drying and death of tender region of the plants. Besides, they also transmit many important plant viruses. For the management, destroy the infested plant parts and Azadirachtin 5 per cent w/w neem extract concentrate @ 80 ml in 160 l of water/acre (Seni, 2018).

Besides these major insects, the incidence of grasshoppers such as *Atractomorpha crenulata* (Pyrgomorphidae: Orthoptera) and *Pyrgomorpha conica* (Pyrgomorphidae: Orthoptera) also found on amaranthus. Both adults and nymphs feed on the leaves and cut the tender shoots.

10.0 Seed Production

The agronomic practices normally followed for producing tender leaves and stems are also usually employed for seed production. In a soil of high alkalinity, the quality of seed will be very much affected. For seed crop, a spacing of 30 x 30 cm is usually maintained. From a study on the effect of sowing time and nitrogen level on seed yield of *A. tricolor*, Das and Ghosh (1999) reported that sowing during winter/rabi was best in terms of seed yield, followed by sowing during pre-kharif/ summer and kharif/rains. They further opined that yield components and seed yield increased with increasing nitrogen rate up to 120 kg/ha. The highest seed yield of 4.5 q/ha was obtained in rabi/November sown crops with 120 kg N/ha. Amaranth responds well to a fertilizer schedule of 50 kg each of N and P and 30 kg K per hectare in respect of seed yield. In addition to the soil application, spraying of urea at the rate of 12 kg N/ha and micronutrients at the rate of 400 g/ha once in 10 days from first flowering increased the yield of good quality seed by 88 per cent (Anon., 1984).

Since it is a cross-pollinated crop, an isolation distance of about 400 m has been recommended between two cultivars (Anon., 1971). It is a quick growing crop and forms seed in about 10-12 weeks. After taking a number of leaf cuttings, only a few last cuttings can be omitted to produce seeds. The seed grower should be well acquainted with the varietal characteristics and rogue out the off-type plants at different stages of growth.

Harvesting of inflorescence starts from 35 days after flowering when the glumes turn brown in colour and seeds turn black. Initial drying of the spikes in sun and drying the seeds to a15 per cent moisture content are usually practised so that the seeds can be threshed with pliable bamboo sticks. Seeds are then strained through 2 mm sieve. The seed yield is about 200 kg/ha and is slightly higher in *Badi Chaulai* (Nath, 1976).

Krishnakumary and Pradeep (1994) opined seed yield of *A. tricolor* was less under partial shade. They observed that seed yields were reduced when the crop was grown under 4-month old banana plants (5.10 g seeds/plant) and even more so when grown under 8-month old banana plants (2.90 g seeds/plant) compared with controls grown in the open (22.80 g seeds/plant).

Root pruning when seedling roots of *A. hypochondriacus* and *A. caudatus* were pruned to remove one-third or two-thirds of the length at transplanting reduced the plant height and leaf number. Plant height was particularly affected when pruning

Seed Production of Amaranthus

was severe. Root pruning stimulated lateral root development but had no net effect on root systems. It increased ethylene production, initially reduced leaf chlorophyll concentration but then increased it as new roots developed and had little effect on seed yield (Chakhatrakan *et al.*, 1994).

Dried seeds of 7 per cent moisture content are stored after treating the seeds with Bavistin at the rate of 2 g/kg of seed. The viability of seeds can be maintained up to 10 months in ordinary containers such as cloth bags and over 18 months in moisture proof containers such as 700 guage polythene bag or polythene laminated aluminium foil bags. The longevity of large and small seeds of *A. gangeticus* (*A. tricolor*) and *A. hybridus* stored at different moisture levels and temperatures was studied by Oladiran and Mumford (1990). Small seeds (<1.00 mm) could be stored for longer than large (>1.41 mm) ones. Within temperature ranges of 25-40 and 22-40°C for *A. hybridus* and *A. tricolor*, respectively, seeds retained viability better at lower moisture contents and storage temperatures. Time required for germination increased with seed age, with the rate of increase higher in more severe environments. Before storage, large seeds germinated faster than small ones, but this trend was reversed as seeds aged. Large seeds produced taller seedlings than did small seeds before storage, but not after storage. Frequency of abnormal seedlings increased with storage period, with the higher frequency generally in large seeds.

11.0 Crop Improvement

Higher leaf yield, longer harvest duration with multiple cutting, uniform growth habit, resistance to lodging, tolerance to low and high temperature, resistance to the insects especially sucking and stem borers, resistance to major diseases like damping off and blights, good nutritional profile that includes high protein and low oxalate content are some of the major breeding objectives in vegetable amaranthus.

Yield potential of most of the agricultural crops has been improved by introduction of hybrid breeding. However, in case of amaranthus, most of the lines are bred as line varieties with only little breeding advances, and enormous heterotic effects have been identified in interspecific crosses (Lehman *et al.*, 1991). Creation of interspecific heterotic groups could be successfully used in amaranthus breeding programs. Besides, homozygous parental lines are needed for hybrid production that can be produced by selfing or through double haploid technology. Further, mutation breeding has been effectively utilized in amaranth for cultivar development and for the creation of genetic variability. The varieties *viz.*, 'Centenario' (improved grain yield) in Peru, 'New Asutake' (early maturity) in Japan, 'Sterk' (tolerance to moisture and heat stress) in Russia, 'Pribina' in Slovakia (Gomez-Pando *et al.*, 2009; Das, 2016) have been developed by utilizing this breeding method (Joshi *et al.*, 2018). Male sterility identified in certain grain species (*A. hypochondriacus*) could be very handy to simplify the hybridization process (Das, 2016). Substantial advances in understanding the genome and modern genetic marker systems has been made in amaranth and these resources can be used to improve breeding methodology for amaranth (Stetter *et al.*, 2016). Many tissue culture-based direct shoot regeneration protocols are also available in Amaranthus (Swain *et al.*, 2010; Pal *et al.*, 2013; Munusamy *et al.*, 2013). In the recent years, genome editing technology has emerged as an effective tool to revolutionize basic research and trait development in crop plants by modifying genomes in a rapid yet precise and predictable manner (Bortesi and Fischer, 2015).

Collection, evaluation and characterization of germplasm or lines is very first step in genetic improvement of crop plants. The remarkable germplasm diversity, adaptability to diverse growing conditions and unique matting behaviour ranging from obligate outcrossing (dioecious species) to greater outcrossing to greater self-pollination has led this crop to exhibits wide genetic variability for many of the agronomical traits. Thus, it offers a substantial scope to identify suitable type for any region. In amaranthus, many of the improved cultivars are results of the selection, and negligible efforts have been made for its genetic improvement through the hybridisation programme due to variability and versatility of inflorescence, ratio and distribution of male and pistillate flowers in inflorescence. Thus, development of efficient crossing methods is important for research and improvement of amaranthus. Stetter *et al.* (2016) evaluated three different crossing methods (open pollination, hot water emasculation and hand emasculation) and reported that all three methods successfully produced F_1 hybrid, but with different success rates. Open pollination had the lowest (10 per cent) and hand emasculation the highest success rate (74 per cent). Hot water emasculation showed an intermediate success rate (26 per cent) with a maximum of 94 per cent success. Wu *et al.* (2000) evaluated 20 species of amaranthus at two location, Wuhan and Beijing of China, and revealed wide diversity in agronomic traits among species and among genotypes within the same species. They further reported that many of these species were sensitive to day length and cultivated genotypes generally had higher yield but were more seriously affected by diseases than non-cultivated species. *A. caudatus* genotypes originating from Asian countries produced more leaves (25.8±8.9), while *A. caudatus* from the US only produced 12.2 leaves. Similarly, *A. cruentus*, genotypes from South America tended to have more leaves (71.2±33.7) than others. Of *A. hybridus* genotypes, those from the US had fewest leaves (30.0±8.) and those from Africa had the most (74.8±11.7). Akaneme and Ani (2013) studied genetic variability in five accessions of the *A. hybridus* and revealed highly significant differences for leaf width, hypocotyls length, days to 50 per cent flowering, 500-seed weight and leaf length. Gerrano *et al.* (2014) studied thirty-two Amaranthus accession from *A. cruentus*, *A. tricolor*, *A. greazicans*, *A. viridus* and *A. caudates* and reported high diversity index (HI) for branching index (0.70), stem pigmentation (0.53); leaf pigmentation (0.63); terminal inflorescence shape (0.70) and inflorescence colour (0.51). Alemu *et al.* (2018) evaluated five genotypes of amaranthus under rainfed and irrigated conditions of Ethiopia, introduced from AVRDC, Taiwan. The total leaf yield ranged from10.08 t ha^{-1} to 15.82 t ha^{-1}. Madiira-II was superior in terms of leaf yield (15.82 t ha^{-1}) among genotypes followed by genotype AC-NL (14.65 t ha^{-1}); whereas AM-25was the least performing genotype (11.18 t ha^{-1}). Madiira -II and AC-NL gave 25.3 per cent and 16 per cent yield advantage over the grand mean (12.63 t ha^{-1}).

In India, Shukla *et al.* (2006) evaluated nine genotypes of *A. tricolor* and reported two genotypes *viz.*, AV-38 (5.06 kg/plot) and AV-31 (5.04 kg/plot) with highest foliage yield, followed by AV-30 (4.78 kg/plot) and AV-23 (4.70 kg/plot). The protein and carotenoid content averaged 1.24±0.03 mg/100 mg and 0.83±0.02 mg/g, respectively. The leaves have considerable quantities of ascorbic acid (112.33±5.00 mg/100g) and fibre (8.39±0.10 per cent). Joshi *et al.* (2011) observed thirty one accessions of amaranthus and found that germplasm showed a wide range of variability in plant height (46.0-149 cm), number of branches (5.5-20.5), leaf length (10.0-24.5 cm), leaf width (5.2-12.7 cm), petiole length (5.7-12.7 cm) leaf weight per plant (7.0-119.0 g), stem weight per plant (28.0-975.0 g) and seed yield per plant (6.0- 58.0 g). Mandal *et al.* (2012) observed wide range of variation for yield ha^{-1} (55.8 to 303.9 q ha^{-1}). Further, two genotypes, Bankura Collection 3 (303.9 q ha^{-1}) and Bolpur Collection 1 (287.0 q ha^{-1}) produced highest yield ha^{-1}, followed by Pusa Kirti (283.5 q ha^{-1}). Kumar *et al.* (2015) evaluated twenty-five genotypes of amaranthus, collected from different part of Chhattisgarh and reported highest leaf yield in genotype IGA-2013-1 (72.50 q ha^{-1}). A wide range in the traits has been reported by the workers. Eighty-nine diverse accessions of amaranthus were evaluated at Thiruvananthapuram, Kerala by Celine *et al.* (2007). They reported leaf yield variation from 120 to 1132.5 g/plant. The top yielder was Am 88 followed by Am 7 and Am 86 and these accessions belonged to *A. dubius*. Among the quality parameter studied, Am 5 had the maximum b-carotene content of 4655.5 µg/100 g while Am 90 had the minimum. Vitamin C was highest in am 58 (151.2 mg/100 g) whereas the protein content was maximum (3.5 per cent) in Am 91. Oxalate was minimum in Am 90 (0.6 per cent) and maximum in Am 76 (2.1 per cent). Nitrate content varied from 0.04 per cent to 1.6 per cent.

11.1.0 Association of Characters

Yield is a complex trait and indirect selection through correlated, less complex and easier measurable traits would be an advisable strategy to increase the yield (Devi *et al.*, 2015). Further, effectiveness of indirect selection depends upon the magnitude of association between various traits. In amaranthus, a positive correlation has been reported between green leaf yield with leaves per plant, stem diameter, stem weight per plant, leaf weight per plant and plant height (Ahammed *et al.*, 2012); leaf yield with plant height, leaves per plant, diameter of stem base, fibre content and leaf area (Sarker *et al.*, 2014); harvest index with days to maturity (Patial *et al.*, 2014); 50 per cent flowering with leaf length and stem diameter (Akaneme and Ani, 2013); plant height with fresh biomass and dry biomass (Abe *et al.*, 2015); grain yield with stem girth, number of leaves per plant, plant height, panicle length and seed weight (Venkatesh *et al.*, 2014); grain yield with inflorescence length and biological yield per plant (Shrivastav *et al.*, 2020). Among the biochemical parameter, a positive correlation is reported between leaf yield and carotenoid content (Shukla *et al.*, 2015), leaf yield with total chlorophyll content, protein content, ascorbic acid, moisture content and folic acid (Tejaswini, 2017). A negative correlation has been reported between the pair of traits *viz.*, protein content with biological yield/plant; grain yield with days to 50 per cent flowering (Shrivastav *et al.*, 2020); potassium with dry matter content and protein with dry matter (Vijayakumar, 1980). Thus, indirect selection for the traits *viz.*, higher leaves number per plant, longer leaves with more leaf area and weight, high stem weight and diameter, more harvest index could result in high leaf yield per plant.

12.0 Biotechnology

Biotechnological research on amaranthus is mainly concentrated on isolation and expression of *AmA 1* gene encoding storage protein besides, tissue culture and genetic transformation for transfer of this gene.

12.1.0 Tissue Culture

12.1.1 Micropropagation

Probably the first report on in vitro growth and morphogenetic response was published by Flores *et al.* (1982). Leaf discs and hypocotyl segments from 2 to 3-week-old seedlings of *A. hypochondriacus, A. cruentus,* and *A. tricolor* were cultured in B_5 and MS media supplemented with 2,4-dichlorophenoxyacetic acid, α-naphthalene acetic acid, benzyladenine and zeatin. Callus and abnormal roots formed on leaf discs of *A. hypochondriacus* and *A. cruentus* in the presence of 0.1-1 0 mg/l 2,4-D. Lower ratios of zeatin/NAA stimulated root formation from hypocotyl segments and viable mesophyll protoplasts were isolated from primary leaves of all three species, with an enzyme solution of 1 per cent Cellulysin and 0.05 per cent Pectolyase Y-23, producing a yield of more than 106 protoplasts/g fresh weight. Micropropagation research on tissue culture of *Amaranthus* has been reported on somatic embryogenesis and protoplast fusion. Bagga *et al.* (1987) induced callus from 4-7 days old seedlings of *A. paniculatus* [*A. cruentus*] hypocotyl segments, 5 to 8 mm in length, on Gamborg's B_5 medium supplemented with kinetin at 0.5 ppm and NAA at 0.1 ppm. About 20 per cent of these cultures produced multiple shoots. In the medium with 1 ppm each of kinetin and NAA shoots were formed at one end of the segment and callus was initiated at the other. The plants obtained in either medium formed roots and were transferred to soil for further growth. *In vitro* flowering was induced by Tisserat and Galletta (1988). Shoot tip-derived plantlets of *A. caudatus* cv. Pan, *A. gangeticus* [*A. tricolor*], *A. hypochondriacus, A. retroflexus* and *A. viridis* flowered *in vitro* following 8 to 32 weeks in culture. Shoot tips were cultured on MS salts and (per litre) 30 g sucrose, 100 mg myoinositol, 0.4 mg thiamine-HCl and 8 g agar. Addition of 0.1 mg/l NAA enhanced inflorescence production but was not necessary for flower induction. The fruits produced by *A. gangeticus* and *A. retroflexus* dehisced and their seeds dropped on the surface of agar medium, where they germinated immediately. The seedlings so derived could, in turn, give rise to inflorescences. Inflorescences could be detached from flowering plantlets and grown successfully for several months on basal nutrient medium containing 0.1 mg/l NAA. According to Comia-Yebron *et al.* (2017) callus formation in *A. tricolor* and *A. spinosus* was faster with a combination of 6-benzylaminopurine (BAP) and 2,4- dichlorophenoxyacetic acid (2,4-D) than with α-naphthaleneacetic acid (NAA). In both species, BAP treatments caused profuse callus growth, but *A. tricolor* favored NAA, while *A. spinosus* was more responsive to 2,4-D. Yaacob *et al.* (2012) carried out the pigment analysis and tissue culture of *Amaranthus cruentus* L. They cultured leaves and stems of amaranthus on Murashige and Skoog (MS) media, supplemented with different hormone combinations and incubated at 24-26°C with 16 h of light and 8 h of dark to produce cream, red and green coloured callus. They reported that most optimum media for the induction of red callus was MS media supplemented with 1.5 mg/L IAA and 1.5 mg/L zeatin, while the most responsive explant was the leaves. Further, the tests confirmed that the red pigments were flavonoids called β-cyanins.

12.1.2 Protoplast Fusion

The electrical fusion technique of Zimmermann and Scheurich was used to fuse mesophyll protoplasts (Bates *et al.*, 1983). Electrical fusion proved a simple, effec-tive, and general fusion techniques that could be controlled to form either dikaryons or large multinucleate fusion bodies. Mastuti *et al.* (1997) obtained intergeneric somatic hybrid calluses by electrofusing protoplasts of *Celosia cristata* (C_3) and *Amaranthus tricolor* (C_4) derived from cell suspensions and cotyledons, respectively. The highest binary fusion frequency (4-5 per cent) was attained when protoplasts were suspended in a fusion solution containing 0.4 M sorbitol and 2.0 mM $CaCl_2$, aligned by an AC field (0.5 MHz at 200 V/cm) for 30s and fused by a single DC square pulse at 1.6 kV/cm for 40 μs. Fusion products were identified by characteristics of both parents in the hybrid cells: they possessed green chloroplasts or pink plastids from cotyledonary protoplasts and a cytoplasm-rich colourless sector derived from cell suspension protoplasts. A total of 14 lines of putative hybrid compact calluses were obtained by screening on the basis of callus morphology. An isozyme analysis of acid phosphatase confirmed the hybridity of the 14 lines and callus morphology was also studied using scanning electron microscopy. Unlike the parent line, the hybrid calli between C_3 and C_4 species of Amaranthaceae have poor division capabiity (Mastuti *et al.*, 1998). It might also be related to the pronounced expansion of central vacuole and the reduced cytoplasm.

12.2.0 Molecular Markers

Molecular markers are crucial tools for genetic characterization of plants with a high level of reproducibility and accuracy and to overcome the limitation of morphological characterization. Molecular genetic diversity and population structure of Amaranthus species accessions were studied by various workers by using RAPD (Ranade *et al.*, 1977; Popa *et al.*, 2010; Stefunova *et al.*, 2015; Pandey *et al.*, 2019), SSR primers (Raut *et al.*, 2014; Suresh *et al.*, 2014; Oduwaye *et al.*, 2014; Stefunova *et al.*, 2015; Gelotar, *et al.*, 2019; Nguyen *et al.*, 2019) and *matK* marker (Nguyen *et al.*, 2019). Gelotar *et al.* (2019) reported 10.36 average bands per primer as compared to Raut *et al.* (2014) and Oduwaye *et al.* (2014) who reported 4.41 and 6.57 average bands per ISSR primers in amaranth genotypes, respectively. However, Ray and Roy (2009) and Ziarovska *et al.* (2013) found 11.36 and 16.00 bands per primer, respectively.

12.3.0 Genetic Transformation

The major research on amaranthus has been the isolation and expression of *AmA1*gene encoding storage protein of *A. hypochondriacus*. Kim and Kim (2000) used this gene to transform the lettuce. Chakraborty *et al.* (2000) could able to

Genetic Transformation in Amaranthus

Induction and Establishment of Transformed Hairy Roots followed by Plant Regeneration.

(A) Emergence of hairy roots at the inoculated site of an internode at day 20 following infection with LBA9402 suspension (with acetosyringone pretreatment), (B) Emergence of hairy roots along leaf midrib at day 20 following infection with LBA9402 suspension (with acetosyringone pre-treatment), (C) A proliferated hairy root clone (HR3B9) at day 30 following transfer of a single root on growth regulator-free medium (MS0), (D) Leafy shoot regeneration from the hairy root clone HR3B9 inMS+ 2.0 mg/l zeatin. (E) A potted plant regenerated from HR3B9 rhizoclone (Swain *et al.,* 2010).

increase the nutritive value of transgenic potato by expressing a non-allergenic seed albumin gene *AmA 1* from *Amaranthus hypochondriacus.* Jofre Garfias *et al.* (1997) used mature embryos of *A. hypochondriacus* (amaranth) to develop in *in vitro* culture system for plant regeneration and genetic transformation. Plants were regenerated from embryo-derived callus cultivated on MS medium supplemented with 10 µM 2,4-D or 3,6-dichloro-2-methoxybenzoic acid and 10 per cent coconut liquid endosperm. Transgenic plants were obtained by inoculation of mature embryo explants with a disarmed *Agrobacterium* strain containing the plasmid pGV2260 (pEsc4), which carried the genes encoding neomycin phosphotransferase type II and β-glucuronidase. The presence of transgenes in the genome of transformed amaranth plants and their progeny was demonstrated by Southern blot hybridization. Tissue specific and light-inducible expression directed by a pea chlorophyll a/b-binding protein promoter was observed in transgenic amaranth plants and their progeny.

Swain *et al.* (2010) developed an efficient protocol for rapid and large scale production of hairy roots of *A. tricolor* via genetic transformation using two selected virulent strains of *A. rhizogenes*. This study has also been able to reveal the key bacterial factors (strain type, growth phase, cell density, vilulence inducer) as well as the plant factors (explant source and type, infection method, co-cultivation period, *etc.*) influencing transformation events. Establishment of renewable hairy root cultures of *A. tricolor* as a result of the present investigation, could not only provide a direct source of phytochemicals of medicinal significance and similarly of other amaranths, but also can indirectly help conserve the natural germplasm of pharmaceutically valuable plant species in general whose roots are being ruthlessly over-exploited. Further, successful plant regeneration from selected rhizoclones and demonstration of the transformed status of regenerated plants open up the possibility towards introgression of agronomically useful transgenes such as those conferring resistance to one or other biotic stress factors (foliar insects, viruses or fungal pathogens) aimed at genetic improvement of crop amaranths of immense horticultural significance.

13.0 References

Abe, S., Gerrano, W.S., Rensburg, J.V., Mavengahama, S., Bairu, M., Venter, S. and Adebola, P.O. (2017) *J. Tropic Agric.*,**55**: 12-20

Abe, S.G., Willem, S. and Patrick, O.A. (2015) *South African J. Plant Soil*, **32:** 39-46.

Adebanjo, A. (1990) *J. Phytopath.*, **128**: 146-152.

Adebanjo, A. (1994) *Ambio.*, **23**: 206-209.

Ahammed, A.U., Rahman, M.M. and Mian, M.A.K. (2012) *Bangladesh J. Plant Breed. Gen.*, **25:** 25-32.

Akaneme, F.I. and Ani, G.O. (2013) *World App. Sci J.*, **28:** 568-577.

Alemu, Y., Tilahun, S., Habtam, S., Selamawit, K. and Melkamu, H. (2018) *J. Biol., Agric. Healthcare.*, **8:** 47-50.

Amelin, A.A. (1999) *Agrokhimiya* No. 9, pp. 29-36.

Andrianova, Yu, E., Safina, N.I., Maksyutova, N.I. and Kadoshinikova, I.G. (1996) *Agrokhimiya*, **8-9**: 117-122.

Anonymous (1971) *Indian Minimum seed Certification Standard*, Central Seed Committee, Min. Food Agric. Comm. Dev. Co-op., New Delhi.

Anonymous (1981) Annual report, Faculty of Horticulture, Tamil Nadu Agric. Univ., Coimbatore.

Anonymous (1984) In: *Seed Technology Scientific Worker's Conference*, Tamil Nadu Agricultural University, Coimbatore, p. 68.

Auwalu, B.M. and Tenebe, V.A. (1997) *Adv. Hort. Sci.*, **11**: 70-72.

Ayodele, V. I., Ferreira, M. I. and Jones, H. G. (2000) *Acta Hortic.*, **537**: 767-772.

Bagga, S., Venkateswarlu, K. and Sopory, S.K. (1987) *Plant Cell Rep.*, **6**: 183-184.

Bass, C., Puinean, A.N., Zimmer, C.T., Denholm, I., Field, L.M., Foster, S.P., Gutboard, O., Nauen, R., Slater, R. and Williamson, M.S. (2014) *Insect Biochem. Mol. Bio.*, **51:** 41–51.

Bates, G.W., Gaynor, J.J. and Shekhawat, N.S. (1983) *Plant Physiol.*, **72**: 1110-1113.

Behara, B. and Patnaik, S.N. (1974) *Cytologia*, **39**: 121-131.

Bobrzecka D., Wojciechowska B., Procyk Z. and Wojtas A. (2000) *Oilseed Crops*, **21**: 301–308.

Bortesi, L. and Fischer, R. (2015) *Biotechnol. Adv.*, **33:** 41–52.

Breus, I.P. (1999) *Agrokhimiya*, No. 3, pp. 40-47.

Breus, I.P. (1999) *Sel'skokhozyaistvennaya Biologiya*, No. 2, pp. 80-89.

Breus, I.P., Avdeeva, A.A., Myasoedov, N.A., Balnokin, Yu. V., Arkhipova, N.S. and Chernov, I.A. (1994) *Russian Agril. Sci.*, No. 3, pp. 16-19.

Celine, V.A., Shankaran, S.S., Seema, S., Deepa, S.N., Sreelathakumary, I. and Abdul Vahab, M. (2007) *Acta Hortic.*, **752:** 447-452.

Celine, V.A., Sindhu, L., Rajamony, L. (2011) *Indian J. Hort.*, **68**: 131-135.

Chakhatrakan, S., Tamai, F. and Motoda, Y. (1994) *J. Agril. Sci.*, Tokyo, **39**: 10-20.

Chakraborty, S., Chakraborty, N. and Datta, A. (2000) *PNAS*, **97**: 3724-3729.

Chandwade, P.M., Desai, U.T., Chaudhuri, S.M. and Kale, P.N. (1994) *J. Maharashtra Agric. Univ.*, **19**: 280-281.

Chen, H.B. and Lin, S.C. (1997) *Chinese J. Agromet.*, **4**: 231-236.

Chen, W. and Swart, W.J. (2000) *Plant Dis.*, **84**: 101.

Chen, Y.K., Lin, W.F. and Chen, M.J. (1995) *Plant Protect. Bull. Taipei*, **37**: 97-106.

Chukwuma, G.O. (1988) *Int. Agrophy.*, **4**: 135-142.

Comia-Yebron. R., Aspuria, E.T. and Bernardo, E.L. (2017) *J. ISSAAS*, **23:** 12-23.

Das, N.R. and Ghosh, N. (1999) *Hort. J.*, **12**: 77-82.

Das, S. (2016) In: *Breeding of Amaranths*, pp. 107–145.

Devi, J., Sharma, A., Singh, Y., Katoch, V. and Sharma, K. C. (2015) *Leg. Res.*, **38:** 149.

Dudai, N, Polijakoff Mayber, A., Mayer, A.M., Putievsky, E. and Lerner, H.R. (1999) *J. Chem. Eco.*, **25**: 1079-1089.

Fasheun, A. (1989) *Res. Bull. Nat. Hort. Res. Inst.*, No. 6, p. 5.

Flores, H.E., Their, A. and Galston, W. (1982) *Amer. J. Bot.*, **69**: 1049–1054.

Gelinas, B., Seguin, P. (2007) *J. Agric. Food Chem.*, **55:** 4789-94.

Gelotar, M.J., Dharajiya, D.T., Solanki, S.D., Prajapati, N.N. and Tiwari, K.K. (2019) *Bull. Natl. Res. Cent.*, **43:** 103.

Gerrano, A.S., Rensburg, W.S.J. and Adebola, P.O. (2014) *Acta Hortic.*, **11:** 183-187.

Ghanti, P., Som, M.G. and Chattopadhyay, P.K. (1983) *The Lal Baugh J.*, **28**: 57-60.

Gomez-Pando, L., Eguiluz, A., Jimenez, J., Falconi, J. and Aguilar, E.H. (2009) Food and Agriculture Organization of the United Nations, Rome, pp. 330–332.

Grant, W.F. (1959) *Can. J. Genet. Cytol.*, **1**: 313-328.

Guo, L., Wang, Y., Bi, X., Duo, K., Sun, Q., Yun, X., Zhang, Y., Fei, P. and Han, J. (2020) *Foods*, **9**: 359-373.

Guo,Y.L. and Lu, J.L. (1991) *Acta Hortic. Sinica*, **18**: 61-66.

Haroon, A.R.M. and Ramulu, U.S. (1990) *Transaction 14th Int. Cong. Soil Sci.*, Kyoto, Japan, 12-18 Aug., 1990, **IV:** 193-197.

Hilman, Y. and Abidin, Z. (1987) *Bull. Penelitian Hortikultura*, **15**: 22-29.

Hu, H. *et al.* (2020) *Int. J. Bio.Macro.*, **159**: 675–684.

Huelsman, M.F., Edwards, C.A., Lawrence, J. and Clarke-Harris, D.O. (2000) *The BCPC Conf., Pests and Diseases*, **3**: 895-900.

Igbokwe, P.E. and Hollins, S. (2000) *J. Veg. Crop Prodn.*, **6**: 75-85.

Ikediugwe, F.E.O., Emoghene, A.O. and Ajiodo, P.O. (1994) *J. Hort. Sci.*, **69**: 351-356.

Inacio, C. A., Dianese, J. C. and Carlos, R. M. E. (1999) *Fitopatol. Bras.*, **24**: 185-189.

Jijiamma, N. C. and Prema, L. (1993) *J. Trop. Agric.*, **31**: 219-226.

Jimenez-Aguilar, D. M. and Grusak, M. A. (2017) *J. Food Compost Anal.*, **58:** 33–39.

Jofre Garfias, A.E., Villegas Sepulveda, N., Cabrera Ponce, J.L., Adame Alvarez, R.M., Herrera Estrella, L. and Simpson, J. (1997) *Plant Cell. Rep.*, **16**: 847-852.

Joshi, D.C., Sood, S., Hosahatti, R., Kant, L., Pattanayak, A., Kumar, A., Yadav, D and Stetter, M.J. (2018) *Theor. Appl. Genet.*, **131:** 1807–1823.

Joshi, V., Vijaya, M., Sireesha, K. and Madhavi, P. (2011) *Veg. Sci.*, **38**: 239-240.

Kader Mohideen, M. and Rajagopal, A. (1975) *South Indian Hort.*, **23**: 87-90.

Kepezynski, J. and Knypl, J.S. (1989) *Acta Hortic.*, **251**: 35-39.

Keshar, B.G., Bhore, D.P., Patil, A.V., Sonone, H.N. and Maslekar, S.R. (1981) *J. Maharashtra Agric. Univ.*, **6**: 68-69.

Khanam, U. K. S., Oba, S., Yanase, E. and Murakami, Y. (2012) *J. Functional Foods*, **4**: 979–987.

Khoshoo, T.N. and Pal, M. (1970) *Chromosomes Today*, **3**: 259-267.

Kim, T.G. and Kim, Y.S. (2000) *J. Korean Soc. Hort. Sci.*, **41**: 495-498.

Kowal, T. (1954) *Monograph Botany*, **2**: 162-193.

Krishnakumary, K. and Pradeep, K.P. (1994) *Madras Agric. J.*, **81**: 675-676.

Kumar, A.S.A. and Pillai, S.J. (1990) *Indian Coconut J.*, **20**: 8-9.

Kumar, S.R. and Yassin M. (2013) *Veg. Sci.*, **38**: 239-240.

Leela, D. (1993) *Indian Hort.*, **38**: 13-15.

Lehman, J.W., Clark, R.L. and Frey, K.J. (1991) *Crop Sci.*, **31**: 1111-1116.

Lingaiah, H. B., Uthaiah, B. C., Herele, P. S. and Nagabhushanam, K. (1997) *Karnataka J. Agric. Sci.*, **10**: 253-256.

Maboko, M. M., and Du-Plooy, C.P. (2012) *Acta Hortic.*, **947:** 415–421.

Makambila, C. and Goma, J.B. (1993) *Cah. Agric.*, **2**: 217-219.

Makus, D. G. (1992) *Acta Hortic.*, **318**: 291-298.

Malik, N.A. and Ahmed, A.J. (1963) *Pakistan J. Sci. Res.*, **15**: 58-60.

Mandal, J. and Dhangrah, V.K. (2012) *Envir and Eco.*, **30:** 1430- 1433.

Mastuti, R., Miyake, H., Taniguchi, T. and Takeoka, Y. (1997) *Japanese J. Crop Sci.*, **66**: 456-465.

Mastuti, R., Miyake, H., Taniguchi, T. and Takeoka, Y. (1998) *Plant Prodn. Sci.*, **1**: 136-144.

Materechera, A. and Medupe, M. L. (2006) *Bio. Agric. Hort.*, **23**: 251-262

Mathai, P.J. (1978) *Indian Fmg.*, **28**: 29-32.

Mihalik, E., Maraz, L., Szocz, Z., Franz, Ch., Seitz, R. and Verlet, N. (1993) *Acta Hortic.*, **333**: 295-298.

Mohanty, K.C. and Das, S.N. (1989) *Indian J. Nematol.*, **18**: 120.

Mosyakin, S.L. and Robertson, K.R. (2003) In: Flora of North America, North of Mexico, vol 4: Magnoliophyta, Caryophillidae, Part I. Oxford University Press, New York, pp. 410–435.

Munusamy, U., Abdullah, S.N.A., Aziz, M.A. and Khazaai, H. (2013) *Adv. Biosci. Biotechnol.*, **4**: 188–192.

Murai, Y. and Goto, M. (1996) *Ann. Phytopathol. Soc. Japan*, **62**: 181-183.

Murray, M.J. (1940) *J. Hered.*, **31**: 477-485.

Murugan, S.B., Reshma, A., Ramamoorthy, D., Balamurugan, S. and Sathishkumar R. (2013). *J. Med. Plant Res.* **7:** 2230-2235.

Musa, A., Oladiran, J.A., Ezenwa, M.I.S., Akanya, H.O. and Ogbadoyi, E.O. (2011) *Amer. J. Food Nutr.* **1:** 147–54.

Nath, P. (1976) In: *Vegetables for the Tropical Region*, ICAR, New Delhi.

Nawawi, M., Santoso, W.I. and Notodimejo (1986) *Agrivita*, **8/9**: 17-18.

Nayar, K., Gokulapalan, C. and Nair, M.C. (1996) *Indian Phytopath.*, **49**: 407.

Nguyen, D.C., Tran, D.S., Tran, T.T.H., Ohsawa, R. and Yoshioka, Y. (2019) *Breed. Sci.*, **69:** 640-650.

Norman, J.C. and Shongwe, V.D. (1993) *Adv. Hort. Sci.*, **7**: 169-172.

Oduwaye, O., Baranek, M., Cechova, J. and Raddova, J. (2014) *African J. Plant Breed. Crop Sci.*, **6:** 48–56.

Ohshiro, M., Amzad, H.M., Nakamura, I., Akamine, H., Masanobu, T., Bhowmik, P.C. and Nose, A. (2016) *Plant Production Sci.*, **19**: 61-72

Oladiran, J.A. and Mumford, P.M. (1990) *Seed Sci.Tech.*, **18**: 499-506.

Olufolaji, A.O. and Dinakin, M.J. (1988) *Tests Agrochem. Culti.*, **9**: 100-101.

Olufolaji, A.O. and Tayo, T.O. (1989a) *Samaru J. Agric. Res.*, **6**: 83-86.

Olufolaji, A.O. and Tayo, T.O. (1989b) *Trop. Agric.*, **66**: 273-276.

Ozhiganova, G.U., Degtyareva, I.A., Chernov, I.A. and Breus, I.P. (1999) *Russian Agril. Sci.*, No. 11, pp. 12-24.

Pal, M. and Khoshoo, T.N. (1972) *J. Hered.*, **63**: 78-82.

Pal, M. and Khoshoo, T.N. (1974) *Evolutionary Studies in World Crops: Diversity and Change in the Indian Subcontinent*, Combridge University Press, pp. 129-137.

Pal. A., Swain, S.S., Das, A.B., Mukherjee, A.K. and Chand, P.K. (2013) *In Vitro Cell Dev. Biol. Plant*, **49:** 114–128.

Panda, J.M., Sahu, S.K. and Nauyak, J.K. (1991) *Orissa J. Hort.*, **19**: 64-68.

Pandey, V.K., Jatt, B.L. and Ratnaker (2019) *J. Pharm. Phytochem.*, **SP3**: 83-89.

Patial, M, Chauhan, A, Singh, K.P. and Sharma, D. (2014) *Int. Agric., Envir. Biotech.*, **6:** 101-106.

Paulraj, C. and Ramulu, U.S. (1994) *J. Indian Soc. Soil Sci.*, **42**: 485-487.

Peter, K., and Gandhi, P. (2017) *Egyptian J. Basic Appl. Sci.*, **4**(3): 196–205.

Popa, G., Cornea, C.P., Ciuca, M., Babeanu, N., Popa, O. and Marin, D. (2010) *Analele Universitatii din Oradea-Fascicula Biologie*, **17:** 280–285.

Purushothaman, V. (1978) *Agric. J.*, **51**: 366-374.

Qin, L., Guo, S., Weidang, Ai., Yongkang, T., Cheng, Q. and Chen, G. (2013) *Adv. Space Res.*, **51**: 476–482.

Rajeshkanna, S., Sivaraga, N. and Mikunthan, G. (2017) *Annals of Sri Lanka Department of Agriculture*, **19:** 258 – 266.

Ranade, S.A., Kumar, A., Goswami, M., Farooqui, N. and Sane, P.V. (1997) *J. Biosci.*, **22**: 457-464.

Rattan, R.K., Datta, S.P., Chhonkar, P.K., Suribabu, K. and Sing, A.K. (2005) *Agric. Ecos. Environ.*, **109:** 310–322

Raut. V.R., Dodake, S.S. and Chimote, V.P. (2014) *Indian J. Agric. Biochem.*, **27:** 60–65.

Ray, T. and Roy S.C. (2009) *J. Hered.*, **100:** 338–347.

Rosskopf, E.N., Charudattan, R., Shabana, Y. and Benny, G. (2000) *Mycologia,* **92:** 114-122.

Saha, S.R., Nazim Uddin, M., Rahman, M.A., Sharifuzzaman, S.M. and Anwar-Habib, A.K.M. (2003) *Asian J. Plant Sci.,* **2**: 431-433.

Said, S. and Lamm, F.R. (1995) *Proc. Fifth Int. Microirrigation Cong., Orlando, Florida, USA, 2-6 April, 1995,* pp. 436-441.

Sarker, U. and Oba, S. (2020) *Sci. Rep.,* 10: 1–9.

Sarker, U., Hossain, M.M. and Oba, S. (2020) *Sci. Rep.*, **10**: 1336.

Sarker, U., Islam, T, Rabbani, G. and Oba, S. (2014) *J. Food Agric. Environ.*, **12:** 168- 174.

Sauer, J.D. (1976) *Evolution of Crop Plants* (Ed. N.W. Simmonds), Longman, London and New York, pp. 4-7.

Sauer, J.D. (1980) *Ann. Mo. Bot. Gdn.*, **54**: 103-137.

Sealy, R.L., Kenerley, C.M. and McWilliams, E.L. (1990) *HortScience*, **25**: 293-294.

Sealy, R.L., McWilliams, E.L., Novak, J., Fong, F. and Kenerley, C.M. (1990) In: *Proc. 1st National Symposium, New Crops: research, development, economics* (Eds. J. Janik, and J.E. Simon), Timber Press, Oregon, USA, pp. 396-398.

Seni A. (2018) *Inter. Environ. Agric. Biotech.*, **3:** 1100-1103.

Seth, A.K. (1968) *J. Post Grad. School*, IARI, **6**: 81-82.

Shrivastav, S.P., Yadav, C.B., Singh V., and Maurya, V. (2020) *Int. J. Curr.Microbiol. App. Sci.*, **9:** 345-360.

Shukla, S., Bhargava, A., Chatterjee, A., Srivastava, A. and Singh, S.P. (2006) *Euphytica,* **151:** 103–110.

Shukla, S., Upadhyay, K.K. and Mishra, B. K. (2015) *Int. J. Veg. Sci.*, **22**: 322–332.

Singh, K. (1976) *Manurial Requirement of Vegetables*, ICAR, New Delhi.

Singh, N. and Singh, P. (2011) In: *Flour and Breads and Their Fortification in Health and Disease Prevention*, pp. 101–111.

Singh, U.C., Sundararajan, S. and Veeraragavathatham, D. (1985) *South Indian Hort.,* **33**: 100-102.

Singh, U.C., sundararajan, S. and Veeraragavathatham, D. (1986a) *South Indian Hort.,* **34**: 150-153.

Singh, U.C., Sundararajan, S. and Veeraragavathatham, D. (1986b) *South Indian Hort.,* **34**: 34-42.

Sreeramulu, K.R., Shetty, Y.V. and Setty, T.K.P. (1996) *Madras Agric. J.*, **83**: 362-364.

Stefunova, V., Milan, B., Jana, Z. and Katarina, R. (2015) *Pak. J. Bot.*, **47:** 1293-1301.

Stetter, M. G., Zeitler, L., Steinhaus, A., Kroener, K., Biljecki, M. and Schmid, K. J. (2016) *Front. Plant Sci.*, **7:** 816.

Subbiah, K. (1979) *M.Sc.(Ag.) Thesis*, Tamil Nadu Agric. Univ., Coimbatore.

Subhan (1989) *Buletin Penelitian Hortikultura*, **17**: 31-40.

Suresh, S., Chung, J.W., Cho, G.T., Sung, J.S., Park, J.H., Gwag, J.G. and Baek, H.J. (2014) *Plant Biosyst.* **148:** 635–644.

Swain, S.S., Sahu, L., Banik, D.P. and Chand, P.K. (2010) *Sci. Hort.*, **125:** 461–468.

Tara, J. S., Azam, M., Ayri, S., Feroz, M. and Ramamurthy, V. (2009) *Munis Ento. Zool.*, **4**: 510-518.

Tejaswini, N., Saidaiah, P., Ravinder, R. K. and Ramesh, T. (2017) *J. Pharm. Phytochem.*, **6:** 2572-2578.

Tisserat, B. and Galletta, P.D. (1988) *HortScience*, **23**: 210-212.

Venkatesh, L., Murthy, N., Nehru, S.D. and Manjappa (2014) *Asian J. Bio Sci.*, **9**: 67-70.

Vijayakumar, M. (1980) *M. Sc. (Ag.)Thesis*, Tamil Nadu Agric. Univ., Coimbatore.

Walton, P.D. (1968) *J. Hered.*, **59**: 17-18.

Waselkov, K. E., Boleda, A. S. and Olsen, K. M. (2018) *Sys. Bot.*, **43**: 439–458.

Wesche-Ebeling, P., Maiti, R., Garcia-Diaz, G., Gonzalez, D.I. and Sosa-Alvarado, F. (1995) *Econ. Bot.*, **49**: 423-430.

Wolosik, K. and Markowska, A. (2019) *Nat. Prod.Commun.*, **14**: 1-10.

Wu, H., Sun, M., Yue, S., Sun, H., Cai, Y., Huang, R., Brenner, D. and Corke, H. (2000) *Genet. Resour. Crop Evol.*, **47:** 43–53.

Xu, Y.B. (1987) *Grassland of China*, No. 3, pp. 48-50.

Yaacob, J.S., Hwei, L.C. and Taha, R.M. (2012) *Acta Hortic.*, **958:** 171-178.

Zaman, S.S., Maleque, M.A. and Basak, N.C. (1999) *Bangladesh J. Trg. Development*, **12**: 121-125.

Ziarovska, J., Razna, V. and Labajova, M. (2013) *Emir. J. Food Agric.*, **25:** 673–681.

SPINACH BEET AND SPINACH

B.K. Singh, P. Karmakar, M.L. Pandita, T.K. Maity and S. Lal

Spinach Beet (Palak) and Spinach

There are two distinct types of spinach (i) spinach beet or beet leaf (desi or common palak) and (ii) spinach (*vilayati palak*). They differ from each other as described below:

Spinach Beet	*Spinach*
Botanical name is *Beta vulgaris* L. ssp. *vulgaris* var. *cicla* (previously *Beta vulgaris* var. *bengalensis* Hort.)	Botanical name is *Spinacia oleracea* L.
Chromosome number 2n = 2x = 18	Chromosome number 2n = 2x = 12
Spinach beet has leaves with entire margin	Spinach has leaves with lobed or wavy leaf margin.
It produces bisexual (hermaphrodite) flowers	It may produce staminate or pistillate and hermaphrodite flowers.
It tolerates high temperature and grows well in hot weather	It is purely a cool season crop and cannot tolerate high temperature. In warm season and long days, it quickly tends to flower
Seeds are monogerm	Seeds are multigerm

These crops belong to an important group of leafy vegetables and are grown for their tender and succulent leaves. They are highly nutritious and available at cheaper rate in the market as compared to other vegetables. These can easily be cooked with other vegetables like potato, onion and brinjal or fried alone. They are very rich sources of vitamins A and C, and minerals like iron, calcium and phosphorus. Appreciable quantity of proteins is also found in both vegetables. These vegetables keep people healthy and help children to grow strong. They provide roughage and have an important place in balanced diet.

SPINACH BEET

1.0 Introduction

Spinach beet or *palak* is one of the most common leafy vegetables of tropical and subtropical regions and is grown and consumed widely in India. The *palak* growing Indian States are Uttar Pradesh, West Bengal, Punjab, Haryana, Delhi, Madhya Pradesh, Bihar, Maharashtra and Gujarat. However, it is less popular in the Southern states like Andhra Pradesh, Telangana, Kerala, Tamil Nadu and Karnataka.

2.0 Composition and Uses

2.1.0 Composition

It is a rich source of vitamins and minerals. Spinach beet is cheap source of vitamin A as compared to spinach and carrot. It also contains high quantity of ascorbic acid and iron, and 100 g of leaves supply as much essential amino acids as 100 g of any non-vegetarian food like meat and fish as reported by Nath and Subramanium (1972). It is also excellent source of fiber and a very good source of protein. Nutritive value of spinach beet is given in Table 1.

Table 1: Composition of Spinach (per 100 g of edible portion) *

Nutrient	*Content*	*Nutrient*	*Content*
Moisture	92.2 g	Iron	2.6 mg
Fat	0.1 g	Sulphur	30 mg
Fibre	3.7 g	Selenium	0.9 mcg
Protein	2.2 g	Thiamin	0.1 mg
Carbohydrates	4.3 g	Niacin	0.4 mg
Magnesium	70 mg	Vitamin A	6326 IU
Phosphorus	41 mg	Riboflavin	0.2 mg
Potassium	206 mg	Folic acid	15 mcg
Sodium	226 mg	Oxalic acid	625 mg
Copper	0.2 mg	Vitamin C	30 mg
Zinc	0.4 mg	Vitamin E	1.5 mg
Calcium	117 mg	Omega-3 fatty acid	4.0 mg
Manganese	0.4 mg	Omega-6 fatty acid	41 mg

* Anonymous (2020).

Khader and Rama (1999) estimated micromineral contents of palak at three different stages of maturity. They reported that as the plant matured from stage I (15 days) to stage II (30 days), Fe and Mn contents increased while Zn and Cu contents decreased as the plant matured. Varietal differences were also observed at different stages of maturity. The results also indicated that the consumption of palak at stage I and stage II potentially provide the greatest amount of minerals.

2.2.0 Uses

Its succulent leaves and stems form a nutritious dish after cooking. Tender leaves are used for making saag, dal mix, palak-paneer, mix-vegetables, *etc.* It is also suitable for making leaf concentrate and dry leaf powder. The herbaceous parts of *palak* are mildly laxative besides other medicinal values.

3.0 Taxonomy, Origin and Domestication

Wild forms of *Beta vulgaris* occur along the shores of the Mediterranean, extending eastwards as far as Indonesia, and westwards along the coasts of the Atlantic up to the Canary Islands and southern Norway. *Beta vulgaris* grown for its leaves *i.e.* beet leaf was taken into cultivation in the eastern Mediterranean or the Middle East and is first mentioned in the literature in Mesopotamia in the 9th century BC (Vaughan and Geissle, 1997). All the cultivated beets likely originated from wild progenitors of *B. vulgaris* ssp. *maritima* (Biancardi *et al.*, 2012). The evolution of *Beta*

Uses of *Palak*

Different Recipes of *Palak* in Indian Cuisine.

crops is illustrated in Figure 1 and distribution of cultivated forms of *Beta vulgaris* subsp. *vulgaris* summarized in Table 2.

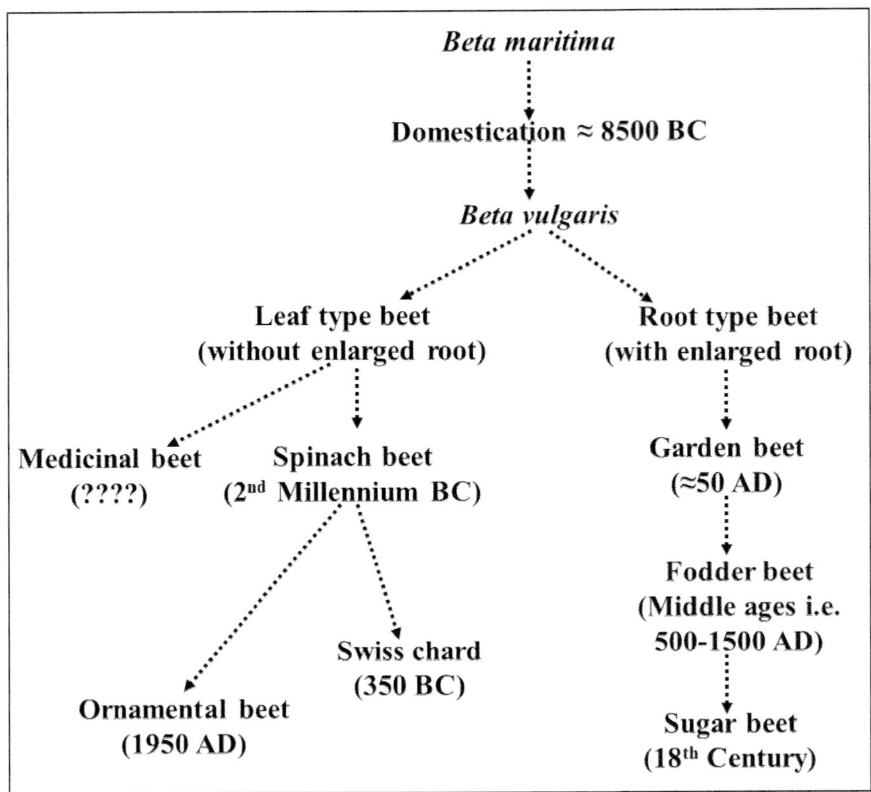

Figure 1: Chronology of *Beta vulgaris* after Domestication (Stevanato *et al.*, 2019; Biancardi *et al.*, 2005).

The genus *Beta* L. (2n=2x=18), of the family Amaranthaceae (formerly Chenopodiaceae) and subfamily Betoideae, is subdivided into four sections *Vulgares*, *Corollinae*, *Nanae* and *Procumbents*, and further section *Vulgares* comprised of 3 species *i.e. vulgaris*, *macrocarpa* and *patura*. The predicted genome size of *Beta vulgaris* is 758 Mb (Arumuganathan and Earle, 1991) and chromosomes at metaphase exhibit similar morphology (Paesold *et al.*, 2012). Three subspecies are typically recognised in *Beta vulgaris*.

A. *Beta vulgaris* subsp. *adanensis* (Pamukç. ex Aellen) Ford-Lloyd and J.T.Williams (Syn.: *Beta adanensis* Pamukç. ex Aellen): occurring in disturbed habitats and steppes of Southeast Europe (Greece) and Western Asia (Cyprus, Irael, western Syria and Turkey).

B. *Beta vulgaris* subsp. *maritima*, Sea beet, the wild ancestor of all cultivated beets (*Beta vulgaris* ssp. *vulgaris*). It is indigenous to European coastal regions, particularly the Mediterranean. The distribution area reaches from the coasts of Western Europe and the Mediterranean Sea to the Near and Middle East.

Subspecies of *Beta vulgaris*

Beta vulgaris* subsp. *adanensis

Beta vulgaris* subsp. *maritima

Beta vulgaris* subsp. *vulgaris

C. *Beta vulgaris* subsp. *vulgaris* (Syn.: *Beta vulgaris* subsp. *cicla* (L.) Arcang., *Beta vulgaris* subsp. *rapacea* (Koch) Döll).: All cultivated beets belong to this subspecies with five Cultivar groups:

I. Spinach beet or beet leaf or desi palak or common palak or chard [*Beta vulgaris* L. ssp. *vulgaris* var. *cicla* or *Beta vulgaris* L. ssp. *vulgaris* (Cicla Group)] is a leafy vegetable. The leaf beet group has a long history dating to the 2nd millennium BC. The first cultivated forms were believed to have been domesticated in the Mediterranean, and were introduced to the Middle East, India, and finally China by 850 AD. These were used as medicinal plants in Ancient Greece and Medieval Europe. Their popularity declined in Europe following the introduction of spinach. This variety is widely cultivated for its leaves, which are usually cooked like spinach. It can be found in many grocery stores around the world.

II. Swiss chard [*Beta vulgaris* L. ssp. *vulgaris* var. *flavescens* or *Beta vulgaris* L. ssp. *vulgaris* (Flavescens Group)] (Syn. *Beta vulgaris* subsp. v. convar. *cicla* var. *flavescens*). Leaves have thick and fleshy midribs. Both the midribs and the leaf blades are used as vegetables, often in separate dishes. Some cultivars are also grown ornamentally for their coloured midribs. The thickened midribs are thought to have arisen from the spinach beet by mutation.

III. Beetroot or garden beet [*Beta vulgaris* L. ssp. *vulgaris* var. *conditiva* or *Beta vulgaris* L. ssp. *vulgaris* (Conditiva Group)] is a red root vegetable that is most typically associated with the word 'beet'. It is especially popular in Eastern Europe where it is the main ingredient of borscht. The modern table beet is one of the progenitors of the sugar beet.

IV. Sugar beet, of greatest importance to produce table sugar [*Beta vulgaris* L. ssp. *vulgaris* var. *altissima* or *Beta vulgaris* L. ssp. *vulgaris* (Altissima Group)]. The sugar beet is a major commercial crop due to its high concentrations of sucrose, which is extracted to produce table sugar. It was developed in Germany in the late 18th century after the roots of beets were found to contain sugar in 1747.

V. Mangelwurzel or fodder spinach [*Beta vulgaris* L. ssp. *vulgaris* var. *crassa* or *Beta vulgaris* L. ssp. *vulgaris* (Crassa Group)] is used as a fodder crop. This variety was developed in the 18th century for its tubers for use as a fodder crop.

Table 2: Distribution and Use of Cultivated Forms of *Beta vulgaris* ssp. *vulgaris* (Mansfeld, 1986)

Cultivated Forms	*Common Name*	*Distribution*	*Usage*
cicla	Spinach beet	Central, western and southern Europe; Asia	Cooking vegetable
flavescens	Swiss chard	Central, western and southern Europe; Asia	Cooking vegetable
conditiva (*vulgaris*)	Garden beet (Red beet, beet root)	Central, western and southern Europe; Asia (Western India)	Cooking and salad vegetable
lutea	Yellow beet	Central, western and southern Europe; Asia	Cooking and salad vegetable
altissima	Sugar beet	Europe; CIS*; China; Asia; North America; South America	Sugar and Molasses
crassa (*rapacea*)	Fodder beet	Europe; CIS*; North America	Fodder

* CIS: Commonwealth of Independent States (formerly the USSR).

4.0 Botany

Spinach beet is an herbaceous annual for the edible leaf production, while it is biennial for seed production. In the early stages of the plant growth (at vegetative phase), it produces rosette, succulent tender edible leaves on a small thick stem. Later, at reproductive phase the stem elongates after about 75 days of sowing which results in bolting. Plant has taproot system along with more number of feeder roots. Leaves are long with entire margin and long petioles. Nath (1969) reported that Flowers of *Beta vulgaris* ssp *vulgaris* are located on the terminal portions of the main axis and on lateral branches subtended from this. The inflorescence in spinach beet is racemose with bracts of cymose type and emerges either in the axil of the leaf or from a spurious terminal bud from the main shoot as well as from the lateral shoots.

Flowers are small sessile, bisexual (hermaphrodite), perfect, bracteate (leafy) and are borne in a group of 1-3 in the axil of short bracts or without bracts in upper portion, 5 stamens opposite to perianth lobes (tepal) on a fleshy disc within unilocular semi-inferior ovary with 2-3 stigmas. The perianths of neighbouring flowers are often fused. Flowers produce abundant, round and small (20 μ) and light pollen grains which are carried by wind leading to wind pollination. The crop is highly cross-pollinated due to lighter pollen, protogyny flower and self-incompatibility. The whole phase of flower bud development takes 35 days to come in bloom. The anthesis takes place between 7 a.m. to 5 p.m. with peak period between 11.00 a.m. to 1.00 p.m. The flowers open mostly during mid-day. In an individual flower, anthesis is completed within 2 hours and hastened by high temperature and low humidity. Anther dehiscence takes place between 8.30 a.m. to 6.30 p.m. with the peak period between 12.30 to 2.30 p.m. depending upon the temperature and humidity. It is protogyny in nature; stigma receptivity begins 8 hours before anthesis, reaches the maximum just after anthesis and continues for another 10 hours. The flowers in the beet crops *Beta vulgaris* are joined in clusters of two or more which develop fruits with multigerm "seed," botanically classified as utricle which formed by the aggregation of as many fruits each containing the true seed.

5.0 Cultivars

Based on pigmentation of midrib and leaf veins, beet leaf cultivars can be divided into two groups (i) with reddish midribs and leaf veins, (ii) with green mid-ribs and leaf veins. In India, several improved cultivars have been developed and released.

All Green

The cultivar is evolved at Indian Agricultural Research Institute (IARI), New Delhi (Singh and Joshi, 1960). Leaves are green, uniform, tender with entire leaf margins and gives about 6-7 cuttings at 15-20 days interval. It yields green leaves about 125 q/ha. During winter season, it produces seed stalks in 75 days.

Pusa Palak

This cultivar was released by IARI, New Delhi during 1967. It is a selection

Flower Development, Anthesis and Pollen Dehiscence in Spinach Beet

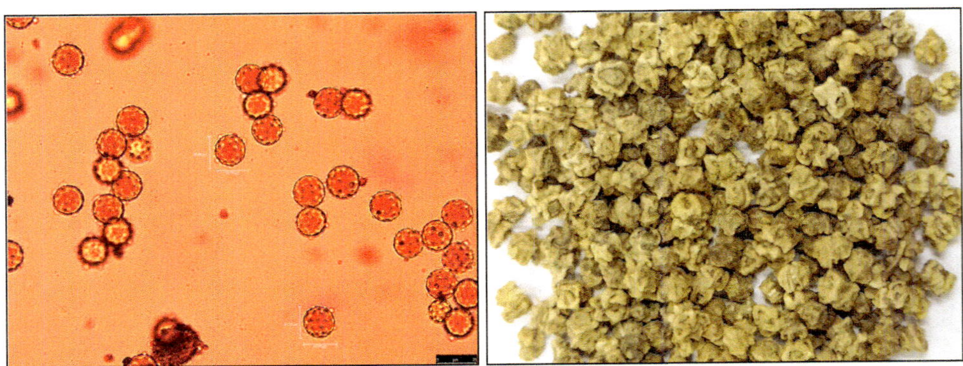

Pollen Grains of Beet Leaf

Seed of Beet Leaf

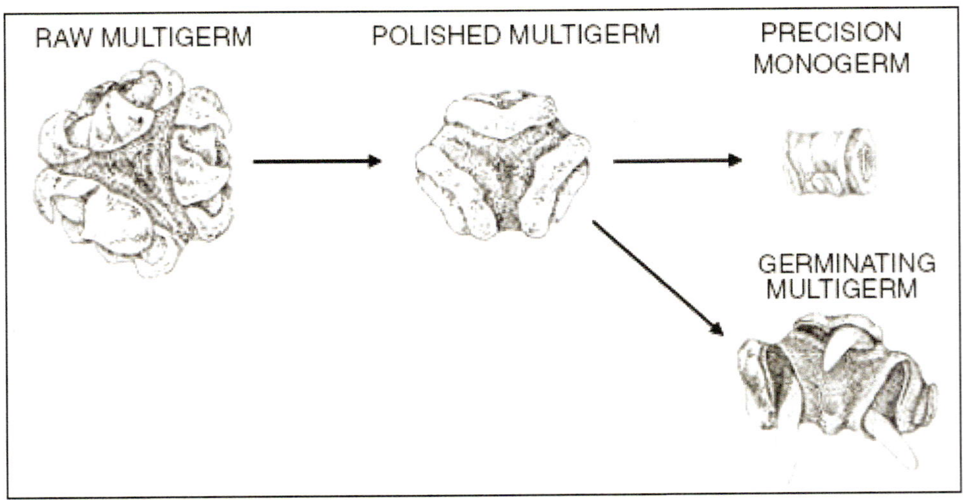

Multigerm Seed of Spinach Beet (Biancardi *et al.*, 2010)

from a cross between Swiss chard and local palak. It produces uniform green leaves without any purple pigmentation and has delayed bolting habit.

Pusa Jyoti

This cultivar is evolved because of treating the seeds of cv. All Green with 2 per cent colchicine solution for 24 hours and selection made from subsequent generations up to 10 years. It is a giant-leaved strain and produces large, green, thick, tender, succulent and crisp leaves of good taste even when eaten raw as salad. Plants are very vigorous, quick growing and regenerate quickly after each cutting, and can be grown throughout the year. It gives 6-8 cuttings and yields about 500 q/ha.

Pusa Harit

This cultivar has been developed by hybridization between sugar beet and local palak at IARI, Regional Research Station, Katrain, HP. It is suitable for cultivation in the hills throughout the year. Plants are upright, vigorous growing with uniformly thick, green, slightly crinkled giant-sized leaves. It is high yielder with greater ability of rejuvenation. It has late bolting habit and wide range of adaptability to varying climates. It can also tolerate alkaline soils.

Pusa Bharati

This cultivar is evolved at IARI, New Delhi. Leaves are long, succulent and flavoured. It is suited for rabi, kharif and summer seasons. It yields about 500 q/ ha of green leaves.

Kashi Baramasi

This cultivar has been developed at ICAR-Indian Institute of Vegetable Research, Varanasi. Suitable for growing round the year, able to tolerate high temperature of 40-45 °C and have green leaf yield ability of 500-800 q/ha in autumn/ winter sowing, 125-175 q/ha in spring sowing, and 175-225 q/ha in summer/rainy sowing. Leaves are ready to harvest in 25-30 days after sowing and thereafter at 15 days interval. Leaves are large, green, smooth, succulent, attractive with entire leaf margin. Plant growth is luxuriant, faster and showed delayed bolting by 20-30 days.

Jobner Green

This cultivar was evolved at the Department of Horticulture, University of Udaipur, Jobner during 1964 because of a spontaneous mutation detected from the population of a local collection S. No. 5 (Nath, 1969). It produces uniform green, large, thick, succulent, tender leaves having entire margin. It is a prolific yielder and produces green leave about 300 q/ha.

HS-23

This cultivar is developed through mass selection and released by the Department of Vegetable Crops, Haryana Agricultural University, Hisar. It produces dark green, thick, tender and juicy leaves. It is a quick growing cultivar, gives first cutting in about 30 days after sowing and a total of 6-8 cuttings at 15 days interval.

Cultivars of Spinach Beet

All Green

Arka Anupama

Kashi Baramasi

Palak No. 51-16

This cultivar has been developed by the Department of Agriculture, Maharashtra State (Kolhe, 1961). It produces green leaves, gives several cuttings and yields about 160 q/ha. It is a late bolting cultivar.

Banerjee's Giant

It is a very popular cultivar evolved by S.P. Banerjee in West Bengal. This cultivar has been developed from a cross between local palak and beet root. It produces large thick leaves with very succulent stem, having yield ability of 50-200 q/ha.

Punjab Selection

This cultivar was released by Punjab Agricultural University, Ludhiana, Punjab. Leaves are light green, long, thin, narrow and smooth with slightly sour taste; along with purple pigmentation on stem. It yields 250 q/ha of green leaves.

Punjab Green

This cultivar is also released by Punjab Agricultural University, Ludhiana, Punjab. Leaves are bright dark green, thick, long, succulent and free from sourness. There is mild purple pigment on the stem. This cultivar has low oxalate content and is slow in bolting. Average yield is 300 q/ha.

Arka Anupama

This cultivar has been developed through hybridization of IIHR-10 and IIHR-8 followed by pedigree method of selection and released from Indian Institute of Horticultural Research, Bangalore. Leaves are medium-large, dark green, succulent, crinkled and attractive. It is a late bolter type with vigorous initial growth and regenerates at a faster rate. It provides first picking at 35 days after sowing and yields 410 q/ha of fresh green leaves from 4 pickings. It shows field tolerance to *Cercospora* leaf spot.

6.0 Soil and Climate

6.1.0 Soil

Palak can be grown in any type of soils having enough fertility and proper drainage system but perform best in sandy loam soil. Mishra *et al.* (1973) working with this crop at Central Soil Salinity Research Institute concluded that spinach beet is a highly salt tolerant vegetable and can be grown successfully in saline sodic soils. Cultivar Jobner Green gave excellent performance in soil with pH 7.0-10.5 as reported by Nath (1969). Repeatedly growing beet leaf on the same field causes a rapid yield drop, mainly due to increased disease inoculants; hence, this practice must be carefully avoided.

6.2.0 Climate

Beet leaf cultivars do best in full sun to part shade. Grow faster in the cool temperatures. The cultivars of beet leaf are tough plants that can tolerate heat, humidity, and drought. Normally this crop is grown during autumn and spring seasons in plains but can be cultivated throughout the year in the regions where relatively mild climate prevails or by adopting heat/humid tolerant varieties. This crop can withstand frost and tolerate warm weather, but high temperature leads to early bolting. During hot weather, leaves mature faster.

7.0 Cultivation

7.1.0 Sowing

The sowing time of this crop varies from region to region. In plains of India, it is grown three times in a year: (i) as main crop during September to November, (ii) in early spring and (iii) in the beginning of rainy season. However, it can be grown throughout the year in regions having mild climatic conditions. In the hills, the sowing time of this crop is from April to June. Seed rate for one hectare is about 25-30 kg. After preparing soil thoroughly, beds of convenient size and irrigation channels are made. The seeds are sown either in row at 25 x 2 cm spacing or broadcasted uniformly at a depth of 1 cm, but previous one facilitates weeding, hoeing and cuttings. Seeds are germinated in 7-10 days after sowing under optimum soil moisture and temperature. If there is insufficient soil moisture, a light irrigation should be done to get proper germination.

7.2.0 Manuring and Fertilization

Spinach beet requires sufficient N fertilizer and irrigation to ensure optimal growth and to meet high quality criteria (*e.g.* dark green leaf color). Various recommendations on manuring and fertilization have been made by different workers. Singh and Joshi (1960) recommended a basal dose of 25 cartloads of farmyard manure and three top dressings of ammonium sulphate, each of 100 kg/ha. According to Choudhury (1967), a basal dressing of farmyard manure @ 35-40 t/ha should be added to the soil at the time of field preparation and a top dressing of N @ 20 kg/ha after every cutting for quick leaf growth. Singh (1969) recommended application of 25 kg nitrogen, 25 kg phosphorus and 50 kg potash per hectare at the time of sowing and thereafter additional N @ 25 kg/ha as top dressing after each cutting. Application of N @ 120 kg/ha, half as broadcast at sowing and half as top dressing 15 days later along with 40 kg P_2O_5 and 40 kg K_2O/ha in cv. All Green produced maximum yield (11.6 t/ha) (Rajput *et al.*, 1989). According to Jana *et al.* (1999), green yield, seed yield and 1000-seed weight of palak cv. All Green was greatly influenced by different nitrogen rates and number of cuttings. Green yield was maximum at 150 kg N/ha. The highest seed yield and 1000-seed weight were obtained with 100 kg N/ha. The maximum green yield was recorded at three cuttings. No cutting gave the highest seed yield and 1000-seed weight. The application of N @ 100 kg/ha and three cuttings gave the highest economic return. Maity *et al.* (1999) opined that crop growth rate and green leaf yield increased as the rate of nitrogen application increased. Total green leaf yield was highest with a combination of nitrogen at 150 kg/ha and 3 cuttings. Seed yield was highest with a combination of 100 kg N/ha and no cutting. Pandita *et al.* (1981) obtained highest leaf yield with application of 0.05 per cent Atonik (sodium mononitroguaiacol) 5 times at weekly interval in combination with seed treatment.

Jabeen *et al.* (2018) studied the effect of organic manures *viz.*, farm yard manure, sheep manure, vermicompost, mustard cake and biofertilizers *Azospirillum* and PSB including RFD (recommended fertilizer dose) on quality of spinach beet. The quality parameters like dry matter content (15.92, 16.47 and 29.26), ascorbic acid content (202.52, 166.10 and 155.12 mg 100/g), crude fibre content (9.50, 8.86 and 7.76 per cent) and leaf chlorophyll content (3.25, 0.98 and 0.76 mg/g) were found higher by application of vermicompost @ 3 t/ha + biofertilizers @ 5 kg/ha. Lowest nitrate content (447.33, 348.33 and 268.33 mg/kg in 1st, 2nd and 3rd cutting, respectively) in leaves was found when applied FYM @ 12.0 t/ha + biofertilizers @ 5 kg/ha) which was statistically at par with treatment comprising vermicompost @ 3 t/ha + biofertilizers @ 5 kg/ha.

7.3.0 Irrigation

Spinach beet is sturdier than spinach and can cope better with water shortages, however one should water regularly to ensure optimum plant growth and prevent bolting. The quantity and frequency of irrigation in spinach beet depends on the soil condition and season. For proper seed germination, if sufficient soil moisture is not available, a light irrigation is done just after sowing. The frequency of irrigation is more in light soils than heavy soils. Further, the spring-summer season crop needs frequent irrigation at 6-7 days interval, whereas autumn-winter crop requires irrigation at about 10-15 days interval. However, rainy season crop does not require much irrigation.

Singh *et al.* (2012) observed that use of the domestic wastewater with fertilizers has shown the improvement in the physico-chemical properties (N, P, K and organic carbon) of the soil, crop yields (Palak, methi and berseem) and also in the nutrient status as compared to that of the resulted from the application of groundwater with fertilizer. The domestic wastewater irrigation applied for a season had no significant effects apart from, slight changes in salt solubility and alkalinity on a clay soil with sewage wastewater irrigation.

7.4.0 Weed Management

The crop of spinach beet is very sensitive to weed competition, so weeding or herbicide treatments immediately manifest their beneficial effects on yield increase. To keep away the weeds from the field and to loosen the soil for proper aeration, 2-3 hoeing-cum-weeding are required. The selective herbicide like Pyrazone at 2.4-2.8 kg/ha can also be used to control the weeds in the crop as pre-emergence application (Mitidieri and Bianchini, 1973).

8.0 Harvesting, Yield and Storage

8.1.0 Harvesting and Yield

The crop becomes ready for first cutting in about 25-30 days after sowing, and thereafter subsequent cuttings are done at 15-20 days interval. Only well grown succulent and tender leaves should be trimmed or cut. Spinach beet plant can regrow and provide multiple harvests when properly cared for. Proper harvesting can also extend the productive life. Harvesting correctly improves the chances of the spinach

Field Growing, Harvesting and Selling of *Palak*

beet re-growth. It regenerates leaves from its growing point, or the crown of the plant, which is where the stems join the root system near the soil surface. For proper harvesting, cut the palak leaves back to within 5 cm of the ground, taking care not to cut into the growing point. Within about 15 days, the leaves should regrow for a next harvest. Winter crop gives more cuttings than spring/summer/rainy crop. Number of cuttings depends upon season and cultivar. The cultivar Jobner Green and Kashi Baramasi gives more cuttings than cv. All Green. On an average, winter crop gives 4-6 cuttings, and spring/summer/rainy crop produces 1-3 cuttings. The total yield of greens depends on cultivars, soil conditions and season. On an average, palak crop yields about 300-500 q/ha. As the weather cools, the leaves are at their tastiest and turn a little bitter in hot weather.

8.2.0 Storage

The green leaves of spinach beet are very tender and succulent. After harvest, the leaves cannot be stored for longer period and hence should be sent to the market immediately. Before marketing leaves are tied in small bundles. All the weeds, diseased and broken leaves should be discarded at the time of bunching. Washing of the leaves is not desirable because it causes rotting and yellowing of the leaves in the centre of the bundle. In the early morning the leaves are crisp and may break during harvesting. The leaves retain their crispness if kept in the coolest part of the fridge for up to 4 days. Wrap in damp paper towels and store unwashed in a perforated plastic bag to extend shelf-life. Deore and Bharud (1989) recorded smallest moisture losses from the leaves 72 hours after treatment with GA_3 at 20 ppm and stored under a wet cloth.

Palak cannot be stored at room temperature but at low temperature (0°C) and high relative humidity (90-95 per cent), leaves can be stored for about 10-14 days (Nath, 1976).

The combination of cool chamber and packaging, which independently had beneficial effects, further enhanced the storage life of palak compared to those in ambient conditions. Palak packed in polyethylene bags (100 or 150 gauge with 1 and 0.5 per cent ventilation, respectively) and stored in a cool chamber had a shelf-life extension of up to 5 days (superior to all other treatments), while those without packaging and stored in ambient condition recorded the shortest shelf-life of one day. Cool chamber storage of palak, whether packed or not, slowed down the cumulative physiological loss in weight, rate of physiological loss in weight, maintained freshness, better colour for a longer time and delayed the symptoms of wilting and rotting, in addition to increasing the shelf-life by 4 days compared to those stored in ambient conditions (Srinivasa and Reddy, 1997).

Freshly harvested local cultivars of palak packed in polyethylene bags (200 guage and 2 per cent ventilation) or non-packed were held at room temperature (17.42-30.12 °C, 24.30-73.00 per cent RH), in a cool chamber (14.45-22.62 °C, 90-95 per cent RH) or in cold storage (18 °C, 90-92 per cent RH). Physiological weight loss was lower in packed than in non-packed samples and was lowest in cold storage followed by cool chamber storage. Storage life was increased from one day (non-packed samples held at room temperature) to 6.0 days in packed palak held in cold storage or a cool chamber (Waskar *et al.*, 1998). Postharvest losses in spinach beet is about 25 per cent which is mainly due to high respiration rate and poor management such as poor handling after harvesting, improper packaging and lack of improved storage techniques, *etc.* Ascorbic acid, citric acid and benzoic acids are the most widely used postharvest chemicals to enhance the storage life and delay in colour change, inhibition of oxidation from cut leafy vegetables and improvement of shelf life in leafy vegetables (Whitaker and Lee, 1995). The maximum shelf life for spinach was observed as 3 days and 14 days when stored in LDPE bags with 5 per cent perforation at ambient and cold storage conditions, respectively (Kakade *et al.*, 2015).

9.0 Diseases and Pests

9.1.0 Diseases

Most common diseases of spinach beet are damping off and leaf spot.

9.1.1 Damping Off

This disease is caused by many fungi, especially *Pythium* sp. Several fungi can cause damping off such as Rhizoctonia, Fusarium and Phytophthora. However, Pythium is most often the culprit. A soil-borne fungal disease that affects seeds and new seedlings, damping off usually refer to the rotting of stems and root tissues at and below the soil surface. In most cases, infected seeds will germinate and come up fine, but within a few days they become water-soaked and mushy, fall over at the base and die. High humidity levels, rich soils and sowing too deeply will also encourage its growth. The best defence is to use treated seeds and practice crop rotation. The seeds treated with ceresan, captan or thiram at the rate of 2.5 g per kg of seeds.

9.1.2 Leaf Spot

This disease is caused by the fungus *Cercospora beticola*. Cercospora leaf spot occurs wherever spinach beet, garden beet, swiss chard, sugar beet and spinach are grown and is one of the most important diseases. It can result in significant losses, particularly in late summer when conditions are favorable (high temperatures, high humidity, long leaf wetness periods at night). Brown dead circular spots are found on the diseased leaves. It can be controlled with the spray of 0.2 per cent coprox or azoxystrobin at 15 days interval.

9.2.0 Pests

There is no serious problem of pests in palak. However, aphids and leaf-eating caterpillar can cause damage to the crop.

9.2.1 Aphids

These insects are small and found in large number on the tender portions of the plants. They suck the sap from the plant tissues causing plant to become weak. Aphids can be controlled by the spray of malathion, imidacloprid, acetamaprid or thiomethoxam.

9.2.2 Flea Beetles

The flea beetle (*Phyllotreta* spp.) also damage plant foliage. The small black adults with over-large back legs feed on the leaves, creating pits and small, irregular holes. The plant may then wilt as a result. They are controlled by spray of cypermethrin 0.75 ml/l.

9.2.3 Leaf-eating Caterpillar

The caterpillars of *Laphygma exiqua* feed on leaves and make holes. They are also controlled by the spray of chlorpyriphos @ 0.25 per cent.

10.0 Seed Production

Spinach beet is more prolific seed bearer than spinach. It gives higher seed yield as compared to spinach because of hermaphroditism (bisexual) flowers and all plants bear seed. For seed crop, field preparation, manurial and cultural requirements are the same as for green leaf crop. Singh *et al.* (1983) obtained maximum yield of seed at a spacing of 60 x 10 cm. Pandita *et al.* (1981) reported the highest seed yield from palak plants sprayed with Atonik @ 0.05 per cent, thrice at fortnightly interval. Usually, after taking 2-3 cuttings, the crop is left for seed production, but it is not desirable as removal of leaves reduces the seed yield. An experiment carried out at the Vegetable Research Farm, Haryana Agricultural University, Hissar revealed that application of nitrogen @ 20 kg/ha and one cutting gave the maximum yield of good quality seed and the highest gross and net return (Lal *et al.*, 1979). Phor and Mangal (1991) observed that seed quality in terms of 1000-seed weight, germination per cent and vigour was greatest from plants grown in rows 20 cm apart, irrigated when the pan evaporation coefficient was 0.9 and leaves harvested once.

10.1.0 Isolation

Spinach beet is a cross-pollinated crop. Pollination takes place by wind because it produces abundant small light pollen grains; so, need to be isolated from other *Beta vulgaris* crops 1200-1600 m depending on physical obstructions and wind conditions. It crosses readily with beet root, sugar beet and Swiss chard but does not cross with spinach.

10.2.0 Roguing

In seed production, roguing refers to the act of identifying and removing plants with undesirable characteristics. Rogues are removed from the fields to preserve the quality of the crop being grown. The plants being removed may be diseased, infected, unwanted variety, or undesirable for other reasons. All off-types, early bolters and host plants are removed carefully from the field. The first roguing is done at vegetative stage (pre-flowering stage) based on foliage characters; and subsequent roguings are done for early bolters, off-type plants and pigweed plants are noticed.

10.3.0 Harvesting and Seed Yield

Crop generally bolts about 75 days after sowing and seeds mature in 150-180 days. The seeds do not shed; hence the harvesting is done after full ripening of the whole crop. After the plant is dried, it is threshed by beating with sticks. Seeds are cleaned by winnowing and dried to 9 per cent moisture before storage. The average seed yield is about 6-7 q/ha. Seed can be stored for 3-4 years under proper storage conditions, *i.e.* low humidity and cool temperature.

11.0 Crop Improvement

The main objective of plant breeding is the development of cultivars with the maximum commercial yield at the lowest economic and environmental cost compared to the existing cultivars. The breeding programme in spinach beet should be taken up with the following objectives:

- The improvement in biomass yield should be given due consideration in breeding programme. The yield in spinach beet is directly correlated with number, size, thickness and succulence of leaves. Hence, a genotype with more foliage, larger and erect leaves, succulent in nature, smaller petiole and faster in growth should be developed.
- The cultivars should be developed with higher nutritive value especially vitamin A, amino acids, iron and calcium; and lower content of oxalic acid and nitrate.
- The yield and quality are affected by number of cuttings and bolting habit. The emphasis should, therefore, be on the development of cultivars with late bolting habit and higher regeneration capacity favour multiple cuttings.
- Varieties must possess good yield stability across locations and seasons, which depend on a broad genetic base, wider adaptability and tolerance to abiotic stresses.
- *Cercospora* is often a serious disease in this crop. Hence, breeding programme should also have emphasis on development of cultivars resistant to diseases.

The improvement of composite traits such as yield, germination ability, bolting, and several disease resistances is more difficult due to their quantitative inheritance and genotype x environment interactions. Spinach beet is a cross-pollinated crop; hence, methods employed for the improvement of cross-pollinated crops can be adopted for this crop also. The basic methods employed are introduction, selection and hybridization (Swarup, 1977).

Introduction is useful in the initial stages of a breeding programme. It is the easiest and quickest method of crop improvement. Some cultivars in spinach are introductions.

Mass selection is commonly used in cross-pollinated crops like spinach beet for crop improvement and maintenance of cultivars. The best individual plants are selected in the population on the basis of phenotypic performance and their seeds composited for raising the following generation. Since there is inbreeding depression, selfing should be avoided and instead sib-mating will be more effective. Large numbers of plants are selected to avoid inbreeding depression. It is very effective for improving qualitative or highly heritable characters but not very useful for improvement of polygenic characters like yield, particularly if it has low heritability. The cultivars HS-23 and Kashi Baramasi have been developed by mass selection. The most useful characters for separating the highly selected cultivars from the primitive leaf beets were plant height, leaf dimension, leaf thickness, petiole length and leaf erectness.

In hybridization method, desirable recombinants are selected in the segregating population of cross involving two or more parents. Pedigree method or bulk population method is employed for the selection but former is mostly used in spinach beet. Several cultivars have been developed by this method. Pusa Palak has been developed from the cross of Swiss chard x Local Palak, Pusa Harit from the cross of Sugar beet x Local Palak to improve the traits like late bolting habit, better regeneration ability and wider adaptability.

Both natural and artificial mutations have also been exploited for the development of cultivars in spinach beet. Jobner Green has been developed because of spontaneous mutation in local material (Nath, 1969). Colchicine-induced polyploids have also been used for developing improved cultivars such as Pusa Jyoti.

Selfing is very important to maintain the purity of progeny after selection or hybridization. For selfing, the entire plant or few branches are enclosed in a thick muslin cloth bag so that pollen does not get blown outside or foreign pollen does not pollinate the flowers. The bag should be shaken once or twice a day to ensure better set. The bags must not be opened in windy days (Pandita and Lal, 1990). It is a wind cross pollinated crop. Efforts were made at IARI, New Delhi, to evolve high yielding varieties, which resulted in the new varieties like 'All Green' and 'Pusa Jyoti. At the University of Udaipur, a high yielding variety 'Jobner Green' suited to alkaline soil condition was evolved. Similarly, HS 23 has been developed by HAU, Hisar.

SPINACH

1.0 Introduction

Spinach or Vilayati palak is not commonly grown in India except in the hilly areas. It is a popular pot-herb or leafy vegetable grown in China, USA, Japan, Turkey, Canada and Europe. In 2018, world production of spinach was 26.3 mt, with China alone accounting for 90 per cent. However, spinach seed production is mainly based in Denmark because of favourable environmental conditions, especially long day length (Deleuran, 2010), covering more than 70 per cent of the world's supply.

2.0 Composition and Uses

2.1.0 Composition

It is a rich source of vitamin A, vitamin C, vitamin K, lutein, zeaxanthin, magnesium, manganese, iron and folate; and good source of the riboflavin, vitamin B6, vitamin E, calcium, potassium, omega-3-fatty acid and dietary fiber. Nutritive value of spinach is given in Table 3.

Table 3: Composition of Spinach (per 100 g of edible portion) *

Nutrient	*Content*	*Nutrient*	*Content*
Moisture	91.3 g	Selenium	1.0 mcg
Fat	0.3 g	Thiamin	0.01 mg
Fibre	2.3 g	Niacin	0.7 mg
Protein	3.0 g	Vitamin A	9377 IU
Carbohydrates	3.7 g	Vitamin K	483 mcg
Magnesium	79 mg	Riboflavin	0.3 mg
Phosphorus	49 mg	Folic acid	194 mcg
Potassium	557 mg	Choline	18 mg
Sodium	79 mg	Oxalic acid	950 mg
Zinc	0.7 mg	Vitamin C	28 mg
Calcium	99 mg	Vitamin E	2.0 mg
Manganese	1.0 mg	Omega-3 fatty acid	138 mg
Iron	2.7 mg	Omega-6 fatty acid	26 mg

* Anonymous (2016).

Samples taken in the evening had higher ascorbic acid, DM and total sugar contents than those taken in the morning. Ascorbic acid was positively correlated with DM and oxalic acid content but was negatively associated with total sugar content. Evening harvesting in the 5th WAP was identified as the best harvesting time with respect to high ascorbic acid contents of leaves (Weerakkody and Ando, 1994). Spinach regulates blood pressure; boosts immune system; supports bone health; good for digestion, skin, hair, eye, brain and heart; and has properties of anti-inflammatory, antioxidant, anti-cancerous and anti-ageing.

2.2.0 Uses

Like palak, it is liked by the common people as a cooked vegetable and also preserved by canning, freezing and dehydration. It has got some medicinal values also. Raw spinach contains high levels of oxalates which block absorption of calcium, magnesium, zinc and iron in the stomach and small intestine as they form insoluble salts. Oxalic acid may also lead to the formation of kidney stones. Moreover, cooked spinach contains lower amounts of oxalates and its nutrients absorbed more readily. Spinach is also rich in nitrate which can be converted into nitrite in the digestive system with the help of oxalic acid, and the latter can react with haemoglobin to form methaemoglobin, potentially causing methaemoglobinaemia (Santamaria, 2006).

3.0 Origin and Taxonomy

Spinach had originated from South-West Asia (Iran) and was first cultivated by the Arabs. This was taken to Spain by Moors from where it spread to other parts of the world (Choudhury, 1967). The spread of spinach followed two separate directions, one to Southern and Eastern Asia; and another one to Africa, the Mediterranean and Northern Europe, from which it was later introduced in the

Americas. A distinction is often made between two main groups of spinach cultivars, namely Asian-type and Western-type cultivars (Simoons, 1990; Van der Vossen, 2004). At the morphological level, Asian cultivars maintain the narrow, hastate and smooth leaf shape and the long petioles of wild spinach; whereas Western cultivars changed the hastate leaf shape to a round form, while leaves were enlarged and the savoy leaf texture appeared (Van der Vossen, 2004). This was known in China as early as the 7th century A.D. as well as to the ancient Greeks and Romans and grown in Persia about 2000 years ago. Spinach was introduced into North America by Arabs (Shoemaker, 1947). As known spinach is of two types, prickly-seeded and smooth-seeded, the prickly-seeded spinach is an old one and was used in Europe in 1351, whereas the smooth seeded was known in 1552 (MacGillivray, 1961).

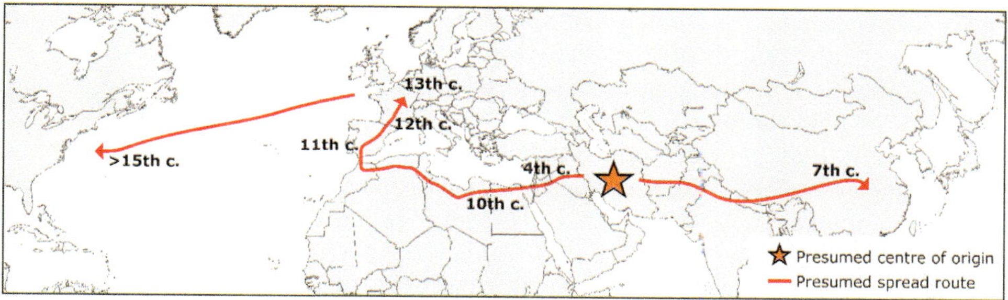

Presumed Centre of Origin and Migration Routes of Spinach. Timelines are indicated by century AD (Ribera *et al.*, 2020).

Spinach is quite different from spinach beet. Spinach (*Spinacia oleracea* L.) belongs to the family Amaranthaceae having chromosome number of 2n = 2x = 12 (Ryder, 1979). The word spinach came from a Spanish word Hispania. The length of six pairs of morphologically distinct chromosomes varies from 4-6 µm. Presently, two species of genus *Spinacia* and 11 species of genus *Blitum* are part of the gene pool of cultivated spinach, respectively (Vincent *et al.*, 2013; Ribera, *et al.*, 2020) such as:

Primary Gene Pool

i. *Spinacia turkestanica* Iljin

Secondary Gene Pool

ii. *Spinacia tetrandra* Steven ex M. Bieb

Tertiary Gene Pool

iii. *Blitum asiaticum* (Fisch. and C.A. Mey.) S. Fuentes *et al.*

iv. *Blitum atriplicinum* F. Mu¨ll.

v. *Blitum bonus-henricus* (L.) Rchb.

vi. *Blitum californicum* S. Wats.

vii. *Blitum capitatum* L.

viii. *Blitum hastatum* Rydb.

ix. *Blitum korshinskyi* Litv.

x. *Blitum litwinowii* (Paulsen) S. Fuentes *et al.*
xi. *Blitum nuttallianum* Roem. and Schult.
xii. *Blitum spathulatum* (A. Gray) S. Fuentes *et al.*
xiii. *Blitum virgatum* L.

Xu *et al.* (2017) studied transcriptome sequencing of 120 accessions of cultivated and wild spinach who revealed >420 K variants and identified 93 domestication sweeps, and suggested *S. turkestanica* as direct progenitor of cultivated spinach.

4.0 Botany

This is also herbaceous annual plant for leaf production and biennial for seed production. Spinach produces rossette leaves from a much shorter stem near the surface of the ground during vegetative phase. Later, stem elongates and forms flower stalks during reproductive phase. The leaves are alternate, simple, ovate to triangular with larger leaves at the base of the plant and small leaves higher on the flowering stem; vary in shape, thickness, and are smooth or savoyed. Flowers are inconspicuous, small, greenish, axillary and borne in clusters on both large stems and small branches. Flowers vary from 6-12 per cluster but all of them do not develop at the same time (Rosa, 1925). Flowers may be staminate, pistillate or hermaphrodite and remain receptive for a week or longer (MacGillivray, 1961). The clusters of ununited flowers, resulting into separate round and slightly flattened fruit cluster containing several seeds. Pistillate flowers have a single ovary with 4 or 5 five styles borne on a 2-4 toothed calyx. Staminate flowers are clustered on a spike. The fruits usually an achene, the fertilized ovary form a one seeded fruit called utricle. Parthenocarpic fruits also occur in spinach.

Spinach plants are usually dioecious but monoecious plants are also found. With reference to sex-expression, Rosa (1925) described 4 types of spinach plants:

1. Extreme males: These type plants produce only staminate flowers with lesser foliage (very little vegetative) tend to bolt quickly; and hence flowers early and generally die soon after flowering. Eliminate the extreme males from commercial strains by selection.
2. Vegetative males: These produce only staminate flowers having more foliage and flower later than extreme males.
3. Female plants: These produce only pistillate flowers, have well developed foliage and remain vegetative for longer time *i.e.* late in bolting. Female and vegetative male types are preferred plants type for commercial cultivation because of higher vegetative growth, greater yield and slower in bolting.
4. Monoecious plants: These produce staminate and pistillate flowers, well developed foliage and are slow in flowering. These plants may be predominantly staminate, predominantly pistillate or purely pistillate early but with some staminate flower later or almost equally staminate and pistillate flowers throughout the season.

The male and female plants are identified based on carbohydrates and pigment contents in the plants. Sugars, chlorophyll and carotenoid contents are higher in female than in male plants. A detailed analysis is made on the combinations of flowering, phenotypic sex expression, flowering dynamics, pollination, fruiting and variability of 1000-seed weight in Smarald spinach over the period 1987-1990 by Stefanescu (1996). The results showed that the male flowers start flowering 3-10 days before the female flowers and continue flowering later, but that peak flowering of the male flowers coincides with that of the female flowers. Flowering and fruit setting proceed acropetally. Phenotype expression of sex is determined by the interaction of the X and Y chromosomes and the autosomal *A* and *G* genes with environmental factors. The optimum flower ratio is 60: 40 (female: male). Seed dormancy of spinach was reduced during the domestication as germination is typically high in cultivated spinach but low in wild type (Van Treuren *et al.*, 2020).

5.0 Cultivars

Generally, spinach cultivars are classified based on morphology of seeds and leaves. There are 2 groups: (i) prickly-seeded, and (ii) round-seeded cultivars; and (i) smooth-leaved, and (ii) savoy-leaved. Spinach cultivars have been classified in various ways:

In the year 1896, Kinney classified cultivars into 4 groups:

Group 1 - Norfolk or Bloomsdale

Group 2 - Round-leaved

Group 3 - Thick-leaved

Group 4 - Prickly-seeded

Later on, in the year 1929, Geise and Fareley classified spinach cultivars into 2 groups: (a) flat-leaved type, and (b) savoy-leaved type. Each class was further subdivided into 3 classes: (i) short-season group, (ii) long-standing group, and (iii) slow-growing group.

According to Drewes (1943) spinach cultivars can be classified as follows:

A. Round-seeded
   1. Smooth-leaved (or slightly savoyed in cool weather)
      (i) Early bolting or not long-standing
      (ii) Slow bolting or long-standing
   2. Savoy-leaved
      (i) Early bolting or not long-standing
         (a) Mosaic resistant
         (b) Not mosaic resistant
      (ii) Slow bolting or long-standing

B. Prickly-seeded
   1. Smooth-leaved (or slightly savoyed in cool weather)
      (i) Early bolting or not long-standing
         (a) Mosaic resistant

(b) Not mosaic resistant

(ii) Slow bolting or long-standing

2. Savoy-leaved

(i) Early bolting or not long-standing

(a) Mosaic resistant

(b) Not mosaic resistant

(ii) Slow bolting or long-standing

On the basis of leaf type, Ryder (1979) classified spinach cultivars into 3 groups:

i) Savoy (wrinkled leaf)

ii) Semi-savoy

iii) Smooth

Popular cultivars grown in India are given below:

Early Smooth Leaf

This is smooth seeded cultivar and produces small light green leaves with a pointed apex.

Virginia Savoy

This is a prickly seeded cultivar having blistered large dark green leaves with round tip. Plants are upright and vigorous growth.

Pusa Vilayati Palak

It is a prickly seeded cultivar, developed by ICAR-IARI, New Delhi, having succulent stem and leaves. Plant is vigorous having completely green stem and leaves. Leaves are smooth, soft having pointed apex with slightly serrated margin. It is short duration crop suitable for 2 cuttings. It recorded high content of ascorbic acid, iron and calcium (65, 8 and 186 mg/100g respectively). It is suitable for autumn winter season for both plain and hilly areas. It gives an average green leaf yield of 12 t/ha.

Cultivar of Spinach

Pusa Vilayati Palak

Virginia Savoy

Cultivars in other countries are as follows:

In a green house trial in in Belgium Jan and Muyldermars (1986) found that the cultivar **Indures** was the earliest followed by **Florisant**. The highest yield was however obtained from **Spark**, followed by **Indures**. In Netherlands **Triathlor, Triade, Hermes, Trio** and **Polka** are recommended for outdoor cultivation. (Canon, 1989).

Goode *et al.* (1989) reported a disease resistant cv. **Fall Green** from Arkansas, USA. It is intended for autumn planting since it has winter hardiness and tolerance of prolonged wet soil conditions.

6.0 Soil and Climate

6.1.0 Soil

The soil requirement of spinach is the same as for spinach beet. Spinach can be grown on a variety of soils. Well drained sandy loams or loams well supplied with organic matter and high in pH are preferred. Good drainage system in the soil is essential. This crop is susceptible to injury by high acidity. Zimmerley (1926) reported soil reactions between pH 6.0 and 7.0 for optimum growth and it was markedly reduced at pH lower than 5.5. Soil temperature also influences the spinach crop. Wendt (1980) found optimum yield of spinach when plants were grown at soil temperature of 20°C. Soil is prepared like that of spinach beet. Well cultivated soil improves the quality like vitamin content of the plants than moderately cultivated soil (Avdonin and Kochubei, 1978).

6.2.0 Climate

Spinach prefers a cool climate and is a long-day crop. The optimum range of temperature for seed germination is 7-24 °C, and 15-20 °C for crop growth with a minimum temperature of 10 °C and a maximum of 32 °C. This crop tolerates frost better than most vegetables. Ample rainfall enhances yields, whereas drought reduces it but raised the leaf dry matter content (Pavlet *et al.*, 1974). The photoperiodic response of this crop can be altered by temperature (Knot, 1939) because temperature has an important role on the life cycle of spinach. A low temperature or cool night tends to make the plants more erect and causes savoying in savoy cultivars, whereas high temperature decreases it. Maximum yield is obtained under short day and moderate to warm temperature. High atmospheric humidity favours succulence and tenderness of foliage. Crop produced in winter generally contains more amount of sugar and total solids as compared to spring-summer crop. The dry weight and size of leaves were affected at high temperature as the plants flowered. Long days and warm weather favour bolting.

Sugiyama (1990) reported that a day length of 16 hours increased leaf number from 8.9 to 16.9 and increased plant weight from 7.2 to 18.5 g. Long days also increased leaf area, leaf dimensions and leaf dry matter/unit area but decreased the percentage dry matter.

Improvements to the environmental conditions in plastic shelters (Japanese style plastic tunnel) for spinach cultivation in the hilly and mountainous areas of midwest Japan were investigated by Inoue *et al.* (2000). UV ray-blocking film used as an outer covering for the plastic shelters decreased solar radiation inside the tunnels by 10-30 per cent, accompanied by a slight decrease in air and soil temperatures. However, this method alone could not guarantee stable spinach growth during July and August, when the average temperature was above 25°C. Shading by 45 per cent improved spinach growth in July and August, while the favourable shading rate in June and September could not be determined because of yearly fluctuation of the climate in the targeted region. The shading treatment caused a decrease in ascorbic acid content of the spinach plants, which became greater as the rate of shading increased. Removal of the covers one week before harvesting helped maintain these qualitative characteristics of the products.

7.0 Cultivation

7.1.0 Sowing

Sowing time of spinach varies from place to place and with cultivars. In plains, winter crop is sown in September-October and in the hills preferably in August-October for supply of green till mid-winter or early spring. Prickly-seeded cultivars are best suited for the autumn-winter crop in the hills, whereas smooth-seeded cultivars are suitable for spring-summer in the hills and autumn sowing in the plains. The growth and yield of the spinach crop are influenced by the dates of sowing and plant population. Moustafa (1969) suggested that early sowing gave larger plants at 20 days, but late sown plants were larger at the stage of market maturity. Moreover, late sown crops tend to bolt earlier.

Sow seed at a depth of 0.5-1.0 cm and spacing of 4-5 cm plant to plant and 20-30 cm line to line. Spinach requires relatively higher seed rate of about 37 to 45 kg/ha (Nath, 1976) than spinach beet because about 50 per cent of the plants come out as male plants with poor growth and is not possible to remove male types except at seed stage. A closer spacing resulted in increased yield (Choudhury *et al.*, 1974; Bradley *et al.*, 1975). Seed should be treated with thiram @3g/kg seed or *Trichoderma* @ 5g/kg seed. The emergence of seedlings (germination) depends upon many factors such as endogenous inhibitors present in the seeds, depth of sowing, environmental as well as soil temperature. Germination of seed may be enhanced by treating the seed with sulphuric acid or a bichromate mixture (Na + K) which destroys the endogenous inhibitors in the outer seed layer (Kerin, 1975).

7.2.0 Manuring and Fertilization

Being a leafy vegetable, spinach has similar nutrient requirement as spinach beet. Higher nitrogen levels increased the yield and dry matter content of the tops (Verma *et al.*, 1969; Choudhury *et al.*, 1974). The yield is closely related to available nitrogen and it was increased by nitrogen in combination with potash as reported by Cervato (1969) and Stanilova *et al.* (1972). A split fertilizer application (half at sowing and half 1 month later) is more effective than a single application at sowing. Leaf N content rose with increasing N rate but leaf Fe and Mn contents decreased (Dhillon *et al.*, 1987).

Among the fertilizers, ammonium sulphate was more effective than calcium ammonium nitrate as suggested by Leskovec and Dobersek (1972). Spinach also responded well to potash and phosphorus (Prummel, 1971). Application of different doses of N as ammonium sulphate, urea or ammonium nitrate revealed that ammonium sulphate at 200 kg/fedan (0.42 ha) gave a good yield and acceptable accumulation of nitrates and nitrites (Farag and Abdel-Aal, 1989). El-Fadaly and Mishriky (1990) noticed that ammonium sulphate and urea enhanced the accumulation of N and P compared with ammonium nitrate while the leaf Ca and Mg contents increased more with ammonium nitrate. In cultivars Dark Green Bloomsdale and Matador, Markovic *et al.* (1988) reported that application of N as calcium ammonium nitrate with P: K basal dressing at 100: 100 kg/ha increased yield with increasing N rates to 150 kg N/ha and then in some cases it declined. Higher dose of nitrogen at crop growing stage enhances nitrate accumulation in the leaf tissue. Foliar applications of magnesium may enhance the production of dark green leaves.

The quality of spinach is also influenced by fertilization. Nitrogen application increased the protein (Kausal *et al.*, 1981; Nicolaisen and Kublen, 1970), β-carotene, reducing sugars, ascorbic acid (Kausal *et al.*, 1981) and oxalic acid (Leskovec and Dobersek, 1972; Macreke, 1973) contents in spinach. Nitrogen in combination with phosphorus and potash increased dry matter, crude protein and vitamin C contents in spinach as reported by Stanilova *et al.* (1972). Phosphorus application decreased the oxalic acid content of spinach (Prummel, 1971). In pot trial with cv. Matador maximum yields were produced by KNO_3 at 200 mg/l or $(NH_4)_2SO_4$ at 400 mg/l. Plants supplied with $(NH_4)_2SO_4$ took up more N from the soil and contained more protein and twice as much nitrates as plants supplied with KNO_3. The addition of N-serve at 5 or 10 ppm did not affect yields, but reduced N uptake and spinach nitrate content and improved protein synthesis. For best results moderate rate of $(NH_4)_2SO_4$ + 10 ppm N-serve are recommended (Kolota *et al.*, 1984).

Effective P management is important to optimize crop yield potential, reduce production costs, and decrease the risk of environmental damage. Plants require adequate P from the very early stages of growth for optimum crop production. Phosphorus supply to the crop is affected by soil P and by other soil and environmental conditions. Therefore, growers can reduce the amount of P that they are currently applying by using sufficient P fertilizer in those growth stages when the plant requirement is greatest and still obtain optimal yields. In addition to effective P management, which is important to reduce production costs, this will decrease the risk of environmental damage (Zahedifar *et al.*, 2011).

Besides macronutrients, micronutrients like zinc (500 and 1000 ppm) in combination with GA_3 increased the fresh and dry weight of spinach (Omran *et al.*, 1973). Spinach is quite susceptible to boron deficiency. Apply boron in the fertilizer and as a foliar spray and/or soil drench. Application of micronutrients such as manganese and boron at 4.5 and 2.2 mg/pot, respectively, enhanced ascorbic acid and carotene contents as reported by Hulewicz and Mokrzecka (1971).

The optimum solution temperature, total ion concentration and Cu and Zn concentrations were determined for the commercial hydroponic cultivation of spinach cv. Lead. All the parameters of plant growth measured were optimized at a controlled temperature of 25°C and an electrical conductivity of 3.6 mS/cm. Optimum concentrations of Zn and Cu in the nutrient solution were 7.5 and 3.2 µmol/l, respectively; growth was significantly retarded at Zn and Cu concentrations above 10.5 and 3.2 µmol/l, respectively (Terazoe and Okano, 1992). Lime should be applied to maintain the soil pH in the range 6.5-6.8 as it is extremely sensitive to soil acidity.

Citak and Sahriye (2010) studied the effect of organic manures like chicken manure (CM), farmyard manure (FM) and blood meal (BM) as a single and as mixtures at different quantities by aiming to receive 150 kg N/ha for each on growth, yield and quality of spinach. In general, autumn season gave the better results in terms of spinach growth, yield and resulted in lower nitrate concentration, whereas the vitamin C concentration was found to be higher in winter season. He advocated reasonable applications of organic manures with regard to the seasons as 3.5 t/ha CM and 0.6 t/ha BM + 0.85 t/ha CM + 4.0 t/ha FM for spinach growth; 3.5 t/ha CM and 5.0 FM + 1.2 CM + 0.4 BM applications for spinach yield; 5.0 t/ha FM + 2.5 t/ha CM and 15.0 t/ha FM for vitamin C and nitrate concentration in the autumn and the winter season, respectively.

7.3.0 Irrigation

Spinach seeds require good moisture to germinate. Therefore, post-sowing light irrigation is necessary, if there is not enough soil moisture in the field. The second irrigation may also be given soon after first irrigation to secure uniform and maximum germination. Subsequent irrigations are required at 10-15 days interval because this is a shallow rooted crop. Ample soil moisture is important near harvest which improves yield and quality of produce (MacGillivray, 1961).

The increasing scarcity of water for irrigation and environmental pollution due to excessive use of fertilizers are the important problems in spinach production. A field experiment with combination of three levels of irrigation and nitrogen fertilization was employed to optimize the irrigation and nitrogen fertilizer usage of spinach (Zhang *et al.*, 2014). The yield was the highest when spinach was grown under the condition of the soil water content at 16.5 per cent combined with 170 kg ha–1 of nitrogen fertilizer, while the lowest yield was recorded for the one under the soil water content at 12.5 per cent with 0 kg ha^{-1}. Nitrate and oxalate contents of spinach were highly dependent on levels of irrigation and nitrogen fertilization. Nitrogen fertilization significantly decreased nitrogen use efficiency. Both water use efficiency and profit responded positively to increased nitrogen fertilizer usage. To optimize the quality and earnings of spinach, and considering the fact that nitrogen fertilizer could degrade the quality of spinach, application of the nitrogen fertilizer at 85 kg ha^{-1} and maintenance of the soil water content at 16.5 per cent could be recommended for spinach cultivation under field conditions.

7.4.0 Weed Control

Annual and perennial weed populations must be reduced prior to planting. To keep the field free from weeds, hoeing is necessary. Weeds can also be checked

Field Growing and Marketing of Spinach

by the application of herbicides as recommended by various workers but apply it before seed germination. Also care must be taken to avoid fields where residual herbicides from previous years persist in the soil. Apply pendimethalin @ 0.3 per cent on soil surface immediately after sowing of seeds Broad-leaved weeds can also be controlled by Phenmedipham applied at 2.2 kg a.i./ha. S-metolachlor @ 1 kg ai.i/ha is safe for pre-emergence use. Lenacil reduced weed densities by 61.5 per cent up to 60 days after application (Silvestri *et al.*, 1997).

Green house studies were conducted to compare the relative tolerance of fresh market spinach to S-metolachlor and metolachlor (Fennimore *et al.*, 2001). At rates above 2.2 kg a.i./ha S-metolachlor is less selective in spinach than metolachlor. S-metolachlor is safe for preemergence use in fresh market spinach at rates up to 1.06 kg/ha.

7.5.0 Use of Growth Substances

Spinach responds well to plant growth regulators. Pee (1970) reported marked improvement in growth with GA_3 applied at 200 ppm. Application of GA_3 at 10 ppm + Tween 20 (surfactant) at 500 ppm markedly enhanced the petiole length, stem height, and leaf blade weight and area (Gonzalez *et al.*, 1979) and GA_3 alone applied at 20 ppm also increased yield, facilitated harvesting and handling, and improved keeping quality of spinach (Gonzalez, 1980). Omran *et al.* (1973) reported that the application of $ZnSO_4$ at 100 ppm + GA_3 gave the highest total fresh yield. Exogenous application of GA_3 at 20 ppm to autumn-grown spinach induced upright growth of leaves that facilitated mechanical harvest in the USA (Gonzalez *et al.*, 1989). Kadam *et al.* (1988) obtained highest yield of leaves with Vipul (triacontanol) at 5 ppm applied 30 days after sowing. A considerable improvement in biochemical and antioxidant attributes of spinach was recorded due to the foliar application of the selected PGRs, and a remarkable effect was exhibited by *Moringa oleifera* leaf extract (MOLE) (Aslam *et al.*, 2016) Their results support the potential uses of the selected PGRs, and especially of MOLE, as effective and safer alternatives to synthetic PGRs for improving the production of valuable nutrients and phenolic antioxidants in spinach and other leafy green vegetables in order to explore their functional food and nutraceutical potential for value addition.

8.0 Harvesting, Yield and Storage

8.1.0 Harvesting and Yield

The crop becomes ready for cutting about 4 weeks after sowing. The crop should be harvested when leaves are lush and tender. Usually cutting is done by hand with the help of a sharp knife. It gives about 3 to 4 cuttings in the season. Like spinach beet, harvesting of this crop is not done early in the morning due to dew because the leaves are crisp and break easily when wet. This crop gives an average yield of about 80-100 q/ha.

8.2.0 Marketing and Storage

Before marketing, all diseased and yellow leaves should be removed from the produce. Leaves are tied in small bundles for easy handling and marketing. Spinach has one of the highest respiration rates among fruits and vegetables, so cooling is essential for longer shelf-life. Storage of spinach is a problem and cannot be stored for a long time except for few hours under ordinary conditions. However, it can be stored for 10-14 days at 0°C and 90-95 per cent relative humidity. Fluctuation in temperatures accelerates the deterioration. Andre *et al.* (1980) reported 40 days' storage with vacuum pre-refrigeration and storage at 1 °C in a modified atmosphere (in plastic packing with a diffusion window).

9.0 Diseases and Pests

Spinach is affected by number of diseases such as leaf spot, white rust, downy mildew and root rot or damping off.

9.1.0 Diseases

Fungal Diseases

9.1.1 Leaf Spot

A complex of fungi is associated with leaf spot disease on spinach, but *Cladosporium variable* is believed to be the primary pathogen under cool and moist conditions (du Toit *et al.*, 2005). This fungus can grow under a wide range of temperatures, but 59-68 °F with relative humidity above 80 per cent is most conducive to disease development. Early symptoms are numerous small, circular white to yellow spots, beginning on older leaves and progressing to younger ones. Spots often join together and then are irregularly shaped. Older spots develop a narrow darker-colored margin with a tan center. Generally, older leaves die, but in severe infections all leaves may die, and seed may be affected. Ripening seed on spinach seed crops also may develop lesions, shrivel, and become infected with the fungus.

Everts and Armentrout (2001) reported the first incidence of leaf spot caused by *Stemphylium botryosum* in Maryland and Delaware, USA. According to them, leaf spot lesions were small (0.2 - 0.7 cm), circular, tan and papery and lacked visual signs fungal infection. A similar disease infestation was occurred in the Salinas Valley, Monterey County, CA, USA (Koike *et al.*, 2001). Bedlan (2001) suggested control measures include varietal resistance and leaving affected areas for 5-7 years before growing spinach again.

9.1.2 White Rust

This is caused by the fungus *Albugo occidentalis*. Symptoms of white rust include yellow lesions on the upper leaf surface and corresponding white blister-like pustules or bumps on the underside of the leaf. Generally, the upper surface of the leaves remains chlorotic, however in advanced stages, white lesions may form on the upper side of the leaf. Infected plants are weak and might collapse if environmental conditions are favorable for disease development. This is controlled by the spray of copper fungicides and by the other methods including crop rotations and keeping the field free from weeds. Smooth leaf cultivars are more resistant than the savoy-leaf cultivars that may be grown (Dainello *et al.*, 1981).

Diseases of Spinach

Cladosporium **Leaf Spot (Left), and White Rust (Right) Infection**

Downy Mildew Infection

Damping Off (Left), and Anthracnose (Right) Infection

9.1.3 Downy Mildew

Downy mildew is still the main and one of the most destructive diseases of spinach at the global scale which is caused by fungus *Peronospora farinosa* (*Peronospora effusa*). It causes indefinite, yellowish areas on the upper leaf surface. On the lower surface a gray mold appears. Affected areas turn black and die. Disease develops best in cool, moist weather. The disease can be controlled by the spray of 0.25 per cent metalaxyl-mancozeb thrice at an interval of 15 days. Cultivars Solomon, Samson and Gone are resistant to the attack of downy mildew (Shimazaki and Uchcyama, 1985). According to Ozaki and Takahashi (1993) the new lines Konbatto, Hokushu, Jhoka and NSP-901 were resistant to race 4 of *Peronospora spinaciae*.

9.1.4 Root Rot and Damping Off

This disease is described in spinach beet and routine control measures are the same. The incidence of the root diseases damping off and *Fusarium* wilt of spinach were suppressed to one-third in a green house with UV-absorbing vinyl film (UVA) compared with disease when a common agricultural vinyl film (CA) green house was used. Suppression of the disease by UVA was evident in the summer cultivation at high temperature from June-September. The fungi isolated from the diseased roots of spinach were mainly *Pythium* spp., *Fusarium* spp. and *Rhizoctonia* spp. Isolation frequency of *Pythium* spp. was the highest, followed by *Fusarium* spp. and the occurrence of *Rhizoctonia* spp. was sporadic.

Wilt caused by *Fusarium oxysporum* and *Verticillium* spp. (*e.g. V. dahliae*) is also threatening spinach cultivation.

9.1.5 *Phoma* Blight (*P. nebulosa*)

A new fungal disease was identified in a spinach seed crop cv. Virginia Savoy growing at Katrain, India during the rabi seasons of 1986-88 (Kapoor and Thakur, 1994). Symptoms of the disease appeared when the crop was 20-35 days old and severity ranged from 12-90 per cent.

Viral Diseases

9.1.6 Cucumber Mosaic Virus

CMV infection of spinach is often referred to as spinach blight. It attacks spinach causing plants to become faintly chlorotic. Chlorosis increases until the entire plant becomes yellow. Crown leaves are narrow, wrinkled and develop an inward rolling of the margins. Plants are stunted. Death may occur in some severely infected plants. The virus is aphid transmitted. CMV resistance in spinach is controlled by a single dominant gene in the variety Virginia Savoy, and this has been incorporated into many current spinach varieties. This resistance, however, is not complete and may break down at temperatures greater than 28 °C.

9.1.7 Tobacco Ringspot Virus

First symptoms are small, indistinct, chlorotic spots which appear on the young foliage. These may coalesce to form large yellow areas. In advanced stages, leaves take on a copper bronze chlorosis. There is no malformation of the foliage. Affected plants rarely die but remain yellow and stunted.

9.1.8 Beet Curly-Top Virus

Infected plants are marked by a rosette of tightly curled, small leaves in the centre of the plant. As the disease develops, the growing point is killed and the plant dies. The virus is transmitted by beet leafhoppers.

9.2.0 Pests

Spinach crop is not damaged so severely by insects. However, aphids and caterpillars damage this crop and the same control measures are followed as given in spinach beet.

10.0 Seed Production

Seed production of this crop is like spinach beet. Spinach plants produce small flowers. Pollens are small and light, produced in abundance and carried away by air currents. This is a cross-pollinated crop and pollination takes place by wind. Unpollinated pistillate flowers remain receptive for 2-3 weeks. Spinach populations are commonly selected based on the following criteria:

1. Seedling vigor and early robust growth are important traits in all production systems and can be improved over several cycles of selection. This includes selecting for the quickest germinating plants.
2. There are various stages at which leaf size can be selected depending on the leaf shape desired by the vegetable market served by the spinach cultivar. As an example, much of the modern "baby leaf" is selected for rounded leaves. Alternately, some prefer a lobed or arrow-shaped spinach leaf for shape variation.
3. Leaf textures are characterized as savoy (crinkle-leaf), semi-savoy or flat, and are selected along with leaf shape. Leaf texture varies with the stage of growth.
4. Darker leaf colour is commonly preferred. Darker leaf colour is also attributed to higher nutritional value.
5. Plant stature: The ability of the plant to hold its leaves in an upright position is very important in vegetable production in general, especially for harvest of leaves cut for salad mix. Upright foliage also lessens the incidence of fungal and bacterial pathogens.

Sex development is governed by many factors. An increase in nitrogen level decreased the proportion of female plants. The female plants have more leaves than male plants (Quagliotti, 1972). However, femaleness can be enhanced by cutting the shoot of young plants and allowed to form adventitious roots, whereas maleness is enhanced by treating the plants with GA_3 25 mg/l after cutting the leaves. Hence, leaves play an essential role in sex expression and this role is related to gibberellin synthesis in the leaves (Chailakhyan and Khryanin, 1979). Cajlakjain *et al.* (1969) reported that treatment with gibberellin and CCC influenced the formation of

Different Stages of Seed Production of Spinach

reproductive organs through regulation of growth processes but sex was not determined directly. Photoperiod and plant growth regulators showed marked effects on the sex-expression. Sex ratio (male to female) is near 1 in long days. The number of female plants flowering in long days can be increased by gibberellins, whereas kinetin reduced flowering in male plants. Abscisic acid and Amo 1618 reduced flowering in famale plants. In short days gibberellins stimulated flowering in female more than in male plants and at higher temperature flowering (maleness) is reduced (Culafic and Neskovic, 1980).

10.1.0 Isolation

Spinach is a cross-pollinated crop. To produce pure and true-to-type seeds, isolation distance should be kept 1600 m between cultivars.

10.2.0 Roguing

For pure seed production roguing is essential; and should be done at following stages: vegetative, bolting and flowering. Early and late bolters are removed. Most of early bolters are males and have no effect on the seed yield.

10.3.0 Harvesting and Seed Yield

Spinach produces seed stalks in about 65 days and seeds in 150 days. Seeds mature slowly and unevenly. Seeds shatter badly from dried plant; hence the seed plants are harvested before they are fully dried. Plants are cut and dried in the field. After complete drying, the plants are threshed, and seeds are cleaned by winnowing. Average seed yield is 7.5-10.0 q/ha (Choudhury, 1967). Late cultivars in the hills produce less seed of about 100 kg/ha as reported by Singh *et al.* (1962). Seed yield is also adversely affected by cuttings (Hussain *et al.*, 1970). However, to utilize some of the leaves of plants grown for a seed crop and to prevent the harmful accumulation of nitrates, following high N rates, the lower leaves were harvested 2 months after sowing and urea at 0.5-2.0 per cent was applied one month later in cv. Long Standing. The seed yield was highest (6.64 q/ha) in plants receiving urea at the lowest rate (Kohli and Saini, 1985). Seed yield can be increased with the spray of $ZnSO_4$ at 500 ppm plus GA_3 at 50 ppm (Omran *et al.*, 1973).

11.0 Crop Improvement

The success of a plant breeding programme is largely dependent on the availability of genetically diverse plant germplasm to allow for cultivar improvement. Spinach is also a cross-pollinated crop and can be improved by breeding procedures as described in spinach beet. The objectives of breeding are also similar to spinach beet. However, spinach is only grown successfully in winter and cultivars should be developed which can withstand higher temperature. Moreover, spinach has higher oxalate content than spinach beet. Leaf dimension has a strong negative association with leaf number. Leaf yield positively correlated with leaf length and width. Oxalic acid content showed no relation to yield or leaf type, being affected by environmental factors rather than genetic ones. Main objectives of the spinach improvement are high yields, erect leaves for mechanical harvesting, small petioles, late bolting, lower oxalic acid and nitrate content, wider adaptability and stress tolerance.

It is a dioecious species with separate male and female plants as well as occasional monoecious plants with both male and female flowers. The sex-expression is controlled by a single pair of sex-chromosomes namely, X and Y. Homozygous XX plants are female and heterozygous XY are male. When XX plants are crossed with XY plants, the progeny will always segregate in 1: 1 as with a monohybrid back-cross. Sometimes modifier genes are present and these may induce the production of hermaphrodite flowers which allow selfing; the XX females give all female progenies and XY males produce male and female plants in the ratio of 3: 1. Sugimoto (1948) reported a single autosomal gene pair in addition of XX mechanism and proposed a hypothesis, *i.e.*, ZZ is female determining, Zz is intermediate (monoecious) and zz is male determining. The variation in degrees of monoecious is directly related to variations in the Zz gene type as induced environmentally. Later this hypothesis was modified by Bemis and Wilson (1953). They postulated a pair of linked autosomal genes Aa (dominant for maleness) and Gg (dominant for femaleness) in addition to XY chromosome pair. When these are in balance, the normal 1: 1 (male: female) ratio is maintained.

Janick (1955 a,b) investigated sex genetics of spinach tetraploids. A single Y factor is required for maleness and only XXXX is female. This indicates that sex is not dependent on a balance of X and Y, *i.e.*, the monoecious expression is not dependent upon such a balance. The 5: 1 (staminate to pistillate) ratio observed in cross XXXX x XXYY is evidence of chromosome segregation. The pistillate plants were homomorphic for the long arm chromosome 1 type, XX and homomorphic for satellited chromosome 3. Staminate plants were heteromorphic for chromosome 1, such that the long arm carried X and the short arm carried Y. It was also homomorphic for the satellited chromosome 3.

Fresh weight has been reported to be positively correlated with the number of leaves/plant, leaf fresh weight/plant and leaf nitrate and nitrite-N concentrations, but negatively correlated with per cent dry matter and leaf/iron concentration (Farag and Abdel-Aal, 1989).

The inheritance of characters, prickly vs smooth seed, smooth vs savoy leaves, light vs dark green leaves and short vs long petiole is either of the single gene type or relatively simple as reported by Sneep (1958). Parlevliet (1968) investigated the components of earliness in spinach. The earliness is determined by rate of seed-stalk formation or rate of flower formation and controlled probably by polygenic inheritance of both growth and development. The savoy forms have lower amount of calcium oxalate than smooth or semi-savoy form as described by Kitchen *et al.* (1964).

Among the breeding procedures, resistant breeding and breeding for F_1 hybrids are most important to improve the spinach crop.

New selections appeared during the 20th century, the general trend was to breed for smooth-seeded rather than for spiny-seeded. Savoy-leaved cultivars, also called Bloomsdale type spinach, followed a separate breeding history. Due to their leaf shape, these were preferred over smooth-leaved cultivars in regions where vegetables needed transportation over long distances (Sneep, 1983; Rubatzky and Yamaguchi, 1997). Crinkles kept spinach less compact during packing and transport, resulting in an extended shelf life.

11.1.0 Resistant Breeding

The disease resistance breeding in spinach is more common because this crop is damaged generally by virus (cucumber virus 1, spinach blight), downy mildew and white rust. Smith (1921) produced Virginia Savoy, a resistant cultivar to cucumber virus 1 by crossing a domestic Savoy with an introduction from Manchuria, China. This was the first documented case of resistance breeding in spinach. In the year 1950, Smith identified dominant single gene immunity from an isolated line PI 140467. This served as the source for race 1 immunity in cvs. Califlay and Dixie Market and parental stocks of Early Hybrid 7 and Early Hybrid 424. Later, Califlay showed infection to downy mildew and a second physiological race was identified. Smith *et al.* (1962) reported that a single dominant gene carried immunity to both races. These two resistant genes are closely linked as reported by Eenink (1976). Eisbein (1982) described a selection method for breeding of spinach for resistance to cucumber mosaic virus. The method consists of artificially inoculating 6-week-old plants twice at an interval of 2-3 weeks followed by elimination of infected plants 2-3 weeks later. In 11 new breeding lines, progeny from plants selected according to this method had lower percentage of susceptible plants than progeny from unselected plants. Cook *et al.* (1947) reported the development of a wilt resistant line by positive selection of 'Virginia Savoy' and commercialized under the name 'Vates Wilt Resistant'. Smooth-leaved downy mildew resistant cultivars were developed, and the first cultivar was commercialized in 1957 under the name Califlay, which resulted from a cross between PI 140467 and Viroflay and the subsequent backcrossing to Viroflay (Smith and Zahara, 1956). In 1987, the improved open-pollinated cultivar Fallgreen with a high resistance to white rust was released by the University of Arkansas (Morelock, 1999).

Wilt caused by *Fusarium oxysporum* and *Verticillium* spp. (*e.g. V. dahliae*) is also threatening spinach cultivation (Koike *et al.*, 2006). Regarding *V. dahliae*, molecular markers associated with resistance to the fungus have been identified (Shi *et al.*, 2016b).

Goode *et al.* (1987) reported polygenic resistance of cultivar Fall Green to white rust, blue mould, Fusarium decline, damping-off and anthracnose.

Screening of 345 accessions of the US spinach collection has shown genetic variability for leaf miner resistance (Mou, 2008) which can be improved further. Spinach has also been investigated for its tolerance to abiotic factors such as water stress, salinity and osmotic stress, heavy metal and temperature stress. Two genes involved in tolerance to osmotic stress have been identified (Burnet *et al.*, 1995; Hibino *et al.*, 2002), but in general the genetics of tolerance against abiotic stresses is rather understudied in spinach. Research on abiotic stress tolerance will most likely gain considerable importance in spinach breeding in the near future.

Breeding methods applicable to the improvement of cross-pollinated vegetable crops may also be applicable to spinach improvement. Generally, population improvement methods such as recurrent selection, mass selection and progeny testing are suitable for the development of new variation. Vegetative male or female plants are more productive than male plants. Therefore, selection should be made for predominantly female plants.

Transgenic tobacco plants which stably overexpressed *spinach oleracea* S°CYP85A1 gene were used as plant materials to enhance high-salt tolerance. Both wild types and transgenic lines were analysed before and after high-salt stress. The results showed that the germination rate of transgenic plants was significantly higher than that of wild type (Lu *et al.*, 2017).

11.2.0 Heterosis Breeding

The F_1 hybrid gives up to 20 per cent higher yield than its parent (Singh and Swarup, 1971). The different sex types in spinach are related to F_1 hybrid production by utilizing female lines in combination with known pollen (male) parent. Generally, hybrids grow faster than OP varieties and accumulate lower amounts of oxalate and nitrate concentration in their leaf blades. The male parent must be good combiner and produces ample pollen. As soon as bolting allows differentiation between male (staminate) and pistillate (female) plants, the male plants are removed from seed (female) parent rows. For production of F_1 hybrids first inbred line is produced from the ultimate parents. Thompson (1955) proposed two methods of producing hybrids using monoecious females or pure females. The first method is single cross method with a highly female monoecious inbred as the seed parent and highly male monoecious inbred as the pollen parent. However, the second method is a three-way cross hybrid in which the female parent is a female x highly female monoecious F_1 or a highly female monoecious x highly female monoecious F_1. Walts (1980) reported that the dark green leaf colour was closely associated with late bolting. The dark leaved types are somewhat curled and contain larger amount of iron. The F_1 between smooth and savoy leaved types was intermediate. Heterosis breeding has also been exploited for development of better quality and resistant cultivars in spinach.

11.3.0 Breeding for Quality

The chemical composition of spinach has been shown to differ significantly between cultivars, including oxalic acid, nitrate, vitamin C, lutein, carotenoid and phenolic content (Murphy and Morelock, 2000; Pandjaitan *et al.*, 2005; Solberg *et al.*, 2015; Wang *et al.*, 2018). Although nutritional composition is known to be influenced by factors such as cultivation method and storage procedures (Lester *et al.*, 2010; Koh *et al.*, 2012), the large variation observed among cultivars indicates a genetic basis, suggesting that nutritional quality can be improved by plant breeding (Howard *et al.*, 2002; Morelock and Correll, 2008; Wang *et al.*, 2018).

Maximizing health-related compounds such as vitamin C, lutein, carotenoid and phenolic content, and minimizing the content of oxalic acid and nitrate are important targets in improving the quality of spinach cultivars. Savoy-leaf cultivars had the highest nitrate and oxalate concentrations and smooth leaved the lowest because later one grows faster and gets lesser time to accumulate. Semi savoy-leaved cultivars had intermediate concentrations. Further, higher levels of dry-matter content have been reported in smooth-leaved cultivars than that in crinkled types; and also, higher iron content was analysed in varieties with dark-green leaves compared with varieties with light-green ones. The identification of SNP markers associated with 13 different mineral elements in spinach (Qin *et al.*, 2017) would assist in breeding of mineral rich cultivars. Additionally, the identification of SNP markers associated with oxalate concentration (Shi *et al.*, 2016a), and QTLs related to nitrogen use efficiency (Chan-Navarrete *et al.*, 2016) could be helpful to reduce oxalate nitrate levels in spinach leaves, respectively. Differences in the concentration of phenolics and flavonoids between modern cultivars differing in the level of resistance against downy mildew and white rust have been detected by Howard *et al.* (2002). In addition, Pandjaitan *et al.* (2005) reported that the level of phenolics and flavonoids in spinach leaves depended on the maturity stage, mid-mature leaves showing the highest levels in comparison with baby and adult leaves. Recently, a red spinach cultivar with an increased antioxidant content was released (Mou, 2019). This cultivar is rich in betacyanin and has been selected from a red-veined spinach cultivar.

11.4.0 Breeding for Abiotic Stress Tolerance

Spinach has been investigated for its tolerance to abiotic factors such as water stress (Zuccarini and Save, 2016), salinity and osmotic stress (Bagheri *et al.*, 2015; Ors and Suarez, 2017; Ferreira *et al.*, 2018), heavy metal stress (Fagioni *et al.*, 2009; Bagheri *et al.*, 2015) and temperature stress (Mogren *et al.*, 2015; Chitwood *et al.*, 2016). Two genes involved in tolerance to osmotic stress have been identified (Hibino *et al.*, 2002), but in general the genetics of tolerance against Abiotic stresses is rather understudied in spinach. Adaptation to low nitrogen availability is an important aspect in the development of sustainable and more efficient agriculture (Witcombe *et al.*, 2008). In this context, the two QTL regions related to growth in nitrogen-poor conditions identified by Chan-Navarrete *et al.* (2016) will contribute to the improvement of nitrogen use efficiency in spinach. To date, abiotic stress improvement is largely unexplored in spinach as the available literature on this topic is rather scarce. As climate change is expected to pose serious challenges to agriculture (Pereira, 2016), research on abiotic stress tolerance will most likely gain considerable importance in spinach breeding in the near future.

12.0 Biotechnology

Biotechnological approaches reported to improve spinach are very meagre. Though a lot of research has been conducted in this area but those are primarily basic in nature. The following is the comprehensive review on biotechnology of spinach.

12.1.0 Tissue Culture

Tissue culture work on spinach has been reported for micropropagation as well as for protoplast cultures.

12.1.1 Micropropagation

Spinach is suitable for micropropagation *in vitro* and plants may be regenerated with roots, shoot tips, leaves and flower stalks in artificial medium. Kondo *et al.*

(1991) induced tissue-cultured shoot primordia from apical domes of shoot tips in MS liquid medium supplemented with 0.2, 2.0 or 4.0 mg IAA and 0.2, 2.0 or 4.0 mg BA/l at pH 5.8 by shaking at 2 cycles/minute on rotary culture equipment (1 m diameter). Multiplication of the shoot primordia was most effective in MS liquid medium supplemented with 0.2 mg IAA and 0.02 mg BA/L. Regeneration of plantlets was accelerated in MS supplemented with 0.2 mg BA and 4.0 mg IAA, 2.0 mg BA and 0.02, 0.2, 1.0 or 4.0 mg IAA or 4.0 mg BA and 0, 0.02, or 0.2 mg IAA/L in combination. Root formation was easily promoted in MS medium supplemented with no hormones or 0.2 mg IAA/l. Komai *et al.* (1996) established an efficient system for plant regeneration from adventitious roots of spinach. The adventitious roots differentiated from calli that had grown from protoplasts. Protoplasts were obtained from the cotyledon of seedlings by partial digestion of cell walls, followed by mechanical maceration with forceps. The protoplasts formed colonies at a frequency of 1.6 per cent in a MS medium containing 1 µM 2,4-D or 10 µM NAA combined with 10 µM zeatin. Calli derived from the colonies produced adventitious roots on MS medium containing 10?µM NAA and 0.1 µM gibberellic acid (GA_3). Root segments excised from these adventitious roots differentiated into many somatic embryos when subcultured on MS medium supplemented with 30 mM fructose instead of sucrose. The somatic embryos developed into flowering mature plants on a growth regulator-free medium.

Okuse (1994) observed differences in callus formation and growth between explants of male and female flower stalks. Callus formation was observed on MS medium supplemented with auxin, best results being obtained with 1 µM 2,4-D and 10 µM IAA and NAA. Callus formation was also observed on MS medium with cytokinin, best results being with 10 µM for all the cytokinins used. However, callus growth was inferior to that on media with auxins. Callus growth was well promoted on media with optimum concentrations of auxins and cytokinins. Thidiazuron was the most effective for callus growth. There was no callus formation on media supplemented with gibberellin alone. Gibberellin promoted callus growth on media with auxins but not with cytokinins. The minimum concentration of auxins and cytokinins for callus formation was higher for female than male explants. Callus growth was better in male than female explants, especially on methods suitable for male and female plants. Micropropagation methods are suitable for male and female plants. Micropropagation of spinach through leaf explants was reported by Al-Khayri *et al.* (1991). Leaf disc explants (5 mm in diameter) were taken from 3-month-old spinach cv. Fall Green plants and cultured with the abaxial side down on MS medium supplemented with 0.4 mg thiamine-HCl, 100 mg myo-inositol, 30 g sucrose and 8 g agar/l. For callus induction, the medium contained kinetin at 1 or 2 mg/L and 2,4-D at 0.1, 0.5 or 1.0 mg/l. Cultivars were held at 20°C either in the dark or in a 10-h photoperiod of cool-white fluorescent light. Best callus growth was obtained in the dark on the medium containing 2 mg kinetin + 0.5 mg 2,4-D/L. After 5 weeks, the most prolific calluses were transferred to the same MS medium supplemented with 2 mg kinetin, 0.01 mg 2,4-D and 0 or 1 mg GA_3/L. A 16 per cent regeneration rate was obtained on the medium containing GA_3 but no shoot production was observed in the GA_3-free medium. To induce rooting shoots 1-3 cm long were transferred to a growth regulator-free MS medium solidified with

8 g phytagel or agar/L. Phytagel gave 80 per cent rooting compared with 10 per cent rooting using agar. Regenerated plantlets were successfully acclimatized to soil conditions.

Nguyen *et al.* (2013) tested different plant growth regulator (PGR) combinations to induce callus with high frequency (73-100 per cent), and the combination of 5 µM α-naphthaleneacetic acid (NAA), 10 µM 6-benzyladenine (BA) and 0.1 µM gibberellic acid (GA_3) had the most significant effect on callus growth in term of weight (120.98 ± 22.56 mg) in spinach. A high auxin-containing medium induced competent callus for shoot formation, while high cytokinin-containing media enhanced callus growth and made callus incompetent for shoot regeneration. Longer periods of callus induction in a high auxin-containing medium were required to form competent callus and led to a high regeneration capacity in spinach.

12.1.2 Protoplast Culture

Spinach has been used as model system and many tried to use this for various basic studies. Goto *et al.* (1996) cultured spinach cv. Jiromaru protoplasts in a modified half-strength MS medium supplemented with KM8p vitamins, 1.0 mg BA/L, 1.0 mg 2,4-D/L and glucose or mannitol. Planting efficiency in the medium containing glucose as an osmoticum was much higher than that in the medium containing mannitol. The addition of KM8p organic acids to the glucose medium promoted the division of protoplasts slightly; the same addition to the medium containing mannitol appeared to inhibit cell division. Several shoots were formed from calluses obtained on MS medium containing 1.0-10.0 mg IAA or IBA/L. Adventitious roots developed most vigorously on MS medium containing 2 per cent sucrose, 0.5 mg IBA/L and 0.8 per cent agar when pH was adjusted to 6.3. Regenerated plantlets were transferred to pots and acclimatized in the green house. Some were dioecious male, some dioecious female and some monoecious. The monoecious ones produced many seeds following self-pollination.

12.2.0 Molecular Markers

Microsatellites are valuable tools as molecular markers in plant breeding. In order to establish genetic linkage maps or for use in population studies, information about the occurrence and usability of microsatellite markers in different species is necessary. Occurrence of microsatellites in spinach sequences from computer databases and development of polymorphic SSR markers has been reported by Groben and Wricke (1998). Sequences of spinach from computer databases were searched for the presence of microsatellites. Sixty simple sequence repeats were found in 237 spinach sequences with a total of 349.4 kb DNA. After removing duplicated sequences, 50 different microsatellites with various motifs remained. Differences between nuclear and chloroplast DNA were not in the number of microsatellites but in their type and length. Chloroplast sequences from spinach contained only short strings of A and AT repeats, whereas nuclear sequences showed a wider variety of motifs. Flanking primers for polymerase chain reaction analysis were designed for 13 or these microsatellites and tested with two different varieties of spinach. Twelve primer pairs gave amplification products; seven of these showed polymorphisms in the variety Wiremona, but only one in the variety Monatol. These markers may be used for linkage analysis or population studies in spinach.

12.3.0 Genetic Transformation

A lot of basic research has been done on spinach which is used as a model system for various purposes. In the following review the relevant research pertaining to spinach improvement is presented.

Al-Khayri *et al.* (1994) transformed leaf discs from 3-month-old plants of cv. High Pack using *Agrobacterium tumefaciens* strain A208 containing the pMON9749 plasmid carrying genes encoding *gus* and *nptii*. Shoots developed from the transformed callus at a rate of 20 per cent and after 4 weeks on rooting medium, 70 per cent of the shoots produced roots. Transformation of the regenerated plantlets was confirmed using a GUS assay. Similarly, Zhang and Zeevaart (1999) used cotyledon explants to develop an efficient transformation and regeneration system for the production of transgenic spinach plants. Cotyledon explants were infected with *A. tumefaciens* strain LBA4404 carrying the selectable marker gene, *nptII*, and the reporter gene *smgfp*, encoding soluble-modified green-fluorescent protein, driven by the cauliflower mosaic virus 35S promoter. The infected explants were cultured on MS medium, containing 1 mg BA/l and 0.4 mg NAA/l. Shoots were regenerated on selection medium containing 50 mg kanamycin/l. Regenerated kanamycin-resistant shoots were rooted on medium containing 1 mg IBA/l and subsequently grown in soil in the green house. Southern blot analysis indicated that the *smgfp* gene had been integrated into the spinach genome. Northern and western blots showed that the *smgfp* gene was expressed in progeny plants.

Yang *et al.* (1997) regenerated transgenic spinach plants expressing cucumber mosaic virus (CMV) coat protein genes. Leaf and hypocotyl explants from two spinach cultivars, Fall Green and High Pack, were transformed with *A. tumefaciens* strain LBA4404 carrying the binary vector pBI121 in which the β-glucuronidase (uidA or GUS) gene was replaced with the coat protein gene of CMV isolate SP103 or SP104. Calluses were induced from both cultivars at rates of 10-27 per cent on MS selective medium containing 86 µm kanamycin, 257 µm carbenicillin and 209 µm cefotaxime. After 12 weeks (three transfers), putatively transformed calluses were transferred to MS medium without kanamycin to promote callus proliferation. Shoots were regenerated at the rates of 23-29 per cent for High Pack and less than 0.5 per cent for Fall Green. Approximately 80 per cent of the regenerated shoots from both cultivars were subsequently induced to establish whole plants. Regenerated plants were screened for transgene incorporation using a polymerase chain reaction (PCR) method using the same primer pair originally used to amplify the CMV coat protein genes for insertion into pBI121. Based on the PCR screening, approximately 12 per cent of the regenerated spinach plants contained a CMV coat protein gene. Southern blot analysis confirmed the presence of viral coat protein genes in the plant genomes. Western blot analysis indicated that the CMV coat protein genes were expressed in transgenic spinach plants.

Chin *et al.* (2009) reported *Agrobacterium*-mediated transformation [strain EHA101 harboring plasmid pIG121-Hm containing the genes for β-glucuronidase (gus), hygromycin phosphotransferase (hpt) and neomycin phosphotransferase (nptII)] combined with plant regeneration from leaf explant at a low temperature of 14°C for producing transgenic spinach plants with useful genes. Moreover, *Agrobacterium*-mediated genetic transformation with *Bacillus thuringiensis* cry1Ac gene was done for resistance to 2 pests, *Trichoplusia ni* and *Autographa nigrisigna* (Bao *et al.*, 2009).

13.0 References

Al-Khayri, J.M., Huang, F.H. and Morelock, T.E. (1991) *Arkansas Farm Res.*, **40**: 7.

Al-Khayri, J.M., Miles, R.W., Huang, F.H., Morelock, T.E. and Stewart, J.M. (1994) *Arkansas Farm Res.*, **43**: 10-11.

Andre, P., Blance, R., Buret, M., Chambroy, Y., Dauple, P., Flanzy, C., Pelisse, C. and Vergniaud, P. (1980) *Revue Horticole, Suisse*, No. 207, pp. 37-42.

Anonymous (2016) http: //nutritiondata.self.com/facts/vegetables-and-vegetable-products/.

Anonymous (2020) https: //nutritiondata.self.com/facts/vegetables-and-vegetable-products/2352/2

Arumuganathan, K. and Earle, E.D. (1991) *Plant Mol. Biol. Report*, **9**: 208-218.

Aslam, M., Sultana, B., Anwar, F. and Munir, H. (2016) *Turk. J. Agric. For.*, **40**: 136-145.

Avdonin, N.S. and Kochubei, I.V. (1978) *Vliyanie Svoistva Pochvi Udobrna Kachestvo Rast*, No. 4, pp. 43-58.

Bagheri, R., Bashir, H., Ahmad, J., Iqbal, M. and Qureshi, M.I. (2015) *Plant Physiol. Biochem.*, **97**: 235–245.

Bao, J.H., Chin, D.P., Fukami, M., Ugaki, M., Nomura, M. and Mii, M. (2009) *Plant Biotech.*, **26**(2): 249-254.

Bedlan, G. (2001) *Genuse Munchen*, **37**: 18-19.

Bemis, W.P. and Wilson, G.B. (1953) *J. Hered.*, **44**: 91-95.

Biancardi E., Mitchell McGrath J., Panella L.W., Lewellen R.T. and Stevanato P. (2010) *Sugar beet*. CRA-Centro per le Colture Industriali, Sede di Rovigo, Italy.

Biancardi, E., Panella, L.W. and Lewellen, R.T. (2012) *Beta maritima: the origin of beets*. Springer, NY.

Biancardi, E., Skaracis, G.N., Steinrücken, G., De Biaggi, M., Panella, L., Lewellen, R.T., Campbell, L.G., Yu, M.H., Stevanato, P. and McGrath, L.G. (2005) In: *Objectives of sugar beet breeding*. Science Publishers Inc, Enfield, NH, pp. 53-168.

Bradley, G.A., Sistrunk, W.A., Baker, E.C. and Cash, J.N. (1975) *J. Amer. Soc. Hort. Sci.*, **100**: 45-48.

Burnet, M., Lafontaine, P.J., Hanson, A.D. (1995) *Plant Physiol.*, **108**(2): 581-588.

Cajlahjain, M.H., Martinovic, L.N. and Kocankov, V.G. (1969) *Doklady Akademii Nauk*, SSSR, pp. 662-665.

Cervato, A. (1969) *Annali della Facolta di Agraria*, Universita Catholica del S. Cuore, Milano, **10**: 389-421.

Chailakhyan, M.Kh. and Khryanin, V.N. (1979) *Planta*, **44**: 205-207.

Chan-Navarrete, R., Dolstra, O., van Kaauwen, M., van Bueren, E.T. and van der Linden, C.G. (2016) *Euphytica*, **208**: 621-636.

Chin, D. P., Bao, J. H. and Mii, M. (2009) *Plant Biotechnol.*, **26**: 243-248.

Chitwood, J., Shi, A., Mou, B., Evans, M., Clark, J., Motes, D., Chen, P. and Hensley, D. (2016) *HortScience*, **51**: 481–486.

Choudhury, A.R., Rashid, M.M. and Hussain, A. (1974) *Horticulture*, **2**: 39-43.

Choudhury, B. (1967) *Vegetables*, National Book Trust, India, New Delhi, pp. 198.

Citak, S., and Sahriye, S. (2010) *Sci. Hort.*, **4**: 415-420.

Cook H.T., Nugent T.J., Parris G.K. and Porter R.F. (1947) *Va Truck Exp. Sta. Bull.* **110**: 1810-1820

Culafic, L. and Neskovic, M. (1980) *Physiol. Plantarum*, **48**: 588-590.

Dainello, F.J., Heineman, R.R. and Jones, R.K. (1981) *Texas Agric. Exp. Sta.*, No. PR-3878, pp. 7.

Deleuran, L.C. (2010) *Renew Agric Food Syst.*, **26**: 149-160.

Deore, B.P. and Bharud, R.W. (1989) *Annals of Plant Physiol.*, **3**: 163-171.

Dhillon, K.S., Dhillon, S.K., Singh, B. and Kansal, B.D. (1987) *J. Res. Punjab Agric. Univ.*, **24**: 31-36.

Drewes, H. (1943) *Spinach Varieties, Mich. Spl. Bull.*, pp.225.

du Toit, L.J., Derie, M.L., and Hernandez-Perez, P. (2005) *Fungicide and Nematicide Tests*, **60**: V044.

Eenink, A.H. (1976) *Euphytica*, **25**: 713-715.

Eisbein, K. (1982) *Archiv fur Zuchtungsforschung*, **12:** 319-324.

El-Fadaly, K.A. and Mishriky, J.F. (1990) *Bull. Fac. Agric. Univ. Cairo*, **41**: 973-988.

Everts, K.L. and Armentrout, D.K. (2001) *Plant Diseases*, **85**: 1209.

Fagioni, M., D'Amici, G.M., Timperio, A.M. and Zolla, L. (2009) *J. Proteome. Res.*, **8**: 310–326.

Farag, I.A. and Abdel-Aal, S.A. (1989) *Assiut. J. Agric. Sci.*, **20**: 105-119.

Fennimore, S. A., Smith, R.F., and Mcgiffen JR., M.E. (2001) *Weed Tech.*, **15**: 511–516.

Ferreira, J.F.S., Sandhu, D., Liu, X. and Halvorson, J.J. (2018) *Agriculture*, **8**: 163.

Geise, F.W. and Farley, H.B. (1929) *Variety Studies of Spinach, Md. Bull.*, pp.312.

Gonzalez, A.R. (1980) *Proc. Ann. Meeting Arkanasas State Hort Soc.*, pp. 68-70.

Gonzalez, A.R., Laniev, M.R. and Edwards, K.M. (1979) *Arkansas Fm. Res.*, **28**: 15.

Gonzalez, A.R., Prokakis, G. and Mays, J. (1989) *Res. Series-Arkansas Agric. Exp. Sta.* No. 388, pp. 20.

Goode, M.J., Morelock, T.E. and Bowers, J.L. (1987) *Arkansas Fm. Res.*, **36:** 3.

Goto, T., Miyazaki, M. and Oku, M. (1996) *J. Japanese Soc. Hort. Sci.*, **65**: 349-354.

Groben, R. and Wricke, G. (1998) *Plant Breed.*, **117**: 271-274.

Hibino, T., Waditee, R., Araki, E., Ishikawa, H., Aoki, K., Tanaka, Y. and Takabe, T. (2002) *J. Biol. Chem.*, **277**: 41352-41360.

Howard, L.R., Pandjaitan, N., Morelock, T. and Gil, M.I. (2002) *J. Agric. Food Chem.*, **50**: 5891–5896.

Hulewicz, D. and Mokrzecka, E. (1971) *Zeitschriftfur Pflanzenernahrung and Bodenkunde*, pp.214-425.

Hussain, A., Safar, A. and Malukera, A.M. (1970) *Punjab Fruit J.*, **32**: 8-9.

Inoue, S., Murakami, K., Kumakura, H. and Araki, Y. (2000) *Bull. Chugoku Nat. Agril. Expt. Stn.*, **21**: 13-40.

Jana, J.C., Thapa, U. and Maity, T.K. (1999) *Veg. Sci.*, **26**: 61-63.

Jabeen, A., Narayan, S., Hussain, K., Ahmed Mir, S. and Ahmed Khan, F. (2018) *Int. J. Curr. Microbiol. App. Sci.*, **7**(9): 1312-1317.

Janick, J. (1955a) *Proc. Amer. Soc. Hort. Sci.*, **66:** 361-363.

Janick, J. (1955b) *J. Hered.*, **46**: 150-156.

Kadam, N.D., Patil, G.D., Chougule, B.A. and Kadam, S.S. (1988) *Indian J. Plant Physiol.*, **31**: 123-125.

Kakade, A., More, P., Jadhav, S. and Bhosle, V. (2015) *Int. J. Agri. Environ. Biotech.*, **8**(3): 609-614.

Kapoor, K.S. and Thakur, P.C. (1994) *Indian Phytopath.*, **47**: 442-443.

Kausal, B.D., Singh, B., Bajaj, K.L. and Kaur, G. (1981) *Qualitas Plantarum*, **31:** 1 63-170.

Kerin, V. (1975) *Gradinarska i Lozarska Nauka*, **12:** 61-66.

Khader, V. and Rama, S. (1999) *Plant Food Human Nutrition*, **53**: 71-81.

Kitchen, J.W., Burns, *E.g.* and Perry, B.A. (1964) *Proc. Amer. Soc. Hort. Sci.*, **84:** 441-445.

Knot, J.E. (1939) *Cornell Memoir.*, pp. 218.

Kohli, U.K. and Saini, S.S. (1985) *Haryana J. Hort. Sci.*, **14:** 141-142.

Koike, S.T., Gladders, P. and Paulus, A.O. (2006) *Vegetable Diseases- A Colour Handbook*. Manson Publishing, London, pp.368-372.

Koike, S.T., Henderson, D.M. and Butler, E.E. (2001) *Plant Disease*, **85**: 126-130.

Kolota, E., Dobromilska, R. and Zwierzykowski, M. (1984) *Zeszyty Naukowe Akademii Rolniczej w Szczecinie, Rolnictwo*, **33**: 51-60.

Komai, F., Masuda, K., Harada, T. and Okuse, I. (1996) *Plant Sci. Limerick*, **120:** 89-94.

Kondo, K., Nadamitsu, S., Tanaka, R. and Taniguchi, K. (1991) *Plant Tissue Cult. Lett.*, **120**: 89-94.

Lal, S., Pandey, U.C., Pandita, M.L. and Singh, K. (1979) *Seed Res.*, **7:** 136-140.

Leskovec, E. and Dobersekurbanc, A. (1972) *Zbornik Biotechniske Fakultete Univerze, v Ljubljani, Kmetijstvo*,**19**: 101-109.

Lu, X. L., Duan, F. M., Wan, F. Y., LI, B.D. and Song, W.W. (2017) *Plant Physiol. J.*, **3**: 18.

MacGillivray, J.H. (1961) *Vegetable Production.* McGraw-HillBook Co., Inc., New York, pp.171-179.

Macreke, D. Van (1973) *Mededelingen van de Faculteit Landbouwetenschappen, Rijksuniversiteit Genet.*, **38:** 173-199.

Maity, T.K., Jana, J.C. and Thapa, U. (1999) *J. Andaman Sci. Asso.*, **15**: 35-38.

Mansfeld, R. (1986) *Verzeichnis landwirtschaftlicher und gärtnerischer Kulturpflanzen (ohne Zierpflanzen)*. Springer Verlag, Berlin, Heidelberg, New York.

Markovic, V., Lozic, B. and Djurovka, M. (1988) *Acta Hortic.*, **220**: 297-302.

Mishra, B., Joshi, Y.C. and Sarin, M.M. (1973) *Indian Hort.*, **18**: 13-14.

Mitidieri, A. and Bianchini, P.R. (1973) *Idia*, No. 301, pp. 54-68.

Morelock, T.E. (1999) *Vegetable cultivar descriptions for North America* List 25, *HortScience*, **34**: 987-988.

Mogren, L., Reade, J. and Monaghan, J. (2015) *Acta Hortic.*, **1099**: 407–412.

Mou B. (2008). *HortScience*, **43:** 1716-1719.

Moustafa, S.A. (1969) *Diss. Absr. Sect. B.*, **29**: 3988-3989.

Murphy, J.B. and Morelock, T.E. (2000) In: Horticultural studies, Research series (Eds. M.D. Richardson, J.R. Clark), vol 475. University of Arkansas, Fayetteville, pp. 36–39.

Nath, P. (1969) *Indian J. Hort.*, **26**: 156-159.

Nath, P. (1976) *Vegetable for the Tropical Region.* ICAR, New Delhi.

Nath, P. and Subramanyan, S. (1972) *Indian Hort.*, **17**: 23.

Nguyen, Q.V., Sun, H.J., Boo, K.H., Lee, D., Lee, J.H., Lim, P.O., Lee, H.Y., Riu, K.Z. and Lee, D.S. (2013) *Plant Biotech. Rep.*, **7**: 99-108.

Nicolaisen, W. and Kuhlen, H. (1970) *Gartenbauwissenschaft*, **35**: 293-301.

Okuse, I. (1994) *Bull. Fac. Agril. Hirosaki Univ.*, **57:** 55-62.

Omran, A. F., Abo-Elasaod, I. A. and Ashour, N. T. (1973) *Gartenbauwissenschaft*, **38**: 441-446.

Ors, S. and Suarez, D.L. (2017) *Agric. Water Manag.*, **190**: 31–41.

Ozaki, K. and Takahashi, K. (1993) *Annual Report Soc. Plant Protect. North Japan*, **44**: 59-60.

Paesold, S., Borchardt, D., Schmidt, T. and Dechyeva, D. (2012) *Plant J.*, **72**: 600-611.

Pandita, M. L., Arora, S. K. and Sidhu, A. S. (1981) *J. Res., Haryana Agric. Univ.*, **11**: 262-265.

Pandita, M.L. and Lal, S. (1990) In: *Vegetable Crops in India* (Eds. T.K. Bose, and M.G. Som), Naya Prokash, Kolkata, India.

Pandjaitan, N., Howard, L.R., Morelock, T. and Gil, M.I. (2005) *J. Agric. Food Chem.*, **53**: 8618–8623.

Parlevliet, J.E. (1968) *Euphytica*, **17**: 21-27.

Pavlet, P., Durman, P., Heneberg, R. and Horgas, D. (1974) *Polijoprivredna Znanstvena Motra*, **33**: 133-141.

Pee, T.Y. (1970) *Malaysia Agric. J.*, **47:** 357-367.

Pereira, A. (2016) *Front. Plant Sci.*, **7**: 1123.

Phor, S.K. and Mangal, J.L. (1991) *Haryana J. Hort. Sci.*, **20**: 129-133.

Prummel, J. (1971) *Bedrifsontwikkeling, Ed. Tuinb.*, **5**: 49-59.

Qin J., Shi A., Mou B., Grusak M.A., Weng Y., Ravelombola W., Bhattarai G., Dong L. and Yang W. (2017) *BMC Genom.*, **18**: 941.

Quagliotti, L. (1972) *Elette*, **18**: 5-19.

Rajput, B.P., Mehta, V.B., Chavan, A.S. and Sananse, S.L. (1989) *J. Indian Soc. Coastal Agricultural Research*, **7**: 5-19.

Ribera, A., Bai, Y., Marie, A., Wolters, A., van Treuren, R. and Kik, C. (2020) *Euphytica*, **216**: 48

Rosa, R. J. Jr. (1925) *Hilgardia*, **1**: 259-274.

Rubatzky, V.E. and Yamaguchi, M. (1997) *Spinach, Table Beets, and Other Vegetable Chenopods*. Springer-Science, Business Media, Dordrecht, pp. 457-465.

Ryder, E. J. (1979) *Leafy Salad Vegetables*, A. V. Publishing Co. Inc. Westport, Connecticut.

Santamaria, P. (2006) *J. Sci. Food Agric.*, **86**: 10-17

Shi, A., Mou, B. and Correll, J.C. (2016a) *Euphytica*, **212**: 17-28.

Shi, A., Mou, B., Correll, J.C., Motes, D., Weng, Y., Qin, J. and Yang, W. (2016b) *Aust. J. Crop Sci.*, **10**: 1188-1196.

Shimazaki, Y. and Uchiyama, F. (1985) *Plant Protection Society*, No. 32, pp. 110.

Shoemaker, J. S. (1947) *Vegetable Growing, John Wiley and Sons, Inc.*, New York.

Silvestri, N., Ciurli, A., Gorreri, L., Sorce, C. and Lorenzi, R. (1997) *Adv. Hort. Sci.*, **11**: 115-119.

Simoons, F.J. (1990) *Food in China: A Cultural and Historical Inquiry*. CRC Press, Boston, pp. 139-140

Singh, H. B. (1969) *Hand Book of Agriculture* (Rev. edn.) ICAR, New Delhi.

Singh, H. B. and Joshi, S. A. (1960) *Our leafy vegetables*, ICAR, Farm Bull, No. 56.

Singh, H. B. and Swarup, V. (1971) *Technical Bull*, No. 33, ICAR, New Delhi.

Singh, H. B., Thakur, M. R. and Bhagchandani, P. M. (1962) *Indian J. Hort.*, **20**: 148-154.

Singh, K. P., Malik, Y. S. and Yadav, P. S. (1983) *J. Res. Haryana Agric. Univ.*, 13.

Singh, P.K., Deshbhratar, P.B. and Ramteke, D.S. (2012) *Agric. Water manage.*, **103**: 100-104.

Smith P.G. and Zahara M.B. (1956) *Calif Agric.* **10**: 15

Smith, L.B. (1921) *Va. Truck. Exp. Sta. Bull.*, 52.

Smith, P.G., Webb, R.E. and Luhn, C.H. (1962) *Phytopathology*, **52**: 597-599.

Sneep, J. (1983) *Euphytica* Supplement **2:** 1-27.

Sneep, J. (1958) *Euphytica*, **7**: 1-8.

Solberg, S.O., Yndgaard, F. and Axelsson, J. (2015) *Emirates J. Food Agric.*, **27**: 698–705.

Srinivasa, V. and Reddy, T.V. (1997) *J. Maharashtra Agril. Univ.*, **21**: 33-35.

Stanilova, O., Boboshevska, D. adn Vitanova, G. (1972) *Nauchni Trudove, Vissh Selskostopanski Institute 'Georgi Dimitrov'*. Obshcho Zemledelie, **24**: 35-46.

Stefanescu, E. (1996) *Anale Institutul de Cercetari pentru Legumicultura si Floricultura*, Vidra, **14**: 385-388.

Stevanato P., Chiodi C., Broccanello C., Concheri G., Biancardi E., Pavli O. and Skaracis G. (2019) *Sugar Tech.*, **21**(5): 703-716

Sugimoto, Y. (1948) *J. Hort. Ass. Japan*, **17**: 17-83.

Sugiyama, S. (1990) *Agriculture and Horticulture*, **65**: 403-408.

Swarup, V. (1977) *Breeding Procedure for Cross-pollinated Vegetable Crops*, ICAR, New Delhi.

Terazoe, H. and Okano, T. (1992) *Proc. 8th Int. Cong. Soilless Cult.*, Hunters Rest, South Africa, 2-9 Oct., 1992, pp. 405-415.

Thompson, A. E. (1955) *Cornell Agric. Exp. Sta. Mem.*, 336.

Van der Vossen, H.A.M. (2004) *Spinacia oleracea*. Backhuys Publishers, Wageningen, pp. 513-515.

Van Treuren, R., de Groot, L., Hisoriev, H., Khassanov, F., Farzaliyev, V., Melyan, G., Gabrielyan, I., van Soest, L., Tulmans, C., Courand, D., de Visser, J., Kimura, R., Boshoven, J.C., Janda, T., Goossens, R., Verhoef, M., Dijkstra, J. and Kik, C. (2020) *Genet. Resour. Crop Evol.*, **67**: 549-559.

Vaughan, J.G. and Geissler, C.A. (1997) *The new Oxford book of food plants*. Oxford University Press, Oxford, UK, p. 239.

Verma, A.N., Srivastava, D.C. and Sharma R.K. (1969) *Plant Sci.*, **1:** 189-191.

Vincent, H., Wiersema, J., Kell, S., Fielder, H., Dobbie, S., Castan˜eda-A lvarez, N.P., Guarina, L., Eastwood, R., Leo´n, B. and Maxted, N. (2013) *Biol. Cons.*, **167**: 265-275.

Walts, L. (1980) *Flower and Vegetable Plant Breeding, Grower Books*, London, p. 118.

Wang, X., Cai, X., Xu, C., Zhao, Q., Ge, C., Dai, S., Wang, Q.H. (2018) *J. Hortic. Sci. Biotechnol.*, **93**: 409–415.

Waskar, D.P., Gaikwad, R.S. and Daman, S.V. (1998) *Orissa J. Hort.*, **26**: 29-30.

Weerakkody, W.A.P. and Ando, T. (1994) *Trop. Agril. Res.*, **6**: 98-104.

Wendt, T. (1980) *Gemuse*, **16**: 396-398.

Whitaker J.R., Lee C.Y. (1995) *American Chemical Society*, Washington DC, pp. 2-7.

Witcombe, J.R., Hollington, P.A., Howarth, C.J., Reader, S. and Steele, K.A. (2008) *Philos. Trans. R. Soc. B Biol. Sci.*, **363**: 703–716.

Xu, C., Jiao, C., Sun, H. *et al.* (2017) *Nat. Commun.*, 8,15275 https: //doi.org/10.1038/ ncomms15275.

Yang, Y., Al-Khayri, J.M. and Anderson, E.J. (1997) *In Vitro Cell. Dev. Biol. Pant.*, **33**: 200-204.

Zahedifar, M., Karimian, N., Ronaghi, A., Yasrebi, J. and Emam, Y. (2011) *Comm. Soil Sci. Plant Analysis*, **42**(15): 1765-1781.

Zhang, Jinxiu, Yue Yubo, Zhimin, Sha, Kirumbaa, George, Zhangb, Yi, Bei, Zhigang and Cao, Linkui (2014) *Acta Agriculturae Scandinavica, Section B – Soil and Plant Science*, **64**: 590–598.

Zhang, H.X. and Zeevaart, J.A.D. (1999) *Plant Cell Rep.*, **18**: 640-645.

Zimmerley, H. H. (1926) *Va. Truck Exp. Sta. Bull.*, p. 57.

Zuccarini, P. and Save, R. (2016) *Plant Biosyst.*, **150**: 851–854.

FENUGREEK

P. Karmakar, B.K. Singh, Sk. M. Islam, T.K. Maity and M.G. Som

1.0 Introduction

Fenugreek (*Trigonella foenum-graecum* Linn.) commonly known as *methi* and its leaves are used as leafy vegetable and seeds are used as spices and condiments. It is an annual plant, widely grown as a food crop in India, Mediterranean region, North Africa, and Yemen. Fenugreek seeds and leaves are famous for their distinct aroma and slightly bitter in taste. Though it is widely used leafy vegetable during winter and spring season in India but it is also valued in the other parts of the globe as leafy vegetable, spices and condiments as well as for medicinal properties and as nutritious fodder. The crop is grown in Russia, United Kingdom, France, Spain, Portugal, Greece, Italy, Sweden, Germany, Switzerland, Austria, Hungary, Poland, Ukraine, Romania, Croatia, Slovenia (Europe); Egypt, Morocco, Tunisia, Algeria, Sudan, Libya (North Africa); Ethiopia, Eritrea (Horn of Africa); Kenya, South Africa (Sub-Saharan Africa); Turkey, Israel, Lebanon, Jordan, Syria, Saudi Arabia, Bahrain, Qatar, UAE, Kuwait, Oman, Yemen, Iraq, Iran (Middle East); Afghanistan, Turkmenistan, Azerbaijan (Central Asia); China, Taiwan (Far East); India, Pakistan, Nepal (South Asia); Australia (Oceania); Canada, United States (North America); Argentina (South America) (Basu *et al.*, 2019). Though many countries in the world are growing fenugreek, but India, Ethiopia, Egypt, and Turkey are the major players for seed production. In present time India is the leading producing country in the world and contributes more than 68 per cent of world production (Malhotra, 2011). Rajasthan, Madhya Pradesh, Gujarat, Uttar Pradesh, Maharashtra and Punjab are the leading States in fenugreek production in India.

2.0 Composition and Uses

2.1.0 Composition

Fenugreek is rich in minerals, protein, vitamin A and C. The nutritive value may differ in different cultivars. However, a general nutritive analysis of leaves is given in Table 1.

Table 1: Composition of Fenugreek Leaves (per 100 g of edible portion)*

Moisture	86.1 g	Thiamine	0.5 mg
Fat	0.9 g	Calcium	360 mg
Protein	4.4 g	Oxalic acid	13 mg
Fibre	1.1 g	Iron	17.2 mg
Minerals	1.5 g	Potassium	51 mg
Carbohydrates	6.0 g	Sulphur	167 mg
Magnesium	67 g	Vitamin A	6450 I.U.
Phosphorus	51 mg	Nicotinic acid	0.7 mg
Sodium	76.1 mg	Vitamin C	54 mg
Chlorine	165 mg		

* Aykroyd (1963).

After boiling in water, and steaming and frying, the fenugreek leaves lose 10.8 and 7.4 per cent of vitamin C, respectively (Pruthi, 1979). The leaves are very rich in protein containing about 18.6 to 40.9 per cent at different stages of growth, on dry weight basis. Reduction in nitrogen and protein content was observed at the flowering stage (Rao and Babu, 1981).

Tender leaves of fenugreek cultivars Kasuri, Pusa Early Bunch (PEB), HM57 and HM46 were collected on the 25th, 35th and 45th days after sowing (DAS) in the field and seed samples were collected at 15-day intervals from the 65th to 95th days after anthesis (DAA) and at maturity and analyzed for proximate principles of the samples. Moisture content of leaves, seeds and pod wall decreased with increasing maturity in all cultivars. Crude protein ranged from 24.31 to 33.91 per cent DW in leaves, 18.13 to 27.12 per cent DW in seeds and 2.69 to 14.13 per cent DW in pod walls. Crude protein in leaves was highest at 35 DAS in all cultivars except Kasuri, in which it was highest at 25 DAS. Crude protein was highest at 95 DAA in seeds and at 65 DAA in pod walls. Ether extract ranged from 2.06 to 5.74 per cent in leaves, 1.13 to 6.73 per cent in seeds and 1.12 to 2.33 per cent in pod walls. Ether extract increased with increasing maturity in leaves and pod walls but initially increased and then decreased with seed maturity in all cultivars. Total ash ranged from 11.26 to 15.66 per cent in leaves, 3.94 to 11.96 per cent in seeds and 7.70 to 15.75 per cent in pod walls. Total ash increased initially and then decreased with leaf maturity. Total ash content of seeds decreased initially then increased with advanced maturity (Singh *et al.*, 1991).

Chhibba *et al.* (2000) compared yield and nutritive values of different varieties of fenugreek tested at Punjab Agricultural University, Ludhiana, India. The results revealed a considerable variation among varieties of both *T. foenum-graecum* and *T. corniculata* in respect of the yield and the mineral content (Tables 2 and 3).

Table 2: Yield and Mineral Contents (on 86 per cent moisture basis) in different Varieties of *T. foenum-graecum**

Variety	*Green Yield+ (q/ha)*	*Mineral content (mg/100 g)*								
		Zn	*Cu*	*Fe*	*Mn*	*Ca*	*Mg*	*S*	*P*	*K*
BDJ 540	181	0.43	0.05	2.5	0.56	151	41	7	73	3.1
C 2502	158	0.55	0.07	2.5	0.43	157	46	-	80	2.9
C 2513	128	0.30	0.03	3.0	0.84	186	50	10	69	3.9
IC 6694	66	0.43	0.05	2.1	0.52	150	43	33	69	2.7
EC 33904	135	0.42	0.08	2.1	0.64	158	50	39	73	3.3
PLM 6	131	0.44	0.06	3.3	0.43	123	43	7	69	3.1
PLM 12	140	0.43	0.06	2.6	0.45	165	49	3	73	3.2
PLM 15	191	0.49	0.06	2.1	0.36	146	46	3	69	3.2
PLM 18	175	0.49	0.08	2.6	0.46	133	56	2	73	3.2
PLM 21	171	0.38	0.08	2.8	0.45	139	45	33	73	2.9
PLM 53	86	0.44	0.07	2.6	0.46	147	49	54	69	3.0

Variety	Green Yield+ (q/ha)	Mineral content (mg/100 g)								
		Zn	Cu	Fe	Mn	Ca	Mg	S	P	K
PLM 54	176	0.39	0.06	2.4	0.41	165	43	54	69	3.1
PLM 59	125	0.38	0.06	2.9	0.43	165	45	62	69	2.9
PLM 64	108	0.28	0.03	2.7	0.63	169	49	3	68	3.9
PLM 70	151	0.39	0.06	2.4	0.57	140	45	23	73	3.1
PLM 74	143	0.38	0.05	2.3	0.52	140	39	10	69	2.5
PLM 83	170	0.35	0.03	2.7	0.41	139	50	23	69	3.2
PLM 102	128	0.21	0.03	4.0	0.74	203	48	3	68	5.2
PLM 107	161	0.41	0.06	2.3	0.63	155	49	23	78	2.5
PLM 109	l48	0.40	0.06	2.3	0.49	151	55	54	73	2.5
U-8-105	145	0.36	0.03	2.6	0.67	147	49	51	69	3.1
Overall Range	66-361	0.21-0.55	0.03-0.08	2.1-4.0	0.36-0.84	123-203	39-56	2-62	68-80	2.5-5.2

+Total of three cuttings.

* Chhibba *et al.* (2000).

Table 3: Yield and Mineral Contents (on 86 per cent moisture basis) in different Varieties of *T. corniculata**

Variety	Green Yield+ (q/ha)	Mineral Content (mg/100 g)								
		Zn	Cu	Fe	Mn	Ca	Mg	S	P	K
IC 6694	275	0.45	0.13	2.9	0.69	140	50	51	61	3.6
IC 8614	280	0.41	0.12	2.7	0.59	120	48	54	69	4.4
IC 8670	248	0.40	0.13	2.7	0.69	127	46	54	66	3.7
IC 59304	285	0.42	0.13	2.8	0.71	148	52	54	64	4.1
Kasuri	247	0.43	0.15	2.9	0.64	134	52	73	69	3.7
NKG 113	361	0.40	0.09	2.9	0.56	181	52	33	51	4.3
NKG 127	335	0.38	0.13	2.8	0.69	119	43	73	66	3.8
PLM 24	350	0.39	0.13	2.9	0.64	119	45	51	59	3.4
PLM 25	342	0.44	0.13	2.7	0.71	132	46	39	60	3.2
PLM 94	360	0.41	0.13	2.7	0.69	130	50	62	64	3.6
PLM 105	274	0.39	0.13	3.0	0.76	146	59	54	60	4.3
PLM 259	318	0.39	0.13	3.0	0.69	134	46	23	59	3.8
PLX 61	325	0.42	0.15	2.9	0.70	130	49	62	66	4.3
Overall range	247-361	0.38-0.45	0.09-0.15	2.7-3.0	0.56-0.76	119-181	43-59	39-73	51-69	3.2-4.4

+Total of three cuttings.

* Chhibba *et al.* (2000).

Nevertheless, the magnitude of variation was relatively higher among of *T. foenum-graecum*. All varieties of *T. corniculata* in general had relatively higher content of all minerals except P. Only a few varieties of *T. foenum-graecum* are promising or *at par* with *T. corniculata* in terms of their nutritive value. In respect of Ca and Mg both species are *at par* and the P content of *T. foenum-graecum* is much more than that of *T. corniculata* irrespective of the variety. Obviously, all varieties of *T. corniculata* can prove to be better source of these minerals as leafy vegetable for human being.

2.2.0 Uses

Fresh tender leaves, pods and stems are consumed as curried vegetable alone or in combination with potato. Fenugreek seeds are mainly used as spice for the preparation of different tasty dishes. It has also high medicinal and industrial importance. It prevents constipation, removes indigestion, stimulates spleen and liver, and is appetizing and diuretic. In industries, seeds are used as dye and for extraction of alkaloids or steroids. It is also used as fodder. Fenugreek is becoming popular around the world day by day and the extract from the seeds is used to flavor cheese in Switzerland, artificial maple syrup and bitter-rum in Germany. Sometime roasted seeds are used as coffee substitute in Africa, seed powder mixed with flour as fortification to make flat-bread in Egypt, and as an anti-diabetic herb in Israel (Malhotra, 2011). The whole seed and dried plant are also used as insect and pest repellents in grain storage, and the oil is used in perfumery in France. Its leaves are used as greens and leafy vegetables and its seeds as spice and as powder in flour in India (Rajagopalan, 2001) and used as forage in Australia, Canada, and India (Acharya *et al.*, 2008). In the recent past, use of fenugreek has been well recognized and demand is increasing due to its multifarious uses as spice, forage, medicinal plant, and a source for natural diosgenin and galactomannan in the pharmaceutical and steroid industries (Malhotra, 2011).

2.3.0 Medicinal Uses and Chemical Constituents

Leaves and seeds of fenugreek are well known for significant medicinal properties and have been reported to considerably reduce both blood glucose and cholesterol levels in human and animal subjects in clinical trials around the globe. Therefore, Fenugreek is highly considered as a chemurgical crop in the local, regional and international pharmaceutical, nutraceutical and functional food industries and markets as a medicinal herb (Basu *et al.*, 2007). Fenugreek seed and leaves are a rich source of a wide diversity of medicinally rich phytochemicals like steroidal saponins (diosgenin), fenugreekine (alkaloid), galactomannan (carbohydrate), 4-hydroxyisoleucine (amino acid) among several others (Basu *et al.*, 2007). More specifically, fenugreek seed itself contain carbohydrates (45–60 per cent) as in mucilaginous fiber (galactomannans), proteins(20–30 per cent) enriched in tryptophan and lysine, lipids (5–10 per cent) or fixed oil, alkaloids of pyridine type (0.2–0.38 per cent) as in trigonelline; choline (0.5 per cent), and other materials including carpaine and gentianine, flavonoids (apigenin, luteolin, orientin, quercetin, vitexin and isovitexin)and 4-hydroxyisoleucine (0.09 per cent), lysine and histidine, arginine, calcium and iron, saponins (Zandi *et al.*, 2017). The plant has been used traditionally in Indian Ayurvedic medicines as well as in traditional Tibetan and

Uses of Fenugreek

Seed (Left) and Powder (Right) from Fenugreek

Different Recipes of Fenugreek in Indian Cuisine

Chinese medication for several centuries. Modern research has also demonstrated that fenugreek seed and leaves are useful in the treatment of a number of diseases including successfully reducing blood sugar and blood cholesterol levels in both animals and human subjects in experimental trials. The crop has the potential to act as a panacea in treatment of diabetic, microbial and cancer disease. The reason behind the rich medicinal properties of fenugreek is due to the presence of a wide diversity of important phytochemicals (diosgenin, trigonelline, fenugreekine, galactomannan and 4-hydroxy isoleucine. Hence, the crop has huge international demand in the associated pharmaceutical, nutraceutical and functional food industries. Being known as a chemurgical crop, fenugreek has a widespread adoption in industrial sectors. Its seeds contain a reliable source of steroid diosgenin, which acts as a supplement in pharmaceutical industry (Zandi *et al.*, 2017).

3.0 Origin and Taxonomy

South-Eastern Europe and West Asia are considered to be the places of origin of fenugreek. It is found growing wild in North Western India.

Vavilov (1926) described the Near East region, extending from Israel to Syria and southern Turkey, Iran and Iraq, and the Mediterranean center including Spain, Morocco, and Turkey are the centers of origin of related genus of *Trigonella*, *Trifolium*, and *Medicago*. Research findings demonstrated that *Trigonella foenum-graecum* and *T. caerulea* accessions from Turkey exhibited more diversity and support Vavilov's hypothesis (Dangi *et al.*, 2004). It was also revealed that *T. foenum-graecum*, the large seeded cultivar, is abundant around the Mediterranean region, whereas the small seeded cultivars are predominant eastward. Species of the genus exist wild in the countries of Europe, Macronesia (Canary Islands), North and South Africa, central Asia, and Australia (Malhotra, 2011). Six species like *T. caelesyriaca*, *T. calliceras*, *T. emodi*, *T. geminiflora*, *T. glabra*, *T. kotschyi*are reported to be indigenous to Asia, *five species viz.*, *T. graeca*, *T. striata*, *T. polycerata*, *T. monspeliaca*, *T. procumbens* are native to Europe while *T. laciniata* and *T. suavissima* Indigenous to Africa and Australia (Allen and Allen, 1981).

Fenugreek belongs to the family tribe Trifolieae Fabaceae (Leguminosae), subfamily Papilionaceae, sub-tribe Trigonellinae and genus *Trigonella*, Different species of the genus have different chromosome numbers. More than 100 species have been reported in the genus *Trigonella* (Vasil'chenko, 1953). *Trigonella* is considered as one the largest genera of the tribe Trifolieae and *Trigonella foenum-graecum* is a widely cultivated species (Balodi and Rao, 1991). Although there is a controversy present in the document about the exact number of species that comprise the genus *Trigonella* a close scrutiny reveals about 97 distinct species (Fazli, 1967). Sixty species of *Trigonella* was reported by Petropoulos and Kouloumbis (2002). There are two species of the genus which are of economic importance. One is *T. foenum-graecum* L. the common *methi* and another is *T. corniculata*, *kasuri* or *champa methi*. These two species differ in their growth habit and yield.

3.1.0. Cytogenetics and Species Relationships

The common *methi* or *Trigonella foenum-graecum*, the most common species of cultivated fenugreek is a diploid with 2n = 16. The basic chromosome number of the genus *Trigonella* is reported to have x = 8, 9, 11, and 14 (Darlington and Wylie, 1945). The majority of species reported are diploid (2n = 2x = 16). The somatic chromosome number of various species presented in the Table 4.

Table 4: Chromosome Number of Various Species of *Trigonella*

Species	*Chromosome Number (2n)*
T. foenum-graecum	16
T. balansae	16
T. corniculata	16
T. spruneranaBoiss., T. monspeliaca L., T. uncata, T. anguina, T. stellata, and T. astroites	16
T. gladiata, T. cariensis, T. berythea, T. macrorrhyncha, T. cassia, and T. foenum-graecum	16
T. homosa	16, 44
T. ornithpodioides	18
T. polycerata	28, 30, 32
T. geminiflora	44
T. grandiflora	44

The cytological studies *Trigonella* species suggest that some species of this genus have experienced a number of rounds of chromosome duplication that occurred followed by diploidization with gene and chromosome elimination. Variation in chromosome number has been reported in fenugreek. An extra B chromosome in some fenugreek lines has been identified (Joshi and Raghuvanshi, 1968). Presence of the B chromosome is responsible for modified growth habit of plants (Petropoulos, 2002). B chromosomes in *T. corniculate* responsible for two types of pollen mother cells (PMCs): one with 2n = 16 and the other with 2n = 16 + B chromosomes.

4.0 Botany

Fenugreek is an annual plant and herbaceous in nature. Generally, the plants are spreading, moderately branched, and weak. The plant has taproot growth initially, followed by a large number of secondary roots (Malhotra, 2011). Being a legume, the formation of root nodules is very common and establishes a symbiotic relation with the Rhizobium bacterium for fixation of atmospheric nitrogen. The fenugreek stems are generally erect having many branches. The leaves are petiolate, alternate, compound, trifoliolate, fragrant, and stipulate. The genus *Trigonella* contains mostly annual and perennial plants, often strongly scented. Fenugreek plants enter into the reproductive phase following a vegetative growth phase. Two types of flowering shoots *i.e.*, indeterminate and determinate, have been occurred in fenugreek. The most common type is the axillary flowers showing indeterminate growth habit while near-determinate or determinate types (blind shoots bearing axillary and terminal flowers) are rare.

The common *methi* is quick growing annual herb which produces erect shoots to a height of 40-70 cm. It has light to dark green leaves with or without pink margin and produces 2-3 small white flowers at the base of each leaf. The pods are slender and straw coloured when ripe, beak-shaped and are about 8-10 cm long with 8-15 yellowish brown coloured and smooth surfaced seeds. The seeds are about 0.3-0.5 cm long. The *kasuri* types are initially slow grower and remain in a rosette condition during most of its vegetative growth period. It produces flowers, bright orange to yellow in colour. The pods are smaller than the common *methi* and are sickle-shaped. *Kasuri* types are high yielded.

There are two types of flowers reported to occur in methi *viz.*, cleistogamous or closed flower type and anisogamous or open flower type. Cleistogamous flowers ensure a high level of self-pollination and higher seed purity. Rate of out crossing in common *methi* ranges from 2.1-7.0 per cent. The inflorescence is racemose and flowers are bracteates, pedicellate, complete, zygomorphic, bisexual, and hypogynous. The five sepals are gamosepalous, green, and valvate. The five petals are polypetalous, papilionaceous, and standard, with two wings; anterior petals form a boat-shaped keel, with descending imbricate aestivation. There are 10 diadelphous stamens; 9 are united to form a tube around the ovary and the 10th is free. Anthers are basified, introse, and enclosed in the keel. The gynoecium is monocarpellary, ovary superior, and unilocular; there are numerous ovules, placentation is marginal, style is long, and stigma is terminal (Malhotra, 2011).

Kasuri Methi

Common methi

The large-seeded or common fenugreek (*T. foenum-graecum*) produces white flowers with straight pods, while Kasuri Methi (*T. corniculata*) having small seeds produce yellow to orange flowers with sickle-shaped pods (Malhotra, 2003).

The floral biology of the crop was described by Thangaraj and Vijayakumaran (2001). They reported that anthesis takes place between 9.00 and 18.00 hrs while dehiscence between 10.30 and 17.30 hrs with peak between 11.30 and 12.30 hrs. Stigma becomes receptive 12.00 hrs before flower opening and remains receptive for about 10 hours after the opening of flowers. The plants flower in about 30-37 days after sowing and the duration of flowering phase is 7-18 days. Pollen fertility ranges from 95-98 per cent in the unopened flower buds and 67-80 per cent in the open flowers. The pods mature within 60-70 days of sowing. It is typically self pollinated and destogamous.

5.0 Cultivars

Non-scented or Deshi

These are quick growing and upright crops with white flowers. The pods are straight, 8-10 cm long, containing 10-15 seeds in each pod.

Scented

These are slow grower and their cultivation is mainly confined to North India. *Kasuri methi, Marwari methi* and *champa methi* are under this group.

5.1.0 Improved Cultivars

Lam Selection 1

Plants are medium in height and bushy with more branches having green yield potentiality of 10-12 t/ha and seed yield of 7.4 q/ha under better management practices.

Pusa Early Bunching

It is an early cultivar evolved at the Indian Agricultural Research Institute, New Delhi. It is a common *methi*. It takes about 125 days for its seed maturity.

Kasuri Selection

It is a late cultivar of *kasuri methi* and selected at the Indian Agricultural Research Institute, New Delhi. It takes about 156 days for its seed maturity.

ML 150

This cultivar derived from the cross T8 x T36, and was released for general cultivation in the Indian Punjab in February, 1995. ML 150 has more pods, matures earlier and in field trials produced 18.9 and 18.2 per cent higher grain and fodder yields, respectively, than T8. ML 150 compared favourably with T8 for disease resistance and quality (Sohoo *et al.*, 1997).

Co. 1 Fenugreek

A dual purpose cultivar evolved at Tamil Nadu Agricultural University, Coimbatore, India which matures in 85-90 days with a seed yield of 6.8 q/ha. The green yield is around 9 t/ha. The protein contents in the leaves and seeds are 15.9 and 20-22 per cent, respectively.

Rajendra Kranti

Plants are medium tall and bushy which matures in 120 days. Seeds contain around 9.5 per cent protein. It is a selection of Rajendra Agricultural University, Dholi, India.

RMT 1

A long duration cultivar (140-150 days) is selected for SKM College of Agriculture, Jobner, Rajasthan, India with a yield potential of 15.0 q/ha. It is tolerant to root rot disease and moderately resistant to powdery mildew disease. Protein content of the seed is 21.0 per cent.

Co.2 Fenugreek

It is selection from CF 390 and crop duration 85-90 days. It is short duration dual purpose variety, field tolerant to Rhizoctonia root rot disease, suitable for both kharif and rabi season. It is an early maturing short duration cultivar.

Hisar Sonali

It is a pureline selection from local germplasm. Plants are tall and bushy vigorous growing cultivar, dual purpose variety, late maturity (140-145 days), suitable for cultivation under irrigated condition. Moderately resistant to root rot and aphids.

Hisar Suvarna

It is a pureline selection from local germplasm. It is a quick growing, erect and tall, dual purpose, medium maturity (130-140days), moderately resistant to cercospora and powdery mildew cultivar. This cultivar reported to have wider adaptability, suitable for cultivation throughout the country.

Hisar Madhavi

It is pureline selection from local germplasm of UP. It is a quick growing, erect and tall, dual purpose, medium maturity (130-140days), moderately resistant to powdery mildew and to downy mildew. A cultivar with under adaptability suitable for both irrigated and rainfed condition.

Hisar Muktha

It is pureline selection natural green seed coated mutant line from UP. A quick growing seed type cultivar, medium maturity (135-140 days), moderately resistant to powdery mildew and to downy mildew. Erect and tall plants. Wide adaptability. Suitable for both irrigated and rainfed condition.

RMt 303

It is developed through mutation breeding from variety RMt 1. Medium maturity cultivar (145- 150 days) seeds bold, with typical yellow colour, less susceptible to powdery mildew.

RMt 305

It is developed through mutation breeding from variety RMt 1. It is first determinant type, multipotent, early maturing, wider adaptability, resistant to powdery mildew and root knot nematodes. Seeds bold, attractive and yellow, duration 120-125 days.

Cultivars of Fenugreek

Pusa Early Bunching

Diversity of Fenugreek Cultivars around the Globe.

(A) Fenugreek grown in Egypt; (B) Field plot showing fenugreek forage and seed yield trial in Iran; (C) Dried fenugreek foliage at a spice shop in Jordan; (D) Fenugreek seeds from China; (E) Fenugreek cultivation in Vietnam; (F) Fenugreek crop production in India; (G) Fenugreek under dryland production system in Pakistan; (H) Fenugreek crop from Australia; and (I) Fenugreek seeds from Brazil.

RMt 143

Moderately resistant to powdery mildew, seeds bold yellow colour, suitable for heavier soils.

Rajendra Abha

It is an improved cultivar of *Kasuri Methi*.

Pant Ragini

A dual purpose tall bushy type resistant to downy mildew and root rots, medium maturity. Seed contain 2-2.5 per cent essential oil.

AM-01-35

It is a selection from local germplasm. Dual purpose, tolerant to powdery mildew.

Guj Methi 1

Recurrent selection based on pureline selection from J. Fenu 102. The first cultivar from Gujarat released for the state. Plant is dwarf.

5.2.0 Other Cultivars

The Maharashtra Agriculture Department released two improved cultivars known as *Methi* No. 47 and *Methi* No. 14 which are high yielding with more vitamin C content in the leaves. The Punjab Agriculture Department also released *Kasuri*, an improved cultivar. Besides these, EC 4911 is also a promising cultivar.

The cultivar UM 50 has been reported to give high green seed yield by Kohli *et al.* (1988).

Basu *et al.* (2019) have gathered all of the reported accessions, cultivars, lines, masses, and varieties (hereinafter called diversification) of fenugreek in various geographical regions. According to the dataset presented, fenugreek crop has a total of 1090 diversification distributed in 45 countries across the continents.

6.0 Soil and Climate

6.1.0 Soil

The ideal soil for fenugreek is clay loam though it can be grown successfully on all types of soil. The optimum soil pH should be between 6.0 and 7.0 for its better growth and development. The crop is fairly tolerant to salinity. However, it is suggested not to grow this crop under saline and acidic soil conditions.

6.2.0 Climate

The cultivation of this crop is confined to areas with moderate or low rainfall and a cool growing season without extreme temperatures. It can tolerate 10-15°C of frost. It has been reported to be grown under a wide range of soil and climatic conditions in many countries of Europe, Asia, Africa, Australia, and America. Being a cool season crop, fenugreek is fairly tolerant to frost and freezing weather. It can also be grown as a hot weather crop. It is mainly cultivated in moderate or low rainfall area, while the areas with heavy rainfall are not suitable for its cultivation. Hot weather is essential during seed maturity stage. *Kasuri methi* requires a relatively cooler climate. It can tolerate frost and freezing weather. The areas having heavy and continuous rainfall, growing *methi* should be avoided.

7.0 Cultivation

7.1.0 Seed and Sowing

For leafy vegetable, common *methi* is sown from the middle of September to middle of March. The sowing time for both the species in the hills is March to April and October. When it is grown as a leafy vegetable, the duration of the crop is only 30-40 days and thus it can successfully be adopted in the multiple cropping systems (Singh and Joshi, 1960). The seed rate per hectare for common *methi* is 25 kg, while 20 kg of seed per hectare is enough for *kasuri* type. The common *methi* may take 5-6 days to germinate and *kasuri* 7-8 days. However, in cool climate complete germination requires a few days more. Seeds should be soaked before sowing for better germination. Soaking seeds in different chemical solutions *viz.*, cycocel (50-100 ppm), EMS (0.6 per cent) is also reported for improving germination, survival of seedlings and better growth. Seeds are also treated with rhizobium culture prior to sowing. According to Varshney and Singh (1989), seed germination was reduced after soaking in culture filtrates of *Aspergillus flavus, A. fumigatus, A. terreus, A. niger, Fusarium moniliforme, Penicillium oxalicum* and *A. nidulans*. Filtrates of *A. flavus* resulted in the maximum reduction of germination. *A. tamarii* and *A. ochraceus* had no effect on germination. In some cases, rate of germination and length of radicle and plumule were also reduced.

Evaluation of best temperature and medium for testing germination of fenugreek seeds in laboratory was made by Singh *et al.* (1990) and they reported that maximum germination of 97 and 95 per cent was observed for Co.1 and Pusa Early Bunching, respectively at 20°C (constant) in between the paper. The lowest values for number of abnormal seedlings and dead seedlings (ungerminated seeds) loss were also observed for the same treatment combination. From the results, they concluded that for seed germination a constant temperature of 20°C and between the paper medium are the ideal requirements.

A thorough soil preparation is necessary before sowing. Beds of convenient sizes should be prepared at the same time. The seed is generally broadcast in beds and surface is raked to cover it. Line sowing in rows 20-25 cm apart facilitates weeding and intercultural operations during the initial stages of crop growth (Choudhury, 1976). From Egypt, Mohamed (1990) reported increase in numbers of branches, average plant weight and pod number by increasing the in-row width from 10 to 30 cm.

7.2.0 Manuring and Fertilization

It is commonly grown on residual fertility of the previous crops. According to Singh and Joshi (1960), more succulent leafy growth could be obtained by ample

Field Growing of Fenugreek

(A) Fenugreek seed; (B) Germinating seeds; (C) Seedlings; (D) Crop under rainfed conditions.

supply of nitrogen and this should be applied as top dressing in three equal split doses, the first at one month after sowing and the rest two after every two cuttings. Katyal (1977) recommended a basal application of 15 tonnes of farmyard manure and a side dressing of 100 kg ammonium sulphate per hectare after the first cutting for early growth of the crop. Rathore and Manohar (1990) noticed the beneficial effects of N (20 kg/ha) and P (0 to 75 kg/ha) with regard to per cent N and per cent P in straw, seed protein content and dry weight of nodules per plant at 75 days after planting. Godara *et al.* (2012) observed significant increase in fenugreek growth and yield parameters through application of chemical fertilizers over sole organic manures such as vermicompost and poultry manures.

The advantage of fenugreek cultivation is that being a legume crop, it is capable of fixing nitrogen from the atmosphere. It has been estimated that each year fenugreek fixes about 283 kg N per hectare (Saxena and Ahmed, 1983). Hardman and Petro-poulos (1975) studied the response of fenugreek to field inoculation with *Rhizobium meliloti* and obtained taller plants with a higher seed yield but the maturity was delayed. Ramkrishnan *et al.* (1976) were of the same view about the initial inoculation of seeds with *R. meliloti.* A study of inoculum load of Rhizobium by Rasal and Patil (1989) revealed that number and dry weight of nodules/plant and plant dry weight increased with increasing weight of inoculum/kg seed but did not differ significantly between 250 g inoculum/5 kg seed and 250 g inoculum/10 kg seed. According to Poi *et al.* (1991), seed inoculation with eight *R. meliloti* strains increased the nodulation, dry matter accumulation and plant nitrogen content and produced more seed yield over control. Inoculation with strain MI-1 was the most effective. Meena *et al.* (2015) reported that plant height, number of branches, number of pods, days to flowering, days to germination, length, yield per plant and days to maturity of fenugreek were maximum with 50 per cent recommended dose of NPK through chemical fertilizers and 50 per cent *Rhizobium* biofertilizers. Vedpathak and Chavan (2016) also observed that growth characteristics such as plant height, fresh and dry weight of fenugreek plant were better with the application of vermicompost but use of straight chemical fertilizer showed highest yield components of fenugreek as leafy vegetable.

The effects of sewage sludge (0, 15, 20 or 25 t/ha) and Mussoorie rock phosphate (MRP; 0, 125, 150 or 175 kg P/ha) on the growth and yield of fenugreek was investigated by Misra and Mani (1994). No harmful effects on growth were observed following sewage sludge treatments. The content of heavy metals in plants, particularly Zn, increased at increasing rates of sewage sludge. In combined sewage sludge + MRP treatments, the concentrations of Cd, Cr and Pb decreased with increasing rate of MRP. No treatment decreased the concentration of Zn in plant tissues to control concentrations.

According to Kumar *et al.* (2000), water use efficiency of the crop was improved by phosphorus application. The optimum doses of phosphorus were determined by them to be 27.8 kg/ha for no irrigation, 44-45.2 kg/ha for one irrigation at 45 or 80 days and 55.6 kg/ha for two irrigations.

Sreeramulu *et al.* (1996) observed, height and leaf number of plants at 40 days after sowing were significantly greater when soil was inoculated with *Glomus fascicula-tum* than when not inoculated. Inoculated plants also had significantly higher shoot and root biomass than corresponding uninoculated control plants. Shoot FW with 50 per cent of the recommended P rate + *G. fasciculatum* inoculation was comparable to that with 100 per cent of the recommended P-rate without mycorrhizal inoculation. Vesicular arbuscular mycorrhizal (VAM) spores were present in greater numbers in the root zones of inoculated plants than in those of uninoculated plants. P-uptake in both shoot and root was significantly higher in inoculated plants than in control plants. They suggested that VAM inoculation of *T. foenum-graecum* reduces P-input required to obtain maximum leaf yields.

Application of Mn at the rate of 0.05 g per 3000 g garden soil to potted plants resulted in a significant increase in fresh weight, dry weight and vitamin C content. However, with higher level of Mn, little or no beneficial effect was obtained (Venkataramani, 1950).

Effect of Cd, Ni, Pb and Zn on growth and chemical composition of fenugreek was studied by Dang *et al.* (1990) when the crop was grown in a glass house on a slightly alkaline clay-loam soil from Northern India. Cd, Ni, Pb and Zn were applied at 0, 50, 100, 200 or 400 mg/kg of soil. The FW and DW was drastically reduced even at the 50 mg addition of Cd, and decreased further at higher levels. There was a slight decrease in the yield at 50 mg Ni but at 100 mg Ni and above the decrease was significant; no growth was observed at 400 mg Ni. The application of 50 mg Zn slightly increased yield, it decreased at higher levels of applied Zn. The threshold concentration, toxicity index and loading rate to produce 10 per cent yield reduction were also calculated. The concentrations of the elements in roots and leaves increased linearly with increasing levels of applied elements. The uptake of Zn was highest, followed by Cd, Ni and then Pb. The roots accumulated higher amounts of these elements than the leaves. Tissue concentrations of Cd, Ni, Pb and Zn associated with 10 per cent yield reduction for fenugreek, the concentrations were 1.5, 7.8, 11.5 and 54.5ppm, respectively. In general, phytotoxicities were found to be in the order: Cd > Ni >Pb> Zn. The DTPA extractable elements in soil, after the harvest of crops, increased with increasing levels of applied elements.

Krishnan *et al.* (1995) observed fenugreek plants, grown in soil amended with partly composted neem (*Azadirachta indica*) leaves (0.1 and 0.2 per cent), were taller and had heavier dry weights than plants grown in unamended soil. These plants also contained higher concentrations of nitrogen and phosphorus compared with control plants. Plant growth was reduced at higher concentrations of *A. indica* (0.5 and 1.0 per cent).

Effect of integrated nutrient management on productivity and economics of fenugreek (*Trigonella foenum-graecum*) studied by Shivran *et al.* (2016) with fourteen combinations of integrated nutrient management practices for application of recommended dose of nitrogen to fenugreek *i.e.*, 40 kg/ha, through different sources, *viz.*, control, inorganic or organic (farmyard manure), poultry manure, vermicompost and neem cake alone or in combinations was laid out in randomized block design with three replications. Based on three years of study the results revealed that application of all nutrient management treatments through organic and inorganic sources alone or in combination brought significant effect on growth, yield and economics of fenugreek. Further the results revealed that integration of 50 per cent recommended doses through vermicompost + 50 per cent recommended doses through inorganic source to fenugreek recorded higher plant height, branches/plant, pods/plant, seeds/pod, test weight, seed yield (1781 kg/ha), stover yield (3392 kg/ha), biological yield (5173 kg/ha) and net returns (Rs. 52151/ha). However, maximum benefit cost ratio was fetched with application of 100 per cent recommended doses through inorganic source.

7.3.0 Irrigation

Frequent irrigation is necessary to obtain quick growth of the crop. First irrigation should be given just after sowing. The second one is to be given after first cutting and then according to necessity. Generally, subsequent irrigations are given at 7-10 days interval.

The optimum doses of phosphorus were worked out as 27.8 kg/ha for no irrigation, 44.0- 45.2 kg/ha for 1 irrigation at 45 or 80 days and 55.6 kg/ha for 2 irrigations by Kumar *et al.* (2000). The responses of phosphorus to yield as well as phosphorus utilization efficiency at optimum dose increased with irrigation. Water-use efficiency was also improved by phosphorus application.

The highest seed yield of fenugreek was obtained with irrigation at IW/CPE ratio of 1.0 (Dutta and Chatarjee, 2006). The application of irrigation at 1.0 IW/CPE ratio along with weed control by pre-emergence application of pendimethalin @ 0.75 kg a.i./ha + inter-culture at 40 DAS is better for getting higher yield and net return from fenugreek (Mehta *et al.*, 2010). The study of Sharma *et al.* (2017) revealed that potential yield and profit from fenugreek cultivation can be obtained by irrigating the crop at 0.6 IW: CPE ratio (21-27 days interval) during vegetative phase and 1.0 IW: CPE ratio (10-12 days interval) during reproductive phase with application of FYM 5 t ha^{-1} + castor cake 1 t ha^{-1}.

7.4.0 Interculture

Weeding is necessary at the early stage of the crop. The crop is slow growing at the initial stage. The growth become vigorous after 4-5 weeks and the crop does not allow weeds to compete with it. Crop-weed competition studies in fenugreek revealed that the yield was reduced by 14.2 per cent due to weed competition during the first 30 days after sowing, and by 69 per cent due to weed competition during the entire cropping season. Weeds emerging after 30 DAS caused a 12 per cent reduction in yields and there was no significant advantage in increasing the weed-free period beyond 30 days. The study led to a conclusion that 30 days is the critical period with respect to crop-weed competition in fenugreek (Tripathi and Singh, 1993). Meena *et al.* (2013) obtained the maximum gross and net returns in weed free treatment followed by pre-emergence application of oxadiargyl @ 75 g/ha + 1 hand weeding at 45 days after sowing but highest B: C ratio (4.38) was recorded with pre-emergence application of oxadiargyl @ 75 g/ha + 1 hand weeding at 45 DAS. Kumar *et al.* (2016) observed that trifluralin as PPI, pendimethalin as pre-emergence (PRE) and imazethapyr at 55 g a.i./ha either applied as PPI or PRE provided excellent control of weeds up to 25 DAS. At 100 DAS and at harvest, post-emergence application of imazethapyr (55 g a.i./ha) significantly reduced the weed population over other herbicidal treatments.

Usually, no thinning is required when it is grown for leafy vegetable but it is necessary for a seed crop. A spacing of 5 to 7.5 cm within the row is generally given for this purpose (Chauhan, 1972).

7.5.0 Effect of Growth Substances

Spraying of ascorbic acid solutions @ 250 or 450 mg per litre on the seedlings resulted in taller stems, more and larger leaves, earlier and more number of flowers and fruits per plant with heavier seeds (Chinoy *et al.*, 1957). Desai and Pathak (1965) reported that spraying of GA_3 (10 to 100 ppm) greatly enhanced internodal length, height and slightly increased the number of nodes and leaves. GA_3 also induced early flowering. Application of maleic hydrazide, FW 450 (sodium 2,3-isobutyrate) and coumarin retarded plants growth and reduced the yield of fenugreek (Faul and Singh, 1967). FW 450 at 0.05 per cent applied before initiation of flower bud produced higher yield with less pollen sterility. The seedling growth was markedly inhibited when seeds were soaked for 16 hours in solutions of 2,4-D at 10, 25 or 50 mg per litre (Balsimha *et al.*, 1976). According to Megha and Laloraya (1978), seeds soaked in different concentrations of ethrel showed stunted growth of the plant. Bharud *et al.* (1988) reported that foliar or seed treatment with 10, 25 or 50 ppm GA_3, NAA or IBA increased the yield of fenugreek grown as vegetable. Krishnaveni *et al.* (2016) obtained the maximum plant height, number of branches, fresh weight, dry weight and number of flowers per plant of fenugreek by foliar spraying of GA_3 50 ppm thrice (25, 45 and 65 DAS) with no or single pinching at 25 DAS. Foliar application of NAA 75 ppm followed by thiourea 250 ppm and ethrel 200 ppm gave the maximum seed yield under alluvial plains of West Bengal, India (Reddy and Hore, 2020).

8.0 Harvesting, Yield and Storage

8.1.0 Harvesting

When common *methi* is used as leafy vegetable, the young shoots are nipped off in about 3 weeks after sowing. A number of cuttings may be taken from the *kasuri* types. The first cutting is ready in about 25-30 days after sowing. The picking of leaves is done by nipping 2 cm above ground level leaving the stubs which produce new shoots and are cut in turn. Subsequent cuttings may be taken after an interval of 12-15 days. If the cuttings are taken late, the leaves develop a bitter taste. After 2-3 cuttings, the plants are left for seeding, otherwise they are uprooted after 4-5 cuttings. According to Mehta *et al.* (1995), seed yield was adversely affected by cutting leaves. Green leaf yield was highest when one leaf harvest at 45 days after sowing was taken and the crop allowed to re-growth for seed. Cultivar NLM gave the highest seed yield (3-year average of 1344 kg/ha) and cv. UM-17 produced the highest leaf yield (3-year average of 3510 kg/ha). The crop after harvest is bundled and marketed as early as possible. Before disposal, the produce should be kept in cool and moist conditions.

Harvesting and Selling of Fenugreek in the Market

8.2.0 Yield

The yield varies with the method of harvesting in both types. Green leaf yield of *kasuri* type varies from 90-100 q/ha, while the common type produces an yield of 70-80 quintals per hectare. Saini (1975) studied the effect of cuttings on the green and seed yield of fenugreek and reported that after taking one cutting and leaving the crop for seed maturity gave the maximum return.

8.3.0 Storage

Well-dried leaves of fenugreek can be stored for one year. *Kasuri methi* is available in this form in packets. Thus, the dried leaves can be used throughout the year. Freshly harvested local cultivars of *methi* packed in polythene bags (200 gauge and 2 per cent ventilation) or non-packed were held at room temperature (17.42-30.12°C, 24.30-73.00 per cent RH), in a cool chamber (14.45-22.62°C, 90-95 per cent RH) or in cold storage (8°C, 90-92 per cent RH). Physiological weight loss was lower in packed than in the non-packed samples, and was lowest in cold storage followed by cool chamber storage. Storage life was increased from one day (non-packed samples held at room temperature) to 6.5 and 3.5 days in packed *methi* held in cold storage or a cool chamber, respectively (Waskar *et al.*, 1998). A field trial was carried out on fenugreek cv. Pusa Early Bunching during 1985, in which treatments were foliar sprays of 20 ppm GA_3, 0.5 per cent ascorbic acid, 20 ppm IAA, 0.25 per cent urea, water and no spray. Leafy vegetable fenugreek was stored in bamboo baskets and sprinkled with water, in perforated plastic bags, in a wet cloth covering or in open bamboo baskets. Physicochemical analysis of fenugreek was carried out after 24, 48 and 72 h of storage. All growth regulators produced higher yields compared with the control; highest yield being obtained with 20 ppm GA_3 (203.33 q/ha). This treatment also resulted in the greatest plant height (22.07 cm), leaf area (75.43 dm^2/plant), leaf: stem ratio (2.77) and fresh weight (15.07 g/plant). Applying 20 ppm IAA resulted in the highest internode length (4.07 cm) and dry weight (1.23 g/plant). Percentage moisture loss increased with increased length of storage period whereas ascorbic acid, chlorophyll and total acid contents decreased. The wet cloth covering was the most effective storage method for increasing the shelf-life of fenugreek. Pre-harvest foliar application of growth regulators increased the shelf-life of fenugreek compared with the controls (Deore and Bharud, 1990).

Fenugreek could be stored well under modified atmospheric packaging (MAP) utilizing macro-perforated MAP. Modified atmosphere had a beneficial effect on the shelf-life and biochemical content of otherwise highly perishable fenugreek stored for 6 days at 15°C (Brar *et al.*, 2013). MAP resulted in better maintenance of chlorophyll as well as antioxidants. Brar *et al.* (2013) observed that packaging of fenugreek in two perforation packets with mustard seeds resulted in best maintenance of chlorophyll, ascorbic acid and aroma. The phenolic content reduction in 2 perforation packet with mustard seeds was also least as compared to other treatments.

The quality of fresh fenugreek leaves in polypropylene film (34 μm) in non-perforated package stored at 15°C and 75 per cent R.H. was found superior on the basis of colour retention, β-carotene, ascorbic acid, 2,2-dipheyl-1-picrylhydrazyl, total phenol content and overall acceptability (Sidhu *et al.*, 2016).

9.0 Diseases and Pests

9.1.0 Diseases

Frequently, the crop is subjected to the attack of a number of diseases like powdery mildew (*Erysiphe polygoni*), rust (*Uromyces anthyllidis*), downy mildew (*Peronospora trigonella*) and leaf spot (*Cercospora traversiana*) (Chattopadhyay, 1967). Damping off, due to *Rhizoctonia solani* (Singh *et al.*, 1973) and blight (Singh and Pavgi, 1979) diseases are also found in fenugreek. *Rhizoctonia solani* can be controlled by soil drenching with 0.1 per cent carbendazim or quintozene first at the onset of symptoms and second one month later (Prakasam and Vinayagamoorthy, 1990).

9.1.1 Powdery Mildew (*Erysiphe polygoni, E. tryphi, Laveillula taurica*)

The symptoms include the appearance of whitish fungal growth on leaves. Sharma (1999) opined that disease intensity at early sowing dates (October 20 and October 30) was comparatively less than later sowing. The disease appeared in the last week of January but was unable to spread until the first week of March because of the low temperature and low relative humidity. Mehta *et al.* (1994) reported from a study on varietal behaviour of fenugreek to powdery mildew and their differential response to fungicides that all 18 lines tested were highly susceptible to this disease. Disease intensity was > 70 per cent in almost all of them. Foliar sprays of wettable Sulphur were superior to dinocan, reducing disease intensity from 80.8 per cent (unsprayed) to 29.2 per cent.

The efficacy of eight fungicides namely, propiconazole (0.025 per cent), penconazole (0.01 per cent), hexaconazole (0.005 per cent), triadimefon (0.025 per cent), tridemorph (0.08 per cent), dinocap (0.048 per cent), sulphur WP (0.2 per cent) and sulphur dust (20 kg/ha) for the control of powdery mildew of fenugreek caused by *Laveillula taurica* and *Erysiphe polygoni* was determined by Dhruj *et al.* (2000). All the fungicides significantly reduced the disease incidence in comparison with control. Minimum disease intensity (19.38 per cent) was recorded with penconazole followed by hexaconazole (21.34 per cent). The highest net profit was obtained with hexaconazole followed by wettable sulphur but with respect to cost benefit ratio, wettable sulphur was the most economic fungicide.

Results of studies on varietal resistance and chemical control of powdery mildew indicated that bayleton (0.1 per cent) [triadimefon] and sulphur dust (25 kg/ha) gave the most effective control of the disease and significantly increased yields over control plots. Of the 10 genotypes under study, none was found to be resistant. However, minimum disease intensity (34.2 per cent) was recorded on cultivar RMM-2 (Rathore and Rathore, 1995). According to Prakash and Saharan (1999), the germplasms lines GC-39 and UM-32 were free from powdery mildew infection and lines GC-7, GC-20 and UM-34 were resistant.

9.1.2 Fusarium Wilt (*F. oxysporum*)

Efficacy of leaf extracts of 7 plants each at 5 concentrations (20, 40, 60, 80 and

100 per cent) were tested against fenugreek wilt (*Fusarium oxysporum*) *in vitro*. Among the different leaf extracts, *Azadirachta indica* was highly toxic to *F. oxysporum* showing complete inhibition of mycelial growth and spore germination at 100 per cent concentration. *Ocimum basilicum* and *Lantana camara* also inhibited mycelial growth and spore germination at all 5 concentrations although they did not exhibit complete inhibition even at 100 per cent concentration (Bansal and Gupta, 2000).

9.1.3 Leaf Spot (*Cercospora traversiana*)

Presence of round spots which are white at the centre and brown at the periphery is the symptoms of the disease. First report of leaf spot in Bulgaria was made by Babev *et al.* (1999). According to them the primary symptoms were leaf spots 2-8 mm in diameter, light brown then becoming grey and slightly zonate, and surrounded by a diffuse chlorotic margin. When single lesions occur the disease normally develops as a typical leaf spot. However, the development of more than one spot is followed rapidly by yellowing and withering of entire leaflets. Stem lesions usually begin at the point of leaf attachment, extend both directions about 15-20 mm, and after encircle the stems.

9.1.4 Root Rot (*Rhizoctonia solani*)

It is a soil borne disease which produces symptoms of varying degrees of rotting of the root resulting to yellowing of leaves followed by withering and drying up of the plants.

In field trials at Tamil Nadu Agricultural University, India, 2 applications of 0.1 per cent carbendazim, once at the initial appearance of the disease, and a second 30 days later, effectively controlled root rot of fenugreek (average disease intensity after treatment was 3.1 per cent). Application of neem cake reduced disease incidence to 4.9 per cent. Seed treatment with *Trichoderma* spp. also reduced infection, but was less effective than carbendazim treatment (Rajan *et al.*, 1991).

Efficacy of *Trichoderma* spp. and *Rhizobium meliloti* in the control of root rot was studied by Haque and Ghaffar (1992). *T. viride, T. hamatum* and *R. meliloti*, used as seed dressing or soil drench, reduced *Macrophomina phaseolina* infection by > 50 per cent on 30-day-old fenugreek seedlings. Combined application of *T. koningii* and *R. meliloti* gave better control on both 30- and 60-day-old plants. *T. hamatum, T. harzianum, T. pseudokoningii* and *R. meliloti*, as seed dressing or soil drench completely controlled infection by *Rhizoctonia solani* on 30- and 60-day-old plants. The combined use of *T. viride* and *T. koningii* with rhizobia was also effective. *T. harzianum, T. koningii* alone or *R. meliloti* with *T. viride, T. hamatum, T. koningii* or *T. pseudokoningii* completely controlled infection by *Fusarium* spp. on 30-day-old seedlings. In 60-day-old plants > 50 per cent reduction of infection was obtained with *Trichoderma* spp. and *R. meliloti* used alone or *R. meliloti* + *Trichoderma* spp. as a seed dressing or a soil drench.

9.1.5 Anthracnose

Anthracnose of *methi* caused by *Colletotrichum gloeosporioides* a new host record from India was mentioned by Kalha *et al.* (1993).

Diseases of Fenugreek

Powdery Mildew Infection

Cercospora Leaf Spot Infection

Root Rot Infection

Pests of Fenugreek

Aphid Infestation

Life Cycle of Spotted Pod Borer

9.1.6 Bacterial Blight

First report of bacterial blight in fenugreek caused by *Pseudomonas syringae* pv. *syringae* from Wimmera regions of Victoria, Australia by McCormick and Hollaway (1999).

9.1.7 Downy Mildew (*Peronospora trifoliarum*)

The disease appears as presence of yellow patches on the upper surface and a cottony-greyish white mycelium at the lower surface in the corresponding areas. Spraying of 1 per cent Bordeaux mixture is found effective for control of the diseases.

Other diseases of minor importance are rust, *methi* mosaic wilt, *etc.*

Shekhawat *et al.* (2016) advocated IDM strategies for diseases and pest control under organic ecosystem for *methi* crop as follows:

Table 5: IDM Strategies for Fenugreek

Treatment	*Powdery Mildew (PDI)*	*Downey Mildew (PDI)*	*Aphid Index*	*Yield (q ha^{-1})*	*B : C Ratio*
Soil application of FYM @ 10.0 t/ha + *Trichoderma viride* @ 2.5 kg/ha+ spray of *Psuedomonas flourescens* @ 10.0 per cent	37.50 (37.75)	27.30 (31.51)	42.50 (40.71)	12.27	2.77
Soil application of vermi-compost @ 3.5 t/ha + *Trichoderma viride* @ 2.5 kg/ ha + spray of *Psuedomonas flourescens* @10.0 per cent	40.00 (39.22)	29.00 (33.17)	46.30 (42.86)	12.14	2.38
Soil application of neem cake @ 1.0 t/ha+ *Trichoderma viride* @ 2.5 kg/ha + spray of neem seed kernel extract @5.0 per cent	26.10 (30.71)	18.60 (25.54)	14.80 (22.65)	12.95	2.78
Soil application of neem cake @ 2.0 t/ha + *Trichoderma viride* @ 2.5 kg/ha + neem seed kernel extract @ 5.0 per cent	29.90 (33.17)	21.00 (27.28)	17.40 (24.68)	12.41	2.64
Check	72.50 (58.40)	63.80 (53.03)	60.00 (50.79)	9.80	1.91

Figures in parentheses are angular transformed values.

9.2.0 Pests

The crop is not prone to severe damage by any insects. However, leaf eating cater-pillars, aphids and spotted pod borers are found to occur and these can effectively be controlled by the spraying of chlorpyriphos 0.25 per cent, imidacloprid 0.03 per cent and indoxacarb 50 g a.i./ha, respectively.

The effect of leaf powders of *Aegle marmelos, Withania somnifera, Hibiscus rosasinensis, Murraya koenigi* and stem of *Cuscuta reflexa* was evaluated for the control of root knot disease. All the plant extracts significantly enhanced plant growth and reduced root galling. Maximum shoot and root growth were observed in plants grown in soil amended with *W. somnifera* and *A. marmelos*, respectively. Maximum reduction in root knot index was seen in plants grown in *C. reflexa* amended soil (Sharma *et al.*, 1992).

9.2.1 Other Pests

Sometimes rabbits cause severe damage to the crop by feeding on young leaves and stems during night. They can be controlled by trapping.

10.0 Seed Production

Both common and *kasuri* types seed produced freely in the plains of India. The practices normally followed for vegetable production, need to be adopted for seed production also. In Pantnagar, India, the seed yield of fenugreek increased with later sowing dates from 15 October to 15 November with closer spacing of 15 cm and application of 45 kg N/ha (Singh and Nand, 1984). In Hisar, India crops sown on 5th October with in rows spacing of 20 cm gave higher seed yield (Baswana and Pandita, 1989). In Egypt, seed yield has been observed to be highest (0.67-0.42 t/ha) when grown in rows 20 cm apart (Mohamed, 1990). Pandita and Randhawa (1996) also recorded highest seed yield with row spacing of 20 cm and without harvesting leaves.

Application of 40 kg P_2O_5 increased the seed yield by 59 kg/ha over control (Maliwal and Gupta, 1989). They also suggested that high rate of pendimethalin (0.75 kg/ha) applied pre-emergence provided the best chemical control of weeds. Highest seed yields were recorded by hoeing twice and pendimethalin at 0.75 kg/ha (1.023 and 0.891 t/ha, respectively compared to 1.056 and 0.421 t/ha for the weed free and unweeded control, respectively) though the benefit: cost ratio was higher for the chemical control. Response of fenugreek cv. Baladi to foliar spray with some growth regulators was studied by Shahine *et al.* (1992). They observed that GA_3 induced earlier flowering and increased flower number per plant. Paclobutrazol and ethrel delayed flowering. Seed yields were generally increased by GA_3 and paclobutrazol and decreased by ethrel. Seed concentrations of amino acids and sugars were increa-sed, and concentration of total carbohydrates decreased, by growth regulators. Seed N and K contents in most cultivars were decreased and P content increased by growth regulators.

A field experiment was conducted by Alagukannan and Vijayakumar (1999) during kharif and rabi in Tamil Nadu, India to investigate the effect of ethrel (50, 100 and 150 ppm), maleic hydrazide (MH; 10, 25 and 50 ppm), 2,4-dichlorophenoxy acetic acid (2,4-D; 1, 2 and 5 ppm), and naphthalene acetic acid (NAA; 10, 15 and 20 ppm) on the yield, quality and yield attributing characteristics of fenugreek. The treatments were applied as foliar sprays at 30, 40 and 50 days after sowing. MH at 50 ppm gave the highest number of branches per plant (4.57), while NAA at 20 ppm gave the highest number of pods per plant (24.22), double pods per plant (0.67), pod length (10.34 cm), seeds per pod (13.25), 1000-seed weight (16.81 g), and harvest index (36.41 per cent). NAA at 20 ppm also gave the highest seed yield (4.03 g/plant), grain yield per plot (390 kg), and grain yield per hectare (702 kg), while

NAA at 15 ppm gave the highest seed protein content (15.31 per cent). Reddy and Hore (2020) recorded the maximum seed yield by the application NAA 75 ppm (16.31q ha^{-1}) followed by thiourea 250 ppm (15.69 q ha^{-1}) and ethrel 200 ppm (15.46 q ha^{-1}) at West Bengal, India.

The crop can be grown as a pure seed crop or after taking 2 to 3 cuttings the plants are left for seed production. Singh and Joshi (1960) reported that in Kasuri Selection cultivar, it is advised to take 2 cuttings and in common type, 3 cuttings before the crop is allowed to produce seed. But Saini (1975) was of the view that one cutting should be taken if the seed is raised from the crop. It has been found, however, that the seed yield is significantly more in pure seed crop. According to Choudhury (1976), common *methi* (cv. Pusa Early Bunching) produced an average seed yield of 12-15 quintals per hectare, while the cultivar Kasuri Selection gave only 6-7 quintals if cuttings were not taken. Kaswan *et al.* (1994) reported that foliage yield increased with rate of nitrogen application whereas seed yield increased with up to 60 kg N. Foliage yields were similar between cultivars whereas seed yield was much higher in Pusa Early Bunching (1.41 t/ha) than Kasuri (0.79 t/ha). Seed yield was not affected by one cut; however, it was decreased by 2 cuts. Seed quality in terms of percentage germination and protein content was increased by nitrogen application (except the highest rate) and was greater in Pusa Early Bunching. Cutting had no effect on seed germination and it decreased seed protein content compared with no cutting. Net returns were highest with 60 kg N/ha, Pusa Early Bunching and 2 cuts. In an evaluation trial of twelve fenugreek cultivars by Sharma and Bhati (1987), highest mean seed yields of 17.17, 17.13 and 15.32 q/ha were obtained from the cultivars NLM (Prabha), UM-34 and UM-35 respectively. The seeds can be stored for two years (Chauhan, 1972).

11.0 Crop Improvement

The work on the improvement of fenugreek has been done only at a few Research Institutes in India. If the crop is used as leafy vegetable, the maximization of green yield is the major objective for its improvement. The breeding objectives of fenugreek differ with specific situations, purpose of cultivation and area of cultivation but main objectives are for high leaf and seed yield; enhanced saponins, diosgenin content; adaptation to different climates; and resistance to pest and diseases. The four fundamental breeding objectives identified are production breeding involve in the leaf and seed yield enhancement, breeding for resistance to biotic and abiotic stresses involve in the resistance toward major insect pest and diseases, breeding for agronomic and physiological characters, and high nutraceutical content (Malhotra, 2011). Shukla (1981) suggested that selection should be made by considering the plant height, days to flowering, number of pods per plant, pod length and grain weight in case of seed crop. Shukla and Sharma (1977) reported that Pusa Early Bunching, Type 226 and C.S. 960 are high yielding cultivars at the Indian Agricultural Research Institute, New Delhi. According to Rao and Babu (1981), Lam Sel. 1 appeared to be a more promising type which has several desirable attributes like bushy growth with more green matter, more number of branches and better yield potential when compared to the local cultivar at Guntur in Andhra Pradesh. Sixteen fenugreek genotypes were evaluated from their source and sink parameters to identify high yielding genotypes at Jabalpur, India during *rabi* 1991-92. Local 3 had significantly higher values for pod length, seed yield/plant, seed yield/plot and seed yield/ha compared to other genotypes. It also had the third highest 100-seed weight and shelling percentage, fourth greatest pod weight/plant and fifth highest number of pods/plant, number of seeds/plant, number of seeds/pod and filling percentage. Local 3 had the second highest number of branches/plant and the fourth largest leaf area/plant (Lowanshi *et al.*, 1996). Fenugreek accessions from Algeria, Morocco, Ethiopia, and India were evaluated for diosgenin content and reported 0.35, 0.25, 0.20, and 0.10 per cent diosgenin, respectively (Bakshi and Hameed, 1971). Different cultivars from Syria, Spain, France, Libya, India, and Hungary also showed significant variation for diosgenin content (Makai *et al.*, 2004). Similarly, Taylor *et al.* (2002) observed 0.98 per cent variability for diosgenin content.

Studies in Poland during 1995-97, Polish and Egyptian cultivars were compared by Berbec and Kolodziej (1998). The Polish population was characterized by favourable quality features, high diosgenin contents (0.2 per cent), and high seed and straw yields (means of 2.15 and 6.71 t/ha, respectively). The Egyptian cultivar gave average yields (means of 1.21 and 4.18 t/ha for seed and straw, respectively) and was conspicuous for its faster rate of growth and development compared with the Polish population, and also its higher seed protein content (29.22 compared with 25.15 per cent, respectively) under Polish conditions. Results of further investigations based on the various plant characters, namely, plant height, days taken for flowering, number of pods per plant, pod length, number of seeds per pod, 1000-seed weight and yield, showed that Lam Sel. 1, C.S. 960 and Pusa Early Bunching were superior to other cultivars (Rao *et al.*, 1983). Research work on improvement of fenugreek taken up at Tamil Nadu Agricultural University, Coimbatore has resulted in the release of CO.1, a high yielding selection suitable for both grains and leaves (Ramadas *et al.*, 1982).

Singh *et al.* (1993) conducted association analysis using 32 genotypes of exotic and indigenous collection of fenugreek under two dates of sowing. Correlation and path analysis indicated that seed yield could be improved by selecting for greater plant height, more tertiary branches, longer pods, more seeds per pod, and greater 100-seed weight. Seed yield per plant is positively correlated with plant height (Mahey *et al.*, 2003), number of primary branches (Banerjee and Kole, 2004), number of tertiary branches, number of pods per plant (Ayanoglu *et al.*, 2004).

Correlation and path coefficient analysis for different morphological and sink para-meter were carried out in 16 varieties of fenugreek by Lowanshi *et al.* (1998). Seed yield/plant had highly significant and positive association with pod length and seed weight/plant and it was also significant and positively correlated with plant height. The number of pods/plant had highest direct effect on seed yield, while low positive indirect effect was observed for number of pods/plant on 1000-seed weight. They suggested that selection on the basis of pod length and number of pods/plant will lead to improvement of seed yield in fenugreek genotypes at Jabalpur, India.

Soil salinity is a widespread and serious problem of arid and semiarid regions. Fenugreek grows well under saline conditions. Association studies of yield with its contributing traits was performed by Mathur and Lal (1999) under saline conditions, including their direct and indirect effects on yield in 14 varieties grown under four salinity levels (0, EC 2.0, 4.0 dSm-1 and field conditions). It was observed that characters changed their magnitude and direction of contribution under saline conditions. At both salinity levels, root length and harvest index exhibited positive direct effects on yield while branches/plant and protein content had negative effects on yield.

Association analysis of seed yield and its components in 15 genotypes of fenugreek was studied by Das and Kole (2000) and they reported that pods per plant, grains per pod, straw yield, biological yield and harvest index showed significantly positive correlations with grain yield per plant at both genotypic and phenotypic levels. Path coefficient analysis indicated direct positive effects of straw yield, test weight, grains per pod and pods per plant on grain yield per plant.

Kohli *et al.* (1988) reported highest heritability for days to flower and highest genetic gain for seed yield per plant. The correlation coefficient for green yield was highest with number of shoots per plant. Path analysis revealed the greatest contribution to green yield through direct effects of number of shoots/plant which was the single most important character contributing to green yield and seed yield. Character association studies in fenugreek by Berwal *et al.* (1996) revealed that high heritability together with high genetic variability was observed for seed yield/plant, 1000-seed weight, pods/plant, pod length and seeds/pod. Plant height, pods/plant, seeds/pod and 1000-seed weight showed positive and significant correlations with seed yield/plant, except for pods/plant which had significant and negative associations. Path analysis revealed that plant height, seeds/pod and 1000-seed weight were the most important characters contributing to seed yield.

Mutation breeding now-a-days is becoming increasingly popular as an effective tool for crop improvement. It very effective in self pollinated crop like legume and has been also utilized for the genetic improvement of fenugreek. The majority of induced mutations in this crop is recessive and segregate in a 3: 1 ratio. Research report suggested that normal plant type was controlled by a single dominant gene and giant mutant plant type by a recessive allele in fenugreek. Treatment of seed with chemical mutagens such as ethyl methane sulfonate (EMS) has the potential to cause chromosome damage, resulting in individual mutations as well as addition or loss of chromosomes from cells (Malhotra, 2011). Basu *et al.* (2007) got success in identification of determinant growth height, early maturity, and high seed yield mutants from generations of fenugreek seed treated with EMS. Floria and Ichim (2006) developed 11 mutant lines from M2 to M4 generations from seeds of fenugreek treated with 25-15 krad gamma rays, EMS (011-0.5 per cent, 2 h), or ethylene imine (0.01-0.1 per cent, 2 h) with better yield potential and higher diosgenin content. Influence of ionizing radiation on antioxidant enzymes in three species (*T. stellata, T. hamosa, T. anguina*) was studied and found that three species differed in their radio sensitivity. *Trigonella stellata* was more radio sensitive than *T. hamosa* and *T. anguina.* It showed higher stimulation in ascorbate peroxidase (APOX), superoxide dismutase (SOD), and glutathione reductase (GR) activities under irradiation stress (Al-Rumaih and Al-Rumaih, 2008).

In fenugreek experimental euploid and aneuploid cytogenetic stocks, primarily in the form of tetraploids, triploids, and simple primary trisomics, have been documented. Natural and chemically induced autoploidy is also reported in this crop. Few double trisomics (2n + 1 + 1 = 18) and primary trisomics (2n + 1) from the progeny of autotetraploids of Trigonella was isolated in methi (Malhotra, 2011). Autotetraploids also developed in this crop by treating fenugreek shoot apices with colchicine. Increased ploidy levels often result in larger seeds and may contain higher levels of chemical constituents and other valuable morphometric traits not present in genotypes with smaller seeds.

12.0 Biotechnology

Major research on biotechnology of fenugreek has been on the production of secondary metabolites. There is scope for attempting to improve fenugreek as a green and a seed spices.

12.1.0 Tissue Culture

Tissue culture is the novel way to create variation in crop improvement. Various techniques *viz.*, somatic embryogenesis, callus regeneration and micropropagation have also been utilized in fenugreek. Malhotra (2011) documented several studies related to tissue culture like *in vitro* regeneration and callus induction on fenugreek. Successful *in vitro* shoot regeneration plants on Murashige Skoog medium (MS) medium containing Thidiazuron (TDZ) also demonstrated in fenugreek. The research reports also suggested that *T. foenum-graecum* L. hypocotyl explants are most responsive to callus induction and proliferation in tissue culture (Aasim *et al.*, 2010).

12.1.1 Callus Culture

Callus has been induced from various explants such as cotyledon, hypocotyls, and leaves. Azam and Biswas (1989) cultured cotyledon and hypocotyls explants on modified MS medium supplemented with NAA, 2,4-D, kinetin and coconut water. Callus tissue formed on cotyledons within 6 days and after 12-15 days with hypocotyl explants. Chromosome number ranged from the normal 2n = 16 to 48 during prolife-ration of cotyledonary callus. Chromosomal aberrations became more frequent with subsequent subculturing. Sen and Gupta (1979) initiated callus from leaf segment of *Trigonella corniculata* and *T. foenum-graecum* in MS medium and White's medium supplemented with NAA (0.5 mg/l) and coconut milk (15 per cent v/v). Root, shoot and isolated leaf development occurred in the calli of both the species on MS medium containing NAA and coconut milk. Omission of coconut milk with subsequent addition of casein hydrolysate increased the number of differentiated organs per callus. In both species higher differentiation rates were observed at low irradiance than at high irradiance or in darkness. Initially the cells were diploid but at the 5th passage the calli became mixoploid with predominantly diploid cell populations. The differentiated organs were diploid.

12.1.2 Protoplast Culture

Mesophyll protoplasts have been used in the protoplast culture and regeneration of fenugreek. Shekhawat and Galston (1983) obtained high yields of

***In vitro* Shoot Regeneration of Fenugreek**

Callus Induction and Shoot Regeneration from Hypocotyl and Cotyledon Node Explant.

(a) Callus induction on hypocotyls explant; (b) shoot regeneration on cotyledon node explant on MS medium supplemented with BAP and NAA; (c) shoot regeneration on cotyledon node explant on MS medium supplemented with kinetin (Aasim *et al.*, 2010).

mesophyll protoplasts after leaves were treated with purified cellulase. Up to 70 per cent planting efficiency could be obtained. The protoplast-derived colonies developed into rapidly growing green calli and produced leafy shoots on a medium containing 0.1 mg/l each of BA and zeatin. Addition of glutamine and asparagine to the medium was essential for rapid cell division, callus growth and differentiation. Christen (2002) reported protoplast culture of fenugreek. They reported isolation of protoplasts from the root apices of 48 h imbibed seeds. They noted first divisions of root fenugreek protoplasts from 3-4 day culture which resulted in cell colonies subsequently. However, the authors failed to regenerate shoots from these and noted development of a few roots on these colonies. Petropoulos (2002), obtained shoot regeneration from fenugreek protoplasts obtained from the root apices of 48h imbibed seeds. The first divisions of protoplasts were observed from 3-day culture followed by cell colonies due to subsequent divisions resulted in roots only. 3-4 folds more trigonelline contents were scored from callus cultures compared to seeds and 12-13 times more than the roots and shoots.

12.1.3 Secondary Metabolites

The tissue culture work on fenugreek is more concentrated on production of secondary metabolites like rotenoid (Kamal *et al.*, 1997), steroidal sapogenins (Ortuno *et al.*, 2000; Oncina *et al.*, 2000), trigonelline (Radwan and Kokate, 1980; Ahmed *et al.*, 2000) and total alkaloids (Sharon and Bhaskare, 1998).

12.2.0 Molecular Markers

Major work on molecular markers in fenugreek has been on DNA content and photo cytometry. Das *et al.* (1997, 2000) studied *in situ* nuclear DNA content estimation, karyotype analysis, and somatic chromosome number (2n = 16) for nine cultivars of fenugreek revealing cultivar specific chromosomal characteristics and minute structural alternation in chromosomes of the genome. The 4C DNA content varied significantly from 3.12 to 4.36 picograms. Significant variations were also observed in total chromosome length, total chromosome volume, and total form percentage at the cultivar level suggesting a genetic drift among the cultivars.

Harish *et al.* (2011) estimated genetic variability among the 10 accessions of fenugreek repository of National Bureau of Plant Genetic Resources at Jodhpur Station, using RAPD primers and ISSR primers. Multilocus genotyping by 10 SPAR (single primer amplification reaction-RAPD and ISSR) primers revealed intraspecific polymorphism in banding patterns. The referenced tree generated after 100 resampling via bootstrap method showed that accession IC-373449, IC-396616 and IC-448830 are distantly placed. Two major clusters were formed from rest of the accessions, one having four accessions *viz*. IC-448828, IC-448832, IC448833 and IC-448834 and other having three accessions *viz*. IC-448835, IC-448837 and IC-396625. Hora *et al.* (2016) studied the diversity and phylogenetic relationships of different varieties of fenugreek (eight varieties and six populations) collected from northern India using RAPD and ISSR markers. The high similarity coefficient values suggested a diverse genetic diversity in fenugreek populations in India. They concluded that these two molecular markers (RAPD and ISSR) can be used effectively to evaluate genetic diversity and assess genetic relationship.

Amiriyan *et al.* (2019) studied the genetic diversity and population structure of Iranian Fenugreek landraces using SRAP Markers. Genetic diversity and population structure of 88 individuals of eight landraces of Iranian fenugreek evaluated based on SRAP markers. Seventy-two bands generated from 6 primers in which 56 (80.11 per cent) band were polymorph. Hamadan landrace showed the lowest values of percentage of polymorphic loci (67.86), Nei's gene diversity index (0.24), number of effective alleles (1.40) and Shannon's Information index (0.36). Nei's genetic distance matrix revealed the highest genetic distance between Hamadan and Yazd (0.203) and the highest genetic similarity between Mahallat and Varamin (0.036) landraces. The most gene flow was between Mahallat and Varamin landraces (Nm = 8.36) and the least was between Shiraz and Hamadan landraces (Nm = 0.66). An extent admixture of alleles between the Iranian fenugreek landraces was observed by the population structure. Mantel test indicated that the genetic differentiation and gene flow is not associated with geographic distance in Iranian fenugreek landraces.

13.0 References

Aasim, M., Hussain, N., Umer, E.M., Zubair, M., Hussain, S.B., Saeed, S., Rafique, T.S. and Sancak, C. (2010) *Afr. J. Biotech.*, **9**: 7174-7179.

Acharya, S.N., Thomas, J.E. and Basu, S.K. (2007) In: *Advances in Medicinal Plant Research* (Eds. S. N. Acharya and J. E. Thomas), Trivandrum, Kerala, India: Research Signpost.

Allen, O. N., and Allen, E. K. (1981) The Leguminosae. London: Macmillan Co.

Ahmed, F.A., Ghanem, S.A., Reda, A.A. and Solaiman, M. (2000) *Bull. Natl. Res. Centre-Cairo*, **25**: 35-46.

Alagukannar, G. and Vijaykumar, M. (1999) *South Indian Hort.*, **47**: 130-133.

Al-Rumaih, M.M., and Al-Rumaih, M.M. (2008) *Amer. J. Environ. Sci.*, **4**(2): 151-156.

Ayanoglu, F., Arslan, M., and Mert, A. (2004) *Turkish J. Field Crops*, **9**(10): 11-15.

Amiriyan, M., Shojaeiyan, A., Yadollahi, A., Maleki, M., and Bahari, Z (2019) *Mol. Bio. Res. Comm.*, **8**(4): 181-190

Aykroyd, W.R. (1963) *ICMR Special Rept. Series*, No. 42.

Azam, M. and Biswas, A.K. (1989) *Current Sci.*, **58**: 844-847.

Babev, S.G., Margina, A.F. and de Gruyter, J. (1999) *Plant Disease*, **83**: 783.

Bakshi, V.M., and Hameed. Y. K. (1971). *Indian J. Pharm.*, **33** (3): 55-56.

Balasimha, D., Ram, C. and Tewari, M.N. (1976) *Biochemic and Physiologiedar Pflanzen*, **169**: 515-518.

Balodi, B., and Rao. R.R. (1991) *J. Econ. Taxon. Bot.*, **5** (1): 11-16.

Bansal, R.K. and Gupta, R.K. (2000) *Indian Phytopath.*, **53**: 107-108.

Banerjee, A., and Kole. P.C. (2004) *J. Spices and Aromatic Crops*, **13**(1): 44-48.

Basu, S.K., Thomas, J.E., Acharya, S.N. (2007) *Australian J. Basic and Appl. Sci.*, **1**(4): 637-649.

Basu, S.K., Zandi, P. and Cetzal-Ix, William (2009) In: *The Role of Functional Food Security In Global Health*, pp. 471-497, DOI: https: //doi.org/10.1016/B978-0-12-813148-0.00028-1.

Baswana, K.S. and Pandita, M.L. (1989) *Seed Res.*, **17**: 109-112.

Berbec, S. and Kolodziej, B. (1998) *HerbaPolonica*, **44**: 386-391.

Berwal, K.K., Singh, J.V., Jhorar, B.S., Lodhi, G.P. and Kishor, C. (1996) *Annals Agri. Bio. Res.*, **1**: 93-99.

Bharud, R.W., Deore, B.P. and Patil, V.A. (1988) *J. Maharashtra Agril. Univ.*, **13**: 340-341.

Brar, Jasleen Kaur, Rai, Deepak Raj, Singh, Amarjit and Kaur, Navreet (2013) J. Food Sci. Technol., **50**(4): 696–704.

Chattopadhyay, S.B. (1967) *Diseases of Plants yielding Drugs,* Dyes and Spices (1st edn.), ICAR, New Delhi, pp. 32-34.

Chauhan, D.V.S. (1972) *Vegetable Production in India* (3rd edn.), Ram Prasad and Sons, Agra, India, pp. 256-260.

Chhibba, I.M., Kanwar, J.S. and Nayyar, V.K. (2000) *Veg. Sci.*, **27**: 176-179.

Chinoy, J.J., Nanda, K.K. and Garg, O.P. (1957) *Physiol. Plant.*, **10**: 869-876.

Choudhury, B. (1976) *Vegetables,* National Book Trust, India (4th edn.), New Delhi, pp. 188-191.

Dang, Y.P., Chhabra, R. and Verma, K.S. (1990) *Communications in Soil Science and Plant Analysis*, **21**: 717-735.

Dangi, R.S., Lagu, M.D., Choudhary, B.L., Ranjekar, P.K., and Gupta, V.S. (2004) *BMC Plant Biology*, **4**: 4-13.

Darlington, C. D., and A. P. Wylie. (1945) Chromosome atlas of flowering plants. London: George Allen and Unwin Ltd.

Das, S.R. and Kole, P.C. (2000) *Crop Res., Hisar*, **20**: 449-452.

Das, A.B., Mohanty, S., Thangaraj, T. and Das, P. (2000) *J. Herbs, Spices and Medicinal Plants*, **7**: 25-32.

Das, A.B., Pattnaik, M., Thangaraj, T. and Das, P. (1997) *Cytobios*, **91**: 366-367.

Deore, B.P. and Bharud, R.W. (1990) *J. Maharashtra Agril. Univ.*, **15**: 208-210.

Desai, S.H. and Pathak, C.H. (1965) *Indian J. Plant Physiol.*, **8**: 57-60.

Dhruj, I.U., Akbari, L.F., Khandar, R.R. and Jadeja, K.B. (2000) *J. Myco. Plant Path.*, **30**: 98-99.

Dutta, S. and Chatarjee, R. (2006) *Journal of Spices and Aromatic Crops*, **15**(2): 125–129.

Fazli, F.R.Y. (1967) *Ph.D. thesis*, University of Nottingham, England.

Faul, C.L. and Sing, S.P. (1967) *Indian J. Plant Physiol.*, **10**: 54-61.

Floria, F., and Ichim. M.C. (2006) *Plant Mutation Reports*, **1**(2): 30-31.

Godara, A.S., Gupta, U.S., Singh, Ravindra and Mehta, R.S. (2012) *International Journal of Seed Spices*, **2**: 34-37.

Haque, S.E. and Ghaffar, A. (1992) *Pakistan J. Bot.*, **24**: 217-221.

Hardman, R. and Petropoulos, G.A. (1975) *PlantaMedica*, **27**: 53-57.

Harish, G.K.A., Ram, K., Singh, B., Phulwaria, M., and Shekhawar, N.S. (2011) *Libyan Agric. Res. Center J. Internat.*, **2**(3): 150-154

Hora, A., Malik, C.P., Kumari, B. (2016) *Int. J. Pharm. Pharmaceutical Sci.*, **8(**1): 179-183.

Joshi, S., and Raghuvanshi, S.S. (1968) *BeitragezurBiologie der Pflanzen*, **44**(1): 161-166

Kalha, C.S., Chib, H.S. and Singh, S.K. (1993) *Plant Disease Res.*, **8**: 80.

Kamal, R., Yadav, R. and Mehra, P. (1997) *J. Medicinal and Aromatic Plant Sci.*, **19**: 988-993.

Kaswan, M.P., Singh, G.R., Thakral, K.K. and Mangal, J.L. (1994) *Haryana Agril. Univ. J. Res.*, **24**: 123-125.

Katyal, S.L. (1977) *Vegetable Growing in India*, (1st edn.), Oxford and IBH Publishing Co., New Delhi, pp. 104-105.

Kohli, U.K., Sharma, O.P. and Jeet Singh (1988) *Indian J. Hort.*, **45**: 119-125.

Krishnan, N., Ravi, D. and Gajapathy, M.K. (1995) *Adv. Plant Sci.*, **8**: 65-70.

Krishnaveni, V., Padmalatha, T., Padma, S.S.V. and Prasad, A.L.N. (2016) *Journal of Spices and Aromatic Crops*, **25**(1): 41-48.

Kumar, V., Yadav, J.S., Singh, J. and Yadav, B.D. (2000) *Indian J. Agric. Sci.*, **70**(8): 515-517.

Kumar, Ravinder, Malik Y.P. and Punia, S.S. (2016) *Indian J. Weed Sci.*, **48**(2): 225–227.

Christen, P., Ebizuka, Y., Nagata, T. (2002) *In vitro* culture and production of Secondary Metabolites. Medicinal and aromatic plants (Vol. 12). *Biotechnology in Agriculture and Forestry. Springer*, pp. 306-348.

Kumar, V., Yadava, J.S., Singh, J. and Yadav, B.D. (2000) *Indian J. Agric. Sci.*, **70**: 515-517.

Lowanshi, R.K., Shukla, R.S., Dubey, R.K., Choubey, P.C. and Singh, M. (1996) *J. Soils Crops*, **6**: 10-14.

Mahey, J., Raje, R.S. and Singhania. D. L. (2003) *J. Spices Aromatic Crops*,**12**(1): 19-28.

Makai, P.S., Makai, S. and Kismanyoky, A. (2004) *J. Central European Agric.*, **5**(4): 259-262.

Malhotra, S. K. (2003) In: Breeding field crops, (Ed. A. K. Sharma), Bikaner, India: Academic Publishers.

Malhotra, S.K. (2011) In: Genetic Resources, Chromosome Engineering, and Crop Improvement (Ed. Ram J Singh) First Edition, Chapter: 24, Publisher: CRC Press.

Maliwal, P.L. and Gupta, O.P. (1989) *Trop. Pest Management*, **35**: 307-310.

Mathur, V.L. and Lal, L. (1999) *Indian J. Agric. Res.*, **33**: 227-232.

McCormick, K.M. and Hollaway, G.J. (1999) *Australian Plant Pathol.*, **28**: 338.

Megha, B.M. and Laloraya, M.M. (1978) *Biochemic and Physiologiedar Pflanzen*, **173**: 229-237.

Mehta, K.G., Patel, P.G., Kachhadia, B.T. and Edison, S. (1995) *Indian Cocoa, Arecanut, Spices J.*, **19**: 91-93.

Mehta, N., Thakral, K.K. and Gandhi, S.K. (1994) *Plant Diseases Res.*, **9**: 112-115.

Mehta, R.S., Patel B.S. and Meena S.S. (2010) *Indian J. Agron.*, **55**(3): 235-239.

Meena, Jitendra Kumar, Kumar, Sanjay, Maji, Sutanu, Kumar, Manoj and Kumar, Devendra (2015) *Int. J. Agric. Sci.*, **11**(1): 198-200.

Meena, S.S., Mehta, R.S., Lal, G. and Anwer, M.M. (2013) *Indian J. Hort.*, **70**(1): 150-153.

Misra, S.G. and Mani, D. (1994) *Curr. Agric.*, **18**: 49-53.

Mohamed, M.A. (1990) *Egyptian J. Agron.*, **15**: 117-123.

Oncina, R., Botia, J.M., Rio, J.A. del and Ortuno, A. (2000) *Food Chem.*, **70**: 489-492.

Ortuno, A., Oncina, R., Botia, J.M. and Rio, J.A. del. (2000) *Rec. Res. Dev. Agril. Fd. Chem.*, **4**: 323-329.

Pandita, V.K. and Randhawa, K.S. (1996) *Seed Res.*, **22**: 127-129.

Petropoulos, G.A., and Kouloumbis.P. (2002) In Fenugreek: The genus Trigonella, medicinal and aromatic plants-Industrial profiles (Ed. G. A. Petropoulos), Taylor and Francis Inc, London.

Poi, S.C., Basu, T.K., Behari, K. and Srivastava, A. (1991) *Environ. Eco.*, **9**: 286-287.

Prakasam, V., Vinayagamoorthy, A., Khader, J.B.M.M.A. and Jeyarajan (1990) *Indian Cocoa, Arecanut and Spices J.*, **14**: 75-77.

Prakash, S. and Saharan, G.S. (1999) *J. Myco. Plant Path.*, **29**: 383-384.

Pruthi, J.S. (1979) *Spices and Condiments*, National Book Trust, India, New Delhi, pp. 118-123.

Radwan, S.S. and Kokate, C.K. (1980) *Planta*, **147**: 340-344.

Rajagopalan, M. S. (2001) *Supplement Industry Executive*, **5**(6): 43-44.

Rajan, F.S., Vedamuthu, P.G.B., Khader, M.D.A. and Jeyarajan, R. (1991) *South Indian Hort.*, **39**: 221-223.

Ramkrishnan, V., Nagarajan, M., Thangaraj, M. and Giridharan, S. (1976) *South Indian Hort.*, **24**: 60-61.

Rao, T.S. and Babu, M.K. (1981) *Indian Cocoa, Arecanut and Spices J.*, **4**: 85-86.

Rao, T.S., Babu, M.K. and Bavaji, J.N. (1983) *Indian Cocoa, Arecanut and Spices J.*, **6**: 72-73.

Rasal, P.H. and Patil, P.L. (1989) *J. Maharashtra Agril. Univ.*, **14**: 234-235.

Rathore, B.S. and Rathore, R.S. (1995) *Indian J. Myco. Plant Path.*, **25**: 260-262.

Rathore, P.S. and Manohar, S.S. (1990) *Indian Cocoa, Arecanut and Spices J.*, **13**: 148.

Reddy, P.P. and Hore, J.K. (2020) *Int. J. Chem. Stud.*, **8**(2): 2566-2569.

Saini, S.S. (1975) *Haryana J. Hort. Sci.*, **4**: 82-85.

Saxena, P. and Ahmed, S.T. (1983) *Intensive Agric.*, **20**: 15-16.

Sen, B. and Gupta, S. (1979) *Physiologia Plant*, **45**: 425-428.

Shekhawat, K.S., Shivran, A.C., Singh, D., Mittal, G.K. and Singh, B. (2016) *Int. J. Seed Spices*, **6**(1): 39-42

Shahine, A.H., El-Desouky, S.A., Abd-El-Dayem, H.M. and Wanas, A.L.I. (1992) *Ann. Agril. Sci.*, Moshtohor, **30**: 755-775.

Sharma, A., Trivedi, P.C., Ooi, F.A.C., Lim, G.S. and Teng, P.S. (1992) *Proc. 3rd Int. Conf. Plant Protection Tropics*, Genting Highlands, Malaysia, March 20-23, 1990, **5**: 25-28.

Sharma, R.K. and Bhati, D.S. (1987) *Indian Cocoa, Arecanut and Spices J.*, **10**: 89-91.

Sharma, S. (1999) *J. Myco. Plant Path.*, **29**: 144-145.

Sharma, S., Patel, R.H., Surve, V.H. and Kumawat, S. (2017) *Int. J. Seed Spices*, **7**(1): 29-34.

Sharon, M. and Bhaskare, M. (1998) *Indian J. Plant Physiol.*, **3**: 163-165.

Shekhawat, N.S. and Gaston, A.W. (1983) *Plant Cell Rep.*, **2**: 119-121.

Shivran, A.C., Jat, N.L., Singh, D., Rajput. S.S and Mittal, G.K. (2016) *Legume Res.*, **39**(2): 279-283

Shukla, G.P. (1981) Indian *Cocoa, Arecanut and Spices J.*, **4**: 121-122.

Shukla, G.P. and Sharma, R.K. (1977) *Indian J. Agric. Sci.*, **47**: 325-327.

Sidhu, G.K., Kaur, P. and Singh, M. (2016) *Nutri. Food Sci.*, **46**(4): 529-541.

Singh, G., Shukla, T.N., Shukla, H.P. and Singh, P.N. (1973) *Indian J. Mycol. Pl. Path.*, **3**: 228.

Singh, H.B. and Joshi, B.S. (1960) *Farm Bull.*, No. 56, Ministry of Food and Agric., New Delhi.

Singh, J., Gupta, K. and Dhindsa, K.S. (1991) *Annals Bio.*, Ludhiana, **7**: 69-73.

Singh, R. J., Boora K.S., and Lodhi, G.P. (1993) *Annals of Biol.*, **9**(1): 123-126.

Singh, J.N. and Nand, K. (1984) *India Drugs*, **21**: 536-540.

Singh, L. and Pavgi, M.S. (1979) *Indian J. Agric. Sci.*, **49**: 731-734.

Singh, T., Singh, V.P. and Singh, R.K. (1990) *Seed Res.*, **18**: 31-33.

Sohoo, M.S., Bhardwaj, B.L. and Beri, S.M. (1997) *J. Res.*, PAU, **32**: 363.

Sreeramulu, K.R., Shetty, Y.V. and Shetty, T.K.P. (1996) *Madras Agric. J.*, **83**: 362-364.

Taylor, W.G., Zulyniak, H.L., Richards K.W., Acharya, S. N., Bittman, S., and Elder, J.L. (2002) *J. Agric. Food Chem.*, **50**: 5994-5997.

Thangaraj, T. and Vijayakumaran, M. (2001) *Textbook of Vegetables,* Tubercrops and Spices (1st edn.), ICAR, New Delhi, pp. 364-371.

Tripathi, S.S. and Singh, G. (1993) *Proc. Int. Symp. Integrated Weed Management Sustainable Agriculture, Hisar*, India, November 18-20, 1993, **II**: 41-43.

Vasil'chenko, I.T. (1953) Berichtuber die Arten der Gattung. Trigonella. Trudy Botany Institute Akademice Nauk. U.S.S.R. 1, 10.

Vavilov, N.I. (1926) *Trends in Practical Botany and General Selection*, **16**: 3-24.

Varshney, P. and Singh, S.K. (1989) *Nat. Acad. Sci. Letter*, **12**: 65-66.

Vedpathak M.M. and Chavan B.L. (2016) *International Journal for Innovative Research in Science and Technology*, **3**: 180-183.

Venkataramani, K.S. (1950) *Proc. Indian Acad. Sci.*, Sec. B., **32**: 112-125.

Waskar, D.P., Damame, S.V. and Gaikwad, R.S. (1998) *Agril. Sci. Digest*, Karnal, **18**: 264-266.

Zandi, P., Basu, S.K., Cetzal-Ix, W., Kordrostami, M., Chalaras, K.S. and Bazrkar, L.K. (2017) Fenugreek (*Trigonellafoenum-graecum* L.): An Important Medicinal and Aromatic Crop, DOI: 10.5772/66506.

BASELLA

V. Sagar, J. Devi and R.S. Dhua

1.0 Introduction

Basella commonly known as *poi* or *Pui Sak*, Malabar Nightshade or Malabar spinach or Indian spinach or red spinach or Ceylon spinach or vine spinach or Chinese spinach or East-Indian spinach is a popular summer leafy vegetable grown in almost all parts of India (Nirmala *et al.*, 2011). The plant is extremely heat, drought tolerant and easily cultivable in variety of soils, thus it is considered as best spinach for tropical region of the world(Palada, 1999). The plant has fleshy stem and leaves, has both trailing and bushy habit. In India, it is cultivated almost all the places, however more frequently in West Bengal, Odisha and Assam. The reports of its cultivation are available from different countries like, Malaya, the Philippines, Fiji, Java, Indonesia, Burma and China. Among the African countries the cultivation has been reported from Ethopia, Mozambique, Sierra Leone, Cameron and Island of Saint Thomas (Deshmukh and Gaikwad, 2014). It is also cultivated in Puerto Rico and Guatemala. It is reported that *Basella* has been introduced to several countries like Europe in 1688, England in 1691 and USA in 1899 (Winters, 1963). *Basella* is native to tropical Southern Asia, probably originated from India or Indonesia (Saroj *et al.*, 2012). The fruits contain a purple red pigment betacyanin which is used as a natural source of food colourant (Huang *et al.*, 2016).

2.0 Composition and Uses

2.1.0 Composition

The leaves and shoots are good source of vitamins, minerals and anti-oxidants and low percentage of soluble oxalates. According to Deshmukh and Gaikwad (2014) and Kumar *et al.* (2015), the plant is reported to contain the following proximate composition in Table 1.

Table 1: Proximate Composition of 100 g Edible Portion of Basella*

Moisture	92.8-93 per cent	Carbohydrate	0.4 g
Energy	31 kcal	Fat	1.9 g
Fibre	0.3 g	Ascorbic acid	74 mg
Vitamin A	3250IU	Thiamine	84 µg
Riboflavin	126 µg	β-carotene	32.42 mg/100 g DW
Iron	5.45mg	Total carotenoids	121.56 mg/100 g DW
Calcium	187 mg	Oxalic acid	60-84 mg
Phosphorus	83.5 mg	Magnesium	103 mg

* Deshmukh and Gaikwad (2014), Kumar *et al.* (2015).

Minerals content compared to spinach especially iron (five folds), magnesium (three folds), calcium (threefolds), zinc (ten folds) and potassium (three folds) in the leaves of *Basella* spp. makes it one of versatile green leafy vegetable (Kumar and Giridhar, 2015). Pragya *et al.* (2018) reported betalain content of 80 mg/100 g in fresh leaves and 200 mg/100 g in fresh fruit.

2.2.0 Uses

The fresh tender leaves and stems are consumed as leafy vegetables after cooking. As half the water-soluble substances may be lost by boiling in water, it is preferable to cook the leaves in soups or stews. The colouring matter present in the red cultivar is reported to have been used in China as a dye. The ripe fruits also contain a deep violet colouring matter which is also used for colouring food. On account of the presence of mucilaginous substances in the leaves and stems, it is used as poultice. The plant has several ethnomedicinal uses in several countries. The juice of leaves is prescribed in cases of constipation, particularly for children and pregnant women (Burkill, 1935). The green tender leaves are used for treating oral diseases and improves the oral hygiene, especially against the aphthae (Hebbar *et al.*, 2004). The leaves were crushed and used to cure poultice sores in India and Malaya. The red pigment from the mature fruits mainly contains betalain, has been used for centuries as dye and edible purposes. The reports are available that during the 3rd–4th century AD the pigment was used to make purple dye for seal impressions for higher officials in India and China (Tsin dynasty)(Laufer, 1919). Seed contains anti-fungal peptides designated α and β-basrubrins with potent anti-fungal properties towards *Botrytis cinerea, Mycosphaerella arachidicola* and *Fusarium oxysporum*(Wang and Ng, 2001, 2004). Continuous consumption of basella improves the vitamin A status of the body reserve (Haskell *et al.*, 2004). It is used to treat the tooth ache in the tribal areas of Karnataka (Hebbar *et al.*, 2004). The red pigment isolated from the fruits has anti-inflammatory and potent anti-oxidant activity (Kachchhava, 2006; Krishna, 2012; Kumar *et al.*, 2011b; Huang *et al.*, 2016; Lin *et al.*, 2010); used for wound healing (Mohammed *et al.*, 2012), anti-microbial activity against *Staphylococcus aureus, Pseudomonas aeruginosa and Escherichia coli* (Oyewole and Kalejaiye, 2012), against *E. coli, Vibrio cholera, Staphylococcus aureus and Staphylococcus typhi* (Sen *et al.*, 2010; Deka *et al.*, 2018); antiviral activity against tobacco mosaic virus (Liu *et al.*, 2006) and herpes simplex virus type 2 (Dong *et al.*, 2012); insecticidal property against *Spodoptera litura* (Aboshi *et al.*, 2018); anti-ulcer activity (Deshpande *et al.*, 2003; Kumar *et al.*, 2012 and Venkatalakshmi and Senthamaraiselvi, 2012); androgenic effect, the plant extract improves the testosterone production in rats (Moundipa *et al.*, 1999; 2005; 2006; *Nantia et al.*, 2012; Chakraborty *et al.*, 2012; Nantia *et al.*, 2017) and leydigs cells (Nantia *et al.*, 2011); hepatoprotective effect (Bamidele *et al.*, 2010; Yanadaiah *et al.*, 2011; Adekilekun *et al.*, 2012; Sonkar *et al.*, 2012; Das *et al.*, 2015); anti-diabetic activity (Nirmala *et al.*, 2009; 2011; Arokoyo *et al.*, 2018a); antioxidant activity (Anusuya *et al.*, 2012; Reshmi *et al.*, 2012b; Arokoyo *et al.*, 2018b; Adenegan-Alakinde and Ojo 2019; Aryal *et al.*, 2019); treatments of boils (Bhat *et al.*, 2014); immune stimulatory activity (Park, 2014); reduction in body cholesterol (Baskaran *et al.*, 2015); anti-cancerous (Kilari *et al.*, 2016; Kumar *et al.*, 2018; Islam *et al.*, 2018; Kilari *et al.*, 2018).

2.2.1 Betalain Pigment

The mature fruit of *Basella* contains this pigment in higher concentration. The major component of this pigment is betacyanin giving it a purple colour. Presence of

Uses of *Basella*

Natural Dye from *Basella*

Different Recipes of *Basella* in Indian Cuisine

betanidin monoglucoside as the major betacyanin and its 4-coumaroyl and feruloyl derivatives as minor components *viz.*,Gomphrenin I (15S betanidin 6-0-figlucoside), gomphrenin II (1SSbetanidin 6-0-[6′-0-(4-coumaroyl)/I-glucoside]) together with small amounts of the respective R forms (isogomphrenin I and II) and gomphrenin III (15S-betanidin 6-0-[6′-0-feruloyl/I-glucoside])(Glässgen *et al.*, 1993; Lin *et al.*, 2010; Kumar *et al.*, 2015).

The pigment had good stability in between pH 3 to 7 and hence can be widely used in the food industry and cosmetic industry for development of product with low pH (Kumar *et al.*, 2015). They also developed strawberry coloured ice-cream from basella fruit juice and tested the storability of the same for six months without losing its colour and taste. The lecithin nano-liposomal encapsulation enhances the activity of betalain and improves its stability (Kumar *et al.*, 2020) and it was used to make vegan gummy candies.

3.0 Origin and Taxonomy

It belongs to the family Basellaceae and genus *Basella*. The generic name *Basella* given by Linnaeus was derived from Malayalam. Later he cited *Hortus malabaricum* in the synonymy for *Basella rubra* L.= *Basella alba* L. (2n = 2x = 24) (Cook, 2010). Roxburgh (1832) first treated *B. alba*, *B. rubra* and *B. lucida* in synonymy and adapted *B. alba* as a correct name according to the article 11.5 of the code (Sidwell, 1999). According to Bailey (1950), the following species are considered as forms of *Basella rubra*: *B. alba* – a white flowered form, *B. caninifolia*, *B. Cordifolia* with heart-shaped leaves, 9.16-12.7 cm long and 5.0-6.25 cm wide, *B. crassifolia*, *B. japonica*, *B. lucida* from India, *B. nigra* Chinese form, *B. ramosa* and *B. volubilis*. Anatomically also, red vine and green vine types are similar, thus indicating they are morphologically different types of same species. According to Sperling, (1987), the genus consists of five species; one is pantropical and remaining four is widely distributed in East and Southeast Africa and Madagascar. *B. alba* is pantropical indistribution, *B. paniculata* is endemic to south-eastern Africa and the other three species *viz.*, *B. excavata*, *B. leandriana* and *B. madagascariensis* are endemic to Madagascar. However, through palynological study of pollen from Kerala, Farooqui *et al.* (2019) suggested that there was a sixth species which is extinct now. They named this species (*B. keralensis* sp. nov.) has intermediate pollen characteristics of two species *B. paniculate* and *B alba*.

4.0 Botany

Basella is a fleshy annual or biennial, twining much branched herb with alternate, broadly entire leaves 7.5-12.5 cm in length and 7.5-9.0 cm in width. Leaves are broadly ovate and pointed at the apex. Flowers are white or pink, small sessile in clusters on elongated thickened peduncles in an open branched inflorescence. The ovule is basal with cauline placentation andthe gynoecium is tricarpellate. Fruit is enclosed in fleshy perianth.

Xu (1986) observed that pollen grains of *B. rubra* (*B. alba*) germinated on stigma 4-5 hours after anthesis and by 16-24 hours double fertilization was usually complete. After 2-8 days the 2-celled proembryo was visible and this finally gave rise to the multicellular globular embryo proper.

5.0 Cultivars

There are mainly two distinct types, one with reddish petioles and stems, and the other with green leaves, petioles and stems. Both the red- and green-leaved cultivars are consumed as vegetables, but green-leaved cultivars are commercially cultivated and belong to var. *alba*. The red-leaved and green-leaved cultivars are equally common in West Bengal, Assam and South India, while the green-leaved ones are more popular in Uttar Pradesh and Punjab. All the cultivars are trained on poles, pandals or trellis or grown on ground. Three varieties namely, Kashi Poi-1, Kashi Poi-2 and Kashi Poi-3 have been released for cultivation in the state of Uttar Pradesh during 2019. Kashi Poi-1 and Kashi Poi-3 are trailing or pole type while the Kashi Poi-2 is bushy in habit and suitable for high density planting.

6.0 Soil and Climate

The crop grows luxuriantly in well manured sandy loam soil provided it is well drained and well aerated. Adequate moisture and partial shade result in better growth of the plant and formation of bigger succulent leaves. Cultivation of this crop should be avoided in region affected by frost. The crop is usually grown during warm and moist seasons.

Respiration and the activity of catalase and peroxidases were determined in leaves of *B. alba* and *B. rubra*, collected in June-September from plants grown in the field or in the green house in Romania. Differences were found between field and green house cultivated plants, and between species. In all cases, respiration was highest during August and was higher in field grown plants. Peroxidase and catalase activities tended to be the lowest in August, and tended to be higher in field grown plants (Busuioc *et al.*, 1998).

7.0 Cultivation

7.1.0 Seeds and Sowing

Basella is commercially propagated through seeds but stem or root cuttings are also practised. In the northern and eastern plains of India, seeds are sown from March to May, while in the southern parts it is grown twice, once sown in July and again in October to November. Late spring or early summer season is the best time for sowing in the hills. A seed rate of about 12.5-16.0 kg will be required to sow a hectare of land. Miura *et al.* (1997) suggested that optimal emergence of *B. alba* seeds can be obtained by sowing in a potting mixture with a moisture content equivalent to 52 g water/500 g mixture (water content of 41-45 per cent), and chilling at 15°C for 10 days. Stem or root cuttings should be planted during monsoon month or in early summer (Chauhan, 1972). Seeds may be sown *in situ* or in seed bed and seedlings are transplanted when they can easily be handled.

Seeds or seedlings of *Basella* are spaced in beds 45 cm apart each way and the plants are allowed to sprawl over the ground (Herklots, 1972). The crop is also raised on bamboo stakes or trained in trellis. In such cases, seeds are sown 20-25 cm apart in rows at the base of bamboo stake or trellis.

Important Species of *Basella*

Basella alba

Basella rubra

Basella cordifolia

Important Cultivars of *Basella*

Kashi Poi-1

Kashi Pai-2

Kashi Poi-3

Field Growing of *Basella*

Demidov *et al.* (1991) made container trials with *B. rubra* (*B. alba*) raised through vegetatively in three different substrates *viz.* (i) natural zeolite, (ii) zeolite + 5 per cent chernozem, and (iii) a 1: 1 mixture of leaf compost and river sand in green house. All substrates had a similar nutrient content and 4-kg containers were used. The plants were grown in 30/20°C day/night temperatures at 65 ± 5 per cent RH and 50 or 70 W/m^2 PAR, under a daily 12 h illumination. The foliage was cut at 15 cm above the substrate level 3 times per week. *B. alba* was found to be a promising salad crop giving a high yield per unit area. The highest productivity of 250-413 g/m^2 in 24 h was obtained on substrate (ii) at 70 W/m^2 PAR.

7.2.0 Manuring and Fertilization

Rich soil is essential for a good crop and only nitrogenous fertilizer application has been found beneficial. While studying the sources and rates of nitrogen fertilizer, Chattopadhyay (1977) reported that among the different sources of nitrogen, *viz.*, urea, ammonium sulphate, mustard cake or calcium ammonium nitrate on *B. alba* seedlings, mustard cake at the rate of 50 kg N/ha produced the highest yield. In glass house experiments in Amsterdam, nitrate application increased the fresh and dry weight of leaf in *B. alba* with an increase in total nitrogen, nitrate, protein, oxalic acid and carotene contents (Vuurmans and Grubben, 1977). Gogai and Rajagopal (1986) reported that the *B. rubra* (*B. alba*), *B. rubra* var. *rubra* and *B. ruba*var. *alba* were grown in the open or under shade with P_2O_5: K_2O at 40: 40 kg/ha + poultry manure at 8 t/ha as basal dressing along with N at 0-40 kg/ha, for the first 4 harvests plants in the open yielded significantly more than shaded plants but in the 5th harvest the shaded plants gave a higher yield. During the first 4 harvests, rising N rates (0-20 kg/ha) had a beneficial effect on leaf yield. The highest yielder was *B. rubra* var. *alba*, followed by *B. rubra* var. *rubra* and *B. rubra*.

According to Donald *et al.* (1983) nutrient per cent in deficient and healthy leaves respectively were N 1.25 and 2.63; P 0.17 and 0.36 and K 0.46 and 3.55; Ca 0.62 and 1.78; Mg 0.37 and 0.80 and S 0.19 and 0.23.

7.3.0 Irrigation

Sufficient moisture is needed to produce rapid and succulent growth. Inadequate moisture may lead to thin, wiry stems and small leaves. The crop, in general requires 5-6 irrigations when grown in summer and the frequency of irrigation depends on the soil type. Intercultural operations are to be done as and when necessary. The field should be kept free from weeds.

7.4.0 Day Length, Temperature and Growth Substances

Long days inhibited flowering in *B. alba* and CCC suppressed flowering under long days, however, MH did not inhibit flowering under short days (Sinnadurai *et al.*, 1971). In another experiment, Veltkamp and Samson (1973) observed that CCC at 500 or 1000 ppm suppressed growth when applied at the early stage. They also observed that *B. alba* plants grown at temperature regimes of 31°/23°, 28°/20° or 25°/17°C and soil applications of CCC at 1200 ppm at 2-leaf stage and subsequently at 2-week intervals resulted in retardation of growth of the main stem 7 weeks later when grown under lower temperature regime, while the total stem length was reduced only at the higher temperature regimes. The ratio of leaf weight: total dry weight was increased by CCC with all the temperature regimes and that of root weight: total dry weight was increased at higher temperatures.

8.0 Harvesting and Yield

Crop raised from seeds will produce edible leaves and stems after 8-10 weeks from sowing. The plant raised from root or stem cuttings will be ready for harvest in about 6 weeks after planting. The yield varies from 140-185 quintals per hectare. A higher yield can be obtained from a well-managed crop (Chauhan, 1972). *B. rubra* var. *alba* grown by direct sowing or by transplanting seedlings produced yields of 25 and 69 t/ha, respectively and *B. alba* produced 69 and 87 t/ha, respectively (Ciofu *et al.*, 1985).

9.0 Diseases and Pests

This crop is subjected to the attack of several diseases. The most important diseases infesting this crop are listed below.

Fungal Disease

Leaf Spot

Leaf spot is the most frequently occurring and severe disease of *Basella* across India and other countries. Diseased plants show large spots lesions with purple halos on leaves and stems. The necrotic lesions (up to 2 mm in diameter) round, semi-circular, or irregular-shaped with greyish centers surrounded by dark brown borders can be seen on the leaves. There is no consensus about the exact causal organism of the disease. Several fungal species were reported to be associated with this disease *viz*., *Pseudocercospora basellae* (Goh and Hsieh, 1989), *Fusarium proliferatum* (Hirei and Takayuki, 2003), *Acrothecium basellae, Fusarium moniliforme* and *Cercospora*sp. (Hosain, 2007), *Bipolaris* sp. (Zhang *et al.*, 2011), *Curvularia lunata, Curvularia pallescen and Fusarium oxysporum* (Akter and Shamsi, 2013), *Cercospora beticola* (Arun *et al.*, 2011). *B. alba* collected from three districts of eastern Uttar Pradesh, India in 1988 were found to be infected by leaf spot (*Alternaria alternata*). This is the first report of *A. alternata* causing leaf spot on this crop in India (Pandey *et al.*, 1990). Bacterial leaf spot disease of *B. rubra* (*B. alba*) caused by *Curtobacterium flaccumfaciens* was discovered in Jiangsu Province, China by Chen *et al.* (2000).

Damping Off

The causal organism of damping off is *Pythium aphanidermatum* and the symptoms includes severe rot of leaves, stems, and roots. Soft, black lesions appeared initially at the base of the stems and enlarged gradually upward and downward within 2 days after infection. During severe infection the plants are completely destroyed, and it can affect the yield by more than half (Tojo *et al.*, 2006).

Leaf Blight

The causal organism of this disease is *Alterneria alternata* was first time reported from India during 2011. The symptoms include leaf lesions, elliptical to irregular oval, yellow brown to dark brown, and sometimes concentrically zonate with diffuse margins frequently surrounded by light-colored haloes. Infection often started at the leaf tips and progressed to the base of leaves as symptoms developed. In severe infections, lesions enlarged and coalesced, causing necrosis, wilting, and ultimately death of leaves (Sankar *et al.*, 2011). However, during 2012 the leaf blight of *Basella* was reported to be caused by another fungus *Rhizoctonia solani* (Baiswar *et al.*, 2012). The symptoms started from the lower leaves and spread to the upper part. Water-soaked lesions covered the whole leaf and gradually the leaves shreddes and get detached from the plants. Whole plants will be defoliated during severe infection. Lesions will be visible on the stem also.

Anthracnose

Colletotrichum truncatum, C. capsica and C. gloeosporioides (*G. cingulata*) were reported to cause anthracnose disease in *Basella* (Yang *et al.*, 2018). The lesions can be observed on young and mature leaves. The necrotic lesions (up to 5 mm in diameter), semi-circular or irregular-shaped with desiccated and grayish centers surrounded by deep red borders are characteristic symptoms of this disease. Symptomatic leaves later withers and finally dies, reducing the economic and edible value.

Diseases of *Basella*

Alternaria* Infected Leaves: A. *Basella alba* and B. *Basella rubra.

***Cercospora* Leaf Spot Infection in Spinach Beet (Leaf) and Chinese Spinach (*Right*).**

Charcoal Rot

The first incidence of charcoal rot caused by *Macrophomina phaseolina*is was reported from India (Meena *et al.*, 2018). The symptoms start from the collar region of the plants, consists of brownish to black discoloration at the collar region of the stem and branches that progressed into wilting and drying of the entire plant.

Verticillium Wilt

During 1996-98, a new disease occurred in this crop in Miyagi Prefecture, Japan. The causal organism was isolated and identified as *Verticillium dahliae* and its pathogenicity was confirmed. This was the first report of *Verticillium* wilt of this crop in Japan (Kanno and Horiuchi, 1999).

Kanno and Ohkubo (1999a) also mentioned *Sclerotinia* rot (*S. sclerotiorum*) in *B. rubra* which was the first report of this disease in Japan. According to Suriachandraselvan and Narayanasamy (1987), *B. rubra* may contain potent antiviral compounds which inhibit potato virus Y infection on chilli.

Viral Diseases

Huang and Chang (2006) reported the association of Basella rugose mosaic virus, new Potyvirus infecting Malabar spinach from Taiwan, where the infected plants exhibited mosaic and rugose symptoms. A new double stranded RNA (dsRNA) virus putatively endornavirus was reported from *Basella* plant. The plant was asymptomatic *i.e.* there was no difference between the healthy and infected plants (Okada *et al.*, 2014). A new leaf curl disease was reported in *Basella*, caused by a begomovirus tomato leaf curl Palampur virus (Kumari *et al.*, 2020).

Nematode

Basella is also reported to be infected with root knot nematode *Meloidogyne incognita*from Bangladesh, India (Singh and Kumar, 2012) and the Phillipines ((Dumaraos and Lizardo, 2019). Hasan *et al.* (2016) recommended that control of root-knot disease of Indian Spinach may be reduced by the use of bio-agent like BAU-Biofungicide instead of chemical nematicide for eco-friendly management of nematode.

The crop is almost free from insect attack though a minor incidence of the omnivorous 'wooly bear' caterpillars of ermine moth is reported by Winters (1963).

10.0 Seed Production

Basella produces seeds freely during spring in North India and during winter in South and East India. The agronomic practices followed for seed production are similar to that for vegetable purpose. Proper isolation distances should be maintained between two cultivars for producing nucleus and foundation seeds. The seeds do not shed from the fruits and hence can be left on plant to become fully mature. The ripe fleshy fruits borne in cluster on elongated peduncles are usually collected and seeds are extracted. The seeds are then dried and kept in storage.

11.0 Crop Improvement

Until recently, research was concentrated mainly on major leafy vegetable species and very little attention has been given to minor but indigenous leafy vegetable species. *Basella* is an indigenous minor leafy vegetable crop of increasing value but underutilized and underexploited. Despite its multiple virtues, *Basella* crop is not included in the priority leafy vegetables for research in India. Until recently no or little attention was paid to its genetic improvement. From the survey of literature, it is evident that genetic studies in *Basella* is limited, this trend is associated with little preference for this crop among researchers in the world, often termed 'orphan crop'. Genetic studies in Basella revealed that phenotypic coefficient of variation was higher than genotypic coefficient of variation for all the traits studied, indicating environmental influence on expression of these characters (Varalakshmi and Devaraju, 2010). Moderate heritability along with high genetic advance was recorded for leaf weight and total plant weight, indicating the presence of additive gene effects. Higher plant weight was found to be significantly and positively associated with branch number, leaf number, leaf weight and stem weight. Thus, for yield improvement in *Basella*, emphasis may be laid on indirect selection through leaf characters like leaf number, leaf length and leaf weight. Hence, selection can be employed for improvement of these characters in *Basella* (Varalakshmi and Devaraju, 2010).

The morphological evaluation and characterization revealed considerable variability among the six landraces (Reddy *et al.*, 2014). Genetic diversity analysis revealed that landraces from the same geographical region differed genetically as well as phenotypically. Based on mean performance, two landraces RNIS-2 and RNIS-4 were superior in terms of leaf-stalk ratio and harvest index as well as in other important attributes related to yield. Upon assessing yield stability through multi-location trials, these genotypes may be used for large-scale cultivation if found suitable. Leaf weight, petiole length, internodal length, weight of tender shoot, stalk yield, leaf-stalk ratio and harvest index had high estimates of heritability coupled with high genetic advance as percent of mean. Hence, these characters need to be given more emphasis in selection as these are expected to be controlled by additive genes. The breeder should adopt suitable breeding methodology to utilize both additive and non-additive gene effects simultaneously, since varietal development will go a long way in the breeding programs especially in case of Indian spinach. The results also indicated the possibility to improve total biomass and quality thereby farmer's income through selection programs. The results indicated that landraces of Indian spinach traits of high biodiversity and genetic resources are suitable for future breeding. There is a need to continue work in the area of nutritional analysis, which is required as an evaluation tool in varietal selection.

Pushpa *et al.* (1975) reported that out of 10 *B. alba* seedlings treated with colchicine, one developed polyploid characteristics. The leaves of the said autotetraploid line were larger and more succulent than those of the normal diploid but pollen fertility was lower. However, larger thick leaf and fleshy stem should be the main objectives for improving this crop.

12.0 References

Aboshi, T., Ishiguri, S., Shiono, Y. and Murayama, T. (2018) *Bioscience, Biotech. Biochem.*, **82**(1): 9–14.

Adenegan-Alakinde, T. A. and Ojo, F. M. (2019) *Int J. Veg. Sci.*, **25**(5): 431–440.

Akter, S. and Shamsi, S. (2014) *Bangladesh J. Scientific Res.*, **26**(1-2): 83-87.

Anun, C., Hidayat I. and J. Meeboon. (2011) *Plant Pathol. Quaran.*, **1**(1): 11-87.

Anusuya N, Gomathi R, Manian S, Sivaram V and Menon A. (2012) *Int. J. Pharm. Pharm. Sci.*, **4**: 714-720.

Arokoyo, Dennis S., Oyeyipo, I. P., Du Plessis, S. S., Chegou, N. N. and Aboua, Y. G. (2018) *Toxicol. Res.*, **34**(4): 325–332.

Arokoyo, Dennis Seyi, Oyeyipo, I. P., Du Plessis, S. S. and Aboua, Y. G. (2018) *Pharmacog. Res.*, **10**(1): 31–36.

Aryal, S., Baniya, M. K., Danekhu, K., Kunwar, P., Gurung, R. and Koirala, N. (2019) *Plants,* **8**(4): 96, doi.org/10.3390/plants8040096.

Bailey, L.H. (1950) The Standard Cyclopedia of Horticulture, The MacMillan Co., New York.

Baiswar, P., Chandra, S., Kumar, R. and Ngachan, S. V. (2012) *Plant Dis.*, **96**(6): 911.

Bamidele, O. and Akinnuga, A.M. (2010) *Afr. J. Biotech.*, **9**: 6952-6955.

Baskaran, G., Salvamani, S., Ahmad, S. A., Shaharuddin, N. A., Pattiram, P. D. and Shukor, M. Y. (2015) *Drug Design, Development and Therapy*, **9**: 509–517.

Baskaran, G., Salvamani, S., Azlan, A., Ahmad, S. A., Yeap, S. K. and Shukor, M. Y. (2015) *Evidence-Based Complementary and Alternative Medicine*, Article ID 751714, doi.org/10.1155/2015/751714.

Bhat, P., Hegde, G. R., Hegde, G. and Mulgund, G.S. (2014) *J. Ethnopharm.*, **151**(1): 493–502.

Burkill, I.H. (1935) A Dictionary of the Economic Products of the Malay Peninsula, Crown Agents, London.

Busuioc, G., Artenie, V.I. and Virtopeanu, D. (1998) *Biologie. Vegetala.*, **44**: 85-88.

Chattopadhyay, T.K. (1977) *Prog. Hort.*, **9**: 71-75.

Chauhan, D.V.S. (1972) Vegetable Production in India (3rd edn.), Ram Prasad and Sons, Agra.

Chen, Y.F., Guo, J.H. and Fang, Z.D. (2000) *Acta Phytopath. Sinica*, **30**: 171-175.

Ciofu, R., Bereciu, I. and Ionescu, E. (1985) *Lucrari Stiintifice, Institutul Agronomic "N Balcescu" B (Horticultura)*, **28:** 25-30.

Cook A. (2010) *Notes and Records of the Royal Society*, **64**: 121-138.

Das, S., Bandyopadhyay, S., Ramasamy, A. and Mondal, S. (2015) *Natural Product Research*, **29**(11): 1059–1064.

Deka, J., Borah, U., Dash, B., Dash, S. and Kalita L. (2018) *Int. J. Cur. Pharma Res.,* **10**(4): 64.

Deka, J., Borah, U., Dash, B., Dash, S., Kalita, L. (2018) *Int. J. Curr. Pharm. Res.,* **10**(4): 64.

Demidov, A.S., Khrzhanovskii, Y.V., Shaidorov, Y.I. and Geodakyan, R.O. (1991) *Rastitel'nye Resursy,* **27**: 124-129.

Deshmukh, S. A. and Gaikwad, D. K. (2014) *J. Appl. Pharm. Sci.,* **4**(1): 153–165.

Deshpande, S., Shah, G.B., Deshpande, I. and Parmar N.S. (2003) *J. Nat. Remed.,* **3/2**: 212 – 214.

Donald, E.L. De. F., Sarruge, J.R. and Sfredo, G.J. (1983) *Anais da Escola Superior de Agricultura "Luiz de Queiroz",* **40**: 553-573.

Dong, C. (2012) *Int. J. Biol. Macromol.,* **50**: 245-259.

Farooqui, A., Ray, J. G. and Garg, A. (2019) *Grana,* **58**(6): pp. 399–407.

Ghosh, S. and Guha, A. (1933) *Indian J. Med. Res.,* **21**: 447.

Glässgen, W. E., Metzger, J. W., Heuer, S. and Strack, D. (1993) *Phytochem.,* **33**(6): 1525–1527.

Gogai, G. and Rajagopal, A. (1986) *South Indian Hort.,* **34**: 271-273.

Goh, T.K and Hsieh, W.H. (1989) *Bot. Bull. Acad. Sin.* **30**: 124-125.

Hasan, M.M., Hasan M.S., Naznin, S., Islam, N.B. and Mustarin, K.E. (2016) *Universal Journal of Agricultural Research,* **4**(6): 247-253.

Haskell, M. J., Jamil, K. M., Hassan, F., Peerson, J. M., Hossain, M. I., Fuchs, G. J. and Brown, K. H. (2004) *Amer. J. Clin. Nutri.,* **80**(3), 705–714.

Hebbar, S.S., Harsha, V.H., Shripathi, V. and Hegde, G.R. (2004) *J. Ethnopharm.,* **94**(2–3): 261–266.

Herklots, G.A.C. (1972) In: *Vegetables in South-East Asia,* George Allen and Unwin, London.

Hirei, K. and Takayuki, A. (2003) Annual Report of the Society of Plant Protection of North Japan J. Code: F0809A. **54**: 54-57.

Huang, C.H. and Chang, Y.C. (2006) *Plant Pathol.,* **55**(6): 819

Huang, F.L., Chiou, R.Y.Y., Chen, W.C., Ko, H.J., Lai, L.J. and Lin, S.M. (2016) *Plant Foods for Human Nutrition,* **71**(3): 322–329.

Islam, M.S., Rahi, M.S., Jahangir, C.A., Rahman, M.H., Jerin, I., Amin, R., Hoque, K.M.F. and Reza, M.A. (2018) *Evidence-Based Complementary and Alternative Medicine,* doi: 10.1155/2018/1537896.

Kachchhava, A.B. (2006) Dissertation work Submitted to the Rajiv Gandhi University of Health Sciences, Bangalore.

Kanno, H. and Ohkubo, H. (1999) *Soc. Plant Protection* North Japan, **50**: 115-119.

Kilari, B.P., Kotakadi, V. S. and Penchalaneni, J. (2016) *Asia Pacific J. Cancer Prev.,* **17**(1): 73–80.

Kilari, B., Netala, V.R., Penchalaneni, J., Kotakadi, V.S. and Tartte, V. (2018) *Process. Biochem.*, **67**: 175–183.

Krishna Chaitanya, B. (2012) *J. Appl. Pharm. Sci.*, **2**: 87-89.

Kumar, V., Bhat, Z.A., Kumar, D., Bohra, P. and Sheela, S. (2011) *Int. Drug Dev. Res.*, **3**: 176179.

Kumar, B.R., Anupam, A., Manchikanti, P., Rameshbabu, A.P., Dasgupta, S. and Dhara, S. (2018) *J. Food Sci. Technol.*, **55**(5): 1675–1684.

Kumar, S. S., Manoj, P. and Giridhar, P. (2015) *Food Sci. Technol.*, **64**(1): 468–474.

Kumar, S. S., Manoj, P., Shetty, N. P., Prakash, M. and Giridhar, P. (2015) *J. Food Sci. Technol.*, **52**(8): 4994–5002.

Kumar, S., Chauhan, A. and Giridhar, P. (2020) *Food Chem.*, **333**: 127442.

Kumar, V., Bhat, Z.A., Kumar, D., Khan, N.A., Chashoo, I.A. and Ara, I. (2012) *Braz. J. Pharm.*, **22**: 657-662.

Kumari, S., Krishnan, N., Dubey, V., Pandey, K. K., and Singh, J. (2020) *J. Plant Pathol.*, **102**(2): 523–527.

Latina, R.A., Dumaraos, J.D.E. and Lizardo, R.C.M. (2019) *Archives of Phytopath. Plant Protec.*, **52**(11–12): 1018–1034.

Laufer, B. and Sino-iranica, A. (1919) *Anthropological Series*, **15**(3).

Lin, S.M., Lin, B.H., Hsieh, W.M., Ko, H.J., Liu, C.D., Chen, L.G. and Chiou, R.Y.Y. (2010) *J. Agric. Food Chem.*, **58**(19): 10364–10372.

Liu, H., Han, J.F., Zhang, Y.F., Meng, F., Wang, F. and Guo, C. (2006) *Acta Agriculture Boreali-Sinica*, **21**: 184-187.

Makambila, C. (1986) *Agronomic Tropicale*, **41**: 69-74.

Meena, B.R., Nagendran, K., Tripathi, A. N., Kumari, S., Sagar, V., Gupta, N., Rai, A. B. and Singh, B. (2018) *Plant Dis.*, **102**(8): 1669.

Miura, H., Yamazaki, H. and Nishijima, T. (1997) *J. Jap. Soc. Hort. Sci.*, **66**: 513-517.

Mohammed, Haneefa, K.P., Abraham, A., Saraswathi, R., Mohanta, G.P. and Nayar, C. (2012) *J. Pharm. Sci. Res.*, **4**: 1642-1648.

Moundipa, P.F., Ngouela, S., Kamtchouing, P., Tsamo, E., Tchouanguep, F. M. and Carreau, S. (2006) *Asian J. Androl.*, **8**(1): 111–114.

Nantia, E.A., Manfo, P.F.T., Beboy, N.E., Travert, C., Carreau, S., Monsees, T.K. and Moundipa, P.F. (2012) *Andrologia*, **44**(1): 38–45.

Nantia, Edouard Akono, Manfo, F.P.T., Beboy, N.S.E. and Moundipa, P.F. (2017) *Advanced Pharmaceutical Bulletin*, **7**(1): 103–108.

Nantia, Edouard Akono, Travert, C., Manfo, F. P. T., Carreau, S., Monsees, T.K. and Moundipa, P. F. (2011) *Int. J. Mol. Sci.*, **12**(1): 376–384.

Nirmala, A. and Saroja, S. (2009) *J. Pharm. Phytotherapy*, **1**: 25-30.

Nirmala, A. and Saroja, S. (2011) *British Biotech. J.*, **1**: 1–9.

Nirmala, A., Saroja, S., Vasanthi, H.R. and Lalitha, G. (2009) *J. Pharm. Phytotherapy*, **1**: 25-30.

Okada, R., Kiyota, E., Moriyama, H., Toshiyuki, F. and Valverde, R.A. (2014) *Archives of Virology*, **159**(4): 807–809.

Oyewole, O.A. and Kalejaiye, O.A. (2012) *Scientific J. Microbiol.*, **1**: 113-118.

Palada, M.C. and Crossman, S.M.A. (1999) In: Perspectives on New Crops and New Uses. (Ed. J. Janick), pp. 388–393.

Pandey, L., Gaur, R. and Pandey, A.K. (1990) *Acta Botanica Indica*, **18**: 108.

Pandey, R.K., Gupta, P.K. Srivastava, M., Singh, M. and Robin, G.S.R. (2011) *Indian Phytopath.*, **64**(2): 207.

Park, H. J. (2014) *Asia Pacific J. Trop. Med.*, 7(11): 849–853.

Pragya, Singh, B., Singh, P.M., Ranjan, J.K., Sagar, V., Bhardwaj, R. and Pandey, C. (2018) *Veg. Sci.*, **45**(1): 116–118.

Pushpa, G., Nayak, K.M.D. and Hanumanthappa, H.S. (1975) *Curr. Sci.*, **44**: 237-238.

Reddy, Medagam Thirupathi, Begum, Hameedunnisa, Sunil, Neelam, Rao, Pandravada Someswara, Sivaraj, Natarajan, Kumar Sashi (2014) *Plant Breed. Biotech.*, **2**(1): 48-63.

Reitsma, J. and Slooff, W.C. (1950a) *Contr. Gen. Agric. Res. Sta., Bogor*, **109**: 3-11.

Reitsma, J. and Slooff, W.C. (1950b) *Contr. Gen. Agric. Res. Sta., Bogor*, **109**: 12-21.

Reshmi, S.K., Aravindhan, K.M. and Suganya Devi, P. (2012) *Tech. Research*, **4**: 900-913.

Robbs, C.F. (1972) *Lavoura*, **74**: 43-45.

Sankar, N.R., Sreeramulu, A., Gopal, D. S. and Bagyanarayana, G. (2011) *Plant Dis.*, **95**(11): 1476.

Sanni, S.B. (1982) *J. Sci. Food Agric.*, **32**: 686-687.

Sanni, S.B. (1983) *J. Plant Foods*, **5**: 75-79.

Saroj, V., Rao, P.S., Rao, S.K. and Krunal, S. (2012) *International Journal of Research in Pharmaceutical and Biological Sciences*, **3**: 1093–1094.

Sen, K., Goel, A., Rawal, S., Mahajan, N., Baboo, S. and Alak, S. (2010) *Int. J. Pharm. Sci. Res.*, **1**: 88-91.

Sidwell, K. (1999) *Novon*, **9**: 562-563.

Singh, A.U. and Kumar, V. (2012) *Arch. Phytopath. Pl. Prot.*, **45**(16): 1968–1971.

Sinnadurai, S., Abu, J.F. and Amuti, S.K. (1971) *J. West African Sci. Assoc.*, **16**: 1-4.

Sonkar, D., Gupta, R. and Saraf, S. (2012) *Pharm. Comm.*, **2**: 10-13.

Sperling, C.R. (1987) *Ph. D. Thesis*, Harvard University, Cambridge, p.279.

Suriachandraselvan, N. and Narayanasami, P. (1987) *Madras Agric. J.*, **74**: 154-156.

Tojo, M., Yonemoto, K. and Kawamura, A. (2006) *Plant Dis.*, **90**(6): 830.

Varalakshmi, B. and Devaraju (2010) *J. Hortic. Sci.*, **5**: 21-24.

Veltkamp, H.J. and Samson, J.A. (1973) *Surinaamse Landbouw*, **21**: 28-34.

Venkatalakshmi, P. and Senthamaraiselvi, V. (2012) *Int. J. Pharm. Sci. Res.*, **3**: 2539-2542.

Vuurmans, J. and Grubben, G.J.H. (1977) *Vegetables for Hot Humid Tropics*, **2**: 40-43.

Wang, H. and Ng, T.B. (2001) *Biochem. Biophy. Res. Comm.*, **288**(4): 765–770.

Wang, H. and Ng, T.B. (2004) *Peptides*, **25**(7): 1209–1214.

Winters, H. F. (1963) *Econ. Bot.*, **17**(3): 195–199.

Xu, H.Q. (1986) *Acta Botanica Sinica*, **28**: 361-367.

Yanadaiah, J.P., Lakshmi, S.M., Jayaveera, K.N. and Sudhakar, Y. (2011) *Int. J. Pharm. and Pharmac Sci.*, **3**: 502-506.

Yang, X.X., Liu, Y., Zhang, L., Huang, X.Q., Wu, W.X. and Xue, L.H. (2018) *Plant Dis.*, **102**(9): 1853.

Zhang, S., Fu, Y., Mersha, Z., Mo, X. and Palmateer, A.J. (2011) *Plant Dis.*, **95**(7): 880.

12

CORIANDER

A.B. Sharangi and A. Chattopadhyay

1.0 Introduction

Coriander (*Coriandrum sativum* L.) is one of the oldest herbs that have been used over 3,000 years for both culinary and medicinal purposes. It is an important vegetable as well as spice crop which is regarded for the unique flavour to impart a taste and aroma in a variety of food. The stems, leaves and fruits are all having a pleasant aromatic odour. The entire young plant is used in chutney and sauce preparation, and the leaves are used for flavouring continental curries and soups. The fruits are exclusively used as a condiment in the preparations of curry powder, sausages and seasonings. Pastry, biscuits, buns, cakes, and liquors, particularly gin are flavoured with coriander. The major growing states are Madhya Pradesh, Rajasthan, Gujarat, Assam and West Bengal (Anonymous, 2020a). The total production of coriander during last four years in these states is given in Table 1. During 2018-19, about 30,500 tonnes of coriander was exported from India. The top coriander seed export destinations are Malayasia, Pakistan and the UAE. In powder form, coriander is mostly exported to South Africa, the UK and the UAE (Anonymous, 2020b).

Table 1: Production of Coriander in Major Producing States of India (in tonnes)*

State	2015-16	2016-17	2017-18	2018-19
Madhya Pradesh	108490	387430	390550	370332
Rajasthan	227200	206960	130200	77020
Gujarat	138800	189520	116810	45418
Assam	29680	29780	30790	21758
West Bengal	14520	14560	14950	14950

* Anonymous (2020a).

2.0 Composition and Uses

2.1.0 Composition

The green leaves and fruits (seeds) of coriander are packed with scores of chemical compounds offering nutritive, antioxidant and medicinal values. Coriander leaves are not so well studied as the fruits. Nevertheless, essential oil, flavonoids, phenolic acids and polyphenols are among the other compounds detected in the leaves (Matasyoh *et al.*, 2009; Nambiar *et al.*, 2010). The green leaves contain 87.9 per cent moisture, 3.3 per cent protein, 0.6 per cent fat, 6.5 per cent carbohydrates and 1.7 per cent mineral matter. Whereas, the dry seeds have 6.3–8.0 per cent moisture, 1.3 per cent protein, 0.3–1.7 per cent volatile oil, 19.6 per cent non-volatile oil, 31.5 per cent ether extract, 24.0 per cent carbohydrates, 5.3 per cent mineral matter and vitamin A 175 IU per 100 g. The pleasant aroma is due to an essential element called d-linalol or coriandrol ($C_{10}H_{17}OH$), a terpene tertiary alcohol.

Dried ripe coriander seeds contain both steam-volatile oil and fixed oil. The volatile oil is a clear, colourless to light yellow liquid. Higher volatile oil content is found in Norwegian coriander (1.4–1.7 per cent) followed by Bulgarian coriander (0.1– 0.5 per cent). Indian seeds are poor in volatile oil content (0.1–0.4 per cent) (Agrawal and Sharma, 1990). Essential oil of coriander contains different alcohol like linalool (60–80 per cent), geraniol (1.2 per cent –4.6 per cent), terpinen-4-ol (3 per cent), α-terpineol (0.5 per cent); hydrocarbons like γ-terpinene (1–8 per cent), r-cymene (3.5 per cent), limonene (0.5 per cent –4.0 per cent), a-pinene (0.2 per cent –8.5 per cent), camphene (1.4 per cent), myrcene (0.2 per cent –2.0 per cent); ketones like Camphor (0.9 per cent –4.9 per cent) and esters like Geranyl acetate (0.1 per cent –4.7 per cent), linalyl acetate (0 per cent –2.7 per cent). Telci *et al.* (2006) investigated two coriander varieties (*vulgare* Alef. and *microcarpum* DC.) from Turkey and found that the essential oil content varied between 0.15 and 0.25 per cent in *vulgare* and 0.31 to 0.43 per cent in *microcarpum*. Nutrient composition of coriander leaf and seeds as per USDA (National Nutrition Data base, 2013) is given in Table 2.

Table 2: Nutrient Composition of Coriander Leaf and Seed

Nutrient	*Amount (per 100 g)*	
	Leaf	*Seed*
Water	7.30 g	8.86 g
Energy	279 kcal	298 kcal
Protein	21.93 g	12.37 g
Total lipid (fat)	4.78 g	17.77 g
Carbohydrate, by difference	52.10 g	54.99 g
Fiber, total dietary	10.40 g	41.90 g
Calcium	1246 mg	709 mg
Iron	42.46 mg	16.32 mg
Magnesium	694 mg	330 mg
Phosphorus	481 mg	409 mg
Potassium	4466 mg	1267 mg
Sodium	211 mg	35 mg
Zinc	4.72 mg	4.70 mg
Vitamin C	566.7 mg	21.0 mg
Thiamin	1.252 mg	0.239 mg
Riboflavin	1.500 mg	0.290 mg
Niacin	10.707 mg	2.130 mg
Vitamin A, RAE	293 μg	0.00 μg
Vitamin A, IU	5850 IU	0.00 IU
Fatty acids, total saturated	0.115 g	0.990 g
Fatty acids, monounsaturated	2.232 g	13.580 g
Fatty acids, total polyunsaturated	0.328 g	1.750 g

* National Nutrition Database (2013).

2.2.0 Uses

It is used as fresh green herb for its specific flavour, which is completely different from that of the ripe fruits. In the Caucasus, Iran, Iraq, Mexico, and in South America, with the exception of Argentina, coriander is mainly used as green herb. The leafy green is also consumed on a large scale in India, China, Thailand, Malaysia, Indonesia, the American Midwest and in the Near East (Prakash, 1990). The Georgian name 'kinza', also used in Russian, refers to the green plant, and the English term 'Chinese parsley' also describes the herb. The same is true for the Spanish name 'cilantro', which denotes the plant, while the fruits are called 'simiente de cilantro'. The name 'cilantro' is frequently used in American English to refer to the green herb or the dried leaves (Lamberts, 1990), probably because the use of the green plant was introduced in the USA by Spanish-speaking Mexicans. The French name 'persil arabe' also means the green herb. It is interesting to note that the ancient Egyptian hieroglyphs 'venshivu' and'ounshavu' refer to the coriander plant, and a different one, 'ounshi', to the fruits of this species (Sinskaja, 1969). Whether the ancient Egyptians also used the green herb cannot be determined from the available literature, and the later Greek and Latin authors only mention the use of the fruits.

The smell of the green plants is often regarded as unpleasant, particularly by the inhabitants of industrialized countries. It has frequently been observed, however, that those who became accustomed to the smell cease to consider it disagreeable. In countries where use of the green herb is widespread, its flavour is highly regarded, and it is used in a variety of dishes. Several conservation techniques, *e.g.* drying, are applied to ensure a constant supply of the herb (Prakash, 1990; Ivanova and Stoletova, 1990). The green plant is used in soups, salads and dressings, and in India, it is very commonly used in chutneys (Ilyas, 1980). In the Transcaucasian area, the green plant is a most important ingredient in cooking (Stoletova, 1930, 1931; Ivanova, 1966). Here, the local varieties are characterized by high production of green matter (Alborisvili, 1984). Although at present the use of the green herb is much more limited in the industrialized countries, its popularity seems likely to increase, as the market for such 'ethnic' foods is growing (Simon, 1990). The characteristic smell of the green plant is caused by the aldehydic contents of the essential oil. During ripening, these decrease, and after ripening and drying, they are no longer found in the fruits (Lörincz and Tyihák, 1965). Lastly, the green herb's high vitamin C (ascorbic acid, up to 160 mg/100 g), vitamin A (carotene, up to 12 mg/100 g) (Girenko, 1982) and vitamin B_2 content (up to 60 mg/100 g) (Prakash, 1990) should be noted. In China, even the root of coriander, which has channels that also contain essential oils, is used as a vegetable.

Coriander seeds have beneficial medicinal properties. They are considered carminative, diuretic, tonic, stomachic, antibilious, refrigerant and aphrodisiac. Coriander extracts and its bioactive phytochemicals have been reported for a wide range of biological activities including antioxidant, anticancer, neuroprotective, anxiolytic, hypnotic, anticonvulsant, analgesic, antiinflammatory, and antidiabetic (Gastón *et al.*, 2016; Laribi *et al.*, 2015). Ayurvedic medicines often use coriander is a frequent ingredient. Now-a-days seeds of coriander undergo extensive value

Uses of Coriander

Leaves and Seeds of Coriander

Powder (Left) and Seed Oil (Right) from Coriander

Different Recipes of Coriander in Indian Cuisine

addition and are in large demand globally. The volatile oil is also used in flavouring liquors and for masking the foul odour of pharmaceutical products.

3.0 Origin and Taxonomy

The origin of the cultivated species *Coriandrum sativum* is still not clear, and no certain information about the wild species exists. Nevertheless, several authors have named coriander as a wild plant. Linnaeus reported long ago in 1780 that coriander also occurs as a weed in cereals. Alefeld (1866) mentioned that coriander was a common weed spread from southeastern Europe to southern Russia. Stoletova (1930) also reports on wild coriander from Armenia.

In the list of centres of origin of cultivated plants Vavilov (1992) mentioned coriander for Central Asia, the Near East and Abyssinia. In each of these areas, distinct forms of coriander can be found. But the discussions since Vavilov have shown that centres of variation do not have to be the centres of origin at the same time (Zeven and de Wet, 1982). Ivanova and Stoletova (1990) speak more cautiously of centres of formation of different types of coriander and name as cradles for distinct types: (i) India; (ii) Northern Africa; (iii) Central Asia, and (iv) Abyssinia. The geographically widespread cultivation of coriander since ancient times has resulted in a wide range of variation.

According to United Nations COMTRADE Database (2020), in the present world, India is the largest exporter of coriander. It is also cultivated in Bulgaria, Morocco, Rumania, France, Spain, Italy, the Netherlands, Myanmar, Pakistan, Turkey, Mexico, Australia, Canada, Argentina and to some extent, in the UK and the USA. The main coriander growing states in India are Andhra Pradesh, Rajasthan, Madhya Pradesh, Karnataka, Tamil Nadu and Uttar Pradesh.

4.0 Morphology

Coriandrum sativum L. ($2n = 22$) belongs to the family Umbelliferae with systematic position as follows:

- ✰ Kingdom – Plantae
- ✰ Division – Magnoliophyta
- ✰ Class – Magnoliopsida
- ✰ Order – Apiales
- ✰ Family – Apiaceae/Umbelliferae
- ✰ Genus – *Coriandrum*
- ✰ Species – *sativum*

Purseglove *et al.* (1981) have specified a comprehensive botanical narrative of coriander. The plants are 30 to 100 cm tall, depending upon the varietal variations. The flowering comes at 45–60 days after sowing and matures in 65–120 days, depending upon the variety and cropping situation. The leaves are lanceolate, green or dark green, glabrous on both surfaces and are variable in shape and lobed. The flowers are borne in small umbels, white or light pink, asymmetrical, with the petals pointing away from the centre. The coriander seed, is almost ovate globular, schizocarp with two mericarps (Coskuner and Karababa, 2011), and multiple longitudinal ridges on the surface possessing a sweet, slightly pungent, citrus like flavor (Yeung and Bowra, 2007). Each branch as well as the main shoot terminates in a compound umbel (determinate growth) bearing 3–10 umbels, each umbel containing 10–50 pentamerous flowers. The flowers are small, protandrous and difficult to manipulate for controlled pollination.

Coriander is a cross-pollinated crop. The extent of cross-pollination varies from 50 per cent (Ramanujam *et al.*, 1964) to 60 per cent (Dimri *et al.*, 1977). Singh and Ramanujam (1973) reported noteworthy differences in male and perfect flower distribution in the umbels of different varieties. Hermaphrodite flowers opened earlier than males.

5.0 Growth and Development

Growth and development of coriander is precisely influenced by microclimate. The crop requires specific microclimate for expressing the vegetative as well as the reproductive phase apart from the initial effect on crop phenophase. Vegetative stage is considered as the initial stage when only shoots with leaves are there. In the second stage, the parts of plants above ground are in full flowering. Aerial parts at full green immature fruits are considered as third stage of growth. The final stage is denoted by aerial parts of plants at brown mature fruits. Some genotypes of coriander form several basal leaves; others start stem elongation immediately or after the second leaf. Flowering is protandrous and starts in the primary umbel, about 50-90 days after sowing. The peripheral florets of the umbellets are the first to flower. Coriander is cross-fertilized by insects; the stigma remains receptive for 5 days, pollen is fertile for 24 hours only; stamens emerge one by one. The length of the flowering period depends on the number of branches and the weather conditions, and extends for up to 30 days. Thus, maturation of the umbels of different orders is a successive process, and ripe fruits of the primary umbel may shatter before those of umbels of a higher degree have reached full maturity. Coriander seeds reach physiological maturity 6-7 weeks after anthesis. During ripening the aldehyde components of the essential oil disappear, and the odour of the fruits changes notably. This process continues after harvest of the fruits and is accelerated by high temperatures combined with dry weather. The time from sowing to harvesting depends very much on the genotype, and is usually between 90 and 140 days. However, since young coriander of some genotypes is frost resistant it can be cultivated as a cold-season crop over a much longer period.

6.0 Cultivars

Guj. Cor.1

Developed through selection from local germplasm. Average yield is 1100 kg/ha (dry), essential oil 0.35 per cent, duration 112 days. Suitable for early sowing, erect plant, round bold grains, moderately tolerant to wilt and powdery mildew.

CO.1

Developed through selection from Koilpatti local. Average yield is 440 kg/ha (dry), essential oil 0.27 per cent, duration 110 days. A cultivar with small statured plant, suitable for rainfed areas and for greens and grains, small grain.

CO.2

Developed through reselection from culture P2 of Gujarat. Average yield is 520 kg/ha(dry), essential oil 0.40 per cent, duration 90-100 days. A dual purpose cultivar, suitable for saline, alkaline and drought prone areas, seeds oblong, medium.

CO.3

Developed through reselection from Acc.695 of IARI, New Delhi type. Average yield is 650 kg/ha, essential oil 0.38-0.41 per cent, duration 85-95 days. A dual purpose cultivar, good yielder, medium sized grains, suitable for both rainfed and irrigated condition, *rabi* as well as *kharif* season. Field tolerant to powdery mildew, wilt and grain mould.

CO (CR).4

Developed through reselection from germplasm ATP77 Guntur collection. Average yield is 600 kg/ha (dry), essential oil 0.4 per cent, duration 65-70 days. Early maturing cultivar suitable for both rainfed and irrigated condition; grains oblong and medium; field tolerant to wilt and grain mould.

Guj.Cor.2

Developed through reselection from Co.2. Average yield is 1450 kg/ha (dry), essential oil 0.40 per cent, duration 110 days. Semi spreading type, suitable for early sowing, moderately tolerant to powdery mildew, grains oblong, lodging and shattering resistant.

Rajendra Swathi

Developed through pureline selection from Muzaffarpur collection. Average yield is 1300 kg/ha (dry), essential oil 0.65 per cent, duration 100 days. Medium sized plant with fine, aromatic round grains, suitable for intercropping, field tolerance to aphids.

Sadhana

Developed through mass selection from local Alur collection. Average yield is 1025 kg/ha (dry), essential oil 0.20 per cent, duration 95-110 days. A dual purpose, semi-erect cultivar, suitable for rainfed condition, field tolerant to white fly, mites and aphids. A mid-late cultivar withstands moisture stress, responded well to input management under optimum moisture level.

Swathi

Developed through mass selection from Nandyal germplasm. Average yield is 855 kg/ha (dry), essential oil 0.30 per cent, duration 82-85 days. Plants are medium size semi-erect type, early maturing cultivar, suitable for rainfed condition, and late sown season. They are field tolerant to white fly, moderately tolerant to disease. Suits well to the areas where the soil moisture retentiveness in comparatively less, being early maturity. It escapes powdery mildew disease.

CS 287

Developed through reselection from Guntur collection. The average yield is 600 kg/ha (dry), duration 79-97 days. Early maturing cultivar, suitable for both rainfed and irrigated condition. Field tolerant to wilt and grain mould.

Sindhu

Developed through mass selection from germplasm Warangal Local. Average yield is 1000 kg/ha (dry), essential oil 0.40 per cent, duration 100-110 days. Oval medium breakable grains, suitable for rainfed areas, tolerant to wilt, powdery mildew as well as drought condition, medium duration.

Hisar Anand

Developed through mass selection from Haryana collection. Average yield is 1400 kg/ha (dry), essential oil 0.35 per cent. A medium tall, dual purpose cultivar, oval medium sized seeds, wider adaptability to different soil conditions. Resistant to lodging due to spreading habit.

Hisar Sugandh

Developed through mass selection from indigenous germplasm. Average yield is 1400 kg/ha (dry). Suitable for irrigated conditions. Resistant to stem gall disease.

Hisar Surabhi

Developed through mass selection from local germplasm. Average yield is 1800 kg/ha (dry), essential oil 0.4-0.5 per cent, duration 130-140days. Bushy, erect plant type, seed medium, oblong, tolerant to frost, less susceptible to aphids, medium duration.

Azad Dhania-1

Developed through mass selection from Kalyanpur germplasm collection. Average yield is 1000 kg/ha (dry), essential oil 0.29 per cent, duration 120-125 days. Erect, early branching, number of umbelletes per umbel 5, tolerant to moisture stress, powdery mildew and aphids.

Pant Haritima

Developed through selection from local type Pant Dhania. Average yield is 1200 kg/ha (dry), essential oil 0.40 per cent, duration 150-160 days. Tall, erect plant, a dual purpose type, good yielder of leaves, smaller seeds with high oil. Resistant to stem gall.

DWA 3

Developed through pureline selection from Karnataka collection. Average yield is 400 kg/ha (dry), essential oil 0.27 per cent. A dual purpose cultivar best suited for black clay soils, moderately tolerant to powdery mildew.

CIMPOS-33

Developed through selection from germplasm introduced from Bulgaria. Average yield is 2100 kg/ha (dry), essential oil 1.3 per cent. Tall, erect, compact,

Cultivars of Coriander for Leaf Production

Arka Isha

Coriander K5 (Baramasi)

Indam Multi-cut

profusely branching and flowering, grains small and bold. Mainly recommended for oil production.

Arka Isha

This cultivar has been developed through mass selection from an exotic introduction from Japan, IIHR ACC. No.19528. It is a multi-cut type, bushy plants, broad leaves, with good aroma, keeping quality (21 days under refrigeration), rich in Vitamin C (167mg/100g FW), yields 12 t/ha in 3 cuttings in 70 days.

ACR-01-256 (NRCSS)

Developed through reselection from EC-467683. Average yield is 1100 kg/ha (dry), essential oil 0.35-5 per cent, long duration crop. Dual purpose cultivar, long duration, resistant to stem gall and wilt.

RCr 20

Developed through recurrent half sib election from Jaipur Local. Average yield is 900 kg/ha (dry), essential oil 0.25 per cent, duration 100-110 days. Medium sized bushy plant suitable for rainfed crop or limited moisture condition and heavy soils of south Rajasthan. Moderately resistant to stem gall, bold grains, early maturity.

RCr 41

Developed through recurrent half sib selection from local type from Kotta. Average yield is 909 kg/ha (dry), essential oil 0.25 per cent, duration 130-140 days. A tall erect plant with thick stem. Grows well under irrigated conditions, resistant to stem gall, wilt and moderately resistant to powdery mildew; small seeds (9.3g/1000 seed), long duration cultivar.

RCr 435

Developed through recurrent selection from local germplasm from Jalore. Average yield is 1000 kg/ha (dry), essential oil 0.33 per cent, duration 110-130 days. Plants are bushy, erect, bold seeds, medium sized, medium maturing variety, adapted for irrigated condition moderately resistant to root knot and powdery mildew.

RCr 436

Developed through recurrent half sib selection from local germplasm from Kota. Average yield is 1100 kg/ha (dry), essential oil 0.33 per cent, duration 90-100 days. Plants semi dwarf, bushy type with quick early growth and bold seeds. Resistant to root rot and root knot nematodes, most suitable for limited moisture condition and heavy soils of south Rajasthan.

RCr 446

Developed through half sib selection from local type from Jaipur Local. Average yield 1200 kg/ha (dry), essential oil 0.33 per cent, duration 110-130 days. Plants tall, are leafy erect with higher number of seeds per umbel. Seeds are medium in size and moderately resistant to stem gall.

RCr 684

Developed through mutation breeding of gamma rays. Induced mutant of Rcr-20. Average yield is 990 kg/ha (dry), essential oil 0.32 per cent, duration 110-120 days. A cultivar, resistant to stem gall and less susceptible the powdery mildew. Adapted to medium heavy textured soil and sandy loam soil under irrigation. Seeds of the cultivar are bold. Plants are tolerant and erect with higher number of seeds per umbel, medium maturity.

Ajmer Coriander-1 (ACr-1)

Dual purpose cultivar suitable for leaves and seeds production, grains are medium to small, having essential oil content up to 0.5 per cent, resistant to stem gall and have tolerance to powdery mildew, average yield is 1250 kg seeds per hectare.

Some private sector bred cultivars **Coriander K 5 (Bramasi)** and **Indam Multi-cut** were found promising for leaf production.

7.0 Soil and Climate

7.1.0 Climate

Being a tropical crop, coriander requires a cool and comparatively dry frost-free climate, chiefly at the time of flowering and seed formation stages for good quality and high yields. Frost at or before the flowering stage considerably reduces the production. High temperature and high wind velocity during anthesis and seed formation enhances sterility and reduces yield. Cloudy weather at the time of flowering increases the number of aphids and disease.

7.2.0 Soil

Coriander prefers loamy to reasonably heavy soils. It is also cultivated as an unirrigated crop with conserved moisture on black cotton or heavy soil types with high moisture retention capacity. Saline, alkaline and even sandy soils are not suitable for coriander cultivation.

8.0 Cultivation

8.1.0 Sowing

Coriander has a specific planting requirement. The field needs to be ploughed 2-3 times to bring the soil to fine tilth. A pre-sowing irrigation before land preparation is essential, whenever optimum moisture for seed germination is not available in the soil.

Generally coriander is cultivated during winter season. The ideal time for sowing of coriander is 15th October to 15th November. In case of early sowing, germination is affected due to excess sun shine. In late sowing, growth of plants is poor with increased chances of disease development. In frost prone areas, sowing time may be adjusted to avoid frost incidence during flowering stage. Seeds should be rubbed to split into two halves during sowing. Generally row spacing is done at 30 cm and for sowing of one hectare area 15 to 20 kg of seed is required. Sharangi and Roychowdhury (2014) reported that the optimum date of sowing in coriander at the gangetic alluvial region is 5th November for important growth parameters, yield attributes and quality. Lal *et al.* (2017) recorded highest number of umbel plant^{-1} (42.33), number of umbellate umbel^{-1} (8.0), seed yield plot^{-1} (1188.33 g) and seed yield (1980.56 kg ha^{-1}) with 23rd October date of sowing.

Table 2: Sowing Time Recommended for different States in India*

State	*Sowing Time*
Rajasthan	Middle of October to Middle of November
Gujarat	Last week of October to first fortnight of November
Tamil Nadu	June-July (kharif) September – October (rabi)
Andhra Pradesh	First fortnight of November
Bihar	First week of October
Uttar Pradesh	Middle of October to first week of November

* Lal *et al.* (2014).

8.2.0 Manuring and Fertilization

At the time of land preparation 15 to 20 MT FYM per hectare should be incorporated in soil. In case of rain fed crop, apart from 20 kg nitrogen, 30 kg phosphorous and 20 kg potash per hectare should also be applied in the soil as basal dose. For irrigated crop; 20 MT FYM, 60 kg of nitrogen, 30 kg phosphorous and 20 kg potash per hectare is recommended. Full dose of FYM should be mixed in soil at the time of land preparation and one third dose of nitrogen (20 kg), full dose of phosphorous and potash should be applied as basal dose and remaining 40 kg nitrogen should be top dressed in two equal split doses at the time of first irrigation and at flowering stage. Dadiga and Jain (2017) studied the effect of different fertilizers and their doses on nutrient uptake in coriander variety JD-1. Among the organic manures, the maximum nitrogen and phosphorus uptake was maximum in Poultry manure @ 5 t ha^{-1} followed by poultry manure @ 2.5 t ha^{-1}, while maximum potash uptake was found in the vermicompost @ 5 t ha^{-1}. Application of sulphur up to 20 kg/ha through gypsum produced significantly higher seed yield of coriander over control under irrigated coarse textural soil (Anonymous, 2002). Study on response of coriander to micronutrients carried out at Jobner indicated that soil application of Mn and Cu, foliar application of Zn and Cu and foliar + soil application of Fe, Mn and Cu produced more seed yield over control under loamy sand soils. The study also indicated that volatile oil content of coriander seed increased with foliar application of $ZnSO_4$ (0.5 per cent) and $FeSO_4$ (0.25 per cent) (Anonymous, 2001). Apart from the major nutrients Zn, Fe, Mn and Cu also play vital role in seed yield of coriander. Effect of some growth regulators on growth, yield and quality of coriander had also been reported by Panda *et al.* (2007). Use of *Azospirillum* + 50 per cent nitrogen+ 5t farm yard manure ha^{-1} produced higher seed yield of coriander (Malhotra *et al.*, 2006).

Seed Sowing and growing of Coriander under Protected Structure for Green Leaves

Field Growing, Harvesting and Selling of Coriander Leaves in the Market

8.3.0 Irrigation

Generally, 4-6 irrigations are given depending on type of soil and climate. For heavy soils only 3-4 irrigations are sufficient whereas for light textured soils about 6 irrigations are required. First irrigation should be given within 30-35 days from the date of sowing, second after 50-60 days, third after 70-80 days, fourth after 90-100 days, fifth after 105-110 days and sixth after 115-125 days. If sowing is done by broadcasting, first irrigation is given immediately after broadcasting the seeds. Second light irrigation is given after another 8-10 days to facilitate germination of seeds. Subsequent irrigations are given according to the requirement of the crop. A good response of sprinkler and drip irrigation system for efficient water use in fennel, fenugreek, coriander, cumin, ajowan, dill and nigella has been studied (Malhotra *et al.*, 2009).

8.4.0 Weed Control and Intercropping

Primarily, coriander plants grow very slow, hence weeding during early growth period is very essential to save the plants from weed competition. The important weed flora generally found in coriander crop are: Bathua (*Chenopodium album*), Kharbathua (*Chenopodium murale*), Sanji (*Melilotus indica*), Piyaji (*Asphodelus tenuiflolius*) and Krishan neel (*Anagallis arvensis*), *etc.* In rainfed crop, first weeding should be done at about 25-30 days after sowing and in irrigated crop about 40-45 days after sowing. Second weeding should be done 50-60 days after sowing, in case of rain fed crop. Thinning of plants should be done before first irrigation and at the time of ist weeding to maintain a spacing of 5-10 cm between plants. Chemical control of weeds can be done with a pre plant application of the herbicides. In India the crop is mostly taken as rainfed crop (70-80 per cent) and for weed control during the vegetative phase pre-plant incorporation of fluchloralin @ 0.75 to 1.0 kg a.i./ ha or per-emergence application of oxyfluorfen @ 0.15 kg a.i./ha or pendimethalin @ 0.75 to 1.0 ka a.i./ha after dissolving in 400-500 litres of water is recommended. After chemical control of weeds, one hand weeding at 50-55 DAS is beneficial to ensure weed free condition till harvesting of the crop. Singh *et al.* (2018) inferred that intercropping of coriander with garlic is better for realizing higher yield, system productivity and profitability.

8.5.0 Harvesting of Green Leaves

Coriander plant has regenerative capacity and hence 2-3 cuttings can be undertaken very easily. Menon *et al.* (1997), Thapa (1999) and Sharangi (2005) suggested that leaf plucking of coriander at early vegetative stage can provide an extra income to the growers. For leaf purpose coriander is grown all the year round. Harvesting at leafy stage has been found to improve seed yields in coriander. A single cut at 30 DAS provides higher yields of seeds (Thakural *et al.*, 1992; Baboo and Rana, 1995, Sharangi *et al.*, 2011) and higher gross returns (Bhati, 1988). The harvested leaves apart from being used raw could also be used for extraction of essential oil. India being the largest producer as well as consumer of green leaves there is always a persistent demand for green leaves.

8.6.0 Protected Cultivation

Protected cultivation offers distinctive advantages of quality, productivity and exciting market price to the growers particularly when compared with that in open condition. In peri-urban areas of Northern India, Singh *et al.* (2007) reported that protected cultivation of vegetables provides the best way to increase the productivity and quality of vegetables as well as biotic and abiotic stress condition compared to open field condition. Dixit *et al.* (2005) found that the average yields of spinach, amaranth, fenugreek and coriander (6.94, 6.92, 8.63 and 2.34 kg/m^2, respectively) were higher under green house conditions than in the open field (3.15, 3.00, 4.95 and 1.68 kg/m^2, respectively). According to Shoba and Rajamani (2009) the shade net houses during off season reduce the temperature up to 5°C and increase the relative humidity, thereby providing optimum environmental conditions for the growth of coriander and concluded 50 per cent shade resulted in higher leaf yield followed by 25 per cent shade. Sarada *et al.* (2011) reported that the afternoon soil temperature if ameliorated and maintained between 28.0°C and 32.5°C, coriander production can be taken up successfully without any modification of other weather factors like air temperature and relative humidity. However, amelioration in ambient air temperature further improves the efficiency of the production system.

9.0 Diseases and Pests

9.1.0 Diseases

Major diseases of coriander with their control measures are described below:

9.1.1 Wilt (*Fusarium oxysporium* F. spp. *coriandri*)

Younger plants are more susceptible to wilt. Yellowing as well as drooping of leaves and terminal branches are symptoms of disease. The infected plant dries up due to wilting in later stage. The fungus also disturbs root system of the plants. To prevent the crop from infection of the fungus, deep ploughing should be done during summer season. Crop rotation may also be followed. Avoid cultivation of coriander crop for 2-3 years where disease has been noticed previously. Seed should be treated with Bavistin or Thiram @ 1.5 g per kg of seed before sowing. Drenching of soil with chlorothalonil (2 g) + carbendazim (1 g)/lt. may be effective.

9.1.2 Powdery Mildew (*Erysiphe polygoni*)

This disease is mostly seen during cloudy weather condition. White powdery growth appears on the leaves and buds during its primary stage. Seed formation may not take place in affected plants due to this disease. To control this disease, dusting of Sulpher dust @ 20-25 kg per hectare should be done or spraying of wettable sulphur or Kerathane can also control this disease.

9.1.3 Blight (*Alternaria poonensis*)

In this fungal disease, dark brown spots appear on the stem and leaves of infected plants. Spray the crop with 0.25 per cent solution of mancozeb effectively controls this disease.

Diseases of Coriander

Wilt Infection

Leaf Blight Infection

Powdery Mildew Infection

Pests of Coriander

Stem Gall Infestation

Aphid Infestation

Root Knot Nematode Infestation

9.2.0 Pests

9.2.1 Stem Gall (*Protomyces macrosporus*)

Galls appear on the leaves and stems of the plants affected by this disease which change the shape of seeds and reduced the quality. Disease is more effective with increasing level of humidity. To control the disease, sowing may be done only after treating the seeds with 1.5 g Thiram and 1.5 g Bavistin (1: 1)/kg seeds. Spray 0.1 per cent solution of Bavistin when the symptoms start appearing and repeat the spraying at an interval of 20 days till the disease is completely controlled.

9.2.2 Aphid (*Hyadaphis coriandri* and *Aphis gossypii*)

Aphid infestation occurs at the time of flowering or after flowering in coriander crop. It sucks the sap from tender parts of the plants resulting in heavy loss. Spraying of 0.03 per cent solution of imidacloprid or 0.15 per cent acetamiprid may be effective to control this insect.

Sowing during last week of October to first week of November can minimize the damage caused by aphid.

9.2.3 Leaf Eating Caterpillar

It is a caterpillar, brownish in colour which cuts the plants from ground level and makes them to fall down. Infestation of this pest starts at the initial stage of plants resulting in heavy loss to the crop. Before last ploughing land should be drenched with chlorpyriphos is also recommended. Spraying of chlorpyriphos @ 0.25 per cent may also be effective.

9.2.4 Nematodes

Meloidogyne incognita attacks the roots of coriander plants, it sets up a feeding location where it deforms the normal root cells and establishes giant cells. The roots become nodulated, forming galls which consequently affect the nutrient absorption capacity of plants negatively and reduce the yield and quality of produce, hence the term "root-knot" nematode. Resistant cultivars RCr-41, Co-1, Co-2 should be given preference especially in nematode infested field to obtain good returns.

Methodology developed and recommended by NRCSS, Ajmer for plant protection is application of mustard residue (2.5 t/ha) + mustard cake (0.5 t/ha) + neem cake (0.5 t/ha) as soil amendment with one irrigation during hot summer, application of *Trichoderma viride* or *A. versicolar* as seed treatment @ 10g/kg seed and soil application @ 2.5 kg/ha mixed with 50 kg of FYM at the time of sowing and adoption of soil solarization in combination with oil cakes has been found highly effective (Malhotra, 2016).

10.0 Seed Production

The seed is a fundamental unit which plays a crucial role in boosting up the productivity and economy of the country. Without the use of good quality seed, the investments incurred on fertilizers, pesticides and water will not pay dividend which ought to be realized. Seed production of coriander is similar to general production procedure. But there are certain general principles and standards which must strictly be adhered to.

10.1.0 Stages of Seed Production

Breeder seed → Foundation seed → Certified seed

10.2.0 Isolation Requirement

Foundation seed – 800 m

Certified seed – 400 m

10.3.0 Roguing

A minimum of 3 inspections shall be made during before flowering, 50 per cent flowering and prior to harvest to verify the true nature of plant.

10.4.0 Field Standards

The following field standards are crucial for seed production of coriander.

Factors	*Foundation Seed*	*Certified Seed*
Isolation distance (m)	10	5
No. of field inspection	2	2
Off types (per cent)	0.10	0.20

10.5.0 Seed Standards

The following seed standards are crucial for seed production of coriander.

Factors	*Foundation*	*Certified*
Pure seed (minimum)	98 per cent	98 per cent
Inert matter (maximum)	2 per cent	2 per cent
Other crop seed (maximum) (no./kg)	10/kg	20/kg
Total weed seed (maximum) (no./kg)	10/kg	20/kg
Germination (minimum)	65 per cent	65 per cent
Moisture (maximum) (normal container)	10 per cent	10 per cent
For VP container (maximum)	8 per cent	8 per cent

10.6.0 Seed Priming

Seed priming (seed invigoration) has gained substantial magnitude in recent times as it emerged as one of the most potential areas of seed quality enhancement procedure. It is a pre-sowing treatment, that involves exposure of seeds to a low external water potential with restricted (controlled) hydration of seed to a level that permits pre-germinative metabolic activity to proceed, but prevents real emergence of the radical. Seed priming is helpful for augmented speed and evenness of seed germination particularly under unfavorable abnormal and/or extreme conditions related to temperature, moisture and salinity. The practice enhances abiotic stress

Different Stages of Seed Production in Coriander

tolerant capacities, increases the antioxidant enzyme activities like peroxidase, catalase, superoxide dismutase (SOD), polyphenol oxidase (PPO), lipoxygenase (LOX) and phenyl alanine ammonia lyase (PAL), breaks the seed dormancy, increases the germination percentage and promotes early growth. Seeds are soaked in different solutions like various inorganic salts, sugars and polyethylene glycol (PEG), *etc.* The temperature recommended during priming is between 10°C – 15°C. The duration of priming varies with the crop from 1 day to 23 day. After priming the seeds are re-dried to original moisture content (Umarani and Vigneshwari, 2010).

Hydropriming is one of the seed germination enhancing techniques which is simple, safe, nonchemical and harmless to the environment. Seeds were aged at 42°C and 100 per cent relative humidity for 0, 24, 48, 72 and 96 h and then primed by soaking in distilled water for 8 h and incubated at 20 degrees C for 24 h, compared to non-primed seeds as control by Rithichai *et al.*(2009). Results obtained showed that primed seeds resulted in earlier germination and higher first count, germination percentage than those of non-primed seeds, irrespective of aging duration Salinity is one of the biggest limiting factors for agriculture in semi-arid areas of the world. For this reason, an experiment was conducted to study the effect of seed priming with NaCl and $CaCl_2$ on growth and yield responses of Tunisian coriander cultivar exposed to five levels of salinity (0, 2, 4, 6, 8 g l^{-1}). Seeds of coriander were primed with aerated solutions of 0.13 M NaCl and $CaCl_2$ for 24 h. Results indicated that with increasing salinity, emergence traits (total emergence, mean emergence time), growth parameters (plant height, shoot fresh and dry weight) and mineral contents (K^+ and Ca^{2+}) decreased, but to a less degree in primed seeds (Aymen *et al.*, 2013). At Tunisia, a study was conducted to evaluate the interactive effect of salinity and seed priming on coriander. Results revealed that seed priming and salinity had significantly ($p<=0.05$) affected all the parameters under study (Ben Fredj *et al.*, 2014).

Plant growth regulators *viz.*, humic acid, Moringa leaf extract, 6 benzyl amino purine in foliar and biofertilizer and humic acid in seed priming mode were employed on coriander (*Coriandum sativum*) to evaluate their possible effects on total phenolic content and antioxidant attributes. Antioxidant activity was found to be enhanced significantly ($P<0.05$) with exogenously applied treatments of plant growth regulators (Aslam *et al.*, 2014).

A study was carried out by Issac and Vijayakumar (2014) to find out the effect of comprehensive seed treatments on storability of coriander cv. CO-4 seeds. The seeds fortified with KH_2PO_4 @ 1 per cent for 16h and treated with *Azospirillum* sp. @100g/kg were found to be good in storage which maintained quality characteristics such as germination (87 per cent), vigour index (2229), dry matter production (32.33 mg/10 seedlings), α-amylase activity (4.60 mg maltose/g/min) and protein content (12.77 per cent) for six months of storage.

10.7.0 Seed Certification

There are six different phases of seed certification, wherein the application requirements for certification is scrutinized, the basic seed source, its class and other seeds used for raising the seed crop are verified. Cultivars evolved by the State Agricultural Universities and ICAR Institutes, India are notified under section 5 of

Seed Act are only considered for seed certification process. Any person/company desires of producing certified seed shall register his name with concerned Seed Certification Agency. He has to undergo following broad phases before the seed is produced and marketed:

- Receipt and scrutiny of application.
- Verification of seed sources, class and other requirement of the seed used for raising the seed crops.
- Field inspection to verify conformity to the prescribed field standards.
- Supervision at postharvest stages.
- Seed sampling and analysis, including genetic purity test and seed health test.
- Grant of certificate and certification tags, tagging and sealing.

10.8.0 Harvesting, Yield and Storage of Seed Crop

Subject to the cultivar and time of sowing, the crop matures within 110 to 140 days. Maturity is determined by the indication when seeds turn to yellowish green colour and seeds become medium-hard while pressing. After harvest, the crop should be dried under partial shade without much exposure to scorching sun so that it retains the green colour and its attractive aroma. When the plants are completely dried, the seeds should be separated by thrashing. Then the seeds are cleaned by winnowing or with the help of sieve or mechanically by vibrator followed by grading. The average yield of rainfed crop is about 500 to 800 kg coriander per hectare and that of the irrigated crop is about 1200 to 2000 kg per hectare.

11.0 Crop Improvement

The breeding methods suitable for coriander depend on the pollination biology. Coriander is a facultative cross-pollinator and after selfing the studied plants did not show obvious inbreeding depression. Thus the species behaves similarly to rape (*Brassica napus* L.), and the described breeding techniques for rape, which are well-known, can be applied to coriander too. A considerable amount of breeding work in coriander has been done in the former Soviet Union. These activities started in the area of Vorone and involved close cooperation with the germplasm collection of the Vavilov Institute (VIR). The VIR has had a department for essential oil plants until 1938, and a number of the fundamental results from their eco-geographical and chemical investigations have been published (Stoletova, 1931).

The Caucasian centre of diversity was the main source for selecting interesting types from the landraces, which have been famous cultivars for many years in the Soviet Union (Lukjanov and Reznikov, 1976). The selection work itself was done close to the town of Alekseevka, 150 km south of Vorone. Later, research on essential oil plants was based in the town of Simferopol on the Crimea. The institute in Simferopol still exists and now belongs to the Ukraine; in 1995, a breeding variety of coriander for autumn sowing that the institute had developed was put on the Ukrainian and Russian markets (Silcenko, 1995). The research and breeding work in this institute concentrated on the quality of the essential oil of the fruits (Cernodubov and Berestovaja, 1986), winter hardiness (Sergeeva and Silcenko, 1984; Romanenko and Nevkrytaja, 1988), resistance to *Ramularia coriandri* (Silcenko *et al.*, 1984; Cislova, 1988; Romanenko *et al.*, 1990b) and drought resistance (Karpov *et al.*, 1990), as well as on the breeding techniques themselves (Silcenko, 1980a, 1980b; Romanenko *et al.*, 1992). A cytoplasmatic and a nuclear type of male sterility in coriander have been found by researchers of this institute, which can be used in breeding (Romanenko and Svecnikova, 1988).

In the former Soviet Union, the coriander for vegetable use has not been subject to breeding programmes, but the characterization data of the germplasm collection of the VIR in St. Petersburg have been published for a total of 139 accessions (Girenko and Cytovic, 1980; Girenko, 1992). Breeding and screening for essential oil use was done by the VNIIEMK4, and the VIR concentrated on genetic resources and also on aspects of the vegetable's use (Alborisvili, 1984; Girenko, 1980). Today, both institutes involved belong to different states, *i.e.* the Ukraine and Russia, and the future of both institutes seems to be uncertain. At present the situation is problematic because the breeding experts and optimal environmental conditions for coriander are located in the Ukraine, while the factories for processing are located in Russia. The cooperation between these facilities has become difficult after the crash of the former Soviet Union (Silcenko, 1996, pers. comm.).

The breeding of coriander has also long been practised in India (Dimri *et al.*, 1977; Suthanthirapandian *et al.*, 1980; Sethi, 1981; Agrawal *et al.*, 1990; Bhandari and Gupta, 1991). Several studies of genetic diversity in coriander have been carried out, and coriander germplasm collections have been screened for resistance breeding. Several East European countries have bred varieties of coriander, but the interest in this crop has apparently diminished, as there are only few recent publications on this topic. In Egypt some breeding research on coriander was done by El-Ballal and El-Nasr (1987). This concentrated on the description of phenological earliness of the plants and the medicinal properties of the fatty oil in the fruits. In Germany, two companies have been active in coriander breeding, who were mainly interested in the fatty oil (Meier zu Beerentrup and Röbbelen, 1987; Schulenburg *et al.*, 1991). At present only one of them is continuing this work, because the petroselinic acid of this crop is of little commercial interest. Breeding in Germany tries to combine fatty oil content with essential oil content and resistance to bacterial blight.

Large variability exists in the yield and the yield contributing characters in coriander. The characters such as days to flowering, days to maturity and number of secondary branches are to be taken into account while selecting parents for hybridization. The methods of breeding that can be attempted are selection, hybridization and mutation breeding.

Crop improvement of coriander basically aims at the following objectives:

- To evolve high yielding varieties
- To induce earliness
- To develop varieties with higher essential oil content
- To enhance biotic stress tolerance (disease and pest)

- To enhance abiotic stress tolerance
- To develop varieties adaptable for different agro-climatic regions

Selection

Early ecotypes are selected in dense culture under stress through short-cut selection. Mother plants selection is practiced in the full-sibbed population in the oscillated environment. This includes several lines promising in the essential oil content, essential oil yield, crude fat and protein contents per cent as well as fruit yield per plant. The selected population exhibits apparent stability in different studied characters measured through stability against micro environment, as well as flexibility in interactions with seasonal variation which enables maximizing fruit yield through agro-mechanics. López *et al.* (2008) reported that special attention should be paid to the positive correlation between essential-oil content and plant yield, because it means that it should be possible to increase the amount of seed produced and the concentration of essential oil simultaneously in a breeding program. Such improvements need a well organized and systematic program which is founded on sound knowledge of flowering and pollination biology.

Hybridization

Coriander is the most important seed spice crop in India. The full potential of the crop could not be exploited due to failure to take up systematic gene transfer, recombination, linkage-breaking, heterosis, maintenance of certain lines, maintenance of diversity in the gene pool, evaluation of parental lines and genetic analysis. Many times the problems associated with the crop were a need for crop improvement interventions through crossing. Despite the identification of several desirable traits and resistance donors, breeding of varieties is restricted to selection alone due to absence of crossing technique. This was mainly due to herbaceous and delicate nature of the umbels, umbellets and florets thus making the crop difficult or less amenable for crossing (Diederichsen, 1996). Giridhar *et al.* (2016) conducted a study to promote crossing as a breeding technique in coriander. It was found that coriander was protandrous and the sex ratio was 1.75 to 2.07. The stigma was receptive from the 3rd day to 6th day of anthesis indicating that artificial pollination on emasculated florets must be done repeatedly on 3rd and 4th day of anthesis. Pollen was viable for two days in the field. Fresh pollen can be stored in incubator at 25°C for three weeks with 88 per cent retention of viability. Emasculation of florets was carried out with the help of binocular loupe. Emasculation either in the morning before anthesis or previous day evening was found to be suitable. For demonstrating the technique, four parents were reciprocally crossed. Only three crosses were successful indicating the differences in combining ability of the parents. Mean success among the crosses was 23 per cent, fruit set among the florets varied from 9.52-83.3 per cent depending on the cross combination.

Mutation Breeding

Variation is the prerequisite for any improvement programme, there can be no improvement in the absence of variation. Mutation breeding is more advisable for coriander because the exploitation of exotic variability through recombination breeding is tedious and hybridization is difficult due to the tiny size of its flowers. In a study conducted with 60 genotypes of coriander, reported wide range of variability for plant height and number of umbels per plant whereas, low and significant variability was recorded for umbel lets per umbel (Suthanthirapandian *et al.*, 1980).

Chaudhary and Ramkrishna (2003) made an attempt on mutation breeding programme with selective mutagenic doses in three coriander varieties of *Coriandrum sativum* L to investigate the polygenic variation in M_4 progenies for yield and yield attributes. Results indicated the existence of influence of genetic constitution of the parents on the fate of their progenies in advanced generation. Most progenies of RCr-20 and RCr 41 seemed to have undergone inbreeding depression whereas those of RCr-436 have frequently out yielded the control. It is remarkable that in case of RCr-436, the parent and M_4 progenies have shown a high magnitude of C.V. for yield per plant and umbels/plant as well.

Kumar (2005) treated the seeds of coriander with different concentrations of EMS (0.2-1.0 per cent). The percentage of germination was recorded after 15 days of treatment. Percentage of germination was also observed for M_2 generation which showed decrease with increasing concentration of mutagen. The decrease in percentage of germination might be due to chromosomal damage resulting in defective enzyme production. The recovery in M_2 generation was probably due to elimination of defective cells.

Salve and More (2014) treated seeds of *Coriandrum sativum*, L. (Var. CS-287) with different doses of gamma rays (10kR, 20kR, 30kR and 40kR). Seed germination percentage revealed gradual decrease from lower doses to higher doses in given treatments of gamma rays. In case of seedling height there was dose dependent reduction after treatment of gamma rays. They concluded that gamma rays was capable in inducing damage to plants at molecular level and was capable of inducing mutation.

EL-Mouhty *et al.* (2014) studied the effect of gamma radiation on the structure and concentration analysis of chlorophyll and carotene of *coriander* and concluded that higher doses (20, 40 Gy) made different effect of radiation on the structure of carotene and chlorophyll while lower doses does not make effect.

Jafri *et al.* (2013) investigated the mutagenic effects of individual and combination treatment of gamma rays and EMS on *Corriandrum sativum* L. Result revealed the meiotic malfunctioning of pollen mother cells (PMCs) that had shared various types of cytological abnormalities such as univalents, multivalents, stickiness, precocious movement, stray bivalent, non-orientation, cytomixis, laggard, bridges, unequal separation, micronuclei and disturbed polarities. Increasing tendency of abnormality percentage along with increasing dose/concentration of mutagens was registered and asymmetrical distribution of chromatin material in PMCs had definitely affected the pollen fertility and resulted in increased pollen sterility.

Sarada *et al.* (2015) studied the nature and extent of genetic variability induced by three doses of gamma irradiation, three doses of EMS and combination of gamma rays with EMS in coriander cultivar Swathi. Of the twelve characters studied, induced variability was higher for number of primary and secondary branches, leaf yield, number of leaves, number of umbels and yield per plant suggesting scope of improvement through selection in these characters.

Breeding for Development of Leaf Type Coriander

Research work at Indian Institute of Horticultural Research has resulted in development of a leafy coriander cultivar Arka Isha with a high yield potential and good aroma. It is a multicut cultivar where the plants are bushy, leaves are broad and leaf lobes are short, and the cultivar is late flowering. Yield is 3.74t ha^{-1} by pulling at 40 days after sowing, and 11.98 t ha^{-1} by cutting. Leaves have 167.05 mg $100g^{-1}$ of vitamin C with good aroma and keeping-quality (Varalakshmi *et al.*, 2012).

12.0 Biotechnology

Biotechnology and genetic engineering have hitherto been unimportant for coriander breeding, but some techniques have been developed. Despite its commercial importance, only a few *in vitro* culture studies have been published. These studies include reports of plant regeneration via somatic embryogenesis and production of petroselinic acid in suspension cultures (Kim *et al.*, 1996). Liu *et al.* (2002) also reported on *in vitro* culture and the production of secondary metabolites in coriander. Methods for tissue culture and micropropagation of coriander have been developed in India (Mujib *et al.*, 1990) and in Russia (Kataeva and Popovic, 1993). In China, somatic embryos of coriander have been used successfully in the creation of artificial seeds (Chen *et al.*, 1992). In the USA, genetic material from coriander has been successfully introduced in tobacco and has affected its fatty acid biosynthesis (Cahoon *et al.*, 1992). In the United Kingdom, rape (*B. napus*) plants have been transformed with genetic material from coriander in order to obtain petroselinic acid from rape (Murphy *et al.*, 1994).

Hypocotyl segments and zygotic embryos of coriander formed embryogenically at frequencies of up to 75 per cent when cultured on MS medium supplemented with 1 mgl^{-1} 2,4-D. Calli were transferred to MS liquid medium with 1 mgl^{-1} 2,4-D to initiate cell suspension cultures. Embryogenic cells became finely dispersible in the medium as the subculture proceeded. Cultures were transferred to a nitrogen compound enriched liquid MS medium containing 2 per cent sucrose and 0.1 mgl^{-1} 2,4-D, and cultured two weeks before plating on MS basal medium. Approximately 75 per cent of cell aggregates (1 to two mm in diameter) underwent development into globular to cotyledonary somatic embryos after two weeks of planting. Most of the embryos were subsequently regenerated into plantlets. Regenerates were successfully transplanted to potting soil and grown to maturity in a phytotron (Suk *et al.*, 1996).

Song *et al.* (2020) reported about a high quality assembly and analysis of its genome sequence, anchored to 11 chromosomes, with total length of 2118.68 Mb and N50 scaffold length of 160.99 Mb. We found that two whole genome duplication events, respectively, dated to ~45–52 and ~54–61 million years ago, were shared by the Apiaceae family after their split from lettuce. Unbalanced gene loss and expression are observed between duplicated copies produced by these two events. Gene retention, expression, metabolomics and comparative genomic analyses of terpene synthase (TPS) gene family, involved in terpenoid biosynthesis pathway contributing to coriander's special flavour, revealed that tandem duplication contributed to coriander TPS gene family expansion, especially compared to their carrot counterparts. Notably, a TPS gene highly expressed in all 4 tissues and 3 development stages studied is likely a major effect gene encoding linalool synthase and myrcene synthase. The present genome sequencing, transcriptome, metabolome and comparative genomic efforts provide valuable insights into the genome evolution and spice trait biology of Apiaceae and other related plants, and facilitated further research into important gene functions and crop improvement.

13.0 References

Agarawal, S. and Sharma, R.K. (1990) *Indian Cocoa, Arecanut and Spices J.*, **13**: 127–129.

Agrawal, S., Sharma, R.K. and Bhatt, B.N. (1990) *Indian Cocoa, Arecanut and Spices J.*, **13**: 137-138.

Alborisvili, C.A. (1984) *Naucn. techn. bull. VIR im. Vavilova*, **141**: 53-56.

Alefeld, Fr. (1866) In: *Landwirthschaftliche Flora*, Wiegandt and Hempel, Berlin. p. 165.

Anonymous (2020a) The Economic Times. https: //m.economictimes.com/ markets/commodities/views/coriander- hits-2-year-high-on-improving-exports-lower-production/articleshow/68854011.cms

Anonymous (2020b) https: //floraspices.com/blogs/market-coriander.

Anonymous (2001) Twenty fifth Annual Report, AICRP, SKN Collage of Agriculture, Jobner Rajasthan.

Anonymous (2002) All India Coordinated Research Project on Spices, Twenty Ninth Annual Report, pp. 47-53.

Aslam, M. Sultana, B. Ali, S. and Khalil, R.U.R. (2014) *Asian J. Chem.*, **26**(16): 5105-5109.

Aymen, E. M. and Cherif, H. (2013) *Acta Agriculturae Slovenica*, **101**(1): 41-47.

Baboo, R. and Rana, N.S. (1995) *Indian J. Agron.*, **40**: 253 – 255.

Ben, Fredj, M., Zhani, K., Hannachi, C., Mehwachi, T. and Belanger, A. (2014) *J. Stress Physiol. Biochem.*, **10**(3): 84-109.

Bhandari, M.M. and Gupta, A. (1991) *Euphytica*, **58**: 1-4.

Bhati, D.S. (1988). *Indian J. Agron.*, **33**: 242 – 244.

Cahoon, E.B., Shanklin, J. and Ohlrogge, J.B. (1992) *Proc. Nat. Acad. Sci.* USA, **89**: 11184-11188.

Cernodubov, A.I. and Berestovaja, M.M. (1986) Sel. Semenovod. (Moskva), pp. 23-24.

Chaudhary, Pradeep and Ramkrishna, K. (2003) *Indian J. Genet.*, **63**(2): 181-182.

Chen, R.R., Zhang, J.T., Li, B.P., Guo, S.S., Hao, J.P. and Zhou, X.M. (1992) *Chin. J. Biotechnol.*, **7**: 127-134.

Cislova, L.S. (1988) *Sel. Semenovod.* (Moskva), p.33.

Coskuner, Y. and Karababa, E. (2007) *J. Food Eng.*, **80**: 408-16.

Dadiga, A. and Jain, P.K. (2017) *Agric. Update*, **12** (TECHSEAR-8): 2202-2206.

Diederichsen, A. (1996) Coriander (*Coriandrum sativum* L.): promoting the conservation and use of underutilized and neglected crops. 3. Institute of Plant Genetics and Crop Plant Research, Gatersleben/International Plant Genetic Resources Institute, Rome.

Dimri, B.P., Khan, M.N.A. and Narayan, M.R. (1977) *Indian Perfumer*, **20**(1A): 14–21.

El-Ballal, A.S.I. and El-Nasr, T.H.Abou (1987) *Acta Hortic.*, **208**: 179-192.

EL-Mouhty, N.R.A., Gad, H.M.H., EL-Naggar, A.Y., Hassan, R.F. and Abdel Latif, N.A. (2014) *Int. J. of Inov. Res. Sc., Engg. Tech.*, **3**: 8101-8105.

Gastón, M.S., Cid, M.P., Vázquez, A.M., Decarlini, M.F., Demmel, G.I., Rossi, L.I., Salvatierra, N. A. (2016) *Pharmaceutical Biol.*, **54**(10): 1954–1961.

Girenko, M.M. (1980) *Bull. VIR im. Vavilova*, **101**: 14-18.

Girenko, M.M. 1992. In: Katalog mirovoj kollekcii VIR, vypusk 642, vidovoe raznoobrazie zelënnych ovo nych kul'tur (Koriandr, reven'), (Ed M.M. Girenko). VIR, St. Petersburg. pp. 4-37.

Girenko, M.M. and Cytovic, K.I. (1980) In: Katalog mirovoj kollekcii VIR, vypusk 287, vidovoe raznoobraznie zelënnych ovo nych kul'tur (Ed. D.D. Breznev), VIR, Leningrad. pp. 121-152.

Giridhar, K., Suryakumari, S. and Sarada, C. and Naidu, L.N. (2016) *J. Spices Aromatic Crops*, **25** (1): 1-6.

Ilyas, M. (1980) *Econ. Bot.*, **34**: 236-259.

Issac, S. and Vijayakumar, A. (2014) *IOSR J. Agric. Veterinary Sci.* **7**(7): 61-63.

Ivanova, K.V. (1966) *Tr. Po prikl. bot., gen. i sel.*, **28**: 178-185.

Ivanova, K.V. and Stoletova E.A. (1990) *Sb. nau n. tr. prikl. bot., gen. i sel.*, **133**: 26-40.

Jafri, I.F., Khan, A.H. and Gulfishan, M. (2013) *Int. J. Bot. Res.*, **3**: 79-85.

Kataeva, N.V. and Popovic, E.A. (1993) *Plant Cell Tissue Organ Cult.*, **34**: 141-148.

Kim, S.W., Park, M.K., Bae, K.S., Rhee, M.S. and Liu, J.R. (1996) *Phytochem.*, **42**: 1581–1582.

Kumar A. (2005) *Int. J. Mendel*, **22**: 29.

Lal, G., Saran, P.L., Devi, G., Bijarnia, D. and Raj R.(2014) In: Advances in vegetable agronomy (Eds. A.K. Chaudhary, K.S. Rana, A. Dass, M. Srivastav), PGS, IARI and DARE, ICAR, New Delhi, pp. 214-222.

Lal, G., Metha, R. S., Singh, R.M., Choudhary, K. and Maheria, S. P. (2017) *Intl. J. Seed Spices*, 7(2): 14-18.

Lamberts, M. (1990) In: *Advances in new crops* (Eds.J.S. Janick and J.E. Simon), Proceedings of the first national symposium on New crops: research, development, economics", Indianapolis, Indiana, USA, 23-26 October 1988. Timber Press, Portland, pp. 378-387.

Laribi, B., Kouki, K., M'Hamdi, M. and Bettaieb, T. (2015) *Fitoterapia*, 103: 9-26.

Liu, J. R., Kim, S.W. and Oh, S.C. (2002) In: *Biotechnology in Agriculture and Forestry, Vol. 51 Medicinal and Aromatic Plants XII* (Eds. T. Nagata and Y. Ebizuka), Springer-Verlag Berlin, Heidelberg, pp. 13-22.

López, P.A., Widrlechner, M.P., Simon, P.W., Rai, S., Boylston, T. D., Isbell, T.A. and Wilson, L.A. (2008) *Genet. Resour. Crop Evol.*, **55**: 247–275.

Lörincz, K. and Tyihák, E. (1965) *Herba Hung.*, **4**: 191-208.

Luk'janov, I.A. and Reznikov, A.R. (1976) In: Efirnomaslicnye kyl'tury. (Eds. A.M. Smoljanova and A.T. Ksendza,), Kolos, Moskva. pp. 9-57.

Malhotra, S.K., Vashishtha, B.B. and Apparao, V.V. (2006) *J. Spices and Aromatic Crops*, **15** (2): 115-117.

Malhotra, S.K., Mehta, R.S., Meena, S.S. and Anwer, M.M. (2009) Paper presented in World Aqua Congress held at India habitat Center, from 2-4, December, 2009.

Malhotra, S.K. (2016) *Indian J. Arecanut, Spices and Medicinal Plants*, **18** (1): 3-19.

Matasyoh, J.C., Maiyo, Z.C., Ngure, R.M. and Chepkorir, R. (2009) *Food Chem.*, **113**(2): 526–529.

Menon Rema, Khader, M.A. and Menon, R. (1997) *Indian Cocoa, Arecanut, Spices J.*, **21**: 74 – 75.

Meier zu Beerentrup, H.M. and Röbbelen, G. (1987) *Angew. Bot.*, **61**: 287-303.

Mujib, A., Bandyopadhyay, S., Banerjee, S. and Gosh, P.D. (1990) *Hortic. J.*, **3**: 59-62.

Murphy, D.J., Richards, D., Taylor, R., Capdevielle, J., Guillemot, J.C., Grison, R., Fairbairn, D. and Bowra, S. (1994) In: Second European symposium on industrial crops and products, Pisa, Italy, 22-23 November 1993. *Indus. Crops Prod.* (special issue) **3**: 17-27.

Nambiar, V.S., Daniel, M. and Guin, P. (2010) *J. Herb. Med. Toxicol.*, **4**(1): 173–177.

Panda, M.R., Chatterjee, R., Pariari, A., Chattopadhyay, P.K., Sharangi, A.B. and Alam, K. (2007) *Indian J. Hort.*, **64**(3): 369-371.

Prakash, V. (1990) In: *Leafy Spices*, CRC Press Inc., Boca Raton. pp. 31-32.

Purseglove, J.W., Brown, *E.g.*, Green, C.L. and Robbins, S.R.J. (1981) *Spices*, Vol. II. Longman, London and New York, pp.736–88.

Ramanujam, S., Joshi, B.S. and Saxena, M.B.L. (1964) *Indian J. Genet.*, **24**(1): 62–7.

Rithichai, P., Sampantharat, P. and Jirakiattikul, Y. (2009) *Asian J. Food and Agro Industry*, (Special Issue), **2**: S217-S221.

Romanenko, L.G. and Nevkrytaja, N.V. (1988) *Sel. Semenovod.* (*Moskva*), pp. 37-39.

Romanenko, L.G., Nevkrytaja N.V. and Serkov L.N. (1990b) In: Simpozium Osnovnye napravlenija nau nych issledovanij po intensifikacii efirnomasli nogo proizvodstva, Kisinev, 17. - 19. sentjabrja 1990 g. Tezisy dokladov, Simferopol', pp. 40-41.

Romanenko, L.G., Nevkrytaja, N.V. and Kuznecova, E.J.U. (1992) *Sel. Semenovod.* (Moskva) pp.25-28.

Romanenko, L.G. and Svesnikova, T.V. (1988) *Tr. VNII efirnomasli n. kul'tur*, **19**: 20-27.

Salve, K.M. and More, A. D. (2014) *Int. J. Life Sci.*, **2**: 223-225.

Sarada, C., Srinivasa Rao, V., Umajyothi, K. and Reddy, P.V. (2015) *Int. J. Pure App. Biosci.*, **3** (2): 440-444.

Schulenburg, H. Graf von der, Meier zu Beerentrup, H.and Toben, H. (1991) In: Schriftenreihe des Bundesministers für Ernährung, Landwirtschaft und Forsten, Reihe A, Angewandte Wissenschaft, Sonderheft. (Eds. Bundesamt für Ernährung und Forstwirtschaft), Landwirtschaftsverlag GmbH, Münster-Hiltrup, pp. 411-421

Sergeeva, D.S. and Sil' enko. V.M. (1984) *Fiziol. Biokhim. Kul't. Rast.*, **16**: 52-55.

Sethi, K.L. (1981) *Indian J. Agric. Sci.*, **51**: 52-53.

Sharangi, A.B. and Roychowdhury, A. (2014) *The Bioscan*, **9**(4): 13-20.

Sharangi, A.B., Chatterjee, R. Nanda, M.K. and Kumar, R. (2011) *J.Plant Nutri.*, **34** (12): 1762-1768.

Sharangi, A.B., Thapa, U., Pariari, A., Mandal, A.R., Chatterjee, R. and Sivakumar, T. (2005) *Leg. Res.*, **28**(3): 184-188.

Silcenko, V.M. (1980a) In: Tez. dokl. 3-ogo simpoz. aktual'n. vopr. izuc. i ispol'z. efirnomasli n. rast. i efirn. masel, 1980, Simferopol, pp. 63-64

Silcenko, V.M. (1980b) *Sel. Semenovod.* (Kiev), **46**: 46-50.

Silcenko, V.M. (1995) In: *Pervyj me dynarodnyi simpozium*,Novye netradicionnye rastenija i perspektivy ich prakti eskovo ispol'zovanija, tezisy dokladov (Eds. Rossiskaja Akad, Selskochoz. Nayk, Rossiskaja Akad, Nauk), Ministerstvo nauki R. F., Puscino, pp. 452-453.

Silcenko, V.M., Zerebcova, V.G., Vlasov, M.I., Levcenko, I.V., Girenko, M.M., Borodnik, A.S. and Cislova, L.S. (1984) *Tr. VNII efirnomasli n. kul'tur*, **16**: 40-46.

Simon, J.E. (1990) In: Advances in New Crops (Eds. J. Janick and J.E. Simon). Proceedings of the first national symposium on New crops: research, development, economics, Indianapolis, Indiana, USA, 23-26 October 1988. Timber Press, Portland, pp. 472-483.

Singh, V.P and Ramanujam, S. (1973). *Euphytica*, **22**: 181–8.

Singh, D. K., Verma, T. C., Khan M. A. and Gautam, K. M. (2018) *Int. J. Seed Spices*, **8**(2): 58-60.

Sinskaja, E.N. (1969) Istori eskaja geografija kul'turnoj flory (na zare zemledelija). Izdatel'stvo ,Kolos, Leningrad. p. 480.

Song *et al.* (2020) *Plant Biotechnology Journal*, **18** (6): 1444-1456.

Stoletova, E.A. (1930) *Tr. po prikl. bot. gen. i sel.*, **23**: 290-291.

Stoletova, E.A. (1931) Koriandr. Vsesojuznaja akademija sel'skogo-chosjajstvennych nauk imeni Lenina izdanie instituta rastenievodstva, Leningrad.

Suthanthirapandian, I.R., Shaw H.A. and Muthuswami, S. (1980) *Madras Agric. J.*, **67**: 450-452.

Telci, I., Gul, T.O. and Sahbaz, N. (2006) *J. Essent Oil Res.*, **18**: 189–193.

Thakral, K. K., Singh, G. R., Pandey, U. C. and Srivastava, V. K. (1992) *Haryana Agric. Univ. J. Res.*, **22** (1): 35-39.

Thapa, U. (1999) *Ph.D. thesis* submitted to Bidhan Chandra Krishi Viswavidyalaya, Mohanpur, Nadia, West Bengal.

Umarani, R. and Vigneshwari, R. (2010) *Principles and Practices*, **1**: 118.

UN COMTRADE (2020) United Nations Commodity Trade Database. https: // comtrade.un.org/ db/syslognin.aspx?ReturnUrl= per cent 2fdb per cent 2fmr per cent 2frfCommoditiesList.aspx per cent 3fpx per cent 3dH1 per cent 26cc per cent 3d090920 and px=H1 and cc=090920.

USDA National Nutrient Database for Standard Reference Release 26 Full Report (All Nutrients) Nutrient data for 2013, Spices, coriander seed.

Vavilov, N.I. (1992) In: Origin and Geography of Cultivated Plants (Ed.V.F. Dorofeev), Cambridge University Press, Cambridge, pp. 316-366.

Yeung, E.C. and Bowra, S. (2011) *Botany*, **89**: 263-73.

Zeven, A.C. and de Wet, J.M.J. (1982) In: *Dictionary of Cultivated Plants and Their Regions of Diversity - Excluding Most Ornamentals, Forest Trees and Lower Plants.* Centre for Agricultural Publishing and Documentation, Wageningen.

PURSLANE

T.K.Maity and P.P. Mohapatra

1.0 Introduction

Purslane or Portulaca (*Portulaca oleracea* L.) belongs to the family Portulacaceae is a dwarf creeping annual herb with small fleshy leaves. It is a highly variable common weed that grows wild during summer in many parts of the world. The plant name Portulaca is derived from the Latin word *portula*, which means a diminutive porta or gate, referring to the covering of the seed capsule that opens like a small gate. It is known by a variety of names in different places, like Pigweed, garden purslane, common purslane, postelijn, pourpier, fatweed, munyeroo, portulat, ma chi xian, Miner's-lettuce, Cuban spinach, Miner's-salad, little hogweed, Pussle *etc.* In India purslane is known by a variety of names including sanhti, paruppu keerai, gangabayala kura, punarva, and kulfa. It is also a highly nutritious vegetable that can be eaten raw or cooked. Many varieties of purslane under different names grow in a wide range of climates and regions. Purslane has wide acceptability as a potherb in Central Europe, Asia, and the Mediterranean region. It is an important component of green salad and its soft stem and leaves are used raw, alone, or with other greens. It is also used for cooking or used as a pickle.

2.0 Composition and Uses

2.1.0 Composition

Purslane deserves special attention from agriculturalists as well as nutritionists. Recently, research demonstrates that purslane has better nutritional quality than the major cultivated vegetables, with higher beta-carotene, ascorbic acid, and alpha-linolenic acid. Additionally, purslane has been described as a power food because of its high nutritive and antioxidant properties. Different varieties, harvesting times, and environmental conditions can contribute to its nutritional composition and benefits. Purslane is a very good source of alpha-linolenic acid. Alpha-linolenic is an Omega-3 fatty acid which plays an important role in human growth and development and in preventing diseases. Purslane has been shown to contain five times higher Omega-3 fatty acids than spinach. It is rich in vitamin A which is a natural antioxidant value. It can play role in vision healthy mucus membranes and to protect from lung and oral cavity cancer. Purslane contains the highest content of vitamin A among green leafy vegetables. It also contains vitamin C and B-complex vitamins like riboflavin, niacin, and pyridoxine. It provides the highest dietary minerals such as potassium followed by magnesium, calcium, phosphorus, and iron. With just 16 calories per 100 g, purslane is an extremely low-calorie food. It contains two types of betalain alkaloid pigments, the reddish betacyanins (visible in the colouration of the stems) and the yellow betaxanthins (noticeable in the flowers and in the slight yellowish cast of the leaves). Both of these pigment types are potent antioxidants and have been found to have antimutagenic properties in laboratory studies. The composition of purslane leaves is given in Table 1.

Table 1: Chemical Composition of *Portulaca oleracea* Leaves*

Constituent	*Content*	*Constituent*	*Content*
Moisture (per cent)	90.5	Carotene (mg/100 g)	2,292.0
Protein (per cent)	2.4	Thiamine (mg/100 g)	0.10
Fat (per cent)	0.6	Riboflavin (mg/100 g)	0.22
Fibre (per cent)	1.3	Niacin (mg/100 g)	0.70
Carbohydrates (per cent)	2.9	Vitamin C (mg/100 g)	29.0
Calcium (mg/100 g)	111.0	Isoleucine (g/g N)	0.35
Phosphorus (mg/100 g)	45.0	Valine (g/g N)	0.28
Iron (mg/100 g)	14.8	Leucine (g/g N)	0.29
Magnesium (mg/100 g)	120.0	Lysine (g/g N)	0.45
Sodium (mg/100 g)	67.2	Trypotophan (g/g N)	0.03
Potassium (mg/100 g)	716.0	Phenylalanine (g/g N)	0.09
Copper (mg/100 g)	0.9	Methionine (g/g N)	0.02
Sulphur (mg/100 g)	63.0	Threonine (g/g N)	0.12
Sodium (mg/100 g)	73.0		

* Gopalan *et al.* (1999).

However, purslane contains oxalates, which can reduce the absorption of some minerals and contribute to the formation of kidney stones. Oxalates also have anti-nutrient properties, meaning that they may interfere with the absorption of minerals like calcium and magnesium. Purslane grown in the shade may have higher levels of oxalates, compared to plants readily exposed to sunlight (https://www.healthline.com/nutrition/purslane).

2.2.0 Uses

The leaves are, thick, fleshy, palatable, have a mild flavour and a mucilaginous quality. Tender stems and leaves can be eaten raw, alone or with other greens. It can also be cooked or pickled. The leaves of purslane can be frozen or dried and stored in jars for year-round usage. Purslane has been used medicinally for thousands of years. It has been used as a remedy for arthritis, and for treatment of respiratory and circulatory problems. It is believed to lower blood pressure and cholesterol levels, and to help avoid blood clots.

It appears to be an excellent candidate for inclusion in saline drainage water reuse systems (Grieve and Suarez, 1997). It is highly tolerant of both chloride- and sulphate-dominated salinities, is a moderate selenium accumulator and a valuable vegetable crop for human consumption (Bianco *et al.*, 1998) and for livestock forage. It is also a source of gum with emulsification properties that can be used in food industry (Garti *et al.*, 1999).

It is sometimes used as fodder and is fed to poultry to reduce egg cholesterol. According to ASPCA (American Society for the Prevention of Cruelty to Animals), purslane contains soluble calcium oxalates, which are poisonous to cats, dogs and

Uses of Purslane

Different Recipes of Purslane in Indian Cuisine

horses, Cats that ingest part of the plant may drool, vomit, or show other signs of digestive stress such as diarrhoea or bloody urine.

3.0 Origin and Distribution

Purslane is a cosmopolitan weed occurring especially in warm areas. The origin of its cultivation is uncertain, possibly western Asia or India. It is one of the oldest leafy vegetables, used from Europe to Japan, Australia and the Americas. It is eaten in many African countries. It is especially popular in Sudan and Egypt, where it is known in Arabic as 'rigla'.

Purslane grows from sea level to 2600 m (Vengris *et al.*, 1972) and is most common in the temperate and subtropical regions, although it extends into the tropics and higher latitudes. Common latitudes are between 45°N and 40°S, with extension to 58°N in North America and 54°N in Europe (Matthews *et al.*, 1993). There is evidence that purslane has been eaten for at least 2,000 years; it was cultivated in ancient Egypt and was enjoyed by the ancient Romans and Greeks. It was known to the Arabs in the medieval period and may have been cultivated in Europe as early as the 13th century. This plant is widespread all over the globe barring some places in Australia. North of Africa to the Middle East sweeping over the Indian subcontinent to Australasia- a wide growth of this plant can be witnessed.

4.0 Taxonomy and Cultivars

4.1.0 Taxonomy

Portulaca oleracea (purslane) is an ancient, cosmopolitan species in which self-fertilization is the rule. Hence, local populations exist which reflect variable morphological and physiological traits expressed as part of the genome of that population. However, purslane has not been split into a series of microspecies.

The two most recent comprehensive monographs, Legrand (1962) and Geesink (1969) provide, respectively, only three and two names as synonyms. Gorske *et al.* (1979) conducted a numerical taxonomic analysis of 36 morphological characters on 44 ecotypes from 18 countries, which included the cultivated form commonly known as *Portulaca oleracea* var. *sativa*. They found three morphological groups that form a clime: (1) cool temperate, (2) warm temperate to sub-tropic and (3) humid sub-tropic to tropic. They did not propose that these groups should receive nomenclatural recognition.

Danin *et al.* (1978) proposed nine subspecies, on the basis of seed size, seed surface pattern and chromosome number. However, it may be that these subspecies could be expanded almost ad infinitum on the basis of the variety of seed surface patterns that can be found in the species (Matthews *et al.*, 1993). Legrand (1962) stated that seed size, sepal wings and the number of stamens are environmentally influenced and that seed surface pattern was not predictable or dependable as a taxonomic character because of the infinite transition patterns that are present.

In a chemotaxonomic study comparing proteins and free amino acids, Prabhakar and Ramayya (1988) found that, within the complex *P. oleracea*, the var. *ophemera* is distinct from vars. *oleracea* and *sativa*.

Matthews *et al.* (1993) concluded that *P. oleracea* exists as a polymorphic species and is not readily divisible into subspecies on the basis of seed surface patterns, chromosome number or other morphological traits that are subject to environmental influences.

Cytologically, *P. oleracea* is characterized by a sequence of polyploids, from a base number of x = 9. There are diploid races (2n = 2x =18) in Africa, Central and North America; tetraploid races (2n = 4x = 36) in India, Central and North America and hexaploid races (2n = 6x = 54) in India, Africa, Europe, North America and Hawaii. There are additional reports of 2n = 45 in India and 2n = 52 in Japan.

The purslane is a succulent herb with stems that may grow erect or prostrate, very branched from the base, developing mats centred on the main root. It can measure 60 cm in diameter.

Roots: Main taproot and adventitious roots which are able to develop from twigs.

Stems: It has smooth fragile stems rising from the ground in multiple forms. Stems display a mix of green and red colours.

Leaves: Its green juicy leaves look like a spoon. They follow a typical pattern of arrangement where they are grouped around the joint of the stem. They are around 1 cm broad and 3 cm long, with a slight tapering at the base.

Flowers: The yellow flowers of purslane occur singly at leaf axils or stem tips. These yellow flowers are self-fertile. They have four to five petals positioned at the centre of the clusters.

Fruit: Fruits are a dehiscent capsule of spherical shape, surmounted of the base of 2 sepals. It is long from 4 to 8 mm and contains numerous seeds.

Seeds: A single plant may produce more than 200,000 seeds per plant within 40 days. The seeds are black, slightly wrinkled and glossy in appearance in small capsules that resemble flower buds.

4.2.0 Cultivars

There are no real cultivars of purslane, but there are different cultivars that can be distinguished by the colour of their leaves. The varieties with strong green leave have a slightly coarser consistency but a better taste. The yellowish-green forms form softer, less aromatic leaves. However, most commonly available cultivars are 'Gruner Red' and 'Goldberg'

Common

The typical garden cultivar, common purslane (species *P. oleracea*) grows low along the ground and spreads up to 18 inches once mature.

Golden Purslane

The Golden Purslane (*Portulaca sativa*) is a cultivar of purslane with yellow leaves. The seeds of an individual plant have been known to produce both green and golden leaved plants. It grows to a mature size of about 10 inches tall.

Goldberg Golden

It is rich in vitamin C and the highest known plant source of omega-3 fatty acid. Succulent leaves and stems that lie close to the ground with a taste similar to watercress or spinach. Harvest young tender leaves and stem tips or cut 2 inches from ground and bunch.

Goldgelber

Goldgelber (*P. oleracea*) matures in 26 days, and spreads up to 12 inches wide. It grows up to 6 inches tall when mature.

Gruner Red

Gruner Red (*P. oleracea*) has pink-tinted stems, much like the common purslane. It has thick, oval-shaped, green leaves of about an inch long, and mature size of up to 12 inches tall.

5.0 Climate and Soil

Purslane can grow in a wide range of climates and regions. It grows well in full sun in almost any soil, from much high in organic matter to heavy clay. It does best in warm weather, and young plants will remain small and stunted when conditions are cool. Although it prefers regular water it can tolerate drought. It is easily dug or hoed out where unwanted but these plants should be removed from the garden as cut stems from larger plants will root readily at the nodes to become re-established, and seeds will mature in the pods even if the plant is pulled and left with its roots turned upward. Seeds can remain viable in the soil for several decades. The plant is frost tender and will be killed by the first freeze in fall. It needs heat to germinate, so summer is the season for growth. It grows in a wide range of temperate and tropical habitats and on various types of grounds but prefers an open one and a rich moist soil. It is able to grow well under poor, low-fertility soil conditions. In the tropics, the plants may flower all the year round. Optimum pH level should be 6.1 - 6.5.

6.0 Cultivation

In tropical countries, purslane seeds can be sown directly into the soil whenever you like but better is early spring or fall season of the year (Kamil *et al.*, 2000). Normally seeds germinate within 1–2 weeks after sowing. Sometimes purslane seeds can show dormancy and remain alive and dormant for 5–40 years in soils (Egley, 1974). Purslane is an annual plant and generally completes the life cycle (seed to seed) within 60–75 days. A single plant can produce over 50,000 seeds. The seeds are capable of being dormant in the soil for 30 to 40 years. Seed can germinate at a high soil temperature of 32°C. Steinmaus *et al.* (2000) reported that common purslane germination rates increased linearly with a temperature between 15 and 30°C. One interesting feature of common purslane plant is that, it can continuously produce flowers as well as seeds and sometimes can live more than a year under favourable environment, specially the cultivated/ornamental ones.

Propagation, Field Growing, and Harvesting of Purslane

Purslane is typically propagated from seed, but it can also grow from stem cuttings, divisions, or transplants. Branches that have been cut off can be planted with the cut-end in damp potting soil or a soilless potting mix. The cutting will begin to grow almost immediately if the soil is moist. The performance of the common purslane plants propagated by cutting is not satisfactory. However, the cultivated/ ornamental purslanes are suitable for propagation from cuttings and perform better compared to the common ones.

It is easily grown in any good soil from seeds sown in early spring, where the plants are to stand. Although the seeds are small, they germinate strongly. Its shoots grow erect when sown thickly. Mostly, it is sown from the middle of April to the middle of September in hills and from middle of March to the end of June in plains of India. Because purslane is shallow-rooted and grown as a densely planted, short-duration crop, the topsoil should be of good fertility. Organic manure is recommended at a rate of 20–30 t/ha at the time of final land preparation. As topdressing 40 kg/ha of urea may be added 3 weeks after planting.

Purslane requires regular watering at short intervals, 3–4 days in dry warm weather. It tolerates irrigation with saline drainage water. In Indonesia, it is sometimes transplanted to a spacing of 30 cm x 30 cm and grown for a longer duration.

Purslane is often listed as one of the world's worst weeds, although others consider it not very harmful because of its shallow rooting. It is, however, an important host plant for root-knot nematodes. When controlling it in other crops, sprayed herbicides such as glyphosate and 2,4-D can be effective, but seeds may mature in the time it takes for a herbicide to kill the plant. It can also be controlled with the more environmentally friendly corn gluten meal. Purslane rarely develops in mulched areas, and mulch placed over purslane will usually smother it.

7.0 Harvesting

Regular harvesting is very important for growing purslane in the home garden because the plants can become invasive. Harvesting purslane before it develops flowers will help cut down on its spread. One more thing to consider while harvesting purslane is the malic acid content of the plant, which is higher in the morning than at night thus making it tarter'. Some people will prefer this, while others may find they enjoy purslane harvested in the evening (when the flavour is milder). Purslane completes its life cycle within 2 to 4 months. Harvesting is initiated 3 to 4 weeks from sowing and 2 or 3 cuts at intervals of 2 to 3 weeks are carried out in commercial plantations.

The leaves should be washed with clean water and stored in a cool, shady place, and ideally eaten soon after picking. Purslane can be stored in plastic boxes for 2–6 days at 0–1°C and high relative humidity. Plants are dried in the sun and sold in markets in the dry season. Leaves and young plants not usually stored, but can be pickled and dried.

8.0 Diseases and Pests

Purslane is a hardy plant. It doesn't typically attract or succumb to too many diseases or pests.

8.1.0 Diseases

8.1.1 Black Stem (*Dichotomophthora portulacae* Mehrlich and Fitzpatrick ex)

The stems of these portulaca plants have been taken over by a black stem rot while the roots are relatively intact and sturdy. Humid or moist conditions encourage the fungus to flourish, so raising plants slightly off of the ground or reducing moisture levels will help with prevention. Once infected, overhead watering or rain could splash spores onto neighbouring plants.

8.1.2 Botrytis Blight or Gray Mold

It is a fungal disease that affects *Portulaca oleracea*. There are many species of the fungus botrytis, which is the cause of the blights. Favourable conditions for the infection are cool, rainy springs and unusually cool summers. Gray mold can be quite damaging if these conditions continue for several days. The disease can affect leaves, stems, flowers, seeds or any part of the plant except the root. Brown spots and gray spores of dead tissue form on the plant. The best defense is to manage by inspection and good hygiene. Faded or blighted flowers, leaves or entire plants should be removed and put in the garbage or burned. Fungicide sprays can also help protect from infections.

8.1.3 Root and Stem Rot

Either too much water or drought can cause poor root health in landscape plants. There are also fungi in the soil that can infect roots and cause root rot. If major roots are affected, the plant can die. Poor root health will appear as dull leaf colour; the leaves may turn yellow. The tiny feeder roots are missing on sick plants. Rotted roots will turn brown and mushy.

The plants need well-drained soil, not to be over-watered and not fertilized if too hot. Fungicide sprays can also help protect from infections.

8.1.4 White Rust (*Wilsoniana portulacae*)

White rust is the common name for several possible single-celled fungi that make up the *Albugo* genus. It is called white rust because of the white, dusty pustules that develop on infected foliage that look similar to those created by rust fungi. As they are spread by airborne sporangia, the initial site of infection will be an upper-level leaf. This leaf will develop thick bumps anywhere between five days to two weeks following infection. If the leaf is isolated from the rest of the purslane before the bumps rupture and disseminate spores, the infection can be halted.

8.1.5 Powdery Mildew

White powdery mildew is a common problem in damp areas. It takes the appearance of a light off-white to grey-coloured powder on the petals and leaves of the purslane.

Diseases and Pest of Purslane

White Rust Infection

Natural and Artificial Disease Symptoms on *Portulaca oleracea* Leaves Caused by ***Dichotomophthora lutea*** CBS 144679.

a, symptoms on leaves collected from purslane fields of Abadan in Khuzestan province, Iran. b, trial pot used in pathogenicity test. c, leaf spot caused by placing a 0.1ml drop of spore suspension on the leaf. d, leaf spot caused by injecting 0.1 ml spore suspension into the leaf (Heideri *et al.,* 2018).

Leaf-mining Weevil Infestation

8.1.6 Leaf Spot

This disease is caused by *Dichotomophthora lutea*. It is a leaf disease showing necrotic, oval to irregular, pale brown spots with darker border which has been observed for the first time on purslane in Khuzestan province, Iran (Heideri *et al.*, 2018).

8.2.0 Pests

8.2.1 Sawfly [*Schizocerella pilicornis* (Hohngren)]

Purslane sawfly is an insect that feeds and reproduces on common purslane. It eats the leaves of common purslane, leaving the plants low in vigour and with the little photosynthetic area. Unfortunately, by the time it develops sufficient numbers to have an impact on the common purslane population, seed development and much of the damage from purslane competition in the garden or crop have already occurred.

8.2.2 Leaf-mining Weevil (*Hypurus bertrandi* Perris)

Larvae of the weevil, like those of the sawfly, mine the leaves of purslane. Weevil adult's damage leaves by actively feeding on the margins and epidermal surfaces. The adults also feed on stems and developing seed capsules, but such feeding does not appear to be as prevalent as leaf feeding.

8.2.3 Aphids

Portulaca grandiflora and *P. oleracea* are both bothered by aphids. They are typically found gathering on young shoots and leaves. They rarely kill a plant, but in large quantities, they can remove large amounts of sap, which reduces the heartiness of the plant and can result in stunted growth. Leaves often become curled, puckered and yellowed. Many aphids suck out more plant sap than they can use. The excess sap is known as "honeydew."

Control can be natural as predators that prey on aphids or an insecticidal spray. Treating the infested crop with neem oil will eradicate this harmful pest.

9.0 Crop Improvement

Alam *et al.* (2014) made an attempt to amalgamate the assorted scientific information on purslane propagation, cultivation, varietal improvement, nutrient analyses, medicinal uses and to describe prospective research especially for genetic improvement of this crop.

Purslane is still considered as a weed in most countries. Therefore, systematic efforts to improve this potential crop and improved lines with enviable characters for definite cropping environment are not available. Due to the limited genetic diversity in purslane, the attempt to genetic improvement of purslane should aim through molecular and biotechnological approaches with more emphasis on high vegetative yield (through increased numbers of vegetative branches and leaf production), delayed in crop maturity, tolerance/resistance to diseases and pests, and more tolerance to salinity and drought stress. By the way, genetic diversity analysis in wild species among the diverse collected samples and selection of the superior genotypes is a potential way of purslane improvement. Several other advanced technological methods such as; mutation breeding, gene transfer through inter-specific hybridization and biotechnological approaches mainly molecular marker assisted selection breeding to bring the changes in desired traits. Comprehensive work on purslane sample collection, categorization and assessment of germplasm for growth, morphology, physiology, seeds or seedling characteristics and yield components are still scanty. To make this novel plant popular among consumers, exploration and documentation of general characteristics and subsequent introduction and demonstration of all novel qualities in comparison to other leafy vegetables are also important in addition to genetic improvement.

Although selection based on seed characters of purslane would simplify the task, there still subsist a number of precincts and uncertainties because of very low or no seed production by the ornamental purslane cultivars and the concerned genes of interest. Understanding of genetic discrepancy in physiological and nutritional especially high antioxidant property contents, salinity and drought tolerance and yield attributes in different purslane germplasm can be of immense prospective in improvement strategies. This is principally significant in selection of genotypes with more vegetative branches and yield as shoot biomass yield is important for purslane and more branching attribute to more biomass yield. Therefore a wide-ranging collection of germplasm from selected cultivars, their maintenance and the assessment program of diverse purslane accession is indispensable to recognize prototypes of inconsistency.

There are various types of purslane available in different locations of the world. Those variations are not only phenotypic or physiological but also genetic. The genetic diversity of purslane has been identified in different studies based on tolerance of drought, heat, salinity and nutritional conditions. Girenko (1980) recorded the intra-specific variability and composition of purslane cultivars in a range of climatic zones of the Canadian Iris Society, together with a different set of data of agricultural significance. Ren *et al.* (2011) conducted experiment with different purslane accessions collected from diverse geographical locations and stated the drought tolerant and AFLP-based genetic diversity within and among purslane accessions. They collected 10 different purslane samples: Keren and Tokombiya (from Eritrea), POS, PO and Turkey (from Turkey), Golden G and Golden T (from Netherlands), Golden E (from UK) and wild accessions Egypt (Egyptium) and Greece (Wild Greece) and these were evaluated using AFLP fingerprinting with dissimilarity among diverse purslane samples regarding drought tolerance at seed germination, seedling and adult stages. Tokombiya was recognized as the most drought tolerant accession at mature stage. Due to global warming, it is a great challenge to identify and develop new crop varieties which are genetically tolerant to drought and heat. In this regard purslane is a very potential and promising vegetable crop that can withstand biotic and abiotic stresses and can therefore be used as the genetic material for crop improvement.

Siriamornpun and Suttajit (2010) conducted an experiment examining the morpho-physiological and nutritional diversity of eight purslane accessions collected from different geographical locations namely *P. oleracea*, *P. sativa*, Golden Ger-ber (GG Dutch), Golden (G England), and wild (W) accessions (Beltsville, Egyptian and Greece) and found significant differences in the percentage of crude protein, total lipids, essential fatty acids, carbohydrates and ash. Crude protein levels in Gn Dutch (25.0 per cent), W Greece (25.1 per cent) and *P. oleracea* (24.6 per cent) were significantly higher than those of W Beltsville (21.6 per cent) over the three-year period. These findings demonstrate that there is significant nutritional diversity among the collected purslane accessions, allowing selection of better parents for a future breeding program.

Another study by Yan *et al.* (2009) examining essential fatty acid variation among different purslane cultivars, showed significant differences in the amount of saturated fatty acids (palmitic and stearic), monounsaturated fatty acid (oleic) and two poly-unsaturated fatty acids; linoleic and linolenic acids among the *P. oleracea* accessions in both leaves and stem. Fatty acid content and genetic diversity were highly variable among *P. oleracea* accessions. Variation in fatty acid composition and other traits in *P. oleracea* suggest that it is possible to select for phenotypes with multiple desirable traits, such as taste, leaf and stem structure, growth rate and tolerance to abiotic and biotic stresses (Yan *et al.*, 2009).

Transferring desirable genes from one species to another is possibly the more expectant and the most useful contribution of inter-specific hybridization. Information on usual hybridization are pretty infrequent, Kim and Carr (1990) stated a recognized hybrid between *P. oleracea* and *P. lutea* in Hawaii, with a chromosome number of 47, which indicated that allopolyploidy is exceptional. The inter-specific hybridization techniques have been used in many crops to solve the normal hybridization problems and this knowledge will be helpful for purslane improvement.

10.0 Biotechnology

The first successful genetic transformation of *P. oleracea* L. callus mediated by *A. tumefaciens* was done by Hong-Yan *et al.* (2011). There the callus was induced from the leaves of *P. oleracea* L., and was used for genetic transformation mediated by *A. tumefaciens*. The resistant callus was screened and confirmed to be a transformant by GUS histochemical assays and PCR. It showed that genetic transformation of *P. oleraceacallus* mediated by *A. tumefaciens* was entirely feasible. Different research findings have proved that Portulaca has many good qualities especially salinity and drought tolerance. Therefore this successful transformation will be the key for further research aimed at transferring the salinity and drought tolerance gene to other crops to get the desired phenotype. The first time from purslane leaves and seeds Teixeira *et al.* (2009) successfully isolated two cDNA clones of Pole-FAD2 and onecDNA clone of Pole-FAD6, encoding x-6 fatty acid desaturases, which are the core enzymes for the conversion of oleic into linoleic acid. They also proved that, the isoforms of the both cDNA clones of Pole-FAD2 were highly similar with other microsomal x-6 desatuarates and then to Pole-FAD6 or other plasmid orthologues, and vice versa. The RT-PCR expression analysis results also confirmed the expression of all the genes in all the tissues of purslane but only in the reproductive organs and cells the higher levels of mRNA accumulation were observed. In purslane leaves the levels of mRNA accumulation of both *FAD2* and *FAD6* genes were affected by wounds and at transcript level only *FAD2* gene were affected by chilling injury. The expression prototypes detected imitate the distinct function of these genes in membrane synthesis for cell division, thylakoid development, and lipid storage or in the biosynthetic pathway for the production of signalling molecules that influence plant development or defence. The purposeful utilization of these two genes can contribute a lot to develop temperature and mechanical injury resistant purslane cultivars as well as to introduce these genes through recombinant technology to other economically valuable crops to improve their resistance to such stresses.

Scientists already have used RAPD and AFLP markers only for diversity analysis of purslane. Further work is needed to develop DNA-based markers linked to α-linolenic acid (ALA) production and molecular markers which could be used to facilitate the breeding of high ALA varieties. Teixeira *et al.* (2009) have identified and isolated two genes (*FAD2* and *FAD6*; *x-3* fatty acid desaturase genes) from leaves and seeds of purslane. They reported the functions of these two genes in cell division, thylakoid development, and lipid storage for the purpose of membrane synthesis. These genes may also be involved in the biosynthetic procedure for the production of signalling molecules that affect plant growth or defence which could contribute to purslane resistance to temperature and mechanical damage.

11.0 References

Alam, M.A., Juraimi, A.S., Rafii M.Y., Hamid, A.A., Kamal Uddin M., Alam M.Z. and Latif, M.A. (2014) *Mol. Biol. Rep.*, **41**: 7395–7411.

Bianco, V.V., Santamaria, P., Elia, A., Rubatzky, V.E., Chen, H. and Peron, J.Y.(1998) *Acta Hortic.*, **467**: 71-87.

Danin, A., Baker, I. and Baker, H.G. (1978) *Israel J. Bot.*, **27**(3/4): 177-211.

Egley, G.H. (1974) *Weed Sci.*, **22**(6): 535–540.

Garti, N., Slavin, Y. and Aserin, A. (1999) *Food Hydrocolloids*, **13**(2): 145-155.

Geesink, R. (1969) *Blumea*, **17**: 274-301.

Girenko, M.M. (1980) *Plant Breed. Abstr.*, **7104**(101): 14–18.

Gopalan, C., Ramasastry, B.V. and Balasubramanium, S.C. (1999) National Institute of Nutrition, Hyderabad, P. 156

Gorske, S.F., Rhodes, A.M. and Hopen, H.J. (1979) *Weed Sci.*, **27**(1): 96-102.

Grieve, C.M. and Suarez, D.L. (1997) *Plant Soil*, **192**(2): 277-283.

Heidari, Kobra, Farokhinejad, Reza and Mehrabi-Koushki, Mehdi (2018) Australasian Plant Disease Notes, 13: 33, doi.org/10.1007/s13314-018-0317-2.

Hong-Yan, Y., Hong-Lei, Y., Hong-Qin, Y., Qun-Ce, H. and An-Yong, Q. (2011) *Chin. Agric. Sci. Bull.*, **27**: 139–143.

https: //www.healthline.com/nutrition/purslane

Kamil, M., Chan, K. and Habibullah, M. (2000) *Aust. J. Med. Herb*, **12**: 43–48.

Kim, I. and Carr, G.D. (1990) *Syst. Bot.*, **15**: 370–377

Legrand, C.D. (1962) Anales del Museo de Historia Natural de Montevideo. II.

Matthews, J.F., Ketron, D.W. and Zane, S.F. (1993) *Rhodora*, **95**(882): 166-183.

Prabhakar, M. and Ramayya, N. (1988) *Indian J. Bot.*, **11**(1): 51-58.

Ren, S., Weeda, S., Akande, O., Guo, Y., Rutto, L., Mebrahtu, T. (2011) *J. Biotech Res.*, **3**: 51–61.

Siriamornpun, S. and Suttajit, M. (2010) *Weed Sci.*, **58**(3): 182–188.

Steinmaus, S.J., Prather, T.S. and Holt, J.S. (2000) *J. Exp. Bot.*, **51**(343): 275-286.

Teixeira, M.C., Coelho, N., Olsson, M.E., Brodelius, P.E., Carvalho, I.S. and Brodelius, M. (2009) *Biotechnol. Lett.*, **31**: 1089–1101.

Vengris, J., Dunn, S. and Stacewicz-Sopuncakis, M. (1972) *Research Bulletin, Agricultural Experiment Station, University of Massachusetts,* **598**: 1-44.

Yan, G., Aryamanesh, N. and Wang, S. (2009) In: Rural Industries Research and Development Corporation (RIRDC), Australia. Publication No. 09/088, RIRDC Project No. PRJ-000348, pp. 1–15.

Index

C

D

E

F

G

M

N

O

P

Q

R

S

T

U

V

W

Y